Molecular Structures and Dimensions

Vol. 7

Solid State Classes 1–86

Molecular Structures and Dimensions

Vol. 7

Bibliography 1974–75
Organic and Organometallic Crystal Structures

Edited by
Olga Kennard, David G. Watson
Frank H. Allen and Stella M. Weeds
University Chemical Laboratory, Cambridge

Springer Science+Business Media, LLC

Originally published by International Union of Crystallography in 1976
Softcover reprint of the hardcover 1st edition 1976

First published 1976 by International Union of Crystallography.

Library of Congress catalogue card number 76–133989

ISBN 978-94-017-2352-7 ISBN 978-94-017-2350-3 (eBook)
DOI 10.1007/978-94-017-2350-3

Contents

1	Introduction	page vii
2	Acknowledgements	ix
3	Classification Rules	xi
4	List of Classes Vol. 7	xiii
5	Entries for Classes 1–86	1
6	Cumulative Formula Index for Vols. 1–7	F1
7	Cumulative Transition Metal Index for Vols. 1–7	M1
8	Cumulative Author Index for Vols. 1–7	A1

Introduction

This volume is the seventh classified bibliography of organic and organometallic crystal structures prepared by the Crystallographic Data Centre, University Chemical Laboratory, Cambridge, and published jointly with the International Union of Crystallography.

The first six volumes covered the years 1935–1974. The present volume provides references principally to compounds whose structures were reported in the literature during 1974–1975. A few structures published prior to 1974 and omitted from the previous volumes are also included.

There are three cumulative indexes in the present volume: formula, transition metal and author indexes. All three cover the period 1935–1975 and give references to entries in Vols. 1–7.

The bibliography and indexes were prepared, checked and printed by computer techniques described in the previous volumes. Magnetic tapes of the seven volumes are available and anyone interested should contact the Centre for further details.

In the present volume we have continued the special arrangement for literature search with the Centre National de la Recherche Scientifique, Paris, France. Under this arrangement reprints of papers containing crystallographic data are sent directly to the Crystallographic Data Centre, Cambridge, at the same time as they are sent out to abstractors preparing material for the Bulletin Signalétique.

In addition to the above arrangement, 11 journals, covering approximately 78% of the crystallographic literature, are scanned directly in Cambridge. The cut-off dates for Volume 7 can be summarised as follows:

Acta Crystallogr., Sect. B part 6, page 1792, 1975
J. Chem. Soc., Dalton Trans., part 11, page 1105, 1975
J. Chem. Soc., Perkin Trans. 2, part 7, page 774, 1975
J. Chem. Soc., Chem. Commun., part 13, page 553, 1975
J. Am. Chem. Soc., part 9, page 2571, 1975
Acta Chem. Scand. Ser. A, part 3, page 374, 1975
Acta Chem. Scand. Ser. B, part 3, page 401, 1975
Inorg. Chem., part 5, page 1147, 1975
Tetrahedron Lett., page 1745, 1975
J. Cryst. Mol. Struct., part 1, page 59, 1975
Cryst. Struct. Commun., part 2, page 403, 1975
Other Journals: complete for 1973
ca. 95% complete for 1974
ca. 30% complete for 1975

The following Conference Abstracts are included in Vol. 7:

Conference Proceedings of the American Crystallographic Association, Summer 1974, Spring 1975.

Conference Proceedings of the Tenth Congress of the International Union of Crystallography, Amsterdam, August 1975, published as a Supplement to Acta Crystallogr. Sect. A, June 1975.

We are grateful to the Organising Committee of the Tenth Congress of Crystallography for their help in obtaining comprehensive information on new structures reported at the Congress by the use of special abstract forms which they forwarded to us. These forms were invaluable when incorporating the Conference Abstracts in the current bibliographic volume.

The work of the Crystallographic Data Centre is supported by the Science Research Council as part of the British contribution to international data activities.

We are greatly indebted to readers who have notified us of mistakes and omissions in Vols. 1–6. We have attempted to modify our procedures and are at present considering further changes including changes in the contents of forthcoming volumes. We would be grateful to readers for any suggestions on how these volumes could be further improved.

Cambridge November 1975

Olga Kennard
David G. Watson
Frank H. Allen
Stella M. Weeds

Acknowledgements

The production of this bibliography was a collaborative effort by members of the Crystallographic Data Centre: Mrs A. Doubleday, Dr W. D. S. Motherwell, Mr J. R. Rodgers, Miss S. A. Stephenson and Mrs K. A. M. Watson.

Mrs Watson has been in charge of the encoding of information and the clerical checking of new material. In the secretarial work of documentation she has been assisted by Miss Stephenson.

Mrs Doubleday has been responsible for scientific checking, registration of new entries and file editing.

Dr Motherwell and Mr Rodgers have written computer programs for the editing and maintenance of the file.

The work of the Centre was guided by the SRC Steering Panel on Crystallographic Data: Professor R. Mason, F.R.S. (Chairman), Professor D. W. J. Cruickshank, Mr O. S. Mills, Dr P. G. Owston, Mr J. H. Richards, Professor G. A. Sim, Professor M. R. Truter, Professor A. J. C. Wilson, F.R.S. and Dr T. Vickers. We are indebted to Professor R. A. Raphael, F.R.S. of the University Chemical Laboratory, Cambridge, for his help and interest in our activities.

We are grateful to the Medical Research Council for allowing a member of their External Scientific Staff (O. Kennard) to participate in this work.

We thank the University of Cambridge for the provision of accommodation in the University Chemical Laboratory and the administrative staff of the Laboratory, particularly Mr R. E. Maxim, who handled financial matters, for all their help.

Our task was greatly facilitated by the excellent organisation of the Centre National de la Recherche Scientifique. We are especially grateful to Madame C. Degen of the CNRS who was responsible for the improved literature searches referred to in the Introduction.

We have used the IBM 370/165 computer of the University of

Cambridge and we are particularly grateful to the staff for their special help with the production of the final tapes. We are grateful to INSPEC (Information Service in Physics, Electrotechnology and Computers & Control) and especially to Mr P. Simmons and Mr M. Farquharson for the use of their computer typesetting programs, which they specially modified for our purposes.

The bibliography was prepared in parallel with the Organic Volume of 'Crystal Data' (National Bureau of Standards, Washington D.C., USA). The supplement to the third edition is now in preparation and both publications are strengthened by this collaboration.

Classification Rules

To regularise the classification procedures it is considered necessary to order the 86 chemical classes according to a precedence rule. Thus, for example, if a compound can be described as belonging to classes 15 and 17 then we will always assign 17 as the basic class with a cross-reference to 15, not vice-versa.

The order of precedence is indicated below running from top to bottom and left to right (i.e. class 61 has highest precedence and class 5 the lowest):

61, 60
71, 72, 73, 74, 75, 76, 77, 78, 79, 80, 81, 82, 86, 85, 84, 83
70, 69, 68, 67, 66, 65, 64, 63, 62
58, 57, 56, 55, 54, 53, 52, 51, 50, 49, 48, 47, 46, 45, 44, 43, 59
42, 41, 40, 39, 38, 37, 36, 35, 34, 32, 33
31, 30, 29, 28, 27, 23, 22, 20, 21, 26, 25, 24
18, 14, 13, 17, 16, 15
2, 1, 3, 4
12, 8, 11, 7, 10, 9, 6, 19, 5

In addition to the above precedence rule the classification conventions have been better defined for classes 1–59. Some notes on specific classes are given below:

Class 1: Cyclic acid derivatives, e.g. anhydrides and lactones, are classified in the appropriate hetero-class. This rule applies also to class 13.

Class 2: In a few cases where the cation is organic we classify the anion in 2.

Class 4: The compound must contain –C–N–S– or –C–S–N–.

Class 9: The compound must contain –C–N–N–.

Class 10: The compound must contain –C–N–O– or –C–O–N–.

Class 24: The compound must be fully unsaturated.
The same rule applies to classes 25 and 26.

Class 44: The ring system must conform to the unmodified pyrimidine or purine skeleton.

Class 48: This class is reserved for peptides and α-amino-acids, whether or not the amino-acid possesses biological properties. Thus a β-amino-acid would be classified in the appropriate acid and amine classes.

Class 50: A cross-reference to a structural class must be provided.
This rule applies also to class 59.

List of Classes

1 Aliphatic Carboxylic Acid Derivatives 1
2 Aliphatic Carboxylic Acid Salts (Ammonium, IA, IIA Metals) 8
3 Aliphatic Amines 11
4 Aliphatic (N and S) Compounds 21
5 Aliphatic Miscellaneous 23
6 Enolates (Aliphatic and Aromatic) 26
7 Nitriles (Aliphatic and Aromatic) 28
8 Urea Compounds (Aliphatic and Aromatic) 31
9 Nitrogen-Nitrogen Compounds (Aliphatic and Aromatic) 35
10 Nitrogen-Oxygen Compounds (Aliphatic and Aromatic) 38
11 Sulphur and Selenium Compounds 40
12 Carbonium Ions, Carbanions, Radicals 45
13 Benzoic Acid Derivatives 47
14 Benzoic Acid Salts (Ammonium, IA, IIA Metals) 51
15 Benzene Nitro Compounds 52
16 Anilines 56
17 Phenols and Ethers 61
18 Benzoquinones 69
19 Benzene Miscellaneous 70
20 Monocyclic Hydrocarbons (3, 4, 5-Membered Rings) 73
21 Monocyclic Hydrocarbons (6-Membered Rings) 77
22 Monocyclic Hydrocarbons (7, 8-Membered Rings) 79
23 Monocyclic Hydrocarbons (9- and Higher-Membered Rings) 80
24 Naphthalene Compounds 81
25 Naphthoquinones 83
26 Anthracene Compounds 84
27 Hydrocarbons (2 Fused Rings) 85
28 Hydrocarbons (3 Fused Rings) 87

29	Hydrocarbons (4 Fused Rings)	89
30	Hydrocarbons (5 or More Fused Rings)	92
31	Bridged Ring Hydrocarbons	94
32	Hetero-Nitrogen (3, 4, 5-Membered Monocyclic)	99
33	Hetero-Nitrogen (6-Membered Monocyclic)	106
34	Hetero-Nitrogen (7- and Higher-Membered Monocyclic)	114
35	Hetero-Nitrogen (2 Fused Rings)	116
36	Hetero-Nitrogen (More than 2 Fused Rings)	124
37	Hetero-Nitrogen (Bridged Ring Systems)	129
38	Hetero-Oxygen	132
39	Hetero-Sulphur and Hetero-Selenium	146
40	Hetero-(Nitrogen and Oxygen)	152
41	Hetero-(Nitrogen and Sulphur)	156
42	Hetero-Mixed Miscellaneous	162
43	Barbiturates	164
44	Pyrimidines and Purines	166
45	Carbohydrates	173
46	Phosphates	184
47	Nucleosides and Nucleotides	187
48	Amino-Acids and Peptides	193
49	Porphyrins and Corrins	202
50	Antibiotics	205
51	Steroids	210
52	Monoterpenes	217
53	Sesquiterpenes	218
54	Diterpenes	222
55	Sesterterpenes	224
56	Triterpenes	225
57	Tetraterpenes	226
58	Alkaloids	227
59	Miscellaneous Natural Products	231
60	Molecular Complexes	235
61	Clathrates	244
62	Boron Compounds	246
63	Silicon Compounds	251
64	Phosphorus Compounds	254
65	Arsenic Compounds	262
66	Antimony and Bismuth Compounds	264
67	Groups IA and IIA Compounds	266
68	Group III Compounds	268
69	Germanium, Tin, Lead Compounds	271
70	Tellurium Compounds	275

71	Transition Metal-C Compounds	276
72	Metal π-Complexes (Open-Chain)	294
73	Metal π-Complexes (Cyclopentadiene)	302
74	Metal π-Complexes (Arene)	313
75	Metal π-Complexes (Miscellaneous Ring Systems)	314
76	Metal Complexes (Ethylenediamine)	322
77	Metal Complexes (Acetylacetone)	331
78	Metal Complexes (Salicylic Derivatives)	334
79	Metal Complexes (Thiourea)	338
80	Metal Complexes (Thiocarbamate or Xanthate)	342
81	Metal Complexes (Carboxylic Acid)	345
82	Metal Complexes (Amino-Acid)	352
83	Metal Complexes (Nitrogen Ligand)	356
84	Metal Complexes (Oxygen Ligand)	379
85	Metal Complexes (Sulphur or Selenium Ligand)	391
86	Metal Complexes (P, As, Sb Ligand)	399

ALIPHATIC CARBOXYLIC ACID DERIVATIVES

1.1 **N,N - Di - iodoformamide**
CHI_2NO
H.Pritzkow *Monatsh. Chem.*, **105,** 621, 1974

1.C **Potassium hydrogen bis(trichloroacetate)**
$C_2HCl_3O_2$, $C_2Cl_3O_2^-$, K^+
For complete entry see 2.1

1.C **Pyridine - N - oxide trichloroacetic acid**
$C_2HCl_3O_2$, C_5H_5NO
For complete entry see 33.11

1.C **Oxotremorine sesquioxalate**
1 - (4 - (2 - Oxopyrrolidin - 1 - yl)but - 2 - ynyl)pyrrolidinium sesquioxalate
$C_2HO_4^-$, $C_{12}H_{19}N_2O^+$, $0.5C_2H_2O_4$
For complete entry see 32.22

1.2 **Oxalic acid (α form)**
$C_2H_2O_4$
J.L.Derissen, P.H.Smit *Acta Crystallogr., Sect. B,* **30,** 2240, 1974

1.3 **Oxalic acid (β form)**
$C_2H_2O_4$
J.L.Derissen, P.H.Smit *Acta Crystallogr., Sect. B,* **30,** 2240, 1974

1.4 **Potassium monothioacetate**
$C_2H_3OS^-$, K^+
M.M.Borel, M.Ledesert *Acta Crystallogr., Sect. B,* **30,** 2777, 1974
Residue 1 also classified in 11

1.C **Sodium hydrogen diacetate (neutron study)**
$C_2H_4O_2$, $C_2H_3O_2^-$, Na^+
For complete entry see 2.4

1.5 **Fluoromalonic acid (at liquid nitrogen temp.)**
$C_3H_3FO_4$
G.Roelofsen, J.A.Kanters, J.Kroon
Acta Crystallogr., Sect. A, **31,** S173, 1975

1.6 **Bromomalonamide**
$C_3H_5BrN_2O_2$
R.F.Picone, M.T.Rogers, M.Neuman *J. Chem. Phys.*, **61**, 4808, 1974

1.7 **Cesium dimethyldithiocarbamate**
$C_3H_6NS_2^-$, Cs^+
P.Jennische, H.Anacker-Eickhoff, A.Wahlberg
Acta Crystallogr., Sect. A, **31**, S143, 1975
Residue 1 also classified in 11

1.8 **Malonic dihydrazide**
$C_3H_8N_4O_2$
C.Miravitlles, J.L.Brianso, F.Plana, M.Font-Altaba
Cryst. Struct. Commun., **4**, 81, 1975
Also classified in 9

1.C **Lithium hydrogen maleate dihydrate**
$C_4H_3O_4^-$, Li^+ , $2H_2O$
For complete entry see 2.5

1.C **Sodium hydrogen maleate trihydrate**
$C_4H_3O_4^-$, Na^+ , $3H_2O$
For complete entry see 2.6

1.C **Imidazolium maleate**
$C_4H_3O_4^-$, $C_3H_5N_2^+$
For complete entry see 32.1

1.C **(+) - Chloropheniramine maleate (absolute configuration)**
(+) - S - 1 - (p - Chlorophenyl) - 1 - (2 - pyridyl) - 3,N,N - dimethylpropylamine maleate
$C_4H_3O_4^-$, $C_{16}H_{20}ClN_2^+$
For complete entry see 33.46

1.C **bis(β - Picoline - N - oxide) - fumaric acid**
$C_4H_4O_4$, $2C_6H_7NO$
For complete entry see 33.20

1.9 **N - Methyl - propiolamide**
C_4H_5NO
M.Tuval, L.Leiserowitz *Acta Crystallogr., Sect. A*, **31**, S176, 1975

1.C **(+) - 2 - Hydroxypropyl - 1 - ammonium (+) - (hydrogen tartrate) dihydrate**
$C_4H_5O_6^-$, $C_3H_{10}NO^+$, $2H_2O$
For complete entry see 2.8

1.C **(−) - 2 - Hydroxypropyl - 1 - ammonium (+) - (hydrogen tartrate) dihydrate**
$C_4H_5O_6^-$, $C_3H_{10}NO^+$, $2H_2O$
For complete entry see 2.9

1.C **(+) - ((−) - 1 - Methyl - 3 - ethyl - 3 - benzoylpiperidine (+) - R,R - bitartrate)**
$C_4H_5O_6^-$, $C_{15}H_{22}NO^+$
For complete entry see 33.43

1.10 **Rubidium N,N - dimethyl - α,α - dinitroacetamide**
$C_4H_6N_3O_5^-$, Rb^+
N.V.Grigor'eva, N.V.Margolis, I.V.Tselinskii, V.V.Mel'nikov, G.V.Makarenko *Zh. Strukt. Khim.*, **15**, 167, 1974
Residue 1 also classified in 12

1.11 **Crotonic acid**
$C_4H_6O_2$
S.Shimizu, S.Kekka, S.Kashino, M.Haisa
Bull. Chem. Soc. Jpn., **47**, 1627, 1974

1.12 **Crotonamide**
C_4H_7NO
S.Shimizu, S.Kekka, S.Kashino, M.Haisa
Bull. Chem. Soc. Jpn., **47**, 1627, 1974

1.13 **trans - cis - Diacetamide**
$C_4H_7NO_2$
Y.Kuroda, Z.Taira, T.Uno, K.Osaki *Cryst. Struct. Commun.*, **4**, 325, 1975

1.14 **trans - trans - Diacetamide**
$C_4H_7NO_2$
Y.Kuroda, Z.Taira, T.Uno, K.Osaki *Cryst. Struct. Commun.*, **4**, 321, 1975

1.15 **DL - γ - Amino - β - hydroxybutyric acid**
$C_4H_9NO_3$
K.-I.Tomita, M.Harada, T.Fujiwara *Bull. Chem. Soc. Jpn.*, **46**, 2854, 1973
Also classified in 3

1.16 **N - Methyl - tetrolamide**
C_5H_7NO
M.Tuval, L.Leiserowitz *Acta Crystallogr., Sect. A*, **31**, S176, 1975

1.C **Dimethylmalonic acid - bis(triphenylphosphine oxide)**
$C_5H_8O_4$, $2C_{18}H_{15}OP$
For complete entry see 60.2

1.C **1,4,7,10,13,16 - Hexaoxacyclo - octadecane dimethylacetylenedicarboxylate (at −160°C)**
$C_6H_6O_4$, $C_{12}H_{24}O_6$
For complete entry see 38.41

1.C **S,S' - Diethyl - dithio - oxalate (at −60°C)**
$C_6H_{10}O_2S_2$
For complete entry see 11.8

1.17 **racemic - Thiodilactic acid**
$C_6H_{10}O_4S$
E.Martuscelli, L.Mazzarella, A.Zagari *Gazz. Chim. Ital.*, **103**, 563, 1973
Also classified in 11

1.18 **4 - Amino - 3 - hydroxy - 2 - methyl - n - valeric acid hemihydrate**
$C_6H_{13}NO_3$, $0.5H_2O$
H.Nakamua, T.Takita, H.Umezawa, Y.Muraoka, Y.Iitaka
J. Antibiot., **27**, 352, 1974
Residue 1 also classified in 3

1.19 **N - Methyl - sorbamide**
$C_7H_{11}NO$
M.Tuval, L.Leiserowitz *Acta Crystallogr., Sect. A*, **31**, S176, 1975

1.20 **3,3 - Dimethylglutaric acid**
$C_7H_{12}O_4$
E.Benedetti, R.Claverini, C.Pedone *Gazz. Chim. Ital.*, **103**, 525, 1973

1.21 **Diethylpropionamide**
$C_7H_{15}NO$
C.Cohen-Addad, G.d'Assenza, G.Taillandier, J.-L.Benoit-Guyod
Acta Crystallogr., Sect. B, **31**, 835, 1975

1.22 **Acetylcholine bromide**
$C_7H_{16}NO_2^+$, Br^-
T.Svinning, H.Sorum *Acta Crystallogr., Sect. B*, **31**, 1581, 1975
Residue 1 also classified in 3

1.23 **DL - Carnitine hydrochloride**
$C_7H_{16}NO_3^+$, Cl^-
K.-I.Tomita, K.Urabe, Y.B.Kim, T.Fujiwara
Bull. Chem. Soc. Jpn., **47**, 1988, 1974
Residue 1 also classified in 3

1.C **2,4 - Dibromo - 3,6 - dihydroxyphenyl - acetamide**
$C_8H_7Br_2NO_3$
For complete entry see 17.7

1.24 **4 - Chlorophenylsulfinyl acetic acid**
$C_8H_7ClO_3S$
L.Leiserowitz, G.Salem, C.-P.Tang, M.Weinstein
Cryst. Struct. Commun., **4**, 89, 1975
Also classified in 11

1.25 **Phenylsulfinyl acetamide**
$C_8H_9NO_2S$
L.Leiserowitz, Z.Berkovitch-Yellin, M.Weinstein
Cryst. Struct. Commun., **4**, 93, 1975
Also classified in 11

1.26 **p - Chloro - trans - cinnamic acid**
$C_9H_7ClO_2$
J.P.Glusker, D.E.Zacharias, H.L.Carrell
J. Chem. Soc., Perkin Trans. 2, 68, 1975
Also classified in 19

1.27 **β - (p - Chlorophenyl)propionic acid**
$C_9H_9ClO_2$
J.P.Glusker, D.E.Zacharias, H.L.Carrell
J. Chem. Soc., Perkin Trans. 2, 68, 1975
Also classified in 19

1.C **N - Succinyl - pyridine**
$C_9H_9NO_4$
For complete entry see 33.31

1.28 **3 - S(−) - Phenyl - lactic acid**
$C_9H_{10}O_3$
M.Cesario, J.Guilhem *Cryst. Struct. Commun.,* **4,** 245, 1975

1.29 **(+) - (2S,3S) - Phenylglyceric acid**
$C_9H_{10}O_4$
M.Cesario, J.Guilhem *Cryst. Struct. Commun.,* **4,** 197, 1975

1.30 **(−) - (2S,3R) - Phenylglyceric acid**
$C_9H_{10}O_4$
M.Cesario, J.Guilhem *Cryst. Struct. Commun.,* **4,** 193, 1975

1.C **D - β - Tyrosine hydrobromide (absolute configuration)**
$C_9H_{12}NO_3^+$, Br^-
For complete entry see 17.13

1.C **D - β - Tyrosine hydrochloride (absolute configuration)**
$C_9H_{12}NO_3^+$, Cl^-
For complete entry see 17.14

1.C **1,1 - Cyclopentane - diacetic acid**
$C_9H_{14}O_4$
For complete entry see 20.8

1.31 **N,N′ - bis(β - Chloroethyl) - glutaramide**
$C_9H_{16}Cl_2N_2O_2$
E.Benedetti, M.R.Ciajolo, P.Corradini *Eur. Polym. J.,* **10,** 1201, 1974

1.32 **Di - n - propylpropionamide**
$C_9H_{19}NO$
C.Cohen-Addad, G.d′Assenza, G.Taillandier, J.-L.Benoit-Guyod
Acta Crystallogr., Sect. B, **31,** 835, 1975

1.33 **Butyryl - thiocholine iodide**
$C_9H_{20}NOS^+$, I^-
T.Svinning, H.Sorum *Acta Crystallogr., Sect. A*, **31**, S101, 1975
Residue 1 also classified in 3

1.C **Potassium trihydrogen bis(O,O′ - catechol diacetate)**
$C_{10}H_9O_6^-$, $C_{10}H_{10}O_6$, K^+
For complete entry see 17.17

1.34 **N - Methyl - cinnamide**
$C_{10}H_{11}NO$
M.Tuval, L.Leiserowitz *Acta Crystallogr., Sect. A*, **31**, S176, 1975

1.C **DL - Dethiobiotin**
$C_{10}H_{18}N_2O_3$
For complete entry see 32.17

1.C **1 - Naphthylsulfinyl acetic acid**
$C_{12}H_{10}O_3S$
For complete entry see 24.7

1.C **2 - Chloro - 2 - phenoxymalonylamide - amidinium hydrochloride monohydrate**
$C_{13}H_{19}ClN_3O_2^+$, Cl^- , H_2O
For complete entry see 17.20

1.C **3,5 - Dibromothyroacetic acid N - diethanolamine solvate**
$C_{14}H_{10}Br_2O_4$, $C_4H_{11}NO_2$
For complete entry see 17.22

1.C **Kavaic acid**
$C_{14}H_{14}O_3$
For complete entry see 59.2

1.35 **Succinyl choline iodide**
$C_{14}H_{30}N_2O_4^{2+}$, $2I^-$
A.L.Mudzhoyan, R.L.Avoyan, A.A.Avetisyan, E.G.Arutunian
Arm. Khim. Zh., **25**, 710, 1972
Residue 1 also classified in 3

1.C **3,5 - Di - iodothyropropionic acid**
$C_{15}H_{12}I_2O_4$
For complete entry see 17.26

1.36 **(±) - 2 - (2 - Fluoro - 4 - biphenyl)propionic acid**
Flurbiprofen
$C_{15}H_{13}FO_2$
J.L.Flippen, R.D.Gilardi *Acta Crystallogr., Sect. B*, **31**, 926, 1975
Also classified in 19

1.C **Tri - iodo - thyropropionic acid ethyl ester**
$C_{17}H_{15}I_3O_4$
For complete entry see 17.29

1.C **Diethyl - (2 - hydroxyethyl) - methylammonium α - cyclopentyl - 2 - thienylglycollate bromide**
Penthienate bromide
$C_{18}H_{30}NO_3S^+$, Br^-
For complete entry see 39.31

1.37 **2 - Diethylaminoethyl - benzilate hydrochloride**
Benactyzine hydrochloride
$C_{20}H_{26}NO_3^+$, Cl^-
T.J.Petcher *J. Chem. Soc., Perkin Trans. 2,* 1151, 1974
Residue 1 also classified in 3

1.38 **β - Diethylaminopropyl α - diphenylacetate hydrobromide**
$C_{21}H_{28}NO_2^+$, Br^-
H.L.Avoyan, E.R.Arakelova, E.G.Arutunian
Arm. Khim. Zh., **26,** 713, 1973
Residue 1 also classified in 3

1.C **Chloramphenicol palmitate (β form)**
$C_{27}H_{42}Cl_2N_2O_6$
For complete entry see 50.21

ALIPHATIC CARBOXYLIC ACID SALTS (AMMONIUM, IA, IIA METALS)

2.1 **Potassium hydrogen bis(trichloroacetate)**

$C_2Cl_3O_2^-$, $C_2HCl_3O_2$, K^+

L.Golic, F.Lazarini *Cryst. Struct. Commun.*, **3,** 645, 1974

Residue 2 classified in 1

2.C **Oxotremorine sesquioxalate**

1 - (4 - (2 - Oxopyrrolidin - 1 - yl)but - 2 - ynyl)pyrrolidinium sesquioxalate

$C_2HO_4^-$, $C_{12}H_{19}N_2O^+$, $0.5C_2H_2O_4$

For complete entry see 32.22

2.2 **Sodium fluoroacetate**

$C_2H_2FO_2^-$, Na^+

K.Vijayan, A.Mani, B.M.Vedavathi, S.Ramaseshan

Acta Crystallogr., Sect. A, **31,** S58, 1975

2.3 **Strontium acetate thioacetate tetrahydrate**

$C_2H_3O_2^-$, $C_2H_3OS^-$, Sr^{2+} , $4H_2O$

M.M.Borel, M.Ledesert *Acta Crystallogr., Sect. B,* **31,** 725, 1975

Residue 2 classified in 11

2.4 **Sodium hydrogen diacetate (neutron study)**

$C_2H_3O_2^-$, $C_2H_4O_2$, Na^+

M.J.Barrow, M.Currie, K.W.Muir, J.C.Speakman, D.N.J.White

J. Chem. Soc., Perkin Trans. 2, 15, 1975

Residue 2 classified in 1

2.C **bis(N - Methylethylenediamine) copper(ii) malonate dihydrate**

$C_3H_2O_4^{2-}$, $C_6H_{20}CuN_4^{2+}$, $2H_2O$

For complete entry see 76.30

2.5 **Lithium hydrogen maleate dihydrate**

$C_4H_3O_4^-$, Li^+ , $2H_2O$

M.P.Gupta, S.M.Prasad, T.N.P.Gupta

Acta Crystallogr., Sect. B, **31,** 37, 1975

Residue 1 also classified in 1

2.6 **Sodium hydrogen maleate trihydrate**
$C_4H_3O_4^-$, Na^+ , $3H_2O$
M.P.Gupta, B.Yadav *Cryst. Struct. Commun.*, **3**, 595, 1974
Residue 1 also classified in 1

2.C **Imidazolium maleate**
$C_4H_3O_4^-$, $C_3H_5N_2^+$
For complete entry see 32.1

2.C **(+) - Chloropheniramine maleate (absolute configuration)**
(+) - S - 1 - (p - Chlorophenyl) - 1 - (2 - pyridyl) - 3,N,N - dimethylpropylamine maleate
$C_4H_3O_4^-$, $C_{16}H_{20}ClN_2^+$
For complete entry see 33.46

2.7 **Ammonium tartrate (neutron study)**
$C_4H_4O_6^{2-}$, $2H_4N^+$
W.M.Padmanabhan, V.S.Yadava, V.K.Wadhawan
Acta Crystallogr., Sect. A, **31,** S122, 1975

2.C **bis(N - Methylethylenediamine) copper(ii) D - tartrate dihydrate**
$C_4H_4O_6^{2-}$, $C_6H_{20}CuN_4^{2+}$, $2H_2O$
For complete entry see 76.31

2.8 **(+) - 2 - Hydroxypropyl - 1 - ammonium (+) - (hydrogen tartrate) dihydrate**
$C_4H_5O_6^-$, $C_3H_{10}NO^+$, $2H_2O$
S.Larsen *Acta Crystallogr., Sect. A,* **31,** S168, 1975
Residue 1 also classified in 1; residue 2 classified in 3

2.9 **(−) - 2 - Hydroxypropyl - 1 - ammonium (+) - (hydrogen tartrate) dihydrate**
$C_4H_5O_6^-$, $C_3H_{10}NO^+$, $2H_2O$
S.Larsen *Acta Crystallogr., Sect. A,* **31,** S168, 1975
Residue 1 also classified in 1; residue 2 classified in 3

2.C **(+) - ((−) - 1 - Methyl - 3 - ethyl - 3 - benzoylpiperidine (+) - R,R - bitartrate)**
$C_4H_5O_6^-$, $C_{15}H_{22}NO^+$
For complete entry see 33.43

2.C **Dextromoramide bitartrate**
$C_4H_5O_6^-$, $C_{25}H_{33}N_2O_2^+$
For complete entry see 40.24

2.10 **Calcium hydrogen citrate trihydrate**
$C_6H_6O_7^{2-}$, Ca^{2+} , $3H_2O$
B.Sheldrick *Acta Crystallogr., Sect. B,* **30,** 2056, 1974

2.C **Potassium trihydrogen bis(O,O′ - catechol diacetate)**
$C_{10}H_9O_6^-$, $C_{10}H_{10}O_6$, K^+
For complete entry see 17.17

2.C **Sodium magnesium ethylenediaminetetra - acetate tetrahydrate**
$C_{10}H_{12}N_2O_8^{4-}$, $2Na^+$, Mg^{2+} , $4H_2O$
For complete entry see 48.47

ALIPHATIC AMINES

3.1 **Methylammonium tetrachloromanganese(ii) (neutron study, at 404°K)**
$2CH_6N^+$, Cl_4Mn^{2-}
G.Heger, D.Mullen, K.Knorr *Acta Crystallogr., Sect. A*, **31**, S189, 1975

3.C **Aminomethyl - phosphonic acid (β form)**
CH_6NO_3P
For complete entry see 64.2

3.C **Dimethylammonium trichloro - tris(dimethylsulfoxide) ruthenium(ii)**
$C_2H_8N^+$, $C_6H_{18}Cl_3O_3RuS_3^-$
For complete entry see 85.6

3.2 **Ethylenediammonium tetrachlorocuprate (neutron study)**
$C_2H_{10}N_2^{2+}$, Cl_4Cu^{2-}
K.Tichy, J.Benes, H.Arend *Acta Crystallogr., Sect. A*, **31**, S84, 1975

3.3 **Ethylenediammonium tetrachloromanganate (neutron study)**
$C_2H_{10}N_2^{2+}$, Cl_4Mn^{2-}
K.Tichy, J.Benes, H.Arend *Acta Crystallogr., Sect. A*, **31**, S84, 1975

3.C **Ethylenediammonium bis(cis - (ethylenediamine - disulfito - aurate(iii)))**
$C_2H_{10}N_2^{2+}$, $2C_2H_8AuN_2O_6S_2^-$
For complete entry see 76.1

3.C **1 - n - Propyl - 3,5 - dicyano - 4 - phenyl - 6 - hydroxy - pyrid - 2 - one n - propylamine**
C_3H_9N , $C_{16}H_{13}N_3O_2$
For complete entry see 33.45

3.C **(+) - 2 - Hydroxypropyl - 1 - ammonium (+) - (hydrogen tartrate) dihydrate**
$C_3H_{10}NO^+$, $C_4H_5O_6^-$, $2H_2O$
For complete entry see 2.8

3.C **(−) - 2 - Hydroxypropyl - 1 - ammonium (+) - (hydrogen tartrate) dihydrate**
$C_3H_{10}NO^+$, $C_4H_5O_6^-$, $2H_2O$
For complete entry see 2.9

3.C **1,1,1 - Trimethylhydrazinium 3 - carbomethoxy - 5 - pyrazolecarboxylate**
$C_3H_{11}N_2^+$, $C_6H_5N_2O_4^-$
For complete entry see 32.5

3.4 **2 - Aminoethyl - isothiouronium bromide hydrobromide**
$C_3H_{11}N_3S^{2+}$, $2Br^-$
K.Vijayan, A.Mani, B.M.Vedavathi, S.Ramaseshan
Acta Crystallogr., Sect. A, **31,** S58, 1975
Residue 1 also classified in 8

3.C **DL - γ - Amino - β - hydroxybutyric acid**
$C_4H_9NO_3$
For complete entry see 1.15

3.5 **N - (2 - Hydroxyethyl)taurine**
$C_4H_{11}NO_4S$
N.Galesic, M.Herceg, B.Matkovic, M.Sljukic, D.Trupcevic, B.Zelenko
Croat. Chem. Acta, **46,** 97, 1974
Also classified in 11

3.C **Sodium tetramethylammonium bis(2 - aminoethylphospho)pentamolybdate pentahydrate**
$C_4H_{12}N^+$, $C_4H_{14}Mo_5N_2O_{21}P_2^{2-}$, Na^+ , $5H_2O$
For complete entry see 64.8

3.C **Tetramethylammonium 1,1,2,4,5,5 - hexacyanopentadienide**
$C_4H_{12}N^+$, $C_{11}HN_6^-$
For complete entry see 12.4

3.C **Tetramethylammonium cis - acetyl - benzoyl - tetracarbonyl - manganate(i)**
$C_4H_{12}N^+$, $C_{13}H_8MnO_6^-$
For complete entry see 71.31

3.C **t - Butylammonium 3,6,9,12,15 - pentaoxa - bicyclo(15.3.1)heneicosa - 1(21),17,19 - triene - 21 - carboxylate**
$C_4H_{12}N^+$, $C_{17}H_{23}O_7^-$
For complete entry see 38.55

3.C **Tetramethylammonium tetraphenylborate**
$C_4H_{12}N^+$, $C_{24}H_{20}B^-$
For complete entry see 62.13

3.6 **Tetramethylammonium tetrachlorocuprate(ii)**
$2C_4H_{12}N^+$, Cl_4Cu^{2-}
R.Clay, J.Murray-Rust, P.Murray-Rust
Acta Crystallogr., Sect. B, **31,** 289, 1975

3.7 **Tetramethylammonium tetrachloroferrate(ii)**
$2C_4H_{12}N^+$, Cl_4Fe^{2-}
J.W.Lauher, J.A.Ibers *Inorg. Chem.,* **14,** 348, 1975

3.8 **Tetramethylammonium hydroxo - triperoxo - vanadium(v) tetrahydrate**
$2C_4H_{12}N^+$, HO_7V^{2-} , $4H_2O$
R.E.Drew, F.W.B.Einstein, J.S.Field, D.Begin
Acta Crystallogr., Sect. A, **31,** S135, 1975

3.9 **Tetramethylammonium trithiocarbonate tetrahydrate**
$2C_4H_{12}N^+$, CS_3^{2-} , $4H_2O$
M.Robineau, D.Zins, M.-C.Perucaud *Rev. Chim. Miner.,* **11,** 229, 1974

3.C **bis(Tetramethylammonium) 4,4′ - commo - bis(decahydro - 1,6 - dimethyl - 1,6 - dicarba - 4 - titana - closo - tridecaborate) acetone solvate (at −160°C)**
$2C_4H_{12}N^+$, $C_8H_{32}B_{20}Ti^{2-}$, $2C_3H_6O$
For complete entry see 71.8

3.10 **Tetramethylammonium carbido - hexadecacarbonyl - hexaferrate**
$2C_4H_{12}N^+$, $C_{17}Fe_6O_{16}^{2-}$
M.R.Churchill, J.Wormald *J. Chem. Soc., Dalton Trans.,* 2410, 1974

3.C **bis(Tetramethylammonium) tetrakis(thiophenolato - iron sulfide)**
$2C_4H_{12}N^+$, $C_{24}H_{20}Fe_4S_8^{2-}$
For complete entry see 85.42

3.C **2 - Ethoxyethylammonium 5,5 - diethylbarbiturate**
$C_4H_{12}NO^+$, $C_8H_{11}N_2O_3^-$
For complete entry see 43.2

3.C **2 - Dimethylamino - ethylammonium 5,5 - diethylbarbiturate**
$C_4H_{13}N_2^+$, $C_8H_{11}N_2O_3^-$
For complete entry see 43.3

3.11 **Putrescine diphosphate**
$C_4H_{14}N_2^{2+}$, $2H_2O_4P^-$
N.Woo, A.Rich
Am. Cryst. Assoc., Abstr. Papers (Spring Meeting), 13, 1975

3.C **1,1′ - Binaphthyl - (2,3),(2′,3′) - bis - 18 - crown - 6 - 1,4 - diammonium - butane di(hexafluorophosphate) complex (at −160°C)**
$C_4H_{14}N_2^{2+}$, $C_{40}H_{50}O_{12}$, $2F_6P^-$
For complete entry see 60.39

3.C **Histamine hydrobromide**
2 - (4 - Imidazolyl)ethylammonium bromide
$C_5H_{10}N_3^+$, Br^-
For complete entry see 32.4

3.12 **Methyl - bis(β - chloroethyl)amine - N - oxide hydrochloride**
Nitromin
$C_5H_{12}Cl_2NO^+$, Cl^-
K.Nimgirawath, V.J.James *Cryst. Struct. Commun.,* **4,** 41, 1975
Residue 1 also classified in 10

3.13 **Cadaverine phosphate hydrate**
1,5 - Diaminopentane phosphate hydrate
$C_5H_{16}N_2^{2+}$, HO_4P^{2-} , $2.5H_2O$
D.L.Anderson
Am. Cryst. Assoc., Abstr. Papers (Spring Meeting), 14, 1975

3.C **3 - Hydroxy - 5 - (3 - aminopropyl)isoxazole monohydrate**
$C_6H_{10}N_2O_2$, H_2O
For complete entry see 40.1

3.C **4 - Amino - 3 - hydroxy - 2 - methyl - n - valeric acid hemihydrate**
$C_6H_{13}NO_3$, $0.5H_2O$
For complete entry see 1.18

3.14 **3 - Chloropropyltrimethylammonium 3 - iodopropyltrimethylammonium iodide**
$0.48C_6H_{15}ClN^+$, $0.52C_6H_{15}IN^+$, I^-
D.J.H.Mallard, D.P.Vaughan, T.A.Hamor
Acta Crystallogr., Sect. B, **30,** 2825, 1974
Residue 2 classified in 3

3.C **3 - Chloropropyltrimethylammonium 3 - iodopropyltrimethylammonium iodide**
$0.52C_6H_{15}IN^+$, $0.48C_6H_{15}ClN^+$, I^-
For complete entry see 3.14

3.C **Triethylammonium uridine - 3′ - O - thiophosphate methyl ester**
$C_6H_{16}N^+$, $C_{10}H_{14}N_2O_8PS^-$
For complete entry see 47.23

3.15 **bis(Triethylammonium chloride) uranyl chloride**
$2C_6H_{16}N^+$, Cl_2O_2U , $2Cl^-$
H.Brusset, N.-Q.Dao, F.Haffner *J. Inorg. Nucl. Chem.,* **36,** 791, 1974

3.16 **bis(Isopropylammonium)tetrachlorocuprate(ii)**
$2C_6H_{16}N^+$, Cl_4Cu^{2-}
D.N.Anderson, R.D.Willett *Inorg. Chim. Acta,* **8,** 167, 1975

3.C **Indenyl lithium tetramethylethylenediamine**
$C_6H_{16}N_2$, $C_9H_7^-$, Li^+
For complete entry see 27.2

3.C **bis(Tetramethylethylenediamine) di - lithium(i) anthracenide**
$2C_6H_{16}N_2$, $C_{14}H_{10}^{2-}$, $2Li^+$
For complete entry see 26.4

3.17 **bis(Dimethyl) - trimethine - cyanine perchlorate**
$C_7H_{15}N_2^+$, ClO_4^-
K.Sieber, L.Kutschabsky, S.Kulpe *Krist. Tech.,* **9,** 1101, 1974

3.C **Acetylcholine bromide**
$C_7H_{16}NO_2^+$, Br^-
For complete entry see 1.22

3.C **DL - Carnitine hydrochloride**
$C_7H_{16}NO_3^+$, Cl^-
For complete entry see 1.23

3.C **(3,4,6 - Trihydroxyphenyl)ethylammonium 2 - (2′ - ethylammonium) - 5 - hydroxy - 1,4 - benzoquinone dichloride**
$C_8H_{10}NO_3^+$, $C_8H_{12}NO_3^+$, $2Cl^-$
For complete entry see 17.12

3.C **(S) - (−) - α - Phenylethylammonium(R) - (−) - O - 2 - butyl - (S) - (−) - ethylphosphonothioate**
$C_8H_{12}N^+$, $C_6H_{14}O_2PS^-$
For complete entry see 64.12

3.C **Tyramine hydrochloride**
$C_8H_{12}NO^+$, Cl^-
For complete entry see 17.10

3.C **Tyramine hydrochloride**
$C_8H_{12}NO^+$, Cl^-
For complete entry see 17.11

3.C **(3,4,6 - Trihydroxyphenyl)ethylammonium 2 - (2′ - ethylammonium) - 5 - hydroxy - 1,4 - benzoquinone dichloride**
$C_8H_{12}NO_3^+$, $C_8H_{10}NO_3^+$, $2Cl^-$
For complete entry see 17.12

3.18 **Tetraethylammonium dichlorobromate**
$C_8H_{20}N^+$, $BrCl_2^-$
W.Gabes, K.Olie *Cryst. Struct. Commun.*, **3**, 753, 1974

3.19 **Tetraethylammonium aluminium hydride**
$C_8H_{20}N^+$, H_4Al^-
K.N.Semenenko, A.L.Dorosinskii, E.B.Lobkovskii
Zh. Strukt. Khim., **14**, 749, 1973

3.20 **Tetraethylammonium tribromo - dihydro - carbonyl - platinum(iv)**
$C_8H_{20}N^+$, $CH_2Br_3OPt^-$
Yu.I.Mironov, L.M.Plyasova, V.V.Bakakin *Kristallografiya*, **19**, 511, 1974

3.C **Tetraethylammonium trichloro(triethylphosphine) platinum(ii)**
$C_8H_{20}N^+$, $C_6H_{15}Cl_3PPt^-$
For complete entry see 86.2

3.C **Tetraethylammonium tris(O - ethylxanthato) tellurium(ii)**
$C_8H_{20}N^+$, $C_9H_{15}O_3S_6Te^-$
For complete entry see 70.5

3.C **Tetraethylammonium bis(2,3 - quinoxalinedithiolato) nickel(ii) dihydrate**
$C_8H_{20}N^+$, $C_{16}H_8N_4NiS_4^-$, $2H_2O$
For complete entry see 85.27

3.C **Tetraethylammonium μ - oxalato - bis(tetrachlorostannate(iv))**
$2C_8H_{20}N^+$, $C_2Cl_8O_4Sn_2^{2-}$
For complete entry see 69.1

3.21 **bis(Tetraethylammonium) di - μ - hydrido - octacarbonylditungstate**
$2C_8H_{20}N^+$, $C_8H_2O_8W_2^{2-}$
M.R.Churchill, S.W.-Y.Chang *Inorg. Chem.*, **13**, 2413, 1974

3.C **Tetraethylammonium bis(o - xylyl - α,α' - thiolato - μ - sulfido iron(iii))**
$2C_8H_{20}N^+$, $C_{16}H_{16}Fe_2S_6^{2-}$
For complete entry see 85.28

3.C **Tetraethylammonium bis(di - (p - tolylthiolato) - μ - sulfido - iron(iii))**
$2C_8H_{20}N^+$, $C_{28}H_{28}Fe_2S_6^{2-}$
For complete entry see 85.46

3.C **D - β - Tyrosine hydrobromide (absolute configuration)**
$C_9H_{12}NO_3^+$, Br^-
For complete entry see 17.13

3.C **D - β - Tyrosine hydrochloride (absolute configuration)**
$C_9H_{12}NO_3^+$, Cl^-
For complete entry see 17.14

3.C **(−) - Adrenaline**
$C_9H_{13}NO_3$
For complete entry see 17.15

3.22 **N - Methylphenethylammonium trichlorocuprate(ii)**
$C_9H_{14}N^+$, Cl_3Cu^-
R.L.Harlow, W.J.Wells III, G.W.Watt, S.H.Simonsen
Inorg. Chem., **13**, 2860, 1974

3.23 **bis(N - Methylphenethylammonium) tetrachlorocuprate(ii) (green form)**
$2C_9H_{14}N^+$, Cl_4Cu^{2-}
R.L.Harlow, W.J.Wells III, G.W.Watt, S.H.Simonsen
Inorg. Chem., **13**, 2106, 1974

3.24 **bis(N - Methylphenethylammonium) tetrachlorocuprate(ii) (yellow form)**
$2C_9H_{14}N^+$, Cl_4Cu^{2-}
R.L.Harlow, W.J.Wells III, G.W.Watt, S.H.Simonsen
Inorg. Chem., **13**, 2106, 1974

3.C **L - Phenylephrine hydrochloride**
$C_9H_{14}NO_2^+$, Cl^-
For complete entry see 17.16

3.25 **bis(Dimethyl) - pentamethine - cyanine chloride dihydrate**
$C_9H_{17}N_2^+$, Cl^- , $2H_2O$
B.Ziemer, S.Kulpe *J. Prakt. Chem.*, **317**, 185, 1975

3.26 **bis(Dimethyl) - pentamethine - cyanine perchlorate**
$C_9H_{17}N_2^+$, ClO_4^-
K.Sieber, L.Kutschabsky, S.Kulpe *Krist. Tech.*, **9**, 1111, 1974

3.C **Butyryl - thiocholine iodide**
$C_9H_{20}NOS^+$, I^-
For complete entry see 1.33

3.C **Sodium magnesium ethylenediaminetetra - acetate tetrahydrate**
$C_{10}H_{12}N_2O_8^{4-}$, $2Na^+$, Mg^{2+} , $4H_2O$
For complete entry see 48.47

3.27 **Diethyldi - isopropylammonium pentacyano - cobaltate(ii)**
$3C_{10}H_{24}N^+$, $C_5CoN_5^{3-}$
L.D.Brown, K.N.Raymond *J. Chem. Soc., Chem. Commun.*, 910, 1974

3.C **(±) - erythro - 2 - (2,5 - Dimethoxyphenyl) - 2 - hydroxy - 1 - methyl - ethylammonium chloride**
$C_{11}H_{18}NO_3^+$, Cl^-
For complete entry see 17.18

3.28 **1,1,3,3 - Tetra(dimethylamino)allyl perchlorate**
$C_{11}H_{25}N_4^+$, ClO_4^-
E.Oeser *Chem. Ber.*, **107**, 627, 1974

3.29 **Tetra - n - propylammonium dicarbonyl - tetraiodo - rhodium(iii) (form i)**
$C_{12}H_{28}N^+$, $C_2I_4O_2Rh^-$
J.J.Daly, F.Sanz, D.Forster *J. Am. Chem. Soc.*, **97**, 2551, 1975

3.30 **bis(Tetra - n - propylammonium) uranyl tetrachloride**
$2C_{12}H_{28}N^+$, $Cl_4O_2U^{2-}$
L.di Sipio, E.Tondello, G.Pelizzi, G.Ingletto, A.Montenero
Cryst. Struct. Commun., **3**, 731, 1974

3.C **2 - (4 - Ethyl - 2,5 - dimethoxyphenyl) - 1 - methyl - ethylamine**
4 - Ethyl - 2,5 - dimethoxy - amphetamine
$C_{13}H_{21}NO_2$
For complete entry see 17.21

3.C **Procaine dihydrogen orthophosphate hemihydrate**
$C_{13}H_{21}N_2O_2^+$, $H_2O_4P^-$, $0.5H_2O$
For complete entry see 13.19

3.C **8 - (4 - (Triethylammonium) - n - butyloxy) - 6 - tricarbonyl - 6 - manganadecaborane**
$C_{13}H_{35}B_9MnNO_4$
For complete entry see 62.5

3.C **2 - Diethylamino - 2',6' - acetoxylidide**
Lidocaine
$C_{14}H_{22}N_2O$
For complete entry see 16.10

3.C **Lidocaine bis(p - nitrophenyl) phosphate**
$C_{14}H_{23}N_2O^+$, $C_{12}H_8N_2O_8P^-$
For complete entry see 16.11

3.C **Sodium cytidine - 5' - diphosphocholine tetrahydrate**
$C_{14}H_{25}N_4O_{11}P_2^-$, Na^+ , $4H_2O$
For complete entry see 47.35

3.C **Succinyl choline iodide**
$C_{14}H_{30}N_2O_4^{2+}$, $2I^-$
For complete entry see 1.35

3.31 **Succinylcholine picrate**
$C_{14}H_{30}N_2O_4^{2+}$, $2C_6H_2N_3O_7^-$
B.Jensen *Acta Chem. Scand. Ser. B*, **29**, 115, 1975
Residue 2 classified in 15, 6

3.C **(+) - 1 - Isopropylamino - 2 - (β - naphthyl) - ethan - 2 - ol hydrochloride**
Pronethanol
$C_{15}H_{20}NO^+$, Cl^-
For complete entry see 24.8

3.C **(+) - Chloropheniramine maleate (absolute configuration)**
(+) - S - 1 - (p - Chlorophenyl) - 1 - (2 - pyridyl) - 3,N,N - dimethylpropylamine maleate
$C_{16}H_{20}ClN_2^+$, $C_4H_3O_4^-$
For complete entry see 33.46

3.C **Sodium tetrabutylammonium tetrakis(β - mercapto - propionato iron sulfide) N - methylpyrrolidone solvate**
$C_{16}H_{36}N^+$, $C_{12}H_{16}Fe_4O_8S_8^{6-}$, $5Na^+$, $5C_5H_9NO$
For complete entry see 85.18

3.C **10 - (3 - (Dimethylamino)propyl) - 2 - (trifluoromethyl) - phenothiazine hydrochloride**
$C_{18}H_{20}F_3N_2S^+$, Cl^-
For complete entry see 41.25

3.C **2 - (3 - (7 - Chloro - 2 - methoxy - 10 - (benzo(b) - 1,5 - naphthyridinyl)amino) propylamino)ethanol**
$C_{18}H_{21}ClN_4O_2$
For complete entry see 36.21

3.C **(−) - 1 - t - Butylamino - 3 - (2 - cyclopentyl - phenoxy) - propan - 2 - ol methyl - sulfonate (absolute configuration)**
$C_{18}H_{30}NO_2^+$, $CH_3O_3S^-$
For complete entry see 20.18

3.C **Diethyl - (2 - hydroxyethyl) - methylammonium α - cyclopentyl - 2 - thienylglycollate bromide**
Penthienate bromide
$C_{18}H_{30}NO_3S^+$, Br^-
For complete entry see 39.31

3.C **(−) - erythro - 1′ - (2,5 - Dimethoxyphenyl) - 3′ - diethylaminobutyl acetate hydrobromide (absolute configuration)**
$C_{18}H_{30}NO_4^+$, Br^-
For complete entry see 17.32

3.32 **p - Bromophenyldiphenylcarbinyl difluoroamine**
$C_{19}H_{14}BrF_2N$
J.R.Surles, C.L.Bumgardner, J.Bordner *J. Fluorine Chem.*, **5**, 467, 1975

3.C **5 - (3 - Dimethylaminopropyl) - 10,11 - dihydro - 5H - dibenzo(b,f)azepine hydrochloride**
Imipramine hydrochloride
$C_{19}H_{25}N_2^+$, Cl^-
For complete entry see 36.22

3.C **Fluorenyl - potassium - tetramethylethylenediamine**
$(C_{19}H_{25}N_2^-)_n$, nK^+
For complete entry see 67.6

3.C **5 - (3 - (Dimethylamino)propyl) - 6,7,8,9,10,11 - hexahydro - 5H - cyclo - oct(b)indole hydrochloride monohydrate**
Iprindole
$C_{19}H_{29}N_2^+$, Cl^- , H_2O
For complete entry see 36.23

3.33 **S - (2 - Diethylaminoethyl) - diphenylthioacetate hydrochloride monohydrate**
Thiphenamil hydrochloride
$C_{20}H_{26}NOS^+$, Cl^- , H_2O
J.J.Guy, T.A.Hamor *Acta Crystallogr., Sect. B,* **30**, 2277, 1974
Residue 1 also classified in 11

3.C **2 - Diethylaminoethyl - benzilate hydrochloride**
Benactyzine hydrochloride
$C_{20}H_{26}NO_3^+$, Cl^-
For complete entry see 1.37

3.C **β - Diethylaminopropyl α - diphenylacetate hydrobromide**
$C_{21}H_{28}NO_2^+$, Br^-
For complete entry see 1.38

3.34 **N - (3 - Diethylaminopropyl) - 2,2 - diphenylacetamide hydrochloride**
Arpenal hydrochloride
$C_{21}H_{29}N_2O^+$, Cl^-
R.L.Avoyan, E.R.Arakelova, E.G.Arutunian *Arm. Khim. Zh.*, **26**, 76, 1973

3.C **(αS(R,R)) - α - ((Di - s - butylamino)methyl) - 1 - (2 - chlorobenzyl) - 1H - 2 - pyrrole - methanol p - hydroxybenzoate (absolute configuration)**
(S(R,R)) - Viminol
$C_{21}H_{32}ClN_2O^+$, $C_7H_5O_3^-$
For complete entry see 32.39

3.35 **N,N - Diethyl - O - diphenylethoxyacetyl - propanolamine hydrobromide**
$C_{23}H_{32}NO_3^+$, Br^-
R.L.Avoyan, E.R.Arakelova, A.H.Avetissian, E.G.Arutunian
Arm. Khim. Zh., **25**, 702, 1972

3.C **trans - 1 - ((p - Diethylaminoethoxy)phenyl) - 1,2 - diphenyl - 2 - chloroethylene**
Clomiphene
$C_{26}H_{29}ClNO^+$, Cl^-
For complete entry see 17.34

3.C **N,N′ - bis((2 - Hydroxylato - 5 - methylphenyl)phenylmethylene) - 4 - azaheptane - 1,7 - diamine copper(ii)**
$C_{34}H_{35}CuN_3O_2$
For complete entry see 84.76, cross-reference should be in Class 83

3.C **1 - (π - Cyclopentadienyl) - 1 - triphenylphosphine - 2 - phenyl - 3,4,5 - tri(methoxycarbonyl) - 1 - cobaltacyclopent - 2 - ene methylene dichloride solvate**
$C_{39}H_{36}CoO_6P$, $0.25CH_2Cl_2$
For complete entry see 71.114, cross-reference should be in Class 73

3.C **tris(Methyldiphenylsilylmethyl)amine**
$C_{42}H_{45}NSi_3$
For complete entry see 63.20

ALIPHATIC (N AND S) COMPOUNDS

4.1 **bis(Methanosulfonyl)imide monohydrate**
$C_2H_7NO_4S_2$, H_2O
R.Attig, D.Mootz *Acta Crystallogr., Sect. B,* **31,** 1212, 1975

4.C **Methanol - iron oxalate monohydrate**
$(C_3H_4FeO_5)_n$, nH_2O
For complete entry see 81.5, cross-reference should be in Class 84

4.2 **Methyl sulfimide**
$C_3H_9N_3O_6S_3$
A.C.Hazell *Acta Crystallogr., Sect. B,* **30,** 2724, 1974

4.3 **bis(Dimethylthionium)amino bromide monohydrate (at −70°C)**
$C_4H_{12}NS_2^+$, Br^- , H_2O
A.M.Griffin, G.M.Sheldrick *Acta Crystallogr., Sect. B,* **31,** 893, 1975

4.4 **N - (Phenylsulfonyl)thiopropionamide**
$C_9H_{11}NO_2S_2$
W.Walter, J.Holst, A.Rohr *Justus Liebigs Ann. Chem.*, 54, 1975

4.5 **Di - t - butyl - 1,2 - (hexasulfane - 1,6 - diyl) - 1,2 - hydrazinedicarboxylate**
$C_{10}H_{18}N_2O_4S_6$
K.-H.Linke, H.G.Kalker, B.Engelen, J.Lex
Z. Naturforsch., Teil B, **29,** 130, 1974
Also classified in 9

4.6 **Benzylidenimine disulfide**
$C_{14}H_{12}N_2S_2$
J.C.Barrick, C.Calvo, F.P.Olsen *Can. J. Chem.*, **52,** 2985, 1974

4.7 **bis(p - Tolylsulfonyl)sulfur di - imide**
$C_{14}H_{14}N_2O_4S_3$
A.Gieren, F.Pertlik *Eur. Cryst. Meeting,* 303, 1974

4.8 **N - p - Tolylsulfonyl - benzylamine**
$C_{14}H_{15}NO_2S$
T.S.Cameron, K.Prout, B.Denton, R.Spagna, E.White
J. Chem. Soc., Perkin Trans. 2, 176, 1975

4.C **Tetrasulfur - tetranitrile - bis(norbornadiene)**
$C_{14}H_{16}N_4S_4$
For complete entry see 41.21

4.C **1 - Benzenesulfenyl - 2,2,6,6 - tetramethyl - 4 - oxopiperidine**
$C_{15}H_{21}NOS$
For complete entry see 33.39

4.C **1 - Benzenesulfinyl - 2,2,6,6 - tetramethyl - 4 - oxopiperidine**
$C_{15}H_{21}NO_2S$
For complete entry see 33.40

4.C **1 - Benzenesulfonyl - 2,2,6,6 - tetramethyl - 4 - oxopiperidine**
$C_{15}H_{21}NO_3S$
For complete entry see 33.41

4.9 **Potassium sulfur - tris(p - tolylsulfonylimide) monohydrate acetonitrile solvate**
$C_{21}H_{21}N_3O_6S_4^{2-}$, $2K^+$, H_2O , C_2H_3N
A.Gieren, P.Narayanan *Acta Crystallogr., Sect. A,* **31,** S120, 1975

4.C **S(+) - N - Phthalimido - p - tolyl - α - naphthyl sulfoximide (absolute configuration)**
$C_{25}H_{18}N_2O_3S$
For complete entry see 35.36, cross-reference should be in Class 24

ALIPHATIC MISCELLANEOUS

5.1 **Iodoform (neutron study)**
CHI_3
Y.Iwata *Annu. Rep. Res. React. Inst., Kyoto Univ.*, **7**, 87, 1974

5.2 **Deuteromethane (cubic form,neutron study, at 30°K)**
CD_4
E.Arzi, E.Sandor *Acta Crystallogr., Sect. A*, **31**, S188, 1975

5.3 **Deuteromethane (tetragonal form,neutron study, at 4.2°K)**
CD_4
E.Arzi, E.Sandor *Acta Crystallogr., Sect. A*, **31**, S188, 1975

5.4 **Calcium nitrate methanol solvate**
$4CH_4O$, Ca^{2+} , $2NO_3^-$
A.Leclaire *Acta Crystallogr., Sect. B*, **30**, 2259, 1974

5.5 **Deuteroacetylene (orthorhombic form, neutron study, at 4.2°K)**
C_2D_2
H.K.Koski, E.Sandor *Acta Crystallogr., Sect. B*, **31**, 350, 1975

5.6 **Deuteroacetylene (neutron study, at 4.2°K)**
C_2D_2
H.K.Koski *Acta Crystallogr., Sect. B*, **31**, 933, 1975

5.7 **Deuteroacetylene (neutron study, at 77°K, isotropic refinement)**
C_2D_2
H.K.Koski *Cryst. Struct. Commun.*, **4**, 337, 1975

5.8 **Deuteroacetylene (neutron study, at 77°K, anisotropic refinement)**
C_2D_2
H.K.Koski *Cryst. Struct. Commun.*, **4**, 337, 1975

5.9 **Deuteroacetylene (neutron study, at 109°K, anisotropic refinement)**
C_2D_2
H.K.Koski *Cryst. Struct. Commun.*, **4**, 337, 1975

5.10 **Deuteroacetylene (neutron study, at 109°K, isotropic refinement)**
C_2D_2
H.K.Koski *Cryst. Struct. Commun.*, **4**, 343, 1975

5.C **Tri - o - thymotide ethanol clathrate**
C_2H_6O , $C_{33}H_{36}O_6$
For complete entry see 61.3

5.C **Fluorescein acetone**
C_3H_6O , $C_{20}H_{12}O_5$
For complete entry see 38.61

5.C **Cyclohexa - amylose 1 - propanol hydrate**
α - Cyclodextrin 1 - propanol hydrate
C_3H_8O , $C_{36}H_{60}O_{30}$, $4.8H_2O$
For complete entry see 61.6

5.11 **meso - (2S,4S,6R) - 2,4,6 - Heptanetriol**
$C_7H_{16}O_3$
S.Kuribayashi *Bull. Chem. Soc. Jpn.*, **47,** 545, 1974

5.C **cis - bis(bis - (Trifluoromethyl) - ethylene - 1,2 - dithiolato) nickel - phenothiazine**
$C_8F_{12}NiS_4$, $C_{12}H_9NS$
For complete entry see 60.9, cross-reference should be in Class 85

5.12 **Deutero - 2,5 - dimethyl - 3 - hexyl - 2,5 - diol**
$C_8H_2D_{12}O_2$
R.B.Helmholdt *Acta Crystallogr., Sect. A,* **31,** S222, 1975

5.13 **trans - 2,5 - Dimethyl - 3 - hexene - 2,5 - diol hemihydrate (at −160°C)**
$C_8H_{16}O_2$, $0.5H_2O$
A.F.J.Ruysink, A.Vos *Acta Crystallogr., Sect. B,* **30,** 1997, 1974

5.14 **2,5 - Dimethyl - 3 - hexene - 2,5 - diol (mixture of cis and trans forms, at −160°C)**
$C_8H_{16}O_2$, $0.5C_8H_{16}O_2$
A.F.J.Ruysink, A.Vos *Acta Crystallogr., Sect. B,* **30,** 1997, 1974

5.15 **Pentamethylglycerol (at −130°C)**
2,3,4 - Trimethylpentane - 2,3,4 - triol
$C_8H_{18}O_3$
S.B.Redjeb, Y.L.Pascal, C.Bois *Acta Crystallogr., Sect. B,* **30,** 2225, 1974

5.16 **Tetra - acetylethane (neutron study)**
$C_{10}H_{14}O_4$
L.F.Power, K.E.Turner, F.H.Moore *J. Cryst. Mol. Struct.*, **5,** 59, 1975

5.17 **2,7 - Dimethyl - 2,7 - octanediol tetrahydrate**
$C_{10}H_{22}O_2$, $4H_2O$
D.Mastropaolo, G.A.Jeffrey
Am. Cryst. Assoc., Abstr. Papers (Summer Meeting), 210, 1974

5.C **Tri - o - thymotide cetyl alcohol clathrate**
$C_{16}H_{34}O$, $0.2C_{33}H_{36}O_6$
For complete entry see 61.4

5.18 **1 - (p - Bromophenyl) - 6 - benzoyl - hexa - 1,3,5 - triene**
$C_{19}H_{15}BrO$
L.G.Vorontsova, A.I.Isakova *Zh. Strukt. Khim.*, **15**, 99, 1974

5.C **2,4 - Hexadiynylene dibenzoate (form A)**
$C_{20}H_{14}O_4$
For complete entry see 13.26

5.C **2,4 - Hexadiynylene dibenzoate (form B)**
$C_{20}H_{14}O_4$
For complete entry see 13.27

5.19 **12 - Tricosanone**
$C_{23}H_{46}O$
V.Malta, G.Cojazzi, R.Zannetti, L.Amati
Gazz. Chim. Ital., **104**, 921, 1974

ENOLATES (ALIPHATIC AND AROMATIC)

6.1 **Thallium hexafluoroacetylacetonate**
$C_5HF_6O_2^-$, Tl^+
S.Tachiyashiki, H.Nakayama, R.Kuroda, S.Sato, Y.Saito
Acta Crystallogr., Sect. B, **31,** 1483, 1975

6.2 **Lithium acetylacetonate**
$C_5H_7O_2^-$, Li^+
F.A.Schroder, H.P.Weber *Acta Crystallogr., Sect. B,* **31,** 1745, 1975

6.C **bis(Ethylenediamine) copper(ii) acetylacetonate dihydrate**
$2C_5H_7O_2^-$, $C_4H_{16}CuN_4^{2+}$, $2H_2O$
For complete entry see 76.8

6.C **bis(Hydronium) nitranilate tetrahydrate**
$C_6N_2O_8^{2-}$, $2H_3O^+$, $4H_2O$
For complete entry see 18.2

6.C **Cytosine picrate**
$C_6H_2N_3O_7^-$, $C_4H_6N_3O^+$
For complete entry see 44.6

6.C **Guanine picrate monohydrate**
$C_6H_2N_3O_7^-$, $C_5H_6N_5O^+$, H_2O
For complete entry see 44.11

6.C **6 - Thioguanine picrate monohydrate**
$C_6H_2N_3O_7^-$, $C_5H_6N_5S^+$, H_2O
For complete entry see 44.12

6.C **1 - Deaza - isotubercidin picrate**
$C_6H_2N_3O_7^-$, $C_{12}H_{16}N_3O_4^+$
For complete entry see 45.33

6.C **8 - Methyl - 5,5a,6,7,8,9 - hexahydro - pyrido(2,1 - b)quinazolinium picrate**
$C_6H_2N_3O_7^-$, $C_{13}H_{17}N_2^+$
For complete entry see 36.14

6.C **Succinylcholine picrate**
$2C_6H_2N_3O_7^-$, $C_{14}H_{30}N_2O_4^{2+}$
For complete entry see 3.31

6.C **Dipotassium 1,4 - benzoquinone - 2,5 - dihydroxylate**
$C_6H_2O_4^{2-}$, $2K^+$
For complete entry see 18.4

6.C **Potassium p - nitrosophenolate monohydrate**
$C_6H_4NO_2^-$, K^+ , H_2O
For complete entry see 17.3

6.C **Potassium o - nitrophenolate hemihydrate**
$C_6H_4NO_3^-$, K^+ , $0.5H_2O$
For complete entry see 15.2

6.C **bis(Hydronium) cyananilate tetrahydrate**
Cyananilic acid hexahydrate
$C_8N_2O_4^{2-}$, $2H_3O^+$, $4H_2O$
For complete entry see 18.6

NITRILES (ALIPHATIC AND AROMATIC)

7.C **Potassium cyanourea**
$C_2H_2N_3O^-$, K^+
For complete entry see 8.10

7.C **Perdeuteropyrene - tetracyanoethylene complex (at 105°K,ordered model)**
C_6N_4 , $C_{16}D_{10}$
For complete entry see 60.23

7.C **Perdeuteropyrene - tetracyanoethylene complex (at 105°K,disordered model)**
C_6N_4 , $C_{16}D_{10}$
For complete entry see 60.24

7.C **2,4,6 - Triphenylpyrylium 1,1,3,3 - tetracyanopropenide**
$C_7HN_4^-$, $C_{23}H_{17}O^+$
For complete entry see 38.67

7.1 **Tetrafluoro - terephthalonitrile**
$C_8F_4N_2$
D.Britton, J.van Rij
Am. Cryst. Assoc., Abstr. Papers (Spring Meeting), 28, 1975

7.C **bis(Hydronium) cyananilate tetrahydrate**
Cyananilic acid hexahydrate
$C_8N_2O_4^{2-}$, $2H_3O^+$, $4H_2O$
For complete entry see 18.6

7.2 **Terephthalonitrile**
$C_8H_4N_2$
D.Britton, J.van Rij
Am. Cryst. Assoc., Abstr. Papers (Spring Meeting), 28, 1975

7.C **Potassium phenyl - nitro - acetonitrile**
$C_8H_5N_2O_2^-$, K^+
For complete entry see 12.3

7.C **3,4 - Dimethoxybenzoic acid**
$C_9H_{10}O_4$
For complete entry see 13.17, cross-reference should be in Class 17

7.3 **3 - (4 - Chlorophenyl) - 2 - cyanopropenonitrile**
$C_{10}H_5ClN_2$
Y.Delugeard *Cryst. Struct. Commun.*, **4**, 289, 1975

7.C **3 - Amino - 3 - chloro - 2 - cyano - acrylic acid anilide**
$C_{10}H_8ClN_3O$
For complete entry see 16.4

7.C **Tetramethylammonium 1,1,2,4,5,5 - hexacyanopentadienide**
$C_{11}HN_6^-$, $C_4H_{12}N^+$
For complete entry see 12.4

7.C **Carbazole - 7,7,8,8 - tetracyanoquinodimethane complex**
$C_{12}H_4N_4$, $C_{12}H_9N$
For complete entry see 60.14

7.C **N - Methylphenothiazine - 7,7,8,8 - tetracyanoquinodimethane complex**
$C_{12}H_4N_4$, $C_{13}H_{11}NS$
For complete entry see 60.18

7.C **Chrysene - 7,7,8,8 - tetracyanoquinodimethane complex**
$C_{12}H_4N_4$, $C_{18}H_{12}$
For complete entry see 60.28

7.C **Tetrathiofulvalinium 7,7,8,8 - tetracyanoquinodimethanide (at 100°K)**
$C_{12}H_4N_4^-$, $C_6H_4S_4^+$
For complete entry see 60.3

7.C **bis(Benzene) chromium(1) 7,7,8,8 - tetracyano - p - quinodimethanide**
$C_{12}H_4N_4^-$, $C_{12}H_{12}Cr^+$
For complete entry see 71.23

7.C **Acridinium - bis(7,7,8,8 - tetracyanoquinodimethanide)**
$C_{12}H_4N_4^-$, $C_{13}H_{10}N^+$, $C_{12}H_4N_4$
For complete entry see 60.17

7.C **(3,3' - Diethylthiazolinocarbocyanine) - (7,7,8,8 - tetracyanoquinodimethanide) - (9 - dicyanomethylene - 2,4,7 - trinitrofluorene) complex**
$C_{12}H_4N_4^-$, $C_{13}H_{21}N_2S_2^+$, $C_{16}H_5N_5O_6$
For complete entry see 60.21

7.C **3,3' - Dimethylthiacyanine - bis(7,7,8,8 - tetracyanoquinodimethane)**
$C_{12}H_4N_4^-$, $C_{17}H_{15}N_2S_2^+$, $C_{12}H_4N_4$
For complete entry see 60.26

7.C **3,3' - Dimethylthiacarbocyanine bis(7,7,8,8 - tetracyanoquinodimethane) complex**
$C_{12}H_4N_4^-$, $C_{19}H_{17}N_2S_2^+$, $C_{12}H_4N_4$
For complete entry see 60.34

7.C **3,3′ - Diethylthiacarbocyanine - bis(7,7,8,8 - tetracyanoquinodimethane) (monoclinic form)**
$C_{12}H_4N_4^-$, $C_{21}H_{21}N_2S_2^+$, $C_{12}H_4N_4$
For complete entry see 60.35

7.C **3,3′ - Diethylthiacarbocyanine - bis(7,7,8,8 - tetracyanoquinodimethane) (triclinic form)**
$C_{12}H_4N_4^-$, $C_{21}H_{21}N_2S_2^+$, $C_{12}H_4N_4$
For complete entry see 60.36

7.C **1 - Methyl - 3,3 - dimethyl - 2 - (p - N - methyl - N - β - chloroethylstyryl)indole - bis(7,7,8,8 - tetracyanoquinodimethane)**
$C_{12}H_4N_4^-$, $C_{22}H_{26}ClN_2^+$, $C_{12}H_4N_4$
For complete entry see 60.38

7.C **(1,4 - Di - (N - pyridinium methyl)benzene) - tetra(7,7,8,8 - tetracyanoquinodimethane)**
$2C_{12}H_4N_4^-$, $C_{18}H_{18}N_2^{2+}$, $2C_{12}H_4N_4$
For complete entry see 60.29

7.C **1,1 - bis(Methylthio) - 2 - p - bromobenzoyl - 2 - cyanoethylene**
$C_{12}H_{10}BrNOS_2$
For complete entry see 11.14

7.C **1,3 - Dimethyl - 2 - (p - bromobenzoyl - cyanomethylene) - imidazolidine**
$C_{14}H_{14}BrN_3O$
For complete entry see 32.25

7.C **(3,3′ - Diethylthiazolinocarbocyanine) - (7,7,8,8 - tetracyanoquinodimethanide) - (9 - dicyanomethylene - 2,4,7 - trinitrofluorene) complex**
$C_{16}H_5N_5O_6$, $C_{12}H_4N_4^-$, $C_{13}H_{21}N_2S_2^+$
For complete entry see 60.21

7.4 **Benzoyl - cyanide - oxime (α - cyano - α - nitro - benzyl) ether**
$C_{16}H_{10}N_4O_3$
N.E.Alexandrou, P.S.Lianis *Tetrahedron Lett.*, 421, 1975
Also classified in 10

7.5 **2,3 - Dichloro - 1,1 - dicyano - 3,3 - diphenylpropene**
$C_{17}H_{10}Cl_2N_2$
R.Schlodder, J.A.Ibers *Acta Crystallogr., Sect. B,* **31,** 708, 1975

UREA COMPOUNDS
(ALIPHATIC AND AROMATIC)

8.C **Quinol - urea**
CH_4N_2O , $C_6H_6O_2$
For complete entry see 60.5

8.1 **tetrakis(Urea) calcium nitrate**
$4CH_4N_2O$, Ca^{2+} , $2NO_3^-$
L.Lebioda *Acta Crystallogr., Sect. A,* **31,** S144, 1975

8.C **Tetra(urea) hexakis(urea) - cobalt nitrate**
$4CH_4N_2O$, $C_6H_{24}CoN_{12}O_6^{2+}$, $2NO_3^-$
For complete entry see 79.12

8.C **Nickel iodide - urea clathrate**
$4CH_4N_2O$, $C_6H_{24}N_{12}NiO_6^{2+}$, $2I^-$
For complete entry see 61.1

8.2 **Aluminium hexa(urea) perchlorate**
$6CH_4N_2O$, Al^{3+} , $3ClO_4^-$
J.H.Mooy, W.Krieger, D.Heijdenrijk, C.H.Stam
Chem. Phys. Lett., **29,** 179, 1974

8.3 **hexakis(Urea) calcium bromide**
$6CH_4N_2O$, Ca^{2+} , $2Br^-$
L.Lebioda *Acta Crystallogr., Sect. A,* **31,** S144, 1975

8.C **bis(Methyl) - thiourea - S - arsenic(iii) chloride - thiourea**
CH_4N_2S , $C_3H_{10}AsN_2S^+$, Cl^-
For complete entry see 65.1

8.C **trans - Dichloro - bis(ethylenediamine) cobalt(iii) chloride tris(thiourea)**
$3CH_4N_2S$, $C_4H_{16}Cl_2CoN_4^+$, Cl^-
For complete entry see 76.6

8.C **tris(Selenourea) dibromide monohydrate**
CH_4N_2Se , $C_2H_8N_4Se_2^{2+}$, $2Br^-$, H_2O
For complete entry see 8.19

8.C **tris(Selenourea) dichloride monohydrate**
CH_4N_2Se , $C_2H_8N_4Se_2^{2+}$, $2Cl^-$, H_2O
For complete entry see 8.20

8.4 **Guanidinium dipicrylaminate**
$CH_6N_3^+$, $C_{12}H_4N_7O_{12}^-$
M.P.Gupta, B.P.Dutta *Acta Crystallogr., Sect. B,* **31,** 1272, 1975
Residue 2 classified in 15

8.5 **Guanidinium dichromate**
$2CH_6N_3^+$, $Cr_2O_7^{2-}$
A.Stepien, E.Wajsman, M.Cygler, M.J.Grabowski
Acta Crystallogr., Sect. A, **31,** S80, 1975

8.6 **Guanidinium trichromate**
$2CH_6N_3^+$, $Cr_3O_{10}^{2-}$
A.Stepien, E.Wajsman, M.Cygler, M.J.Grabowski
Acta Crystallogr., Sect. A, **31,** S80, 1975

8.7 **Guanidinium carbonate**
$2CH_6N_3^+$, CO_3^{2-}
J.M.Adams, R.W.H.Small *Acta Crystallogr., Sect. B,* **30,** 2191, 1974

8.C **Guanidinium tetramolybdo - dimethylarsinate monohydrate**
$2CH_6N_3^+$, $C_2H_7AsMo_4O_{15}^{2-}$, H_2O
For complete entry see 86.1

8.8 **Guanidinium di - μ - peroxo - bis(tricarbonato - cerate(iv)) dihydrate**
$8CH_6N_3^+$, $C_6Ce_2O_{22}^{8-}$, $2H_2O$
L.A.Butman, V.I.Sokol, M.A.Porai-Koshits, L.A.Pospelova
Eur. Cryst. Meeting, 353, 1974

8.9 **Hydroxyguanidinium sulfate monohydrate**
$2CH_6N_3O^+$, O_4S^{2-} , H_2O
I.K.Larsen *Acta Crystallogr., Sect. B,* **31,** 1626, 1975

8.10 **Potassium cyanourea**
$C_2H_2N_3O^-$, K^+
N.S.Magomedova, Z.V.Zvonkova *Zh. Strukt. Khim.,* **15,** 165, 1974
Residue 1 also classified in 7

8.11 **Potassium bis(3 - hydroxybiuret)**
$C_2H_4N_2O_3^-$, $C_2H_5N_2O_3$, K^+
I.K.Larsen *Acta Chem. Scand. Ser. A,* **28,** 787, 1974

8.C **bis(Lupetidinium) cyanoguanidine sulfate**
bis(2,6 - Dimethylpiperidinium) cyanoguanidine sulfate
$C_2H_4N_4$, $2C_7H_{16}N^+$, O_4S^{2-}
For complete entry see 33.28

8.12 **1,1′ - Azo - bis(carbamide)**
$C_2H_4N_4O_2$
D.T.Cromer, A.C.Larson *J. Chem. Phys.,* **60,** 176, 1974
Also classified in 9

8.13 **tetrakis(Biuret) strontium(ii) perchlorate**
$4C_2H_5N_3O_2$, Sr^{2+} , $2ClO_4^-$
S.Haddad, P.S.Gentile *Inorg. Chim. Acta,* **12,** 131, 1975

8.14 **3 - Hydroxybiuret**
$C_2H_5N_3O_3$
I.K.Larsen *Acta Crystallogr., Sect. A,* **31,** S53, 1975
Also classified in 10

8.15 **1 - Hydroxybiuret**
$C_2H_5N_3O_3$
I.K.Larsen *Acta Crystallogr., Sect. A,* **31,** S53, 1975
Also classified in 10

8.16 **3 - Hydroxybiuret hemihydrate**
$C_2H_5N_3O_3$, $0.5H_2O$
I.K.Larsen *Acta Crystallogr., Sect. A,* **31,** S53, 1975
Residue 1 also classified in 10

8.17 **Dithio - biurea**
$C_2H_6N_4S_2$
A.Pignedoli, G.Peyronel, L.Antolini
Acta Crystallogr., Sect. A, **31,** S107, 1975
Also classified in 9

8.18 **bis(Methylguanidinium) monohydrogen orthophosphate**
$2C_2H_8N_3^+$, HO_4P^{2-}
F.A.Cotton, V.W.Day, E.E.Hazen Junior, S.Larsen, S.T.K.Wong
J. Am. Chem. Soc., **96,** 4471, 1974

8.19 **tris(Selenourea) dibromide monohydrate**
$C_2H_8N_4Se_2^{2+}$, CH_4N_2Se , $2Br^-$, H_2O
S.Hauge, D.Opedal, J.Arskog *Acta Chem. Scand. Ser. A,* **29,** 225, 1975
Residue 1 also classified in 11; residue 2 classified in 8

8.20 **tris(Selenourea) dichloride monohydrate**
$C_2H_8N_4Se_2^{2+}$, CH_4N_2Se , $2Cl^-$, H_2O
S.Hauge, D.Opedal, J.Arskog *Acta Chem. Scand. Ser. A,* **29,** 225, 1975
Residue 1 also classified in 11; residue 2 classified in 8

8.C **2 - Aminoethyl - isothiouronium bromide hydrobromide**
$C_3H_{11}N_3S^{2+}$, $2Br^-$
For complete entry see 3.4

8.21 **Acetone semicarbazone**
$C_4H_9N_3O$
D.V.Naik, G.J.Palenik *Acta Crystallogr., Sect. B,* **30,** 2396, 1974
Also classified in 9

8.22 **Acetone thiosemicarbazide**
$C_4H_9N_3S$
G.J.Palenik, D.F.Rendle, W.S.Carter
Acta Crystallogr., Sect. B, **30,** 2390, 1974
Also classified in 9

8.C **5 - Hydroxy - 2 - formylpyridine thiosemicarbazone sesquihydrate**
$C_7H_8N_4OS$, $1.5H_2O$
For complete entry see 33.26

8.23 **Benzaldehyde semicarbazone**
$C_8H_9N_3O$
D.V.Naik, G.J.Palenik *Acta Crystallogr., Sect. B,* **30,** 2396, 1974
Also classified in 9

8.24 **S - Benzyl - isothiouronium tetrachloroaurate(iii)**
$C_8H_{11}N_2S^+$, $AuCl_4^-$
L.E.Pope, J.C.A.Boeyens *J. Cryst. Mol. Struct.,* **5,** 47, 1975

8.25 **S - Benzyl - isothiouronium hexachloroplatinate(iv)**
$2C_8H_{11}N_2S^+$, Cl_6Pt^{2-}
L.E.Pope, J.C.A.Boeyens *J. Cryst. Mol. Struct.,* **5,** 47, 1975

8.C **N,N′ - Dimethyl - N,N′ - di(2,4 - dinitrophenyl)urea**
$C_{15}H_{12}N_6O_9$
For complete entry see 15.5

8.C **N,N′ - Dimethyl - N,N′ - di(p - nitrophenyl)urea**
$C_{15}H_{14}N_4O_5$
For complete entry see 15.6

8.C **Silver lysocellin hemihydrate (absolute configuration)**
$C_{34}H_{59}O_{10}^-$, Ag^+ , $0.5H_2O$
For complete entry see 50.28, cross-reference should be in Class 38

8.C **Jujubogenin p - bromobenzoate ethyl acetate solvate (absolute configuration)**
$C_{37}H_{51}BrO_5$, $0.5C_4H_8O_2$
For complete entry see 59.23, cross-reference should be in Class 38

NITROGEN-NITROGEN COMPOUNDS (ALIPHATIC AND AROMATIC)

9.1 **Diformylhydrazine**
$C_2H_4N_2O_2$
T.Ottersen *Acta Chem. Scand. Ser. A*, **28**, 1145, 1974

9.2 **Diformylhydrazine (at −165°C)**
$C_2H_4N_2O_2$
T.Ottersen *Acta Chem. Scand. Ser. A*, **28**, 1145, 1974

9.C **1,1′ - Azo - bis(carbamide)**
$C_2H_4N_4O_2$
For complete entry see 8.12

9.C **Dithio - biurea**
$C_2H_6N_4S_2$
For complete entry see 8.17

9.C **Malonic dihydrazide**
$C_3H_8N_4O_2$
For complete entry see 1.8

9.3 **Tetraformylhydrazine**
$C_4H_4N_2O_4$
A.Hinderer, H.Hess *Chem. Ber.*, **107**, 492, 1974

9.C **Acetone semicarbazone**
$C_4H_9N_3O$
For complete entry see 8.21

9.C **Acetone thiosemicarbazide**
$C_4H_9N_3S$
For complete entry see 8.22

9.4 **1,1,5,5 - Tetramethylformazanium copper(i) bromide**
$C_5H_{13}N_4^+$, $2Cu^+$, $3Br^-$
J.R.Boehm, A.L.Balch, K.F.Bizot, J.H.Enemark
J. Am. Chem. Soc., **97**, 501, 1975

9.C **3,6 - bis(Methylthio) - 4,5 - diaza - 2,7 - dithiaocta - 3,5 - diene**
$C_6H_{12}N_2S_4$
For complete entry see 11.9

9.C **5 - Hydroxy - 2 - formylpyridine thiosemicarbazone sesquihydrate**
$C_7H_8N_4OS$, $1.5H_2O$
For complete entry see 33.26

9.C **Benzaldehyde semicarbazone**
$C_8H_9N_3O$
For complete entry see 8.23

9.C **6 - (p - Hydroxyphenylazo)uracil**
$C_{10}H_8N_4O_3$
For complete entry see 44.22

9.C **Di - t - butyl - 1,2 - (hexasulfane - 1,6 - diyl) - 1,2 - hydrazinedicarboxylate**
$C_{10}H_{18}N_2O_4S_6$
For complete entry see 4.5

9.C **4 - Bromo - 4′ - hydroxyazobenzene**
$C_{12}H_9BrN_2O$
For complete entry see 17.19

9.5 **cis - Azobenzene dioxide**
$C_{12}H_{10}N_2O_2$
D.A.Dieterich, I.C.Paul, D.Y.Curtin *J. Am. Chem. Soc.*, **96,** 6372, 1974

9.C **p - Nitro - diazoaminobenzene**
$C_{12}H_{10}N_4O_2$
For complete entry see 16.6

9.C **Diphenylformazane**
$C_{13}H_{12}N_4$
For complete entry see 16.8

9.C **trans - 2,2′ - Dicarboxyazobenzene dioxide**
$C_{14}H_{10}N_2O_6$
For complete entry see 13.22

9.C **S - Methyl - dithizone**
$C_{14}H_{14}N_4S$
For complete entry see 16.9

9.C **1,2 - Naphthoquinone 1 - (2 - nitro - 4 - chlorophenylhydrazone)**
$C_{16}H_{10}ClN_3O_3$
For complete entry see 25.4

9.C **trans - α - (4′ - Iodo - 1′ - diazacyclohexane) - β - naphthol**
$C_{16}H_{17}IN_2O$
For complete entry see 21.7

9.C **cis - α - (4′ - Iodo - 1′ - diazacyclohexane) - β - naphthol**
$C_{16}H_{17}IN_3O$
For complete entry see 21.8

9.C **Ethyl p - azoxybenzoate (high temp.form)**
$C_{18}H_{18}N_2O_5$
For complete entry see 13.25

9.C **α,α' - Azobis(4 - bromobenzaldehyde) - bis(O - carboxyoxime) diethyl ester (monoclinic form)**
$C_{20}H_{18}Br_2N_4O_6$
For complete entry see 10.6

9.C **α,α' - Azobis(4 - bromobenzaldehyde) - bis(O - carboxyoxime) diethyl ester (triclinic form)**
$C_{20}H_{18}Br_2N_4O_6$
For complete entry see 10.7

9.C **1 - (2,5 - Dichlorophenylazo) - 2 - hydroxy - 3 - naphthoic acid 4 - chloro - 2 - methoxy - anilide**
$C_{24}H_{16}Cl_3N_3O_3$
For complete entry see 24.11

9.C **Tetraphenylhydrazine**
$C_{24}H_{20}N_2$
For complete entry see 16.14

9.C **Tetraphenylhydrazine (at $-160°C$)**
$C_{24}H_{20}N_2$
For complete entry see 16.15

9.C **2,2' - bis(2 - Bromoacetylaminoanilino) - benzalazine benzene solvate**
$C_{30}H_{26}Br_2N_6O_2$, C_6H_6
For complete entry see 16.16

NITROGEN-OXYGEN COMPOUNDS (ALIPHATIC AND AROMATIC)

10.C **3 - Hydroxybiuret**
$C_2H_5N_3O_3$
For complete entry see 8.14

10.C **1 - Hydroxybiuret**
$C_2H_5N_3O_3$
For complete entry see 8.15

10.C **3 - Hydroxybiuret hemihydrate**
$C_2H_5N_3O_3$, $0.5H_2O$
For complete entry see 8.16

10.C **Methyl - bis(β - chloroethyl)amine - N - oxide hydrochloride**
Nitromin
$C_5H_{12}Cl_2NO^+$, Cl^-
For complete entry see 3.12

10.C **(Ethylenediamine) zinc(ii) benzohydroxamate benzohydroxamic acid monohydrate**
$C_7H_7NO_2$, $C_{16}H_{20}N_4O_4Zn$, H_2O
For complete entry see 76.67

10.C **N,N - Dimethyl - p - nitroso - aniline hydrochloride monohydrate**
$C_8H_{11}N_2O^+$, Cl^- , H_2O
For complete entry see 16.3

10.1 **E - Benzyl - methyl - ketoxime**
$C_9H_{11}NO$
L.D.Thompson, P.J.Nassiff, E.R.Boyko
Acta Crystallogr., Sect. A, **31,** S11, 1975

10.2 **N,O - bis(2 - Chlorobenzoyl) - hydroxylamine**
$C_{14}H_9Cl_2NO_3$
S.Gottlicher, P.Ochsenreiter *Chem. Ber.,* **107,** 398, 1974

10.3 **N,O - Dibenzoyl - hydroxylamine**
$C_{14}H_{11}NO_3$
S.Gottlicher, P.Ochsenreiter *Chem. Ber.,* **107,** 398, 1974

10.4 **N - Benzoyl - O - o - toluoyl - hydroxylamine**
$C_{15}H_{13}NO_3$
S.Gottlicher, P.Ochsenreiter *Chem. Ber.*, **107**, 398, 1974

10.5 **N - Benzoyl - O - p - toluoyl - hydroxylamine**
$C_{15}H_{13}NO_3$
S.Gottlicher, P.Ochsenreiter *Chem. Ber.*, **107**, 398, 1974

10.C **racemic - 2 - (α - p - Bromophenyl - β - nitro)ethyl - 5 - methylcyclohexanone**
$C_{15}H_{18}BrNO_3$
For complete entry see 21.6

10.C **1 - p - (Oximinoethyl) - phenoxyacetyl - piperidine**
$C_{15}H_{20}N_2O_3$
For complete entry see 33.38

10.C **Benzoyl - cyanide - oxime (α - cyano - α - nitro - benzyl) ether**
$C_{16}H_{10}N_4O_3$
For complete entry see 7.4

10.C **Ethyl p - azoxybenzoate (high temp.form)**
$C_{18}H_{18}N_2O_5$
For complete entry see 13.25

10.6 **α,α' - Azobis(4 - bromobenzaldehyde) - bis(O - carboxyoxime) diethyl ester (monoclinic form)**
$C_{20}H_{18}Br_2N_4O_6$
K.T.Go, G.Kartha, C.T.Lu
Am. Cryst. Assoc., Abstr. Papers (Summer Meeting), 255, 1974
Also classified in 9

10.7 **α,α' - Azobis(4 - bromobenzaldehyde) - bis(O - carboxyoxime) diethyl ester (triclinic form)**
$C_{20}H_{18}Br_2N_4O_6$
K.T.Go, G.Kartha, C.T.Lu
Am. Cryst. Assoc., Abstr. Papers (Summer Meeting), 255, 1974
Also classified in 9

SULPHUR AND SELENIUM COMPOUNDS

11.C **(−) - 1 - t - Butylamino - 3 - (2 - cyclopentyl - phenoxy) - propan - 2 - ol methyl - sulfonate (absolute configuration)**
$CH_3O_3S^-$, $C_{18}H_{30}NO_2^+$
For complete entry see 20.18

11.C **3 - Carboethoxy - 1,6 - dimethyl - 4 - oxo - 6,7,8,9 - tetrahydrohomopyrimidazolium methyl sulfate**
$CH_3O_4S^-$, $C_{13}H_{19}N_2O_3^+$
For complete entry see 35.18

11.C **Potassium monothioacetate**
$C_2H_3OS^-$, K^+
For complete entry see 1.4

11.C **Strontium acetate thioacetate tetrahydrate**
$C_2H_3OS^-$, $C_2H_3O_2^-$, Sr^{2+} , $4H_2O$
For complete entry see 2.3

11.1 **Nona - aquo - holmium(iii) tris(ethylsulfate) (neutron study)**
$3C_2H_5O_4S^-$, $H_{18}HoO_9^{3+}$
C.R.Hubbard, C.O.Quicksall, R.A.Jacobson
Acta Crystallogr., Sect. B, **30**, 2613, 1974

11.C **tris(Selenourea) dibromide monohydrate**
$C_2H_8N_4Se_2^{2+}$, CH_4N_2Se , $2Br^-$, H_2O
For complete entry see 8.19

11.C **tris(Selenourea) dichloride monohydrate**
$C_2H_8N_4Se_2^{2+}$, CH_4N_2Se , $2Cl^-$, H_2O
For complete entry see 8.20

11.C **Cesium dimethyldithiocarbamate**
$C_3H_6NS_2^-$, Cs^+
For complete entry see 1.7

11.2 **Methylthio - dibromo - borane trimer**
$C_3H_9B_3Br_6S_3$
F.Zettler, S.Pollitz, D.Forst, H.Hess *Eur. Cryst. Meeting*, 295, 1974

11.3 **Methylthio - dichloro - borane trimer**
$C_3H_9B_3Cl_6S_3$
F.Zettler, S.Pollitz, D.Forst, H.Hess *Eur. Cryst. Meeting,* 295, 1974

11.4 **Selenium di(methylxanthate)**
$C_4H_6O_2S_4Se$
N.J.Brondmo, S.Esperas, S.Husebye
Acta Chem. Scand. Ser. A, **29,** 93, 1975

11.5 **Sulfur di(methylxanthate)**
$C_4H_6O_2S_5$
N.J.Brondmo, S.Esperas, S.Husebye
Acta Chem. Scand. Ser. A, **29,** 93, 1975

11.C **N - (2 - Hydroxyethyl)taurine**
$C_4H_{11}NO_4S$
For complete entry see 3.5

11.6 **Thallium(i) diethyldithiocarbamate**
$C_5H_{10}NS_2^-$, Tl^+
H.Prigzkow, P.Jennische *Acta Chem. Scand. Ser. A,* **29,** 60, 1975

11.7 **Sodium benzene - hexasulfonate octahydrate**
$C_6O_{18}S_6^{6-}$, $6Na^+$, $8H_2O$
L.A.Chetkina, A.N.Sobolev *Acta Crystallogr., Sect. A,* **31,** S125, 1975

11.C **Picryl sulfonic acid tetrahydrate (neutron study)**
$C_6H_2N_3O_9S^-$, $H_5O_2^+$, $2H_2O$
For complete entry see 15.1

11.C **Sodium sulfanilate dihydrate**
$C_6H_6NO_3S^-$, Na^+, $2H_2O$
For complete entry see 16.2

11.8 **S,S' - Diethyl - dithio - oxalate (at −60°C)**
$C_6H_{10}O_2S_2$
G.Kiel, M.Drager, U.Reuter *Chem. Ber.,* **107,** 1483, 1974
Also classified in 1

11.C **racemic - Thiodilactic acid**
$C_6H_{10}O_4S$
For complete entry see 1.17

11.9 **3,6 - bis(Methylthio) - 4,5 - diaza - 2,7 - dithiaocta - 3,5 - diene**
$C_6H_{12}N_2S_4$
A.M.M.Lanfredi, A.Tiripicchio, M.T.Camellini
Cryst. Struct. Commun., **4,** 141, 1975
Also classified in 9

11.C **S - Methyl - L - methionine chloride hydrochloride**
$C_6H_{15}NO_2S^{2+}$, $2Cl^-$
For complete entry see 48.29

11.C **o - Sulfobenzoic acid trihydrate**
$C_7H_6O_5S$, $3H_2O$
For complete entry see 13.6

11.C **5 - Sulfosalicylic acid dihydrate**
$C_7H_6O_6S$, $2H_2O$
For complete entry see 13.7

11.C **5 - Sulphosalicylic acid trihydrate**
$C_7H_6O_6S$, $3H_2O$
For complete entry see 13.8

11.10 **Deuteronium p - toluenesulfonate (neutron study)**
$C_7H_7O_3S^-$, D_3O^+
J.E.Finholt, J.M.Williams *J. Chem. Phys.*, **59,** 514, 1973

11.C **3,3′ - Diethyl - benzothiacarbocyanine p - toluenesulfonate**
$C_7H_7O_3S^-$, $C_{21}H_{21}N_2S_2^+$
For complete entry see 41.27

11.C **5,6 - Dichloro - 1,3 - diethyl - 2 - ((5,6 - dichloro - 1,3 - diethyl - 2 - benzimidazolinylidene) - 1 - propynyl)benzimidazolium toluene - p - sulfonate hydrate**
$C_7H_7O_3S^-$, $C_{25}H_{25}Cl_4N_4^+$, $0.2H_2O$
For complete entry see 35.38

11.11 **4 - Bromophenyl - thioacetate**
C_8H_7BrOS
J.S.Cantrell, J.R.Grunwell, R.D.Sanner, J.Hudgens
Am. Cryst. Assoc., Abstr. Papers (Summer Meeting), 250, 1974

11.C **4 - Chlorophenylsulfinyl acetic acid**
$C_8H_7ClO_3S$
For complete entry see 1.24

11.C **Cyanomethyl - (2 - picolyl) - sulfone**
$C_8H_8N_2O_2S$
For complete entry see 33.29

11.C **Phenylsulfinyl acetamide**
$C_8H_9NO_2S$
For complete entry see 1.25

11.C **1,6 - Di - O - methanesulfonyl - D - mannitol**
$C_8H_{18}O_{10}S_2$
For complete entry see 45.25

11.12 Ammonium 2 - oxamoyl - 3 - methyl - benzenesulfonate
$C_9H_8NO_5S^-$, H_4N^+
M.A.Pellinghelli, A.Tiripicchio, M.T.Camellini
Cryst. Struct. Commun., **3,** 735, 1974

11.13 Thallium(i) di - isobutyldithiocarbamate
$C_9H_{18}NS_2^-$, Tl^+
H.Anacker-Eickhoff, P.Jennische, R.Hesse
Acta Chem. Scand. Ser. A, **29,** 51, 1975

11.C 3,3,3′,3′ - Tetramethyl - D - cystine dihydrochloride (absolute configuration)
D - Penicillamine disulfide dihydrochloride
$C_{10}H_{22}N_2O_4S_2^{2+}$, $2Cl^-$
For complete entry see 48.49

11.14 1,1 - bis(Methylthio) - 2 - p - bromobenzoyl - 2 - cyanoethylene
$C_{12}H_{10}BrNOS_2$
S.Abrahamsson, G.Rehnberg, T.Liljefors, J.Sandstrom
Acta Chem. Scand. Ser. B, **28,** 1109, 1974
Also classified in 7

11.C 1 - Naphthylsulfinyl acetic acid
$C_{12}H_{10}O_3S$
For complete entry see 24.7

11.15 Diphenyl disulfide
$C_{12}H_{10}S_2$
M.Sacerdoti, G.Gilli, P.Domiano *Acta Crystallogr., Sect. B*, **31,** 327, 1975

11.C 2,4 - Dinitrobenzyl p - tolyl sulfone
$C_{14}H_{12}N_2O_6S$
For complete entry see 15.4

11.C S - Methyl - dithizone
$C_{14}H_{14}N_4S$
For complete entry see 16.9

11.C anti - α - Morpholino - β - methyl - α - chloromethanesulfonyl - styrene
$C_{14}H_{18}ClNO_3S$
For complete entry see 40.9

11.C syn - α - Morpholino - β - methyl - α - chloromethanesulfonyl - styrene
$C_{14}H_{18}ClNO_3S$
For complete entry see 40.10

11.16 Methyl - (2,2 - diphenylvinyl) - sulfone
$C_{15}H_{14}O_2S$
D.Tranqui, H.Fillion *Acta Crystallogr., Sect. A*, **31,** S107, 1975

11.17 **(+) - 1 - Bromo - 2 - hydroxy - 2 - phenylethyl p - tolyl sulfoxide**
$C_{15}H_{15}BrO_2S$
F.Iwasaki, S.Mitamura, G.Tsuchihashi
Acta Crystallogr., Sect. A, **31,** S106, 1975

11.18 **2,4,6 - Trimethyl - diphenylsulfone**
$C_{15}H_{16}O_2S$
S.A.Chawdhury *Acta Crystallogr., Sect. A,* **31,** S107, 1975

11.19 **Diacetylmethylene - diphenylselenurane**
$C_{17}H_{16}O_2Se$
K.-T.H.Wei, I.C.Paul, M.-M.Y.Chang, J.I.Musher
J. Am. Chem. Soc., **96,** 4099, 1974

11.C **trans - 4 - t - Butylcyclohexyl toluene - p - sulfonate (neutron study)**
$C_{17}H_{26}O_3S$
For complete entry see 21.10

11.20 **Dicinnamyl disulfide**
$C_{18}H_{18}S_2$
J.Donohue, J.P.Chesick *Acta Crystallogr., Sect. B,* **31,** 986, 1975

11.C **2 - (Cyclohexyl - (phenyl) - acetoxy)ethyl - dimethyl - sulfonium iodide**
Hexasonium iodide
$C_{18}H_{27}O_2S^+$, I^-
For complete entry see 21.11

11.C **4' - (Acridin - 9 - ylamine)methane - sulfonanilide hydrochloride**
$C_{20}H_{18}N_3O_2S^+$, Cl^-
For complete entry see 36.25

11.C **S - (2 - Diethylaminoethyl) - diphenylthioacetate hydrochloride monohydrate**
Thiphenamil hydrochloride
$C_{20}H_{26}NOS^+$, Cl^- , H_2O
For complete entry see 3.33

11.C **S(+) - N - Phthalimido - p - tolyl - α - naphthyl sulfoximide (absolute configuration)**
$C_{25}H_{18}N_2O_3S$
For complete entry see 35.36

11.21 **hexakis(Phenylthio)ethane**
$C_{38}H_{30}S_6$
G.Roelofsen, J.A.Kanters, D.Seebach *Chem. Ber.,* **107,** 253, 1974

CARBONIUM IONS, CARBANIONS, RADICALS

12.1 **Disodium acetylide (neutron study)**
C_2^{2-} , $2Na^+$
M.Atoji *J. Chem. Phys.*, **60**, 3324, 1974

12.2 **Potassium 1,1 - dinitroethanide**
$C_2H_3N_2O_4^-$, K^+
O.V.Frank-Kamenetskaya, N.V.Grigor'eva, N.V.Margolis, I.V.Tselinskii
Kristallografiya, **18**, 178, 1973

12.C **Rubidium N,N - dimethyl - α,α - dinitroacetamide**
$C_4H_6N_3O_5^-$, Rb^+
For complete entry see 1.10

12.C **2,4,6 - Triphenylpyrylium 1,1,3,3 - tetracyanopropenide**
$C_7HN_4^-$, $C_{23}H_{17}O^+$
For complete entry see 38.67

12.C **Potassium 2,4,6 - trinitrophenyl - dinitromethanide**
$C_7H_2N_5O_{10}^-$, K^+
For complete entry see 15.3

12.3 **Potassium phenyl - nitro - acetonitrile**
$C_8H_5N_2O_2^-$, K^+
N.V.Grigor'eva, N.V.Margolis, I.V.Tselinskii, V.V.Mel'nikov,
G.V.Makarenko *Zh. Strukt. Khim.*, **15**, 724, 1974
Residue 1 also classified in 7

12.C **2,2,5,5 - Tetramethyl - 1 - aza - 3 - cyclopentanone - 3 - oxime - 1 - oxyl**
$C_8H_{15}N_2O_2$
For complete entry see 32.9

12.C **Indenyl lithium tetramethylethylenediamine**
$C_9H_7^-$, $C_6H_{16}N_2$, Li^+
For complete entry see 27.2

12.C **racemic - 2,2,5,5 - Tetramethyl - pyrrolidine - 3 - carboxamide 1 - oxyl**
$C_9H_{17}N_2O_2$
For complete entry see 32.13

12.C **2,2,5,5 - Tetramethyl - pyrrolidine - 3 - carboxamide 1 - oxyl (optically active form)**
$C_9H_{17}N_2O_2$
For complete entry see 32.14

12.C **2,2,6,6 - Tetramethyl - piperidine - 1 - oxyl (monoclinic form)**
$C_9H_{18}NO$
For complete entry see 33.32

12.C **2,2,6,6 - Tetramethyl - piperidine - 1 - oxyl (tetragonal form)**
$C_9H_{18}NO$
For complete entry see 33.33

12.C **Spiro(cyclohexane - 1,2′ - (4′,4′ - dimethyloxazolidine - N - oxyl))**
$C_{10}H_{18}NO_2$
For complete entry see 40.3

12.4 **Tetramethylammonium 1,1,2,4,5,5 - hexacyanopentadienide**
$C_{11}HN_6^-$, $C_4H_{12}N^+$
R.L.Sass, T.D.Nichols *Z. Kristallogr.*, **140,** 1, 1974
Residue 1 also classified in 7; residue 2 classified in 3

12.C **Dithieno(2,1 - b.4,5 - b′)tropylium perchlorate**
$C_{11}H_7S_2^+$, ClO_4^-
For complete entry see 39.21

12.C **1 - Bromo - 2,4,6 - tripyrrolidino - benzenium bromide trihydrate**
$C_{18}H_{27}BrN_3^+$, Br^- , $3H_2O$
For complete entry see 32.31

12.C **2,4,6 - Tripyrrolidino - benzenium bromide dihydrate**
$C_{18}H_{28}N_3^+$, Br^- , $2H_2O$
For complete entry see 32.32

12.5 **Triphenylcarbonium tridecachloro - tritellurium**
$C_{19}H_{15}^+$, $Cl_{13}Te_3^-$
B.Krebs, V.Paulat *Eur. Cryst. Meeting,* 238, 1974

12.C **1 - Methyl - 2,4,6 - tripyrrolidino - benzenium iodide**
$C_{19}H_{30}N_3^+$, I^-
For complete entry see 32.35

BENZOIC ACID DERIVATIVES

13.1 **p - Chlorobenzoic acid**
$C_7H_5ClO_2$
R.S.Miller, I.C.Paul, D.Y.Curtin *J. Am. Chem. Soc.*, **96**, 6334, 1974

13.2 **m - Chlorobenzoic acid**
$C_7H_5ClO_2$
J.Z.Gougoutas, L.Lessinger *J. Solid State Chem.*, **12**, 51, 1975

13.3 **o - Fluorobenzoic acid**
$C_7H_5FO_2$
G.Ferguson, K.M.S.Islam *Cryst. Struct. Commun.*, **4**, 389, 1975

13.4 **o - Chlorobenzamide (α form)**
C_7H_6ClNO
Y.Kato, Y.Takaki, K.Sakurai *Acta Crystallogr., Sect. B*, **30**, 2683, 1974

13.5 **o - Chlorobenzamide (β form)**
C_7H_6ClNO
Y.Kato, Y.Takaki, K.Sakurai *Acta Crystallogr., Sect. B*, **30**, 2683, 1974

13.C **Salicylic acid - cytidine complex**
$C_7H_6O_3$, $C_9H_{13}N_3O_5$
For complete entry see 60.7

13.6 **o - Sulfobenzoic acid trihydrate**
$C_7H_6O_5S$, $3H_2O$
R.Attig, D.Mootz *Eur. Cryst. Meeting*, 383, 1974
Residue 1 also classified in 11

13.7 **5 - Sulfosalicylic acid dihydrate**
$C_7H_6O_6S$, $2H_2O$
R.Attig *Acta Crystallogr., Sect. A*, **31**, S167, 1975
Residue 1 also classified in 17, 11

13.8 **5 - Sulphosalicylic acid trihydrate**
$C_7H_6O_6S$, $3H_2O$
R.Attig *Acta Crystallogr., Sect. A*, **31**, S167, 1975
Residue 1 also classified in 17, 11

13.9 **Terephthaloyl bromide**
$C_8H_4Br_2O_2$
J.Leser, D.Rabinovich *Acta Crystallogr., Sect. A,* **31,** S166, 1975

13.C **Lithium hydrogen phthalate dihydrate (neutron study)**
$C_8H_5O_4^-$, Li^+ , $2H_2O$
For complete entry see 14.3

13.C **Lithium hydrogen phthalate dihydrate**
$C_8H_5O_4^-$, Li^+ , $2H_2O$
For complete entry see 14.4

13.10 **Isophthalic acid**
$C_8H_6O_4$
J.L.Derissen *Acta Crystallogr., Sect. B,* **30,** 2764, 1974

13.11 **O - Methyl p - bromobenzamidate**
C_8H_8BrNO
B.Kolakowski *Acta Phys. Pol.,* **A46,** 373, 1974

13.12 **N - Methyl - benzamide**
C_8H_9NO
M.Tuval, L.Leiserowitz *Acta Crystallogr., Sect. A,* **31,** S176, 1975

13.13 **Trimellitic acid**
Benzene - 1,2,4 - tricarboxylic acid
$C_9H_6O_6$
F.Takusagawa, K.Hirotsu, A.Shimada
Bull. Chem. Soc. Jpn., **46,** 2960, 1973

13.14 **Hemimellitic acid dihydrate**
1,2,3 - Benzenetricarboxylic acid dihydrate
$C_9H_6O_6$, $2H_2O$
F.Mo, E.Adman *Acta Crystallogr., Sect. B,* **31,** 192, 1975

13.15 **Hemimellitic acid dihydrate**
Benzene - 1,2,3 - tricarboxylic acid dihydrate
$C_9H_6O_6$, $2H_2O$
F.Takusagawa, A.Shimada *Bull. Chem. Soc. Jpn.,* **46,** 2998, 1973

13.C **2,4 - Dinitro - 5 - ethyleneimino - benzamide**
$C_9H_8N_4O_5$
For complete entry see 32.11

13.16 **N - Methyl - 2 - (N - methylamino) - 3,5 - dinitrobenzamide (at −160°C)**
$C_9H_{10}N_4O_5$
M.Mathew, G.J.Palenik *Acta Crystallogr., Sect. B,* **30,** 2381, 1974
Also classified in 16, 15

13.17 **3,4 - Dimethoxybenzoic acid**
$C_9H_{10}O_4$
S.Swaminathan, T.M.Vimala, L.Lessinger
Also classified in 7

13.18 **3,3 - Dimethylbutanol - 3,5 - dinitrobenzoate**
$C_{13}H_{16}N_2O_6$
M.Sax, M.Rodrigues, G.Blank, M.Wood, J.Pletcher
Am. Cryst. Assoc., Abstr. Papers (Spring Meeting), 10, 1975
Also classified in 15

13.19 **Procaine dihydrogen orthophosphate hemihydrate**
$C_{13}H_{21}N_2O_2^+$, $H_2O_4P^-$, $0.5H_2O$
G.R.Freeman, C.E.Bugg *Acta Crystallogr., Sect. B,* **31,** 96, 1975
Residue 1 also classified in 16, 3

13.20 **2,4 - Dibromo - o - cresyl 3′,5′ - dinitrosalicylate**
$C_{14}H_8Br_2N_2O_7$
V.Pattabhi *Acta Crystallogr., Sect. B,* **31,** 1766, 1975
Also classified in 17, 15

13.C **3 - Oxo - 3H - 1,2 - benzoxiodol - 1 - yl m - chlorobenzoate (α form)**
$C_{14}H_8ClIO_4$
For complete entry see 42.11

13.C **3 - Oxo - 3H - 1,2 - benzoxiodol - 1 - yl m - chlorobenzoate (β form)**
$C_{14}H_8ClIO_4$
For complete entry see 42.12

13.21 **Biphenyl - 2,2′ - dicarbonyl chloride**
$C_{14}H_8Cl_2O_2$
J.Leser, D.Rabinovich *Acta Crystallogr., Sect. A,* **31,** S166, 1975

13.22 **trans - 2,2′ - Dicarboxyazobenzene dioxide**
$C_{14}H_{10}N_2O_6$
D.A.Dieterich, I.C.Paul, D.Y.Curtin *J. Am. Chem. Soc.,* **96,** 6372, 1974
Also classified in 9

13.23 **Ethylene glycol di - p - chlorobenzoate**
$C_{16}H_{12}Cl_2O_4$
F.Brisse, S.Perez *Acta Crystallogr., Sect. A,* **31,** S101, 1975

13.24 **Ethylene glycol dibenzoate**
$C_{16}H_{14}O_4$
F.Brisse, S.Perez *Acta Crystallogr., Sect. A,* **31,** S101, 1975

13.C **1 - (2 - Hydroxy - 1 - methyl)ethyl - 2 - methyl - 2 - (1 - p - bromobenzoyloxy - 2 - methoxycarbonyl)ethyl - ethylene oxide**
$C_{17}H_{21}BrO_6$
For complete entry see 38.54

13.25 **Ethyl p - azoxybenzoate (high temp.form)**
$C_{18}H_{18}N_2O_5$
W.R.Krigbaum, T.Taga *Mol. Cryst. Liq. Cryst.*, **28,** 85, 1974
Also classified in 10, 9

13.26 **2,4 - Hexadiynylene dibenzoate (form A)**
$C_{20}H_{14}O_4$
A.W.Hanson *Acta Crystallogr., Sect. B,* **31,** 831, 1975
Also classified in 5

13.27 **2,4 - Hexadiynylene dibenzoate (form B)**
$C_{20}H_{14}O_4$
A.W.Hanson *Acta Crystallogr., Sect. B,* **31,** 831, 1975
Also classified in 5

BENZOIC ACID SALTS
(AMMONIUM, IA, IIA METALS)

14.C **bis(1,3 - Propanediamine) copper(ii) m - chlorobenzoate**
$2C_7H_4ClO_2^-$, $C_6H_{20}CuN_4^{2+}$
For complete entry see 83.13

14.C **(αS(R,R)) - α - ((Di - s - butylamino)methyl) - 1 - (2 - chlorobenzyl) - 1H - 2 - pyrrole - methanol p - hydroxybenzoate (absolute configuration)**
(S(R,R)) - Viminol
$C_7H_5O_3^-$, $C_{21}H_{32}ClN_2O^+$
For complete entry see 32.39

14.1 **Calcium phthalate monohydrate**
$C_8H_4O_4^{2-}$, Ca^{2+} , H_2O
M.P.Gupta, R.P.Sinha *Cryst. Struct. Commun.*, **4,** 207, 1975

14.2 **Diammonium phthalate**
$C_8H_4O_4^{2-}$, $2H_4N^+$
R.A.Smith *Acta Crystallogr., Sect. B,* **31,** 1773, 1975

14.3 **Lithium hydrogen phthalate dihydrate (neutron study)**
$C_8H_5O_4^-$, Li^+ , $2H_2O$
H.Bartl, H.Kuppers *Acta Crystallogr., Sect. A,* **31,** S174, 1975
Residue 1 also classified in 13

14.4 **Lithium hydrogen phthalate dihydrate**
$C_8H_5O_4^-$, Li^+ , $2H_2O$
W.Gonschorek, H.Kuppers *Acta Crystallogr., Sect. B,* **31,** 1068, 1975
Residue 1 also classified in 13

14.C **Sodium 5 - bromo - lasalocid (form i)**
$C_{34}H_{52}BrO_8^-$, Na^+
For complete entry see 50.26

14.C **Sodium 5 - bromo - lasalocid (form ii)**
$C_{34}H_{52}BrO_8^-$, Na^+
For complete entry see 50.27

BENZENE NITRO COMPOUNDS

15.C **bis(Hydronium) nitranilate tetrahydrate**
$C_6N_2O_8^{2-}$, $2H_3O^+$, $4H_2O$
For complete entry see 18.2

15.C **Fluoroanthene - picryl bromide complex**
$C_6H_2BrN_3O_6$, $C_{16}H_{10}$
For complete entry see 60.22

15.C **Pyrene - picryl bromide complex**
$C_6H_2BrN_3O_6$, $1.5C_{16}H_{10}$
For complete entry see 60.25

15.C **N,N - Difluoro - 2,4,6 - trinitro - aniline**
$C_6H_2F_2N_4O_6$
For complete entry see 16.1

15.C **Cytosine picrate**
$C_6H_2N_3O_7^-$, $C_4H_6N_3O^+$
For complete entry see 44.6

15.C **Guanine picrate monohydrate**
$C_6H_2N_3O_7^-$, $C_5H_6N_5O^+$, H_2O
For complete entry see 44.11

15.C **6 - Thioguanine picrate monohydrate**
$C_6H_2N_3O_7^-$, $C_5H_6N_5S^+$, H_2O
For complete entry see 44.12

15.C **1 - Deaza - isotubercidin picrate**
$C_6H_2N_3O_7^-$, $C_{12}H_{16}N_3O_4^+$
For complete entry see 45.33

15.C **8 - Methyl - 5,5a,6,7,8,9 - hexahydro - pyrido(2,1 - b)quinazolinium picrate**
$C_6H_2N_3O_7^-$, $C_{13}H_{17}N_2^+$
For complete entry see 36.14

15.C **Succinylcholine picrate**
$2C_6H_2N_3O_7^-$, $C_{14}H_{30}N_2O_4^{2+}$
For complete entry see 3.31

15.1 **Picryl sulfonic acid tetrahydrate (neutron study)**
$C_6H_2N_3O_9S^-$, $H_5O_2^+$, $2H_2O$
J.-O.Lundgren, R.Tellgren *Acta Crystallogr., Sect. B,* **30,** 1937, 1974
Residue 1 also classified in 11

15.C **2 - Chloro - 4,6 - dinitrophenol**
$C_6H_3ClN_2O_5$
For complete entry see 17.1

15.C **Benzidine - s - trinitrobenzene complex**
$C_6H_3N_3O_6$, $C_{12}H_{12}N_2$
For complete entry see 60.15

15.C **Benzidine - s - trinitrobenzene complex benzene solvate**
$C_6H_3N_3O_6$, $C_{12}H_{12}N_2$, $0.5C_6H_6$
For complete entry see 60.16

15.2 **Potassium o - nitrophenolate hemihydrate**
$C_6H_4NO_3^-$, K^+ , $0.5H_2O$
E.K.Andersen, I.G.K.Andersen *Acta Crystallogr., Sect. B,* **31,** 391, 1975
Residue 1 also classified in 6

15.C **m - Nitrophenol (monoclinic form)**
$C_6H_5NO_3$
For complete entry see 17.4

15.C **m - Nitrophenol (orthorhombic form)**
$C_6H_5NO_3$
For complete entry see 17.5

15.3 **Potassium 2,4,6 - trinitrophenyl - dinitromethanide**
$C_7H_2N_5O_{10}^-$, K^+
N.V.Grigor'eva, N.V.Margolis, I.V.Tselinskii, V.V.Mel'nikov, G.V.Makarenko *Zh. Strukt. Khim.,* **14,** 751, 1973
Residue 1 also classified in 12

15.C **2,4 - Dinitro - 5 - ethyleneimino - benzamide**
$C_9H_8N_4O_5$
For complete entry see 32.11

15.C **N - Methyl - 2 - (N - methylamino) - 3,5 - dinitrobenzamide (at −160°C)**
$C_9H_{10}N_4O_5$
For complete entry see 13.16

15.C **1 - (N - Fluoro - N - t - butyl)amino - 2,4,6 - trinitrobenzene**
$C_{10}H_{11}FN_4O_6$
For complete entry see 16.5

15.C **Guanidinium dipicrylaminate**
$C_{12}H_4N_7O_{12}^-$, $CH_6N_3^+$
For complete entry see 8.4

15.C **Lidocaine bis(p - nitrophenyl) phosphate**
$C_{12}H_8N_2O_8P^-$, $C_{14}H_{23}N_2O^+$
For complete entry see 16.11

15.C **p - Nitro - diazoaminobenzene**
$C_{12}H_{10}N_4O_2$
For complete entry see 16.6

15.C **N,N - Di - n - propyl - 2,6 - dinitro - 4 - chloroaniline**
$C_{12}H_{16}ClN_3O_4$
For complete entry see 16.7

15.C **3,3 - Dimethylbutanol - 3,5 - dinitrobenzoate**
$C_{13}H_{16}N_2O_6$
For complete entry see 13.18

15.C **2,4 - Dibromo - o - cresyl 3′,5′ - dinitrosalicylate**
$C_{14}H_8Br_2N_2O_7$
For complete entry see 13.20

15.C **Anhydro - 2 - methyl - 4 - (o - nitroanilino) - 1,2,3 - benzotriazinium hydroxide**
$C_{14}H_{11}N_5O_2$
For complete entry see 35.22

15.4 **2,4 - Dinitrobenzyl p - tolyl sulfone**
$C_{14}H_{12}N_2O_6S$
R.L.Harlow, S.H.Simonsen, C.E.Pfluger, M.P.Sammes
Acta Crystallogr., Sect. B, **30,** 2264, 1974
Also classified in 11

15.5 **N,N′ - Dimethyl - N,N′ - di(2,4 - dinitrophenyl)urea**
$C_{15}H_{12}N_6O_9$
U.Lepore, G.C.Lepore, P.Ganis, M.Goodman
Cryst. Struct. Commun., **4,** 351, 1975
Also classified in 8

15.6 **N,N′ - Dimethyl - N,N′ - di(p - nitrophenyl)urea**
$C_{15}H_{14}N_4O_5$
U.Lepore, G.C.Lepore, P.Ganis *Cryst. Struct. Commun.*, **4,** 351, 1975
Also classified in 8

15.C **N - (2,4,6 - Trinitrobenzyl) - N - methyl - p - anisidine**
$C_{15}H_{14}N_4O_7$
For complete entry see 17.27

15.C **N - (2,4,6 - Trinitrobenzyl) - N - methyl - anisidine**
$C_{15}H_{14}N_4O_7$
For complete entry see 17.28

15.C **1,2 - Naphthoquinone 1 - (2 - nitro - 4 - chlorophenylhydrazone)**
$C_{16}H_{10}ClN_3O_3$
For complete entry see 25.4

15.C **2,4 - Dinitro - 4′ - diethylamino - diphenylamine**
$C_{16}H_{18}N_4O_4$
For complete entry see 16.12

15.C **S - 4 - Nitrophenyl O,O - diphenyl thiophosphate**
$C_{18}H_{14}NO_5PS$
For complete entry see 64.45

15.C **Chloramphenicol palmitate (β form)**
$C_{27}H_{42}Cl_2N_2O_6$
For complete entry see 50.21

ANILINES

16.1 **N,N - Difluoro - 2,4,6 - trinitro - aniline**
$C_6H_2F_2N_4O_6$
P.Batail, D.Grandjean *Acta Crystallogr., Sect. B,* **31,** 1367, 1975
Also classified in 15

16.C **α - Cyclodextrin p - iodoaniline trihydrate**
C_6H_6IN , $C_{36}H_{60}O_{30}$, $3H_2O$
For complete entry see 45.58

16.2 **Sodium sulfanilate dihydrate**
$C_6H_6NO_3S^-$, Na^+ , $2H_2O$
J.W.Bats, P.Coppens *Acta Crystallogr., Sect. B,* **31,** 1467, 1975
Residue 1 also classified in 11

16.C **Triphenylmethane aniline**
C_6H_7N , $C_{19}H_{16}$
For complete entry see 60.33

16.C **p - Hydroxyacetanilide (orthorhombic form)**
$C_8H_9NO_2$
For complete entry see 17.9

16.3 **N,N - Dimethyl - p - nitroso - aniline hydrochloride monohydrate**
$C_8H_{11}N_2O^+$, Cl^- , H_2O
O.Drangfelt, C.Romming *Acta Chem. Scand. Ser. A,* **28,** 1101, 1974
Residue 1 also classified in 10

16.C **N - Methyl - 2 - (N - methylamino) - 3,5 - dinitrobenzamide (at $-160°C$)**
$C_9H_{10}N_4O_5$
For complete entry see 13.16

16.C **Trimethylphenylammonium di - μ - iodo - bis(di - iodo - carbonyl - acetyl rhodium(iii))**
$2C_9H_{14}N^+$, $C_6H_6I_6O_4Rh_2^{2-}$
For complete entry see 71.4

16.C **Trimethylanilinium bis(1,1 - dicarboethoxy - 2,2 - ethylenedithiolato) nickel(iii)**
$2C_9H_{14}N^+$, $C_{16}H_{20}NiO_8S_4^{2-}$
For complete entry see 85.31

16.4 **3 - Amino - 3 - chloro - 2 - cyano - acrylic acid anilide**
$C_{10}H_8ClN_3O$
H.D.Block, H.-D.Lockenhoff, R.Allmann
Cryst. Struct. Commun., **4,** 77, 1975
Also classified in 7

16.5 **1 - (N - Fluoro - N - t - butyl)amino - 2,4,6 - trinitrobenzene**
$C_{10}H_{11}FN_4O_6$
P.Batail, D.Grandjean, F.Dudragne, C.Michaud
Acta Crystallogr., Sect. B, **30,** 2653, 1974
Also classified in 15

16.6 **p - Nitro - diazoaminobenzene**
$C_{12}H_{10}N_4O_2$
Yu.D.Kondrashev *Zh. Strukt. Khim.*, **15,** 517, 1974
Also classified in 15, 9

16.C **Benzidine - s - trinitrobenzene complex**
$C_{12}H_{12}N_2$, $C_6H_3N_3O_6$
For complete entry see 60.15

16.C **Benzidine - s - trinitrobenzene complex benzene solvate**
$C_{12}H_{12}N_2$, $C_6H_3N_3O_6$, $0.5C_6H_6$
For complete entry see 60.16

16.7 **N,N - Di - n - propyl - 2,6 - dinitro - 4 - chloroaniline**
$C_{12}H_{16}ClN_3O_4$
R.L.R.Towns, J.N.Brown, R.G.Teller, C.S.Giam
Cryst. Struct. Commun., **3,** 677, 1974
Also classified in 15

16.8 **Diphenylformazane**
$C_{13}H_{12}N_4$
Yu.A.Omel'chenko, Yu.D.Kondrashev, S.L.Ginzburg, M.G.Neigauz
Kristallografiya, **19,** 522, 1974
Also classified in 9

16.C **(N - Piperidino - acetyl) - m - bromoanilide**
$C_{13}H_{17}BrN_2O$
For complete entry see 33.36

16.C **Procaine dihydrogen orthophosphate hemihydrate**
$C_{13}H_{21}N_2O_2^+$, $H_2O_4P^-$, $0.5H_2O$
For complete entry see 13.19

16.C **Anhydro - 2 - methyl - 4 - (o - nitroanilino) - 1,2,3 - benzotriazinium hydroxide**
$C_{14}H_{11}N_5O_2$
For complete entry see 35.22

16.C **1 - Methyl - 4′ - methoxy - 3,5 - di - iodo - diphenylamine**
$C_{14}H_{13}I_2NO$
For complete entry see 17.25

16.9 **S - Methyl - dithizone**
$C_{14}H_{14}N_4S$
J.Preuss, A.Gieren *Acta Crystallogr., Sect. B,* **31,** 1276, 1975
Also classified in 11, 9

16.C **N - Methylpiperidinium - m - chloroacetanilide iodide**
$C_{14}H_{20}ClN_2O^+$, I^-
For complete entry see 33.37

16.10 **2 - Diethylamino - 2′,6′ - acetoxylidide**
Lidocaine
$C_{14}H_{22}N_2O$
A.W.Hanson, D.W.Banner *Acta Crystallogr., Sect. B,* **30,** 2486, 1974
Also classified in 3

16.11 **Lidocaine bis(p - nitrophenyl) phosphate**
$C_{14}H_{23}N_2O^+$, $C_{12}H_8N_2O_8P^-$
C.S.Yoo, E.Abola, M.K.Wood, M.Sax, J.Pletcher
Acta Crystallogr., Sect. B, **31,** 1354, 1975
Residue 1 also classified in 3; residue 2 classified in 46, 15

16.C **N - (2,4,6 - Trinitrobenzyl) - N - methyl - p - anisidine**
$C_{15}H_{14}N_4O_7$
For complete entry see 17.27

16.C **N - (2,4,6 - Trinitrobenzyl) - N - methyl - anisidine**
$C_{15}H_{14}N_4O_7$
For complete entry see 17.28

16.C **1,2 - Naphthoquinone 1 - (2 - nitro - 4 - chlorophenylhydrazone)**
$C_{16}H_{10}ClN_3O_3$
For complete entry see 25.4

16.C **Ammonium 8 - anilino - 1 - naphthalenesulfonate**
$C_{16}H_{12}NO_3S^-$, H_4N^+
For complete entry see 24.9

16.C **2 - n - Propyl - 4 - anilino - 1,2,3 - benzotriazinium iodide**
$C_{16}H_{17}N_4^+$, I^-
For complete entry see 35.29

16.12 **2,4 - Dinitro - 4′ - diethylamino - diphenylamine**
$C_{16}H_{18}N_4O_4$
A.E.Shvets, Ya.Ya.Bleidelis, Ya.F.Freimanis
Zh. Strukt. Khim., **15,** 504, 1974
Also classified in 15

16.13 **N - (Diphenylmethylene)aniline (at −160°C)**
$C_{19}H_{15}N$
P.A.Tucker, A.Hoekstra, J.M.ten Cate, A.Vos
Acta Crystallogr., Sect. B, **31,** 733, 1975

16.C **2 - Chloro - 3 - (N - methyl - N - p - tolyl)aminomethyl - 1,4 - naphthoquinone**
$C_{19}H_{16}ClNO_2$
For complete entry see 25.5

16.C **N,N′ - Diphenyl - 6 - aminopentafulvene - 2 - aldimine**
$C_{19}H_{16}N_2$
For complete entry see 20.21

16.C **4′ - (Acridin - 9 - ylamine)methane - sulfonanilide hydrochloride**
$C_{20}H_{18}N_3O_2S^+$, Cl^-
For complete entry see 36.25

16.C **bis(2,6 - Diacetylpyridinium - bis(phenylhydrazone)) uranyl tetrachloride acetonitrile solvate**
$2C_{21}H_{22}N_5^+$, $Cl_4O_2U^{2-}$, C_2H_3N
For complete entry see 33.52

16.C **2 - (N - Acetyl - 2(2′ - N′ - ethyl - N′ - phenyl)amino)ethylamino - 3 - chloro - 1,4, - naphthoquinone**
$C_{22}H_{21}ClN_2O_3$
For complete entry see 25.6

16.C **1 - Methyl - 3,3 - dimethyl - 2 - (p - N - methyl - N - β - chloroethylstyryl)indole - bis(7,7,8,8 - tetracyanoquinodimethane)**
$C_{22}H_{26}ClN_2^+$, $C_{12}H_4N_4^-$, $C_{12}H_4N_4$
For complete entry see 60.38

16.C **1 - (2,5 - Dichlorophenylazo) - 2 - hydroxy - 3 - naphthoic acid 4 - chloro - 2 - methoxy - anilide**
$C_{24}H_{16}Cl_3N_3O_3$
For complete entry see 24.11

16.14 **Tetraphenylhydrazine**
$C_{24}H_{20}N_2$
A.Hoekstra, A.Vos, P.B.Braun, J.Hornstra
Acta Crystallogr., Sect. B, **31,** 1708, 1975
Also classified in 9

16.15 **Tetraphenylhydrazine (at −160°C)**
$C_{24}H_{20}N_2$
A.Hoekstra, A.Vos, P.B.Braun, J.Hornstra
Acta Crystallogr., Sect. B, **31,** 1708 + , 1975
Also classified in 9

16.C **2,2′ - Dibromo - 4,4′ - bis(p - methoxybenzylideneamine)biphenyl**
$C_{28}H_{22}Br_2N_2O_2$
For complete entry see 17.35

16.16 **2,2′ - bis(2 - Bromoacetylaminoanilino) - benzalazine benzene solvate**
$C_{30}H_{26}Br_2N_6O_2$, C_6H_6
T.Ozawa, Y.Iitaka, M.Hirobe, T.Okamoto
Chem. Pharm. Bull., **22,** 2069, 1974
Residue 1 also classified in 9

16.17 **Diphenylamino - triphenylmethane (at −160°C)**
$C_{31}H_{25}N$
A.Hoekstra, A.Vos *Acta Crystallogr., Sect. B,* **31,** 1716, 1975

PHENOLS AND ETHERS

17.C **Cytosine picrate**
$C_6H_2N_3O_7^-$, $C_4H_6N_3O^+$
For complete entry see 44.6

17.C **Guanine picrate monohydrate**
$C_6H_2N_3O_7^-$, $C_5H_6N_5O^+$, H_2O
For complete entry see 44.11

17.C **6 - Thioguanine picrate monohydrate**
$C_6H_2N_3O_7^-$, $C_5H_6N_5S^+$, H_2O
For complete entry see 44.12

17.C **1 - Deaza - isotubercidin picrate**
$C_6H_2N_3O_7^-$, $C_{12}H_{16}N_3O_4^+$
For complete entry see 45.33

17.C **8 - Methyl - 5,5a,6,7,8,9 - hexahydro - pyrido(2,1 - b)quinazolinium picrate**
$C_6H_2N_3O_7^-$, $C_{13}H_{17}N_2^+$
For complete entry see 36.14

17.C **Dipotassium 1,4 - benzoquinone - 2,5 - dihydroxylate**
$C_6H_2O_4^{2-}$, $2K^+$
For complete entry see 18.4

17.1 **2 - Chloro - 4,6 - dinitrophenol**
$C_6H_3ClN_2O_5$
E.K.Andersen, I.G.K.Andersen *Acta Crystallogr., Sect. B,* **31,** 387, 1975
Also classified in 15

17.2 **2,6 - Dichlorophenol**
$C_6H_4Cl_2O$
C.Bavoux, P.Michel *Acta Crystallogr., Sect. B,* **30,** 2043, 1974

17.3 **Potassium p - nitrosophenolate monohydrate**
$C_6H_4NO_2^-$, K^+ , H_2O
H.J.Talberg *Acta Chem. Scand. Ser. A,* **28,** 593, 1974
Residue 1 also classified in 6

17.C **ε - Caprolactam - 4 - chlororesorcinol**
$C_6H_5ClO_2$, $C_6H_{11}NO$
For complete entry see 60.6

17.4 **m - Nitrophenol (monoclinic form)**
$C_6H_5NO_3$
A.Coda, M.Fumagalli, F.Pandarese, L.Ungaretti
Acta Crystallogr., Sect. A, **31,** S208, 1975
Also classified in 15

17.5 **m - Nitrophenol (orthorhombic form)**
$C_6H_5NO_3$
A.Coda, M.Fumagalli, F.Pandarese, L.Ungaretti
Acta Crystallogr., Sect. A, **31,** S208, 1975
Also classified in 15

17.C **Quinol - urea**
$C_6H_6O_2$, CH_4N_2O
For complete entry see 60.5

17.C **Progesterone - resorcinol**
$C_6H_6O_2$, $C_{21}H_{30}O_2$
For complete entry see 60.37

17.C **(αS(R,R)) - α - ((Di - s - butylamino)methyl) - 1 - (2 - chlorobenzyl) - 1H - 2 - pyrrole - methanol p - hydroxybenzoate (absolute configuration)**
(S(R,R)) - Viminol
$C_7H_5O_3^-$, $C_{21}H_{32}ClN_2O^+$
For complete entry see 32.39

17.C **Salicylic acid - cytidine complex**
$C_7H_6O_3$, $C_9H_{13}N_3O_5$
For complete entry see 60.7

17.C **5 - Sulfosalicylic acid dihydrate**
$C_7H_6O_6S$, $2H_2O$
For complete entry see 13.7

17.C **5 - Sulphosalicylic acid trihydrate**
$C_7H_6O_6S$, $3H_2O$
For complete entry see 13.8

17.6 **p - Cresol (γ form)**
C_7H_8O
M.Perrin, A.Thozet *Cryst. Struct. Commun.,* **3,** 661, 1974

17.7 **2,4 - Dibromo - 3,6 - dihydroxyphenyl - acetamide**
$C_8H_7Br_2NO_3$
G.E.Krejcaerik, R.H.White, L.P.Hager, W.O.McClure, R.D.Johnson, K.L.Rinehart Junior, J.A.McMillan, I.C.Paul, P.D.Shaw, R.C.Brusca
Tetrahedron Lett., 507, 1975
Also classified in 1

17.8 **p - Hydroxyacetophenone**
$C_8H_8O_2$
B.K.Vainshtein, G.M.Lobanova, G.V.Gurskaya
Kristallografiya, **19,** 531, 1974

17.9 **p - Hydroxyacetanilide (orthorhombic form)**
$C_8H_9NO_2$
M.Haisa, S.Kashino, H.Maeda *Acta Crystallogr., Sect. B,* **30,** 2510, 1974
Also classified in 16

17.10 **Tyramine hydrochloride**
$C_8H_{12}NO^+$, Cl^-
K.Tamura, A.Wakahara, T.Fujiwara, K.-I.Tomita
Bull. Chem. Soc. Jpn., **47,** 2682, 1974
Residue 1 also classified in 3

17.11 **Tyramine hydrochloride**
$C_8H_{12}NO^+$, Cl^-
A.Poddar, N.N.Saha, J.K.Dattagupta, W.Saenger
Acta Crystallogr., Sect. A, **31,** S54, 1975
Residue 1 also classified in 3

17.12 **(3,4,6 - Trihydroxyphenyl)ethylammonium 2 - (2′ - ethylammonium) - 5 - hydroxy - 1,4 - benzoquinone dichloride**
$C_8H_{12}NO_3^+$, $C_8H_{10}NO_3^+$, $2Cl^-$
A.M.Andersen, A.Mostad, C.Romming
Acta Chem. Scand. Ser. B, **29,** 45, 1975
Residue 1 also classified in 3; residue 2 classified in 18, 3

17.13 **D - β - Tyrosine hydrobromide (absolute configuration)**
$C_9H_{12}NO_3^+$, Br^-
A.N.Chekhlov, Yu.T.Struchkov, A.I.Kitaigorodskij
Kristallografiya, **19,** 981, 1974
Residue 1 also classified in 1, 3

17.14 **D - β - Tyrosine hydrochloride (absolute configuration)**
$C_9H_{12}NO_3^+$, Cl^-
A.N.Chekhlov, Yu.T.Struchkov, A.I.Kitaigorodskij
Kristallografiya, **19,** 981, 1974
Residue 1 also classified in 1, 3

17.15 **(−) - Adrenaline**
$C_9H_{13}NO_3$
A.M.Andersen *Acta Chem. Scand. Ser. B,* **29,** 239, 1975
Also classified in 3

17.16 **L - Phenylephrine hydrochloride**
$C_9H_{14}NO_2^+$, Cl^-
D.Bhaduri, N.N.Saha *Acta Crystallogr., Sect. A,* **31,** S54, 1975
Residue 1 also classified in 3

17.C **6 - (p - Hydroxyphenylazo)uracil**
$C_{10}H_8N_4O_3$
For complete entry see 44.22

17.17 **Potassium trihydrogen bis(O,O′ - catechol diacetate)**
$C_{10}H_9O_6^-$, $C_{10}H_{10}O_6$, K^+
E.A.Green, W.L.Duax, G.M.Smith, F.Wudl
Am. Cryst. Assoc., Abstr. Papers (Summer Meeting), 227, 1974
Residue 1 also classified in 2, 1

17.18 **(±) - erythro - 2 - (2,5 - Dimethoxyphenyl) - 2 - hydroxy - 1 - methyl - ethylammonium chloride**
$C_{11}H_{18}NO_3^+$, Cl^-
M.V.Gabrielsen, A.M.Sorensen *Acta Chem. Scand. Ser. A,* **28,** 1162, 1974
Residue 1 also classified in 3

17.19 **4 - Bromo - 4′ - hydroxyazobenzene**
$C_{12}H_9BrN_2O$
Yu.D.Kondrashev *Zh. Strukt. Khim.,* **15,** 675, 1974
Also classified in 9

17.C **3 - Phenoxy - 3 - dimethylcarbamoyldimethylamino - 2 - azirine**
$C_{13}H_{17}N_3O_2$
For complete entry see 32.24

17.20 **2 - Chloro - 2 - phenoxymalonylamide - amidinium hydrochloride monohydrate**
$C_{13}H_{19}ClN_3O_2^+$, Cl^- , H_2O
J.Galloy, J.-P.Putzeys, G.Germain, J.-P.Declercq, M.van Meerssche
Acta Crystallogr., Sect. B, **30,** 2460, 1974
Residue 1 also classified in 1

17.21 **2 - (4 - Ethyl - 2,5 - dimethoxyphenyl) - 1 - methyl - ethylamine**
4 - Ethyl - 2,5 - dimethoxy - amphetamine
$C_{13}H_{21}NO_2$
O.Kennard, C.Giacovazzo, A.S.Horn, R.Mongiorgi, L.Riva di Sanseverino
J. Chem. Soc., Perkin Trans. 2, 1160, 1974
Also classified in 3

17.C **2,4 - Dibromo - o - cresyl 3′,5′ - dinitrosalicylate**
$C_{14}H_8Br_2N_2O_7$
For complete entry see 13.20

17.22 **3,5 - Dibromothyroacetic acid N - diethanolamine solvate**
$C_{14}H_{10}Br_2O_4$, $C_4H_{11}NO_2$
V.Cody, M.Erman, E.DeJarnette
Am. Cryst. Assoc., Abstr. Papers (Spring Meeting), 27, 1975
Residue 1 also classified in 1

17.23 **2 - Hydroxy - 4 - methoxy - 4′ - chloro - benzophenone**
$C_{14}H_{11}ClO_3$
B.W.Liebich *Acta Crystallogr., Sect. A,* **31,** S179, 1975

17.24 **2 - Hydroxy - 4 - methoxy - benzophenone**
$C_{14}H_{12}O_3$
B.W.Liebich, E.Parthe *Acta Crystallogr., Sect. B,* **30,** 2522, 1974

17.25 **1 - Methyl - 4′ - methoxy - 3,5 - di - iodo - diphenylamine**
$C_{14}H_{13}I_2NO$
V.Cody *Am. Cryst. Assoc., Abstr. Papers (Summer Meeting),* 209, 1974
Also classified in 16

17.C **2,4 - Diamino - 5 - (3,4,5 - trimethoxybenzyl) - pyrimidine (neutron study)**
$C_{14}H_{18}N_4O_3$
For complete entry see 44.28

17.26 **3,5 - Di - iodothyropropionic acid**
$C_{15}H_{12}I_2O_4$
V.Cody, M.Erman, E.DeJarnette
Am. Cryst. Assoc., Abstr. Papers (Spring Meeting), 27, 1975
Also classified in 1

17.C **3,5,3′ - Tri - iodo - L - thyronine**
$C_{15}H_{12}I_3NO_4$
For complete entry see 48.55

17.27 **N - (2,4,6 - Trinitrobenzyl) - N - methyl - p - anisidine**
$C_{15}H_{14}N_4O_7$
A.E.Shvets, Ya.Ya.Bleidelis, Ya.F.Freimanis
Acta Crystallogr., Sect. A, **31,** S129, 1975
Also classified in 16, 15

17.28 **N - (2,4,6 - Trinitrobenzyl) - N - methyl - anisidine**
$C_{15}H_{14}N_4O_7$
A.E.Shvets, Ya.Ya.Bleidelis
Latv. PSR Zinat. Akad. Vestis, Kim. Ser., 749, 1973
Also classified in 16, 15

17.C **1 - p - (Oximinoethyl) - phenoxyacetyl - piperidine**
$C_{15}H_{20}N_2O_3$
For complete entry see 33.38

17.C **1 - (p - Methoxyphenyl) - 3 - methyl - 4 - (D - arabino - tetrahydroxybutyl) - imidazolidine - 2 - thione**
$C_{15}H_{22}N_2O_5S$
For complete entry see 45.44

17.C **3,5,3′ - Tri - iodo - L - thyronine methyl ester**
$C_{16}H_{14}I_3NO_4$
For complete entry see 48.60

17.C **3 - (3,5 - Dibromo - 4 - hydroxy)benzoyl - 2 - ethyl - benzo(b)furan**
$C_{17}H_{12}Br_2O_3$
For complete entry see 38.53

17.29 **Tri - iodo - thyropropionic acid ethyl ester**
$C_{17}H_{15}I_3O_4$
N.Camerman, A.Camerman *Can. J. Chem.*, **52**, 3048, 1974
Also classified in 1

17.C **2′,3′ - Dimethyl - 3,5 - di - iodo - DL - thyronine hydrochloride hydrate**
$C_{17}H_{18}I_2NO_4^+$, Cl^- , xH_2O
For complete entry see 48.61

17.C **L - Thyronine ethyl ester hydrochloride monohydrate**
$C_{17}H_{20}NO_4^+$, Cl^- , H_2O
For complete entry see 48.62

17.C **3′ - Isopropyl - 3,5 - di - iodo - L - thyronine hydrochloride trihydrate**
$C_{18}H_{20}I_2NO_4^+$, Cl^- , $3H_2O$
For complete entry see 48.66

17.30 **6′,7 - Dimethyl - 7′ - ethyl - stilboestrol monohydrate**
$C_{18}H_{20}O_2$, H_2O
B.Busetta, S.Geoffre, G.Precigoux, C.Courseille, M.Hospital
Eur. Cryst. Meeting, 410, 1974

17.31 **6′,7 - Dimethyl - 7′ - ethyl - stilboestrol monohydrate ethanol solvate**
$C_{18}H_{20}O_2$, H_2O , C_2H_6O
B.Busetta, S.Geoffre, G.Precigoux, C.Courseille, M.Hospital
Eur. Cryst. Meeting, 410, 1974

17.C **2 - (Di - p - anisyl - methyl) - 1,3 - dioxolane**
$C_{18}H_{20}O_4$
For complete entry see 38.58

17.C **(−) - 1 - t - Butylamino - 3 - (2 - cyclopentyl - phenoxy) - propan - 2 - ol methyl - sulfonate (absolute configuration)**
$C_{18}H_{30}NO_2^+$, $CH_3O_3S^-$
For complete entry see 20.18

17.32 **(−) - erythro - 1′ - (2,5 - Dimethoxyphenyl) - 3′ - diethylaminobutyl acetate hydrobromide (absolute configuration)**
$C_{18}H_{30}NO_4^+$, Br^-
Y.Masuda, Y.Iitaka, H.Hamano *Bull. Chem. Soc. Jpn.*, **47**, 825, 1974
Residue 1 also classified in 3

17.C **α(S) - Cyano - 3 - phenoxybenzyl - cis(1R,3R) - 2,2 - dimethyl - 3 - (2,2 - dibromovinyl) - cyclopropanecarboxylate (absolute configuration)**
$C_{22}H_{19}Br_2NO_3$
For complete entry see 20.22

17.C **trans - 1 - (2 - O - Methoxyphenylethyl) - 4β - hydroxy - 2,5α,9β - trimethyl - 1 - octal - 8 - one**
$C_{22}H_{30}O_3$
For complete entry see 27.10

17.C **Streptonigrin ethylacetate solvate**
$C_{25}H_{22}N_4O_8$, $C_4H_8O_2$
For complete entry see 50.19

17.C **4 - O - Ethyl - ascofuranone (absolute configuration)**
$C_{25}H_{33}ClO_5$
For complete entry see 50.20

17.33 **bis(Diphenylmethyl) ether**
$C_{26}H_{22}O$
M.A.Mazid, R.A.Palmer *J. Cryst. Mol. Struct.*, **5**, 35, 1975

17.34 **trans - 1 - ((p - Diethylaminoethoxy)phenyl) - 1,2 - diphenyl - 2 - chloroethylene**
Clomiphene
$C_{26}H_{29}ClNO^+$, Cl^-
G.Hite, S.Ernst, J.S.Cantrell
Am. Cryst. Assoc., Abstr. Papers (Summer Meeting), 243, 1974
Residue 1 also classified in 3

17.35 **2,2′ - Dibromo - 4,4′ - bis(p - methoxybenzylideneamine)biphenyl**
$C_{28}H_{22}Br_2N_2O_2$
D.P.Lesser, A.de Vries, J.W.Reed, G.H.Brown
Acta Crystallogr., Sect. B, **31**, 653, 1975
Also classified in 16

17.C **Novobiocin monohydrate**
$C_{31}H_{36}N_2O_{11}$, H_2O
For complete entry see 50.24

17.C Sodium 5 - bromo - lasalocid (form i)
$C_{34}H_{52}BrO_8^-$, Na^+
For complete entry see 50.26

17.C Sodium 5 - bromo - lasalocid (form ii)
$C_{34}H_{52}BrO_8^-$, Na^+
For complete entry see 50.27

BENZOQUINONES

18.1 **Tetraiodo - p - benzoquinone**
$C_6I_4O_2$
H.Kobayashi, T.Danno, I.Shirotani *Bull. Chem. Soc. Jpn.*, **47,** 2333, 1974

18.2 **bis(Hydronium) nitranilate tetrahydrate**
$C_6N_2O_8^{2-}$, $2H_3O^+$, $4H_2O$
E.K.Andersen, I.G.K.Andersen *Acta Crystallogr., Sect. B*, **31,** 379, 1975
Residue 1 also classified in 15, 6

18.3 **Fluoranilic acid**
$C_6H_2F_2O_4$
E.K.Andersen, I.G.K.Andersen *Acta Crystallogr., Sect. B*, **31,** 384, 1975

18.4 **Dipotassium 1,4 - benzoquinone - 2,5 - dihydroxylate**
$C_6H_2O_4^{2-}$, $2K^+$
S.Kulpe *J. Prakt. Chem.*, **316,** 353, 1974
Residue 1 also classified in 17, 6

18.5 **p - Benzoquinone 4 - oxime**
$C_6H_5NO_2$
H.J.Talberg *Acta Chem. Scand. Ser. A*, **28,** 910, 1974

18.6 **bis(Hydronium) cyananilate tetrahydrate**
Cyananilic acid hexahydrate
$C_8N_2O_4^{2-}$, $2H_3O^+$, $4H_2O$
E.K.Andersen, I.G.K.Andersen *Acta Crystallogr., Sect. B*, **31,** 379, 1975
Residue 1 also classified in 7, 6

18.C **(3,4,6 - Trihydroxyphenyl)ethylammonium 2 - (2′ - ethylammonium) - 5 - hydroxy - 1,4 - benzoquinone dichloride**
$C_8H_{10}NO_3^+$, $C_8H_{12}NO_3^+$, $2Cl^-$
For complete entry see 17.12

18.7 **2,6 - Di - t - butyl - 1,4 - benzoquinone**
$C_{14}H_{20}O_2$
G.G.Aleksandrov, Yu.T.Struchkov, D.I.Kalinin, M.G.Neigauz
Zh. Strukt. Khim., **14,** 852, 1973

18.C **2,5 - bis - Pentamethyleneimino - 1,4 - benzoquinone**
$C_{16}H_{22}N_2O_2$
For complete entry see 33.47

BENZENE MISCELLANEOUS

19.C **p - Xylene - hexafluorobenzene complex**
C_6F_6 , C_8H_{10}
For complete entry see 60.10

19.1 **p - Dichlorobenzene (γ form, at 100°K)**
$C_6H_4Cl_2$
G.L.Wheeler, S.D.Colson *Acta Crystallogr., Sect. B,* **31,** 911, 1975

19.2 **p - Dichlorobenzene (γ form, at 260°K)**
$C_6H_4Cl_2$
R.Fourme, G.Clec'h, P.Figuiere, M.Ghelfenstein, H.Szwarc
Mol. Cryst. Liq. Cryst., **27,** 315, 1974

19.C **Triphenylmethane benzene**
C_6H_6 , $C_{19}H_{16}$
For complete entry see 60.32

19.3 **p - Bromoacetophenone**
C_8H_7BrO
M.P.Gupta, S.M.Prasad, S.R.P.Yadav *Curr. Sci.,* **43,** 509, 1974

19.C **p - Xylene - hexafluorobenzene complex**
C_8H_{10} , C_6F_6
For complete entry see 60.10

19.C **tris(1,8 - Naphthalenedioxy)cyclotriphosphazene p - xylene clathrate**
$0.5C_8H_{10}$, $C_{30}H_{18}N_3O_6P_3$
For complete entry see 61.2

19.C **p - Chloro - trans - cinnamic acid**
$C_9H_7ClO_2$
For complete entry see 1.26

19.C **β - (p - Chlorophenyl)propionic acid**
$C_9H_9ClO_2$
For complete entry see 1.27

19.4 **Durene (neutron data of Prince et al.,Acta Cryst.,B29,184,1973,constrained refinement i)**
$C_{10}H_{14}$
J.-L.Baudour, M.Sanquer *Acta Crystallogr., Sect. B,* **30,** 2371, 1974

19.5 **Durene (neutron data of Prince et al.,Acta Cryst.,B29,184,1973,constrained refinement ii)**
$C_{10}H_{14}$
J.-L.Baudour, M.Sanquer *Acta Crystallogr., Sect. B,* **30,** 2371, 1974

19.6 **2,2′ - Dichloro - biphenyl**
$C_{12}H_8Cl_2$
C.Romming, H.M.Seip, I.-M.A.Oymo
Acta Chem. Scand. Ser. A, **28,** 507, 1974

19.7 **Diphenyliodonium bromide**
$C_{12}H_{10}I^+$, Br^-
N.W.Alcock, R.Countryman *Acta Crystallogr., Sect. A,* **31,** S62, 1975

19.8 **2 - Iodo - 3′ - chloro - dibenzoyl peroxide**
$C_{14}H_8ClIO_4$
J.Z.Gougoutas, L.Lessinger *J. Solid State Chem.,* **7,** 175, 1973

19.9 **trans - Stilbene**
$C_{14}H_{12}$
J.Bernstein *Acta Crystallogr., Sect. B,* **31,** 1268, 1975

19.C **Pyromellitic dianhydride - trans - stilbene complex**
$C_{14}H_{12}$, $C_{10}H_2O_6$
For complete entry see 60.12

19.C **(±) - 2 - (2 - Fluoro - 4 - biphenyl)propionic acid**
Flurbiprofen
$C_{15}H_{13}FO_2$
For complete entry see 1.36

19.C **Pyromellitic dianhydride - trans - 4 - methyl - stilbene complex**
$C_{15}H_{14}$, $2C_{10}H_2O_6$
For complete entry see 60.13

19.10 **1 - (2,6 - Dichlorophenyl) - 4 - phenyl - trans,trans - 1,3 - butadiene (absolute configuration)**
$C_{16}H_{12}Cl_2$
D.Rabinovich, Z.Shakked *Acta Crystallogr., Sect. B,* **31,** 819, 1975

19.11 **1 - (p - Bromophenyl) - 4 - benzoyl - 1,3 - butadiene**
$C_{17}H_{13}BrO$
L.G.Vorontsova, G.S.Kazaryan *Zh. Strukt. Khim.,* **14,** 1089, 1973

19.12 **4,4′ - Dimethylchalcone**
$C_{17}H_{16}O$
D.Rabinovich, Z.Shakked *Acta Crystallogr., Sect. B,* **30,** 2829, 1974

19.C **Triphenylmethane thiophene**
$C_{19}H_{16}$, C_4H_4S
For complete entry see 60.30

19.C **Triphenylmethane pyrrole**
$C_{19}H_{16}$, C_4H_5N
For complete entry see 60.31

19.C **Triphenylmethane benzene**
$C_{19}H_{16}$, C_6H_6
For complete entry see 60.32

19.C **Triphenylmethane aniline**
$C_{19}H_{16}$, C_6H_7N
For complete entry see 60.33

19.13 **1,3,5 - Triphenylbenzene**
$C_{24}H_{18}$
Y.C.Lin, D.E.Williams *Acta Crystallogr., Sect. B,* **31,** 318, 1975

19.14 **Tetraphenylmethane**
$C_{25}H_{20}$
A.Robbins, G.A.Jeffrey
Am. Cryst. Assoc., Abstr. Papers (Summer Meeting), 210, 1974

19.15 **Tetraphenylethylene (at −160°C)**
$C_{26}H_{20}$
A.Hoekstra, A.Vos *Acta Crystallogr., Sect. B,* **31,** 1716, 1975

19.16 **Trimesityl - methane**
$C_{28}H_{34}$
J.F.Blount, K.Mislow *Tetrahedron Lett.,* 909, 1975

19.17 **s - Tetramesityl - ethane**
$C_{38}H_{46}$
H.J.Postma, F.van Bolhuis *Acta Crystallogr., Sect. B,* **31,** 1792, 1975

MONOCYCLIC HYDROCARBONS (3, 4, 5-MEMBERED RINGS)

20.1 **1 - Aminocyclobutane carboxylic acid hydrochloride monohydrate**
$C_5H_{10}NO_2^+$, Cl^- , H_2O
K.K.Chacko, R.Zand *Cryst. Struct. Commun.*, **4,** 17, 1975

20.2 **Hexachlorofulvene (room temp. form)**
C_6Cl_6
K.Mano, F.Takusagawa, M.Hinamoto, Y.Kushi *Chem. Lett.*, 1083, 1973

20.3 **Hexachlorofulvene (low temp. form, at −130°C)**
C_6Cl_6
K.Mano, F.Takusagawa, M.Hinamoto, Y.Kushi *Chem. Lett.*, 1083, 1973

20.4 **cis - Cyclobut - 1 - ene - 3,4 - dicarboxylic acid**
$C_6H_6O_4$
E.Benedetti, M.R.Ciajolo, J.P.Declercq, G.Germain
Acta Crystallogr., Sect. B, **30,** 2873, 1974

20.5 **cis,trans,cis - 1,2,3,4 - Tetracyanocyclobutane (at −108°C)**
$C_8H_4N_4$
M.Harel, F.L.Hirshfeld *Acta Crystallogr., Sect. B,* **31,** 162, 1975

20.6 **2,2,4,4 - Tetramethyl - cyclobutan - 1 - one - 3 - thione**
$C_8H_{12}OS$
C.D.Shirrell, D.E.Williams *Acta Crystallogr., Sect. B,* **30,** 1974, 1974

20.7 **2,2,4,4 - Tetramethyl - 3 - methylenecyclobutanone**
$C_9H_{14}O$
C.D.Shirell, D.E.Williams *Acta Crystallogr., Sect. B,* **31,** 199, 1975

20.8 **1,1 - Cyclopentane - diacetic acid**
$C_9H_{14}O_4$
E.Benedetti, R.Claverini, C.Pedone *Gazz. Chim. Ital.*, **103,** 525, 1973
Also classified in 1

20.9 **cis - 2,5 - Dimethylcyclopentane - 1,1 - dicarboxylic acid**
$C_9H_{14}O_4$
R.T.Kops, H.Schenk *Cryst. Struct. Commun.*, **3,** 665, 1974

20.10 **Octabromopentafulvalene**
$C_{10}Br_8$
L.Fallon, H.L.Ammon, R.West, V.N.M.Rao
Acta Crystallogr., Sect. B, **30,** 2407, 1974

20.11 **6,6 - bis(Dimethylamino)fulvene**
$C_{10}H_{16}N_2$
R.Bohme, H.Burzlaff *Chem. Ber.,* **107,** 832, 1974

20.12 **N,N - Dimethyl - 2 - phenylcyclopropylamine hydrochloride**
$C_{11}H_{16}N^+$, Cl^-
D.Carlstrom *Acta Crystallogr., Sect. A,* **31,** S53, 1975

20.C **Monobromopentenomycin triacetate**
$C_{12}H_{13}BrO_7$
For complete entry see 50.3

20.13 **1 - Phenyl - cyclopentane - carboxylic acid**
$C_{12}H_{14}O_2$
T.N.Margulis *Acta Crystallogr., Sect. B,* **31,** 1049, 1975

20.14 **Diphenyl - cyclopropenone**
$C_{15}H_{10}O$
H.Tsukada, H.Shimanouchi, Y.Sasada *Chem. Lett.,* 639, 1974

20.15 **1,1 - Dibromo - 2,2 - diphenyl - cyclopropane**
$C_{15}H_{12}Br_2$
J.W.Lauher, J.A.Ibers *J. Am. Chem. Soc.,* **97,** 561, 1975

20.16 **1,1 - Dichloro - 2,2 - diphenyl - cyclopropane**
$C_{15}H_{12}Cl_2$
J.W.Lauher, J.A.Ibers *J. Am. Chem. Soc.,* **97,** 561, 1975

20.C **1 - (2′,6′ - Dichlorobenzyl) - 4 - cyclopentadienylidene - 1,4 - dihydropyridine**
$C_{17}H_{13}Cl_2N$
For complete entry see 33.48

20.C **1 - Benzyl - 2 - cyclopentadienylidene - 1,2 - dihydropyridine**
$C_{17}H_{15}N$
For complete entry see 33.49

20.17 **1,2 - Diphenylcyclopentene**
$C_{17}H_{16}$
J.Bernstein *Acta Crystallogr., Sect. B,* **31,** 418, 1975

20.18 **(−) - 1 - t - Butylamino - 3 - (2 - cyclopentyl - phenoxy) - propan - 2 - ol methyl - sulfonate (absolute configuration)**
$C_{18}H_{30}NO_2^+$, $CH_3O_3S^-$
D.Kobelt, E.F.Paulus *Z. Kristallogr.,* **139,** 1, 1974
Residue 1 also classified in 17, 3; residue 2 classified in 11

20.C **Diethyl - (2 - hydroxyethyl) - methylammonium α - cyclopentyl - 2 - thienylglycollate bromide**
Penthienate bromide
$C_{18}H_{30}NO_3S^+$, Br^-
For complete entry see 39.31

20.19 **Diethyl 2,4 - bis(diethylamino) - 1,3 - cyclobutadienedicarboxylate**
$C_{18}H_{30}N_2O_4$
H.J.Lindner, B.von Gross *Chem. Ber.*, **107,** 598, 1974

20.20 **Methyl tri - t - butyl(4)annulene - carboxylate**
2,3,4 - Tri - t - butyl - 1 - carboxymethyl - cyclobuta - 1,3 - diene
$C_{18}H_{30}O_2$
L.T.J.Delbaere, M.N.G.James, N.Nakamura, S.Masamune
J. Am. Chem. Soc., **97,** 1973, 1975

20.21 **N,N′ - Diphenyl - 6 - aminopentafulvene - 2 - aldimine**
$C_{19}H_{16}N_2$
H.L.Ammon, U.Mueller-Westerhoff *Tetrahedron,* **30,** 1437, 1974
Also classified in 16

Prostaglandin A_1 (orthorhombic form)
$C_{20}H_{32}O_4$
For complete entry see 59.11

20.C **Prostaglandin A_1 (monoclinic form)**
$C_{20}H_{32}O_4$
For complete entry see 59.12

20.C **2 - (2′,6′ - Dichlorobenzyl) - 1 - cyclopentadienylidene - 1,2 - dihydroisoquinoline**
$C_{21}H_{15}Cl_2N$
For complete entry see 35.33

20.22 **α(S) - Cyano - 3 - phenoxybenzyl - cis(1R,3R) - 2,2 - dimethyl - 3 - (2,2 - dibromovinyl) - cyclopropanecarboxylate (absolute configuration)**
$C_{22}H_{19}Br_2NO_3$
J.D.Owen *J. Chem. Soc., Chem. Commun.*, 859, 1974
Also classified in 17

20.23 **2 - Benzylidene - 4 - t - butyl - 4 - cyano - 3 - phenyl - cyclobutanone**
$C_{22}H_{21}NO$
H.A.Bampfield, P.R.Brook, W.S.McDonald
J. Chem. Soc., Chem. Commun., 132, 1975

20.C **Borrelidin solvate**
$C_{28}H_{43}NO_6$, $C_5H_{12}O$
For complete entry see 50.23

20.C **9 - Oxo - 9,11 - secogorgost - 5 - ene - 3β,11 - diol 11 - acetate p - iodobenzoate (absolute configuration)**
Secogorgosterol 3 - p - iodobenzoate 11 - acetate
$C_{39}H_{55}IO_5$
For complete entry see 51.59

MONOCYCLIC HYDROCARBONS (6-MEMBERED RINGS)

21.C **P^1,P^2 - Di - β - naphthyl - pyrophosphate di(cyclohexylammonium) salt**
$2C_6H_{14}N^+$, $C_{20}H_{14}O_7P_2^{2-}$
For complete entry see 46.6

21.C **1 - Aminocyclohexanecarboxylic acid**
$C_7H_{13}NO_2$
For complete entry see 48.36

21.1 **4 - Di - iodomethyl - 3,4,5 - trimethyl - cyclohexa - 2,5 - dienone**
$C_{10}H_{12}I_2O$
G.G.Christoph, E.B.Fleischer *J. Chem. Soc., Perkin Trans. 2,* 600, 1975

21.2 **Diethyl succinylsuccinate**
1,4 - Diethoxycarbonyl - 2,5 - dihydroxy - cyclohexa - 1,4 - diene
$C_{12}H_{16}O_6$
H.-C.Mez, G.Rihs *Helv. Chim. Acta,* **56,** 2766, 1973

21.C **Tetramethylene selenium dimedonylide**
$C_{12}H_{18}O_2Se$
For complete entry see 39.23

21.3 **$\Delta^{1,1'}$ - Dicyclohexenyl ketone (pink form, at −110°C)**
$C_{13}H_{15}O$
S.R.Holbrook, D.van der Helm *Acta Crystallogr., Sect. B,* **31,** 1689, 1975

21.4 **$\Delta^{1,1'}$ - Dicyclohexenyl ketone (colourless form, at −110°C)**
$C_{13}H_{18}O$
S.R.Holbrook, D.van der Helm *Acta Crystallogr., Sect. B,* **31,** 1689, 1975

21.5 **Methyl 6 - (isobuten - 1′ - yl) - 2 - methyl - 4 - oxo - cyclohex - 2 - ene carboxylate**
$C_{13}H_{18}O_3$
M.J.Begley *Acta Crystallogr., Sect. A,* **31,** S128, 1975

21.6 **racemic - 2 - (α - p - Bromophenyl - β - nitro)ethyl - 5 - methylcyclohexanone**
$C_{15}H_{18}BrNO_3$
S.Bruckner, G.Pitacco *Gazz. Chim. Ital.,* **104,** 693, 1974
Also classified in 10

21.7 **trans - α - (4′ - Iodo - 1′ - diazacyclohexane) - β - naphthol**
$C_{16}H_{17}IN_2O$
S.G.Biswas, A.Mukherjee *Acta Crystallogr., Sect. A,* **31,** S119, 1975
Also classified in 24, 9

21.8 **cis - α - (4′ - Iodo - 1′ - diazacyclohexane) - β - naphthol**
$C_{16}H_{17}IN_3O$
S.G.Biswas, A.Mukherjee *Acta Crystallogr., Sect. A,* **31,** S119, 1975
Also classified in 24, 9

21.9 **1 - Phenyl - 4 - t - butylcyclohexane - 1 - carboxylic acid**
$C_{17}H_{24}O_2$
A.Chiaroni, C.Riche, C.Pascard-Billy
Acta Crystallogr., Sect. B, **30,** 1914, 1974

21.10 **trans - 4 - t - Butylcyclohexyl toluene - p - sulfonate (neutron study)**
$C_{17}H_{26}O_3S$
V.J.James, F.H.Moore *Acta Crystallogr., Sect. B,* **31,** 1053, 1975
Also classified in 11

21.11 **2 - (Cyclohexyl - (phenyl) - acetoxy)ethyl - dimethyl - sulfonium iodide**
Hexasonium iodide
$C_{18}H_{27}O_2S^+$, I^-
J.J.Guy, T.A.Hamor *J. Chem. Soc., Perkin Trans. 2,* 467, 1975
Residue 1 also classified in 11

21.C **2,6 - Di - cis - 4 - hydroxyretinoic acid γ - lactone**
$C_{20}H_{26}O_2$
For complete entry see 38.63

21.12 **α - Acetoxy - α,2 - anti - diphenylmethylene - cyclohexane**
$C_{21}H_{22}O_2$
F.P.van Remoortere, J.J.Flynn *J. Am. Chem. Soc.,* **96,** 6593, 1974

21.13 **trans - 2,5 - Di - (p - bromobenzyl) - 2,5 - dicarboethoxy - cyclohexane - 1,4 - dione**
$C_{26}H_{26}Br_2O_6$
C.-K.Chan, J.C.N.Ma, T.C.W.Mak
Acta Crystallogr., Sect. A, **31,** S124, 1975

21.C **Dimethyl 2 - cyclohexyl - 1,2 - dihydro - 1 - oxo - 6 - phenyl - 2 - benzazocine - 4,5 - dicarboxylate**
$C_{27}H_{27}NO_5$
For complete entry see 35.40

21.C **Ergocalciferol**
Vitamin D2
$C_{28}H_{44}O$
For complete entry see 51.46

MONOCYCLIC HYDROCARBONS
(7, 8-MEMBERED RINGS)

22.1 **Hexafluorotropone (at −50°C)**
C_7F_6O
J.J.Guy, T.A.Hamor, C.M.Jenkins *J. Fluorine Chem.*, **5**, 89, 1975

22.2 **Methoxy - 2,4,5,6,7 - pentafluorotropone**
$C_8H_3F_5O_2$
M.J.Hamor, T.A.Hamor *Acta Crystallogr., Sect. A*, **31**, S102, 1975

22.C **γ - Thujaplicin**
$C_{10}H_{12}O_2$
For complete entry see 52.2

22.3 **4 - Bromo - 2,3 - dicarbomethoxy - 2 - cyclohepten - 1 - one**
$C_{11}H_{13}BrO_5$
J.L.Atwood, M.D.Williams, R.H.Garner, E.J.Cone
Acta Crystallogr., Sect. B, **30**, 2066, 1974

22.4 **trans - 2 - Cyclo - octenyl - 3′,5′ - dinitrobenzoate**
$C_{15}H_{16}N_2O_6$
O.Ermer *Angew. Chem.*, **86**, 672, 1974

22.C **4,6 - Di - t - butyl - 2,3 - di(methoxycarbonyl) - 7 - heptatrienylidene - bicyclo(2.2.1)hept - 2,6 - diene**
$C_{26}H_{32}O_4$
For complete entry see 31.29

MONOCYCLIC HYDROCARBONS (9- AND HIGHER-MEMBERED RINGS)

23.1 **Cyclotetradecanone (at −157°C)**
$C_{14}H_{26}O$
P.Groth *Acta Chem. Scand. Ser. A*, **29**, 374, 1975

23.2 **1,1,9,9 - Tetramethyl - cyclohexadecane**
$C_{20}H_{40}$
P.Groth *Acta Chem. Scand. Ser. A*, **28**, 808, 1974

NAPHTHALENE COMPOUNDS

24.1 **Octafluoronaphthalene**
$C_{10}F_8$
N.A.Akhmed *Zh. Strukt. Khim.*, **14,** 573, 1973

24.2 **1,5 - Difluoronaphthalene**
$C_{10}H_6F_2$
A.Meresse, C.Courseille, F.Leroy, N.B.Chanh
Acta Crystallogr., Sect. B, **31,** 1236, 1975

24.3 **1,8 - Difluoronaphthalene**
$C_{10}H_6F_2$
A.Meresse, C.Courseille, F.Leroy, N.B.Chanh
Acta Crystallogr., Sect. B, **31,** 1236, 1975

24.C **Lumiflavin - bis(naphthalene - 2,3 - diol) trihydrate**
$2C_{10}H_8O_2$, $C_{13}H_{12}N_4O_2$, $3H_2O$
For complete entry see 60.19

24.4 **3 - Hydroxy - 2 - naphthoic acid**
$C_{11}H_8O_3$
M.P.Gupta, B.P.Dutta *Cryst. Struct. Commun.*, **4,** 37, 1975

24.5 **2,3,4 - Trichloro - naphthalene - 1 - glyoxylic acid**
$C_{12}H_5Cl_3O_3$
J.S.Cantrell, R.A.Lunsford, J.L.Pyle
Am. Cryst. Assoc., Abstr. Papers (Spring Meeting), 28, 1975

24.6 **1,8 - Di(bromomethyl)naphthalene**
$C_{12}H_{10}Br_2$
J.-B.Robert, J.S.Sherfinski, R.E.Marsh, J.D.Roberts
J. Org. Chem., **39,** 1152, 1974

24.7 **1 - Naphthylsulfinyl acetic acid**
$C_{12}H_{10}O_3S$
L.Leiserowitz, G.Salem, M.Weinstein *Cryst. Struct. Commun.*, **4,** 85, 1975
Also classified in 1, 11

24.C **1 - (2 - Thiazolylazo) - 2 - naphthol**
$C_{13}H_9N_3OS$
For complete entry see 41.14

24.8 **(+) - 1 - Isopropylamino - 2 - (β - naphthyl) - ethan - 2 - ol hydrochloride**
Pronethanol
$C_{15}H_{20}NO^{+}$, Cl^{-}
M.Gadret, M.Goursolle, J.M.Leger, J.C.Colleter
Acta Crystallogr., Sect. B, **31,** 1522, 1975
Residue 1 also classified in 3

24.9 **Ammonium 8 - anilino - 1 - naphthalenesulfonate**
$C_{16}H_{12}NO_{3}S^{-}$, $H_{4}N^{+}$
V.Cody *Acta Crystallogr., Sect. A,* **31,** S58, 1975
Residue 1 also classified in 16

24.C **trans - α - (4′ - Iodo - 1′ - diazacyclohexane) - β - naphthol**
$C_{16}H_{17}IN_{2}O$
For complete entry see 21.7

24.C **cis - α - (4′ - Iodo - 1′ - diazacyclohexane) - β - naphthol**
$C_{16}H_{17}IN_{3}O$
For complete entry see 21.8

24.C **4 - Bromo - 2 - (4 - methyl - 2 - morpholino - 1H - 5 - imidazolyl - methylidene) - 1,2 - dihydro - 1 - oxonaphthalene**
$C_{19}H_{18}BrN_{3}O_{2}$
For complete entry see 40.22

24.C **P^{1},P^{2} - Di - β - naphthyl - pyrophosphate di(cyclohexylammonium) salt**
$C_{20}H_{14}O_{7}P_{2}^{2-}$, $2C_{6}H_{14}N^{+}$
For complete entry see 46.6

24.C **2,2,6,6 - Tetramethylpiperidine - 1 - iminoxyl - 4 - (N - 2 - hydroxy - 1 - naphthaldehyde imine)**
$C_{20}H_{25}N_{2}O_{2}$
For complete entry see 33.51

24.10 **(−) - 2,2′ - bis(Bromomethyl) - 1,1′ - binaphthyl (absolute configuration)**
$C_{22}H_{16}Br_{2}$
K.Harata, J.Tanaka *Bull. Chem. Soc. Jpn.,* **46,** 2747, 1973

24.11 **1 - (2,5 - Dichlorophenylazo) - 2 - hydroxy - 3 - naphthoic acid 4 - chloro - 2 - methoxy - anilide**
$C_{24}H_{16}Cl_{3}N_{3}O_{3}$
D.Kobelt, E.F.Paulus, W.Kunstmann *Z. Kristallogr.,* **139,** 15, 1974
Also classified in 16, 9

NAPHTHOQUINONES

25.1 **3 - Bromo - 1,2 - naphthoquinone**
$C_{10}H_5BrO_2$
J.Gaultier, C.Hauw, J.Housty, M.Schvoerer
Cryst. Struct. Commun., **4,** 211, 1975

25.2 **2 - Amino - 1,4 - naphthoquinone iminium chloride dihydrate**
$C_{10}H_9N_2O^+$, Cl^- , $2H_2O$
C.Courseille, S.Geoffre, F.Leroy, M.Hospital
Cryst. Struct. Commun., **3,** 583, 1974

25.3 **4 - Diethylamino - 1,2 - naphthoquinone**
$C_{14}H_{15}NO_2$
F.Bechtel, J.Gaultier, S.Geoffre, C.Hauw
Cryst. Struct. Commun., **4,** 221, 1975

25.4 **1,2 - Naphthoquinone 1 - (2 - nitro - 4 - chlorophenylhydrazone)**
$C_{16}H_{10}ClN_3O_3$
L.J.Guggenberger, G.Teufer *Acta Crystallogr., Sect. B,* **31,** 785, 1975
Also classified in 16, 15, 9

25.5 **2 - Chloro - 3 - (N - methyl - N - p - tolyl)aminomethyl - 1,4 - naphthoquinone**
$C_{19}H_{16}ClNO_2$
A.E.Shvets, Ya.Ya.Bleidelis, Ya.F.Freimanis
Acta Crystallogr., Sect. A, **31,** S129, 1975
Also classified in 16

25.6 **2 - (N - Acetyl - 2(2′ - N′ - ethyl - N′ - phenyl)amino)ethylamino - 3 - chloro - 1,4, - naphthoquinone**
$C_{22}H_{21}ClN_2O_3$
A.E.Shvets, Ya.Ya.Bleidelis, Ya.F.Freimanis
Acta Crystallogr., Sect. A, **31,** S129, 1975
Also classified in 16

ANTHRACENE COMPOUNDS

26.1 **Octafluoroanthraquinone**
$C_{14}F_8O_2$
L.A.Chetkina, E.G.Popova *Kristallografiya,* **18,** 1162, 1973

26.2 **3 - Bromo - 1,4 - anthraquinone**
$C_{14}H_7BrO_2$
J.Gaultier, C.Hauw, J.-M.Lago, M.Schvoerer
Cryst. Struct. Commun., **4,** 271, 1975

26.3 **1 - Iodo - anthraquinone**
$C_{14}H_7IO_2$
N.L.Klimasenko, V.K.Bel'skii, V.K.Sarynin, G.A.Gol'der
Kristallografiya, **18,** 725, 1973

26.C **Anthracene - pyromellitic dithioanhydride complex**
$C_{14}H_{10}$, $C_{10}H_2O_4S$
For complete entry see 60.11

26.4 **bis(Tetramethylethylenediamine) di - lithium(i) anthracenide**
$C_{14}H_{10}^{2-}$, $2C_6H_{16}N_2$, $2Li^+$
W.E.Rhine, J.Davis, G.Stucky *J. Am. Chem. Soc.*, **97,** 2079, 1975
Residue 2 classified in 3

26.5 **9,10 - Dimethylanthracene**
$C_{16}H_{14}$
J.Iball, J.N.Low *Acta Crystallogr., Sect. B,* **30,** 2203, 1974

26.6 **9 - Anthryl styryl ketone**
$C_{23}H_{16}O$
R.L.Harlow, R.A.Loghry, H.J.Williams, S.H.Simonsen
Acta Crystallogr., Sect. B, **31,** 1344, 1975

26.7 **9,10 - Anthryl - bis(styryl ketone)**
$C_{32}H_{22}O_2$
R.L.Harlow, R.A.Loghry, H.J.Williams, S.H.Simonsen
Acta Crystallogr., Sect. B, **31,** 1344, 1975

HYDROCARBONS (2 FUSED RINGS)

27.1 **Perchlorobenzocyclobutene (β form)**
C_8Cl_8
J.Haase, P.Widmann *Z. Naturforsch., Teil A,* **29,** 831, 1974

27.2 **Indenyl lithium tetramethylethylenediamine**
$C_9H_7^-$, $C_6H_{16}N_2$, Li^+
W.E.Rhine, G.D.Stucky *J. Am. Chem. Soc.,* **97,** 737, 1975
Residue 1 also classified in 12; residue 2 classified in 3

27.3 **cis - Perfluorodecalin - 9,10 - diol**
$C_{10}H_2F_{16}O_2$
M.J.Hamor, T.A.Hamor *Acta Crystallogr., Sect. A,* **31,** S102, 1975

27.4 **Spirodienone**
Spiro(5.5)undeca - 1,4,7,10 - tetraene - 3,9 - dione
$C_{11}H_8O_2$
B.S.Hass, T.V.Willoughby, C.N.Morimoto, D.L.Cullen, E.F.Meyer
Acta Crystallogr., Sect. B, **31,** 1225, 1975

27.C **Dimethyl 2,5 - dibromo - 7 - phenyl - norcaradiene - 7 - phosphonate**
$C_{15}H_{15}Br_2O_3P$
For complete entry see 64.39

27.C **Dimethyl 2,5 - dichloro - 7 - phenyl - norcaradiene - 7 - phosphonate**
$C_{15}H_{15}Cl_2O_3P$
For complete entry see 64.40

27.5 **9 - Carbomethoxy - 4,6,6 - trimethyl - trans - decal - 3 - one**
$C_{15}H_{24}O_3$
C.S.Huber, E.J.Gabe *Acta Crystallogr., Sect. B,* **30,** 2519, 1974

27.6 **1 - Phenyl - 1,2,3,4 - tetrachlorotetralin**
$C_{16}H_{12}Cl_4$
J.E.Godfrey, J.M.Waters *Cryst. Struct. Commun.,* **4,** 45, 1975

27.C **6α - Bromo - 4α - (3,5 - dimethyl - 4 - isoxazolyl - methyl) - 1β - hydroxy - 7aβ - methyl - 3aα,6,7,7a - tetrahydro - 5(4H) - indanone**
$C_{16}H_{22}BrNO_3$
For complete entry see 40.11

27.7 **2 - (p - Chlorobenzylidene) - tetral - 1 - one**
$C_{17}H_{13}ClO$
M.Bolognesi, A.Coda, A.C.Corsico, G.Desimoni
Acta Crystallogr., Sect. A, **31,** S119, 1975

27.8 **1 - Phenylamino - 2 - phenyl - 3 - indenone**
$C_{21}H_{15}NO$
A.E.Shvets, A.A.Kemme, Ya.Ya.Bleidelis, Ya.F.Freimanis, R.P.Shibaeva
Zh. Strukt. Khim., **14,** 576, 1973

27.9 **endo,endo - 2,6 - bis(Phenylcarbamoyl - oxy) - cis - bicyclo(3.3.0)octane**
$C_{22}H_{24}N_2O_4$
G.Ferguson, S.Phillips, R.J.Restivo
J. Chem. Soc., Perkin Trans. 2, 405, 1975

27.10 **trans - 1 - (2 - O - Methoxyphenylethyl) - 4β - hydroxy - 2,5α,9β - trimethyl - 1 - octal - 8 - one**
$C_{22}H_{30}O_3$
F.Brisse, A.Lectard, C.Schmidt *Can. J. Chem.,* **52,** 1123, 1974
Also classified in 17

27.11 **endo - 6 - Methoxy - 1,3,6 - triphenyl - bicyclo(3.1.0)hex - 3 - ene - 2 - one**
$C_{25}H_{20}O_2$
W.J.Seifert, T.Debaerdemaeker, U.Muller
Acta Crystallogr., Sect. B, **31,** 537, 1975

27.C **Ergocalciferol**
Vitamin D2
$C_{28}H_{44}O$
For complete entry see 51.46

27.C **9 - Oxo - 9,11 - secogorgost - 5 - ene - 3β,11 - diol 11 - acetate p - iodobenzoate (absolute configuration)**
Secogorgosterol 3 - p - iodobenzoate 11 - acetate
$C_{39}H_{55}IO_5$
For complete entry see 51.59

HYDROCARBONS (3 FUSED RINGS)

28.1 **Acenaphthylene**
$C_{12}H_8$
F.Cser *Acta Chim. Acad. Sci. Hung.*, **80,** 317, 1974

28.2 **trans - racemic - 1,2 - Dibromoacenaphthene**
$C_{12}H_8Br_2$
M.-C.Perucaud, J.Canceill, J.Jacques *Bull. Soc. Chim. Fr.*, 1011, 1974

28.3 **1,2,5,6 - Tetracyano - anti - tricyclo(4.2.0.0^{2,5})octane**
$C_{12}H_8N_4$
D.Bellus, H.-C.Mez, G.Rihs, H.Sauter *J. Am. Chem. Soc.*, **96,** 5007, 1974

28.4 **7 - Acenaphthenol**
$C_{12}H_{10}O$
M.P.Gupta, T.N.P.Gupta *Acta Crystallogr., Sect. B*, **31,** 7, 1975

28.C **8β,9β - Difluoromethylene - 10 - methyl - 2 - decalone 2 - ethylene acetal**
$C_{14}H_{20}F_2O_2$
For complete entry see 38.47

28.C **(3,3' - Diethylthiazolinocarbocyanine) - (7,7,8,8 - tetracyanoquinodimethanide) - (9 - dicyanomethylene - 2,4,7 - trinitrofluorene) complex**
$C_{16}H_5N_5O_6$, $C_{12}H_4N_4^-$, $C_{13}H_{21}N_2S_2^+$
For complete entry see 60.21

28.5 **5,6,11,12 - Tetradehydro - dibenzo(a,e)cyclo - octene**
Dibenzo - 1,5 - cyclo - octadiene - 3,7 - diyne
$C_{16}H_8$
R.Destro, T.Pilati, M.Simonetta *J. Am. Chem. Soc.*, **97,** 658, 1975

28.C **2,7 - Pentamethylene - 4,5 - benzotropone**
$C_{16}H_{16}O$
For complete entry see 31.12

28.6 **DL - 1,1,4aα - Trimethyl - 2α - hydroxy - 8β - methoxy - 1,2,3,4,4a,5,6,7,8,9 - decahydrophenanthrene**
$C_{18}H_{28}O_2$
A.Furusaki, N.Hamanaka, T.Matsumoto
Bull. Chem. Soc. Jpn., **47,** 537, 1974

28.C **Fluorenyl - potassium - tetramethylethylenediamine**
$(C_{19}H_{25}N_2^-)_n$, nK^+
For complete entry see 67.6

28.C **Isocilludin S triacetate (absolute configuration)**
$C_{21}H_{26}O_7$
For complete entry see 50.11

28.7 **anti - Tetraethoxycarbonyl - 2,5,8,11 - tetrahydroxy - tricyclo(6.4.0.0^{2,7})dodeca - 4,10 - diene**
$C_{24}H_{32}O_{12}$
H.-C.Mez, G.Rihs *Helv. Chim. Acta,* **56,** 2772, 1973

28.8 **12,13 - bis(t - Butyl) - 9,9,10,10 - tetracyano - tricyclo(9.3.0.0^{2,8})tetradeca - 2,4,6,12 - tetraene**
$C_{26}H_{28}N_4$
R.E.Davis, W.Henslee, A.Garza, H.Knofel, H.Prinzbach
Tetrahedron Lett., 2823, 1974

28.9 **6,12 - Dicarbomethoxy - 5,11 - di(isobuten - 1′ - yl) - 1,7 - dimethyl - tricyclo(6.4.0.0^{2,7})dodeca - 3,9 - dione**
$C_{26}H_{36}O_6$
M.J.Begley *Acta Crystallogr., Sect. A,* **31,** S128, 1975

28.C **1 - Phenyl - 3,3 - biphenylene - allene dimer**
$C_{42}H_{28}$
For complete entry see 29.12

HYDROCARBONS (4 FUSED RINGS)

29.1 1,3,6,8 - Tetrafluoro - 2,4,5,7,9,10 - hexachloro - pyrene
$C_{16}Cl_6F_4$
A.Hazell, S.Jagner, A.Weigelt *Acta Crystallogr., Sect. A,* **31,** S126, 1975

29.2 Decachloropyrene
$C_{16}Cl_{10}$
A.Hazell, S.Jagner, A.Weigelt *Acta Crystallogr., Sect. A,* **31,** S126, 1975

29.C Fluoroanthene - picryl bromide complex
$C_{16}H_{10}$, $C_6H_2BrN_3O_6$
For complete entry see 60.22

29.C Perdeuteropyrene - tetracyanoethylene complex (at 105°K,ordered model)
$C_{16}D_{10}$, C_6N_4
For complete entry see 60.23

29.C Perdeuteropyrene - tetracyanoethylene complex (at 105°K,disordered model)
$C_{16}D_{10}$, C_6N_4
For complete entry see 60.24

29.C Pyrene - picryl bromide complex
$1.5C_{16}H_{10}$, $C_6H_2BrN_3O_6$
For complete entry see 60.25

29.3 9H,10H - Indeno(1,2 - a)indene
$C_{16}H_{12}$
T.Matsuzaki *Acta Crystallogr., Sect. B,* **30,** 2060, 1974

29.C Benz(a)anthracene - pyromellitic dianhydride complex
$C_{18}H_{12}$, $C_{10}H_2O_6$
For complete entry see 60.27

29.C Chrysene - 7,7,8,8 - tetracyanoquinodimethane complex
$C_{18}H_{12}$, $C_{12}H_4N_4$
For complete entry see 60.28

29.C (±) - 3 - Methoxy - β - nor - 9β - estra - 1,3,5(10) - trien - 17 - one
$C_{18}H_{22}O_2$
For complete entry see 51.2

29.4 **7 - Chloromethyl - benz(a)anthracene**
$C_{19}H_{13}Cl$
J.P.Glusker, H.L.Carrell, D.E.Zacharias, R.Harvey
Acta Crystallogr., Sect. A, **31**, S122, 1975

29.5 **1 - Methyl - benz(a)anthracene**
$C_{19}H_{14}$
D.W.Jones, J.M.Sowden *Acta Crystallogr., Sect. A*, **31**, S126, 1975

29.6 **12 - Methyl - benz(a)anthracene**
$C_{19}H_{14}$
D.W.Jones, J.M.Sowden *Acta Crystallogr., Sect. A*, **31**, S126, 1975

29.C **5β - Cyano - A - nor - 6 - androstene - 3,17 - dione**
$C_{19}H_{23}NO_2$
For complete entry see 51.4

29.7 **7 - Chloromethyl - 12 - methyl - benz(a)anthracene**
$C_{20}H_{15}Cl$
J.P.Glusker, H.L.Carrell, D.E.Zacharias, R.Harvey
Acta Crystallogr., Sect. A, **31**, S122, 1975

29.8 **1,12 - Dimethylbenz(a)anthracene**
$C_{20}H_{16}$
D.W.Jones, J.M.Sowden *Acta Crystallogr., Sect. A*, **31**, S126, 1975

29.9 **5,6 - Dihydro - 7,12 - dimethylbenz(a)anthracene - 5,6 - cis - diol**
$C_{20}H_{18}O_2$
J.P.Glusker, H.L.Carrell, D.E.Zacharias, R.Harvey
Acta Crystallogr., Sect. A, **31**, S122, 1975

29.C **(10S) - 17β - Acetoxy - 3,10 - cyclo - 3,4 - seco - 4,9(11) - estradien - 1 - one**
$C_{20}H_{26}O_3$
For complete entry see 51.10

29.C **16α - Acetyl - 3β - methoxy - CD - cis - D - nor - androstane**
$C_{21}H_{34}O_2$
For complete entry see 51.21

29.C **16β - Acetyl - 3β - methoxy - CD - cis - D - nor - androstane**
$C_{21}H_{34}O_2$
For complete entry see 51.22

29.C **Dipotassium oxytetracycline tetrahydrate (at −150°C)**
$C_{22}H_{22}N_2O_9^{2-}$, $2K^+$, $4H_2O$
For complete entry see 50.14

29.C **Tetracycline hexahydrate (at −150°C)**
$C_{22}H_{24}N_2O_8$, $6H_2O$
For complete entry see 50.15

29.C **Oxytetracycline mercury(ii) chloride dihydrate (at −150°C)**
$C_{22}H_{24}N_2O_9$, Cl_2Hg , $2H_2O$
For complete entry see 50.16

29.10 **1,1,7,7 - Tetramethyl - 1,2,3,4,7,8,9,10 - octahydrochrysen - 4,10 - dione**
$C_{22}H_{24}O_2$
A.Drusiani, L.Plessi, E.Foresti Serantoni, R.Mongiorgi
Atti Accad. Naz. Lincei, Cl. Sci. Fis., Mat. Nat., Rend., **55,** 711, 1973

29.11 **N,N - dimethylspiro(5H - dibenzo(a,d)cycloheptene - 5,1′ - cyclohexan) - 4′ - amine**
$C_{22}H_{25}N$
O.Kennard, M.L.Post, J.R.Rodgers, A.S.Horn
Acta Crystallogr., Sect. A, **31,** S54, 1975

29.C **Triol Q acetonide p - iodobenzoate**
$C_{30}H_{41}IO_4$
For complete entry see 59.19

29.C **1(10 - 6)abeo - Cholesta - 5,7,9 - trien - 3 - yl p - bromobenzoate**
$C_{34}H_{45}BrO_2$
For complete entry see 51.52

29.C **Toxisterol$_2$ - A 3,5 - dinitrobenzoate**
$C_{35}H_{46}N_2O_6$
For complete entry see 51.56

29.12 **1 - Phenyl - 3,3 - biphenylene - allene dimer**
$C_{42}H_{28}$
W.Dreissig, P.Luger, D.Rewicki
Acta Crystallogr., Sect. B, **30,** 2037, 1974
Also classified in 28

HYDROCARBONS (5 OR MORE FUSED RINGS)

30.C **4,5 - Dihydro - benzo(a)pyrene - 4,5 - oxide**
$C_{20}H_{12}O$
For complete entry see 38.60

30.C **5,6 - Dihydro - 7,12 - dimethyl - benz(a)anthracene - 5,6 - oxide**
$C_{20}H_{16}O$
For complete entry see 38.62

30.1 **Anthanthrene**
$C_{22}H_{12}$
G.W.Smith, J.C.Stalley *Acta Crystallogr., Sect. A,* **31,** S122, 1975

30.2 **Dihydroanthanthrene**
$C_{22}H_{14}$
G.W.Smith, J.C.Stalley *Acta Crystallogr., Sect. A,* **31,** S122, 1975

30.3 **6,12 - Dimethyl - dibenzo(def,mno)chrysene**
6,12 - Dimethylanthanthrene
$C_{24}H_{16}$
J.Iball, S.N.Scrimgeour *J. Chem. Soc., Perkin Trans. 2,* 1445, 1974

30.4 **6aα,6bβ,12bβ,12cα - Tetrahydro - 5,5,8,8 - tetramethyl - dinaphtho(1,2 - a.1′,2′ - c)cyclobutene - 6(5H),7(8H) - dione**
$C_{24}H_{24}O_2$
J.Iball, J.N.Low *J. Chem. Soc., Perkin Trans. 2,* 1423, 1974

30.5 **6,6 - Dimethyl - 2,3 - benzo - 2,4 - cycloheptadienone dimer A**
$C_{26}H_{28}O_2$
C.G.Biefeld, B.L.Barnett *Acta Crystallogr., Sect. B,* **30,** 2411, 1974

30.6 **6,6 - Dimethyl - 2,3 - benzo - 2,4 - cycloheptadienone dimer B**
$C_{26}H_{28}O_2$
C.G.Biefeld, B.L.Barnett *Acta Crystallogr., Sect. B,* **30,** 2411, 1974

30.7 **Dinaphtho(1,2 - a.1′,2′ - h)anthracene**
$C_{30}H_{18}$
B.G.M.C.Hummelink-Peters, T.E.M.van den Hark, J.H.Noordik, P.T.Beurskens *Cryst. Struct. Commun.,* **4,** 281, 1975

30.8 **9 - Cyano - 10 - methoxyphenanthrene head - to - tail syn photodimer**
$C_{32}H_{22}N_2O_2$
C.Courseille, A.Castellan, B.Busetta, M.Hospital
Cryst. Struct. Commun., **4,** 1, 1975

30.9 **Dimethyl - 8b,8c,16b,16c - tetrahydrocyclobuta(1,2 - e,3,4 - e′)diphenanthrene 8b,16b - cis - dicarboxylate**
$C_{32}H_{24}O_4$
R.S.Harvey, B.V.McNally, C.J.Timmons, S.C.Wallwork
Cryst. Struct. Commun., **3,** 747, 1974

BRIDGED RING HYDROCARBONS

31.1 **1 - Cyano - tricyclo(3.3.0.0^{2,8})octa - 3,6 - diene (at −45°C)**
C_9H_7N
L.A.Paquette, W.E.Volz, M.A.Beno, G.G.Christoph
J. Am. Chem. Soc., **97**, 2562, 1975

31.C **2,2,5 - endo,6 - exo - 8,9,10 - Heptachloro - bornane**
$C_{10}H_{11}Cl_7$
For complete entry see 52.1

31.2 **(−) - 3,3,4 - Trimethyl - 1,7 - dibromonorbornan - 2 - one**
$C_{10}H_{14}Br_2O$
C.A.Bear, J.Trotter *Acta Crystallogr., Sect. B*, **31**, 904, 1975

31.3 **(4.4.2)Propella - 3,8 - diene - 11,12 - dione**
$C_{12}H_{12}O_2$
R.Fink, D.van der Helm, S.C.Neely
Acta Crystallogr., Sect. B, **31**, 1299, 1975

31.4 **Adamantane - 1,3 - dicarbonyl chloride**
$C_{12}H_{14}Cl_2O_2$
J.Leser, D.Rabinovich *Acta Crystallogr., Sect. A*, **31**, S166, 1975

31.5 **4,10 - Dibromo - 1,7 - methano(12)annulene**
$C_{13}H_{10}Br_2$
A.Mugnoli, M.Simonetta *Acta Crystallogr., Sect. A*, **31**, S121, 1975

31.6 **2 - Adamantylidene malonodinitrile**
$C_{13}H_{14}N_2$
A.Kutoglu, R.Allmann *Cryst. Struct. Commun.*, **4**, 57, 1975

31.7 **5,5,6,6 - Tetracyano - tricyclo(5.2.1.0^{4,8})dec - 2 - ene**
$C_{14}H_{10}N_4$
M.A.Battiste, J.M.Coxon, R.G.Posey, R.W.King, M.Mathew, G.J.Palenik
J. Am. Chem. Soc., **97**, 945, 1975

31.C **Tetrasulfur - tetranitrile - bis(norbornadiene)**
$C_{14}H_{16}N_4S_4$
For complete entry see 41.21

31.8 **3 - Carboxy(7)paracyclophane**
$C_{14}H_{18}O_2$
N.L.Allinger, T.J.Walter, M.G.Newton *J. Am. Chem. Soc.*, **96**, 4588, 1974

31.9 **2,6,6 - Trimethyl - 9 - bromomethyl - tricyclo(5.4.0.0^{2,9})undecan - 3,4 - dione**
$C_{15}H_{21}BrO_2$
G.Mehta, S.K.Kapoor, T.N.G.Row, K.Venkatesan
Tetrahedron Lett., 2653, 1974

31.10 **1,8 - Dicarboxymethyl - tricyclo(4.3.1.1^{3,8})undecane**
$C_{15}H_{22}O_4$
J.Murray-Rust, P.Murray-Rust, R.S.Henry
Acta Crystallogr., Sect. B, **31**, 585, 1975

31.11 **4,4′ - Dibromo(2,2)paracyclophane**
$C_{16}H_{14}Br_2$
P.Goldstein
Am. Cryst. Assoc., Abstr. Papers (Summer Meeting), 209, 1974

31.C **1 - (2′ - (N - Benzamidinato) - 1′,1′,1′,3′,3′,3′ - hexafluoro - prop - 2 - yl) - 1 - phospha - 2,8,9 - trioxa - adamantane benzene solvate**
$C_{16}H_{14}F_6NO_4P$, $0.5C_6H_6$
For complete entry see 64.43

31.12 **2,7 - Pentamethylene - 4,5 - benzotropone**
$C_{16}H_{16}O$
K.Ibata, H.Shimanouchi, Y.Sasada
Acta Crystallogr., Sect. B, **31**, 482, 1975
Also classified in 28

31.13 **1,5 - Dimethyl - 5 - chloro - cyclohexa - 1,3 - diene - 6 - one dimer**
$C_{16}H_{18}Cl_2O_2$
B.Karlsson, A.-M.Pilotti, A.-C.Wiehager *Eur. Cryst. Meeting*, 564, 1974

31.14 **4 - Acetyl - 11 - hydroxy - 1,7,11 - trimethyl - tricyclo(4.3.2^{2,5}.0)undec - 3 - ene - 8,10 - dione**
$C_{16}H_{20}O_4$
J.L.Flippen *Am. Cryst. Assoc., Abstr. Papers (Spring Meeting)*, 12, 1975

31.15 **6,12 - Dihydroxy - 2,6,9,12 - tetramethyl - tetracyclo(6.2.2.0^{2,7},0^{4,9})dodecane - 5,11 - dione**
$C_{16}H_{22}O_4$
K.-I.Hirao, T.Iwakuma, M.Taniguchi, E.Abe, O.Yonemitsu, T.Date, K.Kotera *J. Chem. Soc., Chem. Commun.*, 691, 1974

31.16 **1,6.8,13 - Cyclopropanediylidene(14)annulene**
$C_{17}H_{12}$
A.Mugnoli, M.Simonetta *Acta Crystallogr., Sect. B*, **30**, 2896, 1974

31.17 **3,4 - Dichloro - tricyclo(4.3.1.$0^{2,5}$)dec - 8 - en - 10 - yl p - nitrobenzoate**
$C_{17}H_{15}Cl_2NO_4$
G.Subrahmanyam, R.Srinivasan, S.J.LaPlaca, J.E.Weidenborner
J. Chem. Soc., Chem. Commun., 231, 1975

31.18 **Diketo - peristylane acetate**
Peristylane - 94
$C_{17}H_{18}O_4$
F.E.Scarbrough, W.Nowacki *Z. Kristallogr.*, **139**, 395, 1974

31.19 **2 - Hydroxy(4.2.2)propellane p - nitrobenzoate**
$C_{17}H_{19}NO_4$
J.V.Silverton, G.W.A.Milne, P.E.Eaton, K.Nyi, G.H.Temme III
J. Am. Chem. Soc., **96**, 7429, 1974

31.20 **15,16 - Dimethyl - 1,6.8,13 - ethanediylidene(14)annulene**
$C_{18}H_{16}$
R.Bianchi, G.Casalone, M.Simonetta
Acta Crystallogr., Sect. B, **31**, 1207, 1975

31.21 **(2.2.2)(1,2,4) - Cyclophane**
$C_{18}H_{18}$
R.Weiss, K.N.Trueblood
Am. Cryst. Assoc., Abstr. Papers (Summer Meeting), 251, 1974

31.22 **1 - Carbomethoxy - tetracyclo(9,2,2,$1^{4,11}$,$0^{8,16}$)hexadeca - 5,7,12,14,16(4) - pentaene**
$C_{18}H_{18}O_2$
E.Maverick, S.Smith, L.Kozerski, F.A.L.Anet, K.N.Trueblood
Acta Crystallogr., Sect. B, **31**, 805, 1975

31.23 **1,2,2,5,12,12 - Hexamethyl - pentacyclo (6.$2^{1,4}$.$2^{8,5}$.0.$0^{3,11}$.$0^{7,10}$) dodeca - 6,9 - dione**
$C_{18}H_{24}O_2$
J.L.Flippen *Am. Cryst. Assoc., Abstr. Papers (Spring Meeting)*, 12, 1975

31.24 **1,6.8,13 - Pentane - 1,5 - diylidene - (14)annulene**
$C_{19}H_{18}$
R.Bianchi, A.Mugnoli, M.Simonetta
Acta Crystallogr., Sect. B, **31**, 1283, 1975

31.C **Hexacyclo (9.3.2.$2^{4,7}$.$0^{2,9}$.$0^{3,8}$.$0^{10,12}$) octadeca - 13,15,17 - triene - 5,6 - carbonate**
$C_{19}H_{18}O_3$
For complete entry see 38.59

31.C **Phlebianorkauranol**
$C_{19}H_{28}O_5$
For complete entry see 50.9

31.25 **1,5 - Dimethyl - 5 - acetoxy - cyclohexa - 1,3 - diene - 6 - one dimer**
$C_{20}H_{24}O_6$
B.Karlsson, A.-M.Pilotti, A.-C.Wiehager *Eur. Cryst. Meeting,* 564, 1974

31.26 **9 - (p - Bromobenzoyloxy) - 11 - methylene - tetracyclo(8.2.1^{2,5}.0.0^{1,6}) tridecane**
$C_{21}H_{23}BrO_2$
M.Przybylska, F.R.Ahmed *Acta Crystallogr., Sect. B,* **30,** 2338, 1974

31.C **Phlebiakauranol**
$C_{21}H_{32}O_7$
For complete entry see 50.13

31.27 **(+) - 2,5 - Dimethoxy - 8 - chloro - triptycene**
$C_{22}H_{17}ClO_2$
T.Kaneda, N.Sakabe, J.Tanaka *Bull. Chem. Soc. Jpn.,* **47,** 1858, 1974

31.28 **1,4,4aα,4bβ - 5,8,8aβ,9aα - Octahydro - anti,anti - 10,11 - di - t - butoxy - 1,4.5,8 - dimethanofluorene - 9 - one**
$C_{23}H_{32}O_3$
S.E.Ealick, D.van der Helm *Cryst. Struct. Commun.,* **4,** 369, 1975

31.29 **4,6 - Di - t - butyl - 2,3 - di(methoxycarbonyl) - 7 - heptatrienylidene - bicyclo(2.2.1)hept - 2,6 - diene**
$C_{26}H_{32}O_4$
W.Henslee, R.E.Davis *Acta Crystallogr., Sect. B,* **31,** 1511, 1975
Also classified in 22

31.30 **9,9′,10,10′ - Tetradehydro - dianthracene**
$C_{28}H_{16}$
R.L.Viavattene, F.D.Greene, L.D.Cheung, R.Majeste, L.M.Trefonas
J. Am. Chem. Soc., **96,** 4342, 1974

31.31 **15 - Bromoacetoxymethyl - 15 - methyl - 8 - methoxy - 5 - isopropyl - heptacyclo(12.4.1^{1,5}.1^{7,10}.0.0^{2,11}. 9^{4,9}.0^{6,11})dodecan - 20 - one**
$C_{28}H_{39}BrO_4$
J.A.Turner, R.S.McEwen *Acta Crystallogr., Sect. B,* **30,** 2151, 1974

31.32 **(2_4) - Metacyclophane (form i)**
$C_{32}H_{32}$
I.Ueda, W.Nowacki *Z. Kristallogr.,* **139,** 70, 1974

31.33 **(2_6) - Metacyclophane n - hexane solvate**
$C_{48}H_{48}$, $0.33C_6H_{14}$
I.Ueda, F.E.Scarbrough, W.Nowacki *Z. Kristallogr.,* **140,** 169, 1974

31.34 **cis - (Resorcinol p - bromobenzaldehyde condensation tetramer) octa - acetate**
$C_{68}H_{52}Br_4O_{16}$
K.J.Palmer, R.Y.Wong, L.Jurd, K.Stevens
Am. Cryst. Assoc., Abstr. Papers (Spring Meeting), 59, 1974

31.35 trans - (Resorcinol p - bromobenzaldehyde condensation tetramer) octa - acetate

$C_{68}H_{52}Br_4O_{16}$

K.J.Palmer, R.Y.Wong, L.Jurd, K.Stevens

Am. Cryst. Assoc., Abstr. Papers (Spring Meeting), 59, 1974

HETERO-NITROGEN (3, 4, 5-MEMBERED MONOCYCLIC)

32.1 **Imidazolium maleate**
$C_3H_5N_2^+$, $C_4H_3O_4^-$
M.Matsushima, M.N.G.James
Am. Cryst. Assoc., Abstr. Papers (Summer Meeting), 254, 1974
Residue 2 classified in 2, 1

32.2 **5 - Amino - 1H - 1,2,3 - triazole - 4 - carboxamide**
$C_3H_5N_5O$
A.Kalman, K.Simon, J.Schawartz, G.Horvath
J. Chem. Soc., Perkin Trans. 2, 1849, 1974

32.C **Triphenylmethane pyrrole**
C_4H_5N , $C_{19}H_{16}$
For complete entry see 60.31

32.3 **5 - Amino - 3H - imidazole - 4 - carboxamide monohydrate**
$C_4H_6N_4O$, H_2O
A.Kalman, K.Simon *Acta Crystallogr., Sect. A,* **31,** S175, 1975

32.4 **Histamine hydrobromide**
2 - (4 - Imidazolyl)ethylammonium bromide
$C_5H_{10}N_3^+$, Br^-
K.Prout, S.R.Critchley, C.R.Ganellin
Acta Crystallogr., Sect. B, **30,** 2884, 1974
Residue 1 also classified in 3

32.5 **1,1,1 - Trimethylhydrazinium 3 - carbomethoxy - 5 - pyrazolecarboxylate**
$C_6H_5N_2O_4^-$, $C_3H_{11}N_2^+$
R.L.Harlow, S.H.Simonsen *Acta Crystallogr., Sect. B,* **30,** 2505, 1974
Residue 2 classified in 3

32.6 **1,3 - Dimethyl - pyrazole - 5 - carboxylic acid**
$C_6H_8N_2O_2$
E.Foresti Serantoni, R.Mongiorgi, L.Riva di Sanseverino
Atti Accad. Naz. Lincei, Cl. Sci. Fis., Mat. Nat., Rend., **54,** 787, 1973

32.C **3R - (1′(S) - Aminocarboxymethyl) - 2 - pyrrolidone - 5(S) - carboxylic acid**
$C_7H_{10}N_2O_5$
For complete entry see 48.34

32.7 **4,4 - Diethyl - 3,5 - dioxo - pyrazolidine**
$C_7H_{12}N_2O_2$
O.Dideberg, L.Dupont, J.Toussaint
Acta Crystallogr., Sect. B, **30,** 2444, 1974

32.8 **N,N′ - Dithio - di(succinimide)**
$C_8H_8N_2O_4S_2$
M.-ul-Haque, M.Behforouz *J. Chem. Soc., Perkin Trans. 2,* 1459, 1974

32.C **1 - Methyl - 4 - (β - D - erythrofuranosyl) - 4 - imidazoline - 2 - thione**
$C_8H_{12}N_2O_3S$
For complete entry see 45.19

32.9 **2,2,5,5 - Tetramethyl - 1 - aza - 3 - cyclopentanone - 3 - oxime - 1 - oxyl**
$C_8H_{15}N_2O_2$
B.Chion, M.Thomas *Acta Crystallogr., Sect. B,* **31,** 472, 1975
Also classified in 12

32.10 **4 - Phenyl - 3 - azidopyrazole**
$C_9H_7N_5$
P.Domiano, A.Musatti *Cryst. Struct. Commun.,* **3,** 713, 1974

32.C **6 - ((1 - Methyl - 4 - nitroimidazol - 5 - yl)mercapto)purine dihydrate**
$C_9H_7N_7O_2S$, $2H_2O$
For complete entry see 44.21

32.11 **2,4 - Dinitro - 5 - ethyleneimino - benzamide**
$C_9H_8N_4O_5$
J.Iball, S.N.Scrimgeour, B.C.Williams
Acta Crystallogr., Sect. B, **31,** 1121, 1975
Also classified in 13, 15

32.C **5 - Amino - 1 - (3 - deoxy - β - D - xylofuranosyl)imidazole - 4 - carboxamidinium - N⁵ - 3′ - cyclonucleoside formate**
$C_9H_{14}N_5O_3^+$, CHO_2^-
For complete entry see 47.15

32.12 **N - (4 - Butyloxycarbonyl) - L - azetidine - 2 - carboxylic acid**
$C_9H_{15}NO_4$
M.Cesari, L.d'Ilario, E.Giglio, G.Perego
Acta Crystallogr., Sect. B, **31,** 49, 1975

32.13 **racemic - 2,2,5,5 - Tetramethyl - pyrrolidine - 3 - carboxamide 1 - oxyl**
$C_9H_{17}N_2O_2$
B.Chion, J.Lajzerowicz *Acta Crystallogr., Sect. B,* **31,** 1430, 1975
Also classified in 12

32.14 **2,2,5,5 - Tetramethyl - pyrrolidine - 3 - carboxamide 1 - oxyl (optically active form)**
$C_9H_{17}N_2O_2$
B.Chion, J.Lajzerowicz *Acta Crystallogr., Sect. B,* **31,** 1430, 1975
Also classified in 12

32.15 **5,5 - Dimethyl - 3,3 - bis(trifluoromethyl) - Δ^1 - pyrazolin - 2 - yl - (1′,1′,1′,3′,3′,3′ - hexafluoro - 2′ - propanide)**
$C_{10}H_8F_{12}N_2$
A.Gieren, P.Narayanan, K.Burger, W.Thenn *Angew. Chem.,* **86,** 482, 1974

32.16 **3 - Methyl - 5 - phenylpyrazole**
$C_{10}H_{10}N_2$
E.N.Maslen, J.R.Cannon, A.H.White, A.C.Willis
J. Chem. Soc., Perkin Trans. 2, 1298, 1974

32.17 **DL - Dethiobiotin**
$C_{10}H_{18}N_2O_3$
C.S.Chen, R.Parthasarathy
Am. Cryst. Assoc., Abstr. Papers (Summer Meeting), 240, 1974
Also classified in 1

32.18 **DL - Methylene - bis(N - pyrrolid - 2 - one - 5 - carboxylic acid)**
$C_{11}H_{14}N_2O_6$
F.Baert, M.Lobry, V.Warin *C. R. Acad. Sci., Ser. C,* **280,** 743, 1975

32.19 **Dipyrrolidyl - trimethinecyanine perchlorate**
$C_{11}H_{19}N_2^+$, ClO_4^-
A.Zedler, S.Kulpe *J. Prakt. Chem.,* **317,** 38, 1975

32.20 **1 - (4 - Imidazolylsulfonyl) - 4 - phenylimidazole**
$C_{12}H_{10}N_4O_2S$
L.J.Guggenberger *Acta Crystallogr., Sect. B,* **31,** 13, 1975

32.21 **2,5 - bis(Pyrrolidin - 2 - on - 5 - yl)pyrrole monohydrate**
$C_{12}H_{15}N_3O_2$, H_2O
G.D.Andreetti, G.Bocelli, L.Cavalca, P.Sgarabotto
Gazz. Chim. Ital., **104,** 331, 1974

32.22 **Oxotremorine sesquioxalate**
1 - (4 - (2 - Oxopyrrolidin - 1 - yl)but - 2 - ynyl)pyrrolidinium sesquioxalate
$C_{12}H_{19}N_2O^+$, $C_2HO_4^-$, $0.5C_2H_2O_4$
P.J.Clarke, P.J.Pauling, T.J.Petcher
J. Chem. Soc., Perkin Trans. 2, 774, 1975
Residue 2 classified in 2, 1

32.C **1 - p - Chlorophenyl - 4 - (α - D - erythrofuranosyl) - 4 - imidazoline - 2 - thione**
$C_{13}H_{13}ClN_2O_3S$
For complete entry see 45.38

32.23 **N - Cinnamoyl - pyrrolidine**
$C_{13}H_{15}NO$
P.Engel, W.Nowacki *Z. Kristallogr.*, **139**, 207, 1974

32.C **Aminopyrine - cyclobarbital**
$C_{13}H_{17}N_3O$, $C_{12}H_{16}N_2O_3$
For complete entry see 60.20

32.24 **3 - Phenoxy - 3 - dimethylcarbamoyldimethylamino - 2 - azirine**
$C_{13}H_{17}N_3O_2$
J.Galloy, J.-P.Putzeys, G.Germain, J.-P.Declercq, M.van Meerssche
Acta Crystallogr., Sect. B, **30**, 2462, 1974
Also classified in 17

32.25 **1,3 - Dimethyl - 2 - (p - bromobenzoyl - cyanomethylene) - imidazolidine**
$C_{14}H_{14}BrN_3O$
S.Abrahamsson, G.Rehnberg, T.Liljefors, J.Sandstrom
Acta Chem. Scand. Ser. B, **28**, 1109, 1974
Also classified in 7

32.C **1 - p - Tolyl - 4 - (α - D - erythrofuranosyl)imidazoline - 2 - thione**
$C_{14}H_{16}N_2O_3S$
For complete entry see 45.42

32.C **1 - (Tri - O - acetyl - α - D - xylopyranosyl) - imidazole**
$C_{14}H_{18}N_2O_7$
For complete entry see 45.43

32.C **2,5 - Diethyl - 7 - (1 - methyl - 2 - imidazolyl) - 1H - pyrrolo(3,4 - c)pyridine - 1,3,6(2H,5H) - trione**
$C_{15}H_{16}N_4O_3$
For complete entry see 35.23

32.C **6 - Bromo - 1 - (1 - pyrrolidinyl) - 3H - pyrroline - 2,3 - dicarboxylic acid dimethyl ester**
$C_{15}H_{17}BrN_2O_4$
For complete entry see 35.24

32.C **1 - (p - Methoxyphenyl) - 3 - methyl - 4 - (D - arabino - tetrahydroxybutyl) - imidazolidine - 2 - thione**
$C_{15}H_{22}N_2O_5S$
For complete entry see 45.44

32.26 **1 - (bis - (p - Chlorophenyl)methylene) - 3 - oxo - 4,4 - dimethyl - 1,2 - diazetidinium hydroxide inner salt**
$C_{17}H_{14}Cl_2N_2O$
C.Calvo, P.C.Ip, N.Krishnamachari, J.Warkentin
Can. J. Chem., **52**, 2613, 1974

32.27 **3 - Cyclohexylimino - 2 - (p - fluorophenyl) - 4,4 - bis(trifluoromethyl) - 1 - azetidine**
$C_{17}H_{15}F_7N_2$
A.Gieren, K.Burger, W.Thenn *Z. Naturforsch., Teil B,* **29,** 399, 1974

32.28 **cis - 3,4 - Diphenyl - 3 - (methoxycarbonyl) - 1 - pyrazoline**
$C_{17}H_{16}N_2O_2$
B.Dewulf, J.Meunier-Piret, J.P.Putzeys, M.van Meerssche
Cryst. Struct. Commun., **4,** 175, 1975

32.29 **trans - 3 - Phenyl - 4 - (p - nitrophenyl) - 3 - (carbomethoxy) - anti - 5 - methyl - 1 - pyrazoline**
$C_{18}H_{17}N_3O_4$
B.Dewulf, J.P.Putzeys, M.van Meerssche
Cryst. Struct. Commun., **4,** 181, 1975

32.30 **cis - 3,4 - Diphenyl - 3 - (methoxycarbonyl) - syn - 5 - methyl - 1 - pyrazoline**
$C_{18}H_{18}N_2O_2$
S.A.Chawdhury *Cryst. Struct. Commun.,* **4,** 145, 1975

32.C **3 - Hydroxy - 1,1 - dimethylpyrrolidinium α - phenyl - 2 - thiophene - glycollate bromide**
$C_{18}H_{22}NO_3S^+$, Br^-
For complete entry see 39.30

32.31 **1 - Bromo - 2,4,6 - tripyrrolidino - benzenium bromide trihydrate**
$C_{18}H_{27}BrN_3^+$, Br^- , $3H_2O$
J.J.Stezowski *Eur. Cryst. Meeting,* 333, 1974
Residue 1 also classified in 12

32.32 **2,4,6 - Tripyrrolidino - benzenium bromide dihydrate**
$C_{18}H_{28}N_3^+$, Br^- , $2H_2O$
J.J.Stezowski *Eur. Cryst. Meeting,* 333, 1974
Residue 1 also classified in 12

32.C **4 - Bromo - 2 - (4 - methyl - 2 - morpholino - 1H - 5 - imidazolyl - methylidene) - 1,2 - dihydro - 1 - oxonaphthalene**
$C_{19}H_{18}BrN_3O_2$
For complete entry see 40.22

32.33 **4 - Butyl - 1,2 - diphenyl - pyrazolidine - 3,5 - dione**
Phenylbutazone
$C_{19}H_{20}N_2O_2$
T.P.Singh, M.Vijayan *Acta Crystallogr., Sect. A,* **31,** S51, 1975

32.34 **Triprolidine hydrochloride monohydrate**
trans - 1 - (p - Tolyl) - 1 - (2 - pyridyl) - 3 - (1 - pyrrolidino)prop - 1 - ene hydrochloride monohydrate
$C_{19}H_{23}N_2^+$, Cl^- , H_2O
M.N.G.James, G.J.B.Williams *Can. J. Chem.*, **52,** 1880, 1974
Residue 1 also classified in 33

32.35 **1 - Methyl - 2,4,6 - tripyrrolidino - benzenium iodide**
$C_{19}H_{30}N_3^+$, I^-
J.J.Stezowski *Eur. Cryst. Meeting,* 333, 1974
Residue 1 also classified in 12

32.36 **cis - 1 - Acetyl - 4 - (1 - (p - bromophenyl) - 2 - phenyl) - vinyl - 3 - pyrrolin - 2 - one**
$C_{20}H_{16}BrNO_2$
L.H.Weaver, Y.N.Hwang, B.W.Matthews
Acta Crystallogr., Sect. B, **30,** 2775, 1974

32.37 **2 - Phenyl - 4,5 - di(anilino) - 1,2,3 - triazole (at −30°C)**
$C_{20}H_{17}N_5$
R.L.Harlow, S.H.Simonsen
Am. Cryst. Assoc., Abstr. Papers (Spring Meeting), 30, 1975

32.38 **3 - Phenylazo - 5 - diethylamino - 4 - methyl - 1 - phenyl - pyrazole**
$C_{20}H_{23}N_5$
G.V.Boyd, T.Norris, P.F.Lindley
J. Chem. Soc., Chem. Commun., 639, 1974

32.C **bis(3,5 - Dimethylpyrazolyl)borane dimer**
$C_{20}H_{30}B_2N_8$
For complete entry see 62.10

32.C **2 - Bromo - 9 - (1 - pyrrolidinyl) - 5H - pyrrolo(1,2 - a)azepine - 5,6,7,8 - tetracarboxylic acid tetramethyl ester**
$C_{21}H_{23}BrN_2O_8$
For complete entry see 35.34

32.39 **(αS(R,R)) - α - ((Di - s - butylamino)methyl) - 1 - (2 - chlorobenzyl) - 1H - 2 - pyrrole - methanol p - hydroxybenzoate (absolute configuration)**
(S(R,R)) - Viminol
$C_{21}H_{32}ClN_2O^+$, $C_7H_5O_3^-$
J.V.Silverton, H.A.Lloyd *Acta Crystallogr., Sect. B,* **31,** 1576, 1975
Residue 1 also classified in 3; residue 2 classified in 14, 17

32.40 **1 - (p - Chlorophenyl) - 2 - Z - (α - hydroximinobenzyl) - 4 - phenyl - Δ^3 - imidazoline - 3 - oxide**
$C_{22}H_{18}ClN_3O_2$
K.Deckardt, R.Kreher *Z. Naturforsch., Teil B,* **29,** 237, 1974

32.C **Dextromoramide bitartrate**
$C_{25}H_{33}N_2O_2^+$, $C_4H_5O_6^-$
For complete entry see 40.24

32.C **3′ - Phenyl - 5′ - phenylazo - 2 - pyrrolidino - spiro(1H - indene - 1,2′(3′H) - 1,3,4 - thiadiazole)**
$C_{26}H_{23}N_5S$
For complete entry see 41.31

32.41 **1 - (α - o - Bromobenzoyloxy - o′ - bromobenzylideneamino) - 4,5 - diphenyl - 1,2,3 - triazole**
$C_{28}H_{18}Br_2N_4O_2$
S.C.Kokkou, P.J.Rentzeperis *Acta Crystallogr., Sect. B,* **31,** 1564, 1975

HETERO-NITROGEN
(6-MEMBERED MONOCYCLIC)

33.1 **Melamine**
2,4,6 - Triamino - s - triazine
$C_3H_6N_6$
A.C.Larson, D.T.Cromer *J. Chem. Phys.*, **60**, 185, 1974

33.2 **Cyclo - tetramethylenediamine hexahydrate**
$C_4H_{10}N_2$, $6H_2O$
G.A.Jeffrey, D.Mastropaolo, M.S.Shen
Acta Crystallogr., Sect. A, **31,** S177, 1975

33.3 **Silver iodide piperazinium tetra(dimethylformamide) solvate**
$C_4H_{12}N_2^{2+}$, $4C_3H_7NO$, $10Ag^+$, $12I^-$
J.Coetzer *Acta Crystallogr., Sect. B,* **31,** 622, 1975

33.4 **Hexahydro - 1,4 - dimethyl - s - tetrazine**
$C_4H_{12}N_4$
G.B.Ansell, J.L.Erickson *J. Chem. Soc., Perkin Trans. 2,* 270, 1975

33.5 **3,5 - Dinitropyridine**
$C_5H_3N_3O_4$
R.Destro, T.Pilati, M.Simonetta *Acta Crystallogr., Sect. B,* **30,** 2071, 1974

33.6 **3 - Bromopyridine - 1 - nitroimide**
$C_5H_4BrN_3O_2$
J.Arriau, J.Deschamps, J.R.C.Duke, J.Epsztajn, A.R.Katritzky, E.Lunt, J.W.Mitchell, S.Q.A.Rizvi, G.Roch *Tetrahedron Lett.*, 3865, 1974

33.7 **Pyrazinic acid**
$C_5H_4N_2O_2$
F.Takusagawa, T.Higuchi, A.Shimada, C.Tamura, Y.Sasada
Bull. Chem. Soc. Jpn., **47,** 1409, 1974

33.8 **p - Nitro - pyridine - N - oxide (neutron study, at 30°K)**
$C_5H_4N_2O_3$
Y.Wang, R.H.Blessing, P.Coppens, F.K.Ross, J.M.Williams
Am. Cryst. Assoc., Abstr. Papers (Summer Meeting), 204, 1974

33.9 **2 - Amino - 5 - chloro - pyridine (neutron study)**
$C_5H_5ClN_2$
A.Kvick, R.Thomas, T.F.Koetzle
Am. Cryst. Assoc., Abstr. Papers (Spring Meeting), 21, 1975

33.10 **Pyridine tri - iodo - nitrogen**
C_5H_5N , I_3N
H.Hartl, D.Ullrich *Z. Anorg. Allg. Chem.*, **409,** 228, 1974

33.C **Tri - o - thymotide pyridine clathrate**
C_5H_5N , $2C_{33}H_{36}O_6$
For complete entry see 61.5

33.11 **Pyridine - N - oxide trichloroacetic acid**
C_5H_5NO , $C_2HCl_3O_2$
L.Golic, F.Lazarini *Vestn. Slov. Kem. Drus.*, **21,** 17, 1974
Residue 2 classified in 1

33.12 **Pyridine - 1 - nitroimide**
$C_5H_5N_3O_2$
J.Arriau, J.Deschamps, J.R.C.Duke, J.Epsztajn, A.R.Katritzky, E.Lunt, J.W.Mitchell, S.Q.A.Rizvi, G.Roch *Tetrahedron Lett.*, 3865, 1974

33.13 **Pyridinium tetrachloroiodate(iii)**
$C_5H_6N^+$, Cl_4I^-
U.Gerlach, H.Hartl *Acta Crystallogr., Sect. A,* **31,** S172, 1975

33.14 **Pyridinium iodide**
$C_5H_6N^+$, I^-
H.Hartl *Acta Crystallogr., Sect. B,* **31,** 1781, 1975

33.15 **bis(Pyridinium) pentachloro - nitroso - ruthenate**
$2C_5H_6N^+$, Cl_5NORu^{2-}
V.S.Sergienko, T.S.Khodashova, M.A.Porai-Koshits, L.G.Andrutskaya
Zh. Strukt. Khim., **14,** 945, 1973

33.16 **1 - Methyl - 3,6 - pyridazinedione**
$C_5H_6N_2O_2$
T.Ottersen *Acta Chem. Scand. Ser. A,* **28,** 661, 1974

33.17 **Pyrazine - 2,3 - dicarboxylic acid dihydrate**
$C_6H_4N_2O_4$, $2H_2O$
F.Takusagawa, A.Shimada *Chem. Lett.*, 1121, 1973

33.18 **Nicotinic acid**
$C_6H_5NO_2$
M.P.Gupta, P.Kumar *Cryst. Struct. Commun.*, **4,** 365, 1975

33.19 **Picolinic acid**
$C_6H_5NO_2$
F.Takusagawa, A.Shimada *Chem. Lett.*, 1089, 1973

33.20 **bis(β - Picoline - N - oxide) - fumaric acid**
$2C_6H_7NO$, $C_4H_4O_4$
B.T.Gorres, E.R.McAfee, R.A.Jacobson
Acta Crystallogr., Sect. B, **31,** 158, 1975
Residue 2 classified in 1

33.21 **Isonicotinic acid hydrazide**
$C_6H_7N_3O$
T.N.Bhat, T.P.Singh, M.Vijayan
Acta Crystallogr., Sect. B, **30,** 2921, 1974

33.C **Picolinium p - tolyl - tetrachloro - tellurium**
$C_6H_8N^+$, $C_7H_7Cl_4Te^-$
For complete entry see 70.2

33.22 **1 - Methyl - 3 - methoxy - 6 - pyridazone (at −165°C)**
$C_6H_8N_2O_2$
T.Ottersen *Acta Chem. Scand. Ser. A,* **28,** 666, 1974

33.23 **2 - Amino - 5 - methylpyridine hydrochloride**
$C_6H_9N_2^+$, Cl^-
J.S.Sherfinski, R.E.Marsh *Acta Crystallogr., Sect. B,* **31,** 1073, 1975

33.24 **Pyridine 3,4 - dicarboxylic acid**
Cinchomeronic acid
$C_7H_5NO_4$
F.Takusagawa, K.Hirotsu, A.Shimada
Bull. Chem. Soc. Jpn., **46,** 2669, 1973

33.25 **Pyridine - 2,3 - dicarboxylic acid (neutron study)**
Quinolinic acid
$C_7H_5NO_4$
A.Kvick, T.F.Koetzle, R.Thomas, F.Takusagawa
J. Chem. Phys., **60,** 3866, 1974

33.26 **5 - Hydroxy - 2 - formylpyridine thiosemicarbazone sesquihydrate**
$C_7H_8N_4OS$, $1.5H_2O$
G.J.Palenik, D.F.Rendle, W.S.Carter
Acta Crystallogr., Sect. B, **30,** 2390, 1974
Residue 1 also classified in 8, 9

33.27 **2,3 - Dichloro - 5 - ethylamino - 6 - methoxypyrazine**
$C_7H_9Cl_2N_3O$
D.R.Carter, F.P.Boer *J. Chem. Soc., Perkin Trans. 2,* 1841, 1974

33.28 **bis(Lupetidinium) cyanoguanidine sulfate**
bis(2,6 - Dimethylpiperidinium) cyanoguanidine sulfate
$2C_7H_{16}N^+$, $C_2H_4N_4$, O_4S^{2-}
H.P.Weber *Helv. Chim. Acta,* **57,** 623, 1974
Residue 2 classified in 8

33.29 **Cyanomethyl - (2 - picolyl) - sulfone**
$C_8H_8N_2O_2S$
R.L.Harlow, M.P.Sammes, S.H.Simonsen
Acta Crystallogr., Sect. B, **30,** 2903, 1974
Also classified in 11

33.C **6 - Azacytidine**
$C_8H_{12}N_4O_5$
For complete entry see 47.1

33.30 **5 - (p - Chlorophenyl) - 1,2,4 - triazine**
$C_9H_6ClN_3$
J.L.Atwood, D.K.Krass, W.W.Paudler *J. Heterocycl. Chem.,* **11,** 743, 1974

33.31 **N - Succinyl - pyridine**
$C_9H_9NO_4$
M.Matsushima, M.N.G.James
Am. Cryst. Assoc., Abstr. Papers (Summer Meeting), 254, 1974
Also classified in 1

33.32 **2,2,6,6 - Tetramethyl - piperidine - 1 - oxyl (monoclinic form)**
$C_9H_{18}NO$
D.Bordeaux, J.Lajzerowicz-Bonneteau, R.Briere, H.Lemaire, A.Rassat
Org. Magn. Reson., **5,** 47, 1973
Also classified in 12

33.33 **2,2,6,6 - Tetramethyl - piperidine - 1 - oxyl (tetragonal form)**
$C_9H_{18}NO$
A.Capiomont, J.Lajzerowicz-Bonneteau
Acta Crystallogr., Sect. B, **30,** 2160, 1974
Also classified in 12

33.C **2,2′ - Anhydro - 2 - hydroxy - 1 - (β - D - arabinopentofuranosyl) - 4 - pyridine**
$C_{10}H_{11}NO_5$
For complete entry see 47.17

33.C **3 - Deaza - cytidine**
$C_{10}H_{14}N_2O_5$
For complete entry see 47.22

33.34 **N,N′ - Dimethyl - 4,4′ - bipyridylium hexachloro - dicopper(ii)**
Paraquat hexachloro - dicopper(ii)
$C_{12}H_{14}N_2^{2+}$, $Cl_6Cu_2^{2-}$
P.Murray-Rust *Acta Crystallogr., Sect. B,* **31,** 1771, 1975

33.35 **2 - (Diethoxycarbonyl)amino - 3 - carbamoyl - 5 - methylpyrazine - 1 - oxide**
$C_{12}H_{16}N_4O_6$
W.C.Schmidt, C.D.Duncan, R.F.Bryan
Am. Cryst. Assoc., Abstr. Papers (Spring Meeting), 27, 1975

33.36 **(N - Piperidino - acetyl) - m - bromoanilide**
$C_{13}H_{17}BrN_2O$
L.Kutschabsky, P.Leibnitz, J.Wenzel *Krist. Tech.*, **9**, 605, 1974
Also classified in 16

33.37 **N - Methylpiperidinium - m - chloroacetanilide iodide**
$C_{14}H_{20}ClN_2O^+$, I^-
G.Reck, P.Leibnitz, J.-P.Wenzel *Krist. Tech.*, **9**, 345, 1974
Residue 1 also classified in 16

33.38 **1 - p - (Oximinoethyl) - phenoxyacetyl - piperidine**
$C_{15}H_{20}N_2O_3$
D.Tranqui, D.T.Cromer, A.Boucherle
Acta Crystallogr., Sect. B, **30**, 2237, 1974
Also classified in 17, 10

33.39 **1 - Benzenesulfenyl - 2,2,6,6 - tetramethyl - 4 - oxopiperidine**
$C_{15}H_{21}NOS$
S.Sato, T.Yoshioka, C.Tamura *Acta Crystallogr., Sect. B,* **31**, 1385, 1975
Also classified in 4

33.40 **1 - Benzenesulfinyl - 2,2,6,6 - tetramethyl - 4 - oxopiperidine**
$C_{15}H_{21}NO_2S$
S.Sato, T.Yoshioka, C.Tamura *Acta Crystallogr., Sect. B,* **31**, 1385, 1975
Also classified in 4

33.41 **1 - Benzenesulfonyl - 2,2,6,6 - tetramethyl - 4 - oxopiperidine**
$C_{15}H_{21}NO_3S$
S.Sato, T.Yoshioka, C.Tamura *Acta Crystallogr., Sect. B,* **31**, 1385, 1975
Also classified in 4

33.42 **(−) - 1 - Methyl - 3 - methoxy - 3 - benzoylpiperidine methiodide chloroform solvate**
$C_{15}H_{22}NO^+$, I^- , $CHCl_3$
J.Ruble, B.Blackmond, G.Hite
Am. Cryst. Assoc., Abstr. Papers (Summer Meeting), 253, 1974

33.43 **(+) - ((−) - 1 - Methyl - 3 - ethyl - 3 - benzoylpiperidine (+) - R,R - bitartrate)**
$C_{15}H_{22}NO^+$, $C_4H_5O_6^-$
J.Ruble, B.Blackmond, G.Hite
Am. Cryst. Assoc., Abstr. Papers (Summer Meeting), 253, 1974
Residue 2 classified in 2, 1

33.44 **4 - Carboethoxy - 1 - methyl - 4 - phenyl - piperidine hydrochloride**
Pethidine hydrochloride
$C_{15}H_{22}NO_2^+$, Cl^-
J.V.Tillack, R.C.Seccombe, C.H.L.Kennard, P.W.T.Oh
Rec. Trav. Chim. Pays-Bas, **93**, 164, 1974

33.45 **1 - n - Propyl - 3,5 - dicyano - 4 - phenyl - 6 - hydroxy - pyrid - 2 - one n - propylamine**
$C_{16}H_{13}N_3O_2$, C_3H_9N
M.-H.Pera, D.Tranqui, H.Fillion, C.L.Duc *Bull. Soc. Chim. Fr.*, 321, 1975
Residue 2 classified in 3

33.46 **(+) - Chloropheniramine maleate (absolute configuration)**
(+) - S - 1 - (p - Chlorophenyl) - 1 - (2 - pyridyl) - 3,N,N - dimethylpropylamine maleate
$C_{16}H_{20}ClN_2^+$, $C_4H_3O_4^-$
M.N.G.James, G.J.B.Williams *Can. J. Chem.*, **52,** 1872, 1974
Residue 1 also classified in 3; residue 2 classified in 2, 1

33.47 **2,5 - bis - Pentamethyleneimino - 1,4 - benzoquinone**
$C_{16}H_{22}N_2O_2$
S.Kulpe, B.Schulz, H.Schrauber *J. Prakt. Chem.*, **316,** 199, 1974
Also classified in 18

33.48 **1 - (2′,6′ - Dichlorobenzyl) - 4 - cyclopentadienylidene - 1,4 - dihydropyridine**
$C_{17}H_{13}Cl_2N$
H.L.Ammon, G.L.Wheeler *J. Am. Chem. Soc.*, **97,** 2326, 1975
Also classified in 20

33.49 **1 - Benzyl - 2 - cyclopentadienylidene - 1,2 - dihydropyridine**
$C_{17}H_{15}N$
H.L.Ammon, G.L.Wheeler *J. Am. Chem. Soc.*, **97,** 2326, 1975
Also classified in 20

33.C **N - (5 - O - Phosphopyridoxyl) - L - tyrosine heptahydrate**
$C_{17}H_{21}N_2O_8P$, $7H_2O$
For complete entry see 48.63

33.C **(1,4 - Di - (N - pyridinium methyl)benzene) - tetra(7,7,8,8 - tetracyanoquinodimethane)**
$C_{18}H_{18}N_2^{2+}$, $2C_{12}H_4N_4^-$, $2C_{12}H_4N_4$
For complete entry see 60.29

33.C **Triprolidine hydrochloride monohydrate**
trans - 1 - (p - Tolyl) - 1 - (2 - pyridyl) - 3 - (1 - pyrrolidino)prop - 1 - ene hydrochloride monohydrate
$C_{19}H_{23}N_2^+$, Cl^- , H_2O
For complete entry see 32.34

33.50 **Diphenyl - pyraline hydrochloride monohydrate**
$C_{19}H_{24}NO^+$, Cl^- , H_2O
G.Precigoux, Y.Barrans, B.Busetta, P.Marsau
Acta Crystallogr., Sect. B, **31,** 1497, 1975

33.51 **2,2,6,6 - Tetramethylpiperidine - 1 - iminoxyl - 4 - (N - 2 - hydroxy - 1 - naphthaldehyde imine)**
$C_{20}H_{25}N_2O_2$
S.D.Mamedov, A.D.Khalilov, M.K.Guseinova
Zh. Strukt. Khim., **15**, 103, 1974
Also classified in 24

33.52 **bis(2,6 - Diacetylpyridinium - bis(phenylhydrazone)) uranyl tetrachloride acetonitrile solvate**
$2C_{21}H_{22}N_5^+$, $Cl_4O_2U^{2-}$, C_2H_3N
R.Graziani, G.Bombieri, E.Forsellini, G.Paolucci
J. Cryst. Mol. Struct., **5**, 1, 1975
Residue 1 also classified in 16

33.53 **1′ - (4 - (4 - Fluorophenyl) - 4 - oxobutyl) - (1,4′ - bipiperidine) - 4′ - carboxamide**
Pipamperone
$C_{21}H_{30}FN_3O_2$
J.P.Declerq, G.Germain, M.H.J.Koch
Acta Crystallogr., Sect. B, **31**, 628, 1975

33.C **4 - (2 - (Methylthio)dibenzo(b,f)thiepin - 11 - yl) - 1 - piperazinyl - propanol hemihydrate**
$C_{22}H_{28}N_2OS_2$, $0.5H_2O$
For complete entry see 39.36

33.54 **1,4 - bis - (2,2,6,6 - Tetramethyl - 1 - hydroxy - 4 - oxy - 4 - piperidyl) - butane**
$C_{22}H_{42}N_2O_4$
L.D.Arutyunian, R.P.Shibaeva *Dokl. Akad. Nauk SSSR,* **215**, 881, 1974

33.C **β - Flupenthixol**
$C_{23}H_{25}F_3N_2OS$
For complete entry see 39.37

33.C **α - Flupenthixol**
$C_{23}H_{25}F_3N_2OS$
For complete entry see 39.38

33.C **Streptonigrin ethylacetate solvate**
$C_{25}H_{22}N_4O_8$, $C_4H_8O_2$
For complete entry see 50.19

33.C **2 - Phenyl - 4 - benzoyl - 4a,5,6,7,8,8a - hexahydro - 8a - piperidinyl - 4H - 1,3,4 - benzo - oxadiazine**
$C_{25}H_{29}N_3O_2$
For complete entry see 40.23

33.55 1 - trans - Cinnamyl - 4 - diphenylmethyl - piperazine
Cinnarizine
$C_{26}H_{28}N_2$
Y.Mouille, M.Cotrait, M.Hospital, P.Marsau
Acta Crystallogr., Sect. B, **31**, 1495, 1975

33.56 1,4 - Di - p - chlorophenyl - 2,6 - diphenyl - 1,4 - dihydropyrazine
$C_{28}H_{20}Cl_2N_2$
J.J.Stezowski *Cryst. Struct. Commun.*, **4**, 21, 1975

33.C Rifampicine pentahydrate
$C_{43}H_{58}N_4O_{12}$, $5H_2O$
For complete entry see 50.33

HETERO-NITROGEN
(7- AND HIGHER-MEMBERED MONOCYCLIC)

34.1 **1,3,5,7 - Tetranitro - 1,3,5,7 - tetra - azacyclo - octane (δ form)**
$C_4H_8N_8O_8$
R.E.Cobbledick, R.W.H.Small *Acta Crystallogr., Sect. B,* **30,** 1918, 1974

34.2 **Caprolactam**
$C_6H_{11}NO$
F.K.Winkler, J.D.Dunitz *Acta Crystallogr., Sect. B,* **31,** 268, 1975

34.3 **ε - Caprolactam**
$C_6H_{11}NO$
Kh.P.Oya, R.M.Myasnikova *Zh. Strukt. Khim.,* **15,** 679, 1974

34.C **ε - Caprolactam - 4 - chlororesorcinol**
$C_6H_{11}NO$, $C_6H_5ClO_2$
For complete entry see 60.6

34.4 **Caprolactam hydrochloride**
$C_6H_{12}NO^+$, Cl^-
F.K.Winkler, J.D.Dunitz *Acta Crystallogr., Sect. B,* **31,** 270, 1975

34.5 **α - Amino - ε - caprolactam hydrobromide**
$C_6H_{13}N_2O^+$, Br^-
S.N.Rao, R.Parthasarathy
Am. Cryst. Assoc., Abstr. Papers (Summer Meeting), 252, 1974

34.6 **α - Amino - ε - caprolactam hydrochloride monohydrate**
$C_6H_{13}N_2O^+$, Cl^- , H_2O
S.N.Rao, R.Parthasarathy
Am. Cryst. Assoc., Abstr. Papers (Summer Meeting), 252, 1974

34.7 **Enantholactam hydrochloride**
$C_7H_{14}NO^+$, Cl^-
F.K.Winkler, J.D.Dunitz *Acta Crystallogr., Sect. B,* **31,** 273, 1975

34.8 **1,5 - Diaza - 6,10 - cyclodecadione**
$C_8H_{14}N_2O_2$
T.Srikrishnan, J.D.Dunitz *Acta Crystallogr., Sect. B,* **31,** 1372, 1975

34.9 **Caprylolactam**
$C_8H_{15}NO$
F.K.Winkler, J.D.Dunitz *Acta Crystallogr., Sect. B,* **31,** 276, 1975

34.10 **Caprylolactam hydrochloride**
$C_8H_{16}NO^+$, Cl^-
F.K.Winkler, J.D.Dunitz *Acta Crystallogr., Sect. B,* **31,** 278, 1975

34.11 **Pelargolactam**
$C_9H_{17}NO$
F.K.Winkler, J.D.Dunitz *Acta Crystallogr., Sect. B,* **31,** 281, 1975

34.12 **Tri(pelargolactam) oxonium chloride**
$C_9H_{17}NO$, H_3O^+ , Cl^-
F.K.Winkler, J.D.Dunitz *Acta Crystallogr., Sect. B,* **31,** 264, 1975

34.13 **Pelargolactam hemihydrochloride**
$C_9H_{18}NO^+$, $C_9H_{17}NO$, Cl^-
F.K.Winkler, J.D.Dunitz *Acta Crystallogr., Sect. B,* **31,** 283, 1975

34.14 **Caprinolactam hemihydrochloride**
$C_{10}H_{20}NO^+$, $C_{10}H_{19}NO$, Cl^-
F.K.Winkler, J.D.Dunitz *Acta Crystallogr., Sect. B,* **31,** 286, 1975

34.15 **1,4,8,11 - Tetra - azacyclotetradecane di(hydroperchlorate) (α form)**
$C_{10}H_{26}N_4^{2+}$, $2ClO_4^-$
C.Nave, M.R.Truter *J. Chem. Soc., Dalton Trans.,* 2351, 1974

34.16 **($\pm$) - 1 - Benzyl - 5 - phenyl - 1 - azacycloheptan - 4 - one hydrochloride**
$C_{19}H_{22}NO^+$, Cl^-
K.Fukuyama, S.Shimizu, S.Kashino, M.Haisa
Bull. Chem. Soc. Jpn., **47,** 1117, 1974

34.C **Antamanide lithium bromide acetonitrile solvate**
$C_{64}H_{78}N_{10}O_{10}$, Li^+ , Br^- , $3C_2H_3N$
For complete entry see 48.74

34.C **(Phe^4,Val^6) - Antamanide sodium bromide ethanol solvate**
$C_{66}H_{82}N_{10}O_{10}$, Na^+ , Br^- , C_2H_6O
For complete entry see 48.75

HETERO-NITROGEN (2 FUSED RINGS)

35.C **7 - Amino - 2H,4H - vic - triazolo(4,5 - c) - 1,2,6 - thiadiazine - 1,1 - dioxide**
$C_3H_4N_6O_2S$
For complete entry see 41.2

35.C **7 - Amino - 2H,4H - v - triazolo(4,5 - c)(1,2,6)thiadiazine 1,1 - dioxide monohydrate**
$C_3H_4N_6O_2S$, H_2O
For complete entry see 41.3

35.1 **8 - Azaguanine hydrobromide monohydrate**
$C_4H_5N_6O^+$, Br^- , H_2O
D.L.Kozlowski, P.Singh, D.J.Hodgson
Acta Crystallogr., Sect. B, **31,** 1751, 1975

35.2 **8 - Azaguanine hydrochloride monohydrate**
$C_4H_5N_6O^+$, Cl^- , H_2O
D.L.Kozlowski, P.Singh, D.J.Hodgson
Acta Crystallogr., Sect. B, **30,** 2806, 1974

35.3 **8 - Aza - 2,6 - diaminopurine sulfate monohydrate**
$2C_4H_6N_7^+$, O_4S^{2-} , H_2O
P.Singh, D.J.Hodgson *Acta Crystallogr., Sect. B,* **31,** 845, 1975

35.4 **Pteridine**
$C_6H_4N_4$
C.D.Shirrell, D.E.Williams *J. Chem. Soc., Perkin Trans. 2,* 40, 1975

35.5 **Indazole**
$C_7H_6N_2$
A.Esande, J.Lappaset *Acta Crystallogr., Sect. B,* **30,** 2009, 1974

35.6 **Benzimidazole benzimidazolium tetrafluoroborate**
$C_7H_6N_2$, $C_7H_7N_2^+$, BF_4^-
A.Quick, D.J.Williams, B.Borah, J.L.Wood
J. Chem. Soc., Chem. Commun., 891, 1974

35.C **trans - 3 - (7 - Dihydrothiazolo(2,3 - e) - 1,2,3 - triazolyl)propenal**
$C_7H_7N_3OS$
For complete entry see 41.6

35.7 **3,6 - Dimethyl - 7,8 - dihydro - 9H - s - triazolo(4,3 - b)(1,2,4)triazepin - 8 - one**
$C_7H_9N_5O$
F.Leroy, J.Housty, S.Geoffre, M.Hospital
Cryst. Struct. Commun., **4,** 317, 1975

35.C **Cyclo - L - prolyl - glycyl (full - data refinement)**
$C_7H_{10}N_2O_2$
For complete entry see 48.32

35.C **Cyclo - L - prolyl - glycyl (high - angle refinement)**
$C_7H_{10}N_2O_2$
For complete entry see 48.33

35.8 **Heptachloro - 5H - 1 - pyrindine**
C_8Cl_7N
D.R.Carter, F.P.Boer *Acta Crystallogr., Sect. B,* **30,** 2762, 1974

35.9 **Indole**
C_8H_7N
P.Roychowdhury, B.S.Basak *Acta Crystallogr., Sect. B,* **31,** 1559, 1975

35.C **6 - Methyl - 5 - thioformyl - pyrrolo(2,1 - b)thiazole**
$C_8H_7NS_2$
For complete entry see 41.7

35.10 **Quinolinium bis - μ - chloro - bis(trichloro - nitroso - platinum)**
$2C_9H_8N^+$, $Cl_8N_2O_2Pt_2^{2-}$
T.S.Khodashova, V.S.Sergienko, A.N.Stetsenko, M.A.Porai-Koshits, L.A.Butman *Zh. Strukt. Khim.*, **15,** 471, 1974

35.C **2,6 - Dimethyl - 5 - thioformyl - pyrrolo(2,1 - b)thiazole**
$C_9H_9NS_2$
For complete entry see 41.8

35.11 **5 - Hydroxy - 5,6,7,8 - tetrahydroquinoline - 1 - oxide**
$C_9H_{11}NO_2$
A.M.M.Lanfredi, A.Tiripicchio, M.T.Camellini
Cryst. Struct. Commun., **4,** 153, 1975

35.C **8 - Aza - adenosine monohydrate**
$C_9H_{12}N_6O_4$, H_2O
For complete entry see 47.9

35.C **2 - Aza - adenosine hemihydrate**
$C_9H_{12}N_6O_4$, $0.5H_2O$
For complete entry see 47.10

35.C **2 - Methyl - 8 - quinolinolato - (tetrafluorophosphorane)**
$C_{10}H_8F_4NOP$
For complete entry see 64.24

35.12 **2,7 - Dimethyl - 5 - acetylaminopyrazolo(1,5 - a)pyrimidine**
$C_{10}H_{12}N_4O$
R.E.Ballard, E.K.Norris, G.M.Sheldrick
Acta Crystallogr., Sect. B, **31,** 295, 1975

35.C **8 - Aza - 9 - deaza - inosine**
Formycin B
$C_{10}H_{12}N_4O_5$
For complete entry see 50.2

35.13 **8 - Acetoxy - 2,4 - dimethyl - 2,4 - diazabicyclo(4.2.0)octa - 3,5 - dione**
$C_{10}H_{14}N_2O_4$
J.S.Swenton, J.A.Hyatt, J.M.Lisy, J.Clardy
J. Am. Chem. Soc., **96,** 4885, 1974

35.14 **cis - 8 - Aza - bicyclo(4.3.0)non - 3 - ene methiodide**
$C_{10}H_{18}N^+$, I^-
G.D.Smith, R.D.Otzenberger, B.P.Mundy, C.N.Caughlan
J. Org. Chem., **39,** 321, 1974

35.C **3 - Methyl - 2,4 - dicarbomethoxy - Δ^3 - cephem**
$C_{11}H_{13}NO_5S$
For complete entry see 41.12

35.15 **1,6 - Dimethyl - 4 - oxo - 1,6,7,8 - tetrahydro - 3 - homopyrimidazole - carboxylic acid**
$C_{11}H_{14}N_2O_3$
K.Simon, Z.Meszaros, K.Sasvari
Acta Crystallogr., Sect. B, **31,** 1702, 1975

35.C **3 - Deaza - adenosine**
$C_{11}H_{14}N_4O_4$
For complete entry see 47.27

35.C **3,4 - Dihydro - 7 - chloro - 6 - diethylamino - 2H,8H - pyrimido(2,1 - b)(1,3)thiazin - 8 - one**
$C_{11}H_{16}ClN_3OS$
For complete entry see 41.13

35.C **6a - Carbomethoxy - 3 - carboethoxy - 4 - oxo - perhydrofuro(3,4 - c)pyrazole**
$C_{11}H_{16}N_2O_6$
For complete entry see 38.18

35.C **(R) - 3 - (2 - Deoxy - β - D - erythro - pentafuranosyl) - 3,6,7,8 - tetrahydroimidazo(4,5 - d)(1,3)diazepin - 8 - ol**
$C_{11}H_{16}N_4O_4$
For complete entry see 47.29

35.C **Coformycin sesquihydrate**
3 - (β - D - Ribofuranosyl) - 6,7,8 - trihydroimidazo(4,5 - d)(1,3 - diazepin - 8(R) - ol) sesquihydrate
$C_{11}H_{16}N_4O_5$, $1.5H_2O$
For complete entry see 47.30

35.C **1 - Deaza - isotubercidin picrate**
$C_{12}H_{16}N_3O_4^+$, $C_6H_2N_3O_7^-$
For complete entry see 45.33

35.16 **2 - Phenylbenzo - 1,2,3 - triazinium - betaine 1 - oxide**
$C_{13}H_9N_3O_2$
R.E.Ballard, E.K.Norris *Acta Crystallogr., Sect. B*, **31**, 626, 1975

35.17 **N - Acetyl - 5 - methoxy - tryptamine**
Melatonin
$C_{13}H_{16}N_2O_2$
A.Mostad, C.Romming *Acta Chem. Scand. Ser. B*, **28**, 564, 1974

35.18 **3 - Carboethoxy - 1,6 - dimethyl - 4 - oxo - 6,7,8,9 - tetrahydrohomopyrimidazolium methyl sulfate**
$C_{13}H_{19}N_2O_3^+$, $CH_3O_4S^-$
K.Simon, K.Sasvari *Acta Crystallogr., Sect. B*, **31**, 1695, 1975
Residue 2 classified in 11

35.19 **2 - n - Propyl - 7 - methyl - trans - decahydroquinoline hydrochloride**
$C_{13}H_{26}N^+$, Cl^-
J.L.Flippen *Acta Crystallogr., Sect. B*, **30**, 2906, 1974

35.20 **N - (p - Chlorophenyl) - phthalimide**
$C_{14}H_8ClNO_2$
B.L.Farmer, J.B.Lando *Z. Naturforsch., Teil B*, **29**, 769, 1974

35.21 **N - (p - Iodophenyl) - phthalimide**
$C_{14}H_8INO_2$
B.Ribar, S.Stankovic, R.Herak, R.Halasi, S.Djuric
Cryst. Struct. Commun., **3**, 669, 1974

35.22 **Anhydro - 2 - methyl - 4 - (o - nitroanilino) - 1,2,3 - benzotriazinium hydroxide**
$C_{14}H_{11}N_5O_2$
C.H.Schwalbe *Acta Crystallogr., Sect. A*, **31**, S52, 1975
Also classified in 16, 15

35.C **8.5′ - Anhydro - 7 - bromo - 8 - hydroxy - 2′,3′ - isopropylidene - tubercidin (absolute configuration)**
$C_{14}H_{15}BrN_4O_4$
For complete entry see 50.6

35.C **Cyclo - (D - phenylalanyl - L - prolyl)**
$C_{14}H_{16}N_2O_2$
For complete entry see 48.54

35.23 **2,5 - Diethyl - 7 - (1 - methyl - 2 - imidazolyl) - 1H - pyrrolo(3,4 - c)pyridine - 1,3,6(2H,5H) - trione**
$C_{15}H_{16}N_4O_3$
H.P.Weber, T.J.Petcher, A.Jaunin, F.Troxler
Helv. Chim. Acta, **58,** 552, 1975
Also classified in 32

35.24 **6 - Bromo - 1 - (1 - pyrrolidinyl) - 3H - pyrroline - 2,3 - dicarboxylic acid dimethyl ester**
$C_{15}H_{17}BrN_2O_4$
R.D.Gilardi, J.L.Flippen *Cryst. Struct. Commun.,* **3,** 623, 1974
Also classified in 32

35.25 **10 - (p - Bromophenyl) - 10,11,12 - triaza - bicyclo(7.3.0)dodeca - 1,11 - diene**
$C_{15}H_{18}BrN_3$
R.P.Dodge, Q.Johnson *Cryst. Struct. Commun.,* **3,** 685, 1974

35.C **2 - Methyl - 8 - (trifluoro - phenyl - phosphoroxy) - quinoline**
$C_{16}H_{13}F_3NOP$
For complete entry see 64.42

35.26 **1 - Methyl - 4 - phenyl - 1H - 2,3 - benzodiazepine**
$C_{16}H_{14}N_2$
R.O.Gould, S.E.B.Gould *J. Chem. Soc., Perkin Trans. 2,* 1075, 1974

35.27 **1 - Benzyl - 3,4 - dihydro - isoquinoline hydrochloride hemihydrate**
$C_{16}H_{16}N^+$, Cl^- , $0.5H_2O$
K.Simon, Z.Meszaros, A.Kalman *Cryst. Struct. Commun.,* **4,** 135, 1975

35.28 **1 - Methyl - 3 - p - tolylsulfonylamino - indole**
$C_{16}H_{16}N_2O_2S$
T.S.Cameron, K.Prout, B.Denton, R.Spagna, E.White
J. Chem. Soc., Perkin Trans. 2, 176, 1975

35.29 **2 - n - Propyl - 4 - anilino - 1,2,3 - benzotriazinium iodide**
$C_{16}H_{17}N_4^+$, I^-
C.H.Schwalbe *Acta Crystallogr., Sect. A,* **31,** S52, 1975
Residue 1 also classified in 16

35.30 **1,3,4,7,8,8a - Hexahydro - 2 - methyl - 4a - phenylisoquinolin - 6(8H) - one methobromide**
$C_{17}H_{24}NO^+$, Br^-
N.Finch, L.Blanchard, R.T.Puckett, L.H.Werner
J. Org. Chem., **39,** 1118, 1974

35.C **3 - Phenyl - 8 - phenylthiocarbamoyl - 6,7 - dihydro - 5H - thiazolo(3,2 - a)pyrimidinium - betaine**
$C_{19}H_{17}N_3S_2$
For complete entry see 41.26

35.31 **8 - Chloro - 2,3 - dihydro - 5 - hydroxy - 1 - ((4 - methylphenyl)sulfonyl) - 1H - benzazepine - 4 - carboxylic acid methyl ester**
$C_{19}H_{18}ClNO_5S$
F.D.Sancilio, J.F.Blount, G.R.Proctor
Am. Cryst. Assoc., Abstr. Papers (Spring Meeting), 32, 1975

35.32 **2,3,4,7 - Tetrahydro - 3a,4 - bis(methoxycarbonyl) - 2,6 - dimethyl - 5 - phenyl - indazol - 7 - ene**
$C_{19}H_{20}N_2O_5$
R.C.Gearhart, R.H.Wood, P.C.Thorstenson, J.A.Moore
J. Org. Chem., **39,** 1007, 1974

35.33 **2 - (2′,6′ - Dichlorobenzyl) - 1 - cyclopentadienylidene - 1,2 - dihydroisoquinoline**
$C_{21}H_{15}Cl_2N$
H.L.Ammon, G.L.Wheeler *J. Am. Chem. Soc.*, **97,** 2326, 1975
Also classified in 20

35.C **(−) - 5,6,7,8 - Tetrahydro - 4 - (methylthio) - 6 - phenyl - 6 - benzylphosphorinia(4,3 - d)pyrimidine bromide (absolute configuration)**
$C_{21}H_{22}N_2PS^+$, Br^-
For complete entry see 64.52

35.34 **2 - Bromo - 9 - (1 - pyrrolidinyl) - 5H - pyrrolo(1,2 - a)azepine - 5,6,7,8 - tetracarboxylic acid tetramethyl ester**
$C_{21}H_{23}BrN_2O_8$
J.L.Flippen, R.D.Gilardi *Cryst. Struct. Commun.*, **3,** 627, 1974
Also classified in 32

35.C **7α - Methoxy - 7 - phenyl - acetamido - deacetoxy - cephalosporanic acid t - butyl ester**
$C_{21}H_{26}N_2O_5S$
For complete entry see 59.13

35.35 **N - Phthalimido - dibenzylsulfoximide**
$C_{22}H_{18}N_2O_3S$
G.D.Andreetti, G.Bocelli, L.Coghi, P.Sgarabotto
Cryst. Struct. Commun., **3,** 765, 1974

35.C **1 - Methyl - 3,3 - dimethyl - 2 - (p - N - methyl - N - β - chloroethylstyryl)indole - bis(7,7,8,8 - tetracyanoquinodimethane)**
$C_{22}H_{26}ClN_2^+$, $C_{12}H_4N_4^-$, $C_{12}H_4N_4$
For complete entry see 60.38

35.36 **S(+) - N - Phthalimido - p - tolyl - α - naphthyl sulfoximide (absolute configuration)**
$C_{25}H_{18}N_2O_3S$
G.D.Andreetti, G.Bocelli, L.Coghi, P.Sgarabotto
Cryst. Struct. Commun., **4,** 393, 1975
Also classified in 4, 11

35.C **Streptonigrin ethylacetate solvate**
$C_{25}H_{22}N_4O_8$, $C_4H_8O_2$
For complete entry see 50.19

35.37 **10 - Chloro - 2,3 - dihydro - 7 - hydroxy - 1 - ((4 - methylphenyl)sulfonyl) - 1H - 1 - benzazonine - 4,5,6 - tricarboxylic acid trimethyl ester**
$C_{25}H_{24}ClNO_9S$
F.D.Sancilio, J.F.Blount, G.R.Proctor
Am. Cryst. Assoc., Abstr. Papers (Spring Meeting), 32, 1975

35.C **(+) - 7 - Benzylideneamino - 7 - carbomethoxy - 4 - carbo(p - methoxy - benzyloxy) - Δ^3 - cephem (absolute configuration)**
$C_{25}H_{24}N_2O_6S$
For complete entry see 41.30

35.38 **5,6 - Dichloro - 1,3 - diethyl - 2 - ((5,6 - dichloro - 1,3 - diethyl - 2 - benzimidazolinylidene) - 1 - propynyl)benzimidazolium toluene - p - sulfonate hydrate**
$C_{25}H_{25}Cl_4N_4^+$, $C_7H_7O_3S^-$, $0.2H_2O$
D.L.Smith, H.R.Luss *Acta Crystallogr., Sect. B,* **31,** 402, 1975
Residue 2 classified in 11

35.39 **1,2,3,4 - Tetrahydro - 1,2,4 - trimethyl - 4 - p - tolylsulfonylamino - 3 - p - tolylsulfonylimino - quinoline**
$C_{26}H_{28}N_3O_4S_2$
T.S.Cameron, K.Prout, B.Denton, R.Spagna, E.White
J. Chem. Soc., Perkin Trans. 2, 176, 1975

35.40 **Dimethyl 2 - cyclohexyl - 1,2 - dihydro - 1 - oxo - 6 - phenyl - 2 - benzazocine - 4,5 - dicarboxylate**
$C_{27}H_{27}NO_5$
A.Padwa, E.Vega *J. Org. Chem.*, **40,** 175, 1975
Also classified in 21

35.41 **2 - Phenyl - 2 - (1 - methyl - 2 - phenyl - 3 - indolyl) - 3 - indolinone**
$C_{29}H_{22}N_2O$
E.Foresti Serantoni, A.Krajewski, R.Mongiorgi, L.Riva di Sanseverino
Cryst. Struct. Commun., **3,** 641, 1974

35.C **β - Lactam - fused heterocyclic ylide compound from reaction of chloroamine T with methyl 6β - phenylacetamidopenicillanate**
$C_{31}H_{34}N_4O_8S_3$
For complete entry see 41.33

35.42 **2 - Phenyl - 2 - (2 - phenyl - 3 - indolyl) - 3 - (p - bromophenyl)iminoindoline**
$C_{34}H_{24}BrN_3$
E.Foresti Serantoni, A.Krajewski, R.Mongiorgi, L.Riva di Sanseverino
Cryst. Struct. Commun., **3**, 637, 1974

35.43 **2 - (1 - Methyl - 2 - phenyl - 3 - indolyl) - 3 - phenyl - 3 - phenylamino - 3H - indole**
$C_{35}H_{27}N_3$
E.Foresti Serantoni, A.Krajewski, R.Mongiorgi, L.Riva di Sanseverino
Cryst. Struct. Commun., **4**, 403, 1975

35.44 **1 - Ethyl - 2 - phenyl - 2 - (1 - methyl - 2 - phenyl - 3 - indolyl) - 3 - indolinon - O - benzoyl - oxime**
$C_{38}H_{31}N_3O_2$
E.Foresti Serantoni, A.Krajewski, R.Mongiorgi, L.Riva di Sanseverino
Eur. Cryst. Meeting, S6, 1974

HETERO-NITROGEN (MORE THAN 2 FUSED RINGS)

36.1 **1H - 4,6 - Dimethylimidazo(1,2 - a)purine - 9 - one**
$C_9H_9N_5O$
G.Nygjerd, J.McAlister, M.Sundaralingam, S.Matsuura
Acta Crystallogr., Sect. B, **31,** 413, 1975

36.2 **4,9 - Dioxo - 5,8 - diazatricyclo(6.3.0.0^{1,5})undecane**
$C_9H_{12}N_2O_2$
D.van der Helm, S.E.Ealick, D.Washecheck
Acta Crystallogr., Sect. A, **31,** S104, 1975

36.C **1,3,4,5,6,8 - Hexachlorothieno(2,3 - c.5,4 - c')dipyridine**
$C_{10}Cl_6N_2S$
For complete entry see 39.16

36.3 **5,10 - Dioxo - 6,9 - diazatricyclo(7.3.0.0^{1,6})dodecane**
$C_{10}H_{14}N_2O_2$
D.van der Helm, S.E.Ealick, D.Washecheck
Acta Crystallogr., Sect. A, **31,** S104, 1975

36.4 **10 - Methylisoalloxazinium perchlorate monohydrate**
$C_{11}H_9N_4O_2^+$, ClO_4^- , H_2O
J.S.Sherfinski, A.J.deArmendi, C.J.Fritchie Junior
Am. Cryst. Assoc., Abstr. Papers (Summer Meeting), 238, 1974

36.5 **5,12 - Dioxo - 1,4 - diazatetracyclo - (5.5.1.0^{4,13}.0^{10,13})tridecane**
$C_{11}H_{14}N_2O_2$
D.van der Helm, S.E.Ealick, D.Washecheck
Acta Crystallogr., Sect. A, **31,** S104, 1975

36.6 **Naphthaloimide**
$C_{12}H_7NO_2$
A.N.Sobolev, L.A.Chetkina, G.A.Gol'der, Yu.G.Fedorov, V.E.Zavodnik
Kristallografiya, **18,** 1157, 1973

36.7 **8b,8c - Diazacyclopent(f,g)acenaphthylene**
$C_{12}H_8N_2$
J.L.Atwood, D.C.Hrncir, C.Wong, W.W.Paudler
J. Am. Chem. Soc., **96,** 6132, 1974

36.C **Carbazole - 7,7,8,8 - tetracyanoquinodimethane complex**
$C_{12}H_9N$, $C_{12}H_4N_4$
For complete entry see 60.14

36.8 **3 - Acetyl - 4 - methyl - as - triazino(4,3 - b)indazole**
$C_{12}H_{10}N_4O$
R.Allmann, T.Debaerdemaeker, W.Grahn, C.Reichardt
Chem. Ber., **107,** 1555, 1974

36.C **De(hydrogen fluoride) (5 - fluoro - 1,3 - dimethyluracil photodimer)**
$C_{12}H_{13}FN_4O_4$
For complete entry see 44.23

36.C **cis - anti - (5 - Fluoro - 1,3 - dimethyluracil photodimer)**
$C_{12}H_{14}F_2N_4O_4$
For complete entry see 44.24

36.9 **6,13 - Dioxo - 1,5 - diazatetracyclo(6.5.1.0^{5,14}.0^{11,14}) tetradecane**
$C_{12}H_{16}N_2O_2$
D.van der Helm, S.E.Ealick, D.Washecheck
Acta Crystallogr., Sect. A, **31,** S104, 1975

36.C **Acridinium - bis(7,7,8,8 - tetracyanoquinodimethanide)**
$C_{13}H_{10}N^+$, $C_{12}H_4N_4^-$, $C_{12}H_4N_4$
For complete entry see 60.17

36.C **Adenylyl - 3′,3′ - uridine 9 - aminoacridine hydrate**
$C_{13}H_{11}N_2^+$, $C_{19}H_{23}N_7O_{12}P^-$, xH_2O
For complete entry see 47.37

36.10 **Proflavine sulfate hydrate**
$2C_{13}H_{12}N_3^+$, O_4S^{2-} , $3.5H_2O$
A.Jones, S.Neidle *Acta Crystallogr., Sect. B,* **31,** 1324, 1975

36.C **Lumiflavin - bis(naphthalene - 2,3 - diol) trihydrate**
$C_{13}H_{12}N_4O_2$, $2C_{10}H_8O_2$, $3H_2O$
For complete entry see 60.19

36.11 **1,2,3,4 - Tetrahydro - 1 - methyl - 2 - oxo - pyrimido(1,2 - a)benzimidazole - 4 - carboxylic acid methyl ester**
$C_{13}H_{13}N_3O_3$
H.P.Weber, F.Troxler *Helv. Chim. Acta,* **57,** 2364, 1974

36.12 **2 - Chloro - 3,4 - dimethyl - 6 - dimethylamino - 3H - 1,3,5,7 - tetra - azacyclopent(f)azulene trihydrate**
$C_{13}H_{14}ClN_5$, $3H_2O$
L.Cariello, S.Crescenzi, G.Prota, S.Capasso, F.Giordano, L.Mazzarella
Tetrahedron, **30,** 3281, 1974

36.13 **6,6 - Dimethyl - 3 - methylthio - 6,7 - dihydro - as - triazino(1,6 - c)quinazolin - 5 - ium - 1 - olate**
$C_{13}H_{14}N_4OS$
A.J.M.Duisenberg, A.Kalman, G.Doleschall, K.Lempert
Cryst. Struct. Commun., **4,** 295, 1975

36.14 **8 - Methyl - 5,5a,6,7,8,9 - hexahydro - pyrido(2,1 - b)quinazolinium picrate**
$C_{13}H_{17}N_2^+$, $C_6H_2N_3O_7^-$
G.Reck, E.Hohne, G.Adam *J. Prakt. Chem.*, **316,** 496, 1974
Residue 2 classified in 17, 15, 6

36.15 **9 - Chloromethylacridine hydrochloride**
$C_{14}H_{11}ClN^+$, Cl^-
D.E.Zacharias, J.P.Glusker *Acta Crystallogr., Sect. B,* **30,** 2046, 1974

36.16 **N - Ethylcarbazole**
$C_{14}H_{13}N$
T.Kimura, Y.Kai, N.Yasuoka, N.Kasai
Acta Crystallogr., Sect. A, **31,** S125, 1975

36.C **Indolo(2,1 - b)quinazoline - 6,12 - dione**
Tryptanthrin
$C_{15}H_8N_2O_2$
For complete entry see 50.7

36.17 **3 - Bromo - N - methyl - 5,6 - dihydro - 7H,12H - dibenz(c,f)azocine**
$C_{16}H_{16}BrN$
F.R.Ahmed *Acta Crystallogr., Sect. B,* **31,** 26, 1975

36.18 **1,2,3,4,4aα,5,11aα - Heptahydro - acetoxy - 11βH - dibenz(b,e)azepine - 6 - one**
$C_{16}H_{19}NO_3$
J.L.Flippen *Acta Crystallogr., Sect. B,* **31,** 610, 1975

36.19 **5,11 - Dicarbomethoxy - 2,8 - dimethoxy - 3,9 - dimethyl - 1,6,7,12 - tetra - aza - tricyclo(7.3.0.0^{3,7})dodecan - 5,11 - diene**
$C_{16}H_{24}N_4O_6$
L.Dupont, J.Toussaint, O.Dideberg, J.N.Braham, A.F.Noels
Acta Crystallogr., Sect. B, **31,** 548, 1975

36.C **2,2′ - Spiro - (1 - ethyl - 3 - hydroxy - 3 - methyl - pyrrolidine) - (1a′,3a′ - dimethyl - 4′ - ethyl - perhydro - furo(3,2 - b)pyrrole)**
$C_{16}H_{30}N_2O_2$
For complete entry see 38.52

36.20 **5,11 - Dimethyl - 6H - pyrido(4,3 - b)carbazole**
Ellipticine
$C_{17}H_{14}N_2$
C.Courseille, B.Busetta, M.Hospital
Acta Crystallogr., Sect. B, **30,** 2628, 1974

36.C **Adenine - riboflavin trihydrate**
$C_{17}H_{20}N_4O_6$, $C_5H_5N_5$, $3H_2O$
For complete entry see 60.1

36.C **(±) - 2,3 - Dimethoxy - 18 - nor - 8,13 - diaza - 1,3,5(10) - estratrien - 17 - one**
$C_{17}H_{22}N_2O_3$
For complete entry see 51.1

36.C **5,6 - Dihydro - 1,3 - dimethyl - 5,6 - di(1′,3′ - dimethyl - 2′,4′,6′ - trioxopyrimid(5′,5′)yl)furo(2,3 - d)uracil**
1,3 - Dimethylbarbituric acid trimer
$C_{18}H_{18}N_6O_9$
For complete entry see 43.7

36.21 **2 - (3 - (7 - Chloro - 2 - methoxy - 10 - (benzo(b) - 1,5 - naphthyridinyl)amino) propylamino)ethanol**
$C_{18}H_{21}ClN_4O_2$
J.P.Glusker, H.L.Carrell, H.M.Berman, B.Gallen
Acta Crystallogr., Sect. B, **31,** 826, 1975
Also classified in 3

36.22 **5 - (3 - Dimethylaminopropyl) - 10,11 - dihydro - 5H - dibenzo(b,f)azepine hydrochloride**
Imipramine hydrochloride
$C_{19}H_{25}N_2^+$, Cl^-
M.L.Post, O.Kennard, A.S.Horn
Acta Crystallogr., Sect. B, **31,** 1008, 1975
Residue 1 also classified in 3

36.23 **5 - (3 - (Dimethylamino)propyl) - 6,7,8,9,10,11 - hexahydro - 5H - cyclo - oct(b)indole hydrochloride monohydrate**
Iprindole
$C_{19}H_{29}N_2^+$, Cl^- , H_2O
J.R.Rodgers, O.Kennard, A.S.Horn, L.Riva di Sanseverino
Acta Crystallogr., Sect. B, **30,** 1970, 1974
Residue 1 also classified in 3

36.24 **2,7 - Diamino - 9 - phenyl - 10 - methyl - phenanthridinium bromide monohydrate**
$C_{20}H_{18}N_3^+$, Br^- , H_2O
C.Courseille, B.Busetta, M.Hospital
Acta Crystallogr., Sect. B, **30,** 2631, 1974

36.25 **4′ - (Acridin - 9 - ylamine)methane - sulfonanilide hydrochloride**
$C_{20}H_{18}N_3O_2S^+$, Cl^-
D.Hall, D.A.Swann, T.N.Waters
J. Chem. Soc., Perkin Trans. 2, 1334, 1974
Residue 1 also classified in 16, 11

36.26 **3,4,10,11 - Dibenzo - 1,8 - diazacyclotetradeca - 1,3,8,10 - tetraene**
$C_{20}H_{22}N_2$
S.K.Arora, J.P.Schaefer *Acta Crystallogr., Sect. B,* **30,** 2474, 1974

36.27 **1 - Methyl - 3 - p - tolylsulfonylimino - indoline - 2 - spirocyclopentane**
$C_{20}H_{22}N_2O_2S$
T.S.Cameron, K.Prout, B.Denton, R.Spagna, E.White
J. Chem. Soc., Perkin Trans. 2, 176, 1975

36.C **1,3,7,9,13,15,19,21 - Octa - azapentacyclo(19,3,1^{3,7},1^{9,13},1^{15,19}) octeicosane benzene solvate**
$C_{20}H_{40}N_8$, C_6H_6
For complete entry see 37.20

36.28 **2,7 - Diamino - 9 - phenyl - 10 - ethyl - phenanthridinium bromide ethanol solvate**
$C_{21}H_{20}N_3^+$, Br^- , C_2H_6O
C.Courseille, B.Busetta, M.Hospital
Acta Crystallogr., Sect. B, **30,** 2631, 1974

36.29 **2,7 - Diamino - 9 - phenyl - 10 - ethyl - phenanthridinium chloride ethanol solvate**
$C_{21}H_{20}N_3^+$, Cl^- , C_2H_6O
C.Courseille, B.Busetta, M.Hospital
Acta Crystallogr., Sect. B, **30,** 2631, 1974

36.C **Naphthyridinomycin**
$C_{21}H_{27}N_3O_7$
For complete entry see 50.12

36.C **bis(10 - Methylisoalloxazine) lead(ii) perchlorate tetrahydrate**
$C_{22}H_{16}N_8O_6Pb^{2+}$, $2ClO_4^-$, $4H_2O$
For complete entry see 69.18

36.30 **3 - Ethoxycarbonyl - 1,2,5 - tri(methoxycarbonyl) - 3H - benzo(c)(1,2,5)triazepino(1,2 -)cinnoline**
$C_{25}H_{23}N_3O_8$
A.F.Cameron, A.A.Freer *Acta Crystallogr., Sect. B,* **30,** 2696, 1974

36.C **Spermidine - N,N″ - bis(6 - chloro - 2 - methoxy - acridine) chloroform solvate**
$C_{35}H_{35}Cl_2N_5O_2$, $2CHCl_3$
For complete entry see 59.22

HETERO-NITROGEN (BRIDGED RING SYSTEMS)

37.1 **Hexamethylenetetramine iodine**
$C_6H_{12}N_4$, I_2
H.Pritzkow *Acta Crystallogr., Sect. B,* **31,** 1589, 1975

37.2 **Hexamethylenetetramine bis(iodine)**
$C_6H_{12}N_4$, $2I_2$
H.Pritzkow *Acta Crystallogr., Sect. B,* **31,** 1589, 1975

37.3 **Hexamethylenetetramine tri - iodo - nitrogen iodine**
$C_6H_{12}N_4$, I_3N , I_2
H.Pritzkow *Z. Anorg. Allg. Chem.,* **409,** 237, 1974

37.4 **bis(Hexamethylenetetramine) calcium dichromate heptahydrate**
$2C_6H_{12}N_4$, $Cr_2O_7^{2-}$, Ca^{2+} , $7H_2O$
F.Dahan *Acta Crystallogr., Sect. B,* **31,** 423, 1975

37.5 **3 - Aza - bicyclo(3.2.2)nonane (low temp.form)**
$C_8H_{15}N$
L.M.Amzel, S.Baggio, R.F.Baggio, L.N.Becka
Acta Crystallogr., Sect. B, **30,** 2494, 1974

37.6 **1,3,6,8 - Tetra - azatricyclo(4.4.1.1^{3,8})dodecane**
$C_8H_{16}N_4$
P.Murray-Rust *J. Chem. Soc., Perkin Trans. 2,* 1136, 1974

37.7 **4 - Azatricyclo(4.4.0.0^{3,8})decan - 5 - one**
$C_9H_{13}NO$
K.Venkatesan, S.Ramakumar, H.P.Weber
Acta Crystallogr., Sect. A, **31,** S103, 1975

37.8 **2 - p - Bromophenyl - 1,3 - diazabicyclo(3.1.0)hexane**
$C_{10}H_{11}BrN_2$
S.A.Hiller, Ya.Ya.Bleidelis, A.A.Kemme, A.V.Eremeyev
J. Chem. Soc., Chem. Commun., 130, 1975

37.9 **2 - Methyl - 2 - azatricyclo(4.3.1.0^{4,9})decan - 10 - one oxime hydrochloride**
$C_{10}H_{17}N_2O^+$, Cl^-
R.W.Lockhart, K.Hanaya, F.W.B.Einstein, Y.L.Chow
J. Chem. Soc., Chem. Commun., 344, 1975

37.C **Bicyclomycin**
$C_{12}H_{18}N_2O_7$
For complete entry see 50.4

37.10 **bis(Hexamethylenetetramine)iodonium tri - iodide**
$C_{12}H_{24}IN_8^+$, I_3^-
H.Pritzkow *Acta Crystallogr., Sect. B,* **31,** 1505, 1975

37.C **4β - Ethyl - 2β - acetoxymethyl - 9 - methyl - 9 - aza - 3 - oxa - bicyclo(3.3.1)nonan - 7 - one**
$C_{13}H_{21}NO_4$
For complete entry see 40.8

37.11 **4 - Phenyl - 2,4,6 - triazatricyclo(5.2.2.0^{2,6})undecane - 3,5 - dione (at −170°C)**
$C_{14}H_{15}N_3O_2$
C.van der Ende, B.Offereins, C.Romers
Acta Crystallogr., Sect. B, **30,** 1947, 1974

37.12 **1,4 - Dimethanodibenzo(d,i) - 1,3,6,8 - tetrazecine**
$C_{16}H_{16}N_4$
P.Murray-Rust, I.Smith *Acta Crystallogr., Sect. B,* **31,** 587, 1975

37.13 **7,8,9,10 - Tetrahydro - 6,10 - propano - 6H - cyclohepta(b)quinoxaline**
$C_{16}H_{18}N_2$
J.Murray-Rust, P.Murray-Rust *Acta Crystallogr., Sect. B,* **31,** 310, 1975

37.14 **1,2,4,4,5,8 - Hexamethyl - 8 - N - acetylamido - 3 - aza - bicyclo(3.3.1)non - 2 - ene hydrobromide**
$C_{16}H_{29}N_2O^+$, Br^-
S.Garcia-Blanco, F.Florencio, P.Smith-Verdier
Acta Crystallogr., Sect. A, **31,** S102, 1975

37.15 **1,2,4,4,5,8 - Hexamethyl - 8 - N - acetamido - 3 - aza - bicyclo(3.3.1)nonane**
$C_{16}H_{30}N_2O$
S.Garcia-Blanco, F.Florencio, P.Smith-Verdier
Acta Crystallogr., Sect. A, **31,** S102, 1975

37.16 **1,2,2a,3,4,9b - Hexahydro - 9b - acetoxy - 1,3 - propano - 5H - cyclobuta(d)(2)benzazepin - 5 - one**
$C_{17}H_{19}NO_3$
Y.Kanaoka, K.Koyama, J.L.Flippen, I.L.Karle, B.Witkop
J. Am. Chem. Soc., **96,** 4719, 1974

37.17 **16 - Aza - pentacyclo (12.3.2^{2,13}.0.0^{3,12}.0^{4,11}) nonadec - 18 - ene - 15,17 - dione**
$C_{18}H_{23}NO_2$
H.M.Tyrrell, A.P.Wolters *Tetrahedron Lett.,* 4193, 1974

37.18 **DL - 1,2,3,4,5,6 - Hexahydro - 6,11,12,12 - tetramethyl - 2,6 - methano - 3,11 - propano - 3 - benzazocine hydrochloride monohydrate**
$C_{19}H_{28}N^+$, Cl^- , H_2O
M.Kimura, T.Nakajima, S.Inaba, H.Yamamoto
Bull. Chem. Soc. Jpn., **47,** 1404, 1974

37.19 **cis - Dihydroquinaldine dimer**
$C_{20}H_{22}N_2$
J.Bordner, I.W.Elliott *Cryst. Struct. Commun.*, **3,** 689, 1974

37.20 **1,3,7,9,13,15,19,21 - Octa - azapentacyclo(19,3,1^{3,7},1^{9,13},1^{15,19}) octeicosane benzene solvate**
$C_{20}H_{40}N_8$, C_6H_6
P.Murray-Rust *Acta Crystallogr., Sect. B,* **31,** 583, 1975
Residue 1 also classified in 36

37.21 **N - Tosyl - 2,12 - ethano - 2 - ethyl - 8 - methoxy - 1,4 - methylene - 1,2,3,4,5,6,12,13 - octahydrophenanthridin - 3 - one**
$C_{27}H_{31}NO_5S$
A.F.Cameron, A.A.Freer, P.Doyle, N.C.A.Wright
Acta Crystallogr., Sect. B, **30,** 1923, 1974

37.C **1,2 - Dihydronaphtho(2,1 - d) - 3H - indano(2,1 - f) - N - phenylsuccinimido(4,3 - d) - 7α - thia - 2β,3β,5α,6β - tetrahydrobicyclo(2.2.1)hepta - 2,5 - diene benzene solvate**
$C_{29}H_{21}NO_2S$, C_6H_6
For complete entry see 39.40

37.C **6,14,22,30 - Tetrabenzyl - 2,6,10,14,18,22,26,30 - octa - aza - pentacyclo(26.4.0.0^{4,9}.0^{12,17}.0^{20,25})do triaconta - 4,7,12,15,20,23,28,31 - octaen - 3,11,19,27 - tetraone**
$C_{52}H_{48}N_8O_4$
For complete entry see 48.73

HETERO-OXYGEN

38.1 **Vinylene carbonate (at $-3°C$)**
$C_3H_2O_3$
F.Cser *Acta Chim. Acad. Sci. Hung.*, **80,** 49, 1974

38.2 **Di - imine - bis(pentacarbonyl chromium) tetrahydrofuran solvate**
$2C_4H_8O$, $C_{10}H_2Cr_2N_2O_{10}$
G.Huttner, W.Gartzke, K.Allinger *Angew. Chem.*, **86,** 860, 1974

38.C **bis - μ - (Tetrahydrofuran sodium) - bis(diphenyl(ethylene) nickel(0)) disodium tris(tetrahydrofuran)**
$3C_4H_8O$, $C_{36}H_{44}Na_2Ni_2O_2$, 2Na
For complete entry see 71.105

38.3 **Dioxan hexa - aquo - magnesium chloride**
$C_4H_8O_2$, $H_{12}MgO_6^{2+}$, $2Cl^-$
J.C.Barnes, L.J.Sesay *Acta Crystallogr., Sect. A,* **31,** S144, 1975

38.4 **bis(Dioxan) dibromo - tetra - aquo - nickel**
$2C_4H_8O_2$, $H_8Br_2NiO_4$
J.C.Barnes, L.J.Sesay *Acta Crystallogr., Sect. A,* **31,** S144, 1975

38.5 **1,3,5,7 - Tetroxocane**
$C_4H_8O_4$
Y.Chatani, T.Yamauchi, Y.Miyake *Bull. Chem. Soc. Jpn.*, **47,** 583, 1974

38.6 **α - Methyltetronic acid (CuKα radiation)**
$C_5H_6O_3$
E.K.Andersen, I.G.K.Andersen *Acta Crystallogr., Sect. B,* **31,** 394, 1975

38.7 **α - Methyltetronic acid (MoKα radiation)**
$C_5H_6O_3$
E.K.Andersen, I.G.K.Andersen *Acta Crystallogr., Sect. B,* **31,** 394, 1975

38.C **1,4.2,5.3,6 - Trianhydro - D - mannitol (at $-100°C$)**
$C_6H_8O_3$
For complete entry see 45.4

38.C **1,6 - Anhydro - β - D - glucopyranose**
Levoglucosan
$C_6H_{10}O_5$
For complete entry see 45.11

38.8 **5 - Tetrahydrofurano - 6 - (tricarbonyl) - 6 - manganadodecahydrononaborane**
$C_7H_{20}B_9MnO_4$
J.W.Lott, D.F.Gaines *Inorg. Chem.*, **13**, 2261, 1974

38.9 **1,4,7,10 - Tetraoxacyclododecane magnesium chloride hexahydrate**
$C_8H_{16}O_4$, Mg^{2+} , $2Cl^-$, $6H_2O$
M.A.Neuman, E.C.Steiner, F.P.van Remoortere, F.P.Boer
Inorg. Chem., **14**, 734, 1975

38.10 **bis(1,4,7,10 - Tetraoxacyclododecane) sodium chloride hemihydrate**
$2C_8H_{16}O_4$, Na^+ , Cl^- , $0.5H_2O$
F.P.van Remoortere, F.P.Boer *Inorg. Chem.*, **13**, 2071, 1974

38.11 **bis(1,4,7,10 - Tetraoxacyclododecane) sodium hydroxide octahydrate**
$2C_8H_{16}O_4$, Na^+ , HO^- , $8H_2O$
F.P.Boer, M.A.Neuman, F.P.van Remoortere, E.C.Steiner
Inorg. Chem., **13**, 2826, 1974

38.12 **Coumarin**
$C_9H_6O_2$
R.M.Myasnikova, T.S.Davydova, V.I.Simonov
Kristallografiya, **18**, 720, 1973

38.C **Methylenomycin A**
$C_9H_{10}O_4$
For complete entry see 50.1

38.13 **cis - (5S) - 2 - t - Butyl - 5 - carboxymethyl - 1,3 - dioxolan - 4 - one**
$C_9H_{14}O_5$
M.V.Gabrielsen *Acta Chem. Scand. Ser. A*, **29**, 7, 1975

38.C **5 - Chloro - 5 - deoxy - 1,2 - O - isopropylidene - 3 - methanesulfonyl - 4 - thio - β - L - arabinofuranose**
$C_9H_{15}ClO_5S_2$
For complete entry see 45.26

38.C **Pyromellitic dianhydride - trans - stilbene complex**
$C_{10}H_2O_6$, $C_{14}H_{12}$
For complete entry see 60.12

38.C **Benz(a)anthracene - pyromellitic dianhydride complex**
$C_{10}H_2O_6$, $C_{18}H_{12}$
For complete entry see 60.27

38.C **Pyromellitic dianhydride - trans - 4 - methyl - stilbene complex**
$2C_{10}H_2O_6$, $C_{15}H_{14}$
For complete entry see 60.13

38.14 **4 - Methyl - umbelliferone**
$C_{10}H_8O_3$
S.Shimizu, S.Kashino, M.Haisa *Acta Crystallogr., Sect. B,* **31,** 1287, 1975

38.15 **α - (2 - Hydroxy - 5 - fluoro - phenyl) - α,α - dimethyl acetic acid lactone**
$C_{10}H_9FO_2$
S.A.Chawdhury *Cryst. Struct. Commun.,* **4,** 73, 1975

38.16 **6 - (2 - Hydroxy - prop - 1 - enyl) - 2,4 - dimethylpyrylium chloride**
$C_{10}H_{13}O_2{}^+$, Cl^-
N.Serpone, P.H.Bird *J. Chem. Soc., Chem. Commun.,* 284, 1975

38.17 **7 - Hydroxy - 5,6 - dimethoxy - coumarin**
$C_{11}H_{10}O_5$
H.Wagner, S.Bladt, D.J.Abraham, H.Lotter *Tetrahedron Lett.,* 3807, 1974

38.18 **6a - Carbomethoxy - 3 - carboethoxy - 4 - oxo - perhydrofuro(3,4 - c)pyrazole**
$C_{11}H_{16}N_2O_6$
W.H.De Camp, S.W.Pelletier
Am. Cryst. Assoc., Abstr. Papers (Summer Meeting), 255, 1974
Also classified in 35

38.19 **5 - Chloromethyl - 4 - oxa - homoadamantan - 5 - ol**
$C_{11}H_{17}ClO_2$
M.J.Begley, G.B.Gill, D.C.Woods
J. Chem. Soc., Perkin Trans. 2, 74, 1975

38.20 **Octachlorodibenzo - p - dioxin**
$C_{12}Cl_8O_2$
F.P.Boer, M.A.Neuman, F.P.van Remoortere, P.P.North, H.W.Rinn
Adv. Chem. Ser., **120,** 14, 1973

38.21 **2,3,7,8 - Tetrachloro - dibenzo - p - dioxin**
$C_{12}H_4Cl_4O_2$
F.P.Boer, M.A.Neuman, F.P.van Remoortere, P.P.North, H.W.Rinn
Adv. Chem. Ser., **120,** 14, 1973

38.22 **2,8 - Dichloro - dibenzo - p - dioxin**
$C_{12}H_6Cl_2O_2$
F.P.Boer, M.A.Neuman, F.P.van Remoortere, P.P.North, H.W.Rinn
Adv. Chem. Ser., **120,** 14, 1973

38.23 **2,7 - Dichloro - dibenzo - p - dioxin**
$C_{12}H_6Cl_2O_2$
F.P.Boer, M.A.Neuman, F.P.van Remoortere, P.P.North, H.W.Rinn
Adv. Chem. Ser., **120,** 14, 1973

38.24 **1,4,5,8 - Tetracyano - 9 - oxa - bicyclo(6.1.0)non - 4 - ene**
$C_{12}H_8N_4O$
D.Bellus, H.-C.Mez, G.Rihs, H.Sauter *J. Am. Chem. Soc.*, **96,** 5007, 1974

38.25 **Dibenzofuran**
$C_{12}H_8O$
A.Banerjee *Z. Kristallogr.*, **139,** 415, 1974

38.26 **bis(Xanthotoxin) potassium tri - iodide**
$2C_{12}H_8O_4$, K^+ , I_3^-
M.Kapon, F.H.Herbstein *Nature (London)*, **249,** 439, 1974

38.C **5′,2 - O - Cyclo - 2′,3′ - O - isopropylidene - uridine**
$C_{12}H_{14}N_2O_5$
For complete entry see 47.31

38.27 **r - 2 - (p - Bromophenyl)cis - 4,cis - 6 - dimethyl - 1,3 - dioxane**
$C_{12}H_{15}BrO_2$
F.W.Nader *Tetrahedron Lett.*, 1207, 1975

38.28 **8 - Acetyl - 1,2,6,7,8 - pentamethyl - 2,4,5,9 - tetraoxa - tricyclo(4.2.1.0^{3,7})nonane**
$C_{12}H_{18}O_5$
H.Henke, H.Keul, E.Osawa *Acta Crystallogr., Sect. A*, **31,** S125, 1975

38.29 **1,4,7,10,13,16 - Hexaoxacyclo - octadecane**
$C_{12}H_{24}O_6$
J.D.Dunitz, P.Seiler *Acta Crystallogr., Sect. B*, **30,** 2739, 1974

38.30 **1,4,7,10,13,16 - Hexaoxacyclo - octadecane (discussion)**
$C_{12}H_{24}O_6$
J.D.Dunitz, M.Dobler, P.Seiler, R.P.Phizackerley
Acta Crystallogr., Sect. B, **30,** 2733, 1974

38.31 **1,4,7,10,13,16 - Hexaoxacyclo - octadecane calcium thiocyanate**
$C_{12}H_{24}O_6$, Ca^{2+} , $2CNS^-$
J.D.Dunitz, P.Seiler *Acta Crystallogr., Sect. B*, **30,** 2750, 1974

38.32 **1,4,7,10,13,16 - Hexaoxacyclo - octadecane calcium thiocyanate (discussion)**
$C_{12}H_{24}O_6$, Ca^{2+} , $2CNS^-$
J.D.Dunitz, P.Seiler, M.Dobler, R.P.Phizackerly
Acta Crystallogr., Sect. B, **30,** 2733, 1974

38.33 **1,4,7,10,13,16 - Hexaoxacyclo - octadecane cesium thiocyanate**
$C_{12}H_{24}O_6$, Cs^+ , CNS^-
M.Dobler, R.P.Phizackerley *Acta Crystallogr., Sect. B*, **30,** 2748, 1974

38.34 **1,4,7,10,13,16 - Hexaoxacyclo - octadecane cesium thiocyanate (discussion)**
$C_{12}H_{24}O_6$, Cs^+ , CNS^-
J.D.Dunitz, M.Dobler, P.Seiler, R.P.Phizackerley
Acta Crystallogr., Sect. B, **30,** 2733, 1974

38.35 **1,4,7,10,13,16 - Hexaoxacyclo - octadecane potassium thiocyanate**
$C_{12}H_{24}O_6$, K^+ , CNS^-
P.Seiler, M.Dobler, J.D.Dunitz *Acta Crystallogr., Sect. B,* **30,** 2744, 1974

38.36 **1,4,7,10,13,16 - Hexaoxacyclo - octadecane potassium thiocyanate (discussion)**
$C_{12}H_{24}O_6$, K^+ , CNS^-
J.D.Dunitz, M.Dobler, P.Seiler, R.P.Phizackerley
Acta Crystallogr., Sect. B, **30,** 2733, 1974

38.37 **1,4,7,10,13,16 - Hexaoxacyclo - octadecane sodium thiocyanate monohydrate**
$C_{12}H_{24}O_6$, Na^+ , CNS^- , H_2O
M.Dobler, J.D.Dunitz, P.Seiler *Acta Crystallogr., Sect. B,* **30,** 2741, 1974

38.38 **1,4,7,10,13,16 - Hexaoxacyclo - octadecane sodium thiocyanate monohydrate (discussion)**
$C_{12}H_{24}O_6$, Na^+ , CNS^- , H_2O
J.D.Dunitz, M.Dobler, P.Seiler, R.P.Phizackerley
Acta Crystallogr., Sect. B, **30,** 2733, 1974

38.39 **1,4,7,10,13,16 - Hexaoxacyclo - octadecane rubidium thiocyanate**
$C_{12}H_{24}O_6$, Rb^+ , CNS^-
M.Dobler, R.P.Phizackerley *Acta Crystallogr., Sect. B,* **30,** 2746, 1974

38.40 **1,4,7,10,13,16 - Hexaoxacyclo - octadecane rubidium thiocyanate (discussion)**
$C_{12}H_{24}O_6$, Rb^+ , CNS^-
J.D.Dunitz, M.Dobler, P.Seiler, R.P.Phizackerley
Acta Crystallogr., Sect. B, **30,** 2733, 1974

38.41 **1,4,7,10,13,16 - Hexaoxacyclo - octadecane dimethylacetylenedicarboxylate (at −160°C)**
$C_{12}H_{24}O_6$, $C_6H_6O_4$
I.Goldberg *Acta Crystallogr., Sect. B,* **31,** 754, 1975
Residue 2 classified in 1

38.C **Goniothalamin**
(+) - (6S) - 5,6 - Dihydro - 6 - styryl - 2 - pyrone
$C_{13}H_{12}O_2$
For complete entry see 59.1

38.42 **r - 2 - (p - Trifluoromethylphenyl) - trans - 4,trans - 6 - dimethyl - 1,3 - dioxan**
$C_{13}H_{15}F_3O_2$
F.W.Nader *Tetrahedron Lett.*, 1591, 1975

38.43 **2,2 - Dimethyl - 3,4 - dihydroxy - 5 - phenylvaleric acid γ - lactone**
$C_{13}H_{16}O_3$
E.Benedetti, P.Ganis, G.Bombieri, L.Caglioti, G.Germain
Acta Crystallogr., Sect. B, **31,** 1097, 1975

38.C **1 - O - Acetyl - 2,3.4,5 - di - O - isopropylidene - D - erythropent - 1 - enitol**
$C_{13}H_{20}O_6$
For complete entry see 45.41

38.C **Actinobolin hydroiodide monohydrate acetonitrile solvate (absolute configuration)**
$C_{13}H_{21}N_2O_6^+$, I^- , H_2O , $0.25C_2H_3N$
For complete entry see 50.5

38.44 **Phenacyl - kojate cesium thiocyanate**
5 - Phenacyl - 2 - (hydroxymethyl) - 4H - pyran - 4 - one cesium thiocyanate
$C_{14}H_{12}O_5$, Cs^+ , CNS^-
S.E.V.Phillips, M.R.Truter *J. Chem. Soc., Dalton Trans.*, 2517, 1974

38.45 **Phenacylkojate monohydrate**
2 - (Hydroxymethyl) - 5 - phenacyl - 4H - pyran - 4 - one monohydrate
$C_{14}H_{12}O_5$, H_2O
S.E.V.Phillips, M.R.Truter *J. Chem. Soc., Dalton Trans.*, 1071, 1975

38.46 **Phenacylkojate sodium iodide dihydrate**
2 - (Hydroxymethyl) - 5 - phenacyl - 4H - pyran - 4 - one sodium iodide dihydrate
$2C_{14}H_{12}O_5$, Na^+ , I^- , $2H_2O$
S.E.V.Phillips, M.R.Truter *J. Chem. Soc., Dalton Trans.*, 1066, 1975

38.47 **8β,9β - Difluoromethylene - 10 - methyl - 2 - decalone 2 - ethylene acetal**
$C_{14}H_{20}F_2O_2$
R.A.Moss, P.Bekiarian *Tetrahedron Lett.*, 993, 1975
Also classified in 28

38.48 **4′ - Bromo - 5 - hydroxyflavone**
$C_{15}H_9BrO_3$
T.Hayashi, S.Kawai, T.Ohno, Y.Iitaka, T.Akimoto
Chem. Pharm. Bull., **22,** 1219, 1974

38.C **Genisteine**
$C_{15}H_{10}O_5$
For complete entry see 59.3

38.C **Phomenone**
$C_{15}H_{20}O_4$
For complete entry see 59.4

38.49 **9 - Cyano - 1,10 - dimethyl - 6 - ethylenedioxy - 1 - octalin**
$C_{15}H_{21}NO_2$
H.Lynton, P.-Y.Siew *Can. J. Chem.*, **53**, 192, 1975

38.50 **11 - Methoxy - 6,10 - dimethyl - 12 - oxa - tricyclo(8.3.0.0^{1,6})tridecan - 3 - one**
$C_{15}H_{24}O_3$
G.Bernardinelli, R.Gerdil *Helv. Chim. Acta,* **57**, 1846, 1974

38.C **1,7,8,9 - Tetramethyl - 2,3,5,6 - tetra(trifluoromethyl) - 4 - thia - 10 - oxa - tetracyclo(5.2.1.0^{2,6}.0^{3,5})dec - 8 - ene**
$C_{16}H_{12}F_{12}OS$
For complete entry see 39.26

38.51 **2 - (4 - Dibenzofuranyloxy) - 2 - methylpropionic acid**
$C_{16}H_{14}O_4$
G.Malmros, A.Wagner *Acta Crystallogr., Sect. A,* **31**, S124, 1975

38.C **2 - Deacylusnic acid (absolute configuration)**
$C_{16}H_{14}O_6$
For complete entry see 59.5

38.52 **2,2′ - Spiro - (1 - ethyl - 3 - hydroxy - 3 - methyl - pyrrolidine) - (1a′,3a′ - dimethyl - 4′ - ethyl - perhydro - furo(3,2 - b)pyrrole)**
$C_{16}H_{30}N_2O_2$
M.D.d'Engenieres, M.Miocquie, J.Maldonado, J.Etienne
Bull. Soc. Chim. Fr., 658, 1974
Also classified in 36

38.53 **3 - (3,5 - Dibromo - 4 - hydroxy)benzoyl - 2 - ethyl - benzo(b)furan**
$C_{17}H_{12}Br_2O_3$
F.Fontaine, O.Dideberg, L.Dupont *Cryst. Struct. Commun.*, **4**, 49, 1975
Also classified in 17

38.54 **1 - (2 - Hydroxy - 1 - methyl)ethyl - 2 - methyl - 2 - (1 - p - bromobenzoyloxy - 2 - methoxycarbonyl)ethyl - ethylene oxide**
$C_{17}H_{21}BrO_6$
A.I.Meyers, C.C.Shaw, D.Horne, L.M.Trefonas, R.J.Majeste
Tetrahedron Lett., 1745, 1975
Also classified in 13

38.55 **t - Butylammonium 3,6,9,12,15 - pentaoxa - bicyclo(15.3.1)heneicosa - 1(21),17,19 - triene - 21 - carboxylate**
$C_{17}H_{23}O_7^-$, $C_4H_{12}N^+$
I.Goldberg *Acta Crystallogr., Sect. A,* **31**, S166, 1975
Residue 2 classified in 3

38.56 **3,6,9,12,15 - Pentaoxa - 21 - carboxy - bicyclo(15.3.1)heneicosa - 1(21),17,19 - triene**
$C_{17}H_{24}O_7$
I.Goldberg *Acta Crystallogr., Sect. A,* **31,** S166, 1975

38.C **7 - Azido - 8 - deoxy - 1,2.3,4 - di - O - isopropylidene - 6,7 - S,S - trimethylene - 6,7 - dithio - α - L - threo - D - galacto - octopyranose**
$C_{17}H_{27}N_3O_5S_2$
For complete entry see 45.52

38.C **Dihydrobotrydial**
$C_{17}H_{28}O_5$
For complete entry see 50.8

38.57 **Gnidicoumarin**
$C_{18}H_8O_5$
S.M.Kupchan, J.G.Sweeney, T.Murae, M.-S.Shen, R.F.Bryan
J. Chem. Soc., Chem. Commun., 94, 1975

38.C **Usnic acid (at −110°C)**
$C_{18}H_{16}O_7$
For complete entry see 59.6

38.C **5,6 - Dihydro - 1,3 - dimethyl - 5,6 - di(1′,3′ - dimethyl - 2′,4′,6′ - trioxopyrimid(5′,5′)yl)furo(2,3 - d)uracil**
1,3 - Dimethylbarbituric acid trimer
$C_{18}H_{18}N_6O_9$
For complete entry see 43.7

38.58 **2 - (Di - p - anisyl - methyl) - 1,3 - dioxolane**
$C_{18}H_{20}O_4$
T.E.M.van den Hark, H.M.Hendriks, P.T.Beurskens
Cryst. Struct. Commun., **3,** 703, 1974
Also classified in 17

38.C **(−) - (S) - Warfarin (absolute configuration)**
$C_{19}H_{16}O_4$
For complete entry see 59.7

38.59 **Hexacyclo (9.3.2.2^{4,7}.0^{2,9}.0^{3,8}.0^{10,12}) octadeca - 13,15,17 - triene - 5,6 - carbonate**
$C_{19}H_{18}O_3$
J.J.Stezowski *Cryst. Struct. Commun.*, **4,** 329, 1975
Also classified in 31

38.C **1 - Hydroxy - 3 - oxo - gibberellin**
$C_{19}H_{22}O_7$
For complete entry see 59.8

38.C **(3S - (3β,3aβ,5aα,5bR,5cR,7R,9aR,10aα, ,11aβ,11bα)) - Tetradecahydro - 3,7 - dihydroxy - 3a - methyl - 1H - 5b,7 - methano - 8H - as - indaceno(3′,2′.4,5)furo(2,3 - b)pyran - 8 - one monohydrate**
$C_{19}H_{26}O_5$, H_2O
For complete entry see 51.5

38.C **2,4 - Dioxa - 5α - androstan - 17β - ol acetate**
$C_{19}H_{30}O_4$
For complete entry see 51.7

38.60 **4,5 - Dihydro - benzo(a)pyrene - 4,5 - oxide**
$C_{20}H_{12}O$
J.P.Glusker, H.L.Carrell, D.E.Zacharias
Acta Crystallogr., Sect. A, **31,** S122, 1975
Also classified in 30

38.61 **Fluorescein acetone**
$C_{20}H_{12}O_5$, C_3H_6O
R.S.Osborn, D.Rogers *Acta Crystallogr., Sect. B,* **31,** 359, 1975
Residue 2 classified in 5

38.62 **5,6 - Dihydro - 7,12 - dimethyl - benz(a)anthracene - 5,6 - oxide**
$C_{20}H_{16}O$
J.P.Glusker, H.L.Carrell, D.E.Zacharias, R.Harvey
Acta Crystallogr., Sect. A, **31,** S122, 1975
Also classified in 30

38.C **trans - 2 - (2,4 - Dihydroxy - 3 - methoxybenzyl) - 3 - (4′ - hydroxy - 3′ - methoxybenzyl) - butyrolactone**
$C_{20}H_{22}O_7$
For complete entry see 59.9

38.C **Vermiculine**
$C_{20}H_{24}O_8$
For complete entry see 50.10

38.63 **2,6 - Di - cis - 4 - hydroxyretinoic acid γ - lactone**
$C_{20}H_{26}O_2$
M.M.Thackeray, G.Gafner *Acta Crystallogr., Sect. B,* **31,** 335, 1975
Also classified in 21

38.C **Lobophytolide**
$C_{20}H_{28}O_3$
For complete entry see 54.3

38.C **Sarcophine**
$C_{20}H_{28}O_3$
For complete entry see 59.10

38.64 Dicyclohexyl - 18 - crown - 6 (form A)
$C_{20}H_{36}O_6$
N.K.Dalley, J.S.Smith, S.B.Larson, J.J.Christensen, R.M.Izatt
J. Chem. Soc., Chem. Commun., 43, 1975

38.65 Dicyclohexyl - 18 - crown - 6 (form B′)
$C_{20}H_{36}O_6$
N.K.Dalley, J.S.Smith, S.B.Larson, J.J.Christensen, R.M.Izatt
J. Chem. Soc., Chem. Commun., 43, 1975

38.C Peucephyllin
$C_{21}H_{28}O_6$
For complete entry see 59.14

38.C 16α,17α - Epoxy - 3β - hydroxy - 5 - pregnen - 20 - one
16α,17α - Epoxypregnenolone
$C_{21}H_{30}O_3$
For complete entry see 51.19

38.C 17β - Hydroxy - 3 - oxo - 17α - pregna - 4,6 - diene - 21 - carboxylic acid lactone
Canrenone
$C_{22}H_{28}O_3$
For complete entry see 51.24

38.C Δ⁹ - Tetrahydrocannabinolic acid B
$C_{22}H_{30}O_4$
For complete entry see 59.15

38.66 1,1,1a,7a - Tetrachloro - 1a,2,7,7a - tetrahydro - 2,7 - diphenyl - 2,7 - epoxy - 1H - cyclopropa(b)naphthalene
$C_{23}H_{14}Cl_4O$
J.Bordner, G.R.Howard *Cryst. Struct. Commun.*, **4,** 131, 1975

38.67 2,4,6 - Triphenylpyrylium 1,1,3,3 - tetracyanopropenide
$C_{23}H_{17}O^+$, $C_7HN_4^-$
T.Tamamura, T.Yamane, N.Yasuoka, N.Kasai
Bull. Chem. Soc. Jpn., **47,** 832, 1974
Residue 2 classified in 12, 7

38.C β - Peltatin A methyl ether
$C_{23}H_{24}O_8$
For complete entry see 59.16

38.68 Tetraphenyleno - tetrafuran
$C_{24}H_8O_4$
A.-M.Pilotti *Acta Crystallogr., Sect. A,* **31,** S123, 1975

38.69 Tetraphenyleno - trifuran
$C_{24}H_{12}O_3$
A.-M.Pilotti *Acta Crystallogr., Sect. A,* **31,** S123, 1975

38.70 **6 - Bromo - 7 - hydroxy - 1,3,4,1′,3′,4′ - hexamethyl - 2,1′ - di - indanyl - 1,7′ - ether**
$C_{24}H_{27}BrO_2$
A.W.King, L.J.McDonald, J.M.Waters, T.N.Waters
Cryst. Struct. Commun., **3**, 681, 1974

38.71 **(7R,9R,18S,20S) - 6,7,9,10,17,18,20,21 - Octahydro - 7,9,18,20 - tetramethyl - dibenzo(b,k)(1,4,7,10,13,16) - hexaoxacyclo - octadecin**
$C_{24}H_{32}O_6$
P.R.Mallinson *J. Chem. Soc., Perkin Trans. 2*, 266, 1975

38.72 **(7R,9R,18S,20S) - 6,7,9,10,17,18,20,21 - Octahydro - 7,9,18,20 - tetramethyl - dibenzo(b,k)(1,4,7,10,13,16)hexaoxacyclo - octadecin cesium thiocyanate**
$C_{24}H_{32}O_6$, Cs^+ , CNS^-
P.R.Mallinson *J. Chem. Soc., Perkin Trans. 2*, 261, 1975

38.73 **bis((7R,9R,18R,20R) - 6,7,9,10,17,18,20,21 - Octahydro - 7,9,18,20 - tetramethyl - dibenzo(b,k)(1,4,7,10,13,16)hexaoxacyclo - octadecin) cesium thiocyanate**
$2C_{24}H_{32}O_6$, Cs^+ , CNS^-
P.R.Mallinson *J. Chem. Soc., Perkin Trans. 2*, 261, 1975

38.C **4 - O - Ethyl - ascofuranone (absolute configuration)**
$C_{25}H_{33}ClO_5$
For complete entry see 50.20

38.C **Podolactone A p - bromobenzoate**
$C_{26}H_{25}BrO_9$
For complete entry see 59.17

38.C **3,20 - Diethylenedioxy - 9β,11β - oxido - 11α - acetoxy - 9,11 - seco - 11,19 - cyclo - 5α,14β,17α - pregnane**
$C_{27}H_{40}O_7$
For complete entry see 51.42

38.C **p - Bromobenzoyl - leukopleurotin**
$C_{28}H_{27}BrO_6$
For complete entry see 50.22

38.C **Borrelidin solvate**
$C_{28}H_{43}NO_6$, $C_5H_{12}O$
For complete entry see 50.23

38.C **Helichrysoside hydrate**
$C_{30}H_{26}O_{14}$, $5.5H_2O$
For complete entry see 59.18

38.C **Triol Q acetonide p - iodobenzoate**
$C_{30}H_{41}IO_4$
For complete entry see 59.19

38.C **Stemphone**
$C_{30}H_{42}O_8$
For complete entry see 59.20

38.74 **(E) - 1 - Phenyl - 1 - (3,5 - diphenyl - 2 - furyl) - 2 - (3 - nitrobenzoyl)ethylene (at −30°C)**
$C_{31}H_{21}NO_4$
R.L.Harlow, S.H.Simonsen *Cryst. Struct. Commun.*, **4**, 311, 1975

38.C **Novobiocin monohydrate**
$C_{31}H_{36}N_2O_{11}$, H_2O
For complete entry see 50.24

38.75 **1,3,7,9 - Tetramethyl - 5,11 - diphenyl - (1) - benzopyran(4,3 - c)(1) - benzopyran**
$C_{32}H_{30}O_2$
G.D.Andreetti, G.Bocelli, L.Coghi, P.Sgarabotto
Cryst. Struct. Commun., **3**, 761, 1974

38.C **5 - Hydroxy - 6 - bromo - 2″,3″,4′,4″,6″,7 - hexa - acetyl - vitexin 5 - hydroxy - 3,6 - dibromo - 2″,3″,4′,4″,6″,7 - hexa - acetyl - vitexin mixture (absolute configuration)**
$0.30C_{33}H_{30}Br_2O_{16}$, $0.70C_{33}H_{31}BrO_{16}$
For complete entry see 59.21

38.C **5 - Hydroxy - 6 - bromo - 2″,3″,4′,4″,6″,7 - hexa - acetyl - vitexin 5 - hydroxy - 3,6 - dibromo - 2″,3″,4′,4″,6″,7 - hexa - acetyl - vitexin mixture (absolute configuration)**
$0.70C_{33}H_{31}BrO_{16}$, $0.30C_{33}H_{30}Br_2O_{16}$
For complete entry see 59.21

38.76 **Tri - o - thymotide**
$C_{33}H_{35}O_6$
D.J.Williams, D.Lawton *Tetrahedron Lett.*, 111, 1975

38.C **Tri - o - thymotide ethanol clathrate**
$C_{33}H_{36}O_6$, C_2H_6O
For complete entry see 61.3

38.C **Tri - o - thymotide cetyl alcohol clathrate**
$0.2C_{33}H_{36}O_6$, $C_{16}H_{34}O$
For complete entry see 61.4

38.C **Tri - o - thymotide pyridine clathrate**
$2C_{33}H_{36}O_6$, C_5H_5N
For complete entry see 61.5

38.C **Sodium 5 - bromo - lasalocid (form i)**
$C_{34}H_{52}BrO_8^-$, Na^+
For complete entry see 50.26

38.C **Sodium 5 - bromo - lasalocid (form ii)**
$C_{34}H_{52}BrO_8^-$, Na^+
For complete entry see 50.27

38.C **Milbemycin beta1 p - bromophenylurethane**
$C_{39}H_{52}BrNO_8$
For complete entry see 59.24

38.C **Anhydro - erythromycin A carbonate methiodide methanol solvate**
$C_{39}H_{66}NO_{13}^+$, I^- , CH_4O
For complete entry see 50.29

38.C **1,1′ - Binaphthyl - (2,3),(2′,3′) - bis - 18 - crown - 6 - 1,4 - diammonium - butane di(hexafluorophosphate) complex (at −160°C)**
$C_{40}H_{50}O_{12}$, $C_4H_{14}N_2^{2+}$, $2F_6P^-$
For complete entry see 60.39

38.C **Nonactin sodium thiocyanate**
$C_{40}H_{64}O_{12}$, Na^+ , CNS^-
For complete entry see 50.30

38.C **Grisorixin monohydrate**
$C_{40}H_{68}O_{10}$, H_2O
For complete entry see 50.31

38.C **des - Boron - des - valine - boromycin monohydrate**
$C_{40}H_{68}O_{14}$, H_2O
For complete entry see 50.32

38.C **Methyl 24,25 - O - isopropylidene - 3α - p - bromobenzoyloxy - 9(10 - 19) - abeo - lanost - 5(10) - ene - 32 - carboxylate**
$C_{41}H_{59}BrO_6$
For complete entry see 56.7

38.C **Rifampicine pentahydrate**
$C_{43}H_{58}N_4O_{12}$, $5H_2O$
For complete entry see 50.33

38.C **Kansuinine B diol p - bromobenzoate**
$C_{45}H_{47}BrO_{17}$
For complete entry see 54.16

38.C **Antibiotic X - 206 monohydrate**
$C_{47}H_{82}O_{14}$, H_2O
For complete entry see 50.34

38.C **Potassium alborixin**
$C_{48}H_{83}O_{14}^-$, K^+
For complete entry see 50.35

38.C Isokidamycin bis(m - bromobenzoate) benzene solvate (absolute configuration)
$C_{53}H_{54}Br_2N_2O_{11}$, $0.5C_6H_6$
For complete entry see 50.36

38.C p - Bromophenacyl - septamycin monohydrate (absolute configuration)
$C_{56}H_{87}BrO_{17}$, H_2O
For complete entry see 50.38

HETERO-SULPHUR AND HETERO-SELENIUM

39.C **Triphenylmethane thiophene**
C_4H_4S , $C_{19}H_{16}$
For complete entry see 60.30

39.1 **bis(Tetramethylenesulfone) trans - tetra - μ - (chloro - dichloro - diaquo - copper(ii)) copper(ii)**
$2C_4H_8O_2S$, $H_4Cl_6Cu_3O_2$
D.D.Swank, R.D.Willett *Inorg. Chim. Acta,* **8,** 143, 1974

39.2 **1,4 - Dithiane - trichlorostibine**
$C_4H_8S_2$, Cl_3Sb
G.Kiel, R.Engler *Chem. Ber.,* **107,** 3444, 1974

39.3 **Tetrathiofulvalene - iodine complex (form i)**
$C_6H_4S_4$, I_2
C.K.Johnson, C.R.Watson Junior, R.J.Warmack
Am. Cryst. Assoc., Abstr. Papers (Spring Meeting), 19, 1975

39.C **Tetrathiofulvalinium 7,7,8,8 - tetracyanoquinodimethanide (at 100°K)**
$C_6H_4S_4^+$, $C_{12}H_4N_4^-$
For complete entry see 60.3

39.4 **Hepta(tetrathiafulvalene) penta - iodide**
$5C_6H_4S_4^+$, $2C_6H_4S_4$, $5I^-$
J.J.Daly, F.Sanz *Acta Crystallogr., Sect. B,* **31,** 620, 1975

39.C **Di(tetrathiofulvalene) bis(dithiolene) nickel**
$2C_6H_4S_4$, $C_4H_4NiS_4$
For complete entry see 60.4

39.5 **2 - Methyl - 6a - thiathiophthene**
$C_6H_6S_3$
L.J.Saethre, A.Hordvik *Acta Chem. Scand. Ser. A,* **29,** 136, 1975

39.6 **3,4,5 - tris(Methylthio) - 1,2 - dithiolium iodide**
$C_6H_9S_5^+$, I^-
G.Kiel, U.Reuter, G.Gattow *Chem. Ber.,* **107,** 2569, 1974

39.7 **1 - Methyl - 1 - thionia - cyclohexane iodide**
$C_6H_{13}S^+$, I^-
R.Gerdil *Helv. Chim. Acta,* **57,** 489, 1974

39.8 **tetrakis(Chloromethyl)hexathia - adamantane**
$C_8H_8Cl_4S_6$
S.Aleby *Acta Crystallogr., Sect. B,* **30,** 2877, 1974

39.9 **1 - (2 - (1,3 - Dithiolanylidene)) - 2,3,6,9 - tetrathiaspiro(4.4)nonane (monoclinic form)**
$C_8H_{10}S_6$
J.F.Blount, D.L.Coffen, F.Wong *J. Org. Chem.,* **39,** 2374, 1974

39.10 **Benzo(1,2 - c.3,4 - c′.5,6 - c″)tris(1,2)dithiole - 1,4,7 - trithione**
C_9S_9
L.K.Hansen, A.Hordvik *J. Chem. Soc., Chem. Commun.,* 800, 1974

39.C **7 - Methylbenzothieno(3,2 - c)furoxan**
$C_9H_6N_2O_2S$
For complete entry see 40.2

39.11 **(S) - α - (p - Chlorobenzenesulfonamido) - β - propiothiolactone**
$C_9H_8ClNO_3S_2$
I.Milinovic, A.Bezjak, D.Fles *Croat. Chem. Acta,* **45,** 551, 1973

39.12 **cis - 3 - p - Bromophenylthietan 1 - oxide**
C_9H_9BrOS
J.H.Barlow, C.R.Hall, D.R.Russell, D.J.H.Smith
J. Chem. Soc., Chem. Commun., 133, 1975

39.13 **trans - 3 - p - Bromophenylthietan 1 - oxide**
C_9H_9BrOS
J.H.Barlow, C.R.Hall, D.R.Russell, D.J.H.Smith
J. Chem. Soc., Chem. Commun., 133, 1975

39.14 **9 - Thiono - 8,10 - dithiabicyclo(5.3.1)undeca - 2,5 - diene (monoclinic form)**
$C_9H_{10}S_3$
E.Cuthbertson, D.D.MacNicol, P.R.Mallinson
Tetrahedron Lett., 1345, 1975

39.C **5 - Fluoro - 4′ - thiouridine (absolute configuration)**
$C_9H_{11}FN_2O_5S$
For complete entry see 47.3

39.C **5 - Chloro - 5 - deoxy - 1,2 - O - isopropylidene - 3 - methanesulfonyl - 4 - thio - β - L - arabinofuranose**
$C_9H_{15}ClO_5S_2$
For complete entry see 45.26

39.C **5 - Desoxy - 3 - C - formyl - β - L - lyxofuranose - trimethylenedithioacetal**
$C_9H_{16}O_4S_2$
For complete entry see 45.27

39.15 **4 - t - Butyl - thiacyclohexane sulfoxide**
$C_9H_{17}ClOS$
F.Robert *C. R. Acad. Sci., Ser. C,* **279,** 737, 1974

39.16 **1,3,4,5,6,8 - Hexachlorothieno(2,3 - c.5,4 - c′)dipyridine**
$C_{10}Cl_6N_2S$
A.D.Redhouse *J. Chem. Soc., Perkin Trans. 2,* 1925, 1974
Also classified in 36

39.C **Anthracene - pyromellitic dithioanhydride complex**
$C_{10}H_2O_4S$, $C_{14}H_{10}$
For complete entry see 60.11

39.17 **6 - Thia - tetracyclo(5.4.0.0^{2,4}.0^{3,5})undeca - 1(7),8,10 - triene**
$C_{10}H_8S$
C.Kabuto, T.Tatsuoka, I.Murata, Y.Kitahara *Angew. Chem.,* **86,** 738, 1974

39.18 **5 - Bromo - 2,3 - dimethylbenzo(b)thiophen**
$C_{10}H_9BrS$
J.H.C.Hogg, H.H.Sutherland *Acta Crystallogr., Sect. B,* **30,** 2058, 1974

39.19 **axial - 2 - Phenyl - 1,3 - dithiane - 1 - oxide**
$C_{10}H_{12}OS_2$
R.F.Bryan, R.J.Maher, P.M.Smith, I.F.Taylor Junior
Acta Crystallogr., Sect. A, **31,** S166, 1975

39.20 **equatorial - 2 - Phenyl - 1,3 - dithiane - 1 - oxide**
$C_{10}H_{12}OS_2$
R.F.Bryan, R.J.Maher, P.M.Smith, I.F.Taylor Junior
Acta Crystallogr., Sect. A, **31,** S166, 1975

39.C **1,4.3,6 - bis - Thioanhydro - 2,5 - O - acetyl - D - iditol disulfoxide**
$C_{10}H_{14}O_6S_2$
For complete entry see 45.28

39.21 **Dithieno(2,1 - b.4,5 - b′)tropylium perchlorate**
$C_{11}H_7S_2^+$, ClO_4^-
B.Aurivillius *Acta Chem. Scand. Ser. B,* **28,** 681, 1974
Residue 1 also classified in 12

39.22 **8,9 - Dihydro - 4H - cyclohepta(1,2 - b.5,4 - b′)dithiophene - 4 - one**
$C_{11}H_8OS_2$
J.-E.Andersson *Acta Crystallogr., Sect. B,* **31,** 1396, 1975

39.23 **Tetramethylene selenium dimedonylide**
$C_{12}H_{18}O_2Se$
V.V.Saatsozov, R.A.Kyandzhetzian, S.I.Kuznetsov, N.N.Magdesieva, T.L.Khotsyanova *Dokl. Akad. Nauk SSSR,* **206,** 1130, 1972
Also classified in 21

39.C **5 - Desoxy - 3 - C - formyl - 1,2 - O - isopropylidene - β - L - lyxofuranose - trimethylenedithioacetal**
$C_{12}H_{20}O_4S_2$
For complete entry see 45.35

39.24 **3,3.6,6 - bis(Pentamethylene) - s - tetrathiane**
$C_{12}H_{20}S_4$
C.H.Bushweller, G.Bhat, L.J.Letendre, J.A.Brunelle, H.S.Bilofsky, H.Ruben, D.H.Templeton, A.Zalkin *J. Am. Chem. Soc.*, **97**, 65, 1975

39.25 **cis - 2,4 - Diphenylthietane - trans - 1 - oxide**
$C_{15}H_{14}OS$
G.L.Hardgrove Junior, J.S.Bratholdt, M.M.Lein
J. Org. Chem., **39**, 246, 1974

39.26 **1,7,8,9 - Tetramethyl - 2,3,5,6 - tetra(trifluoromethyl) - 4 - thia - 10 - oxa - tetracyclo(5.2.1.0^{2,6}.0^{3,5})dec - 8 - ene**
$C_{16}H_{12}F_{12}OS$
N.Kikutani, Y.Iitaka, Y.Kobayashi, I.Kumadaki, A.Ohsawa, Y.Sekine
Acta Crystallogr., Sect. B, **31**, 1478, 1975
Also classified in 38

39.27 **cis - 9 - Isopropyl - thioxanthene 10 - oxide**
$C_{16}H_{16}OS$
S.S.C.Chu *Acta Crystallogr., Sect. B*, **31**, 1082, 1975

39.C **Tricarbonyl - η^4 - (1 - pentafluorophenyl - 2,3,4,5 - tetrakis(trifluoromethyl) thiophen) manganese**
$C_{17}F_{17}MnO_3S$
For complete entry see 75.16

39.C **7 - Azido - 8 - deoxy - 1,2.3,4 - di - O - isopropylidene - 6,7 - S,S - trimethylene - 6,7 - dithio - α - L - threo - D - galacto - octopyranose**
$C_{17}H_{27}N_3O_5S_2$
For complete entry see 45.52

39.28 **Naphthaceno(5,6 - cd.11,12 - c'd')bis - (1,2 - dithiolane)**
Tetrathiotetracene
$C_{18}H_8S_4$
O.Dideberg, J.Toussaint *Acta Crystallogr., Sect. B*, **30**, 2481, 1974

39.29 **Benzo(c)thiopyrylium - 4 - oxide dimer**
$C_{18}H_{12}O_2S_2$
P.Groth *Acta Chem. Scand. Ser. A*, **29**, 298, 1975

39.30 **3 - Hydroxy - 1,1 - dimethylpyrrolidinium α - phenyl - 2 - thiophene - glycollate bromide**
$C_{18}H_{22}NO_3S^+$, Br^-
K.N.Caughenour, D.Dennis *Acta Crystallogr., Sect. B*, **31**, 1229, 1975
Residue 1 also classified in 32

39.31 **Diethyl - (2 - hydroxyethyl) - methylammonium α - cyclopentyl - 2 - thienylglycollate bromide**
Penthienate bromide
$C_{18}H_{30}NO_3S^+$, Br^-
J.J.Guy, T.A.Hamor *J. Chem. Soc., Perkin Trans. 2,* 1126, 1974
Residue 1 also classified in 20, 1, 3

39.32 **2 - (p - Dimethylanilino) - 4 - phenyl - 6,6a - dithiafurophthene**
$C_{19}H_{17}NOS_2$
L.J.Saethre, A.Hordvik *Acta Crystallogr., Sect. B,* **31,** 30, 1975

39.33 **2 - p - Methoxyphenyl - 4,5 - (1 - (1,3 - dithiolan - 2 - ylidene) - tetramethylene) - 1,6,6a - thiathiophthene**
$C_{19}H_{18}OS_5$
J.Sletten *Acta Chem. Scand. Ser. A,* **28,** 499, 1974

39.34 **2,5 - Diphenyl - 3,4 - trimethylene - 6a - thiathiophthene**
$C_{20}H_{16}S_3$
B.Birknes, A.Hordvik, L.J.Saethre
Acta Chem. Scand. Ser. A, **29,** 195, 1975

39.35 **3,3,7,7,10,10,14,14 - Octamethyl - 5,12 - dithiatricyclo(7.5.0.0^{2,8})tetradeca - 1(9),2(8) - diene**
$C_{20}H_{32}S_2$
H.Irngartinger, H.Rodewald *Angew. Chem.,* **86,** 783, 1974

39.C **2 - Triphenylsilyl - 1,3 - dithiane - 1 - oxide**
$C_{22}H_{22}OS_2Si$
For complete entry see 63.15

39.36 **4 - (2 - (Methylthio)dibenzo(b,f)thiepin - 11 - yl) - 1 - piperazinyl - propanol hemihydrate**
$C_{22}H_{28}N_2OS_2$, $0.5H_2O$
M.H.J.Koch, G.Evrard *Acta Crystallogr., Sect. B,* **30,** 2925, 1974
Residue 1 also classified in 33

39.C **2 - Triphenylsilyl - 2 - methyl - 1,3 - dithiane**
$C_{23}H_{24}S_2Si$
For complete entry see 63.16

39.37 **β - Flupenthixol**
$C_{23}H_{25}F_3N_2OS$
O.Kennard, M.L.Post, J.R.Rodgers, A.S.Horn
Acta Crystallogr., Sect. A, **31,** S54, 1975
Also classified in 33

39.38 **α - Flupenthixol**
$C_{23}H_{25}F_3N_2OS$
O.Kennard, M.L.Post, J.R.Rodgers, A.S.Horn
Acta Crystallogr., Sect. A, **31,** S54, 1975
Also classified in 33

39.39 2,5 - Di - t - butyl - 3,6 - di(neopentyl)thioeno(3,2 - b)thiophene
$C_{24}H_{40}S_2$
T.C.McKenzie *Acta Crystallogr., Sect. B,* **31,** 1778, 1975

39.40 1,2 - Dihydronaphtho(2,1 - d) - 3H - indano(2,1 - f) - N - phenylsuccinimido(4,3 - d) - 7α - thia - 2β,3β,5α,6β - tetrahydrobicyclo(2.2.1)hepta - 2,5 - diene benzene solvate
$C_{29}H_{21}NO_2S$, C_6H_6
M.E.Leonowicz, S.K.Obendorf, A.G.Schultz, R.E.Hughes
Am. Cryst. Assoc., Abstr. Papers (Summer Meeting), 206, 1974
Residue 1 also classified in 37

HETERO-(NITROGEN AND OXYGEN)

40.C **7 - Amino - 4H - furazan(3,4 - d)(1,2,6)thiadiazine 1,1 - dioxide**
$C_3H_3N_5O_3S$
For complete entry see 41.1

40.1 **3 - Hydroxy - 5 - (3 - aminopropyl)isoxazole monohydrate**
$C_6H_{10}N_2O_2$, H_2O
L.Brehm, P.Krogsgaard-Larsen *Acta Chem. Scand. Ser. B,* **28,** 625, 1974
Residue 1 also classified in 3

40.2 **7 - Methylbenzothieno(3,2 - c)furoxan**
$C_9H_6N_2O_2S$
M.Calleri, D.Viterbo, A.C.Villa, C.Guastini
Cryst. Struct. Commun., **4,** 13, 1975
Also classified in 39

40.C **α,D - O^2 - 2$'$ - Cyclouridine**
$C_9H_{10}N_2O_5$
For complete entry see 47.2

40.C **2,2$'$ - Anhydro - 1 - β - DL - arabinofuranosyl cytosine hydrochloride**
$C_9H_{12}N_3O_4^+$, Cl^-
For complete entry see 47.7

40.C **2,2$'$ - Anhydro - 1 - β - L - arabinofuranosyl cytosine hydrochloride**
$C_9H_{12}N_3O_4^+$, Cl^-
For complete entry see 47.8

40.C **2,2$'$ - Anhydro - 2 - hydroxy - 1 - (β - D - arabinopentofuranosyl) - 4 - pyridine**
$C_{10}H_{11}NO_5$
For complete entry see 47.17

40.C **(4R,6R) - 3 - Methoxycarbonyl - 9,9 - dimethyl - 8 - oxa - 4 - thia - 1 - azabicyclo(4.3.0)non - 2 - ene 4 - oxide**
$C_{10}H_{15}NO_4S$
For complete entry see 41.10

40.3 **Spiro(cyclohexane - 1,2′ - (4′,4′ - dimethyloxazolidine - N - oxyl))**
$C_{10}H_{18}NO_2$
D.Bordeaux, J.Lajzerowicz-Bonneteau
Acta Crystallogr., Sect. B, **30,** 2130, 1974
Also classified in 12

40.4 **Phenmetrazine hydrochloride (monoclinic form)**
3 - Methyl - 2 - phenyl - morpholine hydrochloride
$C_{11}H_{16}NO^+$, Cl^-
D.Carlstrom, I.Hacksell *Acta Crystallogr., Sect. B,* **30,** 2477, 1974

40.C **cis - bis(bis - (Trifluoromethyl) - ethylene - 1,2 - dithiolato) nickel - phenoxazine**
$C_{12}H_9NO$, $C_8F_{12}NiS_4$
For complete entry see 60.8

40.5 **2 - Acetyl - 1 - methyl - 8 - nitro - 1,2,4,5 - tetrahydro - 3,2 - benzoxazepine**
$C_{12}H_{14}N_2O_4$
A.Mangia, G.Pelizzi *Cryst. Struct. Commun.,* **3,** 673, 1974

40.C **5′,2 - O - Cyclo - 2′,3′ - O - isopropylidene - uridine**
$C_{12}H_{14}N_2O_5$
For complete entry see 47.31

40.C **Bicyclomycin**
$C_{12}H_{18}N_2O_7$
For complete entry see 50.4

40.6 **1,8 - Diaza - 10,15 - dioxa - 9,16 - dioxo - cyclohexadecane**
$C_{12}H_{22}N_2O_4$
L.E.Alexander, J.J.Beres
Am. Cryst. Assoc., Abstr. Papers (Summer Meeting), 252, 1974

40.7 **N - (3 - Phenyl - 5 - (1,2,3,4 - oxatriazolio))phenylamide (at 105°K)**
$C_{13}H_{10}N_4O$
T.Ottersen, C.Christophersen, S.Treppendahl
Acta Chem. Scand. Ser. A, **29,** 45, 1975

40.8 **4β - Ethyl - 2β - acetoxymethyl - 9 - methyl - 9 - aza - 3 - oxa - bicyclo(3.3.1)nonan - 7 - one**
$C_{13}H_{21}NO_4$
T.Masamune, H.Matsue, S.Numata, A.Furusaki
Tetrahedron Lett., 3933, 1974
Also classified in 37

40.C **8.5′ - Anhydro - 7 - bromo - 8 - hydroxy - 2′,3′ - isopropylidene - tubercidin (absolute configuration)**
$C_{14}H_{15}BrN_4O_4$
For complete entry see 50.6

40.9 **anti - α - Morpholino - β - methyl - α - chloromethanesulfonyl - styrene**
$C_{14}H_{18}ClNO_3S$
G.D.Andreetti, G.Bocelli, L.Cavalca, P.Sgarabotto
Eur. Cryst. Meeting, 297, 1974
Also classified in 11

40.10 **syn - α - Morpholino - β - methyl - α - chloromethanesulfonyl - styrene**
$C_{14}H_{18}ClNO_3S$
G.D.Andreetti, G.Bocelli, L.Cavalca, P.Sgarabotto
Eur. Cryst. Meeting, 297, 1974
Also classified in 11

40.C **3,1′ - Anhydro - 2 - (4′,6′ - di - O - acetyl - 2′,3′ - dideoxy - α - D - ribo - hexopyranose) - 3 - hydroxy - 5 - methyl - 2H - 1,2,6 - thiadiazine - 1,1 - dioxide**
$C_{14}H_{18}N_2O_8S$
For complete entry see 47.34

40.11 **6α - Bromo - 4α - (3,5 - dimethyl - 4 - isoxazolyl - methyl) - 1β - hydroxy - 7aβ - methyl - 3aα,6,7,7a - tetrahydro - 5(4H) - indanone**
$C_{16}H_{22}BrNO_3$
T.C.McKenzie *J. Org. Chem.,* **39,** 629, 1974
Also classified in 27

40.12 **cis - 1,5 - Diphenyl - 6 - oxa - 4 - aza - spiro(2,4)hept - 4 - en - 7 - one**
$C_{17}H_{13}NO_2$
M.L.Martinez-Garcia, F.H.Cano, S.Garcia-Blanco
Acta Crystallogr., Sect. A, **31,** S103, 1975

40.13 **trans - 1,5 - Diphenyl - 6 - oxa - 4 - aza - spiro(2,4)hept - 4 - en - 7 - one**
$C_{17}H_{13}NO_2$
M.L.Martinez-Garcia, F.H.Cano, S.Garcia-Blanco
Acta Crystallogr., Sect. A, **31,** S103, 1975

40.14 **5 - (Hydroxy(phenyl)amino) - 3,3,5 - trimethyl - 2 - phenylisoxazolidine (form B)**
$C_{18}H_{22}N_2O_2$
R.Foster, J.Iball, R.Nash *J. Chem. Soc., Perkin Trans. 2,* 1210, 1974

40.15 **5 - (Hydroxy(phenyl)amino) - 3,3,5 - trimethyl - 2 - phenylisoxazolidine (form A)**
$C_{18}H_{22}N_2O_2$
R.Foster, J.Iball, R.Nash *J. Chem. Soc., Perkin Trans. 2,* 1210, 1974

40.16 **5,8 - Di - t - butyl - 3,3 - dimethyl - 9 - isopropylidene - 5,8 - diaza - 4,7 - dioxa - bicyclo(4.2.1)nonan - 2 - one**
$C_{18}H_{32}N_2O_3$
J.Murray-Rust, P.Murray-Rust *Acta Crystallogr., Sect. B,* **31,** 589, 1975

40.17 **4,7,13,16,21,24 - Hexaoxa - 1,10 - diazabicyclo(8.8.8)hexacosane tetra(silver(i) thiocyanate)**
$C_{18}H_{36}N_2O_6$, $4Ag^+$, $4CNS^-$
B.Metz, D.Moras, R.Weiss *Eur. Cryst. Meeting*, 376, 1974

40.18 **Sodium(2,2,2) - cryptate**
$C_{18}H_{36}N_2O_6$, Na^+ , Na^-
F.J.Tehan, B.L.Barnett, J.L.Dye *J. Am. Chem. Soc.*, **96**, 7203, 1974

40.19 **(4,7,13,16,21,24 - Hexaoxa - 1,10 - diazabicyclo(8.8.8)hexacosane) lead(ii) thiocyanate**
$C_{18}H_{36}N_2O_6$, Pb^{2+} , $2CNS^-$
B.Metz, R.Weiss *Inorg. Chem.*, **13**, 2094, 1974

40.20 **bis(4,7,13,16,21,24 - Hexaoxa - 1,10 - diazabicyclo(8.8.8)hexacosane) tetra(silver(i) thiocyanate)**
$2C_{18}H_{36}N_2O_6$, $4Ag^+$, $4CNS^-$
B.Metz, D.Moras, R.Weiss *Eur. Cryst. Meeting*, 376, 1974

40.21 **5 - Diacetylamino - 3,4 - diphenylisoxazole**
$C_{19}H_{16}N_2O_3$
K.Simon, K.Sasvari, P.Dvortsak, K.Horvath, K.Harsanyi
J. Chem. Soc., Perkin Trans. 2, 1409, 1974

40.22 **4 - Bromo - 2 - (4 - methyl - 2 - morpholino - 1H - 5 - imidazolyl - methylidene) - 1,2 - dihydro - 1 - oxonaphthalene**
$C_{19}H_{18}BrN_3O_2$
K.Kijima, T.Sakaguchi, Y.Iitaka *Chem. Pharm. Bull.*, **21**, 2529, 1973
Also classified in 32, 24

40.C **Naphthyridinomycin**
$C_{21}H_{27}N_3O_7$
For complete entry see 50.12

40.C **Griseoviridin methanol solvate**
$C_{22}H_{27}N_3O_7S$, CH_4O
For complete entry see 50.17

40.23 **2 - Phenyl - 4 - benzoyl - 4a,5,6,7,8,8a - hexahydro - 8a - piperidinyl - 4H - 1,3,4 - benzo - oxadiazine**
$C_{25}H_{29}N_3O_2$
L.Marchetti, E.Foresti Serantoni, R.Mongiorgi, L.Riva di Sanseverino
Gazz. Chim. Ital., **103**, 615, 1973
Also classified in 33

40.24 **Dextromoramide bitartrate**
$C_{25}H_{33}N_2O_2^+$, $C_4H_5O_6^-$
E.Bye *Acta Chem. Scand. Ser. B*, **29**, 22, 1975
Residue 1 also classified in 32; residue 2 classified in 2

HETERO-(NITROGEN AND SULPHUR)

41.1 **7 - Amino - 4H - furazan(3,4 - d)(1,2,6)thiadiazine 1,1 - dioxide**
$C_3H_3N_5O_3S$
C.Foces, F.H.Cano, S.Garcia-Blanco
Acta Crystallogr., Sect. A, **31,** S122, 1975
Also classified in 40

41.2 **7 - Amino - 2H,4H - vic - triazolo(4,5 - c) - 1,2,6 - thiadiazine - 1,1 - dioxide**
$C_3H_4N_6O_2S$
C.Foces-Foces, F.H.Cano, S.Garcia-Blanco
Acta Crystallogr., Sect. B, **31,** 1427, 1975
Also classified in 35

41.3 **7 - Amino - 2H,4H - v - triazolo(4,5 - c)(1,2,6)thiadiazine 1,1 - dioxide monohydrate**
$C_3H_4N_6O_2S$, H_2O
C.Foces, F.H.Cano, S.Garcia-Blanco
Acta Crystallogr., Sect. A, **31,** S122, 1975
Residue 1 also classified in 35

41.4 **2 - Imino - 4 - oxo - 1,3 - thiazolidine hydrochloride**
$C_3H_5N_2OS^+$, Cl^-
R.V.A.Murthy, B.V.R.Murthy *Cryst. Struct. Commun.,* **4,** 117, 1975

41.5 **2,4 - Dimethyl - 1,2,4 - thiazolidine - 3,5 - dithione**
$C_4H_6N_2S_3$
C.L.Raston, A.H.White, A.C.Willis, J.N.Varghese
J. Chem. Soc., Perkin Trans. 2, 1096, 1974

41.C **(5 - (2 - Hydroxyethyl) - 4 - methylthiazolium) tribromo - (5 - (2 - hydroxyethyl) - 4 - methyl - 3 - thiazolo) copper(ii) (absolute configuration)**
$C_6H_{10}NOS^+$, $C_6H_9Br_3CuNOS^-$
For complete entry see 83.8

41.6 **trans - 3 - (7 - Dihydrothiazolo(2,3 - e) - 1,2,3 - triazolyl)propenal**
$C_7H_7N_3OS$
C.Romming *Acta Chem. Scand. Ser. A,* **29,** 282, 1975
Also classified in 35

41.7 **6 - Methyl - 5 - thioformyl - pyrrolo(2,1 - b)thiazole**
$C_8H_7NS_2$
R.C.G.Killean, J.L.Lawrence, J.U.Cameron, A.Sharma
Acta Crystallogr., Sect. B, **31,** 1217, 1975
Also classified in 35

41.8 **2,6 - Dimethyl - 5 - thioformyl - pyrrolo(2,1 - b)thiazole**
$C_9H_9NS_2$
A.Sharma, R.C.G.Killean *Acta Crystallogr., Sect. B,* **30,** 2869, 1974
Also classified in 35

41.9 **2 - Methyl - 3 - ethyl - benzothiazolium tetrafluoroborate**
$C_{10}H_{12}NS^+$, BF_4^-
E.M.Srenger *Acta Crystallogr., Sect. B,* **30,** 1911, 1974

41.10 **(4R,6R) - 3 - Methoxycarbonyl - 9,9 - dimethyl - 8 - oxa - 4 - thia - 1 - azabicyclo(4.3.0)non - 2 - ene 4 - oxide**
$C_{10}H_{15}NO_4S$
J.J.Guy, T.A.Hamor *J. Chem. Soc., Perkin Trans. 2,* 1132, 1974
Also classified in 40

41.11 **D - (−) - Luciferin (for full structural details see Blank et al., Biochem.Biophys.Res.Comm.,42,583,1971)**
$C_{11}H_8N_2O_3S_2$
G.E.Blank, J.Pletcher, M.Sax *Acta Crystallogr., Sect. B,* **30,** 2525, 1974

41.12 **3 - Methyl - 2,4 - dicarbomethoxy - Δ^3 - cephem**
$C_{11}H_{13}NO_5S$
E.F.Paulus *Acta Crystallogr., Sect. B,* **30,** 2915, 1974
Also classified in 35

41.13 **3,4 - Dihydro - 7 - chloro - 6 - diethylamino - 2H,8H - pyrimido(2,1 - b)(1,3)thiazin - 8 - one**
$C_{11}H_{16}ClN_3OS$
H.J.Talberg *Acta Chem. Scand. Ser. A,* **28,** 903, 1974
Also classified in 35

41.C **(3 - Benzothiazolo - dichlorogermano)pentacarbonyl molybdenum**
$C_{12}H_5Cl_2GeMoNO_5S$
For complete entry see 69.9

41.C **cis - bis(bis - (Trifluoromethyl) - ethylene - 1,2 - dithiolato) nickel - phenothiazine**
$C_{12}H_9NS$, $C_8F_{12}NiS_4$
For complete entry see 60.9

41.C **bis(Thiaminium) magnesium chloride decahydrate**
$2C_{12}H_{18}N_4OS^{2+}$, Mg^{2+} , $6Cl^-$, $10H_2O$
For complete entry see 44.26

41.C **Thiamine pyrophosphate (form i)**
$C_{12}H_{18}N_4O_7P_2S$
For complete entry see 46.3

41.C **Thiamine pyrophosphate (form ii)**
$C_{12}H_{18}N_4O_7P_2S$
For complete entry see 46.4

41.14 **1 - (2 - Thiazolylazo) - 2 - naphthol**
$C_{13}H_9N_3OS$
M.Kurahashi *Chem. Lett.*, 181, 1974
Also classified in 24

41.15 **N - Methylphenothiazine**
$C_{13}H_{11}NS$
S.S.C.Chu, D.van der Helm *Acta Crystallogr., Sect. B,* **30,** 2489, 1974

41.C **N - Methylphenothiazine - 7,7,8,8 - tetracyanoquinodimethane complex**
$C_{13}H_{11}NS$, $C_{12}H_4N_4$
For complete entry see 60.18

41.16 **Dimethyl 4 - formyl - 2,3 - dihydrobenzothiazine - 2,3 - dicarboxylate**
$C_{13}H_{13}NO_5S$
H.Ogura, H.Takayanagi, K.Furuhata, Y.Iitaka
J. Chem. Soc., Chem. Commun., 759, 1974

41.C **(3,3′ - Diethylthiazolinocarbocyanine) - (7,7,8,8 - tetracyanoquinodimethanide) - (9 - dicyanomethylene - 2,4,7 - trinitrofluorene) complex**
$C_{13}H_{21}N_2S_2^+$, $C_{12}H_4N_4^-$, $C_{16}H_5N_5O_6$
For complete entry see 60.21

41.17 **4 - Chloro - anthraquinonethiadiazole**
$C_{14}H_5ClN_2O_2S$
O.A.Mikhno, Z.I.Ezhkova, G.S.Zhdanov *Kristallografiya,* **18,** 99, 1973

41.18 **Phenanthro(9,10 - c) - 1,2,5 - thiadiazole 1 - oxide monohydrate**
$C_{14}H_8N_2OS$, H_2O
S.K.Arora *Acta Crystallogr., Sect. B,* **30,** 2923, 1974

41.19 **3 - Phenyl - 5 - phenylazo - Δ^4,1,3,4 - thiadiazoline - 2 - one**
$C_{14}H_{10}N_4OS$
P.N.Preston, N.J.Robinson, K.Turnbull, T.J.King
J. Chem. Soc., Chem. Commun., 998, 1974

41.20 **N - Ethylphenothiazine**
$C_{14}H_{13}NS$
S.S.C.Chu, D.van der Helm *Acta Crystallogr., Sect. B,* **31,** 1179, 1975

41.21 **Tetrasulfur - tetranitrile - bis(norbornadiene)**
$C_{14}H_{16}N_4S_4$
A.M.Griffin, G.M.Sheldrick *Acta Crystallogr., Sect. B,* **31,** 895, 1975
Also classified in 31, 4

41.22 **3 - (2 - (2 - Chloroethoxy)ethyl) - 5 - methyl - 4 - phenyl - Δ^4 - thiazoline 1,1 - dioxide**
$C_{14}H_{18}ClNO_3S$
P.del Buttero, S.Maiorana, D.Pocar, G.D.Andreetti, G.Bocelli, P.Sgarabotto *J. Chem. Soc., Perkin Trans. 2,* 1483, 1974

41.C **3,1′ - Anhydro - 2 - (4′,6′ - di - O - acetyl - 2′,3′ - dideoxy - α - D - ribo - hexopyranose) - 3 - hydroxy - 5 - methyl - 2H - 1,2,6 - thiadiazine - 1,1 - dioxide**
$C_{14}H_{18}N_2O_8S$
For complete entry see 47.34

41.23 **3,4 - Ethano - 2,3,4,5 - tetrahydro - 2,5 - bis(phenylimino) - 1,6,6a - trithia - 3,4 - diazapentalene**
$C_{17}H_{14}N_4S_3$
K.-T.Wei, I.C.Paul, R.J.S.Beer, A.Naylor
J. Chem. Soc., Chem. Commun., 264, 1975

41.C **3,3′ - Dimethylthiacyanine - bis(7,7,8,8 - tetracyanoquinodimethane)**
$C_{17}H_{15}N_2S_2^+$, $C_{12}H_4N_4^-$, $C_{12}H_4N_4$
For complete entry see 60.26

41.24 **3,5 - bis(N,N - Di - isopropylthiocarbamoylimino) - 4 - methyl - 1,2,4 - dithiazolidine**
$C_{17}H_{31}N_5S_4$
J.Sletten *Acta Chem. Scand. Ser. A,* **28,** 989, 1974

41.25 **10 - (3 - (Dimethylamino)propyl) - 2 - (trifluoromethyl) - phenothiazine hydrochloride**
$C_{18}H_{20}F_3N_2S^+$, Cl^-
D.W.Phelps, A.W.Cordes *Acta Crystallogr., Sect. B,* **30,** 2812, 1974
Residue 1 also classified in 3

41.C **3,3′ - Dimethylthiacarbocyanine bis(7,7,8,8 - tetracyanoquinodimethane) complex**
$C_{19}H_{17}N_2S_2^+$, $C_{12}H_4N_4^-$, $C_{12}H_4N_4$
For complete entry see 60.34

41.26 **3 - Phenyl - 8 - phenylthiocarbamoyl - 6,7 - dihydro - 5H - thiazolo(3,2 - a)pyrimidinium - betaine**
$C_{19}H_{17}N_3S_2$
W.Ried, W.Merkel, S.-W.Park, M.Drager
Justus Liebigs Ann. Chem., 79, 1975
Also classified in 35

41.C **2 - (α - Hydroxybenzyl) - thiamine chloride hydrochloride trihydrate**
$C_{19}H_{24}N_4O_2S^{2+}$, $2Cl^-$, $3H_2O$
For complete entry see 44.29

41.27 **3,3′ - Diethyl - benzothiacarbocyanine p - toluenesulfonate**
$C_{21}H_{21}N_2S_2^+$, $C_7H_7O_3S^-$
B.Ziemer, S.Kulpe *J. Prakt. Chem.*, **317**, 199, 1975
Residue 2 classified in 11

41.C **3,3′ - Diethylthiacarbocyanine - bis(7,7,8,8 - tetracyanoquinodimethane) (monoclinic form)**
$C_{21}H_{21}N_2S_2^+$, $C_{12}H_4N_4^-$, $C_{12}H_4N_4$
For complete entry see 60.35

41.C **3,3′ - Diethylthiacarbocyanine - bis(7,7,8,8 - tetracyanoquinodimethane) (triclinic form)**
$C_{21}H_{21}N_2S_2^+$, $C_{12}H_4N_4^-$, $C_{12}H_4N_4$
For complete entry see 60.36

41.C **7α - Methoxy - 7 - phenyl - acetamido - deacetoxy - cephalosporanic acid t - butyl ester**
$C_{21}H_{26}N_2O_5S$
For complete entry see 59.13

41.28 **3,5 - bis(N,N - Di - isopropylthiocarbamoylimino) - 4 - (4 - nitrophenyl) - 1,2,4 - dithiazolidine**
$C_{22}H_{32}N_6O_2S_4$
J.Sletten *Acta Chem. Scand. Ser. A*, **29**, 317, 1975

41.29 **5,5′ - Dichloro - 3,3′,9 - triethylthiacarbocyanine bromide acetic acid solvate**
$C_{23}H_{23}Cl_2N_2S_2^+$, Br^- , $C_2H_4O_2$
J.Potenza, D.Mastropaolo *Acta Crystallogr., Sect. B*, **30**, 2353, 1974

41.30 **(+) - 7 - Benzylideneamino - 7 - carbomethoxy - 4 - carbo(p - methoxy - benzyloxy) - Δ^3 - cephem (absolute configuration)**
$C_{25}H_{24}N_2O_6S$
E.F.Paulus *Acta Crystallogr., Sect. B*, **30**, 2918, 1974
Also classified in 35

41.31 **3′ - Phenyl - 5′ - phenylazo - 2 - pyrrolidino - spiro(1H - indene - 1,2′(3′H) - 1,3,4 - thiadiazole)**
$C_{26}H_{23}N_5S$
G.V.Boyd, T.Norris, P.F.Lindley
J. Chem. Soc., Chem. Commun., 100, 1975
Also classified in 32

41.32 **5,5′,7,7′ - Tetramethyl - 3,3′,9 - triethyl - thiacarbocyanine perchlorate**
$C_{27}H_{33}N_2S_2^+$, ClO_4^-
H.Stoeckli-Evans *Helv. Chim. Acta*, **57**, 1, 1974

41.33 **β - Lactam - fused heterocyclic ylide compound from reaction of chloroamine T with methyl 6β - phenylacetamidopenicillanate**
$C_{31}H_{34}N_4O_8S_3$
M.M.Campbell, G.Johnson, A.F.Cameron, I.R.Cameron
J. Chem. Soc., Chem. Commun., 868, 1974
Also classified in 35

HETERO-MIXED MISCELLANEOUS

42.1 **1 - Oxa - 3,5 - diselenane**
$C_3H_6OSe_2$
G.Valle, G.Zanotti, A.Del Pra *Cryst. Struct. Commun.*, **4**, 349, 1975

42.2 **Potassium 6 - methyl - 1,2,3 - oxathiazin - 4 - one - 2,2 - dioxide**
$C_4H_4NO_4S^-$, K^+
E.F.Paulus *Acta Crystallogr., Sect. B,* **31**, 1191, 1975

42.3 **9 - Oxa - 3,7 - dithiabicyclo(3.3.1)nonane**
$C_6H_{10}OS_2$
A.V.Goucharov, E.N.Kurkutova, V.V.Ilyukhin, N.V.Belov
Dokl. Akad. Nauk SSSR, **214**, 810, 1974

42.4 **3 - Oxa - 7,9 - dithiabicyclo(3.3.1)nonane**
$C_6H_{10}OS_2$
A.N.Goucharov, E.V.Kurkutova, V.V.Ilyukhin, N.V.Belov
Dokl. Akad. Nauk SSSR, **214**, 565, 1974

42.5 **1,4,7 - Trithio - (12 - crown - 4)**
$C_8H_{16}OS_3$
N.K.Dalley, J.S.Smith, S.B.Larson, K.L.Matheson, J.J.Christensen, R.M.Izatt *J. Chem. Soc., Chem. Commun.*, 84, 1975

42.6 **1,4 - Dithio - (15 - crown - 5)**
$C_{10}H_{20}O_3S_2$
N.K.Dalley, J.S.Smith, S.B.Larson, K.L.Matheson, J.J.Christensen, R.M.Izatt *J. Chem. Soc., Chem. Commun.*, 84, 1975

42.7 **Phenoselenazine**
$C_{12}H_9NSe$
P.Villares, A.Conde, R.Marquez *Eur. Cryst. Meeting,* 338, 1974

42.8 **Diethylene tetraxanthogen**
1,6,9,12,17,20 - Hexaoxa - 3,4,14,15 - tetrathia - cyclodocosane - 2,5,13,16 - tetrathione
$C_{12}H_{16}O_6S_8$
B.F.Hoskins, M.R.Hunt *Aust. J. Chem.*, **27**, 27, 1974

42.9 **1,10 - Dithio - (18 - crown - 6)**
$C_{12}H_{24}O_4S_2$
N.K.Dalley, J.S.Smith, S.B.Larson, K.L.Matheson, J.J.Christensen, R.M.Izatt *J. Chem. Soc., Chem. Commun.*, 84, 1975

42.10 **Anthraquinone - selenadiazole**
$C_{14}H_6N_2O_2Se$
N.L.Klimasenko, L.A.Chetkina, G.A.Gol'der
Zh. Strukt. Khim., **14,** 515, 1973

42.11 **3 - Oxo - 3H - 1,2 - benzoxiodol - 1 - yl m - chlorobenzoate** (α **form)**
$C_{14}H_8ClIO_4$
J.Z.Gougoutas, L.Lessinger *J. Solid State Chem.*, **9,** 155, 1974
Also classified in 13

42.12 **3 - Oxo - 3H - 1,2 - benzoxiodol - 1 - yl m - chlorobenzoate** (β **form)**
$C_{14}H_8ClIO_4$
J.Z.Gougoutas, L.Lessinger *J. Solid State Chem.*, **9,** 155, 1974
Also classified in 13

42.13 **Phenyl - phenoxtinium iodide**
$C_{18}H_{13}OS^+$, I^-
A.I.Gusev, Yu.T.Struchkov *Kristallografiya,* **18,** 525, 1973

42.C **Griseoviridin methanol solvate**
$C_{22}H_{27}N_3O_7S$, CH_4O
For complete entry see 50.17

42.14 **2,2' - bis(1,1,1,3,3,3 - Hexafluoro - 2 - hydroxy - 2 - propyl) - 4,4' - di - t - butyl - diphenyl spirosulfurane**
$C_{26}H_{24}F_{12}O_2S$
E.F.Perozzi, J.C.Martin, I.C.Paul *J. Am. Chem. Soc.*, **96,** 6735, 1974

42.15 **2,2' - bis(1,1,1,3,3,3 - Hexafluoro - 2 - hydroxy - 2 - propyl) - 4,4' - di - t - butyl - diphenyl spirosulfurane oxide**
$C_{26}H_{24}F_{12}O_3S$
E.F.Perozzi, J.C.Martin, I.C.Paul *J. Am. Chem. Soc.*, **96,** 6735, 1974

BARBITURATES

43.1 **Potassium 5 - ethylbarbiturate hydrate**
$C_6H_7N_2O_3^-$, K^+ , $1.67H_2O$
G.L.Gartland, B.M.Gatehouse, B.M.Craven
Acta Crystallogr., Sect. B, **31,** 203, 1975

43.2 **2 - Ethoxyethylammonium 5,5 - diethylbarbiturate**
$C_8H_{11}N_2O_3^-$, $C_4H_{12}NO^+$
I.-N.Hsu, D.P.Lesser, B.M.Craven
Acta Crystallogr., Sect. B, **31,** 882, 1975
Residue 2 classified in 3

43.3 **2 - Dimethylamino - ethylammonium 5,5 - diethylbarbiturate**
$C_8H_{11}N_2O_3^-$, $C_4H_{13}N_2^+$
I.-N.Hsu, D.P.Lesser, B.M.Craven
Acta Crystallogr., Sect. B, **31,** 882, 1975
Residue 2 classified in 3

43.4 **1,3 - Diethyl - barbituric acid**
$C_8H_{12}N_2O_3$
S.Toure, J.P.Bideau, S.Geoffre *Cryst. Struct. Commun.,* **4,** 171, 1975

43.5 **5,5 - Diallyl - barbituric acid**
$C_{10}H_{12}N_2O_3$
C.Escobar *Acta Crystallogr., Sect. B,* **31,** 1059, 1975

43.C **Aminopyrine - cyclobarbital**
$C_{12}H_{16}N_2O_3$, $C_{13}H_{17}N_3O$
For complete entry see 60.20

43.6 **N - Cyclohexyl - 5,5 - diallyl - barbituric acid**
$C_{16}H_{22}N_2O_3$
O.Dideberg, L.Dupont, D.Pyzalska
Acta Crystallogr., Sect. B, **31,** 685, 1975

43.7 **5,6 - Dihydro - 1,3 - dimethyl - 5,6 - di(1′,3′ - dimethyl - 2′,4′,6′ - trioxopyrimid(5′,5′)yl)furo(2,3 - d)uracil**
1,3 - Dimethylbarbituric acid trimer
$C_{18}H_{18}N_6O_9$
S.Kato, M.Poling, D.van der Helm, G.Dryhurst
J. Am. Chem. Soc., **96**, 5255, 1974
Also classified in 38, 36

43.8 **N,N′ - Dicyclohexyl - 5,5 - diallyl - barbituric acid**
$C_{22}H_{32}N_2O_3$
L.Dupont, D.Dideberg, D.Pyzalska
Acta Crystallogr., Sect. B, **30**, 2447, 1974

PYRIMIDINES AND PURINES

44.1 **5 - Bromouracil**
$C_4H_3BrN_2O_2$
H.Sternglanz, C.E.Bugg *Biochim. Biophys. Acta,* **378,** 1, 1975

44.2 **5 - Chloro - pyrimidin - 2 - one**
$C_4H_3ClN_2O$
S.Furberg, J.Solbakk *Acta Chem. Scand. Ser. A,* **28,** 435, 1974

44.3 **5 - Chlorouracil**
$C_4H_3ClN_2O_2$
H.Sternglanz, C.E.Bugg *Biochim. Biophys. Acta,* **378,** 1, 1975

44.4 **5 - Iodouracil**
$C_4H_3IN_2O_2$
H.Sternglanz, G.R.Freeman, C.E.Bugg
Acta Crystallogr., Sect. B, **31,** 1393, 1975

44.5 **2 - Aminopyrimidine**
$C_4H_5N_3$
J.Scheinbeim, E.Schempp
Am. Cryst. Assoc., Abstr. Papers (Summer Meeting), 204, 1974

44.6 **Cytosine picrate**
$C_4H_6N_3O^+$, $C_6H_2N_3O_7^-$
H.M.Einspahr, G.L.Gartland, G.R.Freeman, H.Schenk
Am. Cryst. Assoc., Abstr. Papers (Summer Meeting), 208, 1974
Residue 2 classified in 17, 15, 6

44.7 **Cytosinium tetrachloropalladate(ii)**
$2C_4H_6N_3O^+$, Cl_4Pd^{2-}
B.L.Kindberg, E.L.Amma *Acta Crystallogr., Sect. B,* **31,** 1492, 1975

44.8 **5 - Nitro - 6 - methyluracil**
$C_5H_5N_3O_4$
T.Srikrishnan, R.Parthasarathy
Am. Cryst. Assoc., Abstr. Papers (Spring Meeting), 31, 1975

44.9 **Hypoxanthine gold(iii) tetrachloride dihydrate**
$C_5H_5N_4O^+$, $AuCl_4^-$, $2H_2O$
M.R.Caira, L.R.Nassimbeni, A.L.Rodgers
Acta Crystallogr., Sect. B, **31,** 1112, 1975

44.C **Adenine - riboflavin trihydrate**
$C_5H_5N_5$, $C_{17}H_{20}N_4O_6$, $3H_2O$
For complete entry see 60.1

44.10 **Adeninium trichloro - mercury(ii) sesquihydrate**
$C_5H_6N_5^+$, Cl_3Hg^- , $1.5H_2O$
A.L.Beauchamp, M.A.Martin, C.Gagnon, P.Lavertue
Acta Crystallogr., Sect. A, **31,** S45, 1975

44.11 **Guanine picrate monohydrate**
$C_5H_6N_5O^+$, $C_6H_2N_3O_7^-$, H_2O
C.E.Bugg, U.Thewalt *Acta Crystallogr., Sect. B,* **31,** 121, 1975
Residue 2 classified in 17, 15, 6

44.12 **6 - Thioguanine picrate monohydrate**
$C_5H_6N_5S^+$, $C_6H_2N_3O_7^-$, H_2O
C.E.Bugg, U.Thewalt *Acta Crystallogr., Sect. B,* **31,** 121, 1975
Residue 2 classified in 17, 15, 6

44.13 **Adenine dihydrochloride (space group Pnam or Pna2$_1$)**
$C_5H_7N_5^{2+}$, $2Cl^-$
H.Iwasaki *Chem. Lett.*, 409, 1974

44.14 **6 - Methyl - 5,6 - dihydrouracil**
$C_5H_8N_2O_2$
W.H.Kou, R.Parthasarathy
Am. Cryst. Assoc., Abstr. Papers (Spring Meeting), 31, 1975

44.15 **6 - Methylmercaptopurine trihydrate**
$C_6H_6N_4S$, $3H_2O$
W.J.Cook, C.E.Bugg *J. Pharm. Sci.*, **64,** 221, 1975

44.C **(9 - Methyladenine) silver(i) nitrate dihydrate**
$(C_6H_7AgN_5^+)_n$, nNO_3^- , $2nH_2O$
For complete entry see 83.7

44.16 **7 - Methylxanthine hydrochloride monohydrate**
$C_6H_7N_4O_2^+$, Cl^- , H_2O
T.J.Kistenmacher, T.Sorrell *Acta Crystallogr., Sect. B,* **31,** 489, 1975

44.17 **1 - Methyl - thymine (neutron study)**
$C_6H_8N_2O_2$
A.Kvick, T.F.Koetzle, R.Thomas *J. Chem. Phys.*, **61,** 2711, 1974

44.18 **7 - Methyladenine dihydrochloride**
$C_6H_9N_5^{2+}$, $2Cl^-$
T.J.Kistenmacher, T.Shigematsu *Acta Crystallogr., Sect. B,* **31,** 211, 1975

44.19 **3 - Ethyladenine**
$C_7H_9N_5$
C.S.Petersen, S.Furberg *Acta Chem. Scand. Ser. B,* **29,** 37, 1975

44.20 **tetrakis(1 - Propylthymine) dipotassium hexachlorodipalladium**
$4C_8H_{12}N_2O_2$, $2K^+$, $Cl_6Pd_2^{2-}$
B.L.Kindberg, E.H.Griffith, E.L.Amma
J. Chem. Soc., Chem. Commun., 195, 1975

44.21 **6 - ((1 - Methyl - 4 - nitroimidazol - 5 - yl)mercapto)purine dihydrate**
$C_9H_7N_7O_2S$, $2H_2O$
W.J.Cook, C.E.Bugg *J. Pharm. Sci.*, **64,** 221, 1975
Residue 1 also classified in 32

44.C **α,D - O² - 2′ - Cyclouridine**
$C_9H_{10}N_2O_5$
For complete entry see 47.2

44.C **5 - Fluoro - 4′ - thiouridine (absolute configuration)**
$C_9H_{11}FN_2O_5S$
For complete entry see 47.3

44.C **Disodium uridine - 5′ - phosphate heptahydrate**
$C_9H_{11}N_2O_9P^{2-}$, $2Na^+$, $7H_2O$
For complete entry see 47.4

44.C **Uridine**
$C_9H_{12}N_2O_6$
For complete entry see 47.5

44.C **Dipotassium uridine - 5′ - phosphate dihydrate**
$C_9H_{12}N_2O_{12}P_2^{2-}$, $2K^+$, $2H_2O$
For complete entry see 47.6

44.C **2,2′ - Anhydro - 1 - β - DL - arabinofuranosyl cytosine hydrochloride**
$C_9H_{12}N_3O_4^+$, Cl^-
For complete entry see 47.7

44.C **2,2′ - Anhydro - 1 - β - L - arabinofuranosyl cytosine hydrochloride**
$C_9H_{12}N_3O_4^+$, Cl^-
For complete entry see 47.8

44.C **2′ - Deoxycytidine**
$C_9H_{13}N_3O_4$
For complete entry see 47.11

44.C **2 - Selenocytosine monohydrate**
$C_9H_{13}N_3O_4Se$, H_2O
For complete entry see 47.12

44.C **Salicylic acid - cytidine complex**
$C_9H_{13}N_3O_5$, $C_7H_6O_3$
For complete entry see 60.7

44.C **5,6 - Dihydro - 2,4 - dithiouridine**
$C_9H_{14}N_2O_4S_2$
For complete entry see 47.13

44.C **Cytidine - N - carbobenzoxyglutamic acid dihydrate**
$C_9H_{14}N_3O_5^+$, $C_{13}H_{14}NO_6^-$, $2H_2O$
For complete entry see 47.14

44.C **Cytidine - 5′ - diphosphoric acid monohydrate**
$C_9H_{15}N_3O_{11}P_2$, H_2O
For complete entry see 47.16

44.C **Dichloro - bis(6 - mercaptopurine) mercury(ii)**
$C_{10}H_8Cl_2HgN_8S_2$
For complete entry see 85.12

44.22 **6 - (p - Hydroxyphenylazo)uracil**
$C_{10}H_8N_4O_3$
C.L.Coulter, N.R.Cozzarelli *Acta Crystallogr., Sect. B,* **30,** 2176, 1974
Also classified in 17, 9

44.C **6 - Mercaptopurine copper(i) chloride hydrochloride dimer dihydrate**
$C_{10}H_{10}Cl_4Cu_2N_8S_2$, $2H_2O$
For complete entry see 85.13

44.C **Potassium N - (purin - 6 - ylcarbamoyl) - L - threoninate tetrahydrate (absolute configuration)**
$C_{10}H_{11}N_6O_4^-$, K^+ , $4H_2O$
For complete entry see 48.45

44.C **Rubidium N - (purin - 6 - ylcarbamoyl) - L - threoninate tetrahydrate**
$C_{10}H_{11}N_6O_4^-$, Rb^+ , $4H_2O$
For complete entry see 48.46

44.C **8 - Azido - adenosine - 3′,5′ - cyclic phosphate**
$C_{10}H_{11}N_8O_6P$
For complete entry see 47.18

44.C **Thymidine 5′ - carboxylic acid**
$C_{10}H_{12}N_2O_6$
For complete entry see 47.19

44.C **Disodium deoxyguanosine - 5′ - phosphate tetrahydrate**
$C_{10}H_{12}N_5O_7P^{2-}$, $2Na^+$, $4H_2O$
For complete entry see 47.20

44.C **Sodium deoxyadenosine - 5′ - phosphate hexahydrate**
$C_{10}H_{13}N_5O_6P^-$, Na^+ , $6H_2O$
For complete entry see 47.21

44.C **Triethylammonium uridine - 3′ - O - thiophosphate methyl ester**
$C_{10}H_{14}N_2O_8PS^-$, $C_6H_{16}N^+$
For complete entry see 47.23

44.C **Arabinofuranosyl - adenine hydrochloride**
$C_{10}H_{14}N_5O_4^+$, Cl^-
For complete entry see 47.24

44.C **9 - β - D - Arabinofuranosyl - adenine hydrochloride**
$C_{10}H_{14}N_5O_4^+$, Cl^-
For complete entry see 47.25

44.C **2′ - O - Methyl - cytidine**
$C_{10}H_{15}N_3O_5$
For complete entry see 47.26

44.C **6 - Methyl - 9 - β - D - ribofuranosylpurine**
$C_{11}H_{14}N_4O_4$
For complete entry see 47.28

44.C **1 - Deoxy - 1 - S - ethyl - 1 - (5 - fluorouracil - 1 - yl) - thio - D - arabinose aldehydrol**
$C_{11}H_{17}FN_2O_6S$
For complete entry see 45.31

44.23 **De(hydrogen fluoride) (5 - fluoro - 1,3 - dimethyluracil photodimer)**
$C_{12}H_{13}FN_4O_4$
J.L.Flippen
Am. Cryst. Assoc., Abstr. Papers (Summer Meeting), 207, 1974
Also classified in 36

44.24 **cis - anti - (5 - Fluoro - 1,3 - dimethyluracil photodimer)**
$C_{12}H_{14}F_2N_4O_4$
J.L.Flippen
Am. Cryst. Assoc., Abstr. Papers (Summer Meeting), 207, 1974
Also classified in 36

44.25 **5 - Ethyl - 5 - phenyl - hexahydro - pyrimidine - 4,6 - dione**
Primidone
$C_{12}H_{14}N_2O_2$
D.G.R.Yeates, R.A.Palmer *Acta Crystallogr., Sect. B,* **31,** 1077, 1975

44.C **5′,2 - O - Cyclo - 2′,3′ - O - isopropylidene - uridine**
$C_{12}H_{14}N_2O_5$
For complete entry see 47.31

44.C **Uridine - 5 - oxyacetic acid methyl ester monohydrate**
$C_{12}H_{16}N_2O_9$, H_2O
For complete entry see 47.32

44.26 **bis(Thiaminium) magnesium chloride decahydrate**
$2C_{12}H_{18}N_4OS^{2+}$, Mg^{2+} , $6Cl^-$, $10H_2O$
G.Blank, M.Rodrigues, J.Pletcher, M.Sax
Am. Cryst. Assoc., Abstr. Papers (Summer Meeting), 211, 1974
Residue 1 also classified in 41

44.C **Thiamine pyrophosphate (form i)**
$C_{12}H_{18}N_4O_7P_2S$
For complete entry see 46.3

44.C **Thiamine pyrophosphate (form ii)**
$C_{12}H_{18}N_4O_7P_2S$
For complete entry see 46.4

44.C **Tetra - aquo - bis(9 - methyladenine) copper(ii) dichloride dihydrate**
$C_{12}H_{22}CuN_{10}O_4^{2+}$, $2Cl^-$, $2H_2O$
For complete entry see 83.46

44.C **Tetra - aquo - bis(9 - methyladenine) copper(ii) dichloride dihydrate**
$C_{12}H_{22}CuN_{10}O_4^{2+}$, $2Cl^-$, $2H_2O$
For complete entry see 83.47

44.27 **Benzyl 6 - aminopurine7 - carboxylate**
$C_{13}H_{11}N_5O_2$
J.M.Ohrt, S.P.Dutta, G.B.Chheda
Acta Crystallogr., Sect. A, **31,** S43, 1975

44.C **3′,5′ - Diacetyl - 2′ - deoxy - 2′ - fluorouridine**
$C_{13}H_{15}FN_2O_7$
For complete entry see 47.33

44.C **N - (Purin - 6 - yl - carbamoyl) - glycine riboside**
$C_{13}H_{16}N_6O_7$
For complete entry see 48.52

44.C **1 - (6 - Chloropurin - 9 - yl) - 1 - deoxy - 1 - ethyl - 1 - thio - aldehydo - D - glucose aldehydrol**
$C_{13}H_{19}ClN_4O_5S$
For complete entry see 45.39

44.28 **2,4 - Diamino - 5 - (3,4,5 - trimethoxybenzyl) - pyrimidine (neutron study)**
$C_{14}H_{18}N_4O_3$
G.J.B.Williams, T.F.Koetzle
Am. Cryst. Assoc., Abstr. Papers (Spring Meeting), 26, 1975
Also classified in 17

44.C **Sodium cytidine - 5′ - diphosphocholine tetrahydrate**
$C_{14}H_{25}N_4O_{11}P_2^-$, Na^+ , $4H_2O$
For complete entry see 47.35

44.C **(4 - Nitrobenzyl)thioinosine**
$C_{17}H_{27}N_5O_6S$
For complete entry see 47.36

44.C **Adenylyl - 3′,3′ - uridine 9 - aminoacridine hydrate**
$C_{19}H_{23}N_7O_{12}P^-$, $C_{13}H_{11}N_2^+$, xH_2O
For complete entry see 47.37

44.29 **2 - (α - Hydroxybenzyl) - thiamine chloride hydrochloride trihydrate**
$C_{19}H_{24}N_4O_2S^{2+}$, $2Cl^-$, $3H_2O$
J.Pletcher, M.Sax, G.Blank, M.Wood
Am. Cryst. Assoc., Abstr. Papers (Spring Meeting), 28, 1975
Residue 1 also classified in 41

44.C **Calcium guanosine - 3′,5′ - cytidine monophosphate hydrate**
$2C_{19}H_{24}N_8O_{12}P^-$, Ca^{2+} , $18H_2O$
For complete entry see 47.38

44.C **2,3,4,5 - Tetra - O - acetyl - 1 - deoxy - 1 - S - ethyl - 1 - (5 - fluorouracil - 1 - yl) - thio - D - arabinose aldehydrol**
$C_{19}H_{25}FN_2O_{10}S$
For complete entry see 45.53

44.C **Adenosine bis(pyridine) osmate(vi)**
$C_{20}H_{21}N_7O_6Os$
For complete entry see 84.49

44.C **Sodium thymidyl - (5 - 3) - thymidylate tridecahydrate**
$C_{20}H_{25}N_4O_{15}P_2^{3-}$, $3Na^+$, $13H_2O$
For complete entry see 47.39

44.C **(1R) - 2,3,4,5 - Tetra - O - acetyl - 1 - S - ethyl - 1 - (1,6 - dihydro - 6 - thioxopurin - 9 - yl) - 1 - thio - D - arabinitol**
$C_{20}H_{26}N_4O_8S_2$
For complete entry see 47.40

44.C **Adenylyl - (3′,5′) - adenylyl - (3,5′) - adenosine (form i)**
$C_{30}H_{37}N_{15}O_{16}P_2$
For complete entry see 47.41

44.C **Adenylyl - (3′,5′) - adenylyl - (3,5′) - adenosine (form ii)**
$C_{30}H_{37}N_{15}O_{16}P_2$
For complete entry see 47.42

CARBOHYDRATES

45.1 **meso - Erythritol (neutron study)**
$C_4H_{10}O_4$
A.Shimada, T.Higuchi, M.Fukuyo, K.Hirotsu, F.Takusagawa, I.Shibuya, Y.Iwata, N.Koyano *Acta Crystallogr., Sect. A,* **31,** S178, 1975

45.2 **Calcium ascorbate dihydrate**
$2C_6H_7O_6^-$, Ca^{2+} , 2HO
J.Hvoslef, K.E.Kjellevold *Acta Crystallogr., Sect. B,* **30,** 2711, 1974

45.3 **Calcium ascorbate dihydrate**
$2C_6H_7O_6^-$, Ca^{2+} , $2H_2O$
R.A.Hearn, C.E.Bugg *Acta Crystallogr., Sect. B,* **30,** 2705, 1974

45.4 **1,4.2,5.3,6 - Trianhydro - D - mannitol (at −100°C)**
$C_6H_8O_3$
F.W.B.Einstein, K.N.Slessor *Acta Crystallogr., Sect. B,* **31,** 552, 1975
Also classified in 38

45.5 **D - Glucaro - 1,4 - lactone monohydrate**
$C_6H_8O_7$, H_2O
M.E.Gress, G.A.Jeffrey
Am. Cryst. Assoc., Abstr. Papers (Spring Meeting), 9, 1975

45.6 **Calcium 2 - keto - D - gluconate trihydrate**
$2C_6H_9O_7^-$, Ca^{2+} , $3H_2O$
M.A.Mazid, R.A.Palmer, A.A.Balchin
Acta Crystallogr., Sect. A, **31,** S58, 1975

45.7 **Calcium α - D - galacturonate tetrahydrate**
$2C_6H_9O_7^-$, Ca^{2+} , $4H_2O$
S.Thanomkul, J.Hjortas *Eur. Cryst. Meeting,* 336, 1974

45.8 **Calcium sodium α - D - galacturonate hexahydrate**
$3C_6H_9O_7^-$, Ca^{2+} , Na^+ , $6H_2O$
J.Hjortas, B.Larsen, S.Thanomkul
Acta Chem. Scand. Ser. B, **28,** 689, 1974

45.9 **Calcium sodium α - D - galacturonate hexahydrate**
$3C_6H_9O_7^-$, Ca^{2+} , Na^+ , $6H_2O$
S.E.B.Gould, R.O.Gould, D.A.Rees, W.E.Scott
J. Chem. Soc., Perkin Trans. 2, 237, 1975

45.10 **Strontium sodium α - D - galacturonate hexahydrate**
$3C_6H_9O_7^-$, Sr^{2+} , Na^+ , $6H_2O$
S.E.B.Gould, R.O.Gould, D.A.Rees, W.E.Scott
J. Chem. Soc., Perkin Trans. 2, 237, 1975

45.11 **1,6 - Anhydro - β - D - glucopyranose**
Levoglucosan
$C_6H_{10}O_5$
K.B.Lindberg *Acta Chem. Scand. Ser. A*, **28**, 1181, 1974
Also classified in 38

45.12 **2 - Keto - L - gulonic acid monohydrate**
$C_6H_{10}O_7$, H_2O
J.Hvoslef, B.Bergen *Acta Crystallogr., Sect. B*, **31**, 697, 1975

45.13 **2 - Deoxy - 2 - fluoro - β - D - mannopyranose**
$C_6H_{11}FO_5$
W.Choong, D.C.Craig, N.C.Stephenson, J.D.Stevens
Cryst. Struct. Commun., **4**, 111, 1975

45.C **Sodium D - fructose 1,6 - diphosphate heptahydrate**
$C_6H_{11}O_{12}P_2^{3-}$, $3Na^+$, $7H_2O$
For complete entry see 46.1

45.14 **1,6 - Dichloro - 1,6 - dideoxy - D - mannitol**
$C_6H_{12}Cl_2O_4$
J.C.Wallace
Am. Cryst. Assoc., Abstr. Papers (Summer Meeting), 241, 1974

45.15 **α - D - Galactose**
$C_6H_{12}O_6$
B.Sheldrick *Acta Crystallogr., Sect. A*, **31**, S109, 1975

45.16 **β - D - Galactose**
$C_6H_{12}O_6$
B.Sheldrick *Acta Crystallogr., Sect. A*, **31**, S109, 1975

45.17 **1 - Bromo - 1 - deoxy - D - mannitol**
$C_6H_{13}BrO_5$
J.C.Wallace
Am. Cryst. Assoc., Abstr. Papers (Summer Meeting), 241, 1974

45.18 **1 - Chloro - 1 - deoxy - D - mannitol**
$C_6H_{13}ClO_5$
J.C.Wallace
Am. Cryst. Assoc., Abstr. Papers (Summer Meeting), 241, 1974

45.C **myo - Inositol - 2 - phosphate monohydrate**
$C_6H_{13}O_9P$, H_2O
For complete entry see 46.2

45.19 **1 - Methyl - 4 - (β - D - erythrofuranosyl) - 4 - imidazoline - 2 - thione**
$C_8H_{12}N_2O_3S$
A.Conde, E.Moreno, R.Marquez *Acta Crystallogr., Sect. B,* **31,** 648, 1975
Also classified in 32

45.C **6 - Azacytidine**
$C_8H_{12}N_4O_5$
For complete entry see 47.1

45.20 **Ethyl 3 - cyano - 3,4 - dideoxy - α - DL - threo - pentapyranoside**
$C_8H_{13}NO_3$
B.P.Biryukov, B.V.Unkovskii, B.V.Mochalin, A.N.Kornilov
Zh. Strukt. Khim., **14,** 580, 1973

45.21 **N - Acetyl - α - D - galactosamine**
$C_8H_{15}NO_6$
R.D.Gilardi, J.L.Flippen *Acta Crystallogr., Sect. B,* **30,** 2931, 1974

45.22 **N - Acetyl - α - D - galactosamine**
$C_8H_{15}NO_6$
A.Neuman, H.Gillier-Pandraud, F.Longchambon, D.Rabinovich
Acta Crystallogr., Sect. B, **31,** 474, 1975

45.23 **N - Acetylmannosamine**
$C_8H_{15}NO_6$
A.Neuman, H.Gillier-Pandraud, F.Longchambon
Eur. Cryst. Meeting, 324, 1974

45.C **O - (β - D - Xylopyranosyl) - L - serine**
$C_8H_{15}NO_7$
For complete entry see 48.38

45.24 **N - (2 - Chloroethyl) - D - gluconamide**
$C_8H_{16}ClNO_6$
L.O.G.Satzke, M.F.Mackay *Acta Crystallogr., Sect. B,* **31,** 1128, 1975

45.25 **1,6 - Di - O - methanesulfonyl - D - mannitol**
$C_8H_{18}O_{10}S_2$
J.C.Wallace
Am. Cryst. Assoc., Abstr. Papers (Summer Meeting), 241, 1974
Also classified in 11

45.C **α,D - O² - 2′ - Cyclouridine**
$C_9H_{10}N_2O_5$
For complete entry see 47.2

45.C **5 - Fluoro - 4′ - thiouridine (absolute configuration)**
$C_9H_{11}FN_2O_5S$
For complete entry see 47.3

45.C **Disodium uridine - 5′ - phosphate heptahydrate**
$C_9H_{11}N_2O_9P^{2-}$, $2Na^+$, $7H_2O$
For complete entry see 47.4

45.C **Uridine**
$C_9H_{12}N_2O_6$
For complete entry see 47.5

45.C **Dipotassium uridine - 5′ - phosphate dihydrate**
$C_9H_{12}N_2O_{12}P_2^{2-}$, $2K^+$, $2H_2O$
For complete entry see 47.6

45.C **2,2′ - Anhydro - 1 - β - DL - arabinofuranosyl cytosine hydrochloride**
$C_9H_{12}N_3O_4^+$, Cl^-
For complete entry see 47.7

45.C **2,2′ - Anhydro - 1 - β - L - arabinofuranosyl cytosine hydrochloride**
$C_9H_{12}N_3O_4^+$, Cl^-
For complete entry see 47.8

45.C **8 - Aza - adenosine monohydrate**
$C_9H_{12}N_6O_4$, H_2O
For complete entry see 47.9

45.C **2 - Aza - adenosine hemihydrate**
$C_9H_{12}N_6O_4$, $0.5H_2O$
For complete entry see 47.10

45.C **2′ - Deoxycytidine**
$C_9H_{13}N_3O_4$
For complete entry see 47.11

45.C **2 - Selenocytosine monohydrate**
$C_9H_{13}N_3O_4Se$, H_2O
For complete entry see 47.12

45.C **Salicylic acid - cytidine complex**
$C_9H_{13}N_3O_5$, $C_7H_6O_3$
For complete entry see 60.7

45.C **5,6 - Dihydro - 2,4 - dithiouridine**
$C_9H_{14}N_2O_4S_2$
For complete entry see 47.13

45.C **Cytidine - N - carbobenzoxyglutamic acid dihydrate**
$C_9H_{14}N_3O_5^+$, $C_{13}H_{14}NO_6^-$, $2H_2O$
For complete entry see 47.14

45.C **5 - Amino - 1 - (3 - deoxy - β - D - xylofuranosyl)imidazole - 4 - carboxamidinium - N⁵ - 3′ - cyclonucleoside formate**
$C_9H_{14}N_5O_3^+$, CHO_2^-
For complete entry see 47.15

45.26 **5 - Chloro - 5 - deoxy - 1,2 - O - isopropylidene - 3 - methanesulfonyl - 4 - thio - β - L - arabinofuranose**
$C_9H_{15}ClO_5S_2$
W.Clegg, N.A.Hughes, C.J.Wood
J. Chem. Soc., Chem. Commun., 300, 1975
Also classified in 39, 38

45.C **Cytidine - 5′ - diphosphoric acid monohydrate**
$C_9H_{15}N_3O_{11}P_2$, H_2O
For complete entry see 47.16

45.27 **5 - Desoxy - 3 - C - formyl - β - L - lyxofuranose - trimethylenedithioacetal**
$C_9H_{16}O_4S_2$
W.Depmeier, O.H.Jarchow *Acta Crystallogr., Sect. B*, **31**, 945, 1975
Also classified in 39

45.C **2,2′ - Anhydro - 2 - hydroxy - 1 - (β - D - arabinopentofuranosyl) - 4 - pyridine**
$C_{10}H_{11}NO_5$
For complete entry see 47.17

45.C **8 - Azido - adenosine - 3′,5′ - cyclic phosphate**
$C_{10}H_{11}N_8O_6P$
For complete entry see 47.18

45.C **Thymidine 5′ - carboxylic acid**
$C_{10}H_{12}N_2O_6$
For complete entry see 47.19

45.C **8 - Aza - 9 - deaza - inosine**
Formycin B
$C_{10}H_{12}N_4O_5$
For complete entry see 50.2

45.C **Disodium deoxyguanosine - 5′ - phosphate tetrahydrate**
$C_{10}H_{12}N_5O_7P^{2-}$, $2Na^+$, $4H_2O$
For complete entry see 47.20

45.C **Sodium deoxyadenosine - 5′ - phosphate hexahydrate**
$C_{10}H_{13}N_5O_6P^-$, Na^+ , $6H_2O$
For complete entry see 47.21

45.C **3 - Deaza - cytidine**
$C_{10}H_{14}N_2O_5$
For complete entry see 47.22

45.C **Triethylammonium uridine - 3′ - O - thiophosphate methyl ester**
$C_{10}H_{14}N_2O_8PS^-$, $C_6H_{16}N^+$
For complete entry see 47.23

45.C **Arabinofuranosyl - adenine hydrochloride**
$C_{10}H_{14}N_5O_4^+$, Cl^-
For complete entry see 47.24

45.C **9 - β - D - Arabinofuranosyl - adenine hydrochloride**
$C_{10}H_{14}N_5O_4^+$, Cl^-
For complete entry see 47.25

45.28 **1,4.3,6 - bis - Thioanhydro - 2,5 - O - acetyl - D - iditol disulfoxide**
$C_{10}H_{14}O_6S_2$
K.B.Lindberg, A.Wagner *Acta Crystallogr., Sect. A,* **31,** S107, 1975
Also classified in 39

45.C **2′ - O - Methyl - cytidine**
$C_{10}H_{15}N_3O_5$
For complete entry see 47.26

45.C **Penta - aquo - nickel(ii) guanosine - 5′ - monophosphate trihydrate**
$C_{10}H_{22}N_5NiO_{13}P$, $3H_2O$
For complete entry see 83.33

45.C **3 - Deaza - adenosine**
$C_{11}H_{14}N_4O_4$
For complete entry see 47.27

45.C **6 - Methyl - 9 - β - D - ribofuranosylpurine**
$C_{11}H_{14}N_4O_4$
For complete entry see 47.28

45.29 **2,3,4 - Tri - O - acetyl - β - D - xylopyranosyl chloride**
$C_{11}H_{15}ClO_7$
G.Kothe, P.Luger, H.Paulsen *Carbohydr. Res.*, **37,** 283, 1974

45.30 **Tri - O - acetyl - α - D - arabinopyranosylazide**
$C_{11}H_{15}N_3O_7$
P.Luger, H.Paulsen *Chem. Ber.*, **107,** 1579, 1974

45.C **(R) - 3 - (2 - Deoxy - β - D - erythro - pentafuranosyl) - 3,6,7,8 - tetrahydroimidazo(4,5 - d)(1,3)diazepin - 8 - ol**
$C_{11}H_{16}N_4O_4$
For complete entry see 47.29

45.C **Coformycin sesquihydrate**
3 - (β - D - Ribofuranosyl) - 6,7,8 - trihydroimidazo(4,5 - d)(1,3 - diazepin 8(R) - ol) sesquihydrate
$C_{11}H_{16}N_4O_5$, $1.5H_2O$
For complete entry see 47.30

45.31 **1 - Deoxy - 1 - S - ethyl - 1 - (5 - fluorouracil - 1 - yl) - thio - D - arabinose aldehydrol**
$C_{11}H_{17}FN_2O_6S$
A.Ducruix, C.Pascard *Acta Crystallogr., Sect. A,* **31,** S108, 1975
Also classified in 44

45.32 **trans - O - β - D - Glucopyranosyl methyl acetoacetate**
$C_{11}H_{18}O_8$
J.Ruble, G.A.Jeffrey *Carbohydr. Res.,* **38,** 61, 1974

45.C **5′,2 - O - Cyclo - 2′,3′ - O - isopropylidene - uridine**
$C_{12}H_{14}N_2O_5$
For complete entry see 47.31

45.C **Uridine - 5 - oxyacetic acid methyl ester monohydrate**
$C_{12}H_{16}N_2O_9$, H_2O
For complete entry see 47.32

45.33 **1 - Deaza - isotubercidin picrate**
$C_{12}H_{16}N_3O_4^+$, $C_6H_2N_3O_7^-$
A.Ducruix, C.Riche, C.Pascard *Acta Crystallogr., Sect. A,* **31,** S44, 1975
Residue 1 also classified in 35; residue 2 classified in 17, 15, 6

45.34 **Levoglucosan triacetate**
1,6 - Anhydro - 2,3,4 - triacetyl - glucose
$C_{12}H_{16}O_8$
F.Leung, R.H.Marchessault *Can. J. Chem.,* **52,** 2516, 1974

45.35 **5 - Desoxy - 3 - C - formyl - 1,2 - O - isopropylidene - β - L - lyxofuranose - trimethylenedithioacetal**
$C_{12}H_{20}O_4S_2$
W.Depmeier, O.H.Jarchow *Acta Crystallogr., Sect. B,* **31,** 939, 1975
Also classified in 39

45.36 **β - Lactose**
$C_{12}H_{22}O_{11}$
K.Hirotsu, A.Shimada *Bull. Chem. Soc. Jpn.,* **47,** 1872, 1974

45.37 **Sucrose**
$C_{12}H_{22}O_{11}$
H.Hope, M.A.Poling
Am. Cryst. Assoc., Abstr. Papers (Spring Meeting), 21, 1975

45.38 **1 - p - Chlorophenyl - 4 - (α - D - erythrofuranosyl) - 4 - imidazoline - 2 - thione**
$C_{13}H_{13}ClN_2O_3S$
S.Perez-Garrido, A.Conde, R.Marquez
Acta Crystallogr., Sect. B, **30,** 2348, 1974
Also classified in 32

45.C **3′,5′ - Diacetyl - 2′ - deoxy - 2′ - fluorouridine**
$C_{13}H_{15}FN_2O_7$
For complete entry see 47.33

45.C **N - (Purin - 6 - yl - carbamoyl) - glycine riboside**
$C_{13}H_{16}N_6O_7$
For complete entry see 48.52

45.39 **1 - (6 - Chloropurin - 9 - yl) - 1 - deoxy - 1 - ethyl - 1 - thio - aldehydo - D - glucose aldehydrol**
$C_{13}H_{19}ClN_4O_5S$
A.Ducruix, C.Pascard *Acta Crystallogr., Sect. A,* **31,** S108, 1975
Also classified in 44

45.40 **Methyl 4 - desoxy - 2,3 - O - isopropylidene - 6 - O - methyl - 4 - cis - cyanomethylene - α - D - lyxohexopyranoside**
$C_{13}H_{19}NO_5$
G.Bernardinelli, R.Gerdil *Helv. Chim. Acta,* **57,** 1459, 1974

45.41 **1 - O - Acetyl - 2,3.4,5 - di - O - isopropylidene - D - erythropent - 1 - enitol**
$C_{13}H_{20}O_6$
A.Ducruix, C.Pascard *Acta Crystallogr., Sect. A,* **31,** S108, 1975
Also classified in 38

45.C **8.5′ - Anhydro - 7 - bromo - 8 - hydroxy - 2′,3′ - isopropylidene - tubercidin (absolute configuration)**
$C_{14}H_{15}BrN_4O_4$
For complete entry see 50.6

45.42 **1 - p - Tolyl - 4 - (α - D - erythrofuranosyl)imidazoline - 2 - thione**
$C_{14}H_{16}N_2O_3S$
I.Barragan, A.Lopez-Castro, R.Marquez
Acta Crystallogr., Sect. A, **31,** S108, 1975
Also classified in 32

45.43 **1 - (Tri - O - acetyl - α - D - xylopyranosyl) - imidazole**
$C_{14}H_{18}N_2O_7$
P.Luger, G.Kothe, H.Paulsen *Chem. Ber.,* **107,** 2626, 1974
Also classified in 32

45.C **3,1′ - Anhydro - 2 - (4′,6′ - di - O - acetyl - 2′,3′ - dideoxy - α - D - ribo - hexopyranose) - 3 - hydroxy - 5 - methyl - 2H - 1,2,6 - thiadiazine - 1,1 - dioxide**
$C_{14}H_{18}N_2O_8S$
For complete entry see 47.34

45.C **Sodium cytidine - 5′ - diphosphocholine tetrahydrate**
$C_{14}H_{25}N_4O_{11}P_2^-$, Na^+ , $4H_2O$
For complete entry see 47.35

45.44 **1 - (p - Methoxyphenyl) - 3 - methyl - 4 - (D - arabino - tetrahydroxybutyl) - imidazolidine - 2 - thione**
$C_{15}H_{22}N_2O_5S$
R.Jimenez-Garay, A.Lopez-Castro, R.Marquez
Eur. Cryst. Meeting, 313, 1974
Also classified in 32, 17

45.45 **Methyl 2,3,4,6 - tetra - O - acetyl - α - D - mannopyranoside**
$C_{15}H_{22}O_{10}$
J.Hjortas *Acta Crystallogr., Sect. A,* **31,** S109, 1975

45.46 **1,2,3,4,6 - Penta - O - acetyl - α - D - altropyranose**
$C_{16}H_{22}O_{11}$
J.Ollis, V.J.James, J.D.Stevens *Cryst. Struct. Commun.,* **4,** 215, 1975

45.47 **3,4,6 - Tri - O - acetyl - 1,2 - O - (1 - (exo - ethoxy)ethylidene) - α - D - glucopyranose (at −193°C)**
$C_{16}H_{24}O_{10}$
J.A.Heitmann, G.F.Richards, L.R.Schroeder
Acta Crystallogr., Sect. B, **30,** 2322, 1974

45.C **bis(O - (β - D - Xylopyranosyl) - L - serinato) copper(ii)**
$C_{16}H_{28}CuN_2O_{14}$
For complete entry see 82.20

45.48 **α - N,N - Diacetyl - chitobiose monohydrate**
$C_{16}H_{28}N_2O_{11}$, H_2O
F.Mo, L.H.Jensen *Acta Crystallogr., Sect. A,* **31,** S110, 1975

45.49 **β - N,N - Diacetyl - chitobiose trihydrate**
$C_{16}H_{28}N_2O_{11}$, $3H_2O$
F.Mo, L.H.Jensen *Acta Crystallogr., Sect. A,* **31,** S110, 1975

45.50 **1 - Decyl α - D - glucopyranoside**
$C_{16}H_{32}O_6$
P.C.Moews, J.R.Knox
Am. Cryst. Assoc., Abstr. Papers (Spring Meeting), 13, 1975

45.51 **D - Ribose diphenyl dithioacetal**
$C_{17}H_{20}O_4S_2$
A.Ducruix, C.Pascard *Acta Crystallogr., Sect. A,* **31,** S108, 1975

45.52 **7 - Azido - 8 - deoxy - 1,2.3,4 - di - O - isopropylidene - 6,7 - S,S - trimethylene - 6,7 - dithio - α - L - threo - D - galacto - octopyranose**
$C_{17}H_{27}N_3O_5S_2$
A.Gateau-Olesker, S.D.Gero, C.Pascard-Billy, C.Riche, A.-M.Sepulchre, G.Vass, N.A.Hughes *J. Chem. Soc., Chem. Commun.,* 811, 1974
Also classified in 39, 38

45.C **(4 - Nitrobenzyl)thioinosine**
$C_{17}H_{27}N_5O_6S$
For complete entry see 47.36

45.C **Adenylyl - 3′,3′ - uridine 9 - aminoacridine hydrate**
$C_{19}H_{23}N_7O_{12}P^-$, $C_{13}H_{11}N_2^+$, xH_2O
For complete entry see 47.37

45.C **Calcium guanosine - 3′,5′ - cytidine monophosphate hydrate**
$2C_{19}H_{24}N_8O_{12}P^-$, Ca^{2+} , $18H_2O$
For complete entry see 47.38

45.53 **2,3,4,5 - Tetra - O - acetyl - 1 - deoxy - 1 - S - ethyl - 1 - (5 - fluorouracil - 1 - yl) - thio - D - arabinose aldehydrol**
$C_{19}H_{25}FN_2O_{10}S$
A.Ducruix, C.Pascard *Acta Crystallogr., Sect. A,* **31,** S108, 1975
Also classified in 44

45.C **Adenosine bis(pyridine) osmate(vi)**
$C_{20}H_{21}N_7O_6Os$
For complete entry see 84.49

45.C **Sodium thymidyl - (5 - 3) - thymidylate tridecahydrate**
$C_{20}H_{25}N_4O_{15}P_2^{3-}$, $3Na^+$, $13H_2O$
For complete entry see 47.39

45.C **(1R) - 2,3,4,5 - Tetra - O - acetyl - 1 - S - ethyl - 1 - (1,6 - dihydro - 6 - thioxopurin - 9 - yl) - 1 - thio - D - arabinitol**
$C_{20}H_{26}N_4O_8S_2$
For complete entry see 47.40

45.54 **Tri - O - benzoyl - β - D - xylopyranosyl bromide**
$C_{26}H_{21}BrO_7$
P.Luger, P.L.Durette, H.Paulsen *Chem. Ber.,* **107,** 2615, 1974

45.55 **2 - (m - Bromobenzyl) - 4,7,8,9 - tetra - O - acetyl - N - acetyl - α - D - neuraminic acid**
$C_{26}H_{32}BrNO_{13}$
H.Wawra *Z. Naturforsch., Teil C,* **29,** 317, 1974

45.C **Mascaroside**
$C_{26}H_{36}O_{11}$
For complete entry see 54.13

45.C **Helichrysoside hydrate**
$C_{30}H_{26}O_{14}$, $5.5H_2O$
For complete entry see 59.18

45.C **Adenylyl - (3′,5′) - adenylyl - (3,5′) - adenosine (form i)**
$C_{30}H_{37}N_{15}O_{16}P_2$
For complete entry see 47.41

45.C **Adenylyl - (3′,5′) - adenylyl - (3,5′) - adenosine (form ii)**
$C_{30}H_{37}N_{15}O_{16}P_2$
For complete entry see 47.42

45.C **Novobiocin monohydrate**
$C_{31}H_{36}N_2O_{11}$, H_2O
For complete entry see 50.24

45.56 **α - Cyclodextrin krypton hydrate (complex ii)**
$C_{36}H_{60}O_{30}$, $5.28H_2O$, $0.76Kr$
W.Saenger, M.Noltemeyer *Angew. Chem.*, **86,** 594, 1974

45.57 **α - Cyclodextrin krypton hydrate (complex i)**
$C_{36}H_{60}O_{30}$, $5.78H_2O$, $0.47Kr$
W.Saenger, M.Noltemeyer *Angew. Chem.*, **86,** 594, 1974

45.C **Cyclohexa - amylose 1 - propanol hydrate**
α - Cyclodextrin 1 - propanol hydrate
$C_{36}H_{60}O_{30}$, C_3H_8O , $4.8H_2O$
For complete entry see 61.6

45.58 **α - Cyclodextrin p - iodoaniline trihydrate**
$C_{36}H_{60}O_{30}$, C_6H_6IN , $3H_2O$
K.Harata, H.Uedaira *Nature (London)*, **253,** 190, 1975
Residue 2 classified in 16

45.59 **bis(α - Cyclodextrin) lithium tri - iodide iodine complex octahydrate**
$2C_{36}H_{60}O_{30}$, Li^+ , I_3^- , I_2 , $8H_2O$
W.Saenger, M.Noltemeyer *Acta Crystallogr., Sect. A,* **31,** S126, 1975

PHOSPHATES

46.1 **Sodium D - fructose 1,6 - diphosphate heptahydrate**
$C_6H_{11}O_{12}P_2^{3-}$, $3Na^+$, $7H_2O$
G.A.Clegg, L.C.G.Goaman *Acta Crystallogr., Sect. A*, **31,** S108, 1975
Residue 1 also classified in 45

46.2 **myo - Inositol - 2 - phosphate monohydrate**
$C_6H_{13}O_9P$, H_2O
C.S.Yoo, G.Blank, J.Pletcher, M.Sax
Acta Crystallogr., Sect. B, **30,** 1983, 1974
Residue 1 also classified in 45

46.C **Disodium uridine - 5′ - phosphate heptahydrate**
$C_9H_{11}N_2O_9P^{2-}$, $2Na^+$, $7H_2O$
For complete entry see 47.4

46.C **Dipotassium uridine - 5′ - phosphate dihydrate**
$C_9H_{12}N_2O_{12}P_2^{2-}$, $2K^+$, $2H_2O$
For complete entry see 47.6

46.C **Cytidine - 5′ - diphosphoric acid monohydrate**
$C_9H_{15}N_3O_{11}P_2$, H_2O
For complete entry see 47.16

46.C **8 - Azido - adenosine - 3′,5′ - cyclic phosphate**
$C_{10}H_{11}N_8O_6P$
For complete entry see 47.18

46.C **Disodium deoxyguanosine - 5′ - phosphate tetrahydrate**
$C_{10}H_{12}N_5O_7P^{2-}$, $2Na^+$, $4H_2O$
For complete entry see 47.20

46.C **Sodium deoxyadenosine - 5′ - phosphate hexahydrate**
$C_{10}H_{13}N_5O_6P^-$, Na^+ , $6H_2O$
For complete entry see 47.21

46.C **bis(5,5 - Dimethyl - 2 - oxo - 1,3,2 - P - dioxaphosphorinanyl) oxide**
$C_{10}H_{20}O_7P_2$
For complete entry see 64.29

46.C **Penta - aquo - nickel(ii) guanosine - 5′ - monophosphate trihydrate**
$C_{10}H_{22}N_5NiO_{13}P$, $3H_2O$
For complete entry see 83.33

46.C **Lidocaine bis(p - nitrophenyl) phosphate**
$C_{12}H_8N_2O_8P^-$, $C_{14}H_{23}N_2O^+$
For complete entry see 16.11

46.3 **Thiamine pyrophosphate (form i)**
$C_{12}H_{18}N_4O_7P_2S$
G.Blank, M.Wood, J.Pletcher, M.Sax
Am. Cryst. Assoc., Abstr. Papers (Spring Meeting), 27, 1975
Also classified in 44, 41

46.4 **Thiamine pyrophosphate (form ii)**
$C_{12}H_{18}N_4O_7P_2S$
G.Blank, M.Wood, J.Pletcher, M.Sax
Am. Cryst. Assoc., Abstr. Papers (Spring Meeting), 27, 1975
Also classified in 44, 41

46.C **Sodium cytidine - 5′ - diphosphocholine tetrahydrate**
$C_{14}H_{25}N_4O_{11}P_2^-$, Na^+ , $4H_2O$
For complete entry see 47.35

46.5 **trans - Methyl - meso - hydrobenzoin phosphate**
$C_{15}H_{15}O_4P$
M.G.Newton, B.S.Campbell *J. Am. Chem. Soc.,* **96,** 7790, 1974

46.C **N - (5 - O - Phosphopyridoxyl) - L - tyrosine heptahydrate**
$C_{17}H_{21}N_2O_8P$, $7H_2O$
For complete entry see 48.63

46.C **Adenylyl - 3′,3′ - uridine 9 - aminoacridine hydrate**
$C_{19}H_{23}N_7O_{12}P^-$, $C_{13}H_{11}N_2^+$, xH_2O
For complete entry see 47.37

46.C **Calcium guanosine - 3′,5′ - cytidine monophosphate hydrate**
$2C_{19}H_{24}N_8O_{12}P^-$, Ca^{2+} , $18H_2O$
For complete entry see 47.38

46.6 **P^1,P^2 - Di - β - naphthyl - pyrophosphate di(cyclohexylammonium) salt**
$C_{20}H_{14}O_7P_2^{2-}$, $2C_6H_{14}N^+$
M.K.Wood, M.Sax, J.Pletcher *Acta Crystallogr., Sect. B,* **31,** 76, 1975
Residue 1 also classified in 24; residue 2 classified in 21

46.C **Sodium thymidyl - (5 - 3) - thymidylate tridecahydrate**
$C_{20}H_{25}N_4O_{15}P_2^{3-}$, $3Na^+$, $13H_2O$
For complete entry see 47.39

46.C Adenylyl - (3′,5′) - adenylyl - (3,5′) - adenosine (form i)
$C_{30}H_{37}N_{15}O_{16}P_2$
For complete entry see 47.41

46.C Adenylyl - (3′,5′) - adenylyl - (3,5′) - adenosine (form ii)
$C_{30}H_{37}N_{15}O_{16}P_2$
For complete entry see 47.42

NUCLEOSIDES AND NUCLEOTIDES

47.1 **6 - Azacytidine**
$C_8H_{12}N_4O_5$
P.Singh, D.J.Hodgson *Biochemistry,* **13,** 5445, 1974
Also classified in 45, 33

47.2 **α,D - O^2 - 2′ - Cyclouridine**
$C_9H_{10}N_2O_5$
K.C.Lee, G.Kartha
Am. Cryst. Assoc., Abstr. Papers (Summer Meeting), 273, 1974
Also classified in 45, 44, 40

47.3 **5 - Fluoro - 4′ - thiouridine (absolute configuration)**
$C_9H_{11}FN_2O_5S$
R.Parthasarathy
Am. Cryst. Assoc., Abstr. Papers (Spring Meeting), 13, 1975
Also classified in 45, 44, 39

47.4 **Disodium uridine - 5′ - phosphate heptahydrate**
$C_9H_{11}N_2O_9P^{2-}$, $2Na^+$, $7H_2O$
T.P.Seshadri, B.S.Reddy, M.A.Viswamitra, G.Kartha
Curr. Sci., **43,** 339, 1974
Residue 1 also classified in 46, 45, 44

47.5 **Uridine**
$C_9H_{12}N_2O_6$
E.A.Green, R.D.Rosenstein, R.Shiono, D.J.Abraham, B.L.Trus, R.E.Marsh *Acta Crystallogr., Sect. B,* **31,** 102, 1975
Also classified in 45, 44

47.6 **Dipotassium uridine - 5′ - phosphate dihydrate**
$C_9H_{12}N_2O_{12}P_2{}^{2-}$, $2K^+$, $2H_2O$
M.A.Viswamitra, T.P.Seshadri, M.V.Hosur, M.L.Post, O.Kennard
Acta Crystallogr., Sect. A, **31,** S45, 1975
Residue 1 also classified in 46, 45, 44

47.7 **2,2′ - Anhydro - 1 - β - DL - arabinofuranosyl cytosine hydrochloride**
$C_9H_{12}N_3O_4{}^+$, Cl^-
G.Kartha, T.Phillips II, G.Ambady
Acta Crystallogr., Sect. A, **31,** S52, 1975
Residue 1 also classified in 45, 44, 40

47.8 **2,2′ - Anhydro - 1 - β - L - arabinofuranosyl cytosine hydrochloride**
$C_9H_{12}N_3O_4^+$, Cl^-
G.Kartha, T.Phillips II, G.Ambady
Acta Crystallogr., Sect. A, **31,** S52, 1975
Residue 1 also classified in 45, 44, 40

47.9 **8 - Aza - adenosine monohydrate**
$C_9H_{12}N_6O_4$, H_2O
P.Singh, D.J.Hodgson *J. Am. Chem. Soc.,* **96,** 5276, 1974
Residue 1 also classified in 45, 35

47.10 **2 - Aza - adenosine hemihydrate**
$C_9H_{12}N_6O_4$, $0.5H_2O$
P.Singh, D.L.Kozlowski, D.J.Hodgson
Am. Cryst. Assoc., Abstr. Papers (Summer Meeting), 275, 1974
Residue 1 also classified in 45, 35

47.11 **2′ - Deoxycytidine**
$C_9H_{13}N_3O_4$
D.W.Young, H.R.Wilson *Acta Crystallogr., Sect. B,* **31,** 961, 1975
Also classified in 45, 44

47.12 **2 - Selenocytosine monohydrate**
$C_9H_{13}N_3O_4Se$, H_2O
M.K.Wood, J.Abola
Am. Cryst. Assoc., Abstr. Papers (Spring Meeting), 11, 1975
Residue 1 also classified in 45, 44

47.C **Salicylic acid - cytidine complex**
$C_9H_{13}N_3O_5$, $C_7H_6O_3$
For complete entry see 60.7

47.13 **5,6 - Dihydro - 2,4 - dithiouridine**
$C_9H_{14}N_2O_4S_2$
B.Kojic-Prodic, A.Kvick, Z.Ruzic-Toros
Acta Crystallogr., Sect. A, **31,** S45, 1975
Also classified in 45, 44

47.14 **Cytidine - N - carbobenzoxyglutamic acid dihydrate**
$C_9H_{14}N_3O_5^+$, $C_{13}H_{14}NO_6^-$, $2H_2O$
T.Hata, M.Yoshikawa, S.Sato, C.Tamura
Acta Crystallogr., Sect. B, **31,** 312, 1975
Residue 1 also classified in 45, 44; residue 2 classified in 48

47.15 **5 - Amino - 1 - (3 - deoxy - β - D - xylofuranosyl)imidazole - 4 - carboxamidinium - N^5 - 3′ - cyclonucleoside formate**
$C_9H_{14}N_5O_3^+$, CHO_2^-
M.N.G.James
Am. Cryst. Assoc., Abstr. Papers (Summer Meeting), 275, 1974
Residue 1 also classified in 45, 32

47.16 **Cytidine - 5′ - diphosphoric acid monohydrate**
$C_9H_{15}N_3O_{11}P_2$, H_2O
M.A.Viswamitra, T.P.Seshadri, M.V.Hosur, M.L.Post, O.Kennard
Acta Crystallogr., Sect. A, **31,** S45, 1975
Residue 1 also classified in 46, 45, 44

47.17 **2,2′ - Anhydro - 2 - hydroxy - 1 - (β - D - arabinopentofuranosyl) - 4 - pyridine**
$C_{10}H_{11}NO_5$
W.L.B.Hutcheon, M.N.G.James
Am. Cryst. Assoc., Abstr. Papers (Summer Meeting), 274, 1974
Also classified in 45, 40, 33

47.18 **8 - Azido - adenosine - 3′,5′ - cyclic phosphate**
$C_{10}H_{11}N_8O_6P$
G.Ambady, G.Kartha
Am. Cryst. Assoc., Abstr. Papers (Summer Meeting), 274, 1974
Also classified in 46, 44, 45

47.19 **Thymidine 5′ - carboxylic acid**
$C_{10}H_{12}N_2O_6$
D.Suck, W.Saenger, W.Rohde *Biochim. Biophys. Acta,* **361,** 1, 1974
Also classified in 45, 44

47.C **8 - Aza - 9 - deaza - inosine**
Formycin B
$C_{10}H_{12}N_4O_5$
For complete entry see 50.2

47.20 **Disodium deoxyguanosine - 5′ - phosphate tetrahydrate**
$C_{10}H_{12}N_5O_7P^{2-}$, $2Na^+$, $4H_2O$
D.W.Young, P.Tollin, H.R.Wilson
Acta Crystallogr., Sect. B, **30,** 2012, 1974
Residue 1 also classified in 46, 45, 44

47.21 **Sodium deoxyadenosine - 5′ - phosphate hexahydrate**
$C_{10}H_{13}N_5O_6P^-$, Na^+ , $6H_2O$
B.S.Reddy, M.A.Viswamitra *Acta Crystallogr., Sect. B,* **31,** 19, 1975
Residue 1 also classified in 46, 45, 44

47.22 **3 - Deaza - cytidine**
$C_{10}H_{14}N_2O_5$
W.L.B.Hutcheon, M.N.G.James
Am. Cryst. Assoc., Abstr. Papers (Summer Meeting), 274, 1974
Also classified in 45, 33

47.23 **Triethylammonium uridine - 3′ - O - thiophosphate methyl ester**
$C_{10}H_{14}N_2O_8PS^-$, $C_6H_{16}N^+$
W.Saenger, D.Suck, F.Eckstein *Eur. J. Biochem.,* **46,** 559, 1974
Residue 1 also classified in 45, 44; residue 2 classified in 3

47.24 Arabinofuranosyl - adenine hydrochloride
$C_{10}H_{14}N_5O_4^+$, Cl^-
T.Hata, S.Sato, M.Kaneko, B.Shimizu, C.Tamura
Bull. Chem. Soc. Jpn., **47,** 2758, 1974
Residue 1 also classified in 45, 44

47.25 9 - β - D - Arabinofuranosyl - adenine hydrochloride
$C_{10}H_{14}N_5O_4^+$, Cl^-
A.K.Chwang, M.Sundaralingam, S.Hanessian
Acta Crystallogr., Sect. B, **30,** 2273, 1974
Residue 1 also classified in 45, 44

47.26 2′ - O - Methyl - cytidine
$C_{10}H_{15}N_3O_5$
B.Hingerty, P.J.Bond, R.Langridge, F.Rottman
Biochem. Biophys. Res. Commun., **61,** 875, 1974
Also classified in 45, 44

47.C Penta - aquo - nickel(ii) guanosine - 5′ - monophosphate trihydrate
$C_{10}H_{22}N_5NiO_{13}P$, $3H_2O$
For complete entry see 83.33

47.27 3 - Deaza - adenosine
$C_{11}H_{14}N_4O_4$
P.Singh, D.J.Hodgson
Am. Cryst. Assoc., Abstr. Papers (Spring Meeting), 13, 1975
Also classified in 45, 35

47.28 6 - Methyl - 9 - β - D - ribofuranosylpurine
$C_{11}H_{14}N_4O_4$
T.Takeda, Y.Ohashi, Y.Sasada, M.Kakudo
Acta Crystallogr., Sect. B, **31,** 1202, 1975
Also classified in 45, 44

47.29 (R) - 3 - (2 - Deoxy - β - D - erythro - pentafuranosyl) - 3,6,7,8 - tetrahydroimidazo(4,5 - d)(1,3)diazepin - 8 - ol
$C_{11}H_{16}N_4O_4$
P.W.K.Woo, H.W.Dion, S.M.Lange, L.F.Dahl
J. Heterocycl. Chem., **11,** 641, 1974
Also classified in 45, 35

47.30 Coformycin sesquihydrate
3 - (β - D - Ribofuranosyl) - 6,7,8 - trihydroimidazo(4,5 - d)(1,3 - diazepin - 8(R) - ol) sesquihydrate
$C_{11}H_{16}N_4O_5$, $1.5H_2O$
H.Nakamura, G.Koyama, Y.Iitaka, M.Ohno, N.Yagisawa, S.Kondon, K.Maeda, H.Umezawa *J. Am. Chem. Soc.*, **96,** 4327, 1974
Residue 1 also classified in 45, 35

47.31 **5′,2 - O - Cyclo - 2′,3′ - O - isopropylidene - uridine**
$C_{12}H_{14}N_2O_5$
P.C.Manor, W.Saenger, D.B.Davies, K.Jankowski, A.Rabczenko
Biochim. Biophys. Acta, **340,** 472, 1974
Also classified in 45, 44, 40, 38

47.32 **Uridine - 5 - oxyacetic acid methyl ester monohydrate**
$C_{12}H_{16}N_2O_9$, H_2O
K.Morikawa, K.Torii, Y.Iitaka, M.Tsuboi
Acta Crystallogr., Sect. B, **31,** 1004, 1975
Residue 1 also classified in 45, 44

47.33 **3′,5′ - Diacetyl - 2′ - deoxy - 2′ - fluorouridine**
$C_{13}H_{15}FN_2O_7$
D.Suck, W.Saenger, P.Main, G.Germain, J.P.Declercq
Biochim. Biophys. Acta, **361,** 257, 1974
Also classified in 45, 44

47.C **N - (Purin - 6 - yl - carbamoyl) - glycine riboside**
$C_{13}H_{16}N_6O_7$
For complete entry see 48.52

47.C **8.5′ - Anhydro - 7 - bromo - 8 - hydroxy - 2′,3′ - isopropylidene - tubercidin (absolute configuration)**
$C_{14}H_{15}BrN_4O_4$
For complete entry see 50.6

47.34 **3,1′ - Anhydro - 2 - (4′,6′ - di - O - acetyl - 2′,3′ - dideoxy - α - D - ribo - hexopyranose) - 3 - hydroxy - 5 - methyl - 2H - 1,2,6 - thiadiazine - 1,1 - dioxide**
$C_{14}H_{18}N_2O_8S$
C.Foces-Foces, P.Smith-Verdier, F.Florencio-Sabate, S.Garcia-Blanco
Acta Crystallogr., Sect. B, **31,** 140, 1975
Also classified in 45, 41, 40

47.35 **Sodium cytidine - 5′ - diphosphocholine tetrahydrate**
$C_{14}H_{25}N_4O_{11}P_2^-$, Na^+ , $4H_2O$
M.A.Viswamitra, T.P.Seshadri, M.V.Hosur, M.L.Post, O.Kennard
Acta Crystallogr., Sect. A, **31,** S45, 1975
Residue 1 also classified in 46, 45, 44, 3

47.36 **(4 - Nitrobenzyl)thioinosine**
$C_{17}H_{27}N_5O_6S$
M.Soriano-Garcia, R.Parthasarathy, B.Paul, A.R.P.Paterson
Am. Cryst. Assoc., Abstr. Papers (Summer Meeting), 276, 1974
Also classified in 45, 44

47.37 **Adenylyl - 3′,3′ - uridine 9 - aminoacridine hydrate**
$C_{19}H_{23}N_7O_{12}P^-$, $C_{13}H_{11}N_2^+$, xH_2O
N.C.Seeman, R.O.Day, A.Rich *Nature (London),* **253,** 324, 1975
Residue 1 also classified in 46, 45, 44; residue 2 classified in 36

47.38 **Calcium guanosine - 3′,5′ - cytidine monophosphate hydrate**
$2C_{19}H_{24}N_8O_{12}P^-$, Ca^{2+} , $18H_2O$
B.Hingerty, E.Subramanian, S.D.Stellman, S.B.Broyde, T.Sato, R.Langridge *Biopolymers,* **14,** 227, 1975
Residue 1 also classified in 46, 45, 44

47.C **Adenosine bis(pyridine) osmate(vi)**
$C_{20}H_{21}N_7O_6Os$
For complete entry see 84.49

47.39 **Sodium thymidyl - (5 - 3) - thymidylate tridecahydrate**
$C_{20}H_{25}N_4O_{15}P_2^{3-}$, $3Na^+$, $13H_2O$
N.Camerman, J.K.Fawcett, A.Camerman *Science,* **182,** 1142, 1973
Residue 1 also classified in 46, 45, 44

47.40 **(1R) - 2,3,4,5 - Tetra - O - acetyl - 1 - S - ethyl - 1 - (1,6 - dihydro - 6 - thioxopurin - 9 - yl) - 1 - thio - D - arabinitol**
$C_{20}H_{26}N_4O_8S_2$
D.C.Baker, A.Ducruix, D.Horton, C.Pascard-Billy
J. Chem. Soc., Chem. Commun., 729, 1974
Also classified in 45, 44

47.41 **Adenylyl - (3′,5′) - adenylyl - (3,5′) - adenosine (form i)**
$C_{30}H_{37}N_{15}O_{16}P_2$
D.Suck, P.C.Manor, G.Germain, C.H.Schwalbe, G.Weimann, W.Saenger
Eur. Cryst. Meeting, 430, 1974
Also classified in 46, 45, 44

47.42 **Adenylyl - (3′,5′) - adenylyl - (3,5′) - adenosine (form ii)**
$C_{30}H_{37}N_{15}O_{16}P_2$
D.Suck, P.C.Manor, G.Germain, C.H.Schwalbe, G.Weimann, W.Saenger
Eur. Cryst. Meeting, 430, 1974
Also classified in 46, 45, 44

AMINO-ACIDS AND PEPTIDES

48.1 **Triglycine sulfate (ferroelectic form, at 37°C)**
$C_2H_5NO_2$, $2C_2H_6NO_2^+$, O_4S^{2-}
K.Itoh, T.Mitsui *Ferroelectrics,* **5,** 235, 1973

48.2 **Triglycine sulfate (paraelectric form, at 57°C)**
$C_2H_5NO_2$, $2C_2H_6NO_2^+$, O_4S^{2-}
K.Itoh, T.Mitsui *Ferroelectrics,* **5,** 235, 1973

48.3 **Triglycine sulfate (ferroelectric form, at 19°C)**
$C_2H_5NO_2$, $2C_2H_6NO_2^+$, O_4S^{2-}
K.Itoh, T.Mitsui *Ferroelectrics,* **5,** 235, 1973

48.4 **Glycine nitrate**
$C_2H_6NO_2^+$, NO_3^-
P.Narayanan, S.Venkatarman *J. Cryst. Mol. Struct.,* **5,** 15, 1975

48.5 **Glycinium ammonium sulfate**
$C_2H_6NO_2^+$, H_4N^+ , O_4S^{2-}
S.Vilminot, E.Philippot, L.Cot *Acta Crystallogr., Sect. B,* **30,** 2602, 1974

48.6 **Diglycinium sulfate monohydrate**
$2C_2H_6NO_2^+$, O_4S^{2-} , H_2O
F.H.Cano, S.Martinez-Carrera *Acta Crystallogr., Sect. B,* **30,** 2729, 1974

48.7 **Diglycine selenate**
$2C_2H_6NO_2^+$, O_4Se^{2-}
S.Olejnik, K.Lukaszewicz, T.Lis
Acta Crystallogr., Sect. B, **31,** 1785, 1975

48.8 **DL - Serine**
$C_3H_7NO_3$
T.J.Kistenmacher, G.A.Rand, R.E.Marsh
Acta Crystallogr., Sect. B, **30,** 2573, 1974

48.9 **L - (−) - Serine**
$C_3H_7NO_3$
T.J.Kistenmacher, G.A.Rand, R.E.Marsh
Acta Crystallogr., Sect. B, **30,** 2573, 1974

48.10 **L - (−) - Serine**
$C_3H_7NO_3$
E.Benedetti, C.Pedone, A.Sirigu *Gazz. Chim. Ital.*, **103,** 555, 1973

48.11 **N - Acetylglycine (neutron study)**
$C_4H_7NO_3$
M.F.Mackay *Cryst. Struct. Commun.*, **4,** 225, 1975

48.12 **Iminodiacetic acid hydrobromide (neutron study)**
$C_4H_8NO_4^+$, Br^-
A.Oskarsson *Acta Crystallogr., Sect. A,* **31,** S221, 1975

48.13 **Glycylglycine (α form)**
$C_4H_8N_2O_3$
J.F.Griffin, P.Coppens
Am. Cryst. Assoc., Abstr. Papers (Summer Meeting), 203, 1974

48.14 **L - Allothreonine**
$C_4H_9NO_3$
P.Swaminathan, R.Srinivasan *Acta Crystallogr., Sect. B,* **31,** 217, 1975

48.15 **L - Threonine (neutron study)**
$C_4H_9NO_3$
M.Ramanadham, S.K.Sikka, R.Chidambaram *Pramana,* **1,** 247, 1973

48.16 **L - Threonine (further refinement of data of Shoemaker et al., J.Am.Chem.Soc.,72,2328,1950)**
$C_4H_9NO_3$
M.Ramanadham, S.K.Sikka, R.Chidambaram *Pramana,* **1,** 247, 1973

48.17 **2 - Hydrazinobutyric acid**
$C_4H_{10}N_2O_2$
A.S.Yuan, T.H.Doyne *Acta Crystallogr., Sect. A,* **31,** S178, 1975

48.18 **Pyroglutamic acid**
5 - Oxoproline
$C_5H_7NO_3$
V.Pattabhi, K.Venkatesan *J. Chem. Soc., Perkin Trans. 2,* 1085, 1974

48.19 **Calcium L - glutamate chloride monohydrate**
$C_5H_8NO_4^-$, Ca^{2+} , Cl^- , H_2O
H.Einspahr, G.L.Gartland, C.E.Bugg
Am. Cryst. Assoc., Abstr. Papers (Spring Meeting), 29, 1975

48.20 **Monosodium glutamate monohydrate**
$C_5H_8NO_4^-$, Na^+ , H_2O
S.T.Rao, M.Mallikarjunan *Acta Crystallogr., Sect. A,* **31,** S48, 1975

48.C **Arginine glutamate monohydrate**
$C_5H_8NO_4^-$, $C_6H_{15}N_4O_2^+$, H_2O
For complete entry see 48.31

48.21 **Acetylglycine - N - methylamide**
$C_5H_{10}N_2O_2$
F.Iwasaki *Acta Crystallogr., Sect. B,* **30,** 2503, 1974

48.22 **L - Valine hydrochloride (neutron study)**
$C_5H_{12}NO_2^+$, Cl^-
T.F.Koetzle, L.Golic, M.S.Lehmann, J.J.Verbist, W.C.Hamilton
J. Chem. Phys., **60,** 4690, 1974

48.23 **Cyclo - L - cystine acetic acid solvate**
$C_6H_8N_2O_2S_2$, $C_2H_4O_2$
H.-C.Mez *Cryst. Struct. Commun.,* **3,** 657, 1974

48.24 **cis - δ - Methyl - L - proline**
$C_6H_{11}NO_2$
S.T.Rao, M.Mallikarjunan
Am. Cryst. Assoc., Abstr. Papers (Summer Meeting), 243, 1974

48.25 **3 - Hydroxy - 4 - methyl - proline**
$C_6H_{11}NO_3$
G.Koyama *Helv. Chim. Acta,* **57,** 2477, 1974

48.26 **Glycyl - DL - threonine monohydrate**
$C_6H_{12}N_2O_4$, H_2O
P.Swaminathan *Acta Crystallogr., Sect. B,* **31,** 1608, 1975

48.27 **DL - Leucine**
$C_6H_{13}NO_2$
B.di Blasio, C.Pedone, A.Sirigu *Acta Crystallogr., Sect. B,* **31,** 601, 1975

48.28 **L - Cystine dihydrobromide dihydrate**
$C_6H_{14}N_2O_4S_2^{2+}$, $2Br^-$, $2H_2O$
R.E.Rosenfield Junior, R.Parthasarathy
Acta Crystallogr., Sect. B, **31,** 816, 1975

48.29 **S - Methyl - L - methionine chloride hydrochloride**
$C_6H_{15}NO_2S^{2+}$, $2Cl^-$
S.Andini, G.Del Re, E.Gavuzzo, E.Giglio, F.Leli, F.Mazza, V.Zappia
Acta Crystallogr., Sect. A, **31,** S106, 1975
Residue 1 also classified in 11

48.30 **DL - Lysine hydrochloride**
$C_6H_{15}N_2O_2^+$, Cl^-
D.Bhaduri, N.N.Saha *Indian J. Phys.,* **48,** 93, 1974

48.31 **Arginine glutamate monohydrate**
$C_6H_{15}N_4O_2^+$, $C_5H_8NO_4^-$, H_2O
T.N.Bhat, M.Vijayan *Acta Crystallogr., Sect. A,* **31,** S48, 1975
Residue 2 classified in 48

48.C **Diaquo - bis(L - serinato) calcium**
$C_6H_{16}N_2O_8^{2-}$, nCa^{2+}
For complete entry see 67.2

48.32 **Cyclo - L - prolyl - glycyl (full - data refinement)**
$C_7H_{10}N_2O_2$
R.B.von Dreele *Acta Crystallogr., Sect. B,* **31,** 966, 1975
Also classified in 35

48.33 **Cyclo - L - prolyl - glycyl (high - angle refinement)**
$C_7H_{10}N_2O_2$
R.B.von Dreele *Acta Crystallogr., Sect. B,* **31,** 966, 1975
Also classified in 35

48.34 **3R - (1′(S) - Aminocarboxymethyl) - 2 - pyrrolidone - 5(S) - carboxylic acid**
$C_7H_{10}N_2O_5$
L.Dupont, O.Dideberg, A.Welter
Acta Crystallogr., Sect. B, **31,** 1018, 1975
Also classified in 32

48.35 **N - Acetyl - L - glutamine**
$C_7H_{12}N_2O_4$
M.R.Narasimhamurthy, K.Venkatesan, F.Winkler
Cryst. Struct. Commun., **3,** 743, 1974

48.36 **1 - Aminocyclohexanecarboxylic acid**
$C_7H_{13}NO_2$
K.I.Varughese, K.K.Chacko, R.Zand
Acta Crystallogr., Sect. B, **31,** 886, 1975
Also classified in 21

48.37 **N - Acetyl - L - norvaline**
$C_7H_{13}NO_3$
G.Lovas, A.Kalman, G.Argay *Acta Crystallogr., Sect. B,* **30,** 2882, 1974

48.38 **O - (β - D - Xylopyranosyl) - L - serine**
$C_8H_{15}NO_7$
L.T.J.Delbaere, B.Kamenar, K.Prout
Acta Crystallogr., Sect. B, **31,** 862, 1975
Also classified in 45

48.39 **Glycyl - L - leucine**
$C_8H_{16}N_2O_3$
V.Pattabhi, K.Venkatesan, S.R.Hall
J. Chem. Soc., Perkin Trans. 2, 1722, 1974

48.40 **L - Cystine dimethyl ester dihydrochloride monohydrate**
$C_8H_{18}N_2O_4S_2^{2+}$, $2Cl^-$, H_2O
B.K.Vijayalakshmi, R.Srinivasan *Acta Crystallogr., Sect. B,* **31,** 993, 1975

48.41 **DL - 2 - Hydroxyphenylalanine**
o - Tyrosine
$C_9H_{11}NO_3$
A.Mostad, C.Romming, L.Tressum
Acta Chem. Scand. Ser. B, **29,** 171, 1975

48.42 **3 - Hydroxyphenylalanine**
m - Tyrosine
$C_9H_{11}NO_3$
A.Byrkjedal, A.Mostad, C.Romming
Acta Chem. Scand. Ser. B, **28,** 750, 1974

48.43 **3 - (3′,4′ - Dihydroxyphenyl) - L - alanine hydrochloride**
L - DOPA hydrochloride
$C_9H_{12}NO_4^+$, Cl^-
A.Mostad, C.Romming *Acta Chem. Scand. Ser. B,* **28,** 1161, 1974

48.44 **L - Alanyl - L - alanyl - L - alanine hemihydrate**
$C_9H_{17}N_3O_4$, $0.5H_2O$
J.K.Fawcett, N.Camerman, A.Camerman
Acta Crystallogr., Sect. B, **31,** 658, 1975

48.45 **Potassium N - (purin - 6 - ylcarbamoyl) - L - threoninate tetrahydrate (absolute configuration)**
$C_{10}H_{11}N_6O_4^-$, K^+ , $4H_2O$
R.Parthasarathy, J.M.Ohrt, G.B.Chheda
J. Am. Chem. Soc., **96,** 8087, 1974
Residue 1 also classified in 44

48.46 **Rubidium N - (purin - 6 - ylcarbamoyl) - L - threoninate tetrahydrate**
$C_{10}H_{11}N_6O_4^-$, Rb^+ , $4H_2O$
D.A.Adamiak, T.L.Blundell, I.J.Tickle, Z.Kosturkiewicz
Acta Crystallogr., Sect. B, **31,** 1242, 1975
Residue 1 also classified in 44

48.47 **Sodium magnesium ethylenediaminetetra - acetate tetrahydrate**
$C_{10}H_{12}N_2O_8^{4-}$, $2Na^+$, Mg^{2+} , $4H_2O$
A.I.Pozhidaev, T.N.Polynova, M.A.Porai-Koshits, V.G.Dudakov
Zh. Strukt. Khim., **15,** 160, 1974
Residue 1 also classified in 2, 3

48.48 **L - Methionyl - L - methionine**
$C_{10}H_{20}N_2O_3S_2$
R.E.Stenkamp, L.H.Jensen *Acta Crystallogr., Sect. B,* **31,** 857, 1975

48.49 **3,3,3′,3′ - Tetramethyl - D - cystine dihydrochloride (absolute configuration)**
D - Penicillamine disulfide dihydrochloride
$C_{10}H_{22}N_2O_4S_2^{2+}$, $2Cl^-$
R.E.Rosenfield Junior, R.Parthasarathy
Acta Crystallogr., Sect. B, **31,** 462, 1975
Residue 1 also classified in 11

48.50 **N - Acetyl - L - prolyl - D - lactyl - methylamide**
$C_{11}H_{18}N_2O_4$
C.Lecomte, A.Aubry, J.Protas, G.Boussard, M.Marraud
Acta Crystallogr., Sect. B, **30,** 2343, 1974

48.51 **N - Acetyl - L - prolyl - L - lactyl - methylamide**
$C_{11}H_{18}N_2O_4$
C.Lecomte, A.Aubry, J.Protas, G.Boussard, M.Marraud
Acta Crystallogr., Sect. B, **30,** 1992, 1974

48.C **Cytidine - N - carbobenzoxyglutamic acid dihydrate**
$C_{13}H_{14}NO_6^-$, $C_9H_{14}N_3O_5^+$, $2H_2O$
For complete entry see 47.14

48.52 **N - (Purin - 6 - yl - carbamoyl) - glycine riboside**
$C_{13}H_{16}N_6O_7$
R.Parthasarathy, J.M.Ohrt, G.B.Chheda
Am. Cryst. Assoc., Abstr. Papers (Summer Meeting), 276, 1974
Also classified in 47, 45, 44

48.53 **DL - Tryptophan ethyl ester hydrochloride**
$C_{13}H_{17}N_2O_2^+$, Cl^-
B.K.Vijayalakshmi, R.Srinivasan *Acta Crystallogr., Sect. B,* **31,** 999, 1975

48.54 **Cyclo - (D - phenylalanyl - L - prolyl)**
$C_{14}H_{16}N_2O_2$
R.Ramani, K.Venkatesan, W.J.Kung, R.E.Marsh
Acta Crystallogr., Sect. A, **31,** S48, 1975
Also classified in 35

48.55 **3,5,3′ - Tri - iodo - L - thyronine**
$C_{15}H_{12}I_3NO_4$
V.Cody *J. Am. Chem. Soc.,* **96,** 6720, 1974
Also classified in 17

48.56 **Cyclo - triprolyl**
$C_{15}H_{21}N_3O_3$
M.E.Druyan, C.L.Coulter
Am. Cryst. Assoc., Abstr. Papers (Spring Meeting), 29, 1975

48.57 **Cyclo - L - prolyl - L - prolyl - L - prolyl**
$C_{15}H_{21}N_3O_3$
G.Kartha, G.Ambady, P.V.Shankar *Nature (London),* **247,** 204, 1974

48.58 **Cyclo - L - prolyl - L - prolyl - L - hydroxyprolyl**
$C_{15}H_{21}N_3O_4$
G.Kartha, G.Ambady, P.V.Shankar *Nature (London)*, **247**, 204, 1974

48.59 **Cyclo - alanyl - tetrasarcosyl hemihydrate**
$C_{15}H_{25}N_5O_5$, $0.5H_2O$
P.Groth *Acta Chem. Scand. Ser. A*, **28**, 449, 1974

48.60 **3,5,3′ - Tri - iodo - L - thyronine methyl ester**
$C_{16}H_{14}I_3NO_4$
V.Cody *J. Med. Chem.*, **18**, 126, 1975
Also classified in 17

48.61 **2′,3′ - Dimethyl - 3,5 - di - iodo - DL - thyronine hydrochloride hydrate**
$C_{17}H_{18}I_2NO_4^+$, Cl^- , xH_2O
J.K.Fawcett, N.Camerman, A.Camerman
Acta Crystallogr., Sect. A, **31**, S49, 1975
Residue 1 also classified in 17

48.62 **L - Thyronine ethyl ester hydrochloride monohydrate**
$C_{17}H_{20}NO_4^+$, Cl^- , H_2O
A.Camerman, N.Camerman *Can. J. Chem.*, **52**, 3042, 1974
Residue 1 also classified in 17

48.63 **N - (5 - O - Phosphopyridoxyl) - L - tyrosine heptahydrate**
$C_{17}H_{21}N_2O_8P$, $7H_2O$
A.Mangia, M.Nardelli, G.Pelizzi, C.B.Voltattorni, A.Orlacchio, C.Turano
J. Chem. Soc., Perkin Trans. 2, 60, 1975
Residue 1 also classified in 46, 33

48.64 **N - t - Butyloxycarbonyl - S - benzyl - cysteinylglycine methyl ester**
$C_{17}H_{24}N_2O_5S$
S.Kashino, T.Ashida, M.Kakudo
Acta Crystallogr., Sect. B, **30**, 2074, 1974

48.65 **Cyclo - (L - O - t - butyl - seryl) - β - alanyl - glycyl - (O - methyl - β - aspartyl)**
$C_{17}H_{28}N_4O_7$
I.L.Karle, B.K.Handa, C.H.Hassall
Acta Crystallogr., Sect. B, **31**, 555, 1975

48.66 **3′ - Isopropyl - 3,5 - di - iodo - L - thyronine hydrochloride trihydrate**
$C_{18}H_{20}I_2NO_4^+$, Cl^- , $3H_2O$
J.K.Fawcett, N.Camerman, A.Camerman
Acta Crystallogr., Sect. A, **31**, S49, 1975
Residue 1 also classified in 17

48.67 **Carbobenzoxy - L - leucyl - p - nitrophenyl ester**
$C_{20}H_{22}N_2O_6$
V.M.Coiro, F.Mazza, G.Mignucci
Acta Crystallogr., Sect. B, **30**, 2607, 1974

48.68 **Benzyloxycarbonyl - glycyl - prolyl - leucine**
$C_{21}H_{29}N_3O_6$
T.Yamane, T.Ashida, M.Kakudo, Y.Sasada
Acta Crystallogr., Sect. A, **31,** S47, 1975

48.69 **Cycloheptasarcosyl monohydrate**
$C_{21}H_{35}N_7O_7$, H_2O
P.Groth *Acta Chem. Scand. Ser. A,* **29,** 38, 1975

48.C **Griseoviridin methanol solvate**
$C_{22}H_{27}N_3O_7S$, CH_4O
For complete entry see 50.17

48.70 **S - Benzyl - L - cysteinyl - L - prolyl - L - leucylglycinamide**
$C_{23}H_{35}N_5O_4S$
A.D.Rudko, B.W.Low *Acta Crystallogr., Sect. B,* **31,** 713, 1975

48.71 **Se - Benzyl - L - seleno - cysteinyl - L - prolyl - L - leucylglycinamide**
$C_{23}H_{35}N_5O_4Se$
A.D.Rudko, B.W.Low *Acta Crystallogr., Sect. B,* **31,** 713, 1975

48.C **Cyclochlorotine dihydrate**
$C_{24}H_{31}Cl_2N_5O_7$, $2H_2O$
For complete entry see 50.18

48.72 **Cyclic(L - leucyl - L - tyrosyl - Δ - aminovaleryl - Δ - aminovaleryl) dimethylsulfoxide solvate monohydrate**
$C_{25}H_{38}N_4O_5$, C_2H_6OS , H_2O
I.L.Karle *Acta Crystallogr., Sect. A,* **31,** S47, 1975

48.C **Tri - N - methylfrangulanine methiodide**
$C_{32}H_{53}N_4O_4^+$, I^-
For complete entry see 58.25

48.C **Enniatin B potassium iodide**
$C_{33}H_{57}N_3O_9$, K^+ , I^-
For complete entry see 50.25

48.73 **6,14,22,30 - Tetrabenzyl - 2,6,10,14,18,22,26,30 - octa - aza - pentacyclo(26.4.0.0^{4,9}.0^{12,17}.0^{20,25})do triaconta - 4,7,12,15,20,23,28,31 - octaen - 3,11,19,27 - tetraone**
$C_{52}H_{48}N_8O_4$
M.Martinez-Ripoll, F.H.Cano, S.Garcia-Blanco, S.Martinez-Carrera, W.H.Gundel *Acta Crystallogr., Sect. A,* **31,** S104, 1975
Also classified in 37

48.C **bis(Valinomycin) potassium tri - iodide penta - iodide**
$2C_{54}H_{90}N_6O_{18}$, $2K^+$, I_3^- , I_5^-
For complete entry see 50.37

48.C **Ilamycin B_1 p - bromobenzoate ethanol solvate monohydrate (absolute configuration)**
$C_{61}H_{80}BrN_9O_{11}$, C_2H_6O , H_2O
For complete entry see 50.39

48.74 **Antamanide lithium bromide acetonitrile solvate**
$C_{64}H_{78}N_{10}O_{10}$, Li^+ , Br^- , $3C_2H_3N$
I.L.Karle, J.Karle, T.Wieland, W.Burgermeister, H.Faulstich, B.Witkop
Proc. Nat. Acad. Sci. U. S. A., **70**, 1836, 1973
Residue 1 also classified in 34

48.75 **(Phe^4,Val^6) - Antamanide sodium bromide ethanol solvate**
$C_{66}H_{82}N_{10}O_{10}$, Na^+ , Br^- , C_2H_6O
I.L.Karle *Biochemistry*, **13**, 2155, 1974
Residue 1 also classified in 34

PORPHYRINS AND CORRINS

49.1 **rac - 15 - Cyano - 1,2,2,7,7,12,12 - heptamethyl - corrin hydrochloride ethanol solvate**
$C_{27}H_{36}N_5^+$, Cl^- , C_2H_6O
E.D.Edmond, D.C.Hodgkin *Helv. Chim. Acta,* **58,** 641, 1975

49.2 **8,12 - Diethyl - 2,3,7,13,17,18 - hexamethyl - corrole hydrobromide chloroform solvate**
$C_{29}H_{35}N_4^+$, Br^- , $1.5CHCl_3$
B.F.Anderson, T.J.Bartczak, D.C.Hodgkin
J. Chem. Soc., Perkin Trans. 2, 977, 1974

49.C **Hemiporphyrazine - germanium - di(ethylene glycol monoethyl ether)**
$C_{34}H_{32}GeN_8O_4$
For complete entry see 69.29

49.3 **Methyl - pyrochlorophyllide A monohydrate**
$C_{34}H_{34}MgN_4O_3$, H_2O
C.Kratky, J.D.Dunitz *Acta Crystallogr., Sect. A,* **31,** S49, 1975

49.4 **Methylchlorophyllide A dihydrate**
$C_{36}H_{36}MgN_4O_5$, $2H_2O$
C.Kratky, J.D.Dunitz *Acta Crystallogr., Sect. A,* **31,** S49, 1975

49.5 **2,3,7,8,12,13,17,18 - Octaethylporphinato cobalt(ii) bis(3 - methylpyridine) solvate**
$C_{36}H_{44}CoN_4$, $2C_6H_7N$
R.G.Little, J.A.Ibers *J. Am. Chem. Soc.,* **96,** 4440, 1974

49.6 **Ethylchlorophyllide A dihydrate**
$C_{37}H_{36}MgN_4O_5$, $2H_2O$
C.Kratky, J.D.Dunitz *Acta Crystallogr., Sect. A,* **31,** S49, 1975

49.7 **Ethylchlorophyllide A dihydrate**
$C_{37}H_{36}MgN_4O_5$, $2H_2O$
C.Kratky, J.D.Dunitz *Acta Crystallogr., Sect. B,* **31,** 1586, 1975

49.C **bis(Trimethylsilyloxy) - (phthalocyanine) - silicon**
$C_{38}H_{34}N_8O_2Si_3$
For complete entry see 63.19

49.8 **μ - (1,2,3,4,5,6,7,8 - Octaethylporphinato) - bis(dicarbonyl rhodium(i))**
$C_{40}H_{44}N_4O_4Rh_2$
A.Takenaka, Y.Sasada, H.Ogoshi, T.Omura, Z.-I.Yoshida
Acta Crystallogr., Sect. B, **31,** 1, 1975

49.9 **2,3,7,8,12,13,17,18 - Octaethylporphinato(1 - methylimidazole) cobalt(ii)**
$C_{40}H_{50}CoN_6$
R.G.Little, J.A.Ibers *J. Am. Chem. Soc.,* **96,** 4452, 1974

49.10 **(p - Nitrobenzenethiolato) iron(iii) protoporphyrin IX dimethyl ester**
$C_{42}H_{40}FeN_5O_6S$
S.Koch, S.C.Tang, R.H.Holm, R.B.Frankel, J.A.Ibers
J. Am. Chem. Soc., **97,** 916, 1975

49.11 **Nitrosyl - ($\alpha,\beta,\gamma,\delta$ - tetraphenylporphinato) iron(ii)**
$C_{44}H_{28}FeN_5O$
W.R.Scheidt, M.E.Frisse *J. Am. Chem. Soc.,* **97,** 17, 1975

49.12 **bis(Pyridine) - octaethylporphinato ruthenium(ii)**
$C_{46}H_{54}N_6Ru$
F.R.Hopf, T.P.O'Brien, W.R.Scheidt, D.G.Whitten
J. Am. Chem. Soc., **97,** 277, 1975

49.13 **Nitrosyl - ($\alpha,\beta,\gamma,\delta$ - tetraphenylporphinato)(1 - methylimidazole) iron chloroform solvate**
$C_{48}H_{34}FeN_7O$, $CHCl_3$
P.L.Piciulo, G.Rupprecht, W.R.Scheidt *J. Am. Chem. Soc.,* **96,** 5293, 1974

49.14 **endo - 21 - Ethoxycarbonyl - 5,10,15,20 - tetraphenyl - 21H - 21 - homoporphinato nickel(ii) dichloromethane solvate**
$C_{48}H_{34}N_4NiO_2$, $0.5CH_2Cl_2$
B.Chevrier, R.Weiss *J. Am. Chem. Soc.,* **97,** 1416, 1975

49.15 **(1,2 - Dimethylimidazole) - ($\alpha,\beta,\gamma,\delta$ - tetraphenylporphinato) cobalt(ii) benzene solvate**
$C_{49}H_{36}CoN_6$, $2C_6H_6$
P.N.Dwyer, P.Madura, W.R.Scheidt *J. Am. Chem. Soc.,* **96,** 4815, 1974

49.16 **meso - Tetraphenylporphinato - bis(imidazole) cobalt(iii) acetate monohydrate chloroform solvate**
$C_{50}H_{36}CoN_8^+$, $C_2H_3O_2^-$, $CHCl_3$, H_2O
J.W.Lauher, J.A.Ibers *J. Am. Chem. Soc.,* **96,** 4447, 1974

49.17 **Nitrosyl - ($\alpha,\beta,\gamma,\delta$ - tetraphenylporphinato)(4 - methylpiperidine) manganese chloroform solvate**
$C_{50}H_{41}MnN_6O$, $CHCl_3$
P.L.Piciulo, G.Rupprecht, W.R.Scheidt *J. Am. Chem. Soc.,* **96,** 5293, 1974

49.18 **bis(Imidazole) - (tetra(4 - N - methylpyridyl)porphinato) nickel(ii) perchlorate acetone solvate**
$C_{50}H_{44}N_{12}Ni^{4+}$, $4ClO_4^-$, $2C_3H_6O$
J.F.Kirner, J.Garofalo Junior, W.R.Scheidt
Inorg. Nucl. Chem. Lett., **11,** 107, 1975

49.19 **Nitro - ($\alpha,\beta,\gamma,\delta$ - tetraphenylporphinato) - (3,5 - lutidine) cobalt(iii)**
$C_{51}H_{37}CoN_6O_2$
J.A.Kaduk, W.R.Scheidt *Inorg. Chem.*, **13,** 1875, 1974

49.20 **bis(Phthalocyanine) thorium**
$C_{64}H_{32}N_{16}Th$
I.S.Kirin, A.B.Kolyadin, A.A.Lychev *Zh. Strukt. Khim.*, **15,** 486, 1974

49.21 **Di(phthalocyanine) uranium**
$C_{64}H_{32}N_{16}U$
I.S.Kirin, A.B.Kolyadin, A.A.Lychev *Zh. Strukt. Khim.*, **15,** 486, 1974

49.22 **(N - Methylimidazole) - (dioxygen) - (meso - $\alpha,\alpha,\alpha,\alpha$ - tetra(o - pivalamidophenyl) - porphinato) iron(ii)**
$C_{68}H_{70}FeN_{10}O_6$
J.P.Collman, R.R.Gagne, C.A.Reed, W.T.Robinson, G.A.Rodley
Proc. Nat. Acad. Sci. U. S. A., **71,** 1326, 1974

49.23 **Tetrahydrothiophene - (dioxygen) - meso - tetra($\alpha,\alpha,\alpha,\alpha$ - o - pivalamidophenyl) - porphinato iron(ii) tetrahydrothiophene solvate**
$C_{68}H_{72}FeN_8O_6S$, $2C_4H_8S$
W.T.Robinson, G.A.Rodley, G.B.Jameson
Acta Crystallogr., Sect. A, **31,** S49, 1975

ANTIBIOTICS

50.1 **Methylenomycin A**
$C_9H_{10}O_4$
T.Haneishi, A.Terahara, M.Arai, T.Hata, C.Tamura
J. Antibiot., **27**, 393, 1974
Also classified in 38

50.2 **8 - Aza - 9 - deaza - inosine**
Formycin B
$C_{10}H_{12}N_4O_5$
P.Singh, D.J.Hodgson
Am. Cryst. Assoc., Abstr. Papers (Spring Meeting), 13, 1975
Also classified in 47, 45, 35

50.3 **Monobromopentenomycin triacetate**
$C_{12}H_{13}BrO_7$
T.Date, K.Aoe, K.Kotera, K.Umino *Chem. Pharm. Bull.*, **22**, 1963, 1974
Also classified in 20

50.4 **Bicyclomycin**
$C_{12}H_{18}N_2O_7$
Y.Tokuma, S.Koda, T.Miyoshi, Y.Morimoto
Bull. Chem. Soc. Jpn., **47**, 18, 1974
Also classified in 40, 37

50.5 **Actinobolin hydroiodide monohydrate acetonitrile solvate (absolute configuration)**
$C_{13}H_{21}N_2O_6^+$, I^- , H_2O , $0.25C_2H_3N$
J.B.Wetherington, J.W.Moncrief *Acta Crystallogr., Sect. B,* **31**, 501, 1975
Residue 1 also classified in 38

50.6 **8.5′ - Anhydro - 7 - bromo - 8 - hydroxy - 2′,3′ - isopropylidene - tubercidin (absolute configuration)**
$C_{14}H_{15}BrN_4O_4$
K.-I.Asahi, K.Anzai, S.Suzuki, H.Iwasaki *Chem. Lett.*, 1197, 1973
Also classified in 47, 45, 40, 35

50.7 **Indolo(2,1 - b)quinazoline - 6,12 - dione**
Tryptanthrin
$C_{15}H_8N_2O_2$
W.Fedeli, F.Mazza *J. Chem. Soc., Perkin Trans. 2,* 1621, 1974
Also classified in 36

50.8 **Dihydrobotrydial**
$C_{17}H_{28}O_5$
H.J.Lindner, B.von Gross *Chem. Ber.*, **107,** 3332, 1974
Also classified in 38

50.9 **Phlebianorkauranol**
$C_{19}H_{28}O_5$
J.M.Lisy, J.Clardy, M.Anchel, S.M.Weinreb
J. Chem. Soc., Chem. Commun., 406, 1975
Also classified in 31

50.10 **Vermiculine**
$C_{20}H_{24}O_8$
R.K.Boeckman Junior, J.Fayos, J.Clardy
J. Am. Chem. Soc., **96,** 5954, 1974
Also classified in 38

50.11 **Isocilludin S triacetate (absolute configuration)**
$C_{21}H_{26}O_7$
A.Furusaki, H.Shirahama, T.Matsumoto *Chem. Lett.*, 1293, 1973
Also classified in 28

50.12 **Naphthyridinomycin**
$C_{21}H_{27}N_3O_7$
J.Sygusch, F.Brisse, S.Hanessian, D.Kluepfel
Tetrahedron Lett., 4021, 1974
Also classified in 40, 36

50.13 **Phlebiakauranol**
$C_{21}H_{32}O_7$
J.M.Lisy, J.Clardy, M.Anchel, S.M.Weinreb
J. Chem. Soc., Chem. Commun., 406, 1975
Also classified in 31

50.14 **Dipotassium oxytetracycline tetrahydrate (at −150°C)**
$C_{22}H_{22}N_2O_9^{2-}$, $2K^+$, $4H_2O$
J.J.Stezowski *Acta Crystallogr., Sect. A,* **31,** S50, 1975
Residue 1 also classified in 29

50.15 **Tetracycline hexahydrate (at −150°C)**
$C_{22}H_{24}N_2O_8$, $6H_2O$
J.J.Stezowski *Acta Crystallogr., Sect. A,* **31,** S50, 1975
Residue 1 also classified in 29

50.16 **Oxytetracycline mercury(ii) chloride dihydrate (at −150°C)**
$C_{22}H_{24}N_2O_9$, Cl_2Hg , $2H_2O$
J.J.Stezowski *Acta Crystallogr., Sect. A,* **31,** S50, 1975
Residue 1 also classified in 29

50.17 **Griseoviridin methanol solvate**
$C_{22}H_{27}N_3O_7S$, CH_4O
G.I.Birnbaum, S.R.Hall *Acta Crystallogr., Sect. A,* **31,** S51, 1975
Residue 1 also classified in 48, 42, 40

50.C **Hexahydrocoriolin p - bromobenzoate (absolute configuration)**
$C_{22}H_{29}BrO_6$
For complete entry see 53.28

50.18 **Cyclochlorotine dihydrate**
$C_{24}H_{31}Cl_2N_5O_7$, $2H_2O$
H.Yoshioka, K.Nakatsu, M.Sato, T.Tatsuno *Chem. Lett.,* 1319, 1973
Residue 1 also classified in 48

50.19 **Streptonigrin ethylacetate solvate**
$C_{25}H_{22}N_4O_8$, $C_4H_8O_2$
Y.-Y.H.Chiu, W.N.Lipscomb *J. Am. Chem. Soc.,* **97,** 2525, 1975
Residue 1 also classified in 35, 33, 17

50.20 **4 - O - Ethyl - ascofuranone (absolute configuration)**
$C_{25}H_{33}ClO_5$
K.Ando, H.Sasaki, T.Hosokawa, Y.Nawata, Y.Iitaka
Tetrahedron Lett., 887, 1975
Also classified in 38, 17

50.21 **Chloramphenicol palmitate (β form)**
$C_{27}H_{42}Cl_2N_2O_6$
Y.Eguchi, Y.Iitaka *Acta Crystallogr., Sect. B,* **30,** 2781, 1974
Also classified in 15, 1

50.22 **p - Bromobenzoyl - leukopleurotin**
$C_{28}H_{27}BrO_6$
M.Dobler *Cryst. Struct. Commun.,* **4,** 253, 1975
Also classified in 38

50.23 **Borrelidin solvate**
$C_{28}H_{43}NO_6$, $C_5H_{12}O$
B.F.Anderson, G.B.Robertson *Acta Crystallogr., Sect. A,* **31,** S50, 1975
Residue 1 also classified in 38, 20

50.24 **Novobiocin monohydrate**
$C_{31}H_{36}N_2O_{11}$, H_2O
M.O.Boles, D.J.Taylor *Acta Crystallogr., Sect. B,* **31,** 1400, 1975
Residue 1 also classified in 45, 38, 17

50.25 **Enniatin B potassium iodide**
$C_{33}H_{57}N_3O_9$, K^+ , I^-
M.Dobler, J.D.Dunitz, J.Krajewski *J. Mol. Biol.*, **42,** 603, 1969
Residue 1 also classified in 48

50.26 **Sodium 5 - bromo - lasalocid (form i)**
$C_{34}H_{52}BrO_8^-$, Na^+
P.G.Schmidt, A.H.-J.Wang, I.C.Paul *J. Am. Chem. Soc.*, **96,** 6189, 1974
Residue 1 also classified in 38, 14, 17

50.27 **Sodium 5 - bromo - lasalocid (form ii)**
$C_{34}H_{52}BrO_8^-$, Na^+
P.G.Schmidt, A.H.-J.Wang, I.C.Paul *J. Am. Chem. Soc.*, **96,** 6189, 1974
Residue 1 also classified in 38, 14, 17

50.28 **Silver lysocellin hemihydrate (absolute configuration)**
$C_{34}H_{59}O_{10}^-$, Ag^+ , $0.5H_2O$
N.Otake, M.Koenuma, H.Kinashi, S.Sato, Y.Saito
J. Chem. Soc., Chem. Commun., 92, 1975
Residue 1 also classified in 8

50.29 **Anhydro - erythromycin A carbonate methiodide methanol solvate**
$C_{39}H_{66}NO_{13}^+$, I^- , CH_4O
A.Hempel, M.Bogucka-Ledochowska, Z.Dauter, E.Borowski,
Z.Kosturkiewicz *Tetrahedron Lett.*, 1599, 1975
Residue 1 also classified in 38

50.30 **Nonactin sodium thiocyanate**
$C_{40}H_{64}O_{12}$, Na^+ , CNS^-
M.Dobler, R.P.Phizackerley *Helv. Chim. Acta,* **57,** 664, 1974
Residue 1 also classified in 38

50.31 **Grisorixin monohydrate**
$C_{40}H_{68}O_{10}$, H_2O
M.Alleaume *Eur. Cryst. Meeting,* 405, 1974
Residue 1 also classified in 38

50.32 **des - Boron - des - valine - boromycin monohydrate**
$C_{40}H_{68}O_{14}$, H_2O
W.C.Marsh, J.D.Dunitz, D.N.J.White *Helv. Chim. Acta,* **57,** 10, 1974
Residue 1 also classified in 38

50.33 **Rifampicine pentahydrate**
$C_{43}H_{58}N_4O_{12}$, $5H_2O$
M.Gadret, M.Goursolle, J.M.Leger, J.C.Colleter
Acta Crystallogr., Sect. B, **31,** 1454, 1975
Residue 1 also classified in 38, 33

50.34 **Antibiotic X - 206 monohydrate**
$C_{47}H_{82}O_{14}$, H_2O
J.F.Blount, J.W.Westley *J. Chem. Soc., Chem. Commun.*, 533, 1975
Residue 1 also classified in 38

50.35 **Potassium alborixin**
$C_{48}H_{83}O_{14}^-$, K^+
M.Alleaume, B.Busetta, C.Farges, P.Gachon, A.Kergomard, T.Staron
J. Chem. Soc., Chem. Commun., 411, 1975
Residue 1 also classified in 38

50.36 **Isokidamycin bis(m - bromobenzoate) benzene solvate (absolute configuration)**
$C_{53}H_{54}Br_2N_2O_{11}$, $0.5C_6H_6$
M.Furukawa, Y.Iitaka *Tetrahedron Lett.*, 3287, 1974
Residue 1 also classified in 38

50.37 **bis(Valinomycin) potassium tri - iodide penta - iodide**
$2C_{54}H_{90}N_6O_{18}$, $2K^+$, I_3^- , I_5^-
K.Neupert-Laves, M.Dobler *Helv. Chim. Acta,* **58,** 432, 1975
Residue 1 also classified in 48

50.38 **p - Bromophenacyl - septamycin monohydrate (absolute configuration)**
$C_{56}H_{87}BrO_{17}$, H_2O
T.J.Petcher, H.-P.Weber *J. Chem. Soc., Chem. Commun.*, 697, 1974
Residue 1 also classified in 38

50.39 **Ilamycin B$_1$ p - bromobenzoate ethanol solvate monohydrate (absolute configuration)**
$C_{61}H_{80}BrN_9O_{11}$, C_2H_6O , H_2O
Y.Iitaka, H.Nakamura, K.Takada, T.Takita
Acta Crystallogr., Sect. B, **30,** 2817, 1974
Residue 1 also classified in 48

STEROIDS

51.1 **(±) - 2,3 - Dimethoxy - 18 - nor - 8,13 - diaza - 1,3,5(10) - estratrien - 17 - one**
$C_{17}H_{22}N_2O_3$
A.J.Olson, J.C.Hanson, C.E.Nordman
Acta Crystallogr., Sect. B, **31,** 496, 1975
Also classified in 36

51.2 **(±) - 3 - Methoxy - β - nor - 9β - estra - 1,3,5(10) - trien - 17 - one**
$C_{18}H_{22}O_2$
J.C.Hanson, C.E.Nordman *Acta Crystallogr., Sect. B,* **31,** 493, 1975
Also classified in 29

51.3 **19 - Nortestosterone**
$C_{18}H_{24}O_2$
G.Precigoux, B.Busetta, C.Courseille, M.Hospital
Acta Crystallogr., Sect. B, **31,** 1527, 1975

51.4 **5β - Cyano - A - nor - 6 - androstene - 3,17 - dione**
$C_{19}H_{23}NO_2$
E.M.Smith, E.L.Shapiro, G.Teutsch, L.Weber, H.L.Herzog, A.T.McPhail, P.-S.W.Tschang, J.Meinwald *Tetrahedron Lett.,* 3519, 1974
Also classified in 29

51.5 **(3S - (3β,3aβ,5aα,5bR,5cR,7R,9aR,10aα, ,11aβ,11bα)) - Tetradecahydro - 3,7 - dihydroxy - 3a - methyl - 1H - 5b,7 - methano - 8H - as - indaceno(3′,2′.4,5)furo(2,3 - b)pyran - 8 - one monohydrate**
$C_{19}H_{26}O_5$, H_2O
D.F.Rendle, J.Trotter *Acta Crystallogr., Sect. B,* **31,** 1678, 1975
Residue 1 also classified in 38

51.6 **16α - Bromo - 3β - hydroxy - 5 - androsten - 17 - one methanol solvate**
$C_{19}H_{27}BrO_2$, CH_4O
W.L.Duax, T.F.Brennan, C.M.Weeks, Y.Osawa
Cryst. Struct. Commun., **4,** 249, 1975

51.7 **2,4 - Dioxa - 5α - androstan - 17β - ol acetate**
$C_{19}H_{30}O_4$
V.Cody, W.L.Duax, C.M.Weeks, M.E.Wolff
Acta Crystallogr., Sect. B, **31,** 292, 1975
Also classified in 38

51.8 **17β - Hydroxy - 19 - nor - pregna - 4,9,11 - triene - 20 - yne - 3 - one**
$C_{20}H_{22}O_2$
G.Lepicard, J.Delettre, J.-P.Mornon
Acta Crystallogr., Sect. B, **30,** 2751, 1974

51.9 **17β - Ethynyl - 19 - nor - androst - 4 - ene - 3 - one**
$C_{20}H_{26}O$
G.Precigoux, C.Courseille, S.Geoffre, B.Busetta
Acta Crystallogr., Sect. B, **30,** 2757, 1974

51.10 **(10S) - 17β - Acetoxy - 3,10 - cyclo - 3,4 - seco - 4,9(11) - estradien - 1 - one**
$C_{20}H_{26}O_3$
W.L.Duax, C.M.Weeks, D.C.Rohrer
Am. Cryst. Assoc., Abstr. Papers (Spring Meeting), 29, 1975
Also classified in 29

51.11 **3 - Methylene - 5α - androstane**
$C_{20}H_{32}$
A.H.-J.Wang, I.C.Paul *Cryst. Struct. Commun.*, **4,** 303, 1975

51.12 **(19S) - 19 - Methyl - 5 - androsten - 3β,17β - 19 - triol monohydrate**
$C_{20}H_{32}O_3$, H_2O
D.C.Rohrer, C.M.Weeks, W.L.Duax, Y.Osawa
Acta Crystallogr., Sect. A, **31,** S115, 1975

51.13 **17β - Hydroxy - 18 - methyl - 19 - nor - pregna - 4,9,11 - triene - 20 - yne - 3 - one**
$C_{21}H_{24}O_2$
J.Delettre, J.-P.Mornon, G.Lepicard
Acta Crystallogr., Sect. B, **31,** 450, 1975

51.14 **D - Norgestrel**
$C_{21}H_{28}O_2$
N.J.deAngelis, T.H.Doyne, R.L.Grob
Am. Cryst. Assoc., Abstr. Papers (Summer Meeting), 242, 1974

51.15 **16α - Hydroxy - prednisolone**
11β,16α,17α,21 - Tetrahydroxy - 1,4 - pregnadiene - 3,20 - dione
$C_{21}H_{28}O_6$
O.Dideberg, L.Dupont, H.Campsteyn
Acta Crystallogr., Sect. B, **30,** 2064, 1974

51.16 **9α - Chlorocortisol**
$C_{21}H_{29}ClO_5$
C.M.Weeks, W.L.Duax, M.E.Wolff
Acta Crystallogr., Sect. B, **30,** 2516, 1974

51.17 **racemic - pseudo - retro - Progesterone**
9α - Methyl - 19 - nor - progesterone
$C_{21}H_{30}O_2$
H.P.Weber, E.Galantay *Helv. Chim. Acta,* **57,** 187, 1974

51.18 **Progesterone (form ii)**
$C_{21}H_{30}O_2$
E.Foresti Serantoni, A.Krajewski, R.Mongiorgi, L.Riva di Sanseverino, R.Cameroni *Cryst. Struct. Commun.*, **4,** 189, 1975

51.C **Progesterone - resorcinol**
$C_{21}H_{30}O_2$, $C_6H_6O_2$
For complete entry see 60.37

51.19 **16α,17α - Epoxy - 3β - hydroxy - 5 - pregnen - 20 - one**
16α,17α - Epoxypregnenolone
$C_{21}H_{30}O_3$
C.M.Weeks, D.C.Rohrer, J.P.Hazel, D.A.Langs, W.L.Duax
Acta Crystallogr., Sect. A, **31,** S116, 1975
Also classified in 38

51.20 **3 - Oxo - 17β - acetoxy - Δ^4 - 14α - methyl - 8α,9β,10α - estrene**
$C_{21}H_{30}O_3$
G.Koyama *Helv. Chim. Acta,* **57,** 370, 1974

51.21 **16α - Acetyl - 3β - methoxy - CD - cis - D - nor - androstane**
$C_{21}H_{34}O_2$
J.Meinwald, A.J.Taggi, P.A.Luhan, A.T.McPhail
Proc. Nat. Acad. Sci. U. S. A., **71,** 78, 1974
Also classified in 29

51.22 **16β - Acetyl - 3β - methoxy - CD - cis - D - nor - androstane**
$C_{21}H_{34}O_2$
J.Meinwald, A.J.Taggi, P.A.Luhan, A.T.McPhail
Proc. Nat. Acad. Sci. U. S. A., **71,** 78, 1974
Also classified in 29

51.23 **5α - Azido - pregnane**
$C_{21}H_{35}N_3$
A.Chiaroni, C.Riche, C.Pascard-Billy
Cryst. Struct. Commun., **4,** 285, 1975

51.24 **17β - Hydroxy - 3 - oxo - 17α - pregna - 4,6 - diene - 21 - carboxylic acid lactone**
Canrenone
$C_{22}H_{28}O_3$
C.M.Weeks, D.C.Rohrer, J.P.Hazel, D.A.Langs, W.L.Duax
Acta Crystallogr., Sect. A, **31,** S116, 1975
Also classified in 38

51.25 **17β - Hydroxy - 11 - methoxy - 18 - methyl - 19 - nor - pregna - 4,9 - diene - 20 - yne**
$C_{22}H_{28}O_3$
J.-P.Mornon, G.Lepicard, J.Delettre
Acta Crystallogr., Sect. B, **30,** 2754, 1974

51.26 **17α,21 - Dimethyl - Δ^9 - norprogesterone**
$C_{22}H_{30}O_2$
B.Busetta, G.Comberton, C.Courseille, M.Hospital
Acta Crystallogr., Sect. B, **30,** 2759, 1974

51.27 **3β - Hydroxy - 20 - oxo - 5 - pregnene - 16α - carbonitrile**
$C_{22}H_{31}NO_2$
A.Terzis, T.Theophanides *Acta Crystallogr., Sect. B,* **31,** 790, 1975

51.28 **9α - Methoxy - 11β,17α,21 - trihydroxy - 4 - pregnene - 3,20 - dione**
9α - Methylcortisol
$C_{22}H_{32}O_6$
C.M.Weeks, D.C.Rohrer, J.P.Hazel, D.A.Langs, W.L.Duax
Acta Crystallogr., Sect. A, **31,** S116, 1975

51.29 **17β - Acetoxy - 7α - methyl - 5α - androstan - 3 - one**
7α - Methyl - dihydrotestosterone acetate
$C_{22}H_{34}O_3$
C.M.Weeks, D.C.Rohrer, J.P.Hazel, D.A.Langs, W.L.Duax
Acta Crystallogr., Sect. A, **31,** S116, 1975

51.30 **Chlormadinone acetate**
6 - Chloro - 17 - hydroxypregna - 4,6 - diene - 3,20 - dione acetate
$C_{23}H_{29}ClO_4$
R.J.Chandross, J.Bordner *Acta Crystallogr., Sect. B,* **31,** 928, 1975

51.31 **17β - Chloroacetoxy - 2β - acetoxy - 4 - androsten - 3 - one**
$C_{23}H_{31}ClO_5$
C.M.Weeks, W.L.Duax, Y.Osawa *Cryst. Struct. Commun.,* **4,** 97, 1975

51.32 **2α - Hydroxytestosterone diacetate**
$C_{23}H_{32}O_5$
C.M.Weeks, W.L.Duax, Y.Osawa
Acta Crystallogr., Sect. B, **31,** 1502, 1975

51.33 **2β,17β - Diacetoxy - 4 - androsten - 3 - one**
$C_{23}H_{32}O_5$
D.C.Rohrer, W.L.Duax *Cryst. Struct. Commun.,* **4,** 265, 1975

51.34 **3β - Hydroxy - 21 - amino - 14β,20(R),21(S) - oxido - nor - cholan - 23 - acid lactam**
$C_{23}H_{34}NO_3$
D.C.Rohrer, W.L.Duax, M.E.Wolff
Am. Cryst. Assoc., Abstr. Papers (Summer Meeting), 242, 1974

51.35 **3β - Hydroxy - 21 - amino - 14β,20(S),21(R) - oxido - nor - cholan - 23 - acid lactam monohydrate**
$C_{23}H_{34}NO_3$, H_2O
D.C.Rohrer, W.L.Duax, M.E.Wolff
Am. Cryst. Assoc., Abstr. Papers (Summer Meeting), 242, 1974

51.36 **Δ^6 - 6 - Azido - betamethasone - 21 - acetate acetone solvate**
$C_{24}H_{28}FN_3O_6$, C_3H_6O
L.R.Nassimbeni, G.M.Sheldrick, O.Kennard
Acta Crystallogr., Sect. B, **30,** 2401, 1974

51.37 **9α - Fluoro - 16α - methyl - 11β,17,21 - trihydroxy - 1,4 - pregnadiene - 3,20 - dione 21 - acetate monohydrate**
$C_{24}H_{31}FO_6$, H_2O
A.Terzis, T.Theophanides *Acta Crystallogr., Sect. B,* **31,** 796, 1975

51.38 **17β - Hydroxy - 10β - methoxy - estr - 4(5) - en - 3 - one p - bromobenzoate**
$C_{26}H_{31}BrO_4$
J.Bordner, L.R.Morrow *Cryst. Struct. Commun.,* **4,** 241, 1975

51.39 **3 - Oxo - 5α - androstan - 17β - ol toluene - p - sulfonate (at −170°C)**
$C_{26}H_{36}O_4S$
R.A.G.de Graaff, C.Romers *Acta Crystallogr., Sect. B,* **30,** 2029, 1974

51.40 **3β - Hydroxy - 8 - methyl - 5α,13α - androst - 9(11) - en - 15 - one p - bromobenzoate**
$C_{27}H_{33}BrO_3$
J.Bordner, L.R.Morrow *Cryst. Struct. Commun.,* **4,** 237, 1975

51.41 **3α,3β - Dimethoxy - 5α - estran - 17β - ol toluene - p - sulfonate (at −170°C)**
$C_{27}H_{39}O_5S$
R.A.G.de Graaff, C.A.M.van der Ende, C.Romers
Acta Crystallogr., Sect. B, **30,** 2034, 1974

51.42 **3,20 - Diethylenedioxy - 9β,11β - oxido - 11α - acetoxy - 9,11 - seco - 11,19 - cyclo - 5α,14β,17α - pregnane**
$C_{27}H_{40}O_7$
I.L.Karle *Acta Crystallogr., Sect. B,* **31,** 1519, 1975
Also classified in 38

51.43 **Sodium cholesterol sulfate**
$C_{27}H_{45}O_4S^-$, Na^+
S.Abrahamsson, K.Larsson, I.Pascher, S.Sundell
Acta Crystallogr., Sect. A, **31,** S38, 1975

51.44 **1α,25 - Dihydroxy - cholesterol**
$C_{27}H_{46}O_3$
T.A.Narwid, J.F.Blount, J.A.Iacobelli, M.R.Uskokovic
Helv. Chim. Acta, **57,** 781, 1974

51.45 **5α - Chloro - cholestane**
$C_{27}H_{47}Cl$
J.F.Griffin, M.G.Erman, D.C.Rohrer, J.P.Hazel, W.L.Duax, F.A.Carey
Am. Cryst. Assoc., Abstr. Papers (Spring Meeting), 29, 1975

51.46 **Ergocalciferol**
Vitamin D2
$C_{28}H_{44}O$
S.E.Hull, I.Leban, P.Main, P.S.White, M.M.Woolfson
Acta Crystallogr., Sect. A, **31,** S16, 1975
Also classified in 27, 21

51.47 **Ergosterol monohydrate**
Pro - vitamin D2 monohydrate
$C_{28}H_{44}O$, H_2O
S.E.Hull, I.Leban, P.Main, P.S.White, M.M.Woolfson
Acta Crystallogr., Sect. A, **31,** S16, 1975

51.48 **Dentrosterone**
$C_{29}H_{44}O_5$
D.Behr, J.-E.Berg, B.Karlsson, K.Leander, A.-M.Pilotti, A.-C.Wiehager
Acta Chem. Scand. Ser. B, **29,** 401, 1975

51.49 **22,23 - Methylene - 5,24(28) - ergostadien - 3β - ol**
$C_{29}H_{46}O$
G.D.Anderson, T.J.Powers, C.Djerassi, J.Fayos, J.Clardy
J. Am. Chem. Soc., **97,** 388, 1975

51.50 **3β - Acetyloxy - 14 - chloro - 5α,14β,17α - cholestane**
$C_{29}H_{49}ClO_2$
M.Anastasia, A.Scala, M.Bolognesi, A.Coda, G.Rossi
Acta Crystallogr., Sect. A, **31,** S114, 1975

51.51 **Methyl 3 - o - iodobenzoyloxy - 12α - hydroxy - 5β - cholan - 24 - oate**
Methyl dioxycholate 3 - (o - iodobenzoate)
$C_{32}H_{45}IO_5$
C.M.Weeks, D.C.Rohrer, J.P.Hazel, D.A.Langs, W.L.Duax
Acta Crystallogr., Sect. A, **31,** S116, 1975

51.52 **1(10 - 6)abeo - Cholesta - 5,7,9 - trien - 3 - yl p - bromobenzoate**
$C_{34}H_{45}BrO_2$
N.Bosworth, J.M.Midgley, C.J.Moore, W.B.Whalley, G.Ferguson,
W.C.Marsh *J. Chem. Soc., Chem. Commun.,* 719, 1974
Also classified in 29

51.53 **5α,17α - Cholest - 14 - en - 3β - ol p - bromobenzoate**
$C_{34}H_{49}BrO_2$
J.F.Griffin, W.L.Duax *Acta Crystallogr., Sect. A,* **31,** S114, 1975

51.54 **3β - p - Bromobenzoyloxy - cholest - 8(14) - en - 15β - ol**
$C_{34}H_{49}BrO_3$
G.N.Phillips Junior, F.A.Quiocho, H.Emery, F.F.Knapp Junior, G.J.Schroepfer Junior *Acta Crystallogr., Sect. A,* **31,** S113, 1975

51.55 **Pyrocalciferol 3,5 - nitrobenzoate (at −170°C)**
$C_{35}H_{46}N_2O_6$
A.J.de Kok, C.Romers *Acta Crystallogr., Sect. B,* **31,** 1535, 1975

51.56 **Toxisterol₂ - A 3,5 - dinitrobenzoate**
$C_{35}H_{46}N_2O_6$
A.G.M.Barrett, D.H.R.Barton, M.H.Pendlebury, L.Phillips, R.A.Russell, D.A.Widdowson, C.H.Carlisle, P.F.Lindley
J. Chem. Soc., Chem. Commun., 101, 1975
Also classified in 29

51.57 **7,22 - Ergostadiene - 3β - ol tosylate**
$C_{35}H_{52}O_3S$
W.H.Watson, I.F.Taylor Junior *Acta Crystallogr., Sect. A,* **31,** S113, 1975

51.58 **23 - Demethyl - gorgosterol p - iodobenzoate (absolute configuration)**
$C_{36}H_{51}IO_2$
I.-N.Hsu, D.van der Helm *Rec. Trav. Chim. Pays-Bas,* **92,** 1134, 1973

51.59 **9 - Oxo - 9,11 - secogorgost - 5 - ene - 3β,11 - diol 11 - acetate p - iodobenzoate (absolute configuration)**
Secogorgosterol 3 - p - iodobenzoate 11 - acetate
$C_{39}H_{55}IO_5$
E.L.Enwall, D.van der Helm *Rec. Trav. Chim. Pays-Bas,* **93,** 53, 1974
Also classified in 27, 20

MONOTERPENES

52.1 **2,2,5 - endo,6 - exo - 8,9,10 - Heptachloro - bornane**
$C_{10}H_{11}Cl_7$
K.J.Palmer, R.Y.Wong, R.E.Lundin, S.Khalifa, J.E.Casida
J. Am. Chem. Soc., **97**, 408, 1975
Also classified in 31

52.2 **γ - Thujaplicin**
$C_{10}H_{12}O_2$
A.-C.Wiehager, B.Karlsson, A.-M.Pilotti *Eur. Cryst. Meeting*, 573, 1974
Also classified in 22

52.3 **(+) - 8 - Bromocamphor (absolute configuration)**
$C_{10}H_{15}BrO$
C.A.Bear, J.Trotter *Acta Crystallogr., Sect. B*, **31**, 903, 1975

52.4 **Carvoxime**
$C_{10}H_{15}NO$
F.Baert, R.Fouret *Cryst. Struct. Commun.*, **4**, 307, 1975

52.5 **o - Chlorophenylimino - camphor**
$C_{16}H_{18}ClNO$
F.Baert, M.Foulon, R.Fouret *Cryst. Struct. Commun.*, **4**, 61, 1975

52.6 **cis - Pinocarvyl p - nitrobenzoate (at $-193°C$)**
$C_{17}H_{19}NO_4$
G.F.Richards, R.A.Moran, J.A.Heitmann, W.E.Scott
J. Org. Chem., **39**, 86, 1974

52.7 **(+) - Isomenthol p - bromophenyl urethane**
$C_{17}H_{24}BrNO_2$
G.Kartha, K.T.Go
Am. Cryst. Assoc., Abstr. Papers (Spring Meeting), 31, 1975

SESQUITERPENES

53.1 **3 - Bromo - norpatchoulol (absolute configuration)**
$C_{14}H_{21}BrO_2$
W.E.Oberhansli, P.Schonholzer *Recherches,* 62, 1974

53.2 **Miscandenin**
$C_{15}H_{14}O_5$
P.J.Cox, G.A.Sim *J. Chem. Soc., Perkin Trans. 2,* 1359, 1974

53.3 **Dihydromikanolide**
$C_{15}H_{16}O_6$
P.J.Cox, G.A.Sim *J. Chem. Soc., Perkin Trans. 2,* 1355, 1974

53.4 **Tetrachloro - α - santonin**
$C_{15}H_{18}Cl_4O_3$
H.Ogura, H.Takayanagi, A.Yoshino, T.Okamoto
Chem. Pharm. Bull., **22,** 1433, 1974

53.5 **Helenalin oxide**
$C_{15}H_{18}O_4$
A.T.McPhail, K.D.Onan *J. Chem. Soc., Perkin Trans. 2,* 496, 1975

53.6 **Costunolide**
$C_{15}H_{19}O_3$
P.J.Cox, C.J.Gilmore, M.H.P.Guy, G.A.Sim, D.N.J.White
Acta Crystallogr., Sect. A, **31,** S112, 1975

53.7 **Arteannuin B**
$C_{15}H_{20}O_3$
D.G.Leppard, M.Rey, A.S.Dreiding, R.Grieb
Helv. Chim. Acta, **57,** 602, 1974

53.8 **Arteannuin B**
$C_{15}H_{20}O_3$
M.R.Uskokovic, T.H.Williams, J.F.Blount
Helv. Chim. Acta, **57,** 600, 1974

53.9 **Crotocol**
$C_{15}H_{20}O_4$
M.Czugler, A.Kalman *Acta Crystallogr., Sect. B,* **31,** 1204, 1975

53.10 Parthemollin
3,3a,4,5,6,8a - Hexahydro - 7 - (1 - hydroxy - 3 - oxobutyl) - 6 - methyl - 3 - methylene - cyclohepta(b)furan - 2 - one
$C_{15}H_{20}O_4$
P.Sundararaman, R.S.McEwen *J. Chem. Soc., Perkin Trans. 2,* 440, 1975

53.11 Carolenalone
$C_{15}H_{20}O_5$
A.T.McPhail, K.D.Onan, H.Furukawa, K.-H.Lee
Tetrahedron Lett., 1229, 1975

53.12 Capsidiol
$C_{15}H_{24}O_2$
G.I.Birnbaum, A.Stoessl, S.H.Grover, J.B.Stothers
Can. J. Chem., **52,** 993, 1974

53.13 3,8 - Dihydroxy - capnell - 9(12) - ene (absolute configuration)
$C_{15}H_{24}O_3$
M.Kasin, Y.M.Sheikh, L.J.Durham, C.Djerassi, B.Tursch, D.Daloze, J.C.Braekman, D.Losman, R.Karlsson *Tetrahedron Lett.*, 2239, 1974

53.14 Cuauhtemone
$C_{15}H_{24}O_3$
R.A.Ivie, W.H.Watson, X.A.Dominguez
Acta Crystallogr., Sect. B, **30,** 2891, 1974

53.15 Flourensadiol
$C_{15}H_{26}O_2$
R.C.Pettersen, D.L.Cullen, T.D.Spittler, D.G.I.Kingston
Acta Crystallogr., Sect. B, **31,** 1124, 1975

53.16 Pseudoivalin bromoacetate (absolute configuration)
$C_{17}H_{21}BrO_4$
R.Gitany, G.D.Anderson, R.S.McEwen
Acta Crystallogr., Sect. B, **30,** 1900, 1974

53.17 Eupaformonin
$C_{17}H_{22}O_5$
A.T.McPhail, K.D.Onan, K.-H.Lee, T.Ibuka, H.-C.Huang
Tetrahedron Lett., 3203, 1974

53.18 Elatol acetate (absolute configuration)
$C_{17}H_{24}BrClO_2$
J.J.Sims, G.H.Y.Lin, R.M.Wing *Tetrahedron Lett.*, 3487, 1974

53.19 Dihydrophytuberin
$C_{17}H_{28}O_4$
D.L.Hughes, D.T.Coxon *J. Chem. Soc., Chem. Commun.*, 822, 1974

53.20 Onopordopicrin
$C_{19}H_{24}O_6$
P.J.Cox, C.J.Gilmore, M.H.P.Guy, G.A.Sim, D.N.J.White
Acta Crystallogr., Sect. A, **31,** S112, 1975

53.21 Alatolide
$C_{19}H_{26}O_6$
P.J.Cox, C.J.Gilmore, M.H.P.Guy, G.A.Sim, D.N.J.White
Acta Crystallogr., Sect. A, **31,** S112, 1975

53.22 Eupatoriopicrin
$C_{20}H_{26}O_6$
P.J.Cox, C.J.Gilmore, M.H.P.Guy, G.A.Sim, D.N.J.White
Acta Crystallogr., Sect. A, **31,** S112, 1975

53.23 Phantomolin - 1,10 - epoxide
$C_{21}H_{28}O_7$
A.T.McPhail, K.D.Onan, K.-H.Lee, T.Ibuka, M.Kozuka, T.Shingu, H.-C.Huang *Tetrahedron Lett.,* 2739, 1974

53.24 Dihydro - desacetoxy - glaucolide - A
$C_{21}H_{28}O_8$
W.H.Watson, I.B.Wu, S.A.Monti, R.E.Davis, T.J.Mabry, W.G.Padolina
Cryst. Struct. Commun., **3,** 697, 1974

53.25 Plenolin p - iodobenzoate (absolute configuration)
$C_{22}H_{23}IO_5$
A.T.McPhail, K.D.Onan *J. Chem. Soc., Perkin Trans. 2,* 487, 1975

53.26 Berlandin
$C_{22}H_{26}O_7$
P.J.Cox, G.A.Sim, W.Herz *J. Chem. Soc., Perkin Trans. 2,* 459, 1975

53.27 α **- Pompene diol mono - p - bromobenzoate (absolute configuration)**
$C_{22}H_{29}BrO_3$
A.Matsuo, H.Nozaki, M.Nakayama, Y.Kushi, S.Hayashi, N.Kamijo
Tetrahedron Lett., 241, 1975

53.28 Hexahydrocoriolin p - bromobenzoate (absolute configuration)
$C_{22}H_{29}BrO_6$
H.Nakamura, T.Takita, H.Umezawa, M.Kunishima, Y.Nakayama, Y.Iitaka
J. Antibiot., **27,** 301, 1974
Also classified in 50

53.29 Glaucolide A
$C_{23}H_{28}O_{10}$
P.J.Cox, G.A.Sim *J. Chem. Soc., Perkin Trans. 2,* 455, 1975

53.30 4 - O - Acetyl - 2 - O - p - iodobenzoyl - florilenalin (absolute configuration)

$C_{24}H_{25}IO_6$

A.T.McPhail, K.D.Onan *J. Chem. Soc., Perkin Trans. 2,* 492, 1975

DITERPENES

54.1 **Utahin**
$C_{20}H_{20}O_5$
A.-C.Wiehager, B.Karlsson, A.-M.Pilotti *Eur. Cryst. Meeting,* 573, 1974

54.2 **Diterpene from Solidago Arguta**
$C_{20}H_{26}O_3$
G.Ferguson, W.C.Marsh, R.McCrindle, E.Nakamura
J. Chem. Soc., Chem. Commun., 299, 1975

54.3 **Lobophytolide**
$C_{20}H_{28}O_3$
B.Tursch, J.C.Braekman, D.Daloze, M.Herin, R.Karlsson
Tetrahedron Lett., 3769, 1974
Also classified in 38

54.4 **Sinulariolide (absolute configuration)**
$C_{20}H_{30}O_4$
B.Tursch, J.C.Braekman, D.Daloze, M.Herin, R.Karlsson, D.Losman
Tetrahedron, **31,** 129, 1975

54.5 **Isovirescenol B**
$C_{20}H_{32}O_2$
E.Corazza *Acta Crystallogr., Sect. B,* **31,** 1445, 1975

54.6 **3 - Acetoxy - 9 - hydroxy - 17 - norkauran - 16 - one**
$C_{21}H_{32}O_4$
G.Ferguson, W.C.Marsh *Acta Crystallogr., Sect. B,* **31,** 1684, 1975

54.7 **Methyl suaveolate methanol solvate**
$C_{21}H_{34}O_3$, CH_4O
P.S.Manchand, J.D.White, J.Fayos, J.Clardy
J. Org. Chem., **39,** 2306, 1974

54.8 **Pharbitic keto - acid monohydrate**
$C_{22}H_{24}O_9S$, H_2O
T.Yokota, S.Yamazaki, N.Takahashi, Y.Iitaka
Tetrahedron Lett., 2957, 1974

54.9 **Eupalmerin acetate dibromide (at −110°C absolute configuration)**
$C_{22}H_{32}Br_2O_5$
S.E.Ealick, D.van der Helm, A.J.Weinheimer
Acta Crystallogr., Sect. B, **31,** 1618, 1975

54.10 **Eupalmerin acetate (at −110°C)**
$C_{22}H_{32}O_5$
S.E.Ealick, D.van der Helm, A.J.Weinheimer
Acta Crystallogr., Sect. B, **31,** 1618, 1975

54.11 **O - Bromoacetyl - pleuromutilin**
$C_{22}H_{33}BrO_4$
M.Dobler, B.G.Durr *Cryst. Struct. Commun.,* **4,** 259, 1975

54.12 **Prerotundifuran**
$C_{22}H_{34}O_4$
K.Hirotsu, A.Shimada *Chem. Lett.,* 1035, 1973

54.13 **Mascaroside**
$C_{26}H_{36}O_{11}$
A.Ducruix, C.Pascard-Billy, M.Hamonniere, J.Poisson
J. Chem. Soc., Chem. Commun., 396, 1975
Also classified in 45

54.14 **Diterpene II bromobenzoate**
$C_{28}H_{37}BrO_4$
W.C.Marsh, G.Ferguson
Am. Cryst. Assoc., Abstr. Papers (Spring Meeting), 32, 1975

54.C **Delphisine hydrochloride (absolute configuration)**
$C_{28}H_{44}NO_8^+$, Cl^-
For complete entry see 58.21

54.15 **Diterpene III bromobenzoate**
$C_{29}H_{39}BrO_4$
W.C.Marsh, G.Ferguson
Am. Cryst. Assoc., Abstr. Papers (Spring Meeting), 32, 1975

54.16 **Kansuinine B diol p - bromobenzoate**
$C_{45}H_{47}BrO_{17}$
D.Uemura, C.Katayama, E.Uno, K.Sasaki, Y.Hirata, Y.-P.Chen, H.-Y.Hsu
Tetrahedron Lett., 1703, 1975
Also classified in 38

SESTERTERPENES

No entries in this volume

TRITERPENES

56.1 **22,29,30 - tris - nor - Hopane**
$C_{27}H_{46}$
G.W.Smith *Acta Crystallogr., Sect. B,* **31,** 522, 1975

56.2 **29 - nor - 17α - H - Hopane**
$C_{29}H_{50}$
G.W.Smith *Acta Crystallogr., Sect. B,* **31,** 526, 1975

56.3 **Holotoxinogenin 25 - methyl ether (absolute configuration)**
$C_{31}H_{47}O_5$
W.L.Tan, C.Djerassi, J.Fayos, J.Clardy *J. Org. Chem.,* **40,** 466, 1975

56.4 **21α - Methoxy - Δ^{13} - serraten - 3β - ol**
$C_{31}H_{52}O_2$
F.H.Allen, O.Kennard, G.M.Sheldrick
Acta Crystallogr., Sect. A, **31,** S112, 1975

56.5 **Oleanolic acid diacetate bromolactone (absolute configuration)**
$C_{34}H_{51}BrO_6$
T.G.D.van Schalkwyk, G.J.Kruger
Acta Crystallogr., Sect. B, **30,** 2261, 1974

56.6 **3 - O - Acetyl - 16 - O - p - bromobenzoyl - pachysandiol B (absolute configuration)**
$C_{39}H_{57}BrO_4$
N.Masaki, M.Niwa, T.Kikuchi *J. Chem. Soc., Perkin Trans. 2,* 610, 1975

56.7 **Methyl 24,25 - O - isopropylidene - 3α - p - bromobenzoyloxy - 9(10 - 19) - abeo - lanost - 5(10) - ene - 32 - carboxylate**
$C_{41}H_{59}BrO_6$
J.Sakakibara, Y.Hotta, M.Yasue, Y.Iitaka, K.Yamazaki
J. Chem. Soc., Chem. Commun., 839, 1974
Also classified in 38

56.8 **Prieurianin p - bromobenzenesulfonate**
$C_{44}H_{53}BrO_{18}S$
V.P.Gullo, I.Miura, K.Nakanishi, A.F.Cameron, J.D.Connolly, F.D.Duncanson, A.E.Harding, R.McCrindle
J. Chem. Soc., Chem. Commun., 345, 1975

TETRATERPENES

57.1 **Capsanthin bis(p - bromobenzoate)**
$C_{54}H_{62}Br_2O_5$
I.Ueda, W.Nowacki *Z. Kristallogr.*, **140,** 190, 1974

ALKALOIDS

58.1 **Arecoline methiodide**
N - Methyl - 2,5,6 - trihydropyridine 3 - methylcarboxylate methiodide
$C_9H_{16}NO_2^+$, I^-
D.J.H.Mallard, D.P.Vaughan, T.A.Hamor
Acta Crystallogr., Sect. B, **31,** 1109, 1975

58.2 **Lepistine hydrobromide (absolute configuration)**
$C_{10}H_{17}N_2O_2^+$, Br^-
M.Laing, F.L.Warren, E.P.White *Tetrahedron Lett.*, 269, 1975

58.3 **Saxitoxin p - bromobenzenesulfonate (absolute configuration)**
$C_{10}H_{19}N_7O_4^{2+}$, $2C_6H_4BrO_3S^-$
E.J.Schantz, V.E.Ghazarossian, H.K.Schnoes, F.M.Strong, J.P.Springer, J.O.Pexxanit, J.Clardy *J. Am. Chem. Soc.*, **97,** 1238, 1975

58.4 **2,9β - Dimethyl - 6,7 - benzomorphan hydrochloride**
$C_{14}H_{20}N^+$, Cl^-
T.G.Cochran, J.E.Abola *Acta Crystallogr., Sect. B,* **31,** 919, 1975

58.5 **(−) - 5 - m - Hydroxyphenyl - 2 - methylmorphan hydrobromide (absolute configuration)**
$C_{15}H_{22}NO^+$, Br^-
T.G.Cochran *J. Med. Chem.*, **17,** 987, 1974

58.6 **α - Isosparteine diperchlorate**
$C_{15}H_{28}N_2^{2+}$, $2ClO_4^-$
M.Przybylska *Acta Crystallogr., Sect. B,* **30,** 2455, 1974

58.7 **Lycopodine hydrochloride (absolute configuration)**
$C_{16}H_{26}NO^+$, Cl^-
M.-ul-Haque, D.Rogers *J. Chem. Soc., Perkin Trans. 2,* 93, 1975

58.8 **Heliotrine**
$C_{16}H_{27}NO_5$
S.J.Wodak *Acta Crystallogr., Sect. B,* **31,** 569, 1975

58.9 **Piperine**
$C_{17}H_{19}NO_3$
J.Bordner, P.Mullins *Cryst. Struct. Commun.*, **3,** 693, 1974

58.10 **Senkirkine**
$C_{19}H_{27}NO_6$
G.I.Birnbaum *J. Am. Chem. Soc.*, **96,** 6165, 1974

58.11 **Akagerine**
$C_{20}H_{24}N_2O_2$
L.Angenot, O.Dideberg, L.Dupont *Tetrahedron Lett.*, 1357, 1975

58.12 **Aristoteline hydrobromide methanol solvate (absolute configuration)**
$C_{20}H_{27}N_2^+$, Br^- , CH_4O
B.F.Anderson, G.B.Robertson, H.P.Avey, W.F.Donovan, I.R.C.Bick, J.B.Bremner, A.J.T.Finney, N.W.Preston, R.T.Gallagher, G.B.Russell
J. Chem. Soc., Chem. Commun., 511, 1975

58.13 **13 - Epihydroxy - 15 - (5′ - hydroxymethylfuryl - 2′ -) - lupanine hydrobromide**
$C_{20}H_{29}N_2O_4^+$, Br^-
J.Garbarczyk, Z.Kaluski, M.D.Bratek-Wiewiorowska, J.Skolik, M.Wiewiorowski *Bull. Acad. Pol. Sci., Ser. Sci. Chim.*, **22,** 651, 1974

58.14 **5 - Bromobrevianamide A acetone solvate (absolute configuration)**
$C_{21}H_{22}BrN_3O_3$, C_3H_6O
J.Coetzer *Acta Crystallogr., Sect. B,* **30,** 2254, 1974

58.15 **bis(Vindolinine) hydrochloride perchlorate**
$2C_{21}H_{25}N_2O_2^+$, Cl^- , ClO_4^-
C.Riche, C.Pascard *Acta Crystallogr., Sect. A,* **31,** S110, 1975

58.16 **(−) - Pseudocopsinine hydrobromide monohydrate (absolute configuration)**
$C_{21}H_{27}N_2O_2^+$, Br^- , H_2O
S.-M.Nasyrov, V.G.Andrianov, Yu.T.Struchkov
J. Chem. Soc., Chem. Commun., 979, 1974

58.17 **Oxaline**
$C_{24}H_{25}N_5O_4$
D.W.Nagel, K.G.R.Pachler, P.S.Steyn, P.L.Wessels, G.Gafner, G.J.Kruger *J. Chem. Soc., Chem. Commun.*, 1021, 1974

58.18 **Cephalotaxine p - bromobenzoate (absolute configuration)**
$C_{25}H_{24}BrNO_5$
S.K.Arora, R.B.Bates, R.A.Grady, R.G.Powell
J. Org. Chem., **39,** 1269, 1974

58.19 **Fumitremorgin B monohydrate**
$C_{27}H_{33}N_3O_5$, H_2O
M.Yamazaki, H.Fujimoto, T.Akiyama, U.Sankawa, Y.Iitaka
Tetrahedron Lett., 27, 1975

58.20 Verruculogen benzene solvate
$C_{27}H_{33}N_3O_7$, C_6H_6
J.Fayos, D.Lokensgard, J.Clardy, R.J.Cole, J.W.Kirksey
J. Am. Chem. Soc., **96,** 6785, 1974

58.21 Delphisine hydrochloride (absolute configuration)
$C_{28}H_{44}NO_8^+$, Cl^-
S.W.Pelletier, W.H.De Camp, S.D.Lajsic, Z.Djarmati, A.H.Kapadi
J. Am. Chem. Soc., **96,** 7815, 1974
Residue 1 also classified in 54

58.22 Usambarensine dihydrobromide dihydrate (absolute configuration)
$C_{29}H_{30}N_4^{2+}$, $2Br^-$, $2H_2O$
O.Dideberg, L.Dupont, L.Angenot
Acta Crystallogr., Sect. B, **31,** 1571, 1975

58.23 Cytochalasin G
$C_{29}H_{34}N_2O_4$
A.F.Cameron, A.A.Freer, B.Hesp, C.J.Strawson
J. Chem. Soc., Perkin Trans. 2, 1741, 1974

58.24 Fumitremorgin A
$C_{32}H_{41}N_3O_7$
N.Eickman, J.Clardy, R.J.Cole, J.W.Kirksey
Tetrahedron Lett., 1051, 1975

58.25 Tri - N - methylfrangulanine methiodide
$C_{32}H_{53}N_4O_4^+$, I^-
M.Takai, K.-I.Kwai, Y.Ogihara, Y.Iitaka, S.Shibata
J. Chem. Soc., Chem. Commun., 653, 1974
Residue 1 also classified in 48

58.26 Lythracine - IV
$C_{33}H_{41}NO_8$
M.J.Barrow, P.D.Cradwick, G.A.Sim
J. Chem. Soc., Perkin Trans. 2, 1812, 1974

58.27 Talatisamine precursor
$C_{33}H_{47}NO_9S$
F.R.Ahmed *Acta Crystallogr., Sect. B,* **30,** 2558, 1974

58.28 Desacetyl - tryptoquivaline p - bromophenylurethane
$C_{34}H_{32}BrN_5O_7$
J.Clardy, J.P.Springer, G.Buchi, K.Matsuo, R.Wightman
J. Am. Chem. Soc., **97,** 663, 1975

58.29 Lythracine - II O - p - bromobenzenesulfonate (absolute configuration)
$C_{35}H_{40}BrNO_8S$
M.J.Barrow, P.D.Cradwick, G.A.Sim
J. Chem. Soc., Perkin Trans. 2, 1812, 1974

58.30 **Hodgkinsine trimethiodide monohydrate (absolute configuration)**
$C_{36}H_{47}N_6^{3+}$, $3I^-$, H_2O
J.Fridrichsons, M.F.Mackay, A.McL.Mathieson *Tetrahedron,* **30,** 85, 1974

58.31 **Tetrindrine (absolute configuration)**
$C_{38}H_{42}N_2O_6$
C.J.Gilmore, R.F.Bryan *Acta Crystallogr., Sect. A,* **31,** S111, 1975

58.32 **Catharine acetone solvate**
$C_{46}H_{54}N_4O_{10}$, C_3H_6O
P.Rasoanaivo, A.Ahond, J.-P.Cosson, N.Langlois, P.Potier, J.Guilhem, A.Ducruix, C.Riche, C.Pascard *C. R. Acad. Sci., Ser. C,* **279,** 75, 1974

MISCELLANEOUS NATURAL PRODUCTS

59.1 Goniothalamin
(+) - (6S) - 5,6 - Dihydro - 6 - styryl - 2 - pyrone
$C_{13}H_{12}O_2$
P.J.Clarke, P.J.Pauling *J. Chem. Soc., Perkin Trans. 2,* 368, 1975
Also classified in 38

59.2 Kavaic acid
$C_{14}H_{14}O_3$
F.E.Scarbrough, W.Nowacki *Z. Kristallogr.,* **139,** 39, 1974
Also classified in 1

59.3 Genisteine
$C_{15}H_{10}O_5$
M.Breton, G.Precigoux, C.Courseille, M.Hospital
Acta Crystallogr., Sect. B, **31,** 921, 1975
Also classified in 38

59.4 Phomenone
$C_{15}H_{20}O_4$
C.Riche, C.Pascard-Billy *Acta Crystallogr., Sect. B,* **31,** 1476, 1975
Also classified in 38

59.5 2 - Deacylusnic acid (absolute configuration)
$C_{16}H_{14}O_6$
A.L.Macdonald, S.J.Rettig, J.Trotter *Can. J. Chem.,* **52,** 723, 1974
Also classified in 38

59.6 Usnic acid (at −110°C)
$C_{18}H_{16}O_7$
R.Norrestam, M.von Glehn, C.A.Wachtmeister
Acta Chem. Scand. Ser. B, **28,** 1149, 1974
Also classified in 38

59.7 (−) - (S) - Warfarin (absolute configuration)
$C_{19}H_{16}O_4$
E.J.Valente, W.F.Trager, L.H.Jensen
Acta Crystallogr., Sect. B, **31,** 954, 1975
Also classified in 38

59.8 **1 - Hydroxy - 3 - oxo - gibberellin**
$C_{19}H_{22}O_7$
E.Hohne, I.Seidel, G.Adam, P.D.Hung *Tetrahedron,* **31,** 81, 1975
Also classified in 38

59.9 **trans - 2 - (2,4 - Dihydroxy - 3 - methoxybenzyl) - 3 - (4′ - hydroxy - 3′ - methoxybenzyl) - butyrolactone**
$C_{20}H_{22}O_7$
M.S.Shen, R.F.Bryan, R.L.Baxter, J.Sweeney, S.M.Kupchan
Am. Cryst. Assoc., Abstr. Papers (Summer Meeting), 207, 1974
Also classified in 38

59.10 **Sarcophine**
$C_{20}H_{28}O_3$
J.Bernstein, U.Shmueli, E.Zadock, Y.Kashman, I.Neeman
Tetrahedron, **30,** 2817, 1974
Also classified in 38

Prostaglandin A_1 (orthorhombic form)
$C_{20}H_{32}O_4$
J.W.Edmonds, W.L.Duax *J. Am. Chem. Soc.,* **97,** 413, 1975
Also classified in 20

59.12 **Prostaglandin A_1 (monoclinic form)**
$C_{20}H_{32}O_4$
J.W.Edmonds, G.T.DeTitta
Am. Cryst. Assoc., Abstr. Papers (Summer Meeting), 221, 1974
Also classified in 20

59.13 **7α - Methoxy - 7 - phenyl - acetamido - deacetoxy - cephalosporanic acid t - butyl ester**
$C_{21}H_{26}N_2O_5S$
H.E.Applegate, J.E.Dolfini, M.S.Puar, W.A.Slusarchyk, B.Toeplitz, J.Z.Gougoutas *J. Org. Chem.,* **39,** 2794, 1974
Also classified in 41, 35

59.14 **Peucephyllin**
$C_{21}H_{28}O_6$
M.J.Begley, G.Pattenden, T.J.Mabry *Tetrahedron Lett.,* 1105, 1975
Also classified in 38

59.15 **Δ^9 - Tetrahydrocannabinolic acid B**
$C_{22}H_{30}O_4$
E.Rosenqvist, T.Ottersen *Acta Chem. Scand. Ser. B,* **29,** 379, 1975
Also classified in 38

59.16 **β - Peltatin A methyl ether**
$C_{23}H_{24}O_8$
S.K.Arora, R.B.Bates, R.A.Grady, G.Germain, J.P.Declerq
J. Org. Chem., **40,** 28, 1975
Also classified in 38

59.17 **Podolactone A p - bromobenzoate**
$C_{26}H_{25}BrO_9$
B.J.Poppleton *Cryst. Struct. Commun.*, **4,** 101, 1975
Also classified in 38

59.18 **Helichrysoside hydrate**
$C_{30}H_{26}O_{14}$, $5.5H_2O$
H.A.Candy, M.Laing, C.M.Weeks *Tetrahedron Lett.*, 1211, 1975
Residue 1 also classified in 45, 38

59.19 **Triol Q acetonide p - iodobenzoate**
$C_{30}H_{41}IO_4$
G.Ferguson, J.W.B.Fulke *Cryst. Struct. Commun.*, **4,** 233, 1975
Also classified in 38, 29

59.20 **Stemphone**
$C_{30}H_{42}O_8$
C.S.Huber *Acta Crystallogr., Sect. B,* **31,** 108, 1975
Also classified in 38

59.C **5 - Hydroxy - 6 - bromo - 2″,3″,4′,4″,6″,7 - hexa - acetyl - vitexin 5 - hydroxy - 3,6 - dibromo - 2″,3″,4′,4″,6″,7 - hexa - acetyl - vitexin mixture (absolute configuration)**
$0.30C_{33}H_{30}Br_2O_{16}$, $0.70C_{33}H_{31}BrO_{16}$
For complete entry see 59.21

59.21 **5 - Hydroxy - 6 - bromo - 2″,3″,4′,4″,6″,7 - hexa - acetyl - vitexin 5 - hydroxy - 3,6 - dibromo - 2″,3″,4′,4″,6″,7 - hexa - acetyl - vitexin mixture (absolute configuration)**
$0.70C_{33}H_{31}BrO_{16}$, $0.30C_{33}H_{30}Br_2O_{16}$
F.A.Jurnak, D.H.Templeton *Acta Crystallogr., Sect. B,* **31,** 1304, 1975
Residue 1 also classified in 38; residue 2 classified in 59, 38

59.22 **Spermidine - N,N″ - bis(6 - chloro - 2 - methoxy - acridine) chloroform solvate**
$C_{35}H_{35}Cl_2N_5O_2$, $2CHCl_3$
C.Courseille, B.Busetta, M.Hospital *Eur. Cryst. Meeting,* 413, 1974
Residue 1 also classified in 36

59.23 **Jujubogenin p - bromobenzoate ethyl acetate solvate (absolute configuration)**
$C_{37}H_{51}BrO_5$, $0.5C_4H_8O_2$
K.-I.Kawai, Y.Iitaka, S.Shibata *Acta Crystallogr., Sect. B,* **30,** 2886, 1974
Residue 1 also classified in 8

59.24 Milbemycin beta1 p - bromophenylurethane
$C_{39}H_{52}BrNO_8$
H.Mishima, M.Kurabayashi, C.Tamura, S.Sato, H.Kuwano, A.Saito, A.Aoki *Tetrahedron Lett.*, 711, 1975
Also classified in 38

MOLECULAR COMPLEXES

60.C **Quinol - urea**
CH_4N_2O , $C_6H_6O_2$
For complete entry see 60.5

60.C **Di(tetrathiofulvalene) bis(dithiolene) nickel**
$C_4H_4NiS_4$, $2C_6H_4S_4$
For complete entry see 60.4

60.C **Triphenylmethane thiophene**
C_4H_4S , $C_{19}H_{16}$
For complete entry see 60.30

60.C **Triphenylmethane pyrrole**
C_4H_5N , $C_{19}H_{16}$
For complete entry see 60.31

60.C **1,1′ - Binaphthyl - (2,3),(2′,3′) - bis - 18 - crown - 6 - 1,4 - diammonium - butane di(hexafluorophosphate) complex (at −160°C)**
$C_4H_{14}N_2^{2+}$, $C_{40}H_{50}O_{12}$, $2F_6P^-$
For complete entry see 60.39

60.1 **Adenine - riboflavin trihydrate**
$C_5H_5N_5$, $C_{17}H_{20}N_4O_6$, $3H_2O$
K.Tomita, S.Fujii, K.Fujiki, T.Fujiwara
Acta Crystallogr., Sect. A, **31,** S43, 1975
Residue 1 also classified in 44; residue 2 classified in 60, 36

60.2 **Dimethylmalonic acid - bis(triphenylphosphine oxide)**
$C_5H_8O_4$, $2C_{18}H_{15}OP$
J.P.Declercq, G.Germain, J.P.Putzeys, S.Rona, M.van Meerssche
Cryst. Struct. Commun., **3,** 579, 1974
Residue 1 also classified in 1; residue 2 classified in 60, 64

60.C **p - Xylene - hexafluorobenzene complex**
C_6F_6 , C_8H_{10}
For complete entry see 60.10

60.C **Perdeuteropyrene - tetracyanoethylene complex (at 105°K,ordered model)**
C_6N_4 , $C_{16}D_{10}$
For complete entry see 60.23

60.C **Perdeuteropyrene - tetracyanoethylene complex (at 105°K,disordered model)**
C_6N_4 , $C_{16}D_{10}$
For complete entry see 60.24

60.C **Fluoroanthene - picryl bromide complex**
$C_6H_2BrN_3O_6$, $C_{16}H_{10}$
For complete entry see 60.22

60.C **Pyrene - picryl bromide complex**
$C_6H_2BrN_3O_6$, $1.5C_{16}H_{10}$
For complete entry see 60.25

60.C **Benzidine - s - trinitrobenzene complex**
$C_6H_3N_3O_6$, $C_{12}H_{12}N_2$
For complete entry see 60.15

60.C **Benzidine - s - trinitrobenzene complex benzene solvate**
$C_6H_3N_3O_6$, $C_{12}H_{12}N_2$, $0.5C_6H_6$
For complete entry see 60.16

60.3 **Tetrathiofulvalinium 7,7,8,8 - tetracyanoquinodimethanide (at 100°K)**
$C_6H_4S_4^+$, $C_{12}H_4N_4^-$
R.H.Blessing, P.Coppens *Solid State Commun.*, **15,** 215, 1974
Residue 1 also classified in 39; residue 2 classified in 60, 7

60.4 **Di(tetrathiofulvalene) bis(dithiolene) nickel**
$2C_6H_4S_4$, $C_4H_4NiS_4$
J.S.Kasper, L.V.Interrante, C.A.Secaur *J. Am. Chem. Soc.*, **97,** 890, 1975
Residue 1 also classified in 39; residue 2 classified in 60, 85

60.C **ε - Caprolactam - 4 - chlororesorcinol**
$C_6H_5ClO_2$, $C_6H_{11}NO$
For complete entry see 60.6

60.C **Triphenylmethane benzene**
C_6H_6 , $C_{19}H_{16}$
For complete entry see 60.32

60.5 **Quinol - urea**
$C_6H_6O_2$, CH_4N_2O
M.M.Mahmoud, S.C.Wallwork *Acta Crystallogr., Sect. B*. **31,** 338, 1975
Residue 1 also classified in 17; residue 2 classified in 60, 8

60.C **Progesterone - resorcinol**
$C_6H_6O_2$, $C_{21}H_{30}O_2$
For complete entry see 60.37

60.C **Triphenylmethane aniline**
C_6H_7N , $C_{19}H_{16}$
For complete entry see 60.33

60.6 **ε - Caprolactam - 4 - chlororesorcinol**
$C_6H_{11}NO$, $C_6H_5ClO_2$
Kh.P.Oya, R.M.Myasnikova *Zh. Strukt. Khim.*, **14,** 1094, 1973
Residue 1 also classified in 34; residue 2 classified in 60, 17

60.7 **Salicylic acid - cytidine complex**
$C_7H_6O_3$, $C_9H_{13}N_3O_5$
C.Tamura, M.Yoshikawa, S.Sato, T.Hata *Chem. Lett.*, 1221, 1973
Residue 1 also classified in 13, 17; residue 2 classified in 64, 47, 45, 44

60.8 **cis - bis(bis - (Trifluoromethyl) - ethylene - 1,2 - dithiolato) nickel - phenoxazine**
$C_8F_{12}NiS_4$, $C_{12}H_9NO$
A.Singhabhandhu, P.D.Robinson, J.H.Fang, W.E.Geiger Junior
Inorg. Chem., **14,** 318, 1975
Residue 1 also classified in 85; residue 2 classified in 60, 40

60.9 **cis - bis(bis - (Trifluoromethyl) - ethylene - 1,2 - dithiolato) nickel - phenothiazine**
$C_8F_{12}NiS_4$, $C_{12}H_9NS$
A.Singhabhandhu, P.D.Robinson, J.H.Fang, W.E.Geiger Junior
Inorg. Chem., **14,** 318, 1975
Residue 1 also classified in 5; residue 2 classified in 60, 41

60.10 **p - Xylene - hexafluorobenzene complex**
C_8H_{10} , C_6F_6
T.Dahl *Acta Chem. Scand. Ser. A*, **29,** 170, 1975
Residue 1 also classified in 19; residue 2 classified in 60, 19

60.C **Salicylic acid - cytidine complex**
$C_9H_{13}N_3O_5$, $C_7H_6O_3$
For complete entry see 60.7

60.11 **Anthracene - pyromellitic dithioanhydride complex**
$C_{10}H_2O_4S$, $C_{14}H_{10}$
I.V.Bulgarovskaya, E.M.Smelyanskaya, Yu.G.Fedorov, Z.V.Avonkova
Kristallografiya, **19,** 260, 1974
Residue 1 also classified in 39; residue 2 classified in 60, 26

60.12 **Pyromellitic dianhydride - trans - stilbene complex**
$C_{10}H_2O_6$, $C_{14}H_{12}$
T.Kodama, S.Kumakura *Bull. Chem. Soc. Jpn.*, **47,** 1081, 1974
Residue 1 also classified in 38; residue 2 classified in 60, 19

60.C **Benz(a)anthracene - pyromellitic dianhydride complex**
$C_{10}H_2O_6$, $C_{18}H_{12}$
For complete entry see 60.27

60.13 **Pyromellitic dianhydride - trans - 4 - methyl - stilbene complex**
$2C_{10}H_2O_6$, $C_{15}H_{14}$
T.Kodama, S.Kumakura *Bull. Chem. Soc. Jpn.*, **47**, 2146, 1974
Residue 1 also classified in 38; residue 2 classified in 60, 19

60.C **Lumiflavin - bis(naphthalene - 2,3 - diol) trihydrate**
$2C_{10}H_8O_2$, $C_{13}H_{12}N_4O_2$, $3H_2O$
For complete entry see 60.19

60.C **Carbazole - 7,7,8,8 - tetracyanoquinodimethane complex**
$C_{12}H_4N_4$, $C_{12}H_9N$
For complete entry see 60.14

60.C **N - Methylphenothiazine - 7,7,8,8 - tetracyanoquinodimethane complex**
$C_{12}H_4N_4$, $C_{13}H_{11}NS$
For complete entry see 60.18

60.C **Chrysene - 7,7,8,8 - tetracyanoquinodimethane complex**
$C_{12}H_4N_4$, $C_{18}H_{12}$
For complete entry see 60.28

60.C **Tetrathiofulvalinium 7,7,8,8 - tetracyanoquinodimethanide (at 100°K)**
$C_{12}H_4N_4^-$, $C_6H_4S_4^+$
For complete entry see 60.3

60.C **Acridinium - bis(7,7,8,8 - tetracyanoquinodimethanide)**
$C_{12}H_4N_4^-$, $C_{13}H_{10}N^+$, $C_{12}H_4N_4$
For complete entry see 60.17

60.C **(3,3′ - Diethylthiazolinocarbocyanine) - (7,7,8,8 - tetracyanoquinodimethanide) - (9 - dicyanomethylene - 2,4,7 - trinitrofluorene) complex**
$C_{12}H_4N_4^-$, $C_{13}H_{21}N_2S_2^+$, $C_{16}H_5N_5O_6$
For complete entry see 60.21

60.C **3,3′ - Dimethylthiacyanine - bis(7,7,8,8 - tetracyanoquinodimethane)**
$C_{12}H_4N_4^-$, $C_{17}H_{15}N_2S_2^+$, $C_{12}H_4N_4$
For complete entry see 60.26

60.C **3,3′ - Dimethylthiacarbocyanine bis(7,7,8,8 - tetracyanoquinodimethane) complex**
$C_{12}H_4N_4^-$, $C_{19}H_{17}N_2S_2^+$, $C_{12}H_4N_4$
For complete entry see 60.34

60.C **3,3′ - Diethylthiacarbocyanine - bis(7,7,8,8 - tetracyanoquinodimethane) (monoclinic form)**
$C_{12}H_4N_4^-$, $C_{21}H_{21}N_2S_2^+$, $C_{12}H_4N_4$
For complete entry see 60.35

60.C **3,3′ - Diethylthiacarbocyanine - bis(7,7,8,8 - tetracyanoquinodimethane) (triclinic form)**
$C_{12}H_4N_4^-$, $C_{21}H_{21}N_2S_2^+$, $C_{12}H_4N_4$
For complete entry see 60.36

60.C **1 - Methyl - 3,3 - dimethyl - 2 - (p - N - methyl - N - β - chloroethylstyryl)indole - bis(7,7,8,8 - tetracyanoquinodimethane)**
$C_{12}H_4N_4^-$, $C_{22}H_{26}ClN_2^+$, $C_{12}H_4N_4$
For complete entry see 60.38

60.C **(1,4 - Di - (N - pyridinium methyl)benzene) - tetra(7,7,8,8 - tetracyanoquinodimethane)**
$2C_{12}H_4N_4^-$, $C_{18}H_{18}N_2^{2+}$, $2C_{12}H_4N_4$
For complete entry see 60.29

60.14 **Carbazole - 7,7,8,8 - tetracyanoquinodimethane complex**
$C_{12}H_9N$, $C_{12}H_4N_4$
H.Kobayashi *Bull. Chem. Soc. Jpn.*, **46**, 2675, 1973
Residue 1 also classified in 36; residue 2 classified in 60, 7

60.C **cis - bis(bis - (Trifluoromethyl) - ethylene - 1,2 - dithiolato) nickel - phenoxazine**
$C_{12}H_9NO$, $C_8F_{12}NiS_4$
For complete entry see 60.8

60.C **cis - bis(bis - (Trifluoromethyl) - ethylene - 1,2 - dithiolato) nickel - phenothiazine**
$C_{12}H_9NS$, $C_8F_{12}NiS_4$
For complete entry see 60.9

60.15 **Benzidine - s - trinitrobenzene complex**
$C_{12}H_{12}N_2$, $C_6H_3N_3O_6$
N.Tachikawa, K.Yakushi, H.Kuroda
Acta Crystallogr., Sect. B, **30**, 2770, 1974
Residue 1 also classified in 16; residue 2 classified in 60, 15

60.16 **Benzidine - s - trinitrobenzene complex benzene solvate**
$C_{12}H_{12}N_2$, $C_6H_3N_3O_6$, $0.5C_6H_6$
K.Yakushi, N.Tachikawa, I.Ikemoto, H.Kuroda
Acta Crystallogr., Sect. B, **31**, 738, 1975
Residue 1 also classified in 16; residue 2 classified in 60, 15

60.C **Aminopyrine - cyclobarbital**
$C_{12}H_{16}N_2O_3$, $C_{13}H_{17}N_3O$
For complete entry see 60.20

60.17 **Acridinium - bis(7,7,8,8 - tetracyanoquinodimethanide)**
$C_{13}H_{10}N^+$, $C_{12}H_4N_4^-$, $C_{12}H_4N_4$
H.Kobayashi *Bull. Chem. Soc. Jpn.*, **47**, 1346, 1974
Residue 1 also classified in 36; residue 2 classified in 60, 7

60.18 **N - Methylphenothiazine - 7,7,8,8 - tetracyanoquinodimethane complex**
$C_{13}H_{11}NS$, $C_{12}H_4N_4$
H.Kobayashi *Bull. Chem. Soc. Jpn.*, **46,** 2945, 1973
Residue 1 also classified in 41; residue 2 classified in 60, 7

60.19 **Lumiflavin - bis(naphthalene - 2,3 - diol) trihydrate**
$C_{13}H_{12}N_4O_2$, $2C_{10}H_8O_2$, $3H_2O$
C.J.Fritchie Junior, R.M.Johnston *Acta Crystallogr., Sect. B,* **31,** 454, 1975
Residue 1 also classified in 36; residue 2 classified in 60, 24

60.20 **Aminopyrine - cyclobarbital**
$C_{13}H_{17}N_3O$, $C_{12}H_{16}N_2O_3$
S.Kiryu, F.Hirayama, S.Iguchi *Chem. Pharm. Bull.*, **22,** 1588, 1974
Residue 1 also classified in 32; residue 2 classified in 60, 43

60.21 **(3,3′ - Diethylthiazolinocarbocyanine) - (7,7,8,8 - tetracyanoquinodimethanide) - (9 - dicyanomethylene - 2,4,7 - trinitrofluorene) complex**
$C_{13}H_{21}N_2S_2^+$, $C_{12}H_4N_4^-$, $C_{16}H_5N_5O_6$
V.F.Kaminskii, R.P.Shibaeva, L.O.Atovmyan
Zh. Strukt. Khim., **15,** 509, 1974
Residue 1 also classified in 41; residue 2 classified in 60, 7, residue 3 in 60, 28, 7

60.C **Anthracene - pyromellitic dithioanhydride complex**
$C_{14}H_{10}$, $C_{10}H_2O_4S$
For complete entry see 60.11

60.C **Pyromellitic dianhydride - trans - stilbene complex**
$C_{14}H_{12}$, $C_{10}H_2O_6$
For complete entry see 60.12

60.C **Pyromellitic dianhydride - trans - 4 - methyl - stilbene complex**
$C_{15}H_{14}$, $2C_{10}H_2O_6$
For complete entry see 60.13

60.C **(3,3′ - Diethylthiazolinocarbocyanine) - (7,7,8,8 - tetracyanoquinodimethanide) - (9 - dicyanomethylene - 2,4,7 - trinitrofluorene) complex**
$C_{16}H_5N_5O_6$, $C_{12}H_4N_4^-$, $C_{13}H_{21}N_2S_2^+$
For complete entry see 60.21

60.22 **Fluoroanthene - picryl bromide complex**
$C_{16}H_{10}$, $C_6H_2BrN_3O_6$
F.H.Herbstein, M.Kaftory *Acta Crystallogr., Sect. B,* **31,** 60, 1975
Residue 1 also classified in 29; residue 2 classified in 60, 15

60.23 **Perdeuteropyrene - tetracyanoethylene complex (at 105°K,ordered model)**
$C_{16}D_{10}$, C_6N_4
F.K.Larsen, R.G.Little, P.Coppens
Acta Crystallogr., Sect. B, **31,** 430, 1975
Residue 1 also classified in 29; residue 2 classified in 60, 7

60.24 **Perdeuteropyrene - tetracyanoethylene complex (at 105°K,disordered model)**
$C_{16}D_{10}$, C_6N_4
F.K.Larsen, R.G.Little, P.Coppens
Acta Crystallogr., Sect. B, **31,** 430, 1975
Residue 1 also classified in 29; residue 2 classified in 60, 7

60.25 **Pyrene - picryl bromide complex**
$1.5C_{16}H_{10}$, $C_6H_2BrN_3O_6$
F.H.Herbstein, M.Kaftory *Acta Crystallogr., Sect. B,* **31,** 68, 1975
Residue 1 also classified in 29; residue 2 classified in 60, 15

60.26 **3,3′ - Dimethylthiacyanine - bis(7,7,8,8 - tetracyanoquinodimethane)**
$C_{17}H_{15}N_2S_2^+$, $C_{12}H_4N_4^-$, $C_{12}H_4N_4$
R.P.Shibaeva, L.O.Atovmyan, V.I.Ponomarev, O.S.Filipenko, L.P.Rozenberg *Kristallografiya,* **19,** 95, 1974
Residue 1 also classified in 41; residue 2 classified in 60, 7

60.C **Adenine - riboflavin trihydrate**
$C_{17}H_{20}N_4O_6$, $C_5H_5N_5$, $3H_2O$
For complete entry see 60.1

60.27 **Benz(a)anthracene - pyromellitic dianhydride complex**
$C_{18}H_{12}$, $C_{10}H_2O_6$
J.Iball, S.N.Scrimgeour, B.C.Williams
Acta Crystallogr., Sect. A, **31,** S121, 1975
Residue 1 also classified in 29; residue 2 classified in 60, 38

60.28 **Chrysene - 7,7,8,8 - tetracyanoquinodimethane complex**
$C_{18}H_{12}$, $C_{12}H_4N_4$
P.J.Munnoch, J.D.Wright *J. Chem. Soc., Perkin Trans. 2,* 1397, 1974
Residue 1 also classified in 29; residue 2 classified in 60, 7

60.C **Dimethylmalonic acid - bis(triphenylphosphine oxide)**
$2C_{18}H_{15}OP$, $C_5H_8O_4$
For complete entry see 60.2

60.29 **(1,4 - Di - (N - pyridinium methyl)benzene) - tetra(7,7,8,8 - tetracyanoquinodimethane)**
$C_{18}H_{18}N_2^{2+}$, $2C_{12}H_4N_4^-$, $2C_{12}H_4N_4$
G.J.Ashwell, S.C.Wallwork, S.R.Baker, P.I.C.Berthier
Acta Crystallogr., Sect. B, **31,** 1174, 1975
Residue 1 also classified in 33; residue 2 classified in 60, 7

60.30 **Triphenylmethane thiophene**
$C_{19}H_{16}$, C_4H_4S
A.Allemand, R.Gerdil *Acta Crystallogr., Sect. A,* **31,** S130, 1975
Residue 1 also classified in 19; residue 2 classified in 60, 39

60.31 **Triphenylmethane pyrrole**
$C_{19}H_{16}$, C_4H_5N
A.Allemand, R.Gerdil *Acta Crystallogr., Sect. A,* **31,** S130, 1975
Residue 1 also classified in 19; residue 2 classified in 60, 32

60.32 **Triphenylmethane benzene**
$C_{19}H_{16}$, C_6H_6
A.Allemand, R.Gerdil *Acta Crystallogr., Sect. A,* **31,** S130, 1975
Residue 1 also classified in 19; residue 2 classified in 60, 19

60.33 **Triphenylmethane aniline**
$C_{19}H_{16}$, C_6H_7N
A.Allemand, R.Gerdil *Acta Crystallogr., Sect. A,* **31,** S130, 1975
Residue 1 also classified in 19; residue 2 classified in 60, 16

60.34 **3,3′ - Dimethylthiacarbocyanine bis(7,7,8,8 - tetracyanoquinodimethane) complex**
$C_{19}H_{17}N_2S_2^+$, $C_{12}H_4N_4^-$, $C_{12}H_4N_4$
R.P.Shibaeva, V.F.Kaminskii, L.O.Atovmyan
Zh. Strukt. Khim., **15,** 720, 1974
Residue 1 also classified in 41; residue 2 classified in 60, 7

60.35 **3,3′ - Diethylthiacarbocyanine - bis(7,7,8,8 - tetracyanoquinodimethane) (monoclinic form)**
$C_{21}H_{21}N_2S_2^+$, $C_{12}H_4N_4^-$, $C_{12}H_4N_4$
V.F.Kaminskii, R.P.Shibaeva, L.O.Atovmyan
Zh. Strukt. Khim., **14,** 700, 1973
Residue 1 also classified in 41; residue 2 classified in 60, 7

60.36 **3,3′ - Diethylthiacarbocyanine - bis(7,7,8,8 - tetracyanoquinodimethane) (triclinic form)**
$C_{21}H_{21}N_2S_2^+$, $C_{12}H_4N_4^-$, $C_{12}H_4N_4$
V.F.Kaminskii, R.P.Shibaeva, L.O.Atovmyan
Zh. Strukt. Khim., **14,** 1082, 1973
Residue 1 also classified in 41; residue 2 classified in 60, 7

60.37 **Progesterone - resorcinol**
$C_{21}H_{30}O_2$, $C_6H_6O_2$
O.Dideberg, L.Dupont, H.Campsteyn
Acta Crystallogr., Sect. B, **31,** 637, 1975
Residue 1 also classified in 51; residue 2 classified in 60, 17

60.38 **1 - Methyl - 3,3 - dimethyl - 2 - (p - N - methyl - N - β - chloroethylstyryl)indole - bis(7,7,8,8 - tetracyanoquinodimethane)**
$C_{22}H_{26}ClN_2^+$, $C_{12}H_4N_4^-$, $C_{12}H_4N_4$
R.P.Shibaeva, L.P.Rozenberg, L.O.Atovmyan
Kristallografiya, **18,** 518, 1973
Residue 1 also classified in 35, 16; residue 2 classified in 60, 7

60.39 **1,1′ - Binaphthyl - (2,3),(2′,3′) - bis - 18 - crown - 6 - 1,4 - diammonium - butane di(hexafluorophosphate) complex (at −160°C)**
$C_{40}H_{50}O_{12}$, $C_4H_{14}N_2^{2+}$, $2F_6P^-$
I.Goldberg *Am. Cryst. Assoc., Abstr. Papers (Summer Meeting),* 251, 1974
Residue 1 also classified in 38; residue 2 classified in 60, 3

CLATHRATES

61.C **Nickel iodide - urea clathrate**
$4CH_4N_2O$, $C_6H_{24}N_{12}NiO_6^{2+}$, $2I^-$
For complete entry see 61.1

61.C **Cyclohexa - amylose 1 - propanol hydrate**
α - Cyclodextrin 1 - propanol hydrate
C_3H_8O , $C_{36}H_{60}O_{30}$, $4.8H_2O$
For complete entry see 61.6

61.C **Tri - o - thymotide pyridine clathrate**
C_5H_5N , $2C_{33}H_{36}O_6$
For complete entry see 61.5

61.1 **Nickel iodide - urea clathrate**
$C_6H_{24}N_{12}NiO_6^{2+}$, $4CH_4N_2O$, $2I^-$
X.Suleyman, M.A.Porai-Koshits, A.S.Antsyshkina, K.Sulaimankov
Zh. Neorg. Khim., **16,** 3394, 1971
Residue 1 also classified in 79; residue 2 classified in 61, 8

61.C **tris(1,8 - Naphthalenedioxy)cyclotriphosphazene p - xylene clathrate**
$0.5C_8H_{10}$, $C_{30}H_{18}N_3O_6P_3$
For complete entry see 61.2

61.2 **tris(1,8 - Naphthalenedioxy)cyclotriphosphazene p - xylene clathrate**
$C_{30}H_{18}N_3O_6P_3$, $0.5C_8H_{10}$
H.R.Allcock, M.T.Stein, E.C.Bissell *J. Am. Chem. Soc.*, **96,** 4795, 1974
Residue 1 also classified in 64; residue 2 classified in 61, 19

61.3 **Tri - o - thymotide ethanol clathrate**
$C_{33}H_{36}O_6$, C_2H_6O
D.J.Williams, D.Lawton *Tetrahedron Lett.*, 111, 1975
Residue 1 also classified in 38; residue 2 classified in 5

61.4 **Tri - o - thymotide cetyl alcohol clathrate**
$0.2C_{33}H_{36}O_6$, $C_{16}H_{34}O$
D.J.Williams, D.Lawton *Tetrahedron Lett.*, 111, 1975
Residue 1 also classified in 38; residue 2 classified in 5

61.5 **Tri - o - thymotide pyridine clathrate**
$2C_{33}H_{36}O_6$, C_5H_5N
S.Brunie, A.Navaza, G.Tsoucaris
Acta Crystallogr., Sect. A, **31,** S127, 1975
Residue 1 also classified in 38; residue 2 classified in 61, 33

61.6 **Cyclohexa - amylose 1 - propanol hydrate**
α - Cyclodextrin 1 - propanol hydrate
$C_{36}H_{60}O_{30}$, C_3H_8O , $4.8H_2O$
W.Saenger, R.K.McMullan, J.Fayos, D.Mootz
Acta Crystallogr., Sect. B, **30,** 2019, 1974
Residue 1 also classified in 45; residue 2 classified in 61, 5

BORON COMPOUNDS

62.1 **Dimethyl - boron - methylchloride (form ii, at −180°C)**
C_3H_8BCl
J.C.Huffman *Cryst. Struct. Commun.*, **3,** 649, 1974

62.2 **7 - Hydroxy - 6 - methyl - 7,6 - borazarothieno(3,2 - c)pyridine**
$C_6H_7BN_2OS$
B.Aurivillius, I.Lofving *Acta Chem. Scand. Ser. B,* **28,** 989, 1974

62.3 **Boratrane**
$C_6H_{12}BNO_3$
A.A.Kemme, Ya.Ya.Bleidelis
Latv. PSR Zinat. Akad. Vestis, Kim. Ser., 621, 1971

62.4 **1,1,4,4 - Tetramethyl - 1,4 - diazonia - 2,5 - diborata - cyclohexane**
$C_6H_{20}B_2N_2$
T.H.Hseu, L.A.Larsen *Inorg. Chem.*, **14,** 330, 1975

62.C **8 - η - Cyclopentadienyl - 6,7 - dicarba - 8 - cobalta - nido - nonaborate(ii) (at −160°C)**
$C_7H_{16}B_7Co$
For complete entry see 73.1

62.C **2,4,6,7 - Tetramethyl - 2,6,7 - triaza - 1 - phosphabicyclo(2.2.2)octane - 1 - borane**
$C_8H_{21}BN_3P$
For complete entry see 64.23

62.C **closo - 1,1 - bis(Trimethylphosphine) - 1,6,8 - platinadicarbaborane**
$C_8H_{26}B_6P_2Pt$
For complete entry see 86.4

62.C **bis(Tetramethylammonium) 4,4′ - commo - bis(decahydro - 1,6 - dimethyl - 1,6 - dicarba - 4 - titana - closo - tridecaborate) acetone solvate (at −160°C)**
$C_8H_{32}B_{20}Ti^{2-}$, $2C_4H_{12}N^+$, $2C_3H_6O$
For complete entry see 71.8

62.C **closo - 1,1 - bis(Trimethylphosphine) - 6,8 - dimethyl - 1,6,8 - platinadicarbaborane (α form)**
$C_{10}H_{30}B_6P_2Pt$
For complete entry see 86.10

62.C **closo - 1,1 - bis(Trimethylphosphine) - 6,8 - dimethyl - 1,6,8 - platinadicarbaborane (β form)**
$C_{10}H_{30}B_6P_2Pt$
For complete entry see 86.11

62.C **Carbonyl - (hydrido - tris(pyrazol - 1 - yl)borato) - methyl platinum**
$C_{11}H_{13}BN_6OPt$
For complete entry see 83.38

62.C **1,6 - bis(η - Cyclopentadienyl) - 1,6 - diferra - 2,3 - dicarba - closo - decaborane(8)**
$C_{12}H_{18}B_6Fe_2$
For complete entry see 73.16

62.C **Cobaltocenium - carborane**
$C_{12}H_{20}B_9Co$
For complete entry see 73.17

62.5 **8 - (4 - (Triethylammonium) - n - butyloxy) - 6 - tricarbonyl - 6 - manganadecaborane**
$C_{13}H_{35}B_9MnNO_4$
D.F.Gaines, J.W.Lott, J.C.Calabrese *Inorg. Chem.*, **13,** 2419, 1974
Also classified in 3

62.C **Tricarbonyl - (1 - phenylborinato) manganese**
$C_{14}H_{10}BMnO_3$
For complete entry see 75.12

62.6 **B,B - bis(p - Fluorophenyl)boroxazolidine**
$C_{14}H_{14}BF_2NO$
S.J.Rettig, J.Trotter *Acta Crystallogr., Sect. B,* **30,** 2139, 1974

62.7 **4,4 - Dimethyl - 2,2 - diphenyl - 1,3 - dioxa - 4 - azonia - 2 - boranata - cyclopentane**
$C_{15}H_{18}BNO_2$
S.J.Rettig, J.Trotter, W.Kliegel *Can. J. Chem.*, **52,** 2531, 1974

62.C **Di - μ - carbonyl - bis(carbonyl - (1 - methylborinato) iron)**
$C_{16}H_{16}B_2Fe_2O_4$
For complete entry see 75.14

62.C **nido - 6,6 - bis(Triethylphosphine) - 5,8 - dimethyl - 6,5,8 - platinadicarbaborane**
$C_{16}H_{42}B_6P_2Pt$
For complete entry see 71.45

62.8 **Triphenylborane**
$C_{18}H_{15}B$
F.Zettler, H.D.Hausen, H.Hess *J. Organomet. Chem.*, **72,** 157, 1974

62.C **bis((Hydrido - tris(1 - pyrazolyl)borato) copper(i))**
$C_{18}H_{20}B_2Cu_2N_{12}$
For complete entry see 83.83

62.C **(Diethyl - bis(pyrazolyl)borato) - (pyrazolylato) - (trihapto - allyl) - (dicarbonyl) molybdenum**
$C_{18}H_{25}BMoN_6O_2$
For complete entry see 72.15

62.9 **bis(4 - Dithieno(3,2.2′,3′ - f)borepinyl) ether**
$C_{20}H_{12}B_2OS_4$
B.Aurivillius *Acta Chem. Scand. Ser. B*, **28**, 998, 1974

62.10 **bis(3,5 - Dimethylpyrazolyl)borane dimer**
$C_{20}H_{30}B_2N_8$
N.W.Alcock, J.F.Sawyer *Acta Crystallogr., Sect. B*, **30**, 2899, 1974
Also classified in 32

62.C **bis(Diethyl - bis(1 - pyrazolyl)borato) nickel(ii)**
$C_{20}H_{32}B_2N_8Ni$
For complete entry see 83.97

62.C **1,1 - bis(Dimethylphenylphosphine) - 2,4 - dimethyl - 1 - platina - 2,4 - dicarbadodecaborane**
$C_{20}H_{37}B_9P_2Pt$
For complete entry see 71.66

62.11 **15c - Chloro - tri - isoindolo(1,2,3 - cd.1′,2′,3′ - gh.1″,2″,3″ - kl)(2,3a,5,6a,8,9a,9b)hexa - azaboraphenalene**
$C_{24}H_{12}BClN_6$
H.Kietaibl *Monatsh. Chem.*, **105**, 405, 1974

62.12 **Potassium tetraphenylborate**
$C_{24}H_{20}B^-$, K^+
K.Hoffmann, E.Weiss *J. Organomet. Chem.*, **67**, 221, 1974

62.13 **Tetramethylammonium tetraphenylborate**
$C_{24}H_{20}B^-$, $C_4H_{12}N^+$
K.Hoffmann, E.Weiss *J. Organomet. Chem.*, **67**, 221, 1974
Residue 2 classified in 3

62.C **Cyano - (2,2′,2″ - triamino - triethylamine) copper(ii) tetraphenylborate**
$C_{24}H_{20}B^-$, $C_7H_{18}CuN_5{}^+$
For complete entry see 76.33

62.C **Cyanato - (2,2′,2″ - triaminotriethylamine) nickel(ii) tetraphenylborate**
$C_{24}H_{20}B^-$, $C_7H_{18}N_5NiO^+$
For complete entry see 76.34

62.C **bis(N,N′ - Ethylene - bis(salicylaldiminato) cobalt(ii)) bis(tetrahydrofuran) sodium tetraphenylborate**
$C_{24}H_{20}B^-$, $2C_{16}H_{14}CoN_2O_2$, $2C_4H_8O$, Na^+
For complete entry see 78.2

62.C **Carbonyl - bis(dimethylphenylphosphine) - methylbutadiene - iridium(i) tetraphenylborate**
$C_{24}H_{20}B^-$, $C_{22}H_{30}IrOP_2{}^+$
For complete entry see 72.23

62.C **Tricarbonyl - iodo - tris(dimethylphenylphosphino) tungsten(ii) tetraphenylborate**
$C_{24}H_{20}B^-$, $C_{27}H_{33}IO_3P_3W^+$
For complete entry see 86.53

62.C **Bromo - (hexaphenyl - 1,4,7,10 - tetraphosphadecane) iron(ii) tetraphenylborate dichloromethane solvate**
$C_{24}H_{20}B^-$, $C_{42}H_{42}BrFeP_4{}^+$, CH_2Cl_2
For complete entry see 86.81

62.C **(tris(2 - Diphenylarsinoethyl)amine)phenyl nickel(ii) tetraphenylborate**
$C_{24}H_{20}B^-$, $C_{48}H_{47}As_3NNi^+$
For complete entry see 71.130

62.C **μ - Iodo - bis(tris(2 - diphenylarsinoethyl)amine) dinickel(i) tetraphenylborate**
$C_{24}H_{20}B^-$, $C_{84}H_{84}As_6IN_2Ni_2{}^+$
For complete entry see 86.117

62.C **Di - μ - azido - bis(2,2′,2″ - triaminoethylamine) nickel(ii) tetraphenylborate**
$2C_{24}H_{20}B^-$, $C_{12}H_{36}N_{14}Ni_2{}^{2+}$
For complete entry see 76.63

62.C **trans - bis(Acetone hydrazone) - tetrakis(trimethylphosphite) ruthenium(ii) bis(tetraphenylborate)**
$2C_{24}H_{20}B^-$, $C_{18}H_{52}N_4O_{12}P_4Ru^{2+}$
For complete entry see 86.29

62.C **μ - Thio - bis(1,1,1 - tris(diphenylphosphinomethyl)ethane) dinickel(ii) tetraphenylborate**
$2C_{24}H_{20}B^-$, $C_{82}H_{78}Ni_2P_6S^{2+}$
For complete entry see 86.115

62.C **(Diphenyldipyrazolylborato)(2 - methylallyl)dicarbonyl molybdenum**
$C_{24}H_{23}BMoN_4O_2$
For complete entry see 72.28

62.C **α - Carboranyl complex of platinum(ii)**
$C_{26}H_{56}B_{10}P_2Pt$
For complete entry see 71.86

62.14 1,2.3,4.5,6 - tris(o,o′ - Biphenylylene)borazine
$C_{36}H_{24}B_3N_3$

P.J.Roberts, D.J.Brauer, Y.-H.Tsay, C.Kruger
Acta Crystallogr., Sect. B, **30,** 2673, 1974

SILICON COMPOUNDS

63.1 **Trimethylsilyl lithium**
$6C_3H_9Si^-$, Li_6^{6+}
T.F.Schaaf, W.Butler, M.D.Glick, J.P.Oliver
J. Am. Chem. Soc., **96,** 7593, 1974

63.C **(bis(Dimethyl - (dinitrogendisulfido)silyl)oxide) cobalt(ii)**
$C_4H_{12}CoN_4OS_4Si_2$
For complete entry see 83.5

63.2 **Di - μ - (thiodi - imide) - bis(dimethyl silicon)**
$C_4H_{12}N_4S_2Si_2$
G.Ertl, J.Weiss *Z. Naturforsch., Teil B*, **29,** 803, 1974

63.C **(bis(Dimethyl(dinitrogendisulfido)silyl) - ethylamine) nickel(ii)**
$C_6H_{17}N_5NiS_4Si_2$
For complete entry see 83.12

63.3 **Octamethyl - bicyclopentasiloxane**
$C_8H_{24}O_6Si_5$
G.Menczel, J.Kiss *Acta Crystallogr., Sect. B*, **31,** 1214, 1975

63.4 **9,10 - Disila - dihydroanthracene**
$C_{12}H_{12}Si_2$
O.A.D'yachenko, L.O.Atovmyan, S.V.Soboleva, T.Yu.Markova, N.G.Komalenkova, L.N.Shamshin, E.A.Chernyshev
Zh. Strukt. Khim., **15,** 170, 1974

63.5 **1 - Phenyl - silatrane (β form)**
$C_{12}H_{17}NO_3Si$
L.Parkanyi, K.Simon, J.Nagy *Acta Crystallogr., Sect. B*, **30,** 2328, 1974

63.6 **p - bis(Trimethylsilyl)benzene**
$C_{12}H_{22}Si_2$
G.Menczel, J.Kiss *Acta Crystallogr., Sect. B*, **31,** 1787, 1975

63.7 **tetrakis(Trimethylsilyl)tetrazene**
$C_{12}H_{36}N_4Si_4$
M.Veith *Acta Crystallogr., Sect. B*, **31,** 678, 1975

63.8 **9 - Sila - dihydroanthracene**
$C_{13}H_{12}Si$
O.A.D'yachenko, L.O.Atovmyan, S.V.Soboleva, T.Yu.Markova, N.G.Komalenkova, L.N.Shamshin, E.A.Chernyshev
Zh. Strukt. Khim., **15**, 170, 1974

63.9 **1,8 - bis(Trimethylsilyl)octatetrayne**
$C_{14}H_{18}Si_2$
B.F.Coles, P.B.Hitchcock, D.R.M.Walton
J. Chem. Soc., Dalton Trans., 442, 1975

63.10 **(Imino - bis(ethyleneoxy))diphenylsilane**
$C_{16}H_{19}NO_2Si$
J.J.Daly, F.Sanz *J. Chem. Soc., Dalton Trans.*, 2051, 1974

63.11 **9,9,10,10 - Tetramethyl - 9,10 - disiladihydroanthracene**
$C_{16}H_{20}Si_2$
O.A.D'yachenko, L.O.Atovmyan, S.V.Soboleva, T.Yu.Markova, N.G.Komalenkova, L.N.Shamshin, E.A.Chernyshev
Zh. Strukt. Khim., **15**, 667, 1974

63.C **μ - (Trimethylsilyl - cycloheptatrienyl) - pentacarbonyl - trimethylsilyl - di - ruthenium**
$C_{18}H_{24}O_5Ru_2Si_2$
For complete entry see 75.18

63.12 **Triphenyl - silicon isothiocyanate**
$C_{19}H_{15}NSSi$
G.M.Sheldrick, R.Taylor *J. Organomet. Chem.*, **87**, 145, 1975

63.C **Di(bis(trimethylsilyl)methyl) tin - pentacarbonyl - chromium**
$C_{19}H_{38}CrO_5Si_4Sn$
For complete entry see 69.17

63.13 **cis - 1,2,3 - Trimethyl - 1,2,3 - triphenyl - cyclotrisiloxane**
$C_{21}H_{24}O_3Si_3$
V.E.Shklover, N.G.Bokii, Yu.T.Struchkov *Eur. Cryst. Meeting*, 463, 1974

63.14 **trans - 1,2,3 - Trimethyl - 1,2,3 - triphenyl - cyclotrisiloxane**
$C_{21}H_{24}O_3Si_3$
V.E.Shklover, N.G.Bokii, Yu.T.Struchkov, K.A.Andrianov, B.G.Zavin, V.S.Svistunov *Zh. Strukt. Khim.*, **15**, 90, 1974

63.15 **2 - Triphenylsilyl - 1,3 - dithiane - 1 - oxide**
$C_{22}H_{22}OS_2Si$
F.A.Carey, O.Hernandez, I.C.Taylor Junior, R.F.Bryan
Am. Chem. Soc., Abstr. Papers, 261, 1974
Also classified in 39

63.C **η - (1,5 - bis - Trimethylsilyl)pentalene - 1,1,2,2,2,3,3,3 - octacarbonyl - triangulo - triruthenium**
$C_{22}H_{22}O_8Ru_3Si_2$
For complete entry see 75.25

63.16 **2 - Triphenylsilyl - 2 - methyl - 1,3 - dithiane**
$C_{23}H_{24}S_2Si$
R.F.Bryan, R.J.Maher, P.M.Smith, I.F.Taylor Junior
Acta Crystallogr., Sect. A, **31,** S166, 1975
Also classified in 39

63.17 **Tetraphenyl - silane**
$C_{24}H_{20}Si$
L.Parkanyi, K.Sasvari *Chem. Eng. News,* **17,** 271, 1973

63.C **(π - Cyclopentadienyl) - (trans - diphenyl - di(trimethylsilyl) - cyclobutadiene) - cobalt**
$C_{27}H_{33}CoSi_2$
For complete entry see 73.58

63.18 **1,1,2,2 - Tetramethyl - 3,3,4,4 - tetraphenylcyclotetrasiloxane**
$C_{28}H_{32}O_4Si_4$
V.E.Shklover, A.E.Kalinin, A.I.Gusev, N.G.Bokii, Yu.T.Struchkov, K.A.Andrianov, I.M.Petrova *Zh. Strukt. Khim.,* **14,** 692, 1973

63.C **(8 - Triphenylmethyl - 6 - trimethylsilyl - bicyclo(4.2.0)octa - 2,4,7 - triene) tricarbonyl iron**
$C_{33}H_{30}FeO_3Si$
For complete entry see 75.38

63.19 **bis(Trimethylsilyloxy) - (phthalocyanine) - silicon**
$C_{38}H_{34}N_8O_2Si_3$
K.Knox, C.Choy, M.E.Kenney, J.R.Mooney
Am. Cryst. Assoc., Abstr. Papers (Summer Meeting), 224, 1974
Also classified in 49

63.20 **tris(Methyldiphenylsilylmethyl)amine**
$C_{42}H_{45}NSi_3$
J.J.Daly, F.Sanz *Acta Crystallogr., Sect. B,* **30,** 2766, 1974
Also classified in 3

63.21 **Tetra(diphenylketimino) silicon**
$C_{52}H_{40}N_4Si$
N.W.Alcock, M.Pierce-Butler, G.R.Willey, K.Wade
J. Chem. Soc., Chem. Commun., 183, 1975

PHOSPHORUS COMPOUNDS

64.1 **Methylene - bis(phosphonic dichloride)**
$CH_2Cl_4O_2P_2$
W.S.Sheldrick *J. Chem. Soc., Dalton Trans.*, 943, 1975

64.2 **Aminomethyl - phosphonic acid (β form)**
CH_6NO_3P
M.Darriet, J.Darriet, A.Cassaigne, E.Neuzil
Acta Crystallogr., Sect. B, **31,** 469, 1975
Also classified in 3

64.3 **Ammonium bis(methylphospho)pentamolybdate pentahydrate**
$C_2H_6Mo_5O_{21}P_2^{4-}$, $4H_4N^+$, $5H_2O$
J.K.Stalick, C.O.Quicksall
Am. Cryst. Assoc., Abstr. Papers (Spring Meeting), 35, 1975

64.4 **Trimethylphosphine - boron tribromide**
$C_3H_9BBr_3P$
D.L.Black, R.C.Taylor *Acta Crystallogr., Sect. B*, **31,** 1116, 1975

64.5 **Trimethylphosphine - boron trichloride**
$C_3H_9BCl_3P$
D.L.Black, R.C.Taylor *Acta Crystallogr., Sect. B*, **31,** 1116, 1975

64.6 **Trimethylphosphine - boron tri - iodide**
$C_3H_9BI_3P$
D.L.Black, R.C.Taylor *Acta Crystallogr., Sect. B*, **31,** 1116, 1975

64.7 **1,3 - Dimethyl - 2,4 - bis(trichloromethyl) - 2,2,4,4 - tetrafluoro - 1,3 - diaza - 2,4 - diphosphetidine**
$C_4H_6Cl_6F_4N_2P_2$
W.S.Sheldrick, M.J.C.Hewson *Acta Crystallogr., Sect. B*, **31,** 1209, 1975

64.C **Diethyldithiophosphinato thallium(i)**
$C_4H_{10}PS_2Tl$
For complete entry see 68.5

64.8 **Sodium tetramethylammonium bis(2 - aminoethylphospho)pentamolybdate pentahydrate**
$C_4H_{14}Mo_5N_2O_{21}P_2^{2-}$, $C_4H_{12}N^+$, Na^+ , $5H_2O$
J.K.Stalick, C.O.Quicksall
Am. Cryst. Assoc., Abstr. Papers (Spring Meeting), 35, 1975
Residue 2 classified in 3

64.9 **1a - Methyl - 1e - thiono - phosphorinan - 4e - ol**
$C_6H_{13}OPS$
L.D.Quin, A.T.McPhail, S.O.Lee, K.D.Onan
Tetrahedron Lett., 3473, 1974

64.10 **1(e,a) - Methyl - 1(a,e) - thiono - phosphorinan - 4(e,a) - ol (2 - 1 mixture of conformers)**
$C_6H_{13}OPS$
L.D.Quin, A.T.McPhail, S.O.Lee, K.D.Onan
Tetrahedron Lett., 3473, 1974

64.11 **2,5,5 - Trimethyl - 2 - thiono - 1,3,2 - dioxaphosphorinane**
$C_6H_{13}O_2PS$
J.P.Dutasta, A.Grand, J.B.Robert *Tetrahedron Lett.*, 2655, 1974

64.12 **(S) - (−) - α - Phenylethylammonium(R) - (−) - O - 2 - butyl - (S) - (−) - ethylphosphonothioate**
$C_6H_{14}O_2PS^-$, $C_8H_{12}N^+$
G.H.Y.Lin, D.A.Wustner, T.R.Fukuto, R.M.Wing
J. Agric. Food Chem., **22,** 1134, 1974
Residue 2 classified in 3

64.13 **trans - 2,4,6 - Trichloro - 2,4,6 - tris(dimethylamino) - cyclotriphosphazatriene**
$C_6H_{18}Cl_3N_6P_3$
F.R.Ahmed, E.J.Gabe *Acta Crystallogr., Sect. B,* **31,** 1028, 1975

64.14 **Hexamethylcyclotriphosphazene - iodine**
$C_6H_{18}N_3P_3$, I_2
P.L.Markila, J.Trotter *Can. J. Chem.*, **52,** 2197, 1974

64.15 **N,N - bis(2 - Chloroethyl) - N′ - O - propylene phosphoric acid ester diamide monohydrate**
$C_7H_{15}Cl_2N_2O_2P$, H_2O
J.C.Clardy, J.A.Mosbo, J.G.Verkade *Phosphorus,* **4,** 151, 1974

64.16 **1a,4e - Dimethyl - 1e - thiono - phosphorinan - 4a - ol**
$C_7H_{15}OPS$
L.D.Quin, A.T.McPhail, S.O.Lee, K.D.Onan
Tetrahedron Lett., 3473, 1974

64.17 **1e,4e - Dimethyl - 1a - thiono - phosphorinan - 4a - ol**
$C_7H_{15}OPS$
L.D.Quin, A.T.McPhail, S.O.Lee, K.D.Onan
Tetrahedron Lett., 3473, 1974

64.18 **1,3 - Di - t - butyl - 2 - trans - 4 - dichloro - 2,4 - dioxocyclodiphosphazane**
$C_8H_{18}Cl_2N_2O_2P_2$
L.Manojlovic-Muir, K.W.Muir *J. Chem. Soc., Dalton Trans.*, 2395, 1974

64.19 **1,3 - Di - t - butyl - 2,4 - dichlorodiazadiphosphetidine**
$C_8H_{18}Cl_2N_2P_2$
K.W.Muir *J. Chem. Soc., Dalton Trans.*, 259, 1975

64.20 **cis - 2 - t - Butylamino - 2 - seleno - 4 - methyl - 1,3,2 - dioxaphosphorinane**
$C_8H_{18}NO_2PSe$
T.J.Bartczak, A.Christensen, R.Kinas, W.J.Stec
Acta Crystallogr., Sect. A, **31,** S105, 1975

64.21 **trans - 2 - t - Butylamino - 2 - seleno - 4 - methyl - 1,3,2 - dioxaphosphorinane**
$C_8H_{18}NO_2PSe$
T.J.Bartczak, R.Kinas, W.J.Stec
Acta Crystallogr., Sect. A, **31,** S105, 1975

64.22 **2,4,6,7 - Tetramethyl - 2,6,7 - triaza - 1 - phosphabicyclo(2.2.2)octane - 1 - oxide**
$C_8H_{18}N_3OP$
J.C.Clardy, R.L.Kolpa, J.G.Verkade *Phosphorus,* **4,** 133, 1974

64.C **bis(Dimethylene - dimethylphosphonium) - dicopper(i)**
$C_8H_{20}Cu_2P_2$
For complete entry see 71.7

64.23 **2,4,6,7 - Tetramethyl - 2,6,7 - triaza - 1 - phosphabicyclo(2.2.2)octane - 1 - borane**
$C_8H_{21}BN_3P$
J.C.Clardy, R.L.Kolpa, J.G.Verkade *Phosphorus,* **4,** 133, 1974
Also classified in 62

64.24 **2 - Methyl - 8 - quinolinolato - (tetrafluorophosphorane)**
$C_{10}H_8F_4NOP$
K.-P.John, R.Schmutzler, W.S.Sheldrick
J. Chem. Soc., Dalton Trans., 1841, 1974
Also classified in 35

64.25 **bis - (5,5 - Dimethyl - 2 - thiono - 1,3,2 - dioxophosphorinanyl)**
$C_{10}H_{20}O_4P_2S_2$
Z.Galdecki, J.Karolak-Wojciechowska
Acta Crystallogr., Sect. A, **31,** S105, 1975

64.26 **P - Oxo - P′ - thiono - bis(5,5 - dimethyl - 1,3,2 - dioxaphosphorinanyl)**
$C_{10}H_{20}O_5P_2S$
Z.Galdecki, J.Karolak-Wojciechowska
Acta Crystallogr., Sect. A, **31,** S105, 1975

64.27 **bis - (5,5 - Dimethyl - 2 - oxo - 1,3,2 - dioxaphosphorinanyl)**
$C_{10}H_{20}O_6P_2$
Z.Galdecki, J.Karolak-Wojciechowska
Acta Crystallogr., Sect. A, **31,** S105, 1975

64.28 **bis(5,5 - Dimethyl - 2 - oxo - 1,3,2 - P - dioxaphosphorinanyl) sulfide**
$C_{10}H_{20}O_6P_2S$
M.Bukowska-Strzyzewska, W.Dobrowolska, J.Skoweranda
Acta Crystallogr., Sect. A, **31,** S106, 1975

64.29 **bis(5,5 - Dimethyl - 2 - oxo - 1,3,2 - P - dioxaphosphorinanyl) oxide**
$C_{10}H_{20}O_7P_2$
M.Bukowska-Strzyzewska, W.Dobrowolska, J.Skoweranda
Acta Crystallogr., Sect. A, **31,** S106, 1975
Also classified in 46

64.30 **2 - Cyanoethyl - 4 - phenyl - 1,2,3 - phosphadiazole**
$C_{11}H_{10}N_3P$
V.G.Andrianov, Yu.T.Struchkov, N.I.Shvetsov-Shilovskii, N.P.Ignatova, R.G.Bobkova, V.V.Mel′nikov *Dokl. Akad. Nauk SSSR,* **211,** 1101, 1973

64.31 **1 - Methyl - 4 - phenyl - perhydro - 1,4 - azaphosphorine 4 - sulfide**
$C_{11}H_{16}NPS$
B.M.Gatehouse, B.K.Miskin *Acta Crystallogr., Sect. B,* **30,** 2112, 1974

64.32 **(3RS,PSR) - 3 - (Methylphenyl - phosphinoyl) - 2 - methyl - but - 1 - ene**
$C_{12}H_{17}OP$
F.H.Allen, O.Kennard, L.R.Nassimbeni, R.Shepherd, S.Warren
J. Chem. Soc., Perkin Trans. 2, 1530, 1974

64.33 **(2RS,PSR) - 3 - (Methylphenyl - phosphinoyl) - 3 - methylbutan - 2 - ol**
$C_{12}H_{19}O_2P$
F.H.Allen, O.Kennard, L.R.Nassimbeni, R.Shepherd, S.Warren
J. Chem. Soc., Perkin Trans. 2, 1530, 1974

64.34 **2,2 - Dichloro - 4,4,6,6 - tetra(isopropylamino)cyclotriphosphazene**
$C_{12}H_{32}Cl_2N_7P_3$
W.Polder, A.J.Wagner *Acta Crystallogr., Sect. A,* **31,** S63, 1975

64.35 **2,2 - Dichloro - 4,4,6,6 - tetra(isopropylamino)cyclotriphosphazene hydrochloride**
$C_{12}H_{33}Cl_2N_7P_3^+$, Cl^-
W.Polder, A.J.Wagner *Acta Crystallogr., Sect. A,* **31,** S63, 1975

64.36 **bis(hexakis(Dimethylamino) cyclotriphosphonitrilium) tetrachlorocobaltate(ii)**
$2C_{12}H_{37}N_9P_3^+$, Cl_4Co^{2-}
A.L.Macdonald, J.Trotter *Can. J. Chem.*, **52,** 734, 1974

64.C **π - Cyclopentadienyl - (1 - hydroxy - 2,3,4,5 - tetrakis(trifluoromethyl) - phosphole - 1 - oxide) cobalt**
$C_{13}H_6CoF_{12}O_2P$
For complete entry see 73.18

64.37 **P - Phenyl - 2,2,5,5 - tetrakis(trifluoromethyl)1,3,4 - dioxaphospholan - spiro - (1′,3′,2′ - oxathiaphosphorane)**
$C_{14}H_9F_{12}O_3PS$
E.Duff, D.R.Russell, S.Trippett *Phosphorus,* **4,** 203, 1974

64.38 **2 - Phenyl - 1 - phospha - naphthalene**
2 - Phenyl - phosphinoline
$C_{15}H_{11}P$
J.J.Daly, F.Sanz *J. Chem. Soc., Dalton Trans.*, 2388, 1974

64.39 **Dimethyl 2,5 - dibromo - 7 - phenyl - norcaradiene - 7 - phosphonate**
$C_{15}H_{15}Br_2O_3P$
G.Maas, K.Fischer, M.Regitz *Acta Crystallogr., Sect. B,* **30,** 2853, 1974
Also classified in 27

64.40 **Dimethyl 2,5 - dichloro - 7 - phenyl - norcaradiene - 7 - phosphonate**
$C_{15}H_{15}Cl_2O_3P$
G.Maas, K.Fischer, M.Regitz *Acta Crystallogr., Sect. B,* **30,** 2853, 1974
Also classified in 27

64.41 **trans - Methyl - meso - hydrobenzoin phosphite**
$C_{15}H_{15}O_3P$
M.G.Newton, B.S.Campbell *J. Am. Chem. Soc.*, **96,** 7790, 1974

64.42 **2 - Methyl - 8 - (trifluoro - phenyl - phosphoroxy) - quinoline**
$C_{16}H_{13}F_3NOP$
K.-P.John, R.Schmutzler, W.S.Sheldrick
J. Chem. Soc., Dalton Trans., 2466, 1974
Also classified in 35

64.43 **1 - (2′ - (N - Benzamidinato) - 1′,1′,1′,3′,3′,3′ - hexafluoro - prop - 2 - yl) - 1 - phospha - 2,8,9 - trioxa - adamantane benzene solvate**
$C_{16}H_{14}F_6NO_4P$, $0.5C_6H_6$
H.L.Carrell, H.M.Berman, J.S.Ricci Junior, W.C.Hamilton, F.Ramirez, J.F.Marecek, L.Kramer, I.Ugi *J. Am. Chem. Soc.*, **97,** 38, 1975
Residue 1 also classified in 31

64.44 **tris(Pentafluorophenyl)difluorophosphorane**
$C_{18}F_{17}P$
W.S.Sheldrick *Acta Crystallogr., Sect. B,* **31,** 1776, 1975

64.45 **S - 4 - Nitrophenyl O,O - diphenyl thiophosphate**
$C_{18}H_{14}NO_5PS$
R.Gitany, R.S.McEwen *J. Chem. Soc., Perkin Trans. 2,* 57, 1975
Also classified in 15

64.46 **1 - ((Triphenylphosphoranylidene)amino) - 1,3,5 - trithia - 2,4,6 - triazine**
$C_{18}H_{15}N_4PS_3$
E.M.Holt, S.L.Holt *J. Chem. Soc., Dalton Trans.,* 1990, 1974

64.C **Dimethylmalonic acid - bis(triphenylphosphine oxide)**
$2C_{18}H_{15}OP$, $C_5H_8O_4$
For complete entry see 60.2

64.47 **12 - Ethoxy - 2,3 - benzo - 6,5 - naphtho(b) - (7,12) - thiaphosphorin - 7,7,12 - trioxide**
$C_{18}H_{15}O_4PS$
N.N.Dhaneshwar, A.G.Kulkarni, S.S.Tavale, L.M.Pant
Acta Crystallogr., Sect. B, **31,** 750, 1975

64.48 **Di - n - nonylphosphinic acid**
$C_{18}H_{39}O_2P$
P.Bello *Gazz. Chim. Ital.,* **103,** 537, 1973

64.49 **1 - Benzyl - 2 - phenyl - 3 - hydroxy - 4,5 - dimethyl - phosphol - 2 - ene 1 - oxide**
$C_{19}H_{21}O_2P$
D.M.Washecheck, D.van der Helm, W.R.Purdum, K.D.Berlin
J. Org. Chem., **39,** 3305, 1974

64.50 **trans - 1,6 - Diphenyl - 1,6 - diphosphacyclodecan - 1,6 - dione dihydrate**
$C_{20}H_{26}O_2P_2$, $2H_2O$
M.Drager *Chem. Ber.,* **107,** 3246, 1974

64.51 **(−) - Ephedrine dioxadiazaspirophosphorane**
$C_{20}H_{27}N_2O_2P$
M.G.Newton, J.E.Collier, R.Wolf *J. Am. Chem. Soc.,* **96,** 6888, 1974

64.52 **(−) - 5,6,7,8 - Tetrahydro - 4 - (methylthio) - 6 - phenyl - 6 - benzylphosphorinia(4,3 - d)pyrimidine bromide (absolute configuration)**
$C_{21}H_{22}N_2PS^+$, Br^-
S.R.Holbrook, D.van der Helm, M.Poling
Am. Cryst. Assoc., Abstr. Papers (Summer Meeting), 240, 1974
Residue 1 also classified in 35

64.53 **1 - Ethyl - 1,2,3,4 - tetrahydro - 1 - phenyl - benzo(h)phosphinolinium bromide**
$C_{21}H_{22}P^+$, Br^-
S.R.Holbrook, D.van der Helm, M.Poling
Am. Cryst. Assoc., Abstr. Papers (Summer Meeting), 240, 1974

64.C **bis(η - Cyclopentadienyl) - σ - ((2 - dimethylphenylphosphonium)ethyl)methyl - tungsten hexafluorophosphate**
$C_{21}H_{28}PW^{+}$, F_6P^{-}
For complete entry see 71.74

64.54 **1,3 - bis(p - Tolyl) - 2 - phenyl - 4,5 - dihydro - 1,3 - diaza - 2 - phospholidine**
$C_{22}H_{23}N_2P$
J.C.Clardy, R.L.Kolpa, J.G.Verkade, J.J.Zuckerman
Phosphorus, **4,** 145, 1974

64.55 **2,cis - 4,trans - 6,trans - 8 - Tetrachloro - 2,4,6,8 - tetraphenyl - cyclotetraphosphazatetraene**
$C_{24}H_{20}Cl_4N_4P_4$
A.H.Burr, C.H.Carlisle, G.J.Bullen
J. Chem. Soc., Dalton Trans., 1659, 1974

64.56 **2,2,6,6 - Tetrachloro - 4,4,8,8 - tetraphenyl - cyclotetraphosphazatetraene**
$C_{24}H_{20}Cl_4N_4P_4$
G.J.Bullen, P.E.Dann *Acta Crystallogr., Sect. B,* **30,** 2861, 1974

64.57 **Tetraphenylphosphonium tri(thiocyanato) mercury(ii)**
$C_{24}H_{20}P^{+}$, $C_3HgN_3S_3^{-}$
A.Sakhri, A.L.Beauchamp *Inorg. Chem.*, **14,** 740, 1975

64.58 **Tetraphenylphosphonium tetra(thiocyanato) mercury(ii)**
$2C_{24}H_{20}P^{+}$, $C_4HgN_4S_4^{2-}$
A.Sakhri, A.L.Beauchamp *Acta Crystallogr., Sect. B,* **31,** 409, 1975

64.C **tetrakis(Tetraphenylphosphonium) hexakis(1,2 - dithiosquarato) octa - copper(i)**
$4C_{24}H_{20}P^{+}$, $C_{24}Cu_8O_{12}S_{12}^{4-}$
For complete entry see 85.40

64.59 **2,2 - Di(ethoxy)vinylidene - triphenylphosphorane**
$C_{24}H_{25}O_2P$
H.Burzlaff, U.Voll, H.-J.Bestmann *Chem. Ber.*, **107,** 1949, 1974

64.60 **1, - Methyl - 3,3,5,5 - tetraphenyl - 1 - thia - 3,5 - diphospha - 2,6 - diazine**
$C_{26}H_{24}N_2P_2S$
J.Weiss *Acta Crystallogr., Sect. B,* **30,** 2888, 1974

64.61 **Phenyl - dimenthyl - phosphine**
$C_{26}H_{43}P$
C.Kruger, P.J.Roberts *Cryst. Struct. Commun.*, **3,** 707, 1974

64.62 **Trimesityl - phosphine**
$C_{27}H_{33}P$
J.F.Blount, C.A.Maryanoff, K.Mislow *Tetrahedron Lett.*, 913, 1975

64.C **tris(1,8 - Naphthalenedioxy)cyclotriphosphazene p - xylene clathrate**
$C_{30}H_{18}N_3O_6P_3$, $0.5C_8H_{10}$
For complete entry see 61.2

64.63 **bis(Triphenylphosphine)imminium hexacarbonyl - vanadium**
$C_{36}H_{30}NP_2^+$, $C_6O_6V^-$
R.D.Wilson, R.Bau *J. Am. Chem. Soc.*, **96,** 7601, 1974

64.64 **bis(Triphenylphosphine) - imminium cobalt - iron - octacarbonyl**
$C_{36}H_{30}NP_2^+$, $C_8CoFeO_8^-$
H.B.Chin, M.B.Smith, R.D.Wilson, R.Bau
J. Am. Chem. Soc., **96,** 5285, 1974

64.65 **bis(Triphenylphosphine) - imminium di - iron - octacarbonyl acetonitrile solvate**
$2C_{36}H_{30}NP_2^+$, $C_8Fe_2O_8^{2-}$, $2C_2H_3N$
H.B.Chin, M.B.Smith, R.D.Wilson, R.Bau
J. Am. Chem. Soc., **96,** 5285, 1974

64.C **Di(bis(triphenylphosphine)imminium) octachloro - di - niobium acetonitrile adduct chlorobenzene solvate**
$2C_{36}H_{30}NP_2^+$, $C_8H_{12}Cl_8N_4Nb_2^{2-}$, $2C_6H_5Cl$
For complete entry see 83.18

64.66 **Octaphenylcyclotetraphosphazene**
$C_{48}H_{40}N_4P_4$
M.J.Begley, D.B.Sowerby, R.J.Tillott
J. Chem. Soc., Dalton Trans., 2527, 1974

ARSENIC COMPOUNDS

65.1 **bis(Methyl) - thiourea - S - arsenic(iii) chloride - thiourea**
$C_3H_{10}AsN_2S^+$, CH_4N_2S , Cl^-
P.H.Javora, R.A.Zingaro, E.A.Meyers *Cryst. Struct. Commun.*, **4,** 67, 1975
Residue 2 classified in 8

65.2 **5 - Chloro - 1 - oxa - 4,6 - dithia - 5 - arsaocane**
$C_4H_8AsClOS_2$
M.Drager *Z. Anorg. Allg. Chem.*, **411,** 79, 1975

65.3 **2 - Chloro - 1,3,6,2 - trithia - arsa - ocane**
$C_4H_8AsClS_3$
M.Drager *Chem. Ber.*, **107,** 2601, 1974

65.4 **Tetraphenylarsonium chlorohydroxotellurate(iv) monohydrate**
$C_{24}H_{20}As^+$, HCl_4OTe^- , H_2O
P.H.Collins, M.Webster *J. Chem. Soc., Dalton Trans.*, 1545, 1974

65.C **Tetraphenylarsonium bis(η^3 - cyclo - octatetraene) - (η^4 - cyclo - octatetraene) niobium**
$C_{24}H_{20}As^+$, $C_{24}H_{24}Nb^-$
For complete entry see 75.29

65.5 **bis(Tetraphenylarsonium) decachloro - ditellurium**
$2C_{24}H_{20}As^+$, $Cl_{10}Te_2^{2-}$
B.Krebs, V.Paulat *Eur. Cryst. Meeting*, 238, 1974

65.6 **bis(Tetraphenylarsonium) tetranitratocuprate(ii) methylene chloride solvate**
$2C_{24}H_{20}As^+$, $CuN_4O_{12}^{2-}$, CH_2Cl_2
T.J.King, A.Morris *Inorg. Nucl. Chem. Lett.*, **10,** 237, 1974

65.C **bis(Tetraphenylarsonium) uranyl - bis(pyridine - 2,6 - dicarboxylate) hexahydrate**
$2C_{24}H_{20}As^+$, $C_{14}H_6N_2O_{10}U^{2-}$, $6H_2O$
For complete entry see 81.34

65.7 **Octaphenylcyclotetra - arsazene**
$C_{48}H_{40}As_4N_4$
M.J.Begley, D.B.Sowerby, R.J.Tillott
J. Chem. Soc., Dalton Trans., 2527, 1974

65.C **Tri - iodo - bismuth - tri - μ - iodo - tris(triphenylarsine oxide) bismuth(iii)**
$C_{54}H_{45}As_3Bi_2I_6O_3$
For complete entry see 66.12

ANTIMONY AND BISMUTH COMPOUNDS

66.C **Di - μ - chloro - bis(chloro - tri(thiourea) - bismuth(iii)) pentachloro - thiourea - bismuth(iii)**
$CH_4BiCl_5N_2S^{2-}$, $C_6H_{24}Bi_2Cl_4N_{12}S_6^{2+}$
For complete entry see 66.4

66.1 **μ - Oxo - bis(trimethyl - chloro - antimony(v))**
$C_6H_{18}Cl_2O_2Sb_2$
G.Ferguson, F.C.March, D.R.Ridley
Acta Crystallogr., Sect. B, **31,** 1260, 1975

66.2 **μ - Oxo - bis(trimethyl - azido - antimony(v))**
$C_6H_{18}N_6OSb_2$
G.Ferguson, F.C.March, D.R.Ridley
Acta Crystallogr., Sect. B, **31,** 1260, 1975

66.3 **μ - Oxo - bis(trimethyl - antimony(v)) perchlorate**
$C_6H_{18}OSb_2^{2+}$, $2ClO_4^-$
G.Ferguson, F.C.March, D.R.Ridley
Acta Crystallogr., Sect. B, **31,** 1260, 1975

66.4 **Di - μ - chloro - bis(chloro - tri(thiourea) - bismuth(iii)) pentachloro - thiourea - bismuth(iii)**
$C_6H_{24}Bi_2Cl_4N_{12}S_6^{2+}$, $CH_4BiCl_5N_2S^{2-}$
L.P.Battaglia, A.B.Corradi, G.Pelizzi, M.E.V.Tani
Cryst. Struct. Commun., **4,** 399, 1975
Residue 2 classified in 66

66.5 **Potassium antimony tartrate trihydrate (absolute configuration)**
$C_8H_4O_{12}Sb_2^{2-}$, $2K^+$, $3H_2O$
M.E.Gress, R.A.Jacobson *Inorg. Chim. Acta,* **8,** 209, 1974

66.6 **Bismuth(iii) diethyldithiocarbamate**
$C_{15}H_{30}BiN_3S_6$
J.A.Howard, D.R.Russell, W.Scutcher
Acta Crystallogr., Sect. A, **31,** S141, 1975

66.7 **Antimony(ii) diethyldithiocarbamate**
$C_{15}H_{30}N_3S_6Sb$
J.A.Howard, D.R.Russell, W.Scutcher
Acta Crystallogr., Sect. A, **31,** S141, 1975

66.C **Tetracarbonyl - (triphenylstibine) - iron**
$C_{22}H_{15}FeO_4Sb$
For complete entry see 86.39

66.8 **Di - μ - chloro - bis(dichlorodiphenyl antimony)**
$C_{24}H_{20}Cl_6Sb_2$
J.Bordner, G.O.Doak, J.R.Peters Junior
J. Am. Chem. Soc., **96,** 6763, 1974

66.9 **Tetraphenyl antimony acetic nitrosolate**
$C_{26}H_{23}N_2O_2Sb$
J.Kopf, G.Vetter, G.Klar *Z. Anorg. Allg. Chem.*, **409,** 285, 1974

66.10 **8 - Mercapto - quinolinato - antimony**
$C_{27}H_{18}N_3S_3Sb$
L.Pech, J.Ozols, A.Ievins
Latv. PSR Zinat. Akad. Vestis, Kim. Ser., 259, 1973

66.11 **Pentaphenyl - antimony cyclohexane solvate**
$C_{30}H_{25}Sb$, $0.5C_6H_{12}$
C.Brabant, B.Blanck, A.L.Beauchamp *J. Organomet. Chem.*, **82,** 231, 1974

66.12 **Tri - iodo - bismuth - tri - μ - iodo - tris(triphenylarsine oxide) bismuth(iii)**
$C_{54}H_{45}As_3Bi_2I_6O_3$
F.Lazarini, L.Golic, G.Pelizzi *Acta Crystallogr., Sect. A*, **31,** S140, 1975
Also classified in 65

GROUPS IA AND IIA COMPOUNDS

67.1 **Di - cyclopentadienyl calcium**
$2C_5H_5^-$, Ca^{2+}
R.Zerger, G.Stucky *J. Organomet. Chem.*, **80,** 7, 1974

67.2 **Diaquo - bis(L - serinato) calcium**
$C_6H_{16}N_2O_8^{2-}$, nCa^{2+}
A.Sicignano, S.Gandhi, K.Eriks
Am. Cryst. Assoc., Abstr. Papers (Summer Meeting), 227, 1974
Residue 1 also classified in 48

67.3 **Magnesium - methoxy - chloride methanol solvate**
$C_6H_{18}Mg_4O_6^{2+}$, $2Cl^-$, $10CH_4O$
R.I.Bochkova, V.P.Golovachev, E.P.Turevskaya, N.V.Belov
Dokl. Akad. Nauk SSSR, **189,** 1246, 1969

67.4 **Acetylacetonato - beryllium**
$C_{10}H_{14}BeO_4$
J.M.Stewart, B.Morosin *Acta Crystallogr., Sect. B,* **31,** 1164, 1975

67.5 **Ethylenediaminetetra - acetato - (aquo) - magnesium hexa - aquo - magnesium dihydrate**
$C_{10}H_{14}MgN_2O_9^{2-}$, $H_{12}MgO_6^{2+}$, $2H_2O$
A.I.Pozhidaev, T.N.Polynova, M.A.Porai-Koshits, V.A.Logvinenko
Zh. Strukt. Khim., **14,** 746, 1973

67.C **Tri - lithium hexamethyl chromium tri(dioxan)**
$(C_{18}H_{42}CrLi_3O_6)_n$
For complete entry see 71.57

67.6 **Fluorenyl - potassium - tetramethylethylenediamine**
$(C_{19}H_{25}N_2^-)_n$, nK^+
R.Zerger, W.Rhine, G.D.Stucky *J. Am. Chem. Soc.*, **96,** 5441, 1974
Residue 1 also classified in 28, 3

67.7 **Di - μ - bromo - bis(bis(di - isopropyl ether) - ethyl magnesium)**
$C_{28}H_{66}Br_2Mg_2O_4$
A.L.Spek, P.Voorbergen, G.Schat, C.Blomberg, F.Bickelhaupt
J. Organomet. Chem., **77,** 147, 1974

67.8 **Cyclohexyl - lithium hexamer benzene solvate**
$C_{36}H_{66}Li_6$, C_6H_6
R.Zerger, W.Rhine, G.Stucky *J. Am. Chem. Soc.*, **96,** 6048, 1974

67.C **bis - μ - (bis - (η - Cyclopentadienyl)hydride molybdenum) - bis(di - μ - bromo - cyclohexyl magnesium)((diethyl ether magnesium))**
$C_{40}H_{64}Br_4Mg_4Mo_2O_2$
For complete entry see 73.67

GROUP III COMPOUNDS

68.1 **Potassium methyl - trichloro - aluminate**
$CH_3AlCl_3^-$, K^+
J.L.Atwood, D.C.Hrncir, W.R.Newberry III
Cryst. Struct. Commun., **3,** 615, 1974

68.2 **Dimethylthallium chloride**
$C_2H_6Tl^+$, Cl^-
H.-D.Hausen, E.Veigel, H.-J.Guder *Z. Naturforsch., Teil B,* **29,** 269, 1974

68.3 **Cesium azido - trimethyl - aluminium**
$C_3H_9AlN_3^-$, Cs^+
J.L.Atwood, W.R.Newberry III *J. Organomet. Chem.*, **87,** 1, 1975

68.4 **Potassium hydrido - trimethyl - aluminate**
$C_3H_{10}Al^-$, K^+
G.Hencken, E.Weiss *J. Organomet. Chem.*, **73,** 35, 1974

68.5 **Diethyldithiophosphinato thallium(i)**
$C_4H_{10}PS_2Tl$
S.Esperas, S.Husebye *Acta Chem. Scand. Ser. A,* **28,** 1015, 1974
Also classified in 64

68.6 **Di - μ - chloro - bis(dimethyl indium)**
$C_4H_{12}Cl_2In_2$
H.D.Hausen, K.Mertz, E.Veigel, J.Weidlein
Z. Anorg. Allg. Chem., **410,** 156, 1974

68.7 **bis(Dimethylgallium) - oxalate**
$C_6H_{12}Ga_2O_4$
H.D.Hausen, K.Mertz, J.Weidlein *J. Organomet. Chem.*, **67,** 7, 1974

68.8 **Potassium bis(trimethyl - aluminium)fluoride benzene solvate**
$C_6H_{18}Al_2F^-$, K^+ , C_6H_6
J.L.Atwood, W.R.Newberry III *J. Organomet. Chem.*, **66,** 15, 1974

68.9 **Potassium bis(trimethyl - aluminium)azide**
$C_6H_{18}Al_2N_3^-$, K^+
J.L.Atwood, W.R.Newberry III *J. Organomet. Chem.*, **66,** 145, 1974

68.10 **1,2,3,4 - Tetramethyl - 1 - trichloroaluminium - cyclobutadiene**
$C_8H_{12}AlCl_3$
C.Kruger, P.J.Roberts, Y.-H.Tsay, J.B.Koster
J. Organomet. Chem., **78,** 69, 1974

68.11 **N,N - Dimethylethanolamine - gallane dimer**
$C_8H_{24}Ga_2N_2O_2$
S.J.Rettig, A.Storr, J.Trotter *Can. J. Chem.*, **53,** 58, 1975

68.12 **Di - (μ - isopropylideneneamino) - bis(dimethyl - aluminium)**
$C_{10}H_{24}Al_2N_2$
S.K.Seale, J.L.Atwood *J. Organomet. Chem.*, **73,** 27, 1974

68.13 **N - Methyldiethanolamino - gallane dimer**
$C_{10}H_{24}Ga_2N_2O_4$
S.J.Rettig, A.Storr, J.Trotter *Can. J. Chem.*, **52,** 2206, 1974

68.14 **N,N - Dimethylethanolamine - dimethyl - gallium dimer**
$C_{12}H_{32}Ga_2N_2O_2$
S.J.Rettig, A.Storr, J.Trotter *Can. J. Chem.*, **53,** 58, 1975

68.15 **Dichloro - (acetylacetonato) - 2,2′ - bipyridyl indium(iii)**
$C_{15}H_{15}Cl_2InN_2O_2$
J.G.Contreras, F.W.B.Einstein, D.G.Tuck *Can. J. Chem.*, **52,** 3793, 1974

68.16 **Lithium pentakis(isopropylamido) - tris(hydrido aluminium) - tris(dihydrido aluminium) diethyl ether solvate**
$C_{15}H_{44}Al_6N_5^-$, Li^+ , $C_4H_{10}O$
M.Cesari, G.Perego, G.del Piero, M.Corbellini, A.Immirzi
J. Organomet. Chem., **87,** 43, 1975

68.17 **bis(Isopropylamido - hydridoaluminium) - tris(isopropylamino - dihydridoaluminium)**
$C_{15}H_{46}Al_5N_5$
G.Perego, G.del Piero, M.Cesari, A.Zazzetta, G.Dozzi
J. Organomet. Chem., **87,** 53, 1975

68.C **bis(Dimethyl - bis(1 - pyrazolyl) - gallato) copper(ii)**
$C_{16}H_{24}CuGa_2N_8$
For complete entry see 83.72

68.C **bis(Dimethyl - bis(pyrazol - 1 - yl)gallato) nickel(ii)**
$C_{16}H_{24}Ga_2N_8Ni$
For complete entry see 83.73

68.18 **Hexa(hydrido - isopropylamido - aluminium)**
$C_{18}H_{48}Al_6N_6$
M.Cesari, G.Perego, G.del Piero, S.Cucinella, E.Cernia
J. Organomet. Chem., **78,** 203, 1974

68.19 **hexakis(Isopropylamido - hydrido aluminium) trihydrido - aluminium**
$C_{18}H_{51}Al_7N_6$
G.Perego, M.Cesari, G.del Piero, A.Balducci, E.Cernia
J. Organomet. Chem., **87**, 33, 1975

68.C **1 - bis(Cyclopentadienyl) zirconium(iv) - μ - chloro - 2,2 - bis(diethyl aluminium) ethane**
$C_{20}H_{33}Al_2ClZr$
For complete entry see 71.65

68.C **bis(Dimethyl - bis(3,5 - dimethyl - 1 - pyrazolyl) - gallato) copper(ii)**
$C_{24}H_{40}CuGa_2N_8$
For complete entry see 83.102

68.C **(pentahapto - Cyclopentadienyl)hydrido - molybdenum - μ - dimethylaluminium - μ(methylaluminium - di(μ - pentahapto(monohapto) - cyclopentadienyl) - dimethylaluminium) - (pentahapto - cyclopentadienyl)hydrido - molybdenum**
$C_{25}H_{35}Al_3Mo_2$
For complete entry see 73.51

68.C **Dimolybdenum - bis - μ - (methyl aluminium - di(μ - pentahapto(monohapto) - cyclopentadienyl) - dimethylaluminium)**
$C_{26}H_{34}Al_4Mo_2$
For complete entry see 73.55

68.C **Di - μ - carbonyl - bis(carbonyl cyclopentadienyl iron) bis(triethyl aluminium)**
$C_{26}H_{40}Al_2Fe_2O_4$
For complete entry see 73.56

68.C **tris(bis(Triphenylphosphine) silver(i)) tris(dithio - oxalato) aluminium(iii)**
$C_{114}H_{90}Ag_3AlO_6P_6S_6$
For complete entry see 86.118

GERMANIUM, TIN, LEAD COMPOUNDS

69.1 **Tetraethylammonium μ - oxalato - bis(tetrachlorostannate(iv))**
$C_2Cl_8O_4Sn_2^{2-}$, $2C_8H_{20}N^+$
A.C.Skapski, J.-E.Guerchais, J.-Y.Calves
C. R. Acad. Sci., Ser. C, **278,** 1377, 1974
Residue 2 classified in 3

69.2 **N - (Trimethylstannyl) - N - nitromethylamine**
$(C_4H_{12}N_2O_2Sn)_n$
A.M.Domingos, G.M.Sheldrick *J. Organomet. Chem.*, **69,** 207, 1974

69.3 **Trimethyltin - glycinate**
$C_5H_{13}NO_2Sn$
B.Y.K.Ho, J.A.Zubieta, J.J.Zuckerman
J. Chem. Soc., Chem. Commun., 88, 1975

69.4 **π - Benzene lead bis(tetrachloro - aluminium) benzene solvate**
$C_6H_6AlCl_4Pb^+$, $AlCl_4^-$, C_6H_6
A.G.Gash, P.F.Rodesiler, E.L.Amma *Inorg. Chem.*, **13,** 2429, 1974

69.5 **π - Benzene tin bis(tetrachloro - aluminium) benzene solvate**
$C_6H_6AlCl_4Sn^+$, $AlCl_4^-$, C_6H_6
A.G.Gash, P.F.Rodesiler, E.L.Amma *Inorg. Chem.*, **13,** 2429, 1974

69.6 **tris(Dimethyl - tin(iv)) bis(orthophosphate) octahydrate**
$(C_6H_{18}O_8P_2Sn_3)_n$, $8nH_2O$
J.P.Ashmore, T.Chivers, K.A.Kerr, J.H.G.van Roode
J. Chem. Soc., Chem. Commun., 653, 1974

69.7 **bis(N - Methyl - N - acetyl - hydroxylamino) - dimethyl tin(iv)**
$C_8H_{18}N_2O_4Sn$
P.G.Harrison, T.J.King, J.A.Richards
J. Chem. Soc., Dalton Trans., 826, 1975

69.8 **bis - μ - Ethoxy - bis(trichloro - ethanol tin(iv))**
$C_8H_{22}Cl_6O_4Sn_2$
M.Webster, P.H.Collins *Inorg. Chim. Acta*, **9,** 157, 1974

69.9 **(3 - Benzothiazolo - dichlorogermano)pentacarbonyl molybdenum**
$C_{12}H_5Cl_2GeMoNO_5S$
D.J.Brauer, C.Kruger *Eur. Cryst. Meeting*, 435, 1974
Also classified in 41

69.10 **10,10 - Dichloro - 10 - germa - 9 - oxa - 9,10 - dihydroanthracene**
$C_{12}H_8Cl_2GeO$
A.I.Udel′nov, V.E.Shklover, N.G.Bokii, E.A.Chernyshev, T.L.Krasnova, E.F.Shchipanova, Yu.T.Struchkov *Zh. Strukt. Khim.*, **15**, 83, 1974

69.11 **Dodecamethyl - cyclohexagermane**
$C_{12}H_{36}Ge_6$
W.Jensen, R.Jacobson *Cryst. Struct. Commun.*, **4**, 299, 1975

69.C **μ - (Dichlorostannio) - bis(tricarbonyl - π - cyclopentadienyl chromium)**
$C_{16}H_{10}Cl_2Cr_2O_6Sn$
For complete entry see 73.25

69.12 **Dimethylgermyl - acetyl(tetracarbonyl) rhenium dimer**
$C_{16}H_{18}Ge_2O_{10}Re_2$
M.J.Webb, M.J.Bennett, L.Y.Y.Chan, W.A.G.Graham
J. Am. Chem. Soc., **96**, 5931, 1974

69.13 **cis - Dichloro - cis - bis(dimethylsulfoxide) - trans - diphenyl - tin**
$C_{16}H_{22}Cl_2O_2S_2Sn$
L.Coghi, C.Pelizzi, G.Pelizzi *Gazz. Chim. Ital.*, **104**, 873, 1974

69.14 **(2,2′ - Bipyridyl) - bis(trifluoroacetato) - divinyl tin**
$C_{18}H_{14}F_6N_2O_4Sn$
C.D.Garner, B.Hughes, T.J.King *J. Chem. Soc., Dalton Trans.*, 562, 1975

69.15 **Nitrato - triphenylstannyl - tin(ii)**
$C_{18}H_{15}NO_3Sn_2$
M.Nardelli, C.Pelizzi, G.Pelizzi *J. Organomet. Chem.*, **85**, C43, 1975

69.C **Dimethyl tin(iv) chloride - N,N′ - ethylene - bis(salicylideneiminato) nickel(ii)**
$C_{18}H_{20}Cl_2N_2NiO_2Sn$
For complete entry see 78.4

69.C **μ - Oxo - bis((η^5 - cyclopentadienyl - dicarbonyl iron) - dimethyl germanium)**
$C_{18}H_{22}Fe_2Ge_2O_5$
For complete entry see 73.31

69.16 **Triphenyl - tin isothiocyanate**
$C_{19}H_{15}NSSn$
A.M.Domingos, G.M.Sheldrick *J. Organomet. Chem.*, **67**, 257, 1974

69.17 **Di(bis(trimethylsilyl)methyl) tin - pentacarbonyl - chromium**
$C_{19}H_{38}CrO_5Si_4Sn$
J.D.Cotton, P.J.Davison, D.E.Goldberg, M.F.Lappert, K.M.Thomas
J. Chem. Soc., Chem. Commun., 893, 1974
Also classified in 63

69.18 **bis(10 - Methylisoalloxazine) lead(ii) perchlorate tetrahydrate**
$C_{22}H_{16}N_8O_6Pb^{2+}$, $2ClO_4^-$, $4H_2O$
M.W.Yu, C.J.Fritchie Junior *J. Chem. Soc., Dalton Trans.*, 377, 1975
Residue 1 also classified in 36

69.19 **(2,2' - Bipyridyl) - dichloro - diphenyl tin**
$C_{22}H_{18}Cl_2N_2Sn$
P.G.Harrison, T.J.King, J.A.Richards
J. Chem. Soc., Dalton Trans., 1723, 1974

69.20 **Triphenyl - tin - 4 - mercaptopyridine**
$C_{23}H_{19}NSSn$
N.G.Bokii, Yu.T.Struchkov, D.N.Kravtsov, E.M.Rokhlina
Zh. Strukt. Khim., **14,** 502, 1973

69.21 **Triphenyl - tin - (2,6 - dibromo - 4 - fluoro - thiophenolate)**
$C_{24}H_{17}Br_2FSSn$
N.G.Bokii, Yu.T.Struchkov, D.N.Kravtsov, E.M.Rokhlina
Zh. Strukt. Khim., **15,** 497, 1974

69.22 **Diphenyl - (N - (o - hydroxylato)benzylidene - (o - thiolato)aniline) tin**
$C_{25}H_{19}NOSSn$
H.Preut, H.-J.Haupt, F.Huber, R.Cefalu, R.Barbieri
Z. Anorg. Allg. Chem., **407,** 257, 1974

69.23 **Triphenyltin - (2 - methyl - thiophenolate)**
$C_{25}H_{22}SSn$
N.G.Bokii, Yu.T.Struchkov, D.N.Kravtsov, E.M.Rokhlina
Zh. Strukt. Khim., **15,** 497, 1974

69.24 **(Hexamethyltriamidophosphato) - (isothiocyanato) - triphenyl tin(iv)**
$C_{25}H_{33}N_4OPSSn$
L.J.Pazdernik, F.Rochon *Acta Crystallogr., Sect. A,* **31,** S132, 1975

69.25 **Triphenyl germanium p - t - butylphenyl mercaptide**
$C_{28}H_{28}GeS$
M.E.Cradwick, R.D.Taylor, J.L.Wardell
J. Organomet. Chem., **66,** 43, 1974

69.26 **Tetra - (p - tolyl) tin**
$C_{28}H_{28}Sn$
A.Karipides, K.Wolfe *Acta Crystallogr., Sect. B,* **31,** 605, 1975

69.27 **N - Benzoyl - N - phenyl - O - (triphenylstannyl)hydroxylamine**
$C_{31}H_{25}NO_2Sn$
P.G.Harrison, T.J.King *J. Chem. Soc., Dalton Trans.*, 2298, 1974

69.28 **(1,3 - Diphenylpropane - 1,3 - dionato)triphenyl tin(iv)**
$C_{33}H_{26}O_2Sn$
G.M.Bancroft, B.W.Davies, N.C.Payne, T.K.Sham
J. Chem. Soc., Dalton Trans., 973, 1975

69.29 Hemiporphyrazine - germanium - di(ethylene glycol monoethyl ether)
$C_{34}H_{32}GeN_8O_4$
H.-J.Hecht, P.Luger *Acta Crystallogr., Sect. B,* **30,** 2843, 1974
Also classified in 49

69.30 Tetra(diphenylketimino) germanium
$C_{52}H_{40}GeN_4$
N.W.Alcock, M.Pierce-Butler, G.R.Willey, K.Wade
J. Chem. Soc., Chem. Commun., 183, 1975

69.31 Tetra(diphenylketimino) tin
$C_{52}H_{40}N_4Sn$
N.W.Alcock, M.Pierce-Butler, G.R.Willey, K.Wade
J. Chem. Soc., Chem. Commun., 183, 1975

69.32 Nitrato - tris(triphenylstannyl) tin(iv)
$C_{54}H_{45}NO_3Sn_4$
G.Pelizzi *J. Organomet. Chem.*, **87,** C1, 1975

69.C π - Cyclopentadienyl - bis(π - triphenylgermyl - phenyl - acetylene) - carbonyl - niobium
$C_{58}H_{45}Ge_2NbO$
For complete entry see 72.44

TELLURIUM COMPOUNDS

70.1 **Tellurium di(methylxanthate)**
$C_4H_6O_2S_4Te$
H.Graver, S.Husebye *Acta Chem. Scand. Ser. A,* **29,** 14, 1975

70.2 **Picolinium p - tolyl - tetrachloro - tellurium**
$C_7H_7Cl_4Te^-$, $C_6H_8N^+$
B.Krebs, V.Paulat *Eur. Cryst. Meeting,* 238, 1974
Residue 2 classified in 33

70.3 **cis - Dibromo - bis(trimethylenethiourea) tellurium**
$C_8H_{16}Br_2N_4S_2Te$
K.S.Fredin, K.Maroy, S.Slogvik *Acta Chem. Scand. Ser. A,* **29,** 212, 1975

70.4 **cis - Dichloro - bis(trimethylenethiourea) tellurium**
$C_8H_{16}Cl_2N_4S_2Te$
K.S.Fredin, K.Maroy, S.Slogvik *Acta Chem. Scand. Ser. A,* **29,** 212, 1975

70.5 **Tetraethylammonium tris(O - ethylxanthato) tellurium(ii)**
$C_9H_{15}O_3S_6Te^-$, $C_8H_{20}N^+$
B.F.Hoskins, C.D.Pannan *J. Chem. Soc., Chem. Commun.,* 408, 1975
Residue 2 classified in 3

70.6 **trans - Tetrachloro - bis(tetramethylthiourea) tellurium(iv) (monoclinic form)**
$C_{10}H_{24}Cl_4N_4S_2Te$
S.Esperas, J.W.George, S.Husebye, O.Mikalsen
Acta Chem. Scand. Ser. A, **29,** 141, 1975

70.7 **Dibenzotellurophene di - iodide**
$C_{12}H_8I_2Te$
J.D.McCullough *Inorg. Chem.,* **14,** 1142, 1975

TRANSITION METAL-C COMPOUNDS

71.1 **Methyl mercury azide (at 100°K)**
CH_3HgN_3
U.Muller *Z. Naturforsch., Teil B,* **28,** 426, 1973

71.2 **Methyl - dichloro - bis(N - methyl - N - nitrosohydroxylaminato) tantalum(v)**
$C_3H_9Cl_2N_4O_4Ta$
J.D.Wilkins, M.G.B.Drew *J. Organomet. Chem.,* **69,** 111, 1974
Also classified in 84

71.3 **Tetracarbonyl - iodo - (methylmethinyl) chromium**
$C_6H_3CrIO_4$
G.Huttner, H.Lorenz, W.Gartzke *Angew. Chem.,* **86,** 667, 1974

71.4 **Trimethylphenylammonium di - μ - iodo - bis(di - iodo - carbonyl - acetyl rhodium(iii))**
$C_6H_6I_6O_4Rh_2^{2-}$, $2C_9H_{14}N^+$
G.W.Adamson, J.J.Daly, D.Forster *J. Organomet. Chem.,* **71,** C17, 1974
Residue 2 classified in 16

71.5 **Methyl mercury DL - methionine complex**
$C_6H_{13}HgNO_2S$
Y.S.Wong, A.J.Carty, P.C.Chieh, N.J.Taylor
Am. Cryst. Assoc., Abstr. Papers (Summer Meeting), 226, 1974
Also classified in 82

71.6 **Bromo - tricarbonyl - (trimethylphosphine) - (methylmethinyl) chromium**
$C_8H_{12}BrCrO_3P$
G.Huttner, H.Lorenz, W.Gartzke *Angew. Chem.,* **86,** 667, 1974
Also classified in 86

71.7 **bis(Dimethylene - dimethylphosphonium) - dicopper(i)**
$C_8H_{20}Cu_2P_2$
G.Nardin, L.Randaccio, E.Zangrando *J. Organomet. Chem.,* **74,** C23, 1974
Also classified in 64

71.8 **bis(Tetramethylammonium) 4,4′ - commo - bis(decahydro - 1,6 - dimethyl - 1,6 - dicarba - 4 - titana - closo - tridecaborate) acetone solvate (at −160°C)**
$C_8H_{32}B_{20}Ti^{2-}$, $2C_4H_{12}N^+$, $2C_3H_6O$
F.Y.Lo, C.E.Strouse, K.P.Callahan, C.B.Knobler, M.F.Hawthorne
J. Am. Chem. Soc., **97**, 428, 1975
Residue 1 also classified in 62; residue 2 classified in 3

71.9 **tetrakis(Trifluoroacetoxy - mercuri)methane**
$C_9F_{12}Hg_4O_8$
D.Grdenic, B.Kamenar, B.Korpar-Colig, M.Sikirica, G.Jovanovski
J. Chem. Soc., Chem. Commun., 646, 1974
Also classified in 81

71.10 **Phenyl mercury L - cysteine complex**
$C_9H_{11}HgNO_2S$
Y.S.Wong, A.J.Carty, P.C.Chieh, N.J.Taylor
Am. Cryst. Assoc., Abstr. Papers (Summer Meeting), 226, 1974
Also classified in 82, 85

71.11 **tetrakis(Acetoxy - mercuri)methane**
$C_9H_{12}Hg_4O_8$
D.Grdenic, B.Kamenar, B.Korpar-Colig, M.Sikirica, G.Jovanovski
J. Chem. Soc., Chem. Commun., 646, 1974
Also classified in 81

71.12 **Methyl - bis(dimethylglyoximato) - aquo - cobalt(iii)**
$C_9H_{19}CoN_4O_5$
D.Ginderow *Acta Crystallogr., Sect. B*, **31**, 1092, 1975
Also classified in 83

71.13 **Dicarbonyl - (η - cyclopentadienyl) - (3 - aminopropionyl) molybdenum(ii)**
$C_{10}H_{11}MoNO_3$
G.A.Jones, L.J.Guggenberger *Acta Crystallogr., Sect. B*, **31**, 900, 1975
Also classified in 73, 82

71.14 **Tetracarbonyl - iodo - (phenylmethinyl) tungsten**
$C_{11}H_5IO_4W$
G.Huttner, H.Lorenz, W.Gartzke *Angew. Chem.*, **86**, 667, 1974

71.15 **Nonacarbonyl - μ_3 - ethylidyne - tri - μ - hydrido - triruthenium**
$C_{11}H_6O_9Ru_3$
G.M.Sheldrick, J.P.Yesinowski *J. Chem. Soc., Dalton Trans.*, 873, 1975

71.16 **Hydrido - (O - methyl - carbonyl) decacarbonyl tri - iron**
$C_{12}H_4Fe_3O_{11}$
D.F.Shriver, D.Lehman, D.Strope *J. Am. Chem. Soc.*, **97**, 1594, 1975

71.17 **Tetracarbonyl - (2 - methyl - 3 - prop - 1 - ynyl - maleoyl) iron**
$C_{12}H_6FeO_6$
R.C.Pettersen, J.L.Cihonski, F.R.Young III, R.A.Levenson
J. Chem. Soc., Chem. Commun., 370, 1975

71.18 **2 - Chloro - 4 - bromophenolato - (phenyl) - mercury**
$C_{12}H_8BrClHgO$
L.G.Kuz'mina, N.G.Bokii, Yu.T.Struchkov, D.N.Kravtsov, L.S.Golovchenko *Zh. Strukt. Khim.*, **14,** 508, 1973
Also classified in 84

71.19 **(2,3,4,8 - h^4 - Bicyclo(3.2.2)nona - 3,6 - diene - 2,8 - yl - 9 - one)tricarbonyl iron (orthorhombic form)**
$C_{12}H_8FeO_4$
A.H.-J.Wang, I.C.Paul, R.Aumann *J. Organomet. Chem.*, **69,** 301, 1974
Also classified in 75

71.20 **Bicyclo(3.2.2)nona - 6 - ene - 9 - one - (2 - 4)enyl - 8 - yl iron tricarbonyl (triclinic form)**
$C_{12}H_8FeO_4$
F.A.Cotton, J.M.Troup *J. Organomet. Chem.*, **76,** 81, 1974
Also classified in 75

71.21 **3 - Tetracarbonyl - 6,7 - dimethyl - 3 - ferrabicyclo(3.2.0)hept - 6 - ene - 2,4 - dione**
$C_{12}H_8FeO_6$
B.Deppisch *Acta Crystallogr., Sect. A*, **31,** S137, 1975

71.22 **1 - (Difluoromethyl - carbonyl - (cyclopentadienyl) iron) - cyclopentadiene**
$C_{12}H_{10}F_2FeO$
J.L.Davidson, M.Green, F.G.A.Stone, A.J.Welch
J. Chem. Soc., Chem. Commun., 286, 1975
Also classified in 73

71.23 **bis(Benzene) chromium(1) 7,7,8,8 - tetracyano - p - quinodimethanide**
$C_{12}H_{12}Cr^+$, $C_{12}H_4N_4^-$
R.P.Shibaeva, A.E.Shvets, L.O.Atovmyan
Dokl. Akad. Nauk SSSR, **199,** 334, 1971
Residue 2 classified in 7

71.24 **bis(π - Cyclopentadienyl)ethyl - chloro - molybdenum(iv)**
$C_{12}H_{15}ClMo$
K.Prout, T.S.Cameron, R.A.Forder, S.R.Critchley, B.Denton, G.V.Rees
Acta Crystallogr., Sect. B, **30,** 2290, 1974
Also classified in 73

71.25 **hexakis(Methyl - isocyanide) dipalladium(i) hexafluorophosphate acetone solvate**
$C_{12}H_{18}N_6Pd_2^{2+}$, $2F_6P^-$, xC_3H_6O
D.J.Doonan, A.L.Balch, S.Z.Goldberg, R.Eisenberg, J.S.Miller
J. Am. Chem. Soc., **97**, 1961, 1975

71.26 **Di - μ - chloro - bis(chloro - (2 - (hydroxymethyl)pent - 4 - enyl) rhodium(iii)) methanol solvate**
$C_{12}H_{22}Cl_4O_2Rh_2$, CH_4O
J.F.Malone *J. Chem. Soc., Dalton Trans.*, 1699, 1974
Residue 1 also classified in 72, 84

71.27 **Carbonyl - methyl - (6,7,13,14 - tetramethyl - 1,2,4,5,8,9,11,12 - octa - azacyclotetradeca - 1(14),5,7,12 - tetraenyl) iron(ii)**
$C_{12}H_{22}FeN_8O$
V.L.Goedken, S.-M.Peng *J. Am. Chem. Soc.*, **96**, 7826, 1974
Also classified in 83

71.28 **Methyl - (6,7,13,14 - tetramethyl - 1,2,4,5,8,9,11,12 - octa - aza - cyclotetradeca - 5,7,12,14(1) - tetraene - dienyl) - methylhydrazine cobalt(iii)**
$C_{12}H_{23}CoN_{10}$
V.L.Goedken, S.-M.Peng *J. Chem. Soc., Chem. Commun.*, 258, 1975
Also classified in 83

71.29 **Dimethyl - bis(trimethylphosphine) cobalt(iii) - (dimethylphosphonium - bis(methylide))**
$C_{12}H_{34}CoP_3$
D.J.Brauer, C.Kruger, P.J.Roberts, Y.-H.Tsay
Chem. Ber., **107**, 3706, 1974
Also classified in 86

71.30 **Azido - trimethyl platinum tetramer (at −70°C)**
$C_{12}H_{36}N_{12}Pt_4$
M.Atam, U.Muller *J. Organomet. Chem.*, **71**, 435, 1974

71.31 **Tetramethylammonium cis - acetyl - benzoyl - tetracarbonyl - manganate(i)**
$C_{13}H_8MnO_6^-$, $C_4H_{12}N^+$
C.P.Casey, C.A.Bunnell *J. Chem. Soc., Chem. Commun.*, 733, 1974
Residue 2 classified in 3

71.32 **Dimethyl - (3,3′ - (trimethylenedinitrolo) - bis(butan - 2 - one oximato)) cobalt(iii)**
$C_{13}H_{25}CoN_4O_2$
M.Calligaris *J. Chem. Soc., Dalton Trans.*, 1628, 1974
Also classified in 83

71.33 **1 - Tricarbonyl - benzo(c)ferracyclopentadiene iron tricarbonyl**
$C_{14}H_6Fe_2O_6$
R.E.Davis, B.L.Barnett, R.G.Amiet, W.Merk, J.S.McKennis, R.Pettit
J. Am. Chem. Soc., **96,** 7108, 1974
Also classified in 75

71.34 **1 - Tricarbonyl - benzo(b)ferracyclopentadiene iron tricarbonyl**
$C_{14}H_6Fe_2O_6$
R.E.Davis, B.L.Barnett, R.G.Amiet, W.Merk, J.S.McKennis, R.Pettit
J. Am. Chem. Soc., **96,** 7108, 1974
Also classified in 75

71.35 **Methyl(hydro - tris(1 - pyrazolyl)borato)hexafluorobut - 2 - yne platinum(ii)**
$C_{14}H_{13}BF_6N_6Pt$
B.W.Davies, N.C.Payne *Inorg. Chem.*, **13,** 1843, 1974
Also classified in 72, 83

71.36 **Phenyl - mercury - salicylalmethyliminate**
$C_{14}H_{13}HgNO$
L.G.Kuz'mina, N.G.Bokii, Yu.T.Struchkov, V.I.Minkin, L.P.Olekhnovich, I.E.Mikhailov *Zh. Strukt. Khim.*, **15,** 659, 1974
Also classified in 78

71.37 **Phenylmercury - (2,6 - dimethyl - thiophenolate)**
$C_{14}H_{14}HgS$
L.G.Kuz'mina, N.G.Bokii, Yu.T.Struchkov, D.N.Kravtsov, E.M.Rokhlina
Zh. Strukt. Khim., **15,** 491, 1974
Also classified in 85

71.38 **bis(Cyclopentadienyl) - ethyl - (ethylene) niobium**
$C_{14}H_{19}Nb$
L.J.Guggenberger, P.Meakin, F.N.Tebbe
J. Am. Chem. Soc., **96,** 5420, 1974
Also classified in 72, 73

71.39 **1 - (Dicarbonyl - π - cyclopentadienyl - ferrio) - 2 - (phenyl) - ethyne**
$C_{15}H_{10}FeO_2$
R.Goddard, J.Howard, P.Woodward
J. Chem. Soc., Dalton Trans., 2025, 1974
Also classified in 73

71.40 **2 - Chloromercury - tetradymol**
$C_{15}H_{21}ClHgO_2$
P.W.Jennings, S.K.Reeder, J.C.Hurley, C.N.Caughlan, G.D.Smith
J. Org. Chem., **39,** 3392, 1974

71.41 **Trichloro(dimethylaminomethylene) - bis(triethylphosphine) rhodium(iii)**
$C_{15}H_{37}Cl_3NP_2Rh$
B.Cetinkaya, M.F.Lappert, G.M.McLaughlin, K.Turner
J. Chem. Soc., Dalton Trans., 1591, 1974
Also classified in 86

71.42 **bis(Perfluorophenyl) - tetramethyltetrazene zinc(ii)**
$C_{16}H_{12}F_{10}N_4Zn$
V.W.Day, D.H.Campbell, C.J.Michejda
J. Chem. Soc., Chem. Commun., 118, 1975
Also classified in 83

71.43 **Chloro - (2,5 - dithiahexane) - (1 - (1,4 - di - t - butyl - 4 - chloro)butadienyl) palladium**
$C_{16}H_{30}Cl_2PdS_2$
B.E.Mann, P.M.Bailey, P.M.Maitlis *J. Am. Chem. Soc.*, **97,** 1275, 1975
Also classified in 85

71.44 **Di - μ - (dimethylphosphonium - bis(methylide)) - bis(dimethylphosphonium - bis(methylide) nickel)**
$C_{16}H_{40}Ni_2P_4$
D.J.Brauer, C.Kruger, P.J.Roberts, Y.-H.Tsay
Chem. Ber., **107,** 3706, 1974

71.45 **nido - 6,6 - bis(Triethylphosphine) - 5,8 - dimethyl - 6,5,8 - platinadicarbaborane**
$C_{16}H_{42}B_6P_2Pt$
M.Green, J.L.Spencer, F.G.A.Stone, A.J.Welch
J. Chem. Soc., Chem. Commun., 794, 1974
Also classified in 86, 62

71.46 **μ - Carbonyl - μ - dicyanovinylidene - bis(carbonyl - cyclopentadienyl iron(0))**
$C_{17}H_{10}Fe_2N_2O_3$
R.M.Kirchner, J.A.Ibers *J. Organomet. Chem.*, **82,** 243, 1974
Also classified in 73

71.47 **(6,7,8,9 - h^4.2,3,4,11 - h^4 - Bicyclo(4.3.1)undeca - 2,6,8 - triene - 4,11 - yl - 11 - one) - bis(tricarbonyl iron)**
$C_{17}H_{10}Fe_2O_7$
A.H.-J.Wang, I.C.Paul, R.Aumann *J. Organomet. Chem.*, **69,** 301, 1974
Also classified in 75

71.48 **Acetylacetonato - (7,8 - bis(trifluoromethyl) - bicyclo(4.2.2)dec - 7 - ene - 2,5 - diyl) - aquo - rhodium hemihydrate**
$C_{17}H_{21}F_6O_3Rh$, $0.5H_2O$
A.C.Jarvis, R.D.W.Kemmitt, B.Y.Kimura, D.R.Russell, P.A.Tucker
J. Chem. Soc., Chem. Commun., 797, 1974
Residue 1 also classified in 75, 77

71.49 **π - Cyclopentadienyl - π - (hexafluoro - 2 - butyne) - (1,2 - bis(trifluoromethyl) - 2 - cyclopentadienyl - ethylene - 1 - yl) molybdenum**
$C_{18}H_{10}F_{12}Mo$
J.L.Davidson, M.Green, D.W.A.Sharp, F.G.A.Stone, A.J.Welch
J. Chem. Soc., Chem. Commun., 706, 1974
Also classified in 72, 73

71.50 **bis(1,2 - Di(trifluoromethyl) - 2 - (diacetylmethyl) - ethylenyl) palladium**
$C_{18}H_{14}F_{12}O_4Pd$
A.C.Jarvis, R.D.W.Kemmitt, B.Y.Kimura, D.R.Russell, P.A.Tucker
J. Organomet. Chem., **66,** C53, 1974
Also classified in 84

71.51 **cis - (Carbonyl - (π - cyclopentadienyl) iron) - di - μ - carbonyl - (isobutylisocyano - (π - cyclopentadienyl) iron)**
$C_{18}H_{19}Fe_2NO_3$
I.L.C.Campbell, F.S.Stephens *J. Chem. Soc., Dalton Trans.*, 982, 1975
Also classified in 73

71.52 **Di(cyclopentadienyl) - 2,6 - dimethylphenyl titanium**
$C_{18}H_{19}Ti$
G.J.Olthof, H.R.van der Wal *Acta Crystallogr., Sect. A,* **31,** S135, 1975
Also classified in 73

71.53 **bis(π - 2 - Ethyl - 3 - methylthioacrolein iron(0) dicarbonyl)**
$C_{18}H_{20}Fe_2O_4S_2$
C.E.Pfluger *Acta Crystallogr., Sect. A,* **31,** S136, 1975
Also classified in 72, 85

71.54 **(pentahapto - Cyclopentadienyl) - 1,3′,4′ - trihapto - (2 - methyl - 2 - (6′,6′ - dimethyl - bicyclo(3.2.0)hept - 3′ - en - 7′ - on - 2′ - yl)propionyl) nickel(ii)**
$C_{18}H_{22}NiO_2$
M.R.Churchill, B.G.DeBoer, J.J.Hackbarth *Inorg. Chem.*, **13,** 2098, 1974
Also classified in 73, 75

71.55 **4 - ((Diethylamino) - (t - butylamino) - methylene) - 4 - (t - butyl isocyanide) - 2,2,5,5 - tetrakis(trifluoromethyl) - 1,3,4 - dioxapalladolan**
$C_{18}H_{29}F_{12}N_3O_2Pd$
A.Modinos, P.Woodward *J. Chem. Soc., Dalton Trans.*, 2065, 1974
Also classified in 84

71.56 **(π - Pentenyl) - di - (isopropyl - phenyl - phosphine) - methyl nickel(ii)**
$C_{18}H_{31}NiP$
B.L.Barnett, C.Kruger *J. Organomet. Chem.*, **77,** 407, 1974
Also classified in 72, 86

71.57 **Tri - lithium hexamethyl chromium tri(dioxan)**
$(C_{18}H_{42}CrLi_3O_6)_n$
J.Krausse, G.Marx *J. Organomet. Chem.*, **65**, 215, 1974
Also classified in 67

71.58 **Cyclobutadiene - iron - tricarbonyl dimethyl maleate photoadduct**
$C_{19}H_{20}FeO_{11}$
P.E.Riley, R.E.Davis
Am. Cryst. Assoc., Abstr. Papers (Spring Meeting), 11, 1975
Also classified in 75

71.59 **2,3 - bis(π - 1,5 - Cyclo - octadiene) - 4,4 - bis(trifluoromethyl) - 1,2,3 - oxa - diplatina - cyclobutane**
$C_{19}H_{24}F_6OPt_2$
M.Green, J.A.K.Howard, A.Laguna, M.Murray, J.L.Spencer, F.G.A.Stone
J. Chem. Soc., Chem. Commun., 451, 1975
Also classified in 75, 84

71.60 **Iodo - (1,2 - bis(methylphenylarsino)ethane) - trimethyl platinum**
$C_{19}H_{29}As_2IPt$
G.Casalone, R.Mason *Inorg. Chim. Acta*, **7**, 429, 1973
Also classified in 86

71.61 **1 - Carbonyl - 1 - cyclopentadienyl - 1 - (cyclopentadienyl iron) - 2,3,5,6 - tetra - (trifluoromethyl) - 1 - ferra - cyclohexa - 2,5 - diene - 4 - one**
$C_{20}H_{10}F_{12}Fe_2O_2$
J.L.Davidson, M.Green, F.G.A.Stone, A.J.Welch
J. Chem. Soc., Chem. Commun., 286, 1975
Also classified in 73, 75

71.62 **Chloro - (triphenylphosphine) - methylmercaptomethyl palladium(ii)**
$C_{20}H_{20}ClPPdS$
K.Miki, Y.Kai, N.Yasuoka, N.Kasai
Acta Crystallogr., Sect. A, **31**, S138, 1975
Also classified in 86, 85

71.63 **Chloro - (triphenylphosphine) - methylmercaptomethyl palladium(ii) (at −160°C)**
$C_{20}H_{20}ClPPdS$
K.Miki, Y.Kai, N.Yasuoka, N.Kasai
Acta Crystallogr., Sect. A, **31**, S138, 1975
Also classified in 86, 85

71.64 **Di - μ - chloro - bis((3 - allyl - norbornyl) palladium(ii))**
$C_{20}H_{30}Cl_2Pd_2$
M.Zocchi, G.Tieghi *Acta Crystallogr., Sect. A*, **31**, S146, 1975
Also classified in 72

71.65 **1 - bis(Cyclopentadienyl) zirconium(iv) - μ - chloro - 2,2 - bis(diethyl aluminium) ethane**
$C_{20}H_{33}Al_2ClZr$
W.Kaminsky, J.Kopf, G.Thirase *Justus Liebigs Ann. Chem.*, 1531, 1974
Also classified in 73, 68

71.66 **1,1 - bis(Dimethylphenylphosphine) - 2,4 - dimethyl - 1 - platina - 2,4 - dicarbadodecaborane**
$C_{20}H_{37}B_9P_2Pt$
M.Green, J.L.Spencer, F.G.A.Stone, A.J.Welch
J. Chem. Soc., Dalton Trans., 179, 1975
Also classified in 86, 62

71.67 **Iodo - (1 - (p - chlorophenyl)imino - methyl) - bis(triethylphosphine) platinum(ii)**
$C_{20}H_{37}ClINP_2Pt$
K.P.Wagner, P.M.Treichel, J.C.Calabrese
J. Organomet. Chem., **71**, 299, 1974
Also classified in 86

71.68 **1,1,5 - Trichloro - 1,1 - bis(triethylphosphine) - 2 - methylamino - 3 - platina - indole perchlorate**
$C_{20}H_{38}Cl_3N_2P_2Pt^+$, ClO_4^-
R.Walker, K.W.Muir *J. Chem. Soc., Dalton Trans.*, 272, 1975
Residue 1 also classified in 86

71.69 **(Tricyclo(6.6.2^{2,7}.0)hexadeca - 3,5,9,11,13,15 - hexaene) pentacarbonyl di - ruthenium**
$C_{21}H_{16}O_5Ru_2$
R.Goddard, A.P.Humphries, S.A.R.Knox, P.Woodward
J. Chem. Soc., Chem. Commun., 508, 1975
Also classified in 75

71.70 **Hexafluorobut - 2 - yne - tricarbonyl(cyclohexa - 1,3 - diene) - ruthenium adduct phosphite substitution product**
$C_{21}H_{17}F_{12}O_5PRu$
M.Bottrill, R.Goddard, M.Green, R.P.Hughes, M.K.Lloyd, B.Lewis, P.Woodward *J. Chem. Soc., Chem. Commun.*, 253, 1975
Also classified in 75, 86

71.71 **9 - (1',2' - bis(Trifluoromethyl) - 2' - (diacetylmethyl)ethylen - 1' - yl) - 10,11 - bis(trifluoromethyl) - 9 - iridia - bicyclo(6.3.0)undeca - 4,10 - diene**
$C_{21}H_{19}F_{12}IrO_2$
A.C.Jarvis, R.D.W.Kemmitt, B.Y.Kimura, D.R.Russell, P.A.Tucker
J. Chem. Soc., Chem. Commun., 797, 1974
Also classified in 75, 84

71.72 **bis(6,6 - Dimethylpentafulvene) - pentacarbonyl - di - iron**
$C_{21}H_{20}Fe_2O_5$
U.Behrens, E.Weiss *J. Organomet. Chem.*, **73**, C64, 1974
Also classified in 73

71.73 **cis - Chloro - (benzylacetoacetato) - di(pyridine) palladium(ii)**
$C_{21}H_{21}ClN_2O_3Pd$
M.Horike, Y.Kai, N.Yasuoka, N.Kasai
J. Organomet. Chem., **86**, 269, 1975
Also classified in 83

71.74 **bis(η - Cyclopentadienyl) - σ - ((2 - dimethylphenylphosphonium)ethyl)methyl - tungsten hexafluorophosphate**
$C_{21}H_{28}PW^+$, F_6P^-
R.A.Forder, G.D.Gale, K.Prout *Acta Crystallogr., Sect. B*, **31**, 307, 1975
Residue 1 also classified in 73, 64

71.75 **cis - Dichloro(1,3 - diphenylimidazolidin - 2 - ylidene)(triethylphosphine) platinum(ii)**
$C_{21}H_{29}Cl_2N_2PPt$
L.Manojlovic-Muir, K.W.Muir *J. Chem. Soc., Dalton Trans.*, 2427, 1974
Also classified in 86

71.76 **trans - Dichloro - (1,3 - diphenylimidazolidin - 2 - ylidene)(triethylphosphine) platinum(ii)**
$C_{21}H_{29}Cl_2N_2PPt$
L.Manojlovic-Muir, K.W.Muir *J. Chem. Soc., Dalton Trans.*, 2427, 1974
Also classified in 86

71.77 **trans - (Methyl - (2 - oxacyclopentylidene) - bis(dimethylphenylphosphine) platinum(ii)) hexafluorophosphate**
$C_{21}H_{31}OP_2Pt^+$, F_6P^-
R.F.Stepaniak, N.C.Payne *J. Organomet. Chem.*, **72**, 453, 1974
Residue 1 also classified in 86

71.78 **(N,N - dimethylbenzylamine) - (N - phenylsalicylaldiminato) palladium(ii)**
$C_{22}H_{22}N_2OPd$
G.D.Fallon, B.M.Gatehouse *J. Chem. Soc., Dalton Trans.*, 1632, 1974
Also classified in 78, 83

71.79 **((Chloro(methoxycarbonyl)(1,2,3,4,5 - pentakis(methoxycarbonyl)cyclopenta - 2,4 - dienyl)methyl))(pentane - 2,4 - dionato) palladium(ii) chloroform solvate**
$C_{23}H_{25}ClO_{14}Pd$, $CHCl_3$
D.M.Roe, C.Calvo, N.Krishnamachari, P.M.Maitlis
J. Chem. Soc., Dalton Trans., 125, 1975
Residue 1 also classified in 77, 84

71.80 **Tricarbonyl - (cyclohexeno(1,2)ferracyclopentadieno(3,4 - a)phenanthrene) iron tricarbonyl**
$C_{24}H_{14}Fe_2O_6$
H.Irngartinger *Eur. Cryst. Meeting,* 372, 1974
Also classified in 75

71.81 **Tricarbonyl - (2,4,7 - tri - t - butyl - hepta - 1,5 - diene - 3 - one - 1,4,7 - triyl) iron dicarbonyl - iron**
$C_{24}H_{30}Fe_2O_6$
E.Sappa, L.Milone, G.D.Andreetti *Inorg. Chim. Acta,* **13,** 67, 1975
Also classified in 72

71.82 **1,2 - bis(Dimethylphosphino)ethane - (methyl - (2 - dimethylphosphinoethyl) phosphino)methyl ruthenium hydride dimer**
$C_{24}H_{64}P_8Ru_2$
F.A.Cotton, B.A.Frenz, D.L.Hunter
J. Chem. Soc., Chem. Commun., 755, 1974
Also classified in 86

71.83 **Malononitrilato(propane - 1,2 - bis(salicylideneiminato))pyridine cobalt(iii)**
$C_{25}H_{22}CoN_5O_2$
N.A.Bailey, B.M.Higson, E.D.McKenzie
J. Chem. Soc., Dalton Trans., 1105, 1975
Also classified in 78, 83

71.84 **(2,4 - Pentanedionato)(triphenylphosphine)ethyl nickel(ii)**
$C_{25}H_{27}NiO_2P$
F.A.Cotton, B.A.Frenz, D.L.Hunter *J. Am. Chem. Soc.,* **96,** 4820, 1974
Also classified in 77, 86

71.85 **Benzyl - (triphenylmethylthio) mercury**
$C_{26}H_{22}HgS$
R.D.Bach, A.T.Weibel, W.Schmonsees, M.D.Glick
J. Chem. Soc., Chem. Commun., 961, 1974
Also classified in 85

71.86 **α - Carboranyl complex of platinum(ii)**
$C_{26}H_{56}B_{10}P_2Pt$
N.Bresciani, M.Calligaris, P.Delise, G.Nardin, L.Randaccio
J. Am. Chem. Soc., **96,** 5642, 1974
Also classified in 72, 86, 62

71.87 **bis(Dimethylglyoximato) - (pyridine) - (1 - chloro - 2,2 - bis(p - chlorophenyl)vinyl) cobalt(iii)**
$C_{27}H_{27}Cl_3CoN_5O_4$
R.H.Prince, G.M.Sheldrick, D.A.Stotter, R.Taylor
J. Chem. Soc., Chem. Commun., 854, 1974
Also classified in 83

71.88 tris(2′ - (2 - Phenyl - 1,3 - dioxolano)) chromium(iii)
$C_{27}H_{27}CrO_6$
J.J.Daly, F.Sanz, R.P.A.Sneeden, H.H.Zeiss
Helv. Chim. Acta, **57,** 1863, 1974
Also classified in 84

71.89 (π - Pentenyl)(dimenthyl - methyl - phosphine) - methyl nickel(ii)
$C_{27}H_{53}NiP$
B.L.Barnett, C.Kruger *J. Organomet. Chem.*, **77,** 407, 1974
Also classified in 72, 86

71.90 Acetylacetonato - (acetylacetonyl) - triphenylphosphine palladium(ii) benzene solvate
$C_{28}H_{29}O_4PPd$, $0.5C_6H_6$
M.Horike, Y.Kai, N.Yasuoka, N.Kasai
J. Organomet. Chem., **72,** 441, 1974
Residue 1 also classified in 77, 86

71.91 bis - (η - Cyclopentadienyl) - bis(3,5 - dimethylbenzyl) tungsten
$C_{28}H_{32}W$
R.A.Forder, I.W.Jefferson, K.Prout
Acta Crystallogr., Sect. B, **31,** 618, 1975
Also classified in 73

71.92 Tricarbonyl - (2 - (methylsulfidomethyl)phenyl) - triphenylphosphine manganese
$C_{29}H_{24}MnO_3PS$
R.J.Doedens, J.T.Veal, R.G.Little *Inorg. Chem.*, **14,** 1138, 1975
Also classified in 86, 85

71.93 Di - μ - chloro - bis((1 - (dicarbonyl - π - cyclopentadienyl - ferrio) - 2 - phenyl - ethyne) copper(i)) dimer
$C_{30}H_{20}Cl_2Cu_2Fe_2O_4$
R.Clark, J.Howard, P.Woodward
J. Chem. Soc., Dalton Trans., 2027, 1974
Also classified in 72, 73

71.94 μ - Di - ($\eta^5.\eta^1$ - cyclopentadienyl) - di(dicyclopentadienyl thorium)
$C_{30}H_{28}Th_2$
E.C.Baker, K.N.Raymond, T.J.Marks, W.A.Wachter
J. Am. Chem. Soc., **96,** 7586, 1974
Also classified in 73

71.95 Di - iodo - bis(di - (p - tolylamino)carbene) gold(iii) perchlorate diethyl ether solvate
$C_{30}H_{32}AuI_2N_4^+$, ClO_4^- , $C_4H_{10}O$
L.Manojlovic-Muir *J. Organomet. Chem.*, **73,** C45, 1974

71.96 **tris - μ - (t - Butylisocyanide) - tris - (t - butylisocyanide)triangulo - triplatinum**
$C_{30}H_{54}N_6Pt_3$
M.Green, J.A.Howard, J.L.Spencer, F.G.A.Stone
J. Chem. Soc., Chem. Commun., 3, 1975

71.97 **Dicarbonyl - (π - cyclopentadienyl)triphenylphosphine - (1,1,2,2 - tetracyano - prop - 1 - yl) molybdenum**
$C_{32}H_{23}MoN_4O_2P$
M.R.Churchill, S.W.-Y.Chang *Inorg. Chem.*, **14**, 98, 1975
Also classified in 73, 86

71.98 **μ - (1 - Phenyl - 2 - (triethoxyphosphine) - ethylen - 1,2 - diyl) - μ - diphenylphosphino - bis(tricarbonyl iron)**
$C_{32}H_{30}Fe_2O_9P_2$
Y.S.Wong, H.N.Paik, P.C.Chieh, A.J.Carty
J. Chem. Soc., Chem. Commun., 309, 1975
Also classified in 86

71.99 **Methylisocyanide - (2 - (bis(t - butyl)phosphino) - 3 - methoxyphenolato) - (2 - (1,1 - dimethylethyl - (t - butyl)phosphino) - 3 - methoxyphenolato) iridium(iii)**
$C_{32}H_{50}IrNO_4P_2$
S.R.Fletcher, W.S.McDonald, M.C.Norton
Acta Crystallogr., Sect. A, **31**, S137, 1975
Also classified in 86, 84

71.100 **(3,3 - bis(Methoxycarbonyl) - 4 - phenyl - 1 - (1′,2′ - diphenylethenyl) - pyrazolidinyl) - hexacarbonyl - di - iron**
$C_{33}H_{24}Fe_2N_2O_{10}$
C.Kruger, H.Kisch *J. Chem. Soc., Chem. Commun.*, 65, 1975
Also classified in 72, 83

71.101 **7 - (Methoxycarbonyl - methyl) - 1,2,3,4,5,6,7 - heptakis(methoxycarbonyl) - bicyclo(2.2.1)heptadiene - 8 - yl - (chloro - bis(pyridine) palladium)**
$C_{34}H_{35}ClN_2O_{16}Pd$
A.Konietzny, P.M.Bailey, P.M.Maitlis
J. Chem. Soc., Chem. Commun., 78, 1975
Also classified in 75, 83

71.102 **pentakis(Phenyl isocyanide) cobalt(i) perchlorate chloroform solvate**
$C_{35}H_{25}CoN_5^+$, $CHCl_3$, ClO_4^-
L.D.Brown, D.R.Greig, K.N.Raymond *Inorg. Chem.*, **14**, 645, 1975

71.103 **Tetra(nickel) hepta(t - butylisocyanide) benzene solvate**
$C_{35}H_{63}N_7Ni_4$, C_6H_6
V.W.Day, R.O.Day, J.S.Kristoff, F.J.Hirsekorn, E.L.Muetterties
J. Am. Chem. Soc., **97**, 2571, 1975

71.104 **Di - iodo - carbonyl - (triphenylphosphine) - (p - tolyl - isocyanide) - (p - tolyl - methylaminocarbene) ruthenium**
$C_{36}H_{33}I_2N_2OPRu$
D.F.Christian, G.R.Clark, W.R.Roper *J. Organomet. Chem.*, **81,** C7, 1974
Also classified in 86

71.105 **bis - μ - (Tetrahydrofuran sodium) - bis(diphenyl(ethylene) nickel(0)) disodium tris(tetrahydrofuran)**
$C_{36}H_{44}Na_2Ni_2O_2$, $3C_4H_8O$, 2Na
D.J.Brauer, C.Kruger, P.J.Roberts, Y.-H.Tsay
Acta Crystallogr., Sect. A, **31,** S131, 1975
Residue 1 also classified in 72; residue 2 classified in 38

71.106 **Di - μ - ethoxy - bis(dibenzylethoxy) titanium(iv)**
$C_{36}H_{48}O_4Ti_2$
H.Stoeckli-Evans *Helv. Chim. Acta,* **58,** 373, 1975
Also classified in 84

71.107 **μ - Diphenylphosphino - μ - (1,4 - bis(trifluoromethyl) - 2 - diphenylphosphino - 1,4 - butadiene - 1,3,4 - triyl) - tri - iron heptacarbonyl**
$C_{37}H_{20}F_6Fe_3O_7P_2$
A.J.Carty, G.Ferguson, H.N.Paik, R.Restivo
J. Organomet. Chem., **74,** C14, 1974
Also classified in 72, 86

71.108 **Nonacarbonyl - μ - (1,2,3,4 - tetraphenyl - butadiene - 1,4 - diyl) - triangulo - triosmium (monoclinic form)**
$C_{37}H_{20}O_9Os_3$
G.Ferraris, G.Gervasio *J. Chem. Soc., Dalton Trans.*, 1813, 1974
Also classified in 72

71.109 **Nonacarbonyl - μ - (1,2,3,4 - tetraphenyl - butadiene - 1,4 - diyl) - triangulo - triosmium (orthorhombic form)**
$C_{37}H_{20}O_9Os_3$
G.Ferraris, G.Gervasio *J. Chem. Soc., Dalton Trans.*, 1813, 1974
Also classified in 72

71.110 **δ - (trans - 1,2 - bis(o - (Diphenylphosphino)phenyl)ethenyl)chloro - platinum(ii) (absolute configuration)**
$C_{38}H_{29}ClP_2Pt$
G.B.Robertson, P.O.Whimp *Inorg. Chem.*, **13,** 2082, 1974
Also classified in 86

71.111 **Dichloro - (difluoromethyl) - carbonyl - bis(triphenylphosphine) iridium(iii)**
$C_{38}H_{31}Cl_2F_2IrOP_2$
A.J.Schultz, J.V.McArdle, G.P.Khare, R.Eisenberg
J. Organomet. Chem., **72,** 415, 1974
Also classified in 86

71.112 Iodo - (fumaronitrile) - (triphenylphosphite) - bis(p - methoxyphenylisonitrile) rhodium(i)
$C_{38}H_{31}IN_4O_5PRh$
A.P.Gaughan Junior, J.A.Ibers
Am. Cryst. Assoc., Abstr. Papers (Spring Meeting), 35, 1975
Also classified in 72, 86

71.113 Diphenylphosphino - (1 - methoxycarbonyl - 1 - diphenylphosphino - 2,4 - bis(trifluoromethyl) - butadiene - 3,4 - diyl) - tri - iron heptacarbonyl benzene solvate
$C_{39}H_{23}F_6Fe_3O_9P_2$, $2C_6H_6$
H.N.Paik, A.J.Carty, M.Mathew, G.J.Palenik
J. Chem. Soc., Chem. Commun., 946, 1974
Residue 1 also classified in 72, 86

71.114 1 - (π - Cyclopentadienyl) - 1 - triphenylphosphine - 2 - phenyl - 3,4,5 - tri(methoxycarbonyl) - 1 - cobaltacyclopent - 2 - ene methylene dichloride solvate
$C_{39}H_{36}CoO_6P$, $0.25CH_2Cl_2$
Y.Wakatsuki, K.Aoki, H.Yamazaki *J. Am. Chem. Soc.*, **96**, 5284, 1974
Residue 1 also classified in 3, 86

71.115 sym - trans - Di - μ - acetato - bis(o - (t - butyl - o - tolyl - phosphino)benzyl)di - palladium(ii)
$C_{40}H_{50}O_4P_2Pd_2$
G.J.Gainsford, R.Mason *J. Organomet. Chem.*, **80**, 395, 1974
Also classified in 81, 86

71.116 5 - Methyl - 2 - ((dimethylamino)methyl)phenyl copper tetramer (further discussion)
(4 - Methyl - 2 - cupriobenzyl)dimethylamine tetramer
$C_{40}H_{56}Cu_4N_4$
G.van Koten, J.G.Noltes *J. Organomet. Chem.*, **84**, 129, 1975
Also classified in 83

71.117 1,1 - bis(Triphenylarsine) - 3,3,4,4 - tetracyano - 1 - platina - 2 - oxacyclobutane
$C_{42}H_{30}As_2N_4OPt$
R.Schlodder, J.A.Ibers, M.Lenarda, M.Graziani
J. Am. Chem. Soc., **96**, 6893, 1974
Also classified in 86, 84

71.118 trans - bis(Ethoxycarbonyl) - bis(triphenylphosphine) platinum
$C_{42}H_{40}O_4P_2Pt$
P.L.Bellon, M.Manassero, F.Porta, M.Sansoni
J. Organomet. Chem., **80**, 139, 1974
Also classified in 86

71.119 **μ - Oxo - bis(tribenzyl titanium(iv))**
$C_{42}H_{42}OTi_2$
H.Stoeckli-Evans *Helv. Chim. Acta,* **57**, 684, 1974

71.120 **trans - Pentafluorophenyl - bis(triphenylphosphine) iridium(i) carbonyl**
$C_{43}H_{30}F_5IrOP_2$
R.Gopal, A.Clearfield, I.Bernal, M.D.Rausch
Am. Cryst. Assoc., Abstr. Papers (Summer Meeting), 230, 1974
Also classified in 86

71.121 **1,1 - bis(Triphenylphosphine) - 2,2,4,4 - tetracyano - 1 - platinacyclobutane**
$C_{43}H_{32}N_4P_2Pt$
D.J.Yarrow, J.A.Ibers, M.Lenarda, M.Graziani
J. Organomet. Chem., **70,** 133, 1974
Also classified in 86

71.122 **Acetylacetonato - (cis - 1,trans - 3 - tetraphenyl - 4 - ethoxy - butadien - 1 - yl) - (dimethylphenylphosphine) palladium(ii)**
$C_{43}H_{43}O_3PPd$
P.-T.Cheng, S.C.Nyburg *Acta Crystallogr., Sect. A,* **31,** S141, 1975
Also classified in 77, 86

71.123 **(4 - Ethoxy - tetraphenylbuta - 1 - cis,3 - trans - dienyl) - (acetylacetonato) - (dimethylphenylphosphine) palladium(ii)**
$C_{43}H_{43}O_3PPd$
P.-T.Cheng, T.R.Jack, C.J.May, S.C.Nyburg, J.Powell
J. Chem. Soc., Chem. Commun., 369, 1975
Also classified in 77, 86

71.124 **bis(Triphenylphosphine) - (hexafluoroacetonimido) - (N,N' - dimethylimidazolidin - 2 - ylideno) rhodium**
$C_{44}H_{40}F_6N_3P_2Rh$
M.J.Doyle, M.F.Lappert, G.M.McLaughlin, J.McMeeking
J. Chem. Soc., Dalton Trans., 1494, 1974
Also classified in 86, 83

71.125 **1 - Ethoxycyclohex - 1,4 - diyl platinum - bis(triphenylphosphine)**
$C_{44}H_{44}OP_2Pt$
M.E.Jason, J.A.McGinnety, K.B.Wiberg
J. Am. Chem. Soc., **96,** 6531, 1974
Also classified in 86

71.126 **Di - iodo - tetrakis(2,6 - diethylphenylisocyanide) cobalt(ii)**
$C_{44}H_{52}CoI_2N_4$
D.Baumann, H.Endres, H.J.Keller, B.Nuber, J.Weiss
Acta Crystallogr., Sect. B, **31,** 40, 1975

71.127 1,1 - bis(Pentafluorophenyl) - 2,5 - diphenyl - (3 - 4(μ - di(pentafluorophenyl) - phosphino)di - iron hexacarbonyl) - 1 - phosphiniacyclopenta - 2,5 - diene
$C_{46}H_{10}F_{20}Fe_2O_6P_2$
N.J.Taylor, H.N.Paik, P.C.Chieh, A.J.Carty
J. Organomet. Chem., **87,** C31, 1975
Also classified in 86

71.128 Ferrocenyl - di - gold - bis(triphenylphosphine) tetrafluoroborate
$C_{46}H_{39}Au_2FeP_2^+$, BF_4^-
V.G.Andrianov, Yu.T.Struchkov, E.R.Rossinskaya
Zh. Strukt. Khim., **15,** 74, 1974
Residue 1 also classified in 73, 86

71.129 Hydrido - chloro - (phenylazophenyl) - bis(triphenylphosphine) iridium(iii) n - hexane solvate
$C_{48}H_{40}ClIrN_2P_2$, C_6H_{14}
J.F.van Baar, R.Meij, K.Olie *Cryst. Struct. Commun.*, **3,** 587, 1974
Residue 1 also classified in 86, 83

71.130 (tris(2 - Diphenylarsinoethyl)amine)phenyl nickel(ii) tetraphenylborate
$C_{48}H_{47}As_3NNi^+$, $C_{24}H_{20}B^-$
P.Dapporto, L.Sacconi *Inorg. Chim. Acta,* **9,** L2, 1974
Residue 1 also classified in 86, 83; residue 2 classified in 62

71.131 π - Cyclopentadienyl - bis(triphenylphosphine) ruthenium - phenylacetylide chloro - copper(i) acetone solvate
$C_{49}H_{40}ClCuP_2Ru$, C_3H_6O
M.I.Bruce, O.M.A.Salah, R.E.Davis, N.V.Raghavan
J. Organomet. Chem., **66,** C48, 1974
Residue 1 also classified in 72, 73, 86

71.132 bis(o - Diphenylphosphino - phenyl) - (triphenylphosphine) iridium(iii) hydride
$C_{54}H_{44}IrP_3$
G.del Piero, G.Perego, A.Zazzetta, M.Cesari
Cryst. Struct. Commun., **3,** 725, 1974
Also classified in 86

71.133 (Tricarbonyl - triphenylphosphine rhenium - bis(pentafluorophenylacetylide)) - triphenylphosphine copper
$C_{55}H_{30}CuF_{10}O_3P_2Re$
O.M.A.Salah, M.I.Bruce, A.D.Redhouse
J. Chem. Soc., Chem. Commun., 855, 1974
Also classified in 73, 86

71.134 μ - Iodo - bis(iodo - tetrakis(phenylisocyanide) cobalt) iodide
$C_{56}H_{40}Co_2I_3N_8^+$, I^-
D.Baumann, H.Endres, H.J.Keller, B.Nuber, J.Weiss
Acta Crystallogr., Sect. B, **31,** 40, 1975

71.135 Di - μ - hydroxy - bis(2,3,4,5 - tetraphenyl - 1 - aura - cyclopentadiene)
$C_{56}H_{42}Au_2O_2$
M.Bois d'Enghien-Peteau, J.Meunier-Piret, M.van Meerssche
Cryst. Struct. Commun., **4**, 375, 1975

71.136 bis(1,3 - Tetraphenylbutadiene - mercury cyanide) mercury
$C_{58}H_{40}Hg_3N_2$
M.Bois-d'Enghien-Peteau, J.Meunier-Piret, M.van Meerssche
Cryst. Struct. Commun., **4**, 383, 1975

71.137 Chloro - bis(triphenylphosphine) platinum - dithiocarboxylato - (bis(triphenylphosphine) platinum) tetrafluoroborate methylene dichloride solvate
$C_{73}H_{60}ClP_4Pt_2S_2^+$, BF_4^- , $0.2CH_2Cl_2$
J.M.Lisy, E.D.Dobrzynski, R.J.Angelici, J.Clardy
J. Am. Chem. Soc., **97**, 656, 1975
Residue 1 also classified in 86, 85

71.138 Tri - μ - diphenylphosphino - bis(triphenylphosphine platinum) - (phenyl - platinum) benzene solvate
$C_{78}H_{65}P_5Pt_3$, C_6H_6
N.J.Taylor, P.C.Chieh, A.J.Carty
J. Chem. Soc., Chem. Commun., 448, 1975
Residue 1 also classified in 86

71.139 Di(triphenylphosphine - silver) - (triphenylphosphine - rhodium - pentakis(pentafluorophenylacetylide))
$C_{94}H_{45}Ag_2F_{25}P_3Rh$
O.M.A.Salah, M.I.Bruce, M.R.Churchill, B.G.DeBoer
J. Chem. Soc., Chem. Commun., 688, 1974
Also classified in 72, 86

METAL PI-COMPLEXES (OPEN-CHAIN)

72.1 **Tetrafluoroethylene - bis(ethylene) platinum**
$C_6H_8F_4Pt$
M.Green, J.A.K.Howard, J.L.Spencer, F.G.A.Stone
J. Chem. Soc., Chem. Commun., 449, 1975

72.2 **Tricarbonyl - η^4 - (3 - methylene - 4 - vinyldihydrofuran - 2(3H) - one) iron**
$C_{10}H_8FeO_5$
M.Green, R.P.Hughes, A.J.Welch
J. Chem. Soc., Chem. Commun., 487, 1975

72.3 **π - Allyl - dicarbonyl - (trifluoroacetato) - (1,2 - dimethoxyethane) molybdenum**
$C_{11}H_{15}F_3MoO_6$
F.Dawans, J.Dewailly, J.Meunier-Piret, P.Piret
J. Organomet. Chem., **76,** 53, 1974
Also classified in 81, 84

72.4 **π - Allyl - dicarbonyl - (trifluoroacetato) - (1,2 - dimethoxyethane) tungsten**
$C_{11}H_{15}F_3O_6W$
F.Dawans, J.Dewailly, J.Meunier-Piret, P.Piret
J. Organomet. Chem., **76,** 53, 1974
Also classified in 81, 84

72.5 **η^2 - (cis - 2,3 - Dicarbomethoxy - methylene - cyclopropane) iron tetracarbonyl**
$C_{12}H_{10}FeO_8$
T.H.Whitesides, R.W.Slaven, J.C.Calabrese *Inorg. Chem.*, **13,** 1895, 1974

72.6 **($\pm$) - ZZ - 1,2,3 - η - 5,6,7 - η - Heptadienediyl rhodium(iii) hexafluoroacetylacetonate**
$C_{12}H_{11}F_6O_2Rh$
N.W.Alcock, J.M.Brown, J.A.Conneely, J.J.Stofko Junior
J. Chem. Soc., Chem. Commun., 234, 1975
Also classified in 77

72.7 **tris(Butadiene) molybdenum(0)**
$C_{12}H_{18}Mo$
M.M.Yevitz, P.S.Skell
Am. Cryst. Assoc., Abstr. Papers (Summer Meeting), 229, 1974

72.8 **tris(Butadiene) tungsten(0)**
$C_{12}H_{18}W$
M.M.Yevitz, P.S.Skell
Am. Cryst. Assoc., Abstr. Papers (Summer Meeting), 229, 1974

72.C **Di - μ - chloro - bis(chloro - (2 - (hydroxymethyl)pent - 4 - enyl) rhodium(iii)) methanol solvate**
$C_{12}H_{22}Cl_4O_2Rh_2$, CH_4O
For complete entry see 71.26

72.9 **Chloro - π - cyclopentadienyl - bis(π - (hexafluoro - 2 - butyne)) tungsten**
$C_{13}H_5ClF_{12}W$
J.L.Davidson, M.Green, D.W.A.Sharp, F.G.A.Stone, A.J.Welch
J. Chem. Soc., Chem. Commun., 706, 1974
Also classified in 73

72.C **Methyl(hydro - tris(1 - pyrazolyl)borato)hexafluorobut - 2 - yne platinum(ii)**
$C_{14}H_{13}BF_6N_6Pt$
For complete entry see 71.35

72.10 **Dicarbonyl - η^5 - cyclopentadienyl - (η^2 - tetramethylallenyl) iron tetrafluoroborate**
$C_{14}H_{17}FeO_2^+$, BF_4^-
B.M.Foxman *J. Chem. Soc., Chem. Commun.,* 221, 1975
Residue 1 also classified in 73

72.C **bis(Cyclopentadienyl) - ethyl - (ethylene) niobium**
$C_{14}H_{19}Nb$
For complete entry see 71.38

72.11 **(2 - (m - Nitrophenyl - amino) - trans,trans - 3,5 - heptadiene) iron tricarbonyl**
$C_{16}H_{16}FeN_2O_5$
A.Immirzi *J. Organomet. Chem.,* **76,** 65, 1974

72.12 **(2 - (Phenylamino) - cis,trans - 3,5 - heptadiene) iron tricarbonyl**
$C_{16}H_{17}FeNO_3$
A.Immirzi *J. Organomet. Chem.,* **76,** 65, 1974

72.13 **μ - Dimethylarsino - μ - (1,3 - η - (2,3 - bis(dimethylarsino) - 1,1 - difluoro - 3 - trifluoromethylallyl)) - bis(tricarbonyl manganese(i))**
$C_{16}H_{18}As_3F_5Mn_2O_6$
F.W.B.Einstein, J.S.Field *J. Chem. Soc., Dalton Trans.,* 172, 1975
Also classified in 86

72.C **π - Cyclopentadienyl - π - (hexafluoro - 2 - butyne) - (1,2 - bis(trifluoromethyl) - 2 - cyclopentadienyl - ethylene - 1 - yl) molybdenum**
$C_{18}H_{10}F_{12}Mo$
For complete entry see 71.49

72.C **bis(π - 2 - Ethyl - 3 - methylthioacrolein iron(0) dicarbonyl)**
$C_{18}H_{20}Fe_2O_4S_2$
For complete entry see 71.53

72.14 **Di - μ - chloro - bis((π - benzene - π - allyl) molybdenum(ii))**
$C_{18}H_{22}Cl_2Mo_2$
K.Prout, G.V.Rees *Acta Crystallogr., Sect. B,* **30,** 2251, 1974
Also classified in 74

72.15 **(Diethyl - bis(pyrazolyl)borato) - (pyrazolylato) - (trihapto - allyl) - (dicarbonyl) molybdenum**
$C_{18}H_{25}BMoN_6O_2$
F.A.Cotton, B.A.Frenz, A.G.Stanislowski *Inorg. Chim. Acta,* **7,** 503, 1973
Also classified in 83, 62

72.16 **(bis - π - Cyclopentadienyl molybdenum) - bis - μ - methanethiolato - (bis - π - allyl rhodium) hexafluorophosphate**
$C_{18}H_{26}MoRhS_2^+$, F_6P^-
K.Prout, G.V.Rees *Acta Crystallogr., Sect. B,* **30,** 2249, 1974
Residue 1 also classified in 73, 85

72.C **(π - Pentenyl) - di - (isopropyl - phenyl - phosphine) - methyl nickel(ii)**
$C_{18}H_{31}NiP$
For complete entry see 71.56

72.17 **Carbonyl - chloro - bis(dimethylphenylphosphine) - allyl - iridium(iii) hexafluorophosphate**
$C_{20}H_{27}ClIrOP_2^+$, F_6P^-
I.A.Mustafa, J.H.Robertson *Eur. Cryst. Meeting,* 379, 1974
Residue 1 also classified in 86

72.18 **(Tetracyanoethylene) - oxo - bis(di - n - propyldithiocarbamato) molybdenum**
$C_{20}H_{28}MoN_6OS_4$
L.Ricard, R.Weiss *Inorg. Nucl. Chem. Lett.,* **10,** 217, 1974
Also classified in 80

72.C **Di - μ - chloro - bis((3 - allyl - norbornyl) palladium(ii))**
$C_{20}H_{30}Cl_2Pd_2$
For complete entry see 71.64

72.19 **Carbonyl - bis(dimethylphenylphosphine) - butadiene iridium(i) tetrafluoroborate**
$C_{21}H_{28}IrOP_2^+$, BF_4^-
I.A.Mustafa, J.H.Robertson *Eur. Cryst. Meeting,* 379, 1974
Residue 1 also classified in 86

72.20 **Carbonyl - bis(dimethylphenylphosphine) - butadiene iridium(i) perchlorate**
$C_{21}H_{28}IrOP_2^+$, ClO_4^-
I.A.Mustafa, J.H.Robertson *Eur. Cryst. Meeting,* 379, 1974
Residue 1 also classified in 86

72.21 **Carbonyl - chloro - bis(dimethylphenylphosphine) - methallyl - iridium(iii) hexafluorophosphate**
$C_{21}H_{29}ClIrOP_2^+$, F_6P^-
I.A.Mustafa, J.H.Robertson *Eur. Cryst. Meeting,* 379, 1974
Residue 1 also classified in 86

72.22 **π - Allyl - bis(tri - isopropylphosphine) iridium(i)**
$C_{21}H_{47}IrP_2$
G.Perego, G.del Piero, M.Cesari *Cryst. Struct. Commun.,* **3,** 721, 1974
Also classified in 86

72.23 **Carbonyl - bis(dimethylphenylphosphine) - methylbutadiene - iridium(i) tetraphenylborate**
$C_{22}H_{30}IrOP_2^+$, $C_{24}H_{20}B^-$
I.A.Mustafa, J.H.Robertson *Eur. Cryst. Meeting,* 379, 1974
Residue 1 also classified in 86; residue 2 classified in 62

72.24 **6,6 - Diphenylpentafulvene - pentacarbonyl - di - iron**
$C_{23}H_{14}Fe_2O_5$
U.Behrens, E.Weiss *J. Organomet. Chem.,* **73,** C64, 1974
Also classified in 73

72.25 **6,6 - Diphenylpentafulvene - pentacarbonyl - diruthenium (triclinic form)**
$C_{23}H_{14}O_5Ru_2$
U.Behrens, E.Weiss *J. Organomet. Chem.,* **73,** C67, 1974
Also classified in 73

72.26 **π - Allyl - triphenylphosphine - dicarbonyl iron iodide (red form)**
$C_{23}H_{20}FeIO_2P$
M.Kh.Minasyants, Yu.T.Struchkov, V.G.Andrianov
Yerevan Univ. Trans., **3,** 25, 1972
Also classified in 86

72.27 **π - Allyl - triphenylphosphine - dicarbonyl iron iodide (black form)**
$C_{23}H_{20}FeIO_2P$
M.Kh.Minasyants, Yu.T.Struchkov, V.G.Andrianov
Yerevan Univ. Trans., **2,** 27, 1972
Also classified in 86

72.28 **(Diphenyldipyrazolylborato)(2 - methylallyl)dicarbonyl molybdenum**
$C_{24}H_{23}BMoN_4O_2$
F.A.Cotton, B.A.Frenz, C.A.Murillo *J. Am. Chem. Soc.*, **97**, 2118, 1975
Also classified in 83, 62

72.29 **μ - (N,N′ - Ethylene - bis(salicylaldiminato)) - bis(2 - methallyl palladium(ii))**
$C_{24}H_{28}N_2O_2Pd_2$
B.M.Gatehouse, B.E.Reichert, B.O.West
Acta Crystallogr., Sect. B, **30**, 2451, 1974
Also classified in 78

72.C **Tricarbonyl - (2,4,7 - tri - t - butyl - hepta - 1,5 - diene - 3 - one - 1,4,7 - triyl) iron dicarbonyl - iron**
$C_{24}H_{30}Fe_2O_6$
For complete entry see 71.81

72.30 **Trimethylenemethane - chromium - tricarbonyl - triphenylphosphine**
$C_{25}H_{21}CrO_3P$
W.Henslee, R.E.Davis *J. Organomet. Chem.*, **81**, 389, 1974
Also classified in 86

72.C **α - Carboranyl complex of platinum(ii)**
$C_{26}H_{56}B_{10}P_2Pt$
For complete entry see 71.86

72.31 **(Dibenzylideneacetone) - (pentamethylcyclopentadienyl) rhodium(i)**
$C_{27}H_{29}ORh$
J.A.Ibers *J. Organomet. Chem.*, **73**, 389, 1974
Also classified in 73

72.C **(π - Pentenyl)(dimenthyl - methyl - phosphine) - methyl nickel(ii)**
$C_{27}H_{53}NiP$
For complete entry see 71.89

72.C **Di - μ - chloro - bis((1 - (dicarbonyl - π - cyclopentadienyl - ferrio) - 2 - phenyl - ethyne) copper(i)) dimer**
$C_{30}H_{20}Cl_2Cu_2Fe_2O_4$
For complete entry see 71.93

72.32 **(Tetramethylethylene) nickel (1,2 - bis(dicyclohexylphosphino)ethane)**
$C_{32}H_{60}NiP_2$
D.J.Brauer, C.Kruger *J. Organomet. Chem.*, **77**, 423, 1974
Also classified in 86

72.C **(3,3 - bis(Methoxycarbonyl) - 4 - phenyl - 1 - (1′,2′ - diphenylethenyl) - pyrazolidinyl) - hexacarbonyl - di - iron**
$C_{33}H_{24}Fe_2N_2O_{10}$
For complete entry see 71.100

72.C **bis - μ - (Tetrahydrofuran sodium) - bis(diphenyl(ethylene) nickel(0)) disodium tris(tetrahydrofuran)**
$C_{36}H_{44}Na_2Ni_2O_2$, $3C_4H_8O$, 2Na
For complete entry see 71.105

72.C **μ - Diphenylphosphino - μ - (1,4 - bis(trifluoromethyl) - 2 - diphenylphosphino - 1,4 - butadiene - 1,3,4 - triyl) - tri - iron heptacarbonyl**
$C_{37}H_{20}F_6Fe_3O_7P_2$
For complete entry see 71.107

72.C **Nonacarbonyl - μ - (1,2,3,4 - tetraphenyl - butadiene - 1,4 - diyl) - triangulo - triosmium (monoclinic form)**
$C_{37}H_{20}O_9Os_3$
For complete entry see 71.108

72.C **Nonacarbonyl - μ - (1,2,3,4 - tetraphenyl - butadiene - 1,4 - diyl) - triangulo - triosmium (orthorhombic form)**
$C_{37}H_{20}O_9Os_3$
For complete entry see 71.109

72.C **Iodo - (fumaronitrile) - (triphenylphosphite) - bis(p - methoxyphenylisonitrile) rhodium(i)**
$C_{38}H_{31}IN_4O_5PRh$
For complete entry see 71.112

72.33 **bis - μ - (Diphenyl - (t - butylacetylenyl) - phosphine) - di(carbonyl nickel(0))**
$C_{38}H_{38}Ni_2O_2P_2$
H.N.Paik, A.J.Carty, K.Dymock, G.J.Palenik
J. Organomet. Chem., **70,** C17, 1974
Also classified in 86

72.C **Diphenylphosphino - (1 - methoxycarbonyl - 1 - diphenylphosphino - 2,4 - bis(trifluoromethyl) - butadiene - 3,4 - diyl) - tri - iron heptacarbonyl benzene solvate**
$C_{39}H_{23}F_6Fe_3O_9P_2$, $2C_6H_6$
For complete entry see 71.113

72.34 **Iodo - bis(triphenylphosphine)allene rhodium**
$C_{39}H_{34}IP_2Rh$
T.Kashiwagi, N.Yasuoka, N.Kasai, M.Kakudo
Technol. Rep. Osaka Univ., **24,** 355, 1974
Also classified in 86

72.35 **bis(Triphenylphosphine)allene palladium**
$C_{39}H_{34}P_2Pd$
K.Okamoto, G.Kai, N.Yasuoka, N.Kasai
J. Organomet. Chem., **65,** 427, 1974
Also classified in 86

72.36 **bis(Triphenylphosphine)hexafluorobut - 2 - yne platinum(0)**
$C_{40}H_{30}F_6P_2Pt$
B.W.Davies, N.C.Payne *Inorg. Chem.*, **13**, 1848, 1974
Also classified in 86

72.37 **1,1,2 - Trimethylallyl - bis(triphenylphosphine) nickel trichlorozincate**
$C_{42}H_{41}NiP_2^+$, Cl_3Zn^-
M.Zocchi, A.Albinati *J. Organomet. Chem.*, **77**, C40, 1974
Residue 1 also classified in 86

72.38 **bis - μ - (Diphenyl - (phenylacetylenyl) - phosphine) - di(tricarbonyl iron)**
$C_{46}H_{30}Fe_2O_6P_2$
H.N.Paik, A.J.Carty, K.Dymock, G.J.Palenik
J. Organomet. Chem., **70**, C17, 1974
Also classified in 86

72.C **π - Cyclopentadienyl - bis(triphenylphosphine) ruthenium - phenylacetylide chloro - copper(i) acetone solvate**
$C_{49}H_{40}ClCuP_2Ru$, C_3H_6O
For complete entry see 71.131

72.39 **4,4' - Dinitro - trans - stilbene - bis(triphenylphosphine) platinum**
$C_{50}H_{40}N_2O_4P_2Pt$
J.M.Baraban, J.A.McGinnety *Inorg. Chem.*, **13**, 2864, 1974
Also classified in 86

72.40 **tris(Dibenzylidene - acetone) dipalladium(0) chloroform solvate**
$C_{51}H_{42}O_3Pd_2$, $CHCl_3$
T.Ukai, H.Kawazura, Y.Ishii, J.J.Bonnet, J.A.Ibers
J. Organomet. Chem., **65**, 253, 1974

72.41 **tris(Dibenzylideneacetone) dipalladium(0) methylene chloride solvate**
$C_{51}H_{42}O_3Pd_2$, CH_2Cl_2
C.G.Pierpont, M.C.Mazza *Inorg. Chem.*, **13**, 1891, 1974

72.42 **tris(Dibenzylideneacetone) dipalladium(0) benzene solvate**
$C_{51}H_{42}O_3Pd_2$, C_6H_6
C.G.Pierpont, M.C.Mazza *Inorg. Chem.*, **13**, 1891, 1974

72.43 **bis(Tri - p - tolylphosphine) - (trans - stilbene) nickel(0) tetrahydrofuran solvate**
$C_{56}H_{54}NiP_2$, $0.5C_4H_8O$
S.D.Ittel, J.A.Ibers *J. Organomet. Chem.*, **74**, 121, 1974
Residue 1 also classified in 86

72.44 **π - Cyclopentadienyl - bis(π - triphenylgermyl - phenyl - acetylene) - carbonyl - niobium**
$C_{58}H_{45}Ge_2NbO$
N.I.Kirillova, N.E.Kolobova, A.I.Gusev, A.B.Antonova, Yu.T.Struchkov, K.N.Anisimov, O.M.Khitrova *Zh. Strukt. Khim.*, **15**, 651, 1974
Also classified in 73, 69

72.45 **Di - μ - (1 - diphenylphosphino - 2 - trifluoromethyl - acetylene) - bis(triphenylphosphine) dipalladium(0)**
$C_{66}H_{50}F_6P_4Pd_2$
S.Jacobson, A.J.Carty, M.Mathew, G.J.Palenik
J. Am. Chem. Soc., **96**, 4330, 1974
Also classified in 86

72.C **Di(triphenylphosphine - silver) - (triphenylphosphine - rhodium - pentakis(pentafluorophenylacetylide))**
$C_{94}H_{45}Ag_2F_{25}P_3Rh$
For complete entry see 71.139

METAL PI-COMPLEXES (CYCLOPENTADIENE)

73.1 **8 - η - Cyclopentadienyl - 6,7 - dicarba - 8 - cobalta - nido - nonaborate(ii) (at $-160°C$)**
$C_7H_{16}B_7Co$
K.P.Callahan, F.Y.Lo, C.E.Strouse, A.L.Sims, M.F.Hawthorne
Inorg. Chem., **13,** 2842, 1974
Also classified in 62

73.2 **Dicarbonyl - (π - cyclopentadienyl)(bis(trifluoromethyl)phosphino) iron**
$C_9H_5F_6FeO_2P$
M.J.Barrow, G.A.Sim *J. Chem. Soc., Dalton Trans.*, 291, 1975
Also classified in 86

73.3 **Dicarbonyl - (π - cyclopentadienyl)(bis(trifluoromethyl) - oxo - phosphino) iron**
$C_9H_5F_6FeO_3P$
M.J.Barrow, G.A.Sim *J. Chem. Soc., Dalton Trans.*, 291, 1975
Also classified in 86

73.4 **bis(π - Cyclopentadienyl)dibromo - rhenium(v) tetrafluoroborate**
$C_{10}H_{10}Br_2Re^+$, BF_4^-
K.Prout, T.S.Cameron, R.A.Forder, S.R.Critchley, B.Denton, G.V.Rees
Acta Crystallogr., Sect. B, **30,** 2290, 1974

73.5 **bis(π - Cyclopentadienyl)dichloro - molybdenum(iv)**
$C_{10}H_{10}Cl_2Mo$
K.Prout, T.S.Cameron, R.A.Forder, S.R.Critchley, B.Denton, G.V.Rees
Acta Crystallogr., Sect. B, **30,** 2290, 1974

73.6 **bis(π - Cyclopentadienyl)dichloro - molybdenum(v) tetrafluoroborate**
$C_{10}H_{10}Cl_2Mo^+$, BF_4^-
K.Prout, T.S.Cameron, R.A.Forder, S.R.Critchley, B.Denton, G.V.Rees
Acta Crystallogr., Sect. B, **30,** 2290, 1974

73.7 **bis(π - Cyclopentadienyl)dichloro - niobium(iv)**
$C_{10}H_{10}Cl_2Nb$
K.Prout, T.S.Cameron, R.A.Forder, S.R.Critchley, B.Denton, G.V.Rees
Acta Crystallogr., Sect. B, **30,** 2290, 1974

73.8 **bis(π - Cyclopentadienyl)dichloro - zirconium**
$C_{10}H_{10}Cl_2Zr$
K.Prout, T.S.Cameron, R.A.Forder, S.R.Critchley, B.Denton, G.V.Rees
Acta Crystallogr., Sect. B, **30,** 2290, 1974

73.9 **bis - (π - Cyclopentadienyl) molybdenum tetrasulfide**
$C_{10}H_{10}MoS_4$
H.D.Block, R.Allmann *Cryst. Struct. Commun.,* **4,** 53, 1975

73.C **Dicarbonyl - (η - cyclopentadienyl) - (3 - aminopropionyl) molybdenum(ii)**
$C_{10}H_{11}MoNO_3$
For complete entry see 71.13

73.10 **bis(π - Cyclopentadienyl) niobium tetrahydroborate**
$C_{10}H_{14}BNb$
N.I.Kirillova, A.I.Gusev, Yu.T.Struchkov *Zh. Strukt. Khim.,* **15,** 718, 1974

73.11 **Di - μ - carbonyl - (tricarbonylcobaltio) - carbonyl(π - cyclopentadienyl) iron**
$C_{11}H_5CoFeO_6$
I.L.C.Campbell, F.S.Stephens *J. Chem. Soc., Dalton Trans.,* 22, 1975

73.12 **Di - (π - cyclopentadienyl) niobium carbonyl hydrosulfide**
$C_{11}H_{11}NbOS$
N.I.Kirillova, A.I.Gusev, A.A.Pasynskii, Yu.T.Struchkov
Zh. Strukt. Khim., **14,** 868, 1973

73.13 **Iodo - nitrosyl - cyclopentadienyl - (phenylhydrazine) molybdenum tetrafluoroborate**
$C_{11}H_{13}IMoN_3O^+$, BF_4^-
N.A.Bailey, P.D.Frisch, J.A.McCleverty, N.W.J.Walker, J.Williams
J. Chem. Soc., Chem. Commun., 350, 1975
Residue 1 also classified in 83

73.14 **bis(π - Cyclopentadienyl)hydroxy - methylamino molybdenum(iv) hexafluorophosphate**
$C_{11}H_{16}MoNO^+$, F_6P^-
K.Prout, T.S.Cameron, R.A.Forder, S.R.Critchley, B.Denton, G.V.Rees
Acta Crystallogr., Sect. B, **30,** 2290, 1974

73.C **1 - (Difluoromethyl - carbonyl - (cyclopentadienyl) iron) - cyclopentadiene**
$C_{12}H_{10}F_2FeO$
For complete entry see 71.22

73.15 **bis(η - Cyclopentadienyl) - bis(N - cyanato) titanium(iv)**
$C_{12}H_{10}N_2O_2Ti$
S.J.Anderson, D.S.Brown, A.H.Norbury
J. Chem. Soc., Chem. Commun., 996, 1974

73.C **bis(π - Cyclopentadienyl)ethyl - chloro - molybdenum(iv)**
$C_{12}H_{15}ClMo$
For complete entry see 71.24

73.16 **1,6 - bis(η - Cyclopentadienyl) - 1,6 - diferra - 2,3 - dicarba - closo - decaborane(8)**
$C_{12}H_{18}B_6Fe_2$
K.P.Callahan, W.J.Evans, F.Y.Lo, C.E.Strouse, M.F.Hawthorne
J. Am. Chem. Soc., **97**, 296, 1975
Also classified in 62

73.17 **Cobaltocenium - carborane**
$C_{12}H_{20}B_9Co$
M.R.Churchill, B.G.DeBoer *J. Am. Chem. Soc.*, **96**, 6310, 1974
Also classified in 62

73.C **Chloro - π - cyclopentadienyl - bis(π - (hexafluoro - 2 - butyne)) tungsten**
$C_{13}H_5ClF_{12}W$
For complete entry see 72.9

73.18 **π - Cyclopentadienyl - (1 - hydroxy - 2,3,4,5 - tetrakis(trifluoromethyl) - phosphole - 1 - oxide) cobalt**
$C_{13}H_6CoF_{12}O_2P$
M.J.Barrow, A.A.Freer, W.Harrison, G.A.Sim, D.W.Taylor, F.B.Wilson
J. Chem. Soc., Dalton Trans., 197, 1975
Also classified in 75, 64

73.19 **Dicarbonyl - (π - cyclopentadienyl) - manganese - μ(dimethylarsino) - tetracarbonyl - manganese**
$C_{13}H_{11}AsMn_2O_6$
H.Vahrenkamp *Chem. Ber.*, **107**, 3867, 1974
Also classified in 86

73.20 **π - Pentachlorocyclopentadienyl - (cyclo - octadienyl) rhodium**
$C_{13}H_{12}Cl_5Rh$
V.W.Day, K.J.Reimer, A.Shaver
J. Chem. Soc., Chem. Commun., 403, 1975
Also classified in 75

73.21 **Dichloro - (1,1′ - trimethylene - π - dicyclopentadienyl) hafnium**
$C_{13}H_{14}Cl_2Hf$
C.H.Saldarriaga-Molina, A.Clearfield, I.Bernal
Inorg. Chem., **13**, 2880, 1974

73.22 **(1,1′ - Trimethylene - dicyclopentadienyl) - dichloro zirconium**
$C_{13}H_{14}Cl_2Zr$
C.H.Saldarriaga-Molina, A.Clearfield, I.Bernal
J. Organomet. Chem., **80**, 79, 1974

73.C **Dicarbonyl - η^5 - cyclopentadienyl - (η^2 - tetramethylallenyl) iron tetrafluoroborate**
$C_{14}H_{17}FeO_2^+$, BF_4^-
For complete entry see 72.10

73.C **bis(Cyclopentadienyl) - ethyl - (ethylene) niobium**
$C_{14}H_{19}Nb$
For complete entry see 71.38

73.23 **Ammine - bis(η - cyclopentadienyl) - N - (C - methyl - C - ethylketimino) molybdenum bis(hexafluorophosphate)**
$C_{14}H_{22}MoN_2^{2+}$, $2F_6P^-$
R.A.Forder, G.D.Gale, K.Prout *Acta Crystallogr., Sect. B,* **31,** 297, 1975
Residue 1 also classified in 83

73.C **1 - (Dicarbonyl - π - cyclopentadienyl - ferrio) - 2 - (phenyl) - ethyne**
$C_{15}H_{10}FeO_2$
For complete entry see 71.39

73.24 **tris(π - Cyclopentadienyl) dinickel tetrafluoroborate**
$C_{15}H_{15}Ni_2^+$, BF_4^-
E.Dubler, M.Textor, H.-R.Oswald, A.Salzer *Angew. Chem.,* **86,** 125, 1974

73.25 **μ - (Dichlorostannio) - bis(tricarbonyl - π - cyclopentadienyl chromium)**
$C_{16}H_{10}Cl_2Cr_2O_6Sn$
F.S.Stephens *J. Chem. Soc., Dalton Trans.,* 230, 1975
Also classified in 69

73.26 **Zinc - bis(tricarbonyl cyclopentadienyl molybdenum)**
$C_{16}H_{10}Mo_2O_6Zn$
J.St.Denis, W.Butler, M.D.Glick, J.P.Oliver
J. Am. Chem. Soc., **96,** 5427, 1974

73.27 **(Carbonyl - π - norbornadiene cobalt) - di - μ - carbonyl - (carbonyl - π - cyclopentadienyl iron)**
$C_{16}H_{13}CoFeO_4$
I.L.C.Campbell, F.S.Stephens *J. Chem. Soc., Dalton Trans.,* 226, 1975
Also classified in 75

73.28 **μ - Hydrido - μ - dimethylphosphido - bis(π - cyclopentadienyldicarbonyl molybdenum) (neutron study)**
$C_{16}H_{17}Mo_2O_4P$
J.L.Petersen, L.F.Dahl, J.M.Williams *J. Am. Chem. Soc.,* **96,** 6610, 1974
Also classified in 86

73.C **μ - Carbonyl - μ - dicyanovinylidene - bis(carbonyl - cyclopentadienyl iron(0))**
$C_{17}H_{10}Fe_2N_2O_3$
For complete entry see 71.46

73.C **π - Cyclopentadienyl - π - (hexafluoro - 2 - butyne) - (1,2 - bis(trifluoromethyl) - 2 - cyclopentadienyl - ethylene - 1 - yl) molybdenum**
$C_{18}H_{10}F_{12}Mo$
For complete entry see 71.49

73.29 **Phenylacetyl - ferrocene**
$C_{18}H_{16}FeO$
E.Gyepes, F.Hanic *Cryst. Struct. Commun.*, **4,** 229, 1975

73.C **cis - (Carbonyl - (π - cyclopentadienyl) iron) - di - μ - carbonyl - (isobutylisocyano - (π - cyclopentadienyl) iron)**
$C_{18}H_{19}Fe_2NO_3$
For complete entry see 71.51

73.C **Di(cyclopentadienyl) - 2,6 - dimethylphenyl titanium**
$C_{18}H_{19}Ti$
For complete entry see 71.52

73.30 **Neodymium tris(methylcyclopentadienide)**
$C_{18}H_{21}Nd$
J.H.Burns, W.H.Baldwin, F.H.Fink *Inorg. Chem.*, **13,** 1916, 1974

73.31 **μ - Oxo - bis((η^5 - cyclopentadienyl - dicarbonyl iron) - dimethyl germanium)**
$C_{18}H_{22}Fe_2Ge_2O_5$
R.D.Adams, F.A.Cotton, B.A.Frenz *J. Organomet. Chem.*, **73,** 93, 1974
Also classified in 69

73.32 **Tri - μ - formato - tri - μ - hydroxy - tri(cyclopentadienyl niobium) - μ_3 - oxide hydride**
$C_{18}H_{22}Nb_3O_{10}$
N.I.Kirillova, A.I.Gusev, A.A.Pasynskii, Yu.T.Struchkov
Zh. Strukt. Khim., **14,** 1075, 1973
Also classified in 81

73.C **(pentahapto - Cyclopentadienyl) - 1,3′,4′ - trihapto - (2 - methyl - 2 - (6′,6′ - dimethyl - bicyclo(3.2.0)hept - 3′ - en - 7′ - on - 2′ - yl)propionyl) nickel(ii)**
$C_{18}H_{22}NiO_2$
For complete entry see 71.54

73.C **(bis - π - Cyclopentadienyl molybdenum) - bis - μ - methanethiolato - (bis - π - allyl rhodium) hexafluorophosphate**
$C_{18}H_{26}MoRhS_2^+$, F_6P^-
For complete entry see 72.16

73.33 **Di - μ - sulfido - bis(cyclopentadienyl - (t - butylimido) - molybdenum)**
$C_{18}H_{28}Mo_2N_2S_2$
L.F.Dahl, P.D.Frisch, G.R.Gust *J. Less-Common Met.*, **36,** 255, 1974
Also classified in 83

73.34 **t - Butylamido - tris(π - cyclopentadienyl nickel)**
$C_{19}H_{24}NNi_3$
N.Kamijyo, T.Watanabe *Bull. Chem. Soc. Jpn.*, **47**, 373, 1974
Also classified in 83

73.C **1 - Carbonyl - 1 - cyclopentadienyl - 1 - (cyclopentadienyl iron) - 2,3,5,6 - tetra - (trifluoromethyl) - 1 - ferra - cyclohexa - 2,5 - diene - 4 - one**
$C_{20}H_{10}F_{12}Fe_2O_2$
For complete entry see 71.61

73.35 **μ - (1,1′ - Dicyclopentadienyl) - di - μ - chloro - bis(cyclopentadienyl titanium)**
$C_{20}H_{18}Cl_2Ti_2$
G.J.Olthof, H.R.van der Wal *Acta Crystallogr.*, *Sect. A*, **31**, S135, 1975

73.36 **μ - Oxo - bis(bis - (π - cyclopentadienyl)chloro niobium(v)) tetrafluoroborate**
$C_{20}H_{20}Cl_2Nb_2O^{2+}$, $2BF_4^-$
K.Prout, T.S.Cameron, R.A.Forder, S.R.Critchley, B.Denton, G.V.Rees *Acta Crystallogr.*, *Sect. B*, **30**, 2290, 1974

73.37 **bis(Chloro - cyclopentadienyl - zirconium)oxide**
$C_{20}H_{20}Cl_2OZr_2$
J.F.Clarke, M.G.B.Drew *Acta Crystallogr.*, *Sect. B*, **30**, 2267, 1974

73.38 **Tetra(cyclopentadienyl) uranium(iv)**
$C_{20}H_{20}U$
J.H.Burns *J. Organomet. Chem.*, **69**, 225, 1974

73.39 **tetrakis(Cyclopentadienyl)trihydrido - tetranickel**
$C_{20}H_{23}Ni_4$
G.Huttner, H.Lorenz *Chem. Ber.*, **107**, 996, 1974

73.40 **tetrakis(π - Cyclopentadienyl - hydrido cobalt)**
$C_{20}H_{24}Co_4$
G.Huttner, H.Lorenz *Chem. Ber.*, **108**, 973, 1975

73.C **1 - bis(Cyclopentadienyl) zirconium(iv) - μ - chloro - 2,2 - bis(diethyl aluminium) ethane**
$C_{20}H_{33}Al_2ClZr$
For complete entry see 71.65

73.C **bis(6,6 - Dimethylpentafulvene) - pentacarbonyl - di - iron**
$C_{21}H_{20}Fe_2O_5$
For complete entry see 71.72

73.C **bis(η - Cyclopentadienyl) - σ - ((2 - dimethylphenylphosphonium)ethyl)methyl - tungsten hexafluorophosphate**
$C_{21}H_{28}PW^+$, F_6P^-
For complete entry see 71.74

73.41 **η^5 - Tetrahydropentalenyl - η^7 - cyclo - octatrienyl - hexacarbonyl - triruthenium**
$C_{22}H_{18}O_6Ru_3$
R.Bau, B.Ch.-K.Chou, S.A.R.Knox, V.Riera, F.G.A.Stone
J. Organomet. Chem., **82,** C43, 1974
Also classified in 75

73.42 **bis(1,3 - Dimethylindenyl) iron(iii) hexafluorophosphate**
$C_{22}H_{22}Fe^+$, F_6P^-
P.M.Treichel, J.W.Johnson, J.C.Calabrese
J. Organomet. Chem., **88,** 215, 1975

73.43 **bis(π - Cyclopentadienyl) - hexa(dimethylphosphonato) tricobalt**
$C_{22}H_{46}Co_3O_{18}P_6$
V.Harder, E.Dubler, H.Werner *J. Organomet. Chem.*, **71,** 427, 1974
Also classified in 86

73.C **6,6 - Diphenylpentafulvene - pentacarbonyl - di - iron**
$C_{23}H_{14}Fe_2O_5$
For complete entry see 72.24

73.C **6,6 - Diphenylpentafulvene - pentacarbonyl - diruthenium (triclinic form)**
$C_{23}H_{14}O_5Ru_2$
For complete entry see 72.25

73.44 **Dicyclopentadienyl - di - (p - nitrobenzoate) titanium**
$C_{24}H_{18}N_2O_8Ti$
E.A.Gladkikh, T.S.Kuntsevich *Zh. Strukt. Khim.*, **14,** 949, 1973
Also classified in 81

73.45 **π - Cyclopentadienyl - (triphenylphosphine)carbonyl iridium**
$C_{24}H_{20}IrOP$
M.J.Bennett, J.L.Pratt, R.M.Tuggle *Inorg. Chem.*, **13,** 2408, 1974
Also classified in 86

73.46 **bis(1′ - (1 - Acetylferrocenyl))**
$C_{24}H_{22}Fe_2O_2$
Z.Kaluski, A.I.Gusev, Yu.T.Struchkov
Bull. Acad. Pol. Sci., Ser. Sci. Chim., **22,** 739, 1974

73.47 **Di - μ - chloro - bis((tricarbonyl cyclopentadienyl molybdenum) - diethylether zinc)**
$C_{24}H_{30}Cl_2Mo_2O_8Zn_2$
J.St.Denis, W.Butler, M.D.Glick, J.P.Oliver
J. Am. Chem. Soc., **96,** 5427, 1974
Also classified in 84

73.48 **Dicarbonyl - (pentamethylcyclopentadienyl) - chromium dimer**
$C_{24}H_{30}Cr_2O_4$
J.Potenza, P.Giordano, D.Mastropaolo, A.Efraty
Inorg. Chem., **13,** 2540, 1974

73.49 bis(bis - (π - Cyclopentadienyl) molybdenum(iv) - bis - μ - methanethiolato) nickel(ii) tetrafluoroborate
$C_{24}H_{32}Mo_2NiS_4^{2+}$, $2BF_4^-$
K.Prout, S.R.Critchley, G.V.Rees
Acta Crystallogr., Sect. B, **30,** 2305, 1974
Residue 1 also classified in 85

73.50 bis(bis - (π - Cyclopentadienyl) niobium(v) - bis - μ - methanethiolato) nickel(0) tetrafluoroborate dihydrate
$C_{24}H_{32}Nb_7NiS_4^{2+}$, $2BF_4^-$, $2H_2O$
K.Prout, S.R.Critchley, G.V.Rees
Acta Crystallogr., Sect. B, **30,** 2305, 1974
Residue 1 also classified in 85

73.51 (pentahapto - Cyclopentadienyl)hydrido - molybdenum - μ - dimethylaluminium - μ(methylaluminium - di(μ - pentahapto(monohapto) - cyclopentadienyl) - dimethylaluminium) - (pentahapto - cyclopentadienyl)hydrido - molybdenum
$C_{25}H_{35}Al_3Mo_2$
R.A.Forder, K.Prout *Acta Crystallogr., Sect. B*, **30,** 2312, 1974
Also classified in 68

73.52 bis(π - Cyclopentadienyl) tungsten(iv) - (bis - μ - benzenethiolato) chromium(0) tetracarbonyl
$C_{26}H_{20}CrO_4S_2W$
K.Prout, G.V.Rees *Acta Crystallogr., Sect. B*, **30,** 2717, 1974
Also classified in 85

73.53 bis(π - Cyclopentadienyl) tungsten(iv) - (bis - μ - benzenethiolato) molybdenum(0) tetracarbonyl
$C_{26}H_{20}MoO_4S_2W$
K.Prout, G.V.Rees *Acta Crystallogr., Sect. B*, **30,** 2717, 1974
Also classified in 85

73.54 bis(π - Cyclopentadienyl) tungsten(iv) - bis - μ - benzenethiolato - tungsten(0) tetracarbonyl
$C_{26}H_{20}O_4S_2W_2$
K.Prout, G.V.Rees *Acta Crystallogr., Sect. B*, **30,** 2717, 1974
Also classified in 85

73.55 Dimolybdenum - bis - μ - (methyl aluminium - di(μ - pentahapto(monohapto) - cyclopentadienyl) - dimethylaluminium)
$C_{26}H_{34}Al_4Mo_2$
R.A.Forder, K.Prout *Acta Crystallogr., Sect. B*, **30,** 2312, 1974
Also classified in 68

73.56 **Di - μ - carbonyl - bis(carbonyl cyclopentadienyl iron) bis(triethyl aluminium)**
$C_{26}H_{40}Al_2Fe_2O_4$
N.E.Kim, N.J.Nelson, D.F.Shriver *Inorg. Chim. Acta,* **7,** 393, 1973
Also classified in 68

73.57 **(π - Cyclopentadienyl nickel) - di - μ - carbonyl - (dicarbonyl - (tris - p - fluorophenylphosphine) cobalt)**
$C_{27}H_{17}CoF_3NiO_4P$
I.L.C.Campbell, F.S.Stephens *J. Chem. Soc., Dalton Trans.,* 340, 1975
Also classified in 86

73.C **(Dibenzylideneacetone) - (pentamethylcyclopentadienyl) rhodium(i)**
$C_{27}H_{29}ORh$
For complete entry see 72.31

73.58 **(π - Cyclopentadienyl) - (trans - diphenyl - di(trimethylsilyl) - cyclobutadiene) - cobalt**
$C_{27}H_{33}CoSi_2$
M.D.Rausch, I.Bernal, B.R.Davies, A.Siegel, F.A.Higbie, G.F.Westover
J. Coord. Chem., **3,** 149, 1974
Also classified in 75, 63

73.59 **(π - (Methylcyclopentadienyl) nickel) - di - μ - carbonyl - (dicarbonyl - (cyclohexyldiphenylphosphine) cobalt)**
$C_{28}H_{28}CoNiO_4P$
I.L.C.Campbell, F.S.Stephens *J. Chem. Soc., Dalton Trans.,* 337, 1975
Also classified in 86

73.60 **racemic - Chloro - (1 - methyl - 3 - (dimethylphenylmethyl) - π - cyclopentadienyl) - π - cyclopentadienyl - (2,6 - dimethylphenolato) titanium**
$C_{28}H_{31}ClOTi$
C.Lecomte, Y.Dusausoy, J.Protas, J.Tirouflet, A.Dormond
J. Organomet. Chem., **73,** 67, 1974
Also classified in 84

73.C **bis - (η - Cyclopentadienyl) - bis(3,5 - dimethylbenzyl) tungsten**
$C_{28}H_{32}W$
For complete entry see 71.91

73.C **Di - μ - chloro - bis((1 - (dicarbonyl - π - cyclopentadienyl - ferrio) - 2 - phenyl - ethyne) copper(i)) dimer**
$C_{30}H_{20}Cl_2Cu_2Fe_2O_4$
For complete entry see 71.93

73.C **μ - Di - ($\eta^5.\eta^1$ - cyclopentadienyl) - di(dicyclopentadienyl thorium)**
$C_{30}H_{28}Th_2$
For complete entry see 71.94

73.C **Dicarbonyl - (π - cyclopentadienyl)triphenylphosphine - (1,1,2,2 - tetracyano - prop - 1 - yl) molybdenum**
$C_{32}H_{23}MoN_4O_2P$
For complete entry see 71.97

73.61 **(π - 1,2 - Di - iodocyclopentadienyl) - tetraphenylcyclobutadiene cobalt**
$C_{33}H_{23}CoI_2$
A.C.Villa, L.Coghi, A.G.Manfredotti, C.Guastini
Acta Crystallogr., Sect. B, **30,** 2101, 1974
Also classified in 75

73.62 **(π - Iodocyclopentadienyl) - tetraphenylcyclobutadiene cobalt**
$C_{33}H_{24}CoI$
A.C.Villa, L.Coghi, A.G.Manfredotti, C.Guastini
Acta Crystallogr., Sect. B, **30,** 2101, 1974
Also classified in 75

73.63 **(π - Cyanocyclopentadienyl) - tetraphenylcyclobutadiene cobalt**
$C_{34}H_{24}CoN$
A.C.Villa, L.Coghi, A.G.Manfredotti, C.Guastini
Acta Crystallogr., Sect. B, **30,** 2101, 1974
Also classified in 75

73.64 **bis(μ - Diphenylphosphido - μ - carbonyl - π - methylcyclopentadienyl - carbonyl iron) rhodium hexafluorophosphate**
$C_{40}H_{34}Fe_2O_4P_2Rh^+$, F_6P^-
R.Mason, J.A.Zubieta *J. Organomet. Chem.,* **66,** 279, 1974
Residue 1 also classified in 86

73.65 **Cyclotetra - μ - lithio - tetra(hydrido(bis(π - cyclopentadienyl)) molybdenum)**
$C_{40}H_{44}Li_4Mo_4$
R.A.Forder, K.Prout *Acta Crystallogr., Sect. B,* **30,** 2318, 1974

73.66 **Cyclotetra - μ - lithio - tetra(hydrido - bis(π - cyclopentadienyl) tungsten)**
$C_{40}H_{44}Li_4W_4$
R.A.Forder, K.Prout *Acta Crystallogr., Sect. B,* **30,** 2318, 1974

73.67 **bis - μ - (bis - (η - Cyclopentadienyl)hydride molybdenum) - bis(di - μ - bromo - cyclohexyl magnesium)((diethyl ether magnesium))**
$C_{40}H_{64}Br_4Mg_4Mo_2O_2$
K.Prout, R.A.Forder *Acta Crystallogr., Sect. B,* **31,** 852, 1975
Also classified in 67

73.68 **Cyclopentadienyl - carbonyl - bis(triphenylphosphine) manganese benzene solvate**
$C_{42}H_{35}MnOP_2$, C_6H_6
C.Barbeau, R.J.Dubey *Can. J. Chem.,* **52,** 1140, 1974
Residue 1 also classified in 86

73.69 **(π - Cyclopentadienyl) - bis(triphenylphosphine) - niobium carbonyl dihydride**
$C_{42}H_{37}NbOP_2$
N.I.Kirillova, A.I.Gusev, A.A.Pasynskii, Yu.T.Struchkov
Zh. Strukt. Khim., **15**, 288, 1974
Also classified in 86

73.C **Ferrocenyl - di - gold - bis(triphenylphosphine) tetrafluoroborate**
$C_{46}H_{39}Au_2FeP_2^+$, BF_4^-
For complete entry see 71.128

73.C **π - Cyclopentadienyl - bis(triphenylphosphine) ruthenium - phenylacetylide chloro - copper(i) acetone solvate**
$C_{49}H_{40}ClCuP_2Ru$, C_3H_6O
For complete entry see 71.131

73.C **(Tricarbonyl - triphenylphosphine rhenium - bis(pentafluorophenylacetylide)) - triphenylphosphine copper**
$C_{55}H_{30}CuF_{10}O_3P_2Re$
For complete entry see 71.133

73.C **π - Cyclopentadienyl - bis(π - triphenylgermyl - phenyl - acetylene) - carbonyl - niobium**
$C_{58}H_{45}Ge_2NbO$
For complete entry see 72.44

METAL PI-COMPLEXES (ARENE)

74.1 **Anthracene - tetrakis(silver perchlorate) monohydrate**
$C_{14}H_{10}Ag_4^{4+}$, $4ClO_4^-$, H_2O
E.A.H.Griffith, E.L.Amma *J. Am. Chem. Soc.*, **96,** 5407, 1974

74.2 **1,2 - Diphenylethane silver(i) perchlorate**
$(C_{14}H_{14}Ag^+)_n$, $nClO_4^-$
I.F.Taylor Junior, E.L.Amma *Acta Crystallogr., Sect. B,* **31,** 598, 1975

74.3 **4 - t - Butyl - π - (tricarbonyl chromium) - benzoic acid**
$C_{14}H_{14}CrO_5$
F.van Meurs, H.van Koningsveld *J. Organomet. Chem.*, **78,** 229, 1974

74.4 **Benzene - enneacarbonyl - tetracobalt**
$C_{15}H_6Co_4O_9$
P.H.Bird, A.R.Fraser *J. Organomet. Chem.*, **73,** 103, 1974

74.5 **Enneacarbonyl - tetracobalt π - (o - xylene) and π - (m - xylene) (disordered mixture, monoclinic form)**
$C_{17}H_{10}Co_4O_9$, $C_{17}H_{10}Co_4O_9$
P.H.Bird, A.R.Fraser *J. Organomet. Chem.*, **73,** 103, 1974

74.C **Di - μ - chloro - bis((π - benzene - π - allyl) molybdenum(ii))**
$C_{18}H_{22}Cl_2Mo_2$
For complete entry see 72.14

74.6 **bis(π - Duroquinone) nickel**
$C_{20}H_{24}NiO_4$
G.G.Aleksandrov, Yu.T.Struchkov *Zh. Strukt. Khim.*, **14,** 1067, 1973

74.7 **μ - Dinitrogen - bis((π - mesitylene)(1,2 - bis(dimethylphosphino)ethane) molybdenum)**
$C_{30}H_{56}Mo_2N_2P_4$
R.A.Forder, K.Prout *Acta Crystallogr., Sect. B,* **30,** 2778, 1974
Also classified in 86

METAL PI-COMPLEXES (MISCELLANEOUS RING SYSTEMS)

75.1 **π - (1 - Pentafluorophenyl - 2 - trimethylsilyl - cycloheptadienyl) - dicarbonyl - trimethylsilyl ruthenium**
$C_2H_{25}F_5O_2RuSi_2$
J.A.K.Howard, S.A.R.Knox, V.Riera, B.A.Sosinsky, F.G.A.Stone, P.Woodward *J. Chem. Soc., Chem. Commun.*, 673, 1974

75.2 **π - Cycloheptatrienyl molybdenum dicarbonyl bromide**
$C_9H_7BrMoO_2$
M.L.Ziegler, H.E.Sasse, B.Nuber *Z. Naturforsch., Teil B*, **30**, 26, 1975

75.3 **π - Cycloheptatrienyl molybdenum dicarbonyl chloride**
$C_9H_7ClMoO_2$
M.L.Ziegler, H.E.Sasse, B.Nuber *Z. Naturforsch., Teil B*, **30**, 26, 1975

75.4 **Trichlorostannyl - π - cycloheptatrienyl molybdenum dicarbonyl**
$C_9H_7Cl_3MoO_2Sn$
M.L.Ziegler, H.E.Sasse, B.Nuber *Z. Naturforsch., Teil B*, **30**, 22, 1975

75.5 **(1,2 - Dimethyl - 1,2 - dihydropyridazine - 3,6 - dione) iron tricarbonyl**
$C_9H_8FeN_2O_5$
A.N.Nesmeyanov, M.I.Rybinskaya, L.V.Rybin, A.V.Arutunian, L.G.Kuz'mina, Yu.T.Struchkov *J. Organomet. Chem.*, **73**, 365, 1974

75.6 **Tricarbonyl - (8,8 - dibromobicyclo(5.1.0)octa - 2,4 - diene) iron**
$C_{11}H_8Br_2FeO_3$
P.Skarstad, P.Janse-van Vuuren, J.Meinwald, R.E.Hughes *J. Chem. Soc., Perkin Trans. 2*, 88, 1975

75.7 **(Benzo(b)thiophene - 1,1 - dioxide)tetracarbonyl iron**
$C_{12}H_6FeO_6S$
R.Guilard, Y.Dusausoy *J. Organomet. Chem.*, **77**, 393, 1974

75.C **(2,3,4,8 - h^4 - Bicyclo(3.2.2)nona - 3,6 - diene - 2,8 - yl - 9 - one)tricarbonyl iron (orthorhombic form)**
$C_{12}H_8FeO_4$
For complete entry see 71.19

75.C **Bicyclo(3.2.2)nona - 6 - ene - 9 - one - (2 - 4)enyl - 8 - yl iron tricarbonyl (triclinic form)**
$C_{12}H_8FeO_4$
For complete entry see 71.20

75.C **π - Cyclopentadienyl - (1 - hydroxy - 2,3,4,5 - tetrakis(trifluoromethyl) - phosphole - 1 - oxide) cobalt**
$C_{13}H_6CoF_{12}O_2P$
For complete entry see 73.18

75.C **π - Pentachlorocyclopentadienyl - (cyclo - octadienyl) rhodium**
$C_{13}H_{12}Cl_5Rh$
For complete entry see 73.20

75.8 **(Tricyclo(6.2.0.0^{2,7})deca - 3,5 - diene)tricarbonyl iron (absolute configuration)**
$C_{13}H_{12}FeO_3$
F.A.Cotton, J.M.Troup *J. Organomet. Chem.*, **77**, 369, 1974

75.9 **Tetracarbonyl - (7,7 - dimethoxynorborn - 2 - ene) chromium(0)**
$C_{13}H_{14}CrO_6$
P.D.Brotherton, D.Wege, A.H.White, E.N.Maslen
J. Chem. Soc., Dalton Trans., 1876, 1974
Also classified in 84

75.10 **(1,5 - Cyclo - octadiene) - (S - methyl - maleonitriledithiolato) rhodium(i)**
$C_{13}H_{15}N_2RhS_2$
D.G.VanDerveer, R.Eisenberg *J. Am. Chem. Soc.*, **96**, 4994, 1974
Also classified in 85

75.11 **1,5 - Cyclo - octadiene - (acetylacetonato) rhodium(i)**
$C_{13}H_{19}O_2Rh$
P.A.Tucker, W.Scutcher, D.R.Russell
Acta Crystallogr., Sect. B, **31**, 592, 1975
Also classified in 77

75.C **1 - Tricarbonyl - benzo(c)ferracyclopentadiene iron tricarbonyl**
$C_{14}H_6Fe_2O_6$
For complete entry see 71.33

75.C **1 - Tricarbonyl - benzo(b)ferracyclopentadiene iron tricarbonyl**
$C_{14}H_6Fe_2O_6$
For complete entry see 71.34

75.12 **Tricarbonyl - (1 - phenylborinato) manganese**
$C_{14}H_{10}BMnO_3$
G.Huttner, W.Gartzke *Chem. Ber.*, **107**, 3786, 1974
Also classified in 62

75.13 **Tetracarbonyl iron acenaphthylene**
$C_{16}H_8FeO_4$
F.A.Cotton, P.Lahuerta *Inorg. Chem.*, **14,** 116, 1975

75.C **(Carbonyl - π - norbornadiene cobalt) - di - μ - carbonyl - (carbonyl - π - cyclopentadienyl iron)**
$C_{16}H_{13}CoFeO_4$
For complete entry see 73.27

75.14 **Di - μ - carbonyl - bis(carbonyl - (1 - methylborinato) iron)**
$C_{16}H_{16}B_2Fe_2O_4$
G.Huttner, W.Gartzke *Chem. Ber.*, **107,** 3786, 1974
Also classified in 62

75.15 **(1,3,5 - Cyclo - octatriene)(bicyclo(4.2.0)octa - 2,4 - diene) iron(0)**
$C_{16}H_{20}Fe$
G.Huttner, V.Bejenke *Chem. Ber.*, **107,** 156, 1974

75.16 **Tricarbonyl - η^4 - (1 - pentafluorophenyl - 2,3,4,5 - tetrakis(trifluoromethyl) thiophen) manganese**
$C_{17}F_{17}MnO_3S$
M.J.Barrow, A.A.Freer, W.Harrison, G.A.Sim, D.W.Taylor, F.B.Wilson
J. Chem. Soc., Dalton Trans., 197, 1975
Also classified in 39

75.C **(6,7,8,9 - h^4.2,3,4,11 - h^4 - Bicyclo(4.3.1)undeca - 2,6,8 - triene - 4,11 - yl - 11 - one) - bis(tricarbonyl iron)**
$C_{17}H_{10}Fe_2O_7$
For complete entry see 71.47

75.17 **3 - Methoxy - 1 - methyl - 5 - (2 - oxocyclohexyl)cyclohexa - 1,3 - diene iron tricarbonyl**
$C_{17}H_{20}FeO_5$
R.E.Ireland, G.G.Brown Junior, R.H.Stanford Junior, T.C.McKenzie
J. Org. Chem., **39,** 51, 1974

75.C **Acetylacetonato - (7,8 - bis(trifluoromethyl) - bicyclo(4.2.2)dec - 7 - ene - 2,5 - diyl) - aquo - rhodium hemihydrate**
$C_{17}H_{21}F_6O_3Rh$, $0.5H_2O$
For complete entry see 71.48

75.C **(pentahapto - Cyclopentadienyl) - 1,3′,4′ - trihapto - (2 - methyl - 2 - (6′,6′ - dimethyl - bicyclo(3.2.0)hept - 3′ - en - 7′ - on - 2′ - yl)propionyl) nickel(ii)**
$C_{18}H_{22}NiO_2$
For complete entry see 71.54

75.18 **μ - (Trimethylsilyl - cycloheptatrienyl) - pentacarbonyl - trimethylsilyl - di - ruthenium**
$C_{18}H_{24}O_5Ru_2Si_2$
J.Howard, P.Woodward *J. Chem. Soc., Dalton Trans.*, 59, 1975
Also classified in 63

75.19 **(Tetracyclo(8.4.2^{3,8}.0.0^{2,9})hexadeca - 4,6,11,13,15 - pentane) ruthenium tricarbonyl**
$C_{19}H_{16}O_3Ru$
R.Goddard, A.P.Humphries, S.A.R.Knox, P.Woodward
J. Chem. Soc., Chem. Commun., 507, 1975

75.C **Cyclobutadiene - iron - tricarbonyl dimethyl maleate photoadduct**
$C_{19}H_{20}FeO_{11}$
For complete entry see 71.58

75.20 **Dichloro - bis(aniline) - (norbornadiene) ruthenium**
$C_{19}H_{22}Cl_2N_2Ru$
J.-M.Manoli, A.P.Gaughan Junior, J.A.Ibers
J. Organomet. Chem., **72**, 247, 1974
Also classified in 83

75.C **2,3 - bis(π - 1,5 - Cyclo - octadiene) - 4,4 - bis(trifluoromethyl) - 1,2,3 - oxa - diplatina - cyclobutane**
$C_{19}H_{24}F_6OPt_2$
For complete entry see 71.59

75.C **1 - Carbonyl - 1 - cyclopentadienyl - 1 - (cyclopentadienyl iron) - 2,3,5,6 - tetra - (trifluoromethyl) - 1 - ferra - cyclohexa - 2,5 - diene - 4 - one**
$C_{20}H_{10}F_{12}Fe_2O_2$
For complete entry see 71.61

75.21 **Tricarbonyl - (1 - 5,α - η - diphenylfulvene) chromium**
$C_{21}H_{14}CrO_3$
V.W.Day, D.H.Campbell, C.J.Michejda
J. Chem. Soc., Chem. Commun., 117, 1975

75.C **(Tricyclo(6.6.2^{2,7}.0)hexadeca - 3,5,9,11,13,15 - hexaene) pentacarbonyl di - ruthenium**
$C_{21}H_{16}O_5Ru_2$
For complete entry see 71.69

75.C **Hexafluorobut - 2 - yne - tricarbonyl(cyclohexa - 1,3 - diene) - ruthenium adduct phosphite substitution product**
$C_{21}H_{17}F_{12}O_5PRu$
For complete entry see 71.70

75.C **9 - (1′,2′ - bis(Trifluoromethyl) - 2′ - (diacetylmethyl)ethylen - 1′ - yl) - 10,11 - bis(trifluoromethyl) - 9 - iridia - bicyclo(6.3.0)undeca - 4,10 - diene**
$C_{21}H_{19}F_{12}IrO_2$
For complete entry see 71.71

75.22 **tris(Bicyclo(2.2.1)heptene) platinum**
$C_{21}H_{30}Pt$
M.Green, J.A.K.Howard, J.L.Spencer, F.G.A.Stone
J. Chem. Soc., Chem. Commun., 449, 1975

75.23 **Hexafluorobut - 2 - yne - tricarbonyl(cycloheptatriene) - iron adduct phosphite substitution product**
$C_{22}H_{17}F_{12}FeO_5P$
M.Bottrill, R.Goddard, M.Green, R.P.Hughes, M.K.Lloyd, B.Lewis, P.Woodward *J. Chem. Soc., Chem. Commun.*, 253, 1975
Also classified in 86

75.24 **3,3′ - bis(Bicyclo(4.2.0)octa - 2,4 - diene) - bis(tricarbonyl iron)**
$C_{22}H_{18}Fe_2O_6$
F.A.Cotton, J.M.Troup *J. Organomet. Chem.*, **77**, 83, 1974

75.C **η^5 - Tetrahydropentalenyl - η^7 - cyclo - octatrienyl - hexacarbonyl - triruthenium**
$C_{22}H_{18}O_6Ru_3$
For complete entry see 73.41

75.25 **η - (1,5 - bis - Trimethylsilyl)pentalene - 1,1,2,2,2,3,3,3 - octacarbonyl - triangulo - triruthenium**
$C_{22}H_{22}O_8Ru_3Si_2$
J.A.K.Howard, S.A.R.Knox, F.G.A.Stone, A.C.Szary, P.Woodward
J. Chem. Soc., Chem. Commun., 788, 1974
Also classified in 63

75.26 **(Cyclohexeno(1,2)cyclobuta(3,4 - a)phenanthrene) iron tricarbonyl**
$C_{23}H_{16}FeO_3$
H.Irngartinger *Eur. Cryst. Meeting,* 372, 1974

75.27 **(1,3,3,5 - Tetramethyl - 6 - (1′,2′ - naphtho) - bicyclo(3.2.1)octene) chromium(0) tricarbonyl**
$C_{23}H_{24}CrO_3$
R.C.Pettersen, D.L.Cullen, H.L.Pearce, M.J.Shapiro, B.L.Shapiro
Acta Crystallogr., Sect. B, **30,** 2360, 1974

75.28 **(1,2,3,4,5 - Pentamethyl - 6R - phenylbicyclo(3.2.0)hept - 2 - enyl) - (acetylacetonato) palladium**
$C_{23}H_{30}O_2Pd$
D.J.Mabbott, P.M.Bailey, P.M.Maitlis
J. Chem. Soc., Chem. Commun., 521, 1975
Also classified in 77

75.C **Tricarbonyl - (cyclohexeno(1,2)ferracyclopentadieno(3,4 - a)phenanthrene) iron tricarbonyl**
$C_{24}H_{14}Fe_2O_6$
For complete entry see 71.80

75.29 **Tetraphenylarsonium bis(η^3 - cyclo - octatetraene) - (η^4 - cyclo - octatetraene) niobium**
$C_{24}H_{24}Nb^-$, $C_{24}H_{20}As^+$
L.J.Guggenberger, R.R.Schrock *Acta Crystallogr., Sect. A,* **31,** S136, 1975
Residue 2 classified in 65

75.30 **Di - μ - chloro - bis(cyclo - octatetraene - (tetrahydrofuran) titanium)**
$C_{24}H_{32}Cl_2O_2Ti_2$
G.J.Olthof, H.R.van der Wal *Acta Crystallogr., Sect. A,* **31,** S135, 1975
Also classified in 84

75.31 **Di - μ - (3,3,6,6 - tetramethyl - 1 - thiacyclohept - 4 - yne) - bis(dicarbonyl iron)**
$C_{24}H_{32}Fe_2O_4S_2$
H.-J.Schmitt, M.L.Ziegler *Z. Naturforsch., Teil B,* **28,** 508, 1973

75.32 **Di - (N,N,N',N' - tetramethylethylenediamine lithium) - trans,trans,trans - 1,5,9 - cyclododecatriene nickel(0)**
$C_{24}H_{50}Li_2N_4Ni$
D.J.Brauer, C.Kruger, P.J.Roberts, Y.-H.Tsay
Acta Crystallogr., Sect. A, **31,** S131, 1975

75.33 **1 - Bromo - 2 - methylene - 4 - (1' - bromo - 2' - naphthalene - methylene) - dihydronaphthalenyl iron tricarbonyl**
$C_{25}H_{14}Br_2FeO_3$
V.S.Kuz'min, G.P.Zol'nikova, Yu.T.Struchkov, I.I.Kritskaya
Zh. Strukt. Khim., **15,** 162, 1974

75.34 **(1,5 - η - Fluorenyl)(1,3 - η - fluorenyl) - dichloro - zirconium(iv)**
$C_{26}H_{18}Cl_2Zr$
C.Kowala, P.C.Wailes, H.Weigold, J.A.Wunderlich
J. Chem. Soc., Chem. Commun., 993, 1974

75.C **(π - Cyclopentadienyl) - (trans - diphenyl - di(trimethylsilyl) - cyclobutadiene) - cobalt**
$C_{27}H_{33}CoSi_2$
For complete entry see 73.58

75.35 **(Cyclobuta(1,2 - a.3,4 - a')diphenanthrene) iron tricarbonyl**
$C_{31}H_{16}FeO_3$
H.Irngartinger *Eur. Cryst. Meeting,* 372, 1974

75.36 **(Cyclo - octatetraene) - chloro - titanium tetramer**
$C_{32}H_{32}Cl_4Ti_4$
G.J.Olthof, H.R.van der Wal *Acta Crystallogr., Sect. A,* **31,** S135, 1975

75.37 **η^5 - (8 - Phenyl - bicyclo(5.1.0)octadienyl) - (η^4 - cyclo - octatetraene) - (o - phenylene - bis(dimethylarsine)) niobium**
$C_{32}H_{37}As_2Nb$
L.J.Guggenberger, R.R.Schrock *Acta Crystallogr., Sect. A,* **31,** S136, 1975
Also classified in 86

75.C **(π - 1,2 - Di - iodocyclopentadienyl) - tetraphenylcyclobutadiene cobalt**
$C_{33}H_{23}CoI_2$
For complete entry see 73.61

75.C **(π - Iodocyclopentadienyl) - tetraphenylcyclobutadiene cobalt**
$C_{33}H_{24}CoI$
For complete entry see 73.62

75.38 **(8 - Triphenylmethyl - 6 - trimethylsilyl - bicyclo(4.2.0)octa - 2,4,7 - triene) tricarbonyl iron**
$C_{33}H_{30}FeO_3Si$
M.Cooke, J.A.K.Howard, C.R.Russ, F.G.A.Stone, P.Woodward
J. Organomet. Chem., **78,** C43, 1974
Also classified in 63

75.C **(π - Cyanocyclopentadienyl) - tetraphenylcyclobutadiene cobalt**
$C_{34}H_{24}CoN$
For complete entry see 73.63

75.C **7 - (Methoxycarbonyl - methyl) - 1,2,3,4,5,6,7 - heptakis(methoxycarbonyl) - bicyclo(2.2.1)heptadiene - 8 - yl - (chloro - bis(pyridine) palladium)**
$C_{34}H_{35}ClN_2O_{16}Pd$
For complete entry see 71.101

75.39 **3 - Methyl - cyclopropene - bis(triphenylphosphine) platinum(0) (monoclinic form)**
$C_{40}H_{36}P_2Pt$
J.J.de Boer, D.Bright *J. Chem. Soc., Dalton Trans.*, 662, 1975
Also classified in 86

75.40 **1,2 - Dimethyl - cyclopropene - bis(triphenylphosphine) platinum(0)**
$C_{41}H_{38}P_2Pt$
J.J.de Boer, D.Bright *J. Chem. Soc., Dalton Trans.*, 662, 1975
Also classified in 86

75.41 **bis(Triphenylphosphine) - ($\Delta^{1,4}$ - bicyclo(2.2.0)hexene) platinum**
$C_{42}H_{38}P_2Pt$
M.E.Jason, J.A.McGinnety, K.B.Wiberg
J. Am. Chem. Soc., **96,** 6531, 1974
Also classified in 86

75.42 **Cyclohexyne - bis(triphenylphosphine) platinum(0)**
$C_{42}H_{38}P_2Pt$
G.B.Robertson, P.O.Whimp *J. Am. Chem. Soc.*, **97**, 1051, 1975
Also classified in 86

75.43 **Cycloheptyne - bis(triphenylphosphine) platinum(0)**
$C_{43}H_{40}P_2Pt$
G.B.Robertson, P.O.Whimp *J. Am. Chem. Soc.*, **97**, 1051, 1975
Also classified in 86

75.44 **bis(Triphenylphosphine) - (phenyl - methyl - cyclobutene - dione) platinum(0)**
$C_{47}H_{38}O_2P_2Pt$
P.A.Tucker, D.R.Russell *Acta Crystallogr., Sect. A*, **31**, S139, 1975
Also classified in 86

METAL COMPLEXES (ETHYLENEDIAMINE)

76.1 **Ethylenediammonium bis(cis - (ethylenediamine - disulfito - aurate(iii)))**
$2C_2H_8AuN_2O_6S_2^-$, $C_2H_{10}N_2^{2+}$
A.Dunand, R.Gerdil *Acta Crystallogr., Sect. B,* **31,** 370, 1975
Residue 2 classified in 3

76.2 **cis - Dichloro - ethylenediamine - palladium(ii)**
$C_2H_8Cl_2N_2Pd$
J.Iball, M.MacDougall, S.Scrimgeour
Acta Crystallogr., Sect. B, **31,** 1672, 1975

76.3 **cis - Dichloro - ethylenediamine - platinum(ii)**
$C_2H_8Cl_2N_2Pt$
J.Iball, M.MacDougall, S.Scrimgeour
Acta Crystallogr., Sect. B, **31,** 1672, 1975

76.4 **Ethylenediamine - bis(thiocyanato) mercury**
$C_4H_8HgN_4S_2$
D.Grdenic, B.Kamenar, M.Sikirica, T.Duplancic, S.Govedic, A.Hergold, P.Matkovic *Acta Crystallogr., Sect. A,* **31,** S132, 1975

76.5 **mer - Triazido - diethylenetriamine cobalt(iii)**
$C_4H_{13}CoN_{12}$
L.F.Druding, F.D.Sancilio *Acta Crystallogr., Sect. B,* **30,** 2386, 1974

76.6 **trans - Dichloro - bis(ethylenediamine) cobalt(iii) chloride tris(thiourea)**
$C_4H_{16}Cl_2CoN_4^+$, $3CH_4N_2S$, Cl^-
Yu.A.Simonov, L.I.Landa, N.N.Proskina, T.I.Malinovskii, A.V.Ablov
Kristallografiya, **18,** 530, 1973
Residue 2 classified in 8

76.7 **bis(Ethylenediamine) copper(ii) tetracyano - nickel(ii)**
$C_4H_{16}CuN_4^{2+}$, $C_4N_4Ni^{2-}$
M.Dunaj-Jurco, J.Garaj, J.Chomic, F.Valach, V.Haluska
Acta Crystallogr., Sect. A, **31,** S64, 1975

76.8 **bis(Ethylenediamine) copper(ii) acetylacetonate dihydrate**
$C_4H_{16}CuN_4^{2+}$, $2C_5H_7O_2^-$, $2H_2O$
T.Kurauchi, M.Matsui, Y.Nakamura, S.Ooi, S.Kawaguchi, H.Kuroya
Bull. Chem. Soc. Jpn., **47,** 3049, 1974
Residue 2 classified in 6

76.9 **bis(Ethylenediamine) mercury perchlorate**
$C_4H_{16}HgN_4^{2+}$, $2ClO_4^-$
D.Grdenic, B.Kamenar, M.Sikirica, T.Duplancic, S.Govedic, A.Hergold, P.Matkovic *Acta Crystallogr., Sect. A,* **31,** S132, 1975

76.10 **Dioxo - bis(ethylenediamine) rhenium chloride**
$C_4H_{16}N_4O_2Re^+$, Cl^-
V.S.Sergienko, M.A.Porai-Koshits, T.S.Khodashova
Zh. Strukt. Khim., **15,** 275, 1974

76.11 **Nitrito - bis(ethylenediamine) nickel(ii) nitrate**
$C_4H_{16}N_5NiO_2^+$, NO_3^-
A.E.Shvelashvili, M.A.Porai-Koshits, A.I.Kvitashvili, B.M.Shchedrin
Zh. Strukt. Khim., **15,** 307, 1974

76.12 **Nitrito - bis(ethylenediamine) zinc(ii) nitrite**
$C_4H_{16}N_5O_2Zn^+$, NO_2^-
A.E.Shvelashvili, M.A.Porai-Koshits, A.I.Kvitashvili, B.M.Shchedrin
Zh. Strukt. Khim., **15,** 307, 1974

76.13 **Di - nitrato - bis(ethylenediamine) nickel(ii) monohydrate**
$C_4H_{16}N_6NiO_4$, H_2O
A.E.Shvelashvili, M.A.Porai-Koshits, A.I.Kvitashvili, B.M.Shchedrin
Zh. Strukt. Khim., **15,** 307, 1974

76.14 **trans - Aqua - bis(ethylenediamine)sulfito cobalt(iii) perchlorate monohydrate**
$C_4H_{18}CoN_4O_4S^+$, ClO_4^- , H_2O
E.N.Maslen, C.L.Raston, A.H.White, J.K.Yandell
J. Chem. Soc., Dalton Trans., 327, 1975

76.15 **Diaquo - bis(ethylenediamine) nickel(ii) perchlorate**
$C_4H_{20}N_4NiO_2^{2+}$, $2ClO_4^-$
L.Kh.Minacheva, A.S.Antsyshkina, M.A.Porai-Koshits
Zh. Strukt. Khim., **15,** 478, 1974

76.16 **Chloro - isothiocyanato - bis(ethylenediamine) cadmium(ii)**
$C_5H_{16}CdClN_5S$
A.E.Shvelashvili, M.A.Porai-Koshits, A.I.Kvitashvili, B.M.Shchedrin, L.P.Sarishvili *Zh. Strukt. Khim.,* **15,** 315, 1974

76.17 **trans - bis(Ethylenediamine) - nitro - isothiocyanato cobalt(iii) perchlorate**
$C_5H_{16}CoN_6O_2S^+$, ClO_4^-
I.Grenthe, E.Nordin *Eur. Cryst. Meeting,* 362, 1974

76.18 **trans - bis(Ethylenediamine) - nitrito - isothiocyanato cobalt(iii) perchlorate**
$C_5H_{16}CoN_6O_2S^+$, ClO_4^-
I.Grenthe, E.Nordin *Eur. Cryst. Meeting,* 362, 1974

76.19 **Aquo - isothiocyanato - bis(ethylenediamine) nickel(ii) nitrate**
$C_5H_{18}N_5NiOS^+$, NO_3^-
A.E.Shvelashvili, M.A.Porai-Koshits, A.I.Kvitashvili, B.M.Shchedrin, M.G.Tavberidze *Zh. Strukt. Khim.*, **15**, 313, 1974

76.20 **Malonato - (N - methylethylenediamine) copper(ii) monohydrate**
$C_6H_{12}CuN_2O_4$, H_2O
R.Hamalainen, A.Pajunen *Suom. Kemistil. B*, **46**, 285, 1973
Residue 1 also classified in 81

76.21 **Oxalato - (N,N' - dimethylethylenediamine) copper(ii) monohydrate**
$C_6H_{12}CuN_2O_4$, H_2O
J.Korvenranta *Suom. Kemistil. B*, **46**, 296, 1973
Residue 1 also classified in 81

76.22 **Ethylene - bis(biguanide) silver(iii) sulfate hydrogen sulfate monohydrate**
$C_6H_{16}AgN_{10}^{3+}$, O_4S^{2-} , HO_4S^- , H_2O
L.Coghi, G.Pelizzi *Acta Crystallogr., Sect. B*, **31**, 131, 1975
Residue 1 also classified in 79

76.23 **Dibromo - (N,N,N',N' - tetramethylethylenediamine) copper(ii)**
$C_6H_{16}Br_2CuN_2$
E.Luukkonen, A.Pajunen *Suom. Kemistil. B*, **46**, 292, 1973

76.24 **bis(Isothiocyanato) - bis(ethylenediamine) cadmium(ii)**
$C_6H_{16}CdN_6S_2$
A.E.Shvelashvili, M.A.Porai-Koshits, A.I.Kvitashvili, B.M.Shchedrin, L.P.Sarishvili *Zh. Strukt. Khim.*, **15**, 315, 1974

76.25 **Dichloro - (N,N,N',N' - tetramethylethylenediamine) copper(ii)**
$C_6H_{16}Cl_2CuN_2$
E.D.Estes, W.E.Estes, W.E.Hatfield, D.J.Hodgson
Inorg. Chem., **14**, 106, 1975

76.26 **Ammine - (diethylenetriamine) - oxalato cobalt(iii) nitrate**
$C_6H_{16}CoN_4O_4^+$, NO_3^-
M.C.Couldwell, D.A.House, B.R.Penfold *Inorg. Chim. Acta*, **13**, 61, 1975
Residue 1 also classified in 81

76.27 **Dinitrito - (N,N,N',N - tetramethylethylenediamine) copper(ii)**
$C_6H_{16}CuN_4O_4$
E.Luukkonen *Suom. Kemistil. B*, **46**, 302, 1973

76.28 **Dinitrato - (N,N,N',N' - tetramethylethylenediamine) nickel(ii)**
$C_6H_{16}N_4NiO_6$
E.Turpeinen *Suom. Kemistil. B*, **46**, 208, 1973

76.29 **bis(N - Methylethylenediamine) copper(ii) oxalate dihydrate**
$C_6H_{20}CuN_4^{2+}$, $C_2O_4^{2-}$, $2H_2O$
R.Hamalainen *Suom. Kemistil. B*, **46**, 237, 1973

76.30 **bis(N - Methylethylenediamine) copper(ii) malonate dihydrate**
$C_6H_{20}CuN_4^{2+}$, $C_3H_2O_4^{2-}$, $2H_2O$
R.Hamalainen, A.Pajunen *Suom. Kemistil. B,* **46,** 285, 1973
Residue 2 classified in 2

76.31 **bis(N - Methylethylenediamine) copper(ii) D - tartrate dihydrate**
$C_6H_{20}CuN_4^{2+}$, $C_4H_4O_6^{2-}$, $2H_2O$
R.Hamalainen, A.Pajunen *Finn. Chem. Lett.*, 150, 1974
Residue 2 classified in 2

76.32 **Chloro - (ethylenediamine) - (diethylenetriamine) cobalt(iii) chloride**
$C_6H_{21}ClCoN_5^{2+}$, $2Cl^-$
A.V.Ablov, M.D.Mazus, S.E.V.Popa, T.I.Malinovskii, V.N.Byushkin
Dokl. Akad. Nauk SSSR, **194,** 821, 1970

76.33 **Cyano - (2,2′,2″ - triamino - triethylamine) copper(ii) tetraphenylborate**
$C_7H_{18}CuN_5^+$, $C_{24}H_{20}B^-$
D.M.Duggan, D.N.Hendrickson *Inorg. Chem.*, **13,** 1911, 1974
Residue 2 classified in 62

76.34 **Cyanato - (2,2′,2″ - triaminotriethylamine) nickel(ii) tetraphenylborate**
$C_7H_{18}N_5NiO^+$, $C_{24}H_{20}B^-$
D.M.Duggan, D.N.Hendrickson *Inorg. Chem.*, **13,** 2056, 1974
Residue 2 classified in 62

76.35 **trans - Dicyano(triethylenetetramine) cobalt(iii) perchlorate**
$C_8H_{18}CoN_6^+$, ClO_4^-
R.K.Wismer, R.A.Jacobson *Inorg. Chim. Acta,* **7,** 477, 1973

76.36 **bis(Isothiocyanato) - bis(propylenediamine) nickel(ii)**
$C_8H_{20}N_6NiS_2$
A.E.Shvelashvili, M.A.Porai-Koshits, A.I.Kvitashvili, B.M.Shchedrin, M.G.Tavberidze *Zh. Strukt. Khim.*, **15,** 313, 1974

76.37 **tris(2 - Aminoethyl)amine - (glycinato) cobalt(iii) chloride perchlorate**
$C_8H_{22}CoN_5O_2^{2+}$, Cl^- , ClO_4^-
Y.Mitsui, J.-I.Watanabe, Y.Iitaka, E.Kimura
J. Chem. Soc., Chem. Commun., 280, 1975
Residue 1 also classified in 82

76.38 **bis(N,N′ - Dimethylethylenediamine) copper(ii) sulfate tetrahydrate**
$C_8H_{24}CuN_4^{2+}$, O_4S^{2-} , $4H_2O$
J.Korvenranta *Suom. Kemistil. B,* **46,** 240, 1973

76.39 **bis(Diethylenetriamine) zinc(ii) dibromide monohydrate**
$C_8H_{26}N_6Zn^{2+}$, $2Br^-$, H_2O
P.G.Hodgson, B.R.Penfold *J. Chem. Soc., Dalton Trans.*, 1870, 1974

76.40 **(+) - bis - β - Carbonato - (3S,8S - dimethyl - triethylene - tetramine) cobalt(iii) perchlorate (absolute configuration)**
$C_9H_{22}CoN_4O_3^+$, ClO_4^-
K.Toriumi, Y.Saito *Acta Crystallogr., Sect. B,* **31,** 1247, 1975

76.41 **$(-)_{589}$ - tris(R - Propylenediamine) cobalt(iii) bromide**
$C_9H_{30}CoN_6^{3+}$, $3Br^-$
R.Kuroda, N.Shimanouchi, Y.Saito
Acta Crystallogr., Sect. B, **31,** 931, 1975

76.42 **facial - tris - (R - 1,2 - Propylenediamine) cobalt(iii) hexacyanocobaltate(iii) dihydrate (absolute configuration by internal comparison)**
$C_9H_{30}CoN_6^{3+}$, $C_6CoN_6^{3-}$, $2H_2O$
R.Kuroda, Y.Saito *Acta Crystallogr., Sect. B,* **30,** 2126, 1974

76.43 **Potassium ethylenediaminetetra - acetato - copper(ii) trihydrate**
$C_{10}H_{12}CuN_2O_8^{2-}$, $2K^+$, $3H_2O$
M.A.Porai-Koshits, N.V.Novozhilova, T.N.Polynova, T.V.Filippova, L.I.Martynenko *Kristallografiya,* **18,** 89, 1973
Residue 1 also classified in 82

76.44 **Lithium ethylenediaminetetra - acetato - iron(iii) trihydrate**
$C_{10}H_{12}FeN_2O_8^-$, Li^+ , $3H_2O$
N.V.Novozhilova, T.N.Polynova, M.A.Porai-Koshits, N.I.Pechurova, L.I.Martynenko, A.Khadi *Zh. Strukt. Khim.,* **14,** 745, 1973
Residue 1 also classified in 82

76.45 **Cesium di - μ - sulfido - μ - ethylenediaminetetra - acetato - dioxo - dimolybdate(v) dihydrate**
$C_{10}H_{12}Mo_2N_2O_{10}S_2^{2-}$, $2Cs^+$, $2H_2O$
B.Spivack, Z.Dori *J. Less-Common Met.,* **36,** 249, 1974
Residue 1 also classified in 82

76.46 **Sodium ethylenediaminetetra - acetato - aquo - iron(iii) dihydrate**
$C_{10}H_{14}FeN_2O_9^-$, Na^+ , $2H_2O$
N.V.Novozhilova, T.N.Polynova, M.A.Porai-Koshits, L.I.Martynenko
Zh. Strukt. Khim., **15,** 717, 1974
Residue 1 also classified in 82

76.47 **Lithium aquo - (ethylenediaminetetra - acetato) manganese(ii) tetrahydrate**
$C_{10}H_{14}MnN_2O_9^{2-}$, $2Li^+$, $4H_2O$
N.N.Anan'eva, T.N.Polynova, M.A.Porai-Koshits
Zh. Strukt. Khim., **15,** 261, 1974
Residue 1 also classified in 82

76.48 **Diaquo - zirconium ethylenediaminetetra - acetate dihydrate**
$C_{10}H_{16}N_2O_{10}Zr$, $2H_2O$
A.I.Pozhidaev, M.A.Porai-Koshits, T.N.Polynova
Zh. Strukt. Khim., **15**, 644, 1974
Residue 1 also classified in 82

76.49 **Sodium triaquo - samarium(iii) ethylenediaminetetra - acetate pentahydrate (neutron study)**
$C_{10}H_{18}N_2O_{11}Sm^-$, Na^+ , $5H_2O$
T.F.Koetzle *Acta Crystallogr., Sect. A*, **31**, S22, 1975
Residue 1 also classified in 82

76.50 **Tetra - aquo - zinc(ethylenediaminetetra - acetato - zinc) dihydrate**
$(C_{10}H_{20}N_2O_{12}Zn_2)_n$, $2nH_2O$
A.I.Pozhidaev, T.N.Polynova, M.A.Porai-Koshits, N.N.Neronova
Zh. Strukt. Khim., **14**, 570, 1973
Residue 1 also classified in 82

76.51 **Dichloro - (N,N,N′,N′ - tetraethylethylenediamine) copper(ii)**
$C_{10}H_{24}Cl_2CuN_2$
E.D.Estes, D.J.Hodgson
Am. Cryst. Assoc., Abstr. Papers (Spring Meeting), 10, 1975

76.52 **racemic - ((1,8 - Diamino - 2,5 - dimethyl - 5 - hydroxy - 3,6 - diaza - oct - 2 - ene) - ethylenediamine cobalt(iii)) hexachlorothallate(iii) dihydrate**
$C_{10}H_{28}CoN_6O^{3+}$, Cl_6Tl^{3-} , $2H_2O$
J.D.Bell, A.R.Gainsford, B.T.Golding, A.J.Herlt, A.M.Sargeson
J. Chem. Soc., Chem. Commun., 980, 1974
Residue 1 also classified in 83

76.53 **Oxalato - bis(di(ethylenediamine) nickel(ii)) perchlorate**
$C_{10}H_{32}N_8Ni_2O_4^{2+}$, $2ClO_4^-$
A.E.Shvelashvili, M.A.Porai-Koshits, A.I.Kvitashvili, B.M.Shchedrin
Zh. Strukt. Khim., **15**, 310, 1974
Residue 1 also classified in 81

76.54 **Oxalato - bis(di(ethylenediamine) nickel(ii)) nitrate**
$C_{10}H_{32}N_8Ni_2O_4^{2+}$, $2NO_3^-$
A.E.Shvelashvili, M.A.Porai-Koshits, A.I.Kvitashvili, B.M.Shchedrin
Zh. Strukt. Khim., **15**, 310, 1974
Residue 1 also classified in 81

76.55 **Oxalato - bis(di(ethylenediamine) zinc(ii)) thiocyanate**
$C_{10}H_{32}N_8O_4Zn_2^{2+}$, $2CNS^-$
A.E.Shvelashvili, M.A.Porai-Koshits, A.I.Kvitashvili, B.M.Shchedrin
Zh. Strukt. Khim., **15**, 310, 1974
Residue 1 also classified in 81

76.56 **trans(Theophyllinato - chloro - bis(ethylenediamine) cobalt(iii)) chloride dihydrate**
$C_{11}H_{23}ClCoN_8O_2^+$, Cl^- , $2H_2O$
T.J.Kistenmacher *Acta Crystallogr., Sect. B,* **31,** 85, 1975
Residue 1 also classified in 83

76.57 **cis(Theophyllinato - chloro - bis(ethylenediamine) cobalt(iii)) perchlorate**
$C_{11}H_{23}ClCoN_8O_2^+$, ClO_4^-
T.J.Kistenmacher, D.J.Szalda *Acta Crystallogr., Sect. B,* **31,** 90, 1975
Residue 1 also classified in 83

76.58 **Potassium di - μ - hydroxo - μ - acetato - μ - (ethylenediaminetetra - acetato) - bis(molybdenum(iii))**
$C_{12}H_{17}Mo_2N_2O_{12}^-$, K^+
G.G.Kneale, A.J.Geddes *Acta Crystallogr., Sect. B,* **31,** 1233, 1975
Residue 1 also classified in 81

76.59 **(Di - (2 - aminoethyl)amine) di - isothiocyanato - cadmium(ii)**
$(C_{12}H_{26}Cd_2N_{10}S_4)_n$
M.Cannas, G.Carta, A.Cristini, G.Marongiu
Acta Crystallogr., Sect. A, **31,** S140, 1975

76.60 **$(-)_{436}$ - β_2 - ((2S,9S) - 2,9 - Diamino - 4,7 - diazadecane cobalt(iii) amino - methyl - malonato) perchlorate monohydrate (absolute configuration)**
$C_{12}H_{27}CoN_5O_4^+$, ClO_4^- , H_2O
J.P.Glusker, H.L.Carrell, R.Job, T.C.Bruice
J. Am. Chem. Soc., **96,** 5741, 1974
Residue 1 also classified in 82

76.61 **Iodo - (2,2′,2″ - tri(dimethylamino)triethylamine) nickel iodide**
$C_{12}H_{30}IN_4Ni^+$, I^-
P.L.Orioli, N.Nardi *J. Chem. Soc., Chem. Commun.,* 229, 1975

76.62 **bis(Dichloro - (N,N,N′,N′ - tetramethylethylenediamine) copper(ii))**
$C_{12}H_{32}Cl_4CuN_4$
E.D.Estes, D.J.Hodgson
Am. Cryst. Assoc., Abstr. Papers (Spring Meeting), 10, 1975

76.63 **Di - μ - azido - bis(2,2′,2″ - triaminoethylamine) nickel(ii) tetraphenylborate**
$C_{12}H_{36}N_{14}Ni_2^{2+}$, $2C_{24}H_{20}B$
C.G.Pierpont, D.N.Hendrickson, D.M.Duggan, F.Wagner, E.K.Barefield
Inorg. Chem., **14,** 604, 1975
Residue 2 classified in 62

76.64 **Copper(ii) di - hydrogen diethylenetriaminepenta - acetate monohydrate**
$C_{14}H_{21}CuN_3O_{10}$, H_2O
V.V.Fomenko, T.N.Polynova, M.A.Porai-Koshits, G.L.Varlamova, N.I.Pechurova *Zh. Strukt. Khim.*, **14,** 571, 1973
Residue 1 also classified in 82

76.65 **Trihydrogen diethylenetriaminepenta - acetato copper(ii) monohydrate**
$C_{14}H_{21}CuN_3O_{10}$, H_2O
R.C.Seccombe, B.Lee, G.M.Henry *Inorg. Chem.*, **14,** 1147, 1975
Residue 1 also classified in 82

76.66 **Triethylenetetramine - bis(dimethylglyoxaldimine - monoximato) cobalt(iii) dibromide tetrahydrate**
$C_{14}H_{27}CoN_6O_2^{2+}$, $2Br^-$, $4H_2O$
M.D.Mazus, V.N.Biushkin, A.V.Ablov, V.N.Kaftanat, N.I.Belichuk, T.I.Malinovskii *Dokl. Akad. Nauk SSSR,* **208,** 1364, 1973
Residue 1 also classified in 83

76.67 **(Ethylenediamine) zinc(ii) benzohydroxamate benzohydroxamic acid monohydrate**
$C_{16}H_{20}N_4O_4Zn$, $C_7H_7NO_2$, H_2O
S.Gottlicher, P.Ochsenreiter *Chem. Ber.*, **107,** 391, 1974
Residue 1 also classified in 84; residue 2 classified in 10

76.68 **(1,8 - bis(2′ - Pyridyl) - 3,6 - diaza - octane) copper(ii) perchlorate**
$C_{16}H_{22}CuN_4^{2+}$, $2ClO_4^-$
D.A.Wright, J.D.Quinn *Acta Crystallogr., Sect. B,* **30,** 2132, 1974
Residue 1 also classified in 83

76.69 **(bis(Trifluoroacetylacetone)triethylenetetramine) nickel(ii)**
$C_{16}H_{22}F_6N_4NiO_2$
M.F.Richardson *Can. J. Chem.*, **52,** 3716, 1974
Also classified in 77

76.70 **Potassium N,N′ - ethylene - bis(acetylacetoniminato) - trans - di(glycinato) cobalt(iii) hexahydrate**
$C_{16}H_{26}CoN_4O_6^-$, K^+ , $6H_2O$
S.R.Holbrook, D.van der Helm *Acta Crystallogr., Sect. B,* **31,** 1653, 1975
Residue 1 also classified in 77, 82

76.71 **μ - Nitrosyl - di - μ - (N,N′ - dimethyl - N,N′ - bis(β - mercaptoethyl) ethylenediamine) - di - iron hexafluorophosphate acetone solvate**
$C_{16}H_{36}Fe_2N_5OS_4^+$, F_6P^- , C_3H_6O
K.D.Karlin, D.L.Lewis, H.N.Rabinowitz, S.J.Lippard
J. Am. Chem. Soc., **96,** 6519, 1974
Residue 1 also classified in 85

76.72 **μ - Oxo - bis(tetraethylenepenta - amine iron(iii)) iodide**
$C_{16}H_{46}Fe_2N_{10}O^{4+}$, $4I^-$
A.Coda, B.Kamenar, K.Prout, J.R.Carruthers, J.S.Rollett
Acta Crystallogr., Sect. B, **31,** 1438, 1975

76.73 **bis(Isothiocyanato) - (2,6 - di(acetimino)pyridine - triethylenetetramine) zinc(ii) dichloromethane solvate**
$C_{17}H_{23}N_7S_2Zn$, $0.5C_2H_4Cl_2$
M.G.B.Drew, S.H.Nelson *Acta Crystallogr., Sect. A,* **31,** S140, 1975
Residue 1 also classified in 83

76.74 **Diaquo - μ - triethylenetetraminehexa - acetato - dichromium(iii) hexahydrate**
$C_{18}H_{28}Cr_2N_4O_{14}$, $6H_2O$
G.D.Fallon, B.M.Gatehouse *Acta Crystallogr., Sect. B,* **30,** 1987, 1974
Residue 1 also classified in 82

76.75 **bis(N - (6 - Amino - 3,3,6 - trimethyl - 4 - aza - 2 - heptylidenyl)hydroxanato) dinickel(ii) chloride pentahydrate**
$C_{18}H_{40}N_6Ni_2O_2^{2+}$, $2Cl^-$, $5H_2O$
E.O.Schlemper, R.K.Murmann *Inorg. Chem.,* **13,** 2424, 1974
Residue 1 also classified in 84, 83

76.76 **$(+)_{589}$ - tris(($-$) - trans - 1,2 - Diaminocyclohexane) rhodium(iii) nitrate trihydrate**
$C_{18}H_{42}N_6Rh^{3+}$, $3NO_3^-$, $3H_2O$
R.Kuroda, Y.Sasaki, Y.Saito *Acta Crystallogr., Sect. B,* **30,** 2053, 1974

76.77 **N,N,N′ - tris(2 - (2′ - Pyridyl)ethyl)ethane - 1,2 - diamine nickel(ii) perchlorate nitromethane solvate**
$C_{23}H_{29}N_5Ni^{2+}$, $2ClO_4^-$, CH_3NO_2
B.F.Hoskins, F.D.Whillans *J. Chem. Soc., Dalton Trans.,* 657, 1975
Residue 1 also classified in 83

METAL COMPLEXES (ACETYLACETONE)

77.1 **Di - μ - acetylacetonato cadmium(ii)**
$(C_{10}H_{14}CdO_4)_n$
E.N.Maslen, T.M.Greaney, C.L.Raston, A.H.White
J. Chem. Soc., Dalton Trans., 400, 1975

77.2 **Azido - bis(acetylacetonato) manganese(iii)**
$(C_{10}H_{14}MnN_3O_4)_n$
B.R.Stults, R.S.Marianelli, V.W.Day *Inorg. Chem.*, **14,** 722, 1975

77.3 **Dinitrato - bis(acetylacetonato) zirconium(iv)**
$C_{10}H_{14}N_2O_{10}Zr$
R.C.Fay, E.G.Muller, V.W.Day
Am. Cryst. Assoc., Abstr. Papers (Spring Meeting), 33, 1975

77.C **(±) - ZZ - 1,2,3 - η - 5,6,7 - η - Heptadienediyl rhodium(iii) hexafluoroacetylacetonate**
$C_{12}H_{11}F_6O_2Rh$
For complete entry see 72.6

77.C **1,5 - Cyclo - octadiene - (acetylacetonato) rhodium(i)**
$C_{13}H_{19}O_2Rh$
For complete entry see 75.11

77.4 **Potassium tris(acetylacetonato) cadmium(ii) monohydrate**
$C_{15}H_{21}CdO_6^-$, K^+ , H_2O
T.M.Greaney, C.L.Raston, A.H.White, E.N.Maslen
J. Chem. Soc., Dalton Trans., 876, 1975

77.5 **tris(2,4 - Pentanedionato) manganese(iii) (β form)**
$C_{15}H_{21}MnO_6$
J.P.Fackler Junior, A.Avdeef *Inorg. Chem.*, **13,** 1864, 1974

77.6 **Nitrato - tris(acetylacetonato) zirconium(iv)**
$C_{15}H_{21}NO_9Zr$
R.C.Fay, E.G.Muller, V.W.Day
Am. Cryst. Assoc., Abstr. Papers (Spring Meeting), 33, 1975

77.C **(bis(Trifluoroacetylacetone)triethylenetetramine) nickel(ii)**
$C_{16}H_{22}F_6N_4NiO_2$
For complete entry see 76.69

77.C **Potassium N,N′ - ethylene - bis(acetylacetoniminato) - trans - di(glycinato) cobalt(iii) hexahydrate**
$C_{16}H_{26}CoN_4O_6^-$, K^+ , $6H_2O$
For complete entry see 76.70

77.C **Acetylacetonato - (7,8 - bis(trifluoromethyl) - bicyclo(4.2.2)dec - 7 - ene - 2,5 - diyl) - aquo - rhodium hemihydrate**
$C_{17}H_{21}F_6O_3Rh$, $0.5H_2O$
For complete entry see 71.48

77.7 **tetrakis(Acetylacetonato) cerium(iv) (α form)**
$C_{20}H_{28}CeO_8$
H.Titze *Acta Chem. Scand. Ser. A,* **28,** 1079, 1974

77.8 **bis(Acetylacetonato) nickel(ii) dimer isopropanol solvate**
$C_{20}H_{28}Ni_2O_8$, $2C_3H_8O$
K.A.Klanderman, C.E.Pfluger
Am. Cryst. Assoc., Abstr. Papers (Spring Meeting), 11, 1975

77.9 **Di(isothiocyanato) - di(ethoxy) - (1,3 - diphenyl - propane - 1,3 - dionato) niobium(v)**
$C_{21}H_{21}N_2NbO_4S_2$
V.F.Dahan, R.Kergoat, M.C.Tocquer, J.E.Guerchais
Acta Crystallogr., Sect. A, **31,** S133, 1975
Also classified in 84

77.C **((Chloro(methoxycarbonyl)(1,2,3,4,5 - pentakis(methoxycarbonyl)cyclopenta - 2,4 - dienyl)methyl))(pentane - 2,4 - dionato) palladium(ii) chloroform solvate**
$C_{23}H_{25}ClO_{14}Pd$, $CHCl_3$
For complete entry see 71.79

77.C **(1,2,3,4,5 - Pentamethyl - 6R - phenylbicyclo(3.2.0)hept - 2 - enyl) - (acetylacetonato) palladium**
$C_{23}H_{30}O_2Pd$
For complete entry see 75.28

77.10 **bis(Heptanetrionato) - bis(pyridine) copper(ii)**
$C_{24}H_{26}Cu_2N_2O_6$
A.B.Blake, L.R.Fraser *J. Chem. Soc., Dalton Trans.,* 2554, 1974
Also classified in 83

77.C **(2,4 - Pentanedionato)(triphenylphosphine)ethyl nickel(ii)**
$C_{25}H_{27}NiO_2P$
For complete entry see 71.84

77.C **Acetylacetonato - (acetylacetonyl) - triphenylphosphine palladium(ii) benzene solvate**
$C_{28}H_{29}O_4PPd$, $0.5C_6H_6$
For complete entry see 71.90

77.11 **Dibenzoylmethanato - (N,N′ - o - phenylene - bis(salicylideneiminato)) cobalt**
$C_{35}H_{25}CoN_2O_4$
D.Cummins, E.D.McKenzie, H.Milburn *Inorg. Chim. Acta,* **12,** L17, 1975
Also classified in 78

77.12 **Quinuclidine - tris(2,2,6,6 - tetramethylheptan - 3,5 - dionato) europium(iii)**
$C_{40}H_{70}EuNO_6$
E.Bye *Acta Chem. Scand. Ser. A,* **28,** 731, 1974
Also classified in 83

77.13 **cis - Dichloro - (pentane - 2,4 - dionato) - trans - bis(triphenylphosphine) rhenium(iii)**
$C_{41}H_{37}Cl_2O_2P_2Re$
I.D.Brown, C.J.L.Lock, C.Wan *Can. J. Chem.,* **52,** 1704, 1974
Also classified in 86

77.C **Acetylacetonato - (cis - 1,trans - 3 - tetraphenyl - 4 - ethoxy - butadien - 1 - yl) - (dimethylphenylphosphine) palladium(ii)**
$C_{43}H_{43}O_3PPd$
For complete entry see 71.122

77.C **(4 - Ethoxy - tetraphenylbuta - 1 - cis,3 - trans - dienyl) - (acetylacetonato) - (dimethylphenylphosphine) palladium(ii)**
$C_{43}H_{43}O_3PPd$
For complete entry see 71.123

METAL COMPLEXES
(SALICYLIC DERIVATIVES)

78.C **Phenyl - mercury - salicylalmethyliminate**
$C_{14}H_{13}HgNO$
For complete entry see 71.36

78.1 **N - (γ - (Methyl - (3 - aminopropyl)amino)propyl)salicylaldiminato - (nitrato) nickel(ii) hemihydrate**
$C_{14}H_{22}N_4NiO_4$, $0.5H_2O$
M.Nemiroff, P.Ganis, G.Avitabile, S.L.Holt
Cryst. Struct. Commun., **3,** 619, 1974

78.2 **bis(N,N' - Ethylene - bis(salicylaldiminato) cobalt(ii)) bis(tetrahydrofuran) sodium tetraphenylborate**
$2C_{16}H_{14}CoN_2O_2$, $C_{24}H_{20}B^-$, $2C_4H_8O$, Na^+
L.Randaccio *Gazz. Chim. Ital.*, **104,** 991, 1974
Residue 2 classified in 62

78.3 **(N - Methylsalicylaldiminato)(N - methylsalicylaldiminium) nickel(0) toluene solvate**
$C_{16}H_{18}N_2NiO_2$, C_7H_8
M.Matsumoto, K.Nakatsu, K.Tani, A.Nakamura, S.Otsuka
J. Am. Chem. Soc., **96,** 6777, 1974

78.4 **Dimethyl tin(iv) chloride - N,N' - ethylene - bis(salicylideneiminato) nickel(ii)**
$C_{18}H_{20}Cl_2N_2NiO_2Sn$
M.Calligaris, L.Randaccio, R.Barbieri, L.Pellerito
J. Organomet. Chem., **76,** C56, 1974
Also classified in 69

78.5 **Aquo - (N,N' - ethylene - bis(3 - methoxysalicylideneiminato)) cobalt(ii)**
$C_{18}H_{20}CoN_2O_5$
M.Calligaris, G.Nardin, L.Randaccio
J. Chem. Soc., Dalton Trans., 1903, 1974

78.6 **Salicylaldehydato - (N - 2 - dimethylaminoethyl - salicylaldiminato) copper(ii)**
$C_{18}H_{20}CuN_2O_3$
R.C.Srivastava, R.Tewari *Acta Crystallogr., Sect. A*, **31,** S142, 1975

78.7 **meso - (N,N′ - Cyclohexylene - bis(salicylideneiminato) cobalt(iii))**
$C_{20}H_{20}CoN_2O_2$
N.Bresciani, M.Calligaris, G.Nardin, L.Randaccio
J. Chem. Soc., Dalton Trans., 1606, 1974

78.8 **bis(N - iso - Propylidene - N′ - salicyloylhydrazinato) copper(ii)**
$C_{20}H_{22}CuN_4O_4$
P.Domiano, A.Musatti, G.Predieri *Cryst. Struct. Commun.*, **3**, 717, 1974

78.9 **bis(3 - (Salicylideneamino) - 1 - propanolato) - bis(copper(ii))**
$C_{20}H_{22}Cu_2N_2O_4$
P.J.Nassiff, E.R.Boyko, L.D.Thompson
Bull. Chem. Soc. Jpn., **47**, 2321, 1974

78.C **(N,N - dimethylbenzylamine) - (N - phenylsalicylaldiminato) palladium(ii)**
$C_{22}H_{22}N_2OPd$
For complete entry see 71.78

78.10 **Dioxo - uranyl bis(N - (2 - dimethylaminoethyl)salicylaldiminate)**
$C_{22}H_{30}N_4O_4U$
D.A.Clemente, G.Bandoli, F.Bentollo, M.Vidali, P.A.Vigato, U.Casellato
J. Inorg. Nucl. Chem., **36**, 1999, 1974

78.C **μ - (N,N′ - Ethylene - bis(salicylaldiminato)) - bis(2 - methallyl palladium(ii))**
$C_{24}H_{28}N_2O_2Pd_2$
For complete entry see 72.29

78.11 **(Dipropylenetriamine - N,N′ - bis(salicylaldehydato)) - (1 - methylimidazole) cobalt(iii) bromide monohydrate**
$C_{24}H_{29}CoN_5O_2^+$, Br^- , H_2O
T.J.Kistenmacher, L.G.Marzilli, P.A.Marzilli *Inorg. Chem.*, **13**, 2089, 1974

78.C **Malononitrilato(propane - 1,2 - bis(salicylideneiminato))pyridine cobalt(iii)**
$C_{25}H_{22}CoN_5O_2$
For complete entry see 71.83

78.12 **Chloro - (bis(salicylideneiminephenyl)disulfido) iron(iii)**
$C_{26}H_{18}ClFeN_2O_2S_2$
J.A.Bertrand, J.L.Breece *Inorg. Chim. Acta*, **8**, 267, 1974
Also classified in 85

78.13 **bis(N - Picolinylidene - N′ - salicyloylhydrazinato) nickel(ii)**
$C_{26}H_{20}N_6NiO_4$
P.Domiano, A.Musatti, M.Nardelli, C.Pelizzi
J. Chem. Soc., Dalton Trans., 295, 1975
Also classified in 83

78.14 **Salicylaldehyde - thiosemicarbazone - (triphenylphosphine) nickel(ii)**
$C_{26}H_{22}N_3NiOPS$
L.E.Nikolaeva, T.N.Tarkhova, N.V.Belov *Kristallografiya,* **19,** 746, 1974
Also classified in 79, 86

78.15 **Salicylaldehyde - selenosemicarbazonato - (triphenylphosphine) nickel**
$C_{26}H_{22}N_3NiOPSe$
L.E.Nikolaeva, A.A.Shevyrev, T.N.Tarkhova, N.V.Belov
Kristallografiya, **19,** 516, 1974
Also classified in 86, 85

78.16 **bis(N - Cyclohexyl - salicylaldiminato) copper(ii)**
$C_{26}H_{32}CuN_2O_2$
R.P.Kashyap, J.M.Bindlish, S.C.Bhatia, P.C.Jain
Acta Crystallogr., Sect. A, **31,** S146, 1975

78.17 **2,2′,2″ - Tri(5 - chloro - salicylaldiminato)ethylamine iron(iii) trihydrate**
$C_{27}H_{24}Cl_3FeN_4O_3$, $3H_2O$
N.A.Bailey, D.F.Cook, D.Cummins, E.D.McKenzie
Inorg. Nucl. Chem. Lett., **11,** 51, 1975

78.18 **(cis,cis - 1,3,5 - tris(Salicylaldiminato)cyclohexane) cobalt(iii)**
$C_{27}H_{24}CoN_3O_3$
D.A.Rudman, J.C.Huffman, R.F.Childers, W.E.Streib, R.A.D.Wentworth
Inorg. Chem., **14,** 747, 1975

78.19 **bis(Salicylal - N - m - anisidinato) cobalt**
$C_{28}H_{24}CoN_2O_4$
A.N.Knyazeva, L.M.Shkol′nikova *Zh. Strukt. Khim.,* **14,** 1058, 1973

78.20 **Tetra - μ - salicylato - bis(aquo - copper(ii)) dioxan solvate**
$C_{28}H_{24}Cu_2O_{14}$, $C_4H_8O_2$
A.V.Ablov, G.A.Kiosse, G.I.Dimitrova, T.I.Malinovskii, G.A.Popovich
Kristallografiya, **19,** 168, 1974

78.21 **bis(D - N - Phenylethylsalicylaldiminato) copper(ii)**
$C_{30}H_{28}CuN_2O_2$
Z.A.Starikova, M.A.Porai-Koshits, P.M.Zorky
Vestn. Mosk. Univ., Khim., 323, 1972

78.22 **bis(DL - N - Phenylethylsalicylaldiminato) copper(ii)**
$C_{30}H_{28}CuN_2O_2$
Z.A.Starikova, M.A.Porai-Koshits, P.M.Zorky, E.G.Rukhadze,
G.V.Panova *Vestn. Mosk. Univ., Khim.,* 449, 1972

78.23 **bis(N,N′ - Ethylene - bis(salicylaldiminato) copper(ii)) - diaquo - copper(ii) perchlorate monohydrate**
$C_{32}H_{32}Cu_3N_4O_6^{2+}$, $2ClO_4^-$, H_2O
J.M.Epstein, B.N.Figgis, A.H.White, A.C.Willis
J. Chem. Soc., Dalton Trans., 1954, 1974

78.C **Dibenzoylmethanato - (N,N′ - o - phenylene - bis(salicylideneiminato)) cobalt**

$C_{35}H_{25}CoN_2O_4$

For complete entry see 77.11

78.24 **bis(N,N′ - Ethylene - bis(o - hydroxyacetopheniminato) copper(ii)) - aquo - copper(ii) perchlorate**

$C_{36}H_{38}Cu_3N_4O_5^{2+}$, $2ClO_4^-$

J.M.Epstein, B.N.Figgis, A.H.White, A.C.Willis

J. Chem. Soc., Dalton Trans., 1954, 1974

78.25 **bis(n - n - Decyl - o - hydroxy - acetophenoniminato) copper(ii)**

$C_{36}H_{56}CuN_2O_2$

R.P.Kashyap, J.M.Bindlish, S.C.Bhatia

Acta Crystallogr., Sect. A, **31**, S146, 1975

78.26 **mer - tris(N - Benzoylsalicylaldiminato) manganese(iii)**

$C_{42}H_{36}MnN_3O_3$

C.L.Raston, A.H.White, A.C.Willis, K.S.Murray

J. Chem. Soc., Dalton Trans., 1793, 1974

METAL COMPLEXES (THIOUREA)

79.1 **bis(Thiosemicarbazide) iron(ii) sulfate**
$(C_2H_{10}FeN_6O_4S_3)_n$
D.V.Naik, G.J.Palenik *Chem. Phys. Lett.*, **24,** 260, 1974

79.2 **bis(Thiocarbonohydrazide) silver(i) nitrate**
$C_2H_{12}AgN_8S_2^+$, NO_3^-
F.Bigoli, E.Leporati, M.A.Pellinghelli
Cryst. Struct. Commun., **4,** 127, 1975

79.3 **bis(Thiocarbonohydrazide) copper(ii) perchlorate**
$C_2H_{12}CuN_8S_2^{2+}$, $2ClO_4^-$
F.Bigoli, M.A.Pellinghelli, A.Tiripicchio
Cryst. Struct. Commun., **4,** 123, 1975

79.4 **bis(Thiocarbonohydrazide) copper(ii) sulfate tetrahydrate**
$C_2H_{12}CuN_8S_2^{2+}$, O_4S^{2-} , $4H_2O$
F.Bigoli, M.A.Pellinghelli, A.Tiripicchio, M.T.Camellini
Acta Crystallogr., Sect. B, **31,** 55, 1975

79.5 **bis(Thiocarbonohydrazide) copper(ii) oxalate tetrahydrate**
$C_2H_{12}CuN_8S_2^{2+}$, $C_2O_4^{2-}$, $4H_2O$
F.Bigoli, E.Leporati, M.A.Pellinghelli
Cryst. Struct. Commun., **4,** 119, 1975

79.6 **Di(thiocyanato) - bis(thiosemicarbazide) nickel(ii)**
$C_4H_{10}N_9NiS_4$
M.Dunaj-Jurco, J.Garaj, A.Sirota
Collect. Czech. Chem. Commun., **39,** 236, 1974

79.7 **tetrakis(Urea) cobalt(ii) nitrate**
$(C_4H_{16}CoN_8O_4^{2+})_n$, $2nNO_3^-$
P.S.Gentile, J.White, S.Haddad *Inorg. Chim. Acta*, **8,** 97, 1974

79.8 **tetrakis(Thiourea) platinum(ii) chloride**
$C_4H_{16}N_8PtS_4^{2+}$, $2Cl^-$
R.L.Girling, K.K.Chatterjee, E.L.Amma *Inorg. Chim. Acta*, **7,** 557, 1973

79.9 **Potassium tris(biureto) cobalt(iii) hydrate**
$C_6H_9CoN_9O_6^{3-}$, $3K^+$, $6.38H_2O$
P.J.M.W.L.Birker, J.M.M.Smits, J.J.Bour, P.T.Beurskens
Rec. Trav. Chim. Pays-Bas, **92,** 1240, 1973
Residue 1 also classified in 83

79.10 **bis(Ethylenethiourea) zinc(ii) thiosulfate**
$(C_6H_{12}N_4O_3S_4Zn)_n$
S.Baggio, R.F.Baggio, P.K.De Perazzo
Acta Crystallogr., Sect. B, **30,** 2166, 1974

79.C **Ethylene - bis(biguanide) silver(iii) sulfate hydrogen sulfate monohydrate**
$C_6H_{16}AgN_{10}^{3+}$, O_4S^{2-} , HO_4S^- , H_2O
For complete entry see 76.22

79.11 **trans - Di - isothiocyanato - tetra(urea) manganese(ii)**
$C_6H_{16}MnN_{10}O_4$
G.V.Tsintsadze, T.I.Tsivtsivadze, F.V.Orbeladze
Zh. Strukt. Khim., **15,** 306, 1974

79.12 **Tetra(urea) hexakis(urea) - cobalt nitrate**
$C_6H_{24}CoN_{12}O_6^{2+}$, $4CH_4N_2O$, $2NO_3^-$
E.N.Kurkutova, T.F.Rau *Dokl. Akad. Nauk SSSR,* **204,** 342, 1972
Residue 2 classified in 8

79.13 **tris(Thiourea) copper(i) tetrafluoroborate dimer**
$C_6H_{24}Cu_2N_{12}S_6^{2+}$, $2BF_4^-$
I.F.Taylor Junior, M.S.Weininger, E.L.Amma
Inorg. Chem., **13,** 2835, 1974

79.C **Nickel iodide - urea clathrate**
$C_6H_{24}N_{12}NiO_6^{2+}$, $4CH_4N_2O$, $2I^-$
For complete entry see 61.1

79.14 **Barium chromium thiosemicarbazone pentahydrate**
$2C_8H_{10}CrN_6O_4S_2^-$, Ba^{2+} , $5H_2O$
G.F.Volodina, G.A.Kiosse, N.V.Gerely
Dokl. Akad. Nauk SSSR, **200,** 1349, 1971

79.15 **Nitrato - aquo - (o - aminobenzaldehyde - thiosemicarbazone) copper(ii) nitrate monohydrate**
$C_8H_{12}CuN_5O_4S^+$, NO_3^- , H_2O
A.V.Ablov, V.K.Rotaru, G.A.Kiosse, T.I.Malinovskii, M.V.Shopron, N.V.Gerbeleu *Dokl. Akad. Nauk SSSR,* **208,** 353, 1973
Residue 1 also classified in 83

79.16 **3 - Ethoxy - 2 - oxobutyraldehyde - bis(thiosemicarbazonato) copper(ii) (triclinic form)**
$C_8H_{14}CuN_6OS_2$
M.R.Taylor, J.P.Glusker, E.J.Gabe, J.A.Minkin
Bioinorg. Chem., **3,** 189, 1974
Also classified in 83

79.17 **Copper thiosemicarbazide - diacetate dimer**
$C_{10}H_{14}Cu_2N_6O_8S_2$
V.K.Rotaru, G.A.Kiosse, N.V.Gerbeleu, A.V.Ablov, T.I.Malinovskii, V.G.Bodyu *Zh. Strukt. Khim.*, **14,** 948, 1973
Also classified in 82

79.18 **tetrakis - μ - Chloroacetato - bis(urea - copper(ii))**
$C_{10}H_{16}Cl_4Cu_2N_4O_{10}$
Yu.A.Simonov, V.I.Ivanov, T.I.Malinovskii, L.N.Milkova, A.V.Ablov, Yu.V.Yablokov *Acta Crystallogr., Sect. A*, **31,** S139, 1975
Also classified in 81

79.19 **tetrakis - μ - Fluoroacetato - bis(urea - copper(ii))**
$C_{10}H_{16}Cu_2F_4N_4O_{10}$
Yu.A.Simonov, V.I.Ivanov, T.I.Malinovskii, L.N.Milkova, A.V.Ablov, Yu.V.Yablokov *Acta Crystallogr., Sect. A*, **31,** S139, 1975
Also classified in 81

79.20 **Tetra - μ - acetato - di(urea - copper(ii)) dihydrate**
$C_{10}H_{20}Cu_2N_4O_{10}$, $2H_2O$
A.V.Ablov, Yu.V.Yablokov, Yu.A.Simonov, L.I.Landa, T.I.Malinovskii, L.N.Milkova *Dokl. Akad. Nauk SSSR*, **201,** 599, 1971
Residue 1 also classified in 81

79.21 **bis(Pyridinealdehydethiosemicarbazonato) cobalt(iii) nitrate**
$C_{14}H_{14}CoN_8S_2^+$, NO_3^-
K.N.Akatova, T.N.Tarkhova, N.V.Belov *Kristallografiya*, **18,** 263, 1973
Residue 1 also classified in 83

79.22 **Potassium bis(3 - phenylbiuretato) cobalt(iii) dimethylsulfoxide solvate**
$C_{16}H_{14}CoN_6O_4^-$, K^+ , $2C_2H_6OS$
P.J.M.W.L.Birker, P.T.Beurskens
Rec. Trav. Chim. Pays-Bas, **93,** 218, 1974

79.23 **tris(S - Dimethylthiourea) copper(i) tetrafluoroborate dimer**
$C_{18}H_{48}Cu_2N_{12}S_6^{2+}$, $2BF_4^-$
I.F.Taylor Junior, M.S.Weininger, E.L.Amma
Inorg. Chem., **13,** 2835, 1974

79.24 **tetrakis(N,N′ - Diethylthiourea) platinum(ii) chloride monohydrate**
$C_{20}H_{48}N_8PtS_4^{2+}$, $2Cl^-$, H_2O
G.Marcotrigiano, L.P.Battaglia, A.B.Corradi, M.E.V.Tani
Cryst. Struct. Commun., **4,** 361, 1975

79.C **Salicylaldehyde - thiosemicarbazone - (triphenylphosphine) nickel(ii)**
$C_{26}H_{22}N_3NiOPS$
For complete entry see 78.14

79.25 **tetrakis - μ - Phenylacetato - bis(urea - copper(ii))**
$C_{34}H_{36}Cu_2N_4O_{10}$
Yu.A.Simonov, V.I.Ivanov, T.I.Malinovskii, Yu.V.Yablokov
Acta Crystallogr., Sect. A, **31,** S139, 1975
Also classified in 81

79.26 **hexakis(N,N′ - Di - isopropylthiourea) nickel(ii) perchlorate**
$C_{42}H_{96}N_{12}NiS_6^{2+}$, $2ClO_4^-$
G.A.Bentley, J.M.Waters *J. Inorg. Nucl. Chem.,* **36,** 2247, 1974

METAL COMPLEXES (THIOCARBAMATE OR XANTHATE)

80.1 **Copper(ii) bis(N,N - dimethyldithiocarbamate)**
$C_6H_{12}CuN_2S_4$
F.W.B.Einstein, J.S.Field *Acta Crystallogr., Sect. B,* **30,** 2928, 1974

80.2 **Dichloro - bis(O - ethylthiocarbamato) mercury(ii)**
$C_6H_{14}Cl_2HgN_2O_2S_2$
G.Bandoli, D.A.Clemente, L.Sindellari, E.Tondello
J. Chem. Soc., Dalton Trans., 449, 1975

80.3 **cis - bis(N - Isopropyldithiocarbamato) nickel(ii)**
$C_8H_{16}N_2NiS_4$
C.L.Raston, A.H.White *J. Chem. Soc., Dalton Trans.,* 1790, 1974

80.4 **tris(Ethylthioxanthato) chromium(iii)**
$C_9H_{15}CrS_9$
A.C.Villa, A.G.Manfredotti, C.Guastini, M.Nardelli
Acta Crystallogr., Sect. B, **30,** 2788, 1974

80.5 **bis(N,N - Diethyldithiocarbamato) iron(iii) bromide**
$C_{10}H_{20}FeN_2S_4^+$, Br^-
G.E.Chapps, S.W.McCann, H.H.Wickman, R.C.Sherwood
J. Chem. Phys., **60,** 990, 1974

80.6 **bis(Piperidyldithiocarbamato) copper(ii) tetrakis(copper(i) bromide)**
$(C_{12}H_{20}Br_4Cu_5N_2S_4)_n$
R.M.Golding, A.D.Rae, B.J.Ralph, L.Sulligoi
Inorg. Chem., **13,** 2499, 1974

80.7 **bis(Piperidyldithiocarbamato) copper(ii) hexakis(copper(i) bromide)**
$(C_{12}H_{20}Br_6Cu_7N_2S_4)_n$
R.M.Golding, A.D.Rae, B.J.Ralph, L.Sulligoi
Inorg. Chem., **13,** 2499, 1974

80.8 **Di - μ - chloro - dichloro - bis(methyl pyrrolidine - 1 - carbodithioate) dimercury(ii)**
$C_{12}H_{22}Cl_4Hg_2N_2S_4$
P.D.Brotherton, J.M.Epstein, A.H.White, A.C.Willis
J. Chem. Soc., Dalton Trans., 2341, 1974

80.9 **Oxo - bis(di - n - propyldithiocarbamato) molybdenum(iv)**
$C_{14}H_{28}MoN_2OS_4$
L.Ricard, J.Estienne, P.Karagiannidis, P.Toledano, J.Fischer, A.Mitschler, R.Weiss *J. Coord. Chem.*, **3**, 277, 1974

80.10 **Di - oxo - bis(di - n - propyldithiocarbamato) molybdenum(vi)**
$C_{14}H_{28}MoN_2O_2S_4$
L.Ricard, J.Estienne, P.Karagiannidis, P.Toledano, J.Fischer, A.Mitschler, R.Weiss *J. Coord. Chem.*, **3**, 277, 1974

80.11 **tris(4 - Morpholine - carbodithioato) cobalt(iii) dichloromethane solvate**
$C_{15}H_{24}CoN_3O_3S_6$, CH_2Cl_2
P.C.Healy, E.Sinn *Inorg. Chem.*, **14**, 109, 1975

80.12 **tris(4 - Morpholine - carbodithioato) iron(iii) dichloromethane solvate**
$C_{15}H_{24}FeN_3O_3S_6$, CH_2Cl_2
P.C.Healy, E.Sinn *Inorg. Chem.*, **14**, 109, 1975

80.13 **tris(N,N - Diethyldithiocarbamato) ruthenium(iii)**
$C_{15}H_{30}N_3RuS_6$
L.H.Pignolet *Inorg. Chem.*, **13**, 2051, 1974

80.14 **pentakis(N,N - Dimethylcarbamato) niobium(v)**
$C_{15}H_{30}N_5NbO_{10}$
M.H.Chisholm, M.Extine *J. Am. Chem. Soc.*, **97**, 1623, 1975

80.15 **tris(Dimethylaminato) - tris(N,N - dimethylcarbamato) tungsten(vi)**
$C_{15}H_{36}N_6O_6W$
M.H.Chisholm, M.Extine *J. Am. Chem. Soc.*, **96**, 6214, 1974
Also classified in 84, 83

80.16 **Gold(i) dibutyldithiocarbamate**
$C_{18}H_{36}Au_2N_2S_4$
P.Jennische, H.Anacker-Eickhoff, A.Wahlberg
Acta Crystallogr., Sect. A, **31**, S143, 1975

80.17 **cis - Di - iodo - bis(N,N - di - n - butyldithiocarbamato) platinum(iv)**
$C_{18}H_{36}I_2N_2PtS_4$
J.Willemse, J.A.Cras, J.G.Wijnhoven, P.T.Beurskens
Rec. Trav. Chim. Pays-Bas, **92**, 1199, 1973

80.C **(Tetracyanoethylene) - oxo - bis(di - n - propyldithiocarbamato) molybdenum**
$C_{20}H_{28}MoN_6OS_4$
For complete entry see 72.18

80.18 **tetrakis(N,N - Diethyldithiocarbamato) molybdenum(iv)**
$C_{20}H_{40}MoN_4S_8$
J.G.M.van der Aalsvoort, P.T.Beurskens
Cryst. Struct. Commun., **3**, 653, 1974

80.19 **Tri - μ - (N,N - diethyldithiocarbamato) - bis(N,N - diethyldithiocarbamato - ruthenium(iii)) tetrafluoroborate acetone solvate**
$C_{25}H_{50}N_5Ru_2S_{10}^+$, BF_4^- , C_3H_6O
L.H.Pignolet, B.M.Mattson *J. Chem. Soc., Chem. Commun.*, 49, 1975

80.20 **μ - Oxo - bis(oxo - di(di - n - propyldithiocarbamato) molybdenum(v))**
$C_{28}H_{56}Mo_2N_4O_3S_8$
L.Ricard, J.Estienne, P.Karagiannidis, P.Toledano, J.Fischer, A.Mitschler, R.Weiss *J. Coord. Chem.*, **3**, 277, 1974

80.21 **Di - n - butyl - monothiocarbamato nickel(ii) hexamer**
$C_{108}H_{216}N_{12}Ni_6O_{12}S_{12}$
B.F.Hoskins, C.D.Pannan *Inorg. Nucl. Chem. Lett.*, **10**, 229, 1974

METAL COMPLEXES (CARBOXYLIC ACID)

81.1 **Potassium diperoxo - oxalato - vanadium(v) monohydrate**
$C_2O_9V^{3-}$, $3K^+$, H_2O
R.E.Drew, F.W.B.Einstein, J.S.Field, D.Begin
Acta Crystallogr., Sect. A, **31,** S135, 1975

81.2 **Ammonium di - μ - formato - hexachloro - dirhenium**
$C_2H_2Cl_6O_4Re_2^{2-}$, $2H_4N^+$
P.A.Koz'min, M.D.Surazhskaya, T.B.Larina
Zh. Strukt. Khim., **15,** 64, 1974

81.3 **Aquo - cadmium(ii) malonate**
$(C_3H_4CdO_5)_n$
M.L.Post, J.Trotter *J. Chem. Soc., Dalton Trans.,* 1922, 1974

81.4 **Methanol - cobalt(ii) oxalate monohydrate**
$(C_3H_4CoO_5)_n$, nH_2O
E.Canals, C.Berro, R.Deyrieux *Rev. Chim. Miner.,* **11,** 498, 1974
Residue 1 also classified in 84

81.5 **Methanol - iron oxalate monohydrate**
$(C_3H_4FeO_5)_n$, nH_2O
E.Canals, C.Berro, R.Deyrieux *Rev. Chim. Miner.,* **11,** 498, 1974
Residue 1 also classified in 4

81.6 **Methanol - nickel(ii) oxalate monohydrate**
$(C_3H_4NiO_5)_n$, nH_2O
E.Canals, C.Berro, R.Deyrieux *Rev. Chim. Miner.,* **11,** 498, 1974
Residue 1 also classified in 84

81.7 **Methanol - zinc oxalate monohydrate**
$(C_3H_4O_5Zn)_n$, nH_2O
E.Canals, C.Berro, R.Deyrieux *Rev. Chim. Miner.,* **11,** 498, 1974
Residue 1 also classified in 84

81.8 **Dicesium bis(oxalato) copper(ii) dihydrate**
$C_4CuO_8^{2-}$, $2Cs^+$, $2H_2O$
W.Pannhorst, J.Lohn *Z. Kristallogr.,* **139,** 236, 1974

81.9 **Dipotassium bis(oxalato) copper(ii) dihydrate**
$C_4CuO_8^{2-}$, $2K^+$, $2H_2O$
T.Weichert, J.Lohn *Z. Kristallogr.,* **139,** 223, 1974

81.10 **Tripotassium bis(oxalato)dioxo - vanadate(v) trihydrate**
$C_4O_{10}V^{3-}$, $3K^+$, $3H_2O$
R.E.Drew, F.W.B.Einstein, S.E.Gransden *Can. J. Chem.*, **52,** 2184, 1974

81.11 **Oxydiacetato - copper(ii) hemihydrate**
$C_4H_4CuO_5$, $0.5H_2O$
S.H.Whitlow *Acta Crystallogr., Sect. A,* **31,** S143, 1975

81.12 **Aquo - (thiodiacetato) cadmium(ii)**
$C_4H_6CdO_5S$
S.H.Whitlow *Acta Crystallogr., Sect. A,* **31,** S143, 1975
Also classified in 85

81.13 **Cobalt(ii) acetate hydrate**
$(C_4H_6CoO_4)_{5n}$, $2nH_2O$
R.Alcala, J.F.Garcia *Rev. Acad. Cienc. Exactas, Fis.-Quim. Nat. Zaragoza,* **28,** 303, 1973

81.14 **Aquo - manganese(ii) acetate trihydrate**
$(C_4H_8MnO_5)_n$, $3nH_2O$
E.F.Bertaut, T.Q.Duc, P.Burlet, P.Burlet, M.Thomas, J.M.Moreau *Acta Crystallogr., Sect. B,* **30,** 2234, 1974

81.15 **Triaquo - zinc(ii) thiodiglycollate monohydrate**
$C_4H_{10}O_7SZn$, H_2O
M.G.B.Drew, D.A.Rice, C.W.Timewell *J. Chem. Soc., Dalton Trans.*, 144, 1975
Residue 1 also classified in 85

81.16 **n - Propylmercapto - acetato mercury**
$(C_5H_{10}HgO_2S)_n$
H.Puff, R.Sievers, G.Elsner *Z. Anorg. Allg. Chem.*, **413,** 37, 1975
Also classified in 85

81.17 **Potassium trans,trans,trans - dinitro - bis(β - alaninato) cobalt(iii)**
$C_6H_{12}CoN_4O_8^-$, K^+
B.Prelesnik, M.B.Celap, R.Herak *Inorg. Chim. Acta,* **7,** 569, 1973

81.C **Malonato - (N - methylethylenediamine) copper(ii) monohydrate**
$C_6H_{12}CuN_2O_4$, H_2O
For complete entry see 76.20

81.C **Oxalato - (N,N' - dimethylethylenediamine) copper(ii) monohydrate**
$C_6H_{12}CuN_2O_4$, H_2O
For complete entry see 76.21

81.18 **n - Butylmercapto - acetato mercury**
$(C_6H_{12}HgO_2S)_n$
H.Puff, R.Sievers, G.Elsner *Z. Anorg. Allg. Chem.*, **413,** 37, 1975
Also classified in 85

81.19 **Diaquo - bis(lactato) zinc monohydrate**
$C_6H_{14}O_8Zn$, H_2O
K.D.Singh, S.C.Jain, T.D.Sakore, A.B.Biswas
Acta Crystallogr., Sect. B, **31,** 990, 1975

81.C **Ammine - (diethylenetriamine) - oxalato cobalt(iii) nitrate**
$C_6H_{16}CoN_4O_4^+$, NO_3^-
For complete entry see 76.26

81.20 **trans - Dinitro - (β - alaninato) - (1,3 - propanediamine) cobalt(ii)**
$C_6H_{16}CoN_5O_6$
R.M.Herak, M.B.Celap, I.Krstanovic
Acta Crystallogr., Sect. A, **31,** S142, 1975
Also classified in 83

81.21 **cis(Nitro) - mer(amino) - dinitro - (β - alaninato - 1,3 - propanediamine) cobalt(iii) monohydrate**
$C_6H_{16}CoN_5O_6$, H_2O
R.M.Herak, M.Jeremic, M.B.Celap
Acta Crystallogr., Sect. A, **31,** S143, 1975
Residue 1 also classified in 83

81.22 **(Pyridine - 2,6 - dicarboxylato) - dioxouranium(vi) monohydrate**
$(C_7H_5NO_7U)_n$
A.Immirzi, G.Bombieri, S.Degetto, G.Marangoni
Acta Crystallogr., Sect. B, **31,** 1023, 1975
Also classified in 83

81.23 **Diaquo - peroxotitanium dipicolinate dihydrate (orthorhombic form)**
$C_7H_7NO_8Ti$, $2H_2O$
H.Manohar, D.Schwarzenbach *Helv. Chim. Acta,* **57,** 1086, 1974
Residue 1 also classified in 83

81.24 **Potassium tetra(oxalato) thorium(iv) tetrahydrate**
$C_8O_{16}Th^{4-}$, $4K^+$, $4H_2O$
M.N.Akhtar, A.J.Smith *Acta Crystallogr., Sect. B,* **31,** 1361, 1975

81.25 **Potassium di - μ - oxo - tetra(oxalato) dirhenium(iv) trihydrate**
$C_8O_{18}Re_2^{4-}$, $4K^+$, $3H_2O$
T.Lis *Acta Crystallogr., Sect. B,* **31,** 1594, 1975

81.26 **Ammonium dimolybdomalate hexahydrate**
$C_8H_6Mo_4O_{18}^{4-}$, $4H_4N^+$, $6H_2O$
M.A.Porai-Koshits, L.A.Aslanov, G.V.Ivanova, T.N.Polynova
Zh. Strukt. Khim., **9,** 475, 1968

81.27 **Di - molybdenum tetra - acetate**
$C_8H_{12}Mo_2O_8$
F.A.Cotton, Z.C.Mester, T.R.Webb
Acta Crystallogr., Sect. B, **30,** 2768, 1974

81.C **tetrakis(Trifluoroacetoxy - mercuri)methane**
$C_9F_{12}Hg_4O_8$
For complete entry see 71.9

81.C **tetrakis(Acetoxy - mercuri)methane**
$C_9H_{12}Hg_4O_8$
For complete entry see 71.11

81.C **tetrakis - μ - Chloroacetato - bis(urea - copper(ii))**
$C_{10}H_{16}Cl_4Cu_2N_4O_{10}$
For complete entry see 79.18

81.C **tetrakis - μ - Fluoroacetato - bis(urea - copper(ii))**
$C_{10}H_{16}Cu_2F_4N_4O_{10}$
For complete entry see 79.19

81.C **Tetra - μ - acetato - di(urea - copper(ii)) dihydrate**
$C_{10}H_{20}Cu_2N_4O_{10}$, $2H_2O$
For complete entry see 79.20

81.C **Oxalato - bis(di(ethylenediamine) nickel(ii)) perchlorate**
$C_{10}H_{32}N_8Ni_2O_4^{2+}$, $2ClO_4^-$
For complete entry see 76.53

81.C **Oxalato - bis(di(ethylenediamine) nickel(ii)) nitrate**
$C_{10}H_{32}N_8Ni_2O_4^{2+}$, $2NO_3^-$
For complete entry see 76.54

81.C **Oxalato - bis(di(ethylenediamine) zinc(ii)) thiocyanate**
$C_{10}H_{32}N_8O_4Zn_2^{2+}$, $2CNS^-$
For complete entry see 76.55

81.C **π - Allyl - dicarbonyl - (trifluoroacetato) - (1,2 - dimethoxyethane) molybdenum**
$C_{11}H_{15}F_3MoO_6$
For complete entry see 72.3

81.C **π - Allyl - dicarbonyl - (trifluoroacetato) - (1,2 - dimethoxyethane) tungsten**
$C_{11}H_{15}F_3O_6W$
For complete entry see 72.4

81.28 **Trisodium tris(oxydiacetato) cerium(iii) nonahydrate**
$C_{12}H_{12}CeO_{15}^{3-}$, $3Na^+$, $9H_2O$
J.Albertsson, I.Elding *Acta Crystallogr., Sect. A,* **31,** S167, 1975

81.29 **Pyrazine copper acetate (at 100°K)**
$(C_{12}H_{16}Cu_2N_2O_8)_n$
B.Morosin, R.C.Hughes, Z.G.Soos
Acta Crystallogr., Sect. B, **31,** 762, 1975
Also classified in 83

81.30 **Pyrazine copper acetate**
$(C_{12}H_{16}Cu_2N_2O_8)_n$
B.Morosin, R.C.Hughes, Z.G.Soos
Acta Crystallogr., Sect. B, **31,** 762, 1975
Also classified in 83

81.31 **Tetra - aquo - bis(isonicotinato) europium(iii) nitrate**
$(C_{12}H_{16}EuN_2O_8{}^+)_n$, $nNO_3{}^-$
L.A.Aslanov, I.D.Kiekbaev, I.K.Abdul'minev, M.A.Porai-Koshits
Kristallografiya, **19,** 170, 1974

81.C **Potassium di - μ - hydroxo - μ - acetato - μ - (ethylenediaminetetra - acetato) - bis(molybdenum(iii))**
$C_{12}H_{17}Mo_2N_2O_{12}{}^-$, K^+
For complete entry see 76.58

81.32 **bis(Triacetato - diaquo - erbium) tetrahydrate**
$C_{12}H_{26}Er_2O_{16}$, $4H_2O$
L.A.Aslanov, I.K.Abdul'minev, M.A.Porai-Koshits, V.I.Ivanov
Dokl. Akad. Nauk SSSR, **205,** 343, 1972

81.33 **$(+)_{510}$ - Oxalato - bis(R,S - 2,4 - diaminopentane) cobalt(iii) perchlorate monohydrate (absolute configuration)**
$C_{12}H_{28}CoN_4O_4{}^+$, $ClO_4{}^-$, H_2O
I.Oonishi, S.Sato, Y.Saito *Acta Crystallogr., Sect. B,* **30,** 2256, 1974
Residue 1 also classified in 83

81.34 **bis(Tetraphenylarsonium) uranyl - bis(pyridine - 2,6 - dicarboxylate) hexahydrate**
$C_{14}H_6N_2O_{10}U^{2-}$, $2C_{24}H_{20}As^+$, $6H_2O$
G.Marangoni, S.Degetto, R.Graziani, G.Bombieri, E.Forsellini
J. Inorg. Nucl. Chem., **36,** 1787, 1974
Residue 1 also classified in 83; residue 2 classified in 65

81.35 **bis(2 - Acetato - pyridine) copper(ii) dihydrate**
$C_{14}H_{12}CuN_2O_4$, $2H_2O$
R.Faure, H.Loiseleur *Acta Crystallogr., Sect. B,* **31,** 1472, 1975
Residue 1 also classified in 83

81.36 **Ammonium titanyl oxalate tetramer**
$C_{16}H_8O_{40}Ti_4{}^{8-}$, $8H_4N^+$
G.M.H.van de Velde, S.Harkema, P.J.Gellings
Inorg. Chim. Acta, **11,** 243, 1974

81.37 **Copper(ii) octanoate**
$C_{16}H_{30}CuO_4$
T.R.Lomer, K.Perera *Acta Crystallogr., Sect. B,* **30,** 2913, 1974

81.C **Tri - μ - formato - tri - μ - hydroxy - tri(cyclopentadienyl niobium) - μ_3 - oxide hydride**
$C_{18}H_{22}Nb_3O_{10}$
For complete entry see 73.32

81.38 **bis(1,3 - Propanediamine) nickel(ii) benzoate**
$C_{20}H_{30}N_4NiO_4$
A.Pajunen, U.Turpeinen *Suom. Kemistil. B,* **46,** 282, 1973
Also classified in 83

81.39 **Copper(ii) decanoate**
$C_{20}H_{38}CuO_4$
T.R.Lomer, K.Perera *Acta Crystallogr., Sect. B,* **30,** 2912, 1974

81.C **Dicyclopentadienyl - di - (p - nitrobenzoate) titanium**
$C_{24}H_{18}N_2O_8Ti$
For complete entry see 73.44

81.40 **Pyrazine - 2,3 - dicarbonato - carbonyl - triphenylphosphine rhodium(i) monohydrate**
$C_{25}H_{18}N_2O_5PRh$, H_2O
Z.G.Aliev, L.O.Atovmyan, V.I.Ponomarev
Zh. Strukt. Khim., **14,** 748, 1973
Residue 1 also classified in 86, 83

81.41 **Tetra - μ - trifluoroacetato - bis(quinoline copper(ii))**
$C_{26}H_{14}Cu_2F_{12}N_2O_8$
J.A.Moreland, R.J.Doedens *J. Am. Chem. Soc.,* **97,** 508, 1975
Also classified in 83

81.42 **μ - Acetato - μ - phenyldinitromethanato - bis(2,2' - bipyridine copper(ii))**
$C_{29}H_{25}Cu_2N_6O_6$
V.Fares, G.Dessy, M.Bonamico, L.Scaramuzza
Acta Crystallogr., Sect. A, **31,** S139, 1975
Also classified in 84, 83

81.43 **tetrakis(Acetato - (2 - pyridylmethanolato) copper(ii)) tetrahydrate**
$C_{32}H_{36}Cu_4N_4O_{12}$, $4H_2O$
W.Choong, N.C.Stephenson *Cryst. Struct. Commun.,* **4,** 275, 1975
Residue 1 also classified in 84, 83

81.44 **μ_3 - Oxo - hexakis(μ - trimethyl - acetato) - tris(methanol) tri - iron(iii) chloride**
$C_{33}H_{66}Fe_3O_{16}^+$, Cl^-
A.B.Blake, L.R.Fraser *J. Chem. Soc., Dalton Trans.,* 193, 1975

81.C **tetrakis - μ - Phenylacetato - bis(urea - copper(ii))**
$C_{34}H_{36}Cu_2N_4O_{10}$
For complete entry see 79.25

81.C **sym - trans - Di - μ - acetato - bis(o - (t - butyl - o - tolyl - phosphino)benzyl)di - palladium(ii)**
$C_{40}H_{50}O_4P_2Pd_2$
For complete entry see 71.115

81.45 **Hydrido - formato - tris(triphenylphosphine) ruthenium(ii) (monoclinic form)**
$C_{55}H_{47}O_2P_3Ru$
A.I.Gusev, G.G.Aleksandrov, Yu.T.Struchkov
Zh. Strukt. Khim., **14,** 685, 1973
Also classified in 86

81.46 **bis(Quinoline) - hexa(benzoato) - tricobalt(ii)**
$C_{60}H_{44}Co_3N_2O_{12}$
J.Catterick, M.B.Hursthouse, D.B.New, P.Thornton
J. Chem. Soc., Chem. Commun., 843, 1974
Also classified in 83

81.47 **Benzoato - tris(triphenylphosphine) rhodium(i) benzene solvate**
$C_{61}H_{50}O_2P_3Rh$, $0.5C_6H_6$
A.I.Gusev, Yu.T.Struchkov *Zh. Strukt. Khim.*, **15,** 282, 1974
Residue 1 also classified in 86

81.48 **tris(bis(Triphenylphosphine) silver(i)) tris(dithio - oxalato) iron(iii)**
$C_{114}H_{90}Ag_3FeO_6P_6S_6$
F.J.Hollander, D.Coucouvanis *Inorg. Chem.*, **13,** 2381, 1974
Also classified in 86, 85

METAL COMPLEXES (AMINO-ACID)

82.1 **Diaquo - copper(ii) iminodiacetate**
$(C_4H_9CuNO_6)_n$
F.G.Kramarenko, T.N.Polynova, M.A.Porai-Koshits, V.P.Chalyi, G.N.Kupriyanova, L.I.Martynenko *Zh. Strukt. Khim.*, **14,** 744, 1973

82.2 **L - Cysteinato(methyl)mercury(ii) monohydrate**
$C_4H_9HgNO_2S$, H_2O
N.J.Taylor, Y.S.Wong, P.C.Chieh, A.J.Carty
J. Chem. Soc., Dalton Trans., 438, 1975

82.3 **Glycylglycinato diaquo - copper(ii)**
$C_4H_{10}CuN_2O_5$
T.J.Kistenmacher, D.J.Szalda *Acta Crystallogr., Sect. B,* **31,** 1659, 1975

82.4 **Aquo - glutamato cadmium monohydrate**
$(C_5H_9CdNO_5)_n$, nH_2O
H.Soylu, D.Ulku, J.C.Morrow *Z. Kristallogr.*, **140,** 281, 1974

82.5 **Lithium nitrilotriacetato copper(ii) trihydrate**
$C_6H_6CuNO_6^-$, Li^+ , $3H_2O$
V.V.Fomenko, L.I.Kopaneva, M.A.Porai-Koshits, T.N.Polynova
Zh. Strukt. Khim., **15,** 268, 1974

82.6 **Potassium aquo - (nitrilotriacetato) cobalt(ii) dihydrate**
$C_6H_8CoNO_7^-$, K^+ , $2H_2O$
L.P.Battaglia, A.B.Corradi, M.E.V.Tani
Acta Crystallogr., Sect. B, **31,** 1160, 1975

82.7 **Potassium nitrilotriacetato - aquo - copper dihydrate**
$(C_6H_8CuNO_7^-)_n$, nK^+ , $2nH_2O$
V.V.Fomenko, T.N.Polynova, N.D.Mitrofanova
Zh. Strukt. Khim., **14,** 946, 1973

82.C **Methyl mercury DL - methionine complex**
$C_6H_{13}HgNO_2S$
For complete entry see 71.5

82.8 **Calcium bis(L - aspartato) cobalt(iii) hydrate**
$2C_8H_{10}CoN_2O_8^-$, Ca^{2+} , $7.5H_2O$
I.Oonishi, S.Sato, Y.Saito *Acta Crystallogr., Sect. B,* **31,** 1318, 1975

82.9 **Potassium bis(iminodiacetato) copper(ii) monohydrate**
$C_8H_{10}CuN_2O_8^-$, K^+ , H_2O
F.G.Kramarenko, T.N.Polynova, M.A.Porai-Koshits, V.P.Chalyi, N.D.Mitrofanova *Zh. Strukt. Khim.*, **14**, 1113, 1973

82.10 **Lithium bis(iminoacetato) nickel(ii) tetrahydrate**
$C_8H_{10}N_2NiO_8^{2-}$, $2Li^+$, $4H_2O$
F.G.Kramarenko, T.N.Polynova, M.A.Porai-Koshits, V.P.Chalyi, N.D.Mitrofanova *Zh. Strukt. Khim.*, **15**, 161, 1974

82.11 **bis(Hydrogen iminodiacetato)dioxo uranium(vi)**
$(C_8H_{12}N_2O_{10}U)_n$
G.Bombieri, E.Forsellini, G.Tomat, L.Magon, R.Graziani *Acta Crystallogr., Sect. B*, **30**, 2659, 1974

82.12 **bis - (L - Asparaginato) copper(ii)**
$C_8H_{14}CuN_4O_6$
F.S.Stephens, R.S.Vagg, P.A.Williams *Acta Crystallogr., Sect. B*, **31**, 841, 1975

82.13 **bis(Iminodiacetamide) copper(ii) perchlorate**
$C_8H_{18}CuN_6O_4^{2+}$, $2ClO_4^-$
M.Sekizaki *Bull. Chem. Soc. Jpn.*, **47**, 1447, 1974

82.C **tris(2 - Aminoethyl)amine - (glycinato) cobalt(iii) chloride perchlorate**
$C_8H_{22}CoN_5O_2^{2+}$, Cl^- , ClO_4^-
For complete entry see 76.37

82.C **Phenyl mercury L - cysteine complex**
$C_9H_{11}HgNO_2S$
For complete entry see 71.10

82.C **Dicarbonyl - (η - cyclopentadienyl) - (3 - aminopropionyl) molybdenum(ii)**
$C_{10}H_{11}MoNO_3$
For complete entry see 71.13

82.C **Potassium ethylenediaminetetra - acetato - copper(ii) trihydrate**
$C_{10}H_{12}CuN_2O_8^{2-}$, $2K^+$, $3H_2O$
For complete entry see 76.43

82.C **Lithium ethylenediaminetetra - acetato - iron(iii) trihydrate**
$C_{10}H_{12}FeN_2O_8^-$, Li^+ , $3H_2O$
For complete entry see 76.44

82.C **Cesium di - μ - sulfido - μ - ethylenediaminetetra - acetato - dioxo - dimolybdate(v) dihydrate**
$C_{10}H_{12}Mo_2N_2O_{10}S_2^{2-}$, $2Cs^+$, $2H_2O$
For complete entry see 76.45

82.C **Copper thiosemicarbazide - diacetate dimer**
$C_{10}H_{14}Cu_2N_6O_8S_2$
For complete entry see 79.17

82.C **Sodium ethylenediaminetetra - acetato - aquo - iron(iii) dihydrate**
$C_{10}H_{14}FeN_2O_9^-$, Na^+ , $2H_2O$
For complete entry see 76.46

82.C **Lithium aquo - (ethylenediaminetetra - acetato) manganese(ii) tetrahydrate**
$C_{10}H_{14}MnN_2O_9^{2-}$, $2Li^+$, $4H_2O$
For complete entry see 76.47

82.C **Diaquo - zirconium ethylenediaminetetra - acetate dihydrate**
$C_{10}H_{16}N_2O_{10}Zr$, $2H_2O$
For complete entry see 76.48

82.C **Sodium triaquo - samarium(iii) ethylenediaminetetra - acetate pentahydrate (neutron study)**
$C_{10}H_{18}N_2O_{11}Sm^-$, Na^+ , $5H_2O$
For complete entry see 76.49

82.C **Tetra - aquo - zinc(ethylenediaminetetra - acetato - zinc) dihydrate**
$(C_{10}H_{20}N_2O_{12}Zn_2)_n$, $2nH_2O$
For complete entry see 76.50

82.14 **Copper glycyl - glycyl - L - histidine methyl amine monohydrate**
$C_{11}H_{16}CuN_6O_3$, H_2O
N.Camerman, A.Camerman *Acta Crystallogr., Sect. A,* **31,** S48, 1975

82.15 **2,2′ - Bipyridyl - glycinato - chloro - copper(ii) dihydrate**
$C_{12}H_{12}ClCuN_3O_2$, $2H_2O$
C.J.Neitzel, R.Desiderato *Cryst. Struct. Commun.*, **4,** 333, 1975
Residue 1 also classified in 83

82.16 **Potassium bis(nitrilotriacetato) neodymium hexahydrate**
$C_{12}H_{12}N_2NdO_{12}^{3-}$, $3K^+$, $6H_2O$
C.F.Balyaeva, M.A.Porai-Koshits, T.I.Malinovskii
Eur. Cryst. Meeting, 346, 1974

82.17 **(D - Histidinato) - (L - histidinato) cobalt(iii) bromide**
$C_{12}H_{16}CoN_6O_4^+$, Br^-
N.Thorup *Acta Crystallogr., Sect. A,* **31,** S142, 1975

82.18 **Di - μ - sulfido - bis((L - histidinato)oxo - molybdenum(v)) sesquihydrate**
$C_{12}H_{16}Mo_2N_6O_6S_2$, $1.5H_2O$
B.Spivack, Z.Dori *J. Chem. Soc., Dalton Trans.*, 1077, 1975

82.C **$(-)_{436}$ - β_2 - ((2S,9S) - 2,9 - Diamino - 4,7 - diazadecane cobalt(iii) amino - methyl - malonato) perchlorate monohydrate (absolute configuration)**
$C_{12}H_{27}CoN_5O_4^+$, ClO_4^- , H_2O
For complete entry see 76.60

82.C **Copper(ii) di - hydrogen diethylenetriaminepenta - acetate monohydrate**
$C_{14}H_{21}CuN_3O_{10}$, H_2O
For complete entry see 76.64

82.C **Trihydrogen diethylenetriaminepenta - acetato copper(ii) monohydrate**
$C_{14}H_{21}CuN_3O_{10}$, H_2O
For complete entry see 76.65

82.19 **Aquo - (L - leucyl - L - tyrosinato) copper(ii) dihydrate ethanol solvate**
$C_{15}H_{22}CuN_2O_5$, $2H_2O$, C_2H_6O
D.van der Helm, S.E.Ealick, J.E.Burks
Acta Crystallogr., Sect. B, **31,** 1013, 1975

82.C **Potassium N,N' - ethylene - bis(acetylacetoniminato) - trans - di(glycinato) cobalt(iii) hexahydrate**
$C_{16}H_{26}CoN_4O_6^-$, K^+ , $6H_2O$
For complete entry see 76.70

82.20 **bis(O - (β - D - Xylopyranosyl) - L - serinato) copper(ii)**
$C_{16}H_{28}CuN_2O_{14}$
L.T.J.Delbaere, B.Kamenar, K.Prout
Acta Crystallogr., Sect. B, **31,** 862, 1975
Also classified in 45

82.C **Diaquo - μ - triethylenetetraminehexa - acetato - dichromium(iii) hexahydrate**
$C_{18}H_{28}Cr_2N_4O_{14}$, $6H_2O$
For complete entry see 76.74

82.21 **bis(N - Benzyl - L - prolinato) copper(ii)**
$C_{24}H_{28}CuN_2O_4$
G.G.Aleksandrov, Yu.T.Struchkov, A.A.Kurganov
Zh. Strukt. Khim., **14,** 492, 1973

82.22 **bis(Pyridoxylidene - DL - valinato) nickel(ii) hydrate**
$C_{26}H_{34}N_4NiO_8$, $7.4H_2O$
S.Capasso, F.Giordano, C.Mattia, L.Mazzarella, A.Ripamonti
J. Chem. Soc., Dalton Trans., 2228, 1974

82.23 **bis(Pyridoxylidene - L - valinato) zinc(ii) hydrate**
$C_{26}H_{34}N_4O_8Zn$, $8.1H_2O$
S.Capasso, F.Giordano, C.Mattia, L.Mazzarella, A.Ripamonti
J. Chem. Soc., Dalton Trans., 2228, 1974

METAL COMPLEXES (NITROGEN LIGAND)

83.C **μ - Formamido - bis(pentammine cobalt(iii)) pentachloride monohydrate**
$CH_{32}Co_2N_{11}O^{5+}$, $5Cl^-$, H_2O
For complete entry see 84.1

83.1 **Pyrazine di - mercury(i) nitrate**
$(C_4H_4Hg_2N_2{}^{2+})_n$, $2nNO_3{}^-$
K.Brodersen, N.Hacke, G.Liehr *Z. Anorg. Allg. Chem.*, **409,** 1, 1974

83.2 **Di - iodo - bis(dicyandiamide) cadmium(ii)**
$C_4H_8CdI_2N_8$
A.C.Villa, L.Coghi, A.G.Manfredotti, C.Guastini
Cryst. Struct. Commun., **3,** 739, 1974

83.C **Reaction product of dithiazylylcobaltate with ammonia and formaldehyde**
$C_4H_9CoN_5OS_4$
For complete entry see 85.1

83.3 **cis - Dichloro - bis(ethyleneimine) platinum(ii)**
$C_4H_{10}Cl_2N_2Pt$
J.C.Barnes, J.Iball, T.J.R.Weakley
Acta Crystallogr., Sect. B, **31,** 1435, 1975

83.4 **trans - Dichloro - bis(ethyleneimine) platinum(ii)**
$C_4H_{10}Cl_2N_2Pt$
J.C.Barnes, J.Iball, T.J.R.Weakley
Acta Crystallogr., Sect. B, **31,** 1435, 1975

83.5 **(bis(Dimethyl - (dinitrogendisulfido)silyl)oxide) cobalt(ii)**
$C_4H_{12}CoN_4OS_4Si_2$
U.Thewalt, M.Schlingmann *Z. Anorg. Allg. Chem.*, **406,** 319, 1974
Also classified in 63

83.6 **Chloro - (2,3 - diazabicyclo(2.2.1)hept - 2 - ene) copper(i)**
$(C_5H_8ClCuN_2)_n$
G.S.Chandler, C.L.Raston, G.W.Walker, A.H.White
J. Chem. Soc., Dalton Trans., 1797, 1974

83.7 **(9 - Methyladenine) silver(i) nitrate dihydrate**
$(C_6H_7AgN_5^+)_n$, nNO_3^- , $2nH_2O$
A.L.Beauchamp, M.A.Martin, C.Gagnon, P.Lavertue
Acta Crystallogr., Sect. A, **31,** S45, 1975
Residue 1 also classified in 44

83.8 **(5 - (2 - Hydroxyethyl) - 4 - methylthiazolium) tribromo - (5 - (2 - hydroxyethyl) - 4 - methyl - 3 - thiazolo) copper(ii) (absolute configuration)**
$C_6H_9Br_3CuNOS^-$, $C_6H_{10}NOS^+$
M.M.Thackeray, L.R.Nassimbeni
Acta Crystallogr., Sect. B, **30,** 2469, 1974
Residue 2 classified in 41

83.C **Potassium tris(biureto) cobalt(iii) hydrate**
$C_6H_9CoN_9O_6^{3-}$, $3K^+$, $6.38H_2O$
For complete entry see 79.9

83.9 **Dichloro - bis(acetoxime) palladium**
$C_6H_{14}Cl_2N_2O_2Pd$
Y.Kitano, K.Kobori, M.Tanimura, Y.Kinoshita
Bull. Chem. Soc. Jpn., **47,** 2969, 1974

83.10 **Tetrachloro - bis(2 - (5 - amino - 4 - carboxyamidinium)(1,2,3)triazole) copper(ii) monohydrate**
$C_6H_{14}Cl_4CuN_{12}$, H_2O
L.G.Purnell, J.C.Shepherd, D.J.Hodgson
J. Am. Chem. Soc., **97,** 2376, 1975

83.C **bis(Isothiocyanato) - bis(ethanolamine) nickel(ii)**
$C_6H_{14}N_4NiO_2S_2$
For complete entry see 84.14

83.11 **Tetra - aquo - (9 - methyladenine) copper(ii) sulfate monohydrate**
$C_6H_{15}CuN_5O_4^{2+}$, O_4S^{2-} , H_2O
E.Sletten, B.Thorstensen *Acta Crystallogr., Sect. B,* **30,** 2438, 1974

83.C **trans - Dinitro - (β - alaninato) - (1,3 - propanediamine) cobalt(ii)**
$C_6H_{16}CoN_5O_6$
For complete entry see 81.20

83.C **cis(Nitro) - mer(amino) - dinitro - (β - alaninato - 1,3 - propanediamine) cobalt(iii) monohydrate**
$C_6H_{16}CoN_5O_6$, H_2O
For complete entry see 81.21

83.12 **(bis(Dimethyl(dinitrogendisulfido)silyl) - ethylamine) nickel(ii)**
$C_6H_{17}N_5NiS_4Si_2$
U.Thewalt, M.Schlingmann *Z. Anorg. Allg. Chem.,* **406,** 319, 1974
Also classified in 63

83.13 **bis(1,3 - Propanediamine) copper(ii) m - chlorobenzoate**
$C_6H_{20}CuN_4^{2+}$, $2C_7H_4ClO_2^-$
R.Uggla, O.Orama, M.Sundberg, E.Tirronen, M.Klinga
Finn. Chem. Lett., 185, 1974
Residue 2 classified in 14

83.14 **bis(2 - Hydroxy - 1,3 - propanediamine) copper(ii) chloride**
$C_6H_{20}CuN_4O_2^{2+}$, $2Cl^-$
A.Pajunen, R.Kivekas *Finn. Chem. Lett.*, 39, 1974

83.15 **bis(2 - Hydroxy - 1,3 - propanediamine) copper(ii) thiocyanate**
$C_6H_{20}CuN_4O_2^{2+}$, $2CNS^-$
K.Smolander *Finn. Chem. Lett.*, 199, 1974

83.C **(Pyridine - 2,6 - dicarboxylato) - dioxouranium(vi) monohydrate**
$(C_7H_5NO_7U)_n$
For complete entry see 81.22

83.C **Diaquo - peroxotitanium dipicolinate dihydrate (orthorhombic form)**
$C_7H_7NO_8Ti$, $2H_2O$
For complete entry see 81.23

83.C **(2 - Diethylamino - ethanolato) copper(ii) thiocyanate**
$(C_7H_{14}CuN_2OS)_n$
For complete entry see 84.20

83.C **Tetra - ammine - (4,5 - dihydroxy - 4,5 - dimethyl - Δ^1 - pyrroline - 2 - carboxylato) cobalt(iii) perchlorate monohydrate**
$C_7H_{22}CoN_5O_4^{2+}$, $2ClO_4^-$, H_2O
For complete entry see 84.21

83.16 **bis(Di - iminosuccinonitrilo) platinum(ii) trans - dichloro - bis(benzonitrile) platinum(ii)**
$C_8H_4N_8Pt$, $C_{14}H_{10}Cl_2N_2Pt$
J.W.Lauher, J.A.Ibers *Inorg. Chem.*, **14**, 640, 1975
Residue 2 classified in 83

83.17 **bis(Succinimido) mercury**
$C_8H_8HgN_2O_4$
D.Grdenic, B.Kamenar, M.Sikirica, T.Duplancic, S.Govedic, A.Hergold, P.Matkovic *Acta Crystallogr., Sect. A*, **31**, S132, 1975

83.18 **Di(bis(triphenylphosphine)imminium) octachloro - di - niobium acetonitrile adduct chlorobenzene solvate**
$C_8H_{12}Cl_8N_4Nb_2^{2-}$, $2C_{36}H_{30}NP_2^+$, $2C_6H_5Cl$
P.A.Finn, M.S.King, P.A.Kilty, R.E.McCarley
J. Am. Chem. Soc., **97**, 220, 1975
Residue 2 classified in 64

83.19 **Copper(i) tetra(acetonitrile) perchlorate**
$C_8H_{12}CuN_4^+$, ClO_4^-
I.Csoregh, P.Kierkegaard, R.Norrestam
Acta Crystallogr., Sect. B, **31,** 314, 1975

83.C **Nitrato - aquo - (o - aminobenzaldehyde - thiosemicarbazone) copper(ii) nitrate monohydrate**
$C_8H_{12}CuN_5O_4S^+$, NO_3^- , H_2O
For complete entry see 79.15

83.C **3 - Ethoxy - 2 - oxobutyraldehyde - bis(thiosemicarbazonato) copper(ii) (triclinic form)**
$C_8H_{14}CuN_6OS_2$
For complete entry see 79.16

83.20 **Dichloro - (dimethylglyoxime) - (hydrogen dimethylglyoximato) cobalt(iii)**
$C_8H_{15}Cl_2CoN_4O_4$
Yu.A.Simonov, A.A.Dvorkin, O.A.Bologa, A.V.Ablov, T.I.Malinovskii
Dokl. Akad. Nauk SSSR, **210,** 615, 1973

83.21 **Dichloro - (dimethylglyoxime) - (hydrogen dimethylglyoximato) rhodium(iii)**
$C_8H_{15}Cl_2N_4O_4Rh$
Yu.A.Simonov, A.A.Dvorkin, T.I.Malinovskii, A.V.Ablov, O.A.Bologa
Eur. Cryst. Meeting, 402, 1974

83.22 **Dinitro - (1,4,7,10 - tetra - azacyclododecane) cobalt(iii) chloride monohydrate**
$C_8H_{20}CoN_6O_4^+$, Cl^- , H_2O
Y.Iitaka, M.Shina, E.Kimura *Inorg. Chem.,* **13,** 2886, 1974

83.23 **Di(cyanato) - 2,4 - lutidine copper(ii)**
$(C_9H_9CuN_3O_2)_n$
J.Kohout, F.Valach, M.Quastlerova-Hvastijova, J.Gazo
Z. Phys. Chem. (Leipzig), **255,** 901, 1974

83.24 **tris(Imidazole) copper(ii) sulfate monohydrate**
$(C_9H_{14}CuN_6O_5S)_n$
G.Fransson, B.K.S.Lundberg *Acta Chem. Scand. Ser. A,* **28,** 578, 1974

83.C **Methyl - bis(dimethylglyoximato) - aquo - cobalt(iii)**
$C_9H_{19}CoN_4O_5$
For complete entry see 71.12

83.25 **tris(1,3 - Propanediamine) chromium(iii) pentacyanonickelate(ii) dihydrate (at −80°C)**
$C_9H_{30}CrN_6^{3+}$, $C_5N_5Ni^{3-}$, $2H_2O$
F.A.Jurnak, K.N.Raymond *Inorg. Chem.,* **13,** 2387, 1974

83.26 **Diazido - (2,2′ - bipyridine) copper(ii)**
$C_{10}H_8CuN_8$
G.W.Bushnell, M.A.Khan *Can. J. Chem.*, **52,** 3125, 1974

83.27 **Dibromo - bis(pyridine) copper(ii)**
$C_{10}H_{10}Br_2CuN_2$
B.Morosin *Acta Crystallogr., Sect. B,* **31,** 632, 1975

83.28 **Dichloro - bis(pyridine) cobalt(ii) (α form)**
$C_{10}H_{10}Cl_2CoN_2$
P.J.Clarke, H.J.Milledge *Acta Crystallogr., Sect. B,* **31,** 1543, 1975

83.29 **Dichloro - bis(pyridine) cobalt(ii) (γ form, at 89°K)**
$C_{10}H_{10}Cl_2CoN_2$
P.J.Clarke, H.J.Milledge *Acta Crystallogr., Sect. B,* **31,** 1543, 1975

83.30 **Dichloro - bis(pyridine) copper(ii)**
$C_{10}H_{10}Cl_2CuN_2$
B.Morosin *Acta Crystallogr., Sect. B,* **31,** 632, 1975

83.31 **Diaquo - di(nitrato) - bis(pyridine) mercury**
$C_{10}H_{14}HgN_4O_8$
D.Grdenic, B.Kamenar, M.Sikirica, T.Duplancic, S.Govedic, A.Hergold, P.Matkovic *Acta Crystallogr., Sect. A,* **31,** S132, 1975

83.32 **Dibromo - hydrogen - bis(O - methyl - dimethylglyoximato) cobalt(ii)**
$C_{10}H_{19}Br_2CoN_4O_4$
Yu.A.Simonov, A.A.Dvorkin, T.I.Malinovskii, A.V.Ablov, O.A.Bologa
Eur. Cryst. Meeting, 402, 1974

83.C **(1 - Oxa - 7,10 - dithia - 4,13 - diazacyclopentadecane) palladium(ii) nitrate**
$C_{10}H_{22}N_2OPdS_2^{2+}$, $2NO_3^-$
For complete entry see 85.16

83.33 **Penta - aquo - nickel(ii) guanosine - 5′ - monophosphate trihydrate**
$C_{10}H_{22}N_5NiO_{13}P$, $3H_2O$
P.de Meester, D.M.L.Goodgame, A.C.Skapski, B.T.Smith
Biochim. Biophys. Acta, **340,** 113, 1974
Residue 1 also classified in 47, 46, 45

83.34 **Spermine copper(ii) perchlorate**
$C_{10}H_{26}CuN_4^{2+}$, $2ClO_4^-$
R.Boggs, J.Donohue *Acta Crystallogr., Sect. B,* **31,** 320, 1975

83.35 **bis(cis - 3,5 - Diamino - piperidine) palladium(ii) perchlorate**
$C_{10}H_{26}N_6Pd^{2+}$, $2ClO_4^-$
H.Manohar, D.Schwarzenbach *Helv. Chim. Acta,* **57,** 519, 1974

83.36 **(cis - 3,5 - Diamino - piperidine) - (cis - 3,5 - diaminopiperidinium) palladium(ii) nitrate**
$C_{10}H_{27}N_6Pd^{3+}$, $3NO_3^-$
H.Manohar, D.Schwarzenbach *Helv. Chim. Acta,* **57,** 519, 1974

83.C **racemic - ((1,8 - Diamino - 2,5 - dimethyl - 5 - hydroxy - 3,6 - diaza - oct - 2 - ene) - ethylenediamine cobalt(iii)) hexachlorothallate(iii) dihydrate**
$C_{10}H_{28}CoN_6O^{3+}$, Cl_6Tl^{3-} , $2H_2O$
For complete entry see 76.52

83.37 **bis(cis - 3,5 - Diamino - piperidinium) palladium(ii) perchlorate dihydrate**
$C_{10}H_{28}N_6Pd^{4+}$, $4ClO_4^-$, $2H_2O$
H.Manohar, D.Schwarzenbach *Helv. Chim. Acta,* **57,** 519, 1974

83.38 **Carbonyl - (hydrido - tris(pyrazol - 1 - yl)borato) - methyl platinum**
$C_{11}H_{13}BN_6OPt$
P.E.Rush, J.D.Oliver *J. Chem. Soc., Chem. Commun.,* 996, 1974
Also classified in 62

83.C **Iodo - nitrosyl - cyclopentadienyl - (phenylhydrazine) molybdenum tetrafluoroborate**
$C_{11}H_{13}IMoN_3O^+$, BF_4^-
For complete entry see 73.13

83.39 **Carbonyl - (6,7,13,14 - tetramethyl - 1,2,4,5,8,9,11,12 - octa - azacyclotetradeca - 2,5,7,12,14 - pentaenato) cobalt(i)**
$C_{11}H_{17}CoN_8O$
V.L.Goedken, S.-M.Peng *J. Chem. Soc., Chem. Commun.,* 914, 1974

83.C **Diaquo - (2,6 - diacetylpyridine - bis(semicarbazone)) copper(ii) nitrate monohydrate**
$C_{11}H_{19}CuN_7O_4^{2+}$, $2NO_3^-$, H_2O
For complete entry see 84.30

83.C **Diaquo - (2,6 - diacetylpyridine - bis(semicarbazone)) nickel(ii) nitrate monohydrate**
$C_{11}H_{19}N_7NiO_4^{2+}$, $2NO_3^-$, H_2O
For complete entry see 84.31

83.C **trans(Theophyllinato - chloro - bis(ethylenediamine) cobalt(iii)) chloride dihydrate**
$C_{11}H_{23}ClCoN_8O_2^+$, Cl^- , $2H_2O$
For complete entry see 76.56

83.C **cis(Theophyllinato - chloro - bis(ethylenediamine) cobalt(iii)) perchlorate**
$C_{11}H_{23}ClCoN_8O_2^+$, ClO_4^-
For complete entry see 76.57

83.40 **trans - Di(thiocyanato) - bis(pyridine) platinum(ii)**
$C_{12}H_{10}N_4PtS_2$
M.R.Caira, L.R.Nassimbeni *Acta Crystallogr., Sect. B,* **31,** 581, 1975

83.C **2,2′ - Bipyridyl - glycinato - chloro - copper(ii) dihydrate**
$C_{12}H_{12}ClCuN_3O_2$, $2H_2O$
For complete entry see 82.15

83.C **Di - μ - (N,S - dimethyl - dithiocarbamidinato) - bis(tricarbonyl manganese)**
$C_{12}H_{12}Mn_2N_2O_6S_4$
For complete entry see 85.17

83.41 **Dichloro - bis(4 - methylpyridine) zinc(ii)**
$C_{12}H_{14}Cl_2N_2Zn$
M.Laing *Acta Crystallogr., Sect. A,* **31,** S147, 1975

83.C **Chloro - N,N′ - ethylene - bis(acetylacetoneiminato) manganese(iii)**
$C_{12}H_{16}ClMnN_2O_2$
For complete entry see 84.34

83.42 **Diaquo - bis(9 - methylhypoxanthine) copper(ii) chloride trihydrate**
$C_{12}H_{16}CuN_8O_4^{2+}$, $2Cl^-$, $3H_2O$
E.Sletten *Acta Crystallogr., Sect. B,* **30,** 1961, 1974

83.C **Pyrazine copper acetate (at 100°K)**
$(C_{12}H_{16}Cu_2N_2O_8)_n$
For complete entry see 81.29

83.C **Pyrazine copper acetate**
$(C_{12}H_{16}Cu_2N_2O_8)_n$
For complete entry see 81.30

83.43 **Dibromo - bis(5 - (2 - hydroxyethyl) - 4 - methylthiazole) cobalt(ii)**
$C_{12}H_{18}Br_2CoN_2O_2S_2$
M.R.Caira, L.R.Nassimbeni *Acta Crystallogr., Sect. B,* **30,** 2332, 1974

83.44 **hexakis(Acetonitrile) iron(ii) tetrachloroferrate(ii) (twinned model refined in P - 3)**
$C_{12}H_{18}FeN_6^{2+}$, $2Cl_4Fe^-$
A.D.van Ingen Schenau, G.C.Verschoor, C.Romers
Eur. Cryst. Meeting, 370, 1974

83.45 **hexakis(Acetonitrile) iron(ii) tetrachloroferrate(ii) (disordered model)**
$C_{12}H_{18}FeN_6^{2+}$, $2Cl_4Fe^-$
A.D.van Ingen Schenau, G.C.Verschoor, C.Romers
Eur. Cryst. Meeting, 370, 1974

83.46 Tetra - aquo - bis(9 - methyladenine) copper(ii) dichloride dihydrate
$C_{12}H_{22}CuN_{10}O_4^{2+}$, $2Cl^-$, $2H_2O$
E.Sletten, M.Ruud *Acta Crystallogr., Sect. B,* **31,** 982, 1975
Residue 1 also classified in 44

83.47 Tetra - aquo - bis(9 - methyladenine) copper(ii) dichloride dihydrate
$C_{12}H_{22}CuN_{10}O_4^{2+}$, $2Cl^-$, $2H_2O$
S.W.Hawkinson
Am. Cryst. Assoc., Abstr. Papers (Spring Meeting), 10, 1975
Residue 1 also classified in 44

83.C Carbonyl - methyl - (6,7,13,14 - tetramethyl - 1,2,4,5,8,9,11,12 - octa - azacyclotetradeca - 1(14),5,7,12 - tetraenyl) iron(ii)
$C_{12}H_{22}FeN_8O$
For complete entry see 71.27

83.C Methyl - (6,7,13,14 - tetramethyl - 1,2,4,5,8,9,11,12 - octa - aza - cyclotetradeca - 5,7,12,14(1) - tetraene - dienyl) - methylhydrazine cobalt(iii)
$C_{12}H_{23}CoN_{10}$
For complete entry see 71.28

83.C Bromo - (diethylaminoethanolato) copper(ii) dimer (β form)
$C_{12}H_{28}Br_2Cu_2N_2O_2$
For complete entry see 84.35

83.C $(+)_{510}$ - Oxalato - bis(R,S - 2,4 - diaminopentane) cobalt(iii) perchlorate monohydrate (absolute configuration)
$C_{12}H_{28}CoN_4O_4^+$, ClO_4^- , H_2O
For complete entry see 81.33

83.48 bis(2,4,4 - Trimethyl - 1,5 - diazapent - 1 - ene) nickel(ii) perchlorate
$C_{12}H_{28}N_4Ni^{2+}$, $2ClO_4^-$
P.Domiano, A.Musatti, C.Pelizzi *Cryst. Struct. Commun.,* **4,** 185, 1975

83.49 $(+)_{589}$ - tris(1,4 - Diaminobutane) cobalt(iii) bromide (absolute configuration)
$C_{12}H_{36}CoN_6^{3+}$, $3Br^-$
S.Sato, Y.Saito *Acta Crystallogr., Sect. B,* **31,** 1378, 1975

83.50 bis(hexakis(Dimethylamido) di - tungsten) hexakis(dimethylamido) tungsten
$2C_{12}H_{36}N_6W_2$, $C_{12}H_{36}N_6W$
F.A.Cotton, B.R.Stults, J.M.Troup, M.H.Chisholm, M.Extine
J. Am. Chem. Soc., **97,** 1242, 1975

83.51 μ - Carbonyl - bis - (μ - hexafluoroisopropylideneamido) - hexacarbonyl di - manganese
$C_{13}F_{12}Mn_2N_2O_7$
E.W.Abel, C.A.Burton, M.R.Churchill, K.-K.G.Lin
J. Chem. Soc., Chem. Commun., 917, 1974

83.C **Chloro - (1 - (2 - thiazolylazo) - 2 - naphtholato) palladium(ii) dioxan solvate**
$C_{13}H_8ClN_3OPdS$, $C_4H_8O_2$
For complete entry see 84.38

83.C **(1 - (2 - Thiazolylazo) - 2 - naphtholato) - diaquo - copper(ii) perchlorate**
$C_{13}H_{12}CuN_3O_3S^+$, ClO_4^-
For complete entry see 84.39

83.52 **trans - Azido - (pyridine) - bis(dimethylglyoximato) cobalt(iii)**
$C_{13}H_{19}CoN_8O_4$
A.Clearfield, R.Gopal, R.J.Kline, M.Sipski, L.Urban
Am. Cryst. Assoc., Abstr. Papers (Spring Meeting), 32, 1975

83.C **Dimethyl - (3,3′ - (trimethylenedinitrolo) - bis(butan - 2 - one oximato)) cobalt(iii)**
$C_{13}H_{25}CoN_4O_2$
For complete entry see 71.32

83.C **N,N′ - Di(2 - hydroxyethyl) - 2,4 - pentanedi - iminato - bis(ethanolamine) cobalt(iii)**
$C_{13}H_{29}CoN_4O_4$
For complete entry see 84.40

83.C **bis(Tetraphenylarsonium) uranyl - bis(pyridine - 2,6 - dicarboxylate) hexahydrate**
$C_{14}H_6N_2O_{10}U^{2-}$, $2C_{24}H_{20}As^+$, $6H_2O$
For complete entry see 81.34

83.C **bis(Di - iminosuccinonitrilo) platinum(ii) trans - dichloro - bis(benzonitrile) platinum(ii)**
$C_{14}H_{10}Cl_2N_2Pt$, $C_8H_4N_8Pt$
For complete entry see 83.16

83.53 **Tribromo - (2,9 - dimethyl - 1,10 - phenanthroline) gold(iii)**
$C_{14}H_{12}AuBr_3N_2$
W.T.Robinson, E.Sinn *J. Chem. Soc., Dalton Trans.,* 726, 1975

83.54 **Trichloro - (2,9 - dimethyl - 1,10 - phenanthroline) gold(iii)**
$C_{14}H_{12}AuCl_3N_2$
W.T.Robinson, E.Sinn *J. Chem. Soc., Dalton Trans.,* 726, 1975

83.C **bis(2 - Acetato - pyridine) copper(ii) dihydrate**
$C_{14}H_{12}CuN_2O_4$, $2H_2O$
For complete entry see 81.35

83.C **Methyl(hydro - tris(1 - pyrazolyl)borato)hexafluorobut - 2 - yne platinum(ii)**
$C_{14}H_{13}BF_6N_6Pt$
For complete entry see 71.35

83.C **bis(Pyridinealdehyde - thiosemicarbazonato) cobal(iii) chloride trihydrate**
$C_{14}H_{14}CoN_8S_2^+$, Cl^- , $3H_2O$
For complete entry see 85.22

83.C **bis(Pyridinealdehydethiosemicarbazonato) cobalt(iii) nitrate**
$C_{14}H_{14}CoN_8S_2^+$, NO_3^-
For complete entry see 79.21

83.C **(bis(2 - (2 - Pyridyl)ethyl)disulfide) copper(i) perchlorate**
$(C_{14}H_{16}CuN_2S_2^+)_n$, $nClO_4^-$
For complete entry see 85.23

83.55 **Dichloro - bis(4 - ethylpyridine) zinc(ii)**
$C_{14}H_{18}Cl_2N_2Zn$
M.Laing *Acta Crystallogr., Sect. A,* **31,** S147, 1975

83.C **Ammine - bis(η - cyclopentadienyl) - N - (C - methyl - C - ethylketimino) molybdenum bis(hexafluorophosphate)**
$C_{14}H_{22}MoN_2^{2+}$, $2F_6P^-$
For complete entry see 73.23

83.C **N,N′ - Tetramethylene - bis(thioacetylacetoniminato) zinc**
$C_{14}H_{22}N_2S_2Zn$
For complete entry see 85.25

83.C **Triethylenetetramine - bis(dimethylglyoxaldimine - monoximato) cobalt(iii) dibromide tetrahydrate**
$C_{14}H_{27}CoN_6O_2^{2+}$, $2Br^-$, $4H_2O$
For complete entry see 76.66

83.56 **Diaquo - (2,3,9,10 - tetramethyl - 1,4,8,11 - tetra - azacyclotetradeca - 1,3,8,10 - tetraene) cobalt(ii) perchlorate**
$C_{14}H_{28}CoN_4O_2^{2+}$, $2ClO_4^-$
M.D.Glick, W.G.Schmonsees, J.F.Endicott
J. Am. Chem. Soc., **96,** 5661, 1974

83.C **(2 - Diethylaminoethanolato) copper(ii) thiocyanate dimer**
$C_{14}H_{28}Cu_2N_4O_2S_2$
For complete entry see 84.42

83.57 **5,7,12,14 - Tetramethyl - 1,4,8,11 - tetra - azacyclotetradeca - 4,11 - diene nickel(ii) perchlorate**
$C_{14}H_{28}N_4Ni^{2+}$, $2ClO_4^-$
J.Krajewski, Z.Urbanczyk-Lipkowska, P.Gluzinski
Bull. Acad. Pol. Sci., Ser. Sci. Chim., **22,** 955, 1974

83.58 **3,5,10,12 - Tetramethyl - 1,4,8,11 - tetra - azacyclotetradeca - 4,11 - diene nickel(ii) perchlorate**
$C_{14}H_{28}N_4Ni^{2+}$, $2ClO_4^-$
J.Krajewski, Z.Urbanczyk-Lipkowska, P.Gluzinski
Rocz. Chem., **48,** 1821, 1974

83.59 **Trichloro - bis(N - methyl - 1,4 - diazabicyclo(2.2.2)octane) copper(ii) perchlorate**
$C_{14}H_{30}Cl_3CuN_4^+$, ClO_4^-
J.S.Wood *Acta Crystallogr., Sect. A,* **31,** S145, 1975

83.60 **Trichloro - bis(N - methyl - 1,4 - diazabicyclo(2.2.2)octane) nickel(ii) perchlorate**
$C_{14}H_{30}Cl_3N_4Ni^+$, ClO_4^-
J.S.Wood *Acta Crystallogr., Sect. A,* **31,** S145, 1975

83.61 **Diammine - (2,3,9,10 - tetramethyl - 1,4,8,11 - tetra - azacyclotetradeca - 1,3,8,10 - tetraene) cobalt(iii) bromide**
$C_{14}H_{30}CoN_6^{3+}$, $3Br^-$
M.D.Glick, W.G.Schmonsees, J.F.Endicott
J. Am. Chem. Soc., **96,** 5661, 1974

83.62 **Azido - (1,4,8,11 - tetramethyl - 1,4,8,11 - tetra - azacyclotetradecane) nickel(ii) perchlorate**
$C_{14}H_{32}N_7Ni^+$, ClO_4^-
M.J.D'Aniello Junior, M.T.Mocella, F.Wagner, E.K.Barefield, I.C.Paul
J. Am. Chem. Soc., **97,** 192, 1975

83.63 **Di - μ - thiocyanato - bis(di(3 - aminopropyl)amine copper(ii)) perchlorate**
$C_{14}H_{34}Cu_2N_8S_2^{2+}$, $2ClO_4^-$
M.Cannas, G.Carta, G.Marongiu *Gazz. Chim. Ital.,* **104,** 581, 1974

83.64 **Trichloro - tris(pyridine) molybdenum**
$C_{15}H_{15}Cl_3MoN_3$
J.V.Brencic *Z. Anorg. Allg. Chem.,* **403,** 218, 1974

83.65 **(−) - β - Isosparteine copper(ii) chloride**
$C_{15}H_{26}Cl_2CuN_2$
L.S.Childers, K.Folting, L.L.Merritt Junior, W.E.Streib
Acta Crystallogr., Sect. B, **31,** 924, 1975

83.C **tris(Dimethylaminato) - tris(N,N - dimethylcarbamato) tungsten(vi)**
$C_{15}H_{36}N_6O_6W$
For complete entry see 80.15

83.C **bis(Perfluorophenyl) - tetramethyltetrazene zinc(ii)**
$C_{16}H_{12}F_{10}N_4Zn$
For complete entry see 71.42

83.66 **bis((1,8 - Naphthyridine) mercury(i)) diperchlorate**
$C_{16}H_{12}Hg_2N_4^{2+}$, $2ClO_4^-$
J.C.Dewan, D.L.Kepert, A.H.White
J. Chem. Soc., Dalton Trans., 490, 1975

83.67 Potassium tetrakis(succinimidato) copper(ii) hexahydrate
$C_{16}H_{16}CuN_4O_8^{2-}$, $2K^+$, $6H_2O$
T.Tsukihara, Y.Katsube, K.Fujimori, K.Kawashima, Y.Kan-Nan
Bull. Chem. Soc. Jpn., **47**, 1582, 1974

83.68 Lithium tetrakis(succinimidato) copper(ii) monohydrate
$C_{16}H_{16}CuN_4O_8^{2-}$, $2Li^+$, H_2O
T.Tsukihara, Y.Katsube, K.Fujimori, K.Kawashima, Y.Kan-Nan
Bull. Chem. Soc. Jpn., **47**, 1582, 1974

83.69 bis(2 - (2 - Aminoethyl)pyridine) - di - isothiocyanato copper(ii)
$C_{16}H_{20}CuN_6S_2$
D.L.Kozlowski, D.J.Hodgson *J. Chem. Soc., Dalton Trans.*, 55, 1975

83.70 Dichloro - bis(4 - isopropylpyridine) zinc(ii)
$C_{16}H_{22}Cl_2N_2Zn$
M.Laing *Acta Crystallogr., Sect. A,* **31,** S147, 1975

83.C (1,8 - bis(2′ - Pyridyl) - 3,6 - diaza - octane) copper(ii) perchlorate
$C_{16}H_{22}CuN_4^{2+}$, $2ClO_4^-$
For complete entry see 76.68

83.71 Chloro - (2,7,12 - trimethyl - 3,7,11,17 - tetra - azabicyclo(11.3.1)heptadeca - 1(17),2,11,13,15 - pentaene) copper(ii) nitrate dihydrate
$C_{16}H_{24}ClCuN_4^+$, NO_3^- , $2H_2O$
M.R.Caira, L.R.Nassimbeni, P.R.Woolley
Acta Crystallogr., Sect. B, **31,** 1334, 1975

83.72 bis(Dimethyl - bis(1 - pyrazolyl) - gallato) copper(ii)
$C_{16}H_{24}CuGa_2N_8$
D.J.Patmore, D.F.Rendle, A.Storr, J.Trotter
J. Chem. Soc., Dalton Trans., 718, 1975
Also classified in 68

83.73 bis(Dimethyl - bis(pyrazol - 1 - yl)gallato) nickel(ii)
$C_{16}H_{24}Ga_2N_8Ni$
D.F.Rendle, A.Storr, J.Trotter *J. Chem. Soc., Dalton Trans.*, 176, 1975
Also classified in 68

83.74 Tripotassium μ - oxo - di - μ - (dimethylglyoximato - cobalt(iii)) hydroxide dodecahydrate
$C_{16}H_{26}Co_2N_8O_9^{2-}$, $3K^+$, HO^- , $12H_2O$
T.I.Malinovskii, Yu.A.Simonov, S.T.Malinovskii, A.V.Ablov, O.A.Bologa
Acta Crystallogr., Sect. A, **31,** S137, 1975

83.75 Trichloro - bis(N,N′ - di - isopropylacetamidinato) tantalum(v)
$C_{16}H_{34}Cl_3N_4Ta$
M.G.B.Drew, J.D.Wilkins *Acta Crystallogr., Sect. B,* **31,** 177, 1975

83.76 **Dichloro - (5,5,7,12,12,14 - hexamethyl - 1,4,8,11 - tetra - azacyclotetradecane) manganese(iii) chloride trihydrate**
$C_{16}H_{36}Cl_2MnN_4^+$, Cl^- , $3H_2O$
E.Brackett, C.Pfluger
Am. Cryst. Assoc., Abstr. Papers (Summer Meeting), 228, 1974

83.77 **Morpholine copper(i) iodide tetramer**
$C_{16}H_{36}Cu_4I_4N_4O_4$
V.Schramm, K.F.Fischer *Naturwissenschaften,* **61,** 500, 1974

83.C **Di - μ - (bis(2 - (N,N - dimethylamino)ethyl)disulfide) dicopper(i) tetrafluoroborate**
$C_{16}H_{40}Cu_2N_4S_4^{2+}$, $2BF_4^-$
For complete entry see 85.32

83.78 **bis(Dimethylamino) titanium difluoride tetramer**
$C_{16}H_{48}F_8N_8Ti_4$
W.S.Sheldrick *J. Fluorine Chem.,* **4,** 415, 1975

83.79 **Dichloro - (trans - 2 - (2 - quinolyl)methylenequinuclidin - 3 - one) cobalt(ii)**
$C_{17}H_{16}Cl_2CoN_2O$
G.J.Long, E.O.Schlemper *J. Chem. Soc., Dalton Trans.,* 96, 1975

83.80 **trans - Azido - (pyridine) - bis(acetylacetonato)ethylenedi - imine cobalt(iii)**
$C_{17}H_{23}CoN_6O_2$
A.Clearfield, R.Gopal, R.J.Kline
Am. Cryst. Assoc., Abstr. Papers (Spring Meeting), 32, 1975
Also classified in 84

83.C **bis(Isothiocyanato) - (2,6 - di(acetimino)pyridine - triethylenetetramine) zinc(ii) dichloromethane solvate**
$C_{17}H_{23}N_7S_2Zn$, $0.5C_2H_4Cl_2$
For complete entry see 76.73

83.C **bis(8 - Mercaptoquinolinato) palladium**
$C_{18}H_{12}N_2PdS_2$
For complete entry see 85.34

83.C **bis(8 - Mercaptoquinolinato) platinum**
$C_{18}H_{12}N_2PtS_2$
For complete entry see 85.35

83.81 **hexakis(Isoxazole) iron(ii) perchlorate**
$C_{18}H_{18}FeN_6O_6^{2+}$, $2ClO_4^-$
A.D.van Ingen Schenau, G.C.Verschoor, C.Romers
Eur. Cryst. Meeting, 370, 1974

83.82 **(7,8,15,16,17,18 - Hexahydro - dibenzo(e,m)(1,4,8,11)tetra - azacyclotetradecinato) nickel(ii)**
$C_{18}H_{18}N_4Ni$
E.N.Maslen, L.M.Engelhardt, A.H.White
J. Chem. Soc., Dalton Trans., 1799, 1974

83.83 **bis((Hydrido - tris(1 - pyrazolyl)borato) copper(i))**
$C_{18}H_{20}B_2Cu_2N_{12}$
C.S.Arcus, J.L.Wilkinson, C.Mealli, T.J.Marks, J.A.Ibers
J. Am. Chem. Soc., **96**, 7564, 1974
Also classified in 62

83.84 **Hydrogen - bis(dimethylglyoximato) - bis(pyridine) cobalt(iii) dihydrate**
$C_{18}H_{23}CoN_6O_4$, $2H_2O$
Yu.A.Simonov, A.A.Dvorkin, T.I.Malinovskii, A.V.Ablov, O.A.Bologa
Eur. Cryst. Meeting, 402, 1974

83.85 **hexakis(Imidazole) copper(ii) nitrate**
$C_{18}H_{24}CuN_{12}^{2+}$, $2NO_3^-$
D.L.McFadden, A.T.McPhail, C.D.Garner, F.E.Mabbs
J. Chem. Soc., Dalton Trans., 263, 1975

83.C **bis(N - (Picolinoyl) - 3 - amino - 1 - propoxido - aquo - copper(ii)) dihydrate**
$C_{18}H_{24}Cu_2N_4O_6$, $2H_2O$
For complete entry see 84.44

83.C **(Diethyl - bis(pyrazolyl)borato) - (pyrazolylato) - (trihapto - allyl) - (dicarbonyl) molybdenum**
$C_{18}H_{25}BMoN_6O_2$
For complete entry see 72.15

83.86 **Dichloro - bis(4 - t - butylpyridine) zinc(ii)**
$C_{18}H_{26}Cl_2N_2Zn$
M.Laing *Acta Crystallogr., Sect. A*, **31**, S147, 1975

83.C **Di - μ - sulfido - bis(cyclopentadienyl - (t - butylimido) - molybdenum)**
$C_{18}H_{28}Mo_2N_2S_2$
For complete entry see 73.33

83.C **Chloro - bis(triethylphosphine) - (fluorophenylazo) platinum**
$C_{18}H_{34}ClFN_2P_2Pt$
For complete entry see 86.26

83.C **μ - (N,N′ - bis(2 - Dimethylaminoethyl) - oxamidato) - di(isothiocyanato) - bis(dimethylformamide) dicopper(ii)**
$C_{18}H_{34}Cu_2N_8O_4S_2$
For complete entry see 84.45

83.C **trans - Chloro - bis(triethylphosphine)(p - fluorophenyldiazene) platinum(ii) perchlorate**
$C_{18}H_{35}ClFN_2P_2Pt^+$, ClO_4^-
For complete entry see 86.27

83.C **trans - Chloro - bis(triethylphosphine) - (p - fluorophenylhydrazine) platinum(ii) tetrafluoroborate**
$C_{18}H_{37}ClFN_2P_2Pt^+$, BF_4^-
For complete entry see 86.28

83.C **Bromo - (tris(2 - t - butylthioethyl)amine) cobalt(ii) hexafluorophosphate**
$C_{18}H_{39}BrCoNS_3^+$, F_6P^-
For complete entry see 85.37

83.C **bis(N - (6 - Amino - 3,3,6 - trimethyl - 4 - aza - 2 - heptylidenyl)hydroxanato) dinickel(ii) chloride pentahydrate**
$C_{18}H_{40}N_6Ni_2O_2^{2+}$, $2Cl^-$, $5H_2O$
For complete entry see 76.75

83.C **trans - bis(Acetone hydrazone) - tetrakis(trimethylphosphite) ruthenium(ii) bis(tetraphenylborate)**
$C_{18}H_{52}N_4O_{12}P_4Ru^{2+}$, $2C_{24}H_{20}B^-$
For complete entry see 86.29

83.C **Dichloro - (2 - (2′ - pyridyl) - 3 - (N - 2 - picolylimino) - 4 - oxo - 1,2,3,4 - tetrahydroquinazoline) manganese(ii)**
$C_{19}H_{15}Cl_2MnN_5O$
For complete entry see 84.47

83.C **Dichloro - bis(aniline) - (norbornadiene) ruthenium**
$C_{19}H_{22}Cl_2N_2Ru$
For complete entry see 75.20

83.87 **Diaquo - (2,6 - diacetylpyridine - bis(2 - pyridylhydrazone)) cobalt(ii) chloride**
$C_{19}H_{23}CoN_7O_2^{2+}$, $2Cl^-$
D.Wester, G.J.Palenik *J. Chem. Soc., Chem. Commun.*, 74, 1975

83.88 **Diaquo - (2,6 - diacetylpyridine) - bis(2 - pyridylhydrazone) zinc(ii) chloride**
$C_{19}H_{23}N_7O_2Zn^{2+}$, $2Cl^-$
D.Wester, G.J.Palenik *J. Chem. Soc., Chem. Commun.*, 74, 1975

83.C **t - Butylamido - tris(π - cyclopentadienyl nickel)**
$C_{19}H_{24}NNi_3$
For complete entry see 73.34

83.89 **Di - μ - phenyldiazo - bis(tetracarbonyl manganese)**
$C_{20}H_{10}Mn_2N_4O_8$
M.R.Churchill, K.-K.G.Lin *Inorg. Chem.*, **14**, 1133, 1975

83.90 **bis(2,2′ - Bipyridyl) copper(ii) bis(dichlorocuprate(i))**
$(C_{20}H_{16}Cl_2Cu_2N_4{}^{+})_n$, nCl_2Cu^{-}
J.Kaiser, G.Brauer, F.A.Schroder, I.F.Taylor, S.E.Rasmussen
J. Chem. Soc., Dalton Trans., 1490, 1974

83.91 **Aqua - (2,2′.6′,2″.6″,2‴ - quaterpyridyl) - sulfito cobalt(iii) nitrate monohydrate**
$C_{20}H_{16}CoN_4O_4S^{+}$, $NO_3{}^{-}$, H_2O
E.N.Maslen, C.L.Raston, A.H.White
J. Chem. Soc., Dalton Trans., 323, 1975

83.C **μ - (2 - Mercaptoethanolato) - μ - oxo - bis(oxo - (8 - hydroxyquinolinato) - molybdenum(v))**
$C_{20}H_{16}Mo_2N_2O_6S$
For complete entry see 85.38

83.92 **Di - μ - hydroxy - bis(2,2′ - bipyridine copper(ii)) diperchlorate**
$C_{20}H_{18}Cu_2N_4O_2{}^{2+}$, $2ClO_4{}^{-}$
M.-ul-Haque, M.Toofan, A.Boushehri
Acta Crystallogr., Sect. A, **31**, S83, 1975

83.93 **(7,8,15,17,18,20 - Hexahydrodibenzo(e,m)pyrazino(2,3 - b)(1,4,8,11)tetra - azacyclotetradecinato) nickel(ii)**
$C_{20}H_{18}N_6Ni$
E.N.Maslen, L.M.Engelhardt, A.H.White
J. Chem. Soc., Dalton Trans., 1799, 1974

83.94 **Di - iodo - tetrakis(pyridine) nickel(ii)**
$C_{20}H_{20}I_2N_4Ni$
D.J.Hamm, J.Bordner, A.F.Schreiner *Inorg. Chim. Acta*, **7**, 637, 1973

83.C **Adenosine bis(pyridine) osmate(vi)**
$C_{20}H_{21}N_7O_6Os$
For complete entry see 84.49

83.C **bis(Acetophenone thioacetohydrazonato) nickel(ii)**
$C_{20}H_{22}N_4NiS_2$
For complete entry see 85.39

83.95 **Aqua - bis(γ - picoline) - bis(succinimidato) copper(ii)**
$C_{20}H_{24}CuN_4O_5$
N.Latavalya, M.R.Taylor *Cryst. Struct. Commun.*, **4**, 163, 1975

83.96 **bis - μ - Sulfato - bis(bipyridyl - diaquo - nickel)**
$C_{20}H_{24}N_4Ni_2O_{12}S_2$
J.-C.Tedenac, E.Philippot *Acta Crystallogr., Sect. B*, **30**, 2286, 1974

83.C **2 - Acetylpyridine(9 - (2 - pyridyl) - 4,8 - diazadec - 8 - en - 1 - amine) nickel(ii) perchlorate**
$C_{20}H_{29}N_5NiO^{2+}$, $2ClO_4{}^{-}$
For complete entry see 84.50

83.C **bis(1,3 - Propanediamine) nickel(ii) benzoate**
$C_{20}H_{30}N_4NiO_4$
For complete entry see 81.38

83.97 **bis(Diethyl - bis(1 - pyrazolyl)borato) nickel(ii)**
$C_{20}H_{32}B_2N_8Ni$
H.M.Echols, D.Dennis *Acta Crystallogr., Sect. B,* **30,** 2173, 1974
Also classified in 62

83.98 **10,11,12,13 - Tetrahydrodibenzo(b,k)pyrido(g,f)(1,4,7,10,13) penta - azacyclopentadecin - di(perchlorato) manganese**
$C_{21}H_{19}Cl_2MnN_5O_8$
N.W.Alcock, D.C.Liles, M.McPartlin, P.A.Tasker
J. Chem. Soc., Chem. Commun., 727, 1974

83.99 **2,6 - bis(Phenyliminoethyl)pyridine - (dinitrato) nickel(ii)**
$C_{21}H_{19}N_5NiO_6$
E.C.Alyea, G.Ferguson, R.J.Restivo, P.H.Merrell
J. Chem. Soc., Chem. Commun., 269, 1975

83.C **cis - Chloro - (benzylacetoacetato) - di(pyridine) palladium(ii)**
$C_{21}H_{21}ClN_2O_3Pd$
For complete entry see 71.73

83.C **bis(10 - Methylisoalloxazine) copper(i) perchlorate formic acid solvate**
$C_{22}H_{16}CuN_8O_4^+$, ClO_4^- , CH_2O_2
For complete entry see 84.53

83.C **(N,N - dimethylbenzylamine) - (N - phenylsalicylaldiminato) palladium(ii)**
$C_{22}H_{22}N_2OPd$
For complete entry see 71.78

83.100 **bis(2,6 - Diacetylpyridine - hydrazinedi - imine) - bis(acetonitrile) iron(ii) perchlorate**
$C_{22}H_{24}FeN_8^{2+}$, $2ClO_4^-$
V.L.Goedken, Y.-A.Park, S.-M.Peng, J.M.Norris
J. Am. Chem. Soc., **96,** 7693, 1974

83.C **bis(N,N′ - bis(3 - Propanolato)acetylacetondi - imine) tricopper**
$C_{22}H_{38}Cu_3N_4O_4$
For complete entry see 84.59

83.C **bis(2 - Nitroacetophenonato) - (2,6 - dimethylpyridine) copper(ii)**
$C_{23}H_{21}CuN_3O_6$
For complete entry see 84.60

83.C **N,N,N′ - tris(2 - (2′ - Pyridyl)ethyl)ethane - 1,2 - diamine nickel(ii) perchlorate nitromethane solvate**
$C_{23}H_{29}N_5Ni^{2+}$, $2ClO_4^-$, CH_3NO_2
For complete entry see 76.77

83.C **Chloro - (bis(2 - ((2 - pyridylmethyl)imino)phenyl)disulfide) nickel(ii) perchlorate**
$C_{24}H_{18}ClN_4NiS_2^+$, ClO_4^-
For complete entry see 85.41

83.101 **tris(1,8 - Naphthyridine) mercury(ii) bis(perchlorate)**
$C_{24}H_{18}HgN_6^{2+}$, $2ClO_4^-$
J.M.Epstein, J.C.Dewan, D.L.Kepert, A.H.White
J. Chem. Soc., Dalton Trans., 1949, 1974

83.C **bis - (2 - Methoxy - 4 - nitrophenolato) - bis(pyridine) copper(ii)**
$C_{24}H_{22}CuN_4O_8$
For complete entry see 84.61

83.C **(Diphenyldipyrazolylborato)(2 - methylallyl)dicarbonyl molybdenum**
$C_{24}H_{23}BMoN_4O_2$
For complete entry see 72.28

83.C **bis(Heptanetrionato) - bis(pyridine) copper(ii)**
$C_{24}H_{26}Cu_2N_2O_6$
For complete entry see 77.10

83.C **bis(N - Cyclohexyl - thiopicolinamidato) copper(ii)**
$C_{24}H_{30}CuN_4S_2$
For complete entry see 85.43

83.C **cis - Chloro - bis(triethylphosphine) - (1,10 - phenanthroline) platinum(ii) tetrafluoroborate**
$C_{24}H_{38}ClN_2P_2Pt^+$, BF_4^-
For complete entry see 86.42

83.102 **bis(Dimethyl - bis(3,5 - dimethyl - 1 - pyrazolyl) - gallato) copper(ii)**
$C_{24}H_{40}CuGa_2N_8$
D.J.Patmore, D.F.Rendle, A.Storr, J.Trotter
J. Chem. Soc., Dalton Trans., 718, 1975
Also classified in 68

83.C **Chloro - (2 - diethylaminoethanolato) copper(ii) tetramer**
$C_{24}H_{56}Cl_4Cu_4N_4O_4$
For complete entry see 84.63

83.C **Chloro - (2 - diethylaminoethanolato) zinc(ii) tetramer**
$C_{24}H_{56}Cl_4N_4O_4Zn_4$
For complete entry see 84.64

83.103 **Cyano - bis(1,10 - phenanthroline) copper(ii) nitrate monohydrate**
$C_{25}H_{16}CuN_5^+$, NO_3^- , H_2O
O.P.Anderson *Inorg. Chem.*, **14**, 730, 1975

83.C **Pyrazine - 2,3 - dicarbonato - carbonyl - triphenylphosphine rhodium(i) monohydrate**
$C_{25}H_{18}N_2O_5PRh$, H_2O
For complete entry see 81.40

83.C **Malononitrilato(propane - 1,2 - bis(salicylideneiminato))pyridine cobalt(iii)**
$C_{25}H_{22}CoN_5O_2$
For complete entry see 71.83

83.C **Tetra - μ - trifluoroacetato - bis(quinoline copper(ii))**
$C_{26}H_{14}Cu_2F_{12}N_2O_8$
For complete entry see 81.41

83.C **bis(1 - (2 - Thiazolylazo) - 2 - naphtholato) cobalt(iii) perchlorate**
$C_{26}H_{16}CoN_6O_2S_2^+$, ClO_4^-
For complete entry see 84.65

83.104 **cis - Di(thiocyanato) - bis(1,10 - phenanthroline) mercury(ii)**
$C_{26}H_{16}HgN_6S_2$
A.L.Beauchamp, B.Saperas, R.Rivest *Can. J. Chem.*, **52,** 2923, 1974

83.C **bis(1 - (2 - Thiazolylazo) - 2 - naphtholato) nickel(ii)**
$C_{26}H_{16}N_6NiO_2S_2$
For complete entry see 84.66

83.C **bis(2 - Picolyl - phenylketonato) copper(ii)**
$C_{26}H_{20}CuN_2O_2$
For complete entry see 84.67

83.C **bis(N - Picolinylidene - N' - salicyloylhydrazinato) nickel(ii)**
$C_{26}H_{20}N_6NiO_4$
For complete entry see 78.13

83.C **bis(2 - Hydroxy - 4' - methylazobenzene) nickel(ii)**
$C_{26}H_{22}N_4NiO_2$
For complete entry see 84.68

83.C **bis(2 - Mercapto - 4' - methylazobenzene) nickel(ii)**
$C_{26}H_{22}N_4NiS_2$
For complete entry see 85.44

83.C **Chlorobis - (dimethylglyoximato) triphenylstibine cobalt(iii)**
$C_{26}H_{29}ClCoN_4OSb$
For complete entry see 86.48

83.C **bis(Dimethylglyoximato) - (pyridine) - (1 - chloro - 2,2 - bis(p - chlorophenyl)vinyl) cobalt(iii)**
$C_{27}H_{27}Cl_3CoN_5O_4$
For complete entry see 71.87

83.C **bis - μ - Mercaptobenzothiazolato - bis(dicarbonyl - pyridine ruthenium)**
$C_{28}H_{18}N_4O_4Ru_2S_4$
For complete entry see 85.45

83.105 **bis(3,4 - Di - 2 - pyridylpyridazine)dinitrato - manganese(ii)**
$C_{28}H_{20}MnN_{10}O_6$
J.E.Andrew, A.B.Blake, L.R.Fraser
J. Chem. Soc., Dalton Trans., 800, 1975

83.106 **bis(5,5 - Diethylbarbiturato) - bis(picoline) copper(ii) dihydrate**
$C_{28}H_{36}CuN_6O_6$, $2H_2O$
G.V.Fazakerley, P.W.Linder, L.R.Nassimbeni, A.L.Rodgers
Inorg. Chim. Acta, **9**, 193, 1974

83.107 **bis(5,5 - Diethylbarbiturato) - bis(picoline) zinc(ii)**
$C_{28}H_{36}N_6O_6Zn$
L.R.Nassimbeni, A.Rodgers *Acta Crystallogr., Sect. B*, **30**, 1953, 1974

83.C **bis(5 - Ethyl - 5 - isoamylbarbiturato) - bis(imidazole) nickel(ii)**
$C_{28}H_{42}N_8NiO_6$
For complete entry see 84.72

83.C **μ - Acetato - μ - phenyldinitromethanato - bis(2,2′ - bipyridine copper(ii))**
$C_{29}H_{25}Cu_2N_6O_6$
For complete entry see 81.42

83.108 **Dichloro - bis(N,N′ - dicyclohexyl - acetamidinato)methyl tantalum(v)**
$C_{29}H_{53}Cl_2N_4Ta$
M.G.B.Drew, J.D.Wilkins *J. Chem. Soc., Dalton Trans.*, 1973, 1974

83.109 **bis(2,2′.6′,2″ - Terpyridyl) cobalt(ii) bromide trihydrate**
$C_{30}H_{22}CoN_6^{2+}$, $2Br^-$, $3H_2O$
E.N.Maslen, C.L.Raston, A.H.White
J. Chem. Soc., Dalton Trans., 1803, 1974

83.110 **tetrakis(1,8 - Naphthyridine) cadmium(ii) bis(perchlorate)**
$C_{32}H_{24}CdN_8^{2+}$, $2ClO_4^-$
J.M.Epstein, J.C.Dewan, D.L.Kepert, A.H.White
J. Chem. Soc., Dalton Trans., 1949, 1974

83.C **tetrakis(Acetato - (2 - pyridylmethanolato) copper(ii)) tetrahydrate**
$C_{32}H_{36}Cu_4N_4O_{12}$, $4H_2O$
For complete entry see 81.43

83.C **(3,3 - bis(Methoxycarbonyl) - 4 - phenyl - 1 - (1′,2′ - diphenylethenyl) - pyrazolidinyl) - hexacarbonyl - di - iron**
$C_{33}H_{24}Fe_2N_2O_{10}$
For complete entry see 71.100

83.111 **μ - Cyano - bis(5,7,7,12,14,14 - hexamethyl - 1,4,8,11 - tetra - azacyclotetradeca - 4,11 - diene) dicopper(ii) perchlorate**
$C_{33}H_{64}Cu_2N_9^{3+}$, $3ClO_4^-$
R.Jungst, G.Stucky *Inorg. Chem.*, **13**, 2404, 1974

83.C **7 - (Methoxycarbonyl - methyl) - 1,2,3,4,5,6,7 - heptakis(methoxycarbonyl) - bicyclo(2.2.1)heptadiene - 8 - yl - (chloro - bis(pyridine) palladium)**
$C_{34}H_{35}ClN_2O_{16}Pd$
For complete entry see 71.101

83.C **N,N′ - bis((2 - Hydroxylato - 5 - methylphenyl)phenylmethylene) - 4 - azaheptane - 1,7 - diamine nickel(ii)**
$C_{34}H_{35}N_3NiO_2$
For complete entry see 84.77

83.112 **tris(o - Phenanthroline) iron(iii) perchlorate monohydrate**
$C_{36}H_{24}FeN_6^{3+}$, $3ClO_4^-$, H_2O
J.Baker, L.M.Engelhardt, B.N.Figgis, A.H.White
J. Chem. Soc., Dalton Trans., 530, 1975

83.C **tetrakis(8 - Quinolinolato) zirconium(iv) toluene solvate**
$C_{36}H_{24}N_4O_4Zr$, $3C_7H_8$
For complete entry see 84.79

83.C **Iodo - bis(2 - (diphenylphosphinomethyl)pyridine) nickel(ii) iodide**
$C_{36}H_{32}IN_2NiP_2^+$, I^-
For complete entry see 86.69

83.113 **bis(2,6 - Diacetyl - pyridine - bis(2 - pyridylhydrazonato)) di - zinc**
$C_{38}H_{34}N_{14}Zn_2$
D.Wester, G.J.Palenik *J. Chem. Soc., Chem. Commun.*, 74, 1975

83.C **5 - Methyl - 2 - ((dimethylamino)methyl)phenyl copper tetramer (further discussion)**
(4 - Methyl - 2 - cupriobenzyl)dimethylamine tetramer
$C_{40}H_{56}Cu_4N_4$
For complete entry see 71.116

83.C **Quinuclidine - tris(2,2,6,6 - tetramethylheptan - 3,5 - dionato) europium(iii)**
$C_{40}H_{70}EuNO_6$
For complete entry see 77.12

83.C **bis(Dichloro - (2,6 - diacetylpyridine) - bis(picolinoylhydrazonato) dicopper(ii)) dihydrate**
$C_{42}H_{34}Cl_4Cu_4N_{14}O_4$, $2H_2O$
For complete entry see 84.81

83.C **Chloro - (tris(2 - diphenylphosphinoethyl)amine) nickel(ii) hexafluorophosphate**
$C_{42}H_{42}ClNNiP_3^+$, F_6P^-
For complete entry see 86.82

83.C **Chloro - (bis(3 - diphenylphosphinopropyl)phenyl - phosphine) - phenyldiazo - rhodium hexafluorophosphate methylene chloride solvate**
$C_{42}H_{42}ClN_2P_3Rh^+$, F_6P^- , CH_2Cl_2
For complete entry see 86.83

83.C **(tris(2 - Diphenylphosphinoethyl)amine) cobalt tetrafluoroborate**
$C_{42}H_{42}CoNP_3^+$, BF_4^-
For complete entry see 86.84

83.C **Hydrido - (tris(diphenylphosphinoethyl)amine) nickel tetrafluoroborate**
$C_{42}H_{43}NNiP_3^+$, $C_{42}H_{42}NNiP_3^+$, $2BF_4^-$
For complete entry see 86.87

83.C **bis(Triphenylphosphine) - (hexafluoroacetonimido) - (N,N′ - dimethylimidazolidin - 2 - ylideno) rhodium**
$C_{44}H_{40}F_6N_3P_2Rh$
For complete entry see 71.124

83.114 **Di - μ - hydroxo - bis(bis(1,10 - phenanthroline) chromium(iii)) iodide tetrahydrate**
$C_{48}H_{34}Cr_2N_8O_2^{4+}$, $4I^-$, $4H_2O$
R.P.Scaringe, P.Singh, R.P.Eckberg, W.E.Hatfield, D.J.Hodgson
Inorg. Chem., **14**, 1127, 1975

83.C **Hydrido - chloro - (phenylazophenyl) - bis(triphenylphosphine) iridium(iii) n - hexane solvate**
$C_{48}H_{40}ClIrN_2P_2$, C_6H_{14}
For complete entry see 71.129

83.C **(tris(2 - Diphenylarsinoethyl)amine)phenyl nickel(ii) tetraphenylborate**
$C_{48}H_{47}As_3NNi^+$, $C_{24}H_{20}B^-$
For complete entry see 71.130

83.C **trans - Chloro - (1,3 - di - p - tolyltriazenido) - bis(triphenylphosphine) palladium(ii)**
$C_{50}H_{44}ClN_3P_2Pd$
For complete entry see 86.93

83.C **cis - Chloro - (1,3 - di - p - tolyltriazenido) - bis(triphenylphosphine) platinum(ii)**
$C_{50}H_{44}ClN_3P_2Pt$
For complete entry see 86.94

83.C **Iodo - methyldiazenido - di((bis - 1,2 - diphenylphosphino)ethane) molybdenum(0)**
$C_{53}H_{51}IMoN_2P_4$
For complete entry see 86.101

83.C **Iodo - cyclohexyldiazenido - di((bis - 1,2 - diphenylphosphino)ethane) molybdenum(0) benzene solvate**
$C_{58}H_{59}IMoN_2P_4$, $0.5C_6H_6$
For complete entry see 86.108

83.C **bis(Quinoline) - hexa(benzoato) - tricobalt(ii)**
$C_{60}H_{44}Co_3N_2O_{12}$
For complete entry see 81.46

83.C **μ - (trans - 1,1,2,3,4,4 - Hexacyanobutene - di - ido) - bis((carbonyl) - bis(triphenylphosphine) rhodium)**
$C_{84}H_{60}N_6O_2P_4Rh_2$
For complete entry see 86.116

METAL COMPLEXES (OXYGEN LIGAND)

84.1 **μ - Formamido - bis(pentammine cobalt(iii)) pentachloride monohydrate**
$CH_{32}Co_2N_{11}O^{5+}$, $5Cl^-$, H_2O
R.J.Balahura, G.Ferguson, M.L.Schneider
J. Chem. Soc., Dalton Trans., 603, 1975
Residue 1 also classified in 83

84.2 **Dichloro - (1,2 - ethanediol) copper(ii)**
$C_2H_6Cl_2CuO_2$
B.-M.Antti *Acta Chem. Scand. Ser. A*, **29,** 76, 1975

84.C **Methanol - cobalt(ii) oxalate monohydrate**
$(C_3H_4CoO_5)_n$, nH_2O
For complete entry see 81.4

84.C **Methanol - nickel(ii) oxalate monohydrate**
$(C_3H_4NiO_5)_n$, nH_2O
For complete entry see 81.6

84.C **Methanol - zinc oxalate monohydrate**
$(C_3H_4O_5Zn)_n$, nH_2O
For complete entry see 81.7

84.C **Methyl - dichloro - bis(N - methyl - N - nitrosohydroxylaminato) tantalum(v)**
$C_3H_9Cl_2N_4O_4Ta$
For complete entry see 71.2

84.3 **Diaquo - nickel(ii) squarate**
$(C_4H_4NiO_6)_n$
M.Habenschuss, B.C.Gerstein *J. Chem. Phys.*, **61,** 852, 1974

84.4 **cis - Dioxo - bis(2 - hydroxy - ethanolato) molybdenum(vi)**
$C_4H_{10}MoO_6$
F.A.Schroder, J.Scherle, R.G.Hazell
Acta Crystallogr., Sect. B, **31,** 531, 1975

84.5 **cis - Dichloro - tetramethoxy - tungsten(vi)**
$C_4H_{12}Cl_2O_4W$
L.B.Handy *Acta Crystallogr., Sect. B*, **31,** 300, 1975

84.6 Di - μ - dimethylsulfoxide - bis(dichloro - mercury) dichloro - mercury
$C_4H_{12}Cl_4Hg_2O_2S_2$, Cl_2Hg
P.Biscarini, L.Fusina, G.D.Nivellini, A.Mangia, G.Pelizzi
J. Chem. Soc., Dalton Trans., 1846, 1974

84.7 Sodium tetrachloro - bis(dimethylsulfoxide) rhodium(iii)
$C_4H_{12}Cl_4O_2RhS_2^-$, Na^+
V.I.Sokol, N.D.Rubtsova, A.Yu.Gribenyuk
Zh. Strukt. Khim., **15,** 318, 1974

84.8 trans - Dichloro - tetra(methanol) chromium(iii) chloride
$C_4H_{16}Cl_2CrO_4^+$, Cl^-
K.I.Hardcastle, D.O.Skovlin, A.-H.Eidawad
J. Chem. Soc., Chem. Commun., 190, 1975

84.9 (Pyridine - N - oxide) - di - iodo - cadmium
$(C_5H_5CdI_2NO)_n$
G.Sawitzki, H.G.von Schnering *Chem. Ber.*, **107,** 3266, 1974

84.10 (Pyridine - N - oxide)dichloro - mercury
$(C_5H_5Cl_2HgNO)_n$
G.Sawitzki, H.G.von Schnering *Chem. Ber.*, **107,** 3266, 1974

84.11 Ammonium μ - oxo - μ - mannitolato - tetraoxodimolybdate(vi) monohydrate
$C_6H_{11}Mo_2O_{11}^-$, H_4N^+ , H_2O
J.E.Godfrey, J.M.Waters *Cryst. Struct. Commun.*, **4,** 5, 1975

84.12 Dihydrogen mannitolo - dimolybdate trihydrate
$C_6H_{12}Mo_2O_{11}$, $3H_2O$
B.Hedman *Acta Chem. Scand. Ser. A*, **28,** 591, 1974

84.13 tris(Ethylene - 1,2 - dioxo) tungsten(vi)
$C_6H_{12}O_6W$
J.Scherle, F.A.Schroder *Acta Crystallogr., Sect. B*, **30,** 2772, 1974

84.14 bis(Isothiocyanato) - bis(ethanolamine) nickel(ii)
$C_6H_{14}N_4NiO_2S_2$
A.E.Shvelashvili, M.A.Porai-Koshits, A.I.Kvitashvili, B.M.Shchedrin, M.G.Tavberidze *Zh. Strukt. Khim.*, **15,** 313, 1974
Also classified in 83

84.C Trichloro - tris(dimethylsulfoxide) rhodium(iii)
$C_6H_{18}Cl_3O_3RhS_3$
For complete entry see 85.5

84.15 tris(1,2 - Ethanediol) cobalt(ii) sulfate
$C_6H_{18}CoO_6^{2+}$, O_4S^{2-}
B.-M.Antti *Acta Chem. Scand. Ser. A*, **29,** 76, 1975

84.16 **tris(Nitrato) - tris(dimethylsulfoxide) lutetium**
$C_6H_{18}LuN_3O_{12}S_3$
L.A.Aslanov, L.I.Soleva, M.A.Porai-Koshits
Zh. Strukt. Khim., **14,** 1064, 1973

84.17 **tris(Dimethylsulfoxide) - tri(nitrato) ytterbium**
$C_6H_{18}N_3O_{12}S_3Yb$
K.K.Bhandary, H.Manohar, K.Venkatesan
J. Chem. Soc., Dalton Trans., 288, 1975

84.18 **tris(1,2 - Ethanediol) zinc sulfate**
$C_6H_{18}O_6Zn^{2+}$, O_4S^{2-}
B.-M.Antti *Acta Chem. Scand. Ser. A*, **29,** 76, 1975

84.19 **Chloro - aquo - (2,6 - dimethyl - 4 - pyrone) lanthanum(iii)**
$(C_7H_{10}Cl_3LaO_3)_n$
E.C.Bisi, M.Gorio, E.Cannillo, A.Coda, V.Tazzoli
Acta Crystallogr., Sect. A, **31,** S134, 1975

84.20 **(2 - Diethylamino - ethanolato) copper(ii) thiocyanate**
$(C_7H_{14}CuN_2OS)_n$
A.Pajunen, K.Smolander *Finn. Chem. Lett.*, 99, 1974
Also classified in 83

84.21 **Tetra - ammine - (4,5 - dihydroxy - 4,5 - dimethyl - Δ^1 - pyrroline - 2 - carboxylato) cobalt(iii) perchlorate monohydrate**
$C_7H_{22}CoN_5O_4^{2+}$, $2ClO_4^-$, H_2O
J.M.B.Harrowfield, G.B.Robertson, A.M.Sargeson, P.O.Whimp
J. Chem. Soc., Chem. Commun., 109, 1975
Residue 1 also classified in 83

84.22 **catena - Di - μ - dioxan - (dibromo - nickel)**
$(C_8H_{16}Br_2NiO_4)_n$
J.C.Barnes, L.J.Sesay *Acta Crystallogr., Sect. A*, **31,** S144, 1975

84.23 **Dichloro - (1,4,7,10 - tetraoxacyclododecane) copper(ii)**
$C_8H_{16}Cl_2CuO_4$
F.P.van Remoortere, F.P.Boer, E.C.Steiner
Acta Crystallogr., Sect. B, **31,** 1420, 1975

84.24 **Di - μ - chloro - bis(di(1,2 - ethanediol) cobalt(ii)) chloride**
$C_8H_{24}Cl_2Co_2O_8^{2+}$, $2Cl^-$
B.-M.Antti *Acta Chem. Scand. Ser. A*, **29,** 76, 1975

84.25 **Di - μ - chloro - bis(di(1,2 - ethanediol) nickel(ii)) chloride**
$C_8H_{24}Cl_2Ni_2O_8^{2+}$, $2Cl^-$
B.-M.Antti *Acta Chem. Scand. Ser. A*, **29,** 76, 1975

84.26 **Dicarbonyl - tropolonato rhodium(i)**
$C_9H_5O_4Rh$
L.Manojlovic-Muir, K.W.Muir *Acta Crystallogr., Sect. A*, **31,** S132, 1975

84.27 **tris(1,3 - Propanedionato) chromium(iii)**
$C_9H_9CrO_6$
M.D.Glick, B.Andrelczyk, R.L.Lintvedt
Acta Crystallogr., Sect. B, **31,** 916, 1975

84.28 **bis(Pyridine - N - oxide) - di - iodo - zinc**
$C_{10}H_{10}I_2N_2O_2Zn$
G.Sawitzki, H.G.von Schnering *Chem. Ber.*, **107,** 3266, 1974

84.29 **Dichloro - bis(2 - methylpyrazine di - N - oxide) copper(ii)**
$C_{10}H_{12}Cl_2CuN_4O_4$
N.R.Stemple, W.H.Watson *Cryst. Struct. Commun.*, **4,** 355, 1975

84.C **π - Allyl - dicarbonyl - (trifluoroacetato) - (1,2 - dimethoxyethane) molybdenum**
$C_{11}H_{15}F_3MoO_6$
For complete entry see 72.3

84.C **π - Allyl - dicarbonyl - (trifluoroacetato) - (1,2 - dimethoxyethane) tungsten**
$C_{11}H_{15}F_3O_6W$
For complete entry see 72.4

84.30 **Diaquo - (2,6 - diacetylpyridine - bis(semicarbazone)) copper(ii) nitrate monohydrate**
$C_{11}H_{19}CuN_7O_4^{2+}$, $2NO_3^-$, H_2O
D.Wester, G.J.Palenik *J. Am. Chem. Soc.*, **96,** 7565, 1974
Residue 1 also classified in 83

84.31 **Diaquo - (2,6 - diacetylpyridine - bis(semicarbazone)) nickel(ii) nitrate monohydrate**
$C_{11}H_{19}N_7NiO_4^{2+}$, $2NO_3^-$, H_2O
D.Wester, G.J.Palenik *J. Am. Chem. Soc.*, **96,** 7565, 1974
Residue 1 also classified in 83

84.C **2 - Chloro - 4 - bromophenolato - (phenyl) - mercury**
$C_{12}H_8BrClHgO$
For complete entry see 71.18

84.32 **Potassium bis(catecholato) - dioxo - molybdenum(vi) dihydrate**
$C_{12}H_8MoO_6^{2-}$, $2K^+$, $2H_2O$
L.O.Atovmyan, Yu.A.Sokolova, V.V.Tkachev
Dokl. Akad. Nauk SSSR, **195,** 1355, 1970

84.33 **Ammonium (μ - oxo - di - μ - catecholato) bis(dioxo - molybdenum(vi)) dihydrate**
$C_{12}H_8Mo_2O_9^{2-}$, $2H_4N^+$, $2H_2O$
L.O.Atovmyan, V.V.Tkachev, T.G.Shishova
Dokl. Akad. Nauk SSSR, **205,** 609, 1972

84.34 **Chloro - N,N′ - ethylene - bis(acetylacetoneiminato) manganese(iii)**
$C_{12}H_{16}ClMnN_2O_2$
V.W.Day, L.J.Boucher
Am. Cryst. Assoc., Abstr. Papers (Spring Meeting), 34, 1975
Also classified in 83

84.C **Di - μ - chloro - bis(chloro - (2 - (hydroxymethyl)pent - 4 - enyl) rhodium(iii)) methanol solvate**
$C_{12}H_{22}Cl_4O_2Rh_2$, CH_4O
For complete entry see 71.26

84.35 **Bromo - (diethylaminoethanolato) copper(ii) dimer (β form)**
$C_{12}H_{28}Br_2Cu_2N_2O_2$
R.Mergehenn, L.Merz, W.Haase *Z. Naturforsch., Teil B*, **30,** 14, 1975
Also classified in 83

84.36 **μ - Oxo - bis(chloro - bis(N,N - dimethylmethanolamine) dioxo - molybdenum(vi))**
$C_{12}H_{32}Cl_2Mo_2N_4O_9$
L.O.Atovmyan, Yu.A.Sokolova, V.V.Tkachev
Dokl. Akad. Nauk SSSR, **195,** 1355, 1970

84.37 **Dichloro - hexakis(dimethyl sulfoxide) uranium hexachlorouranate**
$C_{12}H_{36}Cl_2O_6S_6U^{2+}$, Cl_6U^{2-}
G.Bombieri, K.W.Bagnall *J. Chem. Soc., Chem. Commun.*, 188, 1975

84.38 **Chloro - (1 - (2 - thiazolylazo) - 2 - naphtholato) palladium(ii) dioxan solvate**
$C_{13}H_8ClN_3OPdS$, $C_4H_8O_2$
M.Kurahashi *Bull. Chem. Soc. Jpn.*, **47,** 2045, 1974
Residue 1 also classified in 83

84.39 **(1 - (2 - Thiazolylazo) - 2 - naphtholato) - diaquo - copper(ii) perchlorate**
$C_{13}H_{12}CuN_3O_3S^+$, ClO_4^-
M.Kurahashi *Chem. Lett.*, 63, 1974
Residue 1 also classified in 83

84.C **Tetracarbonyl - (7,7 - dimethoxynorborn - 2 - ene) chromium(0)**
$C_{13}H_{14}CrO_6$
For complete entry see 75.9

84.40 **N,N′ - Di(2 - hydroxyethyl) - 2,4 - pentanedi - iminato - bis(ethanolamine) cobalt(iii)**
$C_{13}H_{29}CoN_4O_4$
J.A.Bertrand, F.T.Helm, L.J.Carpenter *Inorg. Chim. Acta*, **9,** 69, 1974
Also classified in 83

84.41 **Aquo - (bis(2 - hydroxyphenylimino)ethanato)dioxo - uranium**
Aquo - glyoxal - bis(2 - hydroxyanil)dioxo - uranium
$C_{14}H_{12}N_2O_5U$
G.Bandoli, D.A.Clemente *J. Chem. Soc., Dalton Trans.*, 612, 1975

84.42 **(2 - Diethylaminoethanolato) copper(ii) thiocyanate dimer**
$C_{14}H_{28}Cu_2N_4O_2S_2$
A.Pajunen, K.Smolander *Finn. Chem. Lett.*, 99, 1974
Also classified in 83

84.C **tris(Dimethylaminato) - tris(N,N - dimethylcarbamato) tungsten(vi)**
$C_{15}H_{36}N_6O_6W$
For complete entry see 80.15

84.C **(Ethylenediamine) zinc(ii) benzohydroxamate benzohydroxamic acid monohydrate**
$C_{16}H_{20}N_4O_4Zn$, $C_7H_7NO_2$, H_2O
For complete entry see 76.67

84.C **trans - Azido - (pyridine) - bis(acetylacetonato)ethylenedi - imine cobalt(iii)**
$C_{17}H_{23}CoN_6O_2$
For complete entry see 83.80

84.C **bis - (cis - 1 - Mercapto - 2 - p - bromobenzoyl - ethylene) nickel(ii)**
$C_{18}H_{12}Br_2NiO_2S_2$
For complete entry see 85.33

84.C **bis(1,2 - Di(trifluoromethyl) - 2 - (diacetylmethyl) - ethylenyl) palladium**
$C_{18}H_{14}F_{12}O_4Pd$
For complete entry see 71.50

84.43 **Pentachloro - (triphenylphosphine oxide) uranium**
$C_{18}H_{15}Cl_5OPU$
G.Bombieri, E.Forsellini, D.Brown, B.Whittaker, C.Mealli
Acta Crystallogr., Sect. A, **31**, S134, 1975

84.44 **bis(N - (Picolinoyl) - 3 - amino - 1 - propoxido - aquo - copper(ii)) dihydrate**
$C_{18}H_{24}Cu_2N_4O_6$, $2H_2O$
J.A.Bertrand, E.Fujita, P.G.Eller *Inorg. Chem.*, **13**, 2067, 1974
Residue 1 also classified in 83

84.C **4 - ((Diethylamino) - (t - butylamino) - methylene) - 4 - (t - butyl isocyanide) - 2,2,5,5 - tetrakis(trifluoromethyl) - 1,3,4 - dioxapalladolan**
$C_{18}H_{29}F_{12}N_3O_2Pd$
For complete entry see 71.55

84.45 **μ - (N,N' - bis(2 - Dimethylaminoethyl) - oxamidato) - di(isothiocyanato) - bis(dimethylformamide) dicopper(ii)**
$C_{18}H_{34}Cu_2N_8O_4S_2$
A.Yoshino, W.Nowacki *Z. Kristallogr.*, **139**, 337, 1974
Also classified in 83

84.C **bis(N - (6 - Amino - 3,3,6 - trimethyl - 4 - aza - 2 - heptylidenyl)hydroxanato) dinickel(ii) chloride pentahydrate**
$C_{18}H_{40}N_6Ni_2O_2^{2+}$, $2Cl^-$, $5H_2O$
For complete entry see 76.75

84.46 **Chloro - hexakis(trimethylphosphine oxide) uranium(iv) chloride**
$C_{18}H_{54}ClO_6P_6U^{3+}$, $3Cl^-$
G.Bombieri, E.Forsellini, D.Brown, B.Whittaker, C.Mealli
Acta Crystallogr., Sect. A, **31,** S134, 1975

84.47 **Dichloro - (2 - (2′ - pyridyl) - 3 - (N - 2 - picolylimino) - 4 - oxo - 1,2,3,4 - tetrahydroquinazoline) manganese(ii)**
$C_{19}H_{15}Cl_2MnN_5O$
C.Pelizzi, G.Pelizzi *Acta Crystallogr., Sect. B,* **30,** 2421, 1974
Also classified in 83

84.48 **Dioxo - (pyridine) - bis(tropolonato) uranium(vi)**
$C_{19}H_{15}NO_6U$
S.Degetto, G.Marangoni, G.Bombieri, E.Forsellini, L.Baracco, R.Gaziani
J. Chem. Soc., Dalton Trans., 1933, 1974

84.C **2,3 - bis(π - 1,5 - Cyclo - octadiene) - 4,4 - bis(trifluoromethyl) - 1,2,3 - oxa - diplatina - cyclobutane**
$C_{19}H_{24}F_6OPt_2$
For complete entry see 71.59

84.C **μ - (2 - Mercaptoethanolato) - μ - oxo - bis(oxo - (8 - hydroxyquinolinato) - molybdenum(v))**
$C_{20}H_{16}Mo_2N_2O_6S$
For complete entry see 85.38

84.49 **Adenosine bis(pyridine) osmate(vi)**
$C_{20}H_{21}N_7O_6Os$
J.F.Conn, J.J.Kim, F.L.Suddath, P.Blattmann, A.Rich
J. Am. Chem. Soc., **96,** 7152, 1974
Also classified in 83, 47, 45, 44

84.50 **2 - Acetylpyridine(9 - (2 - pyridyl) - 4,8 - diazadec - 8 - en - i - amine) nickel(ii) perchlorate**
$C_{20}H_{29}N_5NiO^{2+}$, $2ClO_4^-$
G.M.Sheldrick, D.A.Stotter *J. Chem. Soc., Dalton Trans.,* 666, 1975
Residue 1 also classified in 83

84.51 **cis - Chloro - dicarbonyl - (tricyclohexylphosphine oxide) rhodium(i)**
$C_{20}H_{33}Cl_3O_3PRh$
G.Bandoli, D.A.Clemente, G.Deganello, G.Carturan, P.Uguagliati, U.Belluco *J. Organomet. Chem.,* **71,** 125, 1974

84.52 **tris(Tropolonato) manganese(iii) toluene solvate**
$C_{21}H_{15}MnO_6$, $0.25C_7H_8$
A.Avdeef, J.A.Costamagna, J.P.Fackler Junior
Inorg. Chem., **13,** 1854, 1974

84.C **9 - (1',2' - bis(Trifluoromethyl) - 2' - (diacetylmethyl)ethylen - 1' - yl) - 10,11 - bis(trifluoromethyl) - 9 - iridia - bicyclo(6.3.0)undeca - 4,10 - diene**
$C_{21}H_{19}F_{12}IrO_2$
For complete entry see 71.71

84.C **Di(isothiocyanato) - di(ethoxy) - (1,3 - diphenyl - propane - 1,3 - dionato) niobium(v)**
$C_{21}H_{21}N_2NbO_4S_2$
For complete entry see 77.9

84.53 **bis(10 - Methylisoalloxazine) copper(i) perchlorate formic acid solvate**
$C_{22}H_{16}CuN_8O_4^{+}$, ClO_4^{-} , CH_2O_2
M.W.Yu, C.J.Fritchie Junior *J. Biol. Chem.*, **250,** 946, 1975
Residue 1 also classified in 83

84.54 **Diaquo - bis(10 - methylisoalloxazine) iron(ii) perchlorate tetrahydrate**
$C_{22}H_{20}FeN_8O_6^{2+}$, $2ClO_4^{-}$, $4H_2O$
C.J.Fritchie Junior, T.D.Wade
Am. Cryst. Assoc., Abstr. Papers (Summer Meeting), 239, 1974

84.55 **Dinitrato - bis(antipyrine) cobalt(ii)**
$C_{22}H_{24}CoN_6O_8$
C.Brassy, J.-P.Mornon, J.Delettre
Acta Crystallogr., Sect. B, **30,** 2243, 1974

84.56 **Dinitrato - bis(antipyrine) copper(ii) (form i)**
$C_{22}H_{24}CuN_6O_8$
C.Brassy, A.Renaud, J.Delettre, J.-P.Mornon
Acta Crystallogr., Sect. B, **30,** 2246, 1974

84.57 **Dinitrato - bis(antipyrine) copper(ii) (form ii)**
$C_{22}H_{24}CuN_6O_8$
C.Brassy, J.-P.Mornon, J.Delettre, G.Lepicard
Acta Crystallogr., Sect. B, **30,** 2500, 1974

84.58 **Dinitrato - bis(antipyrine) zinc(ii)**
$C_{22}H_{24}N_6O_8Zn$
C.Brassy, M.-C.Michaud, J.Delettre, J.-P.Mornon
Acta Crystallogr., Sect. B, **30,** 2848, 1974

84.59 **bis(N,N' - bis(3 - Propanolato)acetylacetondi - imine) tricopper**
$C_{22}H_{38}Cu_3N_4O_4$
W.A.Baker Junior, F.T.Helm *J. Am. Chem. Soc.*, **97,** 2295, 1975
Also classified in 83

84.60 **bis(2 - Nitroacetophenonato) - (2,6 - dimethylpyridine) copper(ii)**
$C_{23}H_{21}CuN_3O_6$
M.Bonamico, G.Dessy, V.Fares, L.Scaramuzza
Cryst. Struct. Commun., **3,** 633, 1974
Also classified in 83

84.C **((Chloro(methoxycarbonyl)(1,2,3,4,5 - pentakis(methoxycarbonyl)cyclopenta - 2,4 - dienyl)methyl))(pentane - 2,4 - dionato) palladium(ii) chloroform solvate**
$C_{23}H_{25}ClO_{14}Pd$, $CHCl_3$
For complete entry see 71.79

84.61 **bis - (2 - Methoxy - 4 - nitrophenolato) - bis(pyridine) copper(ii)**
$C_{24}H_{22}CuN_4O_8$
J.I.Bullock, R.J.Hobson, D.C.Povey
J. Chem. Soc., Dalton Trans., 2037, 1974
Also classified in 83

84.62 **Di - μ - (4 - methoxypyridine - 1 - oxide) - bis(dinitrato - copper(ii))**
$C_{24}H_{28}Cu_2N_8O_{20}$
W.H.Watson, N.R.Stemple, L.F.Mercer, G.Beal
Cryst. Struct. Commun., **4,** 31, 1975

84.C **Di - μ - chloro - bis((tricarbonyl cyclopentadienyl molybdenum) - diethylether zinc)**
$C_{24}H_{30}Cl_2Mo_2O_8Zn_2$
For complete entry see 73.47

84.C **Di - μ - chloro - bis(cyclo - octatetraene - (tetrahydrofuran) titanium)**
$C_{24}H_{32}Cl_2O_2Ti_2$
For complete entry see 75.30

84.63 **Chloro - (2 - diethylaminoethanolato) copper(ii) tetramer**
$C_{24}H_{56}Cl_4Cu_4N_4O_4$
E.D.Estes, D.J.Hodgson *Inorg. Chem.*, **14,** 334, 1975
Also classified in 83

84.64 **Chloro - (2 - diethylaminoethanolato) zinc(ii) tetramer**
$C_{24}H_{56}Cl_4N_4O_4Zn_4$
W.Haase, R.Mergehenn, R.Allmann
Acta Crystallogr., Sect. B, **31,** 1184, 1975
Also classified in 83

84.65 **bis(1 - (2 - Thiazolylazo) - 2 - naphtholato) cobalt(iii) perchlorate**
$C_{26}H_{16}CoN_6O_2S_2^+$, ClO_4^-
M.Kurahashi *Chem. Lett.*, 1271, 1974
Residue 1 also classified in 83

84.66 **bis(1 - (2 - Thiazolylazo) - 2 - naphtholato) nickel(ii)**
$C_{26}H_{16}N_6NiO_2S_2$
M.Kurahashi *Bull. Chem. Soc. Jpn.*, **47,** 2067, 1974
Also classified in 83

84.67 **bis(2 - Picolyl - phenylketonato) copper(ii)**
$C_{26}H_{20}CuN_2O_2$
J.Sieler, R.Richter, J.Kaiser, W.Schmidt *Eur. Cryst. Meeting,* 380, 1974
Also classified in 83

84.68 **bis(2 - Hydroxy - 4′ - methylazobenzene) nickel(ii)**
$C_{26}H_{22}N_4NiO_2$
O.A.Dyachenko, L.O.Atovmyan, S.M.Aldoshin
J. Chem. Soc., Chem. Commun., 105, 1975
Also classified in 83

84.C **tris(2′ - (2 - Phenyl - 1,3 - dioxolano)) chromium(iii)**
$C_{27}H_{27}CrO_6$
For complete entry see 71.88

84.C **Dichloro(nitrosyl) - (methanol) - bis(methyldiphenylphosphine) rhenium**
$C_{27}H_{30}Cl_2NO_2P_2Re$
For complete entry see 86.52

84.69 **tetrakis(Tropolonato) hafnium(iv) N,N - dimethylformamide solvate**
$C_{28}H_{20}HfO_8$, C_3H_7NO
D.Tranqui, R.Laugier, A.Tissier *Acta Crystallogr., Sect. A,* **31,** S135, 1975

84.70 **Hydrogen tetrakis(tropolonato) scandium(iii)**
$C_{28}H_{21}O_8Sc$
A.R.Davis, F.W.B.Einstein *Inorg. Chem.*, **13,** 1880, 1974

84.71 **Hydrogen tetrakis(tropolonato) scandium(iii)**
$C_{28}H_{21}O_8Sc$
T.J.Anderson, M.A.Neuman, G.A.Melson *Inorg. Chem.*, **13,** 1884, 1974

84.C **racemic - Chloro - (1 - methyl - 3 - (dimethylphenylmethyl) - π - cyclopentadienyl) - π - cyclopentadienyl - (2,6 - dimethylphenolato) titanium**
$C_{28}H_{31}ClOTi$
For complete entry see 73.60

84.72 **bis(5 - Ethyl - 5 - isoamylbarbiturato) - bis(imidazole) nickel(ii)**
$C_{28}H_{42}N_8NiO_6$
L.R.Nassimbeni, A.Rodgers *Acta Crystallogr., Sect. B,* **30,** 2593, 1974
Also classified in 83

84.C **μ - Acetato - μ - phenyldinitromethanato - bis(2,2′ - bipyridine copper(ii))**
$C_{29}H_{25}Cu_2N_6O_6$
For complete entry see 81.42

84.C **Chloro - (diphenylethoxyphosphine) - (1 - diphenylphosphino - 3,3,3 - trifluoropropen - 2 - olato) palladium(ii) dichloromethane solvate**
$C_{29}H_{26}ClF_3O_2P_2Pd$, CH_2Cl_2
For complete entry see 86.56

84.C **(bis - (2 - Diphenylphosphinoethyl)ether) - carbonyl - rhodium(i) hexafluorophosphate**
$C_{29}H_{28}O_2P_2Rh^+$, F_6P^-
For complete entry see 86.58

84.73 **Di - μ - (azobispyridine - 4,4′ - di - N - oxide) - bis(dichloro(azobispyridine - 4,4′ - di - N - oxide) copper(ii))**
$(C_{30}H_{24}Cl_4Cu_2N_{12}O_6)_n$
N.R.Stemple, W.H.Watson *Cryst. Struct. Commun.*, **4**, 25, 1975

84.74 **hexakis(Pyridine - N - oxide) cobalt(ii) perchlorate**
$C_{30}H_{30}CoN_6O_6^{2+}$, $2ClO_4^-$
T.J.Bergendahl, J.S.Wood *Inorg. Chem.*, **14**, 338, 1975

84.C **tetrakis(Acetato - (2 - pyridylmethanolato) copper(ii)) tetrahydrate**
$C_{32}H_{36}Cu_4N_4O_{12}$, $4H_2O$
For complete entry see 81.43

84.C **Methylisocyanide - (2 - (bis(t - butyl)phosphino) - 3 - methoxyphenolato) - (2 - (1,1 - dimethylethyl - (t - butyl)phosphino) - 3 - methoxyphenolato) iridium(iii)**
$C_{32}H_{50}IrNO_4P_2$
For complete entry see 71.99

84.75 **tris(Antipyrine) - tris(nitrato) neodymium**
$C_{33}H_{36}N_9NdO_{12}$
K.K.Bhandary, H.Manohar, K.Venkatesan
Acta Crystallogr., Sect. A, **31**, S143, 1975

84.C **(Chloro - tetrakis(dimethylphenylphosphine) rhenium(i)) - μ - dinitrogen - (tetrachloro - methoxy molybdenum(v)) methanol hydrochloric acid solvate**
$C_{33}H_{47}Cl_5MoN_2OP_4Re$, CH_4O , HCl
For complete entry see 86.63

84.76 **N,N′ - bis((2 - Hydroxylato - 5 - methylphenyl)phenylmethylene) - 4 - azaheptane - 1,7 - diamine copper(ii)**
$C_{34}H_{35}CuN_3O_2$
P.C.Healy, G.M.Mockler, D.P.Freyberg, E.Sinn
J. Chem. Soc., Dalton Trans., 691, 1975
Also classified in 3

84.77 **N,N′ - bis((2 - Hydroxylato - 5 - methylphenyl)phenylmethylene) - 4 - azaheptane - 1,7 - diamine nickel(ii)**
$C_{34}H_{35}N_3NiO_2$
P.C.Healy, G.M.Mockler, D.P.Freyberg, E.Sinn
J. Chem. Soc., Dalton Trans., 691, 1975
Also classified in 83

84.78 **Di - μ - tetrachloro - 1,2 - benzoquinone - di(bis(tetrachloro - 1,2 - benzoquinone) - molybdenum) benzene solvate**
$C_{36}Cl_{24}Mo_2O_{12}$, $1.5C_6H_6$
C.G.Pierpont, H.H.Downs *J. Am. Chem. Soc.*, **97**, 2123, 1975

84.79 **tetrakis(8 - Quinolinolato) zirconium(iv) toluene solvate**
$C_{36}H_{24}N_4O_4Zr$, $3C_7H_8$
D.F.Lewis, R.C.Fay *J. Chem. Soc., Chem. Commun.*, 1046, 1974
Residue 1 also classified in 83

84.80 **Tetrachloro - bis(triphenylphosphine oxide) uranium**
$C_{36}H_{30}Cl_4O_2P_2U$
G.Bombieri, R.Graziani, D.Brown, P.T.Moseley
Eur. Cryst. Meeting, 347, 1974

84.C **Di - μ - ethoxy - bis(dibenzylethoxy) titanium(iv)**
$C_{36}H_{48}O_4Ti_2$
For complete entry see 71.106

84.C **1,1 - bis(Triphenylarsine) - 3,3,4,4 - tetracyano - 1 - platina - 2 - oxacyclobutane**
$C_{42}H_{30}As_2N_4OPt$
For complete entry see 71.117

84.C **Tetrachlorocatecholato - bis(triphenylphosphine) palladium**
$C_{42}H_{30}Cl_4O_2P_2Pd$
For complete entry see 86.76

84.81 **bis(Dichloro - (2,6 - diacetylpyridine) - bis(picolinoylhydrazonato) dicopper(ii)) dihydrate**
$C_{42}H_{34}Cl_4Cu_4N_{14}O_4$, $2H_2O$
A.Mangia, C.Pelizzi, G.Pelizzi *Acta Crystallogr., Sect. B,* **30,** 2146, 1974
Residue 1 also classified in 83

84.82 **Hepta(2,6 - dimethyl - 4 - pyrone) erbium(iii) perchlorate**
$C_{49}H_{56}ErO_{14}^{3+}$, $3ClO_4^-$
E.C.Bisi, M.Gorio, E.Cannillo, A.Coda, V.Tazzoli
Acta Crystallogr., Sect. A, **31,** S134, 1975

84.83 **Tri - μ - dimethylformamide - bis(tri(3 - trifluoroacetyl - d - camphorato) - praseodymium) (absolute configuration)**
$C_{81}H_{105}F_{18}N_3O_{15}Pr_2$
J.A.Cunningham, R.E.Sievers *J. Am. Chem. Soc.*, **97,** 1586, 1975

METAL COMPLEXES
(SULPHUR OR SELENIUM LIGAND)

85.C **Di(tetrathiofulvalene) bis(dithiolene) nickel**
$C_4H_4NiS_4$, $2C_6H_4S_4$
For complete entry see 60.4

85.C **Aquo - (thiodiacetato) cadmium(ii)**
$C_4H_6CdO_5S$
For complete entry see 81.12

85.1 **Reaction product of dithiazylylcobaltate with ammonia and formaldehyde**
$C_4H_9CoN_5OS_4$
U.Thewalt *Z. Naturforsch., Teil B,* **29,** 308, 1974
Also classified in 83

85.2 **Cadmium(ii) bis(thioglycollate)**
$(C_4H_{10}CdO_2S_2)_n$
H.-B.Burgi *Helv. Chim. Acta,* **57,** 513, 1974

85.C **Triaquo - zinc(ii) thiodiglycollate monohydrate**
$C_4H_{10}O_7SZn$, H_2O
For complete entry see 81.15

85.3 **Tetramethyldiphosphine - disulfide copper(ii) chloride**
$(C_4H_{12}Cl_2CuP_2S_2)_n$
F.A.Cotton, B.A.Frenz, D.L.Hunter, Z.C.Mester
Inorg. Chim. Acta, **11,** 119, 1974

85.C **n - Propylmercapto - acetato mercury**
$(C_5H_{10}HgO_2S)_n$
For complete entry see 81.16

85.4 **Calcium potassium (+)$_{589}$ - tris(dithio - oxalato) cobalt(iii) tetrahydrate**
$C_6CoO_6S_6^{3-}$, Ca^{2+} , K^+ , $4H_2O$
K.R.Butler, M.R.Snow *Acta Crystallogr., Sect. B,* **31,** 354, 1975

85.C **n - Butylmercapto - acetato mercury**
$(C_6H_{12}HgO_2S)_n$
For complete entry see 81.18

85.5 **Trichloro - tris(dimethylsulfoxide) rhodium(iii)**
$C_6H_{18}Cl_3O_3RhS_3$
V.I.Sokol, N.D.Rubtsova, A.Yu.Gribenyuk, M.A.Porai-Koshits
Zh. Strukt. Khim., **15,** 716, 1974
Also classified in 84

85.6 **Dimethylammonium trichloro - tris(dimethylsulfoxide) ruthenium(ii)**
$C_6H_{18}Cl_3O_3RuS_3^-$, $C_2H_8N^+$
R.S.McMillan, A.Mercer, B.R.James, J.Trotter
J. Chem. Soc., Dalton Trans., 1006, 1975
Residue 2 classified in 3

85.C **cis - bis(bis - (Trifluoromethyl) - ethylene - 1,2 - dithiolato) nickel - phenoxazine**
$C_8F_{12}NiS_4$, $C_{12}H_9NO$
For complete entry see 60.8

85.7 **(Thiazolidine - 2 - thione)pentacarbonyl tungsten**
$C_8H_5NO_5S_2W$
M.Cannas, G.Carta, G.Marongiu, E.F.Trogu
Acta Crystallogr., Sect. B, **30,** 2252, 1974

85.8 **bis(Diethyl - diseleno - phosphato) nickel(ii)**
$C_8H_{20}NiO_4P_2Se_4$
P.J.H.A.M.van der Leemput, T.W.Hummelink, J.H.Noordik, P.T.Beurskens *Cryst. Struct. Commun.*, **4,** 167, 1975

85.9 **bis(Tetramethyldiphosphine - disulfide) - bis(copper(i) chloride)**
$C_8H_{24}Cl_2Cu_2P_4S_4$
F.A.Cotton, B.A.Frenz, D.L.Hunter, Z.C.Mester
Inorg. Chim. Acta, **11,** 111, 1974

85.10 **(Thiomorpholin - 3 - one) - pentacarbonyl - tungsten**
$C_9H_7NO_6SW$
M.Cannas, G.Carta, D.de Filippo, G.Marongiu, E.F.Trogu
Inorg. Chim. Acta, **10,** 145, 1974

85.C **Phenyl mercury L - cysteine complex**
$C_9H_{11}HgNO_2S$
For complete entry see 71.10

85.11 **tris(S - Methyl - ethylene - 1,2 - dithiolato) rhodium**
$C_9H_{15}RhS_6$
R.Richter, J.Kaiser, J.Sieler, L.Kutschabsky
Acta Crystallogr., Sect. B, **31,** 1642, 1975

85.12 **Dichloro - bis(6 - mercaptopurine) mercury(ii)**
$C_{10}H_8Cl_2HgN_8S_2$
A.L.Beauchamp, M.A.Martin, C.Gagnon, P.Lavertue
Acta Crystallogr., Sect. A, **31,** S45, 1975
Also classified in 44

85.13 **6 - Mercaptopurine copper(i) chloride hydrochloride dimer dihydrate**
$C_{10}H_{10}Cl_4Cu_2N_8S_2$, $2H_2O$
M.R.Caira, L.R.Nassimbeni *Acta Crystallogr., Sect. B,* **31,** 1339, 1975
Residue 1 also classified in 44

85.14 **Di - iodo - (1,3,5,7 - tetramethyl - 2,4,6,8 - tetrathia - adamantane) platinum**
$C_{10}H_{16}I_2PtS_4$
H.A.Levy, J.R.Long
Am. Cryst. Assoc., Abstr. Papers (Spring Meeting), 11, 1975

85.15 **1,4,8,11 - Tetrathiacyclotetradecane bis(pentachloro - niobium) benzene solvate**
$C_{10}H_{20}Cl_{10}Nb_2S_4$, C_6H_6
R.E.DeSimone, M.D.Glick *J. Am. Chem. Soc.,* **97,** 942, 1975

85.16 **(1 - Oxa - 7,10 - dithia - 4,13 - diazacyclopentadecane) palladium(ii) nitrate**
$C_{10}H_{22}N_2OPdS_2^{2+}$, $2NO_3^-$
R.Louis, D.Pellisard, R.Weiss *Acta Crystallogr., Sect. B,* **30,** 1889, 1974
Residue 1 also classified in 83

85.17 **Di - μ - (N,S - dimethyl - dithiocarbamidinato) - bis(tricarbonyl manganese)**
$C_{12}H_{12}Mn_2N_2O_6S_4$
S.R.Finnimore, R.Goddard, S.D.Killops, S.A.R.Knox, P.Woodward
J. Chem. Soc., Chem. Commun., 391, 1975
Also classified in 83

85.18 **Sodium tetrabutylammonium tetrakis(β - mercapto - propionato iron sulfide) N - methylpyrrolidone solvate**
$C_{12}H_{16}Fe_4O_8S_8^{6-}$, $C_{16}H_{36}N^+$, $5Na^+$, $5C_5H_9NO$
H.L.Carrell, J.P.Glusker, T.C.Bruice, R.Maskiewicz, R.Job
Am. Cryst. Assoc., Abstr. Papers (Spring Meeting), 12, 1975
Residue 2 classified in 3

85.19 **bis(O - Ethyl thioacetatothioacetato) zinc(ii)**
$C_{12}H_{18}O_2S_4Zn$
R.Beckett, B.F.Hoskins *J. Chem. Soc., Dalton Trans.,* 908, 1975

85.20 **bis(1,5 - Dithiacyclo - octane) - dichloro - nickel(ii)**
$(C_{12}H_{24}Cl_2NiS_4)_n$
N.L.Hill, H.Hope *Inorg. Chem.,* **13,** 2079, 1974

85.21 **cis - Dichloro - (4,7,13,16 - tetraoxa - 1,10 - dithia - cyclo - octadecane) palladium**
$C_{12}H_{24}Cl_2O_4PdS_2$
B.Metz, D.Moras, R.Weiss *J. Inorg. Nucl. Chem.,* **36,** 785, 1974

85.C **(1,5 - Cyclo - octadiene) - (S - methyl - maleonitriledithiolato) rhodium(i)**
$C_{13}H_{15}N_2RhS_2$
For complete entry see 75.10

85.22 **bis(Pyridinealdehyde - thiosemicarbazonato) cobal(iii) chloride trihydrate**
$C_{14}H_{14}CoN_8S_2^+$, Cl^- , $3H_2O$
K.N.Akatova, T.N.Tarkhova, S.L.Ginzburg, M.G.Neigauz,
L.A.Novakovskaya *Kristallografiya,* **19,** 383, 1974
Residue 1 also classified in 83

85.C **Phenylmercury - (2,6 - dimethyl - thiophenolate)**
$C_{14}H_{14}HgS$
For complete entry see 71.37

85.23 **(bis(2 - (2 - Pyridyl)ethyl)disulfide) copper(i) perchlorate**
$(C_{14}H_{16}CuN_2S_2^+)_n$, $nClO_4^-$
L.G.Warner, T.Ottersen, K.Seff *Inorg. Chem.,* **13,** 2819, 1974
Residue 1 also classified in 83

85.24 **bis(Tricarbonyl(diethyldithiophosphinato) rhenium(i))**
$C_{14}H_{20}O_6P_2Re_2S_4$
G.Thiele, G.Liehr, E.Lindner *Chem. Ber.,* **107,** 442, 1974

85.25 **N,N' - Tetramethylene - bis(thioacetylacetoniminato) zinc**
$C_{14}H_{22}N_2S_2Zn$
M.J.E.Hewlins *J. Chem. Soc., Dalton Trans.,* 429, 1975
Also classified in 83

85.26 **Dithio - tri(iron) - tetra(1,2 - bis(trifluoromethyl) - ethylene - 1,2 - dithiolate) cyclo - octasulfur**
$C_{16}F_{24}Fe_3S_{10}$, $0.5S_8$
K.Gerst, C.E.Nordman
Am. Cryst. Assoc., Abstr. Papers (Summer Meeting), 225, 1974

85.27 **Tetraethylammonium bis(2,3 - quinoxalinedithiolato) nickel(ii) dihydrate**
$C_{16}H_8N_4NiS_4^-$, $C_8H_{20}N^+$, $2H_2O$
A.Pignedoli, G.Peyronel, L.Antolini
Acta Crystallogr., Sect. B, **30,** 2181, 1974
Residue 2 classified in 3

85.28 **Tetraethylammonium bis(o - xylyl - α,α' - thiolato - μ - sulfido iron(iii))**
$C_{16}H_{16}Fe_2S_6^{2-}$, $2C_8H_{20}N^+$
J.J.Mayerle, S.E.Denmark, B.V.DePamphilis, J.A.Ibers, R.H.Holm
J. Am. Chem. Soc., **97,** 1032, 1975
Residue 2 classified in 3

85.29 **Octacarbonyl - di(tetrahydrothiophene)tri - iron**
$C_{16}H_{16}Fe_3O_8S_2$
F.A.Cotton, J.M.Troup *J. Am. Chem. Soc.,* **96,** 5070, 1974

85.30 Potassium bis(1,1 - dicarboethoxy - 2,2 - ethylenedithiolato) copper(iii) diethylether solvate
$C_{16}H_{20}CuO_8S_4^-$, K^+ , $C_4H_{10}O$
F.J.Hollander, M.L.Caffery, D.Coucouvanis
J. Am. Chem. Soc., **96,** 4682, 1974

85.31 Trimethylanilinium bis(1,1 - dicarboethoxy - 2,2 - ethylenedithiolato) nickel(iii)
$C_{16}H_{20}NiO_8S_4^{2-}$, $2C_9H_{14}N^+$
F.J.Hollander, M.L.Caffery, D.Coucouvanis
J. Am. Chem. Soc., **96,** 4682, 1974
Residue 2 classified in 16

85.C Chloro - (2,5 - dithiahexane) - (1 - (1,4 - di - t - butyl - 4 - chloro)butadienyl) palladium
$C_{16}H_{30}Cl_2PdS_2$
For complete entry see 71.43

85.C μ - Nitrosyl - di - μ - (N,N′ - dimethyl - N,N′ - bis(β - mercaptoethyl) ethylenediamine) - di - iron hexafluorophosphate acetone solvate
$C_{16}H_{36}Fe_2N_5OS_4^+$, F_6P^- , C_3H_6O
For complete entry see 76.71

85.32 Di - μ - (bis(2 - (N,N - dimethylamino)ethyl)disulfide) dicopper(i) tetrafluoroborate
$C_{16}H_{40}Cu_2N_4S_4^{2+}$, $2BF_4^-$
T.Ottersen, L.G.Warner, K.Seff *Inorg. Chem.*, **13,** 1904, 1974
Residue 1 also classified in 83

85.33 bis - (cis - 1 - Mercapto - 2 - p - bromobenzoyl - ethylene) nickel(ii)
$C_{18}H_{12}Br_2NiO_2S_2$
L.Kutschabsky *Z. Anorg. Allg. Chem.*, **404,** 239, 1974
Also classified in 84

85.34 bis(8 - Mercaptoquinolinato) palladium
$C_{18}H_{12}N_2PdS_2$
A.Ozola, J.Ozols, A.Ievins
Latv. PSR Zinat. Akad. Vestis, Kim. Ser., 662, 1973
Also classified in 83

85.35 bis(8 - Mercaptoquinolinato) platinum
$C_{18}H_{12}N_2PtS_2$
J.Ozols, A.Ozola, A.Ievins
Latv. PSR Zinat. Akad. Vestis, Kim. Ser., 648, 1973
Also classified in 83

85.36 Oxo - trichloro - (triphenylphosphine sulfide) molybdenum(v)
$C_{18}H_{15}Cl_3MoOPS$
P.M.Boorman, C.D.Garner, F.E.Mabbs, T.J.King
J. Chem. Soc., Chem. Commun., 663, 1974

85.C **bis(π - 2 - Ethyl - 3 - methylthioacrolein iron(0) dicarbonyl)**
$C_{18}H_{20}Fe_2O_4S_2$
For complete entry see 71.53

85.C **(bis - π - Cyclopentadienyl molybdenum) - bis - μ - methanethiolato - (bis - π - allyl rhodium) hexafluorophosphate**
$C_{18}H_{26}MoRhS_2^+$, F_6P^-
For complete entry see 72.16

85.37 **Bromo - (tris(2 - t - butylthioethyl)amine) cobalt(ii) hexafluorophosphate**
$C_{18}H_{39}BrCoNS_3^+$, F_6P^-
G.Fallani, R.Morassi, F.Zanobini *Inorg. Chim. Acta,* **12,** 147, 1975
Residue 1 also classified in 83

85.38 **μ - (2 - Mercaptoethanolato) - μ - oxo - bis(oxo - (8 - hydroxyquinolinato) - molybdenum(v))**
$C_{20}H_{16}Mo_2N_2O_6S$
J.I.Gelder, J.H.Enemark, G.Wolterman, D.A.Boston, G.P.Haight
J. Am. Chem. Soc., **97,** 1616, 1975
Also classified in 84, 83

85.C **Chloro - (triphenylphosphine) - methylmercaptomethyl palladium(ii)**
$C_{20}H_{20}ClPPdS$
For complete entry see 71.62

85.C **Chloro - (triphenylphosphine) - methylmercaptomethyl palladium(ii) (at −160°C)**
$C_{20}H_{20}ClPPdS$
For complete entry see 71.63

85.39 **bis(Acetophenone thioacetohydrazonato) nickel(ii)**
$C_{20}H_{22}N_4NiS_2$
S.Larsen *Acta Chem. Scand. Ser. A,* **28,** 779, 1974
Also classified in 83

85.40 **tetrakis(Tetraphenylphosphonium) hexakis(1,2 - dithiosquarato) octa - copper(i)**
$C_{24}Cu_8O_{12}S_{12}^{4-}$, $4C_{24}H_{20}P^+$
F.J.Hollander, D.Coucouvanis *J. Am. Chem. Soc.,* **96,** 5646, 1974
Residue 2 classified in 64

85.41 **Chloro - (bis(2 - ((2 - pyridylmethyl)imino)phenyl)disulfide) nickel(ii) perchlorate**
$C_{24}H_{18}ClN_4NiS_2^+$, ClO_4^-
L.G.Warner, T.Ottersen, K.Seff *Inorg. Chem.,* **13,** 2529, 1974
Residue 1 also classified in 83

85.42 bis(Tetramethylammonium) tetrakis(thiophenolato - iron sulfide)
$C_{24}H_{20}Fe_4S_8^{2-}$, $2C_4H_{12}N^+$
L.Que Junior, M.A.Bobrik, J.A.Ibers, R.H.Holm
J. Am. Chem. Soc., **96,** 4168, 1974
Residue 2 classified in 3

85.43 bis(N - Cyclohexyl - thiopicolinamidato) copper(ii)
$C_{24}H_{30}CuN_4S_2$
B.F.Hoskins, F.D.Whillans *J. Chem. Soc., Dalton Trans.*, 2112, 1974
Also classified in 83

85.C bis(bis - (π - Cyclopentadienyl) molybdenum(iv) - bis - μ - methanethiolato) nickel(ii) tetrafluoroborate
$C_{24}H_{32}Mo_2NiS_4^{2+}$, $2BF_4^-$
For complete entry see 73.49

85.C bis(bis - (π - Cyclopentadienyl) niobium(v) - bis - μ - methanethiolato) nickel(0) tetrafluoroborate dihydrate
$C_{24}H_{32}Nb_7NiS_4^{2+}$, $2BF_4^-$, $2H_2O$
For complete entry see 73.50

85.C cis - bis(Diethylphosphinodithioato) - bis(dimethylphenylphosphine) ruthenium(ii)
$C_{24}H_{42}P_4RuS_4$
For complete entry see 86.43

85.C Chloro - (bis(salicylideneiminephenyl)disulfido) iron(iii)
$C_{26}H_{18}ClFeN_2O_2S_2$
For complete entry see 78.12

85.C bis(π - Cyclopentadienyl) tungsten(iv) - (bis - μ - benzenethiolato) chromium(0) tetracarbonyl
$C_{26}H_{20}CrO_4S_2W$
For complete entry see 73.52

85.C bis(π - Cyclopentadienyl) tungsten(iv) - (bis - μ - benzenethiolato) molybdenum(0) tetracarbonyl
$C_{26}H_{20}MoO_4S_2W$
For complete entry see 73.53

85.C bis(π - Cyclopentadienyl) tungsten(iv) - bis - μ - benzenethiolato - tungsten(0) tetracarbonyl
$C_{26}H_{20}O_4S_2W_2$
For complete entry see 73.54

85.C Benzyl - (triphenylmethylthio) mercury
$C_{26}H_{22}HgS$
For complete entry see 71.85

85.C **Salicylaldehyde - selenosemicarbazonato - (triphenylphosphine) nickel**
$C_{26}H_{22}N_3NiOPSe$
For complete entry see 78.15

85.44 **bis(2 - Mercapto - 4′ - methylazobenzene) nickel(ii)**
$C_{26}H_{22}N_4NiS_2$
O.A.Dyachenko, L.O.Atovmyan, S.M.Aldoshin
J. Chem. Soc., Chem. Commun., 105, 1975
Also classified in 83

85.C **Triphenylphosphine - bis(diethyldithiophosphato) nickel(ii)**
$C_{26}H_{35}NiO_4P_3S_4$
For complete entry see 86.49

85.45 **bis - μ - Mercaptobenzothiazolato - bis(dicarbonyl - pyridine ruthenium)**
$C_{28}H_{18}N_4O_4Ru_2S_4$
S.Jeannin, Y.Jeannin, G.Lavigne *C. R. Acad. Sci., Ser. C*, **279**, 447, 1974
Also classified in 83

85.46 **Tetraethylammonium bis(di - (p - tolylthiolato) - μ - sulfido - iron(iii))**
$C_{28}H_{28}Fe_2S_6^{2-}$, $2C_8H_{20}N^+$
J.J.Mayerle, S.E.Denmark, B.V.dePamphilis, J.A.Ibers, R.H.Holm
J. Am. Chem. Soc., **97**, 1032, 1975
Residue 2 classified in 3

85.C **Tricarbonyl - (2 - (methylsulfidomethyl)phenyl) - triphenylphosphine manganese**
$C_{29}H_{24}MnO_3PS$
For complete entry see 71.92

85.C **bis(Dithioformate) - bis(triphenylphosphine) ruthenium**
$C_{38}H_{32}P_2RuS_4$
For complete entry see 86.72

85.C **Chloro - bis(triphenylphosphine) platinum - dithiocarboxylato - (bis(triphenylphosphine) platinum) tetrafluoroborate methylene dichloride solvate**
$C_{73}H_{60}ClP_4Pt_2S_2^+$, BF_4^- , $0.2CH_2Cl_2$
For complete entry see 71.137

85.C **tris(bis(Triphenylphosphine) silver(i)) tris(dithio - oxalato) aluminium(iii)**
$C_{114}H_{90}Ag_3AlO_6P_6S_6$
For complete entry see 86.118

85.C **tris(bis(Triphenylphosphine) silver(i)) tris(dithio - oxalato) iron(iii)**
$C_{114}H_{90}Ag_3FeO_6P_6S_6$
For complete entry see 81.48

METAL COMPLEXES (P, AS, SB LIGAND)

86.1 **Guanidinium tetramolybdo - dimethylarsinate monohydrate**
$C_2H_7AsMo_4O_{15}^{2-}$, $2CH_6N_3^+$, H_2O
K.M.Barkigia, C.O.Quicksall
Am. Cryst. Assoc., Abstr. Papers (Spring Meeting), 36, 1975
Residue 2 classified in 8

86.2 **Tetraethylammonium trichloro(triethylphosphine) platinum(ii)**
$C_6H_{15}Cl_3PPt^-$, $C_8H_{20}N^+$
G.W.Bushnell, A.Pidcock, M.A.R.Smith
J. Chem. Soc., Dalton Trans., 572, 1975
Residue 2 classified in 3

86.3 **Tetrachloro - (1,2 - bis(dimethylarsino) - 3,3,4,4 - tetrafluorocyclobut - 1 - ene) rhenium(iv)**
$C_8H_{12}As_2Cl_4F_4Re$
E.N.Maslen, J.C.Dewan, D.L.Kepert, K.R.Trigwell, A.H.White
J. Chem. Soc., Dalton Trans., 2128, 1974

86.C **Bromo - tricarbonyl - (trimethylphosphine) - (methylmethinyl) chromium**
$C_8H_{12}BrCrO_3P$
For complete entry see 71.6

86.4 **closo - 1,1 - bis(Trimethylphosphine) - 1,6,8 - platinadicarbaborane**
$C_8H_{26}B_6P_2Pt$
M.Green, J.L.Spencer, F.G.A.Stone, A.J.Welch
J. Chem. Soc., Chem. Commun., 794, 1974
Also classified in 62

86.C **Dicarbonyl - (π - cyclopentadienyl)(bis(trifluoromethyl)phosphino) iron**
$C_9H_5F_6FeO_2P$
For complete entry see 73.2

86.C **Dicarbonyl - (π - cyclopentadienyl)(bis(trifluoromethyl) - oxo - phosphino) iron**
$C_9H_5F_6FeO_3P$
For complete entry see 73.3

86.5 **trans - bis(Trimethylphosphite) - tricarbonyl iron**
$C_9H_{18}FeO_9P_2$
D.Ginderow *Acta Crystallogr., Sect. B,* **30,** 2798, 1974

86.6 **Dibromo - tris(trimethylphosphine) nickel(ii)**
$C_9H_{27}Br_2NiP_3$
J.W.Dawson, T.J.McLennan, W.Robinson, A.Merle, M.Dartiguenave, Y.Dartiguenave, H.B.Gray *J. Am. Chem. Soc.*, **96**, 4428, 1974

86.7 **Di - iodo - tris(trimethylphosphite) nickel(ii)**
$C_9H_{27}I_2NiO_9P_3$
L.J.V.Griend, J.C.Clardy, J.G.Verkade *Inorg. Chem.*, **14**, 710, 1975

86.8 **cis - Tetracarbonyl - 1,4 - h^2 - 1,2 - bis(dimethylphosphino - 1,2 - dimethyl) - diarsine molybdenum(0)**
$C_{10}H_{18}As_2MoO_4P_2$
W.S.Sheldrick *Acta Crystallogr., Sect. B*, **31**, 1789, 1975

86.9 **(Hexamethyl - cyclohexaphosphine) tungsten tetracarbonyl**
$C_{10}H_{18}O_4P_6W$
P.S.Elmes, B.M.Gatehouse, B.O.West
J. Organomet. Chem., **82**, 235, 1974

86.10 **closo - 1,1 - bis(Trimethylphosphine) - 6,8 - dimethyl - 1,6,8 - platinadicarbaborane (α form)**
$C_{10}H_{30}B_6P_2Pt$
M.Green, J.L.Spencer, F.G.A.Stone, A.J.Welch
J. Chem. Soc., Chem. Commun., 794, 1974
Also classified in 62

86.11 **closo - 1,1 - bis(Trimethylphosphine) - 6,8 - dimethyl - 1,6,8 - platinadicarbaborane (β form)**
$C_{10}H_{30}B_6P_2Pt$
M.Green, J.L.Spencer, F.G.A.Stone, A.J.Welch
J. Chem. Soc., Chem. Commun., 794, 1974
Also classified in 62

86.12 **μ - Carbonyl - bis - μ - (bis(bis(trifluoromethyl)phosphino)sulfur) - bis(carbonyl nickel)**
$C_{11}F_{24}Ni_2O_3P_4S_2$
H.Einspahr, J.Donohue *Inorg. Chem.*, **13**, 1839, 1974

86.13 **(Tricarbonyl - nitrosyl - trimethylphosphite tungsten) - μ - hydrido - (pentacarbonyl tungsten)**
$C_{11}H_{10}NO_{12}PW_2$
R.Bau, H.B.Chin, R.A.Love, S.W.Kirtley, B.R.Whittlesey, T.F.Koetzle
Acta Crystallogr., Sect. A, **31**, S131, 1975

86.14 **(Tricarbonyl - nitrosyl - trimethylphosphite tungsten) - μ - hydrido - (pentacarbonyl tungsten) (neutron study)**
$C_{11}H_{10}NO_{12}PW_2$
R.Bau, H.B.Chin, R.A.Love, S.W.Kirtley, B.R.Whittlesey, T.F.Koetzle
Acta Crystallogr., Sect. A, **31**, S131, 1975

86.15 **(2,3 - bis(Dimethylarsino) - 1,1,1,4,4,4 - hexafluorobut - 2 - ene)tricarbonyl - di - iodo - tungsten(ii)**
$C_{11}H_{12}As_2F_6I_2O_3W$
A.Mercer, J.Trotter *Can. J. Chem.*, **52,** 3331, 1974

86.C **Dimethyl - bis(trimethylphosphine) cobalt(iii) - (dimethylphosphonium - bis(methylide))**
$C_{12}H_{34}CoP_3$
For complete entry see 71.29

86.16 **Di - μ - nitrato - bis(di(trimethyl phosphite) silver(i))**
$C_{12}H_{36}Ag_2N_2O_{18}P_4$
J.H.Meiners, J.C.Clardy, J.G.Verkade *Inorg. Chem.*, **14,** 632, 1975

86.C **Dicarbonyl - (π - cyclopentadienyl) - manganese - μ(dimethylarsino) - tetracarbonyl - manganese**
$C_{13}H_{11}AsMn_2O_6$
For complete entry see 73.19

86.17 **Tetracarbonyl - iodo - (o - phenylenebis(dimethylarsino)) tungsten(ii) tri - iodide**
$C_{14}H_{16}As_2IO_4W^+$, I_3^-
M.G.B.Drew, J.D.Wilkins *J. Organomet. Chem.*, **69,** 271, 1974

86.18 **Cyclonona(methylarsine) - bis(tricarbonyl chromium)**
$C_{15}H_{27}As_9Cr_2O_6$
P.S.Elmes, B.M.Gatehouse, D.J.Lloyd, B.O.West
J. Chem. Soc., Chem. Commun., 953, 1974

86.19 **Nitroso - tris(1,3,7 - trioxa - 2 - phospha - 5 - methylbicyclo(2.2.2)octane) nickel(ii) tetrafluoroborate chloroform solvate**
$C_{15}H_{27}NNiO_{10}P_3^+$, BF_4^- , $CHCl_3$
J.H.Meiners, C.J.Rix, J.C.Clardy, J.G.Verkade
Inorg. Chem., **14,** 705, 1975

86.C **Trichloro(dimethylaminomethylene) - bis(triethylphosphine) rhodium(iii)**
$C_{15}H_{37}Cl_3NP_2Rh$
For complete entry see 71.41

86.20 **tetrakis(2 - Thienyldifluorophosphine) nickel**
$C_{16}H_{12}F_8NiP_4S_4$
W.S.Sheldrick *Acta Crystallogr., Sect. B,* **31,** 305, 1975

86.21 **(cis - 1 - bis(Trifluoromethyl)phosphino - 2 - (diphenylphosphino)ethane) dichloro palladium(ii)**
$C_{16}H_{14}Cl_2F_6P_2Pd$
L.Manojlovic-Muir, D.Millington, K.W.Muir, D.W.A.Sharp, W.E.Hill, J.V.Quagliano, L.M.Vallarino *J. Chem. Soc., Chem. Commun.*, 999, 1974

86.C **μ - Hydrido - μ - dimethylphosphido - bis(π - cyclopentadienyldicarbonyl molybdenum) (neutron study)**
$C_{16}H_{17}Mo_2O_4P$
For complete entry see 73.28

86.C **μ - Dimethylarsino - μ - (1,3 - η - (2,3 - bis(dimethylarsino) - 1,1 - difluoro - 3 - trifluoromethylallyl)) - bis(tricarbonyl manganese(i))**
$C_{16}H_{18}As_3F_5Mn_2O_6$
For complete entry see 72.13

86.22 **Di - μ - tetramethyldiarsine - bis(tetracarbonyl chromium)**
$C_{16}H_{24}As_4Cr_2O_8$
F.A.Cotton, T.R.Webb *Inorg. Chim. Acta,* **10,** 127, 1974

86.C **nido - 6,6 - bis(Triethylphosphine) - 5,8 - dimethyl - 6,5,8 - platinadicarbaborane**
$C_{16}H_{42}B_6P_2Pt$
For complete entry see 71.45

86.23 **Bromo - (triphenylarsine) gold(i)**
$C_{18}H_{15}AsAuBr$
F.W.B.Einstein, R.Restivo *Acta Crystallogr., Sect. B,* **31,** 624, 1975

86.24 **Dinitrato(triphenylphosphine) mercury(ii)**
$(C_{18}H_{15}HgN_2O_6P)_n$
S.H.Whitlow *Can. J. Chem.,* **52,** 198, 1974

86.25 **Di - μ - chloro - bis(carbonyl - (dimethylphenylphosphine) rhodium(i))**
$C_{18}H_{22}Cl_2O_2P_2Rh_2$
J.J.Bonnet, Y.Jeannin, P.Kalck, A.Maisonnat, R.Poilblanc
Inorg. Chem., **14,** 743, 1975

86.C **(π - Pentenyl) - di - (isopropyl - phenyl - phosphine) - methyl nickel(ii)**
$C_{18}H_{31}NiP$
For complete entry see 71.56

86.26 **Chloro - bis(triethylphosphine) - (fluorophenylazo) platinum**
$C_{18}H_{34}ClFN_2P_2Pt$
S.Krogsrud, S.D.Ittel, J.A.Ibers
Am. Cryst. Assoc., Abstr. Papers (Summer Meeting), 225, 1974
Also classified in 83

86.27 **trans - Chloro - bis(triethylphosphine)(p - fluorophenyldiazene) platinum(ii) perchlorate**
$C_{18}H_{35}ClFN_2P_2Pt^+$, ClO_4^-
S.D.Ittel, J.A.Ibers *J. Am. Chem. Soc.,* **96,** 4804, 1974
Residue 1 also classified in 83

86.28 **trans - Chloro - bis(triethylphosphine) - (p - fluorophenylhydrazine) platinum(ii) tetrafluoroborate**
$C_{18}H_{37}ClFN_2P_2Pt^+$, BF_4^-
S.D.Ittel, J.A.Ibers *Inorg. Chem.*, **14**, 636, 1975
Residue 1 also classified in 83

86.29 **trans - bis(Acetone hydrazone) - tetrakis(trimethylphosphite) ruthenium(ii) bis(tetraphenylborate)**
$C_{18}H_{52}N_4O_{12}P_4Ru^{2+}$, $2C_{24}H_{20}B^-$
M.J.Nolte, E.Singleton *J. Chem. Soc., Dalton Trans.*, 2406, 1974
Residue 1 also classified in 83; residue 2 classified in 62

86.30 **cis - Dichloro - carbonyl - (triphenylphosphine) platinum(ii)**
$C_{19}H_{15}Cl_2OPPt$
L.Manojlovic-Muir, K.W.Muir, R.Walker
J. Organomet. Chem., **66**, C21, 1974

86.31 **Di(nitrosyl) - carbonyl - (triphenylphosphine) iron**
$C_{19}H_{15}FeN_2O_3P$
V.G.Albano, A.Araneo, P.L.Bellon, G.Ciani, M.Manassero
J. Organomet. Chem., **67**, 413, 1974

86.C **Iodo - (1,2 - bis(methylphenylarsino)ethane) - trimethyl platinum**
$C_{19}H_{29}As_2IPt$
For complete entry see 71.60

86.C **Chloro - (triphenylphosphine) - methylmercaptomethyl palladium(ii)**
$C_{20}H_{20}ClPPdS$
For complete entry see 71.62

86.C **Chloro - (triphenylphosphine) - methylmercaptomethyl palladium(ii) (at −160°C)**
$C_{20}H_{20}ClPPdS$
For complete entry see 71.63

86.C **Carbonyl - chloro - bis(dimethylphenylphosphine) - allyl - iridium(iii) hexafluorophosphate**
$C_{20}H_{27}ClIrOP_2^+$, F_6P^-
For complete entry see 72.17

86.32 **Tetrabromo - bis(o - phenylene - bis(dimethylarsine)) tantalum(v) hexabromotantalate(v)**
$C_{20}H_{32}As_4Br_4Ta^+$, Br_6Ta^-
M.G.B.Drew, A.P.Wolters, J.D.Wilkins
Acta Crystallogr., Sect. B, **31**, 324, 1975

86.33 **Tetrachloro - bis(o - phenylene - bis(dimethylarsino)) molybdenum(v) tri - iodide**
$C_{20}H_{32}As_4Cl_4Mo^+$, I_3^-
M.G.B.Drew, G.M.Egginton, J.D.Wilkins
Acta Crystallogr., Sect. B, **30**, 1895, 1974

86.34 **Nitrosyl - bis(o - phenylenebis(dimethylarsine)) cobalt perchlorate**
$C_{20}H_{32}As_4CoNO^{2+}$, $2ClO_4^-$
J.H.Enemark, R.D.Feltham, J.Riker-Nappier, K.F.Bizot
Inorg. Chem., **14,** 624, 1975

86.35 **trans - Dichloro - bis(o - phenylene - bis(dimethylphosphine)) manganese(ii)**
$C_{20}H_{32}Cl_2MnP_4$
M.A.Bennett, G.B.Robertson, J.M.Rosalky, L.F.Warren
Acta Crystallogr., Sect. A, **31,** S136, 1975

86.36 **trans - Dichloro - bis(o - phenylene - bis(dimethylphosphine)) manganese(iii) perchlorate**
$C_{20}H_{32}Cl_2MnP_4^+$, ClO_4^-
M.A.Bennett, G.B.Robertson, J.M.Rosalky, L.F.Warren
Acta Crystallogr., Sect. A, **31,** S136, 1975

86.37 **trans - Dichloro - bis(o - phenylene - bis(dimethylphosphine)) manganese(iv) bis(hydrogen dinitrate)**
$C_{20}H_{32}Cl_2MnP_4^{2+}$, $2HN_2O_8^-$
M.A.Bennett, G.B.Robertson, J.M.Rosalky, L.F.Warren
Acta Crystallogr., Sect. A, **31,** S136, 1975

86.C **1,1 - bis(Dimethylphenylphosphine) - 2,4 - dimethyl - 1 - platina - 2,4 - dicarbadodecaborane**
$C_{20}H_{37}B_9P_2Pt$
For complete entry see 71.66

86.C **Iodo - (1 - (p - chlorophenyl)imino - methyl) - bis(triethylphosphine) platinum(ii)**
$C_{20}H_{37}ClINP_2Pt$
For complete entry see 71.67

86.C **1,1,5 - Trichloro - 1,1 - bis(triethylphosphine) - 2 - methylamino - 3 - platina - indole perchlorate**
$C_{20}H_{38}Cl_3N_2P_2Pt^+$, ClO_4^-
For complete entry see 71.68

86.C **Hexafluorobut - 2 - yne - tricarbonyl(cyclohexa - 1,3 - diene) - ruthenium adduct phosphite substitution product**
$C_{21}H_{17}F_{12}O_5PRu$
For complete entry see 71.70

86.C **Carbonyl - bis(dimethylphenylphosphine) - butadiene iridium(i) tetrafluoroborate**
$C_{21}H_{28}IrOP_2^+$, BF_4^-
For complete entry see 72.19

86.C **Carbonyl - bis(dimethylphenylphosphine) - butadiene iridium(i) perchlorate**
$C_{21}H_{28}IrOP_2^+$, ClO_4^-
For complete entry see 72.20

86.C **Carbonyl - chloro - bis(dimethylphenylphosphine) - methallyl - iridium(iii) hexafluorophosphate**
$C_{21}H_{29}ClIrOP_2^+$, F_6P^-
For complete entry see 72.21

86.C **cis - Dichloro(1,3 - diphenylimidazolidin - 2 - ylidene)(triethylphosphine) platinum(ii)**
$C_{21}H_{29}Cl_2N_2PPt$
For complete entry see 71.75

86.C **trans - Dichloro - (1,3 - diphenylimidazolidin - 2 - ylidene)(triethylphosphine) platinum(ii)**
$C_{21}H_{29}Cl_2N_2PPt$
For complete entry see 71.76

86.C **trans - (Methyl - (2 - oxacyclopentylidene) - bis(dimethylphenylphosphine) platinum(ii)) hexafluorophosphate**
$C_{21}H_{31}OP_2Pt^+$, F_6P^-
For complete entry see 71.77

86.38 **Isothiocyanato - nitrosyl - bis(o - phenylenebis(dimethylarsine)) cobalt isothiocyanate**
$C_{21}H_{32}As_4CoN_2OS^+$, CNS^-
J.H.Enemark, R.D.Feltham, J.Riker-Nappier, K.F.Bizot
Inorg. Chem., **14,** 624, 1975

86.C **π - Allyl - bis(tri - isopropylphosphine) iridium(i)**
$C_{21}H_{47}IrP_2$
For complete entry see 72.22

86.39 **Tetracarbonyl - (triphenylstibine) - iron**
$C_{22}H_{15}FeO_4Sb$
R.F.Bryan, W.C.Schmidt Junior *J. Chem. Soc., Dalton Trans.*, 2337, 1974
Also classified in 66

86.C **Hexafluorobut - 2 - yne - tricarbonyl(cycloheptatriene) - iron adduct phosphite substitution product**
$C_{22}H_{17}F_{12}FeO_5P$
For complete entry see 75.23

86.C **Carbonyl - bis(dimethylphenylphosphine) - methylbutadiene - iridium(i) tetraphenylborate**
$C_{22}H_{30}IrOP_2^+$, $C_{24}H_{20}B^-$
For complete entry see 72.23

86.C **bis(π - Cyclopentadienyl) - hexa(dimethylphosphonato) tricobalt**
$C_{22}H_{46}Co_3O_{18}P_6$
For complete entry see 73.43

86.40 **meso - Tricarbonyl - di - iodo - (o - phenylene - bis(methylphenylarsine)) molybdenum(ii)**
$C_{23}H_{20}As_2I_2MoO_3$
J.C.Dewan, K.Henrick, D.L.Kepert, K.R.Trigwell, A.H.White, S.B.Wild
J. Chem. Soc., Dalton Trans., 546, 1975

86.41 **racemic - Tricarbonyl - di - iodo(o - phenylenebis(methylphenylarsine)) molybdenum(ii) chloroform solvate**
$C_{23}H_{20}As_2I_2MoO_3$, $CHCl_3$
J.C.Dewan, K.Henrick, D.L.Kepert, K.R.Trigwell, A.H.White, S.B.Wild
J. Chem. Soc., Dalton Trans., 546, 1975

86.C **π - Allyl - triphenylphosphine - dicarbonyl iron iodide (red form)**
$C_{23}H_{20}FeIO_2P$
For complete entry see 72.26

86.C **π - Allyl - triphenylphosphine - dicarbonyl iron iodide (black form)**
$C_{23}H_{20}FeIO_2P$
For complete entry see 72.27

86.C **π - Cyclopentadienyl - (triphenylphosphine)carbonyl iridium**
$C_{24}H_{20}IrOP$
For complete entry see 73.45

86.42 **cis - Chloro - bis(triethylphosphine) - (1,10 - phenanthroline) platinum(ii) tetrafluoroborate**
$C_{24}H_{38}ClN_2P_2Pt^+$, BF_4^-
G.W.Bushnell, K.R.Dixon, M.A.Khan *Can. J. Chem.*, **52,** 1367, 1974
Residue 1 also classified in 83

86.43 **cis - bis(Diethylphosphinodithioato) - bis(dimethylphenylphosphine) ruthenium(ii)**
$C_{24}H_{42}P_4RuS_4$
J.D.Owen, D.J.Cole-Hamilton *J. Chem. Soc., Dalton Trans.*, 1867, 1974
Also classified in 85

86.44 **Triethylarsine copper(i) iodide tetramer**
$C_{24}H_{60}As_4Cu_4I_4$
M.R.Churchill, K.L.Kalra *Inorg. Chem.*, **13,** 1899, 1974

86.45 **Triethylphosphine copper(i) iodide tetramer**
$C_{24}H_{60}Cu_4I_4P_4$
M.R.Churchill, K.L.Kalra *Inorg. Chem.*, **13,** 1899, 1974

86.C **1,2 - bis(Dimethylphosphino)ethane - (methyl - (2 - dimethylphosphinoethyl) phosphino)methyl ruthenium hydride dimer**
$C_{24}H_{64}P_8Ru_2$
For complete entry see 71.82

86.C **Pyrazine - 2,3 - dicarbonato - carbonyl - triphenylphosphine rhodium(i) monohydrate**
$C_{25}H_{18}N_2O_5PRh$, H_2O
For complete entry see 81.40

86.C **Trimethylenemethane - chromium - tricarbonyl - triphenylphosphine**
$C_{25}H_{21}CrO_3P$
For complete entry see 72.30

86.C **(2,4 - Pentanedionato)(triphenylphosphine)ethyl nickel(ii)**
$C_{25}H_{27}NiO_2P$
For complete entry see 71.84

86.C **Salicylaldehyde - thiosemicarbazone - (triphenylphosphine) nickel(ii)**
$C_{26}H_{22}N_3NiOPS$
For complete entry see 78.14

86.C **Salicylaldehyde - selenosemicarbazonato - (triphenylphosphine) nickel**
$C_{26}H_{22}N_3NiOPSe$
For complete entry see 78.15

86.46 **Tetrachloro - (1,2 - bis(diphenylphosphino)ethane) rhenium(iv) carbon tetrachloride solvate**
$C_{26}H_{24}Cl_4P_2Re$, $0.75CCl_4$
J.A.Jaecker, W.R.Robinson, R.A.Walton
Inorg. Nucl. Chem. Lett., **10**, 93, 1974

86.47 **Trichloro(nitrosyl) - bis(methyldiphenylphosphine) rhenium**
$C_{26}H_{26}Cl_3NOP_2Re$
K.W.Muir, R.Walker *Acta Crystallogr., Sect. A*, **31**, S132, 1975

86.48 **Chlorobis - (dimethylglyoximato) triphenylstibine cobalt(iii)**
$C_{26}H_{29}ClCoN_4OSb$
M.M.Botoshanskii, Yu.A.Simonov, T.I.Malinovskii
Izv. Akad. Nauk Mold. SSR, 39, 1973
Also classified in 83

86.49 **Triphenylphosphine - bis(diethyldithiophosphato) nickel(ii)**
$C_{26}H_{35}NiO_4P_3S_4$
H.W.Chen, J.P.Fackler
Am. Cryst. Assoc., Abstr. Papers (Summer Meeting), 229, 1974
Also classified in 85

86.C **α - Carboranyl complex of platinum(ii)**
$C_{26}H_{56}B_{10}P_2Pt$
For complete entry see 71.86

86.50 **μ - (Carbonyl(triphenylphosphine) platinio)octacarbonyl - di - iron**
$C_{27}H_{15}Fe_2O_9PPt$
R.Mason, J.A.Zubieta *J. Organomet. Chem.*, **66**, 289, 1974

86.C **(π - Cyclopentadienyl nickel) - di - μ - carbonyl - (dicarbonyl - (tris - p - fluorophenylphosphine) cobalt)**
$C_{27}H_{17}CoF_3NiO_4P$
For complete entry see 73.57

86.51 **Dithiocyanato - (bis(diphenylphosphino)methane) palladium(ii)**
$C_{27}H_{22}N_2P_2PdS_2$
G.J.Palenik, M.Mathew, W.L.Steffen, G.Beran
J. Am. Chem. Soc., **97**, 1059, 1975

86.52 **Dichloro(nitrosyl) - (methanol) - bis(methyldiphenylphosphine) rhenium**
$C_{27}H_{30}Cl_2NO_2P_2Re$
K.W.Muir, R.Walker *Acta Crystallogr., Sect. A*, **31**, S132, 1975
Also classified in 84

86.53 **Tricarbonyl - iodo - tris(dimethylphenylphosphino) tungsten(ii) tetraphenylborate**
$C_{27}H_{33}IO_3P_3W^+$, $C_{24}H_{20}B^-$
M.G.B.Drew, J.D.Wilkins *J. Chem. Soc., Dalton Trans.*, 1654, 1974
Residue 2 classified in 62

86.C **(π - Pentenyl)(dimenthyl - methyl - phosphine) - methyl nickel(ii)**
$C_{27}H_{53}NiP$
For complete entry see 71.89

86.54 **Isothiocyanato - thiocyanato - (1,2 - bis(diphenylphosphino)ethane) palladium(ii)**
$C_{28}H_{24}N_2P_2PdS_2$
G.J.Palenik, M.Mathew, W.L.Steffen, G.Beran
J. Am. Chem. Soc., **97**, 1059, 1975

86.C **(π - (Methylcyclopentadienyl) nickel) - di - μ - carbonyl - (dicarbonyl - (cyclohexyldiphenylphosphine) cobalt)**
$C_{28}H_{28}CoNiO_4P$
For complete entry see 73.59

86.55 **Di - iodo - bis(diphenylphosphinoethyl)sulfido - mercury(ii)**
$C_{28}H_{28}HgI_2P_2S$
K.Aurivillius, L.Falth *Chem. Scr.*, **4**, 215, 1973

86.C **Acetylacetonato - (acetylacetonyl) - triphenylphosphine palladium(ii) benzene solvate**
$C_{28}H_{29}O_4PPd$, $0.5C_6H_6$
For complete entry see 71.90

86.C **Tricarbonyl - (2 - (methylsulfidomethyl)phenyl) - triphenylphosphine manganese**
$C_{29}H_{24}MnO_3PS$
For complete entry see 71.92

86.56 **Chloro - (diphenylethoxyphosphine) - (1 - diphenylphosphino - 3,3,3 - trifluoropropen - 2 - olato) palladium(ii) dichloromethane solvate**
$C_{29}H_{26}ClF_3O_2P_2Pd$, CH_2Cl_2
S.Jacobson, N.J.Taylor, A.J.Carty
J. Chem. Soc., Chem. Commun., 668, 1974
Residue 1 also classified in 84

86.57 **Di - isothiocyanato - (1,3 - bis(diphenylphosphino)propane) palladium(ii)**
$C_{29}H_{26}N_2P_2PdS_2$
G.J.Palenik, M.Mathew, W.L.Steffen, G.Beran
J. Am. Chem. Soc., **97,** 1059, 1975

86.58 **(bis - (2 - Diphenylphosphinoethyl)ether) - carbonyl - rhodium(i) hexafluorophosphate**
$C_{29}H_{28}O_2P_2Rh^+$, F_6P^-
N.W.Alcock, J.M.Brown, J.C.Jeffery
J. Chem. Soc., Chem. Commun., 829, 1974
Residue 1 also classified in 84

86.59 **Octa(n - propylarsine) - bis(tricarbonyl molybdenum)**
$C_{30}H_{56}As_8Mo_2O_6$
P.S.Elmes, B.M.Gatehouse, D.J.Lloyd, B.O.West
J. Chem. Soc., Chem. Commun., 953, 1974

86.C **μ - Dinitrogen - bis((π - mesitylene)(1,2 - bis(dimethylphosphino)ethane) molybdenum)**
$C_{30}H_{56}Mo_2N_2P_4$
For complete entry see 74.7

86.60 **Dichloro - ((−) - 2,3 - O - isopropylidene - 2,3 - dihydroxy - 1,4 - bis(diphenylphosphino)butane) nickel(ii)**
$C_{31}H_{32}Cl_2NiO_2P_2$
V.Gramlich, C.Salomon *J. Organomet. Chem.*, **73,** C61, 1974

86.61 **(Tetraphenyldiphosphinomethane)heptacarbonyl di - iron**
$C_{32}H_{22}Fe_2O_7P_2$
F.A.Cotton, J.M.Troup *J. Am. Chem. Soc.*, **96,** 4422, 1974

86.C **Dicarbonyl - (π - cyclopentadienyl)triphenylphosphine - (1,1,2,2 - tetracyano - prop - 1 - yl) molybdenum**
$C_{32}H_{23}MoN_4O_2P$
For complete entry see 71.97

86.C **μ - (1 - Phenyl - 2 - (triethoxyphosphine) - ethylen - 1,2 - diyl) - μ - diphenylphosphino - bis(tricarbonyl iron)**
$C_{32}H_{30}Fe_2O_9P_2$
For complete entry see 71.98

86.C **η^5 - (8 - Phenyl - bicyclo(5.1.0)octadienyl) - (η^4 - cyclo - octatetraene) - (o - phenylene - bis(dimethylarsine)) niobium**
$C_{32}H_{37}As_2Nb$
For complete entry see 75.37

86.C **Methylisocyanide - (2 - (bis(t - butyl)phosphino) - 3 - methoxyphenolato) - (2 - (1,1 - dimethylethyl - (t - butyl)phosphino) - 3 - methoxyphenolato) iridium(iii)**
$C_{32}H_{50}IrNO_4P_2$
For complete entry see 71.99

86.C **(Tetramethylethylene) nickel (1,2 - bis(dicyclohexylphosphino)ethane)**
$C_{32}H_{60}NiP_2$
For complete entry see 72.32

86.62 **(bis(2 - (2 - Diphenylphosphinoethoxy)ethyl)ether) - carbonyl - aquo rhodium(i) hexafluorophosphate**
$C_{33}H_{38}O_5P_2Rh^+$, F_6P^-
N.W.Alcock, J.M.Brown, J.C.Jeffery
J. Chem. Soc., Chem. Commun., 829, 1974

86.63 **(Chloro - tetrakis(dimethylphenylphosphine) rhenium(i)) - μ - dinitrogen - (tetrachloro - methoxy molybdenum(v)) methanol hydrochloric acid solvate**
$C_{33}H_{47}Cl_5MoN_2OP_4Re$, CH_4O , HCl
M.Mercer *J. Chem. Soc., Dalton Trans.*, 1637, 1974
Residue 1 also classified in 84

86.64 **Chloro - sulfato - nitrosyl - bis(triphenylphosphine) ruthenium(ii)**
$C_{36}H_{30}ClNO_5P_2RuS$
J.Reed, S.L.Soled, R.Eisenberg *Inorg. Chem.*, **13**, 3001, 1974

86.65 **μ - Oxido - bis(chloro - triphenylphosphine - nitrosyl - iridium(i))**
$C_{36}H_{30}Cl_2Ir_2N_2O_3P_2$
P.-T.Cheng, S.C.Nyburg *Inorg. Chem.*, **14**, 327, 1975

86.66 **Di(nitrosyl) - bis(triphenylphosphine) iron**
$C_{36}H_{30}FeN_2O_2P_2$
V.G.Albano, A.Araneo, P.L.Bellon, G.Ciani, M.Manassero
J. Organomet. Chem., **67**, 413, 1974

86.67 **Dinitrosyl - bis(triphenylphosphine) ruthenium**
$C_{36}H_{30}N_2O_2P_2Ru$
S.Bhaduri, G.M.Sheldrick *Acta Crystallogr., Sect. B*, **31**, 897, 1975

86.68 **Hydroxo - di(nitrosyl) - bis(triphenylphosphine) osmium(ii)**
$C_{36}H_{31}N_2O_3OsP_2{}^+$, F_6P^-
G.R.Clark, J.M.Waters, K.R.Whittle
J. Chem. Soc., Dalton Trans., 463, 1975

86.69 **Iodo - bis(2 - (diphenylphosphinomethyl)pyridine) nickel(ii) iodide**
$C_{36}H_{32}IN_2NiP_2{}^+$, I^-
W.Haase *Z. Anorg. Allg. Chem.*, **404,** 273, 1974
Residue 1 also classified in 83

86.C **Di - iodo - carbonyl - (triphenylphosphine) - (p - tolyl - isocyanide) - (p - tolyl - methylaminocarbene) ruthenium**
$C_{36}H_{33}I_2N_2OPRu$
For complete entry see 71.104

86.70 **trans - Di - iodo - bis(tricyclohexylphosphine) platinum(ii)**
$C_{36}H_{66}I_2P_2Pt$
N.W.Alcock, P.G.Leviston *J. Chem. Soc., Dalton Trans.*, 1834, 1974

86.71 **Hydrido - bis(tricyclohexylphosphine) cobalt tetrahydroborate**
$C_{36}H_{67}CoP_2{}^+$, H_4B^-
M.Nakajima, H.Moriyama, A.Kobayashi, T.Saito, Y.Sasaki
J. Chem. Soc., Chem. Commun., 80, 1975

86.C **μ - Diphenylphosphino - μ - (1,4 - bis(trifluoromethyl) - 2 - diphenylphosphino - 1,4 - butadiene - 1,3,4 - triyl) - tri - iron heptacarbonyl**
$C_{37}H_{20}F_6Fe_3O_7P_2$
For complete entry see 71.107

86.C **δ - (trans - 1,2 - bis(o - (Diphenylphosphino)phenyl)ethenyl)chloro - platinum(ii) (absolute configuration)**
$C_{38}H_{29}ClP_2Pt$
For complete entry see 71.110

86.C **Dichloro - (difluoromethyl) - carbonyl - bis(triphenylphosphine) iridium(iii)**
$C_{38}H_{31}Cl_2F_2IrOP_2$
For complete entry see 71.111

86.C **Iodo - (fumaronitrile) - (triphenylphosphite) - bis(p - methoxyphenylisonitrile) rhodium(i)**
$C_{38}H_{31}IN_4O_5PRh$
For complete entry see 71.112

86.72 **bis(Dithioformate) - bis(triphenylphosphine) ruthenium**
$C_{38}H_{32}P_2RuS_4$
A.E.Kalinin, A.I.Gusev, Yu.T.Struchkov *Zh. Strukt. Khim.*, **14,** 859, 1973
Also classified in 85

86.C **bis - μ - (Diphenyl - (t - butylacetylenyl) - phosphine) - di(carbonyl nickel(0))**
$C_{38}H_{38}Ni_2O_2P_2$
For complete entry see 72.33

86.C **Diphenylphosphino - (1 - methoxycarbonyl - 1 - diphenylphosphino - 2,4 - bis(trifluoromethyl) - butadiene - 3,4 - diyl) - tri - iron heptacarbonyl benzene solvate**
$C_{39}H_{23}F_6Fe_3O_9P_2$, $2C_6H_6$
For complete entry see 71.113

86.73 **Dibromo - tris(5 - methyl - 5H - dibenzophosphole) platinum(ii) bromobenzene solvate**
$C_{39}H_{33}Br_2P_3Pt$, C_6H_5Br
K.M.Chui, H.M.Powell *J. Chem. Soc., Dalton Trans.*, 1879, 1974

86.C **Iodo - bis(triphenylphosphine)allene rhodium**
$C_{39}H_{34}IP_2Rh$
For complete entry see 72.34

86.C **bis(Triphenylphosphine)allene palladium**
$C_{39}H_{34}P_2Pd$
For complete entry see 72.35

86.C **1 - (π - Cyclopentadienyl) - 1 - triphenylphosphine - 2 - phenyl - 3,4,5 - tri(methoxycarbonyl) - 1 - cobaltacyclopent - 2 - ene methylene dichloride solvate**
$C_{39}H_{36}CoO_6P$, $0.25CH_2Cl_2$
For complete entry see 71.114

86.C **bis(Triphenylphosphine)hexafluorobut - 2 - yne platinum(0)**
$C_{40}H_{30}F_6P_2Pt$
For complete entry see 72.36

86.C **bis(μ - Diphenylphosphido - μ - carbonyl - π - methylcyclopentadienyl - carbonyl iron) rhodium hexafluorophosphate**
$C_{40}H_{34}Fe_2O_4P_2Rh^+$, F_6P^-
For complete entry see 73.64

86.C **3 - Methyl - cyclopropene - bis(triphenylphosphine) platinum(0) (monoclinic form)**
$C_{40}H_{36}P_2Pt$
For complete entry see 75.39

86.C **sym - trans - Di - μ - acetato - bis(o - (t - butyl - o - tolyl - phosphino)benzyl)di - palladium(ii)**
$C_{40}H_{50}O_4P_2Pd_2$
For complete entry see 71.115

86.74 **trans - Dichloro - tetrakis(diethylphenylphosphonite) technetium(ii)**
$C_{40}H_{60}Cl_2O_8P_4Tc$
G.Bandoli, D.A.Clemente, L.Magon, U.Mazzi
Eur. Cryst. Meeting, 343, 1974

86.C **cis - Dichloro - (pentane - 2,4 - dionato) - trans - bis(triphenylphosphine) rhenium(iii)**
$C_{41}H_{37}Cl_2O_2P_2Re$
For complete entry see 77.13

86.C **1,2 - Dimethyl - cyclopropene - bis(triphenylphosphine) platinum(0)**
$C_{41}H_{38}P_2Pt$
For complete entry see 75.40

86.75 **Iodo - (1,1,1 - tris(diphenylphosphinomethyl)ethane) nickel(i)**
$C_{41}H_{39}INiP_3$
P.Dapporto, G.Fallani, L.Sacconi *Inorg. Chem.*, **13,** 2847, 1974

86.C **1,1 - bis(Triphenylarsine) - 3,3,4,4 - tetracyano - 1 - platina - 2 - oxacyclobutane**
$C_{42}H_{30}As_2N_4OPt$
For complete entry see 71.117

86.76 **Tetrachlorocatecholato - bis(triphenylphosphine) palladium**
$C_{42}H_{30}Cl_4O_2P_2Pd$
C.G.Pierpont, H.H.Downs *Inorg. Chem.*, **14,** 343, 1975
Also classified in 84

86.C **Cyclopentadienyl - carbonyl - bis(triphenylphosphine) manganese benzene solvate**
$C_{42}H_{35}MnOP_2$, C_6H_6
For complete entry see 73.68

86.C **(π - Cyclopentadienyl) - bis(triphenylphosphine) - niobium carbonyl dihydride**
$C_{42}H_{37}NbOP_2$
For complete entry see 73.69

86.C **bis(Triphenylphosphine) - ($\Delta^{1,4}$ - bicyclo(2.2.0)hexene) platinum**
$C_{42}H_{38}P_2Pt$
For complete entry see 75.41

86.C **Cyclohexyne - bis(triphenylphosphine) platinum(0)**
$C_{42}H_{38}P_2Pt$
For complete entry see 75.42

86.77 **Dibromo - tris(2 - phenylisophosphindoline) palladium(ii) (red form)**
$C_{42}H_{39}Br_2P_3Pd$
K.M.Chui, H.M.Powell *J. Chem. Soc., Dalton Trans.*, 2117, 1974

86.78 **Dibromo - tris(2 - phenylisophosphindoline) palladium(ii) acetone solvate (orange form)**
$C_{42}H_{39}Br_2P_3Pd$, C_3H_6O
K.M.Chui, H.M.Powell *J. Chem. Soc., Dalton Trans.*, 2117, 1974

86.79 **Dibromo - tris(5 - ethyl - 5H - dibenzophosphole) palladium(ii) chlorobenzene solvate**
$C_{42}H_{39}Br_2P_3Pd$, C_6H_5Cl
K.M.Chui, H.M.Powell *J. Chem. Soc., Dalton Trans.*, 1879, 1974

86.80 **Dibromo - tris(5 - ethyl - 5H - dibenzophosphole) platinum(ii) bromobenzene solvate**
$C_{42}H_{39}Br_2P_3Pt$, C_6H_5Br
K.M.Chui, H.M.Powell *J. Chem. Soc., Dalton Trans.*, 1879, 1974

86.C **trans - bis(Ethoxycarbonyl) - bis(triphenylphosphine) platinum**
$C_{42}H_{40}O_4P_2Pt$
For complete entry see 71.118

86.C **1,1,2 - Trimethylallyl - bis(triphenylphosphine) nickel trichlorozincate**
$C_{42}H_{41}NiP_2^+$, Cl_3Zn^-
For complete entry see 72.37

86.81 **Bromo - (hexaphenyl - 1,4,7,10 - tetraphosphadecane) iron(ii) tetraphenylborate dichloromethane solvate**
$C_{42}H_{42}BrFeP_4^+$, $C_{24}H_{20}B^-$, CH_2Cl_2
M.Bacci, C.A.Ghilardi *Inorg. Chem.*, **13,** 2398, 1974
Residue 2 classified in 62

86.82 **Chloro - (tris(2 - diphenylphosphinoethyl)amine) nickel(ii) hexafluorophosphate**
$C_{42}H_{42}ClNNiP_3^+$, F_6P^-
M.Di Vaira, L.Sacconi *J. Chem. Soc., Dalton Trans.*, 493, 1975
Residue 1 also classified in 83

86.83 **Chloro - (bis(3 - diphenylphosphinopropyl)phenyl - phosphine) - phenyldiazo - rhodium hexafluorophosphate methylene chloride solvate**
$C_{42}H_{42}ClN_2P_3Rh^+$, F_6P^- , CH_2Cl_2
A.P.Gaughan Junior, J.A.Ibers *Inorg. Chem.*, **14,** 352, 1975
Residue 1 also classified in 83

86.84 **(tris(2 - Diphenylphosphinoethyl)amine) cobalt tetrafluoroborate**
$C_{42}H_{42}CoNP_3^+$, BF_4^-
L.Sacconi, A.Orlandini, S.Midollini *Inorg. Chem.*, **13,** 2850, 1974
Residue 1 also classified in 83

86.85 **Hydrido - tris(2 - diphenylphosphinoethyl)phosphine cobalt(i)**
$C_{42}H_{43}CoP_4$
C.A.Ghilardi, L.Sacconi *Cryst. Struct. Commun.*, **4,** 149, 1975

86.86 **Hydrido - tris(2 - diphenylphosphinoethyl)phosphine cobalt(ii) tetrafluoroborate (absolute configuration)**
$C_{42}H_{43}CoP_4^+$, BF_4^-
A.Orlandini, L.Sacconi *Cryst. Struct. Commun.*, **4,** 157, 1975

86.87 **Hydrido - (tris(diphenylphosphinoethyl)amine) nickel tetrafluoroborate**
$C_{42}H_{43}NNiP_3^+$, $C_{42}H_{42}NNiP_3^+$, $2BF_4^-$
L.Sacconi, A.Orlandini, S.Midollini *Inorg. Chem.*, **13,** 2850, 1974
Residue 1 also classified in 83

86.88 **cis - Dicarbonyl - tetrakis(diethyl - phenyl - phosphonite) technetium(i) perchlorate**
$C_{42}H_{60}O_{10}P_4Tc^+$, ClO_4^-
M.B.Cingi, D.A.Clemente, L.Magon, U.Mazzi
Inorg. Chim. Acta, **13,** 47, 1975

86.C **trans - Pentafluorophenyl - bis(triphenylphosphine) iridium(i) carbonyl**
$C_{43}H_{30}F_5IrOP_2$
For complete entry see 71.120

86.C **1,1 - bis(Triphenylphosphine) - 2,2,4,4 - tetracyano - 1 - platinacyclobutane**
$C_{43}H_{32}N_4P_2Pt$
For complete entry see 71.121

86.C **Cycloheptyne - bis(triphenylphosphine) platinum(0)**
$C_{43}H_{40}P_2Pt$
For complete entry see 75.43

86.C **Acetylacetonato - (cis - 1,trans - 3 - tetraphenyl - 4 - ethoxy - butadien - 1 - yl) - (dimethylphenylphosphine) palladium(ii)**
$C_{43}H_{43}O_3PPd$
For complete entry see 71.122

86.C **(4 - Ethoxy - tetraphenylbuta - 1 - cis,3 - trans - dienyl) - (acetylacetonato) - (dimethylphenylphosphine) palladium(ii)**
$C_{43}H_{43}O_3PPd$
For complete entry see 71.123

86.C **bis(Triphenylphosphine) - (hexafluoroacetonimido) - (N,N′ - dimethylimidazolidin - 2 - ylideno) rhodium**
$C_{44}H_{40}F_6N_3P_2Rh$
For complete entry see 71.124

86.C **1 - Ethoxycyclohex - 1,4 - diyl platinum - bis(triphenylphosphine)**
$C_{44}H_{44}OP_2Pt$
For complete entry see 71.125

86.89 **Di - μ - diethylphosphinosulfide - bis(triphenylphosphite platinum)**
$C_{44}H_{50}O_6P_4Pt_2S_2$
K.P.Wagner, R.W.Hess, P.M.Treichel, J.C.Calabrese
Inorg. Chem., **14**, 1121, 1975

86.90 **(Dicyano - tris(dimethylphenylphosphine) cobalt) - μ - cyano(dioxocyano - bis(dimethylphenylphosphine) cobalt) benzene solvate**
$C_{44}H_{55}Co_2N_4O_2P_5$, $0.5C_6H_6$
J.Halpern, B.L.Goodall, G.P.Khare, H.S.Lin, J.J.Pluth
J. Am. Chem. Soc., **97**, 2301, 1975

86.C **1,1 - bis(Pentafluorophenyl) - 2,5 - diphenyl - (3 - 4(μ - di(pentafluorophenyl) - phosphino)di - iron hexacarbonyl) - 1 - phosphiniacyclopenta - 2,5 - diene**
$C_{46}H_{10}F_{20}Fe_2O_6P_2$
For complete entry see 71.127

86.C **bis - μ - (Diphenyl - (phenylacetylenyl) - phosphine) - di(tricarbonyl iron)**
$C_{46}H_{30}Fe_2O_6P_2$
For complete entry see 72.38

86.C **Ferrocenyl - di - gold - bis(triphenylphosphine) tetrafluoroborate**
$C_{46}H_{39}Au_2FeP_2^+$, BF_4^-
For complete entry see 71.128

86.C **bis(Triphenylphosphine) - (phenyl - methyl - cyclobutene - dione) platinum(0)**
$C_{47}H_{38}O_2P_2Pt$
For complete entry see 75.44

86.C **Hydrido - chloro - (phenylazophenyl) - bis(triphenylphosphine) iridium(iii) n - hexane solvate**
$C_{48}H_{40}ClIrN_2P_2$, C_6H_{14}
For complete entry see 71.129

86.C **(tris(2 - Diphenylarsinoethyl)amine)phenyl nickel(ii) tetraphenylborate**
$C_{48}H_{47}As_3NNi^+$, $C_{24}H_{20}B^-$
For complete entry see 71.130

86.91 **Tetrachloro - di - μ - chloro - tetrakis(tri - n - butylphosphine) dirhodium(iii)**
$C_{48}H_{108}Cl_6P_4Rh_2$
J.A.Muir, M.M.Muir, A.J.Rivera
Acta Crystallogr., Sect. B, **30**, 2062, 1974

86.C **π - Cyclopentadienyl - bis(triphenylphosphine) ruthenium - phenylacetylide chloro - copper(i) acetone solvate**
$C_{49}H_{40}ClCuP_2Ru$, C_3H_6O
For complete entry see 71.131

86.C **4,4′ - Dinitro - trans - stilbene - bis(triphenylphosphine) platinum**
$C_{50}H_{40}N_2O_4P_2Pt$
For complete entry see 72.39

86.92 **bis(bis(Diphenylphosphino)methane) tetra(copper(i) bromide)**
$C_{50}H_{44}Br_4Cu_4P_4$
A.Camus, G.Nardin, L.Randaccio *Inorg. Chim. Acta,* **12,** 23, 1975

86.93 **trans - Chloro - (1,3 - di - p - tolyltriazenido) - bis(triphenylphosphine) palladium(ii)**
$C_{50}H_{44}ClN_3P_2Pd$
G.Bombieri, A.Immirzi, L.Toniolo
Acta Crystallogr., Sect. A, **31,** S141, 1975
Also classified in 83

86.94 **cis - Chloro - (1,3 - di - p - tolyltriazenido) - bis(triphenylphosphine) platinum(ii)**
$C_{50}H_{44}ClN_3P_2Pt$
G.Bombieri, A.Immirzi, L.Toniolo
Acta Crystallogr., Sect. A, **31,** S141, 1975
Also classified in 83

86.95 **bis(bis(Diphenylphosphino)methane) tetra(copper(i) chloride) acetone solvate**
$C_{50}H_{44}Cl_4Cu_4P_4$, $2C_3H_6O$
G.Nardin, L.Randaccio *Cryst. Struct. Commun.,* **3,** 607, 1974

86.96 **bis(bis(Diphenylphosphino)methane) tetra(copper(i) iodide)**
$C_{50}H_{44}Cu_4I_4P_4$
A.Camus, G.Nardin, L.Randaccio *Inorg. Chim. Acta,* **12,** 23, 1975

86.97 **Di - μ - chloro - bis((bis(diphenylphosphino)ethane)dichloro - rhenium) acetonitrile solvate**
$C_{52}H_{48}Cl_6P_4Re_2$, $2C_2H_3N$
J.A.Jaecker, W.R.Robinson, R.A.Walton
J. Chem. Soc., Dalton Trans., 698, 1975

86.98 **trans - bis(Dinitrogen) - bis(1,2 - bis(diphenylphosphino)ethane) molybdenum(0)**
$C_{52}H_{48}MoN_4P_4$
T.Uchida, Y.Uchida, M.Hidai, T.Kodama
Acta Crystallogr., Sect. B, **31,** 1197, 1975

86.99 **tetrakis(Methyldiphenylphosphine) iridium(i) tetrafluoroborate cyclohexane solvate**
$C_{52}H_{52}IrP_4^+$, BF_4^- , C_6H_{12}
G.R.Clark, B.W.Skelton, T.N.Waters *J. Organomet. Chem.,* **85,** 375, 1975

86.100 Di - μ - carbonato - bis(carbonyl - tris(dimethylphenylphosphine) molybdenum)
$C_{52}H_{66}Mo_2O_8P_6$
J.Chatt, M.Kubota, G.J.Leigh, F.C.March, R.Mason, D.J.Yarrow
J. Chem. Soc., Chem. Commun., 1033, 1974

86.101 Iodo - methyldiazenido - di((bis - 1,2 - diphenylphosphino)ethane) molybdenum(0)
$C_{53}H_{51}IMoN_2P_4$
S.D.A.Iske, T.A.George, V.W.Day
Am. Cryst. Assoc., Abstr. Papers (Spring Meeting), 34, 1975
Also classified in 83

86.102 Hydrido - tris(2 - diphenylphosphinophenyl)phosphine cobalt(i)
$C_{54}H_{43}CoP_4$
A.Orlandini, L.Sacconi *Cryst. Struct. Commun.*, **4,** 107, 1975

86.C bis(o - Diphenylphosphino - phenyl) - (triphenylphosphine) iridium(iii) hydride
$C_{54}H_{44}IrP_3$
For complete entry see 71.132

86.103 mer - Trihydrido - tris(triphenylphosphine) iridium(iii) benzene solvate
$C_{54}H_{48}IrP_3$, $0.5C_6H_6$
G.R.Clark, B.W.Skelton, T.N.Waters *Inorg. Chim. Acta*, **12,** 235, 1975

86.C (Tricarbonyl - triphenylphosphine rhenium - bis(pentafluorophenylacetylide)) - triphenylphosphine copper
$C_{55}H_{30}CuF_{10}O_3P_2Re$
For complete entry see 71.133

86.104 Tricarbonyl manganese bis - μ - (bis(diphenylphosphino)methane) - dicarbonyl manganese methylene dichloride n - hexane solvate
$C_{55}H_{44}Mn_2O_5P_4$, CH_2Cl_2 , C_6H_{14}
R.Colton, C.J.Commons, B.F.Hoskins
J. Chem. Soc., Chem. Commun., 363, 1975

86.C Hydrido - formato - tris(triphenylphosphine) ruthenium(ii) (monoclinic form)
$C_{55}H_{47}O_2P_3Ru$
For complete entry see 81.45

86.C bis(Tri - p - tolylphosphine) - (trans - stilbene) nickel(0) tetrahydrofuran solvate
$C_{56}H_{54}NiP_2$, $0.5C_4H_8O$
For complete entry see 72.43

86.105 Di - μ - chloro - di - μ - (bis(diphenylphosphinoethyl)sulfido) disilver
$C_{56}H_{56}Ag_2Cl_2P_4S_2$
K.Aurivillius, A.Cassel, L.Falth *Chem. Scr.*, **5,** 9, 1974

86.106 Iodo - (bis(diphenylphosphinoethyl)sulfido) silver dimer
$C_{56}H_{56}Ag_2I_2P_4S_2$
A.Cassel *Acta Crystallogr., Sect. B,* **31,** 1194, 1975

86.107 Tetracarbonyliron - carbonyl - tris(triphenylphosphite) diplatinum
$C_{58}H_{45}FeO_{13}P_3Pt_2$
V.G.Albano, G.Ciani *J. Organomet. Chem.,* **66,** 311, 1974

86.108 Iodo - cyclohexyldiazenido - di((bis - 1,2 - diphenylphosphino)ethane) molybdenum(0) benzene solvate
$C_{58}H_{59}IMoN_2P_4$, $0.5C_6H_6$
S.D.A.Iske, T.A.George, V.W.Day
Am. Cryst. Assoc., Abstr. Papers (Spring Meeting), 34, 1975
Residue 1 also classified in 83

86.109 Di - μ - diphenylphosphino - bis(triphenylphosphine platinum)
$C_{60}H_{50}P_4Pt_2$
N.J.Taylor, P.C.Chieh, A.J.Carty
J. Chem. Soc., Chem. Commun., 448, 1975

86.C Benzoato - tris(triphenylphosphine) rhodium(i) benzene solvate
$C_{61}H_{50}O_2P_3Rh$, $0.5C_6H_6$
For complete entry see 81.47

86.110 Tetrachloro - molybdenum - di(μ - dinitrogen - tetrakis(phenyldimethylphosphine) - chloro - rhenium)
$C_{64}H_{88}Cl_6MoN_4P_8Re_2$
P.D.Cradwick, J.Chatt, R.H.Crabtree, R.L.Richards
J. Chem. Soc., Chem. Commun., 351, 1975

86.C Di - μ - (1 - diphenylphosphino - 2 - trifluoromethyl - acetylene) - bis(triphenylphosphine) dipalladium(0)
$C_{66}H_{50}F_6P_4Pd_2$
For complete entry see 72.45

86.111 tetrakis(Chloro - triphenylphosphine silver(i))
$C_{72}H_{60}Ag_4Cl_4P_4$
B.-K.Teo, J.C.Calabrese *J. Am. Chem. Soc.,* **97,** 1256, 1975

86.112 tetrakis(Iodo - triphenylphosphine silver(i)) methylene dichloride solvate
$C_{72}H_{60}Ag_4I_4P_4$, xCH_2Cl_2
B.-K.Teo, J.C.Calabrese *J. Am. Chem. Soc.,* **97,** 1256, 1975

86.113 Triphenylphosphine - copper(i) iodide tetramer
$C_{72}H_{60}Cu_4I_4P_4$
M.R.Churchill, B.G.DeBoer, D.J.Donovan *Inorg. Chem.,* **14,** 617, 1975

86.114 μ - Decahydrodecaborato - bis(di(triphenylphosphine) copper(i)) chloroform solvate
$C_{72}H_{70}B_{10}Cu_2P_4$, $CHCl_3$
J.T.Gill, S.J.Lippard *Inorg. Chem.,* **14,** 751, 1975

86.C **Chloro - bis(triphenylphosphine) platinum - dithiocarboxylato - (bis(triphenylphosphine) platinum) tetrafluoroborate methylene dichloride solvate**
$C_{73}H_{60}ClP_4Pt_2S_2^+$, BF_4^- , $0.2CH_2Cl_2$
For complete entry see 71.137

86.C **Tri - μ - diphenylphosphino - bis(triphenylphosphine platinum) - (phenyl - platinum) benzene solvate**
$C_{78}H_{65}P_5Pt_3$, C_6H_6
For complete entry see 71.138

86.115 **μ - Thio - bis(1,1,1 - tris(diphenylphosphinomethyl)ethane) dinickel(ii) tetraphenylborate**
$C_{82}H_{78}Ni_2P_6S^{2+}$, $2C_{24}H_{20}B^-$
P.Dapporto, C.Mealli, S.Midollini, L.Sacconi, P.Stoppioni
Acta Crystallogr., Sect. A, **31,** S138, 1975
Residue 2 classified in 62

86.116 **μ - (trans - 1,1,2,3,4,4 - Hexacyanobutene - di - ido) - bis((carbonyl) - bis(triphenylphosphine) rhodium)**
$C_{84}H_{60}N_6O_2P_4Rh_2$
R.Schlodder, J.A.Ibers *Inorg. Chem.,* **13,** 2870, 1974
Also classified in 83

86.117 **μ - Iodo - bis(tris(2 - diphenylarsinoethyl)amine) dinickel(i) tetraphenylborate**
$C_{84}H_{84}As_6IN_2Ni_2^+$, $C_{24}H_{20}B^-$
P.Dapporto, C.Mealli, S.Midollini, L.Sacconi, P.Stoppioni
Acta Crystallogr., Sect. A, **31,** S138, 1975
Residue 2 classified in 62

86.C **Di(triphenylphosphine - silver) - (triphenylphosphine - rhodium - pentakis(pentafluorophenylacetylide))**
$C_{94}H_{45}Ag_2F_{25}P_3Rh$
For complete entry see 71.139

86.118 **tris(bis(Triphenylphosphine) silver(i)) tris(dithio - oxalato) aluminium(iii)**
$C_{114}H_{90}Ag_3AlO_6P_6S_6$
F.J.Hollander, D.Coucouvanis *Inorg. Chem.,* **13,** 2381, 1974
Also classified in 85, 68

86.C **tris(bis(Triphenylphosphine) silver(i)) tris(dithio - oxalato) iron(iii)**
$C_{114}H_{90}Ag_3FeO_6P_6S_6$
For complete entry see 81.48

FORMULA INDEX

C_1

$CBrCl_3Hg$ 71.1 **2**
$CBrN_2O_4^-$, K^+ 12.1 **3**
CBr_4 , C_8H_{10} 60.1 **2**
CCl_4 5.1 **5**
$CF_3O_3S^-$, H_3O^+ 11.1+ **6**
CF_4 5.2 **5**
CIN , C_5H_5N 60.65 **2**
CN_2^{2-} , Sr^{2+} 7.1 **1**
$CN_3O_6^-$, Cs^+ 10.1 **1**
$CN_3O_6^-$, I^+ 10.2 **1**
$CN_3O_6^-$, K^+ 10.3 **1**
$CN_3O_6^-$, K^+ 12.1 **5**
$CN_3O_6^-$, Rb^+ 10.4 **1**
$CN_3O_6^-$, $H_5N_2^+$ 12.2 **3**
CN_9^+ , Cl_6Sb^- 12.2 **5**
CO_5P^{3-} , $3Na^+$, $6H_2O$ 64.1+ **4**
$CHBrCl_2$, $C_4H_{10}O$ 60.61 **2**
$CHBrN_4$ 32.1 **6**
$CHBr_3$ 5.1 **4**
$2CHBr_3$, $C_6H_{12}N_4$ 60.1 **4**
$CHCl_2O_2^-$, $CH_2Cl_2O_2$, K^+ 2.1 **6**
$CHCl_3$ 5.1 **1**
$CHCl_3$, $C_{16}H_{14}CuN_2O_2$ 60.1 **3**
CHI_2NO 1.1 **7**
CHI_3 5.2 **1**
CHI_3 5.1 **7**
CHI_3 , $3S_8$ 60.2 **2**
CHI_3 , $C_4H_8O_2$ 60.2 **3**
CHI_3 , $C_4H_8S_2$ 60.53 **2**
CHI_3 , $C_6H_{12}N_4$ 60.20 **3**
CHI_3 , $3C_9H_7N$ 60.3 **2**
$2CHI_3$, $C_4H_8Se_2$ 60.4 **2**
$CHN_2O_4^-$, Rb^+ 12.1 **1**
CHN_4^- , Na^+ , H_2O 32.1 **1**
CHO_2^- , Li^+ , H_2O 2.1+ **5**
CHO_2^- , Na^+ 2.1 **1**
CHO_2^- , H_4N^+ 2.2 **1**
CHO_2^- , CH_2O_2 , K^+ 2.48 **1**
CHO_2^- , $C_{18}H_{36}N_2O_6Tl^+$, H_2O 68.24 **5**
$2CHO_2^-$, Ba^{2+} 2.3 **1**
$2CHO_2^-$, Ca^{2+} 2.4 **1**
$2CHO_2^-$, Mg^{2+} , $2H_2O$ 2.5 **1**
$2CHO_2^-$, Sr^{2+} 2.6 **1**
$2CHO_2^-$, Sr^{2+} , $2H_2O$ 2.8 **1**
$2CHO_2^-$, Sr^{2+} , $2H_2O$ 2.1 **3**
$2CHO_2^-$, Sr^{2+} , $2H_2O$ 2.1 **4**
$6CHO_2^-$, $16CH_4N_2S$, $3Pb^{2+}$ 8.2+ **4**
CHS_2^- , K^+ 2.2 **6**
CH_2Br_2 5.1 **6**
CH_2Cl_2 5.2 **6**
$CH_2Cl_2O_2$, $CHCl_2O_2^-$, K^+ 2.1+ **6**
$CH_2Cl_4O_2P_2$ 64.1 **7**
CH_2I_2 5.3 **6**
$CH_2NO_2^-$, H_4N^+ 2.3 **6**
$CH_2NS_2^-$, H_4N^+ 11.1 **1**
CH_2N_4 32.2 **6**
$CH_2N_4O_4^{2-}$, $2K^+$ 10.6 **1**
$CH_2N_5^-$, $H_5N_2^+$ 32.2 **1**
CH_2O_2 1.1 **1**
CH_2O_2 , CHO_2^- , K^+ 2.48 **1**
CH_2O_2 , CH_3NO 60.3 **3**
$CH_2O_6S_2^{2-}$, $2K^+$ 11.2 **1**
CH_2S_3 11.3 **1**
$CH_3AlCl_3^-$, K^+ 68.1 **7**
CH_3AsI_2 65.1 **2**
CH_3Br 5.4 **6**
CH_3Cl 5.3 **1**
$CH_3CuN_3O_8$ 84.1 **2**
CH_3HgN_3 71.1 **7**
CH_3I 5.5 **6**
CH_3NO 1.2 **1**
CH_3NO , CH_2O_2 60.3 **3**
$CH_3N_2O^-$, K^+ 9.1 **1**
$CH_3N_2S_2^-$, $H_5N_2^+$ 9.2 **1**
$CH_3N_3O_9Sn$ 69.1 **4**
CH_3N_5 , H_2O 32.3 **1**
$CH_3O_2S_2^-$, Na^+ , H_2O 11.4 **1**
$CH_3O_3S^-$, Cs^+ 11.5 **1**
$CH_3O_3S^-$, Na^+ , $2H_2O$ 11.6 **1**
$CH_3O_3S^-$, $C_8H_{18}NO^+$ 32.23 **3**
$CH_3O_3S^-$, $C_{18}H_{30}NO_2^+$ 20.18 **7**
$2CH_3O_3S^-$, $C_{11}H_{26}N_2O^{2+}$, H_2O 38.10 **3**
$CH_3O_4P^{2-}$, $2H_4N^+$, $2H_2O$ 46.1 **4**
$CH_3O_4S^-$, $C_{10}H_{15}N_3O_5^+$, H_2O 47.22 **6**
$CH_3O_4S^-$, $C_{13}H_{19}N_2O_3^+$ 35.18 **7**
$CH_3O_4S^-$, $C_{18}H_{23}N_2S^+$ 41.45 **5**
CD_4 5.8 **5**
CD_4 5.2+ **7**
$CH_4BiCl_5N_2S^{2-}$, $C_6H_{24}Bi_2Cl_4N_{12}S_6^{2+}$ 66.4 **7**
CH_4BrN_2PbS , $2H_2O$ 69.1 **5**
$CH_4Br_2N_2PbS$ 69.2 **5**
$(CH_4Cl_2HgN_2S)_{2n}$, nCl_2Hg 79.1 **6**
CH_4N_2O 8.1+ **1**
CH_4N_2O 10.7 **1**
CH_4N_2O , Ca^{2+} , $2NO_3^-$, $3H_2O$ 67.1 **5**
CH_4N_2O , Na^+ , Cl^- , H_2O 60.5 **2**
CH_4N_2O , H_2O_2 60.8 **2**
CH_4N_2O , H_3O_4P 8.6 **1**
CH_4N_2O , H_3O_4P 8.1 **5**
CH_4N_2O , H_3O_4P 8.2 **5**
CH_4N_2O , H_4N^+ , Br^- 60.6 **2**
CH_4N_2O , H_4N^+ , Cl^- 60.7 **2**
CH_4N_2O , $C_3H_2N_2O_3$ 60.4 **3**

CH_4N_2O , $C_4H_8Cl_2$ 61.1 **5**
CH_4N_2O , $C_5H_4N_2O_4$ 60.2 **4**
CH_4N_2O , $C_5H_{12}NO_2S^+$, Cl^- 60.68 **2**
CH_4N_2O , $C_6H_6O_2$ 60.5 **7**
CH_4N_2O , $C_6H_{12}O_6$ 60.14 **4**
CH_4N_2O , C_7H_9N 60.100 **2**
CH_4N_2O , $C_8H_{12}N_2O_3$ 60.5 **6**
CH_4N_2O , $C_{18}H_{24}O_2$ 60.38 **4**
$2CH_4N_2O$, $C_2H_2O_4$ 60.3 **4**
$2CH_4N_2O$, $C_2H_{12}CaN_4O_{10}P_2$ 67.1 **6**
$3CH_4N_2O$, $C_{12}H_{28}N^+$, Br^- , H_2O 61.7 **6**
$4CH_4N_2O$, Ca^{2+} , $2NO_3^-$ 8.1 **7**
$4CH_4N_2O$, $C_6H_{24}CoN_{12}O_6^{2+}$, $2NO_3^-$ 79.12 **7**
$4CH_4N_2O$, $C_6H_{24}N_{12}NiO_6^{2+}$, $2I^-$ 61.1 **7**
$6CH_4N_2O$, Al^{3+} , $3ClO_4^-$ 8.2 **7**
$6CH_4N_2O$, Ca^{2+} , $2Br^-$ 8.3 **7**
$CH_4N_2O_2$ 8.7+ **1**
$CH_4N_2O_2$ 10.8 **1**
$CH_4N_2O_2$ 8.1 **4**
$CH_4N_2O_2S$ 8.9 **1**
CH_4N_2S 8.11+ **1**
CH_4N_2S , $C_3H_2N_2O_3$ 60.5 **3**
CH_4N_2S , $C_3H_{10}AsN_2S^+$, Cl^- 65.1 **7**
CD_4N_2S 8.14+ **1**
$2CH_4N_2S$, C_2HgN_2 8.3 **5**
$2CH_4N_2S$, $C_5H_6N^+$, Br^- 60.6 **4**
$2CH_4N_2S$, $C_{15}H_{21}CrO_6$ 60.24 **6**
$3CH_4N_2S$, $C_4H_{16}Cl_2CoN_4^+$, Cl^- 76.6 **7**
$4CH_4N_2S$, Cs^+ , Cl^- , H_2O 60.9 **2**
$4CH_4N_2S$, Cs^+ , F^- , $2H_2O$ 60.10 **2**
$4CH_4N_2S$, Tl^+ , ClO_3^- 60.6 **3**
$4CH_4N_2S$, Tl^+ , $H_2O_4P^-$ 60.11 **2**
$4CH_4N_2S$, $C_7H_5O_2^-$, Tl^+ 60.98 **2**
$6CH_4N_2S$, Pb^{2+} , $2ClO_4^-$, $2H_2O$ 8.16 **1**
$16CH_4N_2S$, $6CHO_2^-$, $3Pb^{2+}$ 8.2 **4**
CH_4N_2Se 8.2+ **3**
CH_4N_2Se , $C_2H_8N_4Se_2^{2+}$, $2Br^-$, H_2O 8.19 **7**
CH_4N_2Se , $C_2H_8N_4Se_2^{2+}$, $2Cl^-$, H_2O 8.20 **7**
$CH_4N_4NiS_4$ 85.1 **2**
$CH_4N_4O_2$ 8.17 **1**
$CH_4N_5^+$, Cl^- 9.1 **4**
CH_4O 5.4 **1**
CH_4O , C_2HgN_2 60.7 **3**
CH_4O , $C_{10}H_6BrNO_2$ 60.32 **3**
CH_4O , $C_{36}H_{60}O_{30}$ 61.8 **5**
$2CH_4O$, Br_2 60.12 **2**
$2CH_4O$, H_4N_2 60.13 **2**
$3CH_4O$, Na^+ , I^- 60.14 **2**
$4CH_4O$, Ca^{2+} , $2NO_3^-$ 5.4 **7**
$4CH_4O$, H_4N_2 60.15 **2**
CH_4O_2S 11.7 **1**
CH_5AgClN_3S 85.2 **2**
$(CH_5Cl_2CuN_3O)_n$ 83.1+ **4**
$CH_5Cl_2N_3SZn$ 85.3 **2**
CH_5N 3.1 **1**
CH_5NO_2S 4.1 **1**
$CH_5N_2O^+$, $CdCl_3^-$ 8.18 **1**
$CH_5N_2O^+$, NO_3^- 8.19+ **1**
$CH_5N_2S^+$, NO_3^- 8.22+ **1**
$2CH_5N_3$, $C_8H_{12}Br_2O_8Rh_2$ 81.56 **2**
$2CH_5N_3$, $C_8H_{12}Cl_2O_8Rh_2$ 81.58 **2**
CH_5N_3S 8.4+ **3**
CH_6N^+ , Br^- 3.2 **1**
CH_6N^+ , Cl^- 3.3 **1**
CH_6N^+ , ClO_4^- 3.4 **1**
CH_6N^+ , Cl_3Ni^- 3.5 **1**
CH_6N^+ , $H_{12}AlO_6^{3+}$, $2O_4S^{2-}$, $6H_2O$ 3.6 **1**
CH_6N^+ , $C_2H_7BNO^-$ 62.6 **2**
CH_6N^+ , $C_{12}H_{18}CuN_2O_2$, ClO_4^- 60.39 **3**
$2CH_6N^+$, $Br_2Cl_2Cu^{2-}$ 3.1 **3**
$2CH_6N^+$, Cl_4Mn^{2-} 3.1 **7**
CH_6N^{2+} , Br_4Fe^- , Br^- 3.1 **5**
CH_6NO^+ , Cl^- 10.9 **1**
CH_6NO^+ , Cl^- 10.24 **1**
CH_6NO_3P 64.2 **7**
$CH_6N_3^+$, Cl^- 8.25 **1**
$CH_6N_3^+$, $H_{12}AlO_6^{3+}$, $2O_4S^{2-}$ 8.26 **1**
$CH_6N_3^+$, $H_{12}CrO_6^{3+}$, $2O_4S^{2-}$ 8.27 **1**
$CH_6N_3^+$, $C_8H_{11}N_2O_3^-$, $2H_2O$ 43.1 **6**
$CH_6N_3^+$, $C_{10}H_{14}FeN_2O_9^-$, $2H_2O$ 76.34 **5**
$CH_6N_3^+$, $C_{10}H_{18}N_2NdO_{11}^-$ 76.37 **5**
$CH_6N_3^+$, $C_{12}H_4N_7O_{12}^-$ 8.4 **7**
$0.5CH_6N_3^+$, $C_5H_{14}N_3O_3P$, $0.5Cl^-$ 8.9 **4**
$2CH_6N_3^+$, $Cr_2O_7^{2-}$ 8.5 **7**
$2CH_6N_3^+$, $Cr_3O_{10}^{2-}$ 8.6 **7**
$2CH_6N_3^+$, Ni^{2+} , $2O_4S^{2-}$, $6H_2O$ 8.3 **4**
$2CH_6N_3^+$, Zn^{2+} , $2O_4S^{2-}$ 8.6 **3**
$2CH_6N_3^+$, CO_3^{2-} 8.7 **7**
$2CH_6N_3^+$, $C_2H_7AsMo_4O_{15}^{2-}$, H_2O 86.1 **7**
$2CH_6N_3^+$, $C_{16}H_{28}Ce_2O_{18}^{2-}$, $2H_2O$ 81.43 **4**
$3CH_6N_3^+$, C_4H_9NO , $3Cl^-$ 60.16 **2**
$8CH_6N_3^+$, $C_6Ce_2O_{22}^{8-}$, $2H_2O$ 8.8 **7**
$CH_6N_3O^+$, $Cl-$ 9.3 **1**
$2CH_6N_3O^+$, O_4S^{2-} , H_2O 8.9 **7**
CH_6N_4O 8.4 **5**
CD_6N_4O 8.1 **6**
CH_6N_4S 8.28 **1**
CH_6N_4S , $CH_7N_4S^+$, Cl^- 8.5 **5**
$CH_6O_6P_2$ 64.1 **6**
$CH_7N_4^+$, Cl^- 8.29 **1**
$(CH_8CdN_2O_6S_2)_n$ 79.2 **2**
$CH_8N_4O^{2+}$, O_4S^{2-} 8.2 **6**
$CH_8N_4S^{2+}$, $2Cl^-$, $2H_2O$ 8.4 **4**
$CH_8N_4S^{2+}$, O_4S^{2-} 8.6 **5**
$CH_9N_6^+$, Cl^- 8.30 **1**
$CH_{32}Co_2N_{11}O^{5+}$, $5Cl^-$, H_2O 84.1 **7**

C_2

C_2^{2-} , $2Na^+$ 12.1 **7**
$C_2Ag_2O_4$ 81.2 **2**
$C_2Br_2O_2$ 1.3 **1**
$C_2Br_2O_2$, $C_4H_8O_2$ 60.17 **2**
C_2Br_4 5.5 **1**
C_2Br_4 , $C_4H_4N_2$ 60.38 **2**
C_2Br_6 5.6 **1**
C_2Br_6 5.2 **4**
$C_2Cl_2O_2$ 1.4 **1**
$C_2Cl_2O_2$, $C_4H_8O_2$ 60.18 **2**
$C_2Cl_3O_2^-$, $C_2HCl_3O_2$, K^+ 2.1 **7**
$C_2Cl_3O_2^-$, $C_2HCl_3O_2$, Rb^+ 2.4 **6**
$C_2Cl_4S_2$ 39.1 **1**
C_2Cl_6 5.7 **1**
$C_2Cl_6S_3$ 11.8 **1**
$C_2Cl_8O_4Sn_2^{2-}$, $2C_8H_{20}N^+$ 69.1 **7**
$C_2F_3O_2^-$, H_4N^+ 2.9 **1**

$C_2F_3O_2^-$, $C_2HF_3O_2$, Cs^+ 2.10 **1**
$C_2F_3O_2^-$, $C_2HF_3O_2$, K^+ 2.11 **1**
$C_2F_3O_2^-$, $C_2HF_3O_2$, K^+ 2.2 **4**
$C_2F_3O_2^-$, $C_2DF_3O_2$, K^+ 2.3 **4**
C_2I_2 5.9 **5**
C_2I_2 , $C_4H_8O_2$ 60.50 **2**
C_2I_2 , $C_4H_8S_2$ 60.19 **2**
C_2I_2 , $C_4H_8Se_2$ 60.20 **2**
C_2I_2 , $C_6H_8O_2$ 60.92 **2**
C_2I_4 5.8 **1**
C_2I_4 , $C_4H_4N_2$ 60.39 **2**
C_2I_4 , $C_4H_8Se_2$ 60.58 **2**
$C_2MoO_7^{2-}$, H_4N^+ , Na^+ , $2H_2O$ 81.1 **2**
$C_2MoO_9^{2-}$, $2K^+$ 81.1 **3**
$C_2NS_2^-$, $C_8H_{20}N^+$ 3.19 **6**
C_2N_2 7.2 **1**
$C_2N_3O_4^-$, K^+ 7.1 **3**
$C_2N_3O_4^-$, K^+ 7.1 **5**
$C_2N_3O_4^-$, Rb^+ 7.3 **1**
$C_2N_3O_4^-$, Rb^+ 7.2 **5**
$C_2N_4O_8^{2-}$, $2K^+$ 10.10+ **1**
$C_2NiS_6^{2-}$, $2C_{24}H_{20}As^+$ 65.46 **2**
$C_2O_4^{2-}$, Ba^{2+} , H_2O 2.5 **6**
$C_2O_4^{2-}$, Ca^{2+} , H_2O 2.12 **1**
$C_2O_4^{2-}$, Ca^{2+} , $2H_2O$ 2.13 **1**
$C_2O_4^{2-}$, $2K^+$, H_2O 2.15+ **1**
$C_2O_4^{2-}$, $2K^+$, H_2O 2.2 **3**
$C_2O_4^{2-}$, $2K^+$, H_2O_2 2.17 **1**
$C_2O_4^{2-}$, $2Li^+$ 2.18 **1**
$C_2O_4^{2-}$, $2Li^+$, H_2O_2 2.3 **3**
$C_2O_4^{2-}$, $2Na^+$, H_2O 2.3 **5**
$C_2O_4^{2-}$, $2Na^+$, H_2O_2 2.19 **1**
$C_2O_4^{2-}$, $2Rb^+$, H_2O 2.20 **1**
$C_2O_4^{2-}$, $2Rb^+$, H_2O_2 2.21 **1**
$C_2O_4^{2-}$, Sr^{2+} , $2.17H_2O$ 2.22 **1**
$C_2O_4^{2-}$, $H_4BaO_2^{2+}$ 2.9 **4**
$C_2O_4^{2-}$, $2H_4N^+$, H_2O 2.23+ **1**
$C_2O_4^{2-}$, $2H_4N^+$, H_2O 2.4+ **5**
$C_2O_4^{2-}$, $2H_4N^+$, H_2O_2 2.4 **4**
$C_2O_4^{2-}$, $2D_4N^+$, D_2O 2.8+ **5**
$C_2O_4^{2-}$, $C_2H_2O_4$, Ba^{2+} , $2H_2O$ 2.5 **3**
$C_2O_9V^{3-}$, $3K^+$, H_2O 81.1 **7**
$C_2HCl_3O_2$ 1.5 **1**
$C_2HCl_3O_2$, $C_2Cl_3O_2^-$, K^+ 2.1 **7**
$C_2HCl_3O_2$, $C_2Cl_3O_2^-$, Rb^+ 2.4 **6**
$C_2HCl_3O_2$, C_5H_5NO 33.11 **7**
$C_2HF_3O_2$, $C_2F_3O_2^-$, Cs^+ 2.10 **1**
$C_2HF_3O_2$, $C_2F_3O_2^-$, K^+ 2.11 **1**
$C_2HF_3O_2$, $C_2F_3O_2^-$, K^+ 2.2 **4**
$C_2DF_3O_2$, $C_2F_3O_2^-$, K^+ 2.3 **4**
$C_2HNO_4^{2-}$, $2K^+$ 2.26 **1**
$C_2HO_4^-$, K^+ 2.27 **1**
$C_2HO_4^-$, K^+ 2.5 **4**
$C_2HO_4^-$, K^+ 2.10 **5**
$C_2HO_4^-$, Li^+ , H_2O 2.7 **3**
$C_2HO_4^-$, Li^+ , H_2O 2.11 **5**
$C_2HO_4^-$, Na^+ 2.28 **1**
$C_2HO_4^-$, Na^+ , H_2O 2.6 **4**
$C_2HO_4^-$, H_4N^+ , $0.5H_2O$ 2.12 **5**
$C_2HO_4^-$, $H_5N_2^+$ 2.29+ **1**
$C_2HO_4^-$, $H_5N_2^+$ 2.6 **6**
$C_2HO_4^-$, $C_2H_2O_4$, K^+ , $2H_2O$ 2.31 **1**
$C_2HO_4^-$, $C_2H_2O_4$, H_4N^+ , $2H_2O$ 2.32 **1**
$C_2HO_4^-$, $C_{12}H_{19}N_2O^+$, $0.5C_2H_2O_4$ 32.22 **7**
$C_2DO_4^-$, Li^+ , D_2O 2.13 **5**
$C_2DO_4^-$, $D_5N_2^+$ 2.7 **6**
C_2H_2 5.9 **1**
C_2D_2 5.5+ **7**
$C_2H_2B_{10}Cl_{10}$ 62.1 **2**
C_2H_2Be 67.1 **2**
$C_2H_2ClO_2^-$, $C_2H_3ClO_2$, H_4N^+ 2.7 **4**
$C_2H_2ClO_2^-$, $C_2H_3ClO_2$, H_4N^+ 2.8 **6**
$C_2H_2Cl_2Hg$ 71.2 **2**
$C_2H_2Cl_4N_2Ti$ 83.3 **4**
$C_2H_2Cl_4N_2V$ 83.1 **6**
$C_2H_2Cl_6O_4Re_2^{2-}$, $2H_4N^+$ 81.2 **7**
$(C_2H_2CuO_2)_n$, $4nH_2O$ 81.3+ **2**
$C_2H_2CuO_4$, $4H_2O$ 81.1 **6**
$(C_2H_2CuO_4)_n$ 81.23 **2**
$C_2H_2FO_2^-$, Na^+ 2.2 **7**
$(C_2H_2FeO_2S)_n$, nH_2O 81.2 **5**
$C_2H_2NO_3^-$, H_4N^+ 2.33 **1**
$C_2H_2N_2OS_2$ 41.1 **1**
$C_2H_2N_2S$ 41.1 **6**
$C_2H_2N_2S_3$ 41.2+ **1**
$C_2H_2N_3O^-$, K^+ 8.10 **7**
$C_2H_2N_3O_5^-$, Rb^+ 12.3 **3**
$C_2H_2N_4$ 33.1 **1**
$C_2H_2O_4$ 1.6+ **1**
$C_2H_2O_4$ 1.2+ **7**
$C_2H_2O_4$, $2H_2O$ 1.9+ **1**
$C_2H_2O_4$, $2CH_4N_2O$ 60.3 **4**
$C_2H_2O_4$, $C_2O_4^{2-}$, Ba^{2+} , $2H_2O$ 2.5 **3**
$C_2H_2O_4$, $C_2HO_4^-$, K^+ , $2H_2O$ 2.31 **1**
$C_2H_2O_4$, $C_2HO_4^-$, H_4N^+ , $2H_2O$ 2.32 **1**
$C_2H_2O_4$, C_2H_5NO 60.1 **5**
$C_2H_2O_4$, $2C_5H_5NO_2$ 60.3 **5**
$C_2D_2O_4$, $2D_2O$ 1.16+ **1**
$C_2H_2O_4S$ 11.9 **1**
$(C_2H_2O_7U)_n$, $2nH_2O$ 81.3 **5**
$C_2H_3BCl_3N$ 7.2 **3**
$C_2H_3BF_3N$ 7.4 **1**
$C_2H_3BF_3N$ 7.3 **3**
$(C_2H_3BrCuN)_n$ 83.4 **4**
$C_2H_3Br_4NORe^-$, $C_{24}H_{20}As^+$ 83.1 **2**
$(C_2H_3ClCuN)_n$ 83.2 **2**
$C_2H_3ClHgO_2$ 71.3 **2**
$C_2H_3ClO_2$, $C_2H_2ClO_2^-$, H_4N^+ 2.7 **4**
$C_2H_3ClO_2$, $C_2H_2ClO_2^-$, H_4N^+ 2.8 **6**
$(C_2H_3Cl_2CuN_3)_n$ 83.3 **2**
$C_2H_3Cl_3$ 5.10+ **5**
$C_2H_3Cl_3O_2$ 5.10 **1**
$C_2H_3Cl_3O_2$ 5.12 **5**
$(C_2H_3CuNO_4)_n$ 81.1 **4**
$C_2H_3FO_2$ 1.1 **4**
$C_2H_3F_2NO$ 1.1 **5**
C_2H_3HgN 71.4+ **2**
C_2H_3N , Cl_5Sb 60.21 **2**
$2C_2H_3N$, Br_2 60.22 **2**
$C_2H_3N_2O_4^-$, K^+ 12.2 **7**
$C_2H_3N_3$ 32.4+ **1**
$C_2H_3N_3S_2$ 41.1 **4**
$C_2H_3N_4O_{10}V$ 83.1 **3**
$C_2H_3N_4S^-$, Na^+ , $3H_2O$ 32.1 **5**
$C_2H_3O^+$, $AlCl_4^-$ 12.1 **4**
$C_2H_3O^+$, Cl_6Sb^- 12.2 **4**
$C_2H_3O^+$, F_6Sb^- 12.3 **1**
$C_2H_3OS^-$, K^+ 1.4 **7**
$C_2H_3OS^-$, $C_2H_3O_2^-$, Sr^{2+} , $4H_2O$ 2.3 **7**
$C_2H_3O_2^-$, Li^+ , $2H_2O$ 2.34+ **1**
$C_2H_3O_2^-$, Li^+ , $2H_2O$ 2.9 **3**
$C_2H_3O_2^-$, H_4N^+ 2.36 **1**

$C_2H_3O_2^-$, H_4N^+ , H_3N	2.37 + **1**
$C_2H_3O_2^-$, H_4N^+ , $2H_3N$	2.39 + **1**
$C_2H_3O_2^-$, $H_5N_2^+$	2.10 **3**
$C_2H_3O_2^-$, $C_2H_3OS^-$, Sr^{2+} , $4H_2O$	2.3 **7**
$C_2H_3O_2^-$, $C_2H_4O_2$, K^+	2.8 **4**
$C_2H_3O_2^-$, $C_2H_4O_2$, Na^+	2.42 **1**
$C_2H_3O_2^-$, $C_2H_4O_2$, Na^+	2.4 **7**
$C_2H_3O_2^-$, $C_2H_4O_2$, H_4N^+	2.11 **3**
$C_2H_3O_3^-$, Li^+ , H_2O	2.43 **1**
$C_2H_3O_3^-$, $C_2H_4O_3$, K^+	2.44 **1**
$C_2H_3O_3^-$, $C_2H_4O_3$, Rb^+	2.45 **1**
$C_2H_2DO_3^-$, Li^+	2.46 + **1**
$C_2H_4AgNO_2$	48.1 **1**
$(C_2H_4AgNO_2)_n$	82.1 **4**
$(C_2H_4AgNO_2)_{2n}$, nH_2O	82.2 **4**
$C_2H_4B_2Cl_4$	62.2 **2**
$C_2H_4B_{10}Cl_8$	62.3 + **2**
C_2H_4BrNO	1.1 **3**
$C_2H_4Br_3Pt^-$, K^+ , H_2O	72.1 **2**
$(C_2H_4CdCl_2N_4)_n$	83.5 **4**
C_2H_4ClNO	1.21 + **1**
$C_2H_4ClPS_3$	64.1 **3**
$C_2H_4ClS_2Sb$	66.1 **2**
$C_2H_4Cl_2$	5.11 + **1**
$C_2H_4Cl_2N_6$	9.4 **1**
$C_2H_4Cl_3Pt^-$, K^+ , H_2O	72.1 **3**
$C_2H_4Cl_3Pt^-$, K^+ , H_2O	72.1 **4**
$C_2H_4Cl_3Pt^-$, K^+ , H_2O	72.2 **4**
$C_2H_4Cl_4Sn$	69.3 **5**
$(C_2H_4Cl_4Te)_n$	70.1 **4**
C_2H_4FNO	1.24 **1**
$C_2H_4FeO_6$	81.2 + **3**
$C_2H_4N_2NiS_4$	80.1 **2**
$C_2H_4N_2O_2$	1.25 **1**
$C_2H_4N_2O_2$	9.5 **1**
$C_2H_4N_2O_2$	10.13 **1**
$C_2H_4N_2O_2$	9.1 + **7**
$C_2H_4N_2O_3^-$, $C_2H_5N_2O_3$, K^+	8.11 **7**
$C_2H_4N_2S_2$	4.2 **1**
$C_2H_4N_3S_2^+$, Br^-	41.4 **1**
$C_2H_4N_3S_2^+$, Cl^- , $0.5H_2O$	41.5 **1**
$C_2H_4N_3S_2^+$, I^-	41.6 **1**
$C_2H_4N_3S_2^+$, I^-	41.2 **4**
$C_2H_4N_4$	7.5 + **1**
$C_2H_4N_4$	8.7 **3**
$C_2H_4N_4$, $2C_7H_{16}N^+$, O_4S^{2-}	33.28 **7**
$C_2D_4N_4$	7.7 **1**
$C_2H_4N_4O_2$	9.6 **1**
$C_2H_4N_4O_2$	8.12 **7**
$C_2H_4N_4O_4^{2-}$, $2Na^+$	9.7 **1**
C_2H_4O , $7.2H_2O$	61.1 **2**
$C_2H_4O_2$	1.26 **1**
$C_2H_4O_2$	1.2 + **3**
$C_2H_4O_2$, H_3O_4P	1.2 + **5**
$C_2H_4O_2$, $C_2H_3O_2^-$, K^+	2.8 **4**
$C_2H_4O_2$, $C_2H_3O_2^-$, Na^+	2.42 **1**
$C_2H_4O_2$, $C_2H_3O_2^-$, Na^+	2.4 **7**
$C_2H_4O_2$, $C_2H_3O_2^-$, H_4N^+	2.11 **3**
$C_2H_4O_2$, $C_6H_5N_2^+$, Cl^-	60.82 **2**
$C_2H_4O_2$, $C_{24}H_{40}O_4$	60.42 **4**
$C_2H_4O_3$	1.5 + **3**
$C_2H_4O_3$, $C_2H_3O_3^-$, K^+	2.44 **1**
$C_2H_4O_3$, $C_2H_3O_3^-$, Rb^+	2.45 **1**
$C_2H_4O_3Se_2$	39.2 **1**
$C_2H_4O_4S$	42.1 **1**
$C_2H_4O_6S_2^{2-}$, $2H_3O^+$	11.1 **3**
$C_2H_4O_8^{2-}$, $2Rb^+$	12.3 **4**
$C_2H_5AgN_2O_2^+$, NO_3^-	82.1 **2**
$C_2H_5BrN^+$, Br^-	3.7 **1**
$C_2H_5ClN^+$, Cl^-	3.8 **1**
$(C_2H_5IZn)_n$	71.1 **5**
C_2H_5NO	1.28 **1**
C_2H_5NO	1.2 **4**
C_2H_5NO , $C_2H_2O_4$	60.1 **5**
C_2H_5NO , $C_8H_{12}N_2O_3$	60.6 **6**
$2C_2H_5NO$, Na^+ , Br^-	60.23 **2**
C_2H_5NOS	4.1 **4**
$C_2H_5NO_2$	48.3 + **1**
$C_2H_5NO_2$	48.1 **4**
$C_2H_5NO_2$	48.1 **6**
$C_2H_5NO_2$, $0.5H_2O$	10.1 **3**
$C_2H_5NO_2$, $C_2H_6NO_2^+$, Cl^-	48.6 **1**
$C_2H_5NO_2$, $C_2H_6NO_2^+$, I^-	48.1 **5**
$C_2H_5NO_2$, $C_2H_6NO_2^+$, NO_3^-	48.7 + **1**
$C_2H_5NO_2$, $2C_2H_6NO_2^+$, O_4S^{2-}	48.9 **1**
$C_2H_5NO_2$, $2C_2H_6NO_2^+$, O_4S^{2-}	48.3 **4**
$C_2H_5NO_2$, $2C_2H_6NO_2^+$, O_4S^{2-}	48.7 **5**
$C_2H_5NO_2$, $2C_2H_6NO_2^+$, O_4S^{2-}	48.2 **6**
$C_2H_5NO_2$, $2C_2H_6NO_2^+$, O_4S^{2-}	48.1 + **7**
$2C_2H_5NO_2$, Ba^{2+} , $2Cl^-$, H_2O	48.1 **3**
$2C_2H_5NO_2$, Mn^{2+} , $2Cl^-$	48.2 **3**
$2C_2H_5NO_2$, Na^+ , I^- , H_2O	48.4 **4**
$2C_2H_5NO_2$, Sr^{2+} , $2Cl^-$, $3H_2O$	48.3 **3**
C_2H_5NS	4.3 **1**
$C_2H_5N_3O_2$, $0.6H_2O$	8.31 **1**
$C_2D_5N_3O_2$, $0.77D_2O$	8.3 **6**
$4C_2H_5N_3O_2$, Sr^{2+} , $2ClO_4^-$	8.13 **7**
$C_2H_5N_3O_3$	8.14 **7**
$C_2H_5N_3O_3$	8.15 **7**
$C_2H_5N_3O_3$, $0.5H_2O$	8.16 **7**
$C_2H_5N_3S_2$	8.7 **5**
$C_2H_5N_5$	32.6 **1**
$C_2H_5N_5S$	32.7 **1**
$C_2H_5N_5S$	32.1 **3**
$C_2H_5O^-$, $C_8H_7N_3O_7$, Cs^+	60.103 **2**
$C_2H_5O_2^+$, FO_3S^-	12.4 **3**
$C_2H_5O_2^+$, HO_4S^-	12.4 **1**
$C_2H_5O_4P^{2-}$, $2K^+$, $4H_2O$	46.2 **4**
$C_2H_5O_4S^-$, K^+	11.10 **1**
$2C_2H_5O_4S^-$, $C_{16}H_{18}N_4^{2+}$, $2H_2O$	8.12 **4**
$2C_2H_5O_4S^-$, $C_{16}H_{18}N_4O^{2+}$, $2H_2O$	8.13 **4**
$3C_2H_5O_4S^-$, $H_{18}ErO_9^{3+}$	11.11 **1**
$3C_2H_5O_4S^-$, $H_{18}HoO_9^{3+}$	11.1 **7**
$3C_2H_5O_4S^-$, $H_{18}O_9Pr^{3+}$	11.12 **1**
$3C_2H_5O_4S^-$, $H_{18}O_9Y^{3+}$	11.13 **1**
$C_2H_6Al_2Cl_4$	68.1 **2**
$C_2H_6As_2$	65.1 **3**
$C_2H_6BF_3OS$	62.5 **2**
$(C_2H_6CaN_4O_4)_n$	67.1 **4**
$(C_2H_6CdCl_2N_2O)_n$	84.2 **2**
$C_2H_6CdCl_2N_2O_2$	84.3 **2**
$C_2H_6CdN_4O_4$	83.4 **2**
$C_2H_6CdN_4O_4$, H_2O	83.5 **2**
$(C_2H_6Cl_2CuN_2O)_n$	84.4 **2**
$(C_2H_6Cl_2CuOS)_n$	84.1 **3**
$C_2H_6Cl_2CuO_2$	84.2 **7**
$(C_2H_6Cl_2Cu_2N_2)_n$	83.6 **2**
$C_2H_6Cl_2FeN_2O_2$	84.1 **5**
$C_2H_6Cl_2N_2P_2S_2$	64.1 **2**
$C_2H_6Cl_2Sn$	69.1 + **3**
$C_2H_6Cl_2Te$	70.1 **2**
$C_2H_6Cl_3Sn^-$, $C_{17}H_{17}ClN_3Sn^+$	69.32 **2**

FORMULA INDEX

$C_2H_6Cl_6N_2P_2$ 64.2+ **2**
$C_2H_6Cl_8O_2Sb_2$ 66.1 **4**
$(C_2H_6CoO_6)_n$ 81.5 **2**
$C_2H_6CuN_2O_4$ 81.6 **2**
$(C_2H_6CuN_2O_4)_n$, $2nH_2O$ 81.4 **5**
$(C_2H_6CuO_6)_n$ 81.7+ **2**
$C_2H_6F_2NP$ 64.2 **3**
$(C_2H_6F_2O_6S_2Sn)_n$ 69.2 **4**
$C_2H_6F_2Sn$ 69.1 **2**
$C_2H_6F_6N_4P_4$ 64.3 **3**
$C_2H_6HgS_2$ 85.4 **2**
$C_2H_6I_2Te$ 70.2 **4**
$C_2H_6MnN_4O_4$, $2H_2O$ 83.7 **2**
$(C_2H_6MnO_6)_n$ 81.9+ **2**
$C_2H_6Mo_5O_{21}P_2^{4-}$, $4H_4N^+$, $5H_2O$ 64.3 **7**
$C_2H_6NO^+$, C_2H_5NO , Br^- 1.7 **3**
$C_2H_6NO_2^+$, NO_3^- 48.4 **7**
$C_2H_6NO_2^+$, H_4N^+ , O_4S^{2-} 48.5 **7**
$2C_2H_6NO_2^+$, O_4S^{2-} , H_2O 48.6 **7**
$2C_2H_6NO_2^+$, O_4Se^{2-} 48.7 **7**
$2C_2H_6NO_2^+$, $C_2H_5NO_2$, BeF_4^{2-} 48.3+ **6**
$C_2H_6N_2O_2$ 9.8 **1**
$C_2H_6N_2O_2$ 10.14 **1**
$C_2H_6N_2O_2$ 10.15+ **1**
$C_2H_6N_2O_6Sn$ 69.4 **5**
$C_2H_6N_2S^+$, $C_7H_5ClO_2^-$ 8.32 **1**
$C_2H_6N_4O_4$ 10.17 **1**
$C_2H_6N_4O_4Zn$ 81.2 **4**
$(C_2H_6N_4O_4Zn)_n$ 83.2 **3**
$C_2H_6N_4S_2$ 8.17 **7**
$C_2H_6N_6S$ 32.2 **3**
$C_2H_6N_{10}$, H_2O 32.3+ **3**
$(C_2H_6NiO_6)_n$ 81.11 **2**
C_2H_6O , $C_{24}H_{40}O_5$ 60.42 **5**
C_2H_6O , $C_{33}H_{36}O_6$ 61.3 **7**
$2C_2H_6O$, H_4N_2 60.24 **2**
C_2H_6OS 11.14+ **1**
C_2H_6OS , $C_9H_{11}BrN_2O_6$ 47.4 **1**
$C_2H_6O_2PS_2^-$, K^+ 64.4 **2**
$2C_2H_6O_2PS_2^-$, $C_{36}H_{24}N_6Ni^{2+}$ 83.141 **3**
$C_2H_6O_2S$ 11.16 **1**
$C_2H_6O_2S$ 11.2 **3**
$C_2H_6O_4P^-$, H_4N^+ 46.1+ **6**
$C_2H_6O_4S_4$ 11.17 **1**
$C_2H_6O_4S_4Te$ 70.2 **2**
$C_2H_6P_2S_4$ 64.5 **2**
$C_2H_6Tl^+$, Cl^- 68.2 **7**
$C_2H_6Tl^-$, $C_4H_9AlNS^+$ 68.1 **6**
$C_2H_7AsMo_4O_{15}^{2-}$, $2CH_6N_3^+$, H_2O 86.1 **7**
$C_2H_7AsO_2$ 65.2 **2**
$C_2H_7BNO^-$, CH_6N^+ 62.6 **2**
$C_2H_7Br_2NPt$ 72.3 **2**
$C_2H_7NO_3S$ 48.10+ **1**
$C_2H_7NO_3S_2$ 4.1 **5**
$C_2H_7NO_4S$ 4.4 **1**
$C_2H_7NO_4S_2$, H_2O 4.1 **7**
$C_2H_7N_2O^+$, Cl^- 8.8+ **3**
$C_2H_7N_2O^+$, NO_3^- 8.33 **1**
$2C_2H_7N_2S^+$, O_4S^{2-} 8.34 **1**
$C_2H_7N_5^+$, Br^- 32.2 **5**
$C_2H_7N_{10}^+$, Br^- 32.5 **3**
$C_2H_7O_2P$ 64.6 **2**
$C_2H_7O_4PSn$ 69.1 **6**
$C_2H_8AgClN_4S_2$ 79.3 **2**
$2C_2H_8AuN_2O_6S_2^-$, $C_2H_{10}N_2^{2+}$ 76.1 **7**
C_2H_8BN 62.1 **3**
$C_2H_8B_4$ 62.7+ **2**
$C_2H_8Br_2N_4PbS_2$ 69.5 **5**
$C_2H_8Br_2N_4S_2Te$ 70.3 **2**
$(C_2H_8CaN_4O_5)_n$ 67.2 **4**
$C_2H_8CdCl_2N_4O_2$ 84.5 **2**
$C_2H_8CdCl_2N_4S_2$ 79.5 **2**
$(C_2H_8CdN_2^{2+})_n$, $2nC_6H_6$, $nC_4CdN_4^{2-}$ 61.1 **6**
$C_2H_8CdN_4O_6S_2$ 79.6 **2**
$(C_2H_8ClCuN_4S_2)_n$ 79.8 **3**
$C_2H_8ClHgN_4S_2^+$, Cl^- 79.8 **2**
$C_2H_8ClHgN_4S_2^+$, Cl^- 79.1 **5**
$C_2H_8Cl_2CoN_4S_2$ 79.2 **3**
$C_2H_8Cl_2CoN_4S_2$ 79.1 **4**
$C_2H_8Cl_2CuN_2$ 76.1 **3**
$C_2H_8Cl_2HgO_2$ 84.6 **2**
$C_2H_8Cl_2N_2Pd$ 76.2 **7**
$C_2H_8Cl_2N_2Pt$ 76.3 **7**
$C_2H_8Cl_2N_4O_4Pt$ 76.1 **6**
$C_2H_8Cl_2N_4PbS_2$ 79.9 **2**
$C_2H_8Cl_2N_4S_2Te$ 70.4 **2**
$C_2H_8Cl_2N_4S_2Zn$ 79.10 **2**
$C_2H_8CuN_2O_6S_2$ 84.1 **6**
$C_2H_8HgI_2N_4S_2$ 79.11 **2**
$C_2H_8IN_4S_2^+$, I^- 8.5 **4**
$(C_2H_8LiN_4O_2^+)_n$, nI^- 67.3 **3**
$C_2H_8N^+$, Br^- 3.9 **1**
$C_2H_8N^+$, Cl^- 3.10 **1**
$C_2H_8N^+$, Cl^- 3.2 **5**
$C_2H_8N^+$, Cl_3Cu^- 3.11 **1**
$C_2H_8N^+$, $C_3H_3CuO_6^-$ 81.3 **6**
$C_2H_8N^+$, $C_4HO_4^-$, C_4HO_4 20.2 **6**
$C_2H_8N^+$, $C_6H_{18}Cl_3O_3RuS_3^-$ 85.6 **7**
$2C_2H_8N^+$, Cl_4Cu^{2-} 3.2 **3**
$2C_2H_8N^+$, Cl_4Cu^{2-} 3.3 **3**
$3C_2H_8N^+$, $BiBr_6^{3-}$ 3.12 **1**
$3C_2H_8N^+$, Cl^- , Cl_4Cu^{2-} 3.1 **4**
$C_2H_8NO^+$, Cl^- 3.3 **5**
$C_2H_8NO^+$, ClO_4^- 3.5 **3**
$C_2H_8NO_2Se^+$, Cl^- 11.3 **3**
$C_2H_8NO_3P$ 64.7 **2**
$C_2H_8NO_4P$ 46.1+ **1**
$C_2H_8N_2$ 3.4 **5**
$2C_2H_8N_2$, Li^+ , Br^- 60.25 **2**
$2C_2H_8N_2$, Li^+ , Cl^- 60.26+ **2**
$3C_2H_8N_2$, Li^+ , I^- 3.3 **4**
$C_2H_8N_2O_6PdS_4^{2-}$, $C_4H_{16}N_4Pd^{2+}$ 76.14 **3**
$C_2H_8N_2S$ 4.5 **1**
$C_2H_8N_3^+$, NO_3^- 8.35 **1**
$C_2H_8N_3^+$, $H_2O_4P^-$ 8.4 **6**
$2C_2H_8N_3^+$, HO_4P^{2-} 8.18 **7**
$(C_2H_8N_4O_4S_3Sn)_n$ 69.2 **6**
$C_2H_8N_4S_2^{2+}$, $2Br^-$, H_2O 8.36 **1**
$C_2H_8N_4S_2^{2+}$, $2Cl^-$ 8.6 **4**
$C_2H_8N_4S_2^{2+}$, $2I^-$, H_2O 8.37 **1**
$C_2H_8N_4Se_2^{2+}$, $2Cl^-$ 8.11 **3**
$C_2H_8N_4Se_2^{2+}$, CH_4N_2Se , $2Br^-$, H_2O 8.19 **7**
$C_2H_8N_4Se_2^{2+}$, CH_4N_2Se , $2Cl^-$, H_2O 8.20 **7**
$C_2H_8N_6NiS_2$ 85.5 **2**
$C_2H_8O_7P_2$, H_2O 64.1 **5**
$C_2H_9B_{10}Br_3$ 62.9 **2**
$C_2H_9BrCuN_2O_2$ 81.5 **5**
$C_2H_{10}B_{10}Br_2$ 62.10 **2**
$C_2H_{10}B_{10}Br_2$ 62.11 **2**

$(\mathbf{C_2H_{10}CaO_9P_2})_n$ 67.2 **5**
$\mathbf{C_2H_{10}CrN_2O_5}$, H_2O 76.1 **2**
$\mathbf{C_2H_{10}CuN_6O_2}^{2+}$, $2Cl^-$ 83.8 **2**
$\mathbf{C_2H_{10}CuN_6S_2}^{2+}$, $2NO_3^-$ 79.2 **4**
$\mathbf{C_2H_{10}CuN_6S_2}^{2+}$, O_4S^{2-} 79.3 **4**
$(\mathbf{C_2H_{10}FeN_6O_4S_3})_n$ 79.1 **7**
$\mathbf{C_2H_{10}N_2}^{2+}$, $2Cl^-$ 3.13 **1**
$\mathbf{C_2H_{10}N_2}^{2+}$, Cl_4Cu^{2-} 3.5 **5**
$\mathbf{C_2H_{10}N_2}^{2+}$, Cl_4Cu^{2-} 3.2 **7**
$\mathbf{C_2H_{10}N_2}^{2+}$, Cl_4Mn^{2-} 3.3 **7**
$\mathbf{C_2H_{10}N_2}^{2+}$, O_4S^{2-} 3.15 **1**
$\mathbf{C_2H_{10}N_2}^{2+}$, $2C_2H_8AuN_2O_6S_2^-$ 76.1 **7**
$\mathbf{C_2H_{10}N_2}^{2+}$, $C_4H_4O_6^{2-}$ 3.6 **5**
$\mathbf{C_2H_{10}N_2}^{2+}$, $C_{20}H_{30}Fe_2N_4O_{15}^{2-}$, $6H_2O$ 83.183 **2**
$\mathbf{2C_2H_{10}N_2}^{2+}$, $2Br^-$, Br_4Cu^{2-} 3.4 **4**
$\mathbf{2C_2H_{10}N_2}^{2+}$, $2Cl^-$, Cl_4Co^{2-} 3.1 **6**
$\mathbf{2C_2H_{10}N_2}^{2+}$, $Cl_8Mo_2^{4-}$, $2H_2O$ 3.8 **3**
$\mathbf{C_2H_{10}N_4NiO_6}$ 83.3 **3**
$\mathbf{C_2H_{10}N_6Ni}^{2+}$, $2NO_3^-$ 79.2 **5**
$\mathbf{C_2H_{10}N_6Ni}^{2+}$, $2NO_3^-$ 79.3 **5**
$\mathbf{C_2H_{10}N_6NiS_2}^{2+}$, O_4S^{2-} 85.6 **2**
$\mathbf{C_2H_{10}N_6NiS_2}^{2+}$, O_4S^{2-} , $3H_2O$ 85.7 **2**
$\mathbf{C_2H_{10}N_6O_2Zn}^{2+}$, $2Cl^-$ 83.9 **2**
$\mathbf{C_2H_{10}N_6PtS_2}^{2+}$, O_4S^{2-} , $3H_2O$ 79.4 **5**
$\mathbf{C_2H_{11}B_{10}I}$ 62.12 **2**
$\mathbf{C_2H_{11}CoN_6O_6}$ 76.2+ **3**
$\mathbf{C_2H_{12}AgN_8S_2}^{+}$, NO_3^- 79.2 **7**
$\mathbf{C_2H_{12}CaN_4O_{10}P_2}$, $2CH_4N_2O$ 67.1 **6**
$\mathbf{C_2H_{12}CdCl_2N_8S_2}$ 79.4 **4**
$\mathbf{C_2H_{12}Cl_2CrN_2O_2}^{+}$, Cl^- 76.4 **3**
$\mathbf{C_2H_{12}CuN_8S_2}^{2+}$, $2ClO_4^-$ 79.3 **7**
$\mathbf{C_2H_{12}CuN_8S_2}^{2+}$, O_4S^{2-} , $4H_2O$ 79.4 **7**
$\mathbf{C_2H_{12}CuN_8S_2}^{2+}$, $C_2O_4^{2-}$, $4H_2O$ 79.5 **7**
$\mathbf{C_2H_{13}B_5}$ 62.13 **2**
$\mathbf{C_2H_{13}B_{10}I}$ 62.14 **2**
$\mathbf{C_2H_{13}N_5P_2S_2}$ 64.8 **2**
$\mathbf{C_2H_{14}B_2N_2}$ 3.5 **4**
$\mathbf{C_2H_{14}B_4F_2NP}$ 64.4 **3**
$\mathbf{C_2H_{14}N_6NiO_2S_2}^{2+}$, $2NO_3^-$ 85.8 **2**
$\mathbf{C_2H_{14}N_8O_4Zn}$ 83.10 **2**
$\mathbf{C_2H_{16}B_9N}$ 62.15 **2**
$\mathbf{C_2H_{16}CoN_4O_6}^{2+}$, $2NO_3^-$ 79.5 **5**
$\mathbf{C_2H_{16}N_7P_2}^{+}$, I^- 64.9 **2**
$\mathbf{C_2H_{18}B_{10}}$ 62.16 **2**
$\mathbf{C_2H_{18}CoN_5O_2}^{2+}$, Cl^- , ClO_4^- 81.12 **2**
$\mathbf{C_2H_{20}CoN_6O}^{2+}$, $2ClO_4^-$ 83.2 **6**
$\mathbf{C_2H_{20}N_5Rh}^{2+}$, $2Br^-$ 71.7 **2**
$\mathbf{C_2H_{21}N_5ORuS}^{2+}$, $2F_6P^-$ 85.1 **4**
$\mathbf{C_2H_{23}Co_2N_6O_4}^{3+}$, $3Br^-$, $3H_2O$ 81.2 **6**
$\mathbf{C_2H_{25}F_5O_2RuSi_2}$ 75.1 **7**

C_3

$(\mathbf{C_3AgN_3O})_n$ 84.2 **6**
$(\mathbf{C_3AgN_3O_2})_n$ 83.3 **6**
$\mathbf{C_3BrN}$ 7.8 **1**
$\mathbf{C_3ClN}$ 7.9 **1**
$\mathbf{C_3Cl_6}$ 20.1 **1**
$\mathbf{C_3Cl_6O_3}$ 5.2 **3**
$\mathbf{C_3IN}$ 7.10 **1**
$\mathbf{C_3N_{12}}$ 33.2 **1**
$\mathbf{C_3O_2S_3}$ 39.3 **1**
$\mathbf{C_3HN}$ 7.11 **1**
$\mathbf{C_3H_2F_2O_4}$ 1.3 **4**
$\mathbf{C_3H_2N_2}$ 7.1 **6**
$\mathbf{C_3H_2N_2O_3}$ 32.8 **1**
$\mathbf{C_3H_2N_2O_3}$, CH_4N_2O 60.4 **3**
$\mathbf{C_3H_2N_2O_3}$, CH_4N_2S 60.5 **3**
$\mathbf{C_3H_2N_2S_2}$ 11.4 **3**
$\mathbf{C_3H_2O_3}$ 38.1 **7**
$\mathbf{C_3H_2O_4}^{2-}$, $C_6H_{20}CuN_4^{2+}$, $2H_2O$ 76.30 **7**
$\mathbf{C_3H_3AsN_2}$ 65.3 **2**
$\mathbf{C_3H_3BiO_6}$ 66.1 **3**
$\mathbf{C_3H_3BrO_2}$ 5.13 **1**
$\mathbf{C_3H_3CuO_6}^{-}$, $C_2H_8N^+$ 81.3 **6**
$\mathbf{C_3H_3FO_4}$ 1.8 **3**
$\mathbf{C_3H_3FO_4}$ 1.5 **7**
$\mathbf{C_3H_3GdO_6}$ 81.13 **2**
$\mathbf{C_3H_3NOS_2}$ 41.7 **1**
$\mathbf{C_3H_3N_2O_6P_2}^{3-}$, $1.5Ca^{2+}$, $6H_2O$ 32.9 **1**
$\mathbf{C_3H_3N_3}$ 33.3+ **1**
$\mathbf{C_3H_3N_3O_2}$ 33.1+ **6**
$\mathbf{C_3H_3N_3O_3}$ 33.1+ **3**
$\mathbf{C_3H_3N_3O_4}$ 40.1 **6**
$\mathbf{C_3H_3N_5O_3S}$ 41.1 **7**
$\mathbf{C_3H_3O_3}^{-}$, Li^+ 2.51 **1**
$\mathbf{C_3H_3O_3}^{-}$, Na^+ 2.52 **1**
$\mathbf{C_3H_3O_4}^{-}$, K^+ 2.12+ **3**
$\mathbf{C_3H_3O_4}^{-}$, Na^+ 2.9 **6**
$\mathbf{C_3H_3O_4}^{-}$, $C_3H_4O_4$, K^+ 2.14 **5**
$(\mathbf{C_3H_3O_6Sc})_n$ 81.14 **2**
$\mathbf{C_3H_3O_6Sn}^{-}$, K^+ 69.4 **3**
$(\mathbf{C_3H_4CdO_5})_n$ 81.3 **7**
$\mathbf{C_3H_4Cl_2N_2O_2}$ 1.4 **4**
$\mathbf{C_3H_4Cl_2O}$ 5.6 **6**
$\mathbf{C_3H_4Cl_2O_3S}$ 42.2 **1**
$(\mathbf{C_3H_4CoO_5})_n$, nH_2O 81.4 **7**
$(\mathbf{C_3H_4FeO_5})_n$, nH_2O 81.5 **7**
$\mathbf{C_3H_4N_2}$ 32.11 **1**
$\mathbf{C_3H_4N_2}$ 32.12 **1**
$\mathbf{C_3H_4N_2}$ 32.6 **3**
$\mathbf{C_3H_4N_2}$ 32.7+ **3**
$\mathbf{C_3H_4N_2}$ 32.1 **4**
$\mathbf{C_3H_4N_2}$ 32.3+ **6**
$\mathbf{C_3H_4N_2}$, $C_8H_{12}N_2O_3$ 60.7 **6**
$\mathbf{C_3H_4N_2O}$ 1.10 **3**
$\mathbf{C_3H_4N_2OS}$ 32.13 **1**
$\mathbf{C_3H_4N_2OS}$ 41.1 **5**
$\mathbf{C_3H_4N_3O_5}^{-}$, K^+ 12.4 **4**
$\mathbf{C_3H_4N_4NiOS_4}$ 83.11 **2**
$\mathbf{C_3H_4N_6O_2S}$ 41.2 **7**
$\mathbf{C_3H_4N_6O_2S}$, H_2O 41.3 **7**
$(\mathbf{C_3H_4NiO_5})_n$, nH_2O 81.6 **7**
$\mathbf{C_3H_4OS_3}$ 39.4 **1**
$\mathbf{C_3H_4O_2}$ 1.29+ **1**
$\mathbf{C_3H_4O_2S}$ 39.5 **1**
$\mathbf{C_3H_4O_3}$ 38.1 **1**
$\mathbf{C_3H_4O_3}$ 5.7 **6**
$\mathbf{C_3H_4O_4}$ 1.31 **1**
$\mathbf{C_3H_4O_4}$, $C_3H_3O_4^-$, K^+ 2.14 **5**
$\mathbf{C_3H_4O_5}$ 1.32 **1**
$(\mathbf{C_3H_4O_5Zn})_n$, nH_2O 81.7 **7**
$\mathbf{C_3H_4O_6P}^{-}$, $C_6H_{14}N^+$ 46.2 **6**
$\mathbf{C_3H_4S_3}$ 39.1 **5**
$\mathbf{C_3H_5AlCl_4O}$ 68.1 **4**
$\mathbf{C_3H_5BrN_2O_2}$ 1.6 **7**
$\mathbf{C_3H_5ClO_2S}$ 39.2 **5**
$\mathbf{C_3H_5NO}$ 1.33 **1**

$C_3H_5NO_2$ 40.1 **4**
$C_3H_5NO_4$ 1.34 **1**
$C_3H_5NO_4$ 1.35 **1**
$C_3H_5N_2^+$, $H_2O_4P^-$ 32.2 **4**
$C_3H_5N_2^+$, $C_4H_3O_4^-$ 32.1 **7**
$C_3H_5N_2OS^+$, Cl^- 41.4 **7**
$C_3H_5N_2S_2^+$, I^- 39.6 **1**
$C_3H_5N_3O$ 1.1 **6**
$C_3H_5N_5O$ 32.2 **7**
$C_3H_5O^+$, Cl_4Ga^- 12.5 **4**
$C_3H_5O^+$, Cl_6Sb^- 12.5 **1**
$C_3H_5OS_2^-$, K^+ 11.18 **1**
$C_3H_5OS_2^-$, Na^+ , $2H_2O$ 11.5 **3**
$(C_3H_5O_6Sc)_n$, nH_2O 81.4 **6**
$C_3H_6AgS_3^+$, ClO_4^- , H_2O 85.9 **2**
$C_3H_6AgS_3^+$, NO_3^- 85.10 **2**
$C_3H_6AgS_3^+$, NO_3^- , H_2O 85.11 **2**
C_3H_6AsN 65.4 **2**
C_3H_6ClNO 1.12 **3**
$C_3H_6Cl_2HgS_3$ 85.12 **2**
$(C_3H_6Cl_3Cu_3S_3)_n$ 85.1 **3**
$C_3H_6Cl_3OZn^-$, $C_{52}H_{48}ClMoOP_4^+$, C_3H_6O 86.88 **3**
$C_3H_6Cl_5N_2P$ 64.5 **3**
$(C_3H_6CoO_3)_n$ 84.1 **4**
$(C_3H_6CoO_7)_n$, nH_2O 81.6 **5**
$C_3H_6NS_2^-$, Cs^+ 1.7 **7**
$C_3H_6NS_2Tl$ 68.2 **6**
$C_3H_6N_2O_2$ 1.11 **3**
$C_3H_6N_2S$ 32.14 **1**
$C_3H_6N_4NiO_2S_4$ 83.12 **2**
$C_3H_6N_4O_3$ 8.39+ **1**
$C_3H_6N_6$ 33.8 **1**
$C_3H_6N_6$ 33.1 **7**
$C_3H_6N_6O_6$ 33.2 **5**
$C_3H_6N_6O_6$, $C_4H_8O_2S$ 60.28 **2**
C_3H_6O , $2Br$ 60.29 **2**
C_3H_6O , $C_6H_6O_2$ 60.88 **2**
C_3H_6O , $C_{20}H_{12}O_5$ 38.61 **7**
$3C_3H_6O$, Na^+ , I^- 60.30 **2**
$C_3H_6OSe_2$ 42.1 **7**
$C_3H_6O_2$ 1.36+ **1**
$C_3H_6O_3$ 38.2+ **1**
$C_3H_6O_3S$ 42.3 **1**
$C_3H_6O_3S$ 39.1 **6**
$C_3H_6O_7S_2^{2-}$, $2Na^+$, $4H_2O$ 11.19 **1**
$C_3H_6S_3$ 39.7+ **1**
$C_3H_6Se_3$ 39.9 **1**
$(C_3H_7CdO_8P)_n$, $2nH_2O$ 84.2 **3**
$C_3H_7Cl_3O_2SeW$ 84.3 **6**
$C_3H_7Cl_5NOSb$ 66.2 **2**
C_3H_7NO 1.38 **1**
C_3H_7NO 10.18 **1**
C_3H_7NO , $C_8H_{14}N_6O_{10}$ 60.12 **6**
C_3H_7NO , $C_{15}H_{13}I_2NO_4$ 60.28 **5**
$2C_3H_7NO$, Na^+ , ClO_4^- 60.31 **2**
$3C_3H_7NO$, Na^+ , I^- 60.32 **2**
$4C_3H_7NO$, Li^+ , Cl^- 60.33 **2**
$6C_3H_7NO$, $2K^+$, I^- , I_3^- 60.34 **2**
$C_3H_7NO_2$ 1.39+ **1**
$C_3H_7NO_2$ 48.12 **1**
$C_3H_7NO_2$ 48.13 **1**
$C_3H_7NO_2$ 48.14+ **1**
$C_3H_7NO_2$ 48.5 **4**
$C_3H_7NO_2S$ 48.16 **1**
$C_3H_7NO_2S$ 48.5+ **6**
$C_3H_7NO_3$ 48.17 **1**
$C_3H_7NO_3$ 48.9+ **5**
$C_3H_7NO_3$ 48.8 **7**
$C_3H_7NO_3$ 48.9+ **7**
$C_3H_7NO_3$, H_2O 48.10 **5**
$C_3H_7NO_5S$ 48.18 **1**
$C_3H_7NO_5S$, H_2O 48.5 **3**
$C_3H_7NO_5S$, H_2O 48.11 **5**
$C_3H_7NSe^+$, Br^- 42.4 **1**
$C_3H_7N_2^+$, Cl^- 32.15 **1**
$C_3H_7N_2O_2^+$, Cl^- 40.1 **1**
$C_3H_7N_3O_2$ 1.4 **5**
$C_3H_7N_3O_2$ 48.7 **6**
$C_3H_7N_5NiS_4$ 83.6 **4**
$C_3H_7O_4P$ 46.1 **3**
$C_3H_7O_4P$ 46.2 **3**
$C_3H_7O_6P^{2-}$, $2Na^+$, $5H_2O$ 46.5 **1**
$C_3H_7O_6P^{2-}$, $2Na^+$, $6H_2O$ 46.2 **5**
C_3H_8BCl 62.1 **7**
$C_3H_8B_4Ga$ 62.1 **4**
$C_3H_8NO_2^+$, Cl^- 48.19 **1**
$C_3H_8NO_2S^+$, Cl^- , H_2O 48.20 **1**
$C_3H_8NO_6P$ 48.21 **1**
$C_3H_8NO_6P$ 48.6 **3**
$C_3H_8NO_6P$, H_2O 48.7 **3**
$C_3H_8N_2O_3S$ 8.7 **4**
$C_3H_8N_3O_2^+$, Br^- 1.41 **1**
$C_3H_8N_3O_6PS^{2-}$, $2H_4N^+$ 8.10 **5**
$C_3H_8N_4O_2$ 1.8 **7**
$C_3H_8N_4O_2$, H_2O 9.1 **5**
$C_3H_8N_5^+$, Cl^- 32.16 **1**
C_3H_8O , $C_{36}H_{60}O_{30}$ 61.9 **5**
C_3H_8O , $C_{36}H_{60}O_{30}$, $4.8H_2O$ 61.6 **7**
$C_3H_8O_3$ 5.14 **1**
$C_3H_9AlCl_3N$ 68.2 **2**
$C_3H_9AlN_3^-$, Cs^+ 68.3 **7**
$(C_3H_9AlS)_n$ 68.1 **3**
$C_3H_9AsBr_3$ 65.1 **4**
$C_3H_9AsCl_2$ 65.2 **4**
$C_3H_9AsCl_3N$ 3.9 **3**
$C_3H_9AsO_3$ 65.3 **4**
$C_3H_9As_3Cl_3N_3$ 65.5 **2**
$C_3H_9AuBr_3P$ 86.1 **2**
$C_3H_9BBr_3N$ 62.2 **4**
$C_3H_9BBr_3P$ 64.4 **7**
$C_3H_9BCl_3N$ 62.17 **2**
$C_3H_9BCl_3N$ 62.3 **4**
$C_3H_9BCl_3P$ 64.5 **7**
$C_3H_9BF_3N$ 3.16 **1**
$C_3H_9BI_3N$ 62.4 **4**
$C_3H_9BI_3P$ 64.6 **7**
$C_3H_9B_3Br_6S_3$ 11.2 **7**
$C_3H_9B_3Cl_6S_3$ 11.3 **7**
$C_3H_9B_3N_3$ 62.18 **2**
$C_3H_9Br_2Sb$ 66.4 **2**
$(C_3H_9Cl_2NSiTi)_n$ 63.1 **6**
$C_3H_9Cl_2N_4O_4Ta$ 71.2 **7**
$C_3H_9Cl_2Sb$ 66.5 **2**
$C_3H_9Cl_2Sn^-$, $C_{15}H_{15}Mo_3S_4^+$ 73.14 **4**
$C_3H_9Cl_4GeN$ 69.3 **4**
$C_3H_9Cl_5OPSb$ 66.6 **2**
$C_3H_9Cl_6Ge_3N_3$ 69.4 **4**
C_3H_9FSn 69.2 **2**
$C_3H_9HgI_3S$ 85.13 **2**
$C_3H_9I_2Sb$ 66.3 **2**
C_3H_9In 68.3 **2**

C_3H_9N , ClI 60.35 **2**
C_3H_9N , I_2 60.36 **2**
C_3H_9N , O_2S 60.37 **2**
C_3H_9N , xH_2O 61.2+ **6**
C_3H_9N , $8H_2O$ 61.2 **3**
C_3H_9N , $10.25H_2O$ 61.2 **2**
C_3H_9N , $C_{16}H_{13}N_3O_2$ 33.45 **7**
C_3H_9NO 10.19 **1**
$C_3H_9NO_2S_2$ 4.6+ **1**
$C_3H_9NO_3S$ 4.2 **5**
$C_3H_9N_3O_2$ 3.6 **4**
$C_3H_9N_3O_2S$, $2H_2O$ 8.8 **4**
$C_3H_9N_3O_3S$ 8.11 **5**
$C_3H_9N_3O_6S_3$ 4.2 **7**
$C_3H_9N_6NiO_6$, H_2O 83.4 **3**
$C_3H_9N_6NiO_6^-$, $H_5N_2^+$, H_2O 83.13+ **2**
$C_3H_9N_6O_6Zn^-$, K^+ 83.15 **2**
$C_3H_9OS^+$, BF_4^- 11.20 **1**
$C_3H_9OS^+$, ClO_4^- 11.21 **1**
$C_3H_9O_4ReSi$ 63.1 **3**
C_3H_9PS 64.6 **3**
$C_3H_9S^+$, I^- 11.22 **1**
$2C_3H_9S^+$, HgI_4^{2-} 11.23 **1**
$C_3H_9Se^+$, I^- 11.24 **1**
$6C_3H_9Si^-$, Li_6^{6+} 63.1 **7**
$C_3H_9Te^+$, $CH_3I_4Te^-$ 70.5 **2**
C_3H_9Tl 68.2 **3**
$C_3H_{10}AgN_7S_3$ 85.14 **2**
$C_3H_{10}Al^-$, K^+ 68.4 **7**
$C_3H_{10}AsN_2S^+$, CH_4N_2S , Cl^- 65.1 **7**
$(C_3H_{10}CdCl_2N_2)_n$ 83.7 **4**
$C_3H_{10}N^+$, Cl^- 3.17 **1**
$C_3H_{10}N^+$, I^- 3.10 **3**
$2C_3H_{10}N^+$, Cl_4Cu^{2-} 3.11 **3**
$2C_3H_{10}N^+$, Cl_4Mn^{2-} 3.7 **5**
$2C_3H_{10}N^+$, $Cl_{10}Cu_4^{2-}$ 3.13 **3**
$2C_3H_{10}N^+$, $C_{12}H_4N_4$, $2C_{12}H_4N_4^-$ 3.2 **6**
$3C_3H_{10}N^+$, Cl_3Cu^- , Cl_4Cu^{2-} 3.8 **5**
$C_3H_{10}NO^+$, Cl^- 10.20+ **1**
$C_3H_{10}NO^+$, $C_4H_5O_6^-$, $2H_2O$ 2.8 **7**
$C_3H_{10}NO^+$, $C_4H_5O_6^-$, $2H_2O$ 2.9 **7**
$C_3H_{10}OSn$ 69.3 **2**
$C_3H_{11}B_{10}Br_3$ 62.2 **3**
$C_3H_{11}CoN_6O_4S$ 76.2 **6**
$C_3H_{11}NO_4Sn$ 69.5 **4**
$C_3H_{11}N_2^+$, $C_6H_5N_2O_4^-$ 32.5 **7**
$C_3H_{11}N_3S^{2+}$, $2Br^-$ 3.4 **7**
$C_3H_{12}CdN_6O_4S_4$ 79.4 **3**
$C_3H_{12}Cl_2HgN_6S_3$ 79.12 **2**
$(C_3H_{12}CuN_6S_3^+)_n$, nCl^- 79.13 **2**
$C_3H_{12}GaN$ 68.4 **2**
$C_3H_{12}NO_9P_3$ 64.10 **2**
$C_3H_{12}N_2^{2+}$, $2Cl^-$ 3.18 **1**
$C_3H_{12}N_6O_3S_5Zn$, H_2O 79.14 **2**
$C_3H_{12}N_6O_4S_4Zn$ 79.15 **2**
$C_3H_{12}N_6Se_3^{2+}$, $2Br^-$, H_2O 8.12 **3**
$C_3H_{12}N_6Se_3^{2+}$, $2Cl^-$, H_2O 8.13 **3**
$(C_3H_{15}Ag_2Br_2N_9S_3)_n$ 79.16 **2**
$C_3H_{15}N_9NiS_3^{2+}$, $2Cl^-$, H_2O 79.2 **6**
$C_3H_{15}N_9NiS_3^{2+}$, $2NO_3^-$ 85.15 **2**
$C_3H_{15}N_9NiS_3^{2+}$, $2NO_3^-$, H_2O 85.16 **2**
$C_3H_{16}B_3N$ 62.19 **2**
$C_3H_{16}B_5BrSi$ 63.1 **4**
$C_3H_{18}CoN_5O_2^{2+}$, $2NO_3^-$ 82.2 **2**
$C_3H_{19}Na_2O_{12}P$ 67.3 **4**
$C_3H_{21}AlAsB_3$ 65.6 **2**
$C_3H_{21}AlB_3N$ 68.5+ **2**
$C_3H_{21}B_9OS$ 62.5 **4**

C_4

$C_4Ag_2F_6O_4$ 81.7 **5**
$C_4As_4F_{12}$ 65.2 **3**
C_4BrN_3 7.1 **4**
C_4ClN_3 7.2 **4**
$C_4Cl_2N_2$ 7.3 **5**
$C_4Cl_2O_2$ 20.1 **5**
$C_4Cl_4N_4S_4$ 41.8 **1**
$C_4Cl_4NiO_4S_4Sn^{2-}$, $2C_{25}H_{22}P^+$ 85.1 **6**
C_4Cl_8 20.2 **1**
$C_4Cl_8NiO_4S_4Sn_2^{2-}$, $2C_{25}H_{22}P^+$, H_2O 85.2 **6**
$C_4Cl_{12}N_2W$ 83.8 **4**
$C_4Cl_{18}N_2W_2$ 83.5 **3**
$C_4CuO_8^{2-}$, $2Cs^+$, $2H_2O$ 81.8 **7**
$C_4CuO_8^{2-}$, $2K^+$, $2H_2O$ 81.9 **7**
$C_4CuO_8^{2-}$, $2H_4N^+$, $2H_2O$ 81.15 **2**
$C_4F_6Hg_2O_4$ 81.5 **6**
$C_4F_{12}P_4$ 64.11 **2**
C_4N_2 7.12 **1**
$C_4N_2S_2^{2-}$, $2Rb^+$, H_2O 7.4 **5**
$C_4N_3^-$, Ag^+ 7.13 **1**
$C_4N_3^-$, K^+ 7.14 **1**
$C_4N_3^-$, K^+ 7.3 **4**
$C_4N_3^-$, Na^+ 7.15 **1**
$C_4N_3^-$, H_4N^+ 7.16 **1**
$2C_4N_3^-$, Cu^{2+} 7.17 **1**
$C_4N_4NiS_4^{2-}$, $2C_{24}H_{20}As^+$ 65.47 **2**
$C_4NbO_{12}^{3-}$, $3H_4N^+$, H_2O 81.3 **4**
$C_4NiO_4S_4^{2-}$, $2K^+$ 85.3 **6**
$C_4O_4^{2-}$, $2K^+$, H_2O 6.1 **1**
$C_4O_8Pd^{2-}$, $2K^+$, $4H_2O$ 81.17 **2**
$C_4O_8Pt^{2-}$, $2K^+$, $2H_2O$ 81.18 **2**
$C_4O_8Pt^{2-}$, $0.82Mg^{2+}$, $5.3H_2O$ 81.19 **2**
$C_4O_8Y^-$, $H_5O_2^+$, $2H_2O$ 81.4+ **4**
$(C_4O_{10}U^{2-})_n$, $2nH_4N^+$ 81.6 **6**
$C_4O_{10}V^{3-}$, $3K^+$, $3H_2O$ 81.10 **7**
$C_4O_{10}V^{3-}$, $3H_4N^+$, $2H_2O$ 81.9+ **4**
$C_4HO_3^-$, K^+ , H_2O 59.1 **6**
$C_4HO_4^-$, K^+ 2.15 **5**
$C_4HO_4^-$, K^+ , H_2O 20.1 **6**
$C_4HO_4^-$, Na^+ , $2H_2O$ 2.10 **6**
$C_4HO_4^-$, Rb^+ 2.16+ **5**
$C_4HO_4^-$, H_4N^+ 2.11 **6**
$C_4HO_4^-$, $C_2H_8N^+$, C_4HO_4 20.2 **6**
$C_4H_2ClO_4^-$, K^+ 2.53 **1**
$C_4H_2Cl_2N_2O_2$ 33.3+ **5**
$C_4H_2Cl_4$ 5.4 **4**
$C_4H_2N_2O_3S^{2-}$, $2Na^+$, $6H_2O$ 44.1 **3**
$C_4H_2N_2O_4$ 43.1 **1**
$C_4H_2N_3O_4^-$, K^+ , $2H_2O$ 43.2 **1**
$C_4H_2N_3O_4^-$, Rb^+ 43.3 **1**
$C_4H_2N_4O_2$ 9.2 **5**
$C_4H_2O_3$ 1.42 **1**
$C_4H_2O_4$ 20.1 **3**
$C_4H_2O_4$ 1.5 **5**
$C_4H_2O_4$ 1.2 **6**
$C_4H_2O_4$ 20.3+ **6**
$C_4H_2O_4$, $2H_2O$ 1.43 **1**
$C_4H_2O_4^{2-}$, $2K^+$, $2H_2O$ 2.15 **3**

$C_4H_2O_4^{2-}$, $2Li^+$, $2H_2O$ 2.12 **6**
$C_4H_2O_4^{2-}$, $2Na^+$, H_2O 2.13 **6**
$C_4H_2O_4^{2-}$, $H_8CaO_4^{2+}$ 2.16 **4**
$C_4H_2O_9Y^-$, H_4N^+ 81.20 **2**
$C_4H_2O_{10}V^{2-}$, $2H_4N^+$, H_2O 81.9 **5**
$C_4H_3BrN_2O_2$ 50.1 **1**
$C_4H_3BrN_2O_2$ 44.1 **7**
$C_4H_3BrN_2O_4$ 40.1 **5**
$C_4H_3ClN_2O$ 44.2 **7**
$C_4H_3ClN_2O_2$ 44.3 **7**
$C_4H_3Cl_2N_3$ 44.1 **1**
$C_4H_3Cl_2N_3$ 44.2 **1**
$(C_4H_3ErO_9)_n$, $2nH_2O$ 81.4 **3**
$C_4H_3FN_2O$ 44.1 **4**
$C_4H_3FN_2O_2$ 44.1 **6**
$C_4H_3FN_2O_2$, $C_4H_5N_3O$, H_2O 44.9 **3**
$C_4H_3FN_2O_2$, $C_5H_7N_3O$ 44.4 **3**
$C_4H_3FN_2O_2$, $C_7H_8N_4O$ 44.3 **1**
$C_4H_3IN_2O_2$ 44.4 **7**
$C_4H_3N_2O_3^-$, H_4N^+ 43.5 **1**
$C_4H_3N_3O_4$, H_2O 44.4 **1**
$C_4D_3N_3O_4$, D_2O 43.6 + **1**
$C_4H_3N_3O_5$ 43.8 **1**
$C_4H_3N_3O_5$, $3H_2O$ 43.9 **1**
$C_4H_3N_5O$ 40.2 **4**
$C_4H_3N_5O_2$, H_2O 44.5 **1**
$C_4H_3N_5O_2$, H_2O 44.6 **3**
$C_4H_3N_5S$ 41.3 **4**
$C_4H_3NbO_{11}^{2-}$, $2H^+$, $4H_2O$ 81.11 **4**
$C_4H_3O_4^-$, K^+ 2.55 + **1**
$C_4H_3O_4^-$, Li^+ , $2H_2O$ 2.5 **7**
$C_4H_3O_4^-$, Na^+ 2.16 **3**
$C_4H_3O_4^-$, Na^+ , $3H_2O$ 2.6 **7**
$C_4H_3O_4^-$, Rb^+ 2.17 **3**
$C_4H_3O_4^-$, $C_3H_5N_2^+$ 32.1 **7**
$C_4H_3O_4^-$, $C_{16}H_{20}BrN_2^+$ 33.38 **4**
$C_4H_3O_4^-$, $C_{16}H_{20}ClN_2^+$ 33.46 **7**
$C_4H_3O_4^-$, $C_{18}H_{23}N_2OS^+$ 41.44 **5**
$2C_4H_3O_4^-$, $H_8CuO_4^{2+}$ 2.11 **4**
$2C_4H_3O_4^-$, $C_4H_4O_4$, $2K^+$ 2.19 **5**
$2C_4H_3O_4^-$, $C_4H_4O_4$, $2K^+$ 2.14 **6**
$(C_4H_4AgN_2^+)_n$, nNO_3^- 83.16 **2**
$C_4H_4Ag_2N_2^{2+}$, $2NO_3^-$ 83.17 **2**
$C_4H_4ClNO_2$ 1.44 **1**
$C_4H_4ClN_3$ 33.1 **4**
$C_4H_4Cl_2O_3$ 1.14 **3**
$C_4H_4Cl_4Ge_2$ 69.4 **2**
$C_4H_4Cl_4O_2$ 38.4 **1**
$C_4H_4Cl_5IrN_2^{2-}$, $2K^+$, $0.5H_2O$ 83.4 **6**
$C_4H_4Cl_{12}O_2Sb_2$ 66.2 **4**
$C_4H_4CrO_{10}^-$, K^+ , $3H_2O$ 81.21 **2**
$(C_4H_4CuN_4O_6)_n$ 83.6 **3**
$C_4H_4CuO_5$ 81.12 **4**
$C_4H_4CuO_5$, $0.5H_2O$ 81.11 **7**
$(C_4H_4CuO_5)_{2n}$, nH_2O 81.7 **6**
$(C_4H_4CuO_6)_n$, $3nH_2O$ 81.13 **4**
$(C_4H_4CuO_6)_n$, $3nH_2O$ 81.14 **4**
$C_4H_4CuO_{10}^{2-}$, $2Li^+$, $4H_2O$ 81.5 **3**
$C_4H_4Cu_2O_8$, $C_4H_8O_2$ 81.22 **2**
$C_4H_4F_6Si_3$ 63.2 **4**
$C_4H_4Ge_2I_4$ 69.5 **2**
$(C_4H_4Hg_2N_2^{2+})_n$, $2nNO_3^-$ 83.1 **7**
$C_4H_4MnO_{10}^{2-}$, $2K^+$ 81.8 **6**
$C_4H_4Mo_2O_{14}^{2-}$, H_2BaO^{2+} , $2H_2O$ 81.25 **2**
$C_4H_4NO_4S^-$, K^+ 42.2 **7**
$C_4H_4N_2$ 33.9 **1**
$C_4H_4N_2$ 44.6 **1**
$C_4H_4N_2$, C_2Br_4 60.38 **2**
$C_4H_4N_2$, C_2I_4 60.39 **2**
$C_4H_4N_2$, $C_{10}H_2CuF_{12}O_4$ 60.14 **6**
$C_4H_4N_2O$ 44.7 **3**
$C_4H_4N_2OS$ 44.7 **1**
$C_4H_4N_2OS_3$ 41.2 **5**
$C_4H_4N_2OSe$ 44.8 **1**
$C_4H_4N_2O_2$ 44.9 + **1**
$2C_4H_4N_2O_2$, Cl_2Hg 44.2 **4**
$C_4H_4N_2O_2S$ 43.10 **1**
$C_4H_4N_2O_3$ 43.11 **1**
$C_4H_4N_2O_3$, $2H_2O$ 43.12 **1**
$C_4H_4N_2O_4$ 9.3 **7**
$C_4H_4N_2O_4$, H_2O 43.13 + **1**
$C_4H_4N_2O_5$ 43.16 **1**
$C_4H_4N_2O_5$, $3H_2O$ 43.17 **1**
$C_4H_4N_2S$ 33.10 **1**
$C_4H_4N_2S_2$ 11.25 **1**
$C_4H_4N_2S_2$ 44.11 **1**
$C_4H_4N_2Se_2$ 44.12 **1**
$C_4H_4N_3O_5^-$, K^+ 12.6 **1**
$C_4H_4N_4$ 7.18 **1**
$C_4H_4N_4$ 7.19 **1**
$C_4H_4N_6$, $0.5H_2O$ 32.9 **3**
$C_4H_4N_6O$, H_2O 35.1 + **1**
$C_4H_4NbO_{11}^-$, Cs^+ , $2H_2O$ 81.10 **5**
$(C_4H_4NiO_6)_n$ 84.3 **7**
$C_4H_4NiS_4$ 85.2 + **3**
$C_4H_4NiS_4$ 85.2 **4**
$C_4H_4NiS_4$, $2C_6H_4S_4$ 60.4 **7**
C_4H_4O 38.1 + **5**
$C_4H_4O_2$ 38.7 **1**
$C_4H_4O_2$ 1.6 **5**
$C_4H_4O_2$ 1.7 + **5**
$C_4H_4O_2S_2$ 39.10 **1**
$C_4H_4O_3$ 1.45 + **1**
$C_4H_4O_4$ 1.49 **1**
$C_4H_4O_4$ 1.47 + **1**
$C_4H_4O_4$ 1.3 **6**
$C_4H_4O_4$, $2C_4H_3O_4^-$, $2K^+$ 2.19 **5**
$C_4H_4O_4$, $2C_4H_3O_4^-$, $2K^+$ 2.14 **6**
$C_4H_4O_4$, $2C_6H_7NO$ 33.20 **7**
$C_4H_4O_4^{2-}$, $2Li^+$ 2.20 **5**
$C_4H_4O_5^{2-}$, Ca^{2+} , $2H_2O$ 2.59 **1**
$C_4H_4O_6$, $2H_2O$ 1.50 **1**
$C_4H_4O_6^{2-}$, Ca^{2+} , $4H_2O$ 2.60 **1**
$C_4H_4O_6^{2-}$, K^+ , Na^+ , $4H_2O$ 2.61 + **1**
$C_4H_4O_6^{2-}$, K^+ , Na^+ , $4H_2O$ 2.22 **5**
$C_4H_4O_6^{2-}$, $2K^+$, $2H_2O$ 2.63 **1**
$C_4H_4O_6^{2-}$, Li^+ , H_4N^+ , H_2O 2.64 **1**
$C_4H_4O_6^{2-}$, Li^+ , H_4N^+ , H_2O 2.21 **5**
$C_4H_4O_6^{2-}$, Na^+ , K^+ , $4H_2O$ 2.23 **5**
$C_4H_4O_6^{2-}$, Na^+ , Rb^+ 2.65 **1**
$C_4H_4O_6^{2-}$, $2Na^+$, $2H_2O$ 2.66 **1**
$C_4H_4O_6^{2-}$, $2Rb^+$, $2H_2O$ 2.67 **1**
$C_4H_4O_6^{2-}$, Sr^{2+} , $3H_2O$ 2.68 **1**
$C_4H_4O_6^{2-}$, $2H_4N^+$ 2.24 **5**
$C_4H_4O_6^{2-}$, $2H_4N^+$ 2.7 **7**
$C_4H_4O_6^{2-}$, $C_2H_{10}N_2^{2+}$ 3.6 **5**
$C_4H_4O_6^{2-}$, $C_6H_{20}CuN_4^{2+}$, $2H_2O$ 76.31 **7**
$(C_4H_4O_7U)_n$ 81.9 **6**
$C_4H_4O_8Rh_2$, $0.5H_2O$ 81.26 **2**
C_4H_4S , $C_{19}H_{16}$ 60.30 **7**
$C_4H_4S_2$ 39.11 **1**
$C_4H_4S_3$ 39.12 **1**

$C_4H_4S_6$ 39.13 1
$C_4H_5BrN_4$ 44.13 1
$C_4H_5ClN_4$ 44.14 1
$C_4H_5ClO_2$ 1.51 1
$C_4H_5ClO_4$ 1.10 5
$(C_4H_5CuINOS_2)_n$ 85.3 4
$C_4H_5FO_5$ 1.52 1
C_4H_5N , $C_{19}H_{16}$ 60.31 7
C_4H_5NO 1.9 7
$C_4H_5NO_2$ 1.53 1
$C_4H_5NO_3$ 1.15 3
$C_4H_5NO_3$ 1.11 5
$C_4H_5N_3$ 44.5 7
$C_4H_5N_3O$ 44.15 1
$C_4H_5N_3O$ 44.16 1
$C_4H_5N_3O$ 44.1 5
$C_4H_5N_3O$, H_2O 44.17 1
$C_4H_5N_3O$, H_2O 44.2 5
$C_4H_5N_3O$, $C_4H_3FN_2O_2$, H_2O 44.9 3
$C_4H_5N_3O$, $C_{10}H_{12}N_4O_5$ 47.20 3
$C_4H_5N_3O$, $C_{13}H_9N$, H_2O 60.132. 2
$C_4H_5N_3O^+$, $C_9H_8NO_3^-$, H_2O 44.4 5
$2C_4H_5N_3O$, $C_7H_6O_4$, H_2O 44.2 6
$C_4H_5N_3O_2$ 33.5 5
$C_4H_5N_3O_8$ 1.16 3
$C_4H_5N_3S$ 44.10 3
$C_4H_5N_6O^+$, Br^- , H_2O 35.1 7
$C_4H_5N_6O^+$, Cl^- , H_2O 35.2 7
$C_4H_5O_2^-$, $2C_4H_6O_2$, Na^+ 60.8 3
$C_4H_5O_3^-$, Na^+ 2.69 1
$C_4H_5O_4^-$, Cs^+ , H_2O 2.12 4
$C_4H_5O_4^-$, K^+ 2.13 + 4
$C_4H_5O_4^-$, $C_4H_6O_4$, K^+ 2.15 6
$C_4H_5O_4S_2Sb$ 66.7 2
$C_4H_5O_5^-$, K^+ 2.16 + 6
$C_4H_5O_5^-$, Na^+ 2.18 6
$C_4H_5O_5^-$, Rb^+ 2.19 + 6
$C_4H_5O_6^-$, K^+ 2.15 4
$C_4H_5O_6^-$, H_4N^+ 2.70 1
$C_4H_5O_6^-$, $C_3H_{10}NO^+$, $2H_2O$ 2.8 7
$C_4H_5O_6^-$, $C_3H_{10}NO^+$, $2H_2O$ 2.9 7
$C_4H_5O_6^-$, $C_9H_{14}NO_3^+$ 17.14 5
$C_4H_5O_6^-$, $C_{13}H_{18}NO^+$, H_2O 33.40 6
$C_4H_5O_6^-$, $C_{15}H_{22}NO^+$ 33.43 7
$C_4H_5O_6^-$, $C_{25}H_{33}N_2O_2^+$ 40.24 7
C_4H_6 5.15 1
$C_4H_6BrNO_3$ 1.17 3
$C_4H_6Br_2O_2$ 38.8 1
$C_4H_6Br_2S_2$ 39.14 1
$C_4H_6Br_4N_2Nb$ 83.18 2
$C_4H_6CdO_5S$ 81.12 7
$(C_4H_6CdO_6)_n$ 81.10 6
$C_4H_6Cl_2N_2Zn$ 83.19 2
$C_4H_6Cl_2OS$ 42.5 1
$C_4H_6Cl_2O_2$ 38.9 1
$C_4H_6Cl_2O_2$ 38.13 1
$C_4H_6Cl_2O_2$ 38.10 + 1
$C_4H_6Cl_2S_2$ 39.15 1
$C_4H_6Cl_3NO$ 32.17 1
$C_4H_6Cl_3NOS$ 4.3 5
$C_4H_6Cl_3N_2NbO$ 83.2 5
$C_4H_6Cl_4Cu_2N_2$ 83.20 2
$C_4H_6Cl_4N_2Sn$ 69.6 3
$C_4H_6Cl_6Cu_3N_2$ 83.21 2
$C_4H_6Cl_6F_4N_2P_2$ 64.7 7
$C_4H_6Cl_6Pt_2^{2-}$, $2C_5H_{14}N^+$ 72.3 3
$(C_4H_6CoO_4)_{5n}$, $2nH_2O$ 81.13 7
$C_4H_6CuN_4O_6$ 83.22 2
$C_4H_6CuN_6O_4^{2-}$, $2K^+$, $4H_2O$ 83.23 2
$C_4H_6CuO_5$, H_2O 81.33 2
$C_4H_6CuO_6$ 81.27 2
$C_4H_6Cu_2O_4$ 81.12 5
$C_4H_6Cu_2O_4$ 81.11 6
$C_4H_6GeN_2$ 69.6 4
$C_4H_6HgO_4$ 81.12 6
$C_4H_6N_2O$ 32.3 4
$C_4H_6N_2O_2$ 33.11 + 1
$C_4H_6N_2O_2$ 44.11 3
$C_4H_6N_2O_2$ 40.3 4
$C_4H_6N_2O_2$, $0.5H_2O$ 40.2 5
$2C_4H_6N_2O_2$, Cl_2Hg 44.3 4
$C_4H_6N_2S$ 41.4 4
$C_4H_6N_2S_2Sn$ 69.7 + 3
$C_4H_6N_2S_3$ 41.5 7
$C_4H_6N_2Si$ 63.3 4
$C_4H_6N_2Sn$ 69.7 4
$C_4H_6N_3O^+$, $C_6H_2N_3O_7^-$ 44.6 7
$2C_4H_6N_3O^+$, Cl_4Pd^{2-} 44.7 7
$C_4H_6N_3O_5^-$, Rb^+ 1.10 7
$C_4H_6N_4NiO_4$ 83.24 + 2
$C_4H_6N_4O$, H_2O 32.3 7
$C_4H_6N_4O_3$ 32.18 1
$C_4H_6N_4O_3$ 40.2 6
$C_4H_6N_4O_3$ 40.3 6
$C_4H_6N_4O_3S_2$ 41.2 6
$C_4H_6N_4O_4Pd$ 83.3 5
$C_4H_6N_4O_4Pt$ 83.26 2
$2C_4H_6N_7^+$, O_4S^{2-} , H_2O 35.3 7
$C_4H_6O_2$ 1.103 1
$C_4H_6O_2$ 1.11 7
$2C_4H_6O_2$, $C_4H_5O_2^-$, Na^+ 60.8 3
$C_4H_6O_2S_2$ 39.16 1
$C_4H_6O_2S_4Se$ 11.4 7
$C_4H_6O_2S_4Te$ 70.1 7
$C_4H_6O_2S_5$ 11.5 7
$C_4H_6O_3S$ 39.1 4
$C_4H_6O_3S$ 39.3 5
$C_4H_6O_4$ 1.54 1
$C_4H_6O_4$ 1.55 1
$C_4H_6O_4$ 1.18 3
$C_4H_6O_4$, $C_4H_5O_4^-$, K^+ 2.15 6
$C_4H_6O_4$, $2C_7H_7NO$ 60.9 5
$C_4H_6O_4S$ 1.56 1
$C_4H_6O_5$ 1.4 + 6
$C_4H_6O_6$ 1.60 1
$C_4H_6O_6$ 1.57 + 1
$C_4H_6O_6$ 1.19 3
$C_4H_6O_6$, H_2O 1.61 1
$C_4H_6O_6$, H_2O 1.62 + 1
$C_4H_7AsO_2S_2$ 65.4 4
$C_4H_7Cl_4N_2O_2Sb$ 66.1 6
$(C_4H_7ErO_7)_n$, nH_2O 81.6 3
C_4H_7NO 20.3 1
C_4H_7NO 1.12 7
$C_4H_7NO_2$ 32.11 3
$C_4H_7NO_2$ 1.13 7
$C_4H_7NO_2$ 1.14 7
$2C_4H_7NO_2$, Br^- , Na^+ 60.40 2
$2C_4H_7NO_2$, I^- , K^+ 60.41 2
$C_4H_7NO_2S$ 41.3 5
$C_4H_7NO_2S$ 48.8 6
$C_4H_7NO_3$ 48.23 1

$C_4H_7NO_3$	48.9 **3**
$C_4H_7NO_3$	48.11 **7**
$C_4H_7NO_4$	48.25 **1**
$C_4H_7NO_4$	48.9 **6**
$C_4H_7NO_4$	48.10 **6**
$C_4H_7N_3O$	48.26 **1**
$C_4H_7N_3O_4$	32.6 **6**
$C_4H_7N_3S_2$	32.12 **3**
$C_4H_7N_4O_4^-$, K^+	2.71 **1**
$C_4H_7O^+$, Cl_6Sb^-	12.6 **4**
$C_4H_8AsClOS_2$	65.2 **7**
$C_4H_8AsClS_3$	65.3 **7**
$C_4H_8BrS^+$, Br^-	39.1 **3**
$C_4H_8Br_2I_2S_2$	39.2 **3**
$C_4H_8Br_2SSe$	42.6 **1**
$C_4H_8Br_2STe$	70.2 **5**
$C_4H_8Br_2Se$	39.17 **1**
$C_4H_8CdI_2N_8$	83.2 **7**
$(C_4H_8CdN_4Ni)_n$, $2nC_6H_6$	61.2 **5**
$(C_4H_8CdO_7)_n$, nH_2O	81.13 **6**
$C_4H_8ClHgS^+$, Cl^-	85.17 **2**
$C_4H_8Cl_2$, CH_4N_2O	61.1 **5**
$C_4H_8Cl_2CuN_2O_2$	83.7 **3**
$C_4H_8Cl_2N_8Zn$	83.8 **3**
$C_4H_8Cl_3O_2Pt^-$, $C_{24}H_{20}P^+$	72.4 **3**
$C_4H_8Cl_3O_4P$	64.7 **3**
$C_4H_8Cl_4Pd_2$	72.4 **2**
$C_4H_8Cl_4Se_2$	39.18 **1**
$C_4H_8Cl_8Si_4$	63.1 **2**
$C_4H_8CoN_2O_8$, $2H_2O$	81.15 **4**
$(C_4H_8CoO_7)_n$, nH_2O	81.14 **5**
$C_4H_8CuN_2O_4$, H_2O	82.4+ **2**
$C_4H_8CuN_2O_4$, $2H_2O$	82.3 **2**
$C_4H_8CuN_4S_2$	76.1 **4**
$C_4H_8Cu_2N_6O_6^{2-}$, $2K^+$, $4H_2O$	79.3 **6**
$C_4H_8HgN_2O_2$	83.9 **3**
$C_4H_8HgN_2O_2$	83.5 **6**
$C_4H_8HgN_4S_2$	76.4 **7**
$C_4H_8HgN_6S_4$	79.17 **2**
$C_4H_8I_2OTe$	70.1 **6**
$C_4H_8I_2STe$	70.2 **3**
$C_4H_8I_2S_3$	11.26+ **1**
$(C_4H_8MnO_5)_n$, $3nH_2O$	81.14 **7**
$C_4H_8NOS^+$, Cl^-	48.27+ **1**
$C_4H_8NO_2^+$, Br^-	48.12+ **5**
$C_4H_8NO_4^+$, Br^-	48.11 **6**
$C_4H_8NO_4^+$, Br^-	48.12 **7**
$C_4H_8NO_4^+$, Cl^-	48.29 **1**
$C_4H_8NO_4^+$, Cl^-	48.12 **6**
$C_4H_8NO_4^+$, F^- , H_2O	48.13 **6**
$C_4H_8NO_4^+$, I^-	48.14 **6**
$C_4H_4D_4NO_4^+$, Cl^-	48.15 **6**
$C_4H_8N_2NiO_2S_2$	84.7 **2**
$C_4H_8N_2NiO_2S_2$	85.18 **2**
$C_4H_8N_2NiO_2S_2$	82.3 **4**
$C_4H_8N_2NiO_4$, $2H_2O$	48.2 **1**
$C_4H_8N_2NiO_4$, $2H_2O$	82.6 **2**
$C_4H_8N_2NiS_4$	80.1 **5**
$C_4H_8N_2O$	20.4 **1**
$C_4H_8N_2O_2$	1.64 **1**
$C_4H_8N_2O_2$	9.9 **1**
$C_4H_8N_2O_2$	10.22 **1**
$C_4H_8N_2O_3$	48.30 **1**
$C_4H_8N_2O_3$	48.10 **3**
$C_4H_8N_2O_3$	48.13 **7**
$C_4H_8N_2O_3$, Li^+ , Br^-	48.9 **4**
$C_4H_8N_2O_3$, H_2O	48.31 **1**
$C_4H_8N_2O_3$, H_2O	48.10 **4**
$C_4H_8N_2O_3$, H_2O	48.14+ **5**
$C_4D_8N_2O_3$	48.11 **3**
$2C_4H_8N_2O_3$, Ca^{2+} , $2Cl^-$	48.11 **4**
$C_4H_8N_2O_4Pt$	82.7 **2**
$C_4H_8N_2O_8Zn$	81.16 **4**
$C_4H_8N_2S$	8.41 **1**
$C_4H_8N_2S$	8.42 **1**
$C_4H_8N_3O^+$, $C_{10}H_{13}N_2O^+$, O_4S^{2-} , H_2O	35.28 **1**
$C_4H_8N_3O_5P^{2-}$, $2Na^+$, $4.5H_2O$	64.12 **2**
$C_4H_8N_3O_5P^{2-}$, $2Na^+$, xH_2O	48.16 **5**
$C_4H_8N_4$	8.14 **3**
$C_4H_8N_4O_6$	9.1 **3**
$C_4H_8N_6NiS_4$	79.18 **2**
$C_4H_8N_6NiS_4$	85.19 **2**
$C_4H_8N_6NiS_4$, $C_2H_6O_2$	79.6 **5**
$C_4H_8N_6PdS_4$	79.19 **2**
$C_4H_8N_8O_8$	34.1+ **1**
$C_4H_8N_8O_8$	34.1 **3**
$C_4H_8N_8O_8$	34.1 **7**
C_4H_8O , Br_2Hg	38.1 **4**
$2C_4H_8O$, $C_{10}H_2Cr_2N_2O_{10}$	38.2 **7**
$3C_4H_8O$, $C_{36}H_{44}Na_2Ni_2O_2$, 2Na	71.105 **7**
$8C_4H_8O$, $7.33H_2S$, $136H_2O$	61.3 **2**
$2C_4H_8OS$, $C_6H_{10}N_2O_8$	60.13 **4**
C_4H_8OSe , ClI	60.42 **2**
C_4H_8OSe , I_2	60.43 **2**
$C_4H_8O_2$	1.65 **1**
$C_4H_8O_2$, Br_2	60.44 **2**
$C_4H_8O_2$, Cl_2Ge	60.9 **3**
$C_4H_8O_2$, Cl_2Hg	60.45 **2**
$C_4H_8O_2$, Li^+ , Cl^-	60.46 **2**
$C_4H_8O_2$, Li^+ , Cl^- , H_2O	60.47 **2**
$C_4H_8O_2$, N_2O_4	60.48 **2**
$C_4H_8O_2$, H_2O_4S	60.49 **2**
$C_4H_8O_2$, $H_{12}MgO_6^{2+}$, $2Cl^-$	38.3 **7**
$C_4H_8O_2$, CHI_3	60.2 **3**
$C_4H_8O_2$, $C_2Br_2O_2$	60.17 **2**
$C_4H_8O_2$, $C_2Cl_2O_2$	60.18 **2**
$C_4H_8O_2$, C_2I_2	60.50 **2**
$C_4H_8O_2$, $C_4H_4Cu_2O_8$	81.22 **2**
$C_4H_8O_2$, $C_8H_{14}N_6O_{10}$	60.13 **6**
$C_4H_8O_2$, $2C_{10}H_{14}O_5V$	60.34 **3**
$2C_4H_8O_2$, Br_2Hg	60.10 **3**
$2C_4H_8O_2$, $H_8Br_2NiO_4$	38.4 **7**
$3C_4H_8O_2$, Ag^+ , ClO_4^-	60.51 **2**
$C_4H_8O_2S$	11.28 **1**
$C_4H_8O_2S$, $C_3H_6N_6O_6$	60.28 **2**
$2C_4H_8O_2S$, $H_4Cl_6Cu_3O_2$	39.1 **7**
$C_4H_8O_2S_2$	39.19 **1**
$C_4H_8O_3$, Na^+ , ClO_4^-	45.1+ **6**
$C_4H_8O_4$	38.1 **6**
$C_4H_8O_4$	38.5 **7**
$C_4H_8O_5$	38.14 **1**
$(C_4H_8O_7U)_n$, nH_2O	81.14 **6**
$C_4H_8O_8$	20.5 **1**
$(C_4H_8O_{10}Th)_n$, nH_2O	81.28 **2**
$C_4H_8O_{12}P_4^{4-}$, $2Ca^{2+}$, $10H_2O$	64.2 **5**
$C_4H_8O_{12}P_4^{4-}$, $4H_4N^+$, $2H_2O$	64.2 **6**
$C_4H_8S_2$	39.20 **1**
$C_4H_8S_2$, Cl_3Sb	39.2 **7**
$C_4H_8S_2$, $2I_2$	60.52 **2**
$C_4H_8S_2$, CHI_3	60.53 **2**
$C_4H_8S_2$, C_2I_2	60.19 **2**

Formula	Entry	Vol.
$2C_4H_8S_2$, I_3Sb	60.54	2
$C_4H_8S_4$	39.2	6
C_4H_8Se , I_2	60.55 +	2
$C_4H_8Se_2$	39.21	1
$C_4H_8Se_2$, $2I_2$	60.57	2
$C_4H_8Se_2$, $2CHI_3$	60.4	2
$C_4H_8Se_2$, C_2I_2	60.20	2
$C_4H_8Se_2$, C_2I_4	60.58	2
$C_4H_9AgNO^+$, I^-	83.27	2
$C_4H_9AlN^-$, K^+	68.1	5
$C_4H_9AlNS^+$, $C_2H_6Tl^-$	68.1	6
$(C_4H_9CdClN_2O_3)_n$	82.1	6
C_4H_9Cl	5.16	1
$C_4H_9Cl_2NO_2PdS$, H_2O	82.1	5
$C_4H_9Cl_2N_3SZn$	79.7	5
$C_4H_9CoN_5OS_4$	85.1	7
$(C_4H_9CuNO_6)_n$	82.1	7
$(C_4H_9CuNPS_2)_n$	85.1	5
C_4H_9GeN	69.6	2
$C_4H_9HgNO_2S$, H_2O	82.2	7
$C_4H_9InO_2$	68.2	5
C_4H_9NO , $3CH_6N_3^+$, $3Cl^-$	60.16	2
C_4H_9NO , C_8H_5I	60.59	2
$C_4H_9NO_2$	1.66 +	1
$C_4H_9NO_2$	48.32	1
$C_4H_9NO_2$	48.17	5
$C_4H_9NO_2$	1.6 +	6
$C_4H_9NO_3$	48.33	1
$C_4H_9NO_3$	1.15	7
$C_4H_9NO_3$	48.14	7
$C_4H_9NO_3$	48.15 +	7
$C_4H_9NO_3S$	48.34	1
$(C_4H_9NO_6Zn)_n$, H_2O	82.2	6
C_4H_9NSSn	69.9	3
C_4H_9NSn	69.8	2
$C_4H_9N_2O_3^+$, Cl^- , H_2O	48.35	1
$C_4H_9N_2O_3^+$, Cl^- , H_2O	48.21	5
$C_4H_9N_2O_3^+$, NO_3^-	48.16	6
$C_4H_9N_2O_3^+$, $H_2O_4P^-$, H_2O	48.23	5
$C_4H_9N_3O$	8.21	7
$C_4H_9N_3O_2$	1.8	6
$C_4H_9N_3O_2$, H_2O	48.36 +	1
$C_4H_9N_3O_2$, H_2O	48.24	5
$C_4H_9N_3S$	8.22	7
$C_4H_9N_5NiOS_4$	85.2	5
$C_4H_9O_2P$	64.8	3
$C_4H_9O_3P$	64.3	6
$(C_4H_{10}AgO_4P)_n$	84.2	5
$C_4H_{10}Ag_2N_2O_4^{2+}$, $2NO_3^-$	82.4 +	4
$C_4H_{10}Ag_2N_2O_4^{2+}$, $2NO_3^-$	82.2	5
$C_4H_{10}Ag_3N_2S^+$, NO_3^-	85.4	6
$C_4H_{10}Br_4O_2Ti^-$, $C_8H_{20}Br_2O_4Ti^+$	84.27	2
$C_4H_{10}CdCl_2N_2O_2$	83.28	2
$C_4H_{10}CdCl_2N_6O_4$	84.8	2
$C_4H_{10}CdN_4O_4S_2$	79.20	2
$(C_4H_{10}CdO_2S_2)_n$	85.2	7
$(C_4H_{10}CdO_6)_n$	81.17	4
$C_4H_{10}ClCuS_2$	85.20	2
$C_4H_{10}ClHgO_3P$	86.1	3
$C_4H_{10}ClHgS^+$, Cl^- , Cl_2Hg	85.21	2
$C_4H_{10}Cl_2N_2Pt$	83.3	7
$C_4H_{10}Cl_2N_2Pt$	83.4	7
$C_4H_{10}Cl_2N_4Pd$	71.1	4
$C_4H_{10}Cl_2N_6O_4Zn$	84.9	2
$C_4H_{10}Cl_3NPt$	72.5	3
$C_4H_{10}Cl_3NPt$	72.6	3
$(C_4H_{10}Cl_3NTi)_n$	83.9	4
$C_4H_{10}Cl_6O_3S_2W_2$	84.2	4
$C_4H_{10}Cl_8O_2Sb_2$	66.8	2
$C_4H_{10}Cl_8S_4W_2$	85.5	6
$C_4H_{10}CuN_2O_5$	82.3	7
$C_4H_{10}CuN_6O_4^{2+}$, $2Cl^-$	84.10	2
$C_4H_{10}Fe_2N_4O_4S_2$	85.22	2
$C_4H_{10}HgS_2$	85.23	2
$4C_4H_{10}LiO^+$, $C_{10}H_{32}Cr_2^{4-}$	71.12	4
$C_4H_{10}MoO_6$	84.4	7
$C_4H_{10}NO^+$, $C_{12}H_4N_4^-$	40.5	5
$2C_4H_{10}NO^+$, $2C_{12}H_4N_4^-$, $C_{12}H_4N_4$	40.4	4
$C_4H_{10}NO_2^+$, Cl^-	48.25	5
$C_4H_{10}NO_2^+$, Cl^-	1.9	6
$C_4H_{10}NO_3^+$, Br^-	48.26	5
$C_4H_{10}NO_6P$	48.12	3
$C_4H_{10}NO_6P$	48.12	4
$C_4H_{10}N_2$, $6H_2O$	33.13	1
$C_4H_{10}N_2$, $6H_2O$	33.2	7
$0.5C_4H_{10}N_2$, Ag^+ , I^-	60.60	2
$C_4H_{10}N_2O$	10.1	6
$C_4H_{10}N_2O_2$	48.17	7
$C_4H_{10}N_4NiS_2$	85.4	3
$C_4H_{10}N_4O_2PbS_2$, H_2O	69.6	5
$C_4H_{10}N_6NiS_4^{2+}$, $2ClO_4^-$, C_2H_6O	79.4	6
$C_4H_{10}N_8NiS_4$	83.30	2
$C_4H_{10}N_9NiS_4$	79.6	7
$C_4H_{10}NiO_2S_2$	85.24	2
$C_4H_{10}O$	5.13	5
$C_4H_{10}O$, $CHBrCl_2$	60.61	2
$4C_4H_{10}O$, $Br_6Mg_4O_7$	67.17	2
$C_4H_{10}O_4$	45.1 +	1
$C_4H_{10}O_4$	45.1	7
$C_4H_{10}O_4P^-$, $C_4H_{12}N_3^+$	8.13	5
$C_4H_{10}O_4P^-$, $C_6H_{15}N_4O_2^+$	48.35	6
$2C_4H_{10}O_4P^-$, Ba^{2+}	46.6	1
$2C_4H_{10}O_4P^-$, Mg^{2+}	46.3	6
$2C_4H_{10}O_4P^-$, $C_4H_{14}N_2^{2+}$	3.14	5
$C_4H_{10}O_6S_3$	11.29	1
$C_4H_{10}O_6Zn$	81.29	2
$C_4H_{10}O_7SZn$, H_2O	81.15	7
$C_4H_{10}O_8Zn$	81.30	2
$C_4H_{10}PSSe^-$, Tl^+	64.4	6
$C_4H_{10}PS_2Tl$	68.5	7
$C_4H_{11}Cl_2NPt$	72.5	2
$C_4H_{11}N$, $8.67H_2O$	61.4	2
$C_4H_{11}N$, $9.75H_2O$	61.5	2
$xC_4H_{11}N$, $C_{40}H_{37}EuO_8$	61.10	5
$(C_4H_{11}NNdO_7^+)_n$, nCl^-	81.18	4
$C_4H_{11}NO_2$, $C_{14}H_9I_3O_4$	60.26	5
$C_4H_{11}NO_3S_2$	4.8	1
$C_4H_{11}NO_4S$	3.5	7
$C_4H_{11}N_2O_2^+$, Br^-	1.68	1
$C_4H_{11}N_2O_2^+$, Cl^-	48.13 +	4
$C_4H_{11}N_3O_3S$	8.5	6
$C_4H_{11}N_3O_4S$	8.6	6
$C_4H_{12}As^+$, Br^-	65.7	2
$C_4H_{12}As_2S_2$	65.8	2
$(C_4H_{12}BLi)_n$	62.3	3
$C_4H_{12}BN$	62.4	3
$C_4H_{12}B_2Cl_4N_2$	62.20	2
$C_4H_{12}B_2F_4N_2$	62.21	2
$C_4H_{12}B_2N_8NiO_8$, $4H_2O$	83.10	4
$C_4H_{12}B_4$	62.22	2
$C_4H_{12}B_6$	62.23	2
$C_4H_{12}B_{10}Br_4$	62.24	2

$C_4H_{12}Br_4Pd_2S_2$ 85.26 **2**
$(C_4H_{12}Cd_2O_{12})_n$ 81.15 **6**
$C_4H_{12}Cl_2CuN_8O_2$ 83.10 **3**
$C_4H_{12}Cl_2CuO_2S_2$ 84.3 **3**
$(C_4H_{12}Cl_2CuP_2S_2)_n$ 85.3 **7**
$C_4H_{12}Cl_2In_2$ 68.6 **7**
$C_4H_{12}Cl_2MnO_4$ 84.4 **6**
$C_4H_{12}Cl_2N_2Pd$ 76.2 **4**
$C_4H_{12}Cl_2O_2PdS_2$ 85.28 + **2**
$C_4H_{12}Cl_2O_4W$ 84.5 **7**
$C_4H_{12}Cl_3N_2Pt^+$, Cl^- , $0.5H_2O$ 72.3 **4**
$C_4H_{12}Cl_4Hg_2O_2S_2$, Cl_2Hg 84.6 **7**
$C_4H_{12}Cl_4O_2RhS_2^-$, Na^+ 84.7 **7**
$C_4H_{12}Cl_4O_6Re_2$ 81.7 **3**
$C_4H_{12}Cl_6N_6P_4$ 64.5 **6**
$C_4H_{12}CoN_4OS_4Si_2$ 83.5 **7**
$C_4H_{12}CuN_2O_4$ 81.31 + **2**
$C_4H_{12}F_4N_4P_4$ 64.9 **3**
$C_4H_{12}Ge_4S_6$ 69.7 **5**
$C_4H_{12}HgO_6P_2$ 86.1 **5**
$C_4H_{12}In^-$, Cs^+ 68.3 **5**
$C_4H_{12}In^-$, K^+ 68.4 **5**
$C_4H_{12}Li_4$ 67.5 **3**
$C_4H_{12}N^+$, $Ag_2I_3^-$ 3.19 **1**
$C_4H_{12}N^+$, $AlO_8Si_2^-$, H_2O 3.14 **3**
$C_4H_{12}N^+$, $AlO_{12}Si_5^-$ 3.15 **3**
$C_4H_{12}N^+$, Br_3Hg^- 3.20 **1**
$C_4H_{12}N^+$, Br_3Ni^- 3.21 **1**
$C_4H_{12}N^+$, $CeCl_6^-$ 3.22 **1**
$C_4H_{12}N^+$, Cl^- , $5Cl_2OSe$ 3.23 **1**
$C_4H_{12}N^+$, ClO_4^- 3.24 **1**
$C_4H_{12}N^+$, Cl_2I^- 3.25 **1**
$C_4H_{12}N^+$, Cl_3Mn^- 3.26 **1**
$C_4H_{12}N^+$, Cl_3Ni^- 3.27 **1**
$C_4H_{12}N^+$, Cs^+ , $H_{13}B_{11}^{2-}$ 3.28 **1**
$C_4H_{12}N^+$, F^- , $4H_2O$ 61.6 **2**
$C_4H_{12}N^+$, HgI_3^- 3.29 **1**
$C_4H_{12}N^+$, HCl_2^- 3.16 + **3**
$C_4H_{12}N^+$, HO^- , $5H_2O$ 61.7 **2**
$C_4H_{12}N^+$, $H_4Cl_4O_2Rh^-$ 3.9 **5**
$C_4H_{12}N^+$, $C_2AgN_2O_2^-$ 3.4 **6**
$C_4H_{12}N^+$, $C_4H_8B_3CrO_4^-$ 3.18 **3**
$C_4H_{12}N^+$, $C_4H_{14}Mo_5N_2O_{21}P_2^{2-}$, Na^+ , $5H_2O$ 64.8 **7**
$C_4H_{12}N^+$, $C_4H_{16}B_{18}Br_6Co^-$ 3.31 **1**
$C_4H_{12}N^+$, $C_4H_{17}B_{10}^-$ 62.2 **6**
$C_4H_{12}N^+$, $C_4H_{22}B_{18}Ni^-$ 62.3 **6**
$C_4H_{12}N^+$, $C_6N_6S_6Tc^-$ 3.5 **6**
$C_4H_{12}N^+$, $C_7H_5N_2O_2^-$ 10.3 **6**
$C_4H_{12}N^+$, $C_{11}HN_6^-$ 12.4 **7**
$C_4H_{12}N^+$, $C_{13}H_8MnO_6^-$ 71.31 **7**
$C_4H_{12}N^+$, $C_{14}H_{10}MnO_3^-$ 71.14 **3**
$C_4H_{12}N^+$, $C_{14}H_{21}B_{10}^-$ 62.8 **6**
$C_4H_{12}N^+$, $C_{15}H_{30}N_3O_2S_6U^-$ 80.31 **2**
$C_4H_{12}N^+$, $C_{17}H_{23}O_7^-$ 38.55 **7**
$C_4H_{12}N^+$, $C_{18}H_{13}O_4Si^-$ 3.32 **1**
$C_4H_{12}N^+$, $C_{18}H_{19}O_6Pb^-$ 69.30 **4**
$C_4H_{12}N^+$, $C_{24}H_{20}B^-$ 62.13 **7**
$2C_4H_{12}N^+$, $13Ag^+$, $15I^-$ 3.19 **3**
$2C_4H_{12}N^+$, $Br_4O_2U^{2-}$ 3.6 + **6**
$2C_4H_{12}N^+$, Cl_4Co^{2-} 3.33 **1**
$2C_4H_{12}N^+$, Cl_4Cu^{2-} 3.34 **1**
$2C_4H_{12}N^+$, Cl_4Cu^{2-} 3.6 **7**
$2C_4H_{12}N^+$, Cl_4Fe^{2-} 3.7 **7**
$2C_4H_{12}N^+$, Cl_4Ni^{2-} 3.35 **1**
$2C_4H_{12}N^+$, $Cl_4O_2U^{2-}$ 3.8 **6**
$2C_4H_{12}N^+$, Cl_4Zn^{2-} 3.36 **1**
$2C_4H_{12}N^+$, $Cl_6Ga_2^{2-}$ 3.9 **6**
$2C_4H_{12}N^+$, $Cl_{18}Nb_6^{2-}$ 3.7 **4**
$2C_4H_{12}N^+$, O_4S^{2-} , $4H_2O$ 61.8 **2**
$2C_4H_{12}N^+$, HO_7V^{2-} , $4H_2O$ 3.8 **7**
$2C_4H_{12}N^+$, $H_6B_6^{2-}$ 3.37 **1**
$2C_4H_{12}N^+$, $H_7B_{10}Cl_3^{2-}$ 3.8 **4**
$2C_4H_{12}N^+$, $H_{14}B_{10}^{2-}$ 3.10 **5**
$2C_4H_{12}N^+$, $H_{20}B_{20}FeS_2^{2-}$ 3.11 **5**
$2C_4H_{12}N^+$, $H_{24}B_{20}Ni^{2-}$ 3.9 **4**
$2C_4H_{12}N^+$, CS_3^{2-} , $4H_2O$ 3.9 **7**
$2C_4H_{12}N^+$, $C_4N_4Pd^{2-}$ 3.10 **4**
$2C_4H_{12}N^+$, $C_6H_4Cu_2N_2O_8S_2^{2-}$ 3.38 **1**
$2C_4H_{12}N^+$, $C_8N_4NiS_4^{2-}$ 85.30 **2**
$2C_4H_{12}N^+$, $C_8H_{32}B_{20}Ti^{2-}$, $2C_3H_6O$ 71.8 **7**
$2C_4H_{12}N^+$, $C_{10}H_{12}Cu_2N_2O_8S_2^{2-}$ 3.39 **1**
$2C_4H_{12}N^+$, $C_{12}N_6S_6V^{2-}$ 3.40 **1**
$2C_4H_{12}N^+$, $C_{12}Ni_3O_{12}^{2-}$ 3.10 **6**
$2C_4H_{12}N^+$, $C_{14}Co_4Ni_2O_{14}^{2-}$ 3.11 **6**
$2C_4H_{12}N^+$, $C_{17}Fe_6O_{16}^{2-}$ 71.35 **4**
$2C_4H_{12}N^+$, $C_{17}Fe_6O_{16}^{2-}$ 3.10 **7**
$2C_4H_{12}N^+$, $C_{18}H_{12}S_6Zr^{2-}$ 85.34 **6**
$2C_4H_{12}N^+$, $C_{24}H_{20}Fe_4S_8^{2-}$ 85.42 **7**
$2C_4H_{12}N^+$, $C_{30}O_{30}Rh_{12}^{2-}$ 3.21 **3**
$3C_4H_{12}N^+$, $Br_9Sb_2^{3-}$, Br_2 3.12 **5**
$3C_4H_{12}N^+$, $Cl_{15}Ge_5Pt^{3-}$ 3.12 **6**
$3C_4H_{12}N^+$, $C_{16}O_{16}Rh_7^{3-}$ 3.41 **1**
$4C_4H_{12}N^+$, $Cl_{10}Cu_4O^{4-}$ 3.22 **3**
$nC_4H_{12}N^+$, $(CdCl_3^-)_n$ 3.13 **5**
$C_4H_{12}NO^+$, $C_8H_{11}N_2O_3^-$ 43.2 **7**
$C_4H_{12}NS_2^+$, Br^- , H_2O 4.3 **7**
$C_4H_{12}N_2^{2+}$, $2Cl^-$, H_2O 33.14 **1**
$C_4H_{12}N_2^{2+}$, $2Cl_2I^-$ 60.62 **2**
$C_4H_{12}N_2^{2+}$, MoS_4^{2-} 33.2 **4**
$C_4H_{12}N_2^{2+}$, $4C_3H_7NO$, $10Ag^+$, $12I^-$ 33.3 **7**
$C_4H_{12}N_2^{2+}$, $C_8H_4O_4^{2-}$ 33.3 **6**
$C_4H_{12}N_2NiO_6$ 83.31 **2**
$C_4H_{12}N_2O_2S$ 4.9 **1**
$(C_4H_{12}N_2O_2Sn)_n$ 69.2 **7**
$C_4H_{12}N_2O_8PdS_2$ 85.31 **2**
$C_4H_{12}N_3^+$, $C_4H_{10}O_4P^-$ 8.13 **5**
$C_4H_{12}N_4$ 33.4 + **7**
$C_4H_{12}N_4NiO_6$ 83.29 **2**
$C_4H_{12}N_4O_5PbS_2$ 69.8 **5**
$C_4H_{12}N_4S_2Si_2$ 63.2 **7**
$C_4H_{12}N_4S_4$ 4.1 **6**
$C_4H_{12}N_8O_8Sc^-$, $H_5N_2^+$, $3H_2O$ 81.16 **6**
$(C_4H_{12}NdO_8S^+)_n$, nCl^- 81.17 **6**
$C_4H_{12}NiO_4P_2S_4$ 85.32 **2**
$C_4H_{12}NiP_2S_4$ 85.33 **2**
$(C_4H_{12}OSn)_n$ 69.3 **6**
$C_4H_{12}O_2Si$ 63.2 **2**
$C_4H_{12}O_4P_2S_4Te$ 70.6 **2**
$C_4H_{12}O_4Ti$ 84.11 **2**
$C_4H_{12}P_2S_2$ 64.13 **2**
$C_4H_{12}P_2S_2$ 64.10 **3**
$C_4H_{12}S_6Si_4$ 63.3 **2**
$C_4H_{12}S_6Sn_4$ 69.9 **5**
$C_4H_{12}Sb^+$, $C_{12}H_{36}AlO_4Si_4^-$ 68.21 **2**
$C_4H_{13}B_7$ 62.25 **2**
$C_4H_{13}Cl_3CrN_3$ 76.5 **3**
$C_4H_{13}CoN_6O_6$ 76.2 **2**
$C_4H_{13}CoN_{12}$ 76.5 **7**
$C_4H_{13}MoN_3O_3$ 76.3 **2**
$C_4H_{13}N_2^+$, $C_8H_{11}N_2O_3^-$ 43.3 **7**

$C_4H_{13}N_4O_2Pt^+$, I^- 76.1 **5**
$C_4H_{14}B_8$ 62.5 **3**
$C_4H_{14}B_{10}Br_2$ 62.26+ **2**
$C_4H_{14}B_{10}Cl_2$ 62.1 **6**
$C_4H_{14}CaO_{10}$, H_2O 67.3 **5**
$C_4H_{14}Cl_2MnN_2O_6$ 82.8 **2**
$C_4H_{14}CoO_8$ 81.34 **2**
$C_4H_{14}ErO_{10}^-$, $C_8H_{12}ErO_{12}^+$ 81.25 **4**
$C_4H_{14}Mo_5N_2O_{21}P_2^{2-}$, $C_4H_{12}N^+$, Na^+ , $5H_2O$ 64.8 **7**
$C_4H_{14}N_2^{2+}$, $2Cl^-$ 3.42 **1**
$C_4H_{14}N_2^{2+}$, $2H_2O_4P^-$ 3.11 **7**
$C_4H_{14}N_2^{2+}$, $2C_4H_{10}O_4P^-$ 3.14 **5**
$C_4H_{14}N_2^{2+}$, $C_{40}H_{50}O_{12}$, $2F_6P^-$ 60.39 **7**
$C_4H_{14}N_2O_2S_2^{2+}$, $2Cl^-$, H_2O 4.10 **1**
$C_4H_{14}N_2O_8Sn_2$ 69.4 **6**
$C_4H_{14}N_4O_4S_6Te$ 70.7 **2**
$(C_4H_{14}N_4O_4Zn)_n$ 81.35 **2**
$C_4H_{14}N_{10}Ni^{2+}$, $2Cl^-$, $2H_2O$ 83.32 **2**
$C_4H_{14}NiO_8$ 81.36 **2**
$C_4H_{14}NiO_8$ 81.8 **3**
$C_4H_{15}B_7$ 62.6 **4**
$C_4H_{15}B_9$ 62.28 **2**
$C_4H_{16}AlB_9$ 62.7 **4**
$C_4H_{16}BN_2^+$, Cl^- 62.29 **2**
$C_4H_{16}B_2N_2$ 62.30 **2**
$C_4H_{16}B_{18}Br_6Co^-$, $C_4H_{12}N^+$ 3.31 **1**
$C_4H_{16}Br_2CoN_4^+$, $H_5O_2^+$, $2Br^-$ 76.4 **2**
$C_4H_{16}ClCoN_5O^+$, ClO_4^- 76.6 **3**
$C_4H_{16}ClCoN_5O_5^+$, ClO_4^- 76.3 **6**
$C_4H_{16}ClN_4Ni^+$, Cl^- 76.6 **2**
$C_4H_{16}Cl_2CoN_4^+$, Cl^- 76.7 **2**
$C_4H_{16}Cl_2CoN_4^+$, Cl^- , H_2O 76.9+ **2**
$C_4H_{16}Cl_2CoN_4^+$, Cl^- , H_2O 76.3 **4**
$C_4H_{16}Cl_2CoN_4^+$, $2Cl^-$, H^+ , $2H_2O$ 76.8 **2**
$C_4H_{16}Cl_2CoN_4^+$, $2Cl^-$, H^+ , $2H_2O$ 76.7 **3**
$C_4H_{16}Cl_2CoN_4^+$, Cl_4Tl^- 76.2 **5**
$C_4H_{16}Cl_2CoN_4^+$, NO_3^- 76.11 **2**
$C_4H_{16}Cl_2CoN_4^+$, $3CH_4N_2S$, Cl^- 76.6 **7**
$2C_4H_{16}Cl_2CoN_4^+$, $O_6S_6^{2-}$, H_2O 76.4 **6**
$C_4H_{16}Cl_2CoN_8S_4$ 79.5 **3**
$C_4H_{16}Cl_2CoO_4$ 84.12 **2**
$C_4H_{16}Cl_2CrN_4^+$, $2Cl^-$, $H_5O_2^+$ 76.13 **2**
$C_4H_{16}Cl_2CrO_4^+$, Cl^- 84.8 **7**
$C_4H_{16}Cl_2IrN_4^+$, Cl^- 76.5 **6**
$C_4H_{16}Cl_2N_4Pd$ 76.14 **2**
$C_4H_{16}Cl_2N_4Pt^{2+}$, $2Cl^-$ 76.8 **3**
$C_4H_{16}Cl_2N_8NiS_4$ 79.23 **2**
$C_4H_{16}Cl_4N_4O_3Re_2$ 76.3 **5**
$C_4H_{16}CoN_6O_4^+$, NO_3^- 76.9 **3**
$C_4H_{16}CoN_6O_4^+$, NO_3^- 76.10 **3**
$C_4H_{16}CoN_6O_4^+$, $C_8H_{12}CoN_2O_8^-$ 76.4 **5**
$(C_4H_{16}CoN_8O_4^{2+})_n$, $2nNO_3^-$ 79.7 **7**
$C_4H_{16}CoN_8S_4^{2+}$, $2NO_3^-$ 79.6 **3**
$C_4H_{16}CoN_8S_4^{2+}$, $2NO_3^-$, H_2O 79.7 **3**
$C_4H_{16}CoN_{10}^+$, NO_3^- 76.24 **2**
$C_4H_{16}CuN_4^{2+}$, $2BF_4^-$ 76.15 **2**
$C_4H_{16}CuN_4^{2+}$, $2ClO_4^-$ 76.16 **2**
$C_4H_{16}CuN_4^{2+}$, $2NO_3^-$ 76.17 **2**
$C_4H_{16}CuN_4^{2+}$, O_3SSe^{2-} 76.4 **4**
$C_4H_{16}CuN_4^{2+}$, $O_3S_2^{2-}$ 76.5 **4**
$C_4H_{16}CuN_4^{2+}$, CNS^- , ClO_4^- 76.6 **4**
$C_4H_{16}CuN_4^{2+}$, CNS^- , ClO_4^- 76.5 **5**
$C_4H_{16}CuN_4^{2+}$, $2CNS^-$ 76.18 **2**
$C_4H_{16}CuN_4^{2+}$, $C_4HgN_4S_4^{2-}$ 76.19 **2**
$C_4H_{16}CuN_4^{2+}$, $C_4N_4Ni^{2-}$ 76.7 **7**
$C_4H_{16}CuN_4^{2+}$, $2C_5H_7O_2^-$, $2H_2O$ 76.8 **7**
$4C_4H_{16}CuN_4^{2+}$, $O_{20}Si_8^{8-}$, $38H_2O$ 76.6 **5**
$2C_4H_{16}CuN_8S_4^+$, F_6Si^{2-} 79.5 **6**
$C_4H_{16}HgN_4^{2+}$, $2ClO_4^-$ 76.9 **7**
$C_4H_{16}HgN_8S_4^{2+}$, $2Cl^-$ 79.6+ **6**
$C_4H_{16}HgN_8S_4^{2+}$, $C_4CoN_4S_4^{2-}$ 79.24 **2**
$C_4H_{16}N_3^{3+}$, Cl^- , Cl_4Cu^{2-} 3.11 **4**
$C_4H_{16}N_4Ni^{2+}$, $2AgI_2^-$ 76.12 **3**
$nC_4H_{16}N_4Ni^{2+}$, $(AgBr_4^{2-})_n$ 76.20 **2**
$nC_4H_{16}N_4Ni^{2+}$, $(AgBr_4^{2-})_n$ 76.11 **3**
$C_4H_{16}N_4O_2Re^+$, Cl^- 76.7 **5**
$C_4H_{16}N_4O_2Re^+$, Cl^- 76.10 **7**
$C_4H_{16}N_4Pd^{2+}$, $C_2H_8N_2O_6PdS_4^{2-}$ 76.14 **3**
$C_4H_{16}N_5NiO_2^+$, BF_4^- 76.21 **2**
$C_4H_{16}N_5NiO_2^+$, ClO_4^- 76.22 **2**
$C_4H_{16}N_5NiO_2^+$, I^- 76.7 **4**
$C_4H_{16}N_5NiO_2^+$, NO_3^- 76.11 **7**
$C_4H_{16}N_5O_2Zn^+$, NO_2^- 76.12 **7**
$C_4H_{16}N_6NiO_4$ 76.15 **3**
$C_4H_{16}N_6NiO_4$, H_2O 76.13 **7**
$C_4H_{16}N_8PbS_4^{2+}$, $2C_6H_2N_3O_7^-$ 69.8 **4**
$C_4H_{16}N_8PdS_4^{2+}$, $2Cl^-$ 79.25 **2**
$C_4H_{16}N_8PdS_4^{2+}$, $2Cl^-$ 79.9 **3**
$C_4H_{16}N_8PtS_4^{2+}$, $2Cl^-$ 79.8 **7**
$C_4H_{16}N_8S_4$, Tl^+ , ClO_4^- 68.7 **2**
$C_4H_{16}N_8S_4$, Tl^+ , NO_3^- 68.8 **2**
$C_4H_{16}N_8S_4Te^{2+}$, $2Cl^-$ 70.8 **2**
$C_4H_{16}N_8S_4Te^{2+}$, $2Cl^-$, $2H_2O$ 70.9 **2**
$C_4H_{16}N_8Se_4Te^{2+}$, $2Cl^-$ 70.3 **4**
$C_4H_{16}N_9Ru^+$, F_6P^- 76.16 **3**
$C_4H_{17}B_7$ 62.31 **2**
$C_4H_{17}B_{10}^-$, $C_4H_{12}N^+$ 62.2 **6**
$C_4H_{17}N_4O_2Re^{2+}$, $2ClO_4^-$ 76.25 **2**
$C_4H_{18}AlB_9$ 62.8 **4**
$C_4H_{18}B_2P_2$ 62.32 **2**
$C_4H_{18}B_{14}Co^-$, $C_8H_{20}N^+$ 62.9 **4**
$C_4H_{18}ClCuN_4O^+$, Cl^- 76.26 **2**
$C_4H_{18}CoN_4O_4^+$, ClO_4^- 81.18 **6**
$C_4H_{18}CoN_4O_4S^+$, ClO_4^- , H_2O 76.14 **7**
$C_4H_{18}Co_2N_5O_2^{4+}$, $4NO_3^-$, H_2O 76.17 **3**
$nC_4H_{18}CuN_4O^{2+}$, $(C_4Cu_2N_4^{2-})_n$ 76.8 **4**
$C_4H_{18}CuO_{12}S_2$ 84.4 **3**
$C_4H_{18}FeN_2O_8^{2+}$, $H_{12}FeO_6^{2+}$, $2O_4S^{2-}$ 82.9 **2**
$C_4H_{18}IN_4NiO^+$, I^- 76.8 **5**
$C_4H_{18}N_6Pt^{2+}$, $2Cl^-$, H_2O 83.33 **2**
$C_4H_{18}N_8O_7U^{2+}$, $2NO_3^-$ 79.11 **5**
$C_4H_{19}Co_2N_5O_2^{4+}$, $4NO_3^-$, $2H_2O$ 76.19 **3**
$C_4H_{20}CoN_8O_6^{2+}$, $2NO_3^-$ 79.12 **5**
$C_4H_{20}N_4NiO_2^{2+}$, $2ClO_4^-$ 76.15 **7**
$C_4H_{20}N_4Pt^{2+}$, $2Cl^-$ 83.11 **3**
$C_4H_{20}N_4Pt^{2+}$, Cl_4Pt^{2-} 83.12 **3**
$C_4H_{22}B_{18}Co^-$, Cs^+ 62.33 **2**
$C_4H_{22}B_{18}Cu^-$, $C_{19}H_{18}P^+$ 64.56 **2**
$C_4H_{22}B_{18}Cu^{2-}$, $2C_8H_{20}N^+$ 62.34 **2**
$C_4H_{22}B_{18}Ni$ 62.8 **3**
$C_4H_{22}B_{18}Ni^-$, $C_4H_{12}N^+$ 62.3 **6**
$C_4H_{22}B_{18}Ni^{2-}$, $C_8H_{18}N_2^{2+}$ 37.5 **3**
$C_4H_{22}B_{20}$ 62.35 **2**
$C_4H_{22}B_{20}N_2$, C_2H_3N 62.36 **2**
$C_4H_{22}CuN_2O_6^{2+}$, O_4S^{2-} 84.5 **6**
$C_4H_{22}Cu_2N_4O_2^{2+}$, O_4S^{2-} , H_2O 83.34 **2**
$C_4H_{24}B_8N_2$ 62.37 **2**
$C_4H_{24}B_{10}S_2$ 62.38 **2**
$C_4H_{26}B_{18}FeP_2$ 62.39 **2**
$C_4H_{26}N_8NiO_3S_6^{2+}$, $O_3S_2^{2-}$, H_2O 79.26 **2**

$C_4H_{46}N_{14}O_2Ru_3^{6+}$, $6Cl^-$ 76.9 **4**

C_5

$C_5Cl_3N_2O_4^-$, K^+ 20.1 **4**
C_5Cl_5N 33.4+ **6**
$C_5F_{15}P_5$ 64.14 **2**
C_5N_4 7.2 **6**
$C_5O_5^{2-}$, $2H_4N^+$ 6.2 **1**
C_5HCl_4NO 33.3 **4**
$C_5HF_6O_2^-$, Tl^+ 6.1 **7**
$C_5HO_5^-$, Rb^+ 6.3 **1**
$C_5HO_5^-$, H_4N^+ 6.4 **1**
$C_5H_2FN_2O_4^-$, Rb^+ , H_2O 44.19 **1**
$C_5H_3BrO_2$ 38.3 **5**
$C_5H_3Br_2Cl_2HgNO$ 84.14 **2**
$C_5H_3N_2O_4^-$, H_4N^+ , H_2O 44.4 **4**
$C_5H_3N_3$ 7.4 **4**
$C_5H_3N_3O_4$ 33.5 **7**
$C_5H_3N_4O_2^-$, Na^+ , $4H_2O$ 44.12 **3**
$C_5H_3N_4S$ 35.1 **6**
$C_5H_3N_7$, H_2O 44.20 **1**
$C_5H_4BrN_2O_3^-$, $C_5H_5BrN_2O_3$, K^+ 44.5 **4**
$C_5H_4BrN_2O_3^-$, $C_5H_5BrN_2O_3$, Rb^+ 44.6 **4**
$C_5H_4BrN_3O_2$ 33.6 **7**
C_5H_4ClNO 33.6 **3**
C_5H_4ClNO 33.6 **5**
C_5H_4ClNO , C_5H_5NO 60.5 **4**
$C_5H_4Cl_2N_2O$ 33.7 **5**
$(C_5H_4CuN_3)_n$ 83.11 **4**
$C_5H_4N_2O_2$ 33.7 **7**
$C_5H_4N_2O_3$ 33.15 **1**
$C_5H_4N_2O_3$ 33.4 **4**
$C_5H_4N_2O_3$ 33.8 **7**
$C_5H_4N_2O_3$, $C_6H_6O_2$ 60.2 **5**
$2C_5H_4N_2O_3$, $H_4Cl_2CuO_2$ 33.5 **4**
$C_5H_4N_2O_4$, H_2O 44.3 **6**
$C_5H_4N_2O_4$, CH_4N_2O 60.2 **4**
$C_5H_4N_4$ 44.22+ **1**
$C_5H_4N_4O$ 35.1 **5**
$C_5H_4N_4O_2$, $2H_2O$ 44.7 **4**
$C_5H_4N_4O_3$ 44.24 **1**
$C_5H_4N_4S$, H_2O 44.25+ **1**
$C_5H_4N_5^-$, $C_{24}H_{20}As^+$, $3H_2O$ 44.4 **6**
$C_5H_4O_2S$ 39.22 **1**
$C_5H_4O_2S$ 39.23+ **1**
$C_5H_4O_2Se$ 39.25 **1**
$C_5H_4O_2Te$ 70.4 **4**
$C_5H_4O_3$ 38.15 **1**
$C_5H_4S_2Se$ 42.1 **4**
$C_5H_4S_3$ 39.5 **5**
$C_5H_4Se_3$ 39.2 **4**
$2C_5H_5^-$, Ca^{2+} 67.1 **7**
$C_5H_5BrN_2O$ 32.4 **5**
$C_5H_5BrN_2O_2$ 44.5 **5**
$C_5H_5BrN_2O_2$, $C_6H_7N_5$ 44.21 **1**
$C_5H_5BrN_2O_2$, $C_7H_8BrN_5$ 44.13 **3**
$C_5H_5BrN_2O_2$, $C_7H_9N_5$ 44.56 **1**
$C_5H_5BrN_2O_2$, $C_7H_9N_5$ 44.14 **3**
$(C_5H_5CdI_2NO)_n$ 84.9 **7**
$C_5H_5ClCrN_2O_2$ 73.16 **2**
C_5H_5ClIN 33.8 **5**
$2C_5H_5ClN^+$, Cl_6Sn^{2-} 33.9 **5**
$C_5H_5ClN_2$ 33.6 **6**
$C_5H_5ClN_2$ 33.9 **7**
$C_5H_5ClN_2O$ 44.5 **6**
$C_5H_5ClO_3$ 1.12 **5**
$(C_5H_5Cl_2HgNO)_n$ 84.10 **7**
$2C_5H_5Cl_3Cr^-$, $C_{20}H_{40}Li_2O_6^{2+}$ 67.16 **5**
$C_5H_5CrNO_5$ 83.35 **2**
$C_5H_5FN_2O_2$, $C_7H_9N_5$ 44.57 **1**
$C_5H_5FN_2O_2$, $C_7H_9N_5$ 44.58 **1**
$C_5H_5FN_2O_2$, $C_7H_9N_5$ 44.17 **3**
$C_5H_5F_4NP$ 64.6 **6**
$C_5H_5F_5NP$ 33.7 **6**
$2C_5H_5IN_2O_2$, $C_7H_9N_5$ 44.18 **3**
$2C_5H_5IN_2O_2$, $C_7H_{10}N_6$ 44.19 **3**
$(C_5H_5In)_n$ 68.9 **2**
C_5H_5N , I_3N 33.10 **7**
C_5H_5N , Li^+ , Cl^- 60.64 **2**
C_5H_5N , ClN 60.65 **2**
C_5H_5N , $C_6H_{14}O_6$ 60.23 **3**
C_5H_5N , $2C_{33}H_{36}O_6$ 61.5 **7**
$2C_5H_5N$, Li^+ , Cl^- , H_2O 60.66 **2**
C_5H_5NO 33.16 **1**
C_5H_5NO 10.2 **3**
C_5H_5NO , $C_2HCl_3O_2$ 33.11 **7**
C_5H_5NO , C_5H_4ClNO 60.5 **4**
$C_5H_5NO_2$ 38.16 **1**
$2C_5H_5NO_2$, $C_2H_2O_4$ 60.3 **5**
C_5H_5NS 33.17 **1**
$C_5H_5N_3O$ 33.18 **1**
$C_5H_5N_3O$ 33.6+ **4**
$C_5H_5N_3O$ 44.6 **6**
$C_5H_5N_3O$ 44.7 **6**
$C_5H_5N_3OS_2$ 41.3 **6**
$C_5H_5N_3O_2$ 33.19 **1**
$C_5H_5N_3O_2$ 33.12 **7**
$C_5H_5N_3O_4$ 44.8 **7**
$C_5H_5N_4O^+$, $AuCl_4^-$, $2H_2O$ 44.9 **7**
$C_5H_5N_4O^+$, Cl^- , H_2O 44.28 **1**
$C_5H_5N_5$, $C_{17}H_{20}N_4O_6$, $3H_2O$ 60.1 **7**
$C_5H_5N_5O$ 35.3 **5**
$C_5H_5N_5O$, H_2O 44.8 **4**
$C_5H_5N_5S$ 44.20 **3**
C_5H_6 20.6 **1**
$C_5H_6BrN_3O$ 32.13 **3**
$C_5H_6BrN_3O$, $C_7H_9N_5O$ 44.59 **1**
$C_5H_6ClNO_3S$ 40.5 **4**
$C_5H_6ClN_3$ 44.44 **1**
$C_5H_6Cl_3N_5OZn$ 83.14 **3**
$C_5H_6Cl_3N_5Zn$ 83.4 **5**
$(C_5H_6Cl_4N_2Sn)_n$ 69.10 **2**
$(C_5H_6CuO_8)_n$ 84.15 **2**
$C_5H_6FN_3O$, $C_7H_9N_5O$ 44.60 **1**
$C_5H_6FeN_4O_3$ 83.36 **2**
$C_5H_6MnO_8$ 84.16 **2**
$C_5H_6N^+$, $5Ag^+$, $6I^-$ 33.10+ **5**
$C_5H_6N^+$, AsF_4O^- 33.20 **1**
$C_5H_6N^+$, AsF_6^- 33.21 **1**
$C_5H_6N^+$, $BrCl_2^-$ 33.7 **3**
$C_5H_6N^+$, Cl^- 33.22 **1**
$C_5H_6N^+$, Cl_2I^- 33.8 **6**
$C_5H_6N^+$, Cl_4I^- 33.13 **7**
$C_5H_6N^+$, Cl_4Sb^- 33.8 **3**
$C_5H_6N^+$, Cl_6Sb^- 33.16 **5**
$C_5H_6N^+$, F_6P^- 33.24 **1**
$C_5H_6N^+$, F_6Sb^- 33.25 **1**
$C_5H_6N^+$, I^- 33.14 **7**

$C_5H_6N^+$, I_4Tl^- , C_5H_5N 33.9 **3**
$C_5H_6N^+$, NO_3^- 33.26 **1**
$C_5H_6N^+$, HBr_4Re^- 33.27+ **1**
$C_5H_6N^+$, $H_2Cl_3MnO^-$ 33.9 **6**
$C_5H_6N^+$, $2CH_4N_2S$, Br^- 60.6 **4**
$C_5H_6N^+$, $C_4H_6CrN_6S_4^-$ 33.29 **1**
$C_5H_6N^+$, $C_{24}H_{16}CrN_4O_4^-$ 83.188 **2**
$C_5H_6N^+$, $C_{32}H_{18}CrN_6O_8^-$, $0.5C_5H_5N$ 83.206 **2**
$2C_5H_6N^+$, $As_2F_8O_2^{2-}$ 33.22 **5**
$2C_5H_6N^+$, $As_2F_8O_2^{2-}$ 33.10 **6**
$2C_5H_6N^+$, $Br_{12}Cl_6Nb_6^{2-}$ 33.17 **5**
$2C_5H_6N^+$, $Br_{12}Cl_6Nb_6^{2-}$ 33.18 **5**
$2C_5H_6N^+$, Cl_5NORu^{2-} 33.15 **7**
$2C_5H_6N^+$, $Cl_{18}Nb_6^{2-}$ 33.19 **5**
$2C_5H_6N^+$, $2H^+$, $Cl_8Re_2^{4-}$ 33.30 **1**
$2C_5H_6N^+$, $H_2Mo_2O_{14}^{2-}$ 33.8 **4**
$2C_5H_6N^+$, $H_4Mo_2O_{13}^{2-}$ 33.9 **4**
$2C_5H_6N^+$, $C_{18}H_{12}O_6Si^{2-}$ 63.10 **3**
$5C_5H_6N^+$, $Br_9Sb_2^{3-}$, $2Br^-$ 33.10 **3**
$6C_5H_6N^+$, Br_6Sb^{3-} , $3Br_6Sb^-$ 33.11 **3**
$C_5H_6NO^+$, Cl^- 33.33 **1**
$C_5H_6NO^+$, Cl^- , H_2O 33.12 **3**
$C_5H_6NO_2^+$, Cl^- 33.13 **3**
$2C_5H_6N_2$, $C_8H_{12}N_2O_3$ 60.8 **6**
$C_5H_6N_2OS$, $C_6H_7N_5$ 44.21 **3**
$C_5H_6N_2OS$, $C_6H_7N_5$ 44.8 **6**
$C_5H_6N_2O_2$ 44.30 **1**
$C_5H_6N_2O_2$ 44.31 **1**
$C_5H_6N_2O_2$ 33.16 **7**
$C_5H_6N_2O_2$, H_2O 44.32 **1**
$C_5H_6N_2O_2$, $C_6H_4O_2$ 60.7 **4**
$C_5H_6N_2O_2$, $C_7H_9BrN_6$ 44.27 **3**
$C_5H_6N_2O_2$, $C_7H_9N_5$ 44.55 **1**
$C_5H_6N_2O_2$, $C_7H_{14}N_6$, H_2O 60.4 **5**
$C_5H_6N_2O_3S$ 41.4 **5**
$C_5H_6N_2S$ 7.20 **1**
$C_5H_6N_4O_3$ 44.9 **4**
$C_5H_6N_5^+$, Cl^- , $0.5H_2O$ 44.9 **6**
$C_5H_6N_5^+$, Cl_3Hg^- , $1.5H_2O$ 44.10 **7**
$2C_5H_6N_5^+$, $2Cl^-$, H_2O 44.33+ **1**
$2C_5H_6N_5^+$, $C_{10}H_{18}CoN_{10}O_4^{2+}$, $2O_4S^{2-}$, $6H_2O$ 83.23 **6**
$C_5H_6N_5O^+$, Cl^- , H_2O 44.35 **1**
$C_5H_6N_5O^+$, Cl^- , $2H_2O$ 44.36 **1**
$C_5H_6N_5O^+$, $C_6H_2N_3O_7^-$, H_2O 44.11 **7**
$2C_5H_6N_5O^+$, O_4S^{2-} , H_2O 44.22 **3**
$C_5H_6N_5S^+$, $C_6H_2N_3O_7^-$, H_2O 44.12 **7**
$C_5H_6N_8O_{13}$ 8.15 **3**
$C_5H_6O_2S$ 39.26 **1**
$C_5H_6O_3$ 38.17 **1**
$C_5H_6O_3$ 38.6+ **7**
$C_5H_6O_4$ 20.2 **3**
$C_5H_6O_4$ 20.3 **3**
$C_5H_6O_4$ 1.13 **5**
$C_5H_6O_4$ 1.10 **6**
$C_5H_6O_5$ 1.69 **1**
$(C_5H_6O_8Zn)_n$ 84.17 **2**
$C_5H_7B_3FeO_3$ 75.1 **6**
$C_5H_7ClN_2O_3$ 50.1 **5**
$C_5H_7ClN_4O_7$ 1.14 **5**
$C_5H_7Cl_2O_4W^-$, $C_{10}H_{13}O_2^+$ 38.16 **6**
$C_5H_7CuNO_5$, $2H_2O$ 82.6 **4**
C_5H_7NO 1.16 **7**
$C_5H_7NOS_2$ 41.5 **5**
$C_5H_7NO_2$ 1.5 **4**
$C_5H_7NO_3$ 48.18 **7**
$C_5H_7NO_4^{2-}$, Ca^{2+} , $3H_2O$ 48.17 **6**
$C_5H_7N_2O_2^+$, Br^- 44.38 **1**
$C_5H_7N_3O$ 44.39 **1**
$C_5H_7N_3O$, $C_4H_3FN_2O_2$ 44.4 **3**
$C_5H_7N_3O$, $C_7H_9N_5O$ 44.61 **1**
$C_5H_7N_3OS$ 32.5 **4**
$C_5H_7N_5^{2+}$, $2Cl^-$ 44.10 **6**
$C_5H_7N_5^{2+}$, $2Cl^-$ 44.13 **7**
$C_5H_7N_5O^{2+}$, O_4S^{2-} 44.8 **5**
$C_5H_7O_2^-$, Li^+ 6.2 **7**
$2C_5H_7O_2^-$, $C_4H_{16}CuN_4^{2+}$, $2H_2O$ 76.8 **7**
$C_5H_7O_3^-$, Na^+ 2.72 **1**
$C_5H_7O_4^-$, K^+ 2.17 **4**
$C_5H_7O_4^-$, K^+ 2.25 **5**
$C_5H_7O_4^-$, Rb^+ 2.18 **4**
$C_5H_7O_4^-$, H_4N^+ 2.19 **4**
$2C_5H_7S_2^+$, Cl_4Fe^{2-} 39.3+ **6**
$C_5H_8B_6MnO_3^-$, $C_{19}H_{18}P^+$ 64.33 **6**
$(C_5H_8ClCuN_2)_n$ 83.6 **7**
C_5H_8ClNO 1.70 **1**
$C_5H_8ClNO_3$ 48.13 **3**
$(C_5H_8CuN_2O_3)_n$, nH_2O 82.3+ **5**
C_5H_8INO 32.14 **3**
C_5H_8INO 32.7 **6**
$C_5H_8NO_4^-$, Ca^{2+} , Cl^- , H_2O 48.19 **7**
$C_5H_8NO_4^-$, Na^+ , H_2O 48.20 **7**
$C_5H_8NO_4^-$, $C_6H_{15}N_4O_2^+$, H_2O 48.31 **7**
$C_5H_8N_2OS$ 11.6 **3**
$C_5H_8N_2OS$ 44.11 **4**
$C_5H_8N_2O_2$ 44.40 **1**
$C_5H_8N_2O_2$ 44.41 **1**
$C_5H_8N_2O_2$ 40.4 **6**
$C_5H_8N_2O_2$ 44.14 **7**
$C_5H_8N_2O_2$, H_2O 32.15 **3**
$C_5H_8N_2O_4$ 40.2 **3**
$C_5H_8N_2O_4$ 44.11 **6**
$C_5H_8N_2S$ 32.6+ **4**
$C_5H_8N_3O^+$, Br^- 44.42 **1**
$C_5H_8N_3O^+$, Cl^- 44.12 **4**
$C_5H_8N_4O_{12}$ 45.3 **1**
$C_5H_8N_4S$ 44.43 **1**
$C_5H_8O_2$ 1.71 **1**
$C_5H_8O_2$ 1.72 **1**
$C_5H_8O_2S$ 39.29 **1**
$C_5H_8O_2S$ 39.4 **3**
$C_5H_8O_3S$ 39.30 **1**
$C_5H_8O_4$ 1.73 **1**
$C_5H_8O_4$ 1.74 **1**
$C_5H_8O_4$ 1.15 **5**
$C_5H_8O_4$ 1.11+ **6**
$C_5H_8O_4$, $2C_{18}H_{15}OP$ 60.2 **7**
$C_5H_8O_4S$ 39.6 **5**
$C_5H_9Br_2O_3P$ 64.15 **2**
$C_5H_9Br_2O_3P$ 64.4 **4**
$(C_5H_9CdNO_5)_n$, nH_2O 82.3+ **6**
$(C_5H_9CdNO_5)_n$, nH_2O 82.4 **7**
$C_5H_9ClCuN_4OS$, $0.5H_2O$ 79.13 **5**
$C_5H_9Cl_2NOPdS$ 72.7 **3**
$C_5H_9Cl_3N_2O$ 32.17 **3**
$C_5H_9CuNO_5$, H_2O 82.10 **2**
$(C_5H_9F_3O_2Sn)_n$ 69.5 **6**
$C_5H_9NO_2$ 48.38 **1**
$C_5H_9NO_2PbS$ 69.6 **6**
$C_5H_9NO_3$ 48.39+ **1**
$C_5H_9NO_3$ 48.27 **5**

$C_5H_9NO_3S$ 48.28 5
$C_5H_9NO_4$ 48.41 1
$C_5H_9NO_4$ 48.14 3
$C_5H_9NO_4$ 48.29 5
$C_5H_9NO_4S_2$ 4.11 1
$C_5H_9NO_5$ 40.6 5
$C_5H_9NO_5Zn$, H_2O 82.11 2
$C_5H_9N_2O_3S$ 44.24 3
$C_5H_9N_3$ 32.8 6
$(C_5H_9N_3Sn)_n$ 69.9 4
$C_5H_9O_4P$ 46.7 1
$C_5H_9O_4P$ 46.8 1
$2C_5H_9O_6^-$, Ca^{2+} , $5H_2O$ 45.4 1
$2C_5H_9O_6^-$, Sr^{2+} , $5H_2O$ 45.5 1
$C_5H_9O_8P^{2-}$, Ba^{2+} , $5H_2O$ 45.6 1
$C_5H_{10}BO_3$ 62.4 6
$C_5H_{10}B_{10}MoO_5^{2-}$, $2C_{16}H_{36}N^+$ 62.9 3
$C_5H_{10}ClO_3P$ 64.5 4
$(C_5H_{10}HgO_2S)_n$ 81.16 7
$C_5H_{10}NO_2^+$, Cl^- 48.42 1
$C_5H_{10}NO_2^+$, Cl^- , H_2O 20.1 7
$C_5H_{10}NO_4^+$, Cl^- 48.43 1
$C_5H_{10}NO_4^+$, Cl^- 48.30 5
$C_5H_{10}NS_2^-$, Na^+ 11.30 1
$C_5H_{10}NS_2^-$, Na^+ , $3H_2O$ 4.12 1
$C_5H_{10}NS_2^-$, Tl^+ 11.6 7
$C_5H_{10}N_2OS$ 41.10 1
$C_5H_{10}N_2O_2$ 1.75 1
$C_5H_{10}N_2O_2$ 48.21 7
$C_5H_{10}N_2O_2S$ 4.2 4
$C_5H_{10}N_2O_3$ 48.44 1
$C_5H_{10}N_2O_3$ 48.15 3
$C_5H_{10}N_2O_3$ 48.18 6
$2C_5H_{10}N_2O_3S$, Na^+ , I^- 60.67 2
$C_5H_{10}N_2O_4PbS$ 69.11 2
$C_5H_{10}N_2S$ 34.4 1
$C_5H_{10}N_3^+$, Br^- 32.4 7
$C_5H_{10}N_6O_4$ 37.1+ 6
$C_5H_{10}O_2$ 1.76 1
$C_5H_{10}O_2S$ 39.5 6
$C_5H_{10}O_4$ 45.7 1
$C_5H_{10}O_5$ 45.8 1
$C_5H_{10}O_5$ 45.9 1
$C_5H_{10}O_5$ 45.10 1
$C_5H_{10}O_5$ 45.1 4
$C_5H_{10}O_5$ 45.1+ 5
$C_5H_{10}O_5$ 38.2+ 6
$2C_5H_{10}O_5$, Ca^{2+} , $2Cl^-$, $3H_2O$ 45.2 6
$C_5H_{10}S_5$ 39.3+ 4
$C_5H_{11}AgIN$ 83.16 3
$C_5H_{11}B_9O_3Re^-$, Cs^+ 62.41 2
$C_5H_{11}Cl_2NO_2PdS$ 82.1 3
$C_5H_{11}Cl_2NO_2PtS$ 82.2 3
$C_5H_{11}Cl_2NO_2PtS$ 82.3 3
$C_5H_{11}CuO_4^+$, $C_6H_2N_3O_7^-$ 77.1 2
$C_5H_{11}NO_2$ 48.45 1
$C_5H_{11}NO_2$ 48.16 3
$C_5H_{11}NO_2S$ 41.11 1
$C_5H_{11}NO_2S$ 48.46+ 1
$C_5H_{11}NO_2S$ 48.19 6
$C_5H_{11}N_2O_2^+$, ClO_4^- 48.48 1
$C_5H_{11}N_2O_3^+$, Cl^- 48.16 4
$C_5H_{11}N_2O_3^+$, Cl^- 48.17 4
$C_5H_{11}N_2O_3^+$, Cl^- , H_2O 48.49 1
$C_5H_{11}N_3^{2+}$, $2Cl^-$ 59.1 1
$C_5H_{11}N_3^{2+}$, Cl_4Co^{2-} 32.8 4
$C_5H_{11}N_3^{2+}$, O_4S^{2-} , H_2O 32.9 6
$C_5H_{11}N_3^{2+}$, $2H_2O_4P^-$, H_2O 32.18+ 3
$2C_5H_{11}N_3O_2$, $C_{10}H_{18}CoN_6O_4^-$, K^+ 79.10 4
$C_5H_{11}O_4P$, H_2O 46.3 3
C_5H_{12} 5.17 1
$C_5H_{12}AlN$ 68.3 6
$C_5H_{12}ClNS_2Sn$ 69.11 3
$C_5H_{12}Cl_2NO^+$, Cl^- 3.12 7
$C_5H_{12}Cl_3NPt$ 72.3 5
$C_5H_{12}Cl_3NPt$ 72.1+ 5
$C_5H_{12}CuN_7S^+$, CNS^- 83.12 4
$C_5H_{12}LiN_2O_4^+$, Br^- , H_2O 67.6 3
$C_5H_{12}N^+$, Cl^- 33.34 1
$C_5H_{12}N^+$, Cl^- 33.11 6
$C_5H_{12}N^+$, ClO_4^- 3.44 1
$C_5H_{12}N^+$, HS^- 33.12 6
$C_5H_{12}N^+$, $C_7H_4BrO_2^-$ 33.10 4
$C_5H_{12}N^+$, $C_7H_4ClO_2^-$ 33.11 4
$C_5H_{12}N^+$, $C_7H_5O_3^-$ 33.13 6
$C_5H_{12}N^+$, $C_8H_7O_2^-$ 33.14 6
$C_5H_{12}N^+$, $C_{40}H_{36}EuO_8^-$ 77.21 5
$2C_5H_{12}N^+$, $As_4S_6^{2-}$ 33.12 4
$2C_5H_{12}N^+$, $BiBr_5^{2-}$ 33.35 1
$2C_5H_{12}N^+$, Br_5Sb^{2-} 33.21 5
$3C_5H_{12}N^+$, $Cl_{18}Nb_6^{3-}$ 3.13 6
$C_5H_{12}NO_2^+$, Br^- 48.50 1
$C_5H_{12}NO_2^+$, Cl^- 48.51 1
$C_5H_{12}NO_2^+$, Cl^- 48.52 1
$C_5H_{12}NO_2^+$, Cl^- 48.17 3
$C_5H_{12}NO_2^+$, Cl^- 48.22 7
$C_5H_{12}NO_2^+$, Cl^- , H_2O 48.18 3
$C_5H_{12}NO_2^+$, NO_3^- 48.20 6
$C_5H_{12}NO_2S^+$, Cl^- , H_2O 48.21 6
$C_5H_{12}NO_2S^+$, CH_4N_2O , Cl^- 60.68 2
$C_5H_{12}N_2O_3S$ 48.19 3
$C_5H_{12}N_3O_2^+$, Br^- 1.16 5
$C_5H_{12}N_3O_2^+$, Cl^- 8.14 5
$C_5H_{12}O_2$ 5.3 3
$(C_5H_{12}O_2Sn)_n$ 69.7 6
$C_5H_{12}O_4$ 45.11+ 1
$C_5H_{12}O_5$ 45.14 1
$C_5H_{12}O_5$ 45.15 1
$C_5H_{12}O_5$ 45.16 1
$C_5H_{12}S_4$ 11.31 1
$C_5H_{13}CuN_4^+$, ClO_4^- 76.6 6
$C_5H_{13}CuN_4S+$, ClO_4^- 76.7 6
$C_5H_{13}NO_2Sn$ 69.3 7
$C_5H_{13}NO_4P^-$, Ca^{2+} , Cl^- 59.1 5
$C_5H_{13}NO_4P^-$, Ca^{2+} , Cl^- , $4H_2O$ 46.4 6
$C_5H_{13}NO_4S$ 59.2 1
$C_5H_{13}N_2^+$, $C_{19}H_{19}N_2O_2^-$ 33.15 6
$C_5H_{13}N_2O_2^+$, Br^- 48.20 3
$C_5H_{13}N_2O_2^+$, Br^- 48.19 4
$C_5H_{13}N_2O_2^+$, Cl^- 48.56 1
$C_5H_{13}N_2O_2^+$, Cl^- 48.21 3
$C_5H_{13}N_2O_3P$ 64.6 4
$C_5H_{13}N_4^+$, $2Cu^+$, $3Br^-$ 9.4 7
$C_5H_{13}N_4NiS_2^+$, Cl^- 72.4 4
$(C_5H_{14}CuN_3O_2^+)_n$, $nCHO_2^-$ 76.20 3
$2C_5H_{14}N^+$, $C_4H_6Cl_6Pt_2^{2-}$ 72.3 3
$C_5H_{14}NO^+$, Cl^- 59.3 1
$C_5H_{14}NO^+$, Cl^- 59.1 4
$C_5H_{14}NO^+$, $C_4H_6CrN_6S_4^-$ 59.4 1
$C_5H_{14}NO_6P$, H_2O 46.3 5
$C_5H_{14}N_3O_3P$, $0.5CH_6N_3^+$, $0.5Cl^-$ 8.9 4
$C_5H_{14}N_8^{2+}$, $2Cl^-$, H_2O 8.43+ 1

$C_5H_{15}AlIN$ 68.4 **6**
$C_5H_{15}As_5$ 65.9 **2**
$C_5H_{16}CdClN_5S$ 76.16 **7**
$C_5H_{16}ClN_5NiS$ 76.27 **2**
$C_5H_{16}CoN_5O_3S_2$, $2H_2O$ 76.28 **2**
$C_5H_{16}CoN_6O_2S^+$, ClO_4^- 76.17 **7**
$C_5H_{16}CoN_6O_2S^+$, ClO_4^- 76.18 **7**
$C_5H_{16}N_2^{2+}$, HO_4P^{2-} , $2.5H_2O$ 3.13 **7**
$C_5H_{16}N_5NiS^+$, I^- 76.21 **3**
$C_5H_{16}N_6NiO_2S$ 76.29 **2**
$C_5H_{18}N_5NiOS^+$, NO_3^- 76.19 **7**
$(C_5H_{20}Cu_2N_{10}S_5^{2+})_n$, nO_4S^{2-} , $2nH_2O$ 79.10 **3**
$C_5H_{21}B_{18}CoS_2$ 71.2 **4**
$C_5H_{23}ClN_5Re^{2+}$, $2ClO_4^-$ 83.6 **6**
$C_5H_{25}ClCoN_5^{2+}$, $2NO_3^-$ 83.13 **4**

C_6

C_6BrCl_5 19.1 **1**
$C_6Br_2Cl_4$ 19.2 **1**
$C_6Br_4O_2$ 18.1 **1**
C_6Br_5Cl 19.3 **1**
C_6Br_6 , $C_6H_2Br_4$ 60.69 **2**
$C_6Cl_2O_4^{2-}$, $2H_4N^+$, H_2O 6.5 **1**
$C_6Cl_4O_2$ 18.2+ **1**
$C_6Cl_4O_2$ 18.1 **4**
$C_6Cl_4O_2$, $2C_9H_7NO$ 60.107 **2**
$C_6Cl_4O_2$, $C_{10}H_{16}N_2$ 60.122 **2**
$C_6Cl_4O_2$, $C_{12}H_{10}$ 60.19 **5**
$C_6Cl_4O_2$, $C_{12}H_{18}$ 60.71 **2**
$C_6Cl_4O_2$, $C_{15}H_{12}$ 60.27 **5**
$C_6Cl_4O_2$, $C_{16}H_{10}$ 60.25 **6**
$C_6Cl_4O_2$, $C_{16}H_{20}N_2$ 60.26 **6**
$C_6Cl_4O_2$, $2C_{16}H_{20}N_2$ 60.33 **4**
$C_6Cl_4O_2$, $C_{18}H_{12}N_2O_2Pd$ 60.70 **2**
$C_6Cl_4O_2^-$, K^+ 18.1 **6**
$C_6Cl_5NO_2$ 15.1 **6**
C_6Cl_6 19.4 **1**
C_6Cl_6 19.1 **6**
C_6Cl_6 20.2+ **7**
$C_6CoO_6S_6^{3-}$, Ca^{2+} , K^+ , $4H_2O$ 85.4 **7**
$C_6CoO_{12}^{3-}$, $C_{36}H_{24}N_6Ni^{2+}$, K^+ , $2H_2O$ 83.142 **3**
$C_6CrO_{12}^{3-}$, $3H_4N^+$, $2H_2O$ 81.37 **2**
$C_6F_4O_2$ 18.2 **6**
$C_6F_4O_2$, $C_{20}H_{12}$ 60.152 **2**
C_6F_5NO , $2C_{12}H_{10}N_2O_2$ 10.2 **6**
C_6F_6 19.2 **6**
C_6F_6 , C_8H_{10} 60.10 **7**
C_6F_6 , C_9H_{12} 60.21 **4**
C_6F_6 , $C_{12}H_{18}$ 60.24+ **5**
$C_6I_4O_2$ 18.1 **7**
C_6I_6 19.1+ **3**
$C_6MoN_6S_6^{3-}$, $3K^+$, $C_2H_4O_2$, H_2O 83.38 **2**
$C_6N_2O_8^{2-}$, $2H_3O^+$, $4H_2O$ 18.2 **7**
$C_6N_2O_8^{2-}$, $2H_4N^+$ 6.6 **1**
$C_6N_2O_8^{2-}$, $2H_5O_2^+$, $2H_2O$ 6.1 **3**
C_6N_4 7.21 **1**
C_6N_4 7.5 **4**
C_6N_4 7.3 **6**
C_6N_4 , $C_{10}H_8$ 60.72 **2**
C_6N_4 , $C_{10}H_8$ 60.1 **6**
C_6N_4 , $C_{10}H_{10}Fe$ 60.119 **2**
C_6N_4 , $C_{16}H_{10}$ 60.142 **2**
C_6N_4 , $C_{16}H_{10}$ 60.44 **3**
C_6N_4 , $C_{16}D_{10}$ 60.23+ **7**
C_6N_4 , $C_{18}H_{20}$ 60.37 **4**
C_6N_4 , $2C_{18}H_{20}$ 60.35 **5**
C_6N_4 , $8C_{20}H_{10}N_4$, $2CH_2Cl_2$ 60.52 **3**
C_6N_4 , $C_{20}H_{12}$ 60.53 **3**
C_6N_4O 38.2+ **4**
C_6N_4O 38.3 **6**
$C_6N_4O_4$ 21.1 **3**
$C_6N_6O_3$, $C_6H_{15}OP$ 60.73 **2**
$C_6N_6O_3$, $C_6H_{15}O_4P$ 60.15+ **4**
$C_6N_6O_6$ 40.2+ **1**
$C_6N_6O_6$, $C_{12}H_{12}S_2$ 60.74 **2**
$C_6N_6O_{12}$ 15.1 **1**
$C_6N_9^{3-}$, $3Na^+$, $3H_2O$ 33.36 **1**
$C_6NbO_{13}^{3-}$, $3H_4N^+$, H_2O 81.19 **4**
$C_6O_{12}Rh^{3-}$, $3K^+$, $4.5H_2O$ 81.20 **4**
$C_6O_{12}Sb^{3-}$, $3K^+$, $4H_2O$ 66.2 **3**
$C_6O_{12}Sb^{3-}$, $3H_4N^+$, $4H_2O$ 66.3 **3**
$C_6O_{12}V^{3-}$, $3K^+$, $3H_2O$ 81.39 **2**
$C_6O_{14}U^{4-}$, $4H_4N^+$ 81.19 **6**
$(C_6O_{16}U_2^{2-})_n$, $2nH_4N^+$ 81.20 **6**
$C_6O_{18}S_6^{6-}$, $6Na^+$, $8H_2O$ 11.7 **7**
$C_6HCl_2N_3O_6$ 15.2 **1**
C_6HCl_5O 17.1 **1**
C_6HCl_5O , C_6H_7N 60.17 **3**
$C_6HN_4O_6^-$, K^+ , H_2O 40.6 **4**
$C_6HN_6O_4^-$, $C_{18}H_{16}OP^+$ 64.43 **3**
$C_6H_2BrClO_2$ 18.4 **1**
$C_6H_2BrN_3O_6$, $C_{16}H_{10}$ 60.22 **7**
$C_6H_2BrN_3O_6$, $1.5C_{16}H_{10}$ 60.25 **7**
$C_6H_2Br_2Cl_2O$ 21.1 **1**
$C_6H_2Br_2O_2$ 18.5 **1**
$C_6H_2Br_4$ 19.6+ **1**
$C_6H_2Br_4$, C_6Br_6 60.69 **2**
$C_6H_2ClN_3O_6$ 15.3 **1**
$C_6H_2ClN_3O_6$ 15.1+ **3**
$C_6H_2Cl_2N_2O$ 21.2 **3**
$C_6H_2Cl_2N_2O_2$ 40.7 **4**
$C_6H_2Cl_2O_2$ 18.1 **3**
$C_6H_2Cl_2O_2$ 18.2 **3**
$C_6H_2Cl_2O_2$ 18.3 **3**
$C_6H_2Cl_2O_4$ 18.8 **1**
$C_6H_2Cl_2O_4$, $2H_2O$ 18.9 **1**
$C_6H_2Cl_4$ 19.8 **1**
$C_6H_2Cl_4O$ 21.2 **1**
$C_6H_2Cl_4O_2$ 17.2+ **1**
$C_6H_2Cl_4O_2$, $C_{18}H_{15}AsO$ 60.35 **4**
$C_6H_2Cl_5N$ 16.1 **3**
$C_6H_2F_2N_4O_6$ 16.1 **7**
$C_6H_2F_2O_4$ 18.3 **7**
$C_6H_2IN_3O_6$ 15.4 **1**
$(C_6H_2Mn_2O_{13}^{2-})_n$, $2nCs^+$, $2nH_2O$ 81.15 **5**
$C_6H_2N_2O_4^{2-}$, $H_{12}N_4Pd^{2+}$ 33.16 **6**
$C_6H_2N_2O_8$, $6H_2O$ 18.10 **1**
$C_6H_2N_2O_{10}S_2^{2-}$, $2K^+$, H_2O 15.3 **3**
$C_6H_2N_3O^-$, K^+ 7.6 **4**
$C_6H_2N_3O_7^-$, K^+ 15.5 **1**
$C_6H_2N_3O_7^-$, K^+ 15.1 **4**
$C_6H_2N_3O_7^-$, Li^+ , H_2O 15.1 **5**
$C_6H_2N_3O_7^-$, H_4N^+ 15.6 **1**
$C_6H_2N_3O_7^-$, $C_4H_6N_3O^+$ 44.6 **7**
$C_6H_2N_3O_7^-$, $C_5H_6N_5O^+$, H_2O 44.11 **7**
$C_6H_2N_3O_7^-$, $C_5H_6N_5S^+$, H_2O 44.12 **7**

$C_6H_2N_3O_7^-$, $C_5H_{11}CuO_4^+$	77.1	**2**
$C_6H_2N_3O_7^-$, $C_7H_9N_2O^+$	33.26	**6**
$C_6H_2N_3O_7^-$, $C_{10}H_{10}Fe^+$	73.25	**2**
$C_6H_2N_3O_7^-$, $C_{10}H_{13}N_2^+$	35.15	**6**
$C_6H_2N_3O_7^-$, $C_{10}H_{13}N_2O^+$, H_2O	35.11	**4**
$C_6H_2N_3O_7^-$, $C_{11}H_{13}N_2O_2^+$, CH_4O	48.58	**6**
$C_6H_2N_3O_7^-$, $C_{12}H_{16}N_3O_4^+$	45.33	**7**
$C_6H_2N_3O_7^-$, $C_{13}H_{17}N_2^+$	36.14	**7**
$C_6H_2N_3O_7^-$, $C_{23}H_{19}N_2^+$	34.7	**5**
$2C_6H_2N_3O_7^-$, $C_4H_{16}N_8PbS_4^{2+}$	69.8	**4**
$2C_6H_2N_3O_7^-$, $C_{14}H_{30}N_2O_4^{2+}$	3.31	**7**
$C_6H_2N_3O_9S^-$, $H_5O_2^+$, $2H_2O$	15.2	**4**
$C_6H_2N_3O_9S^-$, $H_5O_2^+$, $2H_2O$	15.1	**7**
$C_6H_2N_3O_9S^-$, $C_{13}H_{27}S^+$	11.11	**6**
$C_6H_2N_4O_4^{2-}$, $2Na^+$, $4H_2O$	35.3	**1**
$C_6H_2N_4O_4^{2-}$, $2Rb^+$, $2H_2O$	35.4	**1**
$C_6H_2N_4O_6$	40.8	**4**
$C_6H_2N_6O_4$, $C_{18}H_{15}OP$	60.75	**2**
$2C_6H_2N_6O_6$, $C_{18}H_{12}CuN_2O_2$	60.76	**2**
$C_6H_2O_4$, $2H_2O$	1.77	**1**
$C_6H_2O_4^{2-}$, $2K^+$	18.4	**7**
$C_6H_2O_8^{2-}$, $2K^+$	2.73	**1**
$C_6H_3BrN_2O_2$	40.9	**4**
$C_6H_3BrN_2O_4$	15.7	**1**
$C_6H_3BrN_2O_4$	15.5	**3**
$C_6H_3Br_2O_3S^-$, $H_7O_3^+$	11.1	**4**
$C_6H_3Br_3$	19.9	**1**
$C_6H_3ClN_2O_2$	40.4	**1**
$C_6H_3ClN_2O_4$	15.8	**1**
$C_6H_3ClN_2O_4$	15.6	**3**
$C_6H_3ClN_2O_5$	17.1	**7**
$C_6H_3ClO_2$	18.4	**3**
$C_6H_3Cl_2O_3S^-$, $H_7O_3^+$	11.2	**4**
$C_6H_3Cl_3$	19.10+	**1**
$C_6H_3Cl_3$	19.1	**4**
$C_6H_3Cl_3O$, C_6H_7N	60.18	**3**
$C_6H_3Cl_9O_6^-$, $C_{10}H_{10}Fe^+$	73.3	**4**
$C_6H_3CrIO_4$	71.3	**7**
$C_6H_3IN_2O_2$	40.10	**4**
$C_6H_3N_3$	20.7	**1**
$C_6H_3N_3O_6$	15.3	**4**
$C_6H_3N_3O_6$, C_6H_6IN	60.86	**2**
$C_6H_3N_3O_6$, $C_6H_9N_3$	60.12	**3**
$C_6H_3N_3O_6$, C_8H_6S	60.5	**5**
$C_6H_3N_3O_6$, C_8H_7N	60.102	**2**
$C_6H_3N_3O_6$, C_9H_6OS	60.8	**4**
$C_6H_3N_3O_6$, C_9H_7NO	60.29	**3**
$C_6H_3N_3O_6$, C_9H_9N	60.108	**2**
$C_6H_3N_3O_6$, $C_9H_{18}N_6$	60.113	**2**
$C_6H_3N_3O_6$, $C_{10}H_8$	60.115+	**2**
$C_6H_3N_3O_6$, $C_{10}H_8CrO_4$	60.118	**2**
$C_6H_3N_3O_6$, $C_{12}H_9NS$	60.38	**3**
$C_6H_3N_3O_6$, $C_{12}H_{12}N_2$	60.15	**7**
$C_6H_3N_3O_6$, $C_{12}H_{12}N_2$, $0.5C_6H_6$	60.16	**7**
$C_6H_3N_3O_6$, $C_{14}H_{10}$	60.133+	**2**
$C_6H_3N_3O_6$, $C_{16}H_{10}$	60.143	**2**
$C_6H_3N_3O_6$, $C_{16}H_{10}$	60.29	**5**
$C_6H_3N_3O_6$, $C_{22}H_{28}CoN_2O_2$	60.9	**4**
$C_6H_3N_3O_6$, $C_{22}H_{28}CuN_2O_2$	60.10	**4**
$C_6H_3N_3O_6$, $C_{22}H_{28}N_2NiO_2$	60.11	**4**
$2C_6H_3N_3O_6$, $C_{16}H_{16}CuN_2O_2$	60.2	**6**
$C_6H_3N_3O_7$, $C_{10}H_8BrN$	60.77	**2**
$0.2C_6H_3N_3O_7$, $C_9H_6O_6$, $1.5H_2O$	13.12	**6**
$C_6H_3N_5O_8$	16.1	**1**
C_6H_3P	64.7	**4**
$C_6H_4As_2Cl_2O$	65.10	**2**
C_6H_4BrCl	19.12	**1**
$C_6H_4BrNO_2$	15.9	**1**
$C_6H_4BrO_3S^-$, $C_9H_{18}N^+$	3.38	**3**
$C_6H_4BrO_3S^-$, $C_{11}H_{15}N_2O_3^+$	48.65	**5**
$C_6H_4Br_3N$	16.2	**1**
$(C_6H_4CdN_2O_4)_n$	81.21	**6**
$C_6H_4ClIO_2$	19.13	**1**
$C_6H_4ClNO_2$	15.10	**1**
$C_6H_4ClNO_2$	15.11	**1**
$C_6H_4ClNO_2$	18.12	**1**
$C_6H_4ClNO_4S$	11.7	**3**
$C_6H_4ClNO_5$	15.12	**1**
$C_6H_4Cl_2$	19.14+	**1**
$C_6H_4Cl_2$	19.2	**4**
$C_6H_4Cl_2$	19.1+	**7**
$C_6H_4Cl_2N_2O_2$	18.13	**1**
$C_6H_4Cl_2N_4O_4$	40.5	**1**
$C_6H_4Cl_2O$	17.1	**5**
$C_6H_4Cl_2O$	17.1	**6**
$C_6H_4Cl_2O$	17.2	**7**
$C_6H_4Cl_2O_2$	1.6	**4**
$C_6H_4Cl_3N$	16.1	**4**
$C_6H_4Cl_4N_2$	16.3	**1**
$C_6H_4Cu_2N_2O_8S_2^{2-}$, $2C_4H_{12}N^+$	3.38	**1**
C_6H_4INO	10.25	**1**
$C_6H_4I_2$, $2C_{24}H_{40}O_4$	60.40	**5**
$C_6H_4NO_2^-$, K^+ , H_2O	17.3	**7**
$C_6H_4NO_3^-$, K^+ , $0.5H_2O$	6.7	**1**
$C_6H_4NO_3^-$, K^+ , $0.5H_2O$	15.2	**7**
$C_6H_4N_2$	7.22	**1**
$C_6H_4N_2$	33.13	**4**
$C_6H_4N_2$	27.1	**5**
$C_6H_4N_2O$	40.6	**1**
$C_6H_4N_2O_2$	40.7	**1**
$C_6H_4N_2O_2$	40.8	**1**
$C_6H_4N_2O_2$	40.9	**1**
$C_6H_4N_2O_3S$	9.2	**3**
$C_6H_4N_2O_3S$	9.2	**4**
$C_6H_4N_2O_4$	15.15	**1**
$C_6H_4N_2O_4$	15.16	**1**
$C_6H_4N_2O_4$, $2H_2O$	33.17	**7**
$C_6H_4N_2O_4S$, H_2O	9.11	**1**
$C_6H_4N_2O_8P^-$, $C_7H_{10}N^+$	46.5	**3**
$C_6H_4N_2S$	41.12	**1**
$C_6H_4N_2Se$	42.7	**1**
$C_6H_4N_4$	35.5	**1**
$C_6H_4N_4$	35.6	**1**
$C_6H_4N_4$	7.7+	**4**
$C_6H_4N_4$	35.4	**7**
$C_6H_4N_4^{2+}$, Cl_4Zn^{2-}	9.12	**1**
$C_6H_4N_4O_2$	9.13	**1**
$C_6H_4N_4O_2$, $1.5H_2O$	35.1	**4**
$C_6H_4N_4O_6$	16.1	**5**
$C_6H_4O_2$	18.14+	**1**
$C_6H_4O_2$	18.1	**5**
$C_6H_4O_2$, $C_5H_6N_2O_2$	60.7	**4**
$C_6H_4O_2$, C_6H_5BrO	60.78	**2**
$C_6H_4O_2$, $2C_6H_5BrO$	60.79	**2**
$C_6H_4O_2$, C_6H_5ClO	60.80	**2**
$C_6H_4O_2$, $2C_6H_5ClO$	60.81	**2**
$C_6H_4O_2$, $2C_6H_6O$	17.17	**1**
$C_6H_4O_2$, $C_6H_6O_2$	17.35+	**1**
$C_6H_4O_2$, $C_6H_6O_2$	60.13	**3**
$2C_6H_4O_2$, $C_6H_6O_3$	60.12	**4**
$C_6H_4O_4S$	42.1	**3**
$C_6H_4O_5$	38.18	**1**

$C_6H_4O_5$ 38.19 **1**
$C_6H_4O_6$, $2H_2O$ 18.16 **1**
$C_6H_4O_6^{2-}$, $2K^+$ 2.20 **4**
$C_6H_4O_7^{2-}$, Ca^{2+} , $4H_2O$ 2.21 **4**
$C_6H_4O_7^{2-}$, Ca^{2+} , $4H_2O$ 2.26 **5**
$C_6H_4S_2$ 39.31 **1**
$C_6H_4S_4$ 39.6 **6**
$C_6H_4S_4$, I_2 39.3 **7**
$C_6H_4S_4^+$, $C_{12}H_4N_4^-$ 60.3 **7**
$C_6H_4S_4^+$, $C_{12}H_8N_4^-$ 60.3 **6**
$5C_6H_4S_4^+$, $2C_6H_4S_4$, $5I^-$ 39.4 **7**
$2C_6H_4S_4$, $C_4H_4NiS_4$ 60.4 **7**
$C_6H_4Se_2$ 39.32 **1**
C_6H_5BrHg 71.8 **2**
C_6H_5BrMg , $2C_4H_{10}O$ 67.14 **2**
C_6H_5BrO , $C_6H_4O_2$ 60.78 **2**
C_6H_5BrO , $C_{19}H_{22}O_2$ 60.14 **3**
C_6H_5BrO , $C_{19}H_{26}O_2$ 60.39 **4**
C_6H_5BrO , $C_{19}H_{28}O_2$ 60.151 **2**
C_6H_5BrO , $C_{19}H_{28}O_2$ 60.15 **3**
$2C_6H_5BrO$, $C_6H_4O_2$ 60.79 **2**
C_6H_5BrOS 39.5 **3**
$C_6H_5Br_2NO$ 33.37 **1**
C_6H_5Cl 19.3 **4**
$C_6H_5ClN_2O_2$ 16.4 **1**
$C_6H_5ClN_2O_3$ 38.4 **6**
$C_6H_5ClN_2O_3$, H_2O 33.17 **6**
C_6H_5ClO 17.2+ **5**
C_6H_5ClO , $C_6H_4O_2$ 60.80 **2**
$2C_6H_5ClO$, $C_6H_4O_2$ 60.81 **2**
$C_6H_5ClO_2$, $C_6H_{11}NO$ 60.6 **7**
$C_6H_5ClO_2Se$ 11.32 **1**
$C_6H_5Cl_2I$ 19.18 **1**
$C_6H_5Cl_2N$ 16.5 **1**
$C_6H_5Cl_4NOP_2S$ 64.3 **5**
$C_6H_5Cl_5$ 21.3 **1**
$C_6H_5CrN_3O_3$ 73.1 **3**
$C_6H_5FO_7^{2-}$, Rb^+ , H_4N^+ , $2H_2O$ 2.27 **5**
$C_6H_5F_2N_3O_3S_3$ 11.8 **3**
$C_6H_5FeIO_3$ 72.7 **2**
$C_6H_5NO_2$ 15.17 **1**
$C_6H_5NO_2$ 33.38 **1**
$C_6H_5NO_2$ 18.5 **7**
$C_6H_5NO_2$ 33.18 **7**
$C_6H_5NO_2$ 33.19 **7**
$C_6H_5NO_3$ 17.5+ **1**
$C_6H_5NO_3$ 33.14 **4**
$C_6H_5NO_3$ 17.4+ **7**
$C_6H_5NO_3$, $C_{16}H_{14}CuN_2O_2$ 60.16 **3**
$C_6H_5N_2^+$, Br_3^- 9.14 **1**
$C_6H_5N_2^+$, $Br_3Cu_2^-$ 9.15 **1**
$C_6H_5N_2^+$, Cl^- 9.16 **1**
$C_6H_5N_2^+$, $C_2H_4O_2$, Cl^- 60.82 **2**
$C_6H_5N_2O_4^-$, $C_3H_{11}N_2^+$ 32.5 **7**
$C_6H_5N_3$ 35.2 **6**
$C_6H_5N_5O_6$ 15.18 **1**
$2C_6H_5O_3S^-$, H_2MgO^{2+} 11.33 **1**
$2C_6H_5O_3S^-$, $H_{12}O_6Zn^{2+}$ 11.34 **1**
$C_6H_5O_4P$ 46.5 **4**
$C_6H_5O_4P^{2-}$, Ba^{2+} , $1.5H_2O$ 46.9 **1**
$C_6H_5O_4P^{2-}$, $2K^+$, $1.5H_2O$ 46.10 **1**
$C_6H_5O_6^-$, K^+ 2.22 **4**
$C_6H_5O_6^-$, K^+ 2.28 **5**
$C_6H_5O_6^-$, K^+ 2.29 **5**
$C_6H_5O_6^-$, $C_6H_4O_6^{2-}$, $3K^+$, $2H_2O$ 2.21 **6**
$2C_6H_5O_7^{3-}$, $2H_2MgO^{2+}$, $H_{12}MgO_6^{2+}$, $2H_2O$ 2.77 **1**
C_6H_6 19.19+ **1**
C_6H_6 19.3+ **3**
C_6H_6 19.4 **4**
C_6H_6 , Al_2Br_6 60.83 **2**
C_6H_6 , Br_2 60.84 **2**
C_6H_6 , N_2O_4 60.85 **2**
C_6H_6 , $C_{12}H_4N_4$ 60.17 **6**
C_6H_6 , $C_{19}H_{16}$ 60.32 **7**
C_6H_6 , $C_{25}H_{19}MoO_4P$ 61.8 **6**
$2C_6H_6$, $H_6MnN_2^{2+}$, $C_4N_4Ni^{2-}$ 61.4 **6**
$2C_6H_6$, $C_4H_6CdHgN_6$ 61.3 **5**
$2C_6H_6$, $C_4H_6CdN_6Ni$ 61.3 **3**
$2C_6H_6$, $C_4H_6CuN_6Ni$ 61.5 **6**
$3C_6H_6$, $C_{30}H_{18}N_3O_6P_3$ 61.9 **6**
$2nC_6H_6$, $(C_2H_8CdN_2^{2+})_n$, $nC_4CdN_4^{2-}$ 61.1 **6**
$2nC_6H_6$, $(C_4H_8CdN_4Ni)_n$ 61.2 **5**
$(C_6H_6Ag^+)_n$, $(ClO_4^-)_n$ 74.1 **2**
$C_6H_6AgAlCl_4$ 74.2 **2**
$C_6H_6AlCl_4Cu$ 74.3 **2**
$C_6H_6AlCl_4Pb^+$, $AlCl_4^-$, C_6H_6 69.4 **7**
$C_6H_6AlCl_4Sn^+$, $AlCl_4^-$, C_6H_6 69.5 **7**
$C_6H_6Al_2Cl_8Sn$, C_6H_6 69.12 **3**
$C_6H_6AsO_2S^-$, $C_6H_{16}N^+$ 65.5 **4**
$C_6H_6BBrO_2$ 62.42 **2**
$C_6H_6Br_2N_4$ 35.7 **1**
C_6H_6ClN 16.6+ **1**
$C_6H_6Cl_2N_2O_6S_2$ 32.20 **3**
$C_6H_6Cl_4Ir_2O_4$ 71.9 **2**
$C_6H_6Cl_5Sb$ 66.9 **2**
$C_6H_6Cl_5Sb$ 66.2 **6**
$C_6H_6Cl_6$ 21.4 **1**
$C_6H_6Cl_6$ 21.5 **1**
$C_6H_6Cl_6$ 21.3 **3**
$C_6H_6Cl_6$ 21.4 **3**
$C_6H_6Cl_6FeOSi_2$ 73.2 **3**
$(C_6H_6CoN_2O_6)_n$ 81.16 **5**
$C_6H_6CuNO_6^-$, Li^+ , $3H_2O$ 82.5 **7**
$C_6H_6CuNO_6^-$, Na^+ , H_2O 82.5 **6**
$C_6H_6F_2N_2O_6S_2$ 44.13 **4**
C_6H_6IN , $C_6H_3N_3O_6$ 60.86 **2**
C_6H_6IN , $C_{36}H_{60}O_{30}$, $3H_2O$ 45.58 **7**
$C_6H_6I_6O_4Rh_2^{2-}$, $2C_9H_{14}N^+$ 71.4 **7**
$C_6H_6MoNO_9^{3-}$, $3K^+$, H_2O 82.6 **6**
$C_6H_6MoS_6$ 85.34 **2**
$C_6H_6NO_2^+$, Cl^- 33.40 **1**
$C_6H_6NO_3S^-$, Na^+ , $2H_2O$ 16.2 **7**
$C_6H_6NO_6^{3-}$, $3Na^+$, H_2O 2.75 **1**
$C_6H_6NO_6P$ 46.4 **5**
$C_6H_6N_2O$ 33.41 **1**
$C_6H_6N_2O$ 33.42 **1**
$C_6H_6N_2O_2$ 16.10 **1**
$C_6H_6N_2O_2$ 16.2 **5**
$C_6H_2D_4N_2O_2$ 15.8 **3**
$2C_6H_6N_2O_2$, Cd_2Cl_4 , $2H_2O$ 60.87 **2**
$4C_6H_6N_2O_2$, $C_{12}H_{28}N_2^{2+}$, $2Cl^-$ 60.40 **3**
$(C_6H_6N_2O_6Zn)_n$, nH_2O 81.22 **6**
$C_6H_6N_2S$ 33.43 **1**
$C_6H_6N_2S$ 33.15 **4**
$C_6H_6N_2S$ 33.23 **5**
$C_6H_6N_2S_2$ 11.3 **6**
$C_6H_6N_4O_4$ 40.10 **1**
$C_6H_6N_4S$, H_2O 44.45 **1**
$C_6H_6N_4S$, $3H_2O$ 44.15 **7**

$C_6H_6N_6O_6$	15.19 **1**
$(C_6H_6N_6Sn)_n$	69.10 **4**
$(C_6H_6NdO_{15})_{2n}$, $9nH_2O$	81.23 **6**
$C_6H_6Nd_2O_{15}$, $4H_2O$	81.41 **2**
C_6H_6O	17.7+ **1**
C_6H_6O , $0.5H_2O$	17.10 **1**
$2C_6H_6O$, $C_6H_4O_2$	17.17 **1**
$3C_6H_6O$, $C_6H_{12}N_4$	60.21 **3**
$C_6H_6O_2$	17.13 **1**
$C_6H_6O_2$	17.11+ **1**
$C_6H_6O_2$	17.14+ **1**
$C_6H_6O_2$	5.5 **4**
$C_6H_6O_2$	17.2 **4**
$C_6H_6O_2$	17.2 **6**
$C_6H_6O_2$, O_2S	61.9 **2**
$C_6H_6O_2$, CH_4N_2O	60.5 **7**
$C_6H_6O_2$, C_3H_6O	60.88 **2**
$C_6H_6O_2$, $C_5H_4N_2O_3$	60.2 **5**
$C_6H_6O_2$, $C_6H_4O_2$	17.35+ **1**
$C_6H_6O_2$, $C_6H_4O_2$	60.13 **3**
$C_6H_6O_2$, $C_{13}H_{13}N_4O_2^+$, Cl^-	36.14 **5**
$C_6H_6O_2$, $C_{17}H_{22}N_4O_6^{2+}$, $2Br^-$, $2H_2O$	60.27 **6**
$C_6H_6O_2$, $C_{21}H_{30}O_2$	60.37 **7**
$1.5C_6H_6O_2$, $C_{13}H_{13}N_2O_2^+$, Br^-	60.30 **4**
$C_6H_6O_2S$	39.6 **3**
$C_6H_6O_2S$	39.7 **3**
$C_6H_6O_2Se$	11.35 **1**
$C_6H_6O_3$	17.18 **1**
$C_6H_6O_3$	17.3 **4**
$C_6H_6O_3$, $2H_2O$	17.19 **1**
$C_6H_6O_3$, $2C_6H_4O_2$	60.12 **4**
$C_6H_6O_3S_2$	39.8 **3**
$C_6H_6O_4$	20.8 **1**
$C_6H_6O_4$	1.17 **5**
$C_6H_6O_4$	20.6 **6**
$C_6H_6O_4$	20.4 **7**
$C_6H_6O_4$, $C_{12}H_{24}O_6$	38.41 **7**
$C_6H_6O_4^{2-}$, $2C_6H_8O_4$, $2Na^+$	20.9 **1**
$C_6H_6O_4P^-$, $C_6H_{14}N^+$	46.6 **3**
$C_6H_6O_7^{2-}$, Ca^{2+} , $3H_2O$	2.10 **7**
$C_6H_6O_7^{2-}$, Li^+ , H_4N^+ , H_2O	2.76 **1**
$C_6H_6O_9S^{2-}$, Ba^{2+} , $2H_2O$	38.5 **6**
$C_6H_6O_{24}P_6^{12-}$, $12Na^+$, $38H_2O$	46.6 **4**
$C_6H_6S_3$	39.5 **7**
$C_6H_6Sn^{2+}$, $2AlCl_4^-$	69.11 **4**
$C_6H_6U^{3+}$, $3AlCl_4^-$	74.1 **4**
$(C_6H_7AgN_5^+)_n$, nNO_3^- , $2nH_2O$	83.7 **7**
$C_6H_7AsO_3$	65.12+ **2**
$C_6H_7BN_2OS$	62.2 **7**
$2C_6H_7BrN^+$, F_6Si^{2-} , $2H_2O$	16.2 **4**
$C_6H_7BrN_2O_2$	44.10+ **5**
$C_6H_7Cl_3O$	21.1+ **5**
$2C_6H_7MnO_8^-$, $H_{12}MnO_6^{2+}$, $2H_2O$	81.17+ **5**
C_6H_7N , Cl_3Sb	60.89 **2**
C_6H_7N , I_2	60.90 **2**
C_6H_7N , C_6HCl_5O	60.17 **3**
C_6H_7N , $C_6H_3Cl_3O$	60.18 **3**
C_6H_7N , $C_{19}H_{16}$	60.33 **7**
$2C_6H_7N$, Cl_3Sb	60.19 **3**
C_6H_7NO	17.20 **1**
C_6H_7NO	17.3 **6**
C_6H_7NO , $C_8H_{12}N_2O_3$	60.9 **6**
$2C_6H_7NO$, $C_4H_4O_4$	33.20 **7**
$C_6H_7NO_2$	33.16 **4**
$C_6H_7NO_3S$	16.11 **1**
$C_6H_7NO_3S$	16.12 **1**
$C_6H_7NO_3S$, H_2O	16.13 **1**
$C_6H_7NO_6^{2-}$, Ca^{2+} , $2H_2O$	2.23 **4**
$C_6H_7N_2O_2S^-$, Na^+ , H_2O	16.14 **1**
$C_6H_7N_2O_3^-$, K^+ , $1.67H_2O$	43.1 **7**
$C_6H_7N_3O$	33.17 **4**
$C_6H_7N_3O$	33.21 **7**
$C_6H_7N_3O_3$	44.46 **1**
$C_6H_7N_3S_2$	39.9 **3**
$C_6H_7N_4O_2^+$, Cl^- , H_2O	44.16 **7**
$C_6H_7N_5$	44.47 **1**
$C_6H_7N_5$	44.12 **5**
$C_6H_7N_5$, $C_5H_5BrN_2O_2$	44.21 **1**
$C_6H_7N_5$, $C_5H_6N_2OS$	44.21 **3**
$C_6H_7N_5$, $C_5H_6N_2OS$	44.8 **6**
$C_6H_7N_5$, $C_6H_8N_2O_2$	44.51 **1**
$C_6H_7N_5$, $C_6H_8N_2O_2$	44.12 **6**
$C_6H_7O_4P$	46.5 **5**
$C_6H_7O_6^-$, Na^+	45.17 **1**
$C_6H_7O_6^-$, Tl^+	45.3 **6**
$2C_6H_7O_6^-$, Ca^{2+} , $2HO$	45.2 **7**
$2C_6H_7O_6^-$, Ca^{2+} , $2H_2O$	45.3 **7**
$C_6H_7O_7^-$, K^+	2.78+ **1**
$C_6H_7O_7^-$, Li^+	2.80 **1**
$C_6H_7O_7^-$, Na^+	2.81 **1**
$C_6H_7O_7^-$, Rb^+	2.82 **1**
$2C_6H_7O_9S^-$, Ba^{2+} , $2H_2O$	45.2 **4**
$C_6H_8AgN_4^+$, NO_3^-	83.15+ **4**
$C_6H_8AsNO_3$	65.14 **2**
$C_6H_8Br_2$	20.3 **5**
$(C_6H_8Br_2N_2Ni)_n$	83.40 **2**
$C_6H_8Br_4O_4$	38.21 **1**
$(C_6H_8CdCl_2N_4)_n$	83.5 **5**
$(C_6H_8CdN_6Ni)_n$, $2nC_6H_6$	61.4 **3**
$C_6H_8Cl_2CoN_4$	83.6 **5**
$C_6H_8Cl_2CuN_4$	83.7 **5**
$(C_6H_8Cl_2HgO_2)_n$	60.91 **2**
$(C_6H_8Cl_2MnN_4)_n$	83.7 **6**
$C_6H_8Cl_2N_4Zn$	83.41 **2**
$C_6H_8Cl_4$	21.6 **1**
$C_6H_8Cl_8Si_6$	63.2 **6**
$C_6H_8CoNO_7^-$, K^+ , $2H_2O$	82.6 **7**
$C_6H_8CrN_2O_8^-$, $C_6H_{16}CrN_4O_4^+$, $2H_2O$	76.24 **3**
$(C_6H_8CuNO_7^-)_n$, nK^+ , $2nH_2O$	82.7 **7**
$C_6H_8F_4Pt$	72.1 **7**
$C_6H_8N^+$, $BiBr_4^-$	33.44 **1**
$C_6H_8N^+$, BiI_4^-	33.45 **1**
$C_6H_8N^+$, Br^-	16.15 **1**
$C_6H_8N^+$, Cl^-	16.16 **1**
$C_6H_8N^+$, I_4Sb^-	33.24 **5**
$C_6H_8N^+$, $C_7H_7Cl_4Te^-$	70.2 **7**
$C_6H_8N^+$, $C_{18}H_{15}Br_3PZn^-$	86.10 **4**
$2C_6H_8N^+$, Br_6Sb^- , Br_3^-	33.46 **1**
$2C_6H_8N^+$, Br_6Sb^- , Br_3^-	33.25 **5**
$2C_6H_8N^+$, Cl_4Cu^{2-}	16.1 **6**
$C_6H_8NO^+$, Cl^-	17.21 **1**
$C_6H_8NO^+$, Cl^-	33.39 **1**
$C_6H_8NO_2^+$, Cl^-	17.4 **4**
$C_6H_8N_2$	9.17 **1**
$C_6H_8N_2$, $C_{10}H_2N_4$	60.7 **5**
$2C_6H_8N_2$, $C_{24}H_{32}N_8Ni^{2+}$, $2Cl^-$	83.73 **6**
$C_6H_8N_2O$, $2H_2O$	44.49 **1**
$C_6H_8N_2O_2$	40.11 **1**
$C_6H_8N_2O_2$	44.50 **1**

$C_6H_8N_2O_2$	33.26	5
$C_6H_8N_2O_2$	32.6	7
$C_6H_8N_2O_2$	33.22	7
$C_6H_8N_2O_2$	44.17	7
$C_6H_8N_2O_2$, $C_6H_7N_5$	44.51	1
$C_6H_8N_2O_2$, $C_6H_7N_5$	44.12	6
$2C_6H_8N_2O_2$, $C_7H_{10}N_6$, H_2O	44.25	3
$C_6H_8N_2O_2S$	16.17 +	1
$C_6H_8N_2O_2S$	16.3	4
$C_6H_8N_2O_2S$	16.3	5
$C_6H_8N_2O_2S$, H_2O	16.22	1
$C_6H_8N_2O_2S$, $C_9H_9N_3O_2S_2$	60.30	3
$C_6H_8N_2O_2S_2$	48.22	6
$C_6H_8N_2O_2S_2$, $C_2H_4O_2$	48.23	7
$C_6H_8N_2O_2S_4$	41.4	6
$C_6H_8N_2O_3$	43.1	4
$C_6H_8N_2O_4$	43.2	4
$C_6H_8N_4O_2S$	41.7	5
$C_6H_8N_4O_4$	32.6	5
$C_6H_8N_5^+$, Cl^-	44.13	6
$C_6H_8N_5O^+$, Br^-	44.52	1
$C_6H_8O_2$	21.7 +	1
$C_6H_8O_2$, C_2I_2	60.92	2
$C_6H_8O_2S$	39.6	4
$C_6H_8O_3$	38.20	1
$C_6H_8O_3$	45.4	7
$C_6H_8O_4$	1.78	1
$C_6H_8O_4$	20.11	1
$C_6H_8O_4$	20.4	3
$C_6H_8O_4$	20.5	3
$C_6H_8O_4$	20.6	3
$C_6H_8O_4$	20.2	4
$C_6H_8O_4$	45.3	4
$C_6H_8O_4$	20.4	5
$C_6H_8O_4^{2-}$, $C_6H_{18}N_2^{2+}$	2.83 +	1
$C_6H_8O_4S_2$	39.35	1
$C_6H_8O_4Se_2$	39.36	1
$C_6H_8O_6$	45.18	1
$C_6H_8O_6$	45.19 +	1
$C_6H_8O_6$	45.3	5
$C_6H_8O_7$	1.79	1
$C_6H_8O_7$, H_2O	1.7	4
$C_6H_8O_7$, H_2O	45.5	7
$(C_6H_8O_{16}Sc_2)_n$, $2nH_2O$	81.19	5
$(C_6H_8O_{16}Yb_2)_n$, $2nH_2O$	81.20	5
$C_6H_9Br_3CuNOS^-$, $C_6H_{10}NOS^+$	83.8	7
$C_6H_9ClN_2O_4$	48.34	5
$C_6H_9Cl_3$	21.3	5
$C_6H_9CoN_9O_6^{3-}$, $3K^+$, $6.38H_2O$	79.9	7
$C_6H_9CuNO_5$, $2H_2O$	82.13	2
$(C_6H_9EuO_9)_n$	81.21	4
$C_6H_9GdO_9$, $0.5H_2O$	81.10	3
$(C_6H_9GdO_9)_n$	81.21	5
$(C_6H_9LaO_9)_n$	81.22	5
C_6H_9NO	1.80	1
$C_6H_9NO_2$, H_2O	48.20	4
$C_6H_9NO_3$	40.4	3
$C_6H_9NO_3$	1.18	5
$C_6H_9NO_3$	20.5	5
$C_6H_9NO_6$	1.81	1
$C_6H_9N_2^+$, Cl^-	9.18	1
$C_6H_9N_2^+$, Cl^-	16.4	5
$C_6H_9N_2^+$, Cl^-	33.23	7
$C_6H_9N_2O_2P$	64.4	5
$C_6H_9N_2O_2S^+$, Cl^-	16.2	6
$C_6H_9N_3$, $C_6H_3N_3O_6$	60.12	3
$C_6H_9N_3O^{2+}$, $2Cl^-$	33.47 +	1
$C_6H_9N_3O_2$	48.21	4
$C_6H_9N_3O_2$	48.35 +	5
$C_6H_9N_3O_2$	48.23	6
$C_6H_9N_3O_3$	40.5	6
$C_6H_9N_3O_3$	40.6	6
$C_6H_9N_5^{2+}$, $2Br^-$	44.48	1
$C_6H_9N_5^{2+}$, $2Cl^-$	44.18	7
$C_6H_9O_3PS$	64.16	2
$C_6H_9O_6P$	46.5	6
$C_6H_9O_7^-$, Ca^{2+} , Br^- , $3H_2O$	45.4	6
$C_6H_9O_7^-$, K^+ , $2H_2O$	45.22	1
$C_6H_9O_7^-$, Rb^+ , $2H_2O$	45.23	1
$2C_6H_9O_7^-$, Ca^{2+} , $2H_2O$	45.24	1
$2C_6H_9O_7^-$, Ca^{2+} , $3H_2O$	45.6	7
$2C_6H_9O_7^-$, Ca^{2+} , $4H_2O$	45.7	7
$3C_6H_9O_7^-$, Ca^{2+} , Na^+ , $6H_2O$	45.8 +	7
$3C_6H_9O_7^-$, Sr^{2+} , Na^+ , $6H_2O$	45.10	7
$C_6H_9O_8U^-$, Na^+	81.42	2
$C_6H_9S_5^+$, I^-	39.6	7
C_6H_{10}	20.13	1
$C_6H_{10}BrCl$	21.9	1
$C_6H_{10}Br_6Te_2$	70.3	5
$(C_6H_{10}CdO_2S_4)_n$	80.2	5
$C_6H_{10}Cl_2$	21.10	1
$C_6H_{10}Cl_2N_2Pd$	83.42	2
$C_6H_{10}Cl_2Pd_2$	72.8 +	2
$C_6H_{10}Cl_3HgN_3O_2$	82.4	3
$C_6H_{10}Cl_3O_2Sb$	66.1	5
$C_6H_{10}Cr$	72.12	2
$C_6H_{10}CuN_8O_6$	83.8	5
$C_6H_{10}DyNO_8$, $2H_2O$	82.6	5
$C_6H_6D_4Ga_2N_4$	68.5	6
$(C_6H_{10}HgO_2S_4)_n$	80.3	5
$C_6H_{10}Mo_2N_2O_6S_2^{2-}$, $2Na^+$, $2H_2O$	82.7	5
$C_6H_{10}Mo_2N_2O_8S_2^{2-}$, $2Na^+$, $5H_2O$	82.14	2
$C_6H_{10}NNdO_8$, H_2O	81.43	2
$C_6H_{10}NOS^+$, $C_6H_9Br_3CuNOS^-$	83.8	7
$C_6H_{10}NO_2^+$, Cl^- , H_2O	48.22	4
$C_6H_{10}NO_8Pr$, H_2O	82.8	5
$C_6H_{10}NS^+$, Br^-	41.8	5
$C_6H_{10}N_2^{2+}$, $2Br^-$	16.5	5
$C_6H_{10}N_2^{2+}$, $2Cl^-$	16.23	1
$C_6H_{10}N_2^{2+}$, $2Cl^-$	33.49	1
$C_6H_{10}N_2^{2+}$, $2Cl^-$	16.3	6
$C_6H_{10}N_2O_2$	21.11	1
$C_6H_{10}N_2O_2$	33.17	3
$C_6H_{10}N_2O_2$	33.15 +	3
$C_6H_{10}N_2O_2$	33.18 +	3
$C_6H_{10}N_2O_2$	48.23	4
$C_6H_{10}N_2O_2$	48.39	5
$C_6H_{10}N_2O_2$, H_2O	40.1	7
$C_6H_{10}N_2O_8$, $2C_4H_8OS$	60.13	4
$C_6H_{10}N_3O_2^+$, Cl^- , H_2O	48.59	1
$C_6H_{10}N_3O_2^+$, Cl^- , H_2O	48.24	4
$C_6H_{10}N_3O_2^+$, Cl^- , $2H_2O$	48.25	3
$C_6H_{10}N_3O_2^+$, $H_2O_4P^-$, H_3O_4P	48.25	4
$C_6H_{10}N_4$, ClI	60.93 +	2
$C_6H_{10}N_4NiS_4$	85.35	2
$C_6H_{10}N_4OS$	41.9	5
$C_6H_{10}N_6$	36.1	5
$C_6H_{10}NiO_2S_4$	80.3	2
$C_6H_{10}OS_2$	42.3	7
$C_6H_{10}OS_2$	42.4	7
$C_6H_{10}O_2$	20.6	5
$C_6H_{10}O_2PbS_4$	69.12	2

$C_6H_{10}O_2S_2$ 11.8 **7**
$C_6H_{10}O_2S_4$ 11.9 **3**
$C_6H_{10}O_2S_4Te$ 70.10 **2**
$C_6H_{10}O_2S_4Zn$ 80.4 **2**
$C_6H_{10}O_4$ 1.86 **1**
$C_6H_{10}O_4$ 1.82 + **1**
$C_6H_{10}O_4$ 38.22 **1**
$C_6H_{10}O_4$ 38.5 **4**
$C_6H_{10}O_4S$ 1.17 **7**
$C_6H_{10}O_4S_2$ 42.1 **5**
$C_6H_{10}O_5$ 45.3 **3**
$C_6H_{10}O_5$ 45.4 **5**
$C_6H_{10}O_5$ 45.5 **6**
$C_6H_{10}O_5$ 45.11 **7**
$C_6H_{10}O_6$ 45.25 **1**
$C_6H_{10}O_6$ 45.4 **3**
$C_6H_{10}O_6$ 45.5 **3**
$C_6H_{10}O_6$ 1.19 **5**
$C_6H_{10}O_7$, H_2O 45.12 **7**
$C_6H_{10}O_{10}P^{3-}$, $3Na^+$, $2H_2O$ 46.6 + **6**
$C_6H_{10}S_3$ 39.37 **1**
$C_6H_{10}S_4$ 39.38 **1**
$C_6H_{10}S_4$ 39.7 **4**
$(C_6H_{11}CaNO_8)_n$ 67.4 **4**
$(C_6H_{11}CeO_7)_n$ 81.44 **2**
$C_6H_{11}Cl_2NOS$ 4.3 **4**
$C_6H_{11}FO_5$ 45.13 **7**
$C_6H_{11}Mo_2O_{11}^-$, H_4N^+ , H_2O 84.11 **7**
$C_6H_{11}NO$ 34.2 + **7**
$C_6H_{11}NO$, $C_6H_5ClO_2$ 60.6 **7**
$C_6H_{11}NO_2$ 48.24 **7**
$C_6H_{11}NO_2$, H_2O 48.26 **4**
$C_6H_{11}NO_2Pd$ 72.1 **6**
$C_6H_{11}NO_3$ 48.25 **7**
$C_6H_{11}NO_4S_2$ 4.13 **1**
$C_6H_{11}NO_5$ 48.24 **6**
$C_6H_{11}NO_6$ 45.6 **6**
$C_6H_{11}N_3^{2+}$, $2Cl^-$, H_2O 35.8 **1**
$C_6H_{11}N_3O_2^{2+}$, $2Cl^-$ 48.25 **6**
$C_6H_{11}N_3O_4$ 48.60 **1**
$C_6H_{11}N_3O_4$, Ca^{2+} , $2Cl^-$, $3H_2O$ 60.95 **2**
$C_6H_{11}N_3O_4$, Li^+ , Br^- 48.27 **4**
$C_6H_{11}N_4O_2^+$, Br^- 59.5 **1**
$C_6H_{11}N_4O_2^+$, Br^- 59.1 **3**
$C_6H_{11}N_6O^+$, Cl^- , H_2O 32.10 **6**
$C_6H_{11}O_4P$ 64.5 **5**
$C_6H_{11}O_5P^{2-}$, $2Na^+$, $3H_2O$ 46.6 **5**
$2C_6H_{11}O_6^-$, Ca^{2+} 45.27 **1**
$2C_6H_{11}O_6^-$, Sr^{2+} 45.28 **1**
$C_6H_{11}O_7^-$, K^+ 45.29 **1**
$C_6H_{11}O_7^-$, K^+ , H_2O 45.6 **5**
$C_6H_{11}O_7^-$, K^+ , H_2O 45.7 + **6**
$C_6H_{11}O_9P^{2-}$, $2K^+$, $2H_2O$ 45.30 **1**
$C_6H_{11}O_9P^{2-}$, $2K^+$, $2H_2O$ 45.6 **3**
$C_6H_{11}O_{12}P_2^{3-}$, $3Na^+$, $7H_2O$ 46.1 **7**
C_6H_{12} 21.4 + **5**
$0.5C_6H_{12}$, $C_{12}H_{18}B_2F_2FeN_6O_6$ 61.5 **5**
$C_6H_{12}AgS_6^+$, NO_3^- 85.36 **2**
$C_6H_{12}BNO_3$ 62.10 **4**
$C_6H_{12}BNO_3$ 62.1 **5**
$C_6H_{12}BNO_3$ 62.5 **6**
$C_6H_{12}BNO_3$ 62.3 **7**
$C_6H_{12}Br_2N_4S_2Te$ 70.11 **2**
$C_6H_{12}Br_2O_4$ 45.7 **3**
$C_6H_{12}Br_2O_4$ 45.9 **6**
$C_6H_{12}CdCl_2N_4S_2$ 85.37 **2**
$(C_6H_{12}Ce_2O_{18})_n$, $4nH_2O$ 81.11 **3**
$(C_6H_{12}ClCuN_3O_5)_{2n}$, nH_2O 82.15 **2**
$C_6H_{12}Cl_2HgN_4O_2$ 84.4 **5**
$C_6H_{12}Cl_2O_4$ 45.8 **3**
$C_6H_{12}Cl_2O_4$ 45.14 **7**
$C_6H_{12}Cl_3N_2NbS_2$ 85.6 **6**
$C_6H_{12}Cl_4Te$ 70.5 **4**
$C_6H_{12}CoN_3OS_4$ 80.5 **2**
$C_6H_{12}CoN_3OS_4$ 80.1 **4**
$C_6H_{12}CoN_4O_8^-$, Ag^+ 82.7 **6**
$C_6H_{12}CoN_4O_8^-$, K^+ 81.17 **7**
$C_6H_{12}CrN_3O_6$, H_2O 82.7 **4**
$C_6H_{12}CuN_2O_4$ 82.17 **2**
$C_6H_{12}CuN_2O_4$ 82.5 **3**
$C_6H_{12}CuN_2O_4$, H_2O 76.20 **7**
$C_6H_{12}CuN_2O_4$, H_2O 76.21 **7**
$(C_6H_{12}CuN_2O_4)_n$ 84.6 **6**
$C_6H_{12}CuN_2O_6$ 82.18 **2**
$C_6H_{12}CuN_2S_4$ 80.1 + **7**
$C_6H_{12}CuO_7$, $0.5H_2O$ 81.45 **2**
$C_6H_{12}Cu_2N_4O_{10}$ 79.11 **3**
$C_6H_{12}FeN_3OS_4$ 80.1 + **3**
$C_6H_{12}Ga_2O_4$ 68.7 **7**
$C_6H_{12}Ga_3N_3$ 68.6 **5**
$C_6H_{12}HgI_2N_2S_4$ 85.4 **4**
$(C_6H_{12}HgO_2S)_n$ 81.18 **7**
$C_6H_{12}I_2N_4S_2Te$ 70.12 **2**
$(C_6H_{12}La_2O_{18})_n$, $4nH_2O$ 81.12 **3**
$C_6H_{12}LiN_2^+$, $C_7H_7^-$ 67.7 **3**
$C_6H_{12}MoNO_5^-$, H^+ 83.17 **3**
$C_6H_{12}Mo_2O_{11}$, $3H_2O$ 84.12 **7**
$C_6H_{12}NO^+$, Cl^- 34.4 **7**
$C_6H_{12}NO_2^+$, Br^- 20.14 **1**
$C_6H_{12}NO_2^+$, Br^- 48.40 **5**
$C_6H_{12}NO_3S^+$, Cl^- , H_2O 41.13 **1**
$C_6H_{12}NO_4^+$, Cl^- 59.2 **4**
$C_6H_{12}NO_4^+$, Cl^- 33.19 **6**
$C_6H_{12}NS^+$, $C_{12}H_4N_4^-$, $C_{12}H_4N_4$ 60.8 **5**
$C_6H_{12}N_2$ 37.1 + **1**
$C_6H_{12}N_2$, $C_{10}H_2CuF_{12}O_4$ 60.12 **5**
$C_6H_{12}N_2O_2$ 1.87 **1**
$C_6H_{12}N_2O_2$ 1.88 **1**
$C_6H_{12}N_2O_2$ 48.28 **6**
$C_6H_{12}N_2O_2$ 48.26 + **6**
$C_6H_{12}N_2O_3$ 48.26 **3**
$C_6H_{12}N_2O_3$ 1.13 **6**
$C_6H_{12}N_2O_4$, H_2O 48.26 **7**
$C_6H_{12}N_2O_4$, $2H_2O$ 48.41 **5**
$C_6H_{12}N_2O_4S_2$ 48.61 **1**
$C_6H_{12}N_2O_4S_2$ 48.29 **6**
$(C_6H_{12}N_2O_6Zn)_n$ 82.6 **3**
$C_6H_{12}N_2S_4$ 4.14 **1**
$C_6H_{12}N_2S_4$ 11.9 **7**
$C_6H_{12}N_3PS$ 64.11 **3**
$C_6H_{12}N_4$ 37.4 + **1**
$C_6H_{12}N_4$ 37.1 **5**
$C_6H_{12}N_4$, I_2 37.1 **7**
$C_6H_{12}N_4$, $2I_2$ 37.2 **7**
$C_6H_{12}N_4$, I_3N , I_2 37.3 **7**
$C_6H_{12}N_4$, $6H_2O$ 61.10 **2**
$C_6H_{12}N_4$, $H_{12}MgO_6^{2+}$, $2Cl^-$, $4H_2O$ 37.10 **1**
$C_6H_{12}N_4$, $2CHBr_3$ 60.1 **4**
$C_6H_{12}N_4$, CHI_3 60.20 **3**
$C_6H_{12}N_4$, $3C_6H_6O$ 60.21 **3**
$2C_6H_{12}N_4$, Ca^{2+} , $2Br^-$, $10H_2O$ 60.97 **2**

$2C_6H_{12}N_4 , Cr_2O_7^{2-} , Ca^{2+} , 7H_2O$ 37.4 **7**
$2C_6H_{12}N_4 , H_{12}MgO_6^{2+} , Cr_2O_7^{2-}$ 37.2 **6**
$2C_6H_{12}N_4 , CH_{12}ErN_2O_9S^{+} , CNS^{-} , 2H_2O$ 37.3 **6**
$(C_6H_{12}N_4O_3S_4Zn)_n$ 79.10 **7**
$C_6H_{12}N_8NiO_8^{2-} , 2K^{+} , 4H_2O$ 83.43 **2**
$(C_6H_{12}Nd_2O_{18})_n , 4nH_2O$ 81.13 **3**
$(C_6H_{12}Nd_2O_{18})_{2n} , 9nH_2O$ 81.14 **3**
$C_6H_{12}NiO_6S_2$ 81.22 **4**
$(C_6H_{12}O_2SSn)_n$ 69.10 **5**
$C_6H_{12}O_2Se$ 42.2 **5**
$C_6H_{12}O_2Si$ 63.4 **2**
$C_6H_{12}O_3S_2$ 45.10 **6**
$C_6H_{12}O_3S_2 , 0.25H_2O$ 45.11 **6**
$C_6H_{12}O_4S$ 42.10 **1**
$C_6H_{12}O_4S$ 45.31 **1**
$C_6H_{12}O_4S$ 42.2 **4**
$C_6H_{12}O_4S$ 45.8 **5**
$C_6H_{12}O_4S$ 45.9 **5**
$C_6H_{12}O_4S$ 45.10 **5**
$C_6H_{12}O_5$ 45.32 **1**
$C_6H_{12}O_5$ 45.33 **1**
$C_6H_{12}O_5$ 45.11 **5**
$C_6H_{12}O_5$ 45.12 **6**
$C_6H_{12}O_5 , H_2O$ 45.34 **1**
$C_6H_{12}O_5 , H_2O$ 45.7 **4**
$C_6H_{12}O_6$ 21.12 **1**
$C_6H_{12}O_6$ 45.35 **1**
$C_6H_{12}O_6$ 45.38 **1**
$C_6H_{12}O_6$ 45.36+ **1**
$C_6H_{12}O_6$ 45.11 **3**
$C_6H_{12}O_6$ 45.12 **3**
$C_6H_{12}O_6$ 45.13 **3**
$C_6H_{12}O_6$ 21.1 **4**
$C_6H_{12}O_6$ 38.6 **6**
$C_6H_{12}O_6$ 45.15 **7**
$C_6H_{12}O_6$ 45.16 **7**
$C_6H_{12}O_6 , Ca^{2+} , 2Br^{-} , 3H_2O$ 45.13 **6**
$C_6H_{12}O_6 , Ca^{2+} , 2Br^{-} , 5H_2O$ 45.14 **6**
$C_6H_{12}O_6 , Ca^{2+} , 2Cl^{-} , 2H_2O$ 45.15 **6**
$C_6H_{12}O_6 , Mg^{2+} , 2Cl^{-} , 4H_2O$ 45.16 **6**
$C_6H_{12}O_6 , H_2O$ 45.39 **1**
$C_6H_{12}O_6 , H_2O$ 45.12 **5**
$C_6H_{12}O_6 , 2H_2O$ 21.13 **1**
$C_6H_{12}O_6 , H_8MgO_4^{2+} , 2Cl^{-}$ 45.17 **6**
$C_6H_{12}O_6 , CH_4N_2O$ 60.14 **4**
$2C_6H_{12}O_6 , Ca^{2+} , 2Cl^{-} , 3H_2O$ 45.18 **6**
$C_6H_{12}O_6W$ 84.13 **7**
$(C_6H_{12}O_{18}Pr_2)_n , 4nH_2O$ 81.15 **3**
$C_6H_{12}O_{18}Ti_2 , 4H_2O$ 81.46 **2**
$C_6H_{12}P_2S_6$ 64.7 **6**
$C_6H_{12}S_2^{2+} , Cl_4Zn^{2-}$ 39.8 **5**
$C_6H_{12}S_3$ 39.39 **1**
$C_6H_{12}Se_3$ 39.9 **5**
$C_6H_{13}BO_8 , H_2O$ 45.8 **4**
$C_6H_{13}B_7Co^{-} , Cs^{+}$ 73.1 **6**
$C_6H_{13}BrO_5$ 45.17 **7**
$C_6H_{13}ClO_5$ 45.18 **7**
$C_6H_{13}CuN_3O_4 , 4H_2O$ 76.22 **3**
$C_6H_{13}CuN_5S_2$ 76.8 **6**
$C_6H_{13}HgNO_2S$ 71.5 **7**
$C_6H_{13}HgNO_2S , H_2O$ 71.2 **6**
$C_6H_{13}InOS$ 68.6 **6**
$C_6H_{13}InO_2$ 68.7 **6**
$C_6H_{13}NO_2$ 1.90 **1**
$C_6H_{13}NO_2$ 48.62 **1**
$C_6H_{13}NO_2$ 48.28 **4**
$C_6H_{13}NO_2$ 1.20 **5**
$C_6H_{13}NO_2$ 48.42 **5**
$C_6H_{13}NO_2$ 48.30 **6**
$C_6H_{13}NO_2$ 48.27 **7**
$C_6H_{13}NO_2 , C_6H_{14}NO_2^{+} , Cl^{-}$ 48.29 **4**
$C_6H_{13}NO_3 , 0.5H_2O$ 1.18 **7**
$C_6H_{13}N_2O^{+} , Br^{-}$ 34.5 **7**
$C_6H_{13}N_2O^{+} , Cl^{-} , H_2O$ 34.6 **7**
$C_6H_{13}N_2O_3^{+} , Cl^{-}$ 48.63 **1**
$C_6H_{13}N_3O$ 32.11 **6**
$C_6H_{13}OPS$ 64.9 **7**
$C_6H_{13}OPS$ 64.10 **7**
$C_6H_{13}O_2PS$ 64.11 **7**
$C_6H_{13}O_9P , H_2O$ 46.2 **7**
$C_6H_{13}O_{10}P$ 45.14 **5**
$C_6H_{13}S^{+} , I^{-}$ 39.7 **7**
C_6H_{14} 5.18 **1**
$C_6H_{14}AgN_{10}^{+} , NO_3^{-}$ 83.44 **2**
$C_6H_{14}BN_3$ 62.43 **2**
$(C_6H_{14}Ba_2CuO_{16})_n$ 81.47 **2**
$C_6H_{14}ClHgN$ 71.2 **5**
$C_6H_{14}ClNO_3Zn$ 83.17 **4**
$C_6H_{14}Cl_2HgN_2O_2S_2$ 80.2 **7**
$C_6H_{14}Cl_2MoN_2O_4$ 84.18 **2**
$C_6H_{14}Cl_2N_2O_2Pd$ 83.9 **7**
$C_6H_{14}Cl_4CuN_{12} , H_2O$ 83.10 **7**
$C_6H_{14}CoN_4O_4^{+} , Br^{-}$ 82.7 **3**
$C_6H_{14}CuN_4S_2$ 85.5 **3**
$C_6H_{14}CuO_8$ 81.48 **2**
$C_6H_{14}KO_3^{+} , C_{16}H_{16}Ce^{-}$ 75.26 **5**
$2C_6H_{14}KO_3^{+} , C_{12}H_{16}^{2-}$ 22.3 **6**
$C_6H_{14}N^{+} , Cl^{-}$ 21.14+ **1**
$C_6H_{14}N^{+} , C_3H_4O_6P^{-}$ 46.2 **6**
$C_6H_{14}N^{+} , C_6H_6O_4P^{-}$ 46.6 **3**
$2C_6H_{14}N^{+} , C_{20}H_{14}O_7P_2^{2-}$ 46.6 **7**
$C_6H_{14}NO_2^{+} , Br^{-}$ 48.64 **1**
$C_6H_{14}NO_2^{+} , Br^{-} , H_2O$ 48.65 **1**
$C_6H_{14}NO_2^{+} , Cl^{-} , H_2O$ 48.44 **5**
$C_6H_{14}NO_2^{+} , Cl^{-} , H_2O$ 48.45 **5**
$C_6H_{14}NO_2^{+} , I^{-}$ 48.28 **3**
$C_6H_{14}NO_4^{+} , Cl^{-}$ 45.40 **1**
$C_6H_{14}NO_5^{+} , Br^{-}$ 45.41 **1**
$C_6H_{14}NO_5^{+} , Br^{-}$ 45.15 **3**
$C_6H_{14}NO_5^{+} , Cl^{-}$ 45.42 **1**
$C_6H_{14}NO_5^{+} , Cl^{-}$ 45.16 **3**
$C_6H_{14}NO_5^{+} , Cl^{-}$ 45.15 **5**
$C_6H_{14}N_2NiS_4$ 80.4 **5**
$C_6H_{14}N_2O_4S^{2+} , 2Cl^{-}$ 48.31 **6**
$C_6H_{14}N_2O_4S_2^{2+} , 2Br^{-}$ 48.66+ **1**
$C_6H_{14}N_2O_4S_2^{2+} , 2Br^{-} , 2H_2O$ 48.28 **7**
$C_6H_{14}N_2O_4S_2^{2+} , 2Cl^{-}$ 48.68 **1**
$C_6H_{14}N_2O_4S_2^{2+} , 2Cl^{-}$ 48.32+ **6**
$C_6H_{14}N_2S_2$ 41.14 **1**
$C_6H_{14}N_3^{+} , I_3^{-}$ 3.14 **6**
$C_6H_{14}N_3O_3^{+} , Cl^{-}$ 48.32+ **4**
$(C_6H_{14}N_3O_6Zn^{+})_{2n} , nO_4S^{2-} , 4nH_2O$ 82.8 **3**
$C_6H_{14}N_4NiO_2S_2$ 84.14 **7**
$C_6H_{14}N_4NiS_4$ 85.7 **6**
$C_6H_{14}N_4O_2 , 2H_2O$ 48.69 **1**
$C_6H_{14}N_4O_2 , 2H_2O$ 48.47 **5**
$C_6H_{14}N_4O_4S_2Zn$ 79.27 **2**
$C_6H_{14}N_4O_{12}U$ 84.5 **5**
$C_6H_{14}N_5O^{+} , Cl^{-}$ 40.11 **4**
$C_6H_{14}NiO_8$ 81.16 **3**
$C_6H_{14}O_2 , 6H_2O$ 61.5 **3**

$C_6H_{14}O_2PS^-$, $C_8H_{12}N^+$ 64.12 **7**
$C_6H_{14}O_3$, $C_8H_8^{2-}$, $2K^+$ 22.1 **6**
$C_6H_{14}O_6$ 45.46 **1**
$C_6H_{14}O_6$ 45.44+ **1**
$C_6H_{14}O_6$ 45.17 **3**
$C_6H_{14}O_6$ 45.9 **4**
$C_6H_{14}O_6$ 45.12 **4**
$C_6H_{14}O_6$ 45.10+ **4**
$C_6H_{14}O_6$, C_5H_5N 60.23 **3**
$C_6H_{14}O_8Zn$, H_2O 81.19 **7**
$C_6H_{15}BN_4$ 37.11 **1**
$0.48C_6H_{15}ClN^+$, $0.52C_6H_{15}IN^+$, I^- 3.14 **7**
$C_6H_{15}Cl_3PPt^-$, $C_8H_{20}N^+$ 86.2 **7**
$C_6H_{15}CuN_5O_4^{2+}$, O_4S^{2-} , H_2O 83.11 **7**
$0.52C_6H_{15}IN^+$, $0.48C_6H_{15}ClN^+$, I^- 3.14 **7**
$(C_6H_{15}NNaO_3^+)_n$, nI^- 67.2 **6**
$C_6H_{15}NO_2S^{2+}$, $2Cl^-$ 48.29 **7**
$C_6H_{15}NS_2Sn$ 69.13+ **3**
$C_6H_{15}N_2O_2^+$, Br^- 3.24 **3**
$C_6H_{15}N_2O_2^+$, Cl^- 59.2 **5**
$C_6H_{15}N_2O_2^+$, Cl^- 48.30 **7**
$C_6H_{15}N_2O_2^+$, Cl^- , $2H_2O$ 48.70 **1**
$C_6H_{15}N_2O_2^+$, Cl^- , $2H_2O$ 48.48+ **5**
$C_6H_{15}N_2O_2^+$, I^- 59.3 **5**
$C_6H_{15}N_3$, $3H_2O$ 33.50 **1**
$C_6H_{15}N_4O_2^+$, Br^- , H_2O 48.71 **1**
$C_6H_{15}N_4O_2^+$, Cl^- 48.34 **6**
$C_6H_{15}N_4O_2^+$, Cl^- , H_2O 48.29 **3**
$C_6H_{15}N_4O_2^+$, $H_2O_4P^-$, H_2O 48.30 **3**
$C_6H_{15}N_4O_2^+$, $H_2O_4P^-$, H_2O 48.50 **5**
$C_6H_{15}N_4O_2^+$, $C_4H_{10}O_4P^-$ 48.35 **6**
$C_6H_{15}N_4O_2^+$, $C_5H_8NO_4^-$, H_2O 48.31 **7**
$C_6H_{15}OP$, $C_6N_6O_3$ 60.73 **2**
$C_6H_{15}O_4P$, $C_6N_6O_3$ 60.15+ **4**
$C_6H_{15}PS$ 64.18 **2**
$C_6H_{15}PSe$ 64.19 **2**
$C_6H_{16}AgN_{10}^{3+}$, $3ClO_4^-$ 76.9 **5**
$C_6H_{16}AgN_{10}^{3+}$, O_4S^{2-} , HO_4S^- , H_2O 76.22 **7**
$C_6H_{16}BO_3P$ 64.12 **3**
$C_6H_{16}B_2N_2$ 62.11 **4**
$(C_6H_{16}Br_2CdN_2)_n$ 76.9 **6**
$C_6H_{16}Br_2CuN_2$ 76.23 **7**
$C_6H_{16}CdN_6S_2$ 76.24 **7**
$C_6H_{16}Cl_2CuN_2$ 76.25 **7**
$C_6H_{16}Cl_2CuN_{10}$, H_2O 83.45 **2**
$C_6H_{16}Cl_2N_2Zn$ 76.10 **5**
$C_6H_{16}Cl_{10}Cu_5O_2$ 84.19+ **2**
$C_6H_{16}CoN_4O_4^+$, Br^- , H_2O 76.10 **6**
$C_6H_{16}CoN_4O_4^+$, NO_3^- 76.26 **7**
$C_6H_{16}CoN_5O_6$ 81.20 **7**
$C_6H_{16}CoN_5O_6$, H_2O 81.21 **7**
$C_6H_{16}CoN_6^+$, Cl^- 76.23 **3**
$C_6H_{16}CoN_6^+$, Cl^- , H_2O 76.10 **4**
$C_6H_{16}CrN_4O_4^+$, $C_6H_8CrN_2O_8^-$, $2H_2O$ 76.24 **3**
$C_6H_{16}CuN_2O_6$, $4H_2O$ 82.21 **2**
$C_6H_{16}CuN_4O_4$ 76.27 **7**
$C_6H_{16}CuN_{10}^{2+}$, $2Cl^-$, H_2O 79.12 **3**
$C_6H_{16}CuN_{10}2+$, O_4S^{2-} , H_2O 83.46 **2**
$C_6H_{16}MnN_{10}O_4$ 79.11 **7**
$C_6H_{16}N^+$, Cl^- 3.45 **1**
$C_6H_{16}N^+$, $C_6H_6AsO_2S^-$ 65.5 **4**
$C_6H_{16}N^+$, $C_9H_9N_2O_7PS^-$ 47.1 **3**
$C_6H_{16}N^+$, $C_9H_{10}N_2O_8P^-$ 47.1 **1**
$C_6H_{16}N^+$, $C_{10}H_{14}N_2O_8PS^-$ 47.23 **7**
$C_6H_{16}N^+$, $C_{11}HFe_3O_{11}^-$ 3.46 **1**
$C_6H_{16}N^+$, $C_{12}H_4N_4^-$, $C_{12}H_4N_4$ 3.25 **3**
$C_6H_{16}N^+$, $C_{18}H_{12}O_6P^-$ 46.11 **5**
$2C_6H_{16}N^+$, Cl_2O_2U , $2Cl^-$ 3.15 **7**
$2C_6H_{16}N^+$, Cl_4Cu^{2-} 3.15 **5**
$2C_6H_{16}N^+$, Cl_4Cu^{2-} 3.16 **7**
$2C_6H_{16}N^+$, $H_{18}B_{20}^{2-}$ 3.26 **3**
$3C_6H_{16}N^+$, $H_{18}B_{20}NO^{3-}$ 3.27 **3**
$C_6H_{16}N_2$ 3.49 **1**
$C_6H_{16}N_2$, $C_9H_7^-$, Li^+ 27.2 **7**
$C_6H_{16}N_2^{2+}$, $2Br^-$ 34.5 **1**
$C_6H_{16}N_2^{2+}$, $2Cl^-$ 21.16 **1**
$2C_6H_{16}N_2$, $C_{14}H_{10}^{2-}$, $2Li^+$ 26.4 **7**
$C_6H_{16}N_2NiO_6$ 82.19 **2**
$C_6H_{16}N_2NiO_6$ 82.8 **6**
$C_6H_{16}N_2NiO_8$ 82.22 **2**
$C_6H_{16}N_2O_3^{2+}$, $2Cl^-$ 21.17 **1**
$C_6H_{16}N_2O_4S_2$ 4.4 **4**
$C_6H_{16}N_2O_8^{2-}$, nCa^{2+} 67.2 **7**
$C_6H_{16}N_4NiO_4$ 76.31 **2**
$C_6H_{16}N_4NiO_6$ 76.28 **7**
$C_6H_{16}N_4O_2^{2+}$, $2I^-$ 48.72 **1**
$C_6H_{16}N_4O_2S_2^{2+}$, $2Cl^-$ 48.58 **1**
$C_6H_{16}N_6NiS_2$ 76.32 **2**
$C_6H_{16}N_8^{2+}$, $2Cl^-$, $2H_2O$ 9.3 **4**
$C_6H_{16}N_{10}Ni^{2+}$, $2Cl^-$, H_2O 79.13+ **3**
$C_6H_{16}N_{10}Ni^{2+}$, $2Cl^-$, H_2O 79.5 **4**
$C_6H_{17}Cl_4NO_4Zn_2$ 83.18 **4**
$C_6H_{17}N_5NiS_4Si_2$ 83.12 **7**
$C_6H_{18}AgClN_6S_3$ 79.14 **5**
$C_6H_{18}Al_2$ 68.10 **2**
$C_6H_{18}Al_2$ 68.3 **3**
$C_6H_{18}Al_2$ 68.2 **4**
$C_6H_{18}Al_2Br_4O_2$ 68.11 **2**
$C_6H_{18}Al_2F^-$, K^+ , C_6H_6 68.8 **7**
$C_6H_{18}Al_2N_3^-$, K^+ 68.9 **7**
$C_6H_{18}Al_3N_3$ 68.10 **5**
$C_6H_{18}As_2Br_4Pd_2$ 86.2 **2**
$C_6H_{18}As_2Cl_4Pt_2$ 86.2 **3**
$C_6H_{18}As_4N_6$ 65.15 **2**
$C_6H_{18}B_2Cl_4N_2$ 3.28 **3**
$C_6H_{18}Br_3N_2Ti$ 83.19 **4**
$(C_6H_{18}CaO_9^{2+})_n$, $2nCl^-$, nH_2O 45.16 **5**
$C_6H_{18}ClN_4Zn^+$, $C_{24}H_{20}B^-$ 76.25 **3**
$C_6H_{18}Cl_2N_2OV$ 83.47 **2**
$C_6H_{18}Cl_2O_2S_2Sn$ 69.15 **3**
$C_6H_{18}Cl_2O_2Sb_2$ 66.1 **7**
$C_6H_{18}Cl_2P_2Pt$ 86.4 **2**
$C_6H_{18}Cl_3CrN_2$ 83.20 **4**
$C_6H_{18}Cl_3N_2V$ 83.18 **3**
$C_6H_{18}Cl_3N_6P_3$ 64.8 **4**
$C_6H_{18}Cl_3N_6P_3$ 64.13 **7**
$C_6H_{18}Cl_3O_3RhS_3$ 85.5 **7**
$C_6H_{18}Cl_3O_3RuS_3^-$, $C_2H_8N^+$ 85.6 **7**
$C_6H_{18}Cl_8N_6P_4$ 64.20 **2**
$C_6H_{18}CoN_2O_8P_2$ 84.21 **2**
$C_6H_{18}CoN_4O_2S^+$, Cl^- , H_2O 76.11 **6**
$C_6H_{18}CoN_4O_4^+$, $C_{10}H_{12}CoN_2O_8^-$, H_2O 82.9 **6**
$C_6H_{18}CoO_6^{2+}$, O_4S^{2-} 84.15 **7**
$C_6H_{18}CoO_6P_3S_6$ 85.5 **4**
$C_6H_{18}CrN_4O_2S^+$, ClO_4^- 76.12 **6**
$C_6H_{18}CuO_6^{2+}$, O_4S^{2-} 84.6 **5**
$C_6H_{18}ErN_3O_{12}S_3$ 84.7 **6**
$C_6H_{18}Ga_3N_3$ 68.3 **4**
$C_6H_{18}IN_4Zn^+$, I^- 76.11 **5**

$C_6H_{18}LuN_3O_{12}S_3$	84.16	**7**
$C_6H_{18}Mg_4O_6^{2+}$, $2Cl^-$, $10CH_4O$	67.3	**7**
$C_6H_{18}N_2^{2+}$, $2Cl^-$	3.50	**1**
$C_6H_{18}N_2^{2+}$, $2I^-$	3.51	**1**
$C_6H_{18}N_2^{2+}$, $C_6H_8O_4^{2-}$	2.83 +	**1**
$C_6H_{18}N_2Si_2$	63.3	**6**
$C_6H_{18}N_2Ti^{3+}$, $3Br^-$	83.49	**2**
$C_6H_{18}N_3NiO_3^-$, $C_6H_{21}CoN_3O_3^{3+}$, $2I^-$	83.21	**4**
$C_6H_{18}N_3OP$, $2C_8H_{12}N_2O_3$	60.10	**6**
$C_6H_{18}N_3O_6P_3$	64.21	**2**
$C_6H_{18}N_3O_{12}S_3Yb$	84.17	**7**
$C_6H_{18}N_3P_3$, I_2	64.14	**7**
$C_6H_{18}N_4Ni^{2+}$, $2ClO_4^-$	76.12	**5**
$C_6H_{18}N_4O_4W$	71.3	**6**
$C_6H_{18}N_6OSb_2$	66.2	**7**
$C_6H_{18}N_6O_2P_2$	64.7	**5**
$C_6H_{18}N_6P_2$	64.9	**4**
$C_6H_{18}OSb_2^{2+}$, $2ClO_4^-$	66.3	**7**
$C_6H_{18}OSb_2^{2+}$, $2N_3^-$	66.4	**4**
$C_6H_{18}O_3Si_3$	63.5	**2**
$C_6H_{18}O_6Zn^{2+}$, O_4S^{2-}	84.18	**7**
$(C_6H_{18}O_8P_2Sn_3)_n$, $8nH_2O$	69.6	**7**
$C_6H_{18}O_8V_4$	84.22	**2**
$C_6H_{19}AlN_2$	68.12	**2**
$C_6H_{19}N_3NdO_8^{3+}$, $3Cl^-$, H_2O	82.9	**3**
$C_6H_{20}B_2N_2$	62.4	**7**
$C_6H_{20}B_{20}O_2$	62.11	**3**
$(C_6H_{20}CaO_{10}^{2+})_n$, $2nBr^-$, nH_2O	67.6	**4**
$C_6H_{20}ClCoN_4O^{2+}$, $2ClO_4^-$	76.27	**3**
$C_6H_{20}ClCoN_6^{2+}$, $2Cl^-$, H_2O	76.11	**4**
$C_6H_{20}ClCoN_6O_4$	76.34	**2**
$C_6H_{20}Cl_2CoN_4^+$, $2Cl^-$, $H_5O_2^+$	76.35	**2**
$C_6H_{20}Cl_2CoN_4^+$, $2Cl^-$, $H_5O_2^+$	76.28	**3**
$C_6H_{20}Cl_2CuN_4O_8$	76.29	**3**
$C_6H_{20}CoN_5O_2^{2+}$, $O_6S_2^{2-}$	82.10	**6**
$C_6H_{20}CoN_6O_4^+$, Cl^-	76.30	**3**
$C_6H_{20}CoN_6O_6^+$, NO_3^-	76.31	**3**
$C_6H_{20}CuN_4^{2+}$, $2ClO_4^-$	76.13	**5**
$C_6H_{20}CuN_4^{2+}$, $Cr_2O_7^{2-}$	76.32	**3**
$C_6H_{20}CuN_4^{2+}$, $2NO_2^-$	76.33	**3**
$C_6H_{20}CuN_4^{2+}$, $2NO_3^-$	83.50	**2**
$C_6H_{20}CuN_4^{2+}$, $2NO_3^-$	76.34	**3**
$C_6H_{20}CuN_4^{2+}$, $2NO_3^-$	76.35	**3**
$C_6H_{20}CuN_4^{2+}$, O_4S^{2-} , H_2O	76.36	**3**
$C_6H_{20}CuN_4^{2+}$, O_4Se^{2-} , H_2O	76.37	**3**
$C_6H_{20}CuN_4^{2+}$, $2CNS^-$	76.12	**4**
$C_6H_{20}CuN_4^{2+}$, $C_2O_4^{2-}$, $2H_2O$	76.29	**7**
$C_6H_{20}CuN_4^{2+}$, $C_3H_2O_4^{2-}$, $2H_2O$	76.30	**7**
$C_6H_{20}CuN_4^{2+}$, $C_4H_4O_6^{2-}$, $2H_2O$	76.31	**7**
$C_6H_{20}CuN_4^{2+}$, $2C_7H_4ClO_2^-$	83.13	**7**
$C_6H_{20}CuN_4^{2+}$, $2C_7H_5O_2^-$	83.9	**5**
$C_6H_{20}CuN_4O_2^{2+}$, $2Cl^-$	83.14	**7**
$C_6H_{20}CuN_4O_2^{2+}$, $2CNS^-$	83.15	**7**
$C_6H_{20}MoN_3O_7P$	84.3	**4**
$C_6H_{20}N_6ORu^+$, F_6P^- , C_2D_6OS	71.4	**6**
$C_6H_{21}AlN_2$	68.13	**2**
$C_6H_{21}BN_6$	62.2	**5**
$C_6H_{21}B_3N_6$	62.12	**4**
$C_6H_{21}ClCoN_5^{2+}$, $2Cl^-$	76.32	**7**
$C_6H_{21}ClCoN_5^{2+}$, Cl_4Zn^{2-}	76.14	**5**
$C_6H_{21}ClCoN_5^{2+}$, Cl_4Zn^{2-}	76.15	**5**
$C_6H_{21}ClCoN_5^{2+}$, $2NO_3^-$	76.38	**3**
$C_6H_{21}CoN_3O_3^{3+}$, $C_6H_{18}N_3NiO_3^-$, $2I^-$	83.21	**4**
$C_6H_{21}CoN_{15}^{3+}$, $3Cl^-$, H_2O	79.8	**6**
$C_6H_{21}CrN_{15}^{3+}$, $3C_{10}H_{15}O_4S^-$	79.16	**3**
$C_6H_{21}N_4^{3+}$, $3Cl^-$	3.52 +	**1**
$C_6H_{22}CoN_5S^{2+}$, $2CNS^-$	76.13	**6**
$C_6H_{23}HgNO_2S$	71.5	**6**
$C_6H_{24}Ag_2N_{12}S_6^{2+}$, $2ClO_4^-$	79.9	**6**
$C_6H_{24}B_3N_3$	62.44	**2**
$C_6H_{24}B_3P_3$	64.22	**2**
$C_6H_{24}Bi_2Cl_4N_{12}S_6^{2+}$, $CH_4BiCl_5N_2S^{2-}$	66.4	**7**
$C_6H_{24}CdN_6^{2+}$, O_3SSe^{2-}	76.16	**5**
$C_6H_{24}CdN_6^{2+}$, $O_3S_2^{2-}$	76.39	**3**
$C_6H_{24}Cd_2N_{12}O_8S_8$	79.17	**3**
$3C_6H_{24}CoN_6^+$, $H_{12}NaO_6^+$, $4Cl^-$	76.36	**2**
$C_6H_{24}CoN_6^{3+}$, $3Br^-$, H_2O	76.37	**2**
$C_6H_{24}CoN_6^{3+}$, $3Cl^-$, H_2O	76.38	**2**
$C_6H_{24}CoN_6^{3+}$, $3NO_3^-$	76.17	**5**
$C_6H_{24}CoN_6^{3+}$, $C_5CrN_6O^{3-}$, $2H_2O$	76.40	**3**
$C_6H_{24}CoN_6^{3+}$, $C_6FeN_6^{3-}$, $2H_2O$	76.18	**5**
$2C_6H_{24}CoN_6^{3+}$, $CdCl_6^{4-}$, $2Cl^-$, $2H_2O$	76.13	**4**
$2C_6H_{24}CoN_6^{3+}$, $Cl_8Cu_2^{4-}$, $2Cl^-$, $2H_2O$	76.41	**3**
$2C_6H_{24}CoN_6^{3+}$, $3HO_4P^{2-}$, $9H_2O$	76.14	**4**
$C_6H_{24}CoN_{12}O_6^{2+}$, $4CH_4N_2O$, $2NO_3^-$	79.12	**7**
$C_6H_{24}CrN_6^{3+}$, $C_5N_5Ni^{3-}$, $1.5H_2O$	76.39	**2**
$C_6H_{24}CrN_6^{3+}$, $C_6CoN_6^{3-}$, $6H_2O$	76.40	**2**
$C_6H_{24}CuN_6^{2+}$, O_4S^{2-}	76.42	**3**
$C_6H_{24}Cu_2N_{12}S_6^{2+}$, $2BF_4^-$	79.13	**7**
$C_6H_{24}Cu_2N_{12}S_6^{2+}$, $2ClO_4^-$	79.7	**4**
$C_6H_{24}Cu_4N_{12}S_6^{4+}$, $4NO_3^-$, $4H_2O$	79.15	**5**
$C_6H_{24}N_4NiO_2^{2+}$, $2Br^-$	83.10	**5**
$C_6H_{24}N_4NiO_2^{2+}$, $2ClO_4^-$	76.43	**3**
$C_6H_{24}N_4NiO_2^{2+}$, $2NO_3^-$	83.51	**2**
$C_6H_{24}N_4NiO_2^{2+}$, $C_9H_{30}N_6Ni^{2+}$, $4Cl^-$, H_2O	83.33	**4**
$C_6H_{24}N_6Ni^{2+}$, K^+ , $3CNSe^-$	76.41	**2**
$C_6H_{24}N_6Ni^{2+}$, $2NO_3^-$	76.42	**2**
$C_6H_{24}N_6Ni^{2+}$, O_4S^{2-}	76.44	**3**
$3C_6H_{24}N_6Ni^{2+}$, $O_{15}Si_6^{6-}$, $26H_2O$	76.43	**2**
$C_6H_{24}N_6Ru^{3+}$, $3Cl^-$, $3.5H_2O$	76.45	**3**
$C_6H_{24}N_6Zn^{2+}$, $O_3S_2^{2-}$	76.19	**5**
$C_6H_{24}N_{12}NiO_6^{2+}$, $4CH_4N_2O$, $2I^-$	61.1	**7**
$C_6H_{24}N_{12}NiS_6^{2+}$, $2Br^-$	79.28	**2**
$C_6H_{24}N_{12}O_6Ti^{3+}$, $3ClO_4^-$	79.8	**4**
$C_6H_{24}N_{12}O_6Ti^{3+}$, $3ClO_4^-$	79.16	**5**
$C_6H_{24}N_{12}O_6Ti^{3+}$, $3I^-$	79.29	**2**
$C_6H_{24}N_{12}O_6Ti^{3+}$, $3I^-$	79.19	**3**
$C_6H_{24}N_{12}O_6V^{3+}$, $3I^-$	79.17 +	**5**
$C_6H_{24}N_{12}O_6Zn^{2+}$, $2NO_3^-$	84.4	**4**
$C_6H_{24}N_{12}PbS_6^{2+}$, $2ClO_4^-$	69.13	**4**
$C_6H_{24}N_{12}S_6Te_2^{4+}$, $4HF_2^-$	70.13	**2**
$C_6H_{26}N_{12}O_{12}U_2^{2+}$, I_4^{2-}	79.30	**2**
$C_6H_{32}B_{26}Co_2^{2-}$, $2Cs^+$, H_2O	62.12	**3**

C_7

C_7F_6O	22.1	**7**
$C_7HF_2MnO_5$	71.10	**2**
$C_7HF_5O_2$	13.1	**5**
$C_7HF_6O_4Rh$	77.2	**2**
$C_7HN_4^-$, $C_{23}H_{17}O^+$	38.67	**7**
$C_7H_2Br_3N$	7.9	**4**
$C_7H_2Cl_3N$	7.23	**1**

$C_7H_2Cl_3N$	7.10	**4**
$C_7H_2N_4$	20.7	**5**
$C_7H_2N_5O_{10}^-$, K^+	15.3	**7**
$C_7H_2N_5O_{10}^-$, Rb^+	12.7	**4**
$C_7H_3Br_2N_3$	9.19	**1**
$C_7H_3Br_5$	19.5	**4**
$C_7H_3Cl_5$	19.1	**5**
$C_7H_3F_2NO_6Ti^{2-}$, $2K^+$, $2H_2O$	81.23	**5**
$C_7H_3F_{12}N_3$	32.12	**6**
$C_7H_3FeNO_4$	72.13	**2**
$C_7H_3NO_4^{2-}$, Ca^{2+} , $3H_2O$	33.51	**1**
C_7H_4BrNS	16.4	**6**
$C_7H_4BrN_3$	9.4	**4**
$C_7H_4BrO_2^-$, $C_5H_{12}N^+$	33.10	**4**
$C_7H_4ClMnO_6$	71.3	**5**
C_7H_4ClNOS	41.15	**1**
$C_7H_4ClNO_4$	13.1	**1**
C_7H_4ClNS	41.6	**4**
$C_7H_4ClN_2O_4^-$, K^+	12.8	**4**
$C_7H_4ClN_3$	9.20	**1**
$C_7H_4ClO_2^-$, $C_5H_{12}N^+$	33.11	**4**
$C_7H_4ClO_2^-$, $C_7H_5ClO_2$, K^+	14.1	**5**
$2C_7H_4ClO_2^-$, $C_6H_{20}CuN_4^{2+}$	83.13	**7**
$C_7H_4CrO_3S$	75.1	**2**
$C_7H_4CuN_3$	83.14	**4**
C_7H_4IN	7.24	**1**
C_7H_4IN	7.4	**6**
$C_7H_4NO_4^-$, $C_7H_5NO_4$, K^+	14.3	**1**
$C_7H_4NO_4^-$, $C_7H_5NO_4$, Rb^+	14.2	**1**
$C_7H_4N_2O_6$, $C_{12}H_9NS$	60.130	**2**
$C_7H_4N_4S$	41.5	**6**
$C_7H_4O_5^{2-}$, Ca^{2+} , $3H_2O$	38.23	**1**
$C_7H_5AlCl_4O$	68.14	**2**
C_7H_5BrOTe	70.2	**6**
$C_7H_5BrO_2$	13.2	**1**
$C_7H_5BrO_2$	13.3	**1**
$C_7H_5BrO_2$	13.2	**5**
$C_7H_5Br_2NO_2$	13.4+	**1**
$C_7H_5Br_3FeO_2Sn$	73.3	**3**
$C_7H_5ClCoHgO_2^+$, Cl^- , $2Cl_2Hg$	73.1	**5**
$2C_7H_5ClFeO_2$, Cl_3Sb	73.2	**6**
C_7H_5ClO	22.1	**4**
$C_7H_5ClO_2$	13.6	**1**
$C_7H_5ClO_2$	13.7	**1**
$C_7H_5ClO_2$	13.1	**7**
$C_7H_5ClO_2$	13.2	**7**
$C_7H_5ClO_2$, $C_7H_4ClO_2^-$, K^+	14.1	**5**
$C_7H_5ClO_2^-$, $C_2H_6N_2S^+$	8.32	**1**
$C_7H_5ClO_3$, $C_7H_8N_4O_2$	60.24	**3**
$C_7H_5ClO_3$, $C_8H_{10}N_4O_2$	60.104	**2**
$2C_7H_5ClO_3$, $C_7H_8N_5O_2$	60.17	**4**
$C_7H_5Cl_2CoHgO_2$	73.2	**5**
$C_7H_5Cl_2NO_2$	13.8	**1**
$C_7H_5Cl_3FeO_2Sn$	73.4	**3**
$C_7H_5Cl_4O_2Sb$	66.4	**3**
$C_7H_5Cl_6OSb$	66.2	**5**
$C_7H_5FO_2$	13.9	**1**
$C_7H_5FO_2$	13.3	**7**
$C_7H_5IO_3$	42.8	**1**
$C_7H_5MnO_4S$	73.3	**6**
C_7H_5NOS	41.2	**3**
C_7H_5NOS	40.7	**5**
$C_7H_5NOS_2$	39.41	**1**
$C_7H_5NO_2$	40.8	**5**
$C_7H_5NO_3$	15.20+	**1**
$C_7H_5NO_3S$	41.16+	**1**
$C_7H_5NO_4$	13.10	**1**
$C_7H_5NO_4$	13.11+	**1**
$C_7H_5NO_4$	13.1	**4**
$C_7H_5NO_4$	13.1	**6**
$C_7H_5NO_4$	13.2	**6**
$C_7H_5NO_4$	33.20	**6**
$C_7H_5NO_4$	33.21	**6**
$C_7H_5NO_4$	33.24	**7**
$C_7H_5NO_4$	33.25	**7**
$C_7H_5NO_4$, H_2O	33.22	**6**
$C_7H_5NO_4$, $C_7H_4NO_4^-$, K^+	14.3	**1**
$C_7H_5NO_4$, $C_7H_4NO_4^-$, Rb^+	14.2	**1**
$C_7H_5NO_5$	13.13	**1**
$C_7H_5NO_5$	13.1	**3**
$(C_7H_5NO_7U)_n$	81.22	**7**
$C_7H_5NO_8V^-$, H_4N^+ , $1.3H_2O$	81.24	**5**
$C_7H_5NS_2$	41.18	**1**
$C_7H_5NS_2$	41.7	**4**
$C_7H_5N_2O_2^-$, $C_4H_{12}N^+$	10.3	**6**
$C_7H_5N_2O_4^-$, K^+	12.5	**3**
$C_7H_5N_3O$	22.2	**4**
$C_7H_5N_3O$	35.3	**6**
$C_7H_5N_5O_8$	16.24	**1**
$C_7H_5O_2^-$, Na^+	22.2	**1**
$C_7H_5O_2^-$, $4CH_4N_2S$, Tl^+	60.98	**2**
$C_7H_5O_2^-$, $C_7H_6O_2$, K^+	14.4	**1**
$C_7H_5O_2^-$, $C_{16}H_{36}N^+$, $39.5H_2O$	61.14	**2**
$C_7H_5O_3$, $C_{10}H_{15}N_2$	60.22	**4**
$C_7H_5O_3^-$, $C_5H_{12}N^+$	33.13	**6**
$C_7H_5O_3^-$, $C_7H_6O_3$, K^+ , H_2O	14.6	**1**
$C_7H_5O_3^-$, $C_7H_6O_3$, H_4N^+ , H_2O	14.5	**1**
$C_7H_5O_3^-$, $C_{12}H_8N_2$, Tl^+	36.7	**5**
$C_7H_5O_3^-$, $C_{21}H_{32}ClN_2O^+$	32.39	**7**
$C_7H_5O_5S^-$, H_4N^+	13.15	**1**
$C_7H_5O_6S^-$, $H_7O_3^+$	13.2	**3**
$C_7H_5O_6S^-$, $H_7O_3^+$	13.2	**4**
$C_7H_6BF_3FeO_5S$	72.9	**3**
$C_7H_6BrMnN_2O_3$	71.4	**5**
C_7H_6BrNO	13.16	**1**
C_7H_6BrNO	13.17	**1**
$C_7H_6BrN_3O$	9.5+	**4**
C_7H_6BrO	13.3	**6**
C_7H_6ClNO	19.21	**1**
C_7H_6ClNO	19.22	**1**
C_7H_6ClNO	10.3	**3**
C_7H_6ClNO	13.4	**6**
C_7H_6ClNO	13.4+	**7**
$C_7H_6ClNO_2$	18.17	**1**
$C_7H_6ClNO_2$	19.23	**1**
C_7H_6FNO	13.18	**1**
C_7H_6FNO	13.19	**1**
$C_7H_6FeO_3$	72.14	**2**
C_7H_6INO	13.5	**5**
$C_7H_6MnNO_4$	72.2	**6**
$C_7H_6NO_2^-$, K^+ , $3H_2O$	14.7	**1**
$C_7H_6NO_2^-$, Na^+ , $5H_2O$	14.8	**1**
$C_7H_6N_2$	35.4	**5**
$C_7H_6N_2$	35.4	**6**
$C_7H_6N_2$	35.5	**7**
$C_7H_6N_2$, $C_7H_7N_2^+$, BF_4^-	35.6	**7**
$C_7H_6N_2O_2$	40.12	**4**
$C_7H_6N_2O_2S_2$, $0.75H_2O$	39.11	**3**
$C_7H_6N_2O_3$	10.2	**5**
$C_7H_6N_2O_3$	13.6	**5**
$C_7H_6N_2O_3$	15.2	**6**
$C_7H_6N_2S$	16.25	**1**

$C_7H_6N_4$ 32.21 **3**
$C_7H_6N_5O_2^-$, $C_7H_9N_2O^+$, $2H_2O$ 33.33 **5**
C_7H_6O 22.1 **5**
$C_7H_6O_2$ 13.20 **1**
$C_7H_6O_2$ 22.2 **5**
$C_7H_6O_2$ 17.4 **6**
$C_7H_6O_2$, $C_7H_5O_2^-$, K^+ 14.4 **1**
$C_7H_6O_2S$ 39.42 + **1**
$C_7H_6O_3$ 13.21 + **1**
$C_7H_6O_3$ 13.5 + **6**
$C_7H_6O_3$, H_2O 13.7 **6**
$C_7H_6O_3$, $C_7H_5O_3^-$, K^+ , H_2O 14.6 **1**
$C_7H_6O_3$, $C_7H_5O_3^-$, H_4N^+ , H_2O 14.5 **1**
$C_7H_6O_3$, $C_9H_{13}N_3O_5$ 60.7 **7**
$C_7H_6O_3$, $C_{11}H_{12}N_2O$ 60.15 **6**
$C_7H_6O_3S$ 42.11 **1**
$C_7H_6O_3S$ 39.10 **5**
$C_7H_6O_3S$ 39.11 **5**
$C_7H_6O_3S$ 39.12 **5**
$C_7H_6O_4$, $2C_4H_5N_3O$, H_2O 44.2 **6**
$C_7H_6O_5S$, $3H_2O$ 13.6 **7**
$C_7H_6O_6S$, $2H_2O$ 13.7 **7**
$C_7H_6O_6S$, $3H_2O$ 13.8 **7**
$C_7H_7^+$, $C_8F_{12}NiS_4^-$ 22.2 **3**
$C_7H_7^-$, $C_6H_{12}LiN_2^+$ 67.7 **3**
$C_7H_7BrN_2O_2$ 16.2 **3**
$C_7H_7BrN_4O$, $C_7H_8BrN_5$ 44.26 **3**
C_7H_7BrO 17.6 **5**
$C_7H_7BrO_4$ 1.22 **3**
$C_7H_7Cl_2NO$ 33.31 **5**
$(C_7H_7Cl_3CuO_4)_n$ 77.1 **3**
$C_7H_7Cl_4Te^-$, C_6H_8N+ 70.2 **7**
$C_7H_7CuNO_6$ 81.25 **5**
C_7H_7I 19.2 **5**
$C_7H_7IrO_4$ 77.2 **3**
C_7H_7NO 13.23 **1**
C_7H_7NO 13.8 **5**
$2C_7H_7NO$, HI_3 60.99 **2**
$2C_7H_7NO$, $C_4H_6O_4$ 60.9 **5**
$C_7H_7NO_2$ 13.28 **1**
$C_7H_7NO_2$ 13.29 **1**
$C_7H_7NO_2$ 13.24 + **1**
$C_7H_7NO_2$ 13.26 + **1**
$C_7H_7NO_2$ 38.2 **3**
$C_7H_7NO_2$ 15.4 **4**
$C_7H_7NO_2$ 17.5 **6**
$C_7H_7NO_2$, $C_{11}H_{18}N_2O_3$ 60.16 **6**
$C_7H_7NO_2$, $C_{16}H_{20}N_4O_4Zn$, H_2O 76.67 **7**
$C_7H_7NO_2^+$, Br^- 58.1 **3**
$C_7H_7NO_3$ 10.26 **1**
$C_7H_7NO_3$ 13.30 **1**
$C_7H_7NO_3S$ 15.22 **1**
$C_7H_7NO_7V$, $2H_2O$ 81.49 **2**
$C_7H_7NO_8Ti$, $2H_2O$ 81.26 **5**
$C_7H_7NO_8Ti$, $2H_2O$ 81.23 **7**
$C_7H_7N_2O^+$, Cl_4Fe^- 9.22 **1**
$C_7H_7N_2O_2S_2^-$, Na^+ , $2.5H_2O$ 39.44 **1**
$C_7H_7N_3OS$ 41.6 **7**
$C_7H_7N_5NiOS_4$ 85.38 **2**
$C_7H_7O_2^+$, Cl^- 22.3 **1**
$C_7H_7O_3S^-$, H_3O^+ 11.3 + **4**
$C_7H_7O_3S^-$, H_3O^+ 11.3 + **5**
$C_7H_7O_3S^-$, D_3O^+ 11.10 **7**
$C_7H_7O_3S^-$, $C_{10}H_{18}N^+$ 37.5 **4**
$C_7H_7O_3S^-$, $C_{11}H_{15}N_2O_3^+$ 48.66 **5**
$C_7H_7O_3S^-$, $C_{21}H_{21}N_2S_2^+$ 41.27 **7**
$C_7H_7O_3S^-$, $C_{25}H_{25}Cl_4N_4^+$, $0.2H_2O$ 35.38 **7**
$2C_7H_7O_3S^-$, $H_{12}O_6Zn^{2+}$ 11.36 **1**
$C_7H_7O_4Rh$ 77.3 **2**
$C_7H_7O_4Rh$ 77.3 **3**
$C_7H_7O_4S^-$, K^+ 11.37 **1**
$C_7H_8Ag_2^{2+}$, $2NO_3^-$ 75.2 **2**
C_7H_8BrN 16.6 **5**
$C_7H_8BrN_5$, $C_5H_5BrN_2O_2$ 44.13 **3**
$C_7H_8BrN_5$, $C_7H_7BrN_4O$ 44.26 **3**
$2C_7H_8BrN_5$, $C_{12}H_{12}N_2O_3$ 60.131 **2**
$C_7H_8Br_2CuN_2$ 83.11 **5**
$C_7H_8ClN_3O_4S_2$ 41.10 **5**
$C_7H_8Cl_2O$ 27.1 **4**
$C_7H_8Cl_2Pd$ 75.3 + **2**
$2C_7H_8N^+$, Cl_6Sn^{2-} 3.54 **1**
$C_7H_8NO^+$, $Cl-$ 33.52 **1**
$C_7H_8NO_2^+$, Cl^- 13.8 + **6**
$C_7H_8N_2O$ 13.3 **4**
$C_7H_8N_2O_3$ 44.53 **1**
$C_7H_8N_2O_3S$ 41.11 **5**
$C_7H_8N_2O_4$ 44.14 **5**
$C_7H_8N_2S$ 33.20 **3**
$C_7H_8N_3O_2SSe^+$, CNS^- 8.15 **5**
$C_7H_8N_4O$, $C_4H_3FN_2O_2$ 44.3 **1**
$C_7H_8N_4OS$, $1.5H_2O$ 33.26 **7**
$C_7H_8N_4O_2$, H_2O 44.54 **1**
$C_7H_8N_4O_2$, $C_7H_5ClO_3$ 60.24 **3**
$C_7H_8N_4O_2$, $C_9H_9N_3O_2S_2$ 60.31 **3**
$C_7H_8N_4S$ 33.21 **3**
$C_7H_8N_4Se$ 33.32 **5**
$C_7H_8N_5O_2$, $2C_7H_5ClO_3$ 60.17 **4**
C_7H_8O 17.24 **1**
C_7H_8O 17.3 **3**
C_7H_8O 17.5 **4**
C_7H_8O 17.7 **5**
C_7H_8O 17.6 **7**
C_7H_8OS 38.24 **1**
$C_7H_8OS_2$ 39.45 **1**
$C_7H_8O_2S$ 11.38 **1**
$C_7H_8O_4$ 1.91 **1**
$C_7H_8O_4$ 1.92 **1**
$C_7H_8S_2Se$ 39.13 **5**
$C_7H_8S_3$ 39.46 + **1**
$C_7H_9AsFeO_4$ 65.6 **4**
$C_7H_9BrN_2STe$ 70.14 **2**
$C_7H_9BrN_6$, $C_5H_6N_2O_2$ 44.27 **3**
$C_7H_9BrN_6$, $C_7H_{10}N_2O_2$ 44.28 **3**
$C_7H_9ClN_2S_2Te$ 70.15 **2**
$C_7H_9ClN_4$ 33.23 **6**
$C_7H_9Cl_2N_3O$ 33.27 **7**
$C_7H_9CuNO_7$ 83.52 **2**
$C_7H_9FeO_5Sb$ 66.5 **4**
$C_7H_9MnN_2O_5$ 71.2 **3**
$C_7H_9MnN_2O_5$ 71.5 **5**
C_7H_9N 16.27 **1**
C_7H_9N , CH_4N_2O 60.100 **2**
$C_7H_9NO_2S$ 16.28 **1**
$C_7H_9NO_3S$ 16.3 **3**
$C_7H_9N_2O^+$, Cl^- 33.24 **6**
$C_7H_9N_2O^+$, I^- 33.54 **1**
$C_7H_9N_2O^+$, I^- 33.25 **6**
$C_7H_9N_2O^+$, $C_6H_2N_3O_7^-$ 33.26 **6**
$C_7H_9N_2O^+$, $C_7H_6N_5O_2^-$, $2H_2O$ 33.33 **5**
$C_7H_9N_2O_2S_2^+$, Br^- 39.48 **1**
$C_7H_9N_3O_2$, $2H_2O$ 35.2 **4**
$C_7H_9N_3O_2S$ 41.6 **6**

Formula	Reference	
$C_7H_9N_3S$	8.16	**5**
$C_7H_9N_3S$	8.7	**6**
$C_7H_9N_4O_9^-$, Rb^+	12.6	**3**
$C_7H_9N_5$	44.14	**6**
$C_7H_9N_5$	44.19	**7**
$C_7H_9N_5$, $C_5H_5BrN_2O_2$	44.56	**1**
$C_7H_9N_5$, $C_5H_5BrN_2O_2$	44.14	**3**
$C_7H_9N_5$, $C_5H_5FN_2O_2$	44.57	**1**
$C_7H_9N_5$, $C_5H_5FN_2O_2$	44.58	**1**
$C_7H_9N_5$, $C_5H_5FN_2O_2$	44.17	**3**
$C_7H_9N_5$, $2C_5H_5IN_2O_2$	44.18	**3**
$C_7H_9N_5$, $C_5H_6N_2O_2$	44.55	**1**
$C_7H_9N_5$, $C_8H_{12}N_2O_3$	60.10	**5**
$C_7H_9N_5$, $C_{10}H_{13}BrN_2O_3$	60.25	**3**
$C_7H_9N_5$, $C_{10}H_{15}BrN_2O$	60.11	**5**
$C_7H_9N_5O$	44.15	**6**
$C_7H_9N_5O$	35.7	**7**
$C_7H_9N_5O$, $C_5H_6BrN_3O$	44.59	**1**
$C_7H_9N_5O$, $C_5H_6FN_3O$	44.60	**1**
$C_7H_9N_5O$, $C_5H_7N_3O$	44.61	**1**
$C_7H_9O_2^+$, Br^- , H_2O	38.25	**1**
$(C_7H_{10}Br_2CuN_2)_n$	83.22	**4**
$C_7H_{10}ClNO_3S$	42.2	**3**
$(C_7H_{10}Cl_2CuN_2)_n$	83.12	**5**
$C_7H_{10}Cl_2O_3S$	39.7	**6**
$(C_7H_{10}Cl_3LaO_3)_n$	84.19	**7**
$C_7H_{10}N^+$, Br_4Fe^-	33.20	**4**
$C_7H_{10}N^+$, Cl^-	16.29	**1**
$C_7H_{10}N^+$, HF_2^-	16.4	**4**
$C_7H_{10}N^+$, HF_2^-	16.5	**6**
$C_7H_{10}N^+$, $C_6H_4N_2O_8P-$	46.5	**3**
$2C_7H_{10}N^+$, Cl_6Re^{2-}	16.30	**1**
$2C_7H_{10}N^+$, F_6Si^{2-} , $2H_2O$	16.5	**4**
$2C_7H_{10}NO^+$, O_4S^{2-}	16.4	**3**
$C_7H_{10}N_2$	7.25	**1**
$C_7H_{10}N_2NiO$	72.15	**2**
$C_7H_{10}N_2O_2$	44.62	**1**
$C_7H_{10}N_2O_2$	48.32+	**7**
$C_7H_{10}N_2O_2$, $C_7H_9BrN_6$	44.28	**3**
$C_7H_{10}N_2O_3S$	38.7	**6**
$C_7H_{10}N_2O_5$	48.34	**7**
$C_7H_{10}N_4O_2S$, H_2O	16.6	**6**
$C_7H_{10}N_5P$	64.8	**5**
$C_7H_{10}N_6$, $2C_5H_5IN_2O_2$	44.19	**3**
$C_7H_{10}N_6$, $2C_6H_8N_2O_2$, H_2O	44.25	**3**
$C_7H_{10}O_2S$	39.12	**3**
$C_7H_{10}O_4$	20.8	**5**
$C_7H_{11}BClN$	33.27	**6**
$C_7H_{11}ClO_3Pt$	72.4	**5**
$(C_7H_{11}CuNO_4)_n$	82.11	**6**
$C_7H_{11}NO$	37.4	**6**
$C_7H_{11}NO$	1.19	**7**
$(C_7H_{11}NO_8Sr)_n$	67.7	**4**
$C_7H_{11}N_2O_2S^+$, $C_9H_8N_3O_2S_2^-$	11.5	**5**
$C_7H_{11}N_3O_2S$	4.5	**4**
$C_7H_{11}N_3O_3$	44.16	**6**
$C_7H_{11}O_3^-$, Na^+	2.85	**1**
$C_7H_{11}O_5^-$, K^+ , xH_2O	2.22	**6**
$(C_7H_{12}CdN_2O_6)_n$, nH_2O	82.12	**6**
$C_7H_{12}Cl_2O_4$	45.47	**1**
$C_7H_{12}Cl_2O_4$	45.18	**3**
$C_7H_{12}CoN_4O_5$	83.20	**3**
$(C_7H_{12}CuN_2O_6)_n$, $2.5H_2O$	82.9	**5**
$C_7H_{12}CuN_4O_4$, $1.5H_2O$	82.23	**2**
$C_7H_{12}NOS^+$, Br^-	41.7	**6**
$C_7H_{12}N_2O_2$	32.7	**7**
$C_7H_{12}N_2O_4$	48.35	**7**
$C_7H_{12}N_4OS$	35.9	**1**
$C_7H_{12}O_4$	1.93	**1**
$C_7H_{12}O_4$	1.97	**1**
$C_7H_{12}O_4$	1.94+	**1**
$C_7H_{12}O_4$	1.20	**7**
$C_7H_{12}O_5$	45.17	**5**
$C_7H_{12}O_5$	45.18	**5**
$C_7H_{12}O_6$, H_2O	45.19	**3**
$C_7H_{13}BrO_5$	45.48	**1**
$C_7H_{13}Br_2O_2Sb$	66.3	**6**
$C_7H_{13}ClHgO$	71.11	**2**
$C_7H_{13}ClHgO$	71.12	**2**
$C_7H_{13}ClN_2O_3PtS$, H_2O	82.10+	**3**
$C_7H_{13}ClO_5$	45.20	**3**
$C_7H_{13}Cl_2N_2O_3P$	64.8	**6**
$C_7H_{13}CrN_3O_3$	76.44	**2**
$C_7H_{13}NO_2$	48.36	**7**
$C_7H_{13}NO_2$, $0.5H_2O$	48.36	**6**
$C_7H_{13}NO_3$	48.37	**7**
$C_7H_{13}NS$, I_2	34.1	**6**
$C_7H_{13}N_2O_2^+$, Br^-	32.21	**1**
$C_7H_{13}N_3O_4$	48.37	**6**
$(C_7H_{14}AuNS_2)_n$	80.2	**2**
$(C_7H_{14}CuN_2OS)_n$	84.20	**7**
$C_7H_{14}N^+$, Cl^-	35.10	**1**
$4C_7H_{14}N^+$, Br_6Sb^{3-} , Br_6Sb^- , $2Br_2$	37.1	**3**
$C_7H_{14}NO^+$, Cl^-	34.7	**7**
$C_7H_{14}NO_2^+$, Cl^-	48.34	**4**
$C_7H_{14}NS_2Tl$	68.4	**3**
$C_7H_{14}N_2O_2$	1.98	**1**
$C_7H_{14}N_2O_2$	48.38	**6**
$C_7H_{14}N_2O_4S$, $4H_2O$	48.52	**5**
$C_7H_{14}N_4NiO_2$, $3H_2O$	76.15	**4**
$C_7H_{14}N_6$, $C_5H_6N_2O_2$, H_2O	60.4	**5**
$C_7H_{14}O_6$	45.49	**1**
$C_7H_{14}O_6$	45.21	**3**
$C_7H_{14}O_6$	45.22	**3**
$C_7H_{14}O_6$	45.19	**6**
$C_7H_{14}O_6$, H_2O	45.23	**3**
$C_7H_{14}O_7$	45.24	**3**
$C_7H_{14}O_7$, H_2O	45.25	**3**
$C_7H_{15}Cl_2N_2O_2P$, H_2O	64.10	**5**
$C_7H_{15}Cl_2N_2O_2P$, H_2O	64.15	**7**
$C_7H_{15}Hg_2NO_2S$	71.6	**6**
$C_7H_{15}NO$	1.21	**7**
$C_7H_{15}NO_4^+$, Br^-	50.2	**5**
$C_7H_{15}NS_3$	11.39	**1**
$C_7H_{15}N_2^+$, ClO_4^-	3.16	**5**
$C_7H_{15}N_2^+$, ClO_4^-	3.17	**7**
$C_7H_{15}N_5S_4$	41.8	**6**
$C_7H_{15}OPS$	64.16	**7**
$C_7H_{15}OPS$	64.17	**7**
$C_7H_{15}PS_2$	64.23	**2**
C_7H_{16}	5.19	**1**
$C_7H_{16}B_7Co$	73.1	**7**
$C_7H_{16}B_9Fe$	73.5	**2**
$C_7H_{16}CoN_4O_3^+$, Cl^- , H_2O	81.17	**3**
$C_7H_{16}N^+$, Br^-	34.6	**1**
$2C_7H_{16}N^+$, $C_2H_4N_4$, O_4S^{2-}	33.28	**7**
$C_7H_{16}NOS^+$, Br^-	3.29	**3**
$C_7H_{16}NOS^+$, Br^-	3.30	**3**
$C_7H_{16}NOS^+$, ClO_4^-	41.19+	**1**
$C_7H_{16}NOSe^+$, I^-	3.15	**6**
$C_7H_{16}NO_2^+$, Br^-	59.6	**1**
$C_7H_{16}NO_2^+$, Br^-	1.22	**7**

$C_7H_{16}NO_2^+$, Cl^-	3.32	**3**
$C_7H_{16}NO_2^+$, ClO_4^-	3.16	**6**
$C_7H_{16}NO_3^+$, Cl^-	1.23	**7**
$2C_7H_{16}N_2^{2+}$, $Cl_8Ni_2^{4-}$	37.2	**3**
$C_7H_{16}N_3O_3^+$, Cl^-	48.35	**4**
$C_7H_{16}O$, $0.5H_2O$	5.20	**1**
$C_7H_{16}O$, $6C_{18}H_{20}O_2$	61.2	**4**
$C_7H_{16}O_3$	5.11	**7**
$C_7H_{16}O_4P$	46.11	**1**
$C_7H_{16}O_6S_3$	11.40	**1**
$C_7H_{17}B_{10}Co$	73.4	**6**
$C_7H_{17}Cl_3N_2NiO$	83.21	**3**
$C_7H_{17}N_2O_2^+$, Br^-	3.33	**3**
$C_7H_{18}ClCoN_3O_2S_2^+$, Cl^-	85.8	**6**
$C_7H_{18}CuN_5^+$, $C_{24}H_{20}B^-$	76.33	**7**
$C_7H_{18}CuN_5S^+$, CNS^-	76.46	**2**
$C_7H_{18}CuN_5S^+$, CNS^-	76.46	**3**
$C_7H_{18}N^+$, Cl^-	3.55	**1**
$C_7H_{18}N_2O_2^{2+}$, O_4Se^{2-} , H_2O , CH_4O	50.3	**5**
$(C_7H_{18}N_2Sn)_n$	69.16	**3**
$C_7H_{18}N_5NiO^+$, $C_{24}H_{20}B^-$	76.34	**7**
$C_7H_{18}N_5SZn^+$, CNS^-	76.47	**3**
$(C_7H_{19}O_2Sn_2)_n$	69.11	**5**
$C_7H_{20}B_9MnO_4$	62.3	**5**
$C_7H_{20}B_9MnO_4$	38.8	**7**
$C_7H_{20}CoN_4O_3^+$, ClO_4^-	83.24	**4**
$C_7H_{20}CoN_6O_4^+$, Cl^-	76.20	**5**
$C_7H_{20}CuN_4S^+$, ClO_4^-	83.8	**6**
$C_7H_{20}CuN_5S^+$, ClO_4^-	76.48	**3**
$C_7H_{21}N_6P_4^+$, I^-	64.9	**6**
$C_7H_{22}CoN_5O_2^{2+}$, $2I^-$, $2H_2O$	82.24	**2**
$C_7H_{22}CoN_5O_4^{2+}$, $2ClO_4^-$, H_2O	84.21	**7**
$C_7H_{22}N_3^{3+}$, $3Cl^-$	59.8	**1**
$2C_7H_{22}N_3^{3+}$, $3HO_4P^{2-}$, $6H_2O$	59.9	**1**
$C_7H_{26}B_5Br_2IrOP_2$	86.1	**6**
$C_7H_{26}CoN_6^{3+}$, $3Br^-$	76.21	**5**

C_8

$C_8Ag_2F_{14}O_4$	84.23	**2**
$C_8AuF_{12}S_4^-$, $C_{18}H_{15}ClP^+$	64.52	**2**
$C_8AuN_4S_4^-$, $C_{18}H_{36}AuN_2S_4^+$	80.12	**4**
$C_8Cl_2N_2O_2$, $C_{10}H_{12}O_4$	60.33	**3**
$C_8Cl_4O_3$	13.3	**3**
C_8Cl_7N	35.8	**7**
C_8Cl_8	20.8	**3**
C_8Cl_8	27.2	**5**
C_8Cl_8	27.1	**6**
C_8Cl_8	27.2	**6**
C_8Cl_8	27.1	**7**
$C_8CoF_{12}O_8^{2-}$, $2C_{24}H_{20}As^+$	81.50	**2**
$C_8CoN_4S_4^{2-}$, $2C_{16}H_{36}N^+$	85.39	**2**
$C_8CuN_4S_4^{2-}$, $2C_{16}H_{36}N^+$	85.40	**2**
$C_8CuN_4S_4^{2-}$, $2C_{16}H_{36}N^+$	85.3	**5**
$C_8Cu_4F_{12}O_8$, $2C_6H_6$	81.24	**6**
$C_8F_4N_2$	7.1	**7**
$C_8F_4O_4^{2-}$, $2K^+$	14.1	**6**
$C_8F_6FeO_4$	72.16	**2**
C_8F_{12}	31.1	**1**
$C_8F_{12}Mo_2O_8$	81.28	**5**
$C_8F_{12}NiS_4$, $C_{12}H_9NO$	60.8	**7**
$C_8F_{12}NiS_4$, $C_{12}H_9NS$	60.9	**7**
$C_8F_{12}NiS_4$, $C_{20}H_{12}$	60.54	**3**
$C_8F_{12}NiS_4^-$, $C_7H_7^+$	22.2	**3**
$C_8F_{24}N_8O_4S_4$	10.1	**4**
$C_8FeN_4S_4^-$, $C_{16}H_{36}N^+$	85.41	**2**
$C_8FeN_5OS_4^-$, $C_8H_{20}N^+$	85.9	**6**
$C_8FeN_5OS_4^{2-}$, $2C_8H_{20}N^+$	85.42	**2**
$C_8FeN_5OS_4^{2-}$, $2C_8H_{20}N^+$	85.10	**6**
$C_8N_2O_4^{2-}$, $2H_3O^+$, $4H_2O$	18.6	**7**
$C_8N_4NiS_4^-$, $C_{19}H_{18}P^+$	85.43	**2**
$C_8N_4NiS_4^{2-}$, $2C_4H_{12}N^+$	85.30	**2**
$C_8N_4NiS_4^{2-}$, $2C_9H_{12}NO_2^+$	85.11	**6**
$C_8N_4NiS_4^{2-}$, $2C_{10}H_{16}N_2^+$	12.11	**5**
$C_8N_4NiS_4^{2-}$, $2C_{36}H_{30}AgP_2^+$	85.12	**6**
C_8N_4S	39.49	**1**
$C_8N_4S_2$	39.51	**1**
$C_8N_{12}P_4S_8$	64.11	**5**
$C_8NiO_4S_4^{2-}$, $2K^+$, $2H_2O$	85.13	**6**
$C_8O_{16}Th^{4-}$, $4K^+$, $4H_2O$	81.24	**7**
$C_8O_{16}Zr^{4-}$, $4Na^+$, $3H_2O$	81.52	**2**
$C_8O_{18}Re_2^{4-}$, $4K^+$, $3H_2O$	81.25	**7**
$C_8HBrN_4O_5S$	39.50	**1**
$C_8H_2Br_2Fe_2O_6$	72.5	**5**
$C_8H_2Cl_8$	19.3	**5**
$C_8H_2F_8FeO_4$	71.13	**2**
$C_8H_2N_4$, $C_{12}H_{18}$	60.101	**2**
$C_8H_3Cl_4NS$	7.5	**5**
$C_8H_3F_5O_2$	22.2	**7**
$C_8H_3O_9^-$, Rb^+	38.26	**1**
$C_8H_4As_2O_{12}^{2-}$, $C_{11}H_{27}CoN_4O_2^{2+}$, H_2O	77.9	**6**
$C_8H_4Br_2O_2$	13.9	**7**
$C_8H_4Cl_2O_2$	13.4	**4**
$C_8H_4CuN_6O_8$, $4H_2O$	83.53	**2**
$C_8H_4Cu_2O_{12}^{4-}$, $4Na^+$, $10H_2O$	81.29	**5**
$C_8H_4F_3N_5O_8$	16.5	**3**
$C_8H_4FeN_2O_4$	83.9	**6**
$C_8H_4FeO_8$	72.17	**2**
$C_8H_4FeO_8$	72.18	**2**
$(C_8H_4Hg_2O_4)_n$	81.53	**2**
$(C_8H_4I_2NiO_2)_n$	71.3	**3**
$C_8H_4NO_3^-$, $C_8H_7NO_2$, K^+	10.2	**4**
$C_8H_4N_2$	7.2	**7**
$C_8H_4N_2Se_2$	11.10	**3**
$C_8H_4N_4$	20.15	**1**
$C_8H_4N_4$	20.5	**7**
$C_8H_4N_5O_6^-$, Li^+ , $2H_2O$	43.1	**5**
$C_8H_4N_6O_8Sr$, $4H_2O$	67.4	**5**
$C_8H_4N_8Pt$, $C_{14}H_{10}Cl_2N_2Pt$	83.16	**7**
$C_8H_4O_2$	27.2	**3**
$C_8H_4O_4$	38.27	**1**
$C_8H_4O_4$	38.28	**1**
$C_8H_4O_4^{2-}$, Ca^{2+} , H_2O	14.1	**7**
$C_8H_4O_4^{2-}$, $2K^+$	14.2	**6**
$C_8H_4O_4^{2-}$, $2H_4N^+$	14.2	**7**
$C_8H_4O_4^{2-}$, $C_4H_{12}N_2^{2+}$	33.3	**6**
$C_8H_4O_{12}Sb_2^{2-}$, $2K^+$, $3H_2O$	66.5	**3**
$C_8H_4O_{12}Sb_2^{2-}$, $2K^+$, $3H_2O$	66.6	**4**
$C_8H_4O_{12}Sb_2^{2-}$, $2K^+$, $3H_2O$	66.5	**7**
$C_8H_4O_{12}Sb_2^{2-}$, $2H_4N^+$, $3H_2O$	66.11	**2**
$C_8H_4O_{12}Sb_2^{2-}$, $2C_8H_{20}CoN_6S_2^+$, $4H_2O$	76.52	**3**
$C_8H_4O_{12}Sb_2^{2-}$, $C_{36}H_{24}FeN_6^{2+}$, $8H_2O$	83.79	**6**
$C_8H_4O_{14}V_2^{4-}$, $4Na^+$, $12H_2O$	81.54	**2**
$C_8H_4O_{14}V_2^{4-}$, $4H_4N^+$, $2H_2O$	81.55	**2**
$C_8H_5BrN_2O_2$	40.12	**1**
$C_8H_5ClMoO_3$	73.6	**2**

$C_8H_5ClN_4$ 9.23 **1**
$C_8H_5F_5IORh$ 73.7 **2**
$C_8H_5FeO_3^+$, F_6P^- 73.5 **6**
C_8H_5I , C_4H_9NO 60.59 **2**
$C_8H_5MnO_3$ 73.8 **2**
$C_8H_5NO_2$ 35.11 **1**
$C_8H_5NO_2$ 35.3 **4**
$C_8H_5NO_2S$ 42.3 **5**
$C_8H_5NO_5S_2W$ 85.7 **7**
$C_8H_5N_2O_2^-$, K^+ 12.3 **7**
$C_8H_5N_3$ 33.55 **1**
$C_8H_5O_4^-$, K^+ 14.9 **1**
$C_8H_5O_4^-$, Li^+ , $2H_2O$ 14.3+ **7**
$C_8H_5O_4^-$, H_4N^+ 14.10 **1**
$C_8H_5O_4^-$, H_4N^+ 14.2 **5**
C_8H_6 5.21 **1**
$C_8H_6Br_2$ 27.1 **1**
$C_8H_6Br_2$ 27.3 **5**
$C_8H_6Br_2N_2O_5$ 28.1 **1**
$C_8H_6Br_4$ 19.24 **1**
$C_8H_6ClNO_3$ 18.18 **1**
$C_8H_6ClNO_3$ 18.19 **1**
$C_8H_6Cl_2$ 27.2 **1**
$C_8H_6Cl_2N_4O_2$ 35.4 **4**
$C_8H_6Cl_4$ 19.4 **5**
$C_8H_6Cl_4$ 19.5 **5**
$C_8H_6Cl_4$ 27.4 **5**
$C_8H_6CuO_5$ 81.23 **4**
$C_8H_6Fe_2N_2O_6$ 83.22 **3**
$C_8H_6IN_3$ 16.6 **3**
$C_8H_6I_2$ 27.3 **1**
$C_8H_6Mo_4O_{18}^{4-}$, $4H_4N^+$, $6H_2O$ 81.26 **7**
$C_8H_6N_2$ 35.5 **4**
$C_8H_6N_2$ 35.5 **5**
$C_8H_6N_2$, $2H_2O$ 35.12 **1**
$C_8H_6N_2$, $2H_2O$ 35.13+ **1**
$C_8H_6N_2O_4S_2$ 44.16 **4**
$C_8H_6N_2O_6$ 13.32 **1**
$C_8H_6N_2O_6$ 27.3 **3**
$C_8H_6N_4O_4^{2-}$, $2Rb^+$, $2H_2O$ 35.15 **1**
$C_8H_6N_4O_8$, $2H_2O$ 43.18 **1**
$C_8H_6N_4S_2$, $2H_2O$ 44.17 **6**
C_8H_6OS 39.52 **1**
$C_8H_6O_2$, Cl_3Sb 19.6 **5**
$C_8H_6O_2S$ 39.8 **6**
$C_8H_6O_2S_2$ 39.13 **3**
$C_8H_6O_3$ 38.29 **1**
$C_8H_6O_4$ 13.33 **1**
$C_8H_6O_4$ 13.34+ **1**
$C_8H_6O_4$ 13.5 **4**
$C_8H_6O_4$ 38.6 **4**
$C_8H_6O_4$ 13.10 **7**
$C_8H_6O_6$ 1.23 **3**
C_8H_6S , $C_6H_3N_3O_6$ 60.5 **5**
$C_8H_6S_2$ 39.53 **1**
$C_8H_6S_2$ 39.54 **1**
$C_8H_6S_2$ 39.55 **1**
$C_8H_6S_2$ 39.14 **3**
$C_8H_7AlCl_4O$ 68.11 **5**
$C_8H_7AlCl_4O$ 68.12 **5**
$C_8H_7AlCl_4O$ 68.13 **5**
$C_8H_7BrN_4O_5$ 43.1 **3**
C_8H_7BrO 19.3 **7**
C_8H_7BrOS 11.11 **7**
$C_8H_7BrO_2$ 22.4 **1**
$C_8H_7BrO_2$ 22.5 **1**
$C_8H_7BrO_2$ 31.1 **5**
$C_8H_7Br_2NO_3$ 17.7 **7**
$C_8H_7ClN_2O_2$ 16.7 **6**
C_8H_7ClO 19.25 **1**
$C_8H_7ClO_3S$ 1.24 **7**
$C_8H_7ClO_5$ 38.7 **4**
$C_8H_7Cl_2NO_4$, H_2O 48.73 **1**
$C_8H_7Cl_6OSb$ 66.3 **5**
$C_8H_7Cl_6OSb$ 66.4 **5**
$C_8H_7CrNO_3$ 75.1 **4**
$C_8H_7CrNO_5$ 71.14 **2**
$C_8H_7MoN_3O_9^-$, H_4N^+ , H_2O 81.25 **6**
C_8H_7N 35.9 **7**
C_8H_7N , $C_6H_3N_3O_6$ 60.102 **2**
C_8H_7NOS 40.9 **5**
$C_8H_7NO_2$ 40.10 **5**
$C_8H_7NO_2$ 38.8 **6**
$C_8H_7NO_2$, $C_8H_4NO_3^-$, K^+ 10.2 **4**
$C_8H_7NO_2$, $C_8H_6NO_2^-$, K^+ 10.3 **4**
$C_8H_7NO_2S$ 41.9 **6**
$C_8H_7NO_3$ 15.2 **5**
$C_8H_7NO_3$, H_2O 13.4+ **3**
$C_8H_7NO_3S_2$ 39.14 **5**
$C_8H_7NO_4$ 15.3 **5**
$C_8H_7NS_2$ 41.21 **1**
$C_8H_7NS_2$ 41.22 **1**
$C_8H_7NS_2$ 41.23 **1**
$C_8H_7NS_2$ 41.7 **7**
$C_8H_7N_3O_6$ 15.5 **4**
$C_8H_7N_3O_7$ 17.25 **1**
$C_8H_7N_3O_7$, $C_2H_5O^-$, Cs^+ 60.103 **2**
$C_8H_7N_4O_7^-$, K^+ 40.5 **3**
$C_8H_7N_5O$ 44.29 **3**
$C_8H_7N_6O_3^-$, K^+ , H_2O 48.39 **6**
$C_8H_7O^+$, Cl_6Sb^- 12.3 **5**
$C_8H_7O^+$, Cl_6Sb^- 12.4 **5**
$C_8H_7O_2^-$, $C_5H_{12}N^+$ 33.14 **6**
$C_8H_7O_2^-$, $C_8H_8O_2$, K^+ 2.86+ **1**
$C_8H_7O_3^-$, Na^+ , $0.5H_2O$ 2.24 **4**
$C_8H_7O_3^-$, $C_8H_8O_3$, K^+ 14.11 **1**
C_8H_8 31.2 **1**
$C_8H_8^{2-}$, $C_6H_{14}O_3$, $2K^+$ 22.1 **6**
$C_8H_8Ag^+$, NO_3^- 75.5 **2**
$C_8H_8BO_{10}^-$, K^+ , H_2O 62.4 **5**
C_8H_8BrNO 16.31 **1**
C_8H_8BrNO 10.4 **6**
C_8H_8BrNO 13.11 **7**
$C_8H_8BrN_3O_4$ 16.34 **1**
$C_8H_8BrO_2Re$ 71.7 **6**
$(C_8H_8ClCu)_n$ 75.6 **2**
$C_8H_8ClCuN_4O_2S_2$, C_3H_7NO 85.14 **6**
C_8H_8ClNO 16.32 **1**
C_8H_8ClNO 19.26 **1**
$C_8H_8ClNO_3$ 18.20 **1**
$C_8H_8ClNO_3$ 18.21 **1**
$(C_8H_8Cl_2CoN_4)_n$ 83.25 **4**
$C_8H_8Cl_2HgN_4O_4$ 84.6 **3**
$C_8H_8Cl_2N_2O_2$ 16.33 **1**
$C_8H_8Cl_2N_2O_2$ 18.2 **4**
$C_8H_8Cl_2N_8Pd$ 83.26 **4**
$C_8H_8Cl_2O_4$ 1.14 **6**
$C_8H_8Cl_2Pd$ 75.7 **2**
$C_8H_8Cl_4S_2$ 39.15 **5**
$C_8H_8Cl_4S_6$ 39.8 **7**
$(C_8H_8CuN_4^+)_n$, $nClO_4^-$ 83.54 **2**
$C_8H_8CuN_6O_{10}$, $2H_2O$ 83.27 **4**

$C_8H_8HgN_2O_4$ 83.17 **7**
$C_8H_8HgO_2$ 71.6 **5**
$C_8H_8N_2O_2$ 13.10 **5**
$C_8H_8N_2O_2$ 16.8 **5**
$C_8H_8N_2O_2S$ 33.29 **7**
$C_8H_8N_2O_4$ 1.99 **1**
$C_8H_8N_2O_4S_2$ 32.8 **7**
$C_8H_8N_2S$ 41.5 **3**
$C_8H_8N_3O_8^-$, K^+ , $2H_2O$ 15.23 **1**
$C_8H_8N_4O_4$ 44.31 **3**
$C_8H_8N_4O_4$ 44.17 **4**
C_8H_8O 19.4 **6**
$C_8H_8O_2$ 13.36 **1**
$C_8H_8O_2$ 18.22 **1**
$C_8H_8O_2$ 18.25 **1**
$C_8H_8O_2$ 18.23 + **1**
$C_8H_8O_2$ 13.6 **4**
$C_8H_8O_2$ 17.8 **7**
$C_8H_8O_2$, $C_8H_7O_2^-$, K^+ 2.86 + **1**
$C_8H_8O_3$ 13.37 **1**
$C_8H_8O_3$ 13.6 **3**
$C_8H_8O_3$ 1.15 **6**
$C_8H_8O_3$ 17.6 **6**
$C_8H_8O_3$, $C_8H_7O_3^-$, K^+ 14.11 **1**
$C_8H_8O_3S$ 42.12 **1**
$C_8H_8O_3S$ 11.6 **5**
$C_8H_8O_3S$ 13.10 **6**
$C_8H_8O_3Se$ 11.7 **5**
$C_8H_8O_5$ 38.8 **4**
$C_8H_8O_{12}Sb_2^{2-}$, $2H_4N^+$, $4H_2O$ 66.12 **2**
$C_8H_8O_{12}U^{2-}$, $2Na^+$, $2H_2O$ 81.30 **5**
$C_8H_8Pd_2S_8$ 85.4 **5**
$C_8H_8Pt_2S_8$ 85.5 **5**
$C_8H_8S_2$ 39.56 **1**
$C_8H_8S_2$ 39.15 **3**
$C_8H_8Se_3$ 39.16 **5**
$C_8H_9BrO_2$ 38.9 **6**
$C_8H_9BrO_2S$ 11.5 + **6**
$C_8H_9CrO_5PS$ 85.15 **6**
$C_8H_9F_6N_5$ 32.7 **5**
$C_8H_9IO_2$ 38.10 **6**
$C_8H_9MnO_5Sn$ 69.14 **2**
C_8H_9NO 13.38 **1**
C_8H_9NO 16.35 **1**
C_8H_9NO 13.12 **7**
C_8H_9NOS 16.36 **1**
$C_8H_9NOS_4$ 41.12 **5**
$C_8H_9NO_2$ 13.39 + **1**
$C_8H_9NO_2$ 13.7 **4**
$C_8H_9NO_2$ 17.9 **7**
$C_8H_9NO_2S$ 33.28 **6**
$C_8H_9NO_2S$ 1.25 **7**
$C_8H_9N_3O$ 8.23 **7**
$C_8H_9O_5PSW$ 85.16 **6**
$C_8H_9PS_3$ 42.3 **4**
C_8H_{10} 20.16 **1**
C_8H_{10} , CBr_4 60.1 **2**
C_8H_{10} , C_6F_6 60.10 **7**
$0.5C_8H_{10}$, $C_{30}H_{18}N_3O_6P_3$ 61.2 **7**
$C_8H_{10}AgNO_3$ 31.1 **4**
$C_8H_{10}Br_2$ 22.3 **5**
$C_8H_{10}Br_2O_4$ 20.17 **1**
$C_8H_{10}Br_2O_4$ 20.18 **1**
$(C_8H_{10}CaO_7)_n$ 67.5 **5**
$C_8H_{10}Cl_2$ 31.2 **5**
$(C_8H_{10}Cl_2CuN_6)_n$ 83.23 **3**
$C_8H_{10}Cl_2CuN_6O_2$ 83.28 **4**
$C_8H_{10}Cl_2O_6Pd_2$, $3H_2O$ 72.10 **3**
$C_8H_{10}CoN_2O_8^-$, K^+ , $2.5H_2O$ 82.10 **5**
$2C_8H_{10}CoN_2O_8^-$, Ca^{2+} , $7.5H_2O$ 82.8 **7**
$2C_8H_{10}CrN_6O_4S_2^-$, Ba^{2+} , $5H_2O$ 79.14 **7**
$C_8H_{10}CuN_2O$ 83.13 **5**
$(C_8H_{10}CuN_2O_4)_n$ 81.19 **3**
$C_8H_{10}CuN_2O_8^-$, K^+ , H_2O 82.9 **7**
$C_8H_{10}CuN_4O_3$, $1.5H_2O$ 82.25 **2**
$C_8H_{10}CuN_4O_5^{2-}$, $2Na^+$, $10H_2O$ 82.26 **2**
$C_8H_{10}NO_3^+$, Cl^- 33.34 **5**
$C_8H_{10}NO_3^+$, $C_8H_{12}NO_3^+$, $2Cl^-$ 17.12 **7**
$C_8H_{10}NO_5PS$ 64.15 **3**
$C_8H_{10}NO_6P$, H_2O 33.35 **5**
$C_8H_{10}N_2NiO_8^{2-}$, $2Li^+$, $4H_2O$ 82.10 **7**
$C_8H_{10}N_2O$ 16.8 **6**
$C_8H_{10}N_2O_2$ 16.37 **1**
$C_8H_{10}N_2O_2$ 16.7 **3**
$C_8H_{10}N_2O_2$ 40.11 **5**
$C_8H_{10}N_2O_2S$ 40.12 **5**
$C_8H_{10}N_2O_2Se$ 42.4 **5**
$C_8H_{10}N_2O_3$ 33.22 **3**
$C_8H_{10}N_2O_4$ 48.53 **5**
$C_8H_{10}N_2S$ 33.29 **6**
$2C_8H_{10}N_3^+$, Cl_4Zn^{2-} 9.7 **4**
$C_8H_{10}N_4NiO_5^{2-}$, $2Na^+$, $8H_2O$ 82.27 **2**
$C_8H_{10}N_4O$ 32.13 **6**
$C_8H_{10}N_4O_2$ 44.18 **6**
$C_8H_{10}N_4O_2$, H_2O 44.64 **1**
$C_8H_{10}N_4O_2$, $C_7H_5ClO_3$ 60.104 **2**
$C_8H_{10}N_4O_2$, $2C_8H_{12}N_2O_3$ 60.11 **6**
$C_8H_{10}N_6O_3^{2+}$, $2Cl^-$ 48.40 **6**
$C_8H_{10}O$ 17.27 **1**
$C_8H_{10}O$ 17.28 **1**
$C_8H_{10}O$ 17.8 **5**
$C_8H_{10}O$ 17.9 **5**
$C_8H_{10}O$ 17.11 **5**
$C_8H_{10}O$ 17.12 **5**
$C_8H_{10}O$ 17.7 **6**
$C_8H_{10}O$ 17.8 **6**
$C_8H_{10}OS$ 11.11 **3**
$C_8H_{10}OS$ 39.8 **4**
$C_8H_{10}O_2$ 17.29 **1**
$C_8H_{10}O_2$ 31.1 **6**
$C_8H_{10}O_4$ 1.100 **1**
$C_8H_{10}O_4$ 1.101 **1**
$C_8H_{10}O_4$ 1.102 **1**
$C_8H_{10}O_4^{2-}$, $2C_8H_{12}O_4$, $2K^+$ 21.2 **4**
$C_8H_{10}Pd$ 72.19 **2**
$C_8H_{10}S_6$ 39.9 **7**
$C_8H_{11}BrN^+$, $C_{12}H_9O_2^-$ 31.10 **5**
$C_8H_{11}Cl_2HgN$ 83.14 **5**
$C_8H_{11}Cl_2N_2^+$, Br^- , $2H_2O$ 32.10 **4**
$C_8H_{11}CoN_3O_8^-$, K^+ , $1.5H_2O$ 76.22 **5**
$2C_8H_{11}IN^+$, Cl^- , I_3^- 60.105 + **2**
$C_8H_{11}N$ 20.7 **6**
$C_8H_{11}NO_2$ 48.31 **3**
$C_8H_{11}NO_3$ 20.10 **5**
$C_8H_{11}N_2O^+$, Cl^- , H_2O 16.3 **7**
$C_8H_{11}N_2O_3^-$, K^+ 43.20 **1**
$C_8H_{11}N_2O_3^-$, Na^+ 43.21 **1**
$C_8H_{11}N_2O_3^-$, Na^+ 43.3 **4**
$C_8H_{11}N_2O_3-$, $CH_6N_3^+$, $2H_2O$ 43.1 **6**
$C_8H_{11}N_2O_3^-$, $C_4H_{12}NO^+$ 43.2 **7**
$C_8H_{11}N_2O_3^-$, $C_4H_{13}N_2^+$ 43.3 **7**
$2C_8H_{11}N_2O_3^-$, Ca^{2+} , $3H_2O$ 43.5 **4**

$C_8H_{11}N_2O_6P$, $2H_2O$ 46.12 **1**
$C_8H_{11}N_2S^+$, $AuCl_4^-$ 8.24 **7**
$C_8H_{11}N_2S^+$, Br^- 33.56 **1**
$C_8H_{11}N_2S^+$, Cl^- 33.57 **1**
$2C_8H_{11}N_2S^+$, Cl_6Pt^{2-} 8.25 **7**
$C_8H_{11}N_3O_3$, H_2O 48.54 **5**
$C_8H_{11}N_3O_6$ 47.1 **5**
$C_8H_{11}O_5S^-$, Ag^+ , H_2O 21.1 **6**
$C_8H_{11}O_5S^-$, Rb^+ 21.2 **6**
$C_8H_{11}S^+$, ClO_4^- 11.41 **1**
$C_8H_{12}AlCl_3$ 68.10 **7**
$C_8H_{12}As_2Cl_4F_4Re$ 86.3 **7**
$C_8H_{12}B_2O_2$ 62.5 **5**
$(C_8H_{12}BaO_5)_n$ 67.8 **4**
$C_8H_{12}BrCrO_3P$ 71.6 **7**
$C_8H_{12}BrNNiO_2$ 72.3 **6**
$C_8H_{12}Br_2O_4$ 38.9 **4**
$C_8H_{12}Br_2O_8Rh_2$, $2CH_5N_3$ 81.56 **2**
$C_8H_{12}Br_2O_8Rh_2^{2-}$, $2Br^-$, $4H_4N^+$ 81.57 **2**
$C_8H_{12}Br_4$ 22.6 **1**
$C_8H_{12}CdN_6S_4$ 79.33 **2**
$C_8H_{12}ClIr$ 72.4 **6**
$C_8H_{12}ClN_6O^+$, Cl^- 32.22 **1**
$C_8H_{12}ClN_6O^+$, Cl^- 32.11 **4**
$C_8H_{12}Cl_2FeO_4S_4^+$, Cl_4Fe^- 84.24 **2**
$C_8H_{12}Cl_2O$ 21.3 **6**
$C_8H_{12}Cl_2O_4$ 1.16 **6**
$C_8H_{12}Cl_2O_8Rh_2$, $2CH_5N_3$ 81.58 **2**
$(C_8H_{12}Cl_4Hg_2S_2)_n$ 85.44 **2**
$C_8H_{12}Cl_8N_4Nb_2^{2-}$, $2C_{36}H_{30}NP_2^+$, $2C_6H_5Cl$ 83.18 **7**
$C_8H_{12}CoN_2O_8^-$, Na^+ , $2H_2O$ 81.24 **4**
$C_8H_{12}CoN_2O_8^-$, $C_4H_{16}CoN_6O_4^+$ 76.4 **5**
$C_8H_{12}CoN_4O_6^-$, H_4N^+ , $2H_2O$ 82.28 **2**
$2C_8H_{12}CoN_4O_6^-$, Ba^{2+} , xH_2O 82.12 **3**
$2C_8H_{12}CoN_4O_6^-$, $H_{12}CoO_6^{2+}$, $12H_2O$ 82.13 **3**
$2C_8H_{12}CoN_4O_6^-$, $H_{12}CoO_6^{2+}$, $6H_2O$ 82.14 **3**
$C_8H_{12}CuN_4^+$, ClO_4^- 83.19 **7**
$C_8H_{12}CuN_4O_6^{2-}$, $2K^+$, $6H_2O$ 82.29 **2**
$C_8H_{12}CuN_5O_4S^+$, NO_3^- , H_2O 79.15 **7**
$C_8H_{12}ErO_{12}^+$, $C_4H_{14}ErO_{10}^-$ 81.25 **4**
$C_8H_{12}F_3FeP$ 72.6 **5**
$C_8H_{12}I_2$ 31.2 **6**
$C_8H_{12}Mo_2O_8$ 81.59 **2**
$C_8H_{12}Mo_2O_8$ 81.27 **7**
$C_8H_{12}N^+$, Cl^- 3.56 **1**
$C_8H_{12}N^+$, $C_6H_{14}O_2PS^-$ 64.12 **7**
$C_8H_{12}N^+$, $C_{10}H_7MnO_5^-$ 3.57 **1**
$C_8H_{12}N^+$, $C_{11}H_7CrO_5^-$ 3.58 **1**
$C_8H_{12}N^+$, $C_{11}H_7CrO_5^-$ 3.59 **1**
$C_8H_{12}NO^+$, Cl^- 17.10 + **7**
$C_8H_{12}NO_2$ 37.3 **3**
$C_8H_{12}NO_2^+$, Cl^- 59.10 **1**
$C_8H_{12}NO_3^+$, Cl^- 59.11 **1**
$C_8H_{12}NO_3^+$, Cl^- 59.12 **1**
$C_8H_{12}NO_3^+$, Cl^- 59.3 **4**
$C_8H_{12}NO_3^+$, Cl^- 59.4 **5**
$C_8H_{12}NO_3^+$, $C_8H_{10}NO_3^+$, $2Cl^-$ 17.12 **7**
$C_8H_{12}NO_7P$ 46.8 **6**
$C_8H_{12}N_2$ 33.58 **1**
$C_8H_{12}N_2$, $C_{10}H_2N_4$ 60.4 **6**
$C_8H_{12}N_2^+$, Br^- 16.39 **1**
$4C_8H_{12}N_2O_2$, $2K^+$, $Cl_6Pd_2^{2-}$ 44.20 **7**
$C_8H_{12}N_2O_2S$ 43.22 **1**
$C_8H_{12}N_2O_3$ 43.25 **1**
$C_8H_{12}N_2O_3$ 43.23 + **1**
$C_8H_{12}N_2O_3$ 43.6 **4**
$C_8H_{12}N_2O_3$ 43.4 **7**
$C_8H_{12}N_2O_3$, CH_4N_2O 60.5 **6**
$C_8H_{12}N_2O_3$, C_2H_5NO 60.6 **6**
$C_8H_{12}N_2O_3$, $C_3H_4N_2$ 60.7 **6**
$C_8H_{12}N_2O_3$, $2C_5H_6N_2$ 60.8 **6**
$C_8H_{12}N_2O_3$, C_6H_7NO 60.9 **6**
$C_8H_{12}N_2O_3$, $C_7H_9N_5$ 60.10 **5**
$C_8H_{12}N_2O_3$, $C_{13}H_{17}N_3O$ 60.31 **4**
$2C_8H_{12}N_2O_3$, K^+ , I_3^- 43.2 **5**
$2C_8H_{12}N_2O_3$, $C_6H_{18}N_3OP$ 60.10 **6**
$2C_8H_{12}N_2O_3$, $C_8H_{10}N_4O_2$ 60.11 **6**
$C_8H_{12}N_2O_3S$ 45.19 **7**
$C_8H_{12}N_2O_4$ 32.8 **5**
$C_8H_{12}N_2O_4^{2+}$, O_4S^{2-} , $1.5H_2O$ 48.41 **6**
$(C_8H_{12}N_2O_{10}U)_n$ 82.11 **7**
$C_8H_{12}N_3O_9P$ 47.2 **5**
$C_8H_{12}N_4$ 9.8 **4**
$C_8H_{12}N_4$ 9.5 + **5**
$C_8D_{12}N_4$ 9.7 **5**
$C_8H_{12}N_4NiO_6^{2-}$, $2Na^+$, $8H_2O$ 82.30 **2**
$C_8H_{12}N_4NiO_6^{2-}$, $2Na^+$, $9H_2O$ 82.31 **2**
$C_8H_{12}N_4O_5$ 47.1 **7**
$C_8H_{12}N_5O^+$, Br^- 35.1 **3**
$(C_8H_{12}N_6NiS_4)_n$ 79.31 **2**
$C_8H_{12}N_6O^+$, Cl^- 32.9 **5**
$C_8H_{12}N_6S_2Se_2Te$ 70.6 **4**
$C_8H_{12}N_6S_4Te$ 70.7 **4**
$C_8H_{12}Ni_3S_9$ 85.6 **5**
$C_8H_{12}OS$ 20.6 **7**
$C_8H_{12}O_2$ 20.19 **1**
$C_8H_{12}O_2$ 20.11 **5**
$C_8H_{12}O_2$ 20.8 **6**
$C_8H_{12}O_2$ 21.4 **6**
$C_8H_{12}O_4$ 21.18 **1**
$C_8H_{12}O_4$ 21.5 **3**
$C_8H_{12}O_4$ 21.6 **3**
$C_8H_{12}O_4$ 21.7 **3**
$C_8H_{12}O_4$ 21.3 **4**
$C_8H_{12}O_8Pb$ 69.15 **2**
$C_8H_{12}O_8Si$ 63.1 **5**
$(C_8H_{12}O_8U)_n$ 81.60 **2**
$C_8H_{12}Rh^+$, Cl^- 72.20 **2**
$C_8H_{12}S_2$ 20.9 **6**
$C_8H_{12}S_8V$ 85.8 **4**
$C_8H_{13}Cl_3O_3S_2SnW$ 85.6 **3**
$C_8H_{13}NO_3$ 45.20 **7**
$C_8H_{13}NO_4$ 1.22 **5**
$C_8H_{13}N_4S_2Te^+$, Cl^- 70.16 **2**
$C_8H_{13}O_3^-$, Na^+ 2.89 **1**
$(C_8H_{14}CdN_4O_6)_n$ 82.13 **6**
$(C_8H_{14}CdO_2S_4)_n$ 80.2 **6**
$C_8H_{14}ClN$, H_2O 37.4 **3**
$C_8H_{14}ClNSn$ 69.16 **2**
$C_8H_{14}Cl_2$ 22.3 **3**
$C_8H_{14}Cl_2OPt$ 72.7 **5**
$C_8H_{14}Cl_2Pd_2$ 72.21 **2**
$C_8H_{14}Cl_3O_2Pt^-$, $C_{24}H_{20}P^+$ 72.5 **6**
$C_8H_{14}CoN_4O_6^+$, ClO_4^- 82.15 **3**
$C_8H_{14}CuN_4O_4$ 83.56 **2**
$C_8H_{14}CuN_4O_4$ 83.24 **3**
$C_8H_{14}CuN_4O_6$ 82.12 **7**
$C_8H_{14}CuN_6OS_2$ 79.16 **7**
$C_8H_{14}CuN_{10}O_2^{2+}$, $2ClO_4^-$ 83.15 **5**

$C_8H_{14}NO^+$, Cl^- 40.14 **1**
$C_8H_{14}NO_2^+$, Br^- 31.2 **4**
$C_8H_{14}NO_2^+$, Br^- 35.6 **4**
$C_8H_{14}N_2NiS_2$ 85.47 **2**
$C_8H_{14}N_2O_2$ 48.33 **3**
$C_8H_{14}N_2O_2$ 34.8 **7**
$C_8H_{14}N_2O_2^{2+}$, $2Cl^-$, H_2O 33.36 **5**
$C_8H_{14}N_2O_4S$ 38.11 **6**
$C_8H_{14}N_2O_5P^+$, Cl^- 33.23 **3**
$C_8H_{14}N_4NiO_4$ 83.57 **2**
$C_8H_{14}N_4O_4Pd$ 83.58+ **2**
$C_8H_{14}N_4O_4Pt$ 83.60 **2**
$C_8H_{14}N_5O^+$, Br^- 40.13 **5**
$C_8H_{14}N_6NiS_2$ 85.7 **3**
$C_8H_{14}N_6O_{10}$ 3.17 **6**
$C_8H_{14}N_6O_{10}$, C_3H_7NO 60.12 **6**
$C_8H_{14}N_6O_{10}$, $C_4H_8O_2$ 60.13 **6**
$C_8H_{14}Ni$ 72.22 **2**
$C_8H_2D_{12}O_2$ 5.12 **7**
$C_8H_{14}O_2S_2$ 39.57 **1**
$C_8H_{14}O_2S_2$ 39.9 **4**
$C_8H_{14}O_4$ 1.104 **1**
$C_8H_{14}O_4$ 1.105 **1**
$C_8H_{14}O_7$, H_2O 45.20 **6**
$C_8H_{14}Pt$ 71.4 **3**
$C_8H_{14}Pt$ 71.4 **4**
$C_8H_{15}Cl_2CoN_4O_4$ 83.20 **7**
$C_8H_{15}Cl_2N_4O_4Rh$ 83.21 **7**
$C_8H_{15}N$ 37.5 **7**
$C_8H_{15}NO$ 58.1 **1**
$C_8H_{15}NO$ 34.9 **7**
$C_8H_{15}NO_2$ 21.20 **1**
$C_8H_{15}NO_3S$ 48.42 **6**
$C_8H_{15}NO_6$ 45.51 **1**
$C_8H_{15}NO_6$ 45.21 **6**
$C_8H_{15}NO_6$ 45.23 **7**
$C_8H_{15}NO_6$ 45.21+ **7**
$C_8H_{15}NO_7$ 48.38 **7**
$C_8H_{15}N_2O_2$ 32.9 **7**
$C_8H_{15}N_5O_2^{2+}$, $2Br^-$, $3H_2O$ 50.1 **3**
$C_8H_{15}O_2^+$, ClO_4^- 38.12 **6**
$2C_8H_{15}O_2^-$, Sr^{2+} , xH_2O 2.90 **1**
$C_8H_{16}Ag_2N_4O_6^{2+}$, $2NO_3^-$ 82.8 **4**
$C_8H_{16}Br_2HgO_4$ 84.5 **4**
$C_8H_{16}Br_2N_4S_2Te$ 70.3 **7**
$(C_8H_{16}Br_2NiO_4)_n$ 84.22 **7**
$C_8H_{16}CaCdO_{10}$, $6H_2O$ 81.61 **2**
$C_8H_{16}CaCuO_{10}$, $6H_2O$ 81.62 **2**
$C_8H_{16}CaF_4O_8P_2$ 67.2 **2**
$(C_8H_{16}CdCl_2N_4O_6)_n$ 82.14 **6**
$C_8H_{16}ClNO_6$ 45.24 **7**
$C_8H_{16}ClOP$ 64.16 **3**
$C_8H_{16}Cl_2CuN_2O_2S$ 85.8 **3**
$C_8H_{16}Cl_2CuO_4$ 84.23 **7**
$C_8H_{16}Cl_2CuO_4S_2$ 84.7 **3**
$C_8H_{16}Cl_2HgO_2S_2$ 85.48 **2**
$C_8H_{16}Cl_2N_4S_2Te$ 70.4+ **7**
$C_8H_{16}Cl_2Rh_2$ 72.23 **2**
$C_8H_{16}Cl_4MnO_8P_2$ 84.25 **2**
$C_8H_{16}Cl_4S_2Sn$ 69.17 **2**
$C_8H_{16}Cl_8O_4Ti_2$ 84.26 **2**
$C_8H_{16}Cr_2O_{10}$ 81.63 **2**
$C_8H_{16}Cr_2O_{10}$ 81.26 **4**
$(C_8H_{16}CuN_2O_3)_n$ 81.32 **5**
$(C_8H_{16}CuN_2O_4)_n$, $2nH_2O$ 81.33 **5**
$C_8H_{16}CuN_5O_8^+$, NO_3^- 83.16 **5**
$C_8H_{16}Cu_2O_{10}$ 81.64 **2**
$C_8H_{16}Cu_2O_{10}$ 81.26+ **6**
$C_8H_{16}Fe_2S_8^{2-}$, $2C_{16}H_{36}N^+$ 85.7 **5**
$C_8H_{16}Ge_2$ 69.18 **2**
$C_8H_{16}Hg_2O_2$ 71.15 **2**
$C_8H_{16}Mo_2N_2O_6S_4$ 82.16 **3**
$C_8H_{16}N^+$, Br^- 37.12 **1**
$C_8H_{16}N^+$, ClO_4^- 35.7 **4**
$C_8H_{16}NO^+$, Cl^- 34.10 **7**
$C_8H_{16}NO_2^+$, Br^- 21.21 **1**
$C_8H_{16}NO_2^+$, Br^- 21.22 **1**
$C_8H_{16}NO_2^+$, Br^- , H_2O 48.37 **4**
$C_8H_{16}NO_2^+$, Cl^- 21.23 **1**
$C_8H_{16}NO_2^+$, I^- 20.9 **3**
$C_8H_{16}N_2^+$, ClO_4^- 32.23 **1**
$C_8H_{16}N_2NiS_4$ 80.3 **7**
$C_8H_{16}N_2O^{2+}$, $2Cl^-$ 58.2 **1**
$C_8H_{16}N_2O_2$ 1.106 **1**
$C_8H_{16}N_2O_2$ 32.24 **1**
$C_8H_{16}N_2O_2S_2$ 40.7 **6**
$C_8H_{16}N_2O_3$ 48.39 **7**
$C_8H_{16}N_2O_3S$ 48.43 **6**
$C_8H_{16}N_4$ 37.6 **7**
$C_8H_{16}N_4O_2$ 40.8 **6**
$C_8H_{16}N_6OS_2$ 4.15 **1**
$(C_8H_{16}N_8NiPd)_n$ 76.49 **3**
$2C_8H_{16}NaO_2^+$, $C_{24}H_{28}Al_2^{2-}$ 68.21 **3**
$2C_8H_{16}NaO_2^+$, $C_{32}H_{32}Al_2^{2-}$ 68.31 **5**
$C_8H_{16}Ni_2S_6$ 85.49+ **2**
$C_8H_{16}OS_3$ 42.5 **7**
$C_8H_{16}O_2$ 20.21 **1**
$C_8H_{16}O_2$, $0.5H_2O$ 5.13 **7**
$C_8H_{16}O_2$, $0.5C_8H_{16}O_2$ 5.14 **7**
$C_8H_{16}O_4$ 38.30 **1**
$C_8H_{16}O_4$, Mg^{2+} , $2Cl^-$, $6H_2O$ 38.9 **7**
$2C_8H_{16}O_4$, Na^+ , Cl^- , $0.5H_2O$ 38.10 **7**
$2C_8H_{16}O_4$, Na^+ , HO^- , $8H_2O$ 38.11 **7**
$C_8H_{16}O_5S$ 45.52 **1**
$C_8H_{16}O_6$, H_2O 45.23 **6**
$C_8H_{16}O_6S$ 45.13 **4**
$C_8H_{16}O_{10}Rh_2$ 81.28 **4**
$C_8H_{16}P_2S_2$ 64.24 **2**
$C_8H_{17}ClCoN_5O_4$ 83.26 **3**
$C_8H_{17}ClCoN_5O_4$, H_2O 83.10 **6**
$C_8H_{17}CuN_5S_2$ 83.11 **6**
$C_8H_{17}GeNO_3$ 69.17 **3**
$C_8H_{17}NO_2Si$ 63.2 **3**
$C_8H_{17}N_2O_3^+$, Br^- 48.34 **3**
$C_8H_{17}O_2P$, H_2O 64.25 **2**
$C_8H_{17}O_3P$ 64.17 **3**
C_8H_{18} 5.22 **1**
$C_8H_{18}BF_4N_4NiO^+$, BF_4^- 76.23 **5**
$(C_8H_{18}Cd_2O_{15})_n$, $2nH_2O$ 81.28 **6**
$C_8H_{18}ClN_6NiS_2^+$, Cl^- , H_2O 79.20 **5**
$C_8H_{18}Cl_2CoN_2O_2$ 84.8 **6**
$C_8H_{18}Cl_2Cu_2N_4O_8$ 82.11 **5**
$C_8H_{18}Cl_2N_2O_2P_2$ 64.18 **7**
$C_8H_{18}Cl_2N_2O_2Zn$ 84.9 **6**
$C_8H_{18}Cl_2N_2P_2$ 64.19 **7**
$C_8H_{18}Cl_2PdSe_2$ 85.10 **3**
$C_8H_{18}CoN_4O_4^+$, Cl^- , $2H_2O$ 76.15 **6**
$C_8H_{18}CoN_6^+$, ClO_4^- 76.35 **7**
$C_8H_{18}Cr_2N_4O_{10}$ 82.12 **5**
$C_8H_{18}CuN_2O_5$, $2H_2O$ 82.15 **6**
$C_8H_{18}CuN_6O_4^{2+}$, $2ClO_4^-$ 82.13 **7**
$C_8H_{18}CuO_8$ 81.66 **2**

$C_8H_{18}CuO_8$ 81.22 **3**
$C_8H_{18}Fe_4N_6O_4S_2$ 83.12 **6**
$C_8H_{18}HgS_2$ 85.51 **2**
$C_8H_{18}NO^+$, Br^- 58.3 **1**
$C_8H_{18}NO^+$, $CH_3O_3S^-$ 32.23 **3**
$C_8H_{18}NOS^+$, I^- 3.34 **3**
$C_8H_{18}NO_2{}^+$, Br^- 3.17 **5**
$C_8H_{18}NO_2{}^+$, I^- 59.13 **1**
$C_8H_{18}NO_2{}^+$, I^- 59.14 **1**
$C_8H_{18}NO_2{}^+$, I^- 38.3 **3**
$C_8H_{18}NO_2PSe$ 64.20 **7**
$C_8H_{18}NO_2PSe$ 64.21 **7**
$C_8H_{18}N_2{}^{2+}$, $C_4H_{22}B_{18}Ni^{2-}$ 37.5 **3**
$C_8H_{18}N_2{}^{2+}$, $C_{10}H_6FeN_6{}^{2-}$ 83.17 **6**
$C_8H_{18}N_2O_2$ 10.30 **1**
$C_8H_{18}N_2O_4S_2{}^{2+}$, $2Cl^-$, H_2O 48.40 **7**
$C_8H_{18}N_2O_4Sn$ 69.7 **7**
$C_8H_{18}N_2O_{11}Zn$ 82.32 **2**
$C_8H_{18}N_2S_4Sn$ 69.12 **5**
$C_8H_{18}N_3OP$ 64.22 **7**
$C_8H_{18}N_4O_4S_6Te$ 70.4 + **5**
$C_8H_{18}N_6NiS_2$ 76.47 **2**
$C_8H_{18}N_6NiS_2$ 76.50 **3**
$C_8H_{18}N_6NiS_2$ 76.51 **3**
$C_8H_{18}N_6Pt^{2+}$, $2Cl^-$, $2H_2O$ 83.29 **4**
$C_8H_{18}N_7NiO_3S_2{}^+$, $NO_3{}^-$, H_2O 79.21 **5**
$C_8H_{18}O_2$, $4H_2O$ 5.14 **5**
$C_8H_{18}O_3$ 5.15 **7**
$C_8H_{18}O_5Sn_2$ 69.20 **2**
$C_8H_{18}O_{10}S_2$ 45.25 **7**
$C_8H_{19}AsO_2$ 65.3 **3**
$C_8H_{19}B_2{}^-$, $C_{16}H_{36}N^+$ 61.4 **5**
$C_8H_{19}N_2O_2P$ 64.11 **4**
$C_8H_{20}AlLi$ 68.15 **2**
$C_8H_{20}Au_2Br_2$ 71.16 **2**
$C_8H_{20}B_2F_4N_2$ 3.35 **3**
$(C_8H_{20}BrCuP_2)_n$ 86.3 **3**
$C_8H_{20}Br_2MgO_2$ 67.3 **2**
$C_8H_{20}Br_2O_4Ti^+$, $C_4H_{10}Br_4O_2Ti^-$ 84.27 **2**
$C_8H_{20}Br_4Pt_2S_2$ 85.52 **2**
$C_8H_{20}Cd_2O_{16}$ 81.29 **6**
$C_8H_{20}ClCuN_4S_4$ 85.53 **2**
$C_8H_{20}Cl_2N_4NiO_4$ 83.30 **4**
$C_8H_{20}Cl_2N_4NiS_4$ 85.8 **5**
$C_8H_{20}Cl_2PdSe_2$ 85.11 **3**
$C_8H_{20}Cl_4O_4Ti_2$ 84.28 **2**
$C_8H_{20}Cl_5Re_2S_4$ 85.54 **2**
$2C_8H_{20}CoN_2O_8{}^-$, Ca^{2+} , $10H_2O$ 82.16 **6**
$C_8H_{20}CoN_6O_4{}^+$, Cl^- , H_2O 83.22 **7**
$C_8H_{20}CoN_6O_4{}^+$, $NO_3{}^-$ 83.61 **2**
$2C_8H_{20}CoN_6S_2{}^+$, $C_8H_4O_{12}Sb_2{}^{2-}$, $4H_2O$ 76.52 **3**
$C_8H_{20}CoS_4{}^{2+}$, $2ClO_4{}^-$ 85.55 **2**
$C_8H_{20}CuN_2O_6$ 83.62 **2**
$C_8H_{20}CuN_3O_2{}^+$, $ClO_4{}^-$ 81.35 **5**
$C_8H_{20}CuN_6S_2$ 76.24 **5**
$C_8H_{20}Cu_2P_2$ 71.7 **7**
$(C_8H_{20}HgO_4PS_4)_n$ 85.9 **5**
$C_8H_{20}I_2N_4Rh^+$, I^- 83.28 **3**
$C_8H_{20}Li_4$ 67.4 **2**
$C_8H_{20}N^+$, $BrCl_2{}^-$ 3.18 **7**
$C_8H_{20}N^+$, Br_6Sb^- 3.36 **3**
$C_8H_{20}N^+$, Cl_3NOPt^- 3.18 **6**
$C_8H_{20}N^+$, Cl_4In^- 3.60 **1**
$C_8H_{20}N^+$, $I_3{}^-$ 3.61 + **1**
$C_8H_{20}N^+$, $H_2Br_4O_2Re^-$ 3.63 **1**
$C_8H_{20}N^+$, H_4Al^- 3.19 **7**
$C_8H_{20}N^+$, $CH_2Br_3OPt^-$ 3.20 **7**
$C_8H_{20}N^+$, $C_2NS_2{}^-$ 3.19 **6**
$C_8H_{20}N^+$, $C_4Br_3O_4W^-$ 3.12 **4**
$C_8H_{20}N^+$, $C_4HCr_2O_4{}^-$ 3.64 **1**
$C_8H_{20}N^+$, $C_4H_{18}B_{14}Co^-$ 62.9 **4**
$C_8H_{20}N^+$, $C_6H_{15}Cl_3PPt^-$ 86.2 **7**
$C_8H_{20}N^+$, $C_8Br_2Co_2InO_8{}^-$ 3.13 **4**
$C_8H_{20}N^+$, $C_8FeN_5OS_4{}^-$ 85.9 **6**
$C_8H_{20}N^+$, $C_8H_{40}B_{40}Co^-$ 62.6 **5**
$C_8H_{20}N^+$, $C_9H_{15}O_3PbS_6{}^-$ 69.15 **4**
$C_8H_{20}N^+$, $C_9H_{15}O_3S_6Te^-$ 70.5 **7**
$C_8H_{20}N^+$, $C_9H_{25}B_{16}Co_2{}^-$ 71.10 **6**
$C_8H_{20}N^+$, $C_9H_{26}B_{17}CoN^-$ 62.8 **5**
$C_8H_{20}N^+$, $C_{10}HCr_2O_{10}{}^-$ 3.37 **3**
$C_8H_{20}N^+$, $C_{13}H_4F_3O_6S_2W^-$ 85.23 **6**
$C_8H_{20}N^+$, $C_{16}H_8N_4NiS_4{}^-$, $2H_2O$ 85.27 **7**
$C_8H_{20}N^+$, $C_{16}H_{16}N_2NiS_2{}^-$ 85.27 **3**
$C_8H_{20}N^+$, $C_{20}H_{40}EuN_4S_8{}^-$ 80.12 **3**
$C_8H_{20}N^+$, $C_{20}H_{40}N_4NpS_8{}^-$ 80.13 + **3**
$C_8H_{20}N^+$, $C_{28}H_{20}O_7V^{2-}$, Na^+ , $2C_3H_8O$ 81.52 **3**
$2C_8H_{20}N^+$, $Br_6Pt_2{}^{2-}$ 3.65 **1**
$2C_8H_{20}N^+$, ClO_4Ti^{2-} 3.66 **1**
$2C_8H_{20}N^+$, Cl_4Ni^{2-} 3.67 **1**
$2C_8H_{20}N^+$, Cl_5In^{2-} 3.68 **1**
$2C_8H_{20}N^+$, Cl_5OPa^{2-} 3.14 **4**
$2C_8H_{20}N^+$, $Cl_8Hg_2Pt^{2-}$ 3.20 **6**
$2C_8H_{20}N^+$, $Cl_{10}Hg_3Pd^{2-}$ 3.21 **6**
$2C_8H_{20}N^+$, $Cl_{10}Hg_3Pt^{2-}$ 3.22 **6**
$2C_8H_{20}N^+$, $C_2Cl_8O_4Sn_2{}^{2-}$ 69.1 **7**
$2C_8H_{20}N^+$, $C_4H_{22}B_{18}Cu^{2-}$ 62.34 **2**
$2C_8H_{20}N^+$, $C_8FeN_5OS_4{}^{2-}$ 85.42 **2**
$2C_8H_{20}N^+$, $C_8FeN_5OS_4{}^{2-}$ 85.10 **6**
$2C_8H_{20}N^+$, $C_8H_2O_8W_2{}^{2-}$ 3.21 **7**
$2C_8H_{20}N^+$, $C_{14}Mo_2Ni_4O_{14}{}^{2-}$ 3.15 **4**
$2C_8H_{20}N^+$, $C_{16}Ni_3O_{16}W_2{}^{2-}$ 3.16 **4**
$2C_8H_{20}N^+$, $C_{16}H_{16}Fe_2S_6{}^{2-}$ 85.28 **7**
$2C_8H_{20}N^+$, $C_{28}H_{28}Fe_2S_6{}^{2-}$ 85.46 **7**
$2C_8H_{20}N^+$, $C_{28}H_{28}Fe_4S_8{}^{2-}$ 85.31 **5**
$3C_8H_{20}N^+$, $C_5CoN_5O_2{}^{3-}$, $5H_2O$ 3.24 **6**
$3C_8H_{20}N^+$, $C_8H_{42}B_{34}Co_3{}^{3-}$ 62.15 **3**
$3C_8H_{20}N^+$, $C_{12}InN_6S_6{}^{3-}$ 68.7 **4**
$4C_8H_{20}N^+$, $C_8N_8S_8U^{4-}$ 3.17 **4**
$C_8H_{20}NO_6P$ 46.13 **1**
$C_8H_{20}NO_6P$, $CdCl_2$, $3H_2O$ 46.14 **1**
$C_8H_{20}NO_6P$, $CdCl_2$, $3H_2O$ 46.9 **6**
$C_8H_{20}N_2NiO_6$, $2H_2O$ 82.33 **2**
$C_8H_{20}N_2NiS_2$ 85.56 **2**
$C_8H_{20}N_4NiS_4{}^{2+}$, $2Br^-$ 85.10 **4**
$C_8H_{20}N_6NiS_2$ 76.36 **7**
$C_8H_{20}NiO_4P_2S_4$ 85.57 + **2**
$C_8H_{20}NiO_4P_2Se_4$ 85.8 **7**
$C_8H_{20}NiP_2S_4$ 85.60 **2**
$C_8H_{20}Ni_2S_8$ 85.12 **3**
$C_8H_{20}O_4P_2PbS_4$ 69.14 **4**
$C_8H_{20}O_4P_2S_4Zn$ 85.61 **2**
$C_8H_{20}P_2S_2$ 64.26 **2**
$C_8H_{20}P_2S_2Se_2$ 64.27 **2**
$C_8H_{20}P_2S_2Se_2Te$ 64.18 **3**
$C_8H_{20}P_2S_4Tl_2$ 68.9 **6**
$C_8H_{20}P_2Se_5$ 64.19 **3**
$C_8H_{21}BN_3P$ 64.23 **7**
$C_8H_{22}B_{18}Fe_2O_4{}^{2-}$, $2Cs^+$, C_3H_6O , H_2O 62.13 **3**
$C_8H_{22}Be_2{}^{2-}$, $2C_4H_{10}NaO^+$ 67.5 **2**

$C_8H_{22}Cl_2CoN_4^{+}$, NO_3^{-} 76.25 **5**
$C_8H_{22}Cl_6O_4Sn_2$ 69.8 **7**
$C_8H_{22}CoN_5O_2^{2+}$, Cl^{-} , ClO_4^{-} 76.37 **7**
$C_8H_{22}CoN_5O_2^{2+}$, $2Cl^{-}$, H_2O 76.17 **4**
$C_8H_{22}CoN_5O_2^{2+}$, $2I^{-}$, $0.5H_2O$ 76.16 **6**
$C_8H_{22}CoN_6O_4^{+}$, Br^{-} 76.19 **4**
$C_8H_{22}CoN_6O_4^{+}$, Br^{-} 76.17 **6**
$C_8H_{22}CoN_6O_4^{+}$, ClO_4^{-} 76.54 **3**
$C_8H_{22}CoN_6O_4^{+}$, ClO_4^{-} 76.20 **4**
$C_8H_{22}CoN_6O_4^{+}$, ClO_4^{-} 76.21 **4**
$C_8H_{22}CuN_{10}O_4^{2+}$, $2Cl^{-}$ 83.29 **3**
$C_8H_{22}N_2S_2^{2+}$, $2Cl^{-}$ 3.25 **6**
$C_8H_{22}N_{10}NiO_4^{2+}$, $2Cl^{-}$, $2H_2O$ 83.30 **3**
$C_8H_{23}ClCoN_5^{2+}$, Cl^{-} , ClO_4^{-} 76.55 **3**
$C_8H_{23}ClCoN_5^{2+}$, $2ClO_4^{-}$ 76.49 **2**
$C_8H_{23}ClCoN_5^{2+}$, $2ClO_4^{-}$ 76.26 **5**
$C_8H_{23}ClCoN_5^{2+}$, $2ClO_4^{-}$ 76.27 **5**
$C_8H_{23}ClCoN_5^{2+}$, Cl_4Zn^{2-} 76.22 **4**
$C_8H_{23}CoN_8^{2+}$, $2NO_3^{-}$, H_2O 76.23 **4**
$C_8H_{24}Al_2Mg$ 67.8 **3**
$C_8H_{24}Al_2N_2$ 68.5 **3**
$C_8H_{24}Al_2N_2$ 68.14 **5**
$C_8H_{24}Al_3Br_5O_6Si_4$ 68.16 **2**
$C_8H_{24}Br_2N_4Ni$ 76.56 **3**
$C_8H_{24}Cl_2Co_2O_8^{2+}$, $2Cl^{-}$ 84.24 **7**
$C_8H_{24}Cl_2Cu_2P_4S_4$ 85.9 **7**
$C_8H_{24}Cl_2Ni_2O_8^{2+}$, $2Cl^{-}$ 84.25 **7**
$C_8H_{24}Cl_4N_8P_4$ 64.12 **5**
$C_8H_{24}CuN_4^{2+}$, $2ClO_4^{-}$ 83.17+ **5**
$C_8H_{24}CuN_4^{2+}$, $2NO_3^{-}$ 76.53 **3**
$C_8H_{24}CuN_4^{2+}$, O_4S^{2-} , $4H_2O$ 76.38 **7**
$C_8H_{24}CuN_4^{2+}$, $2CNS^{-}$ 76.57 **3**
$C_8H_{24}CuN_4^{2+}$, $2CNSe^{-}$ 76.18 **6**
$C_8H_{24}CuN_4O_2^{2+}$, $2Cl^{-}$ 76.28 **5**
$C_8H_{24}CuN_4O_2^{2+}$, $2ClO_4^{-}$ 76.19 **6**
$C_8H_{24}CuP_4S_4^{+}$, BF_4^{-} 85.10 **5**
$C_8H_{24}Cu_4N_{12}$ 83.65 **2**
$C_8H_{24}F_4N_8P_4$ 64.13 **5**
$C_8H_{24}F_4N_8P_4$ 64.11 **6**
$C_8H_{24}FeN_2P_4S_4$ 85.9 **4**
$C_8H_{24}Ga_2N_2O_2$ 68.11 **7**
$C_8H_{24}I_2InO_4S_4^{+}$, I_4In^{-} 68.6 **3**
$C_8H_{24}LaN_3O_{13}S_4$ 84.9 **5**
$C_8H_{24}Mo_2^{4-}$, $4Li^{+}$, $4C_4H_8O$ 71.8 **6**
$C_8H_{24}N_2NiP_4S_4$ 85.14 **3**
$C_8H_{24}N_3NdO_{13}S_4$ 84.10 **5**
$C_8H_{24}N_4NiO_6^{2+}$, $2Cl^{-}$ 84.7 **4**
$C_8H_{24}N_4Ni_3S_4^{2+}$, $2Cl^{-}$ 83.31 **3**
$C_8H_{24}N_4O_8P_4$ 64.12 **4**
$C_8H_{24}N_4P_4$ 64.29 **2**
$C_8H_{24}N_6NiO_4$ 76.50 **2**
$C_8H_{24}O_2Si_4$ 63.6 **2**
$C_8H_{24}O_4Si_4$ 63.7 **2**
$C_8H_{24}O_4Zn_4$ 71.17 **2**
$C_8H_{24}O_6Si_5$ 63.8 **2**
$C_8H_{24}O_6Si_5$ 63.3 **7**
$C_8H_{24}O_8Si_8$ 63.9 **2**
$C_8H_{24}O_{12}Si_8$ 63.10 **2**
$C_8H_{25}ClCoN_5^{2+}$, Cl_4Zn^{2-} 76.58 **3**
$C_8H_{25}ClCoN_5^{2+}$, $2I^{-}$, H_2O 76.24 **4**
$C_8H_{25}Cl_3CuN_4P_4$ 83.32 **3**
$2C_8H_{25}N_4P_4^{+}$, Cl_4Co^{2-} 64.20 **3**
$C_8H_{26}B_6P_2Pt$ 86.4 **7**
$C_8H_{26}CoN_6^{3+}$, $3Br^{-}$ 76.25 **4**
$C_8H_{26}CoN_6^{3+}$, $C_6CoN_6^{3-}$, $2H_2O$ 76.29 **5**
$C_8H_{26}CuN_6^{2+}$, $2Br^{-}$, H_2O 76.59 **3**
$C_8H_{26}CuN_6^{2+}$, $2NO_3^{-}$ 76.51 **2**
$C_8H_{26}N_6Ni^{2+}$, $2Cl^{-}$, H_2O 76.60 **3**
$C_8H_{26}N_6Zn^{2+}$, $2Br^{-}$, H_2O 76.39 **7**
$C_8H_{26}N_6Zn^{2+}$, $2NO_3^{-}$ 76.26 **4**
$C_8H_{28}Au_4O_4$ 71.18 **2**
$C_8H_{28}Br_2N_4Pt^{2+}$, $C_8H_{28}N_4Pt^{2+}$, $4Br^{-}$ 83.67 **2**
$C_8H_{28}Cl_2N_4Pt^{2+}$, $C_8H_{28}N_4Pt^{2+}$, $4Cl^{-}$, $4H_2O$ 83.68 **2**
$C_8H_{28}N_4Pt^{2+}$, Cl_4Pt^{2-} 83.33 **3**
$C_8H_{28}N_4Pt^{2+}$, $C_8H_{28}Br_2N_4Pt^{2+}$, $4Br^{-}$ 83.67 **2**
$C_8H_{28}N_4Pt^{2+}$, $C_8H_{28}Cl_2N_4Pt^{2+}$, $4Cl^{-}$, $4H_2O$ 83.68 **2**
$C_8H_{28}N_4Si_4$ 63.11 **2**
$C_8H_{30}B_{18}Cr^{-}$, Cs^{+} , H_2O 71.5 **4**
$C_8H_{30}B_{18}Ni$ 62.14 **3**
$C_8H_{32}B_4P_4$ 62.45 **2**
$C_8H_{32}B_{20}Ti^{2-}$, $2C_4H_{12}N^{+}$, $2C_3H_6O$ 71.8 **7**
$C_8H_{32}Cl_2N_8Ni_2^{2+}$, $2ClO_4^{-}$ 76.27 **4**
$C_8H_{32}Co_2N_{10}O_6^{2+}$, $2NO_3^{-}$, $4H_2O$ 76.20 **6**
$C_8H_{34}Co_2N_9O_2^{3+}$, $3CNS^{-}$, H_2O 76.30 **5**
$C_8H_{34}Co_2N_9O_2^{4+}$, $4NO_3^{-}$, H_2O 76.28 **4**
$C_8H_{34}Co_2N_9O_4S^{3+}$, $3Br^{-}$ 76.29 **4**
$C_8H_{34}Co_2N_{10}O_2^{4+}$, $4NO_3^{-}$ 76.54+ **2**
$C_8H_{35}Co_2N_9O^{4+}$, $4NO_3^{-}$, H_2O 76.30 **4**
$C_8H_{35}Co_2N_9O_2^{4+}$, $4NO_3^{-}$, $2H_2O$ 76.56 **2**
$C_8H_{40}B_{40}Co^{-}$, $C_8H_{20}N^{+}$ 62.6 **5**
$C_8H_{42}B_{34}Co_3^{3-}$, $3C_8H_{20}N^{+}$ 62.15 **3**

C_9

$C_9FN_6^{-}$, $C_{24}H_{20}As^{+}$ 65.40 **2**
$C_9F_8FeO_3$ 75.8 **2**
$C_9F_{12}Hg_4O_8$ 71.9 **7**
C_9S_9 39.10 **7**
$C_9H_2N_5^{-}$, K^{+} 7.11 **4**
$C_9H_3Cl_3N_2S$ 7.6 **5**
$C_9H_4N_3O_2^{-}$, K^{+} 7.26 **1**
$C_9H_4O_3$ 27.4 **1**
$C_9H_5BrN_4O_4$ 32.25 **1**
C_9H_5BrO 19.27 **1**
$C_9H_5BrO_3$, H_2O 38.32 **1**
C_9H_5ClINO 35.6 **5**
C_9H_5ClINO 35.5 **6**
$C_9H_5ClN_4O_4$ 32.24 **3**
C_9H_5ClO 19.28 **1**
$C_9H_5CoF_6S_2$ 73.9 **2**
$C_9H_5CoN_2S_2$ 73.10 **2**
$C_9H_5F_6FeO_2P$ 73.2 **7**
$C_9H_5F_6FeO_3P$ 73.3 **7**
$C_9H_5FeNO_4$ 83.13 **6**
$C_9H_5NbO_4$ 73.11 **2**
$C_9H_5O_4Rh$ 84.26 **7**
$C_9H_5O_4V$ 73.12 **2**
$C_9H_5O_5S^{-}$, Rb^{+} 27.3 **6**
C_9H_6ClN 35.16 **1**
C_9H_6ClN 35.17 **1**
C_9H_6ClN 35.18 **1**
C_9H_6ClNO 35.6 **6**
$C_9H_6ClN_3$ 33.30 **7**
$C_9H_6Cl_2N_3O_5S_2$ 41.13 **5**
$C_9H_6CrO_3$ 74.4+ **2**

$C_9H_6CrO_3$ 74.1 + **6**
$C_9H_6CrO_{12}^{3-}$, $C_9H_{30}CoN_6^{3+}$, $3H_2O$ 76.32 **4**
$C_9H_6Fe_2N_2O_7$ 83.69 **2**
$C_9H_6INO_4S$ 35.2 **3**
$C_9H_6NO_2S^-$, Na^+ 41.24 **1**
$C_9H_6N_2O_2S$ 40.2 **7**
$C_9H_6N_2O_3$ 8.45 **1**
$C_9H_6N_6S$, C_5H_5N 35.7 **5**
C_9H_6OS 39.58 **1**
C_9H_6OS , $C_6H_3N_3O_6$ 60.8 **4**
$C_9H_6O_2$ 1.107 **1**
$C_9H_6O_2$ 38.33 + **1**
$C_9H_6O_2$ 38.13 **6**
$C_9H_6O_2$ 38.12 **7**
$C_9H_6O_3$, H_2O 38.35 **1**
$C_9H_6O_4$ 27.5 **1**
$C_9H_6O_6$ 13.41 **1**
$C_9H_6O_6$ 13.13 **7**
$C_9H_6O_6$, xH_2O 13.11 **6**
$C_9H_6O_6$, $2H_2O$ 13.8 **4**
$C_9H_6O_6$, $2H_2O$ 13.14 + **7**
$C_9H_6O_6$, $0.2C_6H_3N_3O_7$, $1.5H_2O$ 13.12 **6**
$C_9H_6S_2$ 39.17 **3**
$C_9H_7^-$, $C_6H_{16}N_2$, Li^+ 27.2 **7**
$C_9H_7BrMoO_2$ 75.2 **7**
$C_9H_7BrN_2O_2$ 40.15 **1**
$C_9H_7BrN_2O_2$ 40.16 **1**
$C_9H_7Br_3NNi^-$, $C_{16}H_{36}N^+$ 3.80 **1**
C_9H_7ClHgO 71.7 **5**
$C_9H_7ClMoO_2$ 75.3 **7**
$C_9H_7ClO_2$ 1.23 **5**
$C_9H_7ClO_2$ 1.24 **5**
$C_9H_7ClO_2$ 1.26 **7**
$C_9H_7Cl_2HgNO$ 84.30 **2**
$C_9H_7Cl_2NO_3S$ 42.4 **6**
$C_9H_7Cl_3MoO_2Sn$ 75.4 **7**
$C_9H_7F_2NO_2$ 10.5 **3**
$C_9H_7FeNO_3$ 75.1 **5**
$C_9H_7IO_4$ 42.5 **5**
$C_9H_7MnO_3$ 75.1 **3**
C_9H_7N 31.1 **7**
$3C_9H_7N$, CHI_3 60.3 **2**
C_9H_7NO , $C_6H_3N_3O_6$ 60.29 **3**
$2C_9H_7NO$, $C_6Cl_4O_2$ 60.107 **2**
$C_9H_7NO_2$ 40.17 **1**
$C_9H_7NO_2$ 40.18 **1**
$C_9H_7NO_2$ 40.6 **3**
$C_9H_7NO_2$ 36.1 **4**
$C_9H_7NO_2$ 35.7 **6**
$C_9H_7NO_2Se$ 42.5 **6**
$C_9H_7NO_6SW$ 85.10 **7**
$C_9H_7N_5$ 32.10 **7**
$C_9H_7N_6Pt^+$, $C_{24}H_{20}B^-$ 71.8 **5**
$C_9H_7N_7O_2S$, $2H_2O$ 44.21 **7**
$C_9H_7O_2^-$, $C_9H_8O_2$, H_4N^+ 2.91 **1**
$C_9H_7O_4^-$, $C_9H_8O_4$, K^+ 14.13 + **1**
$C_9H_7O_4^-$, $C_9H_8O_4$, K^+ 14.3 **5**
$C_9H_7O_4^-$, $C_9H_8O_4$, Rb^+ 14.12 **1**
$C_9H_7O_4^-$, $C_9H_8O_4$, Rb^+ 14.4 **5**
$C_9H_7S_2^+$, Br^- 39.16 **3**
$C_9H_7S_2^+$, Cl^- , $0.79H_2O$ 39.10 **4**
$C_9H_7S_2^+$, I^- 39.59 **1**
$C_9H_7S_2^+$, I^- 39.60 **1**
$C_9H_7S_2^+$, CNS^- 39.18 **3**
$C_9H_8Ag_2^{2+}$, $2ClO_4^-$ 75.2 **3**
C_9H_8BrNO 13.7 **3**
C_9H_8BrNO 32.14 **6**
$C_9H_8BrN_3O$ 32.25 **3**
C_9H_8ClNO 1.25 **3**
C_9H_8ClNO 17.9 **6**
$C_9H_8ClNO_3S_2$ 39.11 **7**
$C_9H_8FeN_2O_4$ 71.9 **5**
$C_9H_8FeN_2O_5$ 75.5 **7**
$C_9H_8FeO_4$ 71.5 **3**
$C_9H_8FeO_5$ 72.11 **3**
$C_9H_8N^+$, Cl^- 35.19 **1**
$C_9H_8N^+$, Cl^- , H_2O 35.20 **1**
$C_9H_8N^+$, $C_{12}H_4N_4^-$, $C_{12}H_4N_4$ 35.4 **3**
$C_9H_8N^+$, $C_{32}H_{16}CeF_{12}O_8S_4^-$ 77.19 **6**
$2C_9H_8N^+$, Br_6Sb^- , Br_3^- 35.8 **6**
$2C_9H_8N^+$, $CdCl_4^{2-}$ 35.5 **3**
$2C_9H_8N^+$, $Cl_8N_2O_2Pt_2^{2-}$ 35.10 **7**
$2C_9H_8N^+$, $C_{10}N_6^{2-}$ 60.20 **4**
$C_9H_8NO^+$, Cl_3OSe^- 35.22 **1**
$C_9H_8NO_3^-$, $C_4H_5N_3O^+$, H_2O 44.4 **5**
$C_9H_8NO_5S^-$, H_4N^+ 11.12 **7**
$C_9H_8N_2OS$ 41.25 **1**
$C_9H_8N_2OS$ 41.6 **3**
$C_9H_8N_2OS$ 41.14 **5**
$C_9H_8N_2OS$ 41.15 **5**
$C_9H_8N_2O_2$ 35.23 **1**
$C_9H_8N_2O_4S$ 40.9 **6**
$C_9H_8N_2O_6$ 13.9 **4**
$C_9H_8N_3O_2S_2^-$, $C_7H_{11}N_2O_2S^+$ 11.5 **5**
$C_9H_8N_4O_5$ 32.11 **7**
$C_9H_8N_4O_7$ 16.9 **5**
$C_9H_8N_4S$ 32.10 **5**
$C_9H_8O_2$ 22.7 **1**
$C_9H_8O_2$, $C_9H_7O_2^-$, H_4N^+ 2.91 **1**
$C_9H_8O_3$ 1.108 **1**
$C_9H_8O_3$ 52.1 **1**
$C_9H_8O_3$ 17.5 **3**
$C_9H_8O_3$ 31.3 **5**
$C_9H_8O_3S$ 39.19 **3**
$C_9H_8O_4$ 13.42 **1**
$C_9H_8O_4$ 52.2 **1**
$C_9H_8O_4$ 13.10 **4**
$C_9H_8O_4$ 13.11 **5**
$C_9H_8O_4$, $C_9H_7O_4^-$, K^+ 14.13 + **1**
$C_9H_8O_4$, $C_9H_7O_4^-$, K^+ 14.3 **5**
$C_9H_8O_4$, $C_9H_7O_4^-$, Rb^+ 14.12 **1**
$C_9H_8O_4$, $C_9H_7O_4^-$, Rb^+ 14.4 **5**
$C_9H_8O_6$ 38.4 **3**
$C_9H_8S_2$ 39.20 **3**
C_9H_9BrOS 39.12 **7**
C_9H_9BrOS 39.13 **7**
$C_9H_9BrO_3$ 1.17 **6**
$C_9H_9BrO_3$, $C_{10}H_{12}O_4$ 1.119 **1**
$C_9H_9Br_2NO_3$ 50.2 **3**
$C_9H_9Br_2NO_3$ 50.4 **5**
$C_9H_9ClN_2O_3$ 15.6 **4**
$C_9H_9ClO_2$ 1.27 **7**
$C_9H_9ClO_3$ 59.2 **3**
$C_9H_9ClO_3$ 1.18 **6**
$C_9H_9Cl_2Rh$ 72.8 **5**
$C_9H_9Cl_3$ 19.29 **1**
$C_9H_9Cl_3$ 19.7 + **5**
$C_9H_9Cl_3O$ 27.4 **3**
$C_9H_9Cl_3O_3$ 38.36 **1**
$C_9H_9CrO_6$ 84.27 **7**
$C_9H_9CuNO_4$, $0.5H_2O$ 78.1 **2**

$C_9H_9CuNO_4$, $4H_2O$	78.2	**2**
$(C_9H_9CuN_3O_2)_n$	83.23	**7**
$C_9H_9FO_3$	1.19	**6**
$C_9H_9FO_3$	1.20	**6**
$C_9H_9F_4Rh$	71.6	**4**
$C_9H_9FeO_4{}^+$, F_6P^-	72.9	**5**
$C_9H_9Fe_2O_6S_3{}^+$, $C_{16}F_{24}Fe_2S_8{}^-$	85.12	**5**
$C_9H_9I_2NO_3$, $2H_2O$	48.76	**1**
C_9H_9N , $C_6H_3N_3O_6$	60.108	**2**
$C_9H_9NO_2S$	33.21	**4**
$C_9H_9NO_3$	13.9	**3**
$C_9H_9NO_3$	35.8	**4**
$C_9H_9NO_3$	48.38	**4**
$C_9H_9NO_3$	48.44	**6**
$C_9H_9NO_3S$	42.4	**4**
$C_9H_9NO_4$	13.13	**6**
$C_9H_9NO_4$	33.31	**7**
$C_9H_9NS_2$	41.8	**7**
$C_9H_9N_2OS$	41.8	**4**
$C_9H_9N_2O_7PS^-$, $C_6H_{16}N^+$	47.1	**3**
$C_9H_9N_2S^+$, Br^- , H_2O	41.10	**6**
$C_9H_9N_3O_2S_2$	41.7	**3**
$C_9H_9N_3O_2S_2$	41.9+	**4**
$C_9H_9N_3O_2S_2$, $C_6H_8N_2O_2S$	60.30	**3**
$C_9H_9N_3O_2S_2$, $C_7H_8N_4O_2$	60.31	**3**
$C_9H_9N_4O^+$, Br^-	44.32	**3**
$C_9H_9N_5O$	36.1	**7**
$C_9H_9O_4{}^-$, K^+ , $2H_2O$	52.3+	**1**
$C_9H_{10}BrNO$	13.43	**1**
$C_9H_{10}Cl_4O_2$	20.10	**3**
$C_9H_{10}MoO_2S$	72.12	**3**
$C_9H_{10}NO_6S^-$, K^+ , $2H_2O$	48.35	**3**
$C_9H_{10}N_2OSe$	8.46	**1**
$C_9H_{10}N_2O_3$	15.4	**5**
$C_9H_{10}N_2O_3$	17.10	**6**
$C_9H_{10}N_2O_5$	36.2	**5**
$C_9H_{10}N_2O_5$	47.3	**5**
$C_9H_{10}N_2O_5$	47.2	**6**
$C_9H_{10}N_2O_5$	47.2	**7**
$C_9H_{10}N_2O_8P^-$, $C_6H_{16}N^+$	47.1	**1**
$C_9H_{10}N_4O_5$	13.16	**7**
$C_9H_{10}N_4O_5$, $0.5H_2O$	44.18	**4**
$C_9H_{10}N_6OS$	32.11	**5**
$C_9H_{10}O_2$	13.44	**1**
$C_9H_{10}O_2$	13.11	**4**
$C_9H_{10}O_3$	13.12	**5**
$C_9H_{10}O_3$	1.21	**6**
$C_9H_{10}O_3$	1.28	**7**
$C_9H_{10}O_4$	18.3	**4**
$C_9H_{10}O_4$	31.3+	**6**
$C_9H_{10}O_4$	38.14	**6**
$C_9H_{10}O_4$	1.29	**7**
$C_9H_{10}O_4$	1.30	**7**
$C_9H_{10}O_4$	13.17	**7**
$C_9H_{10}O_4$	50.1	**7**
$C_9H_{10}O_6S_2{}^{2-}$, $2K^+$, $2H_2O$	11.8	**5**
$C_9H_{10}S_3$	39.14	**7**
$C_9H_{11}BrN_2O_5$	47.2	**1**
$C_9H_{11}BrN_2O_6$	47.3	**1**
$C_9H_{11}BrN_2O_6$	47.3	**6**
$C_9H_{11}BrN_2O_6$, C_2H_6OS	47.4	**1**
$C_9H_{11}BrN_2O_6$, $C_{10}H_{13}N_5O_4$, H_2O	47.22	**1**
$C_9H_{11}ClCuN_6$	83.19	**5**
$C_9H_{11}ClN_2O_5$	47.1	**4**
$C_9H_{11}ClN_2O_5$	47.4	**5**
$C_9H_{11}ClN_2O_6$	47.4	**3**
$C_9H_{11}ClN_2O_6$	47.4	**6**
$C_9H_{11}FN_2O_5$	47.5	**1**
$C_9H_{11}FN_2O_5S$	47.3	**7**
$C_9H_{11}F_4O_2Rh$	72.13	**3**
$C_9H_{11}HgNO_2S$	71.10	**7**
$C_9H_{11}IN_2O_5$	47.6	**1**
$C_9H_{11}IN_2O_6$	47.5	**3**
$C_9H_{11}NO$	16.40	**1**
$C_9H_{11}NO$	16.41	**1**
$C_9H_{11}NO$	20.10	**6**
$C_9H_{11}NO$	10.1	**7**
$C_9H_{11}NO_2$	15.24	**1**
$C_9H_{11}NO_2$	10.7	**3**
$C_9H_{11}NO_2$	13.12	**4**
$C_9H_{11}NO_2$	16.6	**4**
$C_9H_{11}NO_2$	13.14	**6**
$C_9H_{11}NO_2$	35.11	**7**
$C_9H_{11}NO_2S_2$	4.4	**7**
$C_9H_{11}NO_3$	18.26	**1**
$C_9H_{11}NO_3$	48.57	**5**
$C_9H_{11}NO_3$	48.58+	**5**
$C_9H_{11}NO_3$	48.41	**7**
$C_9H_{11}NO_3$	48.42	**7**
$C_9H_{11}NO_4$	48.40	**4**
$C_9H_{11}NO_4$	48.45	**6**
$C_9H_{11}NO_6$	50.3	**3**
$C_9H_{11}NS$	4.16	**1**
$C_9H_{11}N_2{}^+$, $C_{12}H_4N_4{}^-$	35.9	**4**
$C_9H_{11}N_2O_3{}^+$, Br^- , H_2O	33.24	**3**
$C_9H_{11}N_2O_9P^{2-}$, Ba^{2+} , $8.9H_2O$	47.7	**1**
$C_9H_{11}N_2O_9P^{2-}$, $2Na^+$, $4H_2O$	47.2	**4**
$C_9H_{11}N_2O_9P^{2-}$, $2Na^+$, $7H_2O$	47.4	**7**
$C_9H_{11}N_3OS_4$	41.26	**1**
$2C_9H_{11}N_3OS_4$, I_2	60.109	**2**
$C_9H_{11}N_3O_2$	9.4	**3**
$C_9H_{11}N_3O_2$	9.8	**5**
$C_9H_{11}N_3O_7P^-$, Na^+ , $2H_2O$	47.5	**5**
$C_9H_{11}O_4P$	46.15	**1**
C_9H_{12}	23.1	**1**
C_9H_{12} , C_6F_6	60.21	**4**
$C_9H_{12}Ag_3{}^{3+}$, $3NO_3{}^-$	75.9	**2**
$C_9H_{12}Cl_2N_2O_4$	37.5	**6**
$C_9H_{12}Cl_2O$	38.5	**5**
$C_9H_{12}CoN_3O_6$	82.34	**2**
$C_9H_{12}FeO$	72.14	**3**
$C_9H_{12}FeO$	72.10	**5**
$C_9H_{12}FeO_9P_4$	86.2	**5**
$C_9H_{12}Hg_4O_8$	71.11	**7**
$C_9H_{12}IN_3O_4$	47.6	**5**
$C_9H_{12}NO^+$, Br^-	16.10	**5**
$C_9H_{12}NO_2{}^+$, Cl^-	48.41	**4**
$C_9H_{12}NO_2{}^+$, $C_{12}H_8N_2O_8P^-$	13.13	**4**
$2C_9H_{12}NO_2{}^+$, $C_8N_4NiS_4{}^{2-}$	85.11	**6**
$C_9H_{12}NO_3{}^+$, Br^-	17.13	**7**
$C_9H_{12}NO_3{}^+$, Cl^-	48.60	**5**
$C_9H_{12}NO_3{}^+$, Cl^-	17.14	**7**
$C_9H_{12}NO_3{}^+$, Cl^- , H_2O	3.18	**4**
$C_9H_{12}NO_3{}^+$, I^-	48.79	**1**
$C_9H_{12}NO_4{}^+$, Cl^-	48.38	**3**
$C_9H_{12}NO_4{}^+$, Cl^-	48.43	**7**
$C_9H_{12}N_2$	22.8	**1**
$C_9H_{12}N_2O$	16.11	**5**
$C_9H_{12}N_2OS_3$	41.27	**1**
$C_9H_{12}N_2O_2$	36.2	**7**
$C_9H_{12}N_2O_4S_2$, H_2O	47.7	**3**
$C_9H_{12}N_2O_4S_2$, H_2O	47.3	**4**

$C_9H_{12}N_2O_5$ 47.7 **5**
$C_9H_{12}N_2O_5S$ 47.8 **3**
$C_9H_{12}N_2O_5S$ 47.4 **4**
$C_9H_{12}N_2O_5S$, H_2O 47.5 **4**
$C_9H_{12}N_2O_5S$, $0.5H_2O$ 47.5 **6**
$C_9H_{12}N_2O_5S$, $1.5H_2O$ 47.9 **3**
$C_9H_{12}N_2O_6$ 40.7 **3**
$C_9H_{12}N_2O_6$ 47.6 **6**
$C_9H_{12}N_2O_6$ 47.5 **7**
$C_9H_{12}N_2O_6$, H_2O 47.11 **3**
$C_9H_{12}N_2O_7$ 47.7 **6**
$C_9H_{12}N_2O_{12}P_2^{2-}$, $2K^+$, $2H_2O$ 47.6 **7**
$C_9H_{12}N_2S$ 33.25 **3**
$C_9H_{12}N_3O_4^+$, Cl^- 47.9 **5**
$C_9H_{12}N_3O_4+$, Cl^- 47.7 **7**
$C_9H_{12}N_3O_4^+$, Cl^- 47.8 **7**
$C_9H_{12}N_3O_6$ 47.8 **6**
$C_9H_{12}N_3O_8P^{2-}$, Ba^{2+} , $8H_2O$ 47.9 **6**
$C_9H_{12}N_4O$ 44.65 **1**
$C_9H_{12}N_4O_3$ 44.66 + **1**
$C_9H_{12}N_4O_3$, $C_{16}H_{10}$ 60.110 **2**
$2C_9H_{12}N_4O_3$, $C_{20}H_{12}$ 60.111 **2**
$2C_9H_{12}N_4O_3$, $C_{24}H_{12}$ 60.112 **2**
$C_9H_{12}N_6O_4$, H_2O 47.9 **7**
$C_9H_{12}N_6O_4$, $0.5H_2O$ 47.10 **7**
$C_9H_{12}O_2$ 27.2 **4**
$C_9H_{12}O_2S_2$ 39.61 **1**
$C_9H_{12}O_4$ 27.5 **5**
$C_9H_{12}O_4S_3$ 39.62 **1**
$C_9H_{13}BrN_3O_4$, $C_{10}H_{13}N_5O_4$ 47.23 **1**
$C_9H_{13}BrO$ 52.4 **1**
$C_9H_{13}ClO$ 31.3 **1**
$C_9H_{13}ClO$ 52.5 **1**
$C_9H_{13}ClO_4$ 38.11 **4**
$C_9H_{13}NO$ 37.7 **7**
$C_9H_{13}NO_2S_2$ 4.2 **6**
$C_9H_{13}NO_3$ 17.15 **7**
$C_9H_{13}NO_3^-$, K^+ , $0.33H_2O$ 12.7 **3**
$C_9H_{13}N_2O_2^+$, Br^- , H_2O 20.20 **1**
$C_9H_{13}N_2O_3^-$, Na^+ 43.7 **4**
$C_9H_{13}N_3O_4$ 47.11 **7**
$C_9H_{13}N_3O_4S$, $2H_2O$ 47.12 **3**
$C_9H_{13}N_3O_4Se$, H_2O 47.12 **7**
$C_9H_{13}N_3O_5$ 47.9 **1**
$C_9H_{13}N_3O_5$ 37.2 **4**
$C_9H_{13}N_3O_5$ 47.10 **6**
$C_9H_{13}N_3O_5$, $C_7H_6O_3$ 60.7 **7**
$C_9H_{13}N_3O_6$ 37.3 **4**
$C_9H_{13}N_3O_6$, H_2O 50.5 **5**
$C_9H_{13}O_2^-$, K^+ , H_2O 27.6 **5**
$C_9H_{13}Rh$ 72.24 **2**
$C_9H_{14}BrN_3O$ 32.26 **3**
$C_9H_{14}Br_2$ 31.2 **3**
$C_9H_{14}CuN_4O_4$, H_2O 82.35 **2**
$C_9H_{14}CuN_4O_6$ 83.15 **6**
$(C_9H_{14}CuN_6O_5S)_n$ 83.24 **7**
$C_9H_{14}IN_3O_4$ 45.26 **3**
$C_9H_{14}N^+$, Cl_3Cu^- 3.22 **7**
$2C_9H_{14}N^+$, Cl_4Cu^{2-} 3.23 + **7**
$2C_9H_{14}N^+$, O_4S^{2-} 3.19 **4**
$2C_9H_{14}N^+$, $C_6H_6I_6O_4Rh_2^{2-}$ 71.4 **7**
$2C_9H_{14}N^+$, $C_{16}H_{20}NiO_8S_4^{2-}$ 85.31 **7**
$3C_9H_{14}N^+$, $Cl_9Rh_2^{3-}$ 16.12 **5**
$4C_9H_{14}N^+$, $C_{24}Cu_8N_{12}S_{12}^{4-}$ 3.69 **1**
$C_9H_{14}NO^+$, $C_{10}H_{14}BrO_4S^-$ 16.7 **4**
$C_9H_{14}NO_2$ 37.6 **6**
$C_9H_{14}NO_2^+$, Cl^- 17.16 **7**
$C_9H_{14}NO_3^+$, $C_4H_5O_6^-$ 17.14 **5**
$C_9H_{14}NO_5P$ 33.37 **5**
$C_9H_{14}NO_7P$ 46.10 **6**
$C_9H_{14}N_2O$ 33.60 **1**
$C_9H_{14}N_2O_3$ 43.3 **5**
$C_9H_{14}N_2O_4S_2$ 47.13 **7**
$C_9H_{14}N_2O_5S$ 47.11 **6**
$C_9H_{14}N_2O_6$, $0.5H_2O$ 47.7 + **4**
$C_9H_{14}N_3O^+$, $H_2O_4P^-$ 33.27 **3**
$C_9H_{14}N_3O_4^+$, Cl^- 47.13 **3**
$C_9H_{14}N_3O_5^+$, Cl^- 47.12 **5**
$C_9H_{14}N_3O_5^+$, $C_{13}H_{14}NO_6^-$, $2H_2O$ 47.14 **7**
$C_9H_{14}N_3O_7P$, H_2O 47.9 + **4**
$C_9H_{14}N_3O_8P$ 47.10 + **1**
$C_9H_{14}N_3O_8P$, $3H_2O$ 47.13 **5**
$C_9H_{14}N_5^+$, Cl^- 35.24 **1**
$C_9H_{14}N_5O_3^+$, CHO_2^- 47.15 **7**
$C_9H_{14}O$ 20.7 **7**
$C_9H_{14}O_2$ 27.7 **5**
$C_9H_{14}O_3$ 38.37 **1**
$C_9H_{14}O_4$ 20.8 **7**
$C_9H_{14}O_4$ 20.9 **7**
$C_9H_{14}O_4S_2$ 39.63 **1**
$C_9H_{14}O_5$ 1.110 **1**
$C_9H_{14}O_5$ 1.9 **4**
$C_9H_{14}O_5$ 45.24 **6**
$C_9H_{14}O_5$ 38.13 **7**
$C_9H_{15}AsO_3S_6$ 65.16 **2**
$C_9H_{15}ClO_5S_2$ 45.26 **7**
$C_9H_{15}CoO_3S_6$ 80.7 **2**
$C_9H_{15}CoS_9$ 80.5 **5**
$C_9H_{15}CoS_9$ 80.3 **6**
$C_9H_{15}Cr$ 72.15 **3**
$C_9H_{15}CrO_3S_6$ 80.2 **4**
$C_9H_{15}CrS_9$ 80.4 **7**
$C_9H_{15}CuN_5O_5$, H_2O 82.36 **2**
$C_9H_{15}FeO_3S_6$ 80.3 + **3**
$C_9H_{15}NO_4$ 32.12 **7**
$C_9H_{15}N_2O_2$ 32.12 **4**
$C_9H_{15}N_3O_{11}P_2$, H_2O 47.16 **7**
$C_9H_{15}N_4O_4^+$, Br^- 32.15 **6**
$C_9H_{15}N_5S_4$ 41.16 **5**
$C_9H_{15}N_7S$ 35.9 **6**
$C_9H_{15}N_7S$ 35.10 **6**
$C_9H_{15}O_2Rh$ 72.16 **3**
$C_9H_{15}O_3PbS_6^-$, $C_8H_{20}N^+$ 69.15 **4**
$C_9H_{15}O_3S_6Sb$ 66.13 **2**
$C_9H_{15}O_3S_6Te^-$, $C_8H_{20}N^+$ 70.5 **7**
$C_9H_{15}O_4^-$, Rb^+ , $1.5H_2O$ 2.92 **1**
$C_9H_{15}RhS_6$ 85.11 **7**
$C_9H_{16}Cl_2N_2O_2$ 1.31 **7**
$C_9H_{16}CuN_2O_2$ 84.10 **6**
$C_9H_{16}NO_2$ 33.30 + **6**
$C_9H_{16}NO_2^+$, I^- 58.1 **7**
$C_9H_{16}NO_3$ 32.16 **6**
$C_9H_{16}N_2$ 33.28 **3**
$C_9H_{16}N_4S$ 32.17 **6**
$C_9H_{16}O$, Cl_2Hg 23.1 **4**
$C_9H_{16}O$, $6C_{18}H_{20}OS$ 61.1 **4**
$C_9H_{16}O_3S$ 39.9 **6**
$C_9H_{16}O_3S_2$ 1.22 **6**
$C_9H_{16}O_4$ 1.113 **1**
$C_9H_{16}O_4$ 1.111 + **1**
$C_9H_{16}O_4S_2$ 45.27 **7**
$C_9H_{16}O_6$ 45.25 **6**

$C_9H_{16}O_6$ 45.26 **6**
$C_9H_{16}O_7$ 45.19 **5**
$C_9H_{16}O_7$ 45.27 **6**
$C_9H_{16}O_{11}P_2$ 46.11 **6**
$C_9H_{17}BBrN_2^+$, F_6P^- 62.16 **3**
$C_9H_{17}ClOS$ 39.15 **7**
$2C_9H_{17}CuN_2O_3^+$, O_4S^{2-} , $2H_2O$ 78.1 **5**
$C_9H_{17}NO$ 34.11 **7**
$C_9H_{17}NO$, H_2O 33.32+ **6**
$C_9H_{17}NO$, H_3O^+ , Cl^- 34.12 **7**
$C_9H_{17}NS$ 34.1 **5**
$C_9H_{17}N_2^+$, Cl^- , $2H_2O$ 3.25 **7**
$C_9H_{17}N_2^+$, ClO_4^- 3.26 **7**
$C_9H_{17}N_2O_2$ 32.13 **7**
$C_9H_{17}N_2O_2$ 32.14 **7**
$C_9H_{17}N_3O_4$, $0.5H_2O$ 48.44 **7**
$C_9H_{17}OS_2^+$, ClO_4^- 39.21 **3**
$C_9H_{17}O_{10}S^-$, Rb^+ 45.54 **1**
$C_9H_{18}AuBr_2NS_2$ 80.8 **2**
$C_9H_{18}BNO_3$, $3H_2O$ 62.7 **5**
$C_9H_{18}Br_2CuNS_2$ 80.9 **2**
$C_9H_{18}Cl_2N_2^{2+}$, $2ClO_4^-$ 37.4 **4**
$2C_9H_{18}CuN_6S_3^+$, O_4S^{2-} 79.22 **5**
$C_9H_{18}FeO_9P_2$ 86.5 **7**
$C_9H_{18}N^+$, I^- 35.25 **1**
$C_9H_{18}N^+$, $C_6H_4BrO_3S^-$ 3.38 **3**
$C_9H_{18}NO$ 33.32+ **7**
$C_9H_{18}NO^+$, Cl^- 33.29 **3**
$C_9H_{18}NO^+$, $C_9H_{17}NO$, Cl^- 34.13 **7**
$C_9H_{18}NO_2$ 12.8 **1**
$C_9H_{18}NO_2$ 12.8 **3**
$C_9H_{18}NO_2^+$, Br^- 48.42 **4**
$C_9H_{18}NO_6^+$, I^- 45.55 **1**
$C_9H_{18}NS_2^-$, Tl^+ 11.13 **7**
$C_9H_{18}N_2O_2$ 48.80 **1**
$C_9H_{18}N_2O_2$ 1.26+ **3**
$C_9H_{18}N_2O_2$ 48.47 **6**
$C_9H_{18}N_2O_2S$ 48.43 **4**
$C_9H_{18}N_3^+$, ClO_4^- 20.4 **4**
$2C_9H_{18}N_3^+$, $Cl_6Pt_2^{2-}$ 20.11 **6**
$C_9H_{18}N_6$ 33.23 **4**
$C_9H_{18}N_6$, $C_6H_3N_3O_6$ 60.113 **2**
$(C_9H_{18}Nd_2O_{18})_n$ 81.30 **6**
$(C_9H_{18}Nd_2O_{18})_n$, $2nH_2O$ 81.31 **6**
$C_9H_{18}O_3$ 1.115 **1**
$C_9H_{18}O_6$ 38.5 **3**
$C_9H_{19}CoN_4O_5$ 71.9 **6**
$C_9H_{19}CoN_4O_5$ 71.12 **7**
$C_9H_{19}NO$ 1.25 **5**
$C_9H_{19}NO$ 1.32 **7**
$C_9H_{19}N_2O_3^+$, Br^- 48.81 **1**
$C_9H_{20}ClCoN_9^+$, Br^- , H_2O 76.21 **6**
$C_9H_{20}CoN_4O_3^+$, ClO_4^- 83.34 **3**
$C_9H_{20}N^+$, Br^- 23.2 **1**
$C_9H_{20}NOS^+$, I^- 1.33 **7**
$C_9H_{20}NO_2^+$, I^- 38.38 **1**
$C_9H_{20}NO_2^+$, I^- 3.41 **3**
$C_9H_{20}NO_2^+$, I^- 3.39+ **3**
$C_9H_{20}N_2O$ 9.25 **1**
$C_9H_{21}BrCoN_5^{2+}$, $2NO_3^-$ 76.61 **3**
$C_9H_{21}INNaO_3$ 67.3 **6**
$C_9H_{21}N_4^{3+}$, $3Cl^-$, $0.5H_2O$ 36.2 **1**
$C_9H_{22}CoN_4O_3^+$, ClO_4^- 76.40 **7**
$C_9H_{23}Cl_2CoN_3$ 76.62 **3**
$C_9H_{23}CoN_5O_4^+$, ClO_4^- 76.63 **3**
$C_9H_{24}B_3N_3$ 62.46 **2**
$C_9H_{24}BrCoN_4^+$, Br^- , $0.5C_2H_6O$ 83.32 **4**
$(C_9H_{24}CaN_3O_6^{2+})_n$, $2nCl^-$ 67.9 **4**
$C_9H_{24}ClCuN_6S_3$ 79.20 **3**
$C_9H_{24}CoN_6O_3^+$, ClO_4^- 82.37 **2**
$C_9H_{25}B_{16}Co_2^-$, $C_8H_{20}N^+$ 71.10 **6**
$C_9H_{25}ClCoN_5^{2+}$, $2ClO_4^-$ 76.31 **5**
$C_9H_{26}B_{17}CoN^-$, $C_8H_{20}N^+$ 62.8 **5**
$C_9H_{27}Al_3N_3$ 68.15 **5**
$C_9H_{27}Al_3N_3$ 68.16 **5**
$C_9H_{27}Br_2NiP_3$ 86.6 **7**
$C_9H_{27}Cl_3Cu_3P_3S_3$ 85.13 **5**
$C_9H_{27}CuP_3S_3^+$, ClO_4^- 85.15 **3**
$C_9H_{27}I_2NiO_9P_3$ 86.7 **7**
$C_9H_{27}N_4P_4^+$, $C_5CrIO_5^-$ 64.12 **6**
$(C_9H_{28}CrO_5Sn)_n$ 69.8 **6**
$C_9H_{30}Al_3N_3$ 68.7+ **3**
$C_9H_{30}CoN_6^{3+}$, $3Br^-$ 76.31 **4**
$C_9H_{30}CoN_6^{3+}$, $3Br^-$ 76.41 **7**
$C_9H_{30}CoN_6^{3+}$, $3Br^-$, H_2O 76.64 **3**
$C_9H_{30}CoN_6^{3+}$, $3Cl^-$, H_2O 83.16 **6**
$C_9H_{30}CoN_6^{3+}$, $C_6CoN_6^{3-}$, $2H_2O$ 76.42 **7**
$C_9H_{30}CoN_6^{3+}$, $C_9H_6CrO_{12}^{3-}$, $3H_2O$ 76.32 **4**
$C_9H_{30}CrN_6^{3+}$, $C_5N_5Ni^{3-}$, $2H_2O$ 76.65 **3**
$C_9H_{30}CrN_6^{3+}$, $C_5N_5Ni^{3-}$, $2H_2O$ 83.25 **7**
$C_9H_{30}CuN_6^{2+}$, $2Br^-$, $2H_2O$ 76.66 **3**
$C_9H_{30}N_6Ni^{2+}$, $C_6H_{24}N_4NiO_2^{2+}$, $4Cl^-$, H_2O 83.33 **4**
$(C_9H_{36}Cu_4N_{18}S_9^{4+})_n$, $4nNO_3^-$ 79.9 **4**

C_{10}

$C_{10}Br_8$ 20.10 **7**
$C_{10}Cl_5MnO_5$ 71.11 **6**
$C_{10}Cl_6N_2S$ 39.16 **7**
$C_{10}Cl_8$ 20.22 **1**
$C_{10}Cl_8$ 24.1 **1**
$C_{10}Cl_8$ 20.12 **6**
$C_{10}Cl_{10}$ 28.1 **6**
$C_{10}Cl_{10}Ru$ 73.1 **4**
$C_{10}Cl_{12}O_3S$ 31.4 **1**
$C_{10}F_6Mn_2O_8Se_2$ 85.14 **5**
$C_{10}F_8$ 24.1 **4**
$C_{10}F_8$ 24.1 **7**
$C_{10}F_{10}FeO_3$ 71.12 **6**
$C_{10}F_{24}Ni_2O_3P_4S_2$ 86.4 **3**
$C_{10}N_6^{2-}$, Ba^{2+} , $6H_2O$ 67.6 **2**
$C_{10}N_6^{2-}$, Ca^{2+} , $6H_2O$ 67.7 **2**
$C_{10}N_6^{2-}$, $2K^+$ 12.9 **4**
$C_{10}N_6^{2-}$, $2C_9H_8N^+$ 60.20 **4**
$C_{10}H_2CuF_{12}O_4$, $C_4H_4N_2$ 60.14 **6**
$C_{10}H_2CuF_{12}O_4$, $C_6H_{12}N_2$ 60.12 **5**
$C_{10}H_2F_{16}O_2$ 27.3 **7**
$C_{10}H_2N_4$ 7.7 **5**
$C_{10}H_2N_4$, $C_6H_8N_2$ 60.7 **5**
$C_{10}H_2N_4$, $C_8H_{12}N_2$ 60.4 **6**
$C_{10}H_2N_4$, $C_{10}H_8$ 60.117 **2**
$C_{10}H_2N_4$, $C_{10}H_{16}N_2$ 60.123 **2**
$C_{10}H_2N_4$, $C_{14}H_{10}$ 60.32 **4**
$C_{10}H_2N_4$, $C_{16}H_{10}$ 60.30+ **5**
$C_{10}H_2N_4$, $C_{18}H_{12}N_2O_2Pd$ 60.148 **2**
$2C_{10}H_2N_4$, $C_{18}H_{12}CuN_2O_2$ 60.146+ **2**
$C_{10}H_2N_4S$ 39.18 **5**
$C_{10}H_2O_4S$, $C_{14}H_{10}$ 60.11 **7**

$C_{10}H_2O_4S_2$	39.19	5
$C_{10}H_2O_6$, $C_{14}H_{10}$	60.135	2
$C_{10}H_2O_6$, $C_{14}H_{12}$	60.12	7
$C_{10}H_2O_6$, $C_{16}H_{10}$	60.45 +	3
$C_{10}H_2O_6$, $C_{18}H_{12}$	60.27	7
$C_{10}H_2O_6$, $C_{20}H_{12}$	60.154	2
$2C_{10}H_2O_6$, $C_{15}H_{14}$	60.13	7
$C_{10}H_4Br_2Cl_2$	24.2	1
$C_{10}H_4Br_2O_2$	25.1	1
$C_{10}H_4Br_4$	24.3	1
$C_{10}H_4Br_5NO$	32.28	1
$C_{10}H_4Cl_2O_2$	25.2	1
$C_{10}H_4Cl_4$	24.4 +	1
$C_{10}H_4Cl_5NO_2$	27.3	4
$C_{10}H_4F_2O_2$	25.1	5
$C_{10}H_4N_4O_8$	24.6	1
$C_{10}H_5BrN_2O$	40.13	4
$C_{10}H_5BrO_2$	25.3	1
$C_{10}H_5BrO_2$	25.1	7
$C_{10}H_5Br_2N$	36.3	1
$C_{10}H_5ClN_2$	7.8	5
$C_{10}H_5ClN_2$	7.3	7
$C_{10}H_5ClO_2$	25.2	3
$C_{10}H_5ClO_2$	25.2	5
$C_{10}H_5ClO_3$	25.4	1
$C_{10}H_5Cl_2FN_2O_2$	32.29	1
$C_{10}H_5Cl_7$	31.5	6
$C_{10}H_5IO_2$	25.1	4
$C_{10}H_5IO_3$	25.2	4
$C_{10}H_6$	19.9	5
$C_{10}H_6BrNO_2$	25.5	1
$C_{10}H_6BrNO_2$	25.3	4
$C_{10}H_6BrNO_2$, H_2O	25.3	3
$C_{10}H_6BrNO_2$, CH_4O	60.32	3
$C_{10}H_6BrNO_2S_2$	39.22	3
$C_{10}H_6BrNO_4$	13.45	1
$C_{10}H_6ClNO_2$	25.6	1
$C_{10}H_6Cl_2$	24.7	1
$C_{10}H_6Cl_2$	24.8	1
$C_{10}H_6Cl_2N_2O_2$	50.2	1
$C_{10}H_6Cl_2O$	20.23	1
$C_{10}H_6CrO_4$	75.3	3
$C_{10}H_6F_2$	24.2	7
$C_{10}H_6F_2$	24.3	7
$C_{10}H_6FeN_6^{2-}$, $C_8H_{18}N_2^{2+}$	83.17	6
$C_{10}H_6FeO_4$	75.10	2
$C_{10}H_6N_2$	22.9	1
$C_{10}H_6N_2$	7.12 +	4
$C_{10}H_6N_2O_2$	24.2	4
$C_{10}H_6N_2O_3S_2$	39.23	3
$C_{10}H_6N_2O_4$	24.12	1
$C_{10}H_6N_2O_4$	24.9 +	1
$C_{10}H_6O_2$	20.24	1
$C_{10}H_6O_2$	25.7	1
$C_{10}H_6O_3$	25.4	4
$C_{10}H_6O_4$	25.8 +	1
$C_{10}H_6O_4$	25.5 +	4
$C_{10}H_6O_8$, $2H_2O$	13.14	4
$C_{10}H_6O_{10}Os_2$	81.29	4
$C_{10}H_7Br$	24.1	6
$C_{10}H_7BrCl_2O_3$	1.29	3
$C_{10}H_7BrF_2O_3$	1.30	3
$C_{10}H_7BrO_3S$	42.5	4
$C_{10}H_7F$	24.2	5
$C_{10}H_7HgI$	71.19	2
$C_{10}H_7MnO_5^-$, $C_8H_{12}N^+$	3.57	1
$C_{10}H_7MnO_6$	71.10	5
$C_{10}H_7MoO_3^+$, BF_4^-	75.2	5
$C_{10}H_7NO_2$	25.13	1
$C_{10}H_7NO_2$, $0.5H_2O$	25.4	3
$C_{10}H_7N_3OS$, H_2O	41.11	6
$C_{10}H_7N_3S$	41.12	6
$C_{10}H_8$	24.15	1
$C_{10}H_8$	24.13 +	1
$C_{10}H_8$	27.6 +	1
$C_{10}H_8$	24.1	3
$C_{10}H_8$, $4Ag^+$, $4ClO_4^-$, $4H_2O$	24.2	6
$C_{10}H_8$, $2Cl_3Sb$	60.114	2
$C_{10}H_8$, C_6N_4	60.72	2
$C_{10}H_8$, C_6N_4	60.1	6
$C_{10}H_8$, $C_6H_3N_3O_6$	60.115 +	2
$C_{10}H_8$, $C_{10}H_2N_4$	60.117	2
$C_{10}D_8$	24.16	1
$C_{10}H_8AsCl_3N_2$	65.7	4
$C_{10}H_8BrN$, $C_6H_3N_3O_7$	60.77	2
$C_{10}H_8BrNO_2$	1.116	1
$C_{10}H_8BrNO_2$	40.19	1
$C_{10}H_8BrN_3O_2$	32.13	4
$C_{10}H_8Br_2MoN_2O_2$	83.35	3
$C_{10}H_8ClN_3O$	16.4	7
$C_{10}H_8Cl_2FeO_4S_2$	73.14	2
$C_{10}H_8Cl_2HgN_8S_2$	85.12	7
$C_{10}H_8Cl_2Hg_2N_2^{2+}$, $2ClO_4^-$	83.20	5
$C_{10}H_8Cl_2N_2Pt$	83.18	6
$C_{10}H_8Cl_2O_6$	13.15 +	4
$C_{10}H_8Cl_3NO_5S$	17.30	1
$C_{10}H_8Cl_4$	24.17	1
$C_{10}H_8Cl_4$	27.4	6
$C_{10}H_8CrN_2O_5$	83.71 +	2
$C_{10}H_8CrO_4$, $C_6H_3N_3O_6$	60.118	2
$C_{10}H_8CuN_4O_4$	83.36	3
$C_{10}H_8CuN_8$	83.26	7
$C_{10}H_8CuN_{10}$, $4H_2O$	83.73	2
$C_{10}H_8F_4GeNi$, CH_3NO_2	69.16	4
$C_{10}H_8F_4NOP$	64.24	7
$C_{10}H_8F_4N_2Si$	63.4	4
$C_{10}H_8F_4N_2Sn$, CH_3NO_2	69.17	4
$C_{10}H_8F_6NiO_2S_2$	77.1	6
$C_{10}H_8F_{12}N_2$	32.15	7
$C_{10}H_8FeI_2$	73.4	5
$C_{10}H_8FeO_5$	72.2	7
$C_{10}H_8FeS_3$	73.5	5
$C_{10}H_8MnN_5O_9$	83.21	5
$C_{10}H_8MoO_3$	75.12	2
$C_{10}H_8MoO_5$	71.6	3
$C_{10}H_8N$	35.11	6
$C_{10}H_8NO_3S^-$, Na^+ , $4H_2O$	24.18	1
$C_{10}H_8N_2$	33.61	1
$C_{10}H_8N_2$	33.39	5
$C_{10}H_8N_2O_4$	35.26	1
$C_{10}H_8N_4O_2$	9.26	1
$C_{10}H_8N_4O_3$	44.22	7
$C_{10}H_8N_4S$	41.13	6
$C_{10}H_8O$	24.19	1
$C_{10}H_8O$	24.20	1
$C_{10}H_8O$	38.39	1
$C_{10}H_8O_2$	24.21	1
$1.5C_{10}H_8O_2$, $C_{11}H_9N_4O_2^+$, Br^- , H_2O	60.35	3
$2C_{10}H_8O_2$, $C_{13}H_{12}N_4O_2$	60.22	6
$2C_{10}H_8O_2$, $C_{13}H_{12}N_4O_2$	60.23	6
$2C_{10}H_8O_2$, $C_{13}H_{12}N_4O_2$, $3H_2O$	60.19	7

$C_{10}H_8O_2S$ 13.46 **1**
$C_{10}H_8O_3$ 38.14 **7**
$C_{10}H_8O_3V$ 75.13 **2**
$C_{10}H_8O_4$ 38.40 **1**
$C_{10}H_8O_4$ 38.41 **1**
$2C_{10}H_8O_4P^-$, Ca^{2+} , $3H_2O$ 46.16 **1**
$C_{10}H_8S$ 39.17 **7**
$C_{10}H_9BF_2O_2$ 62.14 **4**
$C_{10}H_9BrN_2OS$ 32.18 **6**
$C_{10}H_9BrN_4O_2S$ 44.68 **1**
$C_{10}H_9BrO$ 5.7 **4**
$C_{10}H_9BrO_2$ 1.117 **1**
$C_{10}H_9BrO_2$ 1.118 **1**
$C_{10}H_9BrO_2$ 38.42 **1**
$C_{10}H_9BrS$ 39.18 **7**
$C_{10}H_9Br_3O_2$ 22.3 **4**
$C_{10}H_9ClN_4$ 36.4 **1**
$C_{10}H_9Cl_3O$ 31.5 **1**
$C_{10}H_9CrNO_3$ 74.6 **2**
$C_{10}H_9FO_2$ 38.15 **7**
$C_{10}H_9N$ 27.8 **1**
$C_{10}H_9NO_2$ 35.27 **1**
$C_{10}H_9NO_2$ 40.20 **1**
$C_{10}H_9NO_2$ 32.12 **5**
$C_{10}H_9NO_4S$ 41.9 **3**
$C_{10}H_9NO_4S$, H_2O 35.7 **3**
$2C_{10}H_9N_2^+$, $Cl_8I_6Mo_6^{2-}$ 33.30 **3**
$2C_{10}H_9N_2^+$, $Cl_{14}Mo_6^{2-}$ 33.31 + **3**
$C_{10}H_9N_2O^+$, Cl^- , $2H_2O$ 25.2 **7**
$C_{10}H_9N_2S_2^+$, HO^- 41.28 **1**
$C_{10}H_9N_3$ 33.33 **6**
$C_{10}H_9N_5O_6$ 32.14 **4**
$C_{10}H_9O_6^-$, $C_{10}H_{10}O_6$, K^+ 17.17 **7**
$C_{10}H_{10}$ 28.2 **1**
$C_{10}H_{10}$ 31.6 + **1**
$C_{10}H_{10}AgN_2^+$, NO_3^- , H_2O 83.35 **4**
$C_{10}H_{10}AgN_4O_6$ 83.22 **5**
$C_{10}H_{10}Be$ 67.9 **3**
$C_{10}H_{10}Be$ 67.10 **4**
$C_{10}H_{10}Be$ 67.4 **6**
$C_{10}H_{10}BrNO_4S$ 41.11 **4**
$C_{10}H_{10}BrN_3O$ 40.15 **5**
$C_{10}H_{10}Br_2CuN_2$ 83.37 **3**
$C_{10}H_{10}Br_2CuN_2$ 83.27 **7**
$C_{10}H_{10}Br_2O_2$ 22.4 **4**
$C_{10}H_{10}Br_2O_2$ 22.5 **4**
$C_{10}H_{10}Br_2Re^+$, BF_4^- 73.4 **7**
$C_{10}H_{10}Br_4Cu_2N_2O_2$ 84.11 **5**
$C_{10}H_{10}Br_4MoSn$ 73.6 **5**
$C_{10}H_{10}CdCl_2N_2$ 83.74 **2**
$C_{10}H_{10}CdCl_2N_2$ 83.38 **3**
$C_{10}H_{10}CdN_8$ 83.39 **3**
$C_{10}H_{10}Cl_2$ 23.1 **3**
$C_{10}H_{10}Cl_2CoN_2$ 83.75 **2**
$C_{10}H_{10}Cl_2CoN_2$ 83.28 + **7**
$C_{10}H_{10}Cl_2CuN_2$ 83.76 + **2**
$C_{10}H_{10}Cl_2CuN_2$ 83.30 **7**
$C_{10}H_{10}Cl_2HgRu$, $C_{10}H_{10}ClHgRu^+$, Cl^- , Cl_2Hg 73.7 **5**
$(C_{10}H_{10}Cl_2MnN_2)_n$ 83.19 **6**
$C_{10}H_{10}Cl_2Mo$ 73.5 **7**
$C_{10}H_{10}Cl_2Mo^+$, BF_4^- 73.6 **7**
$C_{10}H_{10}Cl_2N_2OSe$ 11.42 **1**
$C_{10}H_{10}Cl_2N_2Zn$ 83.78 **2**
$C_{10}H_{10}Cl_2Nb$ 73.7 **7**
$C_{10}H_{10}Cl_2O_4$ 59.15 **1**
$C_{10}H_{10}Cl_2Pt$ 73.17 **2**
$C_{10}H_{10}Cl_2Ti$ 73.8 **5**
$C_{10}H_{10}Cl_2Zr$ 73.8 **7**
$C_{10}H_{10}Cl_4Cu_2N_2O_2$ 84.31 **2**
$C_{10}H_{10}Cl_4Cu_2N_2O_2$ 84.32 **2**
$C_{10}H_{10}Cl_4Cu_2N_2O_2$ 84.8 **3**
$C_{10}H_{10}Cl_4Cu_2N_8S_2$, $2H_2O$ 85.13 **7**
$C_{10}H_{10}Cl_4OTi_2$ 73.18 **2**
$C_{10}H_{10}Co^+$, ClO_4^- 73.19 **2**
$C_{10}H_{10}Cr$ 73.20 **2**
$C_{10}H_{10}Cr_2N_4O_4$ 73.6 **6**
$C_{10}H_{10}CuN_4$ 83.36 **4**
$C_{10}H_{10}CuN_4O_6$, $0.5C_5H_5N$ 83.37 **4**
$C_{10}H_{10}CuN_6O_2^{2+}$, $2ClO_4^-$ 83.23 **5**
$C_{10}H_{10}CuN_8$ 83.40 **3**
$C_{10}H_{10}F_2Zr$ 73.2 **4**
$C_{10}H_{10}F_4N_2Si$ 63.12 **2**
$C_{10}H_{10}Fe$ 73.21 + **2**
$C_{10}H_{10}Fe$, C_6N_4 60.119 **2**
$C_{10}H_{10}Fe^+$, I_3^- 73.24 **2**
$C_{10}H_{10}Fe^+$, $C_6H_2N_3O_7^-$ 73.25 **2**
$C_{10}H_{10}Fe^+$, $C_6H_3Cl_9O_6^-$ 73.3 **4**
$C_{10}H_{10}Fe_2N_2O_2$ 73.7 **6**
$C_{10}H_{10}Fe_2O_6S_2$ 85.62 **2**
$C_{10}H_{10}INO_3S$ 40.21 **1**
$C_{10}H_{10}IN_2^+$, I_3^- , $2I_2$ 60.120 **2**
$C_{10}H_{10}IRu^+$, I_3^- 73.8 **6**
$C_{10}H_{10}I_2N_2O_2Zn$ 84.28 **7**
$C_{10}H_{10}I_2Zr$ 73.5 **4**
$C_{10}H_{10}Mn_2N_4O_5$ 73.6 **4**
$C_{10}H_{10}MoO_3$ 73.26 **2**
$C_{10}H_{10}MoS_4$ 73.9 **7**
$C_{10}H_{10}Mo_2O_2S_2$ 73.27 **2**
$C_{10}H_{10}NO^+$, $H_2AsO_4^-$ 35.8 **3**
$C_{10}H_{10}NO^+$, $C_{10}H_9NO$, AsF_6^- 35.22 **3**
$C_{10}H_{10}N_2$ 32.16 **7**
$C_{10}H_{10}N_2^{2+}$, $2Br^-$ 33.62 **1**
$C_{10}H_{10}N_2^{2+}$, $2Br^-$ 33.40 **5**
$C_{10}H_{10}N_2^{2+}$, Br_4Co^{2-} 33.24 **4**
$C_{10}H_{10}N_2^{2+}$, $2Cl^-$ 33.41 **5**
$C_{10}H_{10}N_2^{2+}$, Cl_4OSe^{2-} 33.34 **3**
$C_{10}H_{10}N_2^{2+}$, Cl_5Mn^{2-} 33.25 **4**
$C_{10}H_{10}N_2O$ 32.19 **6**
$C_{10}H_{10}N_2O$ 32.20 **6**
$C_{10}H_{10}N_2O_2$ 32.16 **5**
$C_{10}H_{10}N_2O_2$ 32.21 + **6**
$C_{10}H_{10}N_2O_2$ 36.1 **6**
$C_{10}H_{10}N_2O_3$ 35.12 **6**
$C_{10}H_{10}N_2O_4$ 38.15 **6**
$C_{10}H_{10}N_4O_3$, H_2O 44.20 **6**
$C_{10}H_{10}N_4O_4S_2$ 44.33 **3**
$C_{10}H_{10}N_4O_6Zn$ 83.38 **4**
$C_{10}H_{10}N_5O_5PS$, $3H_2O$ 47.15 **5**
$C_{10}H_{10}N_8Zn$ 83.41 **3**
$C_{10}H_{10}O_2$ 20.25 **1**
$C_{10}H_{10}O_2$ 6.1 **4**
$C_{10}H_{10}O_2$ 23.2 **4**
$C_{10}H_{10}O_2$ 31.3 **4**
$C_{10}H_{10}O_2$ 1.26 **5**
$C_{10}H_{10}O_2$, Cl_3Sb 19.10 **5**
$C_{10}H_{10}O_3S$ 39.24 **3**
$C_{10}H_{10}O_3S$ 39.25 **3**
$C_{10}H_{10}O_4$ 31.4 **3**
$(C_{10}H_{10}Pb)_n$ 69.21 **2**
$C_{10}H_{10}Ru$ 73.28 **2**
$C_{10}H_{10}S_4W$ 73.9 **5**

$C_{10}H_{10}S_5Ti$ 73.8 **4**
$C_{10}H_{10}Ti^+$, H_4B^- 73.10 **5**
$C_{10}H_{11}BrN_2$ 37.8 **7**
$C_{10}H_{11}BrN_4O_5$ 47.15 **3**
$C_{10}H_{11}BrN_5O_6P$ 47.12 **6**
$C_{10}H_{11}BrO_3$ 38.43 **1**
$C_{10}H_{11}ClN_2O_2$ 16.9 **6**
$C_{10}H_{11}ClO_2$ 38.6 **3**
$C_{10}H_{11}Cl_7$ 52.1 **7**
$C_{10}H_{11}CrNO_3$ 75.3 **5**
$C_{10}H_{11}CrNO_6$ 71.8 **4**
$C_{10}H_{11}FN_4O_6$ 16.5 **7**
$C_{10}H_{11}F_3N_2O_5$ 47.11 **4**
$C_{10}H_{11}MoNO_3$ 71.13 **7**
$C_{10}H_{11}NO$ 1.34 **7**
$C_{10}H_{11}NO_2$ 7.28 **1**
$C_{10}H_{11}NO_2S$ 41.29 **1**
$C_{10}H_{11}NO_3S$ 42.6 **5**
$C_{10}H_{11}NO_4S_2$ 4.17 **1**
$C_{10}H_{11}NO_5$ 47.17 **7**
$C_{10}H_{11}N_2O^+$, Br^- 35.13 **6**
$C_{10}H_{11}N_3$ 35.9 **3**
$C_{10}H_{11}N_3S_2$ 8.17 **3**
$C_{10}H_{11}N_4O_4^-$, Na^+ , H_2O 44.34 **3**
$C_{10}H_{11}N_4O_8P^{2-}$, Ba^{2+} , $6H_2O$ 47.13 **1**
$C_{10}H_{11}N_4O_8P^{2-}$, $2Na^+$, $7.5H_2O$ 47.14 **1**
$C_{10}H_{11}N_5O_2$ 47.13 **6**
$C_{10}H_{11}N_5O_3S$, H_2O 47.14 **5**
$C_{10}H_{11}N_5O_6P$ 47.15 **1**
$C_{10}H_{11}N_5O_7P^-$, Na^+ , $4H_2O$ 47.14 **6**
$C_{10}H_{11}N_6O_4^-$, K^+ , $4H_2O$ 48.45 **7**
$C_{10}H_{11}N_6O_4^-$, Rb^+ , $4H_2O$ 48.46 **7**
$C_{10}H_{11}N_7$, $2H_2O$ 44.20 **4**
$C_{10}H_{11}N_8O_6P$ 47.18 **7**
$C_{10}H_{12}AgO^+$, BF_4^- 75.16 **2**
$C_{10}H_{12}AlN_2O_8^-$, K^+ , $2H_2O$ 68.9 **3**
$C_{10}H_{12}As_2ClCrF_3O_4$ 86.3 **5**
$C_{10}H_{12}As_4Fe_2O_6$ 86.5 **2**
$C_{10}H_{12}BrNO$ 40.22 **1**
$C_{10}H_{12}BrNO_2$ 52.6 **1**
$C_{10}H_{12}BrNO_4$ 48.48 **6**
$C_{10}H_{12}BrN_5O_3$, $C_{17}H_{20}N_4O_6$, $3H_2O$ 60.34 **4**
$C_{10}H_{12}BrN_5O_5$, $2H_2O$ 47.16 **3**
$C_{10}H_{12}Br_2CuN_{10}^{2+}$, $2Br^-$ 83.24 **5**
$C_{10}H_{12}Br_3Cl_3$ 52.1 **5**
$C_{10}H_{12}ClNO$ 10.27 **1**
$C_{10}H_{12}ClN_5$ 32.27 **3**
$C_{10}H_{12}ClN_5O$, H_2O 32.30 **1**
$C_{10}H_{12}ClPt$ 75.4 **5**
$C_{10}H_{12}Cl_2$ 19.8+ **4**
$C_{10}H_{12}Cl_2CuN_4O_4$ 84.29 **7**
$C_{10}H_{12}Cl_2Pt$ 75.2 **6**
$C_{10}H_{12}Cl_6Cu_2N_{10}O_2$, $2H_2O$ 83.39+ **4**
$C_{10}H_{12}Cl_8Cu_3N_{10}$, $4H_2O$ 83.25 **5**
$C_{10}H_{12}CoN_2O_8^-$, H_4N^+ , H_2O 76.59 **2**
$C_{10}H_{12}CoN_2O_8^-$, $C_6H_{18}CoN_4O_4^+$, H_2O 82.9 **6**
$C_{10}H_{12}Cr_2N_4O_3$ 73.6 **3**
$C_{10}H_{12}CuN_2O_8^{2-}$, $2K^+$, $3H_2O$ 76.43 **7**
$(C_{10}H_{12}CuN_4^+)_n$, nCl_2Cu^- 83.79 **2**
$C_{10}H_{12}Cu_2N_2O_8S_2^{2-}$, $2C_4H_{12}N^+$ 3.39 **1**
$C_{10}H_{12}FeN_2O_8^-$, Li^+ , $3H_2O$ 76.44 **7**
$C_{10}H_{12}FeN_6$ 71.20 **2**
$C_{10}H_{12}FeN_6$, $4CHCl_3$ 71.21 **2**
$C_{10}H_{12}Fe_2I_2O_6P_2$ 86.6 **2**
$C_{10}H_{12}I_2O$ 21.1 **7**
$C_{10}H_{12}Mo$ 73.29+ **2**
$C_{10}H_{12}Mo_2N_2O_{10}S_2^{2-}$, $2Cs^+$, $2H_2O$ 76.45 **7**
$C_{10}H_{12}Mo_2N_2O_{14}^{4-}$, $4Na^+$, $8H_2O$ 76.61 **2**
$C_{10}H_{12}NO^+$, HO_4Se^- 50.4 **4**
$C_{10}H_{12}NS^+$, BF_4^- 41.9 **7**
$C_{10}H_{12}N_2OS_3$ 41.30 **1**
$C_{10}H_{12}N_2O_3$ 40.14 **4**
$C_{10}H_{12}N_2O_3$ 17.12 **6**
$C_{10}H_{12}N_2O_3$ 17.13 **6**
$C_{10}H_{12}N_2O_3$ 43.5 **7**
$C_{10}H_{12}N_2O_6$ 47.19 **7**
$C_{10}H_{12}N_2O_8$, CH_4O , H_2O 47.15 **6**
$C_{10}H_{12}N_2O_8^{4-}$, $2Na^+$, Mg^{2+} , $4H_2O$ 48.47 **7**
$C_{10}H_{12}N_2O_8Sm^-$, Na^+ , $8H_2O$ 76.33 **4**
$(C_{10}H_{12}N_2O_8Sn_2)_n$, $2nH_2O$ 69.18 **4**
$C_{10}H_{12}N_2O_8Tb^-$, Na^+ 76.62 **2**
$C_{10}H_{12}N_2O_8Zn^{2-}$, $H_8MgO_4^{2+}$, $2H_2O$ 76.32 **5**
$C_{10}H_{12}N_2O_{10}V^{3-}$, $3Na^+$, $4H_2O$ 76.34 **4**
$C_{10}H_{12}N_3O_2S_2^+$, Br^- , H_2O 41.31 **1**
$C_{10}H_{12}N_3O_8^-$, Cs^+ 17.31 **1**
$C_{10}H_{12}N_3O_8^-$, K^+ 17.32 **1**
$C_{10}H_{12}N_4O$ 35.12 **7**
$C_{10}H_{12}N_4O_4$ 44.69 **1**
$C_{10}H_{12}N_4O_4$ 35.10 **4**
$C_{10}H_{12}N_4O_4$ 36.3 **5**
$C_{10}H_{12}N_4O_4$ 50.1 **6**
$C_{10}H_{12}N_4O_4$, H_2O 44.70 **1**
$C_{10}H_{12}N_4O_4$, H_2O 44.21 **4**
$C_{10}H_{12}N_4O_4S$ 47.16 **1**
$C_{10}H_{12}N_4O_5$ 47.17 **3**
$C_{10}H_{12}N_4O_5$ 35.9 **5**
$C_{10}H_{12}N_4O_5$ 47.16 **5**
$C_{10}H_{12}N_4O_5$ 50.2 **7**
$C_{10}H_{12}N_4O_5$, $2H_2O$ 47.18+ **3**
$C_{10}H_{12}N_4O_5$, $C_4H_5N_3O$ 47.20 **3**
$C_{10}H_{12}N_4O_8P^-$, Na^+ , $8H_2O$ 47.21 **3**
$C_{10}H_{12}N_5O_5P$, H_2O 47.12 **4**
$C_{10}H_{12}N_5O_7P^{2-}$, $2Na^+$, $4H_2O$ 47.20 **7**
$C_{10}H_{12}N_5O_{10}P_2^{3-}$, $3Rb^+$, $3H_2O$ 47.13 **4**
$2C_{10}H_{12}O$, $HClO_4$ 60.121 **2**
$C_{10}H_{12}OS_2$ 39.19 **7**
$C_{10}H_{12}OS_2$ 39.20 **7**
$C_{10}H_{12}O_2$ 18.27 **1**
$C_{10}H_{12}O_2$ 20.26 **1**
$C_{10}H_{12}O_2$ 28.3 **1**
$C_{10}H_{12}O_2$ 13.12 **3**
$C_{10}H_{12}O_2$ 13.13 **3**
$C_{10}H_{12}O_2$ 23.3 **3**
$C_{10}H_{12}O_2$ 31.5 **3**
$C_{10}H_{12}O_2$ 22.6 **4**
$C_{10}H_{12}O_2$ 13.13 **5**
$C_{10}H_{12}O_2$ 28.2 **6**
$C_{10}H_{12}O_2$ 52.2 **7**
$C_{10}H_{12}O_2S$ 11.9 **5**
$C_{10}H_{12}O_2S_4$ 39.65 **1**
$C_{10}H_{12}O_3$ 31.4 **4**
$C_{10}H_{12}O_3$ 38.6 **5**
$C_{10}H_{12}O_4$ 31.4 **5**
$C_{10}H_{12}O_4$, $C_8Cl_2N_2O_2$ 60.33 **3**
$C_{10}H_{12}O_4$, $C_9H_9BrO_3$ 1.119 **1**
$C_{10}H_{12}O_4S$ 42.6 **6**
$C_{10}H_{12}O_6$ 21.24+ **1**
$C_{10}H_{12}S_2$ 39.66 **1**
$C_{10}H_{13}As_2CrF_3O_4$ 86.4 **5**

$C_{10}H_{13}As_2F_3MoO_4$ 86.1 **4**
$C_{10}H_{13}As_2F_3MoO_4$ 86.5 **5**
$C_{10}H_{13}Br$ 19.30 + **1**
$C_{10}H_{13}BrNO$ 52.2 **5**
$C_{10}H_{13}BrN_2O_3$, $C_7H_9N_5$ 60.25 **3**
$C_{10}H_{13}BrN_2O_4$ 47.18 **1**
$C_{10}H_{13}BrO_3$ 27.4 **4**
$C_{10}H_{13}BrO_4$ 59.16 **1**
$C_{10}H_{13}Cl$ 19.32 **1**
$C_{10}H_{13}ClNO$ 40.23 **1**
$C_{10}H_{13}ClN_5^+$, I^- 44.71 **1**
$C_{10}H_{13}ClO_4$ 59.5 **5**
$C_{10}H_{13}Cl_3Se$ 11.7 **6**
$C_{10}H_{13}CuN_5O_4$, Na^+ , ClO_4^- , H_2O 82.13 **5**
$C_{10}H_{13}CuN_5O_4$, $12H_2O$ 82.14 **5**
$C_{10}H_{13}CuN_5O_6^{2-}$, $2Na^+$, $4.5H_2O$ 82.17 **3**
$C_{10}H_{13}F_3MoO_4P_2$ 86.6 **5**
$C_{10}H_{13}NO_2$ 32.24 **6**
$C_{10}H_{13}NO_2$, $C_{10}H_{14}NO_2^+$, I_3^- , I_2 16.15 **5**
$C_{10}H_{13}NO_3$ 48.49 **6**
$C_{10}H_{13}NO_6$ 33.42 **5**
$C_{10}H_{13}N_2+$, Cl^- 35.14 **6**
$C_{10}H_{13}N_2^+$, $C_6H_2N_3O_7^-$ 35.15 **6**
$C_{10}H_{13}N_2^+$, $C_{12}H_4N_4$, $C_{12}H_4N_4^-$, C_2H_3N 7.9 **5**
$C_{10}H_{13}N_2^+$, $C_{12}H_4N_4^-$ 35.10 **5**
$C_{10}H_{13}N_2O^+$, $C_4H_8N_3O^+$, O_4S^{2-} , H_2O 35.28 **1**
$C_{10}H_{13}N_2O^+$, $C_6H_2N_3O_7^-$, H_2O 35.11 **4**
$C_{10}H_{13}N_2O_8P^{2-}$, Ca^{2+} , $6H_2O$ 47.19 **1**
$C_{10}H_{13}N_4O_8P$, CH_4O , H_2O 47.16 **6**
$C_{10}H_{13}N_5$ 44.16 **5**
$C_{10}H_{13}N_5O_3$, H_2O 47.21 **1**
$C_{10}H_{13}N_5O_3$, H_2O 47.17 **6**
$C_{10}H_{13}N_5O_4$ 47.17 **5**
$C_{10}H_{13}N_5O_4$ 47.18 **6**
$C_{10}H_{13}N_5O_4$, H_2O 50.2 **6**
$C_{10}H_{13}N_5O_4$, $C_9H_{11}BrN_2O_6$, H_2O 47.22 **1**
$C_{10}H_{13}N_5O_4$, $C_9H_{13}BrN_3O_4$ 47.23 **1**
$2C_{10}H_{13}N_5O_4$, $C_{62}H_{86}N_{12}O_{16}$, $12H_2O$ 60.43 + **4**
$C_{10}H_{13}N_5O_4S$, H_2O 47.19 **5**
$C_{10}H_{13}N_5O_5$, $2H_2O$ 47.23 **3**
$C_{10}H_{13}N_5O_6P^-$, Na^+ , $6H_2O$ 47.21 **7**
$C_{10}H_{13}O_2^+$, Cl^- 38.16 **7**
$C_{10}H_{13}O_2^+$, $C_5H_7Cl_2O_4W^-$ 38.16 **6**
$C_{10}H_{14}$ 19.11 + **5**
$C_{10}H_{14}$ 19.4 + **7**
$C_{10}H_{14}As_2CrF_2O_4$ 86.7 **5**
$C_{10}H_{14}BNb$ 73.10 **7**
$C_{10}H_{14}BeO_4$ 67.8 **2**
$C_{10}H_{14}BeO_4$ 67.4 **7**
$C_{10}H_{14}BrO_4S^-$, $C_9H_{14}NO^+$ 16.7 **4**
$C_{10}H_{14}BrO_4S^-$, $C_{19}H_{26}NO_7^+$, H_2O 58.38 **1**
$C_{10}H_{14}Br_2O$ 52.3 **5**
$C_{10}H_{14}Br_2O$ 31.7 **6**
$C_{10}H_{14}Br_2O$ 31.2 **7**
$C_{10}H_{14}Br_2O_2$ 23.4 **3**
$C_{10}H_{14}CdN_2O_9^{2-}$, $H_{12}MgO_6^{2+}$, $3H_2O$ 76.33 **5**
$(C_{10}H_{14}CdO_4)_n$ 77.1 **7**
$C_{10}H_{14}ClFeO_4$ 77.4 **3**
$C_{10}H_{14}ClO_4Pt^-$, K^+ 77.4 **2**
$C_{10}H_{14}Cl_2FeGe$ 72.5 + **4**
$C_{10}H_{14}Cl_2O_3Pt$ 75.17 **2**
$C_{10}H_{14}Cl_2O_4Re$ 77.2 **6**
$C_{10}H_{14}Cl_6O_4Ti$ 77.1 **4**
$C_{10}H_{14}CoN_4O_4$ 81.32 **6**
$C_{10}H_{14}CoN_6O_4S_2^-$, H_4N^+ , $3H_2O$ 83.20 **6**
$C_{10}H_{14}CoN_6O_4Se_2^-$, H_4N^+ , $3H_2O$ 83.21 **6**
$C_{10}H_{14}CoS_4$ 77.3 **6**
$C_{10}H_{14}CuN_2O_2^{2+}$, O_4S^{2-} 83.81 **2**
$C_{10}H_{14}CuO_4$ 77.5 **3**
$C_{10}H_{14}Cu_2N_6O_8S_2$ 79.17 **7**
$C_{10}H_{14}FeN_2O_9^-$, Li^+ , $2H_2O$ 76.63 **2**
$C_{10}H_{14}FeN_2O_9^-$, Na^+ , $2H_2O$ 76.46 **7**
$C_{10}H_{14}FeN_2O_9^-$, Rb^+ , H_2O 76.64 **2**
$C_{10}H_{14}FeN_2O_9^-$, $CH_6N_3^+$, $2H_2O$ 76.34 **5**
$C_{10}H_{14}HgN_4O_8$ 83.31 **7**
$C_{10}H_{14}IMoO_5P$ 73.11 **5**
$C_{10}H_{14}I_2O_4Pt$ 77.1 **5**
$C_{10}H_{14}MgN_2O_9^{2-}$, $2Na^+$, $5H_2O$ 67.5 **6**
$C_{10}H_{14}MgN_2O_9^{2-}$, $H_{12}MgO_6^{2+}$, $2H_2O$ 67.5 **7**
$C_{10}H_{14}MnN_2O_9^{2-}$, $2Li^+$, $4H_2O$ 76.47 **7**
$(C_{10}H_{14}MnN_3O_4)_n$ 77.2 **7**
$C_{10}H_{14}MoO_6$ 77.2 **5**
$C_{10}H_{14}N_2NiO_8$ 83.41 **4**
$C_{10}H_{14}N_2O$ 3.20 **4**
$C_{10}H_{14}N_2O$ 52.4 **5**
$C_{10}H_{14}N_2O_2$ 36.3 **7**
$C_{10}H_{14}N_2O_4$ 35.13 **7**
$C_{10}H_{14}N_2O_4S_2$ 41.15 **6**
$C_{10}H_{14}N_2O_5$ 47.25 **1**
$C_{10}H_{14}N_2O_5$ 47.22 **7**
$C_{10}H_{14}N_2O_5S$ 47.24 **3**
$C_{10}H_{14}N_2O_6$ 47.21 **5**
$C_{10}H_{14}N_2O_6$ 47.19 **6**
$C_{10}H_{14}N_2O_6$, $0.5H_2O$ 47.26 **1**
$C_{10}H_{14}N_2O_8^{2-}$, $2K^+$, H_2O 2.93 **1**
$C_{10}H_{14}N_2O_8^{2-}$, $2Rb^+$, $2H_2O$ 2.23 **3**
$C_{10}H_{14}N_2O_8^{2-}$, $H_{12}MgO_6^{2+}$ 2.95 **1**
$C_{10}H_{14}N_2O_8^{2-}$, $H_{12}MgO_6^{2+}$ 48.50 **6**
$C_{10}H_{14}N_2O_8PS^-$, $C_6H_{16}N^+$ 47.23 **7**
$C_{10}H_{14}N_2O_8Sn$ 69.13 **5**
$C_{10}H_{14}N_2O_9Sn$ 69.19 **4**
$C_{10}H_{14}N_2O_{10}V^-$, H_4N^+ , $3H_2O$ 76.36 **4**
$C_{10}H_{14}N_2O_{10}Zr$ 77.3 **7**
$C_{10}H_{14}N_2S$ 41.12 **4**
$C_{10}H_{14}N_2S$ 33.34 **6**
$C_{10}H_{14}N_4O_2$ 40.15 **4**
$C_{10}H_{14}N_5O_4^+$, Br^- , H_2O 50.3 **6**
$C_{10}H_{14}N_5O_4^+$, Cl^- 47.22 **5**
$C_{10}H_{14}N_5O_4^+$, Cl^- 47.24 **7**
$C_{10}H_{14}N_5O_4^+$, Cl^- 47.25 **7**
$C_{10}H_{14}N_5O_5^+$, Br^- , $0.5H_2O$ 47.20 **6**
$C_{10}H_{14}N_5O_6P$, C_2H_6O 47.15 **4**
$C_{10}H_{14}N_5O_7P$, H_2O 47.27 **1**
$C_{10}H_{14}N_5O_7P$, $2H_2O$ 47.28 **1**
$C_{10}H_{14}N_5O_8P$, $3H_2O$ 47.20 **1**
$C_{10}H_{14}N_5O_{13}P_3^{2-}$, $2Na^+$, $3H_2O$ 47.24 **5**
$C_{10}H_{14}N_6O_3$, H_2O 47.26 **3**
$C_{10}H_{14}N_8Ni$ 83.22 **6**
$C_{10}H_{14}NiO_2S_2$ 77.5 **6**
$C_{10}H_{14}NiS_4$ 77.6 **6**
$C_{10}H_{14}Ni_2$ 72.17 **3**
$C_{10}H_{14}O_2$ 52.1 **6**
$C_{10}H_{14}O_2S$ 11.43 **1**
$C_{10}H_{14}O_3$ 38.7 **5**
$C_{10}H_{14}O_4$ 1.120 + **1**
$C_{10}H_{14}O_4$ 5.23 **1**

$C_{10}H_{14}O_4$ 31.6 **3**
$C_{10}H_{14}O_4$ 38.12 **4**
$C_{10}H_{14}O_4$ 5.16 **7**
$C_{10}H_{14}O_4Pd$ 77.2 **4**
$C_{10}H_{14}O_4S_2$ 11.5 **4**
$C_{10}H_{14}O_4S_3$ 11.12 **3**
$C_{10}H_{14}O_5V$ 77.6+ **2**
$2C_{10}H_{14}O_5V$, $C_4H_8O_2$ 60.34 **3**
$C_{10}H_{14}O_6S_2$ 45.28 **7**
$C_{10}H_{14}O_8$ 1.122 **1**
$C_{10}H_{15}BFeN_2O$ 71.11 **5**
$C_{10}H_{15}BrClNO$ 52.7 **1**
$C_{10}H_{15}BrN_2O$, $C_7H_9N_5$ 60.11 **5**
$C_{10}H_{15}BrO$ 52.8 **1**
$C_{10}H_{15}BrO$ 52.10 **1**
$C_{10}H_{15}BrO$ 52.11 **1**
$C_{10}H_{15}BrO$ 27.5 **3**
$C_{10}H_{15}BrO$ 52.1 **3**
$C_{10}H_{15}BrO$ 52.3 **7**
$C_{10}H_{15}BrO_2$ 31.8 **6**
$C_{10}H_{15}Br_2N_3Zn$ 83.44 **3**
$C_{10}H_{15}Br_3O$ 52.12 **1**
$C_{10}H_{15}CeO_2^{2-}$, $2K^+$, H_2O 81.67 **2**
$C_{10}H_{15}ClO$ 27.6 **3**
$C_{10}H_{15}CoN_5^+$, ClO_4^- 72.25 **2**
$C_{10}H_{15}CrN_2O_9$ 76.66 **2**
$C_{10}H_{15}CuN_6O_3^+$, ClO_4^- 82.38 **2**
$C_{10}H_{15}FeN_2O_9$ 76.37 **4**
$C_{10}H_{15}GaN_2O_9$ 68.4 **4**
$C_{10}H_{15}I$ 31.9+ **6**
$2C_{10}H_{15}MnN_2O_9^-$, $H_8MnO_4^{2+}$, $4H_2O$ 76.65 **2**
$C_{10}H_{15}NO$ 52.4 **7**
$C_{10}H_{15}NO$, $C_{20}H_{28}CuN_2O_2$, $2H_2O$ 83.182 **2**
$C_{10}H_{15}NO_4$ 59.17+ **1**
$C_{10}H_{15}NO_4$, H_2O 59.19 **1**
$C_{10}H_{15}NO_4S$ 41.10 **7**
$C_{10}H_{15}NO_5Zn$, $2H_2O$ 81.68 **2**
$C_{10}H_{15}N_2$, $C_7H_5O_3$ 60.22 **4**
$2C_{10}H_{15}N_2NiO_9^-$, Ni^{2+} , $8H_2O$ 76.67 **3**
$C_{10}H_{15}N_2O_9Rh$ 76.35 **5**
$C_{10}H_{15}N_3O_5$ 47.26 **7**
$C_{10}H_{15}N_3O_5^+$, $CH_3O_4S^-$, H_2O 47.22 **6**
$3C_{10}H_{15}O_4S^-$, $C_6H_{21}CrN_{15}^{3+}$ 79.16 **3**
$C_{10}H_{16}$ 31.9+ **1**
$C_{10}H_{16}Ag^+$, NO_3^- 75.4 **3**
$C_{10}H_{16}Ag_4O_4^{4+}$, $4ClO_4^-$ 74.1 **3**
$C_{10}H_{16}As_2Cl_4OW$ 86.8 **3**
$C_{10}H_{16}BrNO_2$ 31.12 **1**
$C_{10}H_{16}BrNO_2$ 52.13 **1**
$(C_{10}H_{16}Br_2MnO_4)_n$ 84.10 **3**
$C_{10}H_{16}Br_2N_2O_2$ 33.44 **5**
$C_{10}H_{16}Br_2NiO_4$ 77.3 **4**
$C_{10}H_{16}Br_2O$ 21.26 **1**
$C_{10}H_{16}Br_2O$ 52.14 **1**
$C_{10}H_{16}Cl_2N_2O_8Pd$, $5H_2O$ 76.68 **3**
$C_{10}H_{16}Cl_2O$ 21.4 **4**
$C_{10}H_{16}Cl_2O$ 21.6 **5**
$C_{10}H_{16}Cl_2O_4$ 45.28 **6**
$C_{10}H_{16}Cl_2Pt$ 75.18 **2**
$C_{10}H_{16}Cl_4Cu_2N_4O_{10}$ 79.18 **7**
$C_{10}H_{16}Cl_4Pt_2$ 75.3 **6**
$C_{10}H_{16}CoN_2O_8^-$, Na^+ , $2H_2O$ 82.10 **4**
$C_{10}H_{16}CuN_2O_6$, $4H_2O$ 81.23 **3**
$C_{10}H_{16}CuN_2O_9$ 76.69 **3**
$C_{10}H_{16}CuN_2S_4$ 80.6 **5**
$C_{10}H_{16}Cu_2F_4N_4O_{10}$ 79.19 **7**
$C_{10}H_{16}Cu_2N_2O_{10}$, $2H_2O$ 76.22 **6**
$C_{10}H_{16}I_2PtS_4$ 85.14 **7**
$C_{10}H_{16}Mo_2N_2O_{12}S_2^{2-}$, $2Cs^+$, $2H_2O$ 76.36 **5**
$2C_{10}H_{16}N^+$, Cl_4Cu^{2-} 3.70 **1**
$2C_{10}H_{16}N^+$, $C_{16}O_{15}Rh^{2-}$ 71.35 **5**
$C_{10}H_{16}NO^+$, Cl^- 58.4 **1**
$C_{10}H_{16}NO^+$, Cl^- 58.2 **3**
$C_{10}H_{16}NO^+$, $H_2O_4P^-$ 3.18 **5**
$2C_{10}H_{16}NO^+$, HO_4P^{2-} , H_2O 58.1 **6**
$C_{10}H_{16}NO_3^-$, H_4N^+ 2.30 **5**
$C_{10}H_{16}NO_9S_2^-$, K^+ , H_2O 45.27 **3**
$C_{10}H_{16}N_2$ 20.11 **7**
$C_{10}H_{16}N_2$, $C_6Cl_4O_2$ 60.122 **2**
$C_{10}H_{16}N_2$, $C_{10}H_2N_4$ 60.123 **2**
$C_{10}H_{16}N_2$, $C_{12}H_4N_4$ 60.124 **2**
$C_{10}H_{16}N_2$, $2C_{12}H_4N_4$ 60.127 **2**
$C_{10}H_{16}N_2^+$, I^- 16.42 **1**
$2C_{10}H_{16}N_2^+$, $C_8N_4NiS_4^{2-}$ 12.11 **5**
$C_{10}H_{16}N_2^{2+}$, $2I^-$ 58.3 **3**
$C_{10}H_{16}N_2NiO_9$ 76.69 **2**
$C_{10}H_{16}N_2NiS_4$ 80.7 **5**
$C_{10}H_{16}N_2O_2S_4Se$ 40.9 **3**
$C_{10}H_{16}N_2O_2S_4Te$ 70.3 **3**
$C_{10}H_{16}N_2O_2S_5$ 40.16 **5**
$C_{10}H_{16}N_2O_3$ 43.8 **4**
$C_{10}H_{16}N_2O_3S$ 59.2 **6**
$C_{10}H_{16}N_2O_4Pd$ 82.18 **3**
$C_{10}H_{16}N_2O_5$ 47.27 **3**
$C_{10}H_{16}N_2O_8$ 1.123 **1**
$C_{10}H_{16}N_2O_8$ 1.10 **4**
$C_{10}H_{16}N_2O_8$ 1.27 **5**
$C_{10}H_{16}N_2O_8$ 48.51 **6**
$C_{10}H_{16}N_2O_{10}Zr$, $2H_2O$ 76.48 **7**
$C_{10}H_{16}N_4NiO_6$ 83.45 **3**
$C_{10}H_{16}N_6NiO_4$ 82.15 **5**
$C_{10}H_{16}N_6S_2Se_2Te$ 70.8 **4**
$C_{10}H_{16}O_2$ 52.15 **1**
$C_{10}H_{16}O_2$ 52.16 **1**
$C_{10}H_{16}O_3$ 21.5 **4**
$C_{10}H_{16}O_3$ 1.28 **5**
$C_{10}H_{16}O_3S$ 52.17 **1**
$C_{10}H_{16}O_4$ 22.11 **1**
$C_{10}H_{16}O_4$ 22.12 **1**
$C_{10}H_{16}O_4$ 21.8 **3**
$C_{10}H_{16}O_5Zn$ 77.8+ **2**
$C_{10}H_{17}BrO$ 21.27 **1**
$C_{10}H_{17}ClO$ 21.6 **4**
$C_{10}H_{17}NO_3$, Na^+ , Br^- , H_2O 37.6 **3**
$C_{10}H_{17}NO_4$ 48.52 **6**
$C_{10}H_{17}N_2^+$, ClO_4^- 16.8+ **4**
$C_{10}H_{17}N_2O^+$, Cl^- 37.9 **7**
$C_{10}H_{17}N_2O_2^+$, Br^- 58.2 **7**
$C_{10}H_{17}N_3O_6S$ 48.82 **1**
$C_{10}H_{17}N_3O_6S$ 48.39 **3**
$C_{10}H_{18}As_2ClMnO_3$ 86.8 **5**
$C_{10}H_{18}As_2MoO_4P_2$ 86.8 **7**
$C_{10}H_{18}Br_2$ 23.4 **1**
$C_{10}H_{18}Br_2Ni_2O_4$ 72.27 **2**
$(C_{10}H_{18}CdO_2S_4)_n$ 80.10 **2**
$C_{10}H_{18}ClNO_3S$ 39.11 **4**
$C_{10}H_{18}Cl_2Pd_2$ 72.28 **2**
$C_{10}H_{18}CoN_4O_4S_2$ 79.21 **3**
$C_{10}H_{18}CoN_6O_4^-$, $2C_5H_{11}N_3O_2$, K^+ 79.10 **4**
$C_{10}H_{18}CoN_{10}O_4^{2+}$, $2C_5H_6N_5^+$, $2O_4S^{2-}$, $6H_2O$ 83.23 **6**

$C_{10}H_{18}CoO_6$	77.11	**2**
$C_{10}H_{18}CuN_2O_7S_2$, $2H_2O$	81.24	**3**
$C_{10}H_{18}CuN_4O_4$	83.82	**2**
$C_{10}H_{18}CuN_4O_6$, H_2O	82.39	**2**
$C_{10}H_{18}CuN_6^{2+}$, $2BF_4^-$	83.46	**3**
$C_{10}H_{18}CuN_6^{2+}$, $2ClO_4^-$	83.47	**3**
$C_{10}H_{18}I_2N_2Pd$	71.12	**5**
$C_{10}H_{18}KLaN_2O_{11}$, $5H_2O$	76.71	**2**
$C_{10}H_{18}MgO_6$	67.9	**2**
$C_{10}H_{18}MnO_6$	77.12	**2**
$C_{10}H_{18}MnO_6$	77.6	**3**
$C_{10}H_{18}N^+$, I^-	35.14	**7**
$C_{10}H_{18}N^+$, $C_7H_7O_3S^-$	37.5	**4**
$C_{10}H_{18}NO^+$, Cl^-	37.7	**6**
$C_{10}H_{18}NO_2$	40.3	**7**
$C_{10}H_{18}NO_2^+$, I^-	37.2	**5**
$C_{10}H_{18}NS^+$, Br^- , $3H_2O$	41.17	**5**
$C_{10}H_{18}N_2NdO_{11}^-$, $CH_6N_3^+$	76.37	**5**
$C_{10}H_{18}N_2NiO_5$, $2H_2O$	82.11	**4**
$C_{10}H_{18}N_2O_3$	32.17	**7**
$C_{10}H_{18}N_2O_4S_6$	4.5	**7**
$C_{10}H_{18}N_2O_5$	40.10	**3**
$C_{10}H_{18}N_2O_{11}Sm^-$, Na^+ , $5H_2O$	76.49	**7**
$C_{10}H_{18}N_2Rh^+$, BF_4^-	72.6	**6**
$C_{10}H_{18}N_4NiO_2$	83.27	**5**
$C_{10}H_{18}N_4NiO_2$, $2H_2O$	82.16	**5**
$C_{10}H_{18}N_4NiO_4$	83.83	**2**
$C_{10}H_{18}N_4NiO_4$	83.28	**5**
$C_{10}H_{18}N_4O_6S_2$, $2H_2O$	48.83	**1**
$C_{10}H_{18}NiO_6$	77.13	**2**
$C_{10}H_{18}NiO_6$, H_2O	77.7	**6**
$C_{10}H_{18}NiS_4$	85.15	**5**
$C_{10}H_{18}O_2$	52.1	**4**
$C_{10}H_{18}O_2$	5.8	**6**
$C_{10}H_{18}O_2PbS_4$	69.22	**2**
$C_{10}H_{18}O_2Si_2$	63.13	**2**
$C_{10}H_{18}O_4$	1.124 +	**1**
$C_{10}H_{18}O_4$	1.31	**3**
$C_{10}H_{18}O_4$	1.11	**4**
$C_{10}H_{18}O_4P_6W$	86.9	**7**
$C_{10}H_{18}O_7U$	77.10	**2**
$C_{10}H_{19}Br_2CoN_4O_4$	83.32	**7**
$C_{10}H_{19}ClNO_3P$	64.13	**6**
$C_{10}H_{19}NO_8$, $3H_2O$	45.28	**3**
$C_{10}H_{19}N_2O_2$	48.61	**5**
$C_{10}H_{19}N_7O_4^{2+}$, $2C_6H_4BrO_3S^-$	58.3	**7**
$C_{10}H_{19}O_2^-$, K^+	2.96	**1**
$C_{10}H_{20}AsBrN_2S_4$	65.4	**3**
$C_{10}H_{20}Br_2N_2NiO_2$	83.29	**5**
$(C_{10}H_{20}CdN_2O_4S_2)_n$	82.17	**6**
$C_{10}H_{20}CdN_2S_4$	80.11	**2**
$(C_{10}H_{20}Cd_2N_2O_{11})_n$, nH_2O	82.18	**6**
$C_{10}H_{20}ClCoN_4O_2$, H_2O	71.13	**5**
$C_{10}H_{20}ClFeN_2S_4$	80.6	**3**
$C_{10}H_{20}ClN_6NiO_5^+$, ClO_4^-	83.48	**3**
$C_{10}H_{20}Cl_2CoN_2O_2$	83.49	**3**
$C_{10}H_{20}Cl_2N_2NiO_2$	83.30	**5**
$C_{10}H_{20}Cl_{10}Nb_2S_4$, C_6H_6	85.15	**7**
$C_{10}H_{20}CuN_2O_4S_2$	82.41	**2**
$C_{10}H_{20}CuN_2O_6$	82.42	**2**
$C_{10}H_{20}CuN_2O_6$	82.17	**5**
$C_{10}H_{20}CuN_2S_4$	80.13 +	**2**
$C_{10}H_{20}CuN_2Se_4$	80.7	**3**
$C_{10}H_{20}Cu_2N_4O_{10}$, $2H_2O$	79.20	**7**
$C_{10}H_{20}FeIN_2S_4$	80.3	**4**
$C_{10}H_{20}FeN_2S_4^+$, Br^-	80.5	**7**
$C_{10}H_{20}FeN_3OS_4$	80.16	**2**
$C_{10}H_{20}HgN_2S_4$	80.9	**5**
$C_{10}H_{20}HgN_2S_4$	80.4	**6**
$C_{10}H_{20}MoN_2O_2S_4$	80.5	**3**
$C_{10}H_{20}MoN_2O_2S_4$	80.10	**5**
$C_{10}H_{20}Mo_2N_2O_8S_2$	82.19	**3**
$C_{10}H_{20}N^+$, Cl^-	52.18	**1**
$C_{10}H_{20}N^+$, Cl^-	37.3	**5**
$C_{10}H_{20}N^+$, I^-	35.29	**1**
$C_{10}H_{20}NO^+$, Br^-	37.7	**3**
$C_{10}H_{20}NO^+$, $C_{10}H_{19}NO$, Cl^-	34.14	**7**
$C_{10}H_{20}N_2NiS_4$	80.17	**2**
$C_{10}H_{20}N_2NiSe_4$	80.8	**3**
$C_{10}H_{20}N_2O_2$	1.126	**1**
$C_{10}H_{20}N_2O_2$	3.71	**1**
$C_{10}H_{20}N_2O_2$	1.29	**5**
$C_{10}H_{20}N_2O_2$	48.62	**5**
$C_{10}H_{20}N_2O_2$	48.53	**6**
$C_{10}H_{20}N_2O_3S_2$	48.48	**7**
$C_{10}H_{20}N_2O_4S_2$, $2H_2O$	48.54	**6**
$(C_{10}H_{20}N_2O_{12}Zn_2)_n$, $2nH_2O$	76.50	**7**
$C_{10}H_{20}N_2PbS_4$	69.20	**4**
$C_{10}H_{20}N_2PdS_4$	80.4	**4**
$C_{10}H_{20}N_2PtS_4$	80.19	**2**
$C_{10}H_{20}N_2S_3^{2+}$, $Hg_2I_6^{2-}$	39.20	**5**
$C_{10}H_{20}N_2S_4$	4.18	**1**
$C_{10}H_{20}N_2S_4Se$	39.26	**3**
$C_{10}H_{20}N_2S_4Se$	39.21	**5**
$C_{10}H_{20}N_2S_4Sn$	69.9	**6**
$C_{10}H_{20}N_2S_4Te$	70.9	**4**
$C_{10}H_{20}N_2S_4Zn$	80.21	**2**
$C_{10}H_{20}N_2Se_4Zn$	80.9	**3**
$C_{10}H_{20}N_3ReS_4$	80.5	**4**
$C_{10}H_{20}NiO_6^{2+}$, $2ClO_4^-$	77.4	**4**
$C_{10}H_{20}O_2$	21.7	**5**
$C_{10}H_{20}O_2$	23.1 +	**6**
$C_{10}H_{20}O_3S_2$	42.6	**7**
$C_{10}H_{20}O_4$	38.8	**5**
$C_{10}H_{20}O_4P_2S_2$	64.25	**7**
$C_{10}H_{20}O_4S_2$	45.14	**4**
$C_{10}H_{20}O_5P_2S$	64.26	**7**
$C_{10}H_{20}O_6P_2$	64.27	**7**
$C_{10}H_{20}O_6P_2S$	64.28	**7**
$C_{10}H_{20}O_7P_2$	64.29	**7**
$C_{10}H_{20}P_2S_2$	64.21	**3**
$C_{10}H_{21}AgNPS$	85.64	**2**
$C_{10}H_{21}LaN_2O_{12}$, $3H_2O$	76.72	**2**
$C_{10}H_{21}NO$	1.127	**1**
$C_{10}H_{21}NO$	1.23	**6**
$C_{10}H_{21}N_4NiO_{13}P$, $2H_2O$	83.24	**6**
$C_{10}H_{21}OP$	64.13	**4**
$C_{10}H_{21}S^+$, ClO_4^-	39.10	**6**
$C_{10}H_{21}S^+$, ClO_4^-	39.11	**6**
$C_{10}H_{22}AlN$	68.5	**4**
$C_{10}H_{22}Cl_2N_2O_2Pt$	83.50	**3**
$C_{10}H_{22}Cl_2N_2PtS$	72.11	**5**
$C_{10}H_{22}N^+$, Cl^- , $1.5H_2O$	23.5	**1**
$C_{10}H_{22}NO_4^+$, I^-	45.29	**3**
$C_{10}H_{22}N_2OPdS_2^{2+}$, $2NO_3^-$	85.16	**7**
$C_{10}H_{22}N_2O_4Pd$, H_2O	82.19	**6**
$C_{10}H_{22}N_2O_4S_2^{2+}$, $2Cl^-$	48.49	**7**
$C_{10}H_{22}N_4O_4Pd$	82.18	**5**
$C_{10}H_{22}N_5NiO_{13}P$, $3H_2O$	83.33	**7**
$C_{10}H_{22}O_2$, $4H_2O$	5.17	**7**
$C_{10}H_{22}O_5$, Cl_2Hg	5.9	**6**
$C_{10}H_{23}Cl_2CoN_4O_2$	83.51	**3**

$C_{10}H_{23}N_4NiO_2^+$, Cl^- , H_2O 83.84 **2**
$C_{10}H_{23}N_4NiO_2^+$, Cl^- , H_2O 83.42 **4**
$C_{10}H_{23}N_4O_2Pt^+$, Cl^- , H_2O 83.52 **3**
$C_{10}H_{24}Al_2N_2$ 68.12 **7**
$C_{10}H_{24}Br_4N_4S_2Te$ 70.17 **2**
$C_{10}H_{24}ClN_4Ni^+$, Cl^- 83.86 **2**
$C_{10}H_{24}ClN_4Ni^+$, Cl^- 83.25 **6**
$C_{10}H_{24}Cl_2CoN_4S_2$ 79.23 **5**
$C_{10}H_{24}Cl_2CuN_2$ 76.51 **7**
$C_{10}H_{24}Cl_2CuN_4S_2$ 79.22 **3**
$C_{10}H_{24}Cl_2N_4O_3V$ 79.23 **3**
$C_{10}H_{24}Cl_2N_4S_2Zn$ 79.11 **4**
$C_{10}H_{24}Cl_4N_4S_2Te$ 70.18 **2**
$C_{10}H_{24}Cl_4N_4S_2Te$ 70.6 **7**
$C_{10}H_{24}CuN_4^{2+}$, $2NO_3^-$, $0.5H_2O$ 83.87 **2**
$C_{10}H_{24}CuN_4O_4^{2+}$, $2Cl^-$, $2H_2O$ 82.21 **3**
$C_{10}H_{24}Ga_2N_2O_4$ 68.13 **7**
$C_{10}H_{24}N^+$, Cl^- 3.72 **1**
$3C_{10}H_{24}N^+$, $C_5CoN_5^{3-}$ 3.27 **7**
$C_{10}H_{24}N_2^{2+}$, $2Cl^-$ 23.6+ **1**
$C_{10}H_{24}N_2^{2+}$, $2Cl^-$, $2H_2O$ 23.8 **1**
$C_{10}H_{24}N_2NiO_2S_2^{2+}$, $2NO_3^-$ 85.18 **6**
$C_{10}H_{24}N_4O_5P_2$ 64.14 **5**
$C_{10}H_{24}N_4PtS_2^{2+}$, $2F_6P^-$ 71.14 **5**
$C_{10}H_{24}Si_4$ 63.2 **5**
$C_{10}H_{25}BrMgO_2$ 67.10 **2**
$C_{10}H_{26}Al_2O_2$ 68.18 **2**
$C_{10}H_{26}CuN_4^{2+}$, $2ClO_4^-$ 83.34 **7**
$C_{10}H_{26}Cu_2N_6O_4^{2+}$, $2ClO_4^-$ 76.23 **6**
$C_{10}H_{26}Cu_2N_8S_2^{2+}$, $2ClO_4^-$ 76.38 **5**
$C_{10}H_{26}N_2^{2+}$, $2Br^-$, $2H_2O$ 3.21 **4**
$2C_{10}H_{26}N_2^{2+}$, $Cl_{10}Cu_4O^{4-}$ 3.19 **5**
$C_{10}H_{26}N_4^{2+}$, $2ClO_4^-$ 34.15 **7**
$C_{10}H_{26}N_6Pd^{2+}$, $2ClO_4^-$ 83.35 **7**
$C_{10}H_{27}N_6Pd^{3+}$, $3NO_3^-$ 83.36 **7**
$C_{10}H_{28}CoN_6^{3+}$, $C_6CoN_6^{3-}$, $2H_2O$ 76.71 **3**
$C_{10}H_{28}CoN_6O^{3+}$, Cl_6Tl^{3-} , $2H_2O$ 76.52 **7**
$C_{10}H_{28}N_4Zn_2$ 76.72+ **3**
$C_{10}H_{28}N_6Pd^{4+}$, $4ClO_4^-$, $2H_2O$ 83.37 **7**
$C_{10}H_{30}Al_4Cl_4N_6$ 68.6 **4**
$C_{10}H_{30}B_6P_2Pt$ 86.10+ **7**
$C_{10}H_{30}I_2Rh_2S_3$ 85.65 **2**
$C_{10}H_{30}N_2Si_4$ 63.14 **2**
$C_{10}H_{30}N_4^{4+}$, $4Cl^-$ 59.20 **1**
$C_{10}H_{30}N_4^{4+}$, $2HO_4P^{2-}$, $6H_2O$ 59.21 **1**
$C_{10}H_{30}N_5Nb$ 83.53 **3**
$C_{10}H_{32}Cr_2^{4-}$, $4C_4H_{10}LiO^+$ 71.12 **4**
$C_{10}H_{32}N_5P_5^{2+}$, Cl_4Cu^{2-} , H_2O 64.14 **6**
$C_{10}H_{32}N_8Ni_2O_4^{2+}$, $2ClO_4^-$ 76.53 **7**
$C_{10}H_{32}N_8Ni_2O_4^{2+}$, $2NO_3^-$ 76.24 **6**
$C_{10}H_{32}N_8Ni_2O_4^{2+}$, $2NO_3^-$ 76.54 **7**
$C_{10}H_{32}N_8O_4Zn_2^{2+}$, $2CNS^-$ 76.55 **7**
$C_{10}H_{32}N_{10}Ni_2S_2^{2+}$, $2ClO_4^-$ 76.70 **2**
$C_{10}H_{32}N_{10}Ni_2S_2^{2+}$, $2I^-$ 76.74 **2**
$C_{10}H_{33}N_6^{5+}$, $5Cl^-$ 3.73 **1**
$(C_{10}H_{40}Cu_4N_{20}S_{10}^{4+})_n$, $2nF_6Si^{2-}$, nH_2O 79.24 **5**
$C_{10}H_{45}N_5Si_5$ 63.15 **2**

C_{11}

$C_{11}F_{24}Ni_2O_3P_4S_2$ 86.12 **7**
$C_{11}HN_6^-$, $C_4H_{12}N^+$ 12.4 **7**
$C_{11}H_2Co_2O_9$ 71.15 **5**
$C_{11}H_3Co_3O_9$ 71.22 **2**
$C_{11}H_5BrClN_3O$ 40.17 **5**
$C_{11}H_5CoFeHgO_6$ 73.31 **2**
$C_{11}H_5CoFeO_6$ 73.11 **7**
$C_{11}H_5F_7MoO_3$ 73.32 **2**
$C_{11}H_5IO_4W$ 71.14 **7**
$C_{11}H_5O_4ReS_2$ 85.11 **4**
$C_{11}H_6BrN_3O_2$ 40.16 **4**
$C_{11}H_6Br_2O$ 27.5 **4**
$C_{11}H_6Fe_3N_2O_9$ 83.88 **2**
$C_{11}H_6O_9Ru_3$ 71.15 **7**
$C_{11}H_7BrN_2O_2$ 36.4 **5**
$C_{11}H_7BrN_2O_2$ 40.18 **5**
$C_{11}H_7ClN_4O_2S$ 42.4 **3**
$C_{11}H_7ClO$ 27.6 **4**
$C_{11}H_7ClO_2$ 25.15 **1**
$C_{11}H_7CrO_5^-$, $C_8H_{12}N^+$ 3.58 **1**
$C_{11}H_7CrO_5^-$, $C_8H_{12}N^+$ 3.59 **1**
$C_{11}H_7FeO_3$ 71.9 **4**
$C_{11}H_7NO$ 36.5 **1**
$C_{11}H_7NO$ 36.2 **4**
$C_{11}H_7S_2^+$, ClO_4^- 39.21 **7**
$C_{11}H_8$ 28.3 **6**
$C_{11}H_8Br_2FeO_3$ 75.6 **7**
$C_{11}H_8Br_2O_3$ 1.128 **1**
$C_{11}H_8ClN_3O_2S$ 41.10 **3**
$C_{11}H_8Cl_3NO_3$ 35.16 **6**
$C_{11}H_8CrO_4$ 74.2 **3**
$C_{11}H_8CrO_5$ 74.7 **2**
$C_{11}H_8CrO_5$ 74.3 **5**
$C_{11}H_8F_2$ 31.6 **4**
$C_{11}H_8FeO_3$ 75.19 **2**
$C_{11}H_8FeO_3$ 75.5 **5**
$C_{11}H_8Fe_2N_2O_6$ 83.43 **4**
$C_{11}H_8MoO_3$ 75.20 **2**
$C_{11}H_8N_2O_3S_2$ 59.5 **4**
$C_{11}H_8N_2O_3S_2$ 41.18 **5**
$C_{11}H_8N_2O_3S_2$ 41.11 **7**
$C_{11}H_8N_2S$ 41.19 **5**
$C_{11}H_8N_4O_2$ 36.2 **6**
$C_{11}H_8N_4O_2$, Ag^+ , NO_3^- 36.5 **5**
$4C_{11}H_8N_4O_2$, $3Ag^+$, $3ClO_4^-$ 36.6 **5**
$C_{11}H_8O$ 27.10 **1**
$C_{11}H_8O$ 31.10 **6**
$C_{11}H_8OS_2$ 39.22 **7**
$C_{11}H_8O_2$ 24.22 **1**
$C_{11}H_8O_2$ 24.23 **1**
$C_{11}H_8O_2$ 28.4 **1**
$C_{11}H_8O_2$ 27.4 **7**
$C_{11}H_8O_3$ 25.16 **1**
$C_{11}H_8O_3$ 24.3 **4**
$C_{11}H_8O_3$ 24.4 **7**
$C_{11}H_8O_3Ru$ 75.5 **3**
$C_{11}H_9BrO_3$ 38.8 **3**
$C_{11}H_9Br_2NO$ 40.19 **5**
$C_{11}H_9Br_2NO$ 40.20 **5**
$C_{11}H_9Br_2NO$ 40.10 **6**
$C_{11}H_9Br_2NO$ 40.11 **6**
$C_{11}H_9ClN_2O_3$ 40.25 **1**
$C_{11}H_9ClO$ 20.12 **3**
$C_{11}H_9Cl_2NO$ 40.12 **6**
$C_{11}H_9FeNO_4$ 75.6 **3**
$C_{11}H_9FeNO_5$ 75.7 **3**
$C_{11}H_9FeO_3^+$, BF_4^- 75.21 **2**
$C_{11}H_9NO_2$ 40.17 **4**

$C_{11}H_9NO_2$ 40.22 **5**
$C_{11}H_9NO_2$ 40.23 **5**
$C_{11}H_9NO_2$, H_2O 25.5 **3**
$C_{11}H_9NO_2$, $0.25H_2O$ 25.18 **1**
$C_{11}H_9NO_3$ 35.11 **5**
$C_{11}H_9N_3S$ 41.16 **6**
$C_{11}H_9N_4O_2^+$, Br^- , $2H_2O$ 36.6 **1**
$C_{11}H_9N_4O_2^+$, Br^- , $2H_2O$ 36.3 **3**
$C_{11}H_9N_4O_2^+$, ClO_4^- , H_2O 36.4 **7**
$C_{11}H_9N_4O_2^+$, $1.5C_{10}H_8O_2$, Br^- , H_2O 60.35 **3**
$C_{11}H_9O_3Re$ 73.33 **2**
$C_{11}H_9O_5S^-$, Na^+ 11.44 **1**
$C_{11}H_{10}CoN_4^{2+}$, $2Cl^-$ 83.89 **2**
$C_{11}H_{10}CrO_3$ 75.22 **2**
$C_{11}H_{10}FeN_2O_5$ 75.6 **5**
$C_{11}H_{10}FeO_4S$ 71.8 **3**
$C_{11}H_{10}NO_2^+$, Br^- , H_2O 33.45 **5**
$C_{11}H_{10}NO_{12}PW_2$ 86.13+ **7**
$C_{11}H_{10}N_2$ 27.11 **1**
$C_{11}H_{10}N_2O_2S$ 41.32 **1**
$C_{11}H_{10}N_2O_3$ 40.26 **1**
$C_{11}H_{10}N_2O_3$ 43.26 **1**
$C_{11}H_{10}N_3P$ 64.30 **7**
$C_{11}H_{10}O_2$ 24.24 **1**
$C_{11}H_{10}O_2S$ 27.7 **3**
$C_{11}H_{10}O_5$ 38.17 **7**
$C_{11}H_{10}O_6$ 31.7 **4**
$C_{11}H_{11}BrN_2O$ 32.31 **1**
$C_{11}H_{11}BrN_2O_2$ 40.13 **6**
$C_{11}H_{11}IO_5$ 27.7 **4**
$C_{11}H_{11}NO_2$ 40.11 **3**
$C_{11}H_{11}NO_3$ 20.13 **5**
$C_{11}H_{11}N_3O_2$ 35.12 **5**
$C_{11}H_{11}NbO$ 73.12 **5**
$C_{11}H_{11}NbOS$ 73.12 **7**
$C_{11}H_{11}P$ 64.15 **6**
$C_{11}H_{12}As_2F_6I_2O_3W$ 86.15 **7**
$C_{11}H_{12}As_2F_6MoO_4$ 86.9 **3**
$C_{11}H_{12}As_2F_6MoO_4$ 86.9 **5**
$C_{11}H_{12}BrNO$ 1.12 **4**
$C_{11}H_{12}NO^+$, Cl^- 24.25 **1**
$C_{11}H_{12}NS^+$, Br^- 41.11 **3**
$C_{11}H_{12}N_2$ 35.11 **3**
$C_{11}H_{12}N_2O$ 32.28 **3**
$C_{11}H_{12}N_2O$ 32.17 **5**
$C_{11}H_{12}N_2O$, $C_7H_6O_3$ 60.15 **6**
$C_{11}H_{12}N_2O_2$, H_2O 32.25 **6**
$C_{11}H_{12}N_2O_3$ 48.55 **6**
$C_{11}H_{12}N_2O_3$ 48.56 **6**
$C_{11}H_{12}N_2S$ 41.20 **5**
$C_{11}H_{12}OS_2$ 11.6 **4**
$C_{11}H_{12}O_2$ 20.13 **3**
$C_{11}H_{12}O_2$ 29.1 **4**
$C_{11}H_{12}O_3S$ 39.27 **3**
$C_{11}H_{12}O_4$ 1.32 **3**
$C_{11}H_{12}O_6$ 38.9 **3**
$C_{11}H_{13}As_2ClCrF_4O_4$ 86.10 **5**
$C_{11}H_{13}BN_6OPt$ 83.38 **7**
$C_{11}H_{13}BrClO_4P$ 46.7 **5**
$C_{11}H_{13}BrN_2O_4$ 45.57+ **1**
$C_{11}H_{13}BrO_5$ 38.17 **6**
$C_{11}H_{13}BrO_5$ 22.3 **7**
$C_{11}H_{13}ClO_2S$ 20.5 **4**
$C_{11}H_{13}CrNO_3$ 75.4 **6**
$C_{11}H_{13}CrNO_3$ 75.5 **6**
$C_{11}H_{13}CuNO_4$ 78.1 **4**
$C_{11}H_{13}FeO_4^+$, F_6P^- 72.7 **6**
$C_{11}H_{13}IMoN_3O^+$, BF_4^- 73.13 **7**
$C_{11}H_{13}I_2NO_3$ 48.41 **3**
$C_{11}H_{13}MoNO$ 71.10 **4**
$C_{11}H_{13}N$ 31.5 **5**
$C_{11}H_{13}NO$ 35.17 **6**
$C_{11}H_{13}NO_5S$ 41.12 **7**
$2C_{11}H_{13}N_2^+$, I_4Zn^{2-} 35.18 **6**
$C_{11}H_{13}N_2O_2^+$, Br^- 48.84+ **1**
$C_{11}H_{13}N_2O_2^+$, Cl^- 48.57 **6**
$C_{11}H_{13}N_2O_2^+$, Cl_2^- 48.86 **1**
$C_{11}H_{13}N_2O_2^+$, CHO_2^- 48.63 **5**
$C_{11}H_{13}N_2O_2^+$, $C_6H_2N_3O_7^-$, CH_4O 48.58 **6**
$C_{11}H_{13}N_2S^+$, Cl^- 41.21 **5**
$C_{11}H_{13}OP$ 64.14 **4**
$C_{11}H_{13}O_3ReSi$ 73.9 **4**
$C_{11}H_{14}AgP$ 72.29 **2**
$C_{11}H_{14}AuN$ 71.23 **2**
$C_{11}H_{14}ClO_4P$ 46.12 **6**
$C_{11}H_{14}CoN_2O_8^-$, K^+ , $2H_2O$ 81.31 **4**
$C_{11}H_{14}CrNO_5$ 71.24 **2**
$C_{11}H_{14}CuP$ 72.30 **2**
$C_{11}H_{14}MnNO_4S$ 77.8 **6**
$C_{11}H_{14}N_2O$ 34.2 **6**
$C_{11}H_{14}N_2O$ 35.19 **6**
$C_{11}H_{14}N_2O_2$ 16.10 **6**
$C_{11}H_{14}N_2O_2$ 36.5 **7**
$C_{11}H_{14}N_2O_3$ 48.43 **3**
$C_{11}H_{14}N_2O_3$ 35.15 **7**
$C_{11}H_{14}N_2O_3S$ 33.35 **3**
$C_{11}H_{14}N_2O_3S$ 41.13 **4**
$C_{11}H_{14}N_2O_4$, $2H_2O$ 48.59 **6**
$C_{11}H_{14}N_2O_6$ 32.18 **7**
$C_{11}H_{14}N_4O_4$ 35.13+ **5**
$C_{11}H_{14}N_4O_4$ 47.27 **7**
$C_{11}H_{14}N_4O_4$ 47.28 **7**
$C_{11}H_{14}N_5O_5P$, H_2O 47.26 **5**
$C_{11}H_{14}N_6O_4$ 45.29 **6**
$C_{11}H_{14}O$ 27.8 **4**
$C_{11}H_{14}O_2$ 31.8 **3**
$C_{11}H_{14}O_2$ 27.8 **5**
$C_{11}H_{15}Ag_2AsNO_3^+$, NO_3^- 72.8 **6**
$C_{11}H_{15}BrN_2O_4$ 45.59+ **1**
$C_{11}H_{15}BrN_3O_7^+$, Br^- 58.1 **4**
$C_{11}H_{15}BrO_7$ 45.20 **5**
$C_{11}H_{15}Cl$ 19.33 **1**
$C_{11}H_{15}ClO_7$ 45.21 **5**
$C_{11}H_{15}ClO_7$ 45.29 **7**
$C_{11}H_{15}CoN_7O_2^{2+}$, $2Cl^-$, $3H_2O$ 79.11 **6**
$C_{11}H_{15}F_3MoO_6$ 72.3 **7**
$C_{11}H_{15}F_3O_6W$ 72.4 **7**
$C_{11}H_{15}FeN_7O_2^{2+}$, $2Cl^-$, $3H_2O$ 79.12 **6**
$C_{11}H_{15}NO$ 52.19 **1**
$C_{11}H_{15}NO_2S$ 39.12 **4**
$C_{11}H_{15}NO_3$ 48.44 **3**
$C_{11}H_{15}N_2^+$, $C_{12}H_4N_4$, $C_{12}H_4N_4^-$, C_2H_3N 7.10 **5**
$C_{11}H_{15}N_2O_3^+$, Cl^- , H_2O 48.60 **6**
$C_{11}H_{15}N_2O_3^+$, $C_6H_4BrO_3S^-$ 48.65 **5**
$C_{11}H_{15}N_2O_3^+$, $C_7H_7O_3S^-$ 48.66 **5**
$C_{11}H_{15}N_3$ 37.15 **1**
$C_{11}H_{15}N_3O_7$ 45.30 **7**
$C_{11}H_{15}N_5$, I_2 44.21 **6**
$C_{11}H_{15}N_5O_4$ 47.23 **6**
$C_{11}H_{15}N_5O_4$ 50.4 **6**

$C_{11}H_{15}N_5S$ 44.22 **4**
$C_{11}H_{15}N_7O_2Zn^{2+}$, $2Cl^-$, $3H_2O$ 79.13 **6**
$C_{11}H_{15}O_2Tl$ 68.19 **2**
$C_{11}H_{15}O_3P$ 64.23 **3**
$C_{11}H_{15}Rh$ 72.12 **5**
$C_{11}H_{16}Br_2$ 28.1 **3**
$C_{11}H_{16}ClNO_3S$ 37.8 **6**
$C_{11}H_{16}ClN_3OS$ 41.13 **7**
$C_{11}H_{16}CuN_6O_3$, H_2O 82.14 **7**
$C_{11}H_{16}IMoO_5P$ 73.13 **5**
$C_{11}H_{16}MoNO^+$, F_6P^- 73.14 **7**
$C_{11}H_{16}MoN_3OS_2$ 71.13 **6**
$C_{11}H_{16}N^+$, Cl^- 20.12 **7**
$C_{11}H_{16}NO^+$, Cl^- 40.4 **7**
$C_{11}H_{16}NOSe^+$, Cl^- 3.22 **4**
$C_{11}H_{16}NPS$ 64.31 **7**
$C_{11}H_{16}N_2O_3$ 43.27 **1**
$C_{11}H_{16}N_2O_4^+$, Cl^- , H_2O 48.87 **1**
$C_{11}H_{16}N_2O_5S$ 47.28 **3**
$C_{11}H_{16}N_2O_6$ 38.18 **7**
$C_{11}H_{16}N_4NiO_4$ 84.11 **6**
$C_{11}H_{16}N_4O_4$ 47.29 **7**
$C_{11}H_{16}N_4O_5$, $1.5H_2O$ 47.30 **7**
$C_{11}H_{16}N_5O_3^+$, Br^- 50.5 **1**
$C_{11}H_{16}N_5O_6P$, C_2H_6O 47.16 **4**
$C_{11}H_{16}O_4S$ 31.9 **3**
$C_{11}H_{16}O_8$ 45.22 **5**
$C_{11}H_{16}O_8$, H_2O 45.23 **5**
$C_{11}H_{16}P^+$, I^- 64.31 **2**
$C_{11}H_{17}ClN_5^+$, Br^- 8.47 **1**
$C_{11}H_{17}ClN_5^+$, Cl^- 8.48 **1**
$C_{11}H_{17}ClN_5^+$, Cl^- 8.10 **4**
$C_{11}H_{17}ClO_2$ 38.19 **7**
$C_{11}H_{17}CoN_8O$ 83.39 **7**
$C_{11}H_{17}FN_2O_6S$ 45.31 **7**
$C_{11}H_{17}NO_5$ 45.16 **4**
$C_{11}H_{17}N_2^+$, Br^- 58.4 **3**
$C_{11}H_{17}N_2O_2^+$, Cl^- 58.2 **6**
$C_{11}H_{17}N_2O_2^+$, Cl_3Ge^- , $0.5H_2O$ 58.6 **1**
$2C_{11}H_{17}N_3S_4Zn$, C_6H_6 80.22 **2**
$C_{11}H_{17}N_6O_3^+$, I^- , H_2O 47.17 **4**
$C_{11}H_{18}BrN_3O_{10}S$, H_2O 50.4 **1**
$C_{11}H_{18}NO_2^+$, Br^- 48.61 **6**
$C_{11}H_{18}NO_3^+$, Br^- 17.14 **6**
$C_{11}H_{18}NO_3^+$, Cl^- 3.20 **5**
$C_{11}H_{18}NO_3^+$, Cl^- 17.18 **7**
$2C_{11}H_{18}NO_3^+$, O_4S^{2-} , $2H_2O$ 59.3 **3**
$2C_{11}H_{18}NO_3^+$, O_4S^{2-} , C_2H_6O 17.15 **5**
$C_{11}H_{18}N_2O_2$ 48.47 **4**
$C_{11}H_{18}N_2O_3$ 43.28 + **1**
$C_{11}H_{18}N_2O_3$ 43.2 **6**
$C_{11}H_{18}N_2O_3$, $C_7H_7NO_2$ 60.16 **6**
$C_{11}H_{18}N_2O_4$ 48.50 **7**
$C_{11}H_{18}N_2O_4$ 48.51 **7**
$C_{11}H_{18}N_3O_8^+$, Br^- 58.7 **1**
$C_{11}H_{18}N_3O_8^+$, Br^- 58.2 **4**
$C_{11}H_{18}O$ 31.10 **3**
$C_{11}H_{18}O_7$ 59.3 **6**
$C_{11}H_{18}O_8$ 45.32 **7**
$C_{11}H_{19}CuN_7O_4^{2+}$, $2NO_3^-$, H_2O 84.30 **7**
$C_{11}H_{19}NO_2S_4$ 58.5 **3**
$C_{11}H_{19}NO_8$, H_2O 45.30 **6**
$C_{11}H_{19}NO_9$, $2H_2O$ 45.31 **6**
$C_{11}H_{19}N_2^+$, BF_4^- 35.12 **4**
$C_{11}H_{19}N_2^+$, ClO_4^- 32.19 **7**
$C_{11}H_{19}N_2O^+$, I^- 32.26 **6**
$C_{11}H_{19}N_2O_3S^+$, $C_{11}H_{18}N_2O_3S$, I^- 41.17 **6**
$C_{11}H_{19}N_7NiO_4^{2+}$, $2NO_3^-$, H_2O 84.31 **7**
$C_{11}H_{20}CuP$ 73.7 **3**
$C_{11}H_{20}N_2$ 32.16 **4**
$C_{11}H_{20}O$ 23.3 **6**
$C_{11}H_{20}O_2$ 21.7 **4**
$C_{11}H_{20}O_4$ 1.129 **1**
$C_{11}H_{21}As_2ClF_6Pt$ 71.16 **5**
$C_{11}H_{21}BrO_2$ 1.130 **1**
$C_{11}H_{21}N_4Ni^+$, ClO_4^- 83.45 **4**
$C_{11}H_{21}O_3P$ 64.16 **6**
$C_{11}H_{22}O_2S$ 1.131 **1**
$C_{11}H_{23}BrO$ 5.8 **4**
$C_{11}H_{23}ClCoN_8O_2^+$, Cl^- , $2H_2O$ 76.56 **7**
$C_{11}H_{23}ClCoN_8O_2^+$, ClO_4^- 76.57 **7**
$C_{11}H_{23}MoN_4O_6P$ 83.46 **4**
$C_{11}H_{23}NO$ 1.24 **6**
$C_{11}H_{23}NO$ 1.25 **6**
$C_{11}H_{23}O_4P$ 64.24 **3**
$C_{11}H_{24}I_3SZn^-$, $C_{11}H_{25}S^+$ 11.45 **1**
$C_{11}H_{24}MoN_4O_3P_4$ 75.6 **6**
$C_{11}H_{24}N^+$, Cl^- 34.7 **1**
$C_{11}H_{24}NO_2^+$, Br^- , $0.5H_2O$ 1.132 **1**
$2C_{11}H_{24}NO_2^+$, Cl_4Cu^{2-} 3.21 **5**
$C_{11}H_{25}N_4^+$, ClO_4^- 3.28 **7**
$C_{11}H_{25}S^+$, $C_{11}H_{24}I_3SZn^-$ 11.45 **1**
$C_{11}H_{26}CoN_5O_2^{2+}$, Cl_4Zn^{2-} 76.74 **3**
$C_{11}H_{26}CoN_5O_2^{2+}$, $2I^-$, $2H_2O$ 76.75 **3**
$C_{11}H_{26}N_2O^{2+}$, $2CH_3O_3S^-$, H_2O 38.10 **3**
$C_{11}H_{26}N_4Ni^{2+}$, $2ClO_4^-$ 83.31 **5**
$C_{11}H_{27}As_3Br_2Ni$ 86.7 **2**
$C_{11}H_{27}CoN_4O_2^{2+}$, $C_8H_4As_2O_{12}^{2-}$, H_2O 77.9 **6**
$C_{11}H_{28}N_2^{2+}$, $2I^-$, $0.25H_2O$ 3.74 **1**
$C_{11}H_{29}B_2CuN_5$ 76.26 **6**
$C_{11}H_{30}CoN_6^{3+}$, $C_6CoN_6^{3-}$, $3H_2O$ 76.27 **6**

C_{12}

$C_{12}Cl_8O_2$ 38.20 **7**
$C_{12}Cl_{12}$ 28.5 **1**
$C_{12}Cl_{12}$ 31.13 **1**
$C_{12}Cl_{14}$ 22.13 **1**
$C_{12}Co_2F_5O_7$ 73.34 **2**
$C_{12}Co_2F_6O_6$ 75.23 **2**
$C_{12}F_{10}$ 19.34 **1**
$C_{12}F_{10}Hg$ 71.25 **2**
$C_{12}F_{10}Hg$ 71.9 **3**
$C_{12}F_{10}Se$ 11.8 **6**
$C_{12}F_{10}Se_2$ 11.9 **6**
$C_{12}F_{12}FeO_4$ 73.35 **2**
$C_{12}F_{18}MoSe_6$ 85.12 **4**
$C_{12}FeN_6S_6^{2-}$, $2C_{24}H_{20}As^+$ 85.19 **6**
$C_{12}InN_6S_6^{3-}$, $3C_8H_{20}N^+$ 68.7 **4**
$C_{12}MoN_6S_6^{2-}$, $2C_{24}H_{20}As^+$ 85.20 **6**
$C_{12}N_6S_6V^{2-}$, $2C_4H_{12}N^+$ 3.40 **1**
$C_{12}N_6S_6W^{2-}$, $2C_{24}H_{20}As^+$ 85.21 **6**
$C_{12}O_{12}^{6-}$, $H_{30}Al_2O_{15}^{6+}$, H_2O 13.14 **5**
$C_{12}HF_{10}OTl$ 68.10 **3**
$C_{12}H_2Ca_2O_{12}$, $9H_2O$ 67.7 **5**
$C_{12}H_2Cl_6O_2$ 59.22 **1**
$C_{12}H_4ClO_6Re$ 71.10 **3**
$C_{12}H_4Cl_2O_3$ 24.4 **4**

$C_{12}H_4Cl_4N_2$	36.7	**1**
$C_{12}H_4Cl_4N_2$	36.8	**1**
$C_{12}H_4Cl_4O_2$	38.21	**7**
$C_{12}H_4Fe_3O_{11}$	71.16	**7**
$C_{12}H_4N_2O_2$	27.6	**6**
$C_{12}H_4N_2O_7$	24.5	**4**
$C_{12}H_4N_4$	7.29	**1**
$C_{12}H_4N_4$, $2C_3H_{10}N^+$, $2C_{12}H_4N_4^-$	3.2	**6**
$C_{12}H_4N_4$, C_6H_6	60.17	**6**
$C_{12}H_4N_4$, $C_{10}H_{13}N_2^+$, $C_{12}H_4N_4^-$, C_2H_3N	7.9	**5**
$C_{12}H_4N_4$, $C_{10}H_{16}N_2$	60.124	**2**
$C_{12}H_4N_4$, $C_{11}H_{15}N_2^+$, $C_{12}H_4N_4^-$, C_2H_3N	7.10	**5**
$C_{12}H_4N_4$, $C_{12}H_4N_4^-$, $C_6H_{16}N^+$	12.1	**6**
$C_{12}H_4N_4$, $2C_{12}H_4N_4^-$, $2Cs^+$	7.30	**1**
$C_{12}H_4N_4$, $C_{12}H_8N_2$	60.14	**5**
$C_{12}H_4N_4$, $C_{12}H_8N_2$	60.15	**5**
$C_{12}H_4N_4$, $C_{12}H_8O_2$	60.16	**5**
$C_{12}H_4N_4$, $C_{12}H_9N$	60.14	**7**
$C_{12}H_4N_4$, $C_{12}H_9NS$	60.19	**6**
$C_{12}H_4N_4$, $C_{12}H_{10}$	60.20	**5**
$C_{12}H_4N_4$, $C_{12}H_{12}N_2$	60.21	**5**
$C_{12}H_4N_4$, $C_{12}H_{12}N_2$	60.20	**6**
$C_{12}H_4N_4$, $C_{12}H_{12}N_2$, $1.8CH_2Cl_2$	60.22	**5**
$C_{12}H_4N_4$, $C_{12}H_{12}N_2$, C_6H_6	60.21	**6**
$C_{12}H_4N_4$, $C_{12}H_{18}$	60.36	**3**
$C_{12}H_4N_4$, $C_{12}H_{18}$	60.18	**6**
$C_{12}H_4N_4$, $C_{13}H_{11}NS$	60.18	**7**
$C_{12}H_4N_4$, $C_{14}H_{10}$	60.125	**2**
$C_{12}H_4N_4$, $C_{14}H_{14}N_2$	60.17+	**5**
$C_{12}H_4N_4$, $C_{16}H_{10}$	60.32+	**5**
$C_{12}H_4N_4$, $C_{18}H_{12}$	60.28	**7**
$C_{12}H_4N_4$, $C_{18}H_{12}CuN_2O_2$	60.126	**2**
$C_{12}H_4N_4$, $C_{20}H_{12}$	60.38	**5**
$C_{12}H_4N_4^-$, Na^+	12.2	**6**
$C_{12}H_4N_4^-$, Rb^+	12.10	**4**
$C_{12}H_4N_4^-$, Rb^+	12.3	**6**
$C_{12}H_4N_4^-$, $C_4H_{10}NO^+$	40.5	**5**
$C_{12}H_4N_4^-$, $C_6H_4S_4^+$	60.3	**7**
$C_{12}H_4N_4^-$, $C_6H_{12}NS^+$, $C_{12}H_4N_4$	60.8	**5**
$C_{12}H_4N_4^-$, $C_6H_{16}N^+$, $C_{12}H_4N_4$	3.25	**3**
$C_{12}H_4N_4^-$, $C_9H_8N^+$, $C_{12}H_4N_4$	35.4	**3**
$C_{12}H_4N_4^-$, $C_9H_{11}N_2^+$	35.9	**4**
$C_{12}H_4N_4^-$, $C_{10}H_{13}N_2^+$	35.10	**5**
$C_{12}H_4N_4^-$, $C_{12}H_4N_4$, $C_6H_{16}N^+$	12.1	**6**
$C_{12}H_4N_4^-$, $C_{12}H_{12}Cr^+$	71.23	**7**
$C_{12}H_4N_4^-$, $C_{12}H_{14}N^+$, $C_{12}H_4N_4$	60.25	**4**
$C_{12}H_4N_4^-$, $C_{13}H_{10}N^+$, $C_{12}H_4N_4$	60.17	**7**
$C_{12}H_4N_4^-$, $C_{13}H_{11}N_2^+$	36.23	**1**
$C_{12}H_4N_4^-$, $C_{13}H_{21}N_2S_2^+$, $C_{16}H_5N_5O_6$	60.21	**7**
$C_{12}H_4N_4^-$, $C_{14}H_{16}Cr^+$	74.16	**2**
$C_{12}H_4N_4^-$, $C_{14}H_{16}Cr^+$, $C_{12}H_4N_4$	60.140	**2**
$C_{12}H_4N_4^-$, $C_{17}H_{15}N_2S_2^+$, $C_{12}H_4N_4$	60.26	**7**
$C_{12}H_4N_4^-$, $C_{17}H_{19}ClN_2S^+$	41.28	**6**
$C_{12}H_4N_4^-$, $C_{19}H_{17}N_2S_2^+$, $C_{12}H_4N_4$	60.34	**7**
$C_{12}H_4N_4^-$, $C_{19}H_{18}As^+$, $C_{12}H_4N_4$	65.9	**4**
$C_{12}H_4N_4^-$, $C_{19}H_{18}P^+$, $C_{12}H_4N_4$	64.24	**4**
$C_{12}H_4N_4^-$, $C_{19}H_{18}P^+$, $C_{12}H_4N_4$	64.34+	**6**
$C_{12}H_4N_4^-$, $C_{19}H_{19}N_2S_2^+$	41.29	**6**
$C_{12}H_4N_4^-$, $C_{21}H_{21}N_2S_2^+$, $C_{12}H_4N_4$	60.35+	**7**
$C_{12}H_4N_4^-$, $C_{22}H_{26}ClN_2^+$, $C_{12}H_4N_4$	60.38	**7**
$2C_{12}H_4N_4^-$, $2C_4H_{10}NO^+$, $C_{12}H_4N_4$	40.4	**4**
$2C_{12}H_4N_4^-$, $C_{12}H_{12}N_2^{2+}$	36.10	**5**
$2C_{12}H_4N_4^-$, $C_{18}H_{18}N_2^{2+}$, $2C_{12}H_4N_4$	60.29	**7**
$2C_{12}H_4N_4^-$, $C_{24}H_{22}N_2^{2+}$, $2C_{12}H_4N_4$	33.55	**5**
$2C_{12}H_4N_4$, $C_{10}H_{16}N_2$	60.127	**2**
$C_{12}H_4N_7O_{12}^-$, $CH_6N_3^+$	8.4	**7**
$C_{12}H_4N_8O_{12}$	15.6	**5**
$C_{12}H_4N_8O_{12}$	15.3+	**6**
$C_{12}H_5Cl_2GeMoNO_5S$	69.9	**7**
$C_{12}H_5Cl_3O_3$	24.5	**7**
$C_{12}H_5Cl_7O_4$	31.14	**1**
$C_{12}H_5FeMnO_7$	73.36	**2**
$C_{12}H_5Fe_2NO_6S$	85.16	**5**
$C_{12}H_6AsCrMnO_{10}$	86.11	**5**
$C_{12}H_6BrNO_3$	24.6	**4**
$C_{12}H_6Cl_2O_2$	38.22	**7**
$C_{12}H_6Cl_2O_2$	38.23	**7**
$C_{12}H_6FeO_6$	71.17	**7**
$C_{12}H_6FeO_6S$	75.7	**7**
$C_{12}H_6Fe_2O_6$	72.32	**2**
$C_{12}H_6Fe_2O_6$	73.38	**2**
$C_{12}H_6Fe_2O_8$	72.31	**2**
$C_{12}H_6Mn_2O_8$	72.13	**5**
$C_{12}H_6N_2O_2$	40.24	**5**
$C_{12}H_6N_2S$	41.14	**4**
$C_{12}H_6O_2$	28.6	**1**
$C_{12}H_6O_3$	24.7	**4**
$C_{12}H_6O_{12}$	13.47	**1**
$C_{12}H_7Cl_2NSe$	42.7	**6**
$C_{12}H_7IMoO_3$	73.14	**5**
$C_{12}H_7NO_2$	36.6	**7**
$C_{12}H_8$	28.7	**1**
$C_{12}H_8$	28.4+	**6**
$C_{12}H_8$	28.1	**7**
$(C_{12}H_8AgClO_4)_n$	74.3	**4**
$C_{12}H_8AsClO$	65.1	**5**
$C_{12}H_8AsCl_6O_2Sb$	65.2	**5**
$C_{12}H_8AsO_4^-$, K^+	65.17	**2**
$C_{12}H_8As_2O$	65.1	**6**
$C_{12}H_8As_2S$	65.5	**3**
$C_{12}H_8As_2Se$	65.2	**6**
$C_{12}H_8As_2Te$	65.3	**6**
$C_{12}H_8BrCl$	28.8	**1**
$C_{12}H_8BrClHgO$	71.18	**7**
$C_{12}H_8BrNO_4$	40.14	**6**
$C_{12}H_8Br_2$	28.9	**1**
$C_{12}H_8Br_2$	28.1	**4**
$C_{12}H_8Br_2$	28.2	**7**
$C_{12}H_8Br_2N_2$	9.27+	**1**
$C_{12}H_8Br_2N_2O_2$	10.28	**1**
$C_{12}H_8Br_2O$	17.33	**1**
$C_{12}H_8Br_2O_2S$	11.46	**1**
$C_{12}H_8Br_2O_2S_2$	11.47	**1**
$C_{12}H_8Br_2S$	11.48	**1**
$C_{12}H_8Br_2S_2$	11.49	**1**
$C_{12}H_8CdN_2O_4$	81.33	**6**
$C_{12}H_8Cl_2$	28.10	**1**
$C_{12}H_8Cl_2$	28.2	**4**
$C_{12}H_8Cl_2$	28.7	**6**
$C_{12}H_8Cl_2$	19.6	**7**
$C_{12}H_8Cl_2CoN_2^+$, Cl^- , $3H_2O$	83.91	**2**
$C_{12}H_8Cl_2GeO$	69.10	**7**
$C_{12}H_8Cl_2I_2Te$	70.19	**2**
$C_{12}H_8Cl_2N_2$	9.29	**1**
$C_{12}H_8Cl_2N_2$	9.1	**6**
$C_{12}H_8Cl_2N_2S_3$	4.4	**5**
$C_{12}H_8Cl_2N_2Zn$	83.92	**2**
$C_{12}H_8Cl_2O_2S$	11.50+	**1**

$C_{12}H_8Cl_2Se_2$	11.52	1
$C_{12}H_8Cl_2Te_2$	70.20	2
$C_{12}H_8Cl_3N_2Tl$	68.8	4
$C_{12}H_8Cl_4S$	11.13	3
$C_{12}H_8Cl_6$	31.6	5
$C_{12}H_8Cl_6$	31.7	5
$C_{12}H_8Cl_6O$	31.8	5
$C_{12}H_8Cl_6O$	38.18	6
$C_{12}H_8Cl_{14}Pt_4^{2-}$, $2C_{19}H_{15}^+$, $2CH_2Cl_2$	12.10	5
$C_{12}H_8CrN_2O_5$	83.93	2
$C_{12}H_8CuN_2O_4$, K^+ , CNS^-	60.37	3
$C_{12}H_8CuN_2O_4$, xH_2O	81.34	6
$C_{12}H_8CuN_2O_4$, $2H_2O$	81.25	3
$C_{12}H_8FO_4P$	64.17	6
$C_{12}H_8FeO_4$	72.14	5
$C_{12}H_8FeO_4$	71.19+	7
$C_{12}H_8FeO_6$	71.21	7
$C_{12}H_8Fe_2O_8$	72.33	2
$C_{12}H_8HgN_2^{2+}$, $2NO_3^-$	83.94	2
$C_{12}H_8Hg_2N_4^{2+}$, $2ClO_4^-$	83.32	5
$C_{12}H_8I_2OTe$	70.3	6
$C_{12}H_8I_2Te$	70.7	7
$C_{12}H_8MoO_6^{2-}$, $2K^+$, $2H_2O$	84.32	7
$C_{12}H_8Mo_2O_9^{2-}$, $2H_4N^+$, $2H_2O$	84.33	7
$C_{12}H_8N_2$	36.9	1
$C_{12}H_8N_2$	36.10	1
$C_{12}H_8N_2$	36.4	4
$C_{12}H_8N_2$	36.5	4
$C_{12}H_8N_2$	36.7	7
$C_{12}H_8N_2$, I_2	60.128	2
$C_{12}H_8N_2$, $C_7H_5O_3^-$, Tl^+	36.7	5
$C_{12}H_8N_2$, $C_{12}H_4N_4$	60.14	5
$C_{12}H_8N_2$, $C_{12}H_4N_4$	60.15	5
$C_{12}H_8N_2$, $C_{12}H_8N_2O$	36.6	4
$C_{12}H_8N_2$, $C_{12}H_8N_2O$	36.7	4
$C_{12}H_8N_2$, $C_{12}H_8N_2O$	36.8	4
$C_{12}H_8N_2NbO_6^-$, K^+ , $3H_2O$	83.47	4
$C_{12}H_8N_2NbO_6^-$, K^+ , $3H_2O$, H_2O_2	83.48	4
$C_{12}H_8N_2O$	36.11	1
$C_{12}H_8N_2O$	36.9+	4
$C_{12}H_8N_2O_2$	33.63	1
$C_{12}H_8N_2O_2$	36.12+	1
$C_{12}H_8N_2O_2$	33.36	3
$C_{12}H_8N_2O_4$	15.25	1
$C_{12}H_8N_2O_4$	36.14+	1
$(C_{12}H_8N_2O_4Pb)_n$, nH_2O	69.14	5
$C_{12}H_8N_2O_4S_2$	11.14	3
$C_{12}H_8N_2O_4S_2$	11.15	3
$C_{12}H_8N_2O_6$	28.11	1
$C_{12}H_8N_2O_7Te$	70.5	6
$C_{12}H_8N_2O_8P^-$, $C_9H_{12}NO_2^+$	13.13	4
$C_{12}H_8N_2O_8P^-$, $C_{13}H_{21}N_2O_2^+$	3.47	3
$C_{12}H_8N_2O_8P^-$, $C_{14}H_{23}N_2O^+$	16.11	7
$C_{12}H_8N_2O_8P^-$, $C_{18}H_{23}N_2O_2^+$, H_2O	16.17	4
$C_{12}H_8N_4$	36.18	1
$C_{12}H_8N_4$	21.9	3
$C_{12}H_8N_4$	28.3	7
$C_{12}H_8N_4^-$, $C_6H_4S_4^+$	60.3	6
$C_{12}H_8N_4O$	38.24	7
$C_{12}H_8N_4O_6$	16.11	6
$C_{12}H_8O$	38.15	4
$C_{12}H_8O$	38.19	6
$C_{12}H_8O$	38.25	7
$C_{12}H_8OS$	42.13	1
$C_{12}H_8OTe$	70.6	6
$C_{12}H_8O_2$	38.12	5
$C_{12}H_8O_2$	38.20	6
$C_{12}H_8O_2$, $C_{12}H_4N_4$	60.16	5
$C_{12}H_8O_2S$	39.67	1
$C_{12}H_8O_2S_2$	39.68	1
$C_{12}H_8O_2S_2$	39.28	3
$C_{12}H_8O_4$	59.6	5
$2C_{12}H_8O_4$, K^+ , I_3^-	38.26	7
$C_{12}H_8O_4Te$	70.21	2
$C_{12}H_8O_6Os_2$	72.34	2
$C_{12}H_8S$	39.29	3
$C_{12}H_8S_2$	39.69	1
$C_{12}H_8S_2$	39.70+	1
$C_{12}H_8Se$	39.30	3
$C_{12}H_8Se_2$	39.72	1
$C_{12}H_9AsBrN$	65.18	2
$C_{12}H_9AsClN$	65.19+	2
$C_{12}H_9BO_2$	62.6	6
$C_{12}H_9BrClNO_2S$	16.43	1
$C_{12}H_9BrN_2O$	17.19	7
$C_{12}H_9Br_2N_3$	9.30	1
$C_{12}H_9Br_2N_3$	9.31	1
$C_{12}H_9ClN_2O_2$	32.30	3
$C_{12}H_9ClO$	31.9	5
$C_{12}H_9Cl_2N$	16.44	1
$C_{12}H_9Cl_2NO$	33.35	6
$C_{12}H_9Cl_2O_4P$	46.17	1
$C_{12}H_9Cl_3$	31.8	4
$C_{12}H_9N$	36.19	1
$C_{12}H_9N$	36.4	3
$C_{12}H_9N$, $C_{12}H_4N_4$	60.14	7
$C_{12}H_9NO$, $C_8F_{12}NiS_4$	60.8	7
$C_{12}H_9NO_2$	15.8	5
$C_{12}H_9NS$	41.33	1
$C_{12}H_9NS$, $C_6H_3N_3O_6$	60.38	3
$C_{12}H_9NS$, $C_7H_4N_2O_6$	60.130	2
$C_{12}H_9NS$, $C_8F_{12}NiS_4$	60.9	7
$C_{12}H_9NS$, $C_{12}H_4N_4$	60.19	6
$C_{12}H_9NSe$	42.7	7
$C_{12}H_9N_2^+$, HO_4S^-	36.8	5
$C_{12}H_9N_2O_8P$	46.8	5
$C_{12}H_9N_3O_4$	33.64	1
$C_{12}H_9N_3O_4$	33.36	6
$C_{12}H_9N_3O_4$, $2H_2O$	43.3	6
$C_{12}H_9O_2^-$, $C_8H_{11}BrN^+$	31.10	5
$C_{12}H_9O_2P$	64.18	6
$C_{12}H_{10}$	19.35+	1
$C_{12}H_{10}$	28.12	1
$C_{12}H_{10}$	28.1	5
$C_{12}H_{10}$, $C_6Cl_4O_2$	60.19	5
$C_{12}H_{10}$, $C_{12}H_4N_4$	60.20	5
$2C_{12}H_{10}$, $H_6N_2Ni^{2+}$, $C_4N_4Ni^{2-}$	61.6	6
$C_{12}H_{10}Ag^+$, ClO_4^-	75.8	3
$C_{12}H_{10}AsBr$	65.21	2
$C_{12}H_{10}AsCl$	65.22	2
$C_{12}H_{10}AsI$	65.23	2
$C_{12}H_{10}As_2S_3$	65.8	4
$C_{12}H_{10}BrNOS_2$	11.14	7
$C_{12}H_{10}BrNO_2S$	16.45	1
$C_{12}H_{10}BrNO_2S$	34.8	1
$C_{12}H_{10}BrN_3$	9.9+	5
$C_{12}H_{10}Br_2$	24.6	7
$C_{12}H_{10}Br_2Se$	11.53	1
$C_{12}H_{10}Br_2Te$	70.22	2
$(C_{12}H_{10}Cl_2HgOS)_n$	84.12	5
$C_{12}H_{10}Cl_2N_2$	16.46	1

$C_{12}H_{10}Cl_2N_2$ 16.47 **1**
$C_{12}H_{10}Cl_2N_2O_2Pd$ 83.33 **5**
$C_{12}H_{10}Cl_2O_2$ 5.15 **5**
$(C_{12}H_{10}Cl_2Pb)_n$ 69.15 **5**
$C_{12}H_{10}Cl_2Se$ 11.54 **1**
$C_{12}H_{10}Cl_2Sn$ 69.21 **4**
$C_{12}H_{10}Cl_4N_3P_3$ 64.32 **2**
$C_{12}H_{10}Cl_6$ 21.8 **5**
$C_{12}H_{10}Cl_8N_6P_6$ 64.15 **4**
$C_{12}H_{10}CoN_4S_2$ 83.95 **2**
$C_{12}H_{10}CrO_5$ 74.4 **5**
$C_{12}H_{10}CuN_4O_4$ 84.33 **2**
$C_{12}H_{10}F_2FeO$ 71.22 **7**
$C_{12}H_{10}F_4N_3P_3$ 64.25 **3**
$C_{12}H_{10}FeO_2$ 73.39 **2**
$C_{12}H_{10}FeO_4$ 73.8 **3**
$C_{12}H_{10}FeO_8$ 72.5 **7**
$C_{12}H_{10}Hg$ 71.26 **2**
$C_{12}H_{10}HgN_2O_3S$ 85.18 **3**
$C_{12}H_{10}I^+$, BF_4^- 19.38 **1**
$C_{12}H_{10}I^+$, Br^- 19.7 **7**
$C_{12}H_{10}I^+$, NO_3^- 19.10 **4**
$C_{12}H_{10}Mn_2N_2O_4$ 73.9 **6**
$C_{12}H_{10}NO_4S_2^-$, Na^+ 11.16 **3**
$C_{12}H_{10}N_2$ 9.34 **1**
$C_{12}H_{10}N_2$ 9.35+ **1**
$C_{12}H_{10}N_2$ 9.9 **4**
$C_{12}H_{10}N_2$ 22.2 **6**
$C_{12}H_{10}N_2^{2+}$, $2Br^-$, H_2O 36.11 **4**
$C_{12}H_{10}N_2^{2+}$, Cl_5Mn^{2-} 36.9 **5**
$C_{12}H_{10}N_2O_2$ 33.65 **1**
$C_{12}H_{10}N_2O_2$ 16.12 **6**
$C_{12}H_{10}N_2O_2$ 9.5 **7**
$2C_{12}H_{10}N_2O_2$, C_6F_5NO 10.2 **6**
$C_{12}H_{10}N_2O_2Ti$ 73.15 **7**
$C_{12}H_{10}N_2O_5$ 35.30 **1**
$C_{12}H_{10}N_2S_3$ 4.3 **6**
$C_{12}H_{10}N_4NiO_2$, $2H_2O$ 83.96+ **2**
$C_{12}H_{10}N_4O$ 36.8 **7**
$C_{12}H_{10}N_4O_2$ 16.6 **7**
$C_{12}H_{10}N_4O_2S$ 32.20 **7**
$C_{12}H_{10}N_4PtS_2$ 83.40 **7**
$C_{12}H_{10}N_4S$, $0.5H_2O$ 44.22 **6**
$C_{12}H_{10}O$ 31.11 **3**
$C_{12}H_{10}O$ 28.4 **7**
$C_{12}H_{10}OS$ 11.55 **1**
$C_{12}H_{10}O_2$ 25.19 **1**
$C_{12}H_{10}O_2$ 28.13 **1**
$C_{12}H_{10}O_2$ 31.15 **1**
$C_{12}H_{10}O_2$ 17.8 **4**
$C_{12}H_{10}O_2$ 31.11 **6**
$C_{12}H_{10}O_3$ 38.21 **6**
$C_{12}H_{10}O_3S$ 24.7 **7**
$C_{12}H_{10}O_4S_2$ 11.7 **4**
$C_{12}H_{10}O_4S_2Se$ 11.56 **1**
$C_{12}H_{10}O_4S_2W$ 85.19 **3**
$C_{12}H_{10}O_4S_3$ 11.57 **1**
$C_{12}H_{10}O_4S_4Te$ 70.11 **4**
$C_{12}H_{10}S$ 39.22 **5**
$C_{12}H_{10}S_2$ 11.58 **1**
$C_{12}H_{10}S_2$ 11.15 **7**
$C_{12}H_{10}S_3$ 39.13 **4**
$C_{12}H_{10}Se_2$ 11.59 **1**
$C_{12}H_{10}Sn$, $0.33C_8H_{10}$ 69.23 **2**
$C_{12}H_{10}Te_2$ 70.7 **5**
$C_{12}H_{11}AsMnO_4$ 86.12 **5**
$C_{12}H_{11}Br$ 24.26 **1**
$C_{12}H_{11}BrO_2$ 31.12 **3**
$C_{12}H_{11}BrO_2$ 5.16 **5**
$C_{12}H_{11}F_6O_2Rh$ 72.6 **7**
$C_{12}H_{11}FeNO_4$ 75.7 **6**
$C_{12}H_{11}Fe_2O_5P$ 73.10 **6**
$C_{12}H_{11}N$, $C_{13}H_{10}O$ 60.26 **4**
$C_{12}H_{11}NO_3Ru$ 75.9 **5**
$C_{12}H_{11}NO_4S_2$ 11.17 **3**
$C_{12}H_{11}NS_4$ 39.14 **4**
$C_{12}H_{11}N_3$ 9.10 **4**
$C_{12}H_{11}N_3$ 9.11 **5**
$C_{12}H_{11}O_2P$ 64.15 **5**
$C_{12}H_{11}O_4P$ 46.9 **5**
$C_{12}H_{12}$ 24.27 **1**
$C_{12}H_{12}$ 24.3 **5**
$C_{12}H_{12}$ 24.3 **6**
$C_{12}H_{12}$ 24.4 **6**
$C_{12}H_{12}Al_2Cl_8Pd_2$ 74.3 **3**
$C_{12}H_{12}Al_4Cl_{14}Pd_2$ 74.4 **3**
$C_{12}H_{12}BrNO$ 35.31 **1**
$C_{12}H_{12}BrNO_8S$ 45.61 **1**
$C_{12}H_{12}Br_2O_5$ 59.4 **3**
$C_{12}H_{12}Br_6$ 19.39 **1**
$C_{12}H_{12}CeO_{15}^{3-}$, $3Na^+$, $9H_2O$ 81.28 **7**
$C_{12}H_{12}ClCuN_3O_2$, $2H_2O$ 82.15 **7**
$C_{12}H_{12}ClN_3$ 35.13 **4**
$C_{12}H_{12}Cl_2N_2O_4$ 50.1 **4**
$C_{12}H_{12}Cl_2O_2$ 38.13 **5**
$C_{12}H_{12}Cl_2Sn_2^{2+}$, $2AlCl_4^-$ 69.22 **4**
$C_{12}H_{12}Cl_3OSb$ 66.14 **2**
$C_{12}H_{12}Cl_4$ 28.14 **1**
$C_{12}H_{12}CoN_2O_6$, $2H_2O$ 81.36 **5**
$C_{12}H_{12}Co_2O_4$ 72.35 **2**
$C_{12}H_{12}Cr$ 74.9+ **2**
$C_{12}H_{12}Cr^+$, I^- 74.3 **6**
$C_{12}H_{12}Cr^+$, $C_{12}H_4N_4^-$ 71.23 **7**
$C_{12}H_{12}CrO_5$ 74.5 **5**
$C_{12}H_{12}CuN_2O_6$ 81.32 **4**
$(C_{12}H_{12}Cu_2N_8)_n$ 83.39 **2**
$C_{12}H_{12}FeN_2O_5$ 75.4 **4**
$C_{12}H_{12}Fe_2O_8P_2$ 86.8 **2**
$C_{12}H_{12}Fe_2O_8Sn_2$ 69.23 **4**
$C_{12}H_{12}GdO_{15}^{3-}$, $2ClO_4^-$, $5Na^+$, $6H_2O$ 81.69 **2**
$C_{12}H_{12}MgN_2O_6$ 67.8 **5**
$C_{12}H_{12}Mn_2N_2O_6S_4$ 85.17 **7**
$C_{12}H_{12}MoN_2O_5$ 71.27 **2**
$C_{12}H_{12}MoN_8$ 71.11 **4**
$C_{12}H_{12}MoN_8$ 71.18 **5**
$C_{12}H_{12}N_2$, $C_6H_3N_3O_6$ 60.15 **7**
$C_{12}H_{12}N_2$, $C_6H_3N_3O_6$, $0.5C_6H_6$ 60.16 **7**
$C_{12}H_{12}N_2$, $C_{12}H_4N_4$ 60.21 **5**
$C_{12}H_{12}N_2$, $C_{12}H_4N_4$ 60.20 **6**
$C_{12}H_{12}N_2$, $C_{12}H_4N_4$, $1.8CH_2Cl_2$ 60.22 **5**
$C_{12}H_{12}N_2$, $C_{12}H_4N_4$, C_6H_6 60.21 **6**
$C_{12}H_{12}N_2^{2+}$, $2Br^-$, H_2O 36.20 **1**
$C_{12}H_{12}N_2^{2+}$, $2C_{12}H_4N_4^-$ 36.10 **5**
$C_{12}H_{12}N_2NdO_{12}^{3-}$, $3K^+$, $6H_2O$ 82.16 **7**
$C_{12}H_{12}N_2NiO_6$, $2H_2O$ 81.33+ **4**
$C_{12}H_{12}N_2NiO_{12}^{4-}$, $4K^+$, $8H_2O$ 82.19 **5**
$C_{12}H_{12}N_2O_2$ 33.27 **4**
$C_{12}H_{12}N_2O_2S$ 11.18 **3**
$C_{12}H_{12}N_2O_2S$ 16.10 **4**
$C_{12}H_{12}N_2O_2S$ 34.2 **5**
$C_{12}H_{12}N_2O_3$ 35.15 **5**

$C_{12}H_{12}N_2O_3$ 43.4 **6**
$C_{12}H_{12}N_2O_3$, H_2O 43.5 **6**
$C_{12}H_{12}N_2O_3$, $2C_7H_8BrN_5$ 60.131 **2**
$C_{12}H_{12}N_2O_4$ 36.3 **6**
$C_{12}H_{12}N_2O_4S_2$ 41.15 **4**
$C_{12}H_{12}N_2O_6Zn$, $2H_2O$ 81.26 **3**
$C_{12}H_{12}N_2O_6Zn$, $2H_2O$ 81.35 **4**
$C_{12}H_{12}N_2O_{12}Zr^{2-}$, $2K^+$, H_2O 81.70 **2**
$C_{12}H_{12}N_2O_{15}Ti_2^{4-}$, $4Na^+$, $11H_2O$ 81.37 **5**
$C_{12}H_{12}N_2S$ 16.13 **6**
$C_{12}H_{12}N_2S_2$ 11.61 **1**
$C_{12}H_{12}N_2S_2$ 11.19 **3**
$C_{12}H_{12}N_4Ni$ 83.99 **2**
$C_{12}H_{12}N_4O_8$, $2H_2O$ 44.23 **4**
$C_{12}H_{12}NdO_{15}^{3-}$, $2ClO_4^-$, $5Na^+$, $6H_2O$ 81.72 **2**
$C_{12}H_{12}NdO_{15}^{3-}$, $3Na^+$, $6H_2O$ 81.71 **2**
$C_{12}H_{12}NdO_{15}^{3-}$, $5Na^+$, $2ClO_4^-$, $6H_2O$ 81.27 **3**
$C_{12}H_{12}O_2$ 28.8 **6**
$C_{12}H_{12}O_2$ 31.3 **7**
$C_{12}H_{12}O_3$ 19.5+ **6**
$C_{12}H_{12}O_4$ 31.9 **4**
$C_{12}H_{12}O_{12}$ 45.17 **4**
$C_{12}H_{12}O_{15}Yb^{3-}$, $5Na^+$, $2ClO_4^-$, $6H_2O$ 81.28 **3**
$C_{12}H_{12}S_2$, $C_6N_6O_6$ 60.74 **2**
$C_{12}H_{12}S_2Ti$ 73.11 **6**
$C_{12}H_{12}Si_2$ 63.4 **7**
$C_{12}H_{12}Ti$ 73.12 **6**
$C_{12}H_{12}V$ 75.24 **2**
$C_{12}H_{13}BrO$ 38.14 **5**
$C_{12}H_{13}BrO_7$ 50.3 **7**
$C_{12}H_{13}ClN_2O_4$ 45.32 **6**
$C_{12}H_{13}Cl_2N_2NbO_2$ 83.57 **3**
$C_{12}H_{13}FN_4O_4$ 44.23 **7**
$C_{12}H_{13}FeNO$ 73.9 **3**
$C_{12}H_{13}NO$ 35.14 **4**
$C_{12}H_{13}NO_4$ 35.32 **1**
$C_{12}H_{13}N_3O$ 40.15 **6**
$C_{12}H_{13}N_5O_4$, H_2O 50.5 **6**
$C_{12}H_{14}BrNO_2S$ 27.9 **4**
$C_{12}H_{14}Br_2CuN_2$ 83.34 **5**
$C_{12}H_{14}Br_2N_2Zn$ 83.49 **4**
$(C_{12}H_{14}CdN_4OS_2)_n$ 83.50 **4**
$C_{12}H_{14}ClNO_2$ 40.12 **3**
$C_{12}H_{14}ClN_4^+$, Br^- 44.36 **3**
$C_{12}H_{14}ClN_5$, $3H_2O$ 36.16 **5**
$C_{12}H_{14}Cl_2CuN_2$ 83.100 **2**
$C_{12}H_{14}Cl_2CuN_2$ 83.101 **2**
$C_{12}H_{14}Cl_2CuN_2O_2$ 84.11 **3**
$C_{12}H_{14}Cl_2N_2Zn$ 83.51 **4**
$C_{12}H_{14}Cl_2N_2Zn$ 83.41 **7**
$C_{12}H_{14}Cl_2O_2$ 31.4 **7**
$C_{12}H_{14}CuN_4O_6$ 83.52+ **4**
$C_{12}H_{14}F_2N_4O_4$ 44.24 **7**
$C_{12}H_{14}Fe^+$, I_3^- 73.15 **5**
$C_{12}H_{14}HgN_4O_4$ 83.60 **3**
$C_{12}H_{14}INO_2S$ 35.33 **1**
$C_{12}H_{14}MoNO_2^+$, Cl^- , H_2O 73.16 **5**
$C_{12}H_{14}N^+$, I^- 35.12 **3**
$C_{12}H_{14}N^+$, $C_{12}H_4N_4^-$, $C_{12}H_4N_4$ 60.25 **4**
$C_{12}H_{14}N_2^{2+}$, $2Br^-$ 33.28 **4**
$C_{12}H_{14}N_2^{2+}$, $2Cl^-$ 33.29 **4**
$C_{12}H_{14}N_2^{2+}$, $2Cl^-$ 16.17 **5**
$C_{12}H_{14}N_2^{2+}$, Cl_4Co^{2-} 33.38 **3**
$C_{12}H_{14}N_2^{2+}$, Cl_4Cu^{2-} 33.66 **1**
$C_{12}H_{14}N_2^{2+}$, Cl_4Pd^{2-} 33.39 **3**
$C_{12}H_{14}N_2^{2+}$, $Cl_6Cu_2^{2-}$ 33.34 **7**
$C_{12}H_{14}N_2^{2+}$, $2I^-$ 33.30 **4**
$C_{12}H_{14}N_2^{2+}$, $C_4N_4Ni^{2-}$ 33.67 **1**
$nC_{12}H_{14}N_2^{2+}$, $(Cl_2Cu^-)_{2n}$ 33.40 **3**
$C_{12}H_{14}N_2O_2$ 37.16 **1**
$C_{12}H_{14}N_2O_2$ 44.25 **7**
$C_{12}H_{14}N_2O_2$, H_2O 48.62 **6**
$C_{12}H_{14}N_2O_4$ 40.5 **7**
$C_{12}H_{14}N_2O_4$, H_2O 48.63 **6**
$C_{12}H_{14}N_2O_4S$, H_2O 47.24 **6**
$C_{12}H_{14}N_2O_5$ 47.25 **6**
$C_{12}H_{14}N_2O_5$ 47.31 **7**
$C_{12}H_{14}N_3O_3^+$, Br^- 40.25 **5**
$C_{12}H_{14}N_4$ 32.31 **3**
$C_{12}H_{14}N_4OS$, $2H_2O$ 41.18 **6**
$C_{12}H_{14}N_4O_2$ 32.32 **1**
$C_{12}H_{14}N_4O_8S_2$ 48.45 **3**
$C_{12}H_{14}N_6NiS_2$ 85.20 **3**
$C_{12}H_{14}O_2$ 38.22 **6**
$C_{12}H_{14}O_2$ 20.13 **7**
$C_{12}H_{14}O_3$ 31.10 **4**
$C_{12}H_{14}O_3S$ 27.7 **6**
$C_{12}H_{14}O_4$ 13.48 **1**
$C_{12}H_{15}BrN_2O_5$ 45.63+ **1**
$C_{12}H_{15}BrO_2$ 38.27 **7**
$C_{12}H_{15}ClMo$ 71.24 **7**
$C_{12}H_{15}ClO_2$ 24.4 **5**
$C_{12}H_{15}Cl_3$ 19.14 **5**
$C_{12}H_{15}NO_2$ 58.6 **3**
$C_{12}H_{15}NO_3$ 37.17 **1**
$C_{12}H_{15}N_3O_2$, H_2O 32.21 **7**
$C_{12}H_{15}N_5O_5$ 47.30 **3**
$(C_{12}H_{15}NaO_4)_n$ 67.11 **3**
$C_{12}H_{16}$ 22.4 **3**
$C_{12}H_{16}^{2-}$, $2C_6H_{14}KO_3^+$ 22.3 **6**
$(C_{12}H_{16}AgN_4^+)_n$, $nClO_4^-$ 83.61 **3**
$C_{12}H_{16}B_2Co$ 75.10 **5**
$C_{12}H_{16}B_2CoN_8$ 83.35 **5**
$C_{12}H_{16}B_2CoO_2$ 75.5 **4**
$C_{12}H_{16}B_2CoO_2$ 75.11 **5**
$C_{12}H_{16}BrFO_7$ 45.31 **3**
$C_{12}H_{16}Br_2N_8Ni$ 83.102 **2**
$C_{12}H_{16}CdN_3O_8$ 83.36 **5**
$C_{12}H_{16}CdN_6O_4$, H_2O 82.45 **2**
$C_{12}H_{16}ClMnN_2O_2$ 84.34 **7**
$C_{12}H_{16}ClN_3O_4$ 16.7 **7**
$C_{12}H_{16}Cl_2N_2O_2Sn$ 69.24 **2**
$C_{12}H_{16}Cl_2N_5O^+$, Cl^- 33.31 **4**
$C_{12}H_{16}Cl_2OZr$ 75.12 **5**
$C_{12}H_{16}CoN_2O_8$ 83.103 **2**
$C_{12}H_{16}CoN_6O_4$, H_2O 82.46 **2**
$C_{12}H_{16}CoN_6O_4$, $2H_2O$ 82.22 **3**
$C_{12}H_{16}CoN_6O_4^+$, Br^- 82.17 **7**
$C_{12}H_{16}CuN_2O_8$ 81.38 **5**
$C_{12}H_{16}CuN_4O_4^{2+}$, $2Cl^-$ 83.62 **3**
$C_{12}H_{16}CuN_8^{2+}$, $2ClO_4^-$ 83.26 **6**
$C_{12}H_{16}CuN_8^{2+}$, $2I^-$ 83.104 **2**
$C_{12}H_{16}CuN_8^{2+}$, O_4S^{2-} 83.37 **5**
$C_{12}H_{16}CuN_8O_4^{2+}$, $2Cl^-$, $3H_2O$ 83.42 **7**
$(C_{12}H_{16}Cu_2N_2O_8)_n$ 81.29+ **7**
$C_{12}H_{16}Cu_2N_6O_8^{2-}$, $2Na^+$, $2H_2O$ 82.47 **2**
$(C_{12}H_{16}EuN_2O_8^+)_n$, nNO_3^- 81.31 **7**
$C_{12}H_{16}Fe_4O_8S_8^{6-}$, $C_{16}H_{36}N^+$, $5Na^+$, $5C_5H_9NO$ 85.18 **7**

$C_{12}H_{16}I_2Mo_2N_4O_2$ 73.13 **6**
$C_{12}H_{16}MoNS^+$, I^- 73.40 **2**
$C_{12}H_{16}MoN_3O_2S_2$ 72.9 **6**
$C_{12}H_{16}Mo_2N_6O_6S_2$, $1.5H_2O$ 82.18 **7**
$C_{12}H_{16}Mo_2N_6O_8$, $3H_2O$ 82.23 **3**
$C_{12}H_{16}NO_2^+$, Br^- 3.43 **3**
$C_{12}H_{16}NO_2^+$, Cl^- 3.44 **3**
$C_{12}H_{16}NO_3^+$, Br^- 58.3 **4**
$C_{12}H_{16}NO_4^+$, Br^- , $2H_2O$ 48.88 **1**
$C_{12}H_{16}N_2$ 35.16 **5**
$C_{12}H_{16}N_2O$ 35.17 **5**
$C_{12}H_{16}N_2O$ 58.3 **6**
$C_{12}H_{16}N_2O_2$ 48.64 **6**
$C_{12}H_{16}N_2O_2$ 36.9 **7**
$C_{12}H_{16}N_2O_3$ 43.6 **3**
$C_{12}H_{16}N_2O_3$ 43.7 **3**
$C_{12}H_{16}N_2O_3$ 35.18 **5**
$C_{12}H_{16}N_2O_3$ 48.65 **6**
$C_{12}H_{16}N_2O_3$, $C_{13}H_{17}N_3O$ 60.20 **7**
$C_{12}H_{16}N_2O_5S$ 47.18 **4**
$C_{12}H_{16}N_2O_5Si$ 63.3 **3**
$C_{12}H_{16}N_2O_8Zn$ 83.55 **4**
$C_{12}H_{16}N_2O_8Zn$ 83.38 **5**
$C_{12}H_{16}N_2O_9$, H_2O 47.32 **7**
$C_{12}H_{16}N_3O_4^+$, $C_6H_2N_3O_7^-$ 45.33 **7**
$C_{12}H_{16}N_4NiO_4^{2+}$, $2Cl^-$ 83.105 **2**
$C_{12}H_{16}N_4OS$, $2H_2O$ 59.5 **3**
$C_{12}H_{16}N_4O_4$ 44.72 **1**
$C_{12}H_{16}N_4O_4$ 48.66 **6**
$C_{12}H_{16}N_4O_6$ 33.35 **7**
$C_{12}H_{16}N_6NiO_4$, H_2O 82.48 **2**
$C_{12}H_{16}N_6O_4Zn$, $2H_2O$ 82.49 **2**
$C_{12}H_{16}N_6O_4Zn$, $2H_2O$ 82.12 **4**
$C_{12}H_{16}N_6O_4Zn$, $5H_2O$ 82.50 **2**
$C_{12}H_{16}N_8Ni^{2+}$, $2Cl^-$ 83.106 **2**
$C_{12}H_{16}N_8NiO_2^{2+}$, $2NO_3^-$ 83.107 **2**
$C_{12}H_{16}O_2$ 28.9 **6**
$C_{12}H_{16}O_2S_2$ 39.32 **3**
$C_{12}H_{16}O_4$ 31.12 **6**
$C_{12}H_{16}O_6$ 38.12 **3**
$C_{12}H_{16}O_6$ 21.2 **7**
$C_{12}H_{16}O_6S_8$ 42.8 **7**
$C_{12}H_{16}O_8$ 20.6 **4**
$C_{12}H_{16}O_8$ 45.18 **4**
$C_{12}H_{16}O_8$ 45.34 **7**
$C_{12}H_{17}BrN_2O_5$ 45.33 **3**
$C_{12}H_{17}BrOS$ 11.10 **6**
$C_{12}H_{17}CuN_3O_3$ 76.38 **4**
$C_{12}H_{17}IP^+$, I^- 64.16 **4**
$C_{12}H_{17}Mo_2N_2O_{12}^-$, K^+ 76.58 **7**
$C_{12}H_{17}NO$ 1.32 **5**
$C_{12}H_{17}NO_2$ 37.4 **5**
$C_{12}H_{17}NO_3$ 32.17 **4**
$C_{12}H_{17}NO_3Si$ 63.16 **2**
$C_{12}H_{17}NO_3Si$ 63.5 **7**
$C_{12}H_{17}NO_4S$ 41.16 **4**
$C_{12}H_{17}NO_4S$ 1.26 **6**
$C_{12}H_{17}N_2O_3^+$, Cl^- , $2H_2O$ 48.67 **6**
$C_{12}H_{17}N_2O_4P$, CH_4O 58.4 **6**
$C_{12}H_{17}N_4OS^+$, Cl^- , H_2O 41.22 **5**
$C_{12}H_{17}N_5O_5$ 47.28 **5**
$C_{12}H_{17}OP$ 64.32 **7**
$C_{12}H_{18}$ 19.40 **1**
$C_{12}H_{18}$ 23.9 **1**
$C_{12}H_{18}$ 19.6+ **3**
$C_{12}H_{18}$ 20.14 **3**
$C_{12}H_{18}$, $C_6Cl_4O_2$ 60.71 **2**
$C_{12}H_{18}$, C_6F_6 60.24+ **5**
$C_{12}H_{18}$, $C_8H_2N_4$ 60.101 **2**
$C_{12}H_{18}$, $C_{12}H_4N_4$ 60.36 **3**
$C_{12}H_{18}$, $C_{12}H_4N_4$ 60.18 **6**
$C_{12}H_{18}Ag_2^{2+}$, $2NO_3^-$ 75.9 **3**
$C_{12}H_{18}B_2CoF_2N_6O_6$ 83.56 **4**
$C_{12}H_{18}B_2CoF_2N_6O_6^+$, BF_4^- 83.57 **4**
$C_{12}H_{18}B_2F_2FeN_6O_6$, $0.5C_6H_{12}$ 61.5 **5**
$C_{12}H_{18}B_6Co_2$ 73.14 **6**
$C_{12}H_{18}B_6Fe_2$ 73.16 **7**
$C_{12}H_{18}Be_4O_{13}$ 67.11 **2**
$C_{12}H_{18}Br_2CoN_2O_2S_2$ 83.43 **7**
$C_{12}H_{18}Br_2O_2$ 23.4 **6**
$C_{12}H_{18}CaO_{14}$ 45.24 **5**
$C_{12}H_{18}Cl_2O_6Pd_2$ 72.18 **3**
$C_{12}H_{18}Cl_2Ru$ 72.37 **2**
$(C_{12}H_{18}Cl_6Cu_3N_2O_4)_n$ 84.34 **2**
$C_{12}H_{18}CoN_2O_2$, C_6H_6 76.39 **5**
$C_{12}H_{18}CoN_3O_3$ 76.40 **5**
$C_{12}H_{18}CuN_2O_2$ 83.112 **2**
$C_{12}H_{18}CuN_2O_2$, $0.5H_2O$ 83.109 **2**
$C_{12}H_{18}CuN_2O_2$, CH_6N^+ , ClO_4^- 60.39 **3**
$C_{12}H_{18}CuO_4$ 77.14 **2**
$C_{12}H_{18}CuO_6$ 77.15+ **2**
$C_{12}H_{18}FeN_6^{2+}$, $2Cl^-$, $3H_2O$ 71.28 **2**
$C_{12}H_{18}FeN_6^{2+}$, $2Cl_4Fe^-$ 83.39+ **5**
$C_{12}H_{18}FeN_6^{2+}$, $2Cl_4Fe^-$ 71.14 **6**
$C_{12}H_{18}FeN_6^{2+}$, $2Cl_4Fe^-$ 83.44+ **7**
$C_{12}H_{18}Fe_2Ge_3O_6$ 69.20 **3**
$C_{12}H_{18}MgN_6^{2+}$, $2AlCl_4^-$ 67.9 **5**
$C_{12}H_{18}Mo$ 72.7 **7**
$C_{12}H_{18}N_2O_3V$ 76.76 **3**
$C_{12}H_{18}N_2O_7$ 50.4 **7**
$C_{12}H_{18}N_2O_{14}Pt_4$, $2C_2H_4O_2$ 81.39 **5**
$C_{12}H_{18}N_4OS^{2+}$, $2Cl^-$, H_2O 41.34 **1**
$C_{12}H_{18}N_4OS^{2+}$, Cl_4Cu^{2-} 44.23 **6**
$2C_{12}H_{18}N_4OS^{2+}$, $2Cl^-$, $Cl_4O_2U^{2-}$ 44.24 **6**
$2C_{12}H_{18}N_4OS^{2+}$, Mg^{2+} , $6Cl^-$, $10H_2O$ 44.26 **7**
$C_{12}H_{18}N_4O_4PS^+$, $H_2O_4P^-$, $3H_2O$ 46.20 **1**
$C_{12}H_{18}N_4O_7P_2S$ 46.3+ **7**
$C_{12}H_{18}N_4O_7P_2S$, $4H_2O$ 46.18 **1**
$C_{12}H_{18}N_5O_5^+$, I^- 47.26 **6**
$C_{12}H_{18}N_6$ 36.11 **5**
$C_{12}H_{18}N_6O_6$, $0.5H_2O$ 48.89 **1**
$C_{12}H_{18}N_6Pd_2^{2+}$, $2F_6P^-$, xC_3H_6O 71.25 **7**
$C_{12}H_{18}Ni$ 75.13 **5**
$C_{12}H_{18}O_2$ 21.8 **4**
$C_{12}H_{18}O_2$ 31.13 **6**
$C_{12}H_{18}O_2S_4Zn$ 85.19 **7**
$C_{12}H_{18}O_2Se$ 39.23 **7**
$C_{12}H_{18}O_3W$ 71.20 **5**
$C_{12}H_{18}O_4$ 20.7 **4**
$C_{12}H_{18}O_4P_2$ 64.26+ **3**
$C_{12}H_{18}O_5$ 38.28 **7**
$C_{12}H_{18}O_8$ 38.23 **6**
$C_{12}H_{18}O_8$ 45.33 **6**
$C_{12}H_{18}O_{12}Pd_3$, $0.5H_2O$ 81.29 **3**
$C_{12}H_{18}O_{13}Zn_4$ 81.74 **2**
$C_{12}H_{18}W$ 72.8 **7**
$C_{12}H_{19}AuCl_2S$ 71.15 **6**
$C_{12}H_{19}BrO$ 27.12 **1**
$C_{12}H_{19}Cl_2NPt$ 72.38 **2**
$C_{12}H_{19}Cl_2NPt$ 72.19 **3**
$C_{12}H_{19}Cl_2NPt$ 72.8 **4**

FORMULA INDEX

$C_{12}H_{19}Co$	72.39	**2**
$C_{12}H_{19}CuO_8S_4$	85.17	**5**
$C_{12}H_{19}NO_3$	21.10	**3**
$C_{12}H_{19}N_2O^+$, $C_2HO_4^-$, $0.5C_2H_2O_4$	32.22	**7**
$C_{12}H_{19}N_2O_2^+$, Br^-	3.45	**3**
$C_{12}H_{19}N_2O_2^+$, Br^-	16.11	**4**
$C_{12}H_{19}N_4O_7P_2S^+$, Cl^-	41.18	**4**
$C_{12}H_{19}O_2P$	64.33	**7**
$C_{12}H_{20}$	21.28	**1**
$C_{12}H_{20}$	28.10	**6**
$C_{12}H_{20}AgN_9O_3$	83.58	**4**
$C_{12}H_{20}B_8Co_2$	73.15	**6**
$C_{12}H_{20}B_9Co$	73.17	**7**
$C_{12}H_{20}Br_2O$	23.7	**3**
$C_{12}H_{20}Br_2O$	20.14	**6**
$(C_{12}H_{20}Br_4Cu_5N_2S_4)_n$	80.6	**7**
$(C_{12}H_{20}Br_6Cu_7N_2S_4)_n$	80.7	**7**
$C_{12}H_{20}Cl_2Rh_2$	72.40	**2**
$C_{12}H_{20}Cl_4Pt_4$	71.13	**4**
$C_{12}H_{20}Cr_2$	72.20	**3**
$C_{12}H_{20}CuN_2O_3$	77.17	**2**
$C_{12}H_{20}CuN_2O_4$	83.113	**2**
$C_{12}H_{20}Cu_2O_8$	81.30	**3**
$C_{12}H_{20}Mo_2$	72.9	**4**
$C_{12}H_{20}Mo_2N_6O_6S_2$, $3H_2O$	82.13	**4**
$C_{12}H_{20}Mo_2O_4S_8$, $2C_4H_8O$	85.22	**6**
$C_{12}H_{20}Mo_2O_7S_8$	80.23	**2**
$C_{12}H_{20}NO^+$, Br^-	17.15	**6**
$C_{12}H_{20}NO^+$, Br^- , $0.5H_2O$	17.16	**6**
$C_{12}H_{20}NO^+$, ClO_4^-	58.4	**4**
$C_{12}H_{20}N_2O_3$	43.9	**4**
$C_{12}H_{20}N_2O_3$	43.6	**6**
$C_{12}H_{20}N_2S_3$	33.41	**3**
$C_{12}H_{20}N_2S_4$	33.46	**5**
$C_{12}H_{20}N_4$	9.12	**5**
$C_{12}H_{20}N_4O_4$	48.46	**3**
$C_{12}H_{20}N_4O_4$	34.3	**5**
$C_{12}H_{20}O_2$	31.11	**5**
$C_{12}H_{20}O_2S$	39.73	**1**
$C_{12}H_{20}O_4$	38.46	**1**
$C_{12}H_{20}O_4S_2$	45.35	**7**
$C_{12}H_{20}O_4Sn$	69.16	**5**
$C_{12}H_{20}O_6$	38.47	**1**
$C_{12}H_{20}O_6$	45.25	**5**
$C_{12}H_{20}S_4$	39.24	**7**
$C_{12}H_{20}S_6$	39.12	**6**
$C_{12}H_{21}ClPRh$	72.10	**6**
$C_{12}H_{21}Cl_2NOPd_2$	72.10	**4**
$C_{12}H_{21}Cl_2N_2O_6P_2Pd_2S_2$	71.21	**5**
$C_{12}H_{21}Cl_2O_6Re_2^+$, O_4Re^-	81.31	**3**
$C_{12}H_{21}INO_2$	37.9	**6**
$C_{12}H_{21}NO_9$, H_2O	45.34	**6**
$C_{12}H_{21}O_{12}^-$, Ca^{2+} , Br^- , $4H_2O$	45.26	**5**
$C_{12}H_{22}Br_2$	21.5	**6**
$C_{12}H_{22}Br_2N_2O_2$	1.133	**1**
$C_{12}H_{22}Cl_2$	21.6	**6**
$C_{12}H_{22}Cl_2N_2O_2Pd$	83.114	**2**
$C_{12}H_{22}Cl_3O_2Pt^-$, K^+	72.15	**5**
$C_{12}H_{22}Cl_4Hg_2N_2S_4$	80.8	**7**
$C_{12}H_{22}Cl_4O_2Rh_2$, CH_4O	71.26	**7**
$C_{12}H_{22}CuN_6O_6^{2+}$, $2NO_3^-$	82.51	**2**
$C_{12}H_{22}CuN_{10}O_4^{2+}$, $2Cl^-$, $2H_2O$	83.46+	**7**
$C_{12}H_{22}Cu_2N_4O_4^{2+}$, $2ClO_4^-$, $2H_2O$	84.12	**6**
$C_{12}H_{22}FeN_8O$	71.27	**7**
$C_{12}H_{22}N_2O_2$	34.9	**1**
$C_{12}H_{22}N_2O_2$	21.11	**3**
$C_{12}H_{22}N_2O_2$	34.2	**3**
$C_{12}H_{22}N_2O_4$	33.68	**1**
$C_{12}H_{22}N_2O_4$	40.6	**7**
$C_{12}H_{22}O_4$	1.134	**1**
$C_{12}H_{22}O_6$	21.12	**3**
$C_{12}H_{22}O_{11}$	45.69	**1**
$C_{12}H_{22}O_{11}$	45.65+	**1**
$C_{12}H_{22}O_{11}$	45.31	**5**
$C_{12}H_{22}O_{11}$	45.28+	**5**
$C_{12}H_{22}O_{11}$	45.35	**6**
$C_{12}H_{22}O_{11}$	45.36	**7**
$C_{12}H_{22}O_{11}$	45.37	**7**
$C_{12}H_{22}O_{11}$, Ca^{2+} , $2Br^-$, $7H_2O$	45.32	**5**
$C_{12}H_{22}O_{11}$, Ca^{2+} , $2Br^-$, H_2O	45.36	**6**
$C_{12}H_{22}O_{11}$, Ca^{2+} , $2Cl^-$, $7H_2O$	45.33	**5**
$C_{12}H_{22}O_{11}$, Na^+ , Br^- , $2H_2O$	45.70	**1**
$C_{12}H_{22}O_{11}$, H_2O	45.35	**3**
$C_{12}H_{22}O_{11}$, H_2O	45.36	**3**
$C_{12}H_{22}O_{11}$, H_2O	45.20	**4**
$C_{12}H_{22}O_{11}$, H_2O	45.34	**5**
$C_{12}H_{22}O_{11}$, $2H_2O$	45.35+	**5**
$C_{12}H_{22}O_{12}$, Ca^{2+} , $2Br^-$, $2H_2O$	45.39	**5**
$C_{12}H_{22}Si_2$	63.6	**7**
$C_{12}H_{23}B_{10}P$	64.16	**5**
$C_{12}H_{23}CoN_{10}$	71.28	**7**
$C_{12}H_{23}NO$	27.9	**5**
$C_{12}H_{23}N_3O_5^{2+}$, $2Cl^-$	58.8	**1**
$C_{12}H_{24}$	23.10	**1**
$C_{12}H_{24}Au_2F_6P_2$	71.22	**5**
$C_{12}H_{24}Br_2N_8S_4Te_2^{2+}$, $2Br^-$	70.8+	**5**
$C_{12}H_{24}Cl_2MnN_8$, $2H_2O$	83.115	**2**
$C_{12}H_{24}Cl_2N_8NiS_4$	79.34+	**2**
$(C_{12}H_{24}Cl_2NiS_4)_n$	85.20	**7**
$C_{12}H_{24}Cl_2O_4PdS_2$	85.21	**7**
$C_{12}H_{24}Cl_3O_3Sc$	84.13	**6**
$C_{12}H_{24}CoN_4O_3^+$, ClO_4^-	83.41	**5**
$C_{12}H_{24}Cr_3O_{16}^+$, Cl^- , $5H_2O$	81.75	**2**
$C_{12}H_{24}Cr_3O_{16}^+$, Cl^- , $6H_2O$	81.32	**3**
$(C_{12}H_{24}CuN_2O_4)_n$, $2nH_2O$	81.40	**5**
$C_{12}H_{24}CuN_2O_6$, $2H_2O$	81.36	**6**
$C_{12}H_{24}CuN_2O_8$	82.24	**3**
$C_{12}H_{24}Cu_2O_8$	81.76	**2**
$C_{12}H_{24}Fe_3O_{16}^+$, ClO_4^- , $2C_2H_4O_2$	81.34	**3**
$C_{12}H_{24}IN_8^+$, I_3^-	37.10	**7**
$C_{12}H_{24}N^+$, Br^-	52.2	**3**
$C_{12}H_{24}NO^+$, I^-	58.5	**4**
$C_{12}H_{24}N_2O_2$	1.135+	**1**
$C_{12}H_{24}N_4Ni^{2+}$, $2ClO_4^-$, $2H_2O$	83.42	**5**
$C_{12}H_{24}N_4O_2S_4Sn_4$	69.21	**3**
$C_{12}H_{24}N_4S_8Zn_2$	80.24	**2**
$C_{12}H_{24}N_6S_2Se_2Te$	70.12	**4**
$C_{12}H_{24}O_2$	1.137	**1**
$C_{12}H_{24}O_4$	38.16	**4**
$C_{12}H_{24}O_4S_2$	42.9	**7**
$C_{12}H_{24}O_6$	38.29+	**7**
$C_{12}H_{24}O_6$, Ca^{2+} , $2CNS^-$	38.31+	**7**
$C_{12}H_{24}O_6$, Cs^+ , CNS^-	38.33+	**7**
$C_{12}H_{24}O_6$, K^+ , CNS^-	38.35+	**7**
$C_{12}H_{24}O_6$, Na^+ , CNS^- , H_2O	38.37+	**7**
$C_{12}H_{24}O_6$, Rb^+ , CNS^-	38.39+	**7**
$C_{12}H_{24}O_6$, $C_6H_6O_4$	38.41	**7**
$C_{12}H_{24}Pd_3S_9$	85.66	**2**
$C_{12}H_{25}N_3Nd_2O_{16}^{2+}$, $2Cl^-$, $3H_2O$	82.52	**2**
$(C_{12}H_{25}N_3O_{16}Pr_2^{2+})_n$, $2nCl^-$, $3nH_2O$	82.20	**6**
$C_{12}H_{26}BrN_2NbO_2S_4$	80.6	**4**

$C_{12}H_{26}Br_2CuN_2O_4$	84.14	**6**
$(C_{12}H_{26}Cd_2N_{10}S_4)_n$	76.59	**7**
$C_{12}H_{26}ClN_2NbO_2S_4$	80.7	**4**
$C_{12}H_{26}Cl_2CuN_2O_4$	83.43	**5**
$C_{12}H_{26}CuN_2O_5$	82.53	**2**
$C_{12}H_{26}Er_2O_{16}$, $4H_2O$	81.32	**7**
$C_{12}H_{26}KN_2O_4^+$, CNS^-	67.12	**4**
$C_{12}H_{26}N_2O$	1.138	**1**
$C_{12}H_{26}N_2O_2$	34.10	**1**
$C_{12}H_{26}N_2O_4$	40.18	**4**
$C_{12}H_{26}N_2O_6Sn$	69.17	**5**
$C_{12}H_{26}O_4S_3$	45.37	**6**
$C_{12}H_{26}O_5$, Cl_2Hg	5.10	**6**
$C_{12}H_{27}CoN_5O_4^+$, ClO_4^- , H_2O	76.60	**7**
$C_{12}H_{27}S^+$, F^- , $20H_2O$	61.11	**2**
$C_{12}H_{27}S^+$, F^- , $23H_2O$	61.12	**2**
$(C_{12}H_{28}AgP_2PtS_2^+)_n$, nNO_3^-	85.67	**2**
$C_{12}H_{28}Au_2O_4P_2S_4$	85.18	**5**
$C_{12}H_{28}Au_2O_4S_4$	85.19	**5**
$C_{12}H_{28}BrCuN_6$	83.116	**2**
$C_{12}H_{28}Br_2Cu_2N_2O_2$	83.59	**4**
$C_{12}H_{28}Br_2Cu_2N_2O_2$	84.15	**6**
$C_{12}H_{28}Br_2Cu_2N_2O_2$	84.35	**7**
$C_{12}H_{28}ClCoN_4^+$, Cl^- , C_2H_6O	83.27	**6**
$C_{12}H_{28}Cl_2Cu_2N_2O_2$	84.16	**6**
$C_{12}H_{28}CoN_4O_4^+$, ClO_4^- , H_2O	81.33	**7**
$C_{12}H_{28}CoN_{10}O_{10}^+$, NO_3^- , $2H_2O$	82.20	**5**
$C_{12}H_{28}CuN_4^{2+}$, $2NO_3^-$	83.117	**2**
$(C_{12}H_{28}HgO_4P_2S_4)_n$	85.21	**3**
$C_{12}H_{28}N^+$, Br^-	3.75	**1**
$C_{12}H_{28}N^+$, Br^-	3.28	**6**
$C_{12}H_{28}N^+$, $3CH_4N_2O$, Br^- , H_2O	61.7	**6**
$C_{12}H_{28}N^+$, $C_2I_4O_2Rh^-$	3.29	**7**
$C_{12}H_{28}N^+$, $C_{20}H_8CoN_6S_4^-$	85.35	**3**
$2C_{12}H_{28}N^+$, $Br_4O_2U^{2-}$	3.29	**6**
$2C_{12}H_{28}N^+$, $Br_6Ga_2^{2-}$	3.30	**6**
$2C_{12}H_{28}N^+$, $Cl_4O_2U^{2-}$	3.30	**7**
$4C_{12}H_{28}N^+$, $O_{32}W_{10}^{4-}$	3.31	**6**
$C_{12}H_{28}N_2^{2+}$, $2Br^-$	34.11	**1**
$C_{12}H_{28}N_2^{2+}$, $4C_6H_6N_2O_2$, $2Cl^-$	60.40	**3**
$C_{12}H_{28}N_4Ni^{2+}$, $2Br^-$	83.44	**5**
$C_{12}H_{28}N_4Ni^{2+}$, $2ClO_4^-$	83.48	**7**
$C_{12}H_{28}N_4Ni^{2+}$, $2NO_3^-$	83.118	**2**
$C_{12}H_{28}N_4O_2P_2$	64.34	**2**
$C_{12}H_{28}O_4P_2PbS_4$	69.24	**4**
$C_{12}H_{28}O_4P_2S_4$	64.28	**3**
$C_{12}H_{28}Si_4$	63.17	**2**
$C_{12}H_{29}BrCuN_6$	76.41	**5**
$C_{12}H_{29}Cl_2CoN_3$	76.75	**2**
$C_{12}H_{29}Cl_2N_7P_3^+$, Cl^-	64.29	**3**
$C_{12}H_{29}N_{12}Rh$, $0.8C_2H_6O$	76.29	**6**
$C_{12}H_{30}Al_2F^-$, K^+	68.20	**2**
$C_{12}H_{30}Al_3N_3$	68.11	**3**
$C_{12}H_{30}B_3N_3$	62.47	**2**
$C_{12}H_{30}BrCuN_4^+$, Br^-	83.119	**2**
$C_{12}H_{30}BrFeN_4^+$, Br^-	76.76	**2**
$C_{12}H_{30}BrMnN_4^+$, Br^-	76.77	**2**
$C_{12}H_{30}BrN_4Ni^+$, Br^-	83.120	**2**
$C_{12}H_{30}BrN_4Zn^+$, Br^-	76.78	**2**
$C_{12}H_{30}Br_2CoN_4$	76.79	**2**
$C_{12}H_{30}Br_2NiP_2$	86.9	**2**
$C_{12}H_{30}Br_2P_2Pt$	86.10	**2**
$2C_{12}H_{30}CdIN_4^+$, $Cd_2I_6^{2-}$	76.42	**5**
$C_{12}H_{30}Cl_2N_2Pt$	71.23	**5**
$C_{12}H_{30}Cl_2P_2Pt$	86.11	**2**
$C_{12}H_{30}Cl_4P_2Sn$	69.10	**6**
$C_{12}H_{30}Cl_6P_2Re_2$	86.12	**2**
$C_{12}H_{30}CoN_6O_3^{3+}$, $3ClO_4^-$, H_2O	82.54	**2**
$C_{12}H_{30}IN_4Ni^+$, I^-	76.61	**7**
$C_{12}H_{30}InO_6P_3S_6$	68.12	**3**
$C_{12}H_{30}N_2^{2+}$, $2Br^-$, $2H_2O$	3.77	**1**
$C_{12}H_{30}N_2^{2+}$, $2Cl^-$, $2H_2O$	3.78	**1**
$C_{12}H_{30}N_2NiO_6^{2+}$, $2NO_3^-$	83.60	**4**
$C_{12}H_{30}N_2O_6Sr^{2+}$, $2NO_3^-$	67.6	**6**
$C_{12}H_{30}N_2O_{16}P_2U$	84.35	**2**
$C_{12}H_{30}N_3O_6P_3$	64.30	**3**
$C_{12}H_{30}N_3O_6P_3$	64.17	**4**
$C_{12}H_{30}N_{18}Ni_3O_6^{6+}$, $6NO_3^-$, $2H_2O$	83.63	**3**
$C_{12}H_{30}O_6P_3S_6V$	85.22	**3**
$C_{12}H_{31}BrP_2Pt$	86.13	**2**
$C_{12}H_{31}ClP_2Pd$	86.13	**5**
$(C_{12}H_{32}Al_2MgO_6)_n$	67.12	**2**
$C_{12}H_{32}Cl_2Mo_2N_4O_9$	84.36	**7**
$C_{12}H_{32}Cl_2N_7P_3$	64.34	**7**
$C_{12}H_{32}Cl_4CuN_4$	76.62	**7**
$C_{12}H_{32}Ga_2N_2O_2$	68.14	**7**
$C_{12}H_{32}Mg_2N_4$	67.12	**3**
$C_{12}H_{32}N_8S_4Te^{2+}$, $2Br^-$	70.13	**4**
$C_{12}H_{32}N_8S_4Te^{2+}$, $2Cl^-$	70.14	**4**
$C_{12}H_{33}Cl_2N_7P_3^+$, Cl^-	64.35	**7**
$(C_{12}H_{34}CaO_{15}^{2+})_n$, $2nBr^-$, $3nH_2O$	67.13	**4**
$C_{12}H_{34}CoP_3$	71.29	**7**
$C_{12}H_{34}Cu_2N_4O_2^{2+}$, $2Br^-$	76.77	**3**
$C_{12}H_{34}N_6Ni^{2+}$, $2ClO_4^-$	83.64	**3**
$C_{12}H_{36}Ag_2N_2O_{18}P_4$	86.16	**7**
$C_{12}H_{36}AlO_4Si_4^-$, $C_4H_{12}Sb^+$	68.21	**2**
$C_{12}H_{36}As_4Cu_4I_4$	86.14	**2**
$C_{12}H_{36}As_6S_{13}Zn_4$	85.20	**5**
$C_{12}H_{36}Be_3N_6$	67.13	**3**
$C_{12}H_{36}Cl_2N_{10}P_4$	64.19	**6**
$C_{12}H_{36}Cl_2O_6S_6U^{2+}$, Cl_6U^{2-}	84.37	**7**
$C_{12}H_{36}Cl_4Pt_4$	71.29	**2**
$C_{12}H_{36}CoN_6^{3+}$, $3Br^-$	83.49	**7**
$C_{12}H_{36}Co_3N_6O_6^{2+}$, $2C_2H_3O_2^-$	83.65	**3**
$C_{12}H_{36}CrN_4OSi_3$	83.61	**4**
$C_{12}H_{36}Cs_4O_4Si_4$	63.4	**3**
$C_{12}H_{36}CuN_4O_2^{2+}$, $2NO_3^-$	76.78	**3**
$C_{12}H_{36}Ge_6$	69.11	**7**
$C_{12}H_{36}K_4O_4Si_4$	63.5	**3**
$C_{12}H_{36}Mo_2N_6$	83.28	**6**
$C_{12}H_{36}N_4Si_4$	63.7	**7**
$C_{12}H_{36}N_6O_{12}P_6$	64.20	**6**
$C_{12}H_{36}N_6W$	83.123	**2**
$2C_{12}H_{36}N_6W_2$, $C_{12}H_{36}N_6W$	83.50	**7**
$C_{12}H_{36}N_9P_3$	64.21	**6**
$C_{12}H_{36}N_{12}Pt_4$	71.30 +	**7**
$C_{12}H_{36}N_{14}Ni_2^{2+}$, $2C_{24}H_{20}B^-$	76.63	**7**
$C_{12}H_{36}O_4Rb_4Si_4$	63.6	**3**
$C_{12}H_{36}O_6Si_8$	63.18	**2**
$2C_{12}H_{37}N_9P_3^+$, Cl_4Co^{2-}	64.36	**7**
$2C_{12}H_{37}N_9P_3^+$, $Mo_6O_{19}^{2-}$	64.22	**6**
$C_{12}H_{40}O_4Pt_4$	71.30 +	**2**
$C_{12}H_{41}B_9P_2PtS$	86.10	**3**
$C_{12}H_{42}Co_2N_{10}O_2^{4+}$, $4ClO_4^-$	76.30	**6**
$C_{12}H_{42}N_6Pt_2^{2+}$, $2I^-$	76.80	**2**
$C_{12}H_{54}Co_4N_{12}O_6^{6+}$, $3O_6S_2^{2-}$, $8H_2O$	76.39	**4**
$C_{12}H_{54}Cr_4N_{12}O_6^{6+}$, $6N_3^-$, $4H_2O$	76.81	**2**

C_{13}

$C_{13}Cl_{11}$ 12.11 **4**
$C_{13}F_{12}Mn_2N_2O_7$ 83.51 **7**
$C_{13}H_3Co_2F_9O_4$ 71.24 **5**
$C_{13}H_3N_6^-$, H_4N^+ 7.6 **3**
$C_{13}H_4BrN_3O_7$, $C_{14}H_{10}$ 60.42 **3**
$C_{13}H_4BrN_3O_7$, $C_{20}H_{16}$ 60.55 **3**
$C_{13}H_4BrN_3O_7$, $C_{26}H_{16}$ 60.156 **2**
$C_{13}H_4F_3O_6S_2W^-$, $C_8H_{20}N^+$ 85.23 **6**
$C_{13}H_5ClF_{12}W$ 72.9 **7**
$C_{13}H_5Co_2NO_4Pt$ 83.124 **2**
$C_{13}H_5F_5FeO_4S$ 73.16 **6**
$C_{13}H_5MnMoO_8$ 73.41 **2**
$C_{13}H_5MoN_4S_4^-$, $C_{24}H_{20}P^+$ 64.52 **3**
$C_{13}H_5N_3O_7$ 28.2 **5**
$C_{13}H_6Cl_2N_2O_2$ 31.11 **4**
$C_{13}H_6CoF_{12}O_2P$ 73.18 **7**
$C_{13}H_7BrN_2$ 9.7 **3**
$C_{13}H_7BrO$ 28.2 **3**
$C_{13}H_8BrN_3OS$ 41.23 **5**
$C_{13}H_8Br_2O$ 19.15 **5**
$C_{13}H_8Br_4GeN_2O_3W$ 83.66 **3**
$C_{13}H_8ClN_3OPdS$, $C_4H_8O_2$ 84.38 **7**
$C_{13}H_8Cl_4HgN_2$ 71.25 **5**
$C_{13}H_8Co_2O_6$ 75.14 **5**
$C_{13}H_8CrO_3$ 74.13 **2**
$C_{13}H_8CrO_5S$ 71.14 **4**
$C_{13}H_8CrO_6$ 71.32 **2**
$C_{13}H_8FeO_5$ 75.10 **3**
$C_{13}H_8Fe_2O_5$ 75.26 **2**
$C_{13}H_8Fe_2O_6$ 75.6 **4**
$C_{13}H_8MnO_6^-$, $C_4H_{12}N^+$ 71.31 **7**
$C_{13}H_8N_2$ 28.15 **1**
$C_{13}H_8N_2O_3$ 36.21 **1**
$C_{13}H_8N_4O_4$ 15.9 **5**
$C_{13}H_8O$ 28.3 **4**
$C_{13}H_8O_2$ 28.11 **6**
$C_{13}H_9BrN_2O_4$ 35.13 **3**
$C_{13}H_9Br_3O$ 20.29 **1**
$C_{13}H_9ClO_2$ 13.49 **1**
$C_{13}H_9ClO_2$ 13.50 **1**
$C_{13}H_9Cl_2N$ 16.12 **4**
$C_{13}H_9Cl_2N$ 16.13 **4**
$C_{13}H_9CrNO_3$ 74.14 **2**
$C_{13}H_9FeO_4$ 72.21 **3**
$C_{13}H_9Fe_3NO_{10}Si$ 63.5 **4**
$C_{13}H_9IO_2$ 13.14 **3**
$C_{13}H_9IO_2$ 13.15 **5**
$C_{13}H_9N$ 36.22 **1**
$C_{13}H_9N$ 36.5 **3**
$C_{13}H_9N$ 36.12 **5**
$C_{13}H_9N$, $C_4H_5N_3O$, H_2O 60.132 **2**
$C_{13}H_9NO$ 36.6 **3**
$C_{13}H_9NOS$ 41.13 **3**
$C_{13}H_9NO_2$ 33.69 **1**
$C_{13}H_9N_3O$ 35.19 **5**
$C_{13}H_9N_3OS$ 41.14 **7**
$C_{13}H_9N_3O_2$ 35.16 **7**
$C_{13}H_9N_3S$ 9.38 **1**
$C_{13}H_{10}$ 28.16 **1**
$C_{13}H_{10}BrN$ 33.42 **3**
$C_{13}H_{10}BrN$ 16.14 **4**
$C_{13}H_{10}BrNO$ 16.48 **1**
$C_{13}H_{10}BrNO$ 16.49 **1**
$C_{13}H_{10}Br_2$ 31.5 **7**
$C_{13}H_{10}Br_2O_4S_2$ 11.10 **5**
$C_{13}H_{10}Br_2S_2$ 11.21 **3**
$C_{13}H_{10}ClNO$ 16.50 **1**
$C_{13}H_{10}ClNO$ 16.51 **1**
$C_{13}H_{10}ClNO$ 16.52+ **1**
$C_{13}H_{10}ClNO$ 16.14 **6**
$C_{13}H_{10}Cl_2FeO_2Sn$ 69.22 **3**
$C_{13}H_{10}Cl_2HgN_4S$ 85.24 **6**
$C_{13}H_{10}Cl_2O_2$ 17.38 **1**
$C_{13}H_{10}FeN_2$ 73.10 **3**
$C_{13}H_{10}FeO_3$ 72.41 **2**
$C_{13}H_{10}Fe_2O_5S$ 73.18 **6**
$C_{13}H_{10}N^+$, $C_{12}H_4N_4^-$, $C_{12}H_4N_4$ 60.17 **7**
$C_{13}H_{10}N_2O_2$ 28.12 **6**
$C_{13}H_{10}N_2O_4$ 50.6 **1**
$C_{13}H_{10}N_2O_4$ 35.14+ **3**
$C_{13}H_{10}N_2O_4$ 35.15 **4**
$C_{13}H_{10}N_2O_4$ 35.16 **4**
$C_{13}H_{10}N_4O$ 40.7 **7**
$C_{13}H_{10}N_4S$ 41.35 **1**
$C_{13}H_{10}N_4S^+$, HO^- 32.32 **3**
$C_{13}H_{10}O$ 19.42+ **1**
$C_{13}H_{10}O$, $C_{12}H_{11}N$ 60.26 **4**
$C_{13}H_{10}O_2S$ 39.74 **1**
$C_{13}H_{10}O_2S$ 39.13 **6**
$C_{13}H_{10}O_2S$, H_2O 17.39 **1**
$C_{13}H_{10}O_3Rh_2$ 73.42 **2**
$C_{13}H_{10}S$ 39.14 **6**
$C_{13}H_{11}AsMn_2O_6$ 73.19 **7**
$C_{13}H_{11}Cl_2N_2O^+$, I^- , H_2O 33.43 **3**
$C_{13}H_{11}Cl_2N_3O_3$ 35.20 **5**
$C_{13}H_{11}FeN$ 73.10 **4**
$C_{13}H_{11}FeNO_6$ 75.8 **6**
$C_{13}H_{11}I_2IrN_2O_3$ 71.33 **2**
$C_{13}H_{11}N$ 16.9 **3**
$C_{13}H_{11}NO_2$ 37.8 **4**
$C_{13}H_{11}NO_2$ 34.4 **5**
$C_{13}H_{11}NO_2S_2$ 13.51 **1**
$C_{13}H_{11}NO_3$ 10.29 **1**
$C_{13}H_{11}NS$ 39.34 **3**
$C_{13}H_{11}NS$ 41.15 **7**
$C_{13}H_{11}NS$, $C_{12}H_4N_4$ 60.18 **7**
$C_{13}H_{11}N_2^+$, Cl^- , H_2O 36.4 **6**
$C_{13}H_{11}N_2^+$, $C_{12}H_4N_4^-$ 36.23 **1**
$C_{13}H_{11}N_2^+$, $C_{19}H_{23}N_7O_{12}^-$, xH_2O 47.30 **6**
$C_{13}H_{11}N_2^+$, $C_{19}H_{23}N_7O_{12}P^-$, xH_2O 47.37 **7**
$C_{13}H_{11}N_3O_2$ 33.37 **6**
$C_{13}H_{11}N_5O_2$ 44.27 **7**
$C_{13}H_{11}O_2P$ 64.36 **2**
$C_{13}H_{11}O_3^+$, Br^- 59.23 **1**
$C_{13}H_{11}O_4P$ 64.23 **6**
$C_{13}H_{11}PS_4$ 64.18 **5**
$C_{13}H_{12}BrNO_2$ 33.70 **1**
$C_{13}H_{12}BrNO_2S$ 16.54 **1**
$C_{13}H_{12}Br_2N_2SZn$ 83.62 **4**
$C_{13}H_{12}ClNO_2$ 1.33 **5**
$C_{13}H_{12}ClNO_3S$ 39.16 **4**
$C_{13}H_{12}ClN_3O_2S$ 41.20 **4**
$C_{13}H_{12}Cl_2CuN_2S$ 85.14 **4**
$C_{13}H_{12}Cl_5Rh$ 73.20 **7**
$C_{13}H_{12}CrO_3$ 75.11 **3**
$C_{13}H_{12}CuN_3O_3S^+$, ClO_4^- 84.39 **7**
$C_{13}H_{12}Cu_3S_6$ 85.69 **2**
$C_{13}H_{12}FeO$ 73.43 **2**
$C_{13}H_{12}FeO_2$ 75.12 **3**

$C_{13}H_{12}FeO_3$ 75.13 3
$C_{13}H_{12}FeO_3$ 75.15 5
$C_{13}H_{12}FeO_3$ 75.8 7
$C_{13}H_{12}MnNO_4$ 71.26 5
$C_{13}H_{12}N_2$ 22.4 6
$C_{13}H_{12}N_2O_4$ 32.19 5
$C_{13}H_{12}N_2O_4S$ 9.8 3
$2C_{13}H_{12}N_3^+$, O_4S^{2-} , $3.5H_2O$ 36.10 7
$C_{13}H_{12}N_3O_2^+$, Cl^- , $2H_2O$ 36.13 5
$C_{13}H_{12}N_4$ 16.8 7
$C_{13}H_{12}N_4O_2$, $2C_{10}H_8O_2$ 60.22 6
$C_{13}H_{12}N_4O_2$, $2C_{10}H_8O_2$ 60.23 6
$C_{13}H_{12}N_4O_2$, $2C_{10}H_8O_2$, $3H_2O$ 60.19 7
$C_{13}H_{12}O$ 27.8 6
$C_{13}H_{12}O_2$ 59.1 7
$C_{13}H_{12}Si$ 63.8 7
$C_{13}H_{13}BFeN_6O_3$ 71.16 6
$C_{13}H_{13}Br_2NO_5$ 21.13 3
$C_{13}H_{13}Br_2N_5O_2$ 58.7 3
$C_{13}H_{13}ClN_2O_3S$ 45.38 7
$C_{13}H_{13}FNP$ 64.31 3
$C_{13}H_{13}FeO_2^-$, $C_{20}H_{25}N_2O_2^+$, H_2O 73.44 2
$C_{13}H_{13}NO_5S$ 41.16 7
$C_{13}H_{13}N_2O_2^+$, $1.5C_6H_6O_2$, Br^- 60.30 4
$C_{13}H_{13}N_3O_2$ 48.68 6
$C_{13}H_{13}N_3O_3$ 36.11 7
$C_{13}H_{13}N_4O_2^+$, I^- 36.7 3
$C_{13}H_{13}N_4O_2^+$, $C_6H_6O_2$, Cl^- 36.14 5
$C_{13}H_{13}OPS$ 64.38 2
$C_{13}H_{13}OPSe$ 64.39 2
$C_{13}H_{13}P$ 64.32 3
$C_{13}H_{13}Ti$ 75.14 3
$C_{13}H_{14}$ 31.12 5
$C_{13}H_{14}ClNS$, $3H_2O$ 36.15 5
$C_{13}H_{14}ClN_5$, $3H_2O$ 36.12 7
$C_{13}H_{14}Cl_2Hf$ 73.21 7
$C_{13}H_{14}Cl_2N_2O_2$, $2H_2O$ 33.71 1
$C_{13}H_{14}Cl_2Ti$ 73.11 4
$C_{13}H_{14}Cl_2Ti$ 73.21 6
$C_{13}H_{14}Cl_2Zr$ 73.22 7
$C_{13}H_{14}Cl_3NO_3S$ 35.21 5
$C_{13}H_{14}CrO_6$ 75.9 7
$C_{13}H_{14}FeO$ 72.16 5
$C_{13}H_{14}INO_2$ 33.72 1
$C_{13}H_{14}MoN_2O_5^+$, F_6P^- 71.15 4
$C_{13}H_{14}NO_6^-$, $C_9H_{14}N_3O_5^+$, $2H_2O$ 47.14 7
$C_{13}H_{14}N_2$ 31.6 7
$3C_{13}H_{14}N_2$, Na^+ , Cl^- 16.18 5
$C_{13}H_{14}N_2O$ 33.73 1
$C_{13}H_{14}N_2O$ 58.1 5
$C_{13}H_{14}N_2O_3$ 43.9 3
$C_{13}H_{14}N_2O_4S_2$ 50.7 1
$C_{13}H_{14}N_4OS$ 36.13 7
$C_{13}H_{14}N_4S$ 16.15 6
$C_{13}H_{14}O_2$ 27.11 5
$C_{13}H_{14}O_3$ 29.1 6
$C_{13}H_{14}O_5$ 59.6 4
$C_{13}H_{15}BrN_2O_3S$ 41.24 5
$C_{13}H_{15}ClF_3NO$ 16.19 5
$C_{13}H_{15}CuN_3O_4$, $2H_2O$ 82.25 3
$C_{13}H_{15}FN_2O_7$ 47.33 7
$C_{13}H_{15}F_3O_2$ 38.42 7
$C_{13}H_{15}FeNO_2$ 75.15 3
$C_{13}H_{15}MoNO_2S$, $C_{13}H_{16}MoNO_2S^+$, F_6P^- 73.17 5
$C_{13}H_{15}MoNO_2S$, $C_{13}H_{16}MoNO_2S^+$, Cl^- 82.21 5
$C_{13}H_{15}NO$ 37.5 5
$C_{13}H_{15}NO$ 32.23 7
$C_{13}H_{15}NO_2$ 37.9 4
$C_{13}H_{15}NO_2$ 58.5 6
$C_{13}H_{15}N_2RhS_2$ 75.10 7
$C_{13}H_{15}N_3O_3$, $2H_2O$ 48.90 1
$C_{13}H_{15}N_5O_3S$, $0.33H_2O$ 47.32 3
$C_{13}H_{15}O$ 21.3 7
$C_{13}H_{16}As_2FeO_3$ 86.15 2
$C_{13}H_{16}Cl_2N_2Pt$ 71.16 4
$C_{13}H_{16}Cl_2N_2Pt$ 83.63 4
$C_{13}H_{16}Cl_4N_2Pt$, $CHCl_3$ 71.17 4
$C_{13}H_{16}CuN_2O_4$ 82.55 2
$C_{13}H_{16}FeO$ 75.7 4
$C_{13}H_{16}INO_2S$ 35.17 4
$C_{13}H_{16}MoNO_2^+$, Cl^- , CH_4O 73.18 5
$C_{13}H_{16}NO_2^+$, Br^- , $2H_2O$ 58.9 1
$C_{13}H_{16}N_2O_2$ 40.13 3
$C_{13}H_{16}N_2O_2$ 35.22 5
$C_{13}H_{16}N_2O_2$ 35.20 6
$C_{13}H_{16}N_2O_2$ 35.17 7
$C_{13}H_{16}N_2O_2S$ 41.19 6
$C_{13}H_{16}N_2O_3S$ 48.69 5
$C_{13}H_{16}N_2O_6$ 13.18 7
$C_{13}H_{16}N_4O_4$ 44.37 3
$C_{13}H_{16}N_4O_4$ 44.19 5
$C_{13}H_{16}N_4O_6$ 45.38 6
$C_{13}H_{16}N_5O_3^+$, I^- 47.29 1
$C_{13}H_{16}N_6Ni_2O_2S_2$, $2C_3H_7NO$ 85.15 4
$C_{13}H_{16}N_6O_7$ 48.52 7
$C_{13}H_{16}O$ 31.14 5
$C_{13}H_{14}D_2O$ 52.21 1
$C_{13}H_{16}O_2$ 31.15 5
$C_{13}H_{16}O_3$ 38.43 7
$C_{13}H_{17}B_3Co_2$ 62.9 5
$C_{13}H_{17}BrN_2$ 32.27 6
$C_{13}H_{17}BrN_2O$ 33.36 7
$C_{13}H_{17}BrO_7$ 38.13 3
$C_{13}H_{17}ClN_2O$ 33.38 6
$C_{13}H_{17}Cl_2N_2Ta$ 71.17 6
$C_{13}H_{17}NO$ 33.39 6
$C_{13}H_{17}NO_4$, H_2O 48.49 4
$C_{13}H_{17}N_2^+$, $C_6H_2N_3O_7^-$ 36.14 7
$C_{13}H_{17}N_2O_2$ 32.28 6
$C_{13}H_{17}N_2O_2^+$, Cl^- 48.53 7
$C_{13}H_{17}N_3O$, $C_8H_{12}N_2O_3$ 60.31 4
$C_{13}H_{17}N_3O$, $C_{12}H_{16}N_2O_3$ 60.20 7
$C_{13}H_{17}N_3O_2$ 32.24 7
$C_{13}H_{17}N_3O_4$, H_2O 48.91 1
$C_{13}H_{18}AuN_3S_4$ 80.11 5
$C_{13}H_{18}Cl_2O_5$ 45.71 1
$C_{13}H_{18}CuN_3S_4^-$, $C_{16}H_{36}N^+$ 80.5 6
$2C_{13}H_{18}FeN^+$, Cl_4Zn^{2-} , H_2O 71.18 4
$C_{13}H_{18}NO^+$, $C_4H_5O_6^-$, H_2O 33.40 6
$C_{13}H_{18}N_2O_2S$ 41.14 3
$C_{13}H_{18}N_2O_3$ 43.30 1
$C_{13}H_{18}N_4O_5S$, H_2O 47.27 6
$C_{13}H_{18}O$ 21.4 7
$C_{13}H_{18}OPd$ 71.27 5
$C_{13}H_{18}O_2$ 1.27 6
$C_{13}H_{18}O_2Pd$ 75.27 2
$C_{13}H_{18}O_2S$ 39.23 5
$C_{13}H_{18}O_3$ 21.5 7
$C_{13}H_{18}O_3S$ 21.9+ 4

$C_{13}H_{18}O_9$ 45.40 **5**
$C_{13}H_{18}O_9$ 45.42 **5**
$C_{13}H_{18}O_9$ 45.39 **6**
$C_{13}H_{18}O_9$ 45.40 **6**
$C_{13}H_{18}O_9$ 45.41 **6**
$C_{13}H_{19}ClN_3O_2^+$, Cl^- , H_2O 17.20 **7**
$C_{13}H_{19}ClN_4O_5S$ 45.39 **7**
$C_{13}H_{19}ClO_2S$ 39.17 **4**
$C_{13}H_{19}CoN_8O_4$ 83.52 **7**
$(C_{13}H_{19}CuNO_4)_n$ 81.35 **3**
$C_{13}H_{19}NO_2PbS_4$ 69.27 **2**
$C_{13}H_{19}NO_5$ 45.40 **7**
$C_{13}H_{19}NO_5$, $0.5H_2O$ 37.8 **3**
$C_{13}H_{19}N_2^+$, ClO_4^- 32.18 **4**
$C_{13}H_{19}N_2O^+$, Cl^- 35.16 **3**
$2C_{13}H_{19}N_2OS^+$, Cl_4Co^{2-} 41.21 **4**
$2C_{13}H_{19}N_2OS^+$, Cl_4Cu^{2-} 41.22 **4**
$C_{13}H_{19}N_2O_3^+$, $CH_3O_4S^-$ 35.18 **7**
$C_{13}H_{19}N_2O_4^+$, Cl^- 13.15 **6**
$C_{13}H_{19}O_2Rh$ 75.11 **7**
$C_{13}H_{19}Re$ 73.45 **2**
$C_{13}H_{20}Cl_2NPPt$ 71.19 **4**
$C_{13}H_{20}N^+$, Br^- 32.34 **1**
$C_{13}H_{20}N^+$, I^- 32.19 **4**
$C_{13}H_{20}NO^+$, Cl^- 32.29 **6**
$C_{13}H_{20}NO^+$, Cl^- 33.41 **6**
$C_{13}H_{20}NO_3^+$, Br^- 58.6 **4**
$C_{13}H_{20}NO_5^+$, ClO_4^- 31.14 **3**
$C_{13}H_{20}N_2OS_2$ 8.18 **3**
$C_{13}H_{20}N_2O_3$ 35.23 **5**
$C_{13}H_{20}NiO_2$ 75.28 **2**
$C_{13}H_{20}O_6$ 45.41 **7**
$C_{13}H_{20}O_8$ 45.72 **1**
$C_{13}H_{21}^+$, $AlCl_4^-$ 12.9 **1**
$C_{13}H_{21}Cl_2NPt$ 72.11 **4**
$C_{13}H_{21}NO_2$ 17.21 **7**
$C_{13}H_{21}NO_3$ 3.23 **4**
$C_{13}H_{21}NO_4$ 40.8 **7**
$C_{13}H_{21}N_2O_2^+$, Cl^- 3.24 **4**
$C_{13}H_{21}N_2O_2^+$, $H_2O_4P^-$, $0.5H_2O$ 13.19 **7**
$C_{13}H_{21}N_2O_2^+$, $C_{12}H_8N_2O_8P^-$ 3.47 **3**
$C_{13}H_{21}N_2O_6^+$, I^- , H_2O , $0.25C_2H_3N$ 50.5 **7**
$C_{13}H_{21}N_2S_2^+$, I^- 41.25 **5**
$C_{13}H_{21}N_2S_2^+$, $C_{12}H_4N_4^-$, $C_{16}H_5N_5O_6$ 60.21 **7**
$C_{13}H_{21}N_6OP$ 32.35 **1**
$C_{13}H_{21}N_6OP$ 32.36 **1**
$C_{13}H_{21}N_6OP$ 32.37 **1**
$C_{13}H_{22}Br_2N_2NiO$ 83.127 **2**
$C_{13}H_{22}HgO_4$ 71.18 **6**
$C_{13}H_{22}NO^+$, Br^- 3.48 **3**
$C_{13}H_{22}NO^+$, Br^- 17.17 **6**
$C_{13}H_{22}NO_4^+$, Br^- 50.4 **3**
$C_{13}H_{22}N_3O^+$, Br^- 58.7 **4**
$C_{13}H_{22}O_6$ 45.43 **5**
$C_{13}H_{22}O_6$ 38.24 **6**
$C_{13}H_{22}O_6$ 45.42 **6**
$C_{13}H_{22}S_3$ 39.24 **5**
$C_{13}H_{23}CoN_2O_2$ 76.43 **5**
$C_{13}H_{23}FeN_7^{2+}$, $2F_6P^-$ 71.20 **4**
$C_{13}H_{23}NO$, $C_{13}H_{24}NO^+$, Cl^- 58.8 **4**
$C_{13}H_{23}N_3^{2+}$, $2Cl^-$, $2H_2O$ 36.5 **6**
$C_{13}H_{23}N_3O_4$ 48.92 **1**
$C_{13}H_{24}N^+$, Br^- 52.2 **4**
$C_{13}H_{24}NO^+$, Cl^- 58.6 **6**
$C_{13}H_{24}N_2O$ 8.11 **4**
$C_{13}H_{24}O_3$ 1.139 **1**
$C_{13}H_{24}O_4$ 1.140 **1**
$C_{13}H_{24}O_{11}$, H_2O 45.73 **1**
$C_{13}H_{24}O_{11}$, CH_4O 45.38 **3**
$C_{13}H_{25}BrN^+$, Br^- 52.3 **4**
$C_{13}H_{25}CoN_4O_2$ 71.32 **7**
$C_{13}H_{25}NNiO_4P_2S_4$ 85.70 **2**
$C_{13}H_{25}NS$ 34.5 **5**
$C_{13}H_{25}N_4O_3$, $0.5H_2O$ 48.69 **6**
$C_{13}H_{26}N^+$, Cl^- 58.9 **3**
$C_{13}H_{26}N^+$, Cl^- 35.19 **7**
$C_{13}H_{26}N_2^{2+}$, $2I^-$ 36.25 **1**
$C_{13}H_{27}BrMgO_3$ 67.14 **3**
$C_{13}H_{27}CoN_6O_6$ 83.29+ **6**
$C_{13}H_{27}S^+$, $C_6H_2N_3O_9S^-$ 11.11 **6**
$C_{13}H_{28}N^+$, I^- 52.22 **1**
$C_{13}H_{29}CoN_4O_4$ 84.40 **7**
$C_{13}H_{29}N_3O_3S_4U$ 80.12 **5**
$C_{13}H_{30}ClOP_2Pt^+$, F_5Si^- 86.16 **2**
$C_{13}H_{30}N_5NiS^+$, CNS^- , H_2O 76.44 **5**
$C_{13}H_{35}BMnNO_4$ 62.10 **5**
$C_{13}H_{35}B_9MnNO_4$ 62.5 **7**

C_{14}

$C_{14}Cl_{12}$ 22.7 **4**
$C_{14}F_4Fe_2O_8$ 71.19 **6**
$C_{14}F_8O_2$ 26.1+ **7**
$C_{14}H_4Cl_2N_2O_3$ 40.26 **5**
$C_{14}H_5ClN_2O_2S$ 41.17 **7**
$C_{14}H_5ClN_2O_3$ 40.16 **6**
$C_{14}H_5CoF_{12}O$ 73.46 **2**
$C_{14}H_6Br_2O_2$ 26.3 **1**
$C_{14}H_6Br_2O_2$ 26.1+ **1**
$C_{14}H_6Cl_2O_2$ 26.4+ **1**
$C_{14}H_6Cl_2O_2$ 26.1 **5**
$(C_{14}H_6CuN_2S_2^+)_n$, $nClO_4^-$ 85.25 **6**
$C_{14}H_6F_2O_2$ 26.6 **1**
$C_{14}H_6Fe_2O_6$ 71.33 **7**
$C_{14}H_6Fe_2O_6$ 71.34 **7**
$C_{14}H_6Fe_4O_{12}S_3$ 85.71 **2**
$C_{14}H_6I_2O_2$ 26.7+ **1**
$C_{14}H_6I_2O_2$ 26.1 **6**
$C_{14}H_6N_2O_2S$ 41.20 **6**
$C_{14}H_6N_2O_2Se$ 42.10 **7**
$C_{14}H_6N_2O_3$ 40.27 **5**
$C_{14}H_6N_2O_8$ 26.9 **1**
$C_{14}H_6N_2O_{10}U^{2-}$, $2C_{24}H_{20}As^+$, $6H_2O$ 81.34 **7**
$C_{14}H_6N_4$ 7.5 **6**
$C_{14}H_6O_8$ 59.24 **1**
$C_{14}H_7BrO_2$ 26.1 **3**
$C_{14}H_7BrO_2$ 26.2 **7**
$C_{14}H_7ClO_2$ 26.2 **3**
$C_{14}H_7IO_2$ 26.3 **7**
$C_{14}H_7O_{12}Re_3Sn$ 69.25 **4**
$C_{14}H_8AgN_2O_8$, H_2O 81.36 **3**
$C_{14}H_8AgN_2O_8$, $2H_2O$ 81.36 **4**
$C_{14}H_8BrCl$ 26.10 **1**
$C_{14}H_8BrN_5O_2$, $0.5C_7H_8$ 37.18 **1**
$C_{14}H_8Br_2$ 26.11+ **1**
$C_{14}H_8Br_2F_2$ 5.24 **1**

$C_{14}H_8Br_2N_2O_7$ 13.20 **7**
$C_{14}H_8Br_2O$ 26.13 **1**
$C_{14}H_8Br_2O_3$ 13.52 **1**
$C_{14}H_8Br_2O_4$ 13.17 **4**
$C_{14}H_8ClIO_4$ 19.8 **7**
$C_{14}H_8ClIO_4$ 42.11+ **7**
$C_{14}H_8ClNO_2$ 35.36 **1**
$C_{14}H_8ClNO_2$ 35.20 **7**
$C_{14}H_8Cl_2$ 26.14 **1**
$C_{14}H_8Cl_2$ 26.15 **1**
$C_{14}H_8Cl_2N_2O_2$ 40.19 **4**
$C_{14}H_8Cl_2O_2$ 13.21 **7**
$C_{14}H_8Cl_2O_2S_2$ 11.12 **6**
$C_{14}H_8Cl_2O_3$ 13.53 **1**
$C_{14}H_8Cl_2O_4$ 13.54 **1**
$C_{14}H_8Cl_2O_4$ 13.15 **3**
$C_{14}H_8Cl_3NO_5$ 15.10 **5**
$C_{14}H_8Cl_4$ 26.16 **1**
$C_{14}H_8CuN_2O_8$, xH_2O 81.42 **5**
$C_{14}H_8CuN_2O_8$, $3H_2O$ 81.43 **5**
$C_{14}H_8Fe_2O_6$ 75.29 **2**
$C_{14}H_8INO_2$ 35.21 **7**
$C_{14}H_8I_2O_4$ 13.18 **4**
$C_{14}H_8I_2O_4$ 42.8+ **5**
$C_{14}H_8N_2NiO_8$, $3H_2O$ 81.37 **4**
$C_{14}H_8N_2NiO_8$, $3H_2O$ 81.44+ **5**
$C_{14}H_8N_2OS$, H_2O 41.18 **7**
$C_{14}H_8N_2O_4$ 26.17 **1**
$C_{14}H_8N_2O_8S_2$, H_2O 11.62 **1**
$C_{14}H_8N_4$ 28.13 **6**
$C_{14}H_8O_2$ 26.18+ **1**
$C_{14}H_8O_4$ 26.25 **1**
$C_{14}H_8O_4$ 26.28 **1**
$C_{14}H_8O_4$ 26.29 **1**
$C_{14}H_8O_4$ 26.26+ **1**
$C_{14}H_8O_4$, H_2O 28.14 **6**
$C_{14}H_8O_4S$ 42.10 **5**
$C_{14}H_8O_4S_2$ 39.18 **4**
$C_{14}H_8O_4Se$ 42.9 **6**
$C_{14}H_8O_6Ru_2$ 75.30 **2**
$C_{14}H_8S_2$ 39.19 **4**
$C_{14}H_9BrO$ 26.2 **6**
$C_{14}H_9BrO_2$ 38.48 **1**
$C_{14}H_9ClO_3$ 22.8 **4**
$C_{14}H_9Cl_2NO_3$ 10.2 **7**
$C_{14}H_9Cl_5$ 19.16 **5**
$C_{14}H_9Cl_5$ 19.17 **5**
$C_{14}H_9F_{12}O_3PS$ 64.37 **7**
$C_{14}H_9FeN_2$ 73.12 **4**
$C_{14}H_9I_3O_4$, $C_4H_{11}NO_2$ 60.26 **5**
$C_{14}H_9N$ 36.12 **4**
$C_{14}H_9NO_2$ 26.30 **1**
$C_{14}H_9NO_2$ 40.27 **1**
$2C_{14}H_9O_4S_2^-$, $H_{12}CoO_6^{2+}$, $6H_2O$, $4CH_4O$ 13.16 **6**
$C_{14}H_{10}$ 5.25+ **1**
$C_{14}H_{10}$ 23.11 **1**
$C_{14}H_{10}$ 26.31+ **1**
$C_{14}H_{10}$ 28.17+ **1**
$C_{14}H_{10}$ 26.3+ **3**
$C_{14}H_{10}$ 28.3+ **3**
$C_{14}H_{10}$ 28.4 **4**
$C_{14}H_{10}$ 29.2 **4**
$C_{14}H_{10}$, $2Cl_3Sb$ 28.5 **4**
$C_{14}H_{10}$, $C_6H_3N_3O_6$ 60.133+ **2**
$C_{14}H_{10}$, $C_{10}H_2N_4$ 60.32 **4**
$C_{14}H_{10}$, $C_{10}H_2O_4S$ 60.11 **7**
$C_{14}H_{10}$, $C_{10}H_2O_6$ 60.135 **2**
$C_{14}H_{10}$, $C_{12}H_4N_4$ 60.125 **2**
$C_{14}H_{10}$, $C_{13}H_4BrN_3O_7$ 60.42 **3**
$C_{14}H_{10}$, $C_{18}H_{18}Si$ 60.36 **4**
$C_{14}H_{10}$, $3C_{24}H_{40}O_4$ 60.41 **5**
$C_{14}D_{10}$ 26.8 **3**
$C_{14}H_{10}^{2-}$, $2C_6H_{16}N_2$, $2Li^+$ 26.4 **7**
$C_{14}H_{10}Ag_4^{4+}$, $4ClO_4^-$, H_2O 74.1 **7**
$C_{14}H_{10}BMnO_3$ 75.12 **7**
$C_{14}H_{10}BO_4^+$, Br^- , H_2O 62.11 **5**
$C_{14}H_{10}BrN_5O$ 44.24 **4**
$C_{14}H_{10}Br_2N_2$ 9.11 **4**
$C_{14}H_{10}Br_2O_4$, $C_4H_{11}NO_2$ 17.22 **7**
$C_{14}H_{10}CaO_6$ 67.14 **4**
$C_{14}H_{10}ClNO_3$ 13.16 **3**
$C_{14}H_{10}Cl_2CuN_2O_4$ 78.3 **2**
$C_{14}H_{10}Cl_2FeHg_2N_2O_4$ 83.67 **3**
$C_{14}H_{10}Cl_2Fe_2Ge$ 73.47 **2**
$C_{14}H_{10}Cl_2Fe_2O_4Sb^+$, Cl_4Sb^- , Cl_3Sb 73.23 **6**
$C_{14}H_{10}Cl_2Fe_2O_4Sn$ 73.48 **2**
$C_{14}H_{10}Cl_2N_2$ 36.17 **5**
$C_{14}H_{10}Cl_2N_2Pt$ 71.21 **4**
$C_{14}H_{10}Cl_2N_2Pt$, $C_8H_4N_8Pt$ 83.16 **7**
$C_{14}H_{10}Cl_8FeO_2$ 73.19 **5**
$C_{14}H_{10}CrO_3$ 75.31 **2**
$(C_{14}H_{10}CuN_4O_2S)_n$ 78.2 **5**
$C_{14}H_{10}CuN_6O_3$ 83.64 **4**
$C_{14}H_{10}CuO_4$ 78.4+ **2**
$C_{14}H_{10}CuO_4$ 84.36 **2**
$C_{14}H_{10}CuO_4$, $3H_2O$ 81.77 **2**
$C_{14}H_{10}F_3NO_2$ 13.17 **6**
$C_{14}H_{10}F_4$ 19.44 **1**
$C_{14}H_{10}F_{12}NiS_4$ 85.26 **6**
$C_{14}H_{10}Fe_2IO_4^+$, BF_4^- 73.24 **6**
$C_{14}H_{10}Fe_2N_2O_4Sn$ 73.49 **2**
$C_{14}H_{10}Fe_2O_4$ 73.51 **2**
$C_{14}H_{10}Fe_2O_4$ 73.12 **3**
$C_{14}H_{10}Fe_2O_4$ 73.13 **3**
$C_{14}H_{10}Fe_2O_6$ 71.34 **2**
$C_{14}H_{10}Fe_2O_6$ 75.32 **2**
$C_{14}H_{10}Fe_2O_6S$ 73.25 **6**
$C_{14}H_{10}I_4O_4Re_2$ 81.37 **3**
$C_{14}H_{10}MnO_3^-$, $C_4H_{12}N^+$ 71.14 **3**
$C_{14}H_{10}N_2O_2$ 32.38 **1**
$C_{14}H_{10}N_2O_2$ 35.18 **4**
$C_{14}H_{10}N_2O_2$ 32.30 **6**
$C_{14}H_{10}N_2O_4$ 15.10 **3**
$C_{14}H_{10}N_2O_6$ 13.22 **7**
$C_{14}H_{10}N_2O_{15}Ti_2^{2-}$, $2K^+$, $3H_2O$ 81.38 **3**
$C_{14}H_{10}N_2S$ 41.36 **1**
$C_{14}H_{10}N_4$ 33.32 **4**
$C_{14}H_{10}N_4$ 31.7 **7**
$C_{14}H_{10}N_4OS$ 41.19 **7**
$C_{14}H_{10}NiS_4$ 85.16 **4**
$C_{14}H_{10}NiS_6$ 85.17 **4**
$C_{14}H_{10}O$ 26.3 **5**
$C_{14}H_{10}O$ 31.14 **6**
$C_{14}H_{10}O_2$ 19.45 **1**
$C_{14}H_{10}O_2$ 38.49 **1**
$C_{14}H_{10}O_2S_2$ 11.22 **3**
$C_{14}H_{10}O_3$ 13.17 **3**
$C_{14}H_{10}O_3W$ 73.53 **2**
$C_{14}H_{10}O_4$ 13.55 **1**
$C_{14}H_{10}O_4Ru_2$ 73.54 **2**

$C_{14}H_{10}PdS_4$ 85.72 **2**
$C_{14}H_{10}S_4Zn$ 85.21 **5**
$C_{14}H_{10}S_6Zn$ 85.19 **4**
$C_{14}H_{11}BrO$ 19.46 **1**
$C_{14}H_{11}BrO_4$ 38.25 **6**
$C_{14}H_{11}ClN^+$, Cl^- 36.15 **7**
$C_{14}H_{11}ClN_2O_2$ 35.21 **6**
$C_{14}H_{11}ClO$ 19.47 **1**
$C_{14}H_{11}ClO_3$ 17.23 **7**
$C_{14}H_{11}Cl_3MoN_2O_3Sn$ 69.23 **3**
$C_{14}H_{11}FO$ 19.48 **1**
$C_{14}H_{11}NO_2$ 16.10 **3**
$C_{14}H_{11}NO_3$ 15.5 **6**
$C_{14}H_{11}NO_3$ 10.3 **7**
$C_{14}H_{11}N_5O_2$ 35.22 **7**
$C_{14}H_{12}$ 19.49 **1**
$C_{14}H_{12}$ 29.1 **1**
$C_{14}H_{12}$ 22.9 **4**
$C_{14}H_{12}$ 19.8 **6**
$C_{14}H_{12}$ 19.9 **7**
$C_{14}H_{12}$, $4Cl_3Sb$ 60.136 **2**
$C_{14}H_{12}$, $C_{10}H_2O_6$ 60.12 **7**
$C_{14}H_{12}Ag_4O^{4+}$, $4ClO_4^-$ 74.6 **3**
$C_{14}H_{12}As_2Co_2F_4O_6$ 86.11 **3**
$C_{14}H_{12}As_2F_4Fe_2O_6$ 75.33 **2**
$C_{14}H_{12}AuBr_3N_2$ 83.53 **7**
$C_{14}H_{12}AuCl_3N_2$ 83.54 **7**
$C_{14}H_{12}BrClO_2$ 27.8 **3**
$C_{14}H_{12}BrNO_3$ 38.17 **4**
$C_{14}H_{12}Br_2OS$ 11.9 **4**
$C_{14}H_{12}ClN_3O$ 35.17 **3**
$2C_{14}H_{12}Cl_2N_2Ni$, $2CHCl_3$ 83.128 **2**
$C_{14}H_{12}Cl_2N_2Zn$ 83.68 **3**
$C_{14}H_{12}Cl_2O_2$ 17.40 **1**
$C_{14}H_{12}CoS_4^-$, $C_{19}H_{18}As^+$, $0.5C_2H_6O$ 85.73 **2**
$C_{14}H_{12}CrO_3S$ 75.17 **5**
$C_{14}H_{12}CuN_2O_2$ 78.7 **2**
$C_{14}H_{12}CuN_2O_4$ 78.8 **2**
$C_{14}H_{12}CuN_2O_4$, $2H_2O$ 81.35 **7**
$C_{14}H_{12}N^+$, I^- 36.9 **3**
$C_{14}H_{12}N_2$ 9.9 **3**
$C_{14}H_{12}N_2NiO_2$ 78.9 **2**
$C_{14}H_{12}N_2NiO_4$ 78.10 **2**
$C_{14}H_{12}N_2O$ 40.28 **5**
$C_{14}H_{12}N_2OSe$ 8.49 **1**
$C_{14}H_{12}N_2O_2$ 36.26 **1**
$C_{14}H_{12}N_2O_2$ 9.10 **3**
$C_{14}H_{12}N_2O_2$ 16.11 **3**
$C_{14}H_{12}N_2O_3$ 16.20 **5**
$C_{14}H_{12}N_2O_4Pd$ 78.1 **3**
$C_{14}H_{12}N_2O_5U$ 84.41 **7**
$C_{14}H_{12}N_2O_6S$ 15.4 **7**
$C_{14}H_{12}N_2S_2$ 4.6 **7**
$C_{14}H_{12}N_2S_3$ 4.4 **6**
$C_{14}H_{12}N_2S_4$ 4.5 **6**
$C_{14}H_{12}N_4O_4Pd$ 83.66 **4**
$C_{14}H_{12}N_4S_3$ 39.35 **3**
$C_{14}H_{12}OS$ 39.75 **1**
$C_{14}H_{12}O_2$ 31.12 **4**
$C_{14}H_{12}O_2S_2$ 39.15 **6**
$C_{14}H_{12}O_3$ 59.6 **3**
$C_{14}H_{12}O_3$ 17.24 **7**
$C_{14}H_{12}O_4$ 38.50 **1**
$C_{14}H_{12}O_5$ 59.9 **5**
$C_{14}H_{12}O_5$, Cs^+ , CNS^- 38.44 **7**
$C_{14}H_{12}O_5$, H_2O 38.45 **7**
$2C_{14}H_{12}O_5$, K^+ , I^- 38.26 **6**
$2C_{14}H_{12}O_5$, Na^+ , I^- , $2H_2O$ 38.46 **7**
$C_{14}H_{12}O_7$ 25.20 **1**
$C_{14}H_{12}S_2$ 39.20 **4**
$C_{14}H_{12}S_2$ 39.25 **5**
$C_{14}H_{13}BF_6N_6Pt$ 71.35 **7**
$C_{14}H_{13}BrN_4O_2$, H_2O 36.18 **5**
$C_{14}H_{13}BrN_4O_3$, H_2O 43.31 **1**
$C_{14}H_{13}BrO_2$ 52.23 **1**
$C_{14}H_{13}BrO_2$ 31.16 **5**
$C_{14}H_{13}ClO$ 31.16 **1**
$C_{14}H_{13}Cl_2NO_3$ 50.5 **3**
$C_{14}H_{13}HgNO$ 71.36 **7**
$C_{14}H_{13}HgNO_3$ 71.22 **4**
$C_{14}H_{13}I_2NO$ 17.25 **7**
$C_{14}H_{13}MnO_2$ 75.34 **2**
$C_{14}H_{13}MoO_2$ 71.29 **5**
$C_{14}H_{13}N$ 36.16 **7**
$C_{14}H_{13}NO$ 1.141 **1**
$C_{14}H_{13}NS$ 41.20 **7**
$C_{14}H_{13}N_4O_2^+$, I^- , H_2O 36.11 **3**
$C_{14}H_{14}$ 19.50 **1**
$C_{14}H_{14}$ 19.51+ **1**
$C_{14}H_{14}$ 23.1 **5**
$C_{14}H_{14}$, $2Cl_3Sb$ 60.137 **2**
$C_{14}H_{14}$, $4Cl_3Sb$ 60.138 **2**
$C_{14}H_{14}Ag^+$, ClO_4^- 74.7 **3**
$(C_{14}H_{14}Ag^+)_n$, $nClO_4^-$ 74.2 **7**
$C_{14}H_{14}As_2$ 65.24 **2**
$C_{14}H_{14}As_2Br_2$ 65.25 **2**
$C_{14}H_{14}As_2I_2$ 65.26 **2**
$C_{14}H_{14}Au_2Cl_2S_2$ 85.22 **5**
$C_{14}H_{14}BF_2NO$ 62.6 **7**
$C_{14}H_{14}BrNO$ 37.6 **5**
$C_{14}H_{14}BrNO_2S$ 16.56 **1**
$C_{14}H_{14}BrN_3O$ 32.25 **7**
$C_{14}H_{14}Br_2O_4$, $5C_4H_8O_2$ 31.16 **3**
$C_{14}H_{14}Br_2Se$ 11.63 **1**
$C_{14}H_{14}ClN_2O_4Tl$ 68.22 **2**
$C_{14}H_{14}Cl_2CoN_2$ 83.130+ **2**
$C_{14}H_{14}Cl_2CoN_6O_2$ 83.67 **4**
$C_{14}H_{14}Cl_2CuN_2$ 83.132 **2**
$C_{14}H_{14}Cl_2CuN_2O$ 83.69 **3**
$C_{14}H_{14}Cl_2HgN_2O_3$, $0.5H_2O$ 84.37 **2**
$C_{14}H_{14}Cl_2NO^+$, BF_4^- 33.42 **6**
$C_{14}H_{14}Cl_2Se$ 11.64 **1**
$C_{14}H_{14}Cl_4$ 20.30 **1**
$C_{14}H_{14}CoN_8S_2^+$, Cl^- , $3H_2O$ 85.22 **7**
$C_{14}H_{14}CoN_8S_2^+$, NO_3^- 79.21 **7**
$C_{14}H_{14}CrO_3$ 75.9 **4**
$C_{14}H_{14}CrO_5$ 74.3 **7**
$C_{14}H_{14}CuO_8$, $2H_2O$ 81.78 **2**
$C_{14}H_{14}FeO$ 73.55 **2**
$C_{14}H_{14}FeO_2$ 73.14 **3**
$C_{14}H_{14}FeO_3$ 75.9 **6**
$C_{14}H_{14}GdN_7O_9$ 76.31 **6**
$C_{14}H_{14}Hg$ 71.15 **3**
$C_{14}H_{14}HgS$ 71.37 **7**
$C_{14}H_{14}NO_3$ 17.41 **1**
$C_{14}H_{14}N_2$ 9.40+ **1**
$C_{14}H_{14}N_2$, $C_{12}H_4N_4$ 60.17+ **5**
$C_{14}H_{14}N_2O_3$ 9.11+ **3**
$C_{14}H_{14}N_2O_4S_3$ 4.7 **7**
$C_{14}H_{14}N_2S$ 16.12 **3**
$C_{14}H_{14}N_2Tl^+$, ClO_4^- 68.17 **5**

$C_{14}H_{14}N_3O_3S^-$, Na^+ , H_2O , C_2H_6O 9.13 **5**
$C_{14}H_{14}N_4O_2$ 36.13 **4**
$C_{14}H_{14}N_4S$ 16.9 **7**
$C_{14}H_{14}N_6Ni$ 83.71 **3**
$C_{14}H_{14}N_6NiO_2$ 83.45 **5**
$C_{14}H_{14}NiO_6$ 78.11 **2**
$C_{14}H_{14}O_2$ 17.42 **1**
$C_{14}H_{14}O_2$ 24.5 **5**
$C_{14}H_{14}O_2$ 24.5 **6**
$C_{14}H_{14}O_2PS_2^-$, K^+ 64.19 **5**
$C_{14}H_{14}O_2Ru$ 73.57 **2**
$C_{14}H_{14}O_3$ 38.18 **4**
$C_{14}H_{14}O_3$ 38.15 **5**
$C_{14}H_{14}O_3$ 59.2 **7**
$C_{14}H_{14}O_4$ 31.17 **5**
$C_{14}H_{14}O_4$ 38.27 **6**
$C_{14}H_{14}O_4S_2$ 39.21 **4**
$C_{14}H_{14}O_4S_2Zn$ 85.23 **5**
$C_{14}H_{14}O_4S_4Te$ 70.24 **2**
$C_{14}H_{14}O_4Se_2$ 39.22 **4**
$C_{14}H_{14}O_8Zn$ 78.12 **2**
$C_{14}H_{14}S$ 11.65 **1**
$C_{14}H_{14}S$, I_2 60.139 **2**
$C_{14}H_{14}S_2$ 11.66 **1**
$C_{14}H_{14}S_2$ 11.25 **3**
$C_{14}H_{14}S_2$ 11.23+ **3**
$C_{14}H_{14}S_2$ 11.11 **4**
$C_{14}H_{14}Se$ 11.67 **1**
$C_{14}H_{14}Te$ 70.25 **2**
$C_{14}H_{15}BrN_4O_4$ 50.6 **7**
$C_{14}H_{15}BrO_3$ 53.1 **1**
$C_{14}H_{15}BrO_3$ 59.7 **3**
$C_{14}H_{15}BrO_3S$ 31.17 **1**
$C_{14}H_{15}BrO_7$ 59.25 **1**
$C_{14}H_{15}ClN_4O_5$ 44.25 **4**
$C_{14}H_{15}Cl_2N_3O_3$ 35.24 **5**
$C_{14}H_{15}NO_2$ 21.14 **3**
$C_{14}H_{15}NO_2$ 25.3 **7**
$C_{14}H_{15}NO_2S$ 4.8 **7**
$C_{14}H_{15}NO_4S$ 11.11 **5**
$C_{14}H_{15}NS$ 16.16 **6**
$C_{14}H_{15}N_3$ 9.42 **1**
$C_{14}H_{15}N_3O_2$ 37.11 **7**
$4C_{14}H_{15}OP$, $C_{12}H_8O_{16}W_3$ 64.20 **5**
$C_{14}H_{15}O_4P$ 46.21 **1**
$C_{14}H_{16}Ag^+$, NO_3^- 75.35 **2**
$C_{14}H_{16}Ag^+$, NO_3^- 31.13 **4**
$C_{14}H_{16}As_2IO_4W^+$, I_3^- 86.17 **7**
$C_{14}H_{16}BNO$ 62.7 **6**
$C_{14}H_{16}BrNO_3$ 21.29 **1**
$C_{14}H_{16}ClFO_3$ 21.9 **5**
$C_{14}H_{16}Cl_2CuN_4O_{10}$ 83.68 **4**
$C_{14}H_{16}Cr^+$, I^- 74.15 **2**
$C_{14}H_{16}Cr^+$, $C_{12}H_4N_4^-$ 74.16 **2**
$C_{14}H_{16}Cr^+$, $C_{12}H_4N_4^-$, $C_{12}H_4N_4$ 60.140 **2**
$C_{14}H_{16}CrO_4$ 74.4 **6**
$C_{14}H_{16}CuN_2O_4$, C_5H_5N 81.79 **2**
$(C_{14}H_{16}CuN_2S_2^+)_n$, $nClO_4^-$ 85.23 **7**
$C_{14}H_{16}CuN_4O_2^{2+}$, $Cl_5Cu_3^{2-}$ 83.72 **3**
$C_{14}H_{16}CuN_6O_4S_2$ 81.47 **5**
$C_{14}H_{16}F_4N_2P_2$ 64.41 **2**
$C_{14}H_{16}Fe_2O_2S_2^+$, BF_4^- 73.15 **3**
$C_{14}H_{16}N^+$, ClO_4^- 33.74 **1**
$C_{14}H_{16}NOP$ 64.42 **2**
$C_{14}H_{16}N_2$ 16.57 **1**
$C_{14}H_{16}N_2$ 16.58 **1**
$C_{14}H_{16}N_2^{2+}$, $2Br^-$, H_2O , $0.33HBr$ 34.3 **3**
$C_{14}H_{16}N_2^{2+}$, $C_4N_4Ni^{2-}$ 33.44 **3**
$C_{14}H_{16}N_2Ni_2$ 71.30 **5**
$C_{14}H_{16}N_2O_2$ 33.47 **5**
$C_{14}H_{16}N_2O_2$ 48.54 **7**
$C_{14}H_{16}N_2O_3$ 48.70 **6**
$C_{14}H_{16}N_2O_3S$ 45.42 **7**
$C_{14}H_{16}N_2O_6Zn$ 81.38 **4**
$C_{14}H_{16}N_2P_2S_2$ 64.21 **5**
$C_{14}H_{16}N_4S_4$ 41.21 **7**
$C_{14}H_{16}N_7O^+$, Br^- , $3H_2O$ 44.25 **6**
$C_{14}H_{16}Ni_2$ 72.11 **6**
$C_{14}H_{16}O_3$ 38.16 **5**
$C_{14}H_{16}O_8$ 13.18 **3**
$C_{14}H_{16}P_2S_2$ 64.43 **2**
$C_{14}H_{16}S_5$ 39.23 **4**
$C_{14}H_{17}BrO_4$ 38.17 **5**
$C_{14}H_{17}Br_2N$ 36.14 **4**
$C_{14}H_{17}Br_2N_3Ni$ 83.73 **3**
$C_{14}H_{17}CoO$ 73.58 **2**
$C_{14}H_{17}FeO_2^+$, BF_4^- 72.10 **7**
$C_{14}H_{17}IO_4$ 31.15 **6**
$C_{14}H_{17}NO_2$ 32.31 **6**
$C_{14}H_{17}NO_2$ 38.28 **6**
$C_{14}H_{17}NO_7S$ 41.21 **6**
$C_{14}H_{17}N_5S_4$ 41.22 **6**
$C_{14}H_{17}N_7O^{2+}$, $2Br^-$, $2H_2O$ 44.26 **6**
$C_{14}H_{18}$ 31.14 **4**
$C_{14}H_{18}BrNO$ 40.17 **6**
$C_{14}H_{18}BrNS$ 41.23 **6**
$C_{14}H_{18}Br_2CuN_2$ 83.69 **4**
$C_{14}H_{18}ClCuN_4^+$, Cl^- , xH_2O 76.45 **5**
$C_{14}H_{18}ClNOPt$ 71.23 **4**
$C_{14}H_{18}ClNO_3S$ 40.9 **7**
$C_{14}H_{18}ClNO_3S$ 40.10 **7**
$C_{14}H_{18}ClNO_3S$ 41.22 **7**
$C_{14}H_{18}Cl_2CoN_2$ 83.134 **2**
$C_{14}H_{18}Cl_2CuN_2$ 83.74 **3**
$C_{14}H_{18}Cl_2CuN_2$ 83.70 **4**
$C_{14}H_{18}Cl_2CuN_2O_2$ 84.38 **2**
$C_{14}H_{18}Cl_2N_2O_2Zn$ 84.39 **2**
$C_{14}H_{18}Cl_2N_2Zn$ 83.135 **2**
$C_{14}H_{18}Cl_2N_2Zn$ 83.55 **7**
$C_{14}H_{18}CoI_2N_2$ 83.136 **2**
$C_{14}H_{18}FeNO$ 73.59 **2**
$C_{14}H_{18}MnN_2O_8^-$, K^+ , H_2O 76.32 **6**
$C_{14}H_{18}NO^+$, $I-$ 37.10 **4**
$C_{14}H_{18}NOS^+$, Br^- 41.27 **5**
$C_{14}H_{18}NO_2^+$, I^- 58.10 **1**
$C_{14}H_{18}N_2$ 24.6 **6**
$C_{14}H_{18}N_2^{2+}$, $2Cl^-$ 16.59 **1**
$C_{14}H_{18}N_2^{2+}$, $C_4N_4Ni^{2-}$, $3H_2O$ 33.33 **4**
$C_{14}H_{18}N_2O_5$ 48.71 **6**
$C_{14}H_{18}N_2O_7$ 45.43 **7**
$C_{14}H_{18}N_2O_8S$ 47.34 **7**
$C_{14}H_{18}N_4O_3$ 44.28 **7**
$C_{14}H_{18}N_4O_4$, $2H_2O$ 44.38 **3**
$C_{14}H_{18}N_5O_4^+$, Cl^- , H_2O 47.28 **6**
$C_{14}H_{18}OP^+$, I^- 64.18 **4**
$C_{14}H_{18}O_2$ 31.8 **7**
$C_{14}H_{18}O_3S$ 31.15 **4**
$C_{14}H_{18}Si_2$ 63.9 **7**
$C_{14}H_{19}BrO_2$ 38.18 **5**
$C_{14}H_{19}ClO_9$ 45.39 **3**
$C_{14}H_{19}N_2^+$, $C_{15}H_3CuF_{18}O_6^-$ 77.7 **4**
$C_{14}H_{19}N_2^+$, $C_{15}H_3F_{18}MgO_6^-$ 67.17 **4**

$C_{14}H_{19}N_4O_3^+$, Br^-	44.39	**3**
$C_{14}H_{19}Nb$	71.38	**7**
$C_{14}H_{20}$	31.18	**1**
$C_{14}H_{20}As_2ClF_{12}ORh$	71.20	**6**
$C_{14}H_{20}BrCuN_4^+$, Br^-	83.46	**5**
$C_{14}H_{20}ClNO$	16.21	**5**
$C_{14}H_{20}ClN_2O^+$, I^-	33.43	**6**
$C_{14}H_{20}ClN_2O^+$, I^-	33.37	**7**
$C_{14}H_{20}Cl_2Cu_2N_2O_2$	83.71	**4**
$C_{14}H_{20}Cl_2NPt$	71.25+	**4**
$C_{14}H_{20}Cl_2N_8O_4PdS_2$	83.31	**6**
$C_{14}H_{20}CuN_4^{2+}$, $2ClO_4^-$	83.32	**6**
$C_{14}H_{20}CuN_4^{2+}$, $2I^-$	83.33	**6**
$C_{14}H_{20}CuN_6O_6$	83.75	**3**
$C_{14}H_{20}F_2O_2$	38.47	**7**
$C_{14}H_{20}F_6FeN_2S_6$	80.8	**4**
$C_{14}H_{20}FeN_2O_2S_4$	80.13	**5**
$2C_{14}H_{20}FeN_2O_4^-$, Ca^{2+} , $8H_2O$	83.137	**2**
$C_{14}H_{20}Fe_2S_4$	71.21	**6**
$C_{14}H_{20}N^+$, Cl^-	58.4	**7**
$C_{14}H_{20}N_2^{2+}$, $2Br^-$	37.19	**1**
$C_{14}H_{20}N_2^{2+}$, $2Br^-$, $2H_2O$	37.11	**4**
$C_{14}H_{20}N_2NiO_6$	81.48	**5**
$C_{14}H_{20}N_2O_2$	35.25	**5**
$C_{14}H_{20}N_2O_7$	47.29	**6**
$C_{14}H_{20}N_3S^+$, Cl^-	33.34	**4**
$C_{14}H_{20}N_4O_2$	37.20	**1**
$C_{14}H_{20}N_4O_4$	44.73	**1**
$C_{14}H_{20}N_4O_4$	44.40	**3**
$C_{14}H_{20}O_2$	18.7	**7**
$C_{14}H_{20}O_2S_3$	39.24	**4**
$C_{14}H_{20}O_3$	27.12	**5**
$C_{14}H_{20}O_6P_2Re_2S_4$	85.24	**7**
$C_{14}H_{21}B_{10}^-$, $C_4H_{12}N^+$	62.8	**6**
$C_{14}H_{21}BrO$	17.43	**1**
$C_{14}H_{21}BrO$	53.1	**3**
$C_{14}H_{21}BrO$	53.2	**3**
$C_{14}H_{21}BrO$	17.16	**5**
$C_{14}H_{21}BrO$	53.1	**6**
$C_{14}H_{21}BrO_2$	53.1	**7**
$C_{14}H_{21}BrO_3$	38.51	**1**
$C_{14}H_{21}ClO_7$	45.43	**6**
$C_{14}H_{21}CuN_3O_{10}$, H_2O	76.64+	**7**
$C_{14}H_{21}NO$	33.36	**4**
$C_{14}H_{21}NO$	33.37	**4**
$C_{14}H_{21}NO$	33.48	**5**
$C_{14}H_{21}NO$	33.49	**5**
$C_{14}H_{21}NO_7$	38.29	**6**
$C_{14}H_{21}N_3S^+$, Cl^-	39.36	**3**
$C_{14}H_{21}OP$	64.34	**3**
$C_{14}H_{21}OP$	64.35	**3**
$C_{14}H_{21}OP$	64.19	**4**
$C_{14}H_{22}$	52.24	**1**
$C_{14}H_{22}Cl_4Cu_2N_2O_4S_2$	84.8	**4**
$C_{14}H_{22}CoN_5O_4$	71.22	**6**
$C_{14}H_{22}Cr_2N_4O_2$	73.16	**3**
$C_{14}H_{22}Cr_2N_4O_2$	73.17	**3**
$C_{14}H_{22}FeN_8O_4$, $2CH_4O$	83.72	**4**
$C_{14}H_{22}MoN_2^{2+}$, $2F_6P^-$	73.23	**7**
$C_{14}H_{22}Mo_2S_4$	73.18	**3**
$C_{14}H_{22}Mo_2S_4^+$, F_6P^-	73.19	**3**
$C_{14}H_{22}N^+$, ClO_4^-	33.45	**3**
$C_{14}H_{22}NO_3^+$, Cl^-	3.49	**3**
$C_{14}H_{22}N_2Ni_2O_4$	84.12	**3**
$C_{14}H_{22}N_2O$	16.10	**7**
$C_{14}H_{22}N_2O_2$	40.14	**3**
$C_{14}H_{22}N_2S_2Zn$	85.25	**7**
$C_{14}H_{22}N_2S_4$	58.11	**1**
$C_{14}H_{22}N_4NiO_4$, $0.5H_2O$	78.1	**7**
$C_{14}H_{22}N_4O_2S^{2+}$, $2Cl^-$	44.27	**6**
$C_{14}H_{22}N_6O_6$, $3H_2O$	48.47	**3**
$C_{14}H_{22}NaO_6^+$, I^-	38.15	**3**
$C_{14}H_{22}NiO_2S_2$	85.20	**4**
$C_{14}H_{22}O$	53.3	**3**
$C_{14}H_{22}O_7$	45.22	**4**
$C_{14}H_{22}O_8$	45.44	**6**
$C_{14}H_{23}ClO_2$	53.2	**6**
$C_{14}H_{23}CoN_2O_3$	71.31	**5**
$C_{14}H_{23}N_2O^+$, AsF_6^-	16.15	**4**
$C_{14}H_{23}N_2O^+$, Cl^- , H_2O	16.22	**5**
$C_{14}H_{23}N_2O^+$, $C_{12}H_8N_2O_8P^-$	16.11	**7**
$C_{14}H_{23}N_2O_2^+$, Br^-	3.50	**3**
$C_{14}H_{23}N_4O_4S^+$, Cl^-	48.70	**5**
$C_{14}H_{23}N_6OP$	32.39	**1**
$C_{14}H_{23}N_6OP$	32.40	**1**
$C_{14}H_{24}$	28.19	**1**
$C_{14}H_{24}$	31.16	**4**
$C_{14}H_{24}Br_2Pd_2$	75.36	**2**
$C_{14}H_{24}Cl_2CoN_4^+$, F_6P^-	83.34	**6**
$C_{14}H_{24}Cl_4Pt_2$	75.10	**6**
$C_{14}H_{24}Cl_4Pt_2$, $2CCl_4$	72.22	**3**
$C_{14}H_{24}CuN_2S_4$	80.25	**2**
$C_{14}H_{24}CuO_8S_2$	84.40	**2**
$C_{14}H_{24}GeO_4Ru_2$	75.11	**6**
$C_{14}H_{24}N_2NiS_4$	80.14	**5**
$C_{14}H_{24}N_2O_4$	40.18	**6**
$C_{14}H_{24}N_4O_4$	34.3	**6**
$C_{14}H_{24}N_6NiO_4Se_2$	84.13	**5**
$C_{14}H_{24}NaO_6^+$, I^-	67.15	**4**
$C_{14}H_{24}O_4$	38.52	**1**
$C_{14}H_{24}O_4$	21.21	**4**
$C_{14}H_{24}O_5S$	45.44	**5**
$C_{14}H_{24}P_2^+$, Br^-	64.36	**3**
$C_{14}H_{25}ClN_2OPd$	71.23	**6**
$C_{14}H_{25}Co_3O_4S_5$	85.74	**2**
$C_{14}H_{25}MoO_4P_5$	86.17	**2**
$C_{14}H_{25}N_3O_3$	40.19	**6**
$C_{14}H_{25}N_4O_{11}P_2^-$, Na^+ , $4H_2O$	47.35	**7**
$C_{14}H_{26}Cl_2Pd_2$	72.42	**2**
$C_{14}H_{26}N_3O_9^+$, Br^- , H_2O	50.9	**1**
$C_{14}H_{26}N_4O_4PbS_2$	69.18	**5**
$C_{14}H_{26}N_6O_2Pt$	83.49	**5**
$C_{14}H_{26}NiO_6$	77.10	**6**
$C_{14}H_{26}O$	23.1	**7**
$C_{14}H_{26}O_2Si_4$	63.4	**6**
$C_{14}H_{27}BrO_4$	1.142	**1**
$C_{14}H_{27}Cl_2N_2Rh$	75.37	**2**
$C_{14}H_{27}CoN_6O_2^{2+}$, $2Br^-$, $4H_2O$	76.66	**7**
$C_{14}H_{27}MnO_5Si_4$	63.3	**5**
$C_{14}H_{28}Au_2N_2S_4$	80.15	**5**
$C_{14}H_{28}Cl_4N_4O_6P_2$	64.24	**6**
$C_{14}H_{28}CoN_4O_2^{2+}$, $2ClO_4^-$	83.56	**7**
$C_{14}H_{28}CuN_2O_4$	82.26	**3**
$C_{14}H_{28}CuN_2S_4$	80.16	**5**
$C_{14}H_{28}Cu_2N_4O_2S_2$	84.42	**7**
$C_{14}H_{28}LiN_2O_4^+$, I^-	67.10	**5**
$C_{14}H_{28}MoN_2OS_4$	80.9	**7**
$C_{14}H_{28}MoN_2O_2S_4$	80.10	**7**
$C_{14}H_{28}N^+$, I^-	35.37	**1**
$C_{14}H_{28}N_2NiS_4$	80.27+	**2**
$C_{14}H_{28}N_2NiS_4$	80.17	**5**
$C_{14}H_{28}N_2O_2S$	41.28	**5**

$C_{14}H_{28}N_2O_8^{2+}$, $2Br^-$, $5H_2O$ 50.2 **4**
$C_{14}H_{28}N_2S_4Tl_2$ 68.18 **5**
$C_{14}H_{28}N_4Ni^{2+}$, $2ClO_4^-$ 83.57 **7**
$C_{14}H_{28}N_4Ni^{2+}$, $2ClO_4^-$ 83.58 **7**
$C_{14}H_{28}N_6NiO_4Se_2$ 83.138 **2**
$C_{14}H_{28}P_2^{2+}$, $2Br^-$ 64.37 **3**
$C_{14}H_{29}NO$ 1.143 **1**
$C_{14}H_{29}NO$ 1.28 **6**
$C_{14}H_{29}O_4P$ 64.38 **3**
$C_{14}H_{30}Be_2N_2$ 67.16 **4**
$C_{14}H_{30}ClN_4NiO^+$, Cl^- 83.73 **4**
$C_{14}H_{30}Cl_3CuN_4^+$, ClO_4^- 83.59 **7**
$C_{14}H_{30}Cl_3N_4Ni^+$, ClO_4^- 83.60 **7**
$C_{14}H_{30}CoN_6^{3+}$, $3Br^-$ 83.61 **7**
$C_{14}H_{30}CuN_4O_4$, $2H_2O$ 76.33 **6**
$C_{14}H_{30}N_2O_4^{2+}$, $2ClO_4^-$ 1.13 **4**
$C_{14}H_{30}N_2O_4^{2+}$, $2I^-$ 3.51 **3**
$C_{14}H_{30}N_2O_4^{2+}$, $2I^-$ 1.35 **7**
$C_{14}H_{30}N_2O_4^{2+}$, $2C_6H_2N_3O_7^-$ 3.31 **7**
$C_{14}H_{32}N_7Ni^+$, ClO_4^- 83.62 **7**
$C_{14}H_{32}O_6P_2Ru$ 72.17 **5**
$C_{14}H_{33}BaN_2O_8^+$, $C_2H_3O_2^-$ 67.7 **6**
$C_{14}H_{33}O_2P_4Ta$ 86.2 **6**
$C_{14}H_{34}Cu_2N_8S_2^{2+}$, $2ClO_4^-$ 83.63 **7**
$C_{14}H_{34}KNO_4Si_2$ 63.5 **6**
$C_{14}H_{34}N_6O_4Zn_2^{2+}$, $2ClO_4^-$ 76.34 **6**
$C_{14}H_{36}N_{10}Ni_2O_2^{2+}$, $2C_{24}H_{20}B^-$ 76.46 **5**
$C_{14}H_{42}Al_7N_7$ 68.10 **6**

C_{15}

$C_{15}H_3CuF_{18}O_6^-$, $C_{14}H_{19}N_2^+$ 77.7 **4**
$C_{15}H_3F_{18}MgO_6^-$, $C_{14}H_{19}N_2^+$ 67.17 **4**
$C_{15}H_3F_{18}NaO_6^{2-}$, $2Rb^+$ 67.8 **6**
$C_{15}H_6Co_4O_9$ 74.4 **7**
$C_{15}H_6F_4FeO_3$ 75.16 **3**
$C_{15}H_7CoFeO_6$ 73.26 **6**
$C_{15}H_7F_5MoO_2$ 71.16 **3**
$C_{15}H_8Cl_2O$ 20.14 **5**
$C_{15}H_8F_{12}NiS_4$ 85.23 **3**
$C_{15}H_8Fe_2O_5$ 75.38 **2**
$C_{15}H_8INO_4$, H_2O 35.38 **1**
$C_{15}H_8N_2O_2$ 50.7 **7**
$C_{15}H_9BrO_2$ 38.19 **5**
$C_{15}H_9BrO_2$ 27.9 **6**
$C_{15}H_9BrO_2$ 38.30 **6**
$C_{15}H_9BrO_2$, $0.25C_6H_6$ 27.10 **6**
$C_{15}H_9BrO_3$ 38.48 **7**
$C_{15}H_9Br_2N$ 5.27 **1**
$C_{15}H_9ClN_4$ 27.11 **6**
$C_{15}H_9N$ 26.34 **1**
$C_{15}H_9N_3S_3$ 41.29 **5**
$C_{15}H_{10}BrNO$ 40.20 **4**
$C_{15}H_{10}BrNO_3$ 35.39 **1**
$C_{15}H_{10}BrNO_3$ 15.11 **3**
$C_{15}H_{10}Br_2O_2$ 19.54 **1**
$C_{15}H_{10}ClNO$ 27.13 **5**
$C_{15}H_{10}ClNO_3$ 15.12 **3**
$C_{15}H_{10}Cl_2$ 26.35 **1**
$C_{15}H_{10}Cl_2$ 26.9 **3**
$C_{15}H_{10}Cl_2O_2$ 19.55 **1**
$C_{15}H_{10}Cl_4$ 20.15 **5**
$C_{15}H_{10}FeO_2$ 71.39 **7**
$C_{15}H_{10}FeO_3$ 72.23 **3**
$C_{15}H_{10}N_2S_3$ 41.30 **5**
$C_{15}H_{10}N_4$ 27.12 **6**
$C_{15}H_{10}O$ 26.36 **1**
$C_{15}H_{10}O$ 28.20 **1**
$C_{15}H_{10}O$ 20.14 **7**
$C_{15}H_{10}O$, H_2O 20.15 **6**
$C_{15}H_{10}OS_2$ 39.27 **5**
$C_{15}H_{10}O_2$ 27.13 **6**
$C_{15}H_{10}O_5$ 59.3 **7**
$C_{15}H_{10}O_5Re_2$ 73.13 **4**
$C_{15}H_{10}O_9Ru$ 71.27 **4**
$C_{15}H_{10}O_9Ru_3$ 72.12 **6**
$C_{15}H_{10}S$ 20.8 **4**
$C_{15}H_{11}BrN_2O$ 32.41 **1**
$C_{15}H_{11}BrO$ 19.18 **5**
$C_{15}H_{11}BrO_2$ 38.31 **6**
$C_{15}H_{11}ClN_2O_3$ 40.15 **3**
$C_{15}H_{11}ClN_3Pd^+$, Cl^- , $2H_2O$ 83.50 **5**
$2C_{15}H_{11}ClN_3Pd^+$, Cl_4Pd^{2-} 83.51 **5**
$C_{15}H_{11}Cl_2CoN_3$ 83.76 **3**
$C_{15}H_{11}Cl_2N_3Zn$ 83.141 + **2**
$C_{15}H_{11}Cl_3GaN_3$ 68.13 **3**
$C_{15}H_{11}Cl_3InN_3$ 68.11 **6**
$C_{15}H_{11}Cl_3N_3Tl$ 68.12 **6**
$C_{15}H_{11}I_4NO_4$, $5H_2O$ 48.48 **3**
$C_{15}H_{11}N$ 37.10 **6**
$C_{15}H_{11}N_3O_9S_2$ 11.68 **1**
$C_{15}H_{11}N_5O_6$ 35.18 **3**
$C_{15}H_{11}O_4^+$, Cl^- , H_2O 59.4 **6**
$C_{15}H_{11}P$ 64.38 **7**
$2C_{15}H_{11}S_2^+$, Cl_4Fe^{2-} 39.16 **6**
$2C_{15}H_{11}S_2^+$, Cl_4Hg^{2-} 39.17 **6**
$C_{15}H_{12}$ 26.10 **3**
$C_{15}H_{12}$ 26.1 **4**
$C_{15}H_{12}$ 29.1 **5**
$C_{15}H_{12}$, $C_6Cl_4O_2$ 60.27 **5**
$C_{15}H_{12}Br_2$ 20.15 **7**
$C_{15}H_{12}Br_2N_4O_3$ 9.43 **1**
$C_{15}H_{12}Cl_2$ 20.16 **7**
$C_{15}H_{12}Cl_2O_2S$ 39.28 **5**
$C_{15}H_{12}Cl_2S$ 39.29 **5**
$C_{15}H_{12}Cl_4O_2$ 38.19 **4**
$C_{15}H_{12}CuN_3O_2^+$, ClO_4^- 83.140 **2**
$C_{15}H_{12}Fe_2O_6$ 71.28 + **4**
$C_{15}H_{12}Fe_2O_7$ 71.17 **3**
$C_{15}H_{12}I_2O_4$ 17.26 **7**
$C_{15}H_{12}I_3NO_4$ 48.50 **4**
$C_{15}H_{12}I_3NO_4$ 48.55 **7**
$C_{15}H_{12}I_4NO_4^+$, Cl^- , H_2O 48.73 **6**
$C_{15}H_{12}NO_2S^+$, Cl^- 41.24 **4**
$C_{15}H_{12}N_2OS$ 41.32 **5**
$C_{15}H_{12}N_2OS$ 41.24 **6**
$C_{15}H_{12}N_2OS$, $0.5C_2H_6O$ 41.31 **5**
$C_{15}H_{12}N_2O_2$ 32.20 **4**
$C_{15}H_{12}N_2S$ 41.25 **6**
$C_{15}H_{12}N_4$ 9.14 **5**
$C_{15}H_{12}N_4S_3$ 41.26 **4**
$C_{15}H_{12}N_6O_9$ 15.5 **7**
$C_{15}H_{12}O$ 19.8 **3**
$C_{15}H_{12}O$ 26.2 **4**
$C_{15}H_{12}O$ 19.9 **6**
$C_{15}H_{12}OS$ 11.13 **6**
$C_{15}H_{12}O_2$ 19.56 **1**
$C_{15}H_{12}O_2$ 6.1 **6**
$C_{15}H_{12}O_4S$ 42.6 **4**

$C_{15}H_{12}O_5$	59.26+	**1**
$C_{15}H_{12}O_5$, H_2O	59.11	**5**
$C_{15}H_{12}O_6$	59.8	**3**
$C_{15}H_{12}O_8$	38.20	**5**
$C_{15}H_{13}Br$	5.28	**1**
$C_{15}H_{13}BrO_6$	59.28	**1**
$C_{15}H_{13}ClN_2OS$	41.33	**5**
$C_{15}H_{13}FO_2$	1.36	**7**
$C_{15}H_{13}FeO_2^+$, F_6Sb^-	71.18	**3**
$C_{15}H_{13}I_2NO_4$, C_3H_7NO	60.28	**5**
$C_{15}H_{13}I_3NO_4^+$, Cl^- , $3H_2O$	48.74	**6**
$C_{15}H_{13}NO$	36.19	**5**
$C_{15}H_{13}NO_2S$	41.34	**5**
$C_{15}H_{13}NO_3$	38.17	**3**
$C_{15}H_{13}NO_3$	10.4	**7**
$C_{15}H_{13}NO_3$	10.5	**7**
$C_{15}H_{13}O_2Te^+$, NO_3^-	70.7	**6**
$C_{15}H_{13}O_4P$	46.10	**5**
$C_{15}H_{14}$, $2C_{10}H_2O_6$	60.13	**7**
$C_{15}H_{14}Br_2O_8$	59.7	**4**
$C_{15}H_{14}CrO_3S$	75.19	**5**
$C_{15}H_{14}CrO_3S$	75.20	**5**
$C_{15}H_{14}MoO_2$	74.17	**2**
$C_{15}H_{14}N_2$	32.42	**1**
$C_{15}H_{14}N_2$	16.16	**4**
$C_{15}H_{14}N_2O$	40.29	**5**
$C_{15}H_{14}N_4O_5$	15.6	**7**
$C_{15}H_{14}N_4O_7$	17.27	**7**
$C_{15}H_{14}N_4O_7$	17.28	**7**
$C_{15}H_{14}O$	31.19	**5**
$C_{15}H_{14}OS$	39.18	**6**
$C_{15}H_{14}OS$	39.25	**7**
$C_{15}H_{14}OS_2$	39.26	**4**
$C_{15}H_{14}O_2S$	11.16	**7**
$C_{15}H_{14}O_3$	17.44	**1**
$C_{15}H_{14}O_5$	38.21	**5**
$C_{15}H_{14}O_5$	53.2	**7**
$C_{15}H_{15}BrN_2O$	33.44	**6**
$C_{15}H_{15}BrO_2S$	11.17	**7**
$C_{15}H_{15}BrO_5$	59.9	**3**
$C_{15}H_{15}BrO_6$	59.29	**1**
$C_{15}H_{15}BrO_6$	59.30	**1**
$C_{15}H_{15}Br_2O_3P$	64.39	**7**
$C_{15}H_{15}CdN_5O_6$	83.52	**5**
$C_{15}H_{15}ClU$	73.62	**2**
$C_{15}H_{15}Cl_2CoN_3OS$	83.74	**4**
$C_{15}H_{15}Cl_2InN_2O_2$	68.15	**7**
$C_{15}H_{15}Cl_2O_3P$	64.40	**7**
$C_{15}H_{15}Cl_3MoN_3$	83.64	**7**
$C_{15}H_{15}Cl_3N_3Ti$, C_5H_5N	83.54	**5**
$C_{15}H_{15}CoN_5O_6$	83.55	**5**
$C_{15}H_{15}Co_3S_2$	73.22+	**5**
$C_{15}H_{15}Co_3S_2^+$, I^-	73.25	**5**
$C_{15}H_{15}CuN_5O_6$	83.56	**5**
$C_{15}H_{15}Ge_3O_9Ru_3$	69.26	**4**
$(C_{15}H_{15}In)_n$	68.19	**5**
$C_{15}H_{15}Mn_2N_3O_3$	71.24	**6**
$C_{15}H_{15}Mn_3N_4O_4$	73.27	**6**
$C_{15}H_{15}MoNO$	73.20	**3**
$C_{15}H_{15}Mo_3S_4^+$, $C_3H_9Cl_2Sn^-$	73.14	**4**
$C_{15}H_{15}N_2O^+$, I^-	33.46	**3**
$C_{15}H_{15}N_5O_6Zn$	83.57	**5**
$C_{15}H_{15}Ni_2^+$, BF_4^-	73.24	**7**
$C_{15}H_{15}Ni_3S_2$	73.64	**2**
$C_{15}H_{15}O_3P$	64.41	**7**
$C_{15}H_{15}O_4P$	46.5	**7**
$(C_{15}H_{15}Sc)_n$	71.33	**5**
$C_{15}H_{15}Sm$	73.65	**2**
$C_{15}H_{15}Ti$	73.28	**6**
$C_{15}H_{16}BrNO_2$	37.12	**4**
$C_{15}H_{16}BrO_3$	53.3	**1**
$C_{15}H_{16}CrO_3$	75.39	**2**
$C_{15}H_{16}FeO$	73.29	**6**
$C_{15}H_{16}FeO_3$	75.22	**5**
$C_{15}H_{16}FeO_6$	71.25	**6**
$C_{15}H_{16}Fe_2GeO_3$	73.30	**6**
$C_{15}H_{16}N_2O_6$	27.9	**3**
$C_{15}H_{16}N_2O_6$	22.4	**7**
$C_{15}H_{16}N_4O_3$	35.23	**7**
$C_{15}H_{16}N_5^+$, Br^-	50.6	**6**
$C_{15}H_{16}O_2S$	11.18	**7**
$C_{15}H_{16}O_6$	59.5	**6**
$C_{15}H_{16}O_6$	53.3	**7**
$C_{15}H_{17}BMoN_6O_2$	72.13	**6**
$C_{15}H_{17}BrN_2O_4$	35.24	**7**
$C_{15}H_{17}BrN_4O_2$	36.15	**4**
$C_{15}H_{17}BrO_3$	31.19	**1**
$C_{15}H_{17}BrO_3$	53.5	**1**
$C_{15}H_{17}BrO_3$	53.6	**1**
$C_{15}H_{17}BrO_3$	53.7	**1**
$C_{15}H_{17}BrO_3$	53.8	**1**
$C_{15}H_{17}BrO_4$	53.4	**3**
$C_{15}H_{17}BrO_5$, H_2O	53.9	**1**
$C_{15}H_{17}BrO_6$	59.31	**1**
$C_{15}H_{17}ClO_3$	53.5	**3**
$C_{15}H_{17}CoO_2$, $2H_2O$	74.6	**5**
$C_{15}H_{17}CrNO_6$	71.30	**4**
$C_{15}H_{17}IrO_2$	74.7	**5**
$C_{15}H_{17}N_4O_2^+$, I^- , H_2O	36.20	**5**
$C_{15}H_{17}O_2Rh$	74.8+	**3**
$C_{15}H_{17}O_3P$	64.25	**6**
$C_{15}H_{18}$	29.2	**1**
$C_{15}H_{18}BNO_2$	62.7	**7**
$C_{15}H_{18}BrCuN_4O^+$, ClO_4^-	83.78	**3**
$C_{15}H_{18}BrNO_2$	37.21	**1**
$C_{15}H_{18}BrNO_3$	21.6	**7**
$C_{15}H_{18}BrN_3$	35.25	**7**
$C_{15}H_{18}ClN_3O$	38.18	**3**
$C_{15}H_{18}Cl_4O_3$	53.4	**7**
$C_{15}H_{18}CrO_3$	74.18	**2**
$C_{15}H_{18}N_4$	33.45	**6**
$C_{15}H_{18}N_4O_2^{2+}$, $2Cl^-$, H_2O	33.46	**6**
$C_{15}H_{18}O_2$	53.3	**6**
$C_{15}H_{18}O_3$	53.3	**5**
$C_{15}H_{18}O_4$	53.5	**7**
$C_{15}H_{18}O_6Rh_2$	72.43	**2**
$C_{15}H_{19}BrF_3NO$	16.18	**6**
$C_{15}H_{19}BrNO_3^+$, Br^-	33.47	**6**
$C_{15}H_{19}BrO_3$	53.1	**4**
$C_{15}H_{19}BrO_3S$	31.20	**1**
$C_{15}H_{19}ClN_3Pd^+$, Cl^-	83.75	**4**
$C_{15}H_{19}ClO$	21.11	**4**
$C_{15}H_{19}ClO_4Zr$	73.21+	**3**
$C_{15}H_{19}ClO_5$	53.5	**6**
$C_{15}H_{19}Cl_3N_3Ti$	83.76	**4**
$C_{15}H_{19}N_2P$	64.26	**6**
$C_{15}H_{19}O_3$	53.6	**7**
$C_{15}H_{19}O_3Re$	75.40	**2**
$C_{15}H_{19}O_4Rh$	74.10	**3**
$C_{15}H_{19}O_4Rh$	74.5	**4**
$C_{15}H_{19}Ti$	72.44	**2**
$C_{15}H_{20}Ag_2O_2^{2+}$, $2NO_3^-$	53.7	**3**

$C_{15}H_{20}As_2FeI_2NiO$	73.15	**4**
$C_{15}H_{20}BrNO_4$	1.30	**6**
$C_{15}H_{20}Br_2O_2$	59.6	**6**
$C_{15}H_{20}Cl_2Ti$	73.31	**6**
$C_{15}H_{20}CoNiO_4P$	73.32	**6**
$C_{15}H_{20}FeO_5$	72.18	**5**
$C_{15}H_{20}N^+$, Br^-	37.9	**3**
$C_{15}H_{20}NO^+$, Cl^-	31.20	**5**
$C_{15}H_{20}NO^+$, Cl^-	24.8	**7**
$C_{15}H_{20}N_2O_2$	58.2	**5**
$C_{15}H_{20}N_2O_3$	40.21	**4**
$C_{15}H_{20}N_2O_3$	33.38	**7**
$C_{15}H_{20}N_3O_9^+$, $0.5CH_4O$, I^-	58.12	**1**
$C_{15}H_{20}N_6O_7$, H_2O	44.26	**4**
$C_{15}H_{20}N_6O_8$	48.75	**6**
$C_{15}H_{20}O_2$	31.16	**6**
$C_{15}H_{20}O_3$	13.18	**6**
$C_{15}H_{20}O_3$	53.7 +	**7**
$C_{15}H_{20}O_4$	53.9	**7**
$C_{15}H_{20}O_4$	53.10	**7**
$C_{15}H_{20}O_4$	59.4	**7**
$C_{15}H_{20}O_5$	53.8	**3**
$C_{15}H_{20}O_5$	53.11	**7**
$C_{15}H_{20}O_{12}$	45.41	**3**
$C_{15}H_{21}AgFeO_6^+$, ClO_4^-	71.34	**5**
$C_{15}H_{21}Ag_3NiO_6^{2+}$, $2NO_3^-$, H_2O	77.19	**2**
$C_{15}H_{21}AlO_6$	68.20	**5**
$C_{15}H_{21}AlO_6$	68.13	**6**
$C_{15}H_{21}BMoN_4O_2$	72.12	**4**
$C_{15}H_{21}BrO_2$	31.9	**7**
$C_{15}H_{21}BrO_3$	59.8	**6**
$C_{15}H_{21}BrO_5$, $0.5C_3H_6O$	50.3	**4**
$C_{15}H_{21}Br_2ClO_2$	53.2	**4**
$C_{15}H_{21}Br_2ClO_3$	59.8	**4**
$C_{15}H_{21}CdO_6^-$, K^+ , H_2O	77.4	**7**
$C_{15}H_{21}ClHgO_2$	71.40	**7**
$C_{15}H_{21}ClN_3O_5Pr^{2+}$, $2Cl^-$, $3H_2O$	83.77	**4**
$C_{15}H_{21}ClO_6Zr$	77.8	**4**
$C_{15}H_{21}CoO_6$	77.20	**2**
$C_{15}H_{21}CoO_6$	77.8	**3**
$C_{15}H_{21}CoO_6$	77.11 +	**6**
$C_{15}H_{21}CoO_6$, $C_{15}H_{21}AlO_6$	77.9	**4**
$C_{15}H_{21}CoO_6^-$, $C_{16}H_{36}N^+$	77.21	**2**
$C_{15}H_{21}CrO_6$	77.22	**2**
$C_{15}H_{21}CrO_6$	77.9	**3**
$C_{15}H_{21}CrO_6$, $2CH_4N_2S$	60.24	**6**
$C_{15}H_{21}FeO_6$	77.23	**2**
$C_{15}H_{21}FeS_6$	85.24	**3**
$C_{15}H_{21}GaO_6$	68.14	**6**
$C_{15}H_{21}MnO_6$	77.24	**2**
$C_{15}H_{21}MnO_6$	77.13	**6**
$C_{15}H_{21}MnO_6$	77.5	**7**
$C_{15}H_{21}NOS$	33.39	**7**
$C_{15}H_{21}NO_2$	38.49	**7**
$C_{15}H_{21}NO_2S$	33.40	**7**
$C_{15}H_{21}NO_3S$	33.41	**7**
$C_{15}H_{21}NO_9Zr$	77.6	**7**
$C_{15}H_{21}N_2O_4^+$, Br^-	48.49	**3**
$C_{15}H_{21}N_3O_2$	58.7	**6**
$C_{15}H_{21}N_3O_3$	48.56 +	**7**
$C_{15}H_{21}N_3O_4$	48.58	**7**
$C_{15}H_{21}N_5O_5$	47.31	**5**
$C_{15}H_{21}O_6Rh$	77.4	**5**
$C_{15}H_{21}O_6Ru$	77.14	**6**
$C_{15}H_{21}O_6Sc$	77.5	**5**
$C_{15}H_{21}O_6V$	77.26 +	**2**
$C_{15}H_{21}O_6V$	77.10	**4**
$C_{15}H_{21}O_6Yb$	77.28	**2**
$C_{15}H_{22}BrN_4Ni^+$, Br^- , H_2O	83.143	**2**
$C_{15}H_{22}Br_2O_4$, $0.5H_2O$	53.9	**3**
$C_{15}H_{22}CuN_2O_5$, $2H_2O$, C_2H_6O	82.19	**7**
$C_{15}H_{22}NO^+$, Br^-	58.5	**7**
$C_{15}H_{22}NO^+$, I^-	33.48	**6**
$C_{15}H_{22}NO^+$, I^- , $CHCl_3$	33.42	**7**
$C_{15}H_{22}NO^+$, $C_4H_5O_6^-$	33.43	**7**
$C_{15}H_{22}NO_2^+$, Br^-	58.10	**3**
$C_{15}H_{22}NO_2^+$, Cl^-	33.44	**7**
$C_{15}H_{22}N_2O_5S$	45.44	**7**
$C_{15}H_{22}O$	53.6	**6**
$C_{15}H_{22}OZr$	72.19	**5**
$C_{15}H_{22}O_2$	53.7	**6**
$C_{15}H_{22}O_4$	31.10	**7**
$C_{15}H_{22}O_5$	59.9	**4**
$C_{15}H_{22}O_5S$	53.4	**5**
$C_{15}H_{22}O_5S$	53.5	**5**
$C_{15}H_{22}O_7S$	53.8	**6**
$C_{15}H_{22}O_{10}$	45.42	**3**
$C_{15}H_{22}O_{10}$	45.45	**6**
$C_{15}H_{22}O_{10}$	45.45	**7**
$C_{15}H_{23}AgFeO_7^+$, ClO_4^-	71.26	**6**
$C_{15}H_{23}Br$	53.7	**5**
$C_{15}H_{23}Br$	53.8	**5**
$C_{15}H_{23}Br$	53.9	**5**
$C_{15}H_{23}BrO$	53.12	**3**
$C_{15}H_{23}ClMoN_2O_6P_2$	71.27	**6**
$C_{15}H_{23}N_3O_3$	31.17	**6**
$C_{15}H_{23}OP$	64.39	**3**
$C_{15}H_{23}O_2Rh$	72.45	**2**
$C_{15}H_{23}O_7Yb$	77.29	**2**
$C_{15}H_{23}O_7Yb$	77.11	**3**
$C_{15}H_{23}O_7Yb$, $0.5C_6H_6$	77.30	**2**
$C_{15}H_{24}$	29.2	**6**
$C_{15}H_{24}Ag^+$, NO_3^-	53.15	**3**
$C_{15}H_{24}AgNO_3$	53.10	**1**
$C_{15}H_{24}Ag_2^{2+}$, $2NO_3^-$	53.11	**1**
$C_{15}H_{24}BrO_3P$	53.13	**3**
$C_{15}H_{24}Br_2$	53.3	**4**
$C_{15}H_{24}Br_2$	28.15	**6**
$C_{15}H_{24}Br_2$	28.16	**6**
$C_{15}H_{24}ClO_3P$	53.14	**3**
$C_{15}H_{24}Cl_2CuN_6$	83.79	**3**
$C_{15}H_{24}CoN_3O_3S_6$, CH_2Cl_2	80.11	**7**
$C_{15}H_{24}FeN_3O_3S_6$, CH_2Cl_2	80.12	**7**
$C_{15}H_{24}FeN_3S_6$	80.9	**4**
$C_{15}H_{24}FeN_3S_6^+$, ClO_4^-	80.6	**6**
$C_{15}H_{24}IN_2O_3$	53.12	**1**
$C_{15}H_{24}N^+$, Br^-	58.4	**5**
$C_{15}H_{24}NO^+$, Br^-	58.11	**3**
$C_{15}H_{24}NO_2^+$, Br^-	58.5	**5**
$C_{15}H_{24}NO_2^+$, Cl^-	3.22	**5**
$C_{15}H_{24}N_4O_2S_2$	44.42	**3**
$C_{15}H_{24}O$	17.17	**5**
$C_{15}H_{24}O$	53.9	**6**
$C_{15}H_{24}O_2$	53.10	**5**
$C_{15}H_{24}O_2$	53.12	**7**
$C_{15}H_{24}O_3$	53.10	**6**
$C_{15}H_{24}O_3$	27.5	**7**
$C_{15}H_{24}O_3$	38.50	**7**
$C_{15}H_{24}O_3$	53.13	**7**
$C_{15}H_{24}O_3$	53.14	**7**
$C_{15}H_{24}O_4$	27.14	**5**
$C_{15}H_{24}O_4$	31.18	**6**

$C_{15}H_{25}Br$ 53.11 **5**
$C_{15}H_{25}BrO$ 53.13 **1**
$C_{15}H_{25}BrO$ 53.11 **6**
$C_{15}H_{25}Br_2$ 53.14 **1**
$C_{15}H_{25}Cl$ 53.15 **1**
$C_{15}H_{25}Cl$ 53.16 **1**
$C_{15}H_{25}Cl$ 53.17 **1**
$C_{15}H_{25}Cl$ 53.18 **1**
$C_{15}H_{25}ClFeN_5^{2+}$, $2Cl_4Fe^-$ 83.58 **5**
$C_{15}H_{25}Cl_2NOPPt$ 71.38 **2**
$C_{15}H_{25}EuO_8$, H_2O 77.12 **3**
$C_{15}H_{25}HoO_8$, $2H_2O$ 77.11 **4**
$C_{15}H_{25}LaO_8$ 77.32 **2**
$C_{15}H_{25}N_2^+$, ClO_4^- 59.32 **1**
$C_{15}H_{25}N_2O_2^+$, ClO_4^- 58.6 **5**
$C_{15}H_{25}N_5O_5$, $0.5H_2O$ 48.59 **7**
$C_{15}H_{25}N_5O_5$, $2H_2O$ 48.76 **6**
$C_{15}H_{25}N_6OP$ 32.43 **1**
$C_{15}H_{25}N_6OP$, $0.5H_2O$ 32.44 **1**
$C_{15}H_{25}NdO_8$ 77.12 **4**
$C_{15}H_{25}NdO_8$, H_2O 77.33 **2**
$C_{15}H_{25}O_8Y$, H_2O 77.34 **2**
$C_{15}H_{26}Br_2$ 53.19 **1**
$C_{15}H_{26}Cl_2$ 53.20 **1**
$C_{15}H_{26}Cl_2CuN_2$ 83.65 **7**
$C_{15}H_{26}N_2$, H_2O 58.13 **1**
$C_{15}H_{26}N_4Ni^{2+}$, $2ClO_4^-$ 83.80 **3**
$C_{15}H_{26}O$ 53.12 **6**
$C_{15}H_{26}O_2$ 53.15 **7**
$C_{15}H_{27}As_9Cr_2O_6$ 86.18 **7**
$C_{15}H_{27}FeS_9$ 80.10 **4**
$C_{15}H_{27}NNiO_{10}P_3^+$, BF_4^- , $CHCl_3$ 86.19 **7**
$C_{15}H_{27}N_2^+$, ClO_4^- 58.8 **6**
$C_{15}H_{27}N_2O^+$, ClO_4^- 58.14 **1**
$C_{15}H_{27}N_2O^+$, $C_{15}H_{28}N_2O^{2+}$, $3ClO_4^-$ 58.15 **1**
$C_{15}H_{27}N_3O_9Pd_3$, $0.5C_6H_6$ 81.39 **4**
$C_{15}H_{27}N_9Ni^{2+}$, $2ClO_4^-$ 83.59 **5**
$C_{15}H_{28}NO^+$, Br^- 58.7 **5**
$C_{15}H_{28}NO^+$, Br^- 58.9 **6**
$C_{15}H_{28}NO_2^+$, I^- 27.13 **4**
$C_{15}H_{28}N_2^{2+}$, $2ClO_4^-$ 58.6 **7**
$C_{15}H_{28}O_2$ 23.3 **4**
$C_{15}H_{29}BrMgN_2$ 67.11 **5**
$C_{15}H_{29}N_2^+$, I^- 37.10 **3**
$C_{15}H_{30}AsN_3S_6$ 65.27 **2**
$C_{15}H_{30}AuN_3S_6$ 80.18 **5**
$C_{15}H_{30}B_3CrN_3O_3$ 75.23 **5**
$C_{15}H_{30}BiN_3S_6$ 66.6 **7**
$C_{15}H_{30}CoN_3S_6$ 80.29+ **2**
$C_{15}H_{30}CuN_{10}O_5^{2+}$, $2ClO_4^-$ 79.14 **6**
$C_{15}H_{30}FeN_3S_6$ 80.7+ **6**
$C_{15}H_{30}MnN_3S_6$ 80.19 **5**
$C_{15}H_{30}N_3NbOS_6$ 80.9 **6**
$C_{15}H_{30}N_3OS_6V$ 80.10 **6**
$C_{15}H_{30}N_3O_2S_6U^-$, $C_4H_{12}N^+$ 80.31 **2**
$C_{15}H_{30}N_3RuS_6$ 80.32 **2**
$C_{15}H_{30}N_3RuS_6$ 80.13 **7**
$C_{15}H_{30}N_3S_6Sb$ 66.7 **7**
$C_{15}H_{30}N_4ORuS_6$ 80.33 **2**
$C_{15}H_{30}N_5NbO_{10}$ 80.14 **7**
$C_{15}H_{30}Pd_3S_{12}$ 85.27 **6**
$C_{15}H_{31}O_4P$ 64.40 **3**
$C_{15}H_{36}CoN_6^{3+}$, $3Cl^-$, $4H_2O$ 76.40 **4**
$C_{15}H_{36}CuN_6S_3^+$, BF_4^- 79.25 **5**
$C_{15}H_{36}N_6O_6W$ 80.15 **7**
$C_{15}H_{37}Cl_3NP_2Rh$ 71.41 **7**
$C_{15}H_{42}CoN_6^{3+}$, $3Cl^-$, H_2O 83.60 **5**
$C_{15}H_{42}CoN_6^{3+}$, $3Cl^-$, $2H_2O$ 83.35 **6**
$C_{15}H_{44}Al_6N_5^-$, Li^+ , $C_4H_{10}O$ 68.16 **7**
$C_{15}H_{45}As_5NiO_5^{2+}$, $2ClO_4^-$ 84.41 **2**
$C_{15}H_{45}N_3Si_6$ 63.7 **3**
$C_{15}H_{46}Al_5N_5$ 68.17 **7**
$C_{15}H_{54}N_3NiS_6^+$, Br^- 80.20 **5**

C_{16}

$C_{16}As_2F_{10}FeO_4$ 86.2 **4**
$C_{16}Cl_6F_4$ 29.1 **7**
$C_{16}Cl_{10}$ 29.2 **7**
$C_{16}Co_2F_{24}S_8$ 85.45 **2**
$C_{16}F_{24}Fe_2S_8^-$, $C_9H_9Fe_2O_6S_3^+$ 85.12 **5**
$C_{16}F_{24}Fe_2S_8^{2-}$, $2C_{16}H_{36}N^+$ 85.25 **3**
$C_{16}F_{24}Fe_3S_{10}$, $0.5S_8$ 85.26 **7**
$C_{16}O_{15}Rh^{2-}$, $2C_{10}H_{16}N^+$ 71.35 **5**
$C_{16}O_{36}Ti_4^{8-}$, $8H_4N^+$, $4H_2O$ 81.37 **6**
$C_{16}H_4N_6O_8$ 28.17 **6**
$C_{16}H_5Fe_3O_{11}Rh$ 73.16 **4**
$C_{16}H_5N_5O_6$ 28.21 **1**
$C_{16}H_5N_5O_6$, $C_{12}H_4N_4^-$, $C_{13}H_{21}N_2S_2^+$ 60.21 **7**
$C_{16}H_6Br_2N_2$ 28.18 **6**
$C_{16}H_6N_4$ 24.9 **4**
$C_{16}H_6N_4O_4$ 28.22 **1**
$C_{16}H_6O_8Ru_3$ 75.12 **6**
$C_{16}H_8$ 28.5 **7**
$C_{16}H_8Cl_2$ 29.3 **1**
$C_{16}H_8Cl_2N_2O_4$ 40.17 **3**
$C_{16}H_8Cl_4$ 27.13 **1**
$C_{16}H_8F_6O_5Te$ 70.8 **6**
$C_{16}H_8F_8$ 31.17 **4**
$C_{16}H_8FeN_4O_3$ 71.31 **4**
$C_{16}H_8FeO_4$ 75.13 **7**
$C_{16}H_8Mn_2O_6$ 75.42 **2**
$C_{16}H_8Mo_2O_6$ 75.43+ **2**
$C_{16}H_8Mo_2O_6$ 75.24 **5**
$C_{16}H_8N_4$ 36.6 **6**
$C_{16}H_8N_4NiS_4^-$, $C_8H_{20}N^+$, $2H_2O$ 85.27 **7**
$C_{16}H_8O_2S_2$ 39.76+ **1**
$C_{16}H_8O_2Se_2$ 39.78 **1**
$C_{16}H_8O_2Se_2$ 39.30 **5**
$C_{16}H_8O_4$ 38.53 **1**
$C_{16}H_8O_{40}Ti_4^{8-}$, $8H_4N^+$ 81.36 **7**
$C_{16}H_9BrO$ 38.22 **5**
$C_{16}H_9ClN_2O$ 36.16 **4**
$(C_{16}H_9CuF_{12}NO_5)_n$ 77.13 **4**
$C_{16}H_{10}$ 29.4 **1**
$C_{16}H_{10}$ 29.5 **1**
$C_{16}H_{10}$ 29.3 **4**
$C_{16}H_{10}$ 29.2 **5**
$C_{16}H_{10}$, $2Br_3Sb$ 29.3 **5**
$C_{16}H_{10}$, $C_6Cl_4O_2$ 60.25 **6**
$C_{16}H_{10}$, C_6N_4 60.142 **2**
$C_{16}H_{10}$, C_6N_4 60.44 **3**
$C_{16}H_{10}$, $C_6H_2BrN_3O_6$ 60.22 **7**
$C_{16}H_{10}$, $C_6H_3N_3O_6$ 60.143 **2**
$C_{16}H_{10}$, $C_6H_3N_3O_6$ 60.29 **5**
$C_{16}H_{10}$, $C_9H_{12}N_4O_3$ 60.110 **2**
$C_{16}H_{10}$, $C_{10}H_2N_4$ 60.30+ **5**

$C_{16}H_{10}$, $C_{10}H_2O_6$	60.45+	3
$C_{16}H_{10}$, $C_{12}H_4N_4$	60.32+	5
$C_{16}D_{10}$, C_6N_4	60.23+	7
$1.5C_{16}H_{10}$, $C_6H_2BrN_3O_6$	60.25	7
$C_{16}H_{10}BrN_4V$	73.24	3
$C_{16}H_{10}Br_4O_5$	38.20	4
$C_{16}H_{10}ClN_3O_3$	25.4	7
$C_{16}H_{10}Cl_2Cr_2O_6Sn$	73.25	7
$C_{16}H_{10}Cl_2O_2$	19.9	3
$C_{16}H_{10}Cl_4Co_2FeGe_2O_6$	73.17	4
$C_{16}H_{10}Co_4O_{10}$	72.46	2
$C_{16}H_{10}Cr_2O_6$	73.33	6
$C_{16}H_{10}Fe_2O_6$	72.25	3
$C_{16}H_{10}Fe_2O_6$	72.26	3
$C_{16}H_{10}Fe_2O_6$	75.17	3
$C_{16}H_{10}Fe_2O_6$	71.36	5
$C_{16}H_{10}Mo_2O_6$	73.34	6
$C_{16}H_{10}Mo_2O_6Zn$	73.26	7
$C_{16}H_{10}N_2$	5.29	1
$C_{16}H_{10}N_2$	36.21	5
$C_{16}H_{10}N_2O_2$	35.40	1
$C_{16}H_{10}N_2O_2$	35.41	1
$C_{16}H_{10}N_2O_2$	35.19+	3
$C_{16}H_{10}N_4O_3$	7.4	7
$C_{16}H_{10}O$	31.21	1
$C_{16}H_{10}O_2$	25.21	1
$C_{16}H_{10}O_2$	13.19	4
$C_{16}H_{10}O_6W_2$	73.35	6
$C_{16}H_{10}O_7Ru_2$	71.28	6
$C_{16}H_{11}ClN_4$	36.7	6
$C_{16}H_{11}ClO_3$	38.54	1
$C_{16}H_{11}Cl_2NO$	35.22	6
$C_{16}H_{11}FeO_4P$	86.12	3
$C_{16}H_{11}N_3O_3$	9.44	1
$C_{16}H_{11}N_4^+$, Cl^- , H_2O	36.16	3
$C_{16}H_{12}$	31.22	1
$C_{16}H_{12}$	30.1	3
$C_{16}H_{12}$	31.20	3
$C_{16}H_{12}$	24.6	5
$C_{16}H_{12}$	31.21	5
$C_{16}H_{12}$	29.3	7
$0.5C_{16}H_{12}$, $C_{19}H_{12}FeO_3$	72.25	5
$C_{16}H_{12}As_2F_4I_2Mn_2O_8$	86.3	4
$C_{16}H_{12}As_2F_4Mn_2O_8$	86.14	5
$C_{16}H_{12}As_2F_4Mn_2O_8$	71.29	6
$C_{16}H_{12}Br_2$	29.4	4
$C_{16}H_{12}Cl_2$	26.3	4
$C_{16}H_{12}Cl_2$	19.10	7
$C_{16}H_{12}Cl_2CuN_4$	83.81	3
$C_{16}H_{12}Cl_2O_4$	13.23	7
$C_{16}H_{12}Cl_4$	27.6	7
$C_{16}H_{12}CrO_3$	75.46	2
$C_{16}H_{12}CuN_2O_6$	84.14	5
$C_{16}H_{12}F_8NiP_4S_4$	86.20	7
$C_{16}H_{12}F_{10}N_4Zn$	71.42	7
$C_{16}H_{12}F_{12}OS$	39.26	7
$C_{16}H_{12}Fe$	75.13	6
$C_{16}H_{12}Fe_2O_6$	72.20	5
$C_{16}H_{12}Fe_2O_6$	75.25	5
$C_{16}H_{12}Fe_2O_6$	71.31	6
$C_{16}H_{12}Fe_2O_8$	75.18	3
$C_{16}H_{12}Hg_2N_4^{2+}$, $2ClO_4^-$	83.66	7
$C_{16}H_{12}MoN_4O_4$	83.36	6
$C_{16}H_{12}N^+$, ClO_4^-	12.5	5
$C_{16}H_{12}NO_3S^-$, H_4N^+	24.9	7
$C_{16}H_{12}N_4OS$	41.37	1
$C_{16}H_{12}OS_2$	39.27	4
$C_{16}H_{12}O_2$	19.10	3
$C_{16}H_{12}O_2$	26.4	4
$C_{16}H_{12}O_2$	27.14	4
$C_{16}H_{12}O_6$	50.6	3
$C_{16}H_{12}O_{12}Re_4S_4$	85.21	4
$C_{16}H_{12}S_4Si$	63.6	6
$C_{16}H_{13}BiN_2O_2S_2$	66.7	4
$C_{16}H_{13}Br$	26.37	1
$C_{16}H_{13}Br$	28.6	3
$C_{16}H_{13}Cl$	29.4	5
$C_{16}H_{13}Cl$	26.3	6
$C_{16}H_{13}ClN_2O$	35.20	4
$C_{16}H_{13}ClN_4O_4$	15.26	1
$C_{16}H_{13}CoFeO_4$	73.27	7
$C_{16}H_{13}F_3NOP$	64.42	7
$C_{16}H_{13}MoN_3O_2S$	72.27	3
$C_{16}H_{13}NO$	35.23	6
$C_{16}H_{13}NO_3$	1.33	3
$C_{16}H_{13}N_2O_2^+$, Br^- , $2H_2O$	58.12	3
$C_{16}H_{13}N_3O_2$, C_3H_9N	33.45	7
$C_{16}H_{14}$	26.5	7
$0.5C_{16}H_{14}$, $C_{19}H_{14}FeO_3$	60.51	3
$C_{16}H_{14}BrNO$	31.23	1
$C_{16}H_{14}BrN_5O_3$	37.23	1
$C_{16}H_{14}BrO_6$	38.21	4
$C_{16}H_{14}Br_2$	31.11	7
$C_{16}H_{14}CdMn_2O_{13}$	84.17	6
$C_{16}H_{14}ClFeN_2O_2$	78.13	2
$C_{16}H_{14}ClFeN_2O_2$, CH_3NO_2	78.14	2
$C_{16}H_{14}Cl_2CoN_2$	83.61	5
$C_{16}H_{14}Cl_2Cu_2N_2O_2$	78.2	3
$C_{16}H_{14}Cl_2F_6P_2Pd$	86.21	7
$C_{16}H_{14}Cl_2N_2$	35.21	4
$C_{16}H_{14}Cl_2N_2O_2Ti$, C_4H_8O	78.2	4
$C_{16}H_{14}Cl_2O$	31.21	3
$C_{16}H_{14}Cl_2S_2$	11.69	1
$C_{16}H_{14}CoN_2O_2$, $CHCl_3$	78.15	2
$2C_{16}H_{14}CoN_2O_2$, $C_{24}H_{20}B^-$, $2C_4H_8O$, Na^+	78.2	7
$C_{16}H_{14}CoN_6O_4^-$, K^+ , $2C_2H_6OS$	79.22	7
$C_{16}H_{14}CrO_5$	74.11	3
$C_{16}H_{14}CuN_2O_2$	78.16	2
$C_{16}H_{14}CuN_2O_2$, $CHCl_3$	78.17	2
$C_{16}H_{14}CuN_2O_2$, $CHCl_3$	60.1	3
$C_{16}H_{14}CuN_2O_2$, $C_6H_5NO_3$	78.18	2
$C_{16}H_{14}CuN_2O_2$, $C_6H_5NO_3$	60.16	3
$2C_{16}H_{14}CuN_2O_2$, Na^+ , ClO_4^- , C_8H_{10}	60.145	2
$C_{16}H_{14}CuO_{10}$	81.80	2
$C_{16}H_{14}F_2$	5.30	1
$C_{16}H_{14}F_6NO_4P$, $0.5C_6H_6$	64.43	7
$C_{16}H_{14}FeO_7$	75.14	6
$C_{16}H_{14}Fe_2O_4Si$	73.36	6
$C_{16}H_{14}I_3NO_4$	48.60	7
$C_{16}H_{14}MoS_2$	73.26	3
$C_{16}H_{14}N_2$	35.26	7
$C_{16}H_{14}N_2NiO_2$	78.3	4
$C_{16}H_{14}N_2OS$	41.35	5
$C_{16}H_{14}N_2OS$	32.32	6
$C_{16}H_{14}N_2OS$	32.33	6
$C_{16}H_{14}N_2OS$	41.26	6
$C_{16}H_{14}N_2O_2S$	39.31	5
$C_{16}H_{14}O$	38.55	1
$C_{16}H_{14}O_2$	17.6	3
$C_{16}H_{14}O_2$	6.2	4

$C_{16}H_{14}O_2S$	1.34	**3**
$C_{16}H_{14}O_4$	13.24	**7**
$C_{16}H_{14}O_4$	38.51	**7**
$C_{16}H_{14}O_6$	59.5	**7**
$C_{16}H_{14}S_2Ti$	73.27	**5**
$C_{16}H_{14}S_2W$	73.37	**6**
$C_{16}H_{14}S_3$	39.38	**3**
$C_{16}H_{15}Br$	5.31	**1**
$C_{16}H_{15}Br_2N_3O_2Si$, C_3H_6O	63.19	**2**
$C_{16}H_{15}Co$	73.67	**2**
$C_{16}H_{15}Co_3OS$	73.28	**5**
$C_{16}H_{15}Co_3O_2$	73.27	**3**
$C_{16}H_{15}I_2NO_3$	38.32	**6**
$C_{16}H_{15}N$	37.11	**3**
$C_{16}H_{15}N_2O^+$, Br^-	32.45	**1**
$C_{16}H_{15}N_3O^+$, Cl^- , H_2O	35.24	**6**
$C_{16}H_{15}N_4NiS_2^+$, ClO_4^- , H_2O	83.62	**5**
$C_{16}H_{16}$	31.24	**1**
$C_{16}H_{16}$	31.25 +	**1**
$C_{16}H_{16}$	19.11	**3**
$C_{16}H_{16}$	23.8 +	**3**
$C_{16}H_{16}$	31.18	**4**
$C_{16}H_{16}$	31.22	**5**
$C_{16}H_{16}Ag^+$, NO_3^-	75.47	**2**
$C_{16}H_{16}B_2Fe_2O_4$	75.14	**7**
$C_{16}H_{16}BrN$	36.17	**7**
$C_{16}H_{16}Br_2O_2$	17.18	**5**
$C_{16}H_{16}Ce^-$, $C_6H_{14}KO_3^+$	75.26	**5**
$C_{16}H_{16}ClN$	35.25	**6**
$C_{16}H_{16}Cl_2Cu_2N_2O_2$	78.4	**4**
$C_{16}H_{16}Cl_2Cu_2N_2O_2$	78.1	**6**
$C_{16}H_{16}Cl_2MoN_2O_2$	78.2	**6**
$C_{16}H_{16}Cl_4CuN_2O_4$	81.40	**4**
$C_{16}H_{16}Cl_4Pd_2$	72.47	**2**
$C_{16}H_{16}Co_2O_4$	75.27	**5**
$C_{16}H_{16}CuN_2O_2$	78.21	**2**
$C_{16}H_{16}CuN_2O_2$	78.19 +	**2**
$C_{16}H_{16}CuN_2O_2$	78.5	**4**
$C_{16}H_{16}CuN_2O_2$	78.3	**5**
$C_{16}H_{16}CuN_2O_2$, $2C_6H_3N_3O_6$	60.2	**6**
$C_{16}H_{16}CuN_2S_4$	80.21	**5**
$C_{16}H_{16}CuN_4O_8^{2-}$, $2Cs^+$, $2H_2O$	83.63	**5**
$C_{16}H_{16}CuN_4O_8^{2-}$, $2K^+$	83.79	**4**
$C_{16}H_{16}CuN_4O_8^{2-}$, $2K^+$, $6H_2O$	83.67	**7**
$C_{16}H_{16}CuN_4O_8^{2-}$, $2Li^+$	83.80	**4**
$C_{16}H_{16}CuN_4O_8^{2-}$, $2Li^+$, H_2O	83.68	**7**
$C_{16}H_{16}Fe$	75.15	**6**
$C_{16}H_{16}FeN^+$, I^-	73.38	**6**
$C_{16}H_{16}FeN_2O_5$	72.11	**7**
$C_{16}H_{16}Fe_2N_2O_2$	71.32	**6**
$C_{16}H_{16}Fe_2O_4Pb$	73.68	**2**
$C_{16}H_{16}Fe_2O_4Sn$	73.69	**2**
$C_{16}H_{16}Fe_2S_6^{2-}$, $2C_8H_{20}N^+$	85.28	**7**
$C_{16}H_{16}Fe_3O_8S_2$	85.29	**7**
$C_{16}H_{16}HgN_2O_2S_2$	85.75	**2**
$C_{16}H_{16}HgO_2$	71.39	**2**
$C_{16}H_{16}IN$	37.13	**4**
$C_{16}H_{16}MoO_4$	75.16	**6**
$C_{16}H_{16}MoO_4S_2Ti$	73.28	**3**
$C_{16}H_{16}N^+$, Cl^- , $0.5H_2O$	35.27	**7**
$C_{16}H_{16}N_2^{2+}$, $2Br^-$, H_2O	31.23	**5**
$C_{16}H_{16}N_2NiO_2$	78.22 +	**2**
$C_{16}H_{16}N_2NiO_6$	83.144	**2**
$C_{16}H_{16}N_2NiS_2^-$, $C_8H_{20}N^+$	85.27	**3**
$C_{16}H_{16}N_2NiS_4$	80.22	**5**
$C_{16}H_{16}N_2O_2$	9.45	**1**
$C_{16}H_{16}N_2O_2S$	35.28	**7**
$C_{16}H_{16}N_2O_2Zn$	78.24	**2**
$C_{16}H_{16}N_2O_3Zn$	78.25	**2**
$C_{16}H_{16}N_2O_4S$, C_2H_6O	50.6	**5**
$C_{16}H_{16}N_2O_5S$	50.7	**3**
$C_{16}H_{16}N_4$	37.12	**7**
$C_{16}H_{16}N_4S$	41.36	**5**
$C_{16}H_{16}NiO_4$	72.48	**2**
$C_{16}H_{16}Ni_2$	75.28	**5**
$C_{16}H_{16}O$	31.12	**7**
$C_{16}H_{16}OS$	39.27	**7**
$C_{16}H_{16}OS_2$	39.39	**3**
$C_{16}H_{16}O_2$	17.18	**6**
$C_{16}H_{16}O_2$	31.19	**6**
$C_{16}H_{16}O_3$	17.46	**1**
$C_{16}H_{16}O_3S_2$	11.14	**6**
$C_{16}H_{16}O_4$	27.14	**1**
$C_{16}H_{16}O_7U$	84.18	**6**
$C_{16}H_{16}Th$	75.14	**4**
$C_{16}H_{16}Ti$	75.19	**3**
$C_{16}H_{16}U$	75.15	**4**
$C_{16}H_{17}BCo_3NO_{10}$	71.40	**2**
$C_{16}H_{17}BrN_4O_3$	36.17	**4**
$C_{16}H_{17}BrO_2S$	11.26	**3**
$C_{16}H_{17}CoFeO_4$	72.14	**6**
$C_{16}H_{17}FeNO_3$	72.12	**7**
$C_{16}H_{17}IN_2O$	21.7	**7**
$C_{16}H_{17}IN_3O$	21.8	**7**
$C_{16}H_{17}MnO_7$	75.17	**6**
$C_{16}H_{17}Mo_2O_4P$	73.70	**2**
$C_{16}H_{17}Mo_2O_4P$	73.28	**7**
$C_{16}H_{17}N$	36.8	**6**
$C_{16}H_{17}NOS$	41.27	**4**
$C_{16}H_{17}NO_2S_2$, $C_{16}H_{17}NO_2S_2$	11.13	**5**
$C_{16}H_{17}NO_3$	38.23	**5**
$C_{16}H_{17}NO_4$	58.10	**6**
$C_{16}H_{17}N_2O_4S^-$, K^+	50.10	**1**
$C_{16}H_{17}N_2O_4S^-$, Na^+	50.11	**1**
$C_{16}H_{17}N_4^+$, I^-	35.29	**7**
$C_{16}H_{18}$	19.57	**1**
$C_{16}H_{18}As_3F_5Mn_2O_6$	72.13	**7**
$C_{16}H_{18}BrN_2O_2$	21.10	**5**
$C_{16}H_{18}ClNO$	52.5	**7**
$C_{16}H_{18}ClN_3O$	36.18	**4**
$C_{16}H_{18}Cl_2CuN_2O_4$	81.40	**3**
$C_{16}H_{18}Cl_2O_2$	31.13	**7**
$C_{16}H_{18}Cl_4$	31.20	**6**
$C_{16}H_{18}CoO_6$	78.3	**6**
$C_{16}H_{18}CrN_2O_4^+$, Cl^-	78.3	**3**
$C_{16}H_{18}CrO_4$	75.16	**4**
$C_{16}H_{18}CuO_8$	81.81	**2**
$C_{16}H_{18}Fe$	73.71	**2**
$C_{16}H_{18}Ge_2O_{10}Re_2$	69.12	**7**
$C_{16}H_{18}NO_2Sb$	66.4	**6**
$C_{16}H_{18}NO_4^+$, Br^-	58.11	**6**
$C_{16}H_{18}NO_4^+$, Br^- , $2H_2O$	58.16	**1**
$C_{16}H_{18}N_2$	37.13	**7**
$C_{16}H_{18}N_2NiO_2$, C_7H_8	78.3	**7**
$C_{16}H_{18}N_2O_2$	9.14	**3**
$C_{16}H_{18}N_2O_3$	9.15	**3**
$C_{16}H_{18}N_2O_3$	58.12	**4**
$C_{16}H_{18}N_2O_4S$	50.12	**1**
$C_{16}H_{18}N_2O_5S$	50.13	**1**
$C_{16}H_{18}N_2O_6S$, CH_4O	50.8	**3**
$C_{16}H_{18}N_2O_6S_2$	33.49	**6**
$C_{16}H_{18}N_3S^+$, Cl^- , $5H_2O$	41.37	**5**

$C_{16}H_{18}N_3S^+$, CNS^- 41.38 **5**
$C_{16}H_{18}N_4^{2+}$, $2C_2H_5O_4S^-$, $2H_2O$ 8.12 **4**
$C_{16}H_{18}N_4O^{2+}$, $2C_2H_5O_4S^-$, $2H_2O$ 8.13 **4**
$C_{16}H_{18}N_4O_3$ 36.27 **1**
$C_{16}H_{18}N_4O_4$ 16.12 **7**
$C_{16}H_{18}N_6Ni$ 71.20 **3**
$C_{16}H_{18}Ni_2$ 72.13 **4**
$C_{16}H_{19}BrN_2O_2S$ 33.75 **1**
$C_{16}H_{19}F_{12}N_3NiO$ 71.32 **4**
$C_{16}H_{19}GeNO_3$ 69.11 **6**
$C_{16}H_{19}NO_2S_2$ 11.13 **4**
$C_{16}H_{19}NO_2Si$ 63.10 **7**
$C_{16}H_{19}NO_3$ 36.18 **7**
$C_{16}H_{19}N_3O$ 36.9 **6**
$C_{16}H_{19}N_3O_4S$, $3H_2O$ 50.14 **1**
$C_{16}H_{19}N_4O_2^+$, ClO_4^- 36.19 **4**
$C_{16}H_{20}Ag^+$, ClO_4^- 74.12 **3**
$C_{16}H_{20}Ag^+$, ClO_4^- 74.13 **3**
$C_{16}H_{20}Ag^+$, ClO_4^- 74.14 **3**
$C_{16}H_{20}As_2PdS_2$, C_5H_5N 86.5 **4**
$C_{16}H_{20}As_2PtS_2$ 86.13 **3**
$C_{16}H_{20}BrN$ 31.27 **1**
$C_{16}H_{20}BrNO_5$ 50.15 **1**
$C_{16}H_{20}BrN_2^+$, $C_4H_3O_4^-$ 33.38 **4**
$C_{16}H_{20}ClN_2^+$, $C_4H_3O_4^-$ 33.46 **7**
$C_{16}H_{20}Cl_2Sn_2^{2+}$, $2AlCl_4^-$ 69.27 **4**
$C_{16}H_{20}Co_2O_4$ 72.29 **3**
$C_{16}H_{20}CuN_2O_6$, $4H_2O$ 81.41 **4**
$C_{16}H_{20}CuN_5S^+$, CNS^- 83.64 + **5**
$C_{16}H_{20}CuN_6S_2$ 83.69 **7**
$C_{16}H_{20}CuO_8S_4^-$, K^+ , $C_4H_{10}O$ 85.30 **7**
$C_{16}H_{20}CuO_9$ 81.42 **4**
$C_{16}H_{20}Fe$ 73.72 **2**
$C_{16}H_{20}Fe$ 75.15 **7**
$C_{16}H_{20}Fe_2Ge_2O_8$ 69.29 **2**
$C_{16}H_{20}GeO_{10}$, $2H_2O$ 69.30 **2**
$C_{16}H_{20}Mo_2O_8P_2$ 86.3 **6**
$C_{16}H_{20}NO_2^+$, Br^- 58.17 **1**
$C_{16}H_{20}NO_4^+$, Br^- 58.18 **1**
$C_{16}H_{20}N_2$, $C_6Cl_4O_2$ 60.26 **6**
$2C_{16}H_{20}N_2$, $C_6Cl_4O_2$ 60.33 **4**
$3C_{16}H_{20}N_2$, Na^+ , Cl^- 16.19 **6**
$3C_{16}H_{20}N_2$, Na^+ , NO_3^- 16.20 **6**
$3C_{16}H_{20}N_2$, Na^+ , CN^- 16.21 **6**
$C_{16}H_{20}N_2NiO_4P_2S_4$ 85.28 + **3**
$C_{16}H_{20}N_2O_3$ 40.30 **5**
$C_{16}H_{20}N_2P_2S_2$ 64.27 **6**
$C_{16}H_{20}N_2P_2S_2$ 64.28 **6**
$C_{16}H_{20}N_3O_2^+$, Br^- 51.1 **3**
$C_{16}H_{20}N_3O_8S^-$, Na^+ , $2H_2O$ 50.16 **1**
$C_{16}H_{20}N_4O_4Zn$, $C_7H_7NO_2$, H_2O 76.67 **7**
$C_{16}H_{20}NiO_8S_4^{2-}$, $2C_9H_{14}N^+$ 85.31 **7**
$C_{16}H_{20}NiP_2S_4$ 85.76 **2**
$C_{16}H_{20}O_2$ 31.28 **1**
$C_{16}H_{20}O_4$ 31.21 **6**
$C_{16}H_{20}O_4$ 31.22 **6**
$C_{16}H_{20}O_4$ 31.14 **7**
$C_{16}H_{20}O_6$ 31.23 **6**
$C_{16}H_{20}S_5$ 39.28 **4**
$C_{16}H_{20}Si_2$ 63.11 **7**
$C_{16}H_{21}Cl_3CuNOP$ 84.19 **6**
$C_{16}H_{21}NO$ 36.18 **3**
$C_{16}H_{21}NO_2$ 24.7 **6**
$C_{16}H_{21}NO_3$ 58.12 **6**
$C_{16}H_{21}N_3^+$, ClO_3^- 12.12 **4**
$C_{16}H_{21}N_3^+$, ClO_4^- 12.10 **1**
$C_{16}H_{21}N_3^+$, I^- 12.11 **1**
$C_{16}H_{21}O_{11}^-$, Rb^+ , $2H_2O$ 52.25 **1**
$C_{16}H_{22}$ 31.19 **4**
$C_{16}H_{22}Al_2$ 68.21 **5**
$C_{16}H_{22}BrNO_3$ 58.19 **1**
$C_{16}H_{22}BrNO_3$ 40.11 **7**
$C_{16}H_{22}Br_3NiP_2$, $0.5C_{16}H_{22}Br_2NiP_2$, C_6H_6 86.14 **3**
$C_{16}H_{22}ClN_4NiS_2^+$, ClO_4^- 85.24 **5**
$C_{16}H_{22}Cl_2N_2Zn$ 83.70 **7**
$C_{16}H_{22}Cl_2O_2S_2Sn$ 69.13 **7**
$C_{16}H_{22}Cl_2P_2Pd$ 86.8 **4**
$C_{16}H_{22}Cl_2Pd_2$ 71.33 **6**
$C_{16}H_{22}Cl_4HgP_2Pt$ 86.15 **5**
$C_{16}H_{22}Cl_4Hg_2N_2$ 83.145 **2**
$C_{16}H_{22}Co_2O_2Sn_2$ 73.29 **5**
$C_{16}H_{22}CuN_4^{2+}$, $2ClO_4^-$ 76.68 **7**
$C_{16}H_{22}F_6N_4NiO_2$ 76.69 **7**
$C_{16}H_{22}INO_2S$ 36.19 **3**
$C_{16}H_{22}I_2P_2Pd$ 86.18 + **2**
$C_{16}H_{22}NO_2^+$, Cl^- 24.7 **5**
$C_{16}H_{22}NO_2^+$, Cl^- 24.8 **6**
$C_{16}H_{22}NO_3^+$, Br^- 58.20 **1**
$C_{16}H_{22}N_2$ 35.42 **1**
$C_{16}H_{22}N_2$ 35.22 **4**
$C_{16}H_{22}N_2O_2$ 33.47 **7**
$C_{16}H_{22}N_2O_3$ 43.6 **7**
$C_{16}H_{22}NiO_{10}$ 81.42 **3**
$C_{16}H_{22}O_2$ 31.24 **6**
$C_{16}H_{22}O_4$ 31.15 **7**
$C_{16}H_{22}O_8Si_2W_2$ 63.6 **4**
$C_{16}H_{22}O_{11}$ 45.46 **7**
$C_{16}H_{23}Br$ 21.12 **4**
$C_{16}H_{23}IO_2S$ 52.26 **1**
$C_{16}H_{23}MoNO_2$ 73.39 **6**
$C_{16}H_{23}NO$ 52.27 **1**
$C_{16}H_{23}NO_5$ 58.13 **6**
$C_{16}H_{23}O_2S_2$ 39.29 **4**
$C_{16}H_{24}$ 22.10 **4**
$C_{16}H_{24}$ 31.25 **5**
$C_{16}H_{24}As_4Cr_2O_8$ 86.22 **7**
$C_{16}H_{24}Br_2MnN_8$ 83.81 **4**
$C_{16}H_{24}Br_2O_5$ 53.13 **6**
$C_{16}H_{24}ClCuN_4^+$, NO_3^- , $2H_2O$ 83.71 **7**
$C_{16}H_{24}Cl_2Cu_2$ 75.50 **2**
$C_{16}H_{24}Cl_2Rh_2$ 75.51 **2**
$C_{16}H_{24}Cl_3IrSn$ 75.52 **2**
$C_{16}H_{24}Cl_4Ni_2$, C_6H_6 75.53 **2**
$C_{16}H_{24}Cl_4O_4$ 28.4 **5**
$C_{16}H_{24}CoN_5O_6$ 83.146 **2**
$C_{16}H_{24}CoN_9O_3^+$, NO_3^- , $0.5C_2H_6O$ 83.82 **4**
$C_{16}H_{24}CuGa_2N_8$ 83.72 **7**
$C_{16}H_{24}Ga_2N_8Ni$ 83.73 **7**
$C_{16}H_{24}NO_2^+$, Br^- 33.76 **1**
$C_{16}H_{24}NO_2^+$, Br^- 33.50 **6**
$C_{16}H_{24}NO_2^+$, Cl^- 33.77 **1**
$C_{16}H_{24}NO_2^+$, Cl^- 33.78 **1**
$C_{16}H_{24}N_4O_6$ 36.19 **7**
$C_{16}H_{24}Ni$ 75.54 **2**
$C_{16}H_{24}O_4$ 59.10 **4**
$C_{16}H_{24}O_4Pt_2$ 71.33 **4**
$C_{16}H_{24}O_{10}$ 45.47 **7**
$C_{16}H_{24}S_5$ 39.19 **6**
$C_{16}H_{25}AsN_2S_4$ 65.28 **2**
$C_{16}H_{25}Cl_2NPt$ 75.21 **3**
$C_{16}H_{25}Co$ 75.55 **2**

$(C_{16}H_{25}Mn_3O_{17})_n$ 81.43 **3**
$C_{16}H_{25}N_2O_2^+$, ClO_4^- 58.14 **6**
$C_{16}H_{26}As_2CrO_4$ 86.16 **5**
$C_{16}H_{26}Br_2O_8Sn$ 69.31 **2**
$C_{16}H_{26}Br_2O_8Sn$ 69.25 **3**
$C_{16}H_{26}CoN_4O_6^-$, K^+ , $6H_2O$ 76.70 **7**
$C_{16}H_{26}Co_2N_8O_9^{2-}$, $3K^+$, HO^- , $12H_2O$ 83.74 **7**
$C_{16}H_{26}CuN_8O_7^{2-}$, $3H_4N^+$, ClO_4^- , $6H_2O$ 83.66 **5**
$C_{16}H_{26}Cu_2N_2O_4$ 77.14 **3**
$C_{16}H_{26}NO^+$, Cl^- 58.7 **7**
$C_{16}H_{26}NO_2^+$, Br^- 58.13 **3**
$C_{16}H_{26}N_2O_{18}Pt_4$ 81.49 **5**
$C_{16}H_{27}Cl_2NOPt$ 72.15 **6**
$C_{16}H_{27}NO_5$ 58.8 **7**
$C_{16}H_{27}N_2O_2^+$, Br^- 58.21 **1**
$C_{16}H_{28}AuN_{16}^-$, $C_{24}H_{20}As^+$ 71.34 **4**
$C_{16}H_{28}B_3P_2Pt$ 86.17 **5**
$C_{16}H_{28}CaN_4O_8^{2+}$, $2Br^-$ 67.17 **3**
$C_{16}H_{28}Ce_2O_{18}^{2-}$, $2CH_6N_3^+$, $2H_2O$ 81.43 **4**
$C_{16}H_{28}CuN_2O_{14}$ 82.20 **7**
$C_{16}H_{28}Cu_2O_8$ 81.82 **2**
$C_{16}H_{28}Cu_2O_8$ 81.44 **4**
$C_{16}H_{28}N_2O_{11}$, H_2O 45.48 **7**
$C_{16}H_{28}N_2O_{11}$, $3H_2O$ 45.49 **7**
$C_{16}H_{28}N_4Ni^{2+}$, $2ClO_4^-$ 83.83 **4**
$C_{16}H_{28}N_4O_2$ 12.13 **4**
$C_{16}H_{28}N_{12}Ni_2O_4^{4+}$, $4Cl^-$, $2H_2O$ 83.84 **3**
$C_{16}H_{28}O_4$ 38.56 **1**
$C_{16}H_{28}O_4$ 21.11 **5**
$C_{16}H_{28}O_8Re_2^{2+}$, $2O_4Re^-$ 81.44 **3**
$C_{16}H_{28}O_8Ru_2^+$, Cl^- 81.83 **2**
$C_{16}H_{28}P^+$, Br^- 64.24 **5**
$C_{16}H_{29}N_2O^+$, Br^- 37.14 **7**
$C_{16}H_{30}B_2$ 62.9 **6**
$C_{16}H_{30}Cl_2PdS_2$ 71.43 **7**
$C_{16}H_{30}Cl_2Pd_2$ 72.14 **4**
$C_{16}H_{30}Co_2S_{14}$ 85.22 **4**
$C_{16}H_{30}CuO_4$ 81.37 **7**
$C_{16}H_{30}Fe_2HgN_2O_6P_2$ 86.18 **5**
$C_{16}H_{30}Fe_2S_{14}$ 85.31 **3**
$C_{16}H_{30}N_2O$ 37.15 **7**
$C_{16}H_{30}N_2O_2$ 38.52 **7**
$C_{16}H_{30}N_3OReS_6$ 80.11 **6**
$C_{16}H_{30}O_6Ru_2Si_4$ 63.4 **5**
$C_{16}H_{30}O_6Ru_2Sn_4$ 69.26 **3**
$C_{16}H_{30}O_8P_2W_2$ 86.4 **6**
$C_{16}H_{31}O_2^-$, K^+ 2.97 **1**
$C_{16}H_{32}BeN_2$ 67.18 **4**
$C_{16}H_{32}ClFeN_4^+$, I^- 83.38 **6**
$C_{16}H_{32}Cl_2N_8NiS_4$ 79.36 **2**
$C_{16}H_{32}LiO_4^+$, $C_{32}H_{36}Lu^-$ 71.79 **5**
$C_{16}H_{32}MgN_2$ 67.18 **3**
$C_{16}H_{32}N_2O_5$, $C_4CoN_4S_4^{2-}$, Co^{2+} 40.20 **6**
$C_{16}H_{32}N_4Ni^{2+}$, $2ClO_4^-$ 83.85 **3**
$C_{16}H_{32}N_4Ni^{2+}$, $2ClO_4^-$ 83.84 **4**
$C_{16}H_{32}N_4Ni^{2+}$, $2CNS^-$, H_2O 83.67 **5**
$C_{16}H_{32}O_6$ 45.50 **7**
$C_{16}H_{34}Cl_3N_4Ta$ 83.75 **7**
$C_{16}H_{34}O$ 5.32 **1**
$C_{16}H_{34}O$, $0.2C_{33}H_{36}O_6$ 61.4 **7**
$C_{16}H_{34}O_7$, $2Cl_2Hg$ 5.11 **6**
$C_{16}H_{36}As_3NNiP^+$, ClO_4^- 86.20 **2**
$C_{16}H_{36}Br_2F_2NiP_2$ 86.19 **5**
$C_{16}H_{36}Cl_2MnN_4^+$, Cl^- , $3H_2O$ 83.76 **7**
$C_{16}H_{36}Cl_4N_4S_4Zn_4$, $2H_2O$ 76.47 **5**
$C_{16}H_{36}CoN_4O_2^{2+}$, $2BF_4^-$ 83.39 **6**
$C_{16}H_{36}Co_4N_8O_4$ 83.40 **6**
$C_{16}H_{36}Cu_4I_4N_4O_4$ 83.77 **7**
$C_{16}H_{36}Fe_2N_4S_4$ 76.35 **6**
$C_{16}H_{36}Fe_2N_5OS_4^+$, F_6P^- , C_3H_6O 76.71 **7**
$C_{16}H_{36}N^+$, $Ag_3I_4^-$ 3.52 **3**
$C_{16}H_{36}N^+$, F^- , $32.8H_2O$ 61.13 **2**
$C_{16}H_{36}N^+$, $C_2Cl_2O_2Rh^-$ 3.53 **3**
$C_{16}H_{36}N^+$, $C_4H_{10}B_6FeO_4^-$ 3.32 **6**
$C_{16}H_{36}N^+$, $C_7H_5O_2^-$, $39.5H_2O$ 61.14 **2**
$C_{16}H_{36}N^+$, $C_8FeN_4S_4^-$ 85.41 **2**
$C_{16}H_{36}N^+$, $C_8H_{19}B_2^-$ 61.4 **5**
$C_{16}H_{36}N^+$, $C_9H_7Br_3NNi^-$ 3.80 **1**
$C_{16}H_{36}N^+$, $C_{12}H_{16}Fe_4O_8S_8^{6-}$, $5Na^+$, $5C_5H_9NO$ 85.18 **7**
$C_{16}H_{36}N^+$, $C_{13}H_{18}CuN_3S_4^-$ 80.5 **6**
$C_{16}H_{36}N^+$, $C_{15}IO_{15}Rh_6^-$ 3.54 **3**
$C_{16}H_{36}N^+$, $C_{15}H_{21}CoO_6^-$ 77.21 **2**
$C_{16}H_{36}N^+$, $C_{26}H_{15}F_{12}FeOPS_4^-$ 84.20 **3**
$C_{16}H_{36}N^+$, $C_{26}H_{20}CoN_2S_4^-$ 85.45 **3**
$2C_{16}H_{36}N^+$, $Cl_4O_2U^{2-}$ 3.33 **6**
$2C_{16}H_{36}N^+$, $O_{19}W_6^{2-}$ 3.55 **3**
$2C_{16}H_{36}N^+$, $C_5H_{10}B_{10}MoO_5^{2-}$ 62.9 **3**
$2C_{16}H_{36}N^+$, $C_8CoN_4S_4^{2-}$ 85.39 **2**
$2C_{16}H_{36}N^+$, $C_8CuN_4S_4^{2-}$ 85.40 **2**
$2C_{16}H_{36}N^+$, $C_8CuN_4S_4^{2-}$ 85.3 **5**
$2C_{16}H_{36}N^+$, $C_8H_{16}Fe_2S_8^{2-}$ 85.7 **5**
$2C_{16}H_{36}N^+$, $C_{16}F_{24}Fe_2S_8^{2-}$ 85.25 **3**
$2C_{16}H_{36}N^+$, $C_{16}O_{16}Re_4^{2-}$ 3.82 **1**
$2C_{16}H_{36}N^+$, $C_{24}Cl_{16}Co_2S_8^{2-}$ 85.86 **2**
$3C_{16}H_{36}N^+$, $C_6ErN_6S_6^{3-}$ 3.83 **1**
$3C_{16}H_{36}N^+$, $C_8MoN_8^{3-}$ 3.56 **3**
$C_{16}H_{36}N_4Ni^{2+}$, $2ClO_4^-$ 76.36 **6**
$C_{16}H_{36}N_4Ni^{2+}$, $2ClO_4^-$ 76.37 **6**
$C_{16}H_{36}N_4Ni^{2+}$, Cl_4Zn^{2-} , H_2O 76.38 **6**
$(C_{16}H_{36}O_2P_2S_2Zn)_n$ 85.32+ **3**
$C_{16}H_{36}O_4Ti_2$ 71.21 **3**
$C_{16}H_{38}CoN_6^{3+}$, $3ClO_4^-$ 83.41 **6**
$C_{16}H_{38}N_2^{2+}$, $2Br^-$, $2H_2O$ 3.84 **1**
$C_{16}H_{40}Br_2Mg_2N_2$ 67.15 **2**
$C_{16}H_{40}Br_2Mg_2O_4$ 67.16 **2**
$C_{16}H_{40}CrN_2O_2Si_4$ 83.85 **4**
$C_{16}H_{40}Cu_2N_4S_4^{2+}$, $2BF_4^-$ 85.32 **7**
$C_{16}H_{40}Mo_2O_{11}P_4S_8$, $2C_6H_4Cl_2$ 85.78 **2**
$C_{16}H_{40}Ni_2P_4$ 71.44 **7**
$C_{16}H_{40}P_4S_8Zn_2$ 85.34 **3**
$C_{16}H_{42}B_6P_2Pt$ 71.45 **7**
$C_{16}H_{43}B_7P_2Pt$ 86.5 **6**
$C_{16}H_{44}B_4N_8$ 62.48 **2**
$C_{16}H_{44}Cu_4Si_4$ 71.38 **5**
$C_{16}H_{45}B_7NiP_2$ 71.34 **6**
$C_{16}H_{46}Fe_2N_{10}O^{4+}$, $4I^-$ 76.72 **7**
$C_{16}H_{48}Be_4O_4Si_4$ 63.8 **3**
$C_{16}H_{48}CuN_8O_6P_4^{2+}$, $2ClO_4^-$ 84.13 **3**
$C_{16}H_{48}F_8N_8Ti_4$ 83.78 **7**
$C_{16}H_{48}Hg_4O_4Si_4$ 63.9 **3**
$C_{16}H_{48}N_8O_{16}P_8$ 64.45 **2**
$C_{16}H_{48}N_{12}P_4$ 64.46 **2**
$C_{16}H_{48}Si_9$ 63.5 **5**
$C_{16}H_{64}B_{20}Cd_2O_4$ 84.9 **4**

C_{17}

$C_{17}F_{17}MnO_3S$ 75.16 **7**
$C_{17}Fe_6O_{16}^{2-}$, $2C_4H_{12}N^+$ 71.35 **4**
$C_{17}H_5Co_3GeO_{11}$ 69.27 **3**
$C_{17}H_5F_{18}Rh$ 74.19 **2**
$C_{17}H_8FeN_4O_3$ 71.35 **6**
$C_{17}H_8Fe_2N_2O_7$ 83.42 **6**
$C_{17}H_8Fe_2O_5$ 75.22 **3**
$C_{17}H_8Mn_2O_{10}$ 71.39 **5**
$C_{17}H_8N_2O$ 26.38 **1**
$C_{17}H_8O_7Ru_3$ 75.29 **5**
$C_{17}H_{10}BrMnN_2O_3$ 71.36 **6**
$C_{17}H_{10}BrNS_2$ 41.38 **1**
$C_{17}H_{10}Cl_2N_2$ 7.5 **7**
$C_{17}H_{10}Co_4O_9$, $C_{17}H_{10}Co_4O_9$ 74.5 **7**
$C_{17}H_{10}CrO_3$ 74.20 **2**
$C_{17}H_{10}CrO_3$ 74.21 **2**
$C_{17}H_{10}CrO_3$ 74.22 **2**
$C_{17}H_{10}CrO_3$ 74.5 **6**
$C_{17}H_{10}Fe_2N_2O_3$ 71.46 **7**
$C_{17}H_{10}Fe_2O_6$ 71.37 **6**
$C_{17}H_{10}Fe_2O_7$ 71.47 **7**
$C_{17}H_{10}MnNO_4$ 71.40 **5**
$C_{17}H_{11}BrS_3$ 39.40 **3**
$C_{17}H_{11}MnO_5$ 72.16 **6**
$C_{17}H_{11}Mn_2O_9P$ 86.6 **6**
$C_{17}H_{11}NO_2S_2$ 39.41 **3**
$C_{17}H_{12}$ 29.1 **3**
$C_{17}H_{12}$ 31.16 **7**
$C_{17}H_{12}As_2F_4Fe_3O_9$ 71.37 **4**
$C_{17}H_{12}BrN$ 24.11 **4**
$C_{17}H_{12}BrNO_2$ 32.21 **4**
$C_{17}H_{12}BrNS_2$, C_3H_7NO 35.26 **6**
$C_{17}H_{12}BrN_3O_4$ 36.20 **3**
$C_{17}H_{12}Br_2O_3$ 38.53 **7**
$C_{17}H_{12}CrO_3$ 74.23 **2**
$C_{17}H_{12}CrO_3$ 74.6 **6**
$C_{17}H_{12}FeN_4O_2$ 71.38 **6**
$C_{17}H_{12}OS_2$ 39.42 **3**
$C_{17}H_{12}O_2$ 27.10 **3**
$C_{17}H_{12}O_2$ 29.3 **6**
$C_{17}H_{12}O_6$ 59.10 **3**
$C_{17}H_{12}O_6$ 59.11 **4**
$C_{17}H_{12}O_6$ 59.9 **6**
$C_{17}H_{12}O_6$, $CHCl_3$ 59.11 **3**
$C_{17}H_{12}O_7$, $0.5C_4H_3BrS$ 59.33 **1**
$C_{17}H_{12}O_7$, $0.5C_6H_5Br$ 59.34 **1**
$C_{17}H_{12}O_7$, $0.5C_6H_6$ 59.35 **1**
$C_{17}H_{12}S_2Se$ 39.32 **5**
$C_{17}H_{12}S_3$ 39.83 **1**
$C_{17}H_{12}S_3$ 39.43 **3**
$C_{17}H_{12}S_3$ 39.30 **4**
$C_{17}H_{13}BrO$ 19.11 **7**
$C_{17}H_{13}ClN_2O_4$ 35.27 **6**
$C_{17}H_{13}ClN_4$, $C_{17}H_{14}ClN_4^+$, Br^- , $2C_2H_6O$ 36.20 **4**
$C_{17}H_{13}ClO$ 27.7 **7**
$C_{17}H_{13}Cl_2N$ 33.48 **7**
$C_{17}H_{13}FeNO_4$ 75.19 **6**
$C_{17}H_{13}MnO_4P$ 86.20 **5**
$C_{17}H_{13}Mo_2NO_5$ 73.40 **6**
$C_{17}H_{13}NO_2$ 40.12 **7**
$C_{17}H_{13}NO_2$ 40.13 **7**
$C_{17}H_{13}N_3OS$ 41.27 **6**
$C_{17}H_{14}$ 29.7 **1**
$C_{17}H_{14}$ 31.20 **4**
$C_{17}H_{14}ClNO_2$ 40.21 **6**
$C_{17}H_{14}ClN_3$ 35.26 **5**
$C_{17}H_{14}Cl_2N_2O$ 32.26 **7**
$C_{17}H_{14}CoO$ 73.73 **2**
$C_{17}H_{14}Co_2O_{11}Sn$ 69.12 **6**
$C_{17}H_{14}CuN_3O_2^+$, BF_4^- 81.45 **3**
$C_{17}H_{14}CuN_3O_2^+$, F_6P^- 81.46 **3**
$C_{17}H_{14}FeN_2O_8$ 83.43 **6**
$C_{17}H_{14}NO_3S^-$, K^+ 24.4 **3**
$C_{17}H_{14}N_2$ 36.20 **7**
$C_{17}H_{14}N_4S_2$ 41.39 **1**
$C_{17}H_{14}N_4S_3$ 41.23 **7**
$C_{17}H_{14}OS_2$ 39.85 **1**
$C_{17}H_{14}O_2$ 27.14 **6**
$C_{17}H_{14}O_2Ti$ 71.38 **4**
$C_{17}H_{14}O_6$ 59.12 **3**
$C_{17}H_{15}As_2Co_3F_4O_7$ 86.9 **4**
$C_{17}H_{15}BMoN_8O_2$ 83.86 **3**
$C_{17}H_{15}BrN_2$ 5.33 **1**
$C_{17}H_{15}BrO_2$ 31.25 **3**
$C_{17}H_{15}BrO_3S$ 52.28 **1**
$C_{17}H_{15}BrO_3S$ 31.26 **3**
$C_{17}H_{15}Cl_2NO_4$ 31.17 **7**
$C_{17}H_{15}Cl_3N_2O$ 35.43 **1**
$C_{17}H_{15}FSi$ 63.21 **2**
$C_{17}H_{15}F_7N_2$ 32.27 **7**
$C_{17}H_{15}I_3O_4$ 17.29 **7**
$C_{17}H_{15}N$ 33.49 **7**
$C_{17}H_{15}NO_2S$ 39.33 **5**
$C_{17}H_{15}N_2S_2^+$, $C_{12}H_4N_4^-$, $C_{12}H_4N_4$ 60.26 **7**
$C_{17}H_{15}Ni_3O_2$ 73.74 **2**
$C_{17}H_{16}$ 20.17 **7**
$C_{17}H_{16}BrClO_6$ 50.17 **1**
$C_{17}H_{16}BrNO$ 35.28 **6**
$C_{17}H_{16}Br_2N_4O_3$ 32.21 **5**
$C_{17}H_{16}Cl_2CoN_2O$ 83.79 **7**
$C_{17}H_{16}Cl_2Cu_2N_2O_2$ 78.4 **3**
$C_{17}H_{16}CuN_2O_2$, H_2O 78.26 **2**
$C_{17}H_{16}CuN_3O_3^+$, BF_4^- 81.47 **3**
$C_{17}H_{16}Fe_2O_5$ 75.57 **2**
$C_{17}H_{16}Fe_2O_5$ 75.58 **2**
$C_{17}H_{16}Ge$ 69.28 **3**
$C_{17}H_{16}I_3N_2ORh$ 71.39 **6**
$C_{17}H_{16}MoN_2O_4$ 78.4 **6**
$C_{17}H_{16}MoS_2$ 73.75 **2**
$C_{17}H_{16}N_2O_2$ 32.28 **7**
$C_{17}H_{16}N_2O_3V$ 78.5 **3**
$C_{17}H_{16}N_2O_5S_2$ 41.39 **5**
$C_{17}H_{16}O$ 19.12+ **7**
$C_{17}H_{16}O_2$ 20.16 **6**
$C_{17}H_{16}O_2Se$ 11.19 **7**
$C_{17}H_{16}O_3$ 28.19 **6**
$C_{17}H_{16}O_3$ 59.10 **6**
$C_{17}H_{16}O_4$ 26.5 **4**
$C_{17}H_{16}O_6$, H_2O 59.12 **5**
$C_{17}H_{16}O_7$ 59.13 **3**
$C_{17}H_{16}Si$ 63.22 **2**
$C_{17}H_{17}BBrNO_4$ 45.43 **3**
$C_{17}H_{17}ClN_3Sn^+$, $C_2H_6Cl_3Sn^-$ 69.32 **2**
$C_{17}H_{17}Cl_2O_2Sb$ 66.5+ **5**
$C_{17}H_{17}NO$ 35.23 **4**
$C_{17}H_{17}NO_4S$ 13.19 **6**
$C_{17}H_{17}NO_5$ 13.20 **6**
$C_{17}H_{17}NO_5S$ 13.21 **6**

$C_{17}H_{17}NO_6S$ 13.22 **6**
$C_{17}H_{17}N_2^+$, Br^-, $0.5H_2O$ 16.22 **6**
$C_{17}H_{18}$ 26.4 **5**
$C_{17}H_{18}BrNO$ 40.22 **4**
$C_{17}H_{18}BrNO_2$ 13.19+ **3**
$C_{17}H_{18}CuN_2O_8P$ 82.57 **2**
$C_{17}H_{18}CuN_4$ 76.39 **6**
$C_{17}H_{18}I_2NO_4^+$, Cl^-, xH_2O 48.61 **7**
$C_{17}H_{18}NO_2^+$, Br^-, xH_2O 58.16 **6**
$C_{17}H_{18}NO_2^+$, Cl^-, xH_2O 58.17 **6**
$C_{17}H_{18}N_2O$ 58.8 **5**
$C_{17}H_{18}N_2O$ 58.9 **5**
$C_{17}H_{18}N_2O_5S$ 36.21 **3**
$C_{17}H_{18}N_2O_5S$ 50.7 **5**
$C_{17}H_{18}N_2O_5U$ 76.40 **6**
$C_{17}H_{18}N_4Ni$ 83.44 **6**
$C_{17}H_{18}OS$ 11.14 **5**
$C_{17}H_{18}O_2S$ 39.20 **6**
$C_{17}H_{18}O_4$ 31.18 **7**
$C_{17}H_{18}S$ 39.21 **6**
$C_{17}H_{19}BrN_4O_3$ 36.22 **3**
$C_{17}H_{19}BrO$ 52.2 **6**
$C_{17}H_{19}ClN_2S$ 41.40 **1**
$C_{17}H_{19}ClN_2S^+$, $C_{12}H_4N_4^-$ 41.28 **6**
$C_{17}H_{19}NO_2$ 21.13 **4**
$C_{17}H_{19}NO_3$ 37.16 **7**
$C_{17}H_{19}NO_3$ 58.9 **7**
$C_{17}H_{19}NO_3S_2$ 37.12 **3**
$C_{17}H_{19}NO_4$ 31.19 **7**
$C_{17}H_{19}NO_4$ 52.6 **7**
$C_{17}H_{19}NO_5$ 58.22 **1**
$C_{17}H_{19}N_3O_4$, H_2O 58.10 **5**
$C_{17}H_{19}N_3O_5$, $4H_2O$ 59.13 **4**
$C_{17}H_{19}N_5O_4$ 36.10 **6**
$C_{17}H_{20}BrClO_3$ 59.11 **6**
$C_{17}H_{20}ClN_2OP$ 64.21 **4**
$C_{17}H_{20}ClN_2S^+$, Cl^- 41.28 **4**
$C_{17}H_{20}FeO_5$ 75.17 **7**
$C_{17}H_{20}NO_3^+$, $Br-$, H_2O 58.23 **1**
$C_{17}H_{20}NO_3^+$, Cl^-, $3H_2O$ 58.18 **6**
$C_{17}H_{20}NO_3^+$, I^-, $2H_2O$ 58.14 **3**
$C_{17}H_{20}NO_4^+$, Cl^-, H_2O 48.62 **7**
$C_{17}H_{20}N_2O$ 8.18 **5**
$C_{17}H_{20}N_4O_2$ 58.19 **6**
$C_{17}H_{20}N_4O_6$, $C_5H_5N_5$, $3H_2O$ 60.1 **7**
$C_{17}H_{20}N_4O_6$, $C_{10}H_{12}BrN_5O_3$, $3H_2O$ 60.34 **4**
$C_{17}H_{20}N_8O_4$, H_2O 44.43 **3**
$C_{17}H_{20}O_2S$ 11.15 **6**
$C_{17}H_{20}O_3$ 24.9 **6**
$C_{17}H_{20}O_4S_2$ 45.51 **7**
$C_{17}H_{20}O_6$ 17.19 **5**
$C_{17}H_{21}BrO_2$ 53.17 **3**
$C_{17}H_{21}BrO_4$ 53.16 **7**
$C_{17}H_{21}BrO_5$ 53.22 **1**
$C_{17}H_{21}BrO_5$ 53.4 **4**
$C_{17}H_{21}BrO_5$ 53.14 **6**
$C_{17}H_{21}BrO_6$ 38.54 **7**
$C_{17}H_{21}F_6O_3Rh$, $0.5H_2O$ 71.48 **7**
$C_{17}H_{21}IO_2$ 52.29 **1**
$C_{17}H_{21}NO_4S$ 11.15 **5**
$C_{17}H_{21}N_2O_8P$, $7H_2O$ 48.63 **7**
$C_{17}H_{21}N_2S^+$, Br^- 41.40 **5**
$C_{17}H_{21}N_2S^+$, Cl^- 41.42 **1**
$C_{17}H_{21}N_4O_6^+$, Br^-, H_2O 59.15 **3**
$C_{17}H_{22}CuN_6S^{2+}$, $2ClO_4^-$ 76.41 **6**
$C_{17}H_{22}NO_2^+$, Br^- 51.2 **3**
$C_{17}H_{22}NO_2^+$, Br^- 58.20 **6**
$C_{17}H_{22}NO_4^+$, Br^- 3.25 **4**
$C_{17}H_{22}NO_4^+$, Br^- 3.26 **4**
$C_{17}H_{22}NO_4^+$, Br^-, $0.5H_2O$ 58.24 **1**
$C_{17}H_{22}NO_4^+$, Cl^- 58.25 **1**
$C_{17}H_{22}N_2O_3$ 51.1 **7**
$C_{17}H_{22}N_2O_6S$, H_2O 48.93 **1**
$C_{17}H_{22}N_2O_6S$, H_2O 48.52 **4**
$C_{17}H_{22}N_3^+$, ClO_4^- 16.60 **1**
$C_{17}H_{22}N_4O_6^{2+}$, $C_6H_6O_2$, $2Br^-$, $2H_2O$ 60.27 **6**
$C_{17}H_{22}O_5$ 53.17 **7**
$C_{17}H_{22}O_5$, H_2O 59.12 **6**
$C_{17}H_{22}P^+$, Br^- 64.48 **2**
$C_{17}H_{23}BrO_2$ 21.12 **5**
$C_{17}H_{23}BrO_3$ 59.16 **3**
$C_{17}H_{23}BrO_4S$ 31.30 **1**
$C_{17}H_{23}BrO_5$ 53.25 **1**
$C_{17}H_{23}CoN_3O_4$ 76.42 **6**
$C_{17}H_{23}CoN_6O_2$ 83.80 **7**
$C_{17}H_{23}CuN_3O_5$, $0.5C_4H_{10}O$, $4H_2O$ 82.58 **2**
$C_{17}H_{23}FeN_7S_2^+$, ClO_4^- 83.150 **2**
$C_{17}H_{23}NOPd$ 72.21 **5**
$C_{17}H_{23}NO_2$ 51.1 **4**
$C_{17}H_{23}NO_2$ 51.1 **5**
$C_{17}H_{23}N_2^+$, I^- 3.85 **1**
$C_{17}H_{23}N_7S_2Zn$, $0.5C_2H_4Cl_2$ 76.73 **7**
$C_{17}H_{23}O_7^-$, $C_4H_{12}N^+$ 38.55 **7**
$C_{17}H_{24}BrClO_2$ 53.18 **7**
$C_{17}H_{24}BrNO_2$ 52.7 **7**
$C_{17}H_{24}NO^+$, Br^- 35.30 **7**
$C_{17}H_{24}NO^+$, Br^-, H_2O 58.15 **3**
$C_{17}H_{24}NO_3$ 58.12 **5**
$C_{17}H_{24}NO_3^+$, Br^- 58.13 **5**
$C_{17}H_{24}N_2O_5S$ 48.64 **7**
$C_{17}H_{24}O_2$ 1.144 **1**
$C_{17}H_{24}O_2$ 57.1 **4**
$C_{17}H_{24}O_2$ 38.24 **5**
$C_{17}H_{24}O_2$ 21.9 **7**
$C_{17}H_{24}O_2Pd$ 75.23 **3**
$C_{17}H_{24}O_3S$ 27.15 **5**
$C_{17}H_{24}O_5$ 53.15 **6**
$C_{17}H_{24}O_7$ 38.56 **7**
$C_{17}H_{24}O_{12}$ 45.47 **6**
$C_{17}H_{25}Al_2N$ 68.15 **3**
$C_{17}H_{25}BrO_4$ 53.18 **3**
$C_{17}H_{25}ClO_5$ 31.25 **6**
$C_{17}H_{26}BrNO_5$ 45.23 **4**
$C_{17}H_{26}N^+$, Cl^- 33.47 **3**
$C_{17}H_{26}NO_2^+$, Br^- 58.21 **6**
$C_{17}H_{26}NO_3^+$, Br^- 58.22 **6**
$C_{17}H_{26}NO_4^+$, Br^-, $0.5H_2O$ 58.16 **3**
$C_{17}H_{26}O_2$ 21.8 **6**
$C_{17}H_{26}O_3P$ 52.4 **4**
$C_{17}H_{26}O_3S$ 21.15 **3**
$C_{17}H_{26}O_3S$ 21.14 **4**
$C_{17}H_{26}O_3S$ 21.13 **5**
$C_{17}H_{26}O_3S$ 21.14 **5**
$C_{17}H_{26}O_3S$ 21.15 **5**
$C_{17}H_{26}O_3S$ 21.10 **7**
$C_{17}H_{27}Br_2ClO_3$ 53.16 **6**
$C_{17}H_{27}Cl_2NPt$ 72.30 **3**
$C_{17}H_{27}NNiP_2S_4$ 83.87 **3**
$C_{17}H_{27}N_3O_5S_2$ 45.52 **7**
$C_{17}H_{27}N_3O_6$, H_2O 48.53 **4**
$C_{17}H_{27}N_5O_6S$ 47.36 **7**

$C_{17}H_{27}N_8O_5^+$, Br^- 50.18 **1**
$C_{17}H_{27}O_2PS$ 52.5 **4**
$C_{17}H_{28}NO^+$, Br^- 58.16 **4**
$C_{17}H_{28}N_4O_7$ 48.65 **7**
$C_{17}H_{28}O_4$ 53.19 **7**
$C_{17}H_{28}O_5$ 50.8 **7**
$C_{17}H_{28}O_{15}$, $3H_2O$ 45.48 **6**
$C_{17}H_{29}BrO$ 20.9 **4**
$C_{17}H_{31}N_5S_4$ 41.24 **7**
$C_{17}H_{35}CoN_5S_3$, $0.5C_4H_{10}O$ 83.88 **3**
$C_{17}H_{36}Si_8$ 63.7 **6**
$C_{17}H_{39}NNiOP_3^+$, BF_4^- 86.16 **3**
$C_{17}H_{49}N_3Si_7$ 63.7 **4**

C_{18}

$C_{18}F_{17}P$ 64.44 **7**
$C_{18}HCo_5O_{15}$ 71.39 **4**
$C_{18}H_8Fe_2N_2O_6$ 83.89 **3**
$C_{18}H_8O_5$ 38.57 **7**
$C_{18}H_8S_4$ 39.28 **7**
$C_{18}H_9BFN_6NiO_3P^+$, BF_4^- 83.68 **5**
$C_{18}H_9ClO_2$ 29.8 **1**
$C_{18}H_{10}$ 30.1 **1**
$C_{18}H_{10}Cl_2O_2$ 20.15 **3**
$C_{18}H_{10}Cr_2O_6$ 74.24+ **2**
$C_{18}H_{10}F_{12}Mo$ 71.49 **7**
$C_{18}H_{10}F_{12}Mo_2N_2O_8$ 81.50 **5**
$C_{18}H_{10}F_{12}O_8V_2$ 73.29 **3**
$C_{18}H_{10}Fe_2N_2O_6$ 83.69 **5**
$C_{18}H_{10}Fe_2O_6S_2$ 85.25 **5**
$C_{18}H_{10}Fe_2O_8Rh_2$ 73.18 **4**
$C_{18}H_{10}Mn_2N_2O_8Pt$ 83.153 **2**
$C_{18}H_{10}N_2$ 20.17 **6**
$C_{18}H_{12}$ 23.15 **1**
$C_{18}H_{12}$ 29.9 **1**
$C_{18}H_{12}$ 29.10 **1**
$C_{18}H_{12}$ 29.11 **1**
$C_{18}H_{12}$ 29.12 **1**
$C_{18}H_{12}$ 29.14 **1**
$C_{18}H_{12}$ 31.27 **5**
$C_{18}H_{12}$ 29.4 **6**
$C_{18}H_{12}$, $C_{10}H_2O_6$ 60.27 **7**
$C_{18}H_{12}$, $C_{12}H_4N_4$ 60.28 **7**
$C_{18}H_{12}AsO_6^-$, K^+ , $1.5H_2O$ 65.3 **5**
$C_{18}H_{12}As_2F_4Fe_3O_{10}$ 86.17 **3**
$C_{18}H_{12}As_2F_4O_{10}Ru_3$ 86.18 **3**
$C_{18}H_{12}AuCl_3N_2$ 83.154 **2**
$C_{18}H_{12}BCoFN_6O_3P^+$, BF_4^- , C_2H_3N 83.45 **6**
$C_{18}H_{12}BFFeN_6O_3P^+$, BF_4^- , CH_2Cl_2 83.70 **5**
$C_{18}H_{12}BFN_6NiO_3P^+$, BF_4^- 83.90 **3**
$C_{18}H_{12}BFN_6O_3PZn^+$, BF_4^- 83.46 **6**
$C_{18}H_{12}B_3NO_6$ 62.17 **3**
$C_{18}H_{12}Br_2NiO_2S_2$ 85.33 **7**
$C_{18}H_{12}ClNO_3$ 24.28 **1**
$C_{18}H_{12}Cl_2N_2O_4S$ 17.19 **6**
$C_{18}H_{12}CuN_2O_2$ 83.155+ **2**
$C_{18}H_{12}CuN_2O_2$, $2C_6H_2N_6O_6$ 60.76 **2**
$C_{18}H_{12}CuN_2O_2$, $2C_{10}H_2N_4$ 60.146+ **2**
$C_{18}H_{12}CuN_2O_2$, $C_{12}H_4N_4$ 60.126 **2**
$C_{18}H_{12}F_3N$ 16.14 **3**
$C_{18}H_{12}Fe_2O_7$ 75.17+ **4**
$C_{18}H_{12}HgN_2S_2$ 85.30 **6**
$C_{18}H_{12}I_3N$ 16.15 **3**
$C_{18}H_{12}MoN_2O_4$ 84.10 **4**
$C_{18}H_{12}MoS_6$ 85.31 **6**
$C_{18}H_{12}Mo_2O_6$ 75.25 **3**
$C_{18}H_{12}N_2O_2Pd$ 83.159 **2**
$C_{18}H_{12}N_2O_2Pd$, $C_6Cl_4O_2$ 60.70 **2**
$C_{18}H_{12}N_2O_2Pd$, $C_{10}H_2N_4$ 60.148 **2**
$C_{18}H_{12}N_2O_2Zn$, $2H_2O$ 83.160 **2**
$C_{18}H_{12}N_2PbS_2$ 69.28 **4**
$C_{18}H_{12}N_2PdS_2$ 85.34 **7**
$C_{18}H_{12}N_2PtS_2$ 85.35 **7**
$C_{18}H_{12}N_3O_6P_3$, $0.5C_6H_5Br$ 64.49 **2**
$C_{18}H_{12}N_3O_6P_3$, $0.5C_6H_6$ 64.50 **2**
$C_{18}H_{12}N_3O_{10}P$ 46.7 **3**
$C_{18}H_{12}N_4S_2$, C_3H_7NO 41.41 **5**
$C_{18}H_{12}N_5O_6$, C_6H_6 12.12 **1**
$C_{18}H_{12}NbS_6^-$, $C_{24}H_{20}As^+$ 85.33 **6**
$C_{18}H_{12}O_2S_2$ 39.31 **4**
$C_{18}H_{12}O_2S_2$ 39.29 **7**
$C_{18}H_{12}O_6P^-$, $C_6H_{16}N^+$ 46.11 **5**
$C_{18}H_{12}O_6Si^{2-}$, $2C_5H_6N^+$ 63.10 **3**
$C_{18}H_{12}S_2$ 39.36 **5**
$C_{18}H_{12}S_6Zr^{2-}$, $2C_4H_{12}N^+$ 85.34 **6**
$C_{18}H_{13}AgN_2O_2$, C_5H_5N 83.161 **2**
$C_{18}H_{13}As$ 65.29 **2**
$C_{18}H_{13}FeNO_3$ 72.22+ **5**
$C_{18}H_{13}MoN_3O_3$ 83.87 **4**
$C_{18}H_{13}NO_2S$ 39.22 **6**
$C_{18}H_{13}OS^+$, I^- 42.13 **7**
$C_{18}H_{13}O_4Si^-$, $C_4H_{12}N^+$ 3.32 **1**
$C_{18}H_{13}S_3$ 39.32 **4**
$C_{18}H_{14}$ 19.58 **1**
$C_{18}H_{14}$ 19.59 **1**
$C_{18}H_{14}$ 30.3 **1**
$C_{18}H_{14}$ 19.12+ **3**
$C_{18}H_{14}$ 19.12 **4**
$C_{18}H_{14}$, $C_{26}H_{28}N_6NiS_2$ 61.8 **3**
$C_{18}H_{14}BrOP$ 64.22 **4**
$C_{18}H_{14}BrP$ 64.51 **2**
$C_{18}H_{14}BrPS$ 64.25 **5**
$C_{18}H_{14}Br_2N_2$ 9.47 **1**
$C_{18}H_{14}Br_4$ 20.18 **6**
$C_{18}H_{14}Br_4$ 20.19 **6**
$C_{18}H_{14}ClOP$ 64.23 **4**
$C_{18}H_{14}ClPS$ 64.26 **5**
$C_{18}H_{14}CuO_8$ 81.84 **2**
$(C_{18}H_{14}CuO_8)_n$ 81.38 **6**
$C_{18}H_{14}F_6N_2O_4Sn$ 69.14 **7**
$C_{18}H_{14}F_{12}O_4Pd$ 71.50 **7**
$C_{18}H_{14}Fe$ 75.60 **2**
$C_{18}H_{14}Fe_2O_4$ 73.76 **2**
$C_{18}H_{14}Fe_2O_4$ 71.24 **3**
$C_{18}H_{14}Fe_2O_{10}$ 72.24 **5**
$(C_{18}H_{14}Mg)_n$ 67.9 **6**
$C_{18}H_{14}NO_5PS$ 64.45 **7**
$C_{18}H_{14}N_2OS^+$, HO^- 36.11 **6**
$C_{18}H_{14}N_4$ 9.12 **4**
$C_{18}H_{14}O_6$ 59.13 **6**
$C_{18}H_{14}Ru$ 73.78 **2**
$C_{18}H_{15}AsAuBr$ 86.23 **7**
$C_{18}H_{15}AsO$, H_2O 65.30 **2**
$C_{18}H_{15}AsO$, $C_6H_2Cl_4O_2$ 60.35 **4**
$C_{18}H_{15}AuCl_3P$ 86.21 **5**
$C_{18}H_{15}B$ 62.8 **7**

$C_{18}H_{15}Be_3N_3O_9$, H_2O	67.10	**6**
$C_{18}H_{15}Bi$	66.15+	**2**
$C_{18}H_{15}BiCl_2$	66.17	**2**
$C_{18}H_{15}BrO_5$	31.27	**3**
$C_{18}H_{15}Br_3PZn^-$, $C_6H_8N^+$	86.10	**4**
$C_{18}H_{15}ClP^+$, $C_8AuF_{12}S_4^-$	64.52	**2**
$C_{18}H_{15}ClSn$	69.29	**4**
$C_{18}H_{15}Cl_2Sb$	66.18	**2**
$C_{18}H_{15}Cl_3MoOPS$	85.36	**7**
$C_{18}H_{15}Cl_5OPU$	84.43	**7**
$C_{18}H_{15}FeN_6O_6$	84.44	**2**
$C_{18}H_{15}Ga$	68.16	**3**
$(C_{18}H_{15}HgN_2O_6P)_n$	86.24	**7**
$C_{18}H_{15}I_3NiP^-$, $C_{24}H_{20}As^+$	65.42	**2**
$C_{18}H_{15}In$	68.17	**3**
$C_{18}H_{15}NO_3Sn_2$	69.15	**7**
$C_{18}H_{15}N_3Si$	63.8	**6**
$C_{18}H_{15}N_4PS_3$	64.46	**7**
$C_{18}H_{15}N_6O_8U^-$, H_4N^+	84.45	**2**
$C_{18}H_{15}OP$	64.42	**3**
$C_{18}H_{15}OP$	64.29	**6**
$C_{18}H_{15}OP$, Cl_2OSe	60.48	**3**
$C_{18}H_{15}OP$, $C_6H_2N_6O_4$	60.75	**2**
$2C_{18}H_{15}OP$, $C_5H_8O_4$	60.2	**7**
$4C_{18}H_{15}OP$, $C_8H_4Mo_4N_4O_{16}$	60.149	**2**
$5C_{18}H_{15}OP$, I^- , Li^+	61.15	**2**
$C_{18}H_{15}O_3P_3S_3$	64.30	**6**
$C_{18}H_{15}O_3Rh_3$	73.79	**2**
$C_{18}H_{15}O_3Rh_3$	73.80+	**2**
$C_{18}H_{15}O_4P$	46.23+	**1**
$C_{18}H_{15}O_4PS$	64.47	**7**
$C_{18}H_{15}P$	64.53	**2**
$2C_{18}H_{15}PS$, $3I_2$	60.150	**2**
$C_{18}H_{16}$	31.28	**5**
$C_{18}H_{16}$	31.20	**7**
$C_{18}H_{16}Ag_2ClO_4^+$, ClO_4^-	75.19	**4**
$C_{18}H_{16}AsClO$	65.32	**2**
$C_{18}H_{16}AsO^+$, Br^-	65.31	**2**
$2C_{18}H_{16}AsO^+$, $Cl_{18}Nb_6^{2-}$	65.4	**6**
$C_{18}H_{16}BrNO_2$	31.31	**1**
$C_{18}H_{16}CdN_4O_7$	83.47	**6**
$C_{18}H_{16}Cl_2Co_2O_4Sn$	75.20	**4**
$C_{18}H_{16}Cl_2N_2O_2$	31.23	**4**
$(C_{18}H_{16}CoN_3O_2)_n$, nCH_4O	71.25	**3**
$C_{18}H_{16}Cu_3N_6O_5S$, $16.3H_2O$	83.88	**4**
$C_{18}H_{16}ErN_3O_8$	81.39	**6**
$C_{18}H_{16}FeO$	73.29	**7**
$C_{18}H_{16}Fe_2O_6$	71.40	**6**
$(C_{18}H_{16}LaN_3O_8)_n$	81.51	**5**
$C_{18}H_{16}Mn_2O_4$	72.51	**2**
$C_{18}H_{16}N_2O_4Zn$	83.162	**2**
$C_{18}H_{16}N_4O_2$	32.22	**4**
$C_{18}H_{16}OP^+$, $C_6HN_6O_4^-$	64.43	**3**
$C_{18}H_{16}O_2$	31.29	**5**
$C_{18}H_{16}O_3$	38.22	**4**
$C_{18}H_{16}O_7$	59.6	**7**
$C_{18}H_{17}BrN_2O_2$, $2C_2H_6O$	36.22	**5**
$C_{18}H_{17}BrO_3$	59.38	**1**
$C_{18}H_{17}Cl$	26.5	**5**
$C_{18}H_{17}ClN_2$	20.10	**4**
$C_{18}H_{17}ClN_2OS$	41.42	**5**
$C_{18}H_{17}ClS$	28.5	**5**
$(C_{18}H_{17}MnN_2O_4)_n$	78.5	**6**
$C_{18}H_{17}NO_3$	24.29	**1**
$C_{18}H_{17}NO_4S$	39.37	**5**
$C_{18}H_{17}NO_5$	13.23	**6**
$C_{18}H_{17}NS$, $CHCl_3$	39.26	**5**
$C_{18}H_{17}N_3O_4$	32.29	**7**
$C_{18}H_{18}$	23.16	**1**
$C_{18}H_{18}$	28.23+	**1**
$C_{18}H_{18}$	31.21	**7**
$C_{18}H_{18}BN_8Ru^+$, F_6P^-	74.7	**6**
$C_{18}H_{18}BrNO$	40.31	**5**
$C_{18}H_{18}Br_2$	31.33	**1**
$C_{18}H_{18}ClNO$	32.33	**3**
$C_{18}H_{18}ClNS$	39.23	**6**
$C_{18}H_{18}ClN_3O$	35.27	**5**
$C_{18}H_{18}ClN_3O$	40.32	**5**
$C_{18}H_{18}ClN_3O_2$	40.33	**5**
$C_{18}H_{18}Cl_2$	5.34	**1**
$C_{18}H_{18}Cl_2Cu_2N_2O_2$	78.6	**4**
$C_{18}H_{18}Cl_2Cu_2N_2O_2$	76.43	**6**
$C_{18}H_{18}Cl_2Pd_2$	72.32	**3**
$C_{18}H_{18}CoN_2O_2$	78.5	**5**
$C_{18}H_{18}CoN_2O_2$	78.6	**5**
$C_{18}H_{18}Co_2$	75.61	**2**
$C_{18}H_{18}F_{18}NiO_6P_2$	71.41	**5**
$C_{18}H_{18}FeN_6O_6^{2+}$, $2ClO_4^-$	83.81	**7**
$C_{18}H_{18}Fe_2O_7$	75.27	**3**
$C_{18}H_{18}I_2N_2NiO_2$	78.6	**3**
$C_{18}H_{18}N_2^{2+}$, $2C_{12}H_4N_4^-$, $2C_{12}H_4N_4$	60.29	**7**
$C_{18}H_{18}N_2O_2$	32.30	**7**
$C_{18}H_{18}N_2O_3$	32.23	**4**
$C_{18}H_{18}N_2O_5$	13.20	**4**
$C_{18}H_{18}N_2O_5$	13.25	**7**
$C_{18}H_{18}N_2PS_2^+$, ClO_4^-	64.54	**2**
$C_{18}H_{18}N_2S_2$	41.16	**3**
$C_{18}H_{18}N_4Ni$	83.82	**7**
$C_{18}H_{18}N_6$	7.32	**1**
$C_{18}H_{18}N_6NiO_4$	83.48	**6**
$C_{18}H_{18}N_6O_9$	43.7	**7**
$C_{18}H_{18}N_6S_2$	41.44	**1**
$C_{18}H_{18}NiO_4$	77.15	**6**
$C_{18}H_{18}O$	21.17	**3**
$C_{18}H_{18}O_2$	29.15	**1**
$C_{18}H_{18}O_2$	17.9	**4**
$C_{18}H_{18}O_2$	31.22	**7**
$C_{18}H_{18}S_2$	11.15	**4**
$C_{18}H_{18}S_2$	11.20	**7**
$C_{18}H_{18}S_3$	39.33	**4**
$C_{18}H_{18}Si$, $C_{14}H_{10}$	60.36	**4**
$C_{18}H_{19}BrN_4O_4$	36.12+	**6**
$C_{18}H_{19}BrO_3$	31.28	**3**
$C_{18}H_{19}ClN_4$	36.23	**5**
$C_{18}H_{19}ClN_4$	36.24	**5**
$C_{18}H_{19}CoN_2O_2$	71.40	**4**
$C_{18}H_{19}Fe_2NO_3$	71.41	**6**
$C_{18}H_{19}Fe_2NO_3$	71.51	**7**
$(C_{18}H_{19}HoN_3O_8^{3+})_n$, $nC_6CrN_6S_6^{3-}$, $2nH_2O$	81.52	**5**
$C_{18}H_{19}IO_8$	59.17	**3**
$C_{18}H_{19}NO_2$	32.34	**6**
$C_{18}H_{19}NO_4$	48.74	**5**
$C_{18}H_{19}N_2^+$, I^-	35.44	**1**
$C_{18}H_{19}N_3O_4U$	78.7	**3**
$C_{18}H_{19}N_3O_4U$	78.7	**5**
$C_{18}H_{19}N_3O_6S$, $C_2H_4O_2$, H_2O	50.9	**3**
$C_{18}H_{19}N_6NiO_3^{3+}$, $3Br^-$, H_2O	83.92	**3**
$C_{18}H_{19}O_6Pb^-$, $C_4H_{12}N^+$	69.30	**4**
$C_{18}H_{19}Ti$	71.52	**7**
$C_{18}H_{20}$	29.16	**1**
$C_{18}H_{20}$	31.34	**1**

$C_{18}H_{20}$ 31.35 **1**
$C_{18}H_{20}$ 26.7 **4**
$C_{18}H_{20}$ 28.20 **6**
$C_{18}H_{20}$, C_6N_4 60.37 **4**
$2C_{18}H_{20}$, C_6N_4 60.35 **5**
$C_{18}H_{20}B_2CoN_{12}$ 83.93 **3**
$C_{18}H_{20}B_2Cu_2N_{12}$ 83.83 **7**
$C_{18}H_{20}BrNO_{10}S$ 45.44 **3**
$C_{18}H_{20}Br_2Cu_2N_2O_2$ 78.6 **6**
$C_{18}H_{20}Br_2N_2$ 33.40 **4**
$C_{18}H_{20}ClNO_2$ 31.24 **4**
$C_{18}H_{20}ClN_3O_2$ 36.21 **4**
$C_{18}H_{20}ClN_3O_6S_2$, $0.65CH_2Br_2$ 58.27 **1**
$C_{18}H_{20}ClN_3O_6S_4$, $C_4H_{10}O$ 58.23 **6**
$C_{18}H_{20}Cl_2Cu_2N_2O_2$ 78.7 **4**
$C_{18}H_{20}Cl_2N_2$ 9.17 **3**
$C_{18}H_{20}Cl_2N_2NiO_2Sn$ 78.4 **7**
$C_{18}H_{20}Cl_4$ 31.31 **5**
$C_{18}H_{20}CoN_2O_5$ 78.5 **7**
$C_{18}H_{20}CuN_2O_2$ 78.27 **2**
$C_{18}H_{20}CuN_2O_2$ 78.28 **2**
$C_{18}H_{20}CuN_2O_2$ 78.8 **3**
$C_{18}H_{20}CuN_2O_2$ 78.7+ **6**
$C_{18}H_{20}CuN_2O_3$ 78.6 **7**
$C_{18}H_{20}CuN_2O_4$ 78.29 **2**
$C_{18}H_{20}CuN_2O_4$ 78.9 **3**
$C_{18}H_{20}CuN_2O_4$ 82.14 **4**
$C_{18}H_{20}CuN_2O_6$ 82.23 **5**
$C_{18}H_{20}CuN_2O_8$ 82.21 **6**
$C_{18}H_{20}CuN_3O_3$ 78.8 **5**
$C_{18}H_{20}F_3N_2S^+$, Cl^- 41.25 **7**
$C_{18}H_{20}F_{12}N_2Ni_2S_8$ 80.23 **5**
$C_{18}H_{20}Fe$ 73.41 **6**
$C_{18}H_{20}Fe_2O_4S_2$ 71.53 **7**
$C_{18}H_{20}I_2NO_4^+$, Cl^- , $3H_2O$ 48.66 **7**
$C_{18}H_{20}NO_3^+$, Br^- 58.28 **1**
$C_{18}H_{20}N_2NiO_2$ 78.30 **2**
$C_{18}H_{20}N_2O_2Pd$ 78.31 **2**
$C_{18}H_{20}N_2O_2PdS_2$ 83.89 **4**
$C_{18}H_{20}N_2O_2Sn$ 69.19 **5**
$C_{18}H_{20}N_2O_4Pd$, $0.5H_2O$ 82.29 **3**
$C_{18}H_{20}N_2O_6Pd$, $0.5H_2O$ 82.22 **6**
$C_{18}H_{20}N_2S_2$ 41.43 **5**
$C_{18}H_{20}N_4$ 31.29 **3**
$3C_{18}H_{20}OS$, C_2H_6O 61.16 **2**
$6C_{18}H_{20}OS$, $C_9H_{16}O$ 61.1 **4**
$C_{18}H_{20}O_2$ 17.47 **1**
$C_{18}H_{20}O_2$ 17.7+ **3**
$C_{18}H_{20}O_2$, H_2O 17.30 **7**
$C_{18}H_{20}O_2$, H_2O , C_2H_6O 17.31 **7**
$C_{18}H_{20}O_2$, $0.167CHCl_3$ 61.6 **3**
$C_{18}H_{20}O_2$, CH_4O , H_2O 17.20 **6**
$C_{18}H_{20}O_2$, C_2H_6O 17.21 **6**
$C_{18}H_{20}O_2$, $0.33C_2H_6O$ 61.7 **3**
$C_{18}H_{20}O_2$, C_2H_6OS 17.22 **6**
$6C_{18}H_{20}O_2$, $C_7H_{16}O$ 61.2 **4**
$C_{18}H_{20}O_4$ 38.58 **7**
$C_{18}H_{20}O_6$, $2C_3H_7NO$ 31.25 **4**
$(C_{18}H_{20}O_7Zn)_n$ 81.40 **6**
$C_{18}H_{20}S_2$ 39.34 **4**
$C_{18}H_{21}BrN_2OPd$ 71.26 **3**
$C_{18}H_{21}BrO_2$ 51.2+ **1**
$C_{18}H_{21}BrO_7S$ 45.45 **3**
$C_{18}H_{21}ClN_4O_2$ 36.21 **7**
$C_{18}H_{21}ClN_4Pd$ 71.41 **4**
$C_{18}H_{21}CuN_3^+$, ClO_4^- 83.90 **4**
$C_{18}H_{21}NO$, H_2O 12.14 **4**
$C_{18}H_{21}N_2^+$, Br^- 36.14 **6**
$C_{18}H_{21}N_2O_4S^+$, Cl^- 40.34 **5**
$C_{18}H_{21}N_3O$ 36.15 **6**
$C_{18}H_{21}Nd$ 73.30 **7**
$C_{18}H_{22}BrN_3$ 36.22 **4**
$C_{18}H_{22}Br_2O_2$ 51.2+ **4**
$C_{18}H_{22}Cl_2Mo_2$ 72.14 **7**
$(C_{18}H_{22}Cl_2N_2O_2Sn)_n$ 69.13 **6**
$C_{18}H_{22}Cl_2O_2P_2Rh_2$ 86.25 **7**
$C_{18}H_{22}Co_2O_4$ 72.52 **2**
$C_{18}H_{22}CrN_4O_2^+$, I^- 78.8 **4**
$C_{18}H_{22}Cu_2N_2O_8$ 81.85+ **2**
$C_{18}H_{22}FeNO_3^+$, Cl^- 72.33 **3**
$C_{18}H_{22}Fe_2Ge_2O_5$ 73.31 **7**
$C_{18}H_{22}HgN_2$ 71.41 **2**
$C_{18}H_{22}NO_3^+$, Br^- , $2H_2O$ 58.29 **1**
$C_{18}H_{22}NO_3^+$, Cl^- 58.24 **6**
$C_{18}H_{22}NO_3S^+$, Br^- 39.30 **7**
$C_{18}H_{22}NS^+$, Cl^- 39.35 **4**
$C_{18}H_{22}N_2O_2$ 37.14 **4**
$C_{18}H_{22}N_2O_2$ 40.14+ **7**
$C_{18}H_{22}N_2O_2Si_2$ 63.11 **3**
$C_{18}H_{22}N_2S$ 41.17 **3**
$C_{18}H_{22}N_4O_2$ 36.23 **4**
$C_{18}H_{22}N_4O_4S_6Te$ 70.15 **4**
$C_{18}H_{22}N_4O_7$ 48.79 **6**
$C_{18}H_{22}N_4O_{10}S_2$ 47.30 **1**
$C_{18}H_{22}N_4O_{10}S_2$ 47.34 **3**
$C_{18}H_{22}Nb_3O_{10}$ 73.32 **7**
$C_{18}H_{22}Ni$ 75.62 **2**
$C_{18}H_{22}NiO_2$ 71.54 **7**
$C_{18}H_{22}O_2$ 51.6 **4**
$C_{18}H_{22}O_2$ 51.2+ **5**
$C_{18}H_{22}O_2$ 17.23 **6**
$C_{18}H_{22}O_2$ 51.2 **7**
$C_{18}H_{22}O_2S$ 11.27 **3**
$C_{18}H_{23}BrO_2$, CH_4O 51.3 **1**
$C_{18}H_{23}BrO_3$ 52.5 **3**
$C_{18}H_{23}CoN_2O_2$, H_2O 76.81 **3**
$C_{18}H_{23}CoN_6O_4$, $2H_2O$ 83.84 **7**
$C_{18}H_{23}CrNO_5Sn$ 69.14 **6**
$C_{18}H_{23}CuF_{15}N_4O_4$ 76.41 **4**
$C_{18}H_{23}NO_2$ 37.17 **7**
$C_{18}H_{23}N_2O^+$, Br^- 58.17 **3**
$C_{18}H_{23}N_2OS^+$, $C_4H_3O_4^-$ 41.44 **5**
$C_{18}H_{23}N_2O_2^+$, $C_{12}H_8N_2O_8P^-$, H_2O 16.17 **4**
$C_{18}H_{23}N_2S^+$, Cl^- 41.45 **1**
$C_{18}H_{23}N_2S^+$, $CH_3O_4S^-$ 41.45 **5**
$C_{18}H_{24}$ 24.30 **1**
$C_{18}H_{24}$ 31.30 **3**
$C_{18}H_{24}B^-$, Li^+ , $2C_4H_{10}O_2$ 62.10 **6**
$C_{18}H_{24}Br_2O_{12}$ 38.33 **6**
$C_{18}H_{24}CdN_{12}^{2+}$, $2NO_3^-$ 83.91 **4**
$C_{18}H_{24}CdN_{12}^{2+}$, HO^- , NO_3^- , $4H_2O$ 83.92 **4**
$C_{18}H_{24}CdN_{12}^{2+}$, CO_3^{2-} , $5H_2O$ 83.93 **4**
$C_{18}H_{24}Cl_2N_6NiP_2$ 86.21 **2**
$C_{18}H_{24}CoN_{12}^{2+}$, $2NO_3^-$ 83.94+ **4**
$C_{18}H_{24}CoN_{12}^{2+}$, CO_3^{2-} , $5H_2O$ 83.96 **4**
$C_{18}H_{24}CoN_{12}^{2+}$, $2C_2H_3O_2^-$, H_2O 83.49 **6**
$C_{18}H_{24}CuN_2O_4$, $3H_2O$ 81.48 **4**
$C_{18}H_{24}CuN_{12}^{2+}$, $2NO_3^-$ 83.85 **7**
$C_{18}H_{24}Cu_2N_4O_6$, $2H_2O$ 84.44 **7**
$C_{18}H_{24}N^+$, I^- 3.35 **6**
$C_{18}H_{24}NO_2^+$, Br^- 51.7 **4**

$C_{18}H_{24}NO_3^{+}$, I^{-}	58.30 **1**
$C_{18}H_{24}NO_4^{+}$, Cl^{-} , $2H_2O$	58.25 **6**
$C_{18}H_{24}NO_5^{+}$, Br^{-}	58.19 **3**
$C_{18}H_{24}N_2NiO_4P_2S_4$	83.94 **3**
$C_{18}H_{24}N_2O_2Pt$	83.164 **2**
$C_{18}H_{24}N_4O_2$, $3H_2O$	36.24 **4**
$C_{18}H_{24}N_{12}Ni^{2+}$, $2NO_3^{-}$	83.165 **2**
$C_{18}H_{24}N_{12}Ni^{2+}$, $2NO_3^{-}$	83.95 **3**
$C_{18}H_{24}N_{12}Zn^{2+}$, $2Cl^{-}$, $4H_2O$	83.166 **2**
$C_{18}H_{24}NiO_2$	75.63 **2**
$C_{18}H_{24}O_2$	31.23 **7**
$C_{18}H_{24}O_2$	51.3 **7**
$C_{18}H_{24}O_2$, $0.5H_2O$	51.8 **4**
$C_{18}H_{24}O_2$, CH_4N_2O	60.38 **4**
$C_{18}H_{24}O_2$, C_3H_8O	51.9 **4**
$C_{18}H_{24}O_3$	51.5 **1**
$C_{18}H_{24}O_3$	29.2 **3**
$C_{18}H_{24}O_3$	51.10 **4**
$C_{18}H_{24}O_5Ru_2Si_2$	75.18 **7**
$C_{18}H_{25}AlN_2O_2$	68.22 **5**
$C_{18}H_{25}BMoN_6O_2$	72.15 **7**
$C_{18}H_{25}Cl_2NO_3$	40.22 **6**
$C_{18}H_{25}NO$	58.31 **1**
$C_{18}H_{25}NOP^{+}$, Cl_2Cu^{-}	64.31 **6**
$C_{18}H_{26}As_2Cl_3O_2Rh$	86.19 **3**
$C_{18}H_{26}As_2I_2PdS_2$	86.20 **3**
$C_{18}H_{26}BrNO_6$, $0.5C_2H_6O$	58.32 **1**
$C_{18}H_{26}ClN_3$	35.28 **5**
$C_{18}H_{26}Cl_2N_2Zn$	83.86 **7**
$C_{18}H_{26}CoN_3O_2$	71.42 **5**
$C_{18}H_{26}CoN_4$	83.98 **4**
$C_{18}H_{26}CuN_4$	83.71 + **5**
$C_{18}H_{26}CuN_8O_6$	81.49 **4**
$C_{18}H_{26}FeN_8O_6$, $2H_2O$	83.167 **2**
$C_{18}H_{26}Fe_2N_2S_2^{2+}$, $2F_6P^{-}$	73.31 **5**
$C_{18}H_{26}MoRhS_2^{+}$, F_6P^{-}	72.16 **7**
$C_{18}H_{26}NO^{+}$, Br^{-} , H_2O	58.33 **1**
$C_{18}H_{26}NOP$	64.28 **5**
$C_{18}H_{26}NPS$	64.29 **5**
$C_{18}H_{26}N_4Ni$	83.73 **5**
$C_{18}H_{26}O_2$	51.7 **3**
$C_{18}H_{26}O_2$	51.5 **5**
$C_{18}H_{27}BrN_3^{+}$, Br^{-} , $3H_2O$	32.31 **7**
$C_{18}H_{27}BrO_3$	31.26 **4**
$C_{18}H_{27}N$	33.51 **6**
$C_{18}H_{27}N_5Ni$	71.42 **6**
$C_{18}H_{27}O_2S^{+}$, I^{-}	21.11 **7**
$C_{18}H_{28}ClN_3^{2+}$, O_4S^{2-} , H_2O	35.29 **5**
$C_{18}H_{28}ClN_3^{2+}$, $2H_2O_4P^{-}$, H_2O	35.21 **3**
$C_{18}H_{28}Cl_2FeMoS_2$	73.19 **4**
$C_{18}H_{28}CoN_2O_4$, H_2O	78.9 **5**
$C_{18}H_{28}Cr_2N_4O_{14}$, $6H_2O$	76.74 **7**
$C_{18}H_{28}Mo_2N_2S_2$	73.33 **7**
$C_{18}H_{28}NO_2^{+}$, Cl^{-}	20.17 **5**
$C_{18}H_{28}NO_3^{+}$, I^{-}	58.20 **3**
$2C_{18}H_{28}NO_6^{+}$, Cl_6Pt^{2-} , $2H_2O$	58.34 **1**
$C_{18}H_{28}N_3^{+}$, Br^{-} , $2H_2O$	32.32 **7**
$C_{18}H_{28}N_{10}NiS_2$	83.99 **4**
$C_{18}H_{28}O_2$	28.6 **7**
$C_{18}H_{28}O_2$, H_2O	51.6 **5**
$C_{18}H_{28}O_4Si_4$	63.9 **6**
$C_{18}H_{29}F_{12}N_3O_2Pd$	71.55 **7**
$C_{18}H_{30}$	29.19 **1**
$C_{18}H_{30}$	29.20 **1**
$C_{18}H_{30}$	29.17 + **1**
$C_{18}H_{30}$	52.30 **1**
$C_{18}H_{30}$, $0.5CHCl_3$	61.17 **2**
$C_{18}H_{30}$, $0.39C_6H_{12}$	61.18 **2**
$C_{18}H_{30}$, $0.225C_7H_{16}$	61.19 **2**
$C_{18}H_{30}Ag_2^{2+}$, $2ClO_4^{-}$	72.53 **2**
$C_{18}H_{30}Al_2$	68.18 **3**
$C_{18}H_{30}BrNO$	21.31 **1**
$C_{18}H_{30}Br_2O$	54.1 **1**
$C_{18}H_{30}Ca_2O_{12}Sr$	67.18 + **2**
$C_{18}H_{30}ClCo_3F_9O_{16}$	81.50 **4**
$C_{18}H_{30}Cl_2$	5.35 **1**
$C_{18}H_{30}Cl_4O_2Sn_2$	69.34 **2**
$C_{18}H_{30}Co_2HgO_6P_2$	86.23 **2**
$C_{18}H_{30}Cu_2N_4O_2^{2+}$, $2ClO_4^{-}$	83.75 **5**
$C_{18}H_{30}Cu_2N_4O_2^{2+}$, $2ClO_4^{-}$	83.50 **6**
$C_{18}H_{30}Cu_2N_4O_2^{2+}$, $2ClO_4^{-}$	83.51 **6**
$C_{18}H_{30}Cu_2N_4O_{14}$, $5H_2O$	76.44 **6**
$C_{18}H_{30}FeN_6^{2+}$, $2F_6P^{-}$	83.52 **6**
$C_{18}H_{30}FeN_{12}^{2+}$, Cl_4Zn^{2-}	83.53 **6**
$C_{18}H_{30}InN_3S_6$	68.23 **5**
$C_{18}H_{30}MnN_3S_6^{+}$, ClO_4^{-} , $CHCl_3$	80.12 **6**
$C_{18}H_{30}N^{+}$, I^{-}	58.35 **1**
$C_{18}H_{30}NO_2^{+}$, $CH_3O_3S^{-}$	20.18 **7**
$C_{18}H_{30}NO_2S_2^{+}$, BF_4^{-}	39.45 **3**
$C_{18}H_{30}NO_3S^{+}$, Br^{-}	39.31 **7**
$C_{18}H_{30}NO_4^{+}$, Br^{-}	17.32 **7**
$C_{18}H_{30}N_2NiO_4P_2S_4$	85.81 **2**
$C_{18}H_{30}N_2O_4$	20.19 **7**
$C_{18}H_{30}N_3^{3+}$, $3Br^{-}$, $2H_2O$	36.16 **6**
$C_{18}H_{30}O_2$	20.20 **7**
$C_{18}H_{30}O_7S$	59.15 **5**
$C_{18}H_{31}Cl_4P_2RhSi$, $xHCl_3Si$	86.21 **3**
$C_{18}H_{31}NiP$	71.56 **7**
$C_{18}H_{32}$	21.15 **4**
$C_{18}H_{32}CoN_6^{+}$, ClO_4^{-}	83.100 **3**
$C_{18}H_{32}NO_3^{+}$, Br^{-}	53.19 **3**
$C_{18}H_{32}N_2O_3$	40.16 **7**
$C_{18}H_{32}O_2$	28.8 **3**
$C_{18}H_{32}O_{16}$	45.24 **4**
$C_{18}H_{32}O_{16}$, H_2O	45.49 **6**
$C_{18}H_{32}O_{16}$, $2H_2O$	45.25 **4**
$C_{18}H_{32}O_{16}$, $5H_2O$	45.47 **3**
$C_{18}H_{34}ClFN_2P_2Pt$	86.26 **7**
$C_{18}H_{34}ClN_2O_5S^{+}$, Cl^{-} , H_2O	50.10 **3**
$C_{18}H_{34}CoN_4O_2S_2$	80.13 **6**
$C_{18}H_{34}Cu_2N_8O_4S_2$	84.45 **7**
$C_{18}H_{34}O_2$	1.145 **1**
$C_{18}H_{34}O_2$	1.146 **1**
$C_{18}H_{34}O_3$	1.35 **5**
$C_{18}H_{34}O_8$	1.35 **3**
$C_{18}H_{35}BrO_2$	1.147 **1**
$C_{18}H_{35}ClFN_2P_2Pt^{+}$, ClO_4^{-}	86.27 **7**
$C_{18}H_{35}N_2O_6S^{+}$, Cl^{-} , H_2O	50.19 **1**
$C_{18}H_{36}$	21.32 **1**
$C_{18}H_{36}AuN_2S_4^{+}$, $AgBr_2^{-}$	80.11 **4**
$C_{18}H_{36}AuN_2S_4^{+}$, $AuBr_2^{-}$	80.34 **2**
$C_{18}H_{36}AuN_2S_4^{+}$, Br^{-}	80.11 **3**
$C_{18}H_{36}AuN_2S_4^{+}$, $C_8AuN_4S_4^{-}$	80.12 **4**
$C_{18}H_{36}Au_2N_2S_4$	80.16 **7**
$C_{18}H_{36}CdN_{12}O_6^{2+}$, $2ClO_4^{-}$	84.15 **5**
$C_{18}H_{36}Cl_2N_2O_4PdS_2$	85.35 **6**
$C_{18}H_{36}CsN_2O_6^{+}$, CNS^{-} , H_2O	67.12 **5**
$C_{18}H_{36}Cs_2N_2S_4$	67.20 **3**
$C_{18}H_{36}CuN_2S_4$, $2C_{18}H_{36}CuN_2S_4^{+}$, $Br_6Cd_2^{2-}$	80.14 **6**
$C_{18}H_{36}CuN_2S_4^{+}$, I_3^{-}	80.24 **5**
$C_{18}H_{36}Cu_2N_{12}S_6^{2+}$, $2BF_4^{-}$	79.24 **3**

$C_{18}H_{36}I_2N_2PtS_4$	80.17	**7**
$C_{18}H_{36}KN_2O_6^{+}$, I^{-}	67.13	**5**
$C_{18}H_{36}N_2NaO_6^{+}$, I^{-}	67.14	**5**
$C_{18}H_{36}N_2NiSe_4$	80.15	**6**
$C_{18}H_{36}N_2O_6$, $4Ag^{+}$, $4CNS^{-}$	40.17	**7**
$C_{18}H_{36}N_2O_6$, Na^{+} , Na^{-}	40.18	**7**
$C_{18}H_{36}N_2O_6$, Pb^{2+} , $2CNS^{-}$	40.19	**7**
$2C_{18}H_{36}N_2O_6$, $4Ag^{+}$, $4CNS^{-}$	40.20	**7**
$C_{18}H_{36}N_2O_6Rb^{+}$, CNS^{-} , H_2O	67.15	**5**
$C_{18}H_{36}N_2O_6Tl^{+}$, CHO_2^{-} , H_2O	68.24	**5**
$C_{18}H_{36}N_4Ni^{2+}$, $2ClO_4^{-}$	83.101	**4**
$C_{18}H_{36}O_2$	1.148	**1**
$C_{18}H_{36}O_2$	1.36	**3**
$C_{18}H_{36}O_2$	1.36	**5**
$C_{18}H_{36}O_6Pt_2$	77.36	**2**
$C_{18}H_{36}Pd_2S_8$	85.36	**6**
$C_{18}H_{37}ClFN_2P_2Pt^{+}$, BF_4^{-}	86.28	**7**
$C_{18}H_{38}$	5.17	**5**
$C_{18}H_{38}BaN_2O_7^{2+}$, $2CNS^{-}$	67.12	**6**
$C_{18}H_{38}CaN_2O_7^{2+}$, $2Br^{-}$, $2H_2O$	67.13	**6**
$C_{18}H_{38}N_4O_{11}^{2+}$, O_4S^{2-} , H_2O	50.20	**1**
$C_{18}H_{38}N_4O_{11}^{2+}$, O_4Se^{2-} , H_2O	50.21	**1**
$C_{18}H_{39}BrCoNS_3^{+}$, F_6P^{-}	85.37	**7**
$C_{18}H_{39}N_4NiO_2^{+}$, ClO_4^{-}	81.48	**3**
$C_{18}H_{39}O_2P$	64.48	**7**
$C_{18}H_{40}Fe_2N_4S_4$	85.37	**6**
$C_{18}H_{40}N_2O_4Pt_2$	77.37	**2**
$C_{18}H_{40}N_6Ni_2O_2^{2+}$, $2Cl^{-}$, $5H_2O$	76.75	**7**
$C_{18}H_{42}BiO_6P_3S_6$	66.5	**6**
$C_{18}H_{42}Cl_4Mo_4O_{12}$	84.16	**5**
$C_{18}H_{42}Cl_4P_2Pt_2$	86.24	**2**
$C_{18}H_{42}CoN_6^{3+}$, $3Cl^{-}$, H_2O	76.48	**5**
$C_{18}H_{42}CoN_6^{3+}$, $3Cl^{-}$, $5H_2O$	76.82	**3**
$(C_{18}H_{42}CrLi_3O_6)_n$	71.57	**7**
$C_{18}H_{42}CrN_3$	83.102	**3**
$C_{18}H_{42}Cu_2N_4O_2^{2+}$, $2ClO_4^{-}$, H_2O	76.45	**6**
$C_{18}H_{42}N_6Rh^{3+}$, $3NO_3^{-}$, $3H_2O$	76.76	**7**
$C_{18}H_{48}Al_6N_6$	68.18	**7**
$C_{18}H_{48}Cu_2N_{12}S_6^{2+}$, $2BF_4^{-}$	79.23	**7**
$C_{18}H_{48}Fe_3N_6O_{16}^{7+}$, $7ClO_4^{-}$	82.23	**6**
$C_{18}H_{51}Al_7N_6$	68.19	**7**
$C_{18}H_{52}N_4O_{12}P_4Ru^{2+}$, $2C_{24}H_{20}B^{-}$	86.29	**7**
$C_{18}H_{54}AlN_3Si_6$	63.12	**3**
$C_{18}H_{54}B_2N_4Si_6$	63.23	**2**
$C_{18}H_{54}ClO_6P_6U^{3+}$, $3Cl^{-}$	84.46	**7**
$C_{18}H_{54}Cl_3N_9O_3P_3Pr$	84.20	**6**
$C_{18}H_{54}EuN_3Si_6$	83.54	**6**
$C_{18}H_{54}FeN_3Si_6$	63.6	**5**
$C_{18}H_{54}Li_3N_3Si_6$	63.13	**3**
$C_{18}H_{54}N_3ScSi_6$	83.55	**6**

C_{19}

$C_{19}H_9CoF_6N_2O_3$	71.44	**5**
$C_{19}H_{10}Br_2O_6$	38.23	**4**
$C_{19}H_{10}Fe_2N_2O_7$	83.168+	**2**
$C_{19}H_{10}O_{11}Ru_4$	75.30	**5**
$C_{19}H_{11}BrN_4O_7$	17.12	**4**
$C_{19}H_{11}BrN_4O_7$	17.13	**4**
$C_{19}H_{12}Br_2N_2$	35.45	**1**
$C_{19}H_{12}Br_2O$	19.15	**3**
$C_{19}H_{12}FeO_3$, $0.5C_{16}H_{12}$	72.25	**5**
$C_{19}H_{12}N_3O_6$	12.13	**1**
$C_{19}H_{12}O_2$	29.21	**1**
$C_{19}H_{12}O_2$	29.22	**1**
$C_{19}H_{12}O_3$	38.34	**6**
$C_{19}H_{12}O_6$	59.39	**1**
$C_{19}H_{13}BrN_4O_4$	9.15	**5**
$C_{19}H_{13}BrOS_4$	39.46	**3**
$C_{19}H_{13}Cl$	29.4	**7**
$C_{19}H_{13}NO_4$	20.18	**5**
$C_{19}H_{14}$	29.5	**7**
$C_{19}H_{14}$	29.6	**7**
$C_{19}H_{14}BrF_2N$	3.32	**7**
$C_{19}H_{14}BrNO_9$, C_3H_7NO	50.22	**1**
$C_{19}H_{14}Br_2O$	20.31	**1**
$C_{19}H_{14}Br_2O_2$	17.14	**4**
$C_{19}H_{14}ClNO_9$, C_3H_7NO	50.23	**1**
$C_{19}H_{14}CrO_3$	75.28	**3**
$C_{19}H_{14}FeO_3$	72.17	**6**
$C_{19}H_{14}FeO_3$, $0.5C_{16}H_{14}$	60.51	**3**
$C_{19}H_{14}I_2O$	20.32	**1**
$C_{19}H_{14}S_3$	39.38	**5**
$C_{19}H_{15}^{+}$, ClO_4^{-}	12.14	**1**
$C_{19}H_{15}^{+}$, Cl_5Pt^{-} , CH_2Cl_2	12.6	**5**
$C_{19}H_{15}^{+}$, $Cl_{13}Te_3^{-}$	12.5	**7**
$2C_{19}H_{15}^{+}$, $Cl_{10}Pt_2^{2-}$	12.7	**5**
$2C_{19}H_{15}^{+}$, $Cl_{12}Pt_3^{2-}$	12.8	**5**
$2C_{19}H_{15}^{+}$, $Cl_{12}Pt_3^{2-}$, $2C_2H_2Cl_4$	12.9	**5**
$2C_{19}H_{15}^{+}$, $C_{12}H_8Cl_{14}Pt_4^{2-}$, $2CH_2Cl_2$	12.10	**5**
$C_{19}H_{15}AuNP$	86.22	**3**
$C_{19}H_{15}Br$	19.60	**1**
$C_{19}H_{15}Br$	19.16	**3**
$C_{19}H_{15}BrO$	5.18	**7**
$C_{19}H_{15}BrOSi$	63.8	**4**
$C_{19}H_{15}BrO_6$	45.47	**5**
$C_{19}H_{15}Cl_2CuN_5O$	83.56	**6**
$C_{19}H_{15}Cl_2MnN_5O$	84.47	**7**
$C_{19}H_{15}Cl_2OPPt$	86.30	**7**
$C_{19}H_{15}FeN_2O_3P$	86.31	**7**
$C_{19}H_{15}Fe_2NO_5$	72.18	**6**
$C_{19}H_{15}Fe_2NO_{10}S$	72.54	**2**
$C_{19}H_{15}MoN_3O_2S$	71.27	**3**
$C_{19}H_{15}N$	28.9	**3**
$C_{19}H_{15}N$	16.13	**7**
$C_{19}H_{15}NO$	10.5	**5**
$C_{19}H_{15}NO_2S$	39.25	**6**
$C_{19}H_{15}NO_6U$	84.48	**7**
$C_{19}H_{15}NSSi$	63.12	**7**
$C_{19}H_{15}NSSn$	69.16	**7**
$C_{19}H_{15}N_3O$	32.24	**4**
$C_{19}H_{16}$	19.10	**6**
$C_{19}H_{16}$, C_4H_4S	60.30	**7**
$C_{19}H_{16}$, C_4H_5N	60.31	**7**
$C_{19}H_{16}$, C_6H_6	60.32	**7**
$C_{19}H_{16}$, C_6H_7N	60.33	**7**
$C_{19}H_{16}BrN_3O_6$	36.25	**4**
$C_{19}H_{16}ClNO_2$	25.5	**7**
$C_{19}H_{16}ClNO_4$	35.24	**4**
$C_{19}H_{16}Cl_3NO_2S_2$	24.12	**4**
$(C_{19}H_{16}Eu_2O_{17})_n$, $3nH_2O$	81.41	**6**
$C_{19}H_{16}FeO_2$	73.20	**4**
$C_{19}H_{16}FeO_3$	75.64	**2**
$C_{19}H_{16}FeO_3$	75.29	**3**
$C_{19}H_{16}N_2$	20.21	**7**
$C_{19}H_{16}N_2O_3$	40.21	**7**
$C_{19}H_{16}N_4$	35.25	**4**
$C_{19}H_{16}O_3Ru$	75.19	**7**

$C_{19}H_{16}O_3S$ 39.88 **1**
$C_{19}H_{16}O_4$ 59.7 **7**
$C_{19}H_{16}O_4$, CH_4O 59.15 **6**
$C_{19}H_{16}O_5$ 45.48 **5**
$C_{19}H_{16}P^+$, Cl_4Fe^- 12.15 **1**
$C_{19}H_{16}S_3$ 39.39 **5**
$C_{19}H_{17}BMoN_8O_2$ 73.42 **6**
$C_{19}H_{17}BrO_2$ 28.7 **5**
$C_{19}H_{17}ClN_4O_3$ 35.29 **6**
$C_{19}H_{17}CuN_4O_3S^+$, ClO_4^- 81.53 **5**
$C_{19}H_{17}FSi$ 63.10 **6**
$C_{19}H_{17}NO$ 34.6 **5**
$C_{19}H_{17}NOS_2$ 39.32 **7**
$C_{19}H_{17}NO_2$ 25.6 **3**
$C_{19}H_{17}NO_2S$ 25.7 **4**
$C_{19}H_{17}NO_2S_2$ 11.28 **3**
$C_{19}H_{17}NS_3$ 39.40 **5**
$C_{19}H_{17}N_2S_2^+$, $C_{12}H_4N_4^-$, $C_{12}H_4N_4$ 60.34 **7**
$C_{19}H_{17}N_3S_2$ 41.26 **7**
$C_{19}H_{17}N_5S_2Sn$ 69.21 **5**
$C_{19}H_{17}P$ 64.45 **3**
$C_{19}H_{18}$ 31.24 **7**
$C_{19}H_{18}As^+$, $C_{12}H_4N_4^-$, $C_{12}H_4N_4$ 65.9 **4**
$C_{19}H_{18}As^+$, $C_{14}H_{12}CoS_4^-$, $0.5C_2H_6O$ 85.73 **2**
$2C_{19}H_{18}As^+$, Cl_4Ni^{2-} 65.33 **2**
$C_{19}H_{18}BrF_{12}O_2P$ 64.32 **6**
$C_{19}H_{18}BrN_3O_2$ 40.22 **7**
$C_{19}H_{18}BrN_3O_6$ 37.11 **6**
$C_{19}H_{18}ClNO_5S$ 35.31 **7**
$C_{19}H_{18}CrO_4$ 74.8+ **6**
$C_{19}H_{18}N_2Ni$ 71.45 **5**
$C_{19}H_{18}N_3^+$, ClO_4^- 12.15 **4**
$C_{19}H_{18}N_3O_4S_2^+$, Cl^- , H_2O 50.11 **3**
$C_{19}H_{18}OPdS$ 72.34 **3**
$C_{19}H_{18}OS_5$ 39.33 **7**
$C_{19}H_{18}O_3$ 38.59 **7**
$C_{19}H_{18}O_6$ 59.40 **1**
$C_{19}H_{18}O_6$ 59.14 **4**
$C_{19}H_{18}P^+$, $H_{24}B_{20}Zn^-$ 64.46 **3**
$C_{19}H_{18}P^+$, $C_4H_{22}B_{18}Cu^-$ 64.56 **2**
$C_{19}H_{18}P^+$, $C_5H_8B_6MnO_3^-$ 64.33 **6**
$C_{19}H_{18}P^+$, $C_8N_4NiS_4^-$ 85.43 **2**
$C_{19}H_{18}P^+$, $C_{12}H_4N_4^-$, $C_{12}H_4N_4$ 64.24 **4**
$C_{19}H_{18}P^+$, $C_{12}H_4N_4^-$, $C_{12}H_4N_4$ 64.34+ **6**
$C_{19}H_{19}BrO$ 21.16 **4**
$C_{19}H_{19}BrO_5$ 59.41 **1**
$C_{19}H_{19}Cl_2PRu$ 74.8 **5**
$C_{19}H_{19}CuN_3O_4$ 84.21 **6**
$C_{19}H_{19}NO_4$, H_2O 58.17 **4**
$C_{19}H_{19}N_2O_2^-$, $C_5H_{13}N_2^+$ 33.15 **6**
$C_{19}H_{19}N_2O_4P$ 64.36 **6**
$C_{19}H_{19}N_2S_2^+$, Br^- 41.46 **5**
$C_{19}H_{19}N_2S_2^+$, $C_{12}H_4N_4^-$ 41.29 **6**
$C_{19}H_{19}O_2Rh$ 73.21 **4**
$C_{19}H_{20}BrN_5O_2S$ 39.37 **4**
$C_{19}H_{20}BrN_5O_2S$ 39.38 **4**
$C_{19}H_{20}Co_6O_{11}S_5$ 85.82 **2**
$C_{19}H_{20}CuN_2O_2$ 78.10 **3**
$C_{19}H_{20}F_2NO_2^+$, Br^- 33.48 **3**
$C_{19}H_{20}FeO_{11}$ 71.58 **7**
$C_{19}H_{20}NOP$ 64.37 **6**
$C_{19}H_{20}N_2O_2$ 32.33 **7**
$C_{19}H_{20}N_2O_5$ 35.32 **7**
$C_{19}H_{20}OS_5$ 39.41 **5**
$C_{19}H_{20}O_2$ 59.16 **6**
$C_{19}H_{20}O_5$ 54.1 **6**
$C_{19}H_{21}Br$ 31.26 **6**
$C_{19}H_{21}CuNO_4$ 77.15 **3**
$C_{19}H_{21}NO_2$ 36.24 **3**
$C_{19}H_{21}NO_5$ 28.7 **4**
$C_{19}H_{21}O_2P$ 64.49 **7**
$C_{19}H_{22}ClN_3O_2$ 36.26 **4**
$C_{19}H_{22}Cl_2N_2Ru$ 75.20 **7**
$C_{19}H_{22}IrO_3P_2^+$, ClO_4^- 86.7 **6**
$C_{19}H_{22}N^+$, Br^- 36.27 **4**
$C_{19}H_{22}NO^+$, Cl^- 34.16 **7**
$C_{19}H_{22}NO_4^+$, Br^- 58.21 **3**
$C_{19}H_{22}NO_4^+$, Cl^- , $2H_2O$ 58.26 **6**
$C_{19}H_{22}N_2O$ 58.27 **6**
$C_{19}H_{22}N_3PPdS_2$ 86.23 **3**
$C_{19}H_{22}O$ 38.57 **1**
$C_{19}H_{22}O_2$, C_6H_5BrO 60.14 **3**
$C_{19}H_{22}O_2$, C_6H_5BrO 51.1 **6**
$C_{19}H_{22}O_6$ 59.16 **5**
$C_{19}H_{22}O_7$ 59.8 **7**
$C_{19}H_{23}BMoN_4O_2$ 75.31 **5**
$C_{19}H_{23}BMoN_4O_2$ 75.21 **6**
$C_{19}H_{23}BrO$ 21.9 **6**
$C_{19}H_{23}BrO_2$ 27.16 **5**
$C_{19}H_{23}BrO_4$ 54.1 **3**
$C_{19}H_{23}BrO_{10}S$ 45.74 **1**
$C_{19}H_{23}ClO_2$ 28.8 **5**
$C_{19}H_{23}CoN_7O_2^{2+}$, $2Cl^-$ 83.87 **7**
$C_{19}H_{23}N$ 36.17 **6**
$C_{19}H_{23}NO_2$ 51.4 **7**
$C_{19}H_{23}NO_3$ 15.11 **5**
$C_{19}H_{23}N_2^+$, Cl^- , H_2O 32.34 **7**
$C_{19}H_{23}N_2O_3^+$, Br^- 48.54 **4**
$C_{19}H_{23}N_7O_2Zn^{2+}$, $2Cl^-$ 83.88 **7**
$C_{19}H_{23}N_7O_{12}^-$, $C_{13}H_{11}N_2^+$, xH_2O 47.30 **6**
$C_{19}H_{23}N_7O_{12}P^-$, Na^+ , $6.5H_2O$ 47.33 **5**
$C_{19}H_{23}N_7O_{12}P^-$, $C_{13}H_{11}N_2^+$, xH_2O 47.37 **7**
$C_{19}H_{24}$ 28.10 **3**
$C_{19}H_{24}BrNO_3$ 31.27 **6**
$C_{19}H_{24}BrNO_9S$ 45.49 **5**
$C_{19}H_{24}Br_2O_6$ 53.20 **3**
$C_{19}H_{24}ClNO_2$ 27.17 **5**
$C_{19}H_{24}ClNO_6$ 58.36 **1**
$C_{19}H_{24}Cl_2O_7$ 53.13 **5**
$C_{19}H_{24}F_6OPt_2$ 71.59 **7**
$C_{19}H_{24}FeNiO_3$ 75.30 **3**
$C_{19}H_{24}N^+$, Br^- 33.80 **1**
$C_{19}H_{24}N^+$, Br^- 32.34 **3**
$C_{19}H_{24}NNi_3$ 73.82 **2**
$C_{19}H_{24}NNi_3$ 73.34 **7**
$C_{19}H_{24}NO^+$, Cl^- , H_2O 33.50 **7**
$C_{19}H_{24}NO_4^+$, Br^- 58.22 **3**
$C_{19}H_{24}NO_4^+$, I^- 58.23 **3**
$C_{19}H_{24}NO_5^+$, I^- 58.24 **3**
$C_{19}H_{24}NO_5^+$, I^- , CH_4O 58.25 **3**
$C_{19}H_{24}N_2NiO_3$ 78.32 **2**
$C_{19}H_{24}N_4O_2S^{2+}$, $2Cl^-$, $3H_2O$ 44.29 **7**
$C_{19}H_{24}N_7O_{12}P$ 47.20 **4**
$C_{19}H_{24}N_7O_{12}P$, $4H_2O$ 47.31 **1**
$C_{19}H_{24}N_8O_{12}P^-$, Na^+ , $9H_2O$ 47.31 **6**
$2C_{19}H_{24}N_8O_{12}P^-$, Ca^{2+} , $18H_2O$ 47.38 **7**
$C_{19}H_{24}O_2$ 51.8 **3**
$C_{19}H_{24}O_2$ 51.2 **6**
$C_{19}H_{24}O_3$ 24.13 **4**
$C_{19}H_{24}O_3$ 51.11 **4**
$C_{19}H_{24}O_6$ 53.20 **7**

$0.7C_{19}H_{24}O_6$, $0.3C_{20}H_{26}O_6$ 53.17 **6**
$C_{19}H_{25}BN_6Ru^+$, F_6P^- 73.43 **6**
$C_{19}H_{25}BrO_2$ 27.11 **3**
$C_{19}H_{25}BrO_2$ 51.9 **3**
$C_{19}H_{25}Cl_2IrO_3S_2$ 71.28 **3**
$C_{19}H_{25}FN_2O_{10}S$ 45.53 **7**
$C_{19}H_{25}N_2^+$, Br^- 58.26 **3**
$C_{19}H_{25}N_2^+$, Cl^- 36.22 **7**
$(C_{19}H_{25}N_2^-)_n$, nK^+ 67.6 **7**
$C_{19}H_{25}N_2O_2^+$, I^- 58.27 **3**
$C_{19}H_{25}N_2S^+$, Cl^- 41.31 **4**
$C_{19}H_{25}O_2Rh$ 74.16 **3**
$C_{19}H_{25}O_3$, C_2H_3N 51.3 **6**
$C_{19}H_{26}NO_2^+$, Br^- 58.28 **6**
$C_{19}H_{26}NO_4^+$, I^- 58.15 **5**
$C_{19}H_{26}NO_5^+$, I^- 58.28 **3**
$C_{19}H_{26}NO_6^+$, I^- 58.16 **5**
$C_{19}H_{26}NO_7^+$, $C_{10}H_{14}BrO_4S^-$, H_2O 58.38 **1**
$C_{19}H_{26}O_2$ 20.16 **3**
$C_{19}H_{26}O_2$ 51.12 **4**
$C_{19}H_{26}O_2$, C_6H_5BrO 60.39 **4**
$C_{19}H_{26}O_3$ 51.10 **3**
$C_{19}H_{26}O_3$ 51.7 **5**
$C_{19}H_{26}O_3$ 51.4 **6**
$C_{19}H_{26}O_5$, H_2O 51.5 **7**
$C_{19}H_{26}O_6$ 53.21 **7**
$C_{19}H_{27}BrO_2$ 54.3 **1**
$C_{19}H_{27}BrO_2$, CH_4O 51.6 **7**
$C_{19}H_{27}BrO_3$ 51.13 **4**
$C_{19}H_{27}NO_6$ 58.10 **7**
$C_{19}H_{27}N_3O_7$ 16.23 **6**
$C_{19}H_{28}Co_2N_2O$ 73.33 **3**
$C_{19}H_{28}N^+$, Cl^- , H_2O 37.18 **7**
$C_{19}H_{28}NO^+$, Cl^- 58.30 **6**
$C_{19}H_{28}NO_3^+$, Br^- 32.35 **6**
$C_{19}H_{28}O_2$ 51.9 **5**
$C_{19}H_{28}O_2$ 51.5 **6**
$C_{19}H_{28}O_2$ 51.6 **6**
$C_{19}H_{28}O_2$, H_2O 51.11 **5**
$C_{19}H_{28}O_2$, H_2O 51.7 **6**
$C_{19}H_{28}O_2$, C_6H_5BrO 60.151 **2**
$C_{19}H_{28}O_2$, C_6H_5BrO 60.15 **3**
$C_{19}H_{28}O_2$, C_6H_5BrO 51.8 **6**
$C_{19}H_{28}O_3$ 28.21 **6**
$C_{19}H_{28}O_5$ 50.9 **7**
$C_{19}H_{29}As_2IPt$ 71.60 **7**
$C_{19}H_{29}BrO_3$ 51.11 **3**
$C_{19}H_{29}ClO$, $0.5CH_4O$ 51.12 **3**
$C_{19}H_{29}N_2^+$, Cl^- , H_2O 36.23 **7**
$C_{19}H_{30}Br_2$ 51.14 **4**
$C_{19}H_{30}NO^+$, Cl^- 32.25 **4**
$C_{19}H_{30}N_3^+$, I^- 32.35 **7**
$C_{19}H_{30}N_4NiO_2^{2+}$, $2ClO_4^-$ 83.57 **6**
$C_{19}H_{30}O_2$ 51.6 **1**
$C_{19}H_{30}O_2$ 51.15 **4**
$C_{19}H_{30}O_2$ 51.9 **6**
$C_{19}H_{30}O_2$, H_2O 51.16 **4**
$C_{19}H_{30}O_2$, H_2O 51.12 **5**
$C_{19}H_{30}O_3$ 51.10 **6**
$C_{19}H_{30}O_4$ 51.7 **7**
$C_{19}H_{31}Cl_2CrO_3$ 84.46 **2**
$C_{19}H_{31}KO_9S$ 67.19 **4**
$C_{19}H_{31}N_3$ 51.11 **6**
$C_{19}H_{31}O_2Rh$ 72.35 **3**
$C_{19}H_{32}IMoO_2P$ 73.22 **4**
$C_{19}H_{32}NO_2^+$, CH_4O , Br^- 58.39 **1**
$C_{19}H_{32}O_2$ 51.17 **4**
$C_{19}H_{32}O_2$ 51.18 **4**
$C_{19}H_{32}O_2$, H_2O 51.13 **5**
$C_{19}H_{34}ClOP_2Pt^+$, F_6P^- 71.43 **6**
$C_{19}H_{36}O_2$ 1.149 **1**
$C_{19}H_{38}BaN_3O_7S^+$, CNS^- 67.20 **4**
$C_{19}H_{38}CrO_5Si_4Sn$ 69.17 **7**
$C_{19}H_{38}O$ 5.18 **5**
$C_{19}H_{38}O_2$ 1.150 **1**
$C_{19}H_{38}O_2$ 1.37 **3**

C_{20}

$C_{20}Co_6O_{18}$ 71.42 **4**
$C_{20}H_4AmF_{24}O_8^-$, Cs^+ 77.16 **3**
$C_{20}H_4EuF_{24}O_8^-$, Cs^+ 77.17 **3**
$C_{20}H_4F_{24}O_8Y^-$, Cs^+ 77.39 **2**
$C_{20}H_6ClNO_3$ 35.30 **6**
$C_{20}H_8Br_4O_4$ 30.4 **1**
$C_{20}H_8CdMn_2N_2O_{10}$ 83.58 **6**
$C_{20}H_8CoN_6S_4^-$, $C_{12}H_{28}N^+$ 85.35 **3**
$C_{20}H_8F_{12}N_2$ 35.30 **5**
$C_{20}H_8F_{12}N_2$ 36.25 **5**
$C_{20}H_8F_{18}O_6Zr$ 73.34 **3**
$C_{20}H_{10}$ 5.36 **1**
$C_{20}H_{10}$ 30.2 **3**
$C_{20}H_{10}Cl_4$ 20.17 **3**
$C_{20}H_{10}Co_2O_6$ 72.56 **2**
$C_{20}H_{10}CrO_5$ 71.30 **3**
$C_{20}H_{10}CuF_{12}N_2O_4$ 77.18 **3**
$C_{20}H_{10}F_{12}Fe_2O_2$ 71.61 **7**
$C_{20}H_{10}Fe_2O_6$ 72.57 **2**
$C_{20}H_{10}Fe_2O_6S$ 71.31 **3**
$C_{20}H_{10}Fe_2O_6S_2$ 85.83 **2**
$C_{20}H_{10}Fe_2O_8$ 71.32 **3**
$C_{20}H_{10}Mn_2N_4O_8$ 83.89 **7**
$8C_{20}H_{10}N_4$, C_6N_4 , $2CH_2Cl_2$ 60.52 **3**
$C_{20}H_{10}O_4$ 25.22 **1**
$C_{20}H_{11}Mn_2O_8P$ 86.25 **2**
$C_{20}H_{12}$ 30.2 **1**
$C_{20}H_{12}$ 30.5 + **1**
$C_{20}H_{12}$, $C_6F_4O_2$ 60.152 **2**
$C_{20}H_{12}$, C_6N_4 60.53 **3**
$C_{20}H_{12}$, $C_8F_{12}NiS_4$ 60.54 **3**
$C_{20}H_{12}$, $2C_9H_{12}N_4O_3$ 60.111 **2**
$C_{20}H_{12}$, $C_{10}H_2O_6$ 60.154 **2**
$C_{20}H_{12}$, $C_{12}H_4N_4$ 60.38 **5**
$C_{20}H_{12}B_2OS_4$ 62.9 **7**
$C_{20}H_{12}Br_2N_2O$ 40.30 **1**
$C_{20}H_{12}Cl_2Fe_2N_2O_8$ 83.103 **3**
$C_{20}H_{12}CuF_{12}N_2O_4$ 77.6 **5**
$C_{20}H_{12}F_{12}N_2O_4Zn$ 77.7 **5**
$C_{20}H_{12}FeO_4$ 71.44 **6**
$C_{20}H_{12}FeO_6$ 72.26 **5**
$C_{20}H_{12}Fe_4O_{16}Sn_3$ 69.35 **2**
$C_{20}H_{12}IN_3$ 28.25 **1**
$C_{20}H_{12}N_2$ 36.29 **1**
$C_{20}H_{12}N_2O_2$ 36.30 **1**
$C_{20}H_{12}O$ 38.60 **7**
$C_{20}H_{12}O_5$, C_3H_6O 38.61 **7**
$C_{20}H_{12}O_8Re_2Si$ 63.14 **3**
$C_{20}H_{13}Br$ 31.37 **1**
$C_{20}H_{13}Fe_2NO_6$ 83.170 **2**

$C_{20}H_{13}NSe$ 42.14 **1**
$C_{20}H_{14}$ 24.5 **3**
$C_{20}H_{14}$ 31.31 **3**
$C_{20}H_{14}$ 31.27 **4**
$C_{20}H_{14}BrN$ 16.16 **3**
$C_{20}H_{14}Br_2O_2$ 30.1 **4**
$C_{20}H_{14}Cl_2CoN_6S_2$, H_2O 83.60 **6**
$C_{20}H_{14}Cr_2O_6$ 75.22 **6**
$C_{20}H_{14}N_2O$ 24.10 **6**
$C_{20}H_{14}N_4$ 49.1 **1**
$C_{20}H_{14}N_4$ 49.1 **4**
$C_{20}H_{14}O_4$ 13.26+ **7**
$C_{20}H_{14}O_7P_2^{2-}$, $2C_6H_{14}N^+$ 46.6 **7**
$C_{20}H_{15}AsCoNO_3$ 86.8 **6**
$C_{20}H_{15}AsHgN_2S_2$ 86.23 **5**
$C_{20}H_{15}Cl$ 29.7 **7**
$C_{20}H_{15}CoNO_3P$ 86.11 **4**
$C_{20}H_{15}CoNO_3P$ 86.24 **5**
$C_{20}H_{15}CoNO_3Sb$ 86.9 **6**
$C_{20}H_{15}F_{12}O_2P$ 64.25 **4**
$C_{20}H_{15}Fe_2O_7P$ 86.10 **6**
$C_{20}H_{15}N$ 35.31 **5**
$C_{20}H_{15}N_5O_2$ 32.23 **5**
$C_{20}H_{15}OP$ 64.57 **2**
$C_{20}H_{15}PS$ 64.58 **2**
$C_{20}H_{16}$ 29.23 **1**
$C_{20}H_{16}$ 29.24 **1**
$C_{20}H_{16}$ 28.11 **3**
$C_{20}H_{16}$ 29.5 **4**
$C_{20}H_{16}$ 28.10 **5**
$C_{20}H_{16}$ 29.8 **7**
$C_{20}H_{16}$, $C_{13}H_4BrN_3O_7$ 60.55 **3**
$C_{20}H_{16}AgN_4^{2+}$, $2NO_3^-$, H_2O 83.79 **5**
$C_{20}H_{16}BrNO_2$ 32.36 **7**
$C_{20}H_{16}Br_4Hg_2N_4$ 83.61 **6**
$C_{20}H_{16}ClCuN_4^+$, Cl^- , $6H_2O$ 83.62 **6**
$C_{20}H_{16}ClGaN_2O_2$ 68.10 **4**
$C_{20}H_{16}ClGaN_2O_2$ 68.15 **6**
$(C_{20}H_{16}Cl_2Cu_2N_4^+)_n$, nCl_2Cu^- 83.90 **7**
$C_{20}H_{16}Cl_2Cu_2N_{16}O_4^{2+}$, $2Cl^-$, $6H_2O$ 83.104 **3**
$C_{20}H_{16}Cl_2Fe_2$ 73.85 **2**
$C_{20}H_{16}Cl_2GaN_4^+$, Cl_4Ga^- 68.11 **4**
$C_{20}H_{16}CoN_4O_4S^+$, NO_3^- , H_2O 83.91 **7**
$C_{20}H_{16}CoN_5O_3^{2+}$, HO^- , NO_3^- , $4H_2O$ 83.102 **4**
$C_{20}H_{16}CuIN_4^+$, I^- 83.171 **2**
$C_{20}H_{16}CuN_4^{2+}$, $2ClO_4^-$ 83.103 **4**
$C_{20}H_{16}CuN_5O_2^+$, NO_3^- 83.105 **3**
$C_{20}H_{16}CuO_4$ 77.40 **2**
$C_{20}H_{16}F_2N_4O_4V$ 83.80 **5**
$C_{20}H_{16}F_{12}O_8Th$ 77.8 **5**
$C_{20}H_{16}Fe$ 75.31 **3**
$C_{20}H_{16}Fe_2$ 73.35 **3**
$C_{20}H_{16}LaN_4^{3+}$, $3NO_3^-$ 83.172 **2**
$C_{20}H_{16}LaN_7O_9$ 83.81+ **5**
$C_{20}H_{16}Mo_2N_2O_6S$ 85.38 **7**
$C_{20}H_{16}N_2$ 33.41 **4**
$C_{20}H_{16}N_2$ 33.51 **5**
$C_{20}H_{16}N_2O_3V$ 83.104 **4**
$C_{20}H_{16}N_4O_4$ 35.26 **4**
$C_{20}H_{16}N_4Pd^{2+}$, $2NO_3-$, H_2O 83.105+ **4**
$C_{20}H_{16}N_6Pd$ 83.173 **2**
$C_{20}H_{16}N_7O_9Tb$ 83.174 **2**
$C_{20}H_{16}O$ 38.62 **7**
$C_{20}H_{16}O_2$ 24.6 **3**
$C_{20}H_{16}O_4Ru_3$ 75.66 **2**
$C_{20}H_{16}O_5$ 1.38 **3**
$C_{20}H_{16}O_6$ 59.15 **4**
$C_{20}H_{16}O_6Ru_3$ 75.23 **6**
$C_{20}H_{16}O_7$ 38.24 **4**
$C_{20}H_{16}O_7$ 59.17 **5**
$C_{20}H_{16}S_3$ 39.34 **7**
$C_{20}H_{17}AuBr_3P$ 71.43 **4**
$C_{20}H_{17}ClN_2O_3$ 34.3 **4**
$C_{20}H_{17}Cl_3O_5$ 50.25 **1**
$C_{20}H_{17}NO_5$ 58.29 **3**
$C_{20}H_{17}NO_6$ 58.31+ **6**
$C_{20}H_{17}N_4$ 12.12 **5**
$2C_{20}H_{17}N_4^+$, $C_4CoN_4S_4^{2-}$ 32.36 **3**
$C_{20}H_{17}N_5$ 32.37 **7**
$C_{20}H_{18}$ 19.61 **1**
$C_{20}H_{18}$ 19.11 **6**
$C_{20}H_{18}BrNO_2$ 28.22 **6**
$C_{20}H_{18}Br_2$ 21.18 **3**
$C_{20}H_{18}Br_2N_2$ 9.18 **3**
$C_{20}H_{18}Br_2N_4$ 36.28 **4**
$C_{20}H_{18}Br_2N_4$ 36.29 **4**
$C_{20}H_{18}Br_2N_4$ 36.30 **4**
$C_{20}H_{18}Br_2N_4$ 36.26 **5**
$C_{20}H_{18}Br_2N_4O_6$ 10.6+ **7**
$C_{20}H_{18}ClN_3O_4$ 9.19 **3**
$C_{20}H_{18}Cl_2$ 5.37 **1**
$C_{20}H_{18}Cl_2$ 5.38 **1**
$C_{20}H_{18}Cl_2F_{12}N_2Pd$ 83.83 **5**
$C_{20}H_{18}Cl_2Ti_2$ 73.35 **7**
$2C_{20}H_{18}CuIN_6^+$, I^- , ClO_4^- 83.84 **5**
$C_{20}H_{18}CuN_6^{2+}$, $2ClO_4^-$ 83.107 **3**
$C_{20}H_{18}Cu_2N_4O_2^{2+}$, $2ClO_4^-$ 83.92 **7**
$C_{20}H_{18}Cu_2N_4O_2^{2+}$, $2NO_3^-$ 83.108 **3**
$C_{20}H_{18}FeO$ 73.32 **5**
$C_{20}H_{18}Fe_2$ 73.86+ **2**
$C_{20}H_{18}GeO$ 69.36 **2**
$C_{20}H_{18}MoN_3O_2^+$, BF_4^- 72.27 **5**
$C_{20}H_{18}NO_5^+$, Br^- 58.18 **4**
$C_{20}H_{18}N_2O_2Sn$ 69.37 **2**
$C_{20}H_{18}N_3^+$, Br^- , H_2O 36.24 **7**
$C_{20}H_{18}N_3O_2S^+$, Cl^- 36.25 **7**
$C_{20}H_{18}N_4Ni$ 83.85 **5**
$C_{20}H_{18}N_4OPd^{2+}$, $2NO_3^-$ 83.86 **5**
$C_{20}H_{18}N_6Ni$ 83.93 **7**
$C_{20}H_{18}OSi$ 63.15 **3**
$C_{20}H_{18}O_2$ 17.20 **5**
$C_{20}H_{18}O_2$ 29.9 **7**
$C_{20}H_{18}O_4Pd$ 77.41 **2**
$C_{20}H_{18}O_5$ 17.24 **6**
$C_{20}H_{18}O_5V$ 77.43 **2**
$C_{20}H_{18}O_6$, CCl_4 59.16 **4**
$C_{20}H_{18}O_7$ 59.19 **3**
$C_{20}H_{18}O_8$ 27.15 **6**
$C_{20}H_{18}Si$ 63.8 **5**
$C_{20}H_{19}BrN_2O$ 37.24 **1**
$C_{20}H_{19}BrO_2$ 28.11 **5**
$C_{20}H_{19}BrO_7$ 59.42 **1**
$C_{20}H_{19}BrO_8$ 38.35 **6**
$C_{20}H_{19}Cl_3Cu_2N_6$ 83.63 **6**
$C_{20}H_{19}CuN_5^{2+}$, $2BF_4^-$ 83.107 **4**
$C_{20}H_{19}I_2NiPS_2$ 85.36 **3**
$C_{20}H_{19}NO_5$ 58.41 **1**
$C_{20}H_{19}N_2S^+$, I^- 41.46 **1**
$C_{20}H_{20}$ 29.25 **1**
$C_{20}H_{20}$ 31.32 **5**

$C_{20}H_{20}BrF_{12}O_2P$ 64.38 **6**
$C_{20}H_{20}BrNO_3$ 58.30 **3**
$C_{20}H_{20}BrNO_4$ 13.21 **4**
$C_{20}H_{20}Br_2N_2$ 37.15 **4**
$C_{20}H_{20}Br_2N_4Ni$ 83.175 **2**
$C_{20}H_{20}Br_4Cu_2N_4O_4$ 84.11 **4**
$C_{20}H_{20}Br_4Hg_2Ru_2$ 73.33 **5**
$C_{20}H_{20}Br_6Cu_4N_4O$ 83.64 **6**
$C_{20}H_{20}ClN$ 28.26 **1**
$C_{20}H_{20}ClPPdS$ 71.62 + **7**
$C_{20}H_{20}Cl_2CoN_4$ 83.176 **2**
$C_{20}H_{20}Cl_2Cu_2N_{20}^{2+}$, $2Cl^-$, $6H_2O$ 83.108 **4**
$C_{20}H_{20}Cl_2N_4Ni$ 83.177 **2**
$C_{20}H_{20}Cl_2N_4Rh^+$, $HN_2O_6^-$ 83.178 **2**
$C_{20}H_{20}Cl_2N_4V$ 83.65 **6**
$C_{20}H_{20}Cl_2Nb_2O^{2+}$, $2BF_4^-$ 73.36 **7**
$C_{20}H_{20}Cl_2OZr_2$ 73.37 **7**
$C_{20}H_{20}Cl_2Sc_2$ 73.46 **6**
$C_{20}H_{20}Cl_4O_4Ti_4$ 73.36 **3**
$C_{20}H_{20}Cl_4Ti_2Zn$, $2C_6H_6$ 73.44 **6**
$C_{20}H_{20}Cl_6Cu_4N_4O$ 83.87 **5**
$C_{20}H_{20}Cl_6N_4W_2$, xC_3H_6O 83.109 **4**
$C_{20}H_{20}CoN_2O_2$ 78.7 **7**
$C_{20}H_{20}Co_4P_4$ 73.34 **5**
$C_{20}H_{20}Co_4S_4$ 73.35 **5**
$C_{20}H_{20}Co_4S_4^+$, F_6P^- 73.36 **5**
$C_{20}H_{20}Co_4S_6$, $0.5CHCl_3$ 73.37 **3**
$C_{20}H_{20}CuN_2O_2$ 83.179 **2**
$C_{20}H_{20}CuN_4O_4^{2+}$, $2BF_4^-$ 84.47 **2**
$C_{20}H_{20}CuN_4O_4^{2+}$, $2ClO_4^-$ 84.48 **2**
$C_{20}H_{20}CuN_5O_3^+$, NO_3^- , H_2O 83.109 **3**
$C_{20}H_{20}Cu_2N_4O_7S$, $4H_2O$ 83.110 **3**
$C_{20}H_{20}Cu_2N_8O_{12}$, C_5H_5N 83.111 **3**
$C_{20}H_{20}Cu_2N_8O_{16}$ 84.49 **2**
$C_{20}H_{20}Fe_4S_4$ 73.89 + **2**
$C_{20}H_{20}Fe_4S_6$ 73.37 **5**
$C_{20}H_{20}Hf$ 71.47 **5**
$(C_{20}H_{20}Hg_2N_4O_4^{2+})_n$, $2nClO_4^-$ 84.17 **5**
$C_{20}H_{20}I_2N_4Ni$ 83.94 **7**
$C_{20}H_{20}NP$ 64.31 **5**
$C_{20}H_{20}N_2$ 9.20 **3**
$C_{20}H_{20}N_2$ 9.16 **5**
$C_{20}H_{20}N_4O_2Re^+$, Cl^- , $2H_2O$ 83.110 **4**
$C_{20}H_{20}N_5NiO_3^+$, NO_3^- 83.111 **4**
$C_{20}H_{20}Nb_2$ 71.48 **5**
$C_{20}H_{20}Ni_5S_4$ 73.91 **2**
$C_{20}H_{20}O_4$ 59.17 **6**
$C_{20}H_{20}O_5$ 59.18 **6**
$C_{20}H_{20}O_5$ 54.1 **7**
$2C_{20}H_{20}P^+$, $CeN_5O_{15}^{2-}$ 64.39 **6**
$C_{20}H_{20}Pt_2$ 75.23 **4**
$C_{20}H_{20}Ti$ 73.23 **4**
$C_{20}H_{20}U$ 73.38 **7**
$C_{20}H_{20}Zr$ 73.38 **3**
$C_{20}H_{21}BrO_6$ 59.43 **1**
$C_{20}H_{21}NO_4$ 58.18 **5**
$C_{20}H_{21}NO_5$ 36.18 **6**
$C_{20}H_{21}N_7O_6Os$ 84.49 **7**
$C_{20}H_{21}O_2Sb$ 66.19 **2**
$C_{20}H_{21}Rh_3$ 73.92 **2**
$C_{20}H_{22}Al_2O_6W_2$ 73.24 **4**
$C_{20}H_{22}As_2F_8Ni$ 71.49 **5**
$C_{20}H_{22}BrNO_4S_2$ 39.43 **5**
$C_{20}H_{22}Br_2O_5$ 59.44 **1**
$C_{20}H_{22}Cl_2Fe_2N_2O_4$ 78.10 **5**
$C_{20}H_{22}Cl_2Fe_2N_2O_4$, C_7H_8 78.9 **6**
$C_{20}H_{22}Cl_4Cu_2N_2O_8$ 81.49 **3**
$C_{20}H_{22}CuN_4$ 76.46 **6**
$C_{20}H_{22}CuN_4O_4$ 78.8 **7**
$C_{20}H_{22}Cu_2N_2O_4$ 78.11 **5**
$C_{20}H_{22}Cu_2N_2O_4$ 78.9 **7**
$C_{20}H_{22}Fe_2N_2O_4$ 73.39 **3**
$C_{20}H_{22}NO_4^+$, I^- 58.42 **1**
$C_{20}H_{22}NO_5^+$, Br^- 58.31 **3**
$C_{20}H_{22}N_2$ 36.26 **7**
$C_{20}H_{22}N_2$ 37.19 **7**
$C_{20}H_{22}N_2O_2$ 36.25 **3**
$C_{20}H_{22}N_2O_2S$ 36.27 **7**
$C_{20}H_{22}N_2O_3$ 58.43 **1**
$C_{20}H_{22}N_2O_5$ 48.80 **6**
$C_{20}H_{22}N_2O_6$ 48.67 **7**
$C_{20}H_{22}N_4NiS_2$ 85.39 **7**
$C_{20}H_{22}N_4O$ 32.26 **4**
$C_{20}H_{22}NiS_5$ 85.26 **5**
$C_{20}H_{22}NiS_6$ 85.38 **3**
$C_{20}H_{22}O_2$ 51.8 **7**
$C_{20}H_{22}O_3$ 51.12 **6**
$C_{20}H_{22}O_4Pt$ 78.33 **2**
$C_{20}H_{22}O_7$ 59.9 **7**
$C_{20}H_{22}S_6Zn$ 85.27 **5**
$C_{20}H_{23}BrO_2$ 28.12 **5**
$C_{20}H_{23}BrO_4$ 21.33 **1**
$C_{20}H_{23}BrO_4$ 54.1 **4**
$C_{20}H_{23}BrO_4$ 54.2 **4**
$C_{20}H_{23}BrO_6$ 54.4 **1**
$C_{20}H_{23}CoN_2O_3$ 78.12 **3**
$C_{20}H_{23}IN_2O_4$ 59.45 **1**
$C_{20}H_{23}N_2O^+$, Br^- , C_2H_6O 58.32 **3**
$C_{20}H_{23}N_3NiO_2$ 78.9 **4**
$C_{20}H_{23}N_5$ 32.38 **7**
$C_{20}H_{23}Ni_4$ 73.39 **7**
$C_{20}H_{24}BrNO$ 21.17 **4**
$C_{20}H_{24}BrNO_2$ 31.32 **3**
$C_{20}H_{24}Br_2N_2Ni$ 76.83 **3**
$C_{20}H_{24}ClFeN_2O_2$ 78.12 **5**
$C_{20}H_{24}Cl_2Cu_2N_2O_4$ 78.10 **4**
$C_{20}H_{24}Cl_5NO_3$ 37.25 **1**
$C_{20}H_{24}CoN_2O_4$ 77.44 **2**
$C_{20}H_{24}Co_4$ 73.40 **7**
$C_{20}H_{24}CuN_2O_2$ 78.34 **2**
$C_{20}H_{24}CuN_2O_2$ 78.35 **2**
$C_{20}H_{24}CuN_4O_5$ 83.95 **7**
$C_{20}H_{24}CuN_6O_4S_2$ 81.54 **5**
$C_{20}H_{24}Cu_2N_{20}O_2^{4+}$, $4ClO_4^-$, $2H_2O$ 83.88 **5**
$C_{20}H_{24}NO_4^+$, I^- 58.44 **1**
$C_{20}H_{24}NO_4^+$, I^- 58.19 **5**
$C_{20}H_{24}NO_5^+$, Br^- 58.45 **1**
$C_{20}H_{24}NO_5^+$, Br^- , H_2O 36.26 **3**
$C_{20}H_{24}NS^+$, Cl^- , H_2O 39.44 **5**
$C_{20}H_{24}N_2NiO_2$ 78.36 **2**
$C_{20}H_{24}N_2NiO_4$ 77.45 **2**
$C_{20}H_{24}N_2NiO_6$ 77.46 **2**
$C_{20}H_{24}N_2OS$ 42.10 **6**
$C_{20}H_{24}N_2O_2$ 58.20 **5**
$C_{20}H_{24}N_2O_2$ 58.11 **7**
$C_{20}H_{24}N_3O_6^+$, Br^- 48.94 **1**
$C_{20}H_{24}N_4NiO_2$, $6H_2O$ 76.42 **4**
$C_{20}H_{24}N_4Ni_2O_{12}S_2$ 83.96 **7**
$C_{20}H_{24}NaO_6^+$, Br^- , $2H_2O$ 38.21 **3**
$C_{20}H_{24}NiO_4$ 74.6 **7**
$C_{20}H_{24}OZr$ 75.32 **5**
$C_{20}H_{24}O_2$ 54.1 **5**

$C_{20}H_{24}O_3$ 54.2 **6**
$C_{20}H_{24}O_3S$ 51.19 **4**
$C_{20}H_{24}O_4S_8$ 39.41 **4**
$C_{20}H_{24}O_6$ 59.18 **5**
$C_{20}H_{24}O_6$ 31.25 **7**
$2C_{20}H_{24}O_6$, $0.55Rb^+$, $0.45Na^+$, CNS 38.22 **3**
$C_{20}H_{24}O_7$ 59.19 **5**
$C_{20}H_{24}O_8$ 50.10 **7**
$C_{20}H_{24}Si_2Ti_2$ 73.45 **6**
$C_{20}H_{25}BrO_7$ 54.3 **4**
$C_{20}H_{25}Co_5O_{10}S_5$ 85.84 **2**
$C_{20}H_{25}IO_3$ 31.38 **1**
$C_{20}H_{25}NO_3S$, CH_3NO_2 51.15 **3**
$C_{20}H_{25}N_2O_2$ 33.51 **7**
$C_{20}H_{25}N_2O_2^+$, Br^- 58.33 **6**
$C_{20}H_{25}N_4O_{15}P_2^{3-}$, $3Na^+$, $13H_2O$ 47.39 **7**
$C_{20}H_{26}BrNO_2$ 28.15 **5**
$C_{20}H_{26}BrNO_2$ 58.21 **5**
$C_{20}H_{26}BrNaO_7$, $C_{20}H_{28}NaO_8^+$, Br^- , $2H_2O$ 67.21 **4**
$C_{20}H_{26}Br_2O$ 59.46 **1**
$C_{20}H_{26}Br_2O_3$ 54.4 **4**
$C_{20}H_{26}CuN_4O_2$, $2H_2O$ 83.181 **2**
$C_{20}H_{26}HgN_4O_2$ 83.112 **4**
$C_{20}H_{26}N^+$, Cl^- 33.52 **6**
$C_{20}H_{26}NOS^+$, Cl^- , H_2O 3.33 **7**
$C_{20}H_{26}NO_2^+$, Cl^- 3.23 **5**
$C_{20}H_{26}NO_3^+$, Cl^- 58.20 **4**
$C_{20}H_{26}NO_3^+$, Cl^- 1.37 **7**
$C_{20}H_{26}NO_4^+$, Br^- 58.22 **5**
$C_{20}H_{26}NO_4^+$, $C_6H_2BrN_2O_5^-$ 58.34 **3**
$C_{20}H_{26}N_2$ 9.2 **6**
$C_{20}H_{26}N_3O^+$, $C_7H_4IO_2^-$, H_2O 58.23 **5**
$C_{20}H_{26}N_3O_4S^+$, Br^- , H_2O 50.26 **1**
$C_{20}H_{26}N_4O_4S_6Te$ 70.4 **3**
$C_{20}H_{26}N_4O_8S_2$ 47.40 **7**
$C_{20}H_{26}O$ 51.9 **7**
$C_{20}H_{26}O_2$ 54.5 **1**
$C_{20}H_{26}O_2$ 54.3 **6**
$C_{20}H_{26}O_2$ 38.63 **7**
$C_{20}H_{26}O_2P_2$, $2H_2O$ 64.50 **7**
$C_{20}H_{26}O_3$ 59.19 **6**
$C_{20}H_{26}O_3$ 51.10 **7**
$C_{20}H_{26}O_3$ 54.2 **7**
$C_{20}H_{26}O_6$ 53.22 **7**
$0.3C_{20}H_{26}O_6$, $0.7C_{19}H_{24}O_6$ 53.17 **6**
$C_{20}H_{26}S_5$ 39.26 **6**
$C_{20}H_{27}BrO$ 59.20 **6**
$C_{20}H_{27}BrO_2$ 54.4 **6**
$C_{20}H_{27}BrO_3$ 51.7 **1**
$C_{20}H_{27}BrO_4$ 27.15 **4**
$C_{20}H_{27}ClIrOP_2^+$, F_6P^- 72.17 **7**
$C_{20}H_{27}IO_3$ 51.20 **4**
$C_{20}H_{27}NO_3$ 28.23 **6**
$C_{20}H_{27}N_2^+$, Br^- , CH_4O 58.12 **7**
$C_{20}H_{27}N_2^+$, I^- 58.46 **1**
$C_{20}H_{27}N_2O^+$, Br^- 16.61 **1**
$C_{20}H_{27}N_2O^+$, Br^- 58.47 **1**
$C_{20}H_{27}N_2O_2^+$, I^- 58.48 **1**
$C_{20}H_{27}N_2O_2^+$, I^- 58.49 **1**
$C_{20}H_{27}N_2O_2P$ 64.51 **7**
$C_{20}H_{27}N_2O_3^+$, Br^- 58.35 **3**
$C_{20}H_{27}N_4O_{15}P_2^-$, Na^+ , $12H_2O$ 47.35 **5**
$C_{20}H_{28}$ 31.33 **3**
$C_{20}H_{28}CeO_8$ 77.47 **2**
$C_{20}H_{28}CeO_8$ 77.19 **3**
$C_{20}H_{28}CeO_8$ 77.7 **7**
$C_{20}H_{28}Cl_2O_9Ti_2$, $CHCl_3$ 77.48 **2**
$C_{20}H_{28}CoN_6O_4^+$, Cl^- 83.66 **6**
$C_{20}H_{28}CuN_2O_2$, $0.67C_6H_6$ 61.21 **2**
$C_{20}H_{28}CuN_2O_2$, $C_{10}H_{15}NO$, $2H_2O$ 83.182 **2**
$C_{20}H_{28}MoN_6OS_4$ 72.18 **7**
$C_{20}H_{28}N^+$, Cl^- 37.8 **5**
$C_{20}H_{28}NO_3^+$, Br^- 58.50 **1**
$C_{20}H_{28}N_2NiO_4P_2S_4$ 83.113 **4**
$C_{20}H_{28}N_8P_2Pd$ 83.90 **5**
$C_{20}H_{28}Ni_2O_8$, $2C_3H_8O$ 77.8 **7**
$C_{20}H_{28}NpO_8$ 77.9 **5**
$C_{20}H_{28}O$ 54.6+ **4**
$C_{20}H_{28}O$ 54.2 **5**
$C_{20}H_{28}O_2$ 57.1+ **5**
$C_{20}H_{28}O_3$ 51.21 **4**
$C_{20}H_{28}O_3$ 54.3 **7**
$C_{20}H_{28}O_3$ 59.10 **7**
$C_{20}H_{28}O_4$ 54.2 **3**
$(C_{20}H_{28}O_4P_2Zn)_n$ 84.50+ **2**
$C_{20}H_{28}O_6$ 54.8 **4**
$C_{20}H_{28}O_8Th$ 77.49 **2**
$C_{20}H_{28}O_8U$ 77.20 **3**
$C_{20}H_{28}O_8Zr$ 77.50 **2**
$C_{20}H_{28}O_{10}Ti_2$ 77.10 **5**
$C_{20}H_{28}O_{10}Ti_2$, $2C_4H_8O_2$ 77.11 **5**
$C_{20}H_{29}Al_2NO_2$ 68.25 **5**
$C_{20}H_{29}BrO_3$ 51.22 **4**
$C_{20}H_{29}BrO_3$ 54.5 **6**
$C_{20}H_{29}IO_2$ 51.14 **5**
$C_{20}H_{29}NO_7Yb$, H_2O 77.22 **3**
$C_{20}H_{29}N_2O^+$, I^- 58.36 **3**
$C_{20}H_{29}N_2O_4^+$, Br^- 58.13 **7**
$C_{20}H_{29}N_5NiO^{2+}$, $2ClO_4^-$ 84.50 **7**
$C_{20}H_{30}$ 31.39 **1**
$C_{20}H_{30}Al_2NO_2$ 68.19 **3**
$C_{20}H_{30}As_4Cl_2Ni^+$, Cl^- 86.26 **2**
$C_{20}H_{30}B_2N_8$ 62.10 **7**
$C_{20}H_{30}Br_2O_3$ 54.7 **1**
$C_{20}H_{30}ClF_2OP_3Pt$ 86.11 **6**
$C_{20}H_{30}Cl_2Pd_2$ 71.64 **7**
$C_{20}H_{30}Cl_3OP_2Re$ 86.27 **2**
$C_{20}H_{30}Co_2N_{10}^{4+}$, $4ClO_4^-$ 71.44 **2**
$C_{20}H_{30}Fe_2N_4O_{15}^{2-}$, $C_2H_{10}N_2^{2+}$, $6H_2O$ 83.183 **2**
$C_{20}H_{30}Mn_2O_8P_2$ 86.28 **2**
$C_{20}H_{30}NO_3^+$, Br^- 33.53 **6**
$C_{20}H_{30}NO_7Yb$ 77.23 **3**
$C_{20}H_{30}N_4NiO_4$ 81.38 **7**
$C_{20}H_{30}N_8P_2Pd$ 83.114 **4**
$C_{20}H_{30}O$ 59.21 **3**
$C_{20}H_{30}O$ 54.9 **4**
$C_{20}H_{30}O_2$ 54.3 **5**
$C_{20}H_{30}O_2$ 51.13 **6**
$C_{20}H_{30}O_2$, $0.5H_2O$ 51.14 **6**
$C_{20}H_{30}O_3$ 28.13 **5**
$C_{20}H_{30}O_3$ 54.4 **5**
$C_{20}H_{30}O_4$ 54.5 **5**
$C_{20}H_{30}O_4$ 54.6 **6**
$C_{20}H_{30}O_4$ 54.4 **7**
$C_{20}H_{30}O_5$ 50.5 **4**
$C_{20}H_{31}Cl_3Rh_2$ 73.39 **5**
$C_{20}H_{31}NO$ 33.52 **5**
$C_{20}H_{31}OP_2Pt^+$, F_6P^- 71.45 **6**
$C_{20}H_{32}$ 54.8 **1**

Formula	Entry	Vol.
$C_{20}H_{32}$	54.4	3
$C_{20}H_{32}$	51.11	7
$C_{20}H_{32}As_2Cl_3N_2Rh$	86.25	3
$C_{20}H_{32}As_2Cu_2I_2N_2$	86.12	4
$C_{20}H_{32}As_4AuI_2^+$, I^-	86.26	3
$C_{20}H_{32}As_4Br_4Ta^+$, Br_6Ta^-	86.32	7
$C_{20}H_{32}As_4Cl_2Co^+$, Cl^-	86.25	5
$C_{20}H_{32}As_4Cl_2Co^+$, ClO_4^-	86.27	3
$C_{20}H_{32}As_4Cl_2Ni^+$, Cl^-	86.26	5
$C_{20}H_{32}As_4Cl_2Pt$	86.31	2
$C_{20}H_{32}As_4Cl_4Mo^+$, I_3^-	86.33	7
$C_{20}H_{32}As_4Co^{2+}$, $2ClO_4^-$	86.32	2
$C_{20}H_{32}As_4CoNO^{2+}$, $2ClO_4^-$	86.34	7
$C_{20}H_{32}As_4I_2Ni$	86.33	2
$C_{20}H_{32}As_4I_2Pd$	86.34	2
$C_{20}H_{32}As_4I_2Pt$	86.35	2
$C_{20}H_{32}B_2N_8Ni$	83.97	7
$C_{20}H_{32}Cl_2MnP_4$	86.35	7
$C_{20}H_{32}Cl_2MnP_4^+$, ClO_4^-	86.36	7
$C_{20}H_{32}Cl_2MnP_4^{2+}$, $2HN_2O_8^-$	86.37	7
$C_{20}H_{32}Cl_4Ru_2$	72.16	4
$C_{20}H_{32}Co_2O_{10}$	77.51	2
$C_{20}H_{32}N_4O_4S_8Te$, $3C_6H_6$	70.10	5
$C_{20}H_{32}N_8Ni^{2+}$, $2F_6P^-$	83.112	3
$C_{20}H_{32}O_2$	59.21	5
$C_{20}H_{32}O_2$	54.5	7
$C_{20}H_{32}O_3$	51.15	6
$C_{20}H_{32}O_3$, H_2O	51.12	7
$C_{20}H_{32}O_4$	59.21	6
$C_{20}H_{32}O_4$	59.11	7
$C_{20}H_{32}O_4$	59.12	7
$C_{20}H_{32}O_5$	59.22	6
$C_{20}H_{32}O_5$, H_2O	54.6	5
$C_{20}H_{32}S_2$	39.35	7
$C_{20}H_{33}Al_2ClZr$	71.65	7
$C_{20}H_{33}BrO$	51.23	4
$C_{20}H_{33}ClP_2PtSi$	71.46	6
$C_{20}H_{33}Cl_3O_3PRh$	84.51	7
$C_{20}H_{34}N_3^+$, Br^-	58.24	5
$C_{20}H_{35}BrO_2$	54.9	1
$C_{20}H_{35}BrO_2$	54.7	5
$C_{20}H_{35}N_3^{2+}$, $2ClO_4^-$	58.51	1
$C_{20}H_{36}Ag^+$, NO_3^-	75.67	2
$C_{20}H_{36}AgNO_3$	23.4	4
$C_{20}H_{36}AgO_{12}P_4^+$, ClO_4^-	86.36	2
$C_{20}H_{36}B_4N_8S_4$	62.50	2
$C_{20}H_{36}NaO_6^+$, Br^- , $2H_2O$	38.25	5
$C_{20}H_{36}O_6$	38.64 +	7
$C_{20}H_{36}O_6$, Na^+ , Br^- , $2H_2O$	38.36	6
$C_{20}H_{37}B_9P_2Pt$	71.66	7
$C_{20}H_{37}ClINP_2Pt$	71.67	7
$C_{20}H_{38}Cl_3N_2P_2Pt^+$, ClO_4^-	71.68	7
$C_{20}H_{38}Cl_4$	5.39	1
$C_{20}H_{38}CuO_4$	81.39	7
$C_{20}H_{38}FeSi_4$	73.94	2
$C_{20}H_{38}N_3^{3+}$, $3Cl^-$, $2H_2O$	58.34	6
$C_{20}H_{40}$	23.2	7
$C_{20}H_{40}Cd_2N_4S_8$	80.35 +	2
$(C_{20}H_{40}Cl_2Hg_3N_4S_8)_n$	80.25	5
$C_{20}H_{40}Co_4O_4S_8$	85.38	6
$C_{20}H_{40}Cu_2N_4O_{10}S_4$, $7H_2O$	82.24	6
$C_{20}H_{40}Cu_4N_4S_8$	80.37	2
$C_{20}H_{40}EuN_4S_8^-$, $C_8H_{20}N^+$	80.12	3
$C_{20}H_{40}Hg_2N_4S_8$	80.16	6
$C_{20}H_{40}Li_2O_6^{2+}$, $2C_5H_5Cl_3Cr^-$	67.16	5
$C_{20}H_{40}MoN_4S_8$	80.18	7
$C_{20}H_{40}N_4NpS_8^-$, $C_8H_{20}N^+$	80.13	3
$C_{20}H_{40}N_4O_3Re_2S_8$	80.13 +	4
$C_{20}H_{40}N_4Pb_2S_8$	69.38	2
$C_{20}H_{40}N_4S_8Sn$	69.31	4
$C_{20}H_{40}N_4S_8Te$	70.11	5
$C_{20}H_{40}N_4S_8Th$	80.15	3
$C_{20}H_{40}N_4S_8Ti$	80.15	4
$C_{20}H_{40}N_4S_8W^+$, Br^-	80.17	6
$C_{20}H_{40}N_4S_8Zn_2$	80.20	2
$C_{20}H_{40}N_8$, C_6H_6	37.20	7
$C_{20}H_{40}O_2$	1.151 +	1
$C_{20}H_{42}Cl_2N_2P_2Pt_2S_2$	86.28 +	3
$C_{20}H_{42}Li_2N_4$	67.17	5
$C_{20}H_{42}N_2P_2PdS_2$	86.38	2
$C_{20}H_{44}BaN_2O_9^{2+}$, $2CNS^-$	67.14	6
$C_{20}H_{44}Cl_2Cu_2N_4O_{12}^{2+}$, $2ClO_4^-$	81.55	5
$C_{20}H_{44}CuN_4O_8^{2+}$, $2ClO_4^-$	81.56	5
$C_{20}H_{44}N^+$, F^- , $38H_2O$	61.22	2
$C_{20}H_{47}Cl_2P_2Rh$	86.14	4
$C_{20}H_{48}N_8PtS_4^{2+}$, $2Cl^-$, H_2O	79.24	7
$C_{20}H_{48}N_{12}O_4P_4W$	83.67	6
$C_{20}H_{50}Cu_2N_4O_2^{2+}$, $2ClO_4^-$	76.47	6

C_{21}

Formula	Entry	Vol.
$C_{21}Co_6O_{19}$	71.45	2
$C_{21}H_7F_{24}O_2Rh$	71.34	3
$C_{21}H_9N_3NdO_{12}^{3-}$, $3Na^+$, $15H_2O$	81.51	4
$C_{21}H_9N_3O_{12}Yb^{3-}$, ClO_4^- , $4Na^+$, $10H_2O$	81.52	4
$C_{21}H_9N_3O_{12}Yb^{3-}$, $3Na^+$, $14H_2O$	81.50	3
$C_{21}H_9N_3O_{12}Yb^{3-}$, $3Na^+$, $13H_2O$	81.53	4
$C_{21}H_{10}CoMnO_9Sn$	69.39	2
$C_{21}H_{10}F_{10}O_2$	31.28	6
$C_{21}H_{10}Mo_2O_{11}Re_2S_2$	85.24	4
$C_{21}H_{11}Fe_2O_6^+$, BF_4^-	75.24	4
$C_{21}H_{12}N_2O_3$	36.27	3
$C_{21}H_{12}N_4$	36.31	1
$2C_{21}H_{12}N_4^+$, Cl_4Cu^{2-}	36.32	1
$C_{21}H_{13}BrN_2O_3$	40.18	3
$C_{21}H_{13}Co_3O_6$	71.35	3
$C_{21}H_{13}Fe_2NO_6$	72.17	4
$C_{21}H_{13}N$	36.33	1
$C_{21}H_{13}O^+$, Cl^- , $C_2H_4O_2$	38.26	5
$C_{21}H_{14}BrNO_3$	19.62	1
$C_{21}H_{14}BrN_3OS$	41.32	4
$C_{21}H_{14}CrO_3$	75.21	7
$C_{21}H_{14}O$	38.37	6
$C_{21}H_{15}^+$, ClO_4^-	12.18	1
$C_{21}H_{15}AlO_6$	68.26	5
$C_{21}H_{15}AsO_3S_3$	65.5	6
$C_{21}H_{15}BrOSi$	63.9	5
$2C_{21}H_{15}ClFe_3O_6Sb^+$, Cl_4Fe^{2-} , CH_2Cl_2	73.25	4
$C_{21}H_{15}ClO_6Sn$, $CHCl_3$	69.29	3
$C_{21}H_{15}Cl_2N$	35.33	7
$C_{21}H_{15}CoN_7OS_2$	83.68	6
$C_{21}H_{15}CrS_6$	85.85	2
$C_{21}H_{15}FeO_6$	84.52	2
$C_{21}H_{15}MnNO_4P$	86.39	2
$C_{21}H_{15}MnO_6$, $0.25C_7H_8$	84.52	7
$C_{21}H_{15}NO$	27.8	7
$C_{21}H_{15}NS_2$	39.27	6

$C_{21}H_{15}N_3$ 33.81 **1**
$C_{21}H_{15}O_6Sc$ 84.22 **6**
$C_{21}H_{16}$ 30.8 **1**
$C_{21}H_{16}ClNO$ 32.36 **6**
$C_{21}H_{16}CrO_5$ 75.68 **2**
$C_{21}H_{16}O$ 23.5 **4**
$C_{21}H_{16}O_5Ru_2$ 71.69 **7**
$C_{21}H_{16}O_7Sn$, $2.1H_2O$, $0.45CH_4O$ 69.30 **3**
$C_{21}H_{16}O_9Ru_3$ 71.45 **4**
$C_{21}H_{17}BrO_2$ 28.14 **5**
$C_{21}H_{17}ClN_4O_3$ 40.24 **6**
$C_{21}H_{17}F_{12}O_5PRu$ 71.70 **7**
$C_{21}H_{17}N$ 16.17 **3**
$C_{21}H_{17}N_3S_2Zn$ 85.39 **6**
$C_{21}H_{18}$ 29.6 **4**
$C_{21}H_{18}CuN_7S^+$, $C_{22}H_{18}CuN_8S_2$, ClO_4^- 83.117 **4**
$C_{21}H_{18}FeN_3O_6$, $3H_2O$ 84.53 **2**
$C_{21}H_{18}Fe_2O$ 73.95 **2**
$C_{21}H_{18}Mo_2O_6$ 75.69 **2**
$C_{21}H_{18}N_2$ 9.13 **4**
$C_{21}H_{18}N_2$ 9.14 **4**
$C_{21}H_{19}AuBr_3P$ 71.46 **4**
$C_{21}H_{19}BrO_7S$ 53.21 **3**
$C_{21}H_{19}BrS$ 39.90 **1**
$C_{21}H_{19}ClN_2O_4$ 35.28 **4**
$C_{21}H_{19}ClN_2O_4$ 35.29 **4**
$C_{21}H_{19}ClN_2O_4$ 40.23 **4**
$C_{21}H_{19}Cl_2MnN_5O_8$ 83.98 **7**
$C_{21}H_{19}CoN_3O_2$ 78.14 **3**
$C_{21}H_{19}F_{12}IrO_2$ 71.71 **7**
$C_{21}H_{19}N_2O_4V$ 84.23 **6**
$C_{21}H_{19}N_4NiO_5^+$, NO_3^- 83.184 **2**
$C_{21}H_{19}N_5NiO_6$ 83.99 **7**
$C_{21}H_{19}O_2P$ 64.32 **5**
$C_{21}H_{20}BrNO_7S$ 45.48 **3**
$C_{21}H_{20}BrO_8$ 31.40 **1**
$C_{21}H_{20}ClPPd$ 72.36 **3**
$C_{21}H_{20}Cl_3PPdSn$, $0.4C_3H_6O$ 72.37 **3**
$C_{21}H_{20}CoN_4O_3^+$, ClO_4^- , H_2O 83.91 **5**
$C_{21}H_{20}CuN_6S^{2+}$, $2ClO_4^-$ 79.16 **6**
$C_{21}H_{20}Fe_2O_5$ 71.72 **7**
$C_{21}H_{20}N_3^+$, Br^- , H_2O 36.31 **4**
$C_{21}H_{20}N_3^+$, Br^- , CH_4O 36.34 **1**
$C_{21}H_{20}N_3^+$, Br^- , C_2H_6O 36.28 **7**
$C_{21}H_{20}N_3^+$, Cl^- , C_2H_6O 36.29 **7**
$C_{21}H_{20}Si$ 63.10 **5**
$C_{21}H_{21}As$ 65.34 **2**
$C_{21}H_{21}BrO_4$ 24.14 **4**
$C_{21}H_{21}ClN_2O_3Pd$ 71.73 **7**
$C_{21}H_{21}ClN_2O_8^+$, Cl^- , $3H_2O$ 50.8 **5**
$C_{21}H_{21}ClNiPS_3^+$, ClO_4^- 86.31 **3**
$C_{21}H_{21}CoN_2O_4$, $0.7H_2O$ 77.14 **4**
$C_{21}H_{21}N$ 3.24 **5**
$C_{21}H_{21}N_2NbO_4S_2$ 77.9 **7**
$C_{21}H_{21}N_2S_2^+$, $AuCl_2^-$ 41.30 **6**
$C_{21}H_{21}N_2S_2^+$, Br^- 41.47 **1**
$C_{21}H_{21}N_2S_2^+$, Br^- , C_2H_6O 41.48 **1**
$C_{21}H_{21}N_2S_2^+$, $C_7H_7O_3S^-$ 41.27 **7**
$C_{21}H_{21}N_2S_2^+$, $C_{12}H_4N_4^-$, $C_{12}H_4N_4$ 60.35 + **7**
$C_{21}H_{21}N_3O_6S_4^{2-}$, $2K^+$, H_2O , C_2H_3N 4.9 **7**
$C_{21}H_{21}OP$ 64.28 **4**
$C_{21}H_{21}O_4P$ 46.8 **3**
$C_{21}H_{21}P$ 64.29 **4**
$C_{21}H_{21}P$ 64.40 **6**
$C_{21}H_{21}PSe$ 64.30 **4**
$C_{21}H_{22}BrNO_3$ 13.22 **4**
$C_{21}H_{22}BrNO_4$ 53.22 **3**
$C_{21}H_{22}BrN_3O_3$, C_3H_6O 58.14 **7**
$C_{21}H_{22}NO_4^+$, I^- , $2H_2O$ 58.21 **4**
$C_{21}H_{22}NO_6^+$, Br^- 58.52 **1**
$C_{21}H_{22}NO_6^+$, I^- 58.37 **3**
$C_{21}H_{22}NO_6+$, I^- 58.22 **4**
$C_{21}H_{22}N_2O_5S$, $4H_2O$ 58.35 **6**
$C_{21}H_{22}N_2O_8S$ 41.31 **6**
$C_{21}H_{22}N_2PS^+$, Br^- 64.52 **7**
$2C_{21}H_{22}N_5^+$, $Cl_4O_2U^{2-}$, C_2H_3N 33.52 **7**
$C_{21}H_{22}O_2$ 21.19 **4**
$C_{21}H_{22}O_2$ 21.12 **7**
$C_{21}H_{22}O_4$ 51.15 **5**
$C_{21}H_{22}P^+$, Br^- 64.53 **7**
$C_{21}H_{23}AgAs_3CoO_4$ 86.32 **3**
$C_{21}H_{23}BrN_2O_8$ 35.34 **7**
$C_{21}H_{23}BrO_2$ 31.26 **7**
$C_{21}H_{23}BrO_3$ 24.15 **4**
$C_{21}H_{23}BrO_4$ 38.58 **1**
$C_{21}H_{23}ClFNO_2$ 33.54 **6**
$C_{21}H_{23}F_2NO_2$ 33.43 **4**
$C_{21}H_{23}NO_5$ 58.53 **1**
$C_{21}H_{23}NO_5V$ 77.15 **4**
$C_{21}H_{23}N_2O_2^+$, Br^- , $2H_2O$ 58.54 **1**
$C_{21}H_{24}BrN_3O_3$ 58.25 **5**
$C_{21}H_{24}Br_2O_7$ 59.17 **4**
$C_{21}H_{24}F_2NO_2^+$, Cl^- 33.44 **4**
$C_{21}H_{24}FeO_4S$ 73.40 **3**
$C_{21}H_{24}GdO_6^{3+}$, $3Cl^-$ 84.18 **5**
$C_{21}H_{24}NO_3^+$, Br^- 37.26 **1**
$C_{21}H_{24}N_2O_2^{2+}$, O_4S^{2-} , $5H_2O$ 58.55 **1**
$C_{21}H_{24}OSi_2$ 63.25 **2**
$C_{21}H_{24}O_2$ 51.13 **7**
$C_{21}H_{24}O_3S$ 31.33 **5**
$C_{21}H_{24}O_3Si_3$ 63.13 **7**
$C_{21}H_{24}O_3Si_3$ 63.14 **7**
$C_{21}H_{24}O_9$ 53.6 **4**
$C_{21}H_{24}O_9$ 53.14 **5**
$C_{21}H_{25}BMoN_4O_2$ 72.19 **6**
$C_{21}H_{25}BrO_5$, CH_3NO_2 59.18 **4**
$C_{21}H_{25}BrO_5$, $0.5C_6H_6$ 31.29 **6**
$C_{21}H_{25}NO_5$ 58.36 **6**
$C_{21}H_{25}N_2O^+$, I^- 58.56 **1**
$2C_{21}H_{25}N_2O_2^+$, Cl^- , ClO_4^- 58.15 **7**
$C_{21}H_{25}N_2O_2Rh$ 71.49 **6**
$C_{21}H_{25}N_2O_3^+$, I^- 58.23 **4**
$C_{21}H_{25}N_3NiO_2$ 78.15 **3**
$C_{21}H_{25}N_3O_2$ 35.31 **6**
$C_{21}H_{25}N_5O$ 35.30 **4**
$C_{21}H_{26}BrNO$ 53.27 **1**
$C_{21}H_{26}ClNO_7$ 58.57 **1**
$C_{21}H_{26}ClN_3O_2$ 36.32 **4**
$C_{21}H_{26}IrOP_2^+$, BF_4^- 72.28 **5**
$C_{21}H_{26}IrOP_2^+$, ClO_4^- 72.29 **5**
$C_{21}H_{26}N^+$, Br^- , H_2O 35.46 **1**
$C_{21}H_{26}NO_2^+$, Cl^- 33.55 **6**
$C_{21}H_{26}NO_4^+$, I^- 58.37 **6**
$C_{21}H_{26}NO_5^+$, Br^- 58.58 **1**
$C_{21}H_{26}N_2O_3$ 58.38 **6**
$C_{21}H_{26}N_2O_3$, CH_4O 58.39 **6**
$C_{21}H_{26}N_2O_4S_2$ 59.23 **5**
$C_{21}H_{26}N_2O_5S$ 59.13 **7**
$C_{21}H_{26}O_3$ 38.28 **5**
$C_{21}H_{26}O_7$ 50.11 **7**

$C_{21}H_{27}BrN_2O_2$ 53.7 **4**
$C_{21}H_{27}BrO_2$ 51.8 **1**
$C_{21}H_{27}BrO_2$ 27.16 **4**
$C_{21}H_{27}BrO_3$ 54.7 **6**
$C_{21}H_{27}ClO_5$ 51.16 **3**
$C_{21}H_{27}NO$ 58.40 + **6**
$C_{21}H_{27}NO_7$, H_2O 58.28 **5**
$C_{21}H_{27}N_2O_2^+$, Br^- 58.59 **1**
$C_{21}H_{27}N_2O_2^+$, Br^- , H_2O 58.16 **7**
$C_{21}H_{27}N_2O_3^+$, Br^- , CH_4O 58.42 **6**
$C_{21}H_{27}N_2O_3^+$, Cl^- 58.29 **5**
$C_{21}H_{27}N_3O_7$ 50.12 **7**
$C_{21}H_{28}Br_2O_2$ 59.23 **3**
$C_{21}H_{28}IN_3O_2$ 53.28 **1**
$C_{21}H_{28}IrOP_2^+$, BF_4^- 72.19 **7**
$C_{21}H_{28}IrOP_2^+$, ClO_4^- 72.20 **7**
$C_{21}H_{28}N^+$, Cl^- , H_2O 33.56 **6**
$C_{21}H_{28}NO^+$, Br^- 58.60 **1**
$C_{21}H_{28}NO_2^+$, Br^- 1.38 **7**
$C_{21}H_{28}NO_2^+$, I^- 58.38 **3**
$C_{21}H_{28}NO_4^+$, I^- 58.39 **3**
$C_{21}H_{28}O_2$ 17.50 **1**
$C_{21}H_{28}O_2$ 51.17 **3**
$C_{21}H_{28}O_2$ 17.21 **5**
$C_{21}H_{28}O_2$ 51.14 **7**
$C_{21}H_{28}O_5$ 51.25 **4**
$C_{21}H_{28}O_5$, H_2O 51.23 **5**
$C_{21}H_{28}O_6$ 51.15 **7**
$C_{21}H_{28}O_6$ 59.14 **7**
$C_{21}H_{28}O_7$ 53.23 **7**
$C_{21}H_{28}O_8$ 53.24 **7**
$C_{21}H_{28}PW^+$, F_6P^- 71.74 **7**
$C_{21}H_{29}BrN_2O_3S$ 53.29 **1**
$C_{21}H_{29}BrO_2$ 51.10 **1**
$C_{21}H_{29}BrO_3$ 51.9 **1**
$C_{21}H_{29}BrO_3$ 51.11 **1**
$C_{21}H_{29}BrO_3S$ 53.23 **3**
$C_{21}H_{29}BrO_4S$ 53.24 **3**
$C_{21}H_{29}BrO_5$ 51.16 **6**
$C_{21}H_{29}ClIrOP_2^+$, F_6P^- 72.21 **7**
$C_{21}H_{29}ClO_5$ 51.16 **7**
$C_{21}H_{29}Cl_2N_2PPt$ 71.75 **7**
$C_{21}H_{29}Cl_2N_2PPt$ 71.76 **7**
$C_{21}H_{29}FO_5$ 51.26 **4**
$C_{21}H_{29}FO_5$ 51.27 **4**
$C_{21}H_{29}FO_5$ 51.28 **4**
$C_{21}H_{29}FO_5$ 51.17 **5**
$C_{21}H_{29}N_2O^+$, Cl^- 3.34 **7**
$C_{21}H_{29}N_2O_4^+$, I^- 58.62 **1**
$C_{21}H_{29}N_3O_6$ 48.68 **7**
$C_{21}H_{30}FN_3O_2$ 33.53 **7**
$C_{21}H_{30}NO_4^+$, I^- 58.40 **3**
$C_{21}H_{30}NO_4^+$, I^- 58.41 **3**
$C_{21}H_{30}O_2$ 50.9 **5**
$C_{21}H_{30}O_2$ 51.18 **5**
$C_{21}H_{30}O_2$ 51.17 **6**
$C_{21}H_{30}O_2$ 51.17 **7**
$C_{21}H_{30}O_2$ 51.18 **7**
$C_{21}H_{30}O_2$, $C_6H_6O_2$ 60.37 **7**
$C_{21}H_{30}O_3$ 51.18 **3**
$C_{21}H_{30}O_3$ 51.30 **4**
$C_{21}H_{30}O_3$ 51.19 **5**
$C_{21}H_{30}O_3$ 51.20 **5**
$C_{21}H_{30}O_3$ 51.19 **7**
$C_{21}H_{30}O_3$ 51.20 **7**
$C_{21}H_{30}O_4$ 51.21 **5**
$C_{21}H_{30}O_4$ 51.18 **6**
$C_{21}H_{30}O_5$, CH_4O 51.22 **5**
$C_{21}H_{30}O_5$, C_5H_5N 51.19 **6**
$C_{21}H_{30}Pt$ 75.22 **7**
$C_{21}H_{31}BrO_3$ 51.12 **1**
$C_{21}H_{31}BrO_4$ 51.13 **1**
$C_{21}H_{31}BrO_4$ 51.14 **1**
$C_{21}H_{31}ClO_5$ 54.10 **4**
$C_{21}H_{31}IO_2$ 51.24 **5**
$C_{21}H_{31}NO_2^+$, I^- 54.5 **3**
$C_{21}H_{31}N_2^+$, I^- 58.63 **1**
$C_{21}H_{31}OP_2Pt^+$, F_6P^- 71.77 **7**
$C_{21}H_{31}P_2Pt^+$, F_6P^- 71.50 **6**
$C_{21}H_{32}As_4CoN_2OS^+$, CNS^- 86.38 **7**
$C_{21}H_{32}ClN_2O^+$, $C_7H_5O_3^-$ 32.39 **7**
$C_{21}H_{32}O_2$ 51.25 **5**
$C_{21}H_{32}O_4$ 54.6 **7**
$C_{21}H_{32}O_7$ 50.13 **7**
$C_{21}H_{33}BrHgO_5$ 59.48 **1**
$C_{21}H_{33}NO^+$, I^- 58.64 **1**
$C_{21}H_{33}N_3O_6$ 48.81 **6**
$C_{21}H_{33}N_5O_3S$ 44.76 **1**
$C_{21}H_{34}NO^+$, I^- 58.65 **1**
$C_{21}H_{34}NP_2Pt^+$, F_6P^- 71.51 **6**
$C_{21}H_{34}O_2$ 51.21 **7**
$C_{21}H_{34}O_2$ 51.22 **7**
$C_{21}H_{34}O_3$ 59.23 **6**
$C_{21}H_{34}O_3$, CH_4O 54.7 **7**
$C_{21}H_{34}O_3S$ 21.10 **6**
$C_{21}H_{35}N_3$ 58.66 **1**
$C_{21}H_{35}N_3$ 51.23 **7**
$C_{21}H_{35}N_3S_6Te$ 70.12 **5**
$C_{21}H_{35}N_7O_7$, H_2O 48.69 **7**
$C_{21}H_{36}Mo_3O_{13}$ 77.53 **2**
$C_{21}H_{36}N_3^+$, I^- , $0.5CH_4O$ 37.27 **1**
$C_{21}H_{36}O_2$, $0.5H_2O$ 51.20 **6**
$C_{21}H_{36}O_3$ 38.59 **1**
$C_{21}H_{37}N$ 51.21 **6**
$C_{21}H_{40}BNiP$ 72.30 **5**
$C_{21}H_{40}N_8O_{12}$, $1.5H_2O_4Se$, $4H_2O$ 50.27 **1**
$C_{21}H_{42}O_2$ 1.153 **1**
$C_{21}H_{43}N_4NiO_2^+$, ClO_4^- 77.12 **5**
$C_{21}H_{47}IrP_2$ 72.22 **7**

C_{22}

$C_{22}Co_6O_{18}$ 71.38 **3**
$C_{22}H_8CdMn_2N_2O_{10}$ 83.69 **6**
$C_{22}H_{10}$ 5.40 **1**
$C_{22}H_{10}Fe_2O_8$ 71.46 **2**
$C_{22}H_{10}O_2$ 30.3 **3**
$C_{22}H_{12}$ 30.9 + **1**
$C_{22}H_{12}$ 30.1 **7**
$C_{22}H_{12}O_2$ 30.11 **1**
$C_{22}H_{12}S_2$ 39.91 **1**
$C_{22}H_{14}$ 30.12 **1**
$C_{22}H_{14}$ 30.13 **1**
$C_{22}H_{14}$ 30.14 **1**
$C_{22}H_{14}$ 30.2 **7**
$C_{22}H_{14}O_9Ru_4$ 75.32 + **3**
$C_{22}H_{15}AuCoO_4P$ 86.27 **5**
$C_{22}H_{15}ClFe_2MoO_7Sn$ 73.96 **2**
$C_{22}H_{15}ClMnO_4P$ 86.15 **4**

$C_{22}H_{15}ClN_2O$ 32.27 **4**
$C_{22}H_{15}F_6P$ 64.41 **6**
$C_{22}H_{15}FeO_4Sb$ 86.39 **7**
$C_{22}H_{16}$ 30.16+ **1**
$C_{22}H_{16}$ 24.7 **3**
$C_{22}H_{16}$ 30.4 **3**
$C_{22}H_{16}$ 30.2 **4**
$C_{22}H_{16}$ 30.3 **4**
$C_{22}H_{16}AgN_8O_4^+$, NO_2^- , $4H_2O$ 84.24 **6**
$C_{22}H_{16}AgN_8O_4^+$, xNO_2^- , yNO_3^- , $4H_2O$ 84.25 **6**
$C_{22}H_{16}AsF_4FeO_4P$ 86.16 **4**
$C_{22}H_{16}Br_2$ 24.10 **7**
$C_{22}H_{16}CuN_8O_4^+$, ClO_4^- , CH_2O_2 84.53 **7**
$C_{22}H_{16}FeN_6S_2$ 83.115+ **4**
$C_{22}H_{16}N_8O_6Pb^{2+}$, $2ClO_4^-$, $4H_2O$ 69.18 **7**
$C_{22}H_{16}O_{10}Ru_4$ 71.51 **5**
$C_{22}H_{17}Br$ 28.16 **5**
$C_{22}H_{17}Cl$ 31.34 **3**
$C_{22}H_{17}ClN_2O_2$ 16.24 **5**
$C_{22}H_{17}ClO_2$ 31.27 **7**
$C_{22}H_{17}Co_3O_6$ 71.52 **5**
$C_{22}H_{17}F_{12}FeO_5P$ 75.23 **7**
$C_{22}H_{17}IN_2O_5$ 58.67 **1**
$2C_{22}H_{17}N_2O_3^+$, Cl_4Hg^{2-} , $2H_2O$ 58.68 **1**
$C_{22}H_{17}P$ 64.31 **4**
$C_{22}H_{18}ClN_3O_2$ 32.40 **7**
$C_{22}H_{18}Cl_2N_2Sn$ 69.19 **7**
$C_{22}H_{18}Cl_3NO_3S$ 36.28 **5**
$C_{22}H_{18}Cl_4Cu_2N_2O_2$, H_2O 84.19 **5**
$C_{22}H_{18}CuN_8S_2$, $C_{21}H_{18}CuN_7S^+$, ClO_4^- 83.117 **4**
$C_{22}H_{18}Fe$ 73.98 **2**
$C_{22}H_{18}Fe$ 73.41 **3**
$C_{22}H_{18}Fe_2O_6$ 75.24 **7**
$C_{22}H_{18}N_2O_3S$ 35.35 **7**
$C_{22}H_{18}N_8NiS_2$, H_2O 83.113 **3**
$C_{22}H_{18}O_6Ru_3$ 73.41 **7**
$C_{22}H_{19}Br$ 19.14 **4**
$C_{22}H_{19}BrOS$ 39.43 **4**
$C_{22}H_{19}BrO_6$ 50.13 **3**
$C_{22}H_{19}BrO_7$ 53.15 **5**
$C_{22}H_{19}Br_2NO_3$ 20.22 **7**
$C_{22}H_{19}CuN_3O_6$ 83.92 **5**
$C_{22}H_{20}Cr_2N_2O_2S_2$ 73.99 **2**
$C_{22}H_{20}CuN_8O_6^{2+}$, $2ClO_4^-$, $2H_2O$ 84.26 **6**
$C_{22}H_{20}FeN_6S_2$ 83.185 **2**
$C_{22}H_{20}FeN_8O_6^{2+}$, $2ClO_4^-$, $4H_2O$ 84.54 **7**
$C_{22}H_{20}N_2O_5U$ 78.11 **4**
$C_{22}H_{20}N_2O_8S_2$ 59.50 **1**
$C_{22}H_{20}N_6NiS_2$ 83.186 **2**
$C_{22}H_{20}O_2$ 31.42 **1**
$C_{22}H_{20}O_2$ 28.24 **6**
$C_{22}H_{20}O_4$ 29.26+ **1**
$C_{22}H_{20}O_7$ 53.8 **4**
$C_{22}H_{20}Ti$ 71.47 **4**
$C_{22}H_{21}BrO_8$, $0.5C_4H_8O_2$ 59.24 **5**
$C_{22}H_{21}Br_2N_3O_2$ 40.25 **6**
$C_{22}H_{21}ClN_2O_3$ 25.6 **7**
$C_{22}H_{21}CuNO_4$ 77.54 **2**
$C_{22}H_{21}NO$ 20.23 **7**
$C_{22}H_{21}NO_{11}$ 58.30 **5**
$C_{22}H_{21}N_3O_5$ 48.75 **5**
$C_{22}H_{21}O_3PS_4U$ 85.26 **4**
$C_{22}H_{22}BrNO_8$ 58.43 **6**
$C_{22}H_{22}BrN_3O_8S$, $0.5C_6H_6$ 50.28 **1**
$C_{22}H_{22}ClPPd$ 72.59 **2**
$C_{22}H_{22}CoN_2O_2$ 83.93 **5**
$C_{22}H_{22}CoN_3O_3$ 76.48 **6**
$C_{22}H_{22}CoN_3O_3$, CH_4O 78.16 **3**
$C_{22}H_{22}CuO_4$ 77.55 **2**
$C_{22}H_{22}Cu_2N_2O_4$ 83.90 **2**
$C_{22}H_{22}Fe^+$, F_6P^- 73.42 **7**
$C_{22}H_{22}Fe_2$ 73.47 **6**
$C_{22}H_{22}N_2OPd$ 71.78 **7**
$C_{22}H_{22}N_2O_9^{2-}$, $2K^+$, $4H_2O$ 50.14 **7**
$C_{22}H_{22}OS_2Si$ 63.15 **7**
$C_{22}H_{22}OS_4$ 39.28 **6**
$C_{22}H_{22}O_7$ 59.25 **5**
$C_{22}H_{22}O_8$ 59.26 **5**
$C_{22}H_{22}O_8Ru_3Si_2$ 75.25 **7**
$C_{22}H_{22}S_5$ 39.49 **3**
$C_{22}H_{23}As_3CuMnO_5$ 86.41 **2**
$C_{22}H_{23}IO_5$ 53.25 **7**
$C_{22}H_{23}NO_2$ 24.32 **1**
$C_{22}H_{23}N_2O_7^+$, Br^- 50.14 **3**
$C_{22}H_{23}N_2P$ 64.54 **7**
$C_{22}H_{24}BrN_3O$ 36.33 **4**
$C_{22}H_{24}Br_2O_6$ 38.60 **1**
$C_{22}H_{24}ClN_2O_8^+$, Cl^- 50.29 **1**
$C_{22}H_{24}Cl_2N_4O_2Zn$ 84.12 **4**
$C_{22}H_{24}CoN_6O_8$ 84.55 **7**
$C_{22}H_{24}CuN_6O_8$ 84.56+ **7**
$C_{22}H_{24}FN_3O_2$ 35.32 **6**
$C_{22}H_{24}FeN_8^{2+}$, $2ClO_4^-$ 83.100 **7**
$C_{22}H_{24}FeO$ 73.40 **5**
$C_{22}H_{24}NO_4^+$, I^- 58.69 **1**
$C_{22}H_{24}N_2O_4$ 27.9 **7**
$C_{22}H_{24}N_2O_8$, $6H_2O$ 50.15 **7**
$C_{22}H_{24}N_2O_9$, Cl_2Hg , $2H_2O$ 50.16 **7**
$C_{22}H_{24}N_6NiO_8$ 83.94+ **5**
$C_{22}H_{24}N_6O_8Zn$ 84.58 **7**
$C_{22}H_{24}O_2$ 29.10 **7**
$C_{22}H_{24}O_5Zn$ 77.24 **3**
$C_{22}H_{24}O_9S$, H_2O 54.8 **7**
$C_{22}H_{25}BrN_2O_4$ 48.55 **4**
$C_{22}H_{25}BrO_2$ 31.29 **4**
$C_{22}H_{25}BrO_4$ 53.32 **1**
$C_{22}H_{25}BrO_5$ 53.9 **4**
$C_{22}H_{25}BrO_8$, $0.5C_6H_6$ 53.19 **6**
$C_{22}H_{25}ClN_2O_4$ 48.56 **4**
$C_{22}H_{25}N$ 29.11 **7**
$C_{22}H_{25}NO_8^+$, Cl^- 50.6 **4**
$C_{22}H_{25}N_2O_9^+$, Cl^- 50.30 **1**
$C_{22}H_{25}O_6^+$, Cl_4Fe^- 59.51 **1**
$C_{22}H_{26}ClN_2^+$, $C_{12}H_4N_4^-$, $C_{12}H_4N_4$ 60.38 **7**
$C_{22}H_{26}Cl_2NiP_2$ 86.17 **4**
$C_{22}H_{26}N^+$, Br^- 32.28 **4**
$C_{22}H_{26}NPr$ 71.52 **6**
$C_{22}H_{26}N_2O_2Pd$ 78.38+ **2**
$C_{22}H_{26}N_2O_4$, $2H_2O$ 37.12 **6**
$C_{22}H_{26}N_4$ 36.19 **6**
$C_{22}H_{26}N_4O$ 32.29 **4**
$C_{22}H_{26}O_3$ 51.26 **5**
$C_{22}H_{26}O_7$ 53.26 **7**
$C_{22}H_{27}BrO_2$ 53.35 **1**
$C_{22}H_{27}BrO_3$ 51.19 **3**
$C_{22}H_{27}BrO_3$ 51.27 **5**
$C_{22}H_{27}BrO_4$ 53.27 **3**
$C_{22}H_{27}BrO_7$ 54.8 **6**
$C_{22}H_{27}FeNO_2$ 73.41 **5**
$C_{22}H_{27}N_2O_3^+$, I^- 58.71 **1**

$C_{22}H_{27}N_2O_3^+$, I^- , H_2O	58.72 **1**
$C_{22}H_{27}N_2O_4^+$, Br^- , H_2O	58.73 **1**
$C_{22}H_{27}N_2O_6^+$, Br^- , C_2H_6O	58.42 **3**
$C_{22}H_{27}N_3O_7S$, CH_4O	50.17 **7**
$C_{22}H_{28}$	31.43 **1**
$C_{22}H_{28}BrN_3O_6S$	50.15 **3**
$C_{22}H_{28}Br_2O_4$	59.24 **3**
$C_{22}H_{28}CoN_2O_2$, $C_6H_3N_3O_6$	60.9 **4**
$C_{22}H_{28}CoN_4PS_2$	86.28 **5**
$C_{22}H_{28}CuN_2O_2$	78.40 **2**
$C_{22}H_{28}CuN_2O_2$	78.41 **2**
$C_{22}H_{28}CuN_2O_2$, $C_6H_3N_3O_6$	60.10 **4**
$C_{22}H_{28}MnN_6O_2S_2$	84.21 **5**
$C_{22}H_{28}N_2NiO_2$	78.42 **2**
$C_{22}H_{28}N_2NiO_2$	78.43 **2**
$C_{22}H_{28}N_2NiO_2$, $C_6H_3N_3O_6$	60.11 **4**
$C_{22}H_{28}N_2OS_2$, $0.5H_2O$	39.36 **7**
$C_{22}H_{28}N_2O_2Pd$	78.44 **2**
$C_{22}H_{28}N_2O_2Pd$	78.45 **2**
$C_{22}H_{28}N_4^{2+}$, $2Br^-$	58.74 **1**
$C_{22}H_{28}N_4^{2+}$, $2Br^-$, $2H_2O$	58.70 **1**
$C_{22}H_{28}N_4Ni$	71.48 **4**
$C_{22}H_{28}O_3$	51.24 **7**
$C_{22}H_{28}O_3$	51.25 **7**
$C_{22}H_{28}O_7S$	27.17 **4**
$C_{22}H_{29}BrO_2$	53.16 **5**
$C_{22}H_{29}BrO_3$	51.20 **3**
$C_{22}H_{29}BrO_3$	53.27 **7**
$C_{22}H_{29}BrO_6$	53.36 **1**
$C_{22}H_{29}BrO_6$	53.28 **7**
$C_{22}H_{29}ClO_7$	59.52 **1**
$C_{22}H_{29}CoN_4OPS_2$	76.50 **5**
$C_{22}H_{29}FO_4$	51.22 **6**
$C_{22}H_{29}FO_5$	51.28 **5**
$C_{22}H_{29}FO_5$	51.23 **6**
$C_{22}H_{29}IO_6$	54.10 **1**
$C_{22}H_{29}N_2O_4^+$, I^-	58.75 **1**
$C_{22}H_{29}N_2O_4^+$, CH_4O , Br^-	58.76 **1**
$C_{22}H_{29}N_3S_2$	41.20 **3**
$C_{22}H_{30}Cl_2O_2Pt_2$	75.71 **2**
$C_{22}H_{30}CoN_5O_6$	83.96 **5**
$C_{22}H_{30}CoN_8O_6$	83.115 **3**
$C_{22}H_{30}CuN_2O_2$, $2H_2O$	83.187 **2**
$C_{22}H_{30}CuN_4O_2$	78.12 **4**
$C_{22}H_{30}HgO_4$	71.53 **5**
$C_{22}H_{30}IrOP_2^+$, $C_{24}H_{20}B^-$	72.23 **7**
$C_{22}H_{30}NO_2^+$, Cl^-	1.31 **6**
$C_{22}H_{30}NO_5^+$, Br^- , H_2O	58.43 **3**
$C_{22}H_{30}NO_5^+$, I^-	58.77 **1**
$C_{22}H_{30}NO_5^+$, $C_2HO_4^-$	58.27 **4**
$C_{22}H_{30}N_2O_8$	36.29 **3**
$C_{22}H_{30}N_2O_8$	37.13 **6**
$C_{22}H_{30}N_4O_4U$	78.10 **7**
$C_{22}H_{30}N_8NiO_6$	83.118+ **4**
$C_{22}H_{30}N_8O_6Zn$	83.116 **3**
$C_{22}H_{30}O_2$	29.7 **4**
$C_{22}H_{30}O_2$	51.26 **7**
$C_{22}H_{30}O_3$	51.21 **3**
$C_{22}H_{30}O_3$	27.10 **7**
$C_{22}H_{30}O_4$	59.15 **7**
$C_{22}H_{30}O_5$	51.31 **4**
$C_{22}H_{30}O_{15}$	45.50 **6**
$C_{22}H_{31}As_2Br_3OPt$	71.49 **4**
$C_{22}H_{31}BrO_3$	51.22 **3**
$C_{22}H_{31}Cl_2N_3P_2Re^+$, Br^-	86.13 **6**
$C_{22}H_{31}FO_3$	51.23 **3**
$C_{22}H_{31}FO_3$	51.24 **3**
$C_{22}H_{31}IO_5$	54.11 **1**
$C_{22}H_{31}NO_2$	51.27 **7**
$C_{22}H_{31}NO_2$, $0.5C_6H_6$	58.31 **5**
$C_{22}H_{31}NO_5$	58.32 **5**
$C_{22}H_{31}N_2O^+$, I^-	58.79 **1**
$C_{22}H_{31}N_7O_5^{2+}$, $2Cl^-$, $5H_2O$	50.16 **3**
$C_{22}H_{32}As_4ClMoO_2^+$, I_3^- , $2CHCl_3$	86.14 **6**
$C_{22}H_{32}Br_2O_5$	54.9 **7**
$C_{22}H_{32}ClN_2P$	64.42 **6**
$C_{22}H_{32}IO_3$	51.32 **4**
$C_{22}H_{32}NO_3^+$, I^-	58.44 **3**
$C_{22}H_{32}N_2O_3$	43.8 **7**
$C_{22}H_{32}N_2P^+$, ClO_4^-	64.59 **2**
$C_{22}H_{32}N_4O_4$	48.82 **6**
$C_{22}H_{32}N_6O_2S_4$	41.28 **7**
$C_{22}H_{32}N_6O_4S_2Zn$	83.70 **6**
$C_{22}H_{32}O$	29.5 **6**
$C_{22}H_{32}O_2$	51.25 **3**
$C_{22}H_{32}O_2$	54.10 **6**
$C_{22}H_{32}O_5$	54.10 **7**
$C_{22}H_{32}O_6$	51.28 **7**
$C_{22}H_{32}O_8$	23.10 **3**
$C_{22}H_{32}O_8$	27.18 **4**
$C_{22}H_{33}BrO_4$	54.11 **7**
$C_{22}H_{33}BrO_6S$	59.53 **1**
$C_{22}H_{33}ClP_2Pt$	71.53 **6**
$C_{22}H_{33}IO_3$	51.26 **3**
$C_{22}H_{33}NO$	32.30 **4**
$C_{22}H_{33}NO_2$	58.45 **3**
$C_{22}H_{34}NO_2^+$, Br^-	58.44 **6**
$C_{22}H_{34}NO_5^+$, Br^- , H_2O	58.80 **1**
$C_{22}H_{34}N_4O_4S_6Te$	70.9 **6**
$C_{22}H_{34}O_3$	51.29 **7**
$C_{22}H_{34}O_4$	31.28 **4**
$C_{22}H_{34}O_4$	54.12 **7**
$C_{22}H_{36}Fe_2O_{10}Si_4$	71.54 **5**
$C_{22}H_{36}NO_2+$, Br^- , H_2O	58.81 **1**
$C_{22}H_{36}N_8Ni_2$	76.49 **6**
$C_{22}H_{36}O_2Si$	51.33 **4**
$C_{22}H_{36}O_5$	54.12 **6**
$C_{22}H_{36}O_7$	54.6 **3**
$C_{22}H_{38}$	19.19 **5**
$C_{22}H_{38}BaN_2O_7S_2$	67.23 **4**
$C_{22}H_{38}CuN_6O_6^{2+}$, $2NO_3^-$	83.71 **6**
$C_{22}H_{38}Cu_3N_4O_4$	84.59 **7**
$C_{22}H_{38}N_4O_7$	48.83 **6**
$C_{22}H_{38}NiO_2S_2$	77.17 **4**
$C_{22}H_{38}NiO_4$	77.56 **2**
$C_{22}H_{38}O_4Zn$	77.57 **2**
$C_{22}H_{40}Li_2N_4$	67.18 **5**
$C_{22}H_{40}OsP_4$	71.41 **3**
$C_{22}H_{40}P_4Ru$	71.42 **3**
$C_{22}H_{41}NiP$	72.31 **5**
$C_{22}H_{42}Mo_2O_6P_4$	86.43 **2**
$C_{22}H_{42}N_2O_2$	40.35 **5**
$C_{22}H_{42}N_2O_4$	33.54 **7**
$C_{22}H_{42}N_4O_2$, $2H_2O$	33.57 **6**
$C_{22}H_{42}N_4O_2S_2Zn$	80.18 **6**
$C_{22}H_{46}Co_3O_{18}P_6$	73.43 **7**
$C_{22}H_{48}N_{10}NiS_6$	79.26 **3**
$C_{22}H_{49}IrP_2$	71.54 **6**
$C_{22}H_{52}Cl_2P_2Pt_2S_2$	86.29 **5**

C_{23}

$C_{23}H_{10}Fe_3O_9$ 72.61 **2**
$C_{23}H_{14}Cl_4O$ 38.66 **7**
$C_{23}H_{14}Fe_2O_5$ 72.24 **7**
$C_{23}H_{14}O_5Ru_2$ 72.25 **7**
$C_{23}H_{15}As_2Co_2O_5P$ 86.33 **3**
$C_{23}H_{15}CrO_5P$ 86.30 **5**
$C_{23}H_{15}CrO_8P$ 86.31 **5**
$C_{23}H_{15}GeMnO_5$ 69.48 **2**
$C_{23}H_{15}MnO_5Sn$ 69.40 **2**
$C_{23}H_{16}BrNO$ 32.46+ **1**
$C_{23}H_{16}CoF_4O_3P$ 86.44 **2**
$C_{23}H_{16}FeO_3$ 75.26 **7**
$C_{23}H_{16}O$ 27.17 **6**
$C_{23}H_{16}O$ 26.6 **7**
$C_{23}H_{16}S_3$ 39.44 **4**
$C_{23}H_{17}As$ 65.4 **5**
$C_{23}H_{17}ClN_2O_2Ti$ 73.42 **3**
$C_{23}H_{17}NS_2$ 41.47 **5**
$C_{23}H_{17}O^+$, $C_7HN_4^-$ 38.67 **7**
$C_{23}H_{18}CoFeO_5P$ 73.48 **6**
$C_{23}H_{18}MnO_4P$ 71.55 **5**
$C_{23}H_{18}O_4$ 31.35 **5**
$C_{23}H_{19}BrO_2$ 31.36 **5**
$C_{23}H_{19}BrO_7$, C_3H_7NO 50.31 **1**
$C_{23}H_{19}FeO_2PS$ 72.20 **6**
$C_{23}H_{19}IO_2$ 24.16 **4**
$C_{23}H_{19}NSSn$ 69.20 **7**
$C_{23}H_{19}N_2^+$, $C_6H_2N_3O_7^-$ 34.7 **5**
$C_{23}H_{19}P$ 64.43 **6**
$C_{23}H_{20}As_2I_2MoO_3$ 86.40 **7**
$C_{23}H_{20}As_2I_2MoO_3$, $CHCl_3$ 86.41 **7**
$C_{23}H_{20}Br_2N_2O_5$ 35.32 **5**
$C_{23}H_{20}ClNO_2$ 36.32 **3**
$C_{23}H_{20}Cl_3GeNiP$, $0.5C_6H_6$ 73.43 **3**
$C_{23}H_{20}CuP$ 73.44 **3**
$C_{23}H_{20}FeIO_2P$ 72.26+ **7**
$C_{23}H_{20}N_2O_3$ 32.38 **6**
$C_{23}H_{20}U$ 71.56 **5**
$C_{23}H_{21}Cl_2O_2Rh$ 75.24 **6**
$C_{23}H_{21}CoN_3O_2$, C_2H_6O 78.13 **5**
$C_{23}H_{21}CoN_3O_4$ 78.14 **5**
$C_{23}H_{21}CuN_3O_6$ 84.60 **7**
$C_{23}H_{21}FeN_3O_2$ 78.15 **5**
$C_{23}H_{21}NO_3S$ 11.16 **4**
$C_{23}H_{22}BrNO_6$ 58.45 **6**
$C_{23}H_{22}CoN_3O_2$ 71.57 **5**
$C_{23}H_{22}N_2OS_2$ 41.49 **1**
$C_{23}H_{23}BrO_7$ 59.54 **1**
$C_{23}H_{23}Cl_2N_2S_2^+$, Br^- , $C_2H_4O_2$ 41.29 **7**
$C_{23}H_{23}CoN_3O_2$ 78.11 **6**
$C_{23}H_{23}IO_8$ 45.75 **1**
$C_{23}H_{23}N_2^+$, Br^- 35.31 **4**
$C_{23}H_{23}N_2^+$, Br^- 35.32 **4**
$C_{23}H_{23}N_2^+$, Cl^- , H_2O 35.33 **4**
$C_{23}H_{23}N_2S_2^+$, I^- 41.32 **6**
$C_{23}H_{23}P$ 64.44 **6**
$C_{23}H_{24}BrNO_6$ 53.37 **1**
$C_{23}H_{24}Br_2N_4O_3S$ 59.55 **1**
$C_{23}H_{24}ClO_2PSn$ 69.22 **5**
$C_{23}H_{24}CrO_3$ 75.27 **7**
$C_{23}H_{24}FeN_6^{2+}$, $2ClO_4^-$ 83.98 **5**
$C_{23}H_{24}N_6Zn^{2+}$, $2ClO_4^-$ 83.99 **5**
$C_{23}H_{24}O_2Sn$ 69.41 **2**
$C_{23}H_{24}O_4$ 21.16 **5**
$C_{23}H_{24}O_8$ 59.27 **5**
$C_{23}H_{24}O_8$ 59.16 **7**
$C_{23}H_{24}S_2Si$ 63.16 **7**
$C_{23}H_{25}AuNPS_2$ 80.16 **4**
$C_{23}H_{25}BrO_5$ 53.38 **1**
$C_{23}H_{25}BrO_5$ 53.20 **6**
$C_{23}H_{25}ClO_{14}Pd$, $CHCl_3$ 71.79 **7**
$C_{23}H_{25}F_3N_2OS$ 39.37 **7**
$C_{23}H_{25}F_3N_2OS$ 39.38 **7**
$C_{23}H_{25}NO$ 58.46 **6**
$C_{23}H_{25}N_3O_4$, $2H_2O$ 48.84 **6**
$C_{23}H_{26}FN_3O_2$ 35.33 **5**
$C_{23}H_{26}FeN_6O$ 83.100 **5**
$C_{23}H_{26}N_2O_2$ 33.53 **5**
$C_{23}H_{26}OPdS$ 72.32 **5**
$C_{23}H_{26}O_9S$ 59.24 **6**
$C_{23}H_{27}BrN_2O_3S$ 9.17 **5**
$C_{23}H_{27}BrO_3$ 28.12 **3**
$C_{23}H_{27}Cl_2PRu$ 74.9 **5**
$C_{23}H_{28}FeO_4$ 72.38 **3**
$C_{23}H_{28}O_8$ 54.8 **5**
$C_{23}H_{28}O_{10}$ 53.29 **7**
$C_{23}H_{29}BrO_{10}$ 53.17 **5**
$C_{23}H_{29}ClO_4$ 51.30 **7**
$C_{23}H_{29}N_3O_2S_2$ 39.92 **1**
$C_{23}H_{29}N_5Ni^{2+}$, $2ClO_4^-$, CH_3NO_2 76.77 **7**
$C_{23}H_{29}O_5P$ 64.60+ **2**
$C_{23}H_{30}Cl_3IrP_2Sn$ 75.25 **6**
$C_{23}H_{30}NO_6^+$, Br^- , $2H_2O$ 58.33 **5**
$C_{23}H_{30}O_2Pd$ 71.55 **6**
$C_{23}H_{30}O_2Pd$ 75.28 **7**
$C_{23}H_{30}O_6$ 51.34 **4**
$C_{23}H_{31}BrN_4O_7$ 48.95 **1**
$C_{23}H_{31}ClO_5$ 51.31 **7**
$C_{23}H_{31}ClO_5$, CH_4O 51.35 **4**
$C_{23}H_{31}NO_6$, CH_4O 58.47 **6**
$C_{23}H_{31}N_2O_4^+$, I^- 58.82 **1**
$C_{23}H_{32}ClN_3O^{2+}$, $2Cl^-$, $2H_2O$ 36.20 **6**
$C_{23}H_{32}CoN_4PS_2$ 86.32 **5**
$C_{23}H_{32}NO_3^+$, Br^- 3.35 **7**
$C_{23}H_{32}O_3$ 51.28 **3**
$C_{23}H_{32}O_3$ 31.28 **7**
$C_{23}H_{32}O_5$ 51.27 **6**
$C_{23}H_{32}O_5$ 51.32 **7**
$C_{23}H_{32}O_5$ 51.33 **7**
$C_{23}H_{32}O_5$, C_6H_5BrO 51.36 **4**
$C_{23}H_{32}O_6$, $0.5H_2O$ 51.28 **6**
$C_{23}H_{33}BrO_5$ 51.17 **1**
$C_{23}H_{33}BrO_5S$ 28.10 **4**
$C_{23}H_{33}BrO_6$ 54.7 **3**
$C_{23}H_{33}N_2O^+$, I^- 58.47 **3**
$C_{23}H_{33}N_2O_2^+$, I^- 58.84 **1**
$C_{23}H_{33}N_2O_3^+$, I^- 58.85 **1**
$C_{23}H_{34}NO_3$ 51.34 **7**
$C_{23}H_{34}NO_3$, H_2O 51.35 **7**
$C_{23}H_{34}N_2O_4$ 32.24 **5**
$C_{23}H_{34}O_4$ 51.18 **1**
$C_{23}H_{34}O_5$ 51.29 **5**
$C_{23}H_{34}O_5$ 51.29 **6**
$C_{23}H_{34}O_6$ 51.30 **6**
$C_{23}H_{35}BrO_9$ 27.18 **5**
$C_{23}H_{35}IO_3$ 51.30 **3**
$C_{23}H_{35}NO_2$ 58.35 **5**
$C_{23}H_{35}N_5O_4S$ 48.70 **7**
$C_{23}H_{35}N_5O_4Se$ 48.71 **7**

$C_{23}H_{36}NO_2^+$, I^- 37.28 **1**
$C_{23}H_{36}NO_2^+$, I^- 58.48 **3**
$C_{23}H_{36}NO_4^+$, Br^- , $2H_2O$ 58.86 **1**
$C_{23}H_{37}BrN_2$ 51.38 **4**
$C_{23}H_{37}ClCoN_5O_4$, CH_2Cl_2 83.120 **4**
$C_{23}H_{37}NiO_3P$ 71.59 **5**
$C_{23}H_{38}Cu_2N_2O_8$ 81.57 **5**
$C_{23}H_{38}NO_6^+$, Br^- 58.49 **3**
$C_{23}H_{46}O$ 5.19 **7**

C_{24}

$C_{24}Cl_{16}Co_2S_8^{2-}$, $2C_{16}H_{36}N^+$ 85.86 **2**
$C_{24}Cu_8N_{12}S_{12}^{4-}$, $4C_9H_{14}N^+$ 3.69 **1**
$C_{24}Cu_8O_{12}S_{12}^{4-}$, $4C_{24}H_{20}P^+$ 85.40 **7**
$C_{24}F_{20}Ge$ 69.15 **6**
$C_{24}F_{20}P_4$ 64.49 **3**
$C_{24}F_{20}Sn$ 69.16 **6**
$C_{24}H_3AsCo_6O_{18}$ 72.39 **3**
$C_{24}H_8Cu_2F_{24}N_2O_8$ 77.16 **6**
$C_{24}H_8N_8^-$, $C_{24}H_{20}P^+$ 64.66 **2**
$C_{24}H_8O_4$ 38.68 **7**
$C_{24}H_{12}$ 30.18+ **1**
$C_{24}H_{12}$ 29.3+ **3**
$C_{24}H_{12}$, $2C_9H_{12}N_4O_3$ 60.112 **2**
$C_{24}H_{12}AsN_3O_6$ 65.35 **2**
$C_{24}H_{12}BClN_6$ 62.11 **7**
$C_{24}H_{12}F_{24}N_8O_4Th$ 84.27 **6**
$C_{24}H_{12}F_{24}N_8O_4U$ 84.28 **6**
$C_{24}H_{12}O_3$ 38.69 **7**
$C_{24}H_{12}O_{14}Ru_6$ 71.48 **2**
$C_{24}H_{14}Br_2N_2O_2$ 28.13 **3**
$C_{24}H_{14}Fe_2O_6$ 71.80 **7**
$C_{24}H_{15}AuF_5P$ 71.60 **5**
$C_{24}H_{15}BrMnO_4P$ 71.50 **4**
$C_{24}H_{15}Mo_3O_9Tl$ 68.27 **5**
$C_{24}H_{15}N_4O_5P$, C_6H_6 40.19 **3**
$C_{24}H_{15}N_6O_3P$ 40.20 **3**
$C_{24}H_{15}P$ 64.62 **2**
$C_{24}H_{16}$ 31.35 **3**
$C_{24}H_{16}$ 30.3 **7**
$C_{24}H_{16}As_2O_2S$ 65.5 **5**
$C_{24}H_{16}As_4$ 65.6 **3**
$C_{24}H_{16}BrN_3O_4$ 36.31 **3**
$(C_{24}H_{16}Cl_2HgO_2S_2)_n$ 85.39 **3**
$C_{24}H_{16}Cl_3N_3O_3$ 24.11 **7**
$C_{24}H_{16}Cl_6PtS_2$ 85.27 **4**
$C_{24}H_{16}CrN_4O_4^-$, $C_5H_6N^+$ 83.188 **2**
$C_{24}H_{16}Fe_2$ 75.34 **3**
$C_{24}H_{16}N^+$, BF_4^- 12.13 **5**
$C_{24}H_{16}N_2O_2$ 40.32 **1**
$C_{24}H_{16}N_2O_9Te_2$ 70.10 **6**
$C_{24}H_{16}O_2$ 38.29 **5**
$C_{24}H_{17}BrO_2$ 31.44 **1**
$C_{24}H_{17}Br_2FSSn$ 69.21 **7**
$C_{24}H_{17}NO_3Si$ 63.26 **2**
$C_{24}H_{18}$ 19.64 **1**
$C_{24}H_{18}$ 28.25 **6**
$C_{24}H_{18}$ 19.13 **7**
$C_{24}H_{18}ClN_4NiS_2^+$, ClO_4^- 85.41 **7**
$C_{24}H_{18}Cr_2N_4O_2^{4+}$, $4Cl^-$, $6H_2O$ 83.101 **5**
$C_{24}H_{18}FeO_2$ 73.101 **2**
$C_{24}H_{18}HgN_6^{2+}$, $2ClO_4^-$ 83.101 **7**
$C_{24}H_{18}MnO_4P$ 71.61 **5**
$C_{24}H_{18}MnO_5P$ 86.36 **3**
$C_{24}H_{18}N_2O_8Ti$ 73.44 **7**
$C_{24}H_{18}N_2P^+$, Cl^- , CH_4O 64.33 **5**
$C_{24}H_{18}N_8PdS_2$, C_4H_9NO 85.40 **6**
$C_{24}H_{18}O_6$, C_6H_5Br 24.33 **1**
$C_{24}H_{19}BrNP$ 64.50 **3**
$C_{24}H_{19}BrN_2O_5$ 35.34 **5**
$C_{24}H_{19}P_3$ 64.63 **2**
$C_{24}H_{20}$ 31.45 **1**
$C_{24}H_{20}Ag_2Cl_2O_8$ 74.8 **4**
$C_{24}H_{20}As^+$, Cl_4NOs^- 65.6 **6**
$C_{24}H_{20}As^+$, F_5Si^- 65.7 **3**
$C_{24}H_{20}As^+$, $FeN_4O_{12}^-$ 65.8 **3**
$C_{24}H_{20}As^+$, I^- 65.36 **2**
$C_{24}H_{20}As^+$, I_3^- 65.10 **4**
$C_{24}H_{20}As^+$, I_3^- 65.6 **5**
$C_{24}H_{20}As^+$, H^+ , $2NO_3^-$ 65.37 **2**
$C_{24}H_{20}As^+$, HCl_4OTe^- , H_2O 65.4 **7**
$C_{24}H_{20}As^+$, $H_2Br_4MoO_2^-$ 65.38 **2**
$C_{24}H_{20}As^+$, $H_4Cl_4O_2Ru^-$, H_2O 65.39 **2**
$C_{24}H_{20}As^+$, $H_5O_2^+$, $2Cl^-$ 65.7 **5**
$C_{24}H_{20}As^+$, $C_2H_3Br_4NORe^-$ 83.1 **2**
$C_{24}H_{20}As^+$, $C_5H_4N_5^-$, $3H_2O$ 44.4 **6**
$C_{24}H_{20}As^+$, $C_9FN_6^-$ 65.40 **2**
$C_{24}H_{20}As^+$, $C_{12}H_2O_{12}Re_3^-$ 65.41 **2**
$C_{24}H_{20}As^+$, $C_{14}Mn_3O_{14}^-$ 65.7 **6**
$C_{24}H_{20}As^+$, $C_{16}H_{28}AuN_{16}^-$ 71.34 **4**
$C_{24}H_{20}As^+$, $C_{18}H_{12}NbS_6^-$ 85.33 **6**
$C_{24}H_{20}As^+$, $C_{18}H_{15}I_3NiP^-$ 65.42 **2**
$C_{24}H_{20}As^+$, $C_{24}H_{24}Nb^-$ 75.29 **7**
$2C_{24}H_{20}As^+$, $Cl_6Cu_2^{2-}$ 65.8 **6**
$2C_{24}H_{20}As^+$, $Cl_{10}Te_2^{2-}$ 65.5 **7**
$2C_{24}H_{20}As^+$, $Cl_{11}Re_3^{2-}$, H_2O 65.43 **2**
$2C_{24}H_{20}As^+$, $CoN_4O_{12}^{2-}$ 65.44 **2**
$2C_{24}H_{20}As^+$, $CuN_4O_{12}^{2-}$, CH_2Cl_2 65.6 **7**
$2C_{24}H_{20}As^+$, FeN_{15}^{2-} 65.45 **2**
$2C_{24}H_{20}As^+$, $MnN_4O_{12}^{2-}$ 65.9 **3**
$2C_{24}H_{20}As^+$, $N_{18}Pd_2^{2-}$ 65.11 **4**
$2C_{24}H_{20}As^+$, $C_2NiS_6^{2-}$ 65.46 **2**
$2C_{24}H_{20}As^+$, $C_4N_4NiS_4^{2-}$ 65.47 **2**
$2C_{24}H_{20}As^+$, $C_5N_5NbOS_5^{2-}$ 65.10 **3**
$2C_{24}H_{20}As^+$, $C_8CoF_{12}O_8^{2-}$ 81.50 **2**
$2C_{24}H_{20}As^+$, $C_{12}FeN_6S_6^{2-}$ 85.19 **6**
$2C_{24}H_{20}As^+$, $C_{12}MoN_6S_6^{2-}$ 85.20 **6**
$2C_{24}H_{20}As^+$, $C_{12}N_6S_6W^{2-}$ 85.21 **6**
$2C_{24}H_{20}As^+$, $C_{14}H_6N_2O_{10}U^{2-}$, $6H_2O$ 81.34 **7**
$2C_{24}H_{20}As^+$, $C_{24}O_{24}Pt_{12}^{2-}$ 65.9 **6**
$2C_{24}H_{20}As^+$, $C_{30}O_{30}Pt_{15}^{2-}$ 65.10 **6**
$C_{24}H_{20}As_2O$ 65.48 **2**
$C_{24}H_{20}B^-$, K^+ 62.12 **7**
$C_{24}H_{20}B^-$, Rb^+ 62.51 **2**
$C_{24}H_{20}B^-$, $C_4H_{12}N^+$ 62.13 **7**
$C_{24}H_{20}B^-$, $C_6H_{18}ClN_4Zn^+$ 76.25 **3**
$C_{24}H_{20}B^-$, $C_7H_{18}CuN_5^+$ 76.33 **7**
$C_{24}H_{20}B^-$, $C_7H_{18}N_5NiO^+$ 76.34 **7**
$C_{24}H_{20}B^-$, $C_9H_7N_6Pt^+$ 71.8 **5**
$C_{24}H_{20}B^-$, $2C_{16}H_{14}CoN_2O_2$, $2C_4H_8O$, Na^+ 78.2 **7**
$C_{24}H_{20}B^-$, $C_{22}H_{30}IrOP_2^+$ 72.23 **7**
$C_{24}H_{20}B^-$, $C_{26}H_{42}BrN_3NiP^+$ 83.126 **4**
$C_{24}H_{20}B^-$, $C_{27}H_{33}IO_3P_3W^+$ 86.53 **7**
$C_{24}H_{20}B^-$, $C_{27}H_{42}AsN_4NiS^+$ 76.51 **5**
$C_{24}H_{20}B^-$, $C_{32}H_{24}Br_2N_8Ni^+$ 83.112 **5**
$C_{24}H_{20}B^-$, $C_{40}H_{44}Co_2N_4NaO_6^+$ 67.22 **5**

$C_{24}H_{20}B^-$, $C_{42}H_{42}BrFeP_4^+$, CH_2Cl_2 86.81 **7**
$C_{24}H_{20}B^-$, $C_{48}H_{47}As_3NNi^+$ 71.130 **7**
$C_{24}H_{20}B^-$, $C_{52}H_{48}NOP_4Ru^+$, C_3H_6O 86.75 **5**
$C_{24}H_{20}B^-$, $C_{52}H_{50}ClN_2P_4W^+$ 86.63 **6**
$C_{24}H_{20}B^-$, $C_{54}H_{42}ClCoP_4^+$ 86.56 **4**
$C_{24}H_{20}B^-$, $C_{72}H_{57}IrP_5^+$ 86.64 **4**
$C_{24}H_{20}B^-$, $C_{84}H_{84}As_6IN_2Ni_2^+$ 86.117 **7**
$2C_{24}H_{20}B^-$, $C_{12}H_{36}N_{14}Ni_2^{2+}$ 76.63 **7**
$2C_{24}H_{20}B^-$, $C_{14}H_{36}N_{10}Ni_2O_2^{2+}$ 76.46 **5**
$2C_{24}H_{20}B^-$, $C_{18}H_{52}N_4O_{12}P_4Ru^{2+}$ 86.29 **7**
$2C_{24}H_{20}B^-$, $C_{82}H_{78}Ni_2P_6S^{2+}$ 86.115 **7**
$2C_{24}H_{20}B^-$, $C_{126}H_{126}Au_6P_6^{2+}$ 86.75 **6**
$4C_{24}H_{20}B^-$, $xC_{72}H_{62}N_4P_4Pt_2^{2+}$, $yC_{72}H_{64}N_2P_4Pt_2^{2+}$ 86.98 **3**
$C_{24}H_{20}B_2I_4P_2$ 62.15 **4**
$C_{24}H_{20}BrNO_5$ 59.19 **4**
$C_{24}H_{20}Cl_2N_3P_3$ 64.64 **2**
$C_{24}H_{20}Cl_4N_4P_4$ 64.65 **2**
$C_{24}H_{20}Cl_4N_4P_4$ 64.34 **5**
$C_{24}H_{20}Cl_4N_4P_4$ 64.45 **6**
$C_{24}H_{20}Cl_4N_4P_4$ 64.55 **7**
$C_{24}H_{20}Cl_4N_4P_4$ 64.56 **7**
$C_{24}H_{20}Cl_4O_4Ti_2$ 84.54 **2**
$C_{24}H_{20}Cl_6Sb_2$ 66.8 **7**
$(C_{24}H_{20}CoO_2P_2S_2)_n$ 85.40 **3**
$C_{24}H_{20}CuN_2O_2$ 78.46 **2**
$C_{24}H_{20}CuN_2O_2$ 78.12 **6**
$C_{24}H_{20}Cu_2N_6$ 83.180 **2**
$C_{24}H_{20}F_3NiP$ 71.43 **3**
$C_{24}H_{20}Fe_2O_2S_2$ 73.102 **2**
$C_{24}H_{20}Fe_2O_4Sn$ 69.31 **3**
$C_{24}H_{20}Fe_4O_4$ 73.28 **4**
$C_{24}H_{20}Fe_4O_4^+$, F_6P^- 73.27 **4**
$C_{24}H_{20}Fe_4S_8^{2-}$, $2C_4H_{12}N^+$ 85.42 **7**
$C_{24}H_{20}Ge$ 69.32 + **4**
$C_{24}H_{20}Ge$ 69.24 **5**
$C_{24}H_{20}In^-$, Na^+ 68.28 **5**
$C_{24}H_{20}IrOP$ 73.45 **7**
$C_{24}H_{20}NS_2^+$, I^- 41.48 **5**
$C_{24}H_{20}N_2$ 16.14 + **7**
$C_{24}H_{20}N_2P_2S_2$ 64.35 **5**
$C_{24}H_{20}N_2S_2$ 41.49 **5**
$C_{24}H_{20}N_6O_2Pd$ 83.189 **2**
$C_{24}H_{20}N_8O_8Zr$ 84.16 **3**
$C_{24}H_{20}NiP_2S_4$ 85.88 **2**
$C_{24}H_{20}Ni_2$ 72.64 **2**
$C_{24}H_{20}OSb_2$ 66.6 **6**
$C_{24}H_{20}O_4$ 31.37 **5**
$C_{24}H_{20}P^+$, Cl_3Cu^- 64.46 **6**
$C_{24}H_{20}P^+$, $C_3HgN_3S_3^-$ 64.57 **7**
$C_{24}H_{20}P^+$, $C_4H_8Cl_3O_2Pt^-$ 72.4 **3**
$C_{24}H_{20}P^+$, $C_8H_{14}Cl_3O_2Pt^-$ 72.5 **6**
$C_{24}H_{20}P^+$, $C_{13}H_5MoN_4S_4^-$ 64.52 **3**
$C_{24}H_{20}P^+$, $C_{24}H_8N_8^-$ 64.66 **2**
$2C_{24}H_{20}P^+$, $C_4HgN_4S_4^{2-}$ 64.58 **7**
$2C_{24}H_{20}P^+$, $C_{12}O_{12}Pt_6^{2-}$ 64.47 **6**
$2C_{24}H_{20}P^+$, $C_{18}O_{18}Pt_9^{2-}$ 64.48 **6**
$4C_{24}H_{20}P^+$, $C_{24}Cu_8O_{12}S_{12}^{4-}$ 85.40 **7**
$C_{24}H_{20}P_4S$ 64.49 **6**
$C_{24}H_{20}Pb$ 69.25 **5**
$C_{24}H_{20}Se_4$ 39.45 **4**
$C_{24}H_{20}Si$ 63.10 + **4**
$C_{24}H_{20}Si$ 63.17 **7**
$C_{24}H_{20}Sn$ 69.32 **3**
$C_{24}H_{20}Sn$ 69.34 **4**
$C_{24}H_{21}BrPRh$ 72.21 **6**
$C_{24}H_{21}FeS_7$ 85.89 **2**
$C_{24}H_{21}InS_6$ 68.20 **3**
$C_{24}H_{21}NO_2$ 27.12 **3**
$C_{24}H_{21}N_3O_3$ 33.54 **5**
$C_{24}H_{21}OSb$ 66.20 **2**
$C_{24}H_{22}BrNO_5$ 58.50 **3**
$C_{24}H_{22}CuN_4O_8$ 84.61 **7**
$C_{24}H_{22}Fe_2O_2$ 73.46 **7**
$C_{24}H_{22}Fe_2O_4$ 73.104 **2**
$C_{24}H_{22}Fe_2O_4$ 73.105 **2**
$C_{24}H_{22}Fe_2O_4$ 73.49 **6**
$C_{24}H_{22}N_2^{2+}$, $2I^-$ 33.46 **4**
$C_{24}H_{22}N_2^{2+}$, $2C_{12}H_4N_4^-$, $2C_{12}H_4N_4$ 33.55 **5**
$C_{24}H_{22}Nb_2O_{10}$ 72.40 **3**
$C_{24}H_{22}O_7$ 59.25 **3**
$C_{24}H_{22}O_8$ 59.28 **5**
$C_{24}H_{23}BMoN_4O_2$ 72.28 **7**
$C_{24}H_{23}BrO_7$ 59.25 **6**
$C_{24}H_{23}NO_2$ 31.38 **5**
$C_{24}H_{23}NO_3S$ 41.50 **5**
$C_{24}H_{23}N_3O_4S$ 33.47 **4**
$C_{24}H_{24}As_4Co_4F_8O_8$ 86.18 **4**
$C_{24}H_{24}As_4F_8O_8Ru_3$ 86.37 **3**
$C_{24}H_{24}FeN_3S_6$ 80.17 **4**
$C_{24}H_{24}Fe_2$ 73.43 **5**
$C_{24}H_{24}IN_3O_5$, C_2H_6O 58.87 **1**
$C_{24}H_{24}NO_2^+$, Br^- 31.40 **5**
$C_{24}H_{24}N_2O_4$ 35.34 **4**
$C_{24}H_{24}N_3P_2^+$, Cl^- 64.67 **2**
$C_{24}H_{24}N_6Ni^{2+}$, $2ClO_4^-$ 83.102 **5**
$C_{24}H_{24}N_6Zn^{2+}$, $2ClO_4^-$ 83.190 **2**
$C_{24}H_{24}Nb^-$, $C_{24}H_{20}As^+$ 75.29 **7**
$C_{24}H_{24}O_2$ 30.4 **7**
$C_{24}H_{24}Ti_2$ 75.72 **2**
$C_{24}H_{25}IO_6$ 53.30 **7**
$C_{24}H_{25}NO$ 13.16 **5**
$C_{24}H_{25}N_5O_4$ 58.17 **7**
$C_{24}H_{25}OP$ 64.68 **2**
$C_{24}H_{25}O_2P$ 64.59 **7**
$C_{24}H_{25}O_3P$ 64.33 **4**
$C_{24}H_{26}$ 29.8 **4**
$C_{24}H_{26}ClNO_6$ 38.25 **4**
$C_{24}H_{26}Cl_2$ 20.11 **4**
$C_{24}H_{26}Cu_2N_2O_6$ 77.10 **7**
$C_{24}H_{26}Fe_2$ 73.107 **2**
$C_{24}H_{26}Fe_2$ 73.45 **3**
$C_{24}H_{26}N_2O_6$ 37.14 **6**
$C_{24}H_{26}N_2O_8S_2$ 59.56 **1**
$C_{24}H_{26}O_3S$ 11.16 **6**
$C_{24}H_{27}As$ 65.49 **2**
$C_{24}H_{27}BrO_2$ 38.70 **7**
$C_{24}H_{27}FeN_7^{2+}$, $2BF_4^-$ 83.118 **3**
$C_{24}H_{28}Al_2^{2-}$, $2C_8H_{16}NaO_2^+$ 68.21 **3**
$C_{24}H_{28}ClFO_4$ 51.31 **6**
$C_{24}H_{28}Cl_4Cu_2N_4O_4$ 84.17 **3**
$C_{24}H_{28}Cl_6Cu_4N_4O$, xH_2O 83.119 **3**
$C_{24}H_{28}CuN_2O_4$ 82.24 **5**
$C_{24}H_{28}CuN_2O_4$ 82.21 **7**
$C_{24}H_{28}Cu_2N_8O_{20}$ 84.62 **7**
$C_{24}H_{28}Cu_2O_6$ 77.13 **5**
$C_{24}H_{28}FN_3O$ 35.33 **6**
$C_{24}H_{28}FN_3O_6$, C_3H_6O 51.36 **7**
$C_{24}H_{28}MoO_2$ 75.26 **6**
$C_{24}H_{28}N_2Ni$ 72.33 **5**
$C_{24}H_{28}N_2O_2Pd_2$ 72.29 **7**

$C_{24}H_{28}N_4Ni^{2+}$, $2ClO_4^-$ 83.72 **6**
$C_{24}H_{28}N_8Ni^{2+}$, $2ClO_4^-$ 83.120 **3**
$C_{24}H_{28}N_8O_2S_4Zn_2$ 84.29 **6**
$C_{24}H_{28}N_8O_6$, $3H_2O$ 37.16 **4**
$C_{24}H_{29}ClO_4$ 51.32 **6**
$C_{24}H_{29}CoN_5O_2^+$, Br^- , H_2O 78.11 **7**
$C_{24}H_{29}FN_2O_2$ 33.58 **6**
$C_{24}H_{30}As_4ClPd^+$, ClO_4^- , C_6H_6 86.45 **2**
$C_{24}H_{30}BrN_3O_5S$ 8.19 **5**
$C_{24}H_{30}Cl_2Mo_2O_8Zn_2$ 73.47 **7**
$C_{24}H_{30}Cr_2O_4$ 73.48 **7**
$C_{24}H_{30}CuN_4S_2$ 85.43 **7**
$C_{24}H_{30}F_2O_6$ 51.32 **5**
$C_{24}H_{30}F_{18}P_2Pt$ 74.10 **5**
$C_{24}H_{30}FeO_{12}S_6^{2-}$, $2C_{25}H_{22}P^+$ 85.41 **6**
$C_{24}H_{30}Fe_2O_6$ 72.65 **2**
$C_{24}H_{30}Fe_2O_6$ 71.81 **7**
$C_{24}H_{30}N_2O_2Pt_2$ 83.191 **2**
$C_{24}H_{30}Na_2O_8$ 67.24 **3**
$C_{24}H_{31}AlTi_2$ 73.50 **6**
$C_{24}H_{31}BrO_5$ 53.28 **3**
$C_{24}H_{31}BrO_8$ 54.12 **1**
$C_{24}H_{31}Cl_2N_5O_7$, $2H_2O$ 50.18 **7**
$C_{24}H_{31}FO_6$, H_2O 51.37 **7**
$C_{24}H_{31}N_2O_5^+$, I^- 58.51 **3**
$C_{24}H_{31}N_2O_6^+$, I^- 58.89 **1**
$C_{24}H_{32}Ag^+$, ClO_4^- 74.9 **4**
$C_{24}H_{32}Cl_2O_2Ti_2$ 75.30 **7**
$C_{24}H_{32}CuN_2O_2$ 78.18 **3**
$C_{24}H_{32}Fe_2O_4S_2$ 75.31 **7**
$C_{24}H_{32}K_2O_8^{2+}$, $2CNS^-$ 67.15 **6**
$C_{24}H_{32}Mo_2NiS_4^{2+}$, $2BF_4^-$ 73.49 **7**
$C_{24}H_{32}NO_3^+$, Br^- 58.52 **3**
$C_{24}H_{32}NO_6^+$, I^- 58.36 **5**
$C_{24}H_{32}N_2NiO_2$ 78.47 **2**
$C_{24}H_{32}N_2O_2Pd$ 78.48 **2**
$C_{24}H_{32}N_2O_5$ 15.12 **5**
$C_{24}H_{32}N_8Ni^{2+}$, $2C_6H_8N_2$, $2Cl^-$ 83.73 **6**
$C_{24}H_{32}Nb_2NiS_4^{2+}$, $2BF_4^-$, $2H_2O$ 73.29 **4**
$C_{24}H_{32}Nb_7NiS_4^{2+}$, $2BF_4^-$, $2H_2O$ 73.50 **7**
$C_{24}H_{32}O_2Pd$ 71.51 **4**
$C_{24}H_{32}O_4S$ 51.33 **5**
$C_{24}H_{32}O_6$ 38.71 **7**
$C_{24}H_{32}O_6$, Cs^+ , CNS^- 38.72 **7**
$2C_{24}H_{32}O_6$, Cs^+ , CNS^- 38.73 **7**
$C_{24}H_{32}O_{12}$ 28.7 **7**
$C_{24}H_{32}U$ 75.34 **5**
$C_{24}H_{33}B_3N_3O_3$ 62.18 **3**
$C_{24}H_{33}BrO_8$ 54.13 **1**
$C_{24}H_{33}Br_4MoP_3$ 86.33 **5**
$C_{24}H_{33}Cl_2MoOP_3$ 86.19 **4**
$C_{24}H_{33}Cl_4MoP_3$, C_2H_6O 86.34 **5**
$C_{24}H_{34}Cl_2Pt_2$ 75.35 **3**
$C_{24}H_{34}NO_6^+$, I^- 58.37 **5**
$C_{24}H_{34}N_4NiO_2$ 78.49 **2**
$C_{24}H_{35}As_2Br_3OPt$ 86.46 **2**
$C_{24}H_{35}IO_4$ 54.14 **1**
$C_{24}H_{36}Ag_2^{2+}$, $2NO_3^-$, $2H_2O$ 53.29 **3**
$C_{24}H_{36}BrNO_2$ 52.6 **3**
$C_{24}H_{36}CdN_{12}^{2+}$, $2BF_4^-$ 83.104 **5**
$C_{24}H_{36}Cl_2CoN_4^+$, ClO_4^- 76.85 **3**
$C_{24}H_{36}Cl_6Pt_3Sn_2$ 69.43 **2**
$C_{24}H_{36}CoN_7^{2+}$, $2BF_4^-$ 83.121 **3**
$C_{24}H_{36}CuN_7^{2+}$, $2BF_4^-$ 83.122 **3**
$C_{24}H_{36}FeN_7^{2+}$, $2BF_4^-$ 83.123 **3**
$C_{24}H_{36}Fe_2O_4$ 72.41 **3**
$C_{24}H_{36}MnN_7^{2+}$, $2BF_4^-$ 83.124 **3**
$C_{24}H_{36}NO_4^+$, I^- 58.90 **1**
$C_{24}H_{36}NO_8^+$, I^- , $3H_2O$ 58.91 **1**
$C_{24}H_{36}N_4O_9$ 50.7 **6**
$C_{24}H_{36}N_7Ni^{2+}$, $2F_6P^-$ 83.125 **3**
$C_{24}H_{36}N_7Zn^{2+}$, $2BF_4^-$ 83.126 **3**
$C_{24}H_{36}O_8$ 59.26 **6**
$C_{24}H_{36}Ru$ 74.11 **5**
$C_{24}H_{38}ClN_2P_2Pt^+$, BF_4^- 86.42 **7**
$C_{24}H_{39}ClN_2P_2Pd$ 71.44 **3**
$C_{24}H_{40}Cl_2Rh_2$ 72.22 **6**
$C_{24}H_{40}CuGa_2N_8$ 83.102 **7**
$C_{24}H_{40}NO_6^+$, I^- , H_2O 58.92 **1**
$C_{24}H_{40}N_8O_8$, $4H_2O$ 48.85 **6**
$C_{24}H_{40}O_4$ 51.33 **6**
$C_{24}H_{40}O_4$, $C_2H_4O_2$ 60.42 **4**
$2C_{24}H_{40}O_4$, $C_6H_4I_2$ 60.40 **5**
$3C_{24}H_{40}O_4$, $C_{14}H_{10}$ 60.41 **5**
$C_{24}H_{40}O_5$, C_2H_6O 60.42 **5**
$C_{24}H_{40}S_2$ 39.39 **7**
$C_{24}H_{42}Cl_2Cu_2$ 75.36 **3**
$C_{24}H_{42}N_2O_6$ 48.53 **3**
$C_{24}H_{42}O_8Zn_7$ 71.45 **3**
$C_{24}H_{42}P_4RuS_4$ 86.43 **7**
$C_{24}H_{43}NO_5$ 59.57 **1**
$C_{24}H_{43}NiO_2P$ 71.62 **5**
$C_{24}H_{44}AgNO_3$ 75.37 **3**
$C_{24}H_{44}N_4O_4$ 34.5 **3**
$C_{24}H_{44}P_4$ 64.69 **2**
$C_{24}H_{48}Ag_2N_4O_6^{2+}$, Ag^+ , $3NO_3^-$ 84.30 **6**
$C_{24}H_{48}MnO_6S_{12}^{2+}$, $2ClO_4^-$ 84.22 **5**
$C_{24}H_{48}O_4Pt_2$ 77.58 **2**
$C_{24}H_{48}O_4Pt_2$ 71.46 **3**
$C_{24}H_{49}I$ 5.4 **3**
$C_{24}H_{50}Li_2N_4Ni$ 75.32 **7**
$C_{24}H_{52}CuN_4O_8^{2+}$, $2ClO_4^-$ 81.51 **3**
$C_{24}H_{56}Cd_2O_8P_4S_8$ 85.41 **3**
$C_{24}H_{56}Cl_4Cu_4N_4O_4$ 84.63 **7**
$C_{24}H_{56}Cl_4N_4O_4Zn_4$ 84.64 **7**
$C_{24}H_{56}Cr_2Li_4O_4$ 71.47 **3**
$C_{24}H_{56}Cu_4O_8P_4S_8$ 85.28 **4**
$C_{24}H_{56}O_8P_4S_8Zn_2$ 85.42 **3**
$C_{24}H_{58}Cu_3N_6O_7^{2+}$, $C_{24}H_{57}Cu_3N_6O_7^+$, $3ClO_4^-$, $8H_2O$ 84.31 **6**
$C_{24}H_{60}As_4Cu_4I_4$ 86.44 **7**
$C_{24}H_{60}Cl_4P_4Re_2$ 86.15 **6**
$C_{24}H_{60}Cl_4Pt_4$ 71.48 **3**
$C_{24}H_{60}Cu_4I_4P_4$ 86.45 **7**
$C_{24}H_{60}Ni_6O_{12}S_{12}$ 85.43 **3**
$C_{24}H_{60}Ni_6S_{12}$ 85.25 **2**
$C_{24}H_{62}Nb_2Si_6$ 71.52 **4**
$C_{24}H_{64}P_8Ru_2$ 71.82 **7**
$C_{24}H_{66}Mo_2Si_6$ 71.53 **4**
$2C_{24}H_{72}ClCoN_{18}P_6^+$, $Cl_6Co_2^{2-}$, $2CHCl_3$ 83.122 **4**
$2C_{24}H_{72}ClCoN_{18}P_6^+$, $Cl_6Co_2^{2-}$, $2CHCl_3$ 83.105 **5**
$C_{24}H_{72}ClCuN_{18}P_6$ + , Cl_2Cu^- 83.127 **3**
$C_{24}H_{72}CoN_{12}O_9P_6^{2+}$, $2ClO_4^-$ 84.18 **3**
$C_{24}H_{72}CuN_{12}O_9P_6^{2+}$, $2ClO_4^-$ 84.19 **3**
$C_{24}H_{72}MgN_{12}O_9P_6^{2+}$, $2ClO_4^-$ 67.25 **3**
$C_{24}H_{72}N_{18}P_6$ 64.70 **2**

C_{25}

$C_{25}H_{11}CdMn_2N_3O_{10}$ 83.106 5
$C_{25}H_{14}Br_2FeO_3$ 75.33 7
$C_{25}H_{15}BrO_7$ 59.20 4
$C_{25}H_{15}CoO_4$ 75.27 6
$C_{25}H_{16}$ 30.4 4
$C_{25}H_{16}CuN_5^+$, NO_3^- , H_2O 83.103 7
$C_{25}H_{16}Fe_3O_9$ 71.54 4
$C_{25}H_{16}Fe_3O_9$, $0.5C_6H_6$ 71.55 4
$C_{25}H_{17}MoO_4P$ 72.66 2
$C_{25}H_{18}ClN_2O_2Rh$ 83.128 3
$C_{25}H_{18}Cl_3N_3O_4$ 24.17 4
$C_{25}H_{18}FeO_2$ 75.38 3
$C_{25}H_{18}FeO_3$ 71.49 3
$C_{25}H_{18}Fe_2O_8$ 71.63 5
$C_{25}H_{18}Fe_2O_8$ 71.64 5
$C_{25}H_{18}N_2O_3S$ 35.36 7
$C_{25}H_{18}N_2O_5PRh$, H_2O 81.40 7
$C_{25}H_{19}BrO_2$ 27.18 6
$C_{25}H_{19}Fe^+$, BF_4^- 73.51 6
$C_{25}H_{19}IN_2O_4$ 32.25 5
$C_{25}H_{19}MoO_3P$ 75.39 3
$C_{25}H_{19}MoO_4P$, C_6H_6 61.8 6
$C_{25}H_{19}NOSSn$ 69.22 7
$C_{25}H_{20}$ 19.65 1
$C_{25}H_{20}$ 19.14 7
$C_{25}H_{20}IMoO_2P$ 73.46 3
$C_{25}H_{20}MnO_2P$ 73.52 6
$C_{25}H_{20}MnO_4P$ 71.56 4
$C_{25}H_{20}MnO_4P$ 72.23 6
$C_{25}H_{20}N_2Si$ 63.11 5
$C_{25}H_{20}OS$ 39.93 1
$C_{25}H_{20}OW$ 72.24 6
$C_{25}H_{20}O_2$ 27.11 7
$C_{25}H_{20}S_4$ 11.17 4
$C_{25}H_{21}CrO_3P$ 72.30 7
$C_{25}H_{21}CrO_5P$ 86.47 2
$C_{25}H_{21}CrO_5P$ 71.52 3
$C_{25}H_{22}CoN_5O_2$ 71.83 7
$C_{25}H_{22}NO_2PPd$ 71.58 4
$C_{25}H_{22}NO_2PS$ 64.50 6
$C_{25}H_{22}N_4O_8$, $C_4H_8O_2$ 50.19 7
$C_{25}H_{22}O_{10}$, CH_4O 59.26 3
$C_{25}H_{22}P^+$, Cl^- 64.51 6
$2C_{25}H_{22}P^+$, $C_4Cl_4NiO_4S_4Sn^{2-}$ 85.1 6
$2C_{25}H_{22}P^+$, $C_4Cl_8NiO_4S_4Sn_2^{2-}$, H_2O 85.2 6
$2C_{25}H_{22}P^+$, $C_{24}H_{30}FeO_{12}S_6^{2-}$ 85.41 6
$C_{25}H_{22}SSn$ 69.23 7
$C_{25}H_{22}Sn$ 69.35 4
$C_{25}H_{23}IO_8$ 59.58 1
$C_{25}H_{23}NP_2S$ 64.36 5
$C_{25}H_{23}N_3O_8$ 36.30 7
$C_{25}H_{23}OSb$ 66.21 2
$C_{25}H_{23}O_2P$ 64.54 3
$C_{25}H_{23}P$ 64.55 3
$C_{25}H_{24}BrNO_4$ 35.23 3
$C_{25}H_{24}BrNO_5$ 58.18 7
$C_{25}H_{24}Br_2O_6S_2$ 31.39 5
$C_{25}H_{24}ClNO_9S$ 35.37 7
$C_{25}H_{24}Cl_2N_2Pt$, $0.5C_2H_6O$ 71.56 6
$C_{25}H_{24}Co_2$ 73.108 2
$C_{25}H_{24}N_2O_6S$ 41.30 7
$C_{25}H_{25}BrO_8S$ 53.18 5
$C_{25}H_{25}Cl_4N_4^+$, $C_7H_7O_3S^-$, $0.2H_2O$ 35.38 7
$C_{25}H_{26}As_4Co_4F_8O_9$ 75.35 5
$C_{25}H_{26}BrN_3O_5$ 36.34 4
$C_{25}H_{26}BrN_5O_{13}$, $7H_2O$ 59.21 4
$C_{25}H_{26}F_6O_2$ 51.34 5
$C_{25}H_{27}BrO_3$ 51.20 1
$C_{25}H_{27}Cl_4N_4^+$, I^- , $2CH_4O$ 35.35 5
$C_{25}H_{27}Cl_4N_4^+$, I^- , C_2H_3N 35.36 5
$C_{25}H_{27}NiO_2P$ 71.84 7
$C_{25}H_{29}BrN_4O_6$ 59.29 5
$C_{25}H_{29}BrO_4S$ 51.40 4
$C_{25}H_{29}BrO_4S$ 51.41 4
$C_{25}H_{29}BrO_6S$ 53.19 5
$C_{25}H_{29}BrO_8$, $0.8CHCl_3$ 54.15 1
$C_{25}H_{29}N_3O_2$ 40.23 7
$C_{25}H_{30}BrNO_5$ 58.93 1
$C_{25}H_{30}NO_3^+$, I^- 58.94 1
$C_{25}H_{30}N_5Ni^+$, Cl^- , xH_2O , yCH_4O 49.2 1
$C_{25}H_{30}N_5Ni^+$, Cl^- , xCH_4O 49.2 4
$C_{25}H_{31}BrO_2$ 51.31 3
$C_{25}H_{31}BrO_3$ 28.26 6
$C_{25}H_{31}BrO_4S$ 51.35 5
$C_{25}H_{31}LiN_2$ 67.19 5
$C_{25}H_{31}N_3O_8S$ 50.7 4
$C_{25}H_{32}N_3O_4^+$, Br^- , H_2O 58.53 3
$C_{25}H_{32}N_3O_4^+$, I^- , H_2O 58.54 3
$C_{25}H_{33}BrO_3S$ 51.36 5
$C_{25}H_{33}BrO_8$ 50.32 1
$C_{25}H_{33}ClIrOP_3$ 86.35 5
$C_{25}H_{33}ClO_5$ 50.20 7
$C_{25}H_{33}N_2O_2^+$, $C_4H_5O_6^-$ 40.24 7
$C_{25}H_{33}N_4OPSSn$ 69.24 7
$C_{25}H_{33}O_2$ 12.10 3
$C_{25}H_{34}Br_2Cl_2O_8S_2$ 50.33 1
$C_{25}H_{34}F_3P_2Pt^+$, F_6Sb^- 71.57 6
$C_{25}H_{35}Al_3Mo_2$ 73.53 6
$C_{25}H_{35}Al_3Mo_2$ 73.51 7
$C_{25}H_{36}O_4$ 51.37 5
$C_{25}H_{36}O_4$ 51.38 5
$C_{25}H_{37}IrP_2$ 75.36 5
$C_{25}H_{38}BrNO_3$ 58.31 4
$C_{25}H_{38}Br_2O_3$ 51.21 1
$C_{25}H_{38}Br_2O_3$ 51.22 1
$C_{25}H_{38}CoN_6O_5$ 71.58 6
$C_{25}H_{38}N_4O_5$, C_2H_6OS , H_2O 48.72 7
$C_{25}H_{40}NO_4^+$, I^- 58.48 6
$C_{25}H_{44}N_2^{2+}$, $2Br^-$ 58.55 3
$C_{25}H_{44}N_2^{2+}$, $2I^-$ 58.56 3
$C_{25}H_{45}CoN_5O_4P$ 71.59 6
$C_{25}H_{45}N_5O_4Pd_3S_2$, $2C_6H_6$ 71.67 5
$C_{25}H_{46}Br_2O_5$ 1.39 3
$C_{25}H_{46}N_{13}O_{10}^{3+}$, $2Br^-$, Cl^- , $3H_2O$ 50.8 4
$C_{25}H_{48}O_5S_2$ 1.154 1
$C_{25}H_{50}N_5Nb$ 83.129 3
$C_{25}H_{50}N_5Ru_2S_{10}^+$, BF_4^- , C_3H_6O 80.19 7
$C_{25}H_{60}S_5Zn_5$ 71.50 2
$C_{25}H_{62}Si_7$ 63.16 3

C_{26}

$C_{26}H_{14}Cu_2F_{12}N_2O_8$ 81.41 7
$C_{26}H_{14}FeN_2O_9$ 71.61 4
$C_{26}H_{14}N_2$ 30.1 5
$C_{26}H_{14}N_2O_4$ 38.38 6

$C_{26}H_{14}N_8Ni$	49.3	**1**
$C_{26}H_{15}AsF_{12}FeS_4$	85.44	**3**
$C_{26}H_{15}Br$	30.5	**4**
$C_{26}H_{15}F_{12}FeOPS_4^-$, $C_{16}H_{36}N^+$	84.20	**3**
$C_{26}H_{15}Fe_2O_6P$	71.60	**6**
$C_{26}H_{16}$	28.27	**1**
$C_{26}H_{16}$	30.21	**1**
$C_{26}H_{16}$	24.18	**4**
$C_{26}H_{16}$	30.2	**5**
$C_{26}H_{16}$, $C_{13}H_4BrN_3O_7$	60.156	**2**
$C_{26}H_{16}CoCuF_{12}N_2O_6$	76.50	**6**
$C_{26}H_{16}CoN_6O_2S_2^+$, ClO_4^-	84.65	**7**
$C_{26}H_{16}CuN_2O_4$	83.130	**3**
$C_{26}H_{16}FeN_6O_2S_2$, $CHCl_3$, C_2H_6O , H_2O	83.107	**5**
$C_{26}H_{16}HgN_6S_2$	83.104	**7**
$C_{26}H_{16}N_2O_4Pd$	84.57	**2**
$C_{26}H_{16}N_6NiO_2S_2$	84.66	**7**
$C_{26}H_{16}N_8$	49.1	**3**
$C_{26}H_{16}O_2$	38.61 +	**1**
$C_{26}H_{17}CrO_3P$	75.37	**5**
$C_{26}H_{18}$	28.18	**5**
$C_{26}H_{18}$	28.20	**5**
$C_{26}H_{18}ClFeN_2O_2S_2$	78.12	**7**
$C_{26}H_{18}Cl_2N_4O_2Rh_2$	71.68	**5**
$C_{26}H_{18}Cl_2Zr$	75.34	**7**
$C_{26}H_{18}CuN_2O_2$	78.50	**2**
$C_{26}H_{18}N_2$	36.21	**6**
$C_{26}H_{18}N_2O_2$	26.39	**1**
$C_{26}H_{18}N_2O_2$	26.40	**1**
$C_{26}H_{18}N_2S_2$	41.34	**4**
$C_{26}H_{19}BrO_3$	27.19	**6**
$C_{26}H_{20}$	27.20	**6**
$C_{26}H_{20}$	19.15	**7**
$C_{26}H_{20}AuO_3PW$	73.109	**2**
$C_{26}H_{20}ClOP$	64.72	**2**
$C_{26}H_{20}Cl_2O_{10}Re_2$	84.32	**6**
$C_{26}H_{20}CoN_2S_4^-$, $C_{16}H_{36}N^+$	85.45	**3**
$C_{26}H_{20}CrO_4S_2W$	73.52	**7**
$C_{26}H_{20}CuN_2O_2$	78.51	**2**
$C_{26}H_{20}CuN_2O_2$	84.67	**7**
$C_{26}H_{20}Fe_2O_8S_2Sn$	73.110	**2**
$C_{26}H_{20}IOP$	64.73	**2**
$C_{26}H_{20}MoO_4S_2W$	73.53	**7**
$C_{26}H_{20}N_2NiO_2$	78.52	**2**
$C_{26}H_{20}N_2OS$	41.23	**3**
$C_{26}H_{20}N_4$	9.3	**6**
$C_{26}H_{20}N_4S_3$	4.6	**6**
$C_{26}H_{20}N_6^{2+}$, $2Br^-$	35.37	**5**
$C_{26}H_{20}N_6NiO_4$	78.13	**7**
$C_{26}H_{20}N_{10}NiS_2$	83.131	**3**
$C_{26}H_{20}Ni$	75.27	**4**
$C_{26}H_{20}O_4S_2W_2$	73.54	**7**
$C_{26}H_{20}P_2$	64.74	**2**
$C_{26}H_{21}BrO_7$	45.54	**7**
$C_{26}H_{21}CuN_6O_4^+$, ClO_4^- , C_2H_6O	78.18	**5**
$C_{26}H_{21}N_4O_2Rh$	71.69	**5**
$C_{26}H_{22}CuN_2O_2$	83.192	**2**
$C_{26}H_{22}Ge$	69.26	**5**
$C_{26}H_{22}HgN_8S_2$, $2C_5H_5N$	85.90	**2**
$C_{26}H_{22}HgS$	71.85	**7**
$C_{26}H_{22}N_3NiOPS$	78.14	**7**
$C_{26}H_{22}N_3NiOPSe$	78.15	**7**
$C_{26}H_{22}N_4NiO_2$	84.68	**7**
$C_{26}H_{22}N_4NiS_2$	85.44	**7**
$C_{26}H_{22}N_8NiS_2$	85.46	**3**
$C_{26}H_{22}N_8S_2Zn$	85.30	**4**
$C_{26}H_{22}N_8S_2Zn$	85.29	**5**
$C_{26}H_{22}O$	17.33	**7**
$C_{26}H_{22}OP^+$, Cl^-	64.37	**5**
$C_{26}H_{22}Se_2$	11.70	**1**
$C_{26}H_{23}CoN_2O_4$, $1.5H_2O$	78.13	**4**
$C_{26}H_{23}NO_8$	34.8	**5**
$C_{26}H_{23}NO_8$	34.9	**5**
$C_{26}H_{23}N_2O_2Sb$	66.9	**7**
$C_{26}H_{23}N_3P_2$	64.56	**3**
$C_{26}H_{23}N_5S$	41.31	**7**
$C_{26}H_{23}O_2PS$	64.75	**2**
$C_{26}H_{24}Br_2CoN_4S_2$	79.26	**5**
$C_{26}H_{24}Br_2N_4NiS_2$	79.27	**5**
$(C_{26}H_{24}Cl_2CuO_2P_2)_n$	84.13	**4**
$C_{26}H_{24}Cl_4P_2Re$, $0.75CCl_4$	86.46	**7**
$C_{26}H_{24}CoIN_2O_3P_2$	86.17	**6**
$C_{26}H_{24}CuN_2O_6$, H_2O	84.33	**6**
$C_{26}H_{24}F_{12}O_2S$	42.14	**7**
$C_{26}H_{24}F_{12}O_3S$	42.15	**7**
$C_{26}H_{24}N_2NiS_4$	85.30	**5**
$C_{26}H_{24}N_2P_2S$	64.60	**7**
$C_{26}H_{24}N_6NiO_3^{2+}$, $2NO_3^-$, $4H_2O$	84.34	**6**
$C_{26}H_{25}BrO_5$	27.21	**6**
$C_{26}H_{25}BrO_9$	59.17	**7**
$C_{26}H_{25}Cl_2NP_2Pd$	86.18	**6**
$C_{26}H_{25}IO_8$	53.30	**3**
$C_{26}H_{26}Br_2O_6$	21.13	**7**
$C_{26}H_{26}Cl_2CoNOP_2$	86.36	**5**
$C_{26}H_{26}Cl_2Pd_2$	71.61	**6**
$C_{26}H_{26}Cl_3NOP_2Re$	86.47	**7**
$C_{26}H_{26}Cl_3NOP_2Ru$	86.19	**6**
$C_{26}H_{26}CuN_2O_7$	81.54	**4**
$C_{26}H_{26}Cu_2N_2O_8$	81.87	**2**
$C_{26}H_{26}O_2$	31.46	**1**
$C_{26}H_{27}N_2O_{11}$	50.17	**3**
$C_{26}H_{27}OPdS$	71.62	**6**
$C_{26}H_{28}BrNO_3$	51.39	**5**
$C_{26}H_{28}BrNO_7S$	58.32	**4**
$C_{26}H_{28}Br_2N_2O_2$, H_2O	36.29	**5**
$C_{26}H_{28}CoP$, $0.5C_6H_6$	72.67	**2**
$C_{26}H_{28}Fe_2$	73.54	**6**
$C_{26}H_{28}NO_9^+$, Br^-	58.57	**3**
$C_{26}H_{28}N_2$	36.30	**5**
$C_{26}H_{28}N_2$	33.55	**7**
$C_{26}H_{28}N_2O_{11}$	50.9	**4**
$C_{26}H_{28}N_3O_4S_2$	35.39	**7**
$C_{26}H_{28}N_4$	28.8	**7**
$C_{26}H_{28}N_6NiS_2$	83.125	**4**
$C_{26}H_{28}N_6NiS_2$, $C_{18}H_{14}$	61.8	**3**
$C_{26}H_{28}O$	31.30	**6**
$C_{26}H_{28}O_2$	30.5	**7**
$C_{26}H_{28}O_2$	30.6	**7**
$C_{26}H_{28}O_{12}$	31.31	**6**
$C_{26}H_{29}BrO_7$	50.8	**6**
$C_{26}H_{29}ClCoN_4OSb$	86.48	**7**
$C_{26}H_{29}ClCoN_4O_4P$	86.20	**6**
$C_{26}H_{29}ClNO^+$, Cl^-	17.34	**7**
$C_{26}H_{29}ClN_4O_4PRh$	83.132	**3**
$C_{26}H_{29}ClN_4O_4RhSb$	86.37	**5**
$C_{26}H_{29}NOP^+$, Cl^-	64.38	**5**
$C_{26}H_{30}F_{12}O_2P_2Pd$	86.38	**5**
$C_{26}H_{30}NO^+$, Br^-	17.10	**3**
$C_{26}H_{30}N_4Ni$	83.133	**3**
$C_{26}H_{30}N_4O_4PRh^+$, Cl^- , CH_2Cl_2	86.39	**5**
$C_{26}H_{30}O_4$	51.40	**5**

$C_{26}H_{30}O_4$ 31.32 **6**
$C_{26}H_{30}O_8$ 31.31 **4**
$C_{26}H_{30}O_8P_2$ 64.39 **5**
$C_{26}H_{31}BrO_3$ 51.32 **3**
$C_{26}H_{31}BrO_3$ 51.33 **3**
$C_{26}H_{31}BrO_3$ 51.43 + **4**
$C_{26}H_{31}BrO_3$ 51.41 **5**
$C_{26}H_{31}BrO_4$ 51.38 **7**
$C_{26}H_{31}BrO_5$ 51.24 **1**
$C_{26}H_{31}BrO_5S$ 54.11 **4**
$C_{26}H_{31}IrO_2$ 71.70 **5**
$C_{26}H_{31}N_2O_5^+$, Br^- 58.33 **4**
$C_{26}H_{32}BrNO_{13}$ 45.55 **7**
$C_{26}H_{32}CuN_2O_2$ 78.16 **7**
$C_{26}H_{32}CuN_6O_6$ 83.74 **6**
$C_{26}H_{32}K_2N_2O_8S_2$ 67.24 **4**
$C_{26}H_{32}NO_4^+$, Br^- 58.49 **6**
$C_{26}H_{32}NO_5^+$, Br^- 58.34 **4**
$C_{26}H_{32}O_4$ 31.29 **7**
$C_{26}H_{32}O_{10}S_2$ 21.17 **5**
$C_{26}H_{33}BrO_2S$ 51.34 **6**
$C_{26}H_{33}BrO_4$ 51.35 **6**
$C_{26}H_{33}BrO_4S$ 51.25 **1**
$C_{26}H_{33}BrO_7$ 59.27 **3**
$C_{26}H_{33}Cl_2MoO_2P_3$, CH_4O 86.40 **5**
$C_{26}H_{33}N_2NiP_3$ 86.39 **3**
$C_{26}H_{34}Al_4Mo_2$ 73.55 **7**
$C_{26}H_{34}BrNO_2$ 51.45 **4**
$C_{26}H_{34}MnN_4O_8$, $4H_2O$ 82.56 **2**
$C_{26}H_{34}N_4NiO_8$, $7.4H_2O$ 82.22 **7**
$C_{26}H_{34}N_4O_8Zn$, $8.1H_2O$ 82.23 **7**
$C_{26}H_{34}O_8$ 54.8 **3**
$C_{26}H_{35}ClF_4N_2P_3Rh$ 86.21 **4**
$C_{26}H_{35}NiO_4P_3S_4$ 86.49 **7**
$C_{26}H_{36}Cl_2N_4NiO_2$ 78.53 **2**
$C_{26}H_{36}N_6Ni_2S_8$ 80.19 **6**
$C_{26}H_{36}O_4$ 28.19 **5**
$C_{26}H_{36}O_4S$ 51.39 **7**
$C_{26}H_{36}O_6$ 28.9 **7**
$C_{26}H_{36}O_8$ 54.12 **4**
$C_{26}H_{36}O_{11}$ 54.13 **7**
$C_{26}H_{37}BrN_2O_5S$ 54.13 **4**
$C_{26}H_{37}BrO_3S$ 54.14 **4**
$C_{26}H_{38}B_2Cl_2N_2$ 62.19 **3**
$C_{26}H_{38}Br_2O_8$ 54.9 **3**
$C_{26}H_{38}Cl_2NiO_4P_2$ 86.41 **5**
$C_{26}H_{38}Cl_2NiO_4P_2$ 86.42 **5**
$C_{26}H_{38}Fe$ 75.38 **5**
$C_{26}H_{39}BrO_5$ 55.1 **1**
$C_{26}H_{39}FeN_3O$ 71.63 **6**
$C_{26}H_{40}Al_2Fe_2O_4$ 73.56 **7**
$C_{26}H_{40}Cl_4O_2Rh_4$ 71.54 **3**
$C_{26}H_{40}Ni_2O_4$ 71.64 + **6**
$C_{26}H_{40}O_4Pd_2$ 71.71 **5**
$C_{26}H_{42}BrN_3NiP^+$, $C_{24}H_{20}B^-$ 83.126 **4**
$C_{26}H_{42}ClCoN_2P^+$, ClO_4^- 86.43 **5**
$C_{26}H_{42}Fe$ 73.55 **6**
$C_{26}H_{42}Mn_2N_2O_8$ 77.18 **4**
$C_{26}H_{42}O_4$ 50.10 **5**
$C_{26}H_{43}P$ 64.61 **7**
$C_{26}H_{46}N_2O_6$ 1.37 **5**
$C_{26}H_{54}Br_2Ni_2P_2$ 72.34 **5**
$C_{26}H_{56}B_{10}P_2Pt$ 71.86 **7**

C_{27}

$C_{27}H_{15}Fe_2O_9PPt$ 86.50 **7**
$C_{27}H_{16}AsF_4Fe_3O_9P$ 71.72 **5**
$C_{27}H_{17}CoF_3NiO_4P$ 73.57 **7**
$C_{27}H_{18}$ 30.3 **5**
$C_{27}H_{18}N_2$ 36.35 + **1**
$C_{27}H_{18}N_3S_3Sb$ 66.10 **7**
$C_{27}H_{19}Fe_2O_6P$ 72.68 **2**
$C_{27}H_{19}N_3O_5U$ 83.193 **2**
$C_{27}H_{19}N_3O_5U$, $CHCl_3$ 83.194 **2**
$C_{27}H_{20}NOP$, CH_4O 64.35 **4**
$C_{27}H_{21}ClU$ 75.28 **4**
$C_{27}H_{21}Sm$ 75.39 **5**
$C_{27}H_{22}As_2CrO_2$ 74.11 **6**
$C_{27}H_{22}BrN_5O_5$ 48.86 **6**
$C_{27}H_{22}F_6IrN_2OP$ 86.21 **6**
$C_{27}H_{22}N_2P_2PdS_2$ 86.51 **7**
$C_{27}H_{23}MoO_3P$ 71.51 **2**
$C_{27}H_{24}Cl_3FeN_4O_3$, $3H_2O$ 78.17 **7**
$C_{27}H_{24}CoN_3O_3$ 78.18 **7**
$C_{27}H_{26}BrNO_5$ 38.30 **5**
$C_{27}H_{26}BrNO_6$ 58.58 **3**
$C_{27}H_{26}N_4P_2$ 64.57 **3**
$C_{27}H_{26}SSn$ 69.17 **6**
$C_{27}H_{27}AsNiO_2P_2S_2$, $0.5C_6H_6$ 86.23 **6**
$C_{27}H_{27}BrO_7$ 56.1 **3**
$C_{27}H_{27}BrO_{10}$, $2C_2H_6O$ 54.17 **1**
$C_{27}H_{27}Cl_3CoN_5O_4$ 71.87 **7**
$C_{27}H_{27}CrO_6$ 71.88 **7**
$C_{27}H_{27}NO_5$ 35.40 **7**
$C_{27}H_{27}O_2Rh$ 72.19 **4**
$C_{27}H_{29}BrO_4$ 59.30 **5**
$C_{27}H_{29}EuN_2O_6$ 77.14 **5**
$C_{27}H_{29}N_2P$ 64.36 **4**
$C_{27}H_{29}N_3O_4S_2$ 36.35 **4**
$C_{27}H_{29}N_8^+$, I^- , C_3H_8O 36.33 **3**
$C_{27}H_{29}ORh$ 72.31 **7**
$C_{27}H_{30}Cl_2NO_2P_2Re$ 86.52 **7**
$C_{27}H_{30}NO_6^+$, I^- 58.50 **6**
$C_{27}H_{31}BrO_3$ 51.26 **1**
$C_{27}H_{31}BrO_3$ 30.5 **3**
$C_{27}H_{31}BrO_4$ 31.32 **4**
$C_{27}H_{31}BrO_7$ 53.10 **4**
$C_{27}H_{31}NO_5S$ 37.21 **7**
$C_{27}H_{32}NO_5^+$, CH_4O , Br^- 58.100 **1**
$C_{27}H_{33}B$ 62.11 **6**
$C_{27}H_{33}BrO_3$ 51.46 **4**
$C_{27}H_{33}BrO_3$ 31.41 **5**
$C_{27}H_{33}BrO_3$ 51.40 **7**
$C_{27}H_{33}BrO_4$ 54.13 **6**
$C_{27}H_{33}BrO_6$ 54.18 **1**
$C_{27}H_{33}CoSi_2$ 75.41 **5**
$C_{27}H_{33}CoSi_2$ 75.42 **5**
$C_{27}H_{33}CoSi_2$ 73.58 **7**
$C_{27}H_{33}IO_3P_3W^+$, $C_{24}H_{20}B^-$ 86.53 **7**
$C_{27}H_{33}IO_5$ 54.10 **3**
$C_{27}H_{33}N_2S_2^+$, ClO_4^- 41.32 **7**
$C_{27}H_{33}N_3O_5$, H_2O 58.19 **7**
$C_{27}H_{33}N_3O_7$, C_6H_6 58.20 **7**
$C_{27}H_{33}P$ 64.62 **7**
$C_{27}H_{34}NO_6^+$, I^- 58.101 **1**
$C_{27}H_{34}N_4O_8^{2+}$, $C_4H_8O_2$, $2Br^-$, H_2O 50.34 **1**
$C_{27}H_{34}N_5Ni^+$, ClO_4^- , $0.8C_3H_6O_2$ 49.3 **4**

$C_{27}H_{34}N_5Pd^+$, ClO_4^- , $0.6C_3H_6O$ 49.4 **4**
$C_{27}H_{34}N_5Pt^+$, ClO_4^- 49.5 **4**
$C_{27}H_{34}O_2Si$ 63.12 **5**
$C_{27}H_{35}BrO_7$ 54.14 **6**
$C_{27}H_{35}LiN_2$ 67.20 **5**
$C_{27}H_{36}BrNO_2$ 54.15 **6**
$C_{27}H_{36}N_5+$, Cl^- , C_2H_6O 49.1 **7**
$C_{27}H_{36}N_5NiO^+$, ClO_4^- 49.6 **4**
$C_{27}H_{37}BrO_3$ 51.47 **4**
$C_{27}H_{37}BrO_8$ 54.11 **3**
$C_{27}H_{37}Cl_3NOP_2Re$ 83.195 **2**
$C_{27}H_{37}IO$ 30.6 **3**
$C_{27}H_{37}IO_4$ 51.42 **5**
$C_{27}H_{38}NO_7^+$, Br^- 58.102 **1**
$C_{27}H_{38}NO_7^+$, Br^- 58.59 **3**
$C_{27}H_{39}NO$ 13.23 **4**
$C_{27}H_{39}O_5S$ 51.41 **7**
$C_{27}H_{40}O_7$ 51.42 **7**
$C_{27}H_{41}NO_3$ 58.103 **1**
$C_{27}H_{42}AsN_4NiS^+$, $C_{24}H_{20}B^-$ 76.51 **5**
$C_{27}H_{42}Cl_2N_2O_6$ 50.21 **7**
$C_{27}H_{44}Br_2$ 51.27 **1**
$C_{27}H_{44}NO_2^+$, I^- , CH_4O 58.105 **1**
$C_{27}H_{44}NO_8^+$, I^- , $2H_2O$ 58.106 **1**
$C_{27}H_{44}O$, C_2H_6O 51.36 **6**
$C_{27}H_{44}O_6$ 51.28 **1**
$C_{27}H_{45}Cl_2NO$ 51.36 **3**
$C_{27}H_{45}Cl_2NO$, $0.5C_3H_6O$ 36.36 **4**
$C_{27}H_{45}FeN_6O_9$ 84.35 **6**
$C_{27}H_{45}I$ 51.29 **1**
$C_{27}H_{45}O_4S^-$, Na^+ 51.43 **7**
$C_{27}H_{46}$ 56.1 **7**
$C_{27}H_{46}BrCl$ 51.30 **1**
$C_{27}H_{46}BrCl$ 51.31 **1**
$C_{27}H_{46}Br_2$ 51.32 **1**
$C_{27}H_{46}Cl_2$ 51.33 **1**
$C_{27}H_{46}Cl_2$ 51.34 **1**
$C_{27}H_{46}NO^+$, I^- 58.107 **1**
$C_{27}H_{46}NO^+$, I^- , $0.5C_2H_6O$ 58.39 **5**
$C_{27}H_{46}NO_2^+$, Br^- 58.108 **1**
$C_{27}H_{46}NO_2^+$, I^- 58.109 **1**
$C_{27}H_{46}O_3$ 51.44 **7**
$C_{27}H_{47}Cl$ 51.45 **7**
$C_{27}H_{48}NO_2^+$, I^- 58.40 **5**
$C_{27}H_{48}NO_2^+$, I^- 51.37 **6**
$C_{27}H_{48}NO_3^+$, Br^- 51.38 **6**
$C_{27}H_{48}NO_3^+$, I^- , CH_4O 51.48 **4**
$C_{27}H_{49}NiP$ 71.73 **5**
$C_{27}H_{53}NiP$ 71.89 **7**
$C_{27}H_{54}Cu_4N_{18}S_9^{4+}$, $4NO_3^-$, $6H_2O$ 79.17 **6**
$C_{27}H_{54}FeN_3S_6$ 80.40 **2**
$C_{27}H_{54}MoN_4OS_6$ 80.20 **6**
$C_{27}H_{54}N_3NiS_6^+$, Br^- 80.18 **3**
$C_{27}H_{54}N_3NiSe_6^+$, Br^- 80.18 **4**
$C_{27}H_{63}Cl_3O_3P_3Rh$ 86.40 **3**
$C_{27}H_{78}CuN_{12}O_6P_6^{2+}$, $2ClO_4^-$ 84.23 **5**

C_{28}

$C_{28}H_{12}Cl_2N_2O_2Ti$ 83.196 **2**
$C_{28}H_{12}N_2O_2$ 36.37 **1**
$C_{28}H_{14}N_2O_4$ 36.38 **1**
$C_{28}H_{16}$ 30.22 **1**
$C_{28}H_{16}$ 31.30 **7**
$C_{28}H_{16}$, $0.5C_6H_6$ 30.4 **5**
$C_{28}H_{16}Br_2O_4$ 13.56 **1**
$C_{28}H_{16}Cl_2O_4$ 13.57 **1**
$C_{28}H_{16}N_2O_4$ 26.11 **3**
$(C_{28}H_{16}N_4O_{17}Sn)_n$, nC_4H_8O 69.18 **6**
$C_{28}H_{16}O_2$ 28.28 **1**
$C_{28}H_{17}CrO_5P$ 86.24 **6**
$C_{28}H_{18}$ 31.36 **3**
$C_{28}H_{18}Br_2N_4O_2$ 32.41 **7**
$C_{28}H_{18}Co_3O_8P$ 71.55 **3**
$C_{28}H_{18}FeO_4$ 75.75 **2**
$C_{28}H_{18}N_2$ 35.36 **4**
$C_{28}H_{18}N_4O_4Ru_2S_4$ 85.45 **7**
$C_{28}H_{18}O_2$ 28.29 **1**
$C_{28}H_{18}O_4$ 26.41 **1**
$C_{28}H_{19}BrN_2O_3$ 9.49 **1**
$C_{28}H_{19}BrN_2O_3$ 9.18 **5**
$C_{28}H_{19}BrN_2O_3$ 9.19 **5**
$C_{28}H_{20}$ 31.47 **1**
$C_{28}H_{20}$ 30.6 **5**
$C_{28}H_{20}$ 30.1 **6**
$C_{28}H_{20}$ 30.2 **6**
$C_{28}H_{20}$ 30.3 + **6**
$C_{28}H_{20}Br_4Cu_2O_{10}$ 81.58 **5**
$C_{28}H_{20}Cl^+$, Cl_5Sn^- 12.19 **1**
$C_{28}H_{20}Cl_2N_2$ 33.56 **7**
$C_{28}H_{20}Cl_2O_8Re_2$, $2CHCl_3$ 81.88 **2**
$C_{28}H_{20}HfO_8$, C_3H_7NO 84.69 **7**
$C_{28}H_{20}MnN_{10}O_6$ 83.105 **7**
$C_{28}H_{20}N_2O_2$ 40.33 **1**
$C_{28}H_{20}N_2O_3$ 9.20 **5**
$C_{28}H_{20}N_2O_3$ 9.21 **5**
$C_{28}H_{20}N_4Ni^{2+}$, $2BF_4^-$ 83.134 **3**
$C_{28}H_{20}N_4Ni^{2+}$, $2I^-$, H_2O 83.135 **3**
$C_{28}H_{20}N_6Ni$ 83.127 **4**
$C_{28}H_{20}NiS_4$ 85.91 **2**
$C_{28}H_{20}Ni_2O_4P_2$ 86.41 **3**
$C_{28}H_{20}O_4S$ 42.12 **5**
$C_{28}H_{20}O_7V^{2-}$, $C_8H_{20}N^+$, Na^+ , $2C_3H_8O$ 81.52 **3**
$C_{28}H_{20}S_8V$ 85.42 **6**
$C_{28}H_{21}NO_2PRh$ 83.128 **4**
$C_{28}H_{21}O_8Sc$ 84.70 + **7**
$C_{28}H_{22}$ 5.41 **1**
$C_{28}H_{22}Br_2N_2O_2$ 17.35 **7**
$C_{28}H_{22}CoN_2O_2$ 78.19 **3**
$C_{28}H_{22}CrN_3O_4$ 83.197 **2**
$C_{28}H_{22}FeO_3P_2$ 86.45 **5**
$C_{28}H_{22}IN_4NiO_4$ 60.155 **2**
$C_{28}H_{22}Mo_2O_6$ 75.76 **2**
$C_{28}H_{22}N_3^+$, ClO_4^- 36.37 **4**
$C_{28}H_{22}OP^+$, ClO_4^- 64.52 **6**
$C_{28}H_{23}BrO_{13}$ 59.31 **5**
$C_{28}H_{23}FeOPS$ 71.56 **3**
$C_{28}H_{23}FeO_2P$ 72.42 **3**
$C_{28}H_{24}$ 20.34 **1**
$C_{28}H_{24}BrNO_6$ 58.51 **6**
$C_{28}H_{24}Cl_4N_4Ni_2$, $2CHCl_3$ 83.136 **3**
$C_{28}H_{24}CoN_2O_4$ 78.13 **6**
$C_{28}H_{24}CoN_2O_4$ 78.19 **7**
$C_{28}H_{24}Cu_2O_{14}$, $C_4H_8O_2$ 78.20 **7**
$C_{28}H_{24}Ge_2$ 69.45 **2**
$C_{28}H_{24}N_2$ 36.31 **5**
$C_{28}H_{24}N_2O_4Zn$ 78.20 **3**
$C_{28}H_{24}N_2P_2PdS_2$ 86.54 **7**

$C_{28}H_{24}Ni_2O_{10}$ 84.24 **5**
$C_{28}H_{25}ClSiZr$ 73.30 **4**
$C_{28}H_{26}Br_2O_4$ 31.42 **5**
$C_{28}H_{26}CuN_4O_6$ 84.25 **5**
$C_{28}H_{26}O_3S$ 39.29 **6**
$C_{28}H_{26}O_4Sn_2$ 69.33 **3**
$C_{28}H_{27}BrO_6$ 50.22 **7**
$C_{28}H_{27}ClF_5NO$ 33.60 **6**
$C_{28}H_{27}FeN_4$ 71.66 **6**
$C_{28}H_{28}Br_2NNiP_2$ 86.49 **2**
$C_{28}H_{28}Cd_2O_{16}$ 78.19 **5**
$C_{28}H_{28}Cl_2NiOP_2$ 86.43 **3**
$C_{28}H_{28}CoNiO_4P$ 73.59 **7**
$C_{28}H_{28}Fe_2S_6^{2-}$, $2C_8H_{20}N^+$ 85.46 **7**
$C_{28}H_{28}Fe_4S_8^{2-}$, $2C_8H_{20}N^+$ 85.31 **5**
$C_{28}H_{28}GeS$ 69.25 **7**
$C_{28}H_{28}Hf$ 71.62 **4**
$C_{28}H_{28}HgI_2P_2S$ 86.55 **7**
$C_{28}H_{28}I_2Sn_2$ 69.34 **3**
$C_{28}H_{28}P_2^{2+}$, $2Br^-$, H_2O 64.40 **5**
$C_{28}H_{28}SSn$ 69.19 **6**
$C_{28}H_{28}Sn$ 69.36 **4**
$C_{28}H_{28}Sn$ 69.26 **7**
$C_{28}H_{28}Ti$ 71.63 + **4**
$C_{28}H_{28}Zr$ 71.57 **3**
$C_{28}H_{29}Br_2NNiP_2$ 86.44 **3**
$C_{28}H_{29}O_4PPd$, $0.5C_6H_6$ 71.90 **7**
$C_{28}H_{30}NP_2^+$, I^- 64.76 **2**
$C_{28}H_{31}BrO_5$ 51.35 **1**
$C_{28}H_{31}ClOTi$ 73.60 **7**
$C_{28}H_{31}ClP_2Pt$ 86.50 **2**
$C_{28}H_{31}IO_6$ 56.1 **1**
$C_{28}H_{32}Cl_4Cu_4$ 75.77 **2**
$C_{28}H_{32}O_4Si_4$ 63.18 **7**
$C_{28}H_{32}P^+$, Br^- 64.41 **5**
$C_{28}H_{32}W$ 71.91 **7**
$C_{28}H_{33}AgNO_7^+$, NO_3^- , H_2O 58.52 **6**
$C_{28}H_{33}BrO_5S$ 50.35 **1**
$C_{28}H_{33}ClIrOP_2^+$, BF_4^- , CH_2Cl_2 72.35 **5**
$C_{28}H_{33}IO_9$ 56.2 **1**
$C_{28}H_{34}$ 19.16 **7**
$C_{28}H_{34}O_3$ 17.11 **3**
$C_{28}H_{35}AsN_2O_3S_4U$ 80.19 **3**
$C_{28}H_{35}AsN_2O_3Se_4U$ 85.32 **4**
$C_{28}H_{35}ClP_2Pt$ 71.67 **6**
$C_{28}H_{35}N_2O_3PS_4U$ 80.20 **3**
$C_{28}H_{35}N_3O_7$, $C_4H_8O_2$ 50.9 **6**
$C_{28}H_{36}Cl_2Ga_2N_6O_2^{2+}$, $2Cl^-$, H_2O 68.30 **5**
$C_{28}H_{36}CuN_6O_6$, $2H_2O$ 83.106 **7**
$C_{28}H_{36}NO_4^+$, Br^- 58.110 **1**
$C_{28}H_{36}N_4Ni^{2+}$, $2ClO_4^-$ 83.198 **2**
$C_{28}H_{36}N_4Ni^{2+}$, $2ClO_4^-$ 83.199 **2**
$C_{28}H_{36}N_6O_6Zn$ 83.107 **7**
$C_{28}H_{36}N_8P_4$ 64.37 **4**
$C_{28}H_{36}O_2$ 30.5 **6**
$C_{28}H_{36}O_2$ 30.6 **6**
$C_{28}H_{36}O_3$, $0.5H_2O$ 56.2 **5**
$C_{28}H_{36}O_5$ 51.39 **6**
$C_{28}H_{37}BrNO_5^+$, Br^- , C_2H_6O 58.42 **5**
$C_{28}H_{37}BrO_4$ 54.14 **7**
$C_{28}H_{37}Cl_3NOP_2Re$ 83.200 **2**
$C_{28}H_{38}Al_2Ti_2$, C_7H_8 73.56 **6**
$C_{28}H_{38}BrN_5O_8$, $0.75C_4H_8O_2$, H_2O 48.57 **4**
$C_{28}H_{38}Cl_2NiP_2$ 86.46 **5**
$C_{28}H_{38}CoN_2O_2$ 78.20 **5**
$C_{28}H_{38}CrN_4Si_2^+$, I^- 71.68 **6**
$C_{28}H_{38}O_7$, $2H_2O$ 51.49 **4**
$C_{28}H_{39}BrO_4$ 31.31 **7**
$C_{28}H_{39}BrO_5$ 55.2 **1**
$C_{28}H_{39}BrO_9$ 54.12 **3**
$C_{28}H_{39}BrO_{16}$ 52.31 **1**
$C_{28}H_{40}Al_2Ti_2$ 73.111 **2**
$C_{28}H_{40}CuN_2O_2$ 78.14 **6**
$C_{28}H_{40}KO_{10}^+$, I^- 67.25 **4**
$C_{28}H_{40}KO_{10}^+$, I^- 38.31 **5**
$C_{28}H_{40}MgN_4O_6P_2S_2$ 67.16 **6**
$C_{28}H_{40}NiP_2$ 86.51 + **2**
$C_{28}H_{40}O_{10}$ 38.26 **4**
$C_{28}H_{41}BrO_4$ 54.20 **1**
$C_{28}H_{42}AsN_5NiS_2$ 76.86 **3**
$C_{28}H_{42}Br_2O_9$ 54.13 **3**
$C_{28}H_{42}N_8NiO_6$ 84.72 **7**
$C_{28}H_{43}IO_4$ 51.40 **1**
$C_{28}H_{43}NO_6$, $C_5H_{12}O$ 50.23 **7**
$C_{28}H_{44}Br_2O$ 51.37 + **3**
$C_{28}H_{44}Cu_4N_4O_8$ 77.25 **3**
$C_{28}H_{44}IrO_2P_2$ 86.25 **6**
$C_{28}H_{44}NO_8^+$, Cl^- 58.21 **7**
$C_{28}H_{44}O$ 51.46 **7**
$C_{28}H_{44}O$, H_2O 51.47 **7**
$C_{28}H_{44}O_2$ 51.43 **5**
$C_{28}H_{46}Co_4O_{16}$ 77.16 **5**
$C_{28}H_{46}NO_3^+$, Br^- 58.111 **1**
$C_{28}H_{46}NO_3^+$, Br^- 58.60 **3**
$C_{28}H_{46}P_2Pd$ 86.26 **6**
$C_{28}H_{48}Cd_2N_4S_8$ 80.28 **5**
$C_{28}H_{48}N_4S_8Zn_2$ 80.29 **5**
$C_{28}H_{48}N_8NiS_4^{2+}$, $2I^-$ 79.12 **4**
$C_{28}H_{49}NiP$ 71.65 **4**
$C_{28}H_{52}AgNO_3$ 23.6 **4**
$C_{28}H_{52}Ni_2P_2$ 72.36 **5**
$C_{28}H_{55}BrO$ 51.50 **4**
$C_{28}H_{55}BrO$ 51.51 **4**
$C_{28}H_{56}Mo_2N_4O_3S_8$ 80.20 **7**
$C_{28}H_{56}Mo_2N_4S_8$ 71.69 **6**
$C_{28}H_{58}Cl_6Mg_4O_6$ 67.26 **4**
$C_{28}H_{64}N_6Ni_2O_2^{2+}$, $2ClO_4^-$ 76.87 **3**
$C_{28}H_{64}N_{17}Ni_2^+$, I^- 76.51 **6**
$C_{28}H_{66}Br_2Mg_2O_4$ 67.7 **7**
$C_{28}H_{72}O_{16}Ti_4$ 84.58 **2**
$(C_{28}H_{84}Cl_{10}Mn_5O_{14})_n$ 84.36 **6**

C_{29}

$C_{29}H_{15}Fe_3O_{11}P$, $C_{29}H_{15}Fe_3O_{11}P$ 86.53 **2**
$C_{29}H_{18}Br_2O_2S_2$ 39.46 **4**
$C_{29}H_{18}Fe_2N_2O_6$ 83.137 **3**
$C_{29}H_{20}Br$ 20.12 **4**
$C_{29}H_{20}F_5NiP$ 73.112 **2**
$C_{29}H_{20}F_6FeN_2O_2P_2$ 86.45 **3**
$C_{29}H_{20}S_3$ 39.30 **6**
$C_{29}H_{21}BrO_{11}$ 59.60 **1**
$C_{29}H_{21}NO_2S$, C_6H_6 39.40 **7**
$C_{29}H_{22}$ 20.18 **3**
$C_{29}H_{22}MoO_4P_2$ 86.46 **3**
$C_{29}H_{22}N_2O$ 35.41 **7**
$C_{29}H_{23}IrMnO_5P$, $0.5C_6H_6$ 71.70 **6**
$C_{29}H_{23}MnO_3P_2S_2$ 85.32 **5**
$C_{29}H_{24}Br_2MoO_3P_2$, C_3H_6O 86.23 **4**

$C_{29}H_{24}MnO_3PS$ 71.92 **7**
$C_{29}H_{25}BRu$ 73.57 **6**
$C_{29}H_{25}Cu_2N_6O_6$ 81.42 **7**
$C_{29}H_{25}NiP$ 73.113 **2**
$C_{29}H_{25}O_3P$ 64.38 **4**
$C_{29}H_{26}ClF_3O_2P_2Pd$, CH_2Cl_2 86.56 **7**
$C_{29}H_{26}N_2P_2PdS_2$ 86.57 **7**
$C_{29}H_{27}MnO_3P_2$ 86.28 **6**
$C_{29}H_{28}Br_2O_6$ 53.39 **1**
$C_{29}H_{28}O_2P_2Rh^+$, F_6P^- 86.58 **7**
$C_{29}H_{30}BrNO_{11}$, C_3H_6O 50.11 **5**
$C_{29}H_{30}ClIrO_3P_2$ 86.24 **4**
$C_{29}H_{30}N_4^{2+}$, $2Br^-$, $2H_2O$ 58.22 **7**
$C_{29}H_{32}BrClO_6S$ 50.10 **4**
$C_{29}H_{32}BrNO_9$ 59.61 **1**
$C_{29}H_{33}Cl_3NP_2Re$ 86.47 **3**
$C_{29}H_{34}BrFO_5$ 51.39 **3**
$C_{29}H_{34}CoN_7$ 49.2 **3**
$C_{29}H_{34}N_2O_4$ 58.23 **7**
$C_{29}H_{34}N_4$ 49.3 **3**
$C_{29}H_{35}N_4^+$, Br^- , $1.5CHCl_3$ 49.2 **7**
$C_{29}H_{37}AgNO_5^+$, BF_4^- , $2C_3H_8O$ 50.19 **3**
$C_{29}H_{37}N_3O_6$ 50.10 **6**
$C_{29}H_{39}BrN_2O_5S$ 51.36 **1**
$C_{29}H_{39}BrO_4$ 54.15 **7**
$C_{29}H_{39}NO_9^+$, I^- 58.35 **4**
$C_{29}H_{39}NO_{12}$, $1.5H_2O$ 38.39 **6**
$C_{29}H_{43}IO_4$ 51.37 **1**
$C_{29}H_{43}IO_6$, $1.5C_6H_6$ 59.27 **6**
$C_{29}H_{44}Br_2O_4$ 51.38 **1**
$C_{29}H_{44}O_5$ 51.48 **7**
$C_{29}H_{45}BrO_4$ 51.39 **1**
$C_{29}H_{46}FeN_9O_{14}$, H_2O 82.28 **5**
$C_{29}H_{46}O$ 51.49 **7**
$C_{29}H_{47}NO_5$, CH_4O 58.61 **3**
$C_{29}H_{48}O_2S$ 51.41 **1**
$C_{29}H_{49}ClO_2$ 51.50 **7**
$C_{29}H_{50}$ 56.2 **7**
$C_{29}H_{50}NO_3^+$, Br^- , H_2O 58.43 **5**
$C_{29}H_{53}Cl_2N_4Ta$ 83.108 **7**
$C_{29}H_{58}NO_8P$, $C_2H_4O_2$ 46.14 **6**

C_{30-34}

$C_{30}Co_8O_{24}$ 71.59 **3**
$C_{30}Co_8O_{24}$, $0.5C_6H_8$ 71.60 **3**
$C_{30}H_{14}$ 30.23 **1**
$C_{30}H_{14}O_2$ 30.24 **1**
$C_{30}H_{14}O_2$ 30.25 **1**
$C_{30}H_{15}AuClF_{10}P$ 71.52 **2**
$C_{30}H_{16}$ 30.26 **1**
$C_{30}H_{16}$ 30.8 **4**
$C_{30}H_{16}Fe_4O_{10}$, $C_2H_4Cl_2$ 75.43 **3**
$C_{30}H_{18}$ 30.7 **7**
$C_{30}H_{18}Br_2O_7$ 59.62 **1**
$C_{30}H_{18}Cl_2$ 29.28 **1**
$C_{30}H_{18}Fe_2O_6$ 72.69 **2**
$C_{30}H_{18}N_3O_6P_3$, $3C_6H_6$ 61.9 **6**
$C_{30}H_{18}N_3O_6P_3$, $0.5C_8H_{10}$ 61.2 **7**
$C_{30}H_{18}O_2$ 30.27 **1**
$C_{30}H_{20}$ 31.37+ **3**
$C_{30}H_{20}$ 28.27 **6**
$C_{30}H_{20}$ 28.28 **6**
$C_{30}H_{20}Br_2$ 31.43 **5**
$C_{30}H_{20}Cl_2Cu_2Fe_2O_4$ 71.93 **7**
$C_{30}H_{20}F_{12}O_2S$ 11.16 **5**
$C_{30}H_{20}N_5NaO_3$ 67.17 **6**
$C_{30}H_{20}N_5O_3Rb$ 67.18 **6**
$C_{30}H_{20}Ni_2O_6P_2$ 86.55 **2**
$C_{30}H_{20}O_2$ 31.48 **1**
$C_{30}H_{20}S_2$ 39.47 **4**
$C_{30}H_{22}Br_2O_{10}$, $2CH_4O$, H_2O 59.29 **3**
$C_{30}H_{22}Cl_2$ 20.13 **4**
$C_{30}H_{22}CoN_6^{2+}$, $2Br^-$, $3H_2O$ 83.109 **7**
$C_{30}H_{22}CuN_6^{2+}$, $2NO_3^-$ 83.109 **5**
$C_{30}H_{22}CuO_4$ 77.59 **2**
$C_{30}H_{22}CuO_4$ 77.26 **3**
$C_{30}H_{22}MoO_6$ 77.17 **6**
$C_{30}H_{22}O_2PdS_2$ 77.18 **6**
$C_{30}H_{22}O_4Pd$ 77.60 **2**
$C_{30}H_{22}O_{12}V_2$ 73.58 **6**
$C_{30}H_{23}Br_2NO_6$ 58.53 **6**
$C_{30}H_{23}F_3N_2P_2PdS_2$, CH_2Cl_2 86.47 **5**
$(C_{30}H_{24}Cl_4Cu_2N_{12}O_6)_n$ 84.73 **7**
$C_{30}H_{24}CrO_4P_2$ 86.25 **4**
$C_{30}H_{24}CuN_6^{2+}$, $2ClO_4^-$ 83.110 **5**
$C_{30}H_{24}HgN_4^{2+}$, $2ClO_4^-$ 71.67 **4**
$C_{30}H_{24}N_6V$ 83.202 **2**
$C_{30}H_{25}BrNO_4P$ 64.78 **2**
$C_{30}H_{25}Cr^{2-}$, $2Na^+$, $3C_4H_{10}O$, C_4H_8O 71.75 **5**
$C_{30}H_{25}FeOP$ 71.61 **3**
$C_{30}H_{25}MoNO_4P_2$ 86.56 **2**
$C_{30}H_{25}Ni_2O_5S_4$ 85.47 **3**
$C_{30}H_{25}P$ 64.79 **2**
$C_{30}H_{25}P_5$ 64.80 **2**
$C_{30}H_{25}Sb$ 66.22+ **2**
$C_{30}H_{25}Sb$, $0.5C_6H_{12}$ 66.11 **7**
$C_{30}H_{26}^{2+}$, $2ClO_4^-$ 27.16 **1**
$C_{30}H_{26}BrO_9$ 59.32 **5**
$C_{30}H_{26}Br_2N_6O_2$, C_6H_6 16.16 **7**
$C_{30}H_{26}Br_2Sn$ 69.35 **3**
$C_{30}H_{26}Co_2O_4Sn$ 75.29 **4**
$C_{30}H_{26}Fe_3$ 73.115 **2**
$C_{30}H_{26}Ni_2O_5S_4$ 84.60 **2**
$C_{30}H_{26}O_{14}$, $5.5H_2O$ 59.18 **7**
$C_{30}H_{27}IO_{14}$ 50.36 **1**
$C_{30}H_{28}Br_2CuN_6O_6$, $2H_2O$ 83.76 **6**
$C_{30}H_{28}ClFeN_2O_2$ 78.15 **6**
$C_{30}H_{28}CoN_6S_2$ 83.129 **4**
$C_{30}H_{28}CuN_2O_2$ 78.16 **6**
$C_{30}H_{28}CuN_2O_2$ 78.21 **7**
$C_{30}H_{28}CuN_2O_2$ 78.22 **7**
$C_{30}H_{28}FeN_4S_2^{2+}$, $2ClO_4^-$, CH_4O 83.138 **3**
$C_{30}H_{28}Mn_2N_6O_6$, $2CH_4O$ 78.22 **5**
$C_{30}H_{28}N_6O_6S_4$ 50.12 **5**
$C_{30}H_{28}Ni_2S_8$ 85.43 **6**
$C_{30}H_{28}O_4^{2+}$, $2Cl_2I^-$ 12.20 **1**
$C_{30}H_{28}O_6$ 59.22 **4**
$C_{30}H_{28}Th_2$ 71.94 **7**
$C_{30}H_{29}O_7Y$ 77.61 **2**
$C_{30}H_{30}Ag^+$, BF_4^- 75.78 **2**
$C_{30}H_{30}Br_2N_4O_4$ 59.64 **1**
$C_{30}H_{30}CoN_6O_6^{2+}$, $2ClO_4^-$ 84.74 **7**
$C_{30}H_{30}Cu_3N_{20}^{4+}$, $4ClO_4^-$ 83.111 **5**
$C_{30}H_{30}FeN_6^{2+}$, $C_{13}F_5O_{13}^{2-}$ 83.203 **2**
$C_{30}H_{30}HgN_6O_6^{2+}$, $2ClO_4^-$ 84.26 **5**
$C_{30}H_{30}N_6NiO_6^{2+}$, $2BF_4^-$ 84.37+ **6**
$C_{30}H_{30}N_8Ni$ 83.204 **2**

$C_{30}H_{30}O_2P_2Pt$ 86.48 **5**
$C_{30}H_{30}O_4$ 17.22 **5**
$C_{30}H_{31}BrNiP_2$ 72.43 **3**
$C_{30}H_{32}Al_2N_2O_2$ 68.12 **4**
$C_{30}H_{32}AuI_2N_4^+$, ClO_4^- , $C_4H_{10}O$ 71.95 **7**
$C_{30}H_{32}CuN_2O_6$ 84.23 **3**
$C_{30}H_{32}F_{21}LuO_7$ 77.19 **4**
$C_{30}H_{32}I_2NiO_2P_2$ 86.48 **3**
$C_{30}H_{33}BrF_4O_3$ 51.40 **3**
$C_{30}H_{34}CuN_4$ 49.4 **3**
$C_{30}H_{35}CoN_5O_5^{2+}$, $2ClO_4^-$ 84.24 **3**
$C_{30}H_{36}O_2Sn$ 69.47 **2**
$C_{30}H_{36}O_4$ 59.65 **1**
$C_{30}H_{37}BrO_{12}$, $2CH_4O$ 59.30 **3**
$C_{30}H_{37}N_{15}O_{16}P_2$ 47.41+ **7**
$C_{30}H_{37}O_8^-$, Cs^+ , $3H_2O$ 50.13 **5**
$C_{30}H_{38}BO_6P_2Rh$ 74.12 **6**
$C_{30}H_{38}ClIO_8$ 59.66 **1**
$C_{30}H_{38}Cl_2N_2P_3Re$ 86.49 **3**
$C_{30}H_{38}CoN_7O^+$, Br^- , $xC_4H_8O_2$, yCH_4O 49.5 **1**
$C_{30}H_{38}CoN_7O^+$, ClO_4^- 49.6 **1**
$C_{30}H_{39}BrO_4$ 54.22 **1**
$C_{30}H_{39}IO_8$ 56.3 **1**
$C_{30}H_{40}NO_{13}^+$, I^- , $0.5CH_4O$ 58.36 **4**
$C_{30}H_{41}BrO_4S$ 59.23 **4**
$C_{30}H_{41}IO_4$ 59.19 **7**
$C_{30}H_{42}Ni_3O_{12}$ 77.62 **2**
$C_{30}H_{42}O_7$ 51.44 **5**
$C_{30}H_{42}O_8$ 59.20 **7**
$C_{30}H_{42}O_{12}Zn_3$ 77.63 **2**
$C_{30}H_{44}BrNO_3$ 51.52 **4**
$C_{30}H_{44}BrN_3O_3$ 50.37 **1**
$C_{30}H_{44}Br_2O_2$ 51.42 **1**
$C_{30}H_{44}Br_2O_3$ 56.2 **3**
$C_{30}H_{44}Br_2O_3$ 56.1 **6**
$C_{30}H_{44}Co_3O_{13}$ 77.64 **2**
$C_{30}H_{44}N_2O_2S^{2+}$, $2Br^-$, $4H_2O$ 58.112 **1**
$C_{30}H_{44}N_2O_2S^{2+}$, $2Br^-$, $2H_2O$ 58.54 **6**
$C_{30}H_{44}O_7$ 51.40 **6**
$C_{30}H_{45}Cl_2MoOP_3$ 86.26 **4**
$C_{30}H_{45}Cl_2NP_3Re$ 86.58 **2**
$C_{30}H_{45}Cl_3NP_3Ru$ 86.29 **6**
$C_{30}H_{45}Cl_3P_3Rh$ 86.30 **6**
$C_{30}H_{45}Cl_3P_3Ru^-$, $C_{60}H_{90}Cl_3P_6Ru_2^+$ 86.117 **2**
$C_{30}H_{45}Cl_9P_3Re_3$ 86.59 **2**
$C_{30}H_{45}NiO_{15}P_5^{2+}$, $2ClO_4^-$ 86.27 **4**
$C_{30}H_{46}GeOP_2Pt$ 71.68 **4**
$C_{30}H_{47}BrO$ 56.4 **1**
$C_{30}H_{47}BrO_7$, C_6H_6 56.3 **3**
$C_{30}H_{48}IrO_4P_2$ 86.31 **6**
$C_{30}H_{48}O_4$ 56.4 **3**
$C_{30}H_{49}Br$ 51.41 **3**
$C_{30}H_{50}CrO_4$ 53.40 **1**
$C_{30}H_{50}CrO_4$ 53.11 **4**
$C_{30}H_{50}Fe_2N_{10}O_3^{4+}$, $4ClO_4^-$ 83.205 **2**
$C_{30}H_{50}NO_{11}^+$, Br^- , $0.6C_2H_6O$ 50.20 **3**
$C_{30}H_{50}O$ 56.6 **1**
$C_{30}H_{50}O$ 56.4 **5**
$C_{30}H_{50}O$ 56.2 **6**
$C_{30}H_{50}O_2$ 51.41 **6**
$C_{30}H_{50}O_6$ 56.3 **6**
$C_{30}H_{51}BrO_3$ 56.7 **1**
$C_{30}H_{51}CoN_2PSi_4$ 83.130 **4**
$C_{30}H_{52}$ 56.5 **3**
$C_{30}H_{52}$ 56.6 **3**
$C_{30}H_{52}$ 56.5 **5**
$C_{30}H_{52}$ 56.6+ **5**
$C_{30}H_{54}Co_2O_6P_2$ 86.60+ **2**
$C_{30}H_{54}IMoN_6^+$, I^- 71.69 **4**
$C_{30}H_{54}N_6Pt_3$ 71.96 **7**
$C_{30}H_{56}As_8Mo_2O_6$ 86.59 **7**
$C_{30}H_{56}Mo_2N_2P_4$ 74.7 **7**
$C_{30}H_{60}Ag_6N_6S_{12}$ 80.41 **2**
$C_{31}H_{16}FeO_3$ 75.35 **7**
$C_{31}H_{20}AsMnO_3$ 75.43 **5**
$C_{31}H_{20}Cl_2Fe_2O_8P_2Pd_2$, C_7H_8 86.62 **2**
$C_{31}H_{20}FeO_3$ 75.79 **2**
$C_{31}H_{20}Fe_2Ge_2O_7$ 69.36 **3**
$C_{31}H_{21}BrO_{12}$ 59.31 **3**
$C_{31}H_{21}F_{12}PRu$ 71.70 **4**
$C_{31}H_{21}F_{12}PRu$ 71.76 **5**
$C_{31}H_{21}NO_4$ 38.74 **7**
$C_{31}H_{23}BrN_2O_{11}$, C_6H_6 50.14 **5**
$C_{31}H_{25}BrO_2Si$ 63.12 **4**
$C_{31}H_{25}ClN_2Ni$, C_5H_5N 75.30 **4**
$C_{31}H_{25}ClN_3O_4P$ 64.42 **5**
$C_{31}H_{25}FeO_2P$ 71.62 **3**
$C_{31}H_{25}Fe_2O_6P$ 73.59 **6**
$C_{31}H_{25}N$ 16.17 **7**
$C_{31}H_{25}NO_2Sn$ 69.27 **7**
$C_{31}H_{27}NO_9Th$ 84.25 **3**
$C_{31}H_{27}O_4P$ 64.53 **6**
$C_{31}H_{27}O_4P$ 64.54 **6**
$C_{31}H_{27}O_4P$ 64.55 **6**
$C_{31}H_{29}O_4PRhS_2^+$, I^- , $0.5CH_2Cl_2$ 71.71 **6**
$C_{31}H_{30}Cl_2N_2PRh$, $0.61CHCl_3$ 71.63 **3**
$C_{31}H_{31}BrO_{10}$ 54.15 **4**
$C_{31}H_{32}BN$ 62.13 **5**
$C_{31}H_{32}Br_2O_{10}$ 59.32 **3**
$C_{31}H_{32}Cl_2NiO_2P_2$ 86.60 **7**
$C_{31}H_{33}BrO_9$, C_6H_6 54.9 **5**
$C_{31}H_{34}N_4O_8S_3$ 41.33 **7**
$C_{31}H_{35}As_2I_2NNiO$ 86.49 **5**
$C_{31}H_{35}BrO_9$ 54.16 **4**
$C_{31}H_{35}BrO_{10}$ 54.16 **6**
$C_{31}H_{35}ClCoNOP_2^+$, F_6P^- 86.50 **5**
$C_{31}H_{35}INNiOP_2^+$, I^- 86.52 **3**
$C_{31}H_{36}N_2O_{11}$, H_2O 50.24 **7**
$C_{31}H_{36}N_4O_2$, $2CH_4O$ 58.55 **6**
$C_{31}H_{37}IO_{12}$ 59.25 **4**
$C_{31}H_{38}BrNO_6$ 58.113 **1**
$C_{31}H_{38}BrNO_6$ 58.63 **3**
$C_{31}H_{38}BrN_3O_8S$ 50.21 **3**
$C_{31}H_{38}O_{11}$ 54.17 **6**
$C_{31}H_{42}BrNO$ 55.1 **6**
$C_{31}H_{43}IO_5$ 56.9 **1**
$C_{31}H_{45}ClO_5$ 51.45 **5**
$C_{31}H_{45}CuN_3O_2$ 78.58 **2**
$C_{31}H_{46}O_6$ 51.46 **5**
$C_{31}H_{47}O_5$ 56.3 **7**
$C_{31}H_{48}O_3$ 56.9 **5**
$C_{31}H_{51}BrO_2$ 56.7 **3**
$C_{31}H_{52}O_2$ 56.4 **7**
$C_{31}H_{52}O_4$ 51.42 **3**
$C_{32}H_{14}$ 30.28 **1**
$C_{32}H_{14}$ 30.7 **6**
$C_{32}H_{16}$ 30.29 **1**
$C_{32}H_{16}CeF_{12}O_8S_4^-$, $C_9H_8N^+$ 77.19 **6**
$C_{32}H_{16}Cl_2N_8Sn$ 49.7 **4**
$C_{32}H_{16}CuN_8$ 49.7 **1**
$C_{32}H_{16}F_{12}O_8PrS_4^-$, H_4N^+ , H_2O 77.65 **2**

$C_{32}H_{16}MgN_8$, $2C_5H_5N$, H_2O	49.8 **4**
$C_{32}H_{16}N_8Ni$	49.8 **1**
$C_{32}H_{16}N_8Pb$	49.1 **6**
$C_{32}H_{16}N_8Pt$	49.9+ **1**
$C_{32}H_{16}N_8Sn$	49.5 **3**
$C_{32}H_{18}CrN_6O_8^-$, $C_5H_6N^+$, $0.5C_5H_5N$	83.206 **2**
$C_{32}H_{18}MgN_8O$, $2C_5H_5N$	49.6 **3**
$C_{32}H_{18}N_8$	49.12+ **1**
$C_{32}H_{18}N_8$	49.7 **3**
$C_{32}H_{20}FeO_4$	72.20 **4**
$C_{32}H_{20}Mn_2O_8Si_2$	63.14 **5**
$C_{32}H_{22}CuN_4O_2$	83.207 **2**
$C_{32}H_{22}Fe_2O_7P_2$	86.61 **7**
$C_{32}H_{22}N_2O_2$	30.8 **7**
$C_{32}H_{22}O_2$	26.7 **7**
$C_{32}H_{23}MoN_4O_2P$	71.97 **7**
$C_{32}H_{24}$	31.49 **1**
$C_{32}H_{24}Br_2N_8Ni^+$, $C_{24}H_{20}B^-$	83.112 **5**
$C_{32}H_{24}CdN_8^{2+}$, $2ClO_4^-$	83.110 **7**
$C_{32}H_{24}Cu_2F_{12}O_8$	74.11 **4**
$C_{32}H_{24}FeN_8^{2+}$, $2ClO_4^-$	83.113 **5**
$C_{32}H_{24}O_4$	30.9 **7**
$C_{32}H_{25}BrO_2Si$	63.13 **4**
$C_{32}H_{26}BrF_5NiP_2$	71.72 **6**
$C_{32}H_{26}CrN_4^+$, I^-	71.77 **5**
$C_{32}H_{27}F_6O_5PRu$	71.78 **5**
$C_{32}H_{28}ClCu_2N_4NaO_8$, C_8H_{10}	78.17 **6**
$C_{32}H_{28}Cl_4Cu_4N_4O_4$	76.52 **6**
$C_{32}H_{28}CoI_2N_4$	71.64 **3**
$C_{32}H_{28}Co_2N_4O_4$	78.59 **2**
$C_{32}H_{28}Co_2N_4O_4$	78.14 **4**
$C_{32}H_{28}Fe_2N_4O_5$	76.53 **6**
$C_{32}H_{28}Fe_2N_4O_5$, CH_2Cl_2	78.22+ **3**
$C_{32}H_{28}Fe_2N_4O_5$, $2C_5H_5N$	78.24 **3**
$C_{32}H_{28}Ni_2S_8$	85.93 **2**
$C_{32}H_{28}S_8V$	85.44 **6**
$C_{32}H_{29}ClMoOP_2$	73.51+ **3**
$C_{32}H_{30}Fe_2O_9P_2$	71.98 **7**
$C_{32}H_{30}O_2$	38.75 **7**
$C_{32}H_{32}$	31.32 **7**
$C_{32}H_{32}Al_2^{2-}$, $2C_8H_{16}NaO_2^+$	68.31 **5**
$C_{32}H_{32}Cl_4Ti_4$	75.36 **7**
$C_{32}H_{32}CuN_2O_2$	78.18 **6**
$C_{32}H_{32}Cu_3N_4O_6^{2+}$, $2ClO_4^-$, H_2O	78.23 **7**
$C_{32}H_{32}N_{12}Ni_2$	83.131 **4**
$C_{32}H_{32}Si_2$	63.28 **2**
$C_{32}H_{34}N_4OV$, $C_2H_4Cl_2$	49.14 **1**
$C_{32}H_{35}N_2NiOP_2S^+$, F_6P^-	86.53 **3**
$C_{32}H_{36}CuN_4$	49.2 **6**
$C_{32}H_{36}Cu_4N_4O_{12}$, $4H_2O$	81.43 **7**
$C_{32}H_{36}Lu^-$, $C_{16}H_{32}LiO_4^+$	71.79 **5**
$C_{32}H_{36}N_4Ni$	49.15 **1**
$C_{32}H_{37}As_2Nb$	75.37 **7**
$C_{32}H_{37}BrO_{10}$	54.24 **1**
$C_{32}H_{38}BrCoN_2P_2^+$, F_6P^-	76.43 **4**
$C_{32}H_{38}BrN_2NiP_2^+$, Br^- , $0.5C_4H_{10}O$	76.44 **4**
$C_{32}H_{38}N_4$	49.9 **4**
$C_{32}H_{38}O_8Zn_3$	71.73 **6**
$C_{32}H_{39}BrO_{11}$	56.10 **1**
$C_{32}H_{40}Br_2Cu_6N_4$, $1.5C_6H_6$	71.80 **5**
$C_{32}H_{40}Cl_2P_4Ru$	86.63 **2**
$C_{32}H_{40}O_8$	54.15 **3**
$C_{32}H_{40}P_4Pd^{2+}$, $2Cl^-$, C_2H_6O	86.54 **3**
$C_{32}H_{41}N_2O_{12}^+$, Br^-	35.37 **4**
$C_{32}H_{41}N_3O_7$	58.24 **7**
$C_{32}H_{42}Ag_2O_6$, $2H_2O$	81.55 **4**
$C_{32}H_{43}BrO_2$	55.3 **1**
$C_{32}H_{44}ClN_2P_4Re$	86.55 **3**
$C_{32}H_{44}N_4O_7$, H_2O	48.87 **6**
$C_{32}H_{45}IO_5$	51.51 **7**
$C_{32}H_{45}N_2NiO_6P_3$	86.56 **3**
$C_{32}H_{45}N_3Si_5$	63.29 **2**
$C_{32}H_{48}Ce_2Cl_2O_4$	75.31 **4**
$C_{32}H_{48}CuN_2O_2$	78.19 **6**
$C_{32}H_{48}O_3$	56.3 **4**
$C_{32}H_{48}O_6$	56.8 **3**
$C_{32}H_{50}IrNO_4P_2$	71.99 **7**
$C_{32}H_{50}NO_5^+$, Br^-	58.114 **1**
$C_{32}H_{50}NO_8^+$, I^- , $2C_3H_6O$	58.45 **5**
$C_{32}H_{51}IO_2$	56.11 **1**
$C_{32}H_{51}IO_2$	56.12 **1**
$C_{32}H_{52}Br_2O_2$	51.42 **6**
$C_{32}H_{52}Br_2O_3$	51.43 **1**
$C_{32}H_{52}Br_2O_7S$	1.38 **5**
$C_{32}H_{52}NO_3^+$, I^-	58.115 **1**
$C_{32}H_{53}IO_2$	51.44 **1**
$C_{32}H_{53}IO_2$	56.13 **1**
$C_{32}H_{53}N_4O_4^+$, I^-	58.25 **7**
$C_{32}H_{54}Co_2N_2O_8$	77.20 **4**
$C_{32}H_{54}O_3$	56.4 **4**
$C_{32}H_{56}O_8S_{16}Zn_4$	80.19 **4**
$C_{32}H_{60}NiP_2$	72.32 **7**
$C_{32}H_{64}N_4Na_2O_{10}^{2+}$, $2I^-$	67.19 **6**
$C_{32}H_{72}ClCu_2N_8^{3+}$, $3ClO_4^-$	83.114 **5**
$C_{32}H_{80}Cd_{10}O_{16}S_{16}^{4+}$, $2O_4S^{2-}$, $4H_2O$	85.94 **2**
$C_{32}H_{80}O_{16}Ti_4$	84.61 **2**
$C_{32}H_{80}S_8Zn_8$	71.54 **2**
$C_{33}H_{23}CoI_2$	73.61 **7**
$C_{33}H_{24}CoI$	73.62 **7**
$C_{33}H_{24}FeO_9$	73.116 **2**
$C_{33}H_{24}Fe_2N_2O_{10}$	71.100 **7**
$C_{33}H_{25}Rh$	75.46 **5**
$C_{33}H_{26}O_2Sn$	69.28 **7**
$C_{33}H_{28}ClRh$	72.26 **6**
$0.30C_{33}H_{30}Br_2O_{16}$, $0.70C_{33}H_{31}BrO_{16}$	59.21 **7**
$C_{33}H_{30}FeO_3Si$	75.38 **7**
$C_{33}H_{30}Fe_3$	73.53 **3**
$0.70C_{33}H_{31}BrO_{16}$, $0.30C_{33}H_{30}Br_2O_{16}$	59.21 **7**
$C_{33}H_{31}INO_6$, C_3H_6O	54.25 **1**
$C_{33}H_{33}Br_3O_{12}$	54.17 **4**
$C_{33}H_{33}CrNO_3P_2$	86.57 **3**
$C_{33}H_{33}Fe_3O_9P_3$	86.28 **4**
$C_{33}H_{33}O_8P$	64.56 **6**
$C_{33}H_{35}O_6$	38.76 **7**
$C_{33}H_{35}PPtS_4$	85.49 **3**
$C_{33}H_{36}BiCl_3N_{12}O_9S_3$	66.8 **4**
$C_{33}H_{36}N_9NdO_{12}$	84.75 **7**
$C_{33}H_{36}O_6$	38.40 **6**
$C_{33}H_{36}O_6$, C_2H_6O	61.3 **7**
$0.2C_{33}H_{36}O_6$, $C_{16}H_{34}O$	61.4 **7**
$2C_{33}H_{36}O_6$, C_5H_5N	61.5 **7**
$C_{33}H_{37}ClP_2PtSi$	86.32 **6**
$C_{33}H_{37}IO_{11}S$, C_3H_6O	50.39 **1**
$C_{33}H_{38}I_2O_{12}$	59.26 **4**
$C_{33}H_{38}N_4O_2$	49.16 **1**
$C_{33}H_{38}O_5P_2Rh^+$, F_6P^-	86.62 **7**
$C_{33}H_{39}BrO_{15}$	51.54 **4**
$C_{33}H_{40}BrN_3O_7$	50.22 **3**
$C_{33}H_{40}N_2O_9$	58.116 **1**

$C_{33}H_{41}NO_8$ 58.26 **7**
$C_{33}H_{46}N_2O_3^{2+}$, $2Br^-$, H_2O 58.56 **6**
$C_{33}H_{47}Cl_5MoN_2OP_4Re$, CH_4O , HCl 86.63 **7**
$C_{33}H_{47}IN_2O_6$ 40.24 **4**
$C_{33}H_{47}NO_9S$ 58.27 **7**
$C_{33}H_{48}ClIO_7$ 56.10 **5**
$C_{33}H_{48}Cl_2O_7$ 56.11 **5**
$C_{33}H_{51}BrO_4$ 56.14 **1**
$C_{33}H_{51}IO_4$ 56.15 **1**
$C_{33}H_{51}NO_6$ 56.16 **1**
$C_{33}H_{52}NO_4^+$, I^- , C_3H_6O 58.64 **3**
$C_{33}H_{57}ErO_6$ 77.21 **4**
$C_{33}H_{57}N_3O_9$, K^+ , I^- 50.25 **7**
$C_{33}H_{59}DyO_7$ 77.22 **4**
$C_{33}H_{62}O_6$ 1.155 **1**
$C_{33}H_{64}Cu_2N_9^{3+}$, $3ClO_4^-$ 83.111 **7**
$C_{33}H_{66}Fe_3O_{16}^+$, Cl^- 81.44 **7**
$C_{34}H_{16}Fe_2O_6$ 71.81 **5**
$C_{34}H_{16}O_2$ 30.30 **1**
$C_{34}H_{16}O_2$ 30.31 **1**
$C_{34}H_{18}$ 30.32 **1**
$C_{34}H_{20}$ 30.9 **4**
$C_{34}H_{21}Br_2N_7O_3S_4$, $0.5C_7H_8O$ 50.40 **1**
$C_{34}H_{24}$ 24.19 **4**
$C_{34}H_{24}BrN_3$ 35.42 **7**
$C_{34}H_{24}CoN$ 73.63 **7**
$C_{34}H_{25}FeNiO_3P$ 71.82 **5**
$C_{34}H_{25}NbO$ 72.72 **2**
$C_{34}H_{25}OTa$ 72.73 **2**
$C_{34}H_{26}Cl_2Ti$ 73.60 **6**
$C_{34}H_{26}N_4NiO_2$ 83.208 **2**
$C_{34}H_{28}O_4$ 21.22 **4**
$C_{34}H_{28}O_4$ 28.11 **4**
$C_{34}H_{28}O_4$ 28.12 **4**
$C_{34}H_{28}PdS_4$ 85.33 **5**
$C_{34}H_{30}Br_2O_8$ 54.26 **1**
$C_{34}H_{30}Co_2P_2$ 73.117 **2**
$C_{34}H_{30}CrN_4O_2^+$, I^- , H_2O 71.83 **5**
$C_{34}H_{30}N_2NiP_2S_4$ 85.50 **3**
$C_{34}H_{30}Ni_2P_2$ 73.118 **2**
$C_{34}H_{32}BrN_5O_7$ 58.28 **7**
$C_{34}H_{32}ClFeN_4O_4$ 49.17 **1**
$C_{34}H_{32}Cl_4N_4Ni_2O_2$ 83.78 **6**
$C_{34}H_{32}GeN_8O_4$ 69.29 **7**
$C_{34}H_{33}N_2O_2PPdS_2$ 85.33 **4**
$C_{34}H_{34}Br_2Cl_2Cu_2N_2O_2$ 78.20 **6**
$C_{34}H_{34}MgN_4O_3$, H_2O 49.3 **7**
$C_{34}H_{34}Mn_2N_4O_6$, $4C_5H_5N$ 76.54 **6**
$C_{34}H_{35}ClN_2O_{16}Pd$ 71.101 **7**
$C_{34}H_{35}CuN_3O_2$ 84.76 **7**
$C_{34}H_{35}N_3NiO_2$ 84.77 **7**
$C_{34}H_{36}Cu_2N_4O_{10}$ 79.25 **7**
$C_{34}H_{36}N_4NiO_2$, $0.5C_2H_4Cl_2$ 49.10 **4**
$C_{34}H_{38}Al_2N_2O_4$ 68.28 **2**
$C_{34}H_{38}INO_{12}$, C_3H_6O 56.17 **1**
$C_{34}H_{38}IO_9$ 56.18 **1**
$C_{34}H_{38}NO_{10}^+$, I^- 58.38 **4**
$C_{34}H_{40}Al_2N_2O_4$ 68.32 **5**
$C_{34}H_{40}P_2^{2+}$, $2Cl^-$, $0.5H_2O$ 64.43 **5**
$C_{34}H_{42}CoIN_2P_2^+$, I^- 76.52 **5**
$C_{34}H_{42}IN_2NiP_2^+$, I^- 76.53 **5**
$C_{34}H_{45}BrO_2$ 51.52 **7**
$C_{34}H_{46}Cu_2N_6O_{10}$, $8H_2O$, $C_4H_{10}O$ 82.15 **4**
$C_{34}H_{47}BrO_3$ 51.55 **4**
$C_{34}H_{48}O_6Ti$ 77.17 **5**
$C_{34}H_{49}BrO_2$ 51.53 **7**
$C_{34}H_{49}BrO_3$ 51.54 **7**
$C_{34}H_{50}Br_2O_6$ 56.4 **6**
$C_{34}H_{50}Cl_2O_4P_2Ru_2$ 86.51 **5**
$C_{34}H_{50}N_6Si_6$ 63.11 **6**
$C_{34}H_{51}BrO_6$ 56.5 **6**
$C_{34}H_{51}BrO_6$ 56.5 **7**
$C_{34}H_{51}IO_6$ 56.19 **1**
$C_{34}H_{52}BrO_8^-$, Na^+ 50.26 + **7**
$C_{34}H_{53}BrO_8$, $0.5H_2O$ 50.15 **5**
$C_{34}H_{53}O_8^-$, Ag^+ , C_3H_6O 50.23 **3**
$2C_{34}H_{53}O_8^-$, Ba^{2+} , H_2O 50.24 **3**
$C_{34}H_{56}O_6$ 59.29 **6**
$C_{34}H_{59}O_{10}^-$, Ag^+ , $0.5H_2O$ 50.28 **7**
$C_{34}H_{68}$ 23.17 **1**

C_{35-39}

$C_{35}H_{18}F_9N_2PRu$ 71.74 **6**
$C_{35}H_{20}Fe_3N_4O_9$ 83.209 **2**
$C_{35}H_{21}N_5$ 49.11 **4**
$C_{35}H_{24}BrNO$ 40.25 **4**
$C_{35}H_{24}Br_2N_2$ 35.24 **3**
$C_{35}H_{24}I_2N_2$ 34.6 **3**
$C_{35}H_{25}CoN_2O_4$ 77.11 **7**
$C_{35}H_{25}CoN_5^+$, $CHCl_3$, ClO_4^- 71.102 **7**
$C_{35}H_{25}O_2V$ 75.46 **3**
$C_{35}H_{26}BrO_3P$ 64.39 **4**
$C_{35}H_{27}N_3$ 35.43 **7**
$C_{35}H_{33}NNbP$ 72.27 **6**
$C_{35}H_{33}Sb$ 66.8 **6**
$C_{35}H_{34}Fe_3O$ 73.54 **3**
$C_{35}H_{35}Cl_2N_5O_2$, $2CHCl_3$ 59.22 **7**
$C_{35}H_{39}IrP_2$ 71.84 **5**
$C_{35}H_{40}BrNO_8S$ 58.29 **7**
$C_{35}H_{41}Br_2ClO_4$ 51.56 **4**
$C_{35}H_{46}INO_4$ 51.45 **1**
$C_{35}H_{46}INO_4$ 51.46 **1**
$C_{35}H_{46}INO_4$ 51.43 **3**
$C_{35}H_{46}N_2O_6$ 51.55 **7**
$C_{35}H_{46}N_2O_6$ 51.56 **7**
$C_{35}H_{47}BrO_2$ 51.47 **1**
$C_{35}H_{51}BrO_3$ 51.43 **6**
$C_{35}H_{52}O_3S$ 51.57 **7**
$C_{35}H_{60}N_2O_4^{2+}$, $2Br^-$, CH_2Cl_2 , H_2O 51.44 **3**
$C_{35}H_{63}N_7Ni_4$, C_6H_6 71.103 **7**
$C_{36}Cl_{24}Mo_2O_{12}$, $1.5C_6H_6$ 84.78 **7**
$C_{36}H_{20}Br_4N_4O_4W$, C_6H_6 83.132 **4**
$C_{36}H_{20}F_{24}Ni_4$ 73.61 **6**
$C_{36}H_{20}Fe_3O_8$ 72.74 **2**
$C_{36}H_{20}Fe_3O_8$ 72.75 **2**
$C_{36}H_{20}O_8Os_3$ 72.21 **4**
$C_{36}H_{22}O_{10}Re_2$ 77.18 **5**
$C_{36}H_{24}$ 31.40 + **3**
$C_{36}H_{24}$ 31.44 + **5**
$C_{36}H_{24}B_3N_3$ 62.14 **7**
$C_{36}H_{24}CuN_6^{2+}$, $2ClO_4^-$ 83.115 **5**
$C_{36}H_{24}FeN_6^{2+}$, $C_8H_4O_{12}Sb_2^{2-}$, $8H_2O$ 83.79 **6**
$C_{36}H_{24}FeN_6^{3+}$, $3ClO_4^-$, H_2O 83.112 **7**
$C_{36}H_{24}N_3O_6P_3$ 64.40 **4**
$C_{36}H_{24}N_4O_4Zr$, $3C_7H_8$ 84.79 **7**

$C_{36}H_{24}N_6Ni^{2+}$, $2C_2H_6O_2PS_2^-$ 83.141 **3**
$C_{36}H_{24}N_6Ni^{2+}$, $2C_5MnO_5^-$ 83.133 **4**
$C_{36}H_{24}N_6Ni^{2+}$, $C_6CoO_{12}^{3-}$, K^+ , $2H_2O$ 83.142 **3**
$C_{36}H_{26}Br_2CrN_8O_6^-$, $C_5H_6N^+$, $3H_2O$ 83.211 **2**
$C_{36}H_{26}O_6P_2W$ 84.39 **6**
$C_{36}H_{27}N^+$, ClO_4^- 21.19 **3**
$C_{36}H_{28}Fe_2N_2O_6$ 83.212 **2**
$C_{36}H_{28}NiP_2Se_2$ 85.51 **3**
$2C_{36}H_{30}AgP_2^+$, $C_8N_4NiS_4^{2-}$ 85.12 **6**
$C_{36}H_{30}Al_2$ 68.30 **2**
$C_{36}H_{30}Al_2$ 68.33 **5**
$C_{36}H_{30}As_2Cl_2HgO_2$ 84.62 **2**
$C_{36}H_{30}As_2Cl_4Hg_2O_2$ 84.63 **2**
$C_{36}H_{30}As_2N_2O_{10}U$ 84.27 **5**
$C_{36}H_{30}As_3N_3$ 65.11 **6**
$C_{36}H_{30}As_6$ 65.11 **2**
$C_{36}H_{30}AuClP_2$, $0.5C_6H_6$ 86.34 **6**
$C_{36}H_{30}Bi_2O^{2+}$, $2ClO_4^-$ 66.9 **4**
$C_{36}H_{30}BrCuP_2$, $0.5C_6H_6$ 86.52 **5**
$C_{36}H_{30}Br_2NiP_2$ 86.65 **2**
$C_{36}H_{30}CdCl_2P_2$ 86.58 **3**
$C_{36}H_{30}CeN_4O_{14}P_2$ 84.26 **3**
$C_{36}H_{30}ClNO_5P_2RuS$ 86.64 **7**
$C_{36}H_{30}ClN_2O_2P_2Ru^+$, F_6P^- , C_6H_6 86.29 **4**
$C_{36}H_{30}Cl_2CuO_2P_2$ 84.14 **4**
$C_{36}H_{30}Cl_2Cu_2P_2$ 86.35 **6**
$C_{36}H_{30}Cl_2IrNOP_2$ 86.60 **3**
$C_{36}H_{30}Cl_2Ir_2N_2O_3P_2$ 86.65 **7**
$C_{36}H_{30}Cl_2MoN_2O_2P_2$ 86.53 **5**
$C_{36}H_{30}Cl_2NP_2Re$ 86.66 **2**
$C_{36}H_{30}Cl_2NiP_2$ 86.67 **2**
$C_{36}H_{30}Cl_3HgNOOsP_2$ 86.61 **3**
$C_{36}H_{30}Cl_3InP_2$ 86.68 **2**
$C_{36}H_{30}Cl_4Hg_2P_2Se_2$ 85.52 **3**
$C_{36}H_{30}Cl_4O_2P_2U$ 84.80 **7**
$C_{36}H_{30}CoN_9$ 83.116 **5**
$C_{36}H_{30}CoN_9$, C_7H_8 83.213 **2**
$C_{36}H_{30}CrO_4Si_2$ 63.17 **3**
$C_{36}H_{30}CuNO_3P_2$ 86.62 **3**
$C_{36}H_{30}CuP_2^+$, H_4B^- 86.69 **2**
$C_{36}H_{30}FeN_2O_2P_2$ 86.66 **7**
$C_{36}H_{30}Fe_2N_4O_6$ 83.214 **2**
$C_{36}H_{30}Ge_2O$ 69.20 **6**
$C_{36}H_{30}I_2P_2Pd$, $2CH_2Cl_2$ 86.54 **5**
$C_{36}H_{30}IrN_2O_2P_2^+$, ClO_4^- 86.63 **3**
$C_{36}H_{30}Ir_2N_2O_3P_2$, C_6H_6 86.70 **2**
$C_{36}H_{30}Ir_2N_4O_4P_2$ 86.36 **6**
$C_{36}H_{30}MgMo_2N_4O_6$ 73.62 **6**
$C_{36}H_{30}NP_2^+$, $C_4HFeO_4^-$ 64.44 **5**
$C_{36}H_{30}NP_2^+$, $C_5FeNO_4^-$ 64.57 **6**
$C_{36}H_{30}NP_2^+$, $C_6O_6V^-$ 64.63 **7**
$C_{36}H_{30}NP_2^+$, $C_8CoFeO_8^-$ 64.64 **7**
$C_{36}H_{30}NP_2^+$, $C_{10}Cr_2IO_{10}^-$ 64.62 **3**
$2C_{36}H_{30}NP_2^+$, $C_8Fe_2O_8^{2-}$, $2C_2H_3N$ 64.65 **7**
$2C_{36}H_{30}NP_2^+$, $C_8H_{12}Cl_8N_4Nb_2^{2-}$, $2C_6H_5Cl$ 83.18 **7**
$2C_{36}H_{30}NP_2^+$, $C_{10}Cr_2O_{10}^{2-}$, CH_2Cl_2 64.63 **3**
$2C_{36}H_{30}NP_2^+$, $C_{10}Mo_2O_{10}^{2-}$, CH_2Cl_2 64.64 **3**
$2C_{36}H_{30}NP_2^+$, $C_{16}Mo_2Ni_3O_{16}^{2-}$ 64.41 **4**
$2C_{36}H_{30}NP_2^+$, $C_{16}Ni_3O_{16}W_2^{2-}$ 64.42 **4**
$C_{36}H_{30}N_2O_2P_2Ru$ 86.67 **7**
$C_{36}H_{30}N_2O_2P_2Ru$, $0.5C_6H_6$ 86.37 **6**
$C_{36}H_{30}N_3O_6P_3$ 64.65 **3**
$C_{36}H_{30}N_3P_3$ 64.81 **2**
$C_{36}H_{30}N_4NiOP_2$ 86.30 **4**
$C_{36}H_{30}N_4O_{14}P_2Th$ 84.40 **6**
$C_{36}H_{30}N_6OSb_2$ 66.9 **6**
$C_{36}H_{30}N_{10}NiS_2$ 83.143 **3**
$C_{36}H_{30}O_2P_2Pt$, $2CHCl_3$ 86.31 **4**
$C_{36}H_{30}O_2P_2Pt$, $1.5C_6H_6$ 86.71 **2**
$C_{36}H_{30}O_2P_2Pt$, $2C_7H_8$ 86.64 **3**
$C_{36}H_{30}O_3Si_3$ 63.15 **5**
$C_{36}H_{30}P_3Pt$ 86.72 **2**
$C_{36}H_{30}P_6$ 64.82 + **2**
$C_{36}H_{30}Pb_2$ 69.38 **4**
$C_{36}H_{30}Sn_2$ 69.27 **5**
$2C_{36}H_{31}As_2O_2^+$, $Br_6Hg_2^{2-}$ 65.51 **2**
$C_{36}H_{31}BrSn$ 69.37 **3**
$C_{36}H_{31}ClSn$ 69.38 **3**
$C_{36}H_{31}N_2O_3OsP_2^+$, F_6P^- 86.68 **7**
$C_{36}H_{32}AsBrO_2^+$, Br_3^- 65.11 **3**
$C_{36}H_{32}As_2BrO_2^+$, Br_3^- 65.8 **5**
$C_{36}H_{32}IN_2NiP_2^+$, I^- 86.69 **7**
$C_{36}H_{32}La_2N_6O_{16}$ 81.56 **4**
$C_{36}H_{32}N_6O_{16}Pr_2$ 81.59 **5**
$C_{36}H_{32}N_6O_{16}Sm_2$ 81.58 **4**
$C_{36}H_{33}Br_2CoP_3$ 86.73 **2**
$C_{36}H_{33}ClU$ 73.45 **5**
$C_{36}H_{33}Cl_3NOsP_2$ 86.66 **3**
$C_{36}H_{33}I_2N_2OPRu$ 71.104 **7**
$C_{36}H_{34}B_2P_2$ 64.58 **6**
$C_{36}H_{34}Br_2N_2O_{12}$, C_3H_6O 35.38 **4**
$C_{36}H_{34}MgN_2O_6$ 67.21 **5**
$C_{36}H_{36}B_3P_3$ 64.45 **5**
$C_{36}H_{36}MgN_4O_5$, $2H_2O$ 49.4 **7**
$C_{36}H_{36}N_4NiO_6$, $0.5C_6H_6$ 49.18 **1**
$C_{36}H_{37}ClNOP_3Rh^+$, F_6P^- 86.38 **6**
$C_{36}H_{37}ClP_3Rh$ 86.39 **6**
$C_{36}H_{38}B_3CuP_2$ 86.67 **3**
$C_{36}H_{38}Cu_3N_4O_5^{2+}$, $2ClO_4^-$ 78.24 **7**
$C_{36}H_{38}N_4O_5$ 49.12 **4**
$C_{36}H_{38}N_4O_5$ 49.1 **5**
$C_{36}H_{39}AlO_{12}W_3$ 73.31 **4**
$C_{36}H_{40}Cu_2N_4O_{16}$, $4H_2O$ 81.59 **4**
$C_{36}H_{40}N_4O_4Ti$ 78.61 **2**
$C_{36}H_{40}N_4O_4Zr$ 78.25 **3**
$C_{36}H_{41}BrO_{11}$, xC_3H_6O 59.33 **5**
$C_{36}H_{41}Cl_3N_4O_8Pd$ 83.117 **5**
$C_{36}H_{41}IrP_2$ 71.85 **5**
$C_{36}H_{42}BrNO_{17}$, H_2O 58.39 **4**
$C_{36}H_{42}N_{12}Ni_3O_{12}$, C_6H_6 83.215 **2**
$C_{36}H_{44}Al_4Mo_2$ 68.34 **5**
$C_{36}H_{44}ClN_4Tl$, $2CH_2Cl_2$ 49.3 **6**
$C_{36}H_{44}Cl_2N_4Sn$, $2CH_3NO_2$ 49.4 **6**
$C_{36}H_{44}CoN_4$, $2C_6H_7N$ 49.5 **7**
$C_{36}H_{44}N_4Ni$ 49.2 **5**
$C_{36}H_{44}N_4Ni$ 49.5 **6**
$C_{36}H_{44}Na_2Ni_2O_2$, $3C_4H_8O$, 2Na 71.105 **7**
$C_{36}H_{44}P_2Pt$ 71.86 **5**
$C_{36}H_{44}P_2Pt$ 71.87 **5**
$C_{36}H_{44}P_2Pt$ 71.88 **5**
$C_{36}H_{45}Cl_2PPd$, $C_2H_4O_2$ 75.28 **6**
$C_{36}H_{46}N_2$ 21.20 **3**
$C_{36}H_{46}N_4$ 49.6 **6**
$C_{36}H_{47}N_4^+$, I_3^- 49.7 **6**
$C_{36}H_{47}N_6^{3+}$, $3I^-$, H_2O 58.30 **7**
$C_{36}H_{48}N_4^{2+}$, $2C_2Cl_2O_2Rh^-$ 49.8 **6**
$C_{36}H_{48}O_4Ti_2$ 71.106 **7**

$C_{36}H_{50}N_6Rh^{+}$, Cl^{-} , $2H_2O$ 49.9 **6**
$C_{36}H_{51}CrO_6$ 77.27 **3**
$C_{36}H_{51}IO_2$ 51.58 **7**
$C_{36}H_{52}O_{12}$ 38.32 **5**
$C_{36}H_{53}IO_2$ 51.44 **6**
$C_{36}H_{54}Cl_6Nb_3^{+}$, Cl^{-} 74.13 **6**
$C_{36}H_{54}F_{10}P_2PtS_2$ 85.53 **3**
$C_{36}H_{56}CuN_2O_2$ 78.25 **7**
$C_{36}H_{60}O_{30}$, I_2 , $4H_2O$ 45.51 **6**
$C_{36}H_{60}O_{30}$, $2H_2O$ 61.7 **5**
$C_{36}H_{60}O_{30}$, $5.28H_2O$, 0.76Kr 45.56 **7**
$C_{36}H_{60}O_{30}$, $5.78H_2O$, 0.47Kr 45.57 **7**
$C_{36}H_{60}O_{30}$, $6H_2O$ 45.52 **6**
$C_{36}H_{60}O_{30}$, CH_4O 61.8 **5**
$C_{36}H_{60}O_{30}$, C_3H_8O 61.9 **5**
$C_{36}H_{60}O_{30}$, C_3H_8O , $4.8H_2O$ 61.6 **7**
$C_{36}H_{60}O_{30}$, C_6H_6IN , $3H_2O$ 45.58 **7**
$2C_{36}H_{60}O_{30}$, Li^{+} , I_3^{-} , I_2 , $8H_2O$ 45.59 **7**
$C_{36}H_{61}AgO_{11}$, $2H_2O$ 50.25 **3**
$C_{36}H_{62}O_8P_4Ru$ 86.55 **5**
$C_{36}H_{62}O_{11}$, H_2O 50.11 **4**
$C_{36}H_{65}Br_3O_6$ 1.156 **1**
$C_{36}H_{66}Cl_2NiP_2$ 86.74 **2**
$C_{36}H_{66}CuNO_3P_2$ 86.34 **4**
$C_{36}H_{66}I_2P_2Pt$ 86.70 **7**
$C_{36}H_{66}Li_6$, C_6H_6 67.8 **7**
$C_{36}H_{66}P_2Pd$ 86.40 **6**
$C_{36}H_{67}CoP_2^{+}$, H_4B^{-} 86.71 **7**
$C_{36}H_{72}AlLiN_4$ 68.13 **4**
$C_{36}H_{74}$ 5.42+ **1**
$C_{36}H_{84}Pd_6S_{12}$ 85.95 **2**
$C_{37}H_{20}F_6Fe_3O_7P_2$ 71.107 **7**
$C_{37}H_{20}O_9Os_3$ 71.108+ **7**
$C_{37}H_{24}O_7Os_3P_2$ 71.71 **4**
$C_{37}H_{27}AsO_2$ 65.12 **4**
$C_{37}H_{30}ClIrNO_2P_2^{+}$, BF_4^{-} 86.75 **2**
$C_{37}H_{30}ClIrO_3P_2$ 86.76 **2**
$C_{37}H_{30}ClIrO_3P_2S$ 86.77 **2**
$C_{37}H_{30}ClO_3P_2RhS$ 86.68 **3**
$C_{37}H_{30}ClP_2RhS$ 86.78 **2**
$C_{37}H_{30}Cl_2N_2ORh_2$, $2CH_2Cl_2$ 71.75 **6**
$C_{37}H_{30}CoNO_2P_2$ 86.56 **5**
$C_{37}H_{30}FP_2PtS_2^{+}$, HF_2^{-} 85.34 **4**
$C_{37}H_{30}IIrNO_2P_2^{+}$, BF_4^{-} , C_6H_6 86.69 **3**
$C_{37}H_{30}IIrO_3P_2$, CH_2Cl_2 86.79 **2**
$C_{37}H_{30}INO_2P_2Ru$ 86.41 **6**
$C_{37}H_{30}IrNO_2P_2$ 86.57 **5**
$C_{37}H_{30}O_3P_2Pt$ 86.80 **2**
$C_{37}H_{30}P_2$ 64.43 **4**
$C_{37}H_{30}P_2PdS_2$ 86.81 **2**
$C_{37}H_{30}P_2PtS_2$ 86.70 **3**
$C_{37}H_{31}P_2^{+}$, Br^{-} 64.59 **6**
$C_{37}H_{33}IIrNOP_2$ 71.66 **3**
$C_{37}H_{33}IO_2P_2PtS$ 71.72 **4**
$C_{37}H_{33}IP_2Pt$, O_2S 71.89 **5**
$C_{37}H_{33}I_2P_2Rh$, C_6H_6 71.55 **2**
$C_{37}H_{35}MoN_2O_3P_3$ 86.71 **3**
$C_{37}H_{36}MgN_4O_5$, $2H_2O$ 49.6 **7**
$C_{37}H_{37}BrF_2N_2O_5$ 51.45 **3**
$C_{37}H_{38}MgN_4O_6$, $2H_2O$ 49.7 **7**
$C_{37}H_{40}BrNO_7$, C_3H_8O 50.12 **4**
$C_{37}H_{42}N_2O_6^{2+}$, $2Cl^{-}$, $5H_2O$ 58.46 **5**
$C_{37}H_{42}N_4O_7^{2+}$, $2Br^{-}$ 58.65 **3**
$C_{37}H_{43}BrO_8$, $0.5C_4H_8O_2$ 51.48 **1**
$C_{37}H_{43}FeN_4O_5$ 49.19 **1**
$C_{37}H_{44}O_5P_4Pt_4$ 86.72 **3**
$C_{37}H_{51}BrClN_3O_{10}$ 58.47 **5**
$C_{37}H_{51}BrO_5$, $0.5C_4H_8O_2$ 59.23 **7**
$C_{37}H_{51}IO_{10}S$ 59.33 **3**
$C_{37}H_{52}BrNO$, C_2H_6O 56.10 **3**
$C_{37}H_{53}BrO_4$, $0.5H_2O$ 56.20 **1**
$C_{37}H_{55}BrO_3$ 56.5 **4**
$C_{37}H_{58}O_8$, C_6H_6 51.49 **1**
$C_{37}H_{68}NO_{13}^{+}$, I^{-} , $2H_2O$ 50.43 **1**
$C_{38}H_{18}$ 30.33 **1**
$C_{38}H_{20}F_6Fe_3O_8P_2$, C_6H_6 71.76 **6**
$C_{38}H_{20}F_6Fe_3O_8P_2$, $0.5C_6H_6$ 71.90 **5**
$C_{38}H_{24}O_8Os_3P_2$ 71.73 **4**
$C_{38}H_{26}Cl_5F_5NiP_2$ 71.74 **4**
$C_{38}H_{26}F_{10}NiP_2$ 71.75 **4**
$C_{38}H_{28}^{2+}$, $2Cl_6Sb^{-}$ 12.11 **3**
$C_{38}H_{29}ClP_2Pt$ 71.110 **7**
$C_{38}H_{29}IrO_2P_2$, C_4H_8O 71.78 **6**
$C_{38}H_{30}ClF_3P_2Pt$ 72.22 **4**
$C_{38}H_{30}ClF_4P_2Rh$ 86.84 **2**
$C_{38}H_{30}ClIrO_2P_2$, C_6H_6 86.73 **3**
$C_{38}H_{30}ClP_2Rh$ 72.40 **5**
$C_{38}H_{30}Cl_2F_2P_2Pt$ 72.23 **4**
$C_{38}H_{30}Cl_4P_2Pt$ 72.24 **4**
$C_{38}H_{30}FeOSn$ 72.41 **5**
$C_{38}H_{30}HgN_2P_2S_2$ 86.42 **6**
$C_{38}H_{30}MnNO_3P_2$ 86.85 **2**
$C_{38}H_{30}NO_3OsP_2^{+}$, ClO_4^{-} , CH_2Cl_2 86.35 **4**
$C_{38}H_{30}N_2O_6P_2PdS_2$ 86.43 **6**
$C_{38}H_{30}N_4O_5STh$, C_2H_6OS 83.145 **3**
$C_{38}H_{30}NiO_2P_2$ 86.44 **6**
$C_{38}H_{30}S_6$ 11.21 **7**
$C_{38}H_{31}Cl_2F_2IrOP_2$ 71.111 **7**
$C_{38}H_{31}IN_4O_5PRh$ 71.112 **7**
$C_{38}H_{31}IrO_2P_2$ 86.74 **3**
$C_{38}H_{31}N_3O_2$ 35.44 **7**
$C_{38}H_{31}N_9Zn$, $0.5C_6H_{15}N$ 49.13 **4**
$C_{38}H_{32}Fe_2O_{10}S_2Sn_2$ 73.33 **4**
$C_{38}H_{32}P_2RuS_4$ 86.45 **6**
$C_{38}H_{32}P_2RuS_4$ 86.72 **7**
$C_{38}H_{33}Cl_3NP_2Re$ 86.75 **3**
$C_{38}H_{34}Br_2NiP_2$ 86.76 **3**
$C_{38}H_{34}Br_2O_2$ 20.14 **4**
$C_{38}H_{34}F_6FeOP_2Sn$ 73.55 **3**
$C_{38}H_{34}F_6FeOP_2Sn$ 73.32 **4**
$C_{38}H_{34}N_8O_2Si_3$ 63.19 **7**
$C_{38}H_{34}N_{14}Zn_2$ 83.113 **7**
$C_{38}H_{34}NiP_2$ 72.77 **2**
$C_{38}H_{34}NiP_2$ 72.25 **4**
$C_{38}H_{34}P_2Pt$ 72.42 **5**
$C_{38}H_{35}NO_2Si_3$ 63.30 **2**
$C_{38}H_{36}Cl_2CoN_2O_2P_2$ 84.28 **5**
$C_{38}H_{38}N_2P_2Pd_2S_2$ 86.58 **5**
$C_{38}H_{38}N_2P_2PtS_2$ 86.46 **6**
$C_{38}H_{38}Ni_2O_2P_2$ 72.33 **7**
$C_{38}H_{40}B_{10}NiP_2$ 71.79 **6**
$C_{38}H_{41}N_2O_7^{+}$, Br^{-} 58.41 **4**
$C_{38}H_{42}Co_2N_6O_8$ 78.26 **3**
$C_{38}H_{42}N_2O_6$ 58.31 **7**
$C_{38}H_{44}$ 19.20 **5**
$C_{38}H_{46}$ 19.17 **7**
$C_{38}H_{46}BrClO_{13}$, C_2H_6O 54.10 **5**
$C_{38}H_{46}CuN_4O_8$ 83.146 **3**
$C_{38}H_{46}N_4O_8Pd$ 83.147 **3**
$C_{38}H_{48}N_2O_{12}S_2Sn_2$ 69.21 **6**
$C_{38}H_{50}N_4Ni$ 49.10 **6**
$C_{38}H_{52}CoP_2$ 86.86 **2**

$C_{38}H_{53}BrO_5$ 56.6 **4**
$C_{38}H_{56}Cl_2HgO_4$ 84.64 **2**
$C_{38}H_{67}EuO_7S$ 77.19 **5**
$C_{38}H_{67}EuO_7S$ 77.20 **6**
$C_{38}H_{95}O_{24}Ti_7$ 84.15 **4**
$C_{39}H_{23}F_6Fe_3O_9P_2$, $2C_6H_6$ 71.113 **7**
$C_{39}H_{24}IrO_2P_2S^+$, F_6P^- , C_3H_6O 86.59 **5**
$C_{39}H_{27}Fe_2N_2O_5P$ 86.36 **4**
$C_{39}H_{30}ClIrN_2OP_2S$ 86.60 **5**
$C_{39}H_{30}CoGeO_3P$ 69.39 **3**
$C_{39}H_{30}F_4IIrOP_2$ 86.87 **2**
$C_{39}H_{30}F_6NiOP_2$ 71.92 **5**
$C_{39}H_{30}O_3OsP_2$ 86.88 **2**
$C_{39}H_{31}F_7P_2Pt$ 71.80 **6**
$C_{39}H_{33}Br_2P_3Pt$, C_6H_5Br 86.73 **7**
$C_{39}H_{33}ClO_3P_2Pd$, $0.6CH_2Cl_2$ 71.81 **6**
$C_{39}H_{34}IP_2Rh$ 72.34 **7**
$C_{39}H_{34}N_8OSi_2$ 49.14 **4**
$C_{39}H_{34}P_2Pd$ 72.35 **7**
$C_{39}H_{34}P_2Pt$ 72.27 **4**
$C_{39}H_{35}Cl_5O_3P_2Re_2$ 81.89 **2**
$C_{39}H_{35}IP_2Pd_2$, C_6H_6 72.28 **4**
$C_{39}H_{36}CoO_6P$, $0.25CH_2Cl_2$ 71.114 **7**
$C_{39}H_{36}O_3P_2Pt$ 86.89 **2**
$C_{39}H_{39}BrF_2N_2O_6$, $2C_4H_{10}O$ 51.46 **3**
$C_{39}H_{39}BrO_9S$ 59.68 **1**
$C_{39}H_{39}CuNO_3P_3$ 86.61 **5**
$(C_{39}H_{42}N_6Na^+)_n$, nCl^- 67.27+ **4**
$C_{39}H_{44}Br_2N_4O_{10}$ 59.27 **4**
$C_{39}H_{45}BrFeMgO_2P_2$, C_4H_8O 73.63 **6**
$C_{39}H_{51}Br_2NO_{11}$, CH_4O 58.120 **1**
$C_{39}H_{51}Br_2NO_{11}$, C_2H_6O 59.28 **4**
$C_{39}H_{52}BrNO_8$ 59.24 **7**
$C_{39}H_{53}BrO_7$ 51.50 **1**
$C_{39}H_{55}IO_5$ 51.59 **7**
$C_{39}H_{57}BrO_4$ 56.6 **7**
$C_{39}H_{64}LuNO_6$ 77.20 **5**
$C_{39}H_{66}NO_{13}^+$, I^- , CH_4O 50.29 **7**
$C_{39}H_{74}O_6$ 1.157 **1**
$C_{39}H_{90}CuO_{18}P_6^{2+}$, $2ClO_4^-$ 84.29 **5**

C_{40-49}

$C_{40}H_{20}$ 30.34 **1**
$C_{40}H_{20}Co_4F_6O_{10}P_2$ 72.29 **4**
$C_{40}H_{20}N_{10}O_2U$ 83.80 **6**
$C_{40}H_{28}Cu_2N_4O_4$, $2CHCl_3$ 78.27 **3**
$C_{40}H_{30}Al_2$ 68.16 **6**
$C_{40}H_{30}AuMnO_7P_2$ 86.90 **2**
$C_{40}H_{30}Au_2F_6P_2$ 71.77 **4**
$C_{40}H_{30}BrN_3O_3$ 52.32 **1**
$C_{40}H_{30}BrN_3O_3$ 31.33 **6**
$C_{40}H_{30}Cl_2N_2P_2Pt$ 72.30 **4**
$C_{40}H_{30}CrO_{10}P_2$ 86.38 **4**
$C_{40}H_{30}Cr_2O_4P_2$ 74.14 **6**
$C_{40}H_{30}F_6OP_2$ 64.84 **2**
$C_{40}H_{30}F_6P_2Pt$ 72.36 **7**
$C_{40}H_{30}MnO_4PSn$ 86.91 **2**
$C_{40}H_{30}N_2O_2S$ 35.39 **4**
$C_{40}H_{30}N_2O_2S$, $0.5C_6H_6$ 42.13 **5**
$C_{40}H_{30}N_2P_2Pt$ 71.78 **4**
$C_{40}H_{30}Nb_2O_2$ 72.48 **3**
$C_{40}H_{30}O_4OsSn_2$ 69.22 **6**
$C_{40}H_{31}Cl_2F_4IrO_3P_2$, C_6H_6 71.82 **6**
$C_{40}H_{31}F_6N_3P_2Pt$ 86.92 **2**
$C_{40}H_{32}Al_2N_4O_5$ 68.22 **3**
$C_{40}H_{32}As_2F_8Fe_2O_4P_2$ 75.48 **5**
$C_{40}H_{32}Fe_2N_4O_5$, $CHCl_3$ 84.41 **6**
$C_{40}H_{32}Mn_2N_8O_2^{3+}$, $3ClO_4^-$, $3H_2O$ 83.134 **4**
$C_{40}H_{32}N_2P_2Pt$ 72.49 **3**
$C_{40}H_{32}N_8^{6+}$, $6Cl^-$, H_2O 49.21 **1**
$C_{40}H_{32}P_2^{2+}$, $2Br^-$, H_2O 64.46 **5**
$C_{40}H_{33}MnO_4P_2$ 81.60 **5**
$C_{40}H_{34}F_6Fe_2O_4P_3Rh$ 73.46 **5**
$C_{40}H_{34}Fe_2O_4P_2Rh^+$, F_6P^- 73.64 **7**
$C_{40}H_{34}HgN_4$ 71.79 **4**
$C_{40}H_{34}O_3P_2Pd$ 71.80 **4**
$C_{40}H_{36}Cu_2I_2N_{12}^{2+}$, I^- , ClO_4^- 83.135 **4**
$C_{40}H_{36}EuO_8^-$, $C_5H_{12}N^+$ 77.21 **5**
$C_{40}H_{36}F_{12}NO_2Pt$ 72.44 **5**
$C_{40}H_{36}P_2Pt$ 75.39 **7**
$C_{40}H_{37}As_2Cl_2Rh$ 72.50 **3**
$C_{40}H_{37}EuO_8$, $xC_4H_{11}N$ 61.10 **5**
$C_{40}H_{37}GdO_8$, $C_5H_{12}N$ 77.30 **3**
$C_{40}H_{37}IrP_2$, $1.5C_7H_6$ 71.83 **6**
$C_{40}H_{38}P_2Pt$ 71.84 **6**
$C_{40}H_{40}LaN_8O_8^{3+}$, $3ClO_4^-$ 84.17 **4**
$C_{40}H_{41}ClPtSi$ 71.81 **4**
$C_{40}H_{44}$ 31.46 **5**
$C_{40}H_{44}Co_2N_4NaO_6^+$, $C_{24}H_{20}B^-$ 67.22 **5**
$C_{40}H_{44}Li_4Mo_4$ 73.65 **7**
$C_{40}H_{44}Li_4W_4$ 73.66 **7**
$C_{40}H_{44}N_4O^{2+}$, $2I^-$ 58.42 **4**
$C_{40}H_{44}N_4O_2^{2+}$, $2I^-$ 58.121 **1**
$C_{40}H_{44}N_4O_4Rh_2$ 49.8 **7**
$C_{40}H_{44}Pt_2S_8$ 85.34 **5**
$C_{40}H_{46}Co_2N_6O_6$, C_7H_8 78.15 **4**
$C_{40}H_{46}Fe_2N_4O_8$, C_7H_8 78.21 **6**
$C_{40}H_{48}Fe_2N_4O_5$ 78.16 **4**
$C_{40}H_{48}N_2O_6^{2+}$, $2I^-$ 58.48 **5**
$C_{40}H_{48}N_4O_2^{2+}$, $2I^-$, xH_2O 58.122 **1**
$C_{40}H_{50}CoN_6$ 49.9 **7**
$C_{40}H_{50}N_2O_3$ 51.47 **3**
$C_{40}H_{50}O_2$ 57.1 **1**
$C_{40}H_{50}O_4P_2Pd_2$ 71.115 **7**
$C_{40}H_{50}O_{12}$, $C_4H_{14}N_2^{2+}$, $2F_6P^-$ 60.39 **7**
$C_{40}H_{52}Cr$ 71.85 **6**
$C_{40}H_{52}O_2$ 57.2 **1**
$C_{40}H_{53}N_4O_2^+$, I^- , C_3H_6O 49.13 **6**
$C_{40}H_{54}$ 57.3 **1**
$C_{40}H_{56}$ 57.4 **1**
$C_{40}H_{56}Co_4O_{16}$ 77.67 **2**
$C_{40}H_{56}Cu_4N_4$ 71.82 **4**
$C_{40}H_{56}Cu_4N_4$ 71.116 **7**
$C_{40}H_{56}Ni_4O_{16}$ 78.28 **3**
$C_{40}H_{58}$ 57.5 **1**
$C_{40}H_{58}Cl_4N_4O_4Rh_2$ 71.83 **4**
$C_{40}H_{60}Cl_2O_8P_4Tc$ 86.47 **6**
$C_{40}H_{60}Cl_2O_8P_4Tc$ 86.74 **7**
$C_{40}H_{61}CoO_8P_4$ 86.77 **3**
$C_{40}H_{62}FeO_8P_4$ 86.39+ **4**
$C_{40}H_{62}O_8P_4Ru$ 86.62 **5**
$C_{40}H_{64}BO_{14}^-$, Rb^+ , $2CH_4O$ 50.13 **4**
$C_{40}H_{64}Br_4Mg_4Mo_2O_2$ 73.67 **7**
$C_{40}H_{64}O_{12}$ 50.16 **5**
$C_{40}H_{64}O_{12}$, Na^+ , CNS^- 50.30 **7**
$C_{40}H_{64}O_{12}$, CNS^- , K^+ 50.44 **1**
$C_{40}H_{67}AgO_{10}$ 50.26 **3**

$C_{40}H_{67}O_{10}Tl$, H_2O 50.14 **4**
$C_{40}H_{67}O_{11}^-$, Ag^+ 50.45 **1**
$C_{40}H_{67}O_{11}^-$, Ag^+ 50.27 **3**
$C_{40}H_{68}O_{10}$, H_2O 50.31 **7**
$C_{40}H_{68}O_{14}$, H_2O 50.32 **7**
$C_{40}H_{70}EuNO_6$ 77.12 **7**
$C_{40}H_{80}Co_2N_{20}O_{10}^{4+}$, $4ClO_4^-$ 84.30 **5**
$C_{40}H_{100}Nb_8O_{30}$ 84.65 **2**
$C_{41}H_{28}BrNO$, C_2H_6O 33.82 **1**
$C_{41}H_{30}F_6OP_2RuS_2$ 85.36 **5**
$C_{41}H_{30}F_6OP_2RuS_2$ 86.48 **6**
$C_{41}H_{32}OP_2Ru$ 72.45 **5**
$C_{41}H_{33}FeIO_6P_2$ 73.119 **2**
$C_{41}H_{33}IrN_2OP_2$ 86.79 **3**
$C_{41}H_{33}N_2NiP_3$ 86.80 **3**
$C_{41}H_{33}N_2NiP_3$, $0.33H_2O$ 86.81 **3**
$C_{41}H_{33}N_2NiP_3$, CH_4O 86.82 **3**
$C_{41}H_{34}CuF_3O_2P_2$ 86.83 **3**
$C_{41}H_{35}FeIO_6P_2$ 73.120 **2**
$C_{41}H_{36}O_8$ 5.9 **4**
$C_{41}H_{37}Cl_2O_2P_2Re$ 77.13 **7**
$C_{41}H_{38}P_2Pt$ 75.40 **7**
$C_{41}H_{39}INiP_3$ 86.75 **7**
$C_{41}H_{48}N_4O^{2+}$, $2I^-$, H_2O 58.67 **3**
$C_{41}H_{48}N_4O_4$, CH_4O 58.123 **1**
$C_{41}H_{49}N_5Zn$ 49.15 **4**
$C_{41}H_{52}O_{17}$ 56.11 **3**
$C_{41}H_{58}FeN_9O_{20}$, $4H_2O$ 82.59 **2**
$C_{41}H_{59}BrO_6$ 56.7 **7**
$C_{42}H_{18}$ 30.35 **1**
$C_{42}H_{24}$ 30.8 **6**
$C_{42}H_{27}N_4$, C_3H_6O 36.34 **3**
$C_{42}H_{28}$ 29.12 **7**
$C_{42}H_{30}$ 19.67 **1**
$C_{42}H_{30}As_2N_4OPt$ 71.117 **7**
$C_{42}H_{30}Cl_4O_2P_2Pd$ 86.76 **7**
$C_{42}H_{30}CoN_6^{3+}$, $3I^-$, $3H_2O$ 83.148 **3**
$C_{42}H_{30}Co_2O_{12}P_2$ 86.94 **2**
$C_{42}H_{30}F_{12}N_2P_2Pt$, $0.4CH_2Cl_2$ 71.84 **4**
$C_{42}H_{30}N_4P_2Pt$ 72.51 **3**
$C_{42}H_{30}Ni_3S_{12}$ 85.96 **2**
$C_{42}H_{30}ReS_6$ 85.97 **2**
$C_{42}H_{30}S_6V$ 85.98 **2**
$C_{42}H_{33}BrCrMnO_8P_3$ 86.41 **4**
$C_{42}H_{33}BrCrMoO_8P_3$ 86.49 **6**
$C_{42}H_{33}BrCrMoO_8P_3$, CH_2Cl_2 86.50 **6**
$C_{42}H_{34}Cl_4Cu_4N_{14}O_4$, $2H_2O$ 84.81 **7**
$C_{42}H_{34}FeO_2P_2$ 72.28 **6**
$C_{42}H_{34}O_2P_2Ru$ 71.94 **5**
$C_{42}H_{35}MnOP_2$, C_6H_6 73.68 **7**
$C_{42}H_{35}P_3PdS_4$ 85.36 **4**
$C_{42}H_{36}MnN_3O_3$ 78.26 **7**
$C_{42}H_{36}O_4P_2Pd$ 72.29 **6**
$C_{42}H_{37}NbOP_2$ 73.69 **7**
$C_{42}H_{38}P_2Pt$ 75.41 **7**
$C_{42}H_{38}P_2Pt$ 75.42 **7**
$C_{42}H_{39}Br_2P_3Pd$ 86.77 **7**
$C_{42}H_{39}Br_2P_3Pd$, C_3H_6O 86.78 **7**
$C_{42}H_{39}Br_2P_3Pd$, C_6H_5Cl 86.79 **7**
$C_{42}H_{39}Br_2P_3Pt$, C_6H_5Br 86.80 **7**
$C_{42}H_{40}Cl_3O_5P_2Re_2$ 81.53 **3**
$C_{42}H_{40}FeN_5O_6S$ 49.10 **7**
$C_{42}H_{40}N_8O_4Ru_2^{2+}$, $2BF_4^-$ 81.54 **3**
$C_{42}H_{40}O_4P_2Pt$ 71.118 **7**
$C_{42}H_{40}P_2Ru$, C_7H_8 72.46 **5**
$C_{42}H_{41}NiP_2^+$, Cl_3Zn^- 72.37 **7**
$C_{42}H_{42}BrFeP_4^+$, $C_{24}H_{20}B^-$, CH_2Cl_2 86.81 **7**
$C_{42}H_{42}ClCoNP_3^+$, F_6P^- 86.64 **5**
$C_{42}H_{42}ClNNiP_3^+$, F_6P^- 86.82 **7**
$C_{42}H_{42}ClN_2P_3Rh^+$, F_6P^- , CH_2Cl_2 86.83 **7**
$C_{42}H_{42}Cl_2CoO_2P_2$ 84.42 **6**
$C_{42}H_{42}CoINP_3^+$, I^- 86.44 **4**
$C_{42}H_{42}CoNP_3^+$, BF_4^- 86.84 **7**
$C_{42}H_{42}Co_2N_{12}^{4+}$, $2H_2Cl_3OZn^-$, Cl_4Zn^{2-} , $4H_2O$ 83.81 **6**
$C_{42}H_{42}INNiP_3^+$, I^- 86.85 **3**
$C_{42}H_{42}NNiP_3$ 86.52 **6**
$C_{42}H_{42}OTi_2$ 71.119 **7**
$C_{42}H_{43}CoNP_3$ 86.53 **6**
$C_{42}H_{43}CoP_4$ 86.85 **7**
$C_{42}H_{43}CoP_4^+$, BF_4^- 86.86 **7**
$C_{42}H_{43}NNiP_3^+$, $C_{42}H_{42}NNiP_3^+$, $2BF_4^-$ 86.87 **7**
$C_{42}H_{45}NSi_3$ 63.20 **7**
$C_{42}H_{46}N_2O_4Ti$ 83.118 **5**
$C_{42}H_{46}N_4O_5^{2+}$, $2Br^-$, $2H_2O$ 58.43 **4**
$C_{42}H_{47}Br_3O_8$ 59.69 **1**
$C_{42}H_{48}NNiP_2Si_2$ 83.136 **4**
$C_{42}H_{48}N_{12}Ni_4O_6P_4$ 86.97 **2**
$C_{42}H_{52}FeN_8^+$, ClO_4^- , $2CHCl_3$ 49.6 **5**
$C_{42}H_{59}IO_{14}S$ 54.19 **4**
$C_{42}H_{60}O_{10}P_4Tc^+$, ClO_4^- 86.88 **7**
$C_{42}H_{84}Ag_6N_6O_6S_6$ 80.20 **4**
$C_{42}H_{84}Ag_6N_6S_{12}$ 80.21 **3**
$C_{42}H_{84}Co_3O_{24}P_6$ 84.66 **2**
$C_{42}H_{84}Cu_6N_6O_6S_6$ 80.22 **3**
$C_{42}H_{85}NO_4$ 48.79 **5**
$C_{42}H_{96}N_{12}NiS_6^{2+}$, $2ClO_4^-$ 79.26 **7**
$C_{43}H_{28}O_7Os_3P_2$ 71.95 **5**
$C_{43}H_{30}As_2ClIrN_4O$ 72.47 **5**
$C_{43}H_{30}BrIrN_4OP_2$ 86.98 **2**
$C_{43}H_{30}F_5IrOP_2$ 71.120 **7**
$C_{43}H_{30}OW$ 72.32 **4**
$C_{43}H_{30}O_7Os_3P_2$ 71.96 **5**
$C_{43}H_{32}N_4P_2Pt$ 71.121 **7**
$C_{43}H_{33}AsO$ 65.13 **4**
$C_{43}H_{33}OP$ 64.44 **4**
$C_{43}H_{34}AsO^+$, ClO_4^- 65.14 **4**
$C_{43}H_{34}ClFIrN_2OP_2^+$, BF_4^- , C_3H_6O 71.97 **5**
$C_{43}H_{35}As_3BrNiS^+$, ClO_4^- , C_6H_5Cl 85.37 **4**
$C_{43}H_{35}MoNO_2P_2$, $0.5CH_2Cl_2$ 73.56 **3**
$C_{43}H_{37}Cl_2IrN_2OP_2$, $CHCl_3$ 71.86 **6**
$C_{43}H_{37}Cl_3N_2P_2Ru$, C_3H_6O 86.54 **6**
$C_{43}H_{37}MnN_4O_2$ 77.31 **3**
$C_{43}H_{39}F_4NiP_3$ 72.48 **5**
$C_{43}H_{40}P_2Pt$ 75.43 **7**
$C_{43}H_{43}O_3PPd$ 71.122 **7**
$C_{43}H_{43}O_3PPd$ 71.123 **7**
$C_{43}H_{48}Br_2N_4O_6$ 58.57 **6**
$C_{43}H_{50}N_4O_2$ 49.16 **4**
$C_{43}H_{57}IO_6S$ 51.48 **3**
$C_{43}H_{58}N_4O_{12}$, $5H_2O$ 50.33 **7**
$C_{43}H_{67}EuN_2O_6$ 77.22 **5**
$0.54C_{44}H_{28}AgN_4$, $0.46C_{44}H_{30}N_4$ 49.17 **4**
$C_{44}H_{28}ClN_4O_4Zn$ 49.14 **6**
$C_{44}H_{28}Cl_2N_4Sn$ 49.7 **5**
$C_{44}H_{28}CoN_5O$ 49.15 **6**
$C_{44}H_{28}CuN_4$ 49.22 **1**
$C_{44}H_{28}FeN_5O$ 49.11 **7**
$C_{44}H_{28}MnN_7$ 49.16 **6**

$C_{44}H_{28}Mn_2O_8P_2$	71.98	**5**
$C_{44}H_{28}Mn_2O_8P_2$	71.87	**6**
$C_{44}H_{28}N_4Pd$	49.23	**1**
$C_{44}H_{28}N_4Zn$, H_2O	49.24	**1**
$C_{44}H_{30}ClF_{12}RhSb_2$, CH_2Cl_2	71.68	**3**
$C_{44}H_{30}Cu_2N_2O_4$	78.29	**3**
$C_{44}H_{30}F_2$	27.17	**1**
$C_{44}H_{30}F_{12}Ir_2N_2O_2P_2$	71.86	**4**
$C_{44}H_{30}FeN_{10}$, $2C_6H_7N$	49.18	**4**
$C_{44}H_{30}MgN_4O$	49.10	**3**
$C_{44}H_{30}N_4$	49.25+	**1**
$C_{44}H_{30}O_8Os_3P_2$	71.99	**5**
$C_{44}H_{31}FeN_4O_2$	49.28+	**1**
$C_{44}H_{32}Cl_4F_{24}Rh_4$	71.87	**4**
$C_{44}H_{32}N_4^{2+}$, Cl^- , Cl_4Fe^-	49.27	**1**
$C_{44}H_{32}N_4^{2+}$, $2Cl^-$	49.30	**1**
$C_{44}H_{32}N_4O_2Zn$	49.31	**1**
$C_{44}H_{34}O_2P_2Pt$	71.100+	**5**
$C_{44}H_{37}BrP_2Pt$	71.88	**6**
$C_{44}H_{39}N_2NiP_3$	86.99	**2**
$C_{44}H_{39}N_2NiP_3$	86.86	**3**
$C_{44}H_{40}F_6N_3P_2Rh$	71.124	**7**
$C_{44}H_{42}Cl_2O_6P_2Rh_2$	75.47	**3**
$C_{44}H_{42}HgN_4$	71.88	**4**
$C_{44}H_{42}O_{14}P_2U_2$	81.55	**3**
$C_{44}H_{44}OP_2Pt$	71.125	**7**
$C_{44}H_{45}B_{10}P_2Rh$	71.89	**6**
$C_{44}H_{46}NiO_6P_2$	72.49	**5**
$C_{44}H_{48}Br_2O_6$	56.13	**5**
$C_{44}H_{48}Cl_4Hg_2O_4S_4$	84.67	**2**
$C_{44}H_{48}Cr_2O_{12}P_2$	77.69	**2**
$C_{44}H_{50}N_2O_4Ti$	83.119	**5**
$C_{44}H_{50}O_6P_4Pt_2S_2$	86.89	**7**
$C_{44}H_{52}CoI_2N_4$	71.126	**7**
$C_{44}H_{53}BrO_{18}S$	56.8	**7**
$C_{44}H_{55}Co_2N_4O_2P_5$, $0.5C_6H_6$	86.90	**7**
$C_{44}H_{56}Br_2O_6$, $2C_2H_6O$	56.13	**3**
$C_{44}H_{63}N_2O_2^+$, I^- , C_2H_6O	58.44	**4**
$C_{44}H_{72}KO_{12}^+$, CNS^-	50.17+	**5**
$C_{44}H_{72}NaO_{12}^+$, CNS^-	50.19	**5**
$C_{44}H_{72}O_{12}$	50.11	**6**
$C_{44}H_{72}O_{12}Rb^+$, CNS^-	50.22	**5**
$C_{44}H_{76}NbO_8$	77.23	**5**
$C_{45}H_{29}N_9Zn$	49.11	**3**
$C_{45}H_{30}FeN_3O_{12}^-$, Na^+ , $3CCl_4$, $3CH_4O$	83.137	**4**
$C_{45}H_{30}FeN_3O_{12}^-$, Na^+ , $2C_3H_6O$	83.138	**4**
$C_{45}H_{30}O_9Os_3P_2$	71.89	**4**
$C_{45}H_{31}IrN_4OP_2$	71.102	**5**
$C_{45}H_{31}N_4NdO_9$	81.43	**6**
$C_{45}H_{32}MnN_7O$, CH_4O	49.17	**6**
$C_{45}H_{33}EuN_9^{3+}$, $3ClO_4^-$	83.217	**2**
$C_{45}H_{35}HoO_7$	77.32	**3**
$C_{45}H_{45}NNiO_6P_2$	72.50	**5**
$C_{45}H_{47}BrO_{17}$	54.16	**7**
$C_{45}H_{51}IN_2O_{14}$, C_2H_6OS , H_2O	50.28	**3**
$C_{45}H_{53}IN_2O_{13}$, C_3H_6O , $5H_2O$	50.48	**1**
$C_{45}H_{54}B_{20}ClIrOP_2$, $2.5CH_2Cl_2$	71.103	**5**
$C_{45}H_{60}Br_2O_5$	51.49	**3**
$C_{45}H_{71}HoN_2O_6$	77.24	**4**
$C_{45}H_{72}O_{12}$	50.23	**5**
$C_{45}H_{72}O_{12}$, xH_2O , yCH_4O	50.24	**5**
$C_{45}H_{77}AgO_{13}$	50.15	**4**
$C_{46}H_{10}F_{20}Fe_2O_6P_2$	71.127	**7**
$C_{46}H_{28}N_4O_2Ru$	49.19	**4**
$C_{46}H_{30}Br_2N_4O_8$	58.58	**6**
$C_{46}H_{30}Fe_2O_6P_2$	72.38	**7**
$C_{46}H_{30}Ir_4O_{10}P_2$	86.100	**2**
$C_{46}H_{34}Co_2N_2O_8$	81.61	**5**
$C_{46}H_{34}N_2O_2$	35.38	**5**
$C_{46}H_{38}Co_4N_4O_{12}$, $2C_2H_6O$, $2H_2O$	81.62	**5**
$C_{46}H_{38}N_2P_2RuS_2$	85.38	**4**
$C_{46}H_{39}Au_2FeP_2^+$, BF_4^-	71.104	**5**
$C_{46}H_{39}Au_2FeP_2^+$, BF_4^-	71.128	**7**
$C_{46}H_{54}N_4O_{10}$, C_3H_6O	58.32	**7**
$C_{46}H_{54}N_6Ru$	49.12	**7**
$C_{46}H_{58}ClCoN_6O_{13}$, C_3H_6O , $2H_2O$	49.32	**1**
$C_{46}H_{66}CoN_{11}O_9$, $11H_2O$	49.20	**4**
$C_{47}H_{30}Br_2N_6O_4$	36.32	**5**
$C_{47}H_{34}N_4O_2Ru$	49.8	**5**
$C_{47}H_{38}O_2P_2Pt$	75.44	**7**
$C_{47}H_{59}N_4O_{10}^+$, I^- , $2H_2O$	58.125	**1**
$C_{47}H_{76}KO_{14}$	50.16	**4**
$C_{47}H_{82}O_{14}$, H_2O	50.34	**7**
$C_{48}H_{28}Fe_2O_4$	75.48	**3**
$C_{48}H_{30}$, C_6H_5Cl	31.42	**3**
$C_{48}H_{30}Cu_2F_{18}N_4O_{14}$	81.44	**6**
$C_{48}H_{30}Cu_2N_{12}$	83.120	**5**
$C_{48}H_{30}Cu_4F_{18}N_4O_{14}$	81.60	**4**
$C_{48}H_{30}N_8O_4P_2$	40.21	**3**
$C_{48}H_{30}N_{12}Ni_2$	83.121	**5**
$C_{48}H_{30}N_{12}Pd_2$	83.122	**5**
$C_{48}H_{32}Br_2O_2$, $1.5C_6H_6$	17.51	**1**
$C_{48}H_{34}CoN_6$, C_6H_6	83.82	**6**
$C_{48}H_{34}Cr_2N_8O_2^{4+}$, $4I^-$, $4H_2O$	83.114	**7**
$C_{48}H_{34}FeN_7O$, $CHCl_3$	49.13	**7**
$C_{48}H_{34}N_4NiO_2$, $0.5CH_2Cl_2$	49.14	**7**
$C_{48}H_{35}NbO$	72.53	**3**
$C_{48}H_{36}$	27.19	**4**
$C_{48}H_{40}Al_4N_4$	68.14	**4**
$C_{48}H_{40}As_4N_4$	65.12	**3**
$C_{48}H_{40}As_4N_4$	65.7	**7**
$C_{48}H_{40}ClIrN_2P_2$, C_6H_{14}	71.129	**7**
$C_{48}H_{40}Cl_4N_4O_4P_4Ru_4$	86.65	**5**
$C_{48}H_{40}I_2O_{16}$	59.70	**1**
$C_{48}H_{40}MnN_2P_4S_4$	85.45	**6**
$C_{48}H_{40}N_4P_4$	64.67	**3**
$C_{48}H_{40}N_4P_4$	64.66	**7**
$C_{48}H_{40}N_{12}Ni_2$	83.218	**2**
$C_{48}H_{40}OP_2Pt$	71.90	**4**
$C_{48}H_{42}N_6Rh^+$, Cl^- , $CHCl_3$	49.18	**6**
$C_{48}H_{43}BrO_8P_2Pd$	71.105	**5**
$C_{48}H_{44}Cu_4N_8O_8^{4+}$, $4NO_3^-$	78.17	**4**
$C_{48}H_{47}As_3NNi^+$, $C_{24}H_{20}B^-$	71.130	**7**
$C_{48}H_{48}$, $0.33C_6H_{14}$	31.33	**7**
$C_{48}H_{48}N_4O_8P_2Rh_2$, H_2O	81.61	**4**
$C_{48}H_{64}CsO_{12}^+$, CNS^-	67.20	**6**
$C_{48}H_{64}N_2O_{13}^{2+}$, $2I^-$, $4CH_4O$, H_2O	50.25	**5**
$C_{48}H_{83}O_{14}^-$, K^+	50.35	**7**
$C_{48}H_{108}Cl_6P_4Rh_2$	86.91	**7**
$C_{48}H_{144}Cl_5P_4Ru_2$	86.101	**2**
$C_{49}H_{22}As_2F_{20}Hg_2$	71.91	**4**
$C_{49}H_{30}O_7Os_3$	71.90	**6**
$C_{49}H_{31}IrN_8OP_2$, $0.5C_6H_6$	86.45	**4**
$C_{49}H_{36}CoN_6$, $2C_6H_6$	49.15	**7**
$C_{49}H_{38}F_2IrN_4OP_2^+$, BF_4^-	83.123	**5**
$C_{49}H_{40}Br_4Pb_4$	69.40	**3**
$C_{49}H_{40}ClCuP_2Ru$, C_3H_6O	71.131	**7**
$C_{49}H_{40}O_3Sb_2$	66.10	**6**
$C_{49}H_{56}ErO_{14}^{3+}$, $3ClO_4^-$	84.82	**7**
$C_{49}H_{74}INO_{18}$, $3C_4H_8O$, H_2O	50.17	**4**

$C_{49}H_{83}O_{17}^{-}$, Ag^{+} , C_3H_6O 50.26 **5**
$C_{49}H_{83}O_{17}^{-}$, Na^{+} , C_3H_6O 50.27 **5**
$C_{49}H_{84}O_{17}$ 50.28 **5**

C_{50-99}

$C_{50}H_{28}N_4O_6Re_2$ 49.10 **5**
$C_{50}H_{33}ClN_4Rh$, H_2O 49.33 **1**
$C_{50}H_{33}N_5ORu$, $1.5C_7H_8$ 49.19 **6**
$C_{50}H_{35}MnO_4Sn$ 73.121 **2**
$C_{50}H_{36}CoN_8^{+}$, $C_2H_3O_2^{-}$, $CHCl_3$, H_2O 49.16 **7**
$C_{50}H_{36}FeN_8^{+}$, Cl^{-} 49.12 **3**
$C_{50}H_{36}FeN_8^{+}$, Cl^{-} , $4CH_4O$ 49.21 **4**
$C_{50}H_{38}Al_2Br_2N_2$, $2C_6H_6$ 68.32 **2**
$C_{50}H_{40}F_{12}P_6Rh_2$, $C_4H_{10}O$ 72.51 **5**
$C_{50}H_{40}N_2O_2P_2Pt$, C_2H_6O 86.55 **6**
$C_{50}H_{40}N_2O_4P_2Pt$ 72.39 **7**
$C_{50}H_{40}N_2P_2$ 64.45 **4**
$C_{50}H_{40}N_2P_2$ 64.47 **5**
$C_{50}H_{40}N_8P_2Pd$, CH_2Cl_2 86.66 **5**
$C_{50}H_{40}P_2Pt$ 72.83 **2**
$C_{50}H_{41}Cu_4I_4P_4$ 86.67 **5**
$C_{50}H_{41}MnN_6O$, $CHCl_3$ 49.17 **7**
$C_{50}H_{42}N_2P_4Pd_2S_2$ 86.56 **6**
$C_{50}H_{44}Br_4Cu_4P_4$ 86.92 **7**
$C_{50}H_{44}ClN_3P_2Pd$ 86.93 **7**
$C_{50}H_{44}ClN_3P_2Pt$ 86.94 **7**
$C_{50}H_{44}Cl_4Cu_4P_4$, $C_2H_4Cl_2$ 86.57 **6**
$C_{50}H_{44}Cl_4Cu_4P_4$, $2C_3H_6O$ 86.95 **7**
$C_{50}H_{44}Cu_4I_4P_4$ 86.96 **7**
$C_{50}H_{44}N_{12}Ni^{4+}$, $4ClO_4^{-}$, $2C_3H_6O$ 49.18 **7**
$C_{50}H_{46}F_{30}Ni_6O_{24}$, C_7H_8 77.33 **3**
$C_{50}H_{46}N_2O_2P_4Ru_2$ 86.59 **6**
$C_{50}H_{50}CoN_6$ 83.83 **6**
$C_{50}H_{56}Br_2O_{10}S_2$ 59.30 **6**
$C_{50}H_{64}Cs_2N_2O_{12}S_2$ 67.21 **6**
$C_{50}H_{75}Cl_4P_5Ru_2$, $xC_6H_{12}O_2$ 86.102 **2**
$C_{50}H_{75}IO_{12}$ 50.12 **6**
$C_{50}H_{92}NiP_4$ 86.46 **4**
$C_{51}H_{37}CoN_6O_2$ 49.19 **7**
$C_{51}H_{39}O_{12}P_3Ru_2$ 74.12 **5**
$C_{51}H_{42}O_3Pd$, C_6H_6 84.43 **6**
$C_{51}H_{42}O_3Pd_2$, $CHCl_3$ 72.40 **7**
$C_{51}H_{42}O_3Pd_2$, CH_2Cl_2 72.41 **7**
$C_{51}H_{42}O_3Pd_2$, C_6H_6 72.42 **7**
$C_{51}H_{56}Fe_2O_3$ 75.29 **6**
$C_{52}H_{28}As_4O_2Rh^{+}$, BF_4^{-} 86.69 **5**
$C_{52}H_{28}IrO_2P_4^{+}$, BF_4^{-} 86.70 **5**
$C_{52}H_{28}O_2P_4Rh^{+}$, BF_4^{-} 86.71 **5**
$C_{52}H_{36}Cl_4Fe_2N_4O_5$ 78.22 **6**
$C_{52}H_{40}Fe_4O_6P_2$ 73.122 **2**
$C_{52}H_{40}GeN_4$ 69.30 **7**
$C_{52}H_{40}N_4Si$ 63.21 **7**
$C_{52}H_{40}N_4Sn$ 69.31 **7**
$C_{52}H_{44}As_4Br_2MoO_2$ 86.47 **4**
$C_{52}H_{44}As_4Cl_2O_2Rh_2$ 86.87 **3**
$C_{52}H_{44}CoO_2P_4^{+}$, BF_4^{-} , $2C_6H_6$ 86.48 **4**
$C_{52}H_{48}ClCoP_4^{+}$, Cl_3Sn^{-} 86.60 **6**
$C_{52}H_{48}ClCoP_4^{+}$, Cl_3Sn^{-} , C_6H_5Cl 86.61 **6**
$C_{52}H_{48}ClMoOP_4^{+}$, $C_3H_6Cl_3OZn^{-}$, C_3H_6O 86.88 **3**
$C_{52}H_{48}Cl_6P_4Re_2$, $2C_2H_3N$ 86.97 **7**
$C_{52}H_{48}CoO_2P_4^{+}$, BF_4^{-} 86.74 **5**
$C_{52}H_{48}IrO_2P_4^{+}$, F_6P^{-} 86.89 **3**
$C_{52}H_{48}IrP_4S_2^{+}$, Cl^{-} , C_2H_3N 86.51 **4**
$C_{52}H_{48}MoN_4P_4$ 86.98 **7**
$C_{52}H_{48}NOP_4Ru^{+}$, $C_{24}H_{20}B^{-}$, C_3H_6O 86.75 **5**
$C_{52}H_{48}N_8O_4$ 48.73 **7**
$C_{52}H_{48}O_2P_4Rh^{+}$, F_6P^{-} 86.90 **3**
$C_{52}H_{48}P_4Rh^{+}$, ClO_4^{-} 86.91 **3**
$C_{52}H_{50}ClN_2P_4W^{+}$, $C_{24}H_{20}B^{-}$ 86.63 **6**
$C_{52}H_{51}P_4Re$ 86.76 **5**
$C_{52}H_{52}As_4ClCoO_8^{+}$, ClO_4^{-} 84.68 **2**
$C_{52}H_{52}IrP_4^{+}$, BF_4^{-} , C_6H_{12} 86.99 **7**
$C_{52}H_{52}O_2P_2Pt$ 71.91 **6**
$C_{52}H_{52}O_4P_2Rh_2$ 71.92 **6**
$C_{52}H_{56}MoP_4$ 86.64 **6**
$C_{52}H_{58}N_8O_8P_2Rh_2$, C_3H_8O , H_2O 86.54 **4**
$C_{52}H_{59}BBrNO_{17}$, CH_2Cl_2 50.18 **4**
$C_{52}H_{60}I_2O_{14}$, $2H_2O$ 59.34 **5**
$C_{52}H_{66}Mo_2O_8P_6$ 86.100 **7**
$C_{53}H_{44}As_4Br_2O_3W$ 86.55 **4**
$C_{53}H_{48}IrOP_4^{+}$, Cl^{-} 86.104 **2**
$C_{53}H_{51}IMoN_2P_4$ 86.101 **7**
$C_{53}H_{54}Br_2N_2O_{11}$, $0.5C_6H_6$ 50.36 **7**
$C_{54}H_{42}As_4Br_2Hg$, CH_2Cl_2 86.105 **2**
$C_{54}H_{42}As_4Br_2Ru$ 86.106 **2**
$C_{54}H_{42}ClCoP_4^{+}$, $C_{24}H_{20}B^{-}$ 86.56 **4**
$C_{54}H_{43}ClIrO_9P_3$ 86.78 **5**
$C_{54}H_{43}CoP_4$ 86.102 **7**
$C_{54}H_{44}Co_2N_4O_6$, $4C_5H_5N$ 77.24 **5**
$C_{54}H_{44}IrP_3$ 71.132 **7**
$C_{54}H_{44}N_4Ni_2O_6$, $4C_5H_5N$ 77.21 **6**
$C_{54}H_{45}As_3Bi_2I_6O_3$ 66.12 **7**
$C_{54}H_{45}ClP_3Rh$ 86.108 **2**
$C_{54}H_{45}Cl_2Cu_2P_3$ 86.57 **4**
$C_{54}H_{45}Cl_2Cu_2P_3$, C_6H_6 86.92 **3**
$C_{54}H_{45}Cl_2P_3Ru$ 86.109 **2**
$C_{54}H_{45}CuP_3^{+}$, BF_4^{-} 86.65 **6**
$C_{54}H_{45}IrNOP_3$ 86.58 **4**
$C_{54}H_{45}NO_3Sn_4$ 69.32 **7**
$C_{54}H_{45}O_2P_3PtS$, $1.5C_6H_6$ 86.66 **6**
$C_{54}H_{46}ClP_3Ru$, C_6H_6 86.110 **2**
$C_{54}H_{46}CoN_2P_3$ 86.93 **3**
$C_{54}H_{46}CoN_2P_3$, $C_4H_{10}O$ 86.112 **2**
$C_{54}H_{46}IrNOP_3^{+}$, ClO_4^{-} 86.59 **4**
$C_{54}H_{46}IrNOP_3^{+}$, ClO_4^{-} 86.67 **6**
$C_{54}H_{46}NOP_3Ru$ 86.60 **4**
$C_{54}H_{46}P_3Ru^{+}$, BF_4^{-} 74.15 **6**
$C_{54}H_{48}IrP_3$, $0.5C_6H_6$ 86.103 **7**
$C_{54}H_{50}CoN_6^{+}$, $0.5NO_3$, $0.5CHO_2^{-}$, $2C_5H_{11}N$ 49.21 **6**
$C_{54}H_{50}FeN_6$ 49.22 **4**
$C_{54}H_{52}Cu_2N_2P_4S_2$ 86.61 **4**
$C_{54}H_{52}N_2NiP_2$ 72.30 **6**
$C_{54}H_{54}Co_2N_6O_{12}$, $2C_2H_6O_5$, $2H_2O$ 83.150 **3**
$C_{54}H_{62}Br_2O_5$ 57.1 **7**
$C_{54}H_{68}N_8O_{22}Zn^{2+}$, $2ClO_4^{-}$ 83.124 **5**
$C_{54}H_{70}Br_2O_{10}$ 54.18 **6**
$C_{54}H_{90}N_6O_{18}$ 50.29 **5**
$2C_{54}H_{90}N_6O_{18}$, $2K^{+}$, I_3^{-} , I_5^{-} 50.37 **7**
$C_{55}H_{30}CuF_{10}O_3P_2Re$ 71.133 **7**
$C_{55}H_{44}Mn_2O_5P_4$, CH_2Cl_2 , C_6H_{14} 86.104 **7**
$C_{55}H_{45}OP_3Pt$ 86.94 **3**
$C_{55}H_{45}OP_3Pt_2S$ 86.95 **3**
$C_{55}H_{46}BrOOsP_3$ 86.83 **2**
$C_{55}H_{46}OP_3Rh$ 86.114 **2**

$C_{55}H_{47}IrOP_3^{+}$, F_5Si^{-}	86.68	**6**
$C_{55}H_{47}O_2P_3Ru$	81.45	**6**
$C_{55}H_{47}O_2P_3Ru$	81.45	**7**
$C_{56}H_{40}$	22.16+	**1**
$C_{56}H_{40}$	20.20	**6**
$C_{56}H_{40}Co_2I_3N_8^{+}$, I^{-}	71.134	**7**
$C_{56}H_{42}Au_2O_2$	71.135	**7**
$C_{56}H_{45}O_2P_3Pt_2S$	86.115	**2**
$C_{56}H_{48}Cl_3N_{16}Ni^{+}$, Cl^{-} , $4C_3H_6O$	83.219	**2**
$C_{56}H_{49}O_2P_3Ru$	81.46	**6**
$C_{56}H_{54}NiP_2$, $0.5C_4H_8O$	72.43	**7**
$C_{56}H_{56}Ag_2Cl_2P_4S_2$	86.105	**7**
$C_{56}H_{56}Ag_2I_2P_4S_2$	86.106	**7**
$C_{56}H_{64}LaO_{16}^{3+}$, $3ClO_4^{-}$	84.44	**6**
$C_{56}H_{70}Cu_2P_4$	71.71	**3**
$C_{56}H_{87}BrO_{17}$, H_2O	50.38	**7**
$C_{57}H_{45}P_2Pt^{+}$, F_6P^{-} , C_6H_6	75.30	**6**
$C_{58}H_{40}F_{12}P_4Rh^{+}$, $C_2Cl_2O_2Rh^{-}$	86.62	**4**
$C_{58}H_{40}Hg_3N_2$	71.136	**7**
$C_{58}H_{45}FeO_{13}P_3Pt_2$	86.107	**7**
$C_{58}H_{45}Ge_2NbO$	72.44	**7**
$C_{58}H_{52}Br_3NO_{16}$	50.30	**3**
$C_{58}H_{59}IMoN_2P_4$, $0.5C_6H_6$	86.108	**7**
$C_{59}H_{45}FeO_9P_2Pt_2$	86.79	**5**
$C_{59}H_{48}Ir_2O_2P_2S_6$	85.39	**4**
$C_{60}H_{30}F_{20}P_2Pd_2S_4$	85.55	**3**
$C_{60}H_{30}F_{20}P_2Pd_2S_4$, $2C_2H_6O$	85.40	**4**
$C_{60}H_{40}Br_2Mo_2O_4$	75.31	**6**
$C_{60}H_{44}ClO_2P_2Rh$, H_2O , $0.5C_2H_6O$	71.92	**4**
$C_{60}H_{44}Co_3N_2O_{12}$	81.46	**7**
$C_{60}H_{45}Mo_2N_3O_6P_3$, $0.5C_4H_8O$	83.84	**6**
$C_{60}H_{45}N_6O_6P_3$, $0.5C_7H_8$	64.70	**3**
$C_{60}H_{50}Cl_2O_2Pd_2$	75.80	**2**
$C_{60}H_{50}Cl_2O_2Pd_2$	75.81	**2**
$C_{60}H_{50}PPb_2Pt$	69.28	**5**
$C_{60}H_{50}P_4Pt_2$	86.109	**7**
$C_{60}H_{52}O_{10}Ti_2$	84.18	**4**
$C_{60}H_{62}Br_2Ni_2P_4$, $2CHCl_3$	72.53	**5**
$C_{60}H_{64}F_{42}O_{14}Pr_2$, $2H_2O$	77.34	**3**
$C_{60}H_{90}Cl_3P_6Ru_2^{+}$, $C_{30}H_{45}Cl_3P_3Ru^{-}$	86.117	**2**
$C_{60}H_{102}La_2N_4O_4P_2Si_8$	84.45	**6**
$C_{61}H_{43}Mn_2O_7P_3$, CH_2Cl_2	71.94	**6**
$C_{61}H_{50}O_2P_3Rh$, $0.5C_6H_6$	81.47	**7**
$C_{61}H_{80}BrN_9O_{11}$, C_2H_6O , H_2O	50.39	**7**
$C_{62}H_{50}Ir_2O_2P_4$	86.69	**6**
$C_{62}H_{57}P_4Re$	86.118	**2**
$C_{62}H_{86}N_{12}O_{16}$, $2C_{10}H_{13}N_5O_4$, $12H_2O$	60.43+	**4**
$C_{63}H_{45}Ir_4O_9P_3$	86.119	**2**
$C_{63}H_{88}CoN_{14}O_{14}P$, $18H_2O$	49.13	**3**
$C_{63}H_{88}CoN_{14}O_{14}P$, $20H_2O$	49.23	**4**
$C_{63}H_{88}CoN_{14}O_{14}P$, $22H_2O$	49.34	**1**
$C_{63}H_{89}CoN_{14}O_{17}P_2$, $15H_2O$	49.14	**3**
$C_{64}H_{32}N_{16}Sn$	49.11	**5**
$C_{64}H_{32}N_{16}Th$	49.20	**7**
$C_{64}H_{32}N_{16}U$	49.15	**3**
$C_{64}H_{32}N_{16}U$	49.21	**7**
$C_{64}H_{51}P_2Rh$, C_7H_8	71.93	**4**
$C_{64}H_{52}Co_2F_8N_8O_{14}$, $2CHCl_3$, $C_5H_{11}N$	78.30	**3**
$C_{64}H_{78}N_{10}O_{10}$, Li^{+} , Br^{-} , $3C_2H_3N$	48.74	**7**
$C_{64}H_{88}Cl_6MoN_4P_8Re_2$	86.110	**7**
$C_{66}H_{50}F_6P_4Pd_2$	72.45	**7**
$C_{66}H_{60}Br_2O_{12}$, $2C_6H_6$	59.34	**3**
$C_{66}H_{63}O_{13}P_3Ru_3$	81.63	**5**
$C_{66}H_{72}CaN_{12}O_6^{2+}$, $2ClO_4^{-}$	67.20	**2**
$C_{66}H_{72}MgN_{12}O_6^{2+}$, $2ClO_4^{-}$	67.21	**2**
$C_{66}H_{72}N_{12}O_6Pb^{2+}$, $2ClO_4^{-}$	69.49	**2**
$C_{66}H_{72}N_{12}O_6Y^{3+}$, $3I^{-}$	84.46	**6**
$C_{66}H_{74}N_{10}O_{10}$, Na^{+} , Br^{-}	50.13	**6**
$C_{66}H_{81}LiN_{11}O_{10}^{+}$, Br^{-} , $2C_2H_3N$	67.22	**6**
$C_{66}H_{82}N_{10}O_{10}$, Na^{+} , Br^{-} , C_2H_6O	48.75	**7**
$C_{66}H_{114}O_{12}Pr_2$	77.35	**3**
$C_{68}H_{50}Nb_2O_2$	72.84	**2**
$C_{68}H_{52}Br_4O_{16}$	31.34	**7**
$C_{68}H_{52}Br_4O_{16}$	31.35	**7**
$C_{68}H_{70}FeN_{10}O_6$	49.22	**7**
$C_{68}H_{72}FeN_8O_6S$, $2C_4H_8S$	49.23	**7**
$C_{69}H_{84}N_{18}O_{18}S_5$	50.31	**3**
$C_{72}H_{57}IrP_5^{+}$, $C_{24}H_{20}B^{-}$	86.64	**4**
$C_{72}H_{60}Ag_4Cl_4P_4$	86.111	**7**
$C_{72}H_{60}Ag_4I_4P_4$, xCH_2Cl_2	86.112	**7**
$C_{72}H_{60}Br_4Cu_4P_4$, $2CHCl_3$	86.70	**6**
$C_{72}H_{60}Cl_2O_4P_4Rh_2$, CH_2Cl_2	86.65	**4**
$C_{72}H_{60}Cl_4Cu_4P_4$	86.71	**6**
$C_{72}H_{60}Cl_6Cu_4O_5P_4$	84.69	**2**
$C_{72}H_{60}Cu_2N_6P_4$	86.96	**3**
$C_{72}H_{60}Cu_4I_4P_4$	86.113	**7**
$C_{72}H_{60}IMnO_4P_4^{+}$, $C_4I_2MnO_4^{-}$	84.32	**5**
$C_{72}H_{60}N_{18}OZn_4$, C_6H_6	83.151	**3**
$C_{72}H_{61}AsP_3Rh$, $0.5C_6H_6$	86.97	**3**
$C_{72}H_{61}P_4Rh$	86.120	**2**
$xC_{72}H_{62}N_4P_4Pt_2^{2+}$, $4C_{24}H_{20}B^{-}$, $yC_{72}H_{64}N_2P_4Pt_2^{2+}$	86.98	**3**
$C_{72}H_{62}P_4Ru$	86.80	**5**
$C_{72}H_{70}B_{10}Cu_2P_4$, $CHCl_3$	86.114	**7**
$C_{72}H_{84}Nd_4$	71.108	**5**
$C_{72}H_{100}CoN_{18}O_{17}P$, $17H_2O$	49.35	**1**
$C_{72}H_{120}O_{60}$, $3.8C_2H_3O_2^{-}$, $3.8K^{+}$, $19.4H_2O$	45.76	**1**
$C_{72}H_{132}N_2Ni_2P_4$	86.66	**4**
$C_{73}H_{60}ClP_4Pt_2S_2^{+}$, BF_4^{-} , $0.2CH_2Cl_2$	71.137	**7**
$C_{73}H_{60}Cl_4P_4Ru_2S$	86.72	**6**
$C_{74}H_{42}Mn_2N_{18}O$, $2C_5H_5N$	49.36	**1**
$C_{74}H_{60}Ag_2N_2P_4S_2$	86.81	**5**
$C_{74}H_{60}O_2P_4Rh_2$, $2CH_2Cl_2$	86.82	**5**
$C_{74}H_{66}B_2Cu_2N_2P_4$	86.73	**6**
$C_{75}H_{63}Cu_3P_6$, $2C_7H_8$	71.95	**6**
$C_{75}H_{66}NaO_6P_6^{+}$, Br^{-} , $3H_2O$	67.27	**3**
$C_{78}H_{65}P_5Pt_3$, C_6H_6	71.138	**7**
$C_{78}H_{72}Cl_2Cu_2P_6$, $2C_3H_6O$	86.83	**5**
$C_{78}H_{72}Cu_2N_6P_6$	86.68	**4**
$C_{81}H_{105}F_{18}N_3O_{15}Pr_2$	84.83	**7**
$C_{82}H_{72}Cu_2F_6N_8P_6$	86.69	**4**
$C_{82}H_{78}Ni_2P_6S^{2+}$, $2C_{24}H_{20}B^{-}$	86.115	**7**
$C_{84}H_{60}N_6O_2P_4Rh_2$	86.116	**7**
$C_{84}H_{60}Pd_2$	72.55	**5**
$C_{84}H_{84}As_6IN_2Ni_2^{+}$, $C_{24}H_{20}B^{-}$	86.117	**7**
$C_{84}H_{84}Br_4O_{16}$	30.36	**1**
$C_{88}H_{56}Fe_2N_8O$	49.24	**4**
$C_{88}H_{100}Li_{12}N_4Ni_4O_4$	67.23	**6**
$C_{94}H_{45}Ag_2F_{25}P_3Rh$	71.139	**7**

$C_{100-149}$

$C_{100}H_{70}Cu_4Ir_2P_2$	71.96	**6**
$C_{108}H_{90}Hg_2O_6P_6^{2+}$, $2ClO_4^{-}$	84.47	**6**
$C_{108}H_{96}Cu_6P_6$, C_3H_7NO	86.84	**5**

$C_{108}H_{216}N_{12}Ni_6O_{12}S_{12}$ 80.21 **7**
$C_{110}H_{202}Er_8O_{33}$ 77.25 **5**
$C_{114}H_{90}Ag_3AlO_6P_6S_6$ 86.118 **7**
$C_{114}H_{90}Ag_3FeO_6P_6S_6$ 81.48 **7**
$C_{124}H_{100}Ca_4O_{18}$ 67.25 **5**
$C_{126}H_{84}Au_{11}Cl_{21}I_3P_7$ 86.99 **3**
$C_{126}H_{84}Au_{11}F_{21}I_3P_7$ 86.85 **5**
$C_{126}H_{100}O_{18}Sr_4$ 67.26 **5**
$C_{126}H_{126}Au_6P_6^{2+}$, $2C_{24}H_{20}B^-$ 86.75 **6**

$C_{150-199}$

$C_{168}H_{168}Au_9P_8^{3+}$, $3F_6P^-$ 86.71 **4**

Ag

Vol. 1 48.1, 53.10, 53.11 **Vol. 2** 72.29, 72.53, 74.1, 74.2, 75.2, 75.5, 75.9, 75.16, 75.35, 75.47, 75.67, 75.78, 76.20, 77.19, 79.3, 79.16, 80.41, 81.2, 82.1, 83.16, 83.17, 83.27, 83.44, 83.161, 84.23, 85.2, 85.9, 85.10, 85.11, 85.14, 85.36, 85.64, 85.67, 86.36 **Vol. 3** 53.7, 53.15, 53.29, 74.1, 74.6, 74.7, 74.12, 74.13, 74.14, 75.2, 75.4, 75.8, 75.9, 75.37, 76.11, 76.12, 80.21, 81.36, 83.16, 83.61, 86.32 **Vol. 4** 23.4, 23.6, 31.1, 31.13, 74.3, 74.8, 74.9, 75.19, 80.11, 80.20, 81.36, 81.55, 82.1, 82.2, 82.4, 82.5, 82.8, 83.15, 83.16, 83.35, 83.58 **Vol. 5** 71.34, 76.9, 79.14, 81.7, 82.2, 83.22, 83.79, 84.2, 86.81 **Vol. 6** 71.26, 72.8, 79.9, 82.7, 83.3, 84.2, 84.24, 84.25, 84.30, 85.4, 85.12 **Vol. 7** 71.139, 74.1, 74.2, 76.22, 79.2, 81.48, 83.7, 86.16, 86.105, 86.106, 86.111, 86.112, 86.118

Am

Vol. 3 77.16

Au

Vol. 2 64.52, 71.16, 71.18, 71.23, 71.52, 73.109, 80.2, 80.8, 80.34, 83.154, 86.1, 86.90 **Vol. 3** 80.11, 86.22, 86.26, 86.99 **Vol. 4** 71.34, 71.43, 71.46, 71.77, 80.11, 80.12, 80.16, 86.71 **Vol. 5** 71.22, 71.60, 71.104, 80.11, 80.15, 80.18, 85.18, 85.19, 85.22, 86.21, 86.27, 86.85 **Vol. 6** 71.15, 86.34, 86.75 **Vol. 7** 71.95, 71.128, 71.135, 76.1, 80.16, 83.53, 83.54, 86.23

Cd

Vol. 2 79.2, 79.5, 79.6, 79.20, 79.33, 80.10, 80.11, 80.35, 80.36, 81.61, 82.45, 83.4, 83.5, 83.28, 83.74, 84.2, 84.3, 84.5, 84.8, 85.37, 85.94 **Vol. 3** 61.4, 76.39, 79.4, 79.17, 83.38, 83.39, 84.2, 85.41, 86.58 **Vol. 4** 76.13, 79.4, 81.17, 83.5, 83.7, 83.50, 83.91, 83.92, 83.93, 84.9 **Vol. 5** 61.2, 76.16, 76.33, 76.42, 78.19, 80.2, 80.28, 83.5, 83.36, 83.52, 83.104, 83.106, 84.15 **Vol. 6** 61.1, 76.9, 80.2, 80.14, 81.10, 81.13, 81.15, 81.21, 81.28, 81.29, 81.33, 82.1, 82.3, 82.4, 82.12, 82.13, 82.14, 82.17, 82.18, 83.47, 83.58, 83.69, 84.17 **Vol. 7** 76.16, 76.24, 76.59, 77.1, 77.4, 81.3, 81.12, 82.4, 83.2, 83.110, 84.9, 85.2

Ce

Vol. 2 77.47, 81.44, 81.67 **Vol. 3** 77.19, 81.11, 84.26 **Vol. 4** 75.31, 81.43 **Vol. 5** 75.26 **Vol. 6** 77.19 **Vol. 7** 77.7, 81.28

Co

Vol. 1 3.31 **Vol. 2** 62.33, 71.22, 71.40, 71.44, 71.45, 72.25, 72.35, 72.39, 72.46, 72.52, 72.56, 72.67, 73.9, 73.10, 73.19, 73.31, 73.34, 73.46, 73.58, 73.67, 73.73, 73.108, 73.117, 75.23, 75.55, 75.61, 76.2, 76.4, 76.7, 76.8, 76.9, 76.10, 76.11, 76.24, 76.28, 76.34, 76.35, 76.36, 76.37, 76.38, 76.40, 76.49, 76.54, 76.55, 76.56, 76.59, 76.75, 76.79, 77.11, 77.20, 77.21, 77.44, 77.51, 77.64, 77.67, 78.15, 78.59, 79.24, 80.5, 80.7, 80.29, 80.30, 81.5, 81.12, 81.34, 81.50, 82.2, 82.24, 82.28, 82.34, 82.37, 82.46, 82.54, 83.61, 83.75, 83.89, 83.91, 83.95, 83.103, 83.124, 83.130, 83.131, 83.134, 83.136, 83.146, 83.176, 83.213, 84.12, 84.21, 84.66, 84.68, 85.39, 85.45, 85.55, 85.73, 85.74, 85.82, 85.84, 85.86, 86.23, 86.32, 86.44, 86.60, 86.61, 86.73, 86.86, 86.94, 86.112 **Vol. 3** 62.12, 69.39, 71.25, 71.35, 71.38, 71.55, 71.59, 71.60, 71.64, 72.29, 72.39, 73.27, 73.33, 73.37, 76.2, 76.3, 76.6, 76.7, 76.9, 76.10, 76.17, 76.19, 76.23, 76.27, 76.28, 76.30, 76.31, 76.38, 76.40, 76.41, 76.52, 76.54, 76.55, 76.58, 76.61, 76.62, 76.63, 76.64, 76.71, 76.74, 76.75, 76.81, 76.82, 76.85, 77.8, 78.12, 78.14, 78.16, 78.19, 78.26, 78.30, 79.2, 79.5, 79.6, 79.7, 79.21, 81.17, 82.7, 82.12, 82.13, 82.14, 82.15, 82.22, 83.20, 83.26, 83.34, 83.49, 83.51, 83.65, 83.76, 83.88, 83.93, 83.100, 83.115, 83.121, 83.142, 83.148, 83.150, 84.18, 84.24, 85.35, 85.40, 85.45, 86.11, 86.27, 86.32, 86.33, 86.77, 86.93 **Vol. 4** 60.9, 71.2, 71.39, 71.40, 71.42, 72.29, 73.17, 75.5, 75.20, 75.29, 76.3, 76.10, 76.11, 76.13, 76.14, 76.17, 76.19, 76.20, 76.21, 76.22, 76.23, 76.24, 76.25, 76.28, 76.29, 76.30, 76.31, 76.32, 76.39, 76.40, 76.43, 77.9, 77.14, 77.20, 78.13, 78.14, 78.15, 79.1, 79.10, 80.1, 81.15, 81.24, 81.31, 81.50, 82.10, 83.13, 83.21, 83.24, 83.25, 83.32, 83.56, 83.57, 83.67, 83.74, 83.82, 83.94, 83.95, 83.96, 83.98, 83.102, 83.120, 83.122, 83.129, 83.130, 84.1, 85.5, 85.22, 86.9, 86.11, 86.18, 86.44, 86.48, 86.56 **Vol. 5** 62.9, 67.22, 71.13, 71.15, 71.24, 71.31, 71.42, 71.44, 71.52, 71.57, 73.1, 73.2, 73.22, 73.23, 73.24, 73.25, 73.28, 73.29, 73.34, 73.35, 73.36, 74.6, 75.10, 75.11, 75.14, 75.27, 75.35, 75.41, 75.42, 76.2, 76.4, 76.14, 76.15, 76.17, 76.18, 76.20, 76.21, 76.22, 76.25, 76.26, 76.27, 76.29, 76.30, 76.31, 76.39, 76.40, 76.43, 76.48, 76.50, 76.52, 77.16, 77.24, 78.5, 78.6, 78.9,

78.13, 78.14, 78.20, 79.5, 79.12, 79.23, 79.26, 80.5, 81.6, 81.14, 81.16, 81.36, 81.61, 81.62, 82.10, 82.20, 83.6, 83.35, 83.41, 83.55, 83.60, 83.61, 83.91, 83.93, 83.96, 83.105, 83.116, 84.28, 84.30, 86.24, 86.25, 86.27, 86.28, 86.32, 86.36, 86.43, 86.50, 86.56, 86.64, 86.74 **Vol. 6** 71.9, 71.10, 71.22, 71.58, 71.59, 72.14, 73.1, 73.4, 73.14, 73.15, 73.26, 73.32, 73.48, 75.27, 76.2, 76.3, 76.4, 76.10, 76.11, 76.13, 76.15, 76.16, 76.17, 76.20, 76.21, 76.27, 76.30, 76.42, 76.48, 76.50, 77.3, 77.9, 77.11, 77.12, 78.3, 78.11, 78.13, 79.8, 79.11, 80.3, 80.13, 81.2, 81.18, 81.32, 82.7, 82.9, 82.10, 82.16, 83.2, 83.10, 83.16, 83.20, 83.21, 83.23, 83.27, 83.29, 83.30, 83.34, 83.35, 83.39, 83.40, 83.41, 83.45, 83.49, 83.60, 83.66, 83.68, 83.81, 83.82, 83.83, 84.8, 84.42, 85.8, 85.38, 86.8, 86.9, 86.17, 86.20, 86.53, 86.60, 86.61 **Vol. 7** 71.12, 71.28, 71.29, 71.32, 71.83, 71.87, 71.102, 71.114, 71.126, 71.134, 73.1, 73.11, 73.17, 73.18, 73.27, 73.40, 73.43, 73.57, 73.58, 73.59, 73.61, 73.62, 73.63, 74.4, 74.5, 76.5, 76.6, 76.14, 76.17, 76.18, 76.26, 76.32, 76.35, 76.37, 76.40, 76.41, 76.42, 76.52, 76.56, 76.57, 76.60, 76.66, 76.70, 77.11, 78.2, 78.5, 78.7, 78.11, 78.18, 78.19, 79.7, 79.9, 79.12, 79.21, 79.22, 80.11, 81.4, 81.13, 81.17, 81.20, 81.21, 81.33, 81.46, 82.6, 82.8, 82.17, 83.5, 83.20, 83.22, 83.28, 83.29, 83.32, 83.39, 83.43, 83.49, 83.52, 83.56, 83.61, 83.74, 83.79, 83.80, 83.84, 83.87, 83.91, 83.109, 84.1, 84.15, 84.21, 84.24, 84.40, 84.55, 84.65, 84.74, 85.1, 85.4, 85.22, 85.37, 86.34, 86.38, 86.48, 86.71, 86.84, 86.85, 86.86, 86.90, 86.102

Cr

Vol. 1 3.58, 3.59 **Vol. 2** 60.118, 60.140, 71.14, 71.24, 71.32, 72.12, 73.16, 73.20, 73.99, 74.4, 74.5, 74.6, 74.7, 74.9, 74.10, 74.11, 74.12, 74.13, 74.14, 74.15, 74.16, 74.18, 74.20, 74.21, 74.22, 74.23, 74.24, 74.25, 75.1, 75.22, 75.31, 75.39, 75.46, 75.68, 76.1, 76.13, 76.39, 76.40, 76.44, 76.66, 76.81, 77.22, 77.69, 81.21, 81.37, 81.63, 81.75, 83.35, 83.71, 83.72, 83.93, 83.188, 83.197, 83.206, 83.211, 84.46, 85.85, 86.47 **Vol. 3** 71.30, 71.47, 71.52, 72.15, 72.20, 73.1, 73.6, 73.16, 73.17, 74.2, 74.11, 75.3, 75.11, 75.28, 76.4, 76.5, 76.24, 76.32, 76.40, 76.65, 77.9, 77.27, 78.3, 79.16, 81.32, 83.102, 86.57 **Vol. 4** 71.5, 71.8, 71.12, 71.14, 71.30, 75.1, 75.9, 75.16, 76.32, 78.8, 80.2, 81.26, 82.7, 83.20, 83.61, 83.85, 86.25, 86.38, 86.41 **Vol. 5** 67.16, 71.75, 71.77, 71.83, 74.3, 74.4, 74.5, 75.3, 75.17, 75.19, 75.20, 75.23, 75.37, 81.52, 82.12, 83.101, 86.3, 86.4, 86.7, 86.10, 86.11, 86.16, 86.30, 86.31 **Vol. 6** 60.24, 71.68, 71.85, 73.6, 73.33, 74.1, 74.2, 74.3, 74.4, 74.5, 74.6, 74.8, 74.9, 74.11, 74.14, 75.4, 75.5, 75.22, 76.12, 85.15, 86.24, 86.49, 86.50 **Vol. 7** 71.3, 71.6, 71.23, 71.57, 71.88, 72.30, 73.25, 73.48, 73.52, 74.3, 75.9, 75.21, 75.27, 76.74, 79.14, 80.4, 83.25, 83.114, 84.8, 84.27, 86.18, 86.22

Cu

Vol. 1 3.38, 3.39, 3.69 **Vol. 2** 60.76, 60.126, 60.145, 60.146, 60.147, 61.21, 62.34, 72.30, 74.3, 75.6, 75.50, 75.77, 76.15, 76.16, 76.17, 76.18, 76.19, 76.26, 76.46, 76.51, 77.1, 77.14, 77.15, 77.16, 77.17, 77.40, 77.54, 77.55, 77.59, 78.1, 78.2, 78.3, 78.4, 78.5, 78.6, 78.7, 78.8, 78.16, 78.17, 78.18, 78.19, 78.20, 78.21, 78.26, 78.27, 78.28, 78.29, 78.34, 78.35, 78.40, 78.41, 78.46, 78.50, 78.51, 78.58, 79.13, 80.9, 80.13, 80.14, 80.15, 80.25, 80.37, 81.3, 81.4, 81.6, 81.7, 81.8, 81.15, 81.22, 81.23, 81.27, 81.31, 81.32, 81.33, 81.45, 81.47, 81.48, 81.62, 81.64, 81.66, 81.76, 81.77, 81.78, 81.79, 81.80, 81.81, 81.82, 81.84, 81.85, 81.86, 81.87, 82.3, 82.4, 82.5, 82.10, 82.13, 82.15, 82.17, 82.18, 82.21, 82.23, 82.25, 82.26, 82.29, 82.35, 82.36, 82.38, 82.39, 82.41, 82.42, 82.47, 82.51, 82.53, 82.55, 82.57, 82.58, 83.2, 83.3, 83.6, 83.8, 83.20, 83.21, 83.22, 83.23, 83.34, 83.39, 83.45, 83.46, 83.50, 83.52, 83.53, 83.54, 83.56, 83.62, 83.65, 83.73, 83.76, 83.77, 83.79, 83.81, 83.82, 83.87, 83.90, 83.100, 83.101, 83.104, 83.109, 83.112, 83.113, 83.116, 83.117, 83.119, 83.132, 83.140, 83.155, 83.156, 83.157, 83.158, 83.171, 83.179, 83.180, 83.181, 83.182, 83.187, 83.192, 83.207, 84.1, 84.4, 84.10, 84.15, 84.19, 84.20, 84.31, 84.32, 84.33, 84.34, 84.36, 84.38, 84.40, 84.47, 84.48, 84.49, 84.69, 85.20, 85.40, 85.53, 85.69, 86.14, 86.41, 86.69 **Vol. 3** 60.1, 60.16, 60.37, 60.39, 71.71, 73.7, 73.44, 75.36, 76.1, 76.20, 76.22, 76.29, 76.32, 76.33, 76.34, 76.35, 76.36, 76.37, 76.41, 76.42, 76.46, 76.48, 76.53, 76.57, 76.59, 76.66, 76.69, 76.77, 76.78, 77.1, 77.5, 77.14, 77.15, 77.18, 77.25, 77.26, 78.2, 78.4, 78.8, 78.9, 78.10, 78.18, 78.27, 78.29, 79.8, 79.10, 79.11, 79.12, 79.20, 79.22, 79.24, 80.7, 80.22, 81.5, 81.19, 81.22, 81.23, 81.24, 81.25, 81.30, 81.35, 81.40, 81.45, 81.46, 81.47, 81.49, 81.51, 82.5, 82.17, 82.21, 82.24, 82.25, 82.26, 83.6, 83.7, 83.10, 83.23, 83.24, 83.29, 83.32, 83.36, 83.37, 83.40, 83.46, 83.47, 83.62, 83.69, 83.72, 83.74, 83.75, 83.78, 83.79, 83.81, 83.104, 83.105, 83.107, 83.108, 83.109, 83.110, 83.111, 83.119, 83.122, 83.127, 83.130, 83.146, 84.1, 84.3, 84.4, 84.7, 84.8, 84.11, 84.13, 84.17, 84.19, 84.23, 85.1, 85.5, 85.8, 85.15, 86.3, 86.62, 86.67, 86.83, 86.92, 86.96 **Vol. 4** 60.10, 71.82, 74.11, 76.1, 76.4, 76.5, 76.6, 76.8, 76.12, 76.38, 76.41, 77.7, 77.13, 78.1, 78.4, 78.5, 78.6, 78.7, 78.10, 78.12, 78.17, 79.2, 79.3, 79.7, 79.9, 81.1, 81.12,

81.13, 81.14, 81.23, 81.32, 81.40, 81.41, 81.42, 81.44, 81.48, 81.49, 81.54, 81.59, 81.60, 82.6, 82.14, 82.15, 83.1, 83.2, 83.4, 83.11, 83.12, 83.14, 83.22, 83.27, 83.28, 83.36, 83.37, 83.39, 83.40, 83.52, 83.53, 83.59, 83.64, 83.68, 83.69, 83.70, 83.71, 83.79, 83.80, 83.88, 83.90, 83.103, 83.107, 83.108, 83.117, 83.135, 84.8, 84.11, 84.13, 84.14, 85.3, 85.14, 85.28, 86.12, 86.34, 86.57, 86.61, 86.68, 86.69 **Vol. 5** 60.12, 71.38, 71.80, 76.5, 76.6, 76.13, 76.24, 76.28, 76.38, 76.41, 76.45, 77.6, 77.13, 78.1, 78.2, 78.3, 78.8, 78.11, 78.18, 79.13, 79.15, 79.22, 79.24, 79.25, 80.6, 80.16, 80.21, 80.24, 81.4, 81.5, 81.12, 81.25, 81.29, 81.32, 81.33, 81.35, 81.38, 81.40, 81.42, 81.43, 81.47, 81.53, 81.54, 81.55, 81.56, 81.57, 81.58, 82.3, 82.9, 82.11, 82.13, 82.14, 82.17, 82.23, 82.24, 83.7, 83.8, 83.9, 83.11, 83.12, 83.13, 83.15, 83.16, 83.17, 83.18, 83.19, 83.23, 83.24, 83.25, 83.34, 83.37, 83.43, 83.46, 83.56, 83.63, 83.64, 83.65, 83.66, 83.71, 83.72, 83.75, 83.84, 83.87, 83.88, 83.92, 83.109, 83.110, 83.111, 83.114, 83.115, 83.120, 84.6, 84.11, 84.14, 84.19, 84.23, 84.25, 84.29, 85.1, 85.3, 85.10, 85.13, 85.17, 86.52, 86.61, 86.67, 86.83, 86.84 **Vol. 6** 60.2, 60.14, 71.95, 71.96, 76.6, 76.7, 76.8, 76.18, 76.19, 76.22, 76.23, 76.26, 76.33, 76.39, 76.41, 76.43, 76.44, 76.45, 76.46, 76.47, 76.50, 76.52, 77.16, 78.1, 78.6, 78.7, 78.8, 78.12, 78.14, 78.16, 78.17, 78.18, 78.19, 78.20, 79.3, 79.5, 79.14, 79.16, 79.17, 80.5, 80.14, 81.1, 81.3, 81.7, 81.11, 81.24, 81.26, 81.27, 81.34, 81.36, 81.38, 81.44, 82.5, 82.11, 82.15, 82.21, 82.24, 83.8, 83.11, 83.15, 83.26, 83.32, 83.33, 83.50, 83.51, 83.56, 83.62, 83.63, 83.64, 83.71, 83.74, 83.76, 84.1, 84.5, 84.6, 84.10, 84.12, 84.14, 84.15, 84.16, 84.19, 84.21, 84.26, 84.31, 84.33, 85.14, 85.25, 86.35, 86.57, 86.65, 86.70, 86.71, 86.73 **Vol. 7** 71.7, 71.93, 71.116, 71.131, 71.133, 76.7, 76.8, 76.20, 76.21, 76.23, 76.25, 76.27, 76.29, 76.30, 76.31, 76.33, 76.38, 76.43, 76.51, 76.62, 76.64, 76.65, 76.68, 77.10, 78.6, 78.8, 78.9, 78.16, 78.20, 78.21, 78.22, 78.23, 78.24, 78.25, 79.3, 79.4, 79.5, 79.13, 79.15, 79.16, 79.17, 79.18, 79.19, 79.20, 79.23, 79.25, 80.1, 80.6, 80.7, 81.8, 81.9, 81.11, 81.29, 81.30, 81.35, 81.37, 81.39, 81.41, 81.42, 81.43, 82.1, 82.3, 82.5, 82.7, 82.9, 82.12, 82.13, 82.14, 82.15, 82.19, 82.20, 82.21, 83.6, 83.8, 83.10, 83.11, 83.13, 83.14, 83.15, 83.19, 83.23, 83.24, 83.26, 83.27, 83.30, 83.34, 83.42, 83.46, 83.47, 83.59, 83.63, 83.65, 83.67, 83.68, 83.69, 83.71, 83.72, 83.77, 83.83, 83.85, 83.90, 83.92, 83.95, 83.102, 83.103, 83.106, 83.111, 84.2, 84.20, 84.23, 84.29, 84.30, 84.35, 84.39, 84.42, 84.44, 84.45, 84.53, 84.56, 84.57, 84.59, 84.60, 84.61, 84.62, 84.63, 84.67, 84.73, 84.76, 84.81, 85.3, 85.9, 85.13, 85.23, 85.30, 85.32, 85.40, 85.43, 86.44, 86.45, 86.92, 86.95, 86.96, 86.113, 86.114

Dy

Vol. 4 77.22 **Vol. 5** 82.6

Er

Vol. 3 81.4, 81.6 **Vol. 4** 77.21, 81.25 **Vol. 5** 77.25 **Vol. 6** 81.39, 84.7 **Vol. 7** 81.32, 84.82

Eu

Vol. 2 83.217 **Vol. 3** 77.12, 77.17, 80.12 **Vol. 4** 81.21 **Vol. 5** 61.10, 77.14, 77.19, 77.21, 77.22 **Vol. 6** 77.20, 81.41, 83.54 **Vol. 7** 77.12, 81.31

Fe

Vol. 2 60.119, 62.39, 69.35, 71.13, 71.20, 71.21, 71.28, 71.34, 71.46, 72.7, 72.13, 72.14, 72.16, 72.17, 72.18, 72.31, 72.32, 72.33, 72.41, 72.54, 72.57, 72.61, 72.65, 72.68, 72.69, 72.74, 72.75, 73.5, 73.14, 73.21, 73.22, 73.23, 73.24, 73.25, 73.31, 73.35, 73.36, 73.38, 73.39, 73.43, 73.44, 73.47, 73.48, 73.49, 73.51, 73.55, 73.59, 73.68, 73.69, 73.71, 73.72, 73.76, 73.85, 73.86, 73.87, 73.89, 73.90, 73.94, 73.95, 73.96, 73.98, 73.101, 73.102, 73.104, 73.105, 73.107, 73.110, 73.115, 73.116, 73.119, 73.120, 73.122, 75.8, 75.10, 75.19, 75.21, 75.26, 75.29, 75.32, 75.33, 75.38, 75.57, 75.58, 75.60, 75.64, 75.75, 75.79, 76.63, 76.64, 76.76, 77.23, 78.13, 78.14, 80.16, 80.40, 82.9, 82.59, 83.36, 83.69, 83.88, 83.137, 83.150, 83.167, 83.168, 83.169, 83.170, 83.183, 83.185, 83.203, 83.205, 83.209, 83.212, 83.214, 84.24, 84.44, 84.52, 84.53, 85.22, 85.41, 85.42, 85.62, 85.71, 85.83, 85.89, 86.5, 86.6, 86.8, 86.15, 86.53, 86.62 **Vol. 3** 60.51, 69.22, 69.31, 71.5, 71.8, 71.17, 71.18, 71.24, 71.31, 71.32, 71.49, 71.56, 71.61, 71.62, 72.9, 72.11, 72.14, 72.21, 72.23, 72.25, 72.26, 72.33, 72.38, 72.41, 72.42, 73.2, 73.3, 73.4, 73.8, 73.9, 73.10, 73.12, 73.13, 73.14, 73.15, 73.35, 73.39, 73.40, 73.41, 73.45, 73.53, 73.54, 73.55, 75.6, 75.7, 75.10, 75.12, 75.13, 75.15, 75.16, 75.17, 75.18, 75.22, 75.27, 75.29, 75.30, 75.31, 75.34, 75.38, 75.43, 75.48, 77.4, 78.22, 78.23, 78.24, 80.1, 80.2, 80.3, 80.4, 80.6, 81.2, 81.3, 81.34, 83.22, 83.67, 83.89, 83.103, 83.118, 83.123, 83.137, 83.138, 84.20, 85.24, 85.25, 85.31, 85.44, 86.12, 86.17, 86.45 **Vol. 4** 63.5, 71.9, 71.18, 71.20, 71.28, 71.29, 71.31, 71.35, 71.37, 71.54, 71.55, 71.61, 72.5, 72.6, 72.17,

72.20, 73.3, 73.10, 73.12, 73.15, 73.16, 73.17, 73.18, 73.19, 73.20, 73.25, 73.27, 73.28, 73.32, 73.33, 75.4, 75.6, 75.7, 75.17, 75.18, 75.24, 76.37, 78.16, 80.3, 80.8, 80.9, 80.10, 80.17, 83.43, 83.72, 83.115, 83.116, 83.137, 83.138, 85.9, 86.2, 86.16, 86.28, 86.36, 86.39, 86.40 **Vol. 5** 61.5, 63.6, 71.9, 71.11, 71.34, 71.36, 71.54, 71.63, 71.64, 71.72, 71.81, 71.82, 71.90, 71.104, 72.5, 72.6, 72.9, 72.10, 72.14, 72.16, 72.18, 72.20, 72.22, 72.23, 72.24, 72.25, 72.26, 72.41, 73.4, 73.5, 73.15, 73.19, 73.31, 73.32, 73.37, 73.40, 73.41, 73.43, 73.46, 75.1, 75.5, 75.6, 75.15, 75.22, 75.25, 75.38, 75.48, 76.18, 76.34, 78.10, 78.12, 78.15, 80.13, 81.2, 82.28, 83.39, 83.40, 83.58, 83.69, 83.70, 83.98, 83.100, 83.107, 83.113, 84.1, 85.7, 85.12, 85.16, 85.25, 85.31, 86.2, 86.18, 86.45, 86.79 **Vol. 6** 71.12, 71.14, 71.16, 71.19, 71.21, 71.25, 71.26, 71.31, 71.32, 71.35, 71.37, 71.38, 71.40, 71.41, 71.44, 71.60, 71.63, 71.66, 71.76, 72.7, 72.14, 72.17, 72.18, 72.20, 72.28, 73.2, 73.5, 73.7, 73.10, 73.16, 73.18, 73.23, 73.24, 73.25, 73.26, 73.29, 73.30, 73.36, 73.38, 73.41, 73.47, 73.48, 73.49, 73.51, 73.54, 73.55, 73.59, 73.63, 75.1, 75.7, 75.8, 75.9, 75.13, 75.14, 75.15, 75.19, 75.29, 76.35, 76.53, 78.9, 78.15, 78.21, 78.22, 79.12, 80.6, 80.7, 80.8, 82.23, 83.9, 83.12, 83.13, 83.17, 83.38, 83.42, 83.43, 83.52, 83.53, 83.79, 84.35, 84.41, 85.9, 85.10, 85.19, 85.37, 85.41, 86.10 **Vol. 7** 71.16, 71.17, 71.19, 71.20, 71.21, 71.22, 71.27, 71.33, 71.34, 71.39, 71.46, 71.47, 71.51, 71.53, 71.58, 71.61, 71.72, 71.80, 71.81, 71.93, 71.98, 71.100, 71.107, 71.113, 71.127, 71.128, 72.2, 72.5, 72.10, 72.11, 72.12, 72.24, 72.26, 72.27, 72.38, 73.2, 73.3, 73.11, 73.16, 73.27, 73.29, 73.31, 73.42, 73.46, 73.56, 73.64, 75.5, 75.6, 75.7, 75.8, 75.13, 75.14, 75.15, 75.17, 75.23, 75.24, 75.26, 75.31, 75.33, 75.35, 75.38, 76.44, 76.46, 76.71, 76.72, 78.12, 78.17, 79.1, 80.5, 80.12, 81.5, 81.44, 81.48, 83.44, 83.45, 83.81, 83.100, 83.112, 84.54, 85.18, 85.26, 85.28, 85.29, 85.42, 85.46, 86.5, 86.31, 86.39, 86.50, 86.61, 86.66, 86.81, 86.107

Gd

Vol. 2 81.13, 81.69 **Vol. 3** 77.30, 81.10 **Vol. 5** 81.21, 84.18 **Vol. 6** 76.31

Hf

Vol. 4 71.62 **Vol. 5** 71.47 **Vol. 7** 73.21, 84.69

Hg

Vol. 2 60.91, 71.1, 71.2, 71.3, 71.4, 71.5, 71.8, 71.11, 71.12, 71.15, 71.19, 71.25, 71.26, 71.39, 71.41, 73.31, 76.19, 79.8, 79.11, 79.12, 79.17, 79.24, 81.53, 83.94, 83.145, 84.6, 84.14, 84.30, 84.37, 84.62, 84.63, 84.64, 84.67, 85.4, 85.12, 85.13, 85.17, 85.21, 85.23, 85.44, 85.48, 85.51, 85.75, 85.90, 86.23, 86.105 **Vol. 3** 63.9, 71.9, 71.15, 82.4, 83.9, 83.60, 83.67, 84.6, 85.18, 85.21, 85.39, 85.52, 86.1, 86.61 **Vol. 4** 71.22, 71.67, 71.79, 71.88, 71.91, 83.112, 84.5, 85.4 **Vol. 5** 71.2, 71.6, 71.7, 71.25, 71.53, 73.1, 73.2, 73.7, 73.33, 79.1, 80.3, 80.9, 80.25, 83.14, 83.20, 83.32, 84.4, 84.12, 84.17, 84.26, 85.9, 86.1, 86.15, 86.18, 86.23 **Vol. 6** 71.2, 71.5, 71.6, 71.18, 79.1, 79.6, 79.7, 80.4, 80.16, 81.5, 81.12, 83.5, 83.61, 84.47, 85.24, 85.30, 86.42 **Vol. 7** 71.1, 71.5, 71.9, 71.10, 71.11, 71.18, 71.36, 71.37, 71.40, 71.85, 71.136, 76.4, 76.9, 80.2, 80.8, 81.16, 81.18, 82.2, 83.1, 83.17, 83.31, 83.66, 83.101, 83.104, 84.6, 84.10, 85.12, 86.24, 86.55

Ho

Vol. 3 77.32 **Vol. 4** 77.11, 77.24 **Vol. 5** 81.52

Ir

Vol. 2 71.9, 71.33, 75.52, 86.70, 86.75, 86.76, 86.77, 86.79, 86.87, 86.98, 86.100, 86.104, 86.119 **Vol. 3** 71.28, 71.66, 77.2, 86.60, 86.63, 86.69, 86.73, 86.74, 86.79, 86.89 **Vol. 4** 71.86, 85.39, 86.24, 86.45, 86.51, 86.58, 86.59, 86.64 **Vol. 5** 71.70, 71.84, 71.85, 71.97, 71.102, 71.103, 72.28, 72.29, 72.35, 72.47, 74.7, 75.36, 83.123, 86.35, 86.57, 86.59, 86.60, 86.70, 86.78 **Vol. 6** 71.54, 71.70, 71.78, 71.82, 71.83, 71.86, 71.96, 72.4, 75.25, 76.5, 83.4, 86.1, 86.7, 86.21, 86.25, 86.31, 86.36, 86.67, 86.68, 86.69 **Vol. 7** 71.71, 71.99, 71.111, 71.120, 71.129, 71.132, 72.17, 72.19, 72.20, 72.21, 72.22, 72.23, 73.45, 86.65, 86.99, 86.103

La

Vol. 2 76.71, 76.72, 77.32, 83.172 **Vol. 3** 81.12 **Vol. 4** 81.56, 84.17 **Vol. 5** 81.22, 81.51, 83.81, 83.82, 84.9 **Vol. 6** 84.44, 84.45 **Vol. 7** 84.19

Lu

Vol. 4 77.19 **Vol. 5** 71.79, 77.20 **Vol. 7** 84.16

Mn

Vol. 1 3.57 **Vol. 2** 69.14, 69.40, 71.10, 72.51, 73.8, 73.36, 73.41, 73.121, 75.34, 75.42, 76.65, 76.77, 77.12, 77.24, 81.9, 81.10, 82.8, 82.56, 83.7, 83.115, 83.153, 84.16, 84.25, 86.25, 86.28, 86.39, 86.41, 86.85, 86.90, 86.91 **Vol. 3** 71.2, 71.14, 75.1, 77.6, 77.31, 81.43, 83.124, 84.10, 86.36 **Vol. 4** 71.50, 71.56, 73.6, 77.18, 83.81, 83.133, 83.134, 86.3, 86.15, 86.41 **Vol. 5** 71.3, 71.4, 71.5, 71.10, 71.26, 71.39, 71.40, 71.55, 71.61, 71.98, 72.13, 75.43, 78.22, 80.19, 81.15, 81.17, 81.18, 81.60, 83.21, 83.106, 84.21, 84.22, 84.32, 85.14, 85.32, 86.8, 86.11, 86.12, 86.14, 86.20 **Vol. 6** 71.11, 71.24, 71.29, 71.36, 71.70, 71.87, 71.94, 72.2, 72.16, 72.23, 73.3, 73.9, 73.27, 73.52, 75.17, 76.32, 76.54, 77.8, 77.13, 78.5, 80.12, 81.8, 83.7, 83.19, 83.58, 83.69, 84.4, 84.17, 84.36, 85.45, 86.6, 86.28 **Vol. 7** 71.31, 71.92, 72.13, 73.19, 73.68, 75.12, 75.16, 76.47, 77.2, 77.5, 78.26, 79.11, 81.14, 83.51, 83.76, 83.89, 83.98, 83.105, 84.34, 84.47, 84.52, 85.17, 86.35, 86.36, 86.37, 86.104

Mo

Vol. 2 71.27, 71.51, 72.66, 73.6, 73.26, 73.27, 73.29, 73.30, 73.32, 73.40, 73.41, 73.70, 73.75, 73.96, 74.17, 75.12, 75.20, 75.43, 75.44, 75.69, 75.76, 76.3, 76.61, 77.53, 80.23, 81.1, 81.25, 81.59, 82.14, 83.38, 84.18, 85.34, 85.78, 86.17, 86.43, 86.56 **Vol. 3** 64.52, 69.23, 71.6, 71.16, 71.27, 72.12, 72.27, 73.18, 73.19, 73.20, 73.26, 73.28, 73.46, 73.51, 73.52, 73.56, 75.25, 75.39, 80.5, 81.1, 82.16, 82.19, 82.23, 83.17, 83.35, 83.86, 86.9, 86.46, 86.71, 86.88 **Vol. 4** 71.10, 71.11, 71.15, 71.53, 71.69, 72.9, 72.12, 73.14, 73.19, 73.22, 82.13, 83.46, 83.87, 84.3, 84.10, 85.12, 85.24, 86.1, 86.19, 86.23, 86.26, 86.47 **Vol. 5** 68.27, 71.18, 71.29, 72.27, 73.6, 73.11, 73.13, 73.14, 73.16, 73.17, 73.18, 75.2, 75.24, 75.31, 76.36, 77.2, 80.10, 81.28, 81.50, 82.7, 82.21, 84.16, 86.5, 86.6, 86.9, 86.33, 86.34, 86.40, 86.53 **Vol. 6** 61.8, 71.8, 71.13, 71.27, 71.69, 72.9, 72.13, 72.19, 73.13, 73.34, 73.39, 73.40, 73.42, 73.53, 73.62, 75.6, 75.16, 75.21, 75.26, 75.31, 77.17, 78.2, 78.4, 80.20, 81.25, 82.6, 83.28, 83.36, 83.84, 85.20, 85.22, 85.31, 86.3, 86.14, 86.49, 86.50, 86.64 **Vol. 7** 71.13, 71.24, 71.49, 71.97, 72.3, 72.7, 72.14, 72.15, 72.16, 72.18, 72.28, 73.5, 73.6, 73.9, 73.13, 73.14, 73.23, 73.26, 73.28, 73.33, 73.47, 73.49, 73.51, 73.53, 73.55, 73.65, 73.67, 74.7, 75.2, 75.3, 75.4, 76.45, 76.58, 80.9, 80.10, 80.18, 80.20, 81.26, 81.27, 82.18, 83.64, 84.4, 84.11, 84.12, 84.32, 84.33, 84.36, 84.78, 85.36, 85.38, 86.1, 86.8, 86.33, 86.40, 86.41, 86.59, 86.63, 86.98, 86.100, 86.101, 86.108, 86.110

Nb

Vol. 2 72.72, 72.84, 73.11, 83.18, 84.65 **Vol. 3** 72.40, 72.48, 72.53, 83.53, 83.57, 83.129 **Vol. 4** 71.52, 73.29, 80.6, 80.7, 81.3, 81.11, 81.19, 83.47, 83.48 **Vol. 5** 71.48, 73.12, 77.23, 81.10, 83.2 **Vol. 6** 72.27, 74.13, 80.9, 85.6, 85.33 **Vol. 7** 71.38, 72.44, 73.7, 73.10, 73.12, 73.32, 73.36, 73.50, 73.69, 75.29, 75.37, 77.9, 80.14, 83.18, 85.15

Nd

Vol. 2 77.33, 81.41, 81.43, 81.71, 81.72, 82.52 **Vol. 3** 81.13, 81.14, 81.27, 82.9 **Vol. 4** 77.12, 81.18, 81.51 **Vol. 5** 71.108, 76.37, 84.10 **Vol. 6** 81.17, 81.23, 81.30, 81.31, 81.43 **Vol. 7** 73.30, 82.16, 84.75

Ni

Vol. 1 3.80, 48.2 **Vol. 2** 60.155, 65.42, 65.46, 65.47, 72.15, 72.22, 72.27, 72.48, 72.64, 72.77, 73.64, 73.74, 73.82, 73.91, 73.112, 73.113, 73.118, 75.28, 75.53, 75.54, 75.62, 75.63, 76.6, 76.20, 76.21, 76.22, 76.27, 76.29, 76.31, 76.32, 76.39, 76.41, 76.42, 76.43, 76.47, 76.50, 76.69, 76.70, 76.74, 77.13, 77.19, 77.45, 77.46, 77.56, 77.62, 78.9, 78.10, 78.11, 78.22, 78.23, 78.30, 78.32, 78.36, 78.42, 78.43, 78.47, 78.49, 78.52, 78.53, 79.18, 79.23, 79.26, 79.28, 79.31, 79.34, 79.35, 79.36, 80.1, 80.3, 80.17, 80.27, 80.28, 81.11, 81.36, 82.6, 82.19, 82.22, 82.27, 82.30, 82.31, 82.33, 82.48, 83.11, 83.12, 83.13, 83.14, 83.24, 83.25, 83.29, 83.30, 83.31, 83.32, 83.40, 83.43, 83.51, 83.57, 83.83, 83.84, 83.86, 83.96, 83.97, 83.99, 83.102, 83.105, 83.106, 83.107, 83.118, 83.120, 83.127, 83.128, 83.138, 83.143, 83.144, 83.165, 83.175, 83.177, 83.184, 83.186, 83.198, 83.199, 83.204, 83.208, 83.215, 83.218, 83.219, 84.7, 84.41, 84.60, 85.1, 85.5, 85.6, 85.7, 85.8, 85.15, 85.16, 85.18, 85.19, 85.24, 85.25, 85.30, 85.32, 85.33, 85.35, 85.38, 85.43, 85.47, 85.49, 85.50, 85.56, 85.57, 85.58,

85.59, 85.60, 85.70, 85.76, 85.81, 85.88, 85.91, 85.93, 85.96, 86.7, 86.9, 86.20, 86.21, 86.26, 86.33, 86.49, 86.51, 86.52, 86.55, 86.65, 86.67, 86.74, 86.97, 86.99 **Vol. 3** 22.2, 60.54, 61.4, 61.8, 71.3, 71.20, 71.43, 72.17, 72.43, 73.43, 75.30, 76.11, 76.12, 76.15, 76.21, 76.43, 76.44, 76.49, 76.50, 76.51, 76.56, 76.60, 76.65, 76.67, 76.83, 76.86, 76.87, 77.33, 78.6, 78.15, 78.28, 79.13, 79.14, 79.26, 80.8, 80.18, 81.8, 81.16, 81.42, 81.48, 83.3, 83.4, 83.21, 83.30, 83.31, 83.45, 83.48, 83.63, 83.64, 83.71, 83.73, 83.80, 83.84, 83.85, 83.87, 83.90, 83.92, 83.94, 83.95, 83.112, 83.113, 83.120, 83.125, 83.131, 83.133, 83.134, 83.135, 83.136, 83.141, 83.142, 83.143, 84.12, 85.2, 85.3, 85.4, 85.7, 85.12, 85.14, 85.20, 85.23, 85.27, 85.28, 85.29, 85.36, 85.38, 85.43, 85.46, 85.47, 85.50, 85.51, 86.4, 86.14, 86.16, 86.31, 86.39, 86.41, 86.43, 86.44, 86.48, 86.52, 86.53, 86.56, 86.76, 86.80, 86.81, 86.82, 86.85, 86.86 **Vol. 4** 60.11, 71.32, 71.48, 71.65, 71.74, 71.75, 72.4, 72.13, 72.25, 73.15, 73.29, 75.27, 75.30, 76.7, 76.15, 76.27, 76.42, 76.44, 77.3, 77.4, 77.17, 78.3, 78.9, 79.5, 79.12, 80.18, 81.22, 81.33, 81.34, 81.37, 82.3, 82.11, 83.6, 83.10, 83.21, 83.30, 83.33, 83.41, 83.42, 83.45, 83.60, 83.73, 83.83, 83.84, 83.99, 83.101, 83.111, 83.113, 83.118, 83.119, 83.125, 83.126, 83.127, 83.131, 83.133, 83.136, 84.7, 85.2, 85.10, 85.15, 85.16, 85.17, 85.20, 85.37, 86.17, 86.27, 86.30, 86.46, 86.66 **Vol. 5** 12.11, 61.2, 71.30, 71.41, 71.45, 71.49, 71.59, 71.62, 71.73, 71.82, 71.92, 72.30, 72.31, 72.33, 72.34, 72.36, 72.48, 72.49, 72.50, 72.53, 75.13, 75.28, 76.8, 76.12, 76.23, 76.44, 76.46, 76.51, 76.53, 77.12, 79.2, 79.3, 79.6, 79.20, 79.21, 79.27, 80.1, 80.4, 80.7, 80.14, 80.17, 80.20, 80.22, 80.23, 81.44, 81.45, 81.48, 82.15, 82.16, 82.19, 83.10, 83.27, 83.28, 83.29, 83.30, 83.31, 83.42, 83.44, 83.45, 83.59, 83.62, 83.67, 83.68, 83.73, 83.85, 83.94, 83.95, 83.102, 83.112, 83.121, 84.13, 84.24, 85.2, 85.6, 85.8, 85.15, 85.24, 85.26, 85.30, 86.19, 86.26, 86.41, 86.42, 86.46, 86.49 **Vol. 6** 71.34, 71.42, 71.64, 71.65, 71.72, 71.79, 72.3, 72.11, 72.30, 73.32, 73.61, 76.24, 76.36, 76.37, 76.38, 76.49, 76.51, 77.1, 77.5, 77.6, 77.7, 77.10, 77.15, 77.21, 79.2, 79.4, 80.15, 80.19, 82.8, 83.22, 83.24, 83.25, 83.44, 83.48, 83.57, 83.72, 83.73, 83.78, 84.11, 84.34, 84.37, 84.38, 85.1, 85.2, 85.3, 85.7, 85.11, 85.12, 85.13, 85.18, 85.26, 85.43, 86.23, 86.44, 86.52 **Vol. 7** 60.4, 60.8, 61.1, 71.44, 71.54, 71.56, 71.84, 71.89, 71.103, 71.105, 71.130, 72.32, 72.33, 72.37, 72.43, 73.24, 73.34, 73.39, 73.49, 73.50, 73.57, 73.59, 74.6, 75.32, 76.7, 76.11, 76.13, 76.15, 76.19, 76.28, 76.34, 76.36, 76.53, 76.54, 76.61, 76.63, 76.69, 76.75, 76.77, 77.8, 78.1, 78.3, 78.4, 78.13, 78.14, 78.15, 79.6, 79.26, 80.3, 80.21, 81.6, 81.38, 82.10, 82.22, 83.12, 83.25, 83.33, 83.48, 83.57, 83.58, 83.60, 83.62, 83.73, 83.82, 83.93, 83.94, 83.96, 83.97, 83.99, 84.3, 84.14, 84.22, 84.25, 84.31, 84.50, 84.66, 84.68, 84.72, 84.77, 85.8, 85.20, 85.27, 85.31, 85.33, 85.39, 85.41, 85.44, 86.6, 86.7, 86.12, 86.19, 86.20, 86.49, 86.60, 86.69, 86.75, 86.82, 86.87, 86.115, 86.117

Np

Vol. 3 80.13 **Vol. 5** 77.9

Os

Vol. 2 72.34, 86.83, 86.88 **Vol. 3** 71.41, 86.61, 86.66 **Vol. 4** 71.71, 71.73, 71.89, 72.21, 81.29, 86.35 **Vol. 5** 71.95, 71.96, 71.99 **Vol. 6** 71.90 **Vol. 7** 71.108, 71.109, 84.49, 86.68

Pd

Vol. 2 60.70, 60.148, 72.4, 72.8, 72.9, 72.10, 72.11, 72.19, 72.21, 72.28, 72.42, 72.47, 72.59, 75.3, 75.4, 75.7, 75.27, 75.36, 75.80, 75.81, 76.14, 77.41, 77.60, 78.31, 78.38, 78.39, 78.44, 78.45, 78.48, 79.19, 79.25, 81.17, 83.42, 83.58, 83.59, 83.114, 83.159, 83.173, 83.189, 84.57, 85.26, 85.28, 85.29, 85.31, 85.66, 85.72, 85.95, 86.2, 86.18, 86.19, 86.34, 86.38, 86.45, 86.62, 86.81 **Vol. 3** 71.26, 71.44, 72.7, 72.10, 72.18, 72.32, 72.34, 72.36, 72.37, 74.3, 74.4, 75.23, 76.14, 76.49, 76.68, 78.1, 79.9, 81.29, 82.1, 82.18, 82.29, 83.147, 85.10, 85.11, 85.55, 86.20, 86.23, 86.54 **Vol. 4** 71.1, 71.41, 71.51, 71.58, 71.80, 72.10, 72.14, 72.28, 76.2, 77.2, 80.4, 81.39, 83.26, 83.66, 83.75, 83.89, 83.105, 83.106, 83.114, 85.33, 85.36, 85.40, 86.5, 86.8 **Vol. 5** 71.12, 71.21, 71.27, 71.67, 71.71, 71.105, 72.21, 72.32, 72.55, 82.1, 82.18, 83.3, 83.33, 83.50, 83.51, 83.83, 83.86, 83.90, 83.117, 83.122, 85.4, 85.33, 86.13, 86.38, 86.47, 86.54, 86.58, 86.66 **Vol. 6** 71.23, 71.33, 71.55, 71.61, 71.62, 71.81, 72.1, 72.29, 75.28, 77.18, 82.19, 82.22, 83.31, 84.43, 85.27, 85.35, 85.36, 85.40, 86.18, 86.26, 86.40, 86.43, 86.56 **Vol. 7** 71.25, 71.43, 71.50, 71.55, 71.62, 71.63, 71.64, 71.73, 71.78, 71.79, 71.90, 71.101, 71.115, 71.122, 71.123, 72.29, 72.35, 72.40, 72.41, 72.42, 72.45, 75.28, 76.2, 83.9, 83.35, 83.36, 83.37, 84.38, 85.16, 85.21, 85.34, 86.21, 86.51, 86.54, 86.56, 86.57, 86.76, 86.77, 86.78, 86.79, 86.93

Pr

Vol. 2 77.65 **Vol. 3** 77.34, 77.35, 81.15 **Vol. 4** 83.77 **Vol. 5** 81.59, 82.8 **Vol. 6** 71.52, 82.20, 84.20 **Vol. 7** 84.83

Pt

Vol. 2 69.43, 71.29, 71.30, 71.31, 71.38, 72.1, 72.3, 72.5, 72.38, 72.83, 73.17, 75.17, 75.18, 75.71, 76.80, 77.4, 77.36, 77.37, 77.58, 78.33, 80.19, 81.18, 81.19, 82.7, 83.26, 83.33, 83.60, 83.67, 83.68, 83.124, 83.153, 83.164, 83.191, 85.52, 85.67, 86.4, 86.10, 86.11, 86.13, 86.16, 86.24, 86.31, 86.35, 86.46, 86.50, 86.71, 86.72, 86.80, 86.89, 86.92, 86.115 **Vol. 3** 71.4, 71.46, 71.48, 72.1, 72.3, 72.4, 72.5, 72.6, 72.19, 72.22, 72.30, 72.49, 72.51, 75.21, 75.35, 76.8, 82.2, 82.3, 82.10, 82.11, 83.11, 83.12, 83.33, 83.50, 83.52, 85.49, 85.53, 86.2, 86.10, 86.13, 86.28, 86.29, 86.64, 86.70, 86.72, 86.94, 86.95, 86.98 **Vol. 4** 71.4, 71.13, 71.16, 71.17, 71.19, 71.21, 71.23, 71.25, 71.26, 71.33, 71.49, 71.68, 71.72, 71.78, 71.81, 71.84, 71.90, 72.1, 72.2, 72.3, 72.8, 72.11, 72.22, 72.23, 72.24, 72.27, 72.30, 75.23, 83.29, 83.63, 85.27, 85.34, 86.31 **Vol. 5** 12.10, 69.28, 71.8, 71.14, 71.16, 71.23, 71.86, 71.87, 71.88, 71.89, 71.100, 71.101, 72.1, 72.2, 72.3, 72.4, 72.7, 72.11, 72.15, 72.42, 72.44, 74.10, 75.4, 76.1, 77.1, 79.4, 81.39, 81.49, 83.49, 85.5, 85.34, 86.15, 86.17, 86.29, 86.48, 86.79 **Vol. 6** 71.43, 71.45, 71.46, 71.50, 71.51, 71.53, 71.56, 71.57, 71.67, 71.80, 71.84, 71.88, 71.91, 72.5, 72.15, 75.2, 75.3, 75.10, 75.30, 76.1, 83.18, 86.5, 86.11, 86.32, 86.46, 86.55, 86.66 **Vol. 7** 71.30, 71.35, 71.45, 71.59, 71.60, 71.66, 71.67, 71.68, 71.75, 71.76, 71.77, 71.86, 71.96, 71.110, 71.117, 71.118, 71.121, 71.125, 71.137, 71.138, 72.1, 72.36, 72.39, 75.22, 75.39, 75.40, 75.41, 75.42, 75.43, 75.44, 76.3, 79.8, 79.24, 80.17, 83.3, 83.4, 83.16, 83.38, 83.40, 85.14, 85.35, 86.2, 86.4, 86.10, 86.11, 86.26, 86.27, 86.28, 86.30, 86.42, 86.50, 86.70, 86.73, 86.80, 86.89, 86.94, 86.107, 86.109

Re

Vol. 2 62.41, 73.33, 73.45, 75.40, 76.25, 81.88, 81.89, 83.1, 83.195, 83.200, 85.54, 85.97, 86.12, 86.27, 86.58, 86.59, 86.66, 86.118 **Vol. 3** 71.10, 81.7, 81.31, 81.37, 81.44, 81.53, 86.47, 86.49, 86.55, 86.75 **Vol. 4** 73.9, 73.13, 80.5, 80.13, 80.14, 83.110, 85.11, 85.21, 85.24 **Vol. 5** 76.3, 76.7, 77.18, 86.76 **Vol. 6** 71.7, 77.2, 80.11, 83.6, 84.32, 86.13, 86.15 **Vol. 7** 71.133, 73.4, 76.10, 77.13, 81.2, 81.25, 85.24, 86.3, 86.46, 86.47, 86.52, 86.63, 86.97, 86.110

Rh

Vol. 2 71.7, 71.55, 72.20, 72.23, 72.24, 72.40, 72.43, 72.45, 73.7, 73.42, 73.79, 73.80, 73.81, 73.92, 74.19, 75.37, 75.51, 77.2, 77.3, 81.26, 81.56, 81.57, 81.58, 83.178, 85.65, 86.78, 86.84, 86.108, 86.114, 86.120 **Vol. 3** 71.34, 71.54, 71.63, 71.68, 72.13, 72.16, 72.35, 72.50, 74.8, 74.9, 74.10, 74.16, 75.47, 77.3, 83.28, 83.128, 83.132, 86.19, 86.21, 86.25, 86.40, 86.68, 86.87, 86.90, 86.91, 86.97 **Vol. 4** 71.6, 71.83, 71.87, 71.92, 71.93, 72.19, 73.16, 73.18, 73.21, 74.5, 81.20, 81.28, 81.61, 83.128, 86.14, 86.21, 86.54, 86.62, 86.65 **Vol. 5** 71.35, 71.68, 71.69, 72.8, 72.12, 72.40, 72.51, 73.39, 73.46, 75.46, 76.35, 77.4, 86.37, 86.39, 86.69, 86.71, 86.82 **Vol. 6** 71.20, 71.39, 71.49, 71.71, 71.75, 71.89, 71.92, 72.6, 72.10, 72.21, 72.22, 72.26, 74.12, 75.24, 76.29, 86.30, 86.38, 86.39 **Vol. 7** 71.4, 71.26, 71.41, 71.48, 71.112, 71.124, 71.139, 72.6, 72.16, 72.31, 72.34, 73.20, 73.64, 75.10, 75.11, 76.76, 81.40, 81.47, 83.21, 84.7, 84.26, 84.51, 85.5, 85.11, 86.25, 86.58, 86.62, 86.83, 86.91, 86.116

Sc

Vol. 2 81.14 **Vol. 5** 71.33, 77.5, 81.19 **Vol. 6** 73.46, 81.4, 81.16, 83.55, 84.13, 84.22 **Vol. 7** 84.70, 84.71

Ru

Vol. 2 71.48, 72.37, 73.28, 73.54, 73.57, 73.78, 75.30, 75.66, 80.32, 80.33, 81.83, 86.63, 86.101, 86.102, 86.106, 86.109, 86.110, 86.117 **Vol. 3** 71.42, 75.5, 75.32, 75.33, 76.16, 76.45, 81.54, 86.18, 86.37 **Vol. 4** 71.27, 71.45, 71.70, 72.16, 73.1, 76.9, 85.1, 85.38, 86.29, 86.60 **Vol. 5** 71.51, 71.76, 71.78, 71.94, 72.17, 72.45, 72.46, 73.7, 73.33, 74.8, 74.9, 74.11, 74.12, 75.9, 75.29, 75.30, 81.63, 85.36, 86.51, 86.55, 86.62, 86.65, 86.75, 86.80 **Vol. 6** 71.4, 71.28, 71.74, 72.12, 73.8, 73.43, 73.57, 74.7, 74.15, 75.11, 75.12, 75.23, 77.14, 81.45, 81.46, 86.19, 86.29, 86.37, 86.41, 86.45, 86.48, 86.54, 86.59, 86.72 **Vol. 7** 71.15, 71.69, 71.70, 71.82, 71.104, 71.131, 72.25, 73.41, 75.1, 75.18, 75.19, 75.20, 75.25, 80.13, 80.19, 81.45, 85.6, 85.45, 86.29, 86.43, 86.64, 86.67, 86.72

Sm

Vol. 2 73.65 **Vol. 4** 76.33, 81.58 **Vol. 5** 75.39 **Vol. 7** 76.49

Ta

Vol. 2 72.73 **Vol. 6** 71.17, 86.2 **Vol. 7** 71.2, 83.75, 83.108, 86.32

Tb

Vol. 2 76.62, 83.174

Tc

Vol. 6 86.47 **Vol. 7** 86.74, 86.88

Th

Vol. 2 77.49, 81.28 **Vol. 3** 80.15, 83.145, 84.25 **Vol. 4** 75.14 **Vol. 5** 77.8 **Vol. 6** 84.27, 84.40 **Vol. 7** 71.94, 81.24

Ti

Vol. 2 72.44, 73.18, 73.111, 75.72, 77.48, 78.61, 79.29, 81.46, 83.49, 83.196, 84.11, 84.26, 84.27, 84.28, 84.54, 84.58, 84.61 **Vol. 3** 71.21, 73.28, 73.36, 73.42, 75.14, 75.19, 79.19, 81.38 **Vol. 4** 71.38, 71.47, 71.63, 71.64, 73.8, 73.11, 73.23, 77.1, 78.2, 79.8, 80.15, 83.3, 83.9, 83.19, 83.76, 84.15, 84.18 **Vol. 5** 73.8, 73.10, 73.27, 77.10, 77.11, 77.17, 79.16, 81.23, 81.26, 81.37, 83.54, 83.118, 83.119 **Vol. 6** 73.11, 73.12, 73.21, 73.28, 73.31, 73.44, 73.45, 73.50, 73.56, 73.60, 81.37 **Vol. 7** 71.8, 71.52, 71.106, 71.119, 73.15, 73.35, 73.44, 73.60, 75.30, 75.36, 81.23, 81.36, 83.78

U

Vol. 2 73.62, 77.10, 79.30, 80.31, 81.42, 81.60, 83.193, 83.194, 84.35, 84.45 **Vol. 3** 77.20, 78.7, 80.19, 80.20, 81.55 **Vol. 4** 74.1, 75.15, 75.28, 78.11, 85.26, 85.32 **Vol. 5** 71.56, 73.45, 75.34, 78.7, 79.11, 80.12, 81.3, 81.30, 84.5, 84.27 **Vol. 6** 76.40, 81.6, 81.9, 81.14, 81.19, 81.20, 83.80, 84.18, 84.28 **Vol. 7** 73.38, 78.10, 81.22, 81.34, 82.11, 84.37, 84.41, 84.43, 84.46, 84.48, 84.80

V

Vol. 1 3.40 **Vol. 2** 73.12, 75.13, 75.24, 77.6, 77.7, 77.26, 77.27, 77.43, 81.39, 81.49, 81.54, 81.55, 83.47, 83.202, 84.22, 85.98 **Vol. 3** 60.34, 73.24, 73.29, 75.46, 76.76, 78.5, 79.23, 81.52, 83.1, 83.18, 85.22 **Vol. 4** 76.34, 76.36, 77.10, 77.15, 81.9, 81.10, 83.104, 85.8 **Vol. 5** 79.17, 79.18, 81.9, 81.24, 83.80 **Vol. 6** 73.58, 80.10, 83.1, 83.65, 84.23, 85.42, 85.44 **Vol. 7** 81.1, 81.10

W

Vol. 2 73.53, 73.109, 83.123 **Vol. 3** 83.5, 83.66, 85.6, 85.19, 86.8 **Vol. 4** 72.32, 73.24, 73.31, 83.8, 83.109, 83.132, 84.2, 86.55 **Vol. 5** 71.20, 73.9 **Vol. 6** 38.16, 71.3, 72.24, 73.35, 73.37, 80.17, 83.67, 84.3, 84.39, 85.5, 85.16, 85.21, 85.23, 86.4, 86.63 **Vol. 7** 71.14, 71.74, 71.91, 72.4, 72.8, 72.9, 73.52, 73.53, 73.54, 73.66, 80.15, 83.50, 84.5, 84.13, 85.7, 85.10, 86.9, 86.13, 86.14, 86.15, 86.17, 86.53

Y

Vol. 2 77.34, 77.39, 77.61, 81.20 **Vol. 4** 81.4, 81.5, 81.6, 81.7, 81.8 **Vol. 6** 84.46

Yb

Vol. 2 77.28, 77.29, 77.30 **Vol. 3** 77.11, 77.22, 77.23, 81.28, 81.50 **Vol. 4** 81.52, 81.53 **Vol. 5** 81.20 **Vol. 7** 84.17

Zn

Vol. 1 11.45 **Vol. 2** 71.17, 71.50, 71.54, 76.78, 77.8, 77.9, 77.57, 77.63, 78.12, 78.24, 78.25, 79.10, 79.14, 79.15, 79.27, 80.4, 80.20, 80.21, 80.22, 80.24, 81.29, 81.30, 81.35, 81.68, 81.74, 82.11, 82.32, 82.49, 82.50, 83.9, 83.10, 83.15, 83.19, 83.41, 83.78, 83.92, 83.135, 83.139, 83.141, 83.160, 83.162, 83.166, 83.190, 84.9, 84.17, 84.39, 84.50, 84.51, 85.3, 85.61 **Vol. 3** 71.45, 76.25, 76.47, 76.58, 76.72, 76.73, 76.74, 77.24, 78.20, 80.9, 81.26, 82.6, 82.8, 83.2, 83.8, 83.14, 83.41, 83.44, 83.68, 83.116, 83.126, 83.151, 85.32, 85.33, 85.34, 85.42, 86.88 **Vol. 4** 71.18, 76.22, 76.26, 79.11, 80.19, 81.2, 81.16, 81.35, 81.38, 82.12,

83.17, 83.18, 83.38, 83.49, 83.51, 83.55, 83.62, 84.4, 84.12, 85.19, 85.30, 86.10 **Vol. 5** 71.1, 76.10, 76.11, 76.14, 76.15, 76.19, 76.32, 76.47, 77.7, 79.7, 80.29, 83.4, 83.38, 83.57, 83.99, 83.124, 85.20, 85.21, 85.23, 85.27, 85.29 **Vol. 6** 71.73, 73.44, 76.34, 76.38, 79.13, 80.18, 81.22, 81.40, 82.2, 83.46, 83.53, 83.70, 83.81, 84.9, 84.29, 85.39 **Vol. 7** 71.42, 72.37, 73.26, 73.47, 76.12, 76.39, 76.50, 76.55, 76.67, 76.73, 79.10, 81.7, 81.15, 81.19, 82.23, 83.41, 83.55, 83.70, 83.86, 83.88, 83.107, 83.113, 84.18, 84.28, 84.58, 84.64, 85.19, 85.25

Zr

Vol. 2 77.50, 81.52, 81.70 **Vol. 3** 71.57, 73.21, 73.22, 73.34, 73.38, 78.25, 84.16 **Vol. 4** 73.2, 73.5, 73.30, 77.8 **Vol. 5** 72.19, 75.12, 75.32 **Vol. 6** 85.34 **Vol. 7** 71.65, 73.8, 73.22, 73.37, 75.34, 76.48, 77.3, 77.6, 84.79

AUTHOR INDEX

Aalsvoort, J.G.M.van der **7** 80.18
Aarflot, K. **6** 3.4
Aarskog, J. **3** 8.12, 8.13
Aasen, A. **4** 33.21
Aava, U. **3** 67.20, 80.22
Abdel-Rehim, H.A. **5** 33.21, 33.24
Abdul'minev, I.K. **5** 81.59 **6** 81.39, 81.43 **7** 81.31, 81.32
Abdullayev, G.K. **6** 84.5
Abe, E. **7** 31.15
Abe, J. **1** 50.31 **3** 50.20, 59.1
Abel, E.W. **2** 85.25 **5** 75.43 **6** 72.2 **7** 83.51
Abel, H. **1** 54.24
Abel, J.E. **4** 15.3 **5** 34.3
Abidi, S. **4** 35.12
Ablov, A.V. **2** 66.11, 81.32, 81.76, 81.82, 81.87, 83.91, 83.135 **6** 76.2 **7** 76.6, 76.32, 76.66, 78.20, 79.15, 79.17, 79.18, 79.19, 79.20, 83.20, 83.21, 83.32, 83.74, 83.84
Ablov, A.W. **2** 60.87 **5** 79.13
Abola, E. **6** 51.2 **7** 16.11
Abola, E.E. **6** 31.23
Abola, J. **4** 47.8, 47.12, 47.15 **5** 35.14, 47.26 **7** 47.12
Abola, J.E. **6** 50.4 **7** 58.4
Abowitz, G. **1** 32.5
Abraham, D.J. **1** 32.22 **3** 58.23, 59.26 **4** 37.4, 44.9, 44.18, 44.25, 50.2 **6** 50.4 **7** 38.17, 47.5
Abraham, N.A. **3** 51.7
Abrahams, S.C. **1** 11.29, 11.50, 11.55, 11.65, 11.67, 13.54, 24.13 **2** 70.25, 73.30
Abrahamsson, S. **1** 1.131, 1.146, 1.147, 5.32, 16.34, 41.28, 46.13, 50.13, 53.32, 59.62, 59.69 **3** 5.4 **4** 39.1, 39.16 **5** 1.36, 42.1 **6** 1.22, 31.33 **7** 11.14, 32.25, 51.43
Acheson, R.M. **4** 35.34
Acland, C.B. **4** 82.1, 82.2, 82.8, 83.15
Acton, N. **2** 75.35 **3** 75.34 **5** 31.42
Adam, G. **1** 58.105 **4** 51.48, 51.56 **5** 58.40 **6** 51.37 **7** 36.14, 59.8
Adam, V.C. **3** 72.3, 86.88
Adamiak, D.A. **7** 48.46
Adams, J.M. **6** 2.3 **7** 8.7
Adams, M.J. **3** 82.4 **6** 16.15
Adams, R.D. **5** 71.30 **6** 71.41, 73.30, 73.33, 73.34, 73.35, 73.40 **7** 73.31
Adams, W.W. **6** 28.24, 63.11, 71.59
Adamson, G.W. **2** 67.5, 71.50, 71.54 **3** 63.7, 64.31 **4** 71.4 **7** 71.4
Adamson, M.G. **2** 65.39
Adcock, S. **3** 27.4
Addamiano, A. **1** 9.28
Adeoye, S.A. **1** 59.66
Adley, A.D. **4** 63.4, 69.16, 69.17
Adman, E. **1** 16.30, 20.9 **2** 60.119 **3** 20.6, 44.31 **4** 44.16 **5** 41.11 **7** 13.14
Admiraal, L.J. **1** 58.11 **2** 83.130 **3** 58.5
Adolf, W. **3** 54.8, 54.15 **4** 54.12
Afonia, N.N. **6** 64.29
Agami, C. **6** 71.33
Agawa, T. **3** 41.23 **4** 35.39 **5** 42.13
Agayev, A.M. **6** 84.5
Agranat, I. **6** 20.12
Agre, V.M. **2** 80.11, 80.36 **4** 69.28 **5** 80.14, 80.28, 80.29
Agrell, I. **3** 83.39, 83.40, 83.41
Ahlgren, G. **5** 50.9
Ahlsen, E.L. **6** 34.1
Ahmed, F.R. **1** 1.14, 16.61, 24.14, 29.14, 33.76, 33.77, 33.78, 35.46, 36.22, 38.57, 58.19, 58.29, 58.41, 58.53, 58.60 **2** 64.32, 64.64, 64.81 **3** 26.4, 64.56, 64.57 **4** 33.36, 33.37, 51.45, 64.8 **5** 33.48, 33.49 **6** 36.8, 36.17 **7** 31.26, 36.17, 58.27, 64.13
Ahmed, N.A. **4** 33.32 **5** 19.9
Ahmed, N.A.K. **1** 2.29
Ahn, C.-T. **5** 19.2
Ahond, A. **7** 58.32
Ahrens, H. **6** 38.15
Aihara, A. **3** 21.9 **6** 15.1, 17.4
Aime, S. **3** 25.4
Ainalem, I.-B. **2** 83.71, 83.72
Aizawa, H. **6** 59.19
Akatova, K.N. **7** 79.21, 85.22
Aken-Schutz, E.van **6** 40.22
Aken, E.van **5** 72.11
Akhmed, N.A. **4** 17.8, 69.34 **7** 24.1
Akhtar, F. **2** 83.104 **4** 83.82
Akhtar, M.N. **3** 78.7 **5** 78.7 **7** 81.24
Akimoto, H. **1** 24.33
Akimoto, T. **5** 48.17 **7** 38.48
Akiyama, S. **6** 62.10
Akiyama, T. **3** 56.3 **6** 29.1, 59.29 **7** 58.19
Akopyan, Z.A. **1** 15.1, 24.12 **3** 40.2
Al-Karaghouli, A.R. **2** 83.172 **4** 84.17 **5** 83.81 **6** 64.39
Al-Shamma, A.A. **4** 54.10
Albano, V. **1** 40.25, 40.26 **2** 60.149, 86.72, 86.74, 86.100, 86.118, 86.119

Albano, V.G. **1** 3.41 **2** 71.33 **3** 3.21, 3.54, 86.94, 86.99 **4** 86.57, 86.58 **5** 64.20, 71.35, 86.24, 86.48, 86.56, 86.76, 86.79, 86.83 **6** 3.11 **7** 86.31, 86.66, 86.107
Albert, R. **3** 39.1
Albertsson, J. **2** 81.41, 81.69, 81.71, 81.72, 82.52 **3** 81.27, 81.28, 81.50 **4** 81.51, 81.52, 81.53 **6** 2.16, 2.17, 2.18, 2.19, 2.20, 82.20 **7** 81.28
Albinati, A. **4** 35.21, 76.26 **5** 71.71 **6** 31.13, 40.19 **7** 72.37
Albrand, K.-R. **5** 8.1
Albrecht, G. **2** 72.12, 83.202
Alcala, R. **4** 13.5 **7** 81.13
Alcock, N.W. **1** 39.65 **2** 69.41, 69.47, 73.45, 83.208, 86.102 **4** 38.23, 69.30 **5** 35.12, 83.62 **6** 63.1, 81.6, 81.19, 81.20 **7** 19.7, 62.10, 63.21, 69.30, 69.31, 72.6, 83.98, 86.58, 86.62, 86.70
Alden, R.A. **1** 31.39, 52.24, 52.30, 59.65
Alderman, P.R.H. **2** 72.5, 80.5
Aldoshin, S.M. **7** 84.68, 85.44
Aleby, S. **1** 1.150, 1.152, 1.153 **4** 39.45 **7** 39.8
Aleksandrov, G.G. **2** 72.73, 72.84 **3** 73.29, 74.8, 74.9, 74.10, 74.16, 75.46 **4** 69.34, 71.38, 73.21, 74.5 **5** 64.32, 74.7, 82.24 **6** 71.7, 81.45 **7** 18.7, 74.6, 81.45, 82.21
Alexander, E. **1** 31.31
Alexander, L.E. **1** 1.97, 7.25, 34.9, 43.14 **2** 63.11, 63.13, 78.12 **3** 34.2 **7** 40.6
Alexander, R. **2** 75.2
Alexandrou, N.E. **7** 7.4
Alfonso, T.B.d' **6** 26.2
Ali, S.M. **1** 2.24
Aliev, Z.G. **7** 81.40
Alison, J.M.C. **4** 85.36
Allard, B. **5** 77.9
Allard, M. **6** 48.66
Allcock, H.R. **4** 64.40 **5** 46.11 **6** 61.9, 64.22 **7** 61.2
Alleaume, M. **1** 13.26, 16.14, 16.18, 16.19, 16.22 **3** 50.26 **4** 13.3, 16.3, 16.10, 50.14 **5** 16.3 **6** 16.2, 16.6, 33.29, 83.31 **7** 50.31, 50.35
Allegra, G. **1** 23.9, 29.17, 29.18, 29.19, 38.27, 38.28, 38.41 **2** 61.17, 61.18, 61.19, 68.1, 68.20, 71.45, 72.20, 72.39, 72.43, 72.45, 72.67, 73.18, 74.5, 74.24, 74.25, 75.13 **3** 1.31, 21.6, 21.8, 21.10, 39.1, 62.16, 74.3, 74.4 **4** 71.63, 72.16 **5** 1.32, 8.18, 13.16 **6** 71.89, 75.15
Allemand, A. **7** 60.30, 60.31, 60.32, 60.33
Allen, C.W. **3** 64.25
Allen, D.W. **2** 86.99 **3** 65.5, 86.80
Allen, F.H. **1** 53.13 **3** 10.5, 27.2, 27.3, 35.15, 51.49, 52.1, 53.15, 56.10, 73.41, 86.40 **4** 36.27, 69.2 **5** 1.18, 20.10, 56.9 **6** 38.39, 53.6 **7** 56.4, 64.32, 64.33
Allen, M. **5** 76.41
Allentoff, N. **1** 9.7
Alleyne, A.B. **1** 38.17
Allinger, K. **7** 38.2
Allinger, N.L. **5** 36.30 **6** 19.8, 31.16, 36.21 **7** 31.8
Allison, A.J. **3** 54.2
Allison, D.A. **5** 86.2
Allison, G.B. **3** 35.18 **4** 81.49 **5** 73.16, 73.17, 73.18, 82.21
Allmann, R. **1** 17.51, 33.82, 41.44 **2** 60.105, 60.106, 64.54, 64.59 **4** 29.3, 75.4 **5** 34.2, 71.53 **6** 34.3, 35.21, 36.6, 36.13, 41.19, 71.18, 81.12 **7** 16.4, 31.6, 36.8, 73.9, 84.64
Almeida Santos, R.H.F.de **6** 84.42
Almlof, J. **4** 60.5 **6** 48.1
Almodovar, I. **2** 81.3, 81.4, 81.8, 81.10
Alsop, P.A. **3** 85.46 **6** 16.15
Altenburg, H. **4** 36.36
Altenkirk, B. **3** 58.13, 58.52
Altona, C. **1** 38.8, 38.9, 38.10, 38.11, 38.12, 38.13, 42.3, 42.5 **4** 27.2, 31.15 **6** 21.10, 37.5
Alver, E. **2** 64.31 **3** 64.8
Alvino, G. **1** 1.133
Alyea, E.C. **7** 83.99
Amakasu, O. **1** 58.7, 58.12
Amano, Y. **2** 61.21, 83.187
Amanov, A.Z. **2** 80.19
Amati, L. **7** 5.19
Ambady, G. **5** 47.6, 47.13, 58.29 **6** 32.14, 47.12 **7** 47.7, 47.8, 47.18, 48.57, 48.58
Ambady, G.K. **1** 2.60, 2.66, 2.68
Ambats, I. **1** 40.32
Amendola, A. **1** 39.18
Ament, S.S. **6** 32.16
Amiet, R.G. **7** 71.33, 71.34
Amin, A. **4** 17.8
Amirthalingam, V. **1** 2.35 **2** 67.8 **4** 53.11 **5** 41.1, 41.4, 41.5
Amisimov, K.N. **2** 72.72
Amit, A. **1** 31.6
Amit, A.G. **1** 9.27
Amma, E.L. **2** 68.3, 68.10, 74.2, 74.3, 79.3, 79.19, 79.28, 83.16, 85.19, 85.56, 86.4, 86.10, 86.11, 86.51 **3** 69.12, 74.1, 74.6, 74.7, 74.11, 74.12, 74.13, 74.14, 75.2, 75.8, 79.5, 79.6, 79.7, 79.8, 79.9, 79.20, 79.22, 79.24 **4** 41.2, 69.11, 69.22, 69.27, 74.3, 74.8, 74.9, 74.11, 75.19, 79.9, 85.10, 85.27, 86.24, 86.48 **5** 8.7, 79.14, 79.15, 79.22, 79.24, 79.25, 85.8, 86.69, 86.70, 86.71, 86.74 **6** 24.2, 79.5, 81.24, 85.14 **7** 44.7, 44.20, 69.4, 69.5, 74.1, 74.2, 79.8, 79.13, 79.23
Ammon, H.L. **1** 25.22, 27.14, 35.44, 37.20, 39.90 **3** 11.18, 15.1, 15.2, 28.9, 36.27, 39.6, 39.28, 39.32 **4** 8.10, 11.9, 33.31, 37.9, 39.6, 39.43 **5** 28.2, 37.5 **6** 16.12, 20.7, 20.10, 20.12, 20.15, 20.17, 24.7, 24.8, 27.20, 28.12, 33.35, 33.42, 35.6, 35.7, 35.22, 35.23, 40.14, 64.43 **7** 20.10, 20.21, 33.48, 33.49, 35.33
Amzel, L.M. **3** 76.14, 79.26 **5** 36.3, 38.6 **6** 16.14, 32.24 **7** 37.5
Anacker-Eickhoff, H. **7** 1.7, 11.13, 80.16
Anagnostopoulos, A. **2** 83.103
Anan'eva, N.N. **7** 76.47
Ananthakrishnan, N. **1** 48.66

Anastasia, M. 7 51.50
Anastassiou, A.G. 6 34.2
Anca, R. 1 13.44
Anchel, M. 7 50.9, 50.13
Andersen, A.M. 5 59.4 7 17.12, 17.15
Andersen, E.K. 1 6.5, 6.6, 18.8, 18.9, 18.10, 39.13 7 15.2, 17.1, 18.2, 18.3, 18.6, 38.6, 38.7
Andersen, I.G.K. 7 15.2, 17.1, 18.2, 18.3, 18.6, 38.6, 38.7
Andersen, K.K. 6 77.20
Andersen, N.H. 5 27.17
Andersen, P. 1 7.14, 7.15, 12.13 2 60.61, 64.16
Anderson Junior, A.G. 6 27.20
Anderson, B. 3 50.31
Anderson, B.F. 3 54.4 5 35.25 7 49.2, 50.23, 58.12
Anderson, D.L. 7 3.13
Anderson, D.N. 3 3.2 4 3.4 7 3.16
Anderson, G.D. 6 59.12 7 51.49, 53.16
Anderson, J.S. 3 83.11
Anderson, O.P. 3 40.9 4 70.13, 70.14 5 83.110, 83.115 7 83.103
Anderson, S.J. 7 73.15
Anderson, T.J. 5 77.5 6 84.22 7 84.71
Anderson, W.A. 4 86.34
Andersson, J.-E. 7 39.22
Andini, S. 7 48.29
Ando, K. 5 59.17 7 50.20
Ando, O. 1 48.52
Ando, T. 1 17.40, 17.42
Andre, D. 4 19.3, 19.4 5 5.13, 19.7, 21.4, 21.5 6 58.10
Andre, J.M. 4 38.15
Andreetti, G.D. 1 16.31, 35.8, 39.41, 41.21 2 79.14, 79.15, 79.27 3 8.5, 61.8, 76.47, 81.45, 81.46, 81.47 4 9.5, 9.6, 26.5, 35.2, 39.11, 76.12, 83.7, 83.12, 83.33, 83.125, 83.129 5 9.17, 27.11, 27.12, 32.19, 32.25, 38.26, 39.2, 39.6, 39.26, 39.33, 40.11, 42.3 6 11.5, 11.6, 17.24, 31.17, 38.37, 39.1, 39.22, 39.25, 41.9, 41.11, 41.16, 41.24, 59.3, 72.3, 78.7, 78.8, 78.16, 78.18, 78.19 7 32.21, 35.35, 35.36, 38.75, 40.9, 40.10, 41.22, 71.81
Andrelczyk, B. 7 84.27
Andresen, O. 1 9.14
Andrew, J.E. 3 78.28, 83.84 5 77.13 7 83.105
Andrews, B.C. 6 66.6
Andrews, L.C. 6 48.57
Andrianov, K.A. 7 63.14, 63.18
Andrianov, V.G. 2 62.12, 62.14, 71.34, 73.119, 73.120 3 71.56 4 16.1, 72.5, 72.6 5 71.104, 72.26 6 58.20 7 58.16, 64.30, 71.128, 72.26, 72.27
Andrianov, V.I. 1 32.17 3 27.12, 33.28, 64.30 5 45.47
Andrutskaya, L.G. 7 33.15
Aneja, R. 3 58.39
Anet, F.A.L. 7 31.22
Anex, B.G. 6 33.1
Angeli, F.d' 1 33.75
Angelici, R.J. 5 71.21 6 71.81, 82.11 7 71.137
Angelino, N. 6 38.15
Angelis, N.J.de 7 51.14
Angenot, L. 7 58.11, 58.22
Angiuli, R. 5 50.11
Angoletta, M. 6 86.36
Angyal, S.J. 5 21.17
Anisimov, K.N. 2 73.68 3 71.56 7 72.44
Ansell, G.B. 2 60.60, 64.21, 83.27, 85.33 3 21.1, 32.27, 83.16 4 32.6, 32.7, 38.17, 64.12, 83.114 5 83.90 6 32.1 7 33.4
Antolini, L. 5 79.6, 80.16 6 79.4 7 8.17, 85.27
Antona, D. 5 17.11 6 17.7, 17.8
Antonova, A.B. 7 72.44
Antosz, F.J. 4 50.18
Antsyshkina, A.S. 1 8.6 2 76.6, 76.27, 81.5, 81.14, 81.26, 81.28, 83.138, 83.175, 83.186 3 76.15, 76.21 4 76.7, 76.27 5 76.8, 84.13 6 21.1, 21.2, 27.3 7 61.1, 76.15
Antti, B.-M. 4 83.93 5 84.6 6 84.4 7 84.2, 84.15, 84.18, 84.24, 84.25
Antti, C.-J. 4 83.16 5 83.6
Anzai, K. 7 50.6
Anzenhofer, K. 2 62.18, 81.79 3 31.31, 72.50, 81.34 4 77.4
Aoe, K. 7 50.3
Aoki, A. 7 59.24
Aoki, K. 3 48.30 6 71.63 7 71.114
Aoki, T. 3 72.20 5 41.46, 71.67 6 76.10
Aono, T. 5 31.22, 60.35
Aota, K. 4 53.1
Apgar, P.A. 4 31.2
Apinitis, S.K. 6 21.1, 21.2, 27.3
Applegate, H.E. 5 59.23 7 59.13
Apreda, M.C. 6 13.13
Arai, G. 1 58.47
Arai, M. 7 50.1
Arakelova, E.R. 7 1.38, 3.34, 3.35
Araneo, A. 7 86.31, 86.66
Arcamone, F. 5 50.11
Archer, E.M. 1 19.13, 19.18
Archer, R.D. 4 83.132
Arcovito, G. 3 9.10
Arcus, C.S. 7 83.83
Ardrey, R.E. 6 38.23
Arend, H. 7 3.2, 3.3
Argay, G. 4 9.8 5 8.16, 32.12, 32.16, 40.30, 41.42 6 8.7, 32.25, 35.11, 39.27 7 48.37
Argos, P. 3 33.47 5 37.8
Arigone, D. 1 54.7
Arita, M. 4 54.3
Ariyaratne, J.K.P. 3 71.5, 71.6
Armendi, A.J.de 7 36.4
Armitage, D.A. 6 4.3
Armstrong, R.A. 2 81.27, 81.45, 81.48, 81.66, 81.81
Armstrong, V.S. 2 75.22
Arnott, S. 1 56.2
Arora, S.K. 1 13.8 3 47.12, 50.16 4 11.3, 41.14, 50.1 6 13.8, 13.9 7 36.26, 41.18, 58.18, 59.16
Arran, N. 5 3.2
Arriau, J. 7 33.6, 33.12

Arrington, D.E. **3** 11.8
Arrington, W.S. **2** 84.45
Arskog, J. **7** 8.19, 8.20
Artaud, M. **3** 43.7
Arthur, P. **1** 7.30
Arutunian, A.V. **7** 75.5
Arutunian, E.G. **1** 8.6 **2** 81.28 **7** 1.35, 1.38, 3.34, 3.35
Arutyunian, L.D. **7** 33.54
Arutyunyan, A.V. **4** 86.36
Arzi, E. **5** 5.8 **7** 5.2, 5.3
Asahara, T. **3** 1.23
Asahi, K. **1** 50.4
Asahi, K.-I. **7** 50.6
Asahi, Y. **3** 44.42
Asai, M. **1** 48.73 **3** 50.13 **4** 50.6
Asao, T. **5** 38.19 **6** 38.8
Ase, K. **3** 70.4 **4** 70.6, 70.7, 70.8, 70.11, 70.12, 70.15 **6** 3.4, 70.9
Ashbaugh, A.L. **6** 42.4
Asher, J.D.M. **1** 53.6, 53.25, 58.48, 59.70
Ashida, T. **1** 3.13, 3.42, 11.37, 13.3, 20.30, 22.9, 27.13, 31.27, 33.63, 33.65, 33.69, 38.42, 48.18, 48.52, 48.56, 48.63, 48.84, 48.86, 48.88, 48.95, 56.12, 58.42 **2** 63.21, 63.22, 78.1, 78.2, 82.13, 82.29 **3** 33.36, 59.15 **4** 48.33, 48.35, 48.57, 49.13, 49.18, 54.7, 67.9 **5** 44.11, 45.15 **6** 32.9, 48.81 **7** 48.64, 48.68
Ashmore, J.P. **5** 5.15, 5.16, 16.24 **6** 46.6, 46.7, 48.5, 48.6, 69.1, 85.16 **7** 69.6
Ashwell, G.J. **7** 60.29
Ashworth, R.S. **2** 85.9, 85.10, 85.11, 85.36
Ashworth, T. **5** 86.12, 86.20
Askew, W.B. **2** 71.40
Aslanov, L.A. **2** 77.28, 77.33 **3** 77.11, 77.12, 77.22, 77.30, 82.9 **4** 77.11, 77.12 **5** 61.10, 77.21, 81.59, 84.10 **6** 37.3, 81.39, 81.43, 84.7 **7** 81.26, 81.31, 81.32, 84.16
Asperger, R.G. **3** 76.85
Asplund, R.O. **6** 82.23
Assenza, G.d' **7** 1.21, 1.32
Astakhova, I.S. **3** 62.2, 71.10, 71.49, 75.38 **6** 66.2
Atam, M. **7** 71.30
Atoji, M. **1** 3.1, 5.7 **2** 76.42 **7** 12.1
Atovmyan, L.O. **1** 10.2, 10.3, 13.43, 32.17 **2** 60.140, 74.16, 81.1, 83.78 **3** 12.1, 13.7, 69.17, 78.22, 83.17 **4** 81.10, 84.10 **5** 12.1, 27.13, 60.8, 73.8 **6** 33.30, 41.29, 69.11 **7** 60.21, 60.26, 60.34, 60.35, 60.36, 60.38, 63.4, 63.8, 63.11, 71.23, 81.40, 84.32, 84.33, 84.36, 84.68, 85.44
Attig, R. **7** 4.1, 13.6, 13.7, 13.8
Atwood, J.L. **2** 67.12, 68.18 **3** 67.8, 67.13, 68.11 **4** 67.18, 68.5 **5** 68.1, 71.33, 71.56, 75.39, 76.9, 83.79 **6** 67.9, 68.1, 68.3, 68.4, 73.46, 84.13 **7** 22.3, 33.30, 36.7, 68.1, 68.3, 68.8, 68.9, 68.12
Aubagnac, J.-L. **4** 32.18
Aubry, A. **4** 48.23, 48.43 **5** 48.61, 48.62 **6** 2.5, 37.4, 48.38, 48.47, 48.53 **7** 48.50, 48.51
Aumann, R. **7** 71.19, 71.47
Aupers, J. **6** 41.15
Aurivillius, B. **1** 5.13 **4** 39.22 **5** 39.18 **6** 28.1 **7** 39.21, 62.2, 62.9
Aurivillius, K. **7** 86.55, 86.105
Auvray, P. **4** 7.12
Avdeef, A. **3** 80.18, 85.38 **4** 75.14, 75.15 **5** 80.20 **7** 77.5, 84.52
Averill, B.A. **5** 85.31
Avetissian, A.H. **7** 3.35
Avetisyan, A.A. **7** 1.35
Avey, H.P. **5** 36.3 **6** 16.14 **7** 58.12
Avitabile, G. **1** 1.69, 1.110 **3** 1.32, 22.4, 83.86 **4** 1.9, 15.6, 16.6, 50.17, 72.12 **5** 1.32, 8.18, 13.16, 75.4 **6** 31.3, 75.2 **7** 78.1
Avonkova, Z.V. **7** 60.11
Avoyan, H.L. **7** 1.38
Avoyan, R.L. **1** 28.8, 28.9, 28.10, 29.28 **2** 73.87 **7** 1.35, 3.34, 3.35
Axelrod, M. **1** 52.26
Ayer, W.A. **3** 58.13, 58.52 **6** 58.21
Ayerst, E.M. **1** 1.25
Ayres, F.D. **2** 83.40
Ayyar, R.R. **1** 48.20, 48.85
Azarnia, N. **4** 45.9, 45.12 **5** 45.3
Azzoni, G. **2** 84.2

Baar, J.F.van **7** 71.129
Baba, S. **4** 71.58, 71.80
Babeau, A. **3** 3.50
Babushkina, T.A. **1** 16.3, 19.1, 19.2, 19.3 **2** 71.1 **3** 16.1 **5** 19.1, 19.4, 19.5
Bacci, M. **7** 86.81
Bach, R.D. **7** 71.85
Bachechi, F. **5** 16.11 **6** 15.2
Backeus, M. **6** 33.6
Bacon, G.E. **1** 2.86, 2.87, 11.51, 17.15, 19.19, 19.20 **6** 13.5, 13.6, 17.2
Baddley, W.H. **4** 71.78, 71.90 **5** 86.60
Badley, E.M. **2** 71.38
Baenziger, N.C. **1** 6.2, 6.3, 6.4, 12.9, 12.20 **2** 60.93, 60.94, 72.4, 72.47, 73.17, 75.2, 75.3, 75.4, 75.6, 75.18, 75.77 **3** 11.13, 83.8 **5** 19.14, 49.11 **6** 64.41, 73.47, 85.1, 85.2, 85.3, 85.12, 86.34
Baerlocher, C. **3** 3.14, 3.15
Baert, F. **4** 21.13 **7** 32.18, 52.4, 52.5
Bagga, M.M. **2** 83.214
Baggio, R.F. **5** 58.18 **7** 37.5, 79.10
Baggio, S. **2** 76.28 **3** 76.14, 79.17, 79.26 **4** 38.6 **5** 38.6, 58.18 **6** 16.14, 32.24 **7** 37.5, 79.10
Bagnall, K.W. **7** 84.37
Baikie, P.E. **2** 71.14, 75.31, 75.46, 83.170, 83.209, 83.214 **5** 83.69
Bailey, A.S. **2** 60.75, 60.76 **3** 35.18, 40.19
Bailey, M. **1** 3.71, 13.34, 13.35, 13.48, 26.5, 26.9, 26.39, 26.40, 36.38
Bailey, M.F. **2** 74.4, 74.18, 75.1 **3** 81.48, 83.121, 83.122, 83.123, 83.124, 83.125, 83.126 **4** 83.83, 83.84

Bailey, N.A. **1** 23.11, 23.15, 38.39 **2** 65.6, 68.5, 68.6, 71.9, 73.35, 75.23, 77.2, 77.3, 83.182, 86.18, 86.19 **3** 71.3, 83.75, 85.7, 85.20 **4** 12.14, 28.11, 35.23, 78.13, 83.131 **5** 71.12, 76.45, 83.64, 83.65 **6** 71.13, 72.9, 83.44 **7** 71.83, 73.13, 78.17
Bailey, P.M. **5** 71.105 **6** 71.61, 75.28 **7** 71.43, 71.101, 75.28
Baillie, A.C. **4** 32.10
Bain, V.A. **2** 63.12
Baird, H.W. **2** 73.9
Baird, J.W. **6** 8.1
Baird, W. **2** 73.11
Baird, W.C. **2** 75.50
Baiwir, M. **6** 70.2
Bakakin, V.V. **3** 76.39 **4** 76.4, 76.5 **7** 3.20
Baker Junior, W.A. **4** 81.56, 81.58 **5** 83.50, 83.51 **7** 84.59
Baker-Hawkes, M.J. **2** 85.86
Baker, C. **3** 39.2
Baker, D.C. **7** 47.40
Baker, D.J. **2** 85.49
Baker, E.C. **7** 71.94
Baker, E.N. **2** 78.7, 78.17, 78.18, 78.27 **3** 60.1, 60.16, 60.39 **6** 85.15
Baker, J. **7** 83.112
Baker, R.J. **3** 83.72
Baker, R.W. **2** 71.52, 86.120 **3** 83.67, 86.97 **5** 28.4, 37.2, 41.21, 58.23, 71.60, 86.15 **6** 32.26, 33.53, 84.46
Baker, S.R. **7** 60.29
Baker, W.A. **5** 19.6, 19.10
Balaban, A.T. **1** 33.74
Balahura, R.J. **6** 83.2 **7** 84.1
Balasubramanian, R. **2** 76.24
Balch, A. **5** 85.36
Balch, A.L. **3** 84.20 **4** 71.20 **5** 71.8 **7** 9.4, 71.25
Balchin, A.A. **1** 45.24 **7** 45.6
Balducci, A. **7** 68.19
Baldwin, J.E. **1** 37.18, 37.24, 39.61, 41.38 **3** 20.12
Baldwin, W.H. **5** 71.108 **6** 71.52 **7** 73.30
Balkenhol, W.G. **3** 59.32
Ball, A.L. **6** 32.7
Ball, R. **3** 69.27
Ball, R.D. **2** 76.26 **6** 69.12
Ballard, R.E. **5** 41.38 **6** 45.1, 79.2 **7** 35.12, 35.16
Bally, R. **2** 65.28, 80.15 **3** 64.15, 65.4 **4** 41.8 **5** 41.15, 41.32, 41.35 **6** 41.26
Balta, E.Ya. **2** 81.28
Balundgi, R.H. **5** 78.8
Balyaeva, C.F. **7** 82.16
Balyaeva, K.F. **3** 82.9
Bampfield, H.A. **7** 20.23
Ban, E. **5** 72.55
Ban, Y. **3** 58.36
Bancroft, G.M. **7** 69.28
Bandlish, B.K. **5** 36.11
Bando, S. **4** 48.57, 67.9
Bandoli, G. **3** 39.39, 40.15, 64.42, 69.33, 72.49, 72.51, 81.55 **4** 39.27, 40.19, 71.78, 78.11 **5** 69.28, 83.29, 83.30, 86.21 **6** 16.16, 44.24, 76.40, 84.18, 86.47 **7** 78.10, 80.2, 84.41, 84.51, 86.74
Banerjee, A. **6** 38.19 **7** 38.25
Bang, W.B. **1** 23.1
Banks, C.V. **5** 83.28
Banner, B.L. **6** 27.21
Banner, D.W. **7** 16.10
Baraban, J.M. **7** 72.39
Baracco, L. **7** 84.48
Baraniak, E. **3** 83.20
Baranowski, P. **5** 58.6
Barassin, J. **1** 1.116
Barbeau, C. **6** 73.3, 73.52 **7** 73.68
Barber, P.G. **3** 9.11 **4** 13.20
Barbier, M. **6** 51.41
Barbieri, R. **7** 69.22, 78.4
Barclay, G.A. **2** 76.34, 77.15, 81.23, 81.85, 83.90, 83.113, 83.171, 85.50 **3** 76.30, 83.92
Bardi, R. **1** 4.16, 39.45 **2** 83.56
Bardos, T.J. **1** 47.30 **4** 36.18
Barefield, E.K. **6** 76.51 **7** 76.63, 83.62
Barendregt, F. **3** 3.11
Barieau, R.E. **1** 48.17
Barinov, I.V. **5** 71.45
Barker, D.L. **1** 44.15
Barkigia, K.M. **7** 86.1
Barlex, D.M. **3** 71.34
Barlic, B. **6** 81.16
Barlow, J.H. **7** 39.12, 39.13
Barlow, M. **1** 11.20, 39.17, 47.31
Barnes, J.C. **7** 38.3, 38.4, 83.3, 83.4, 84.22
Barnes, W. **6** 11.11
Barnes, W.H. **1** 33.76, 33.77, 33.78, 36.22, 58.13, 58.25, 58.29 **2** 64.32, 64.64, 64.81
Barnet, M.T. **3** 82.12, 82.13, 82.14, 82.15
Barnett, A.J. **4** 39.14
Barnett, B. **4** 71.65
Barnett, B.L. **3** 27.9, 30.1, 31.29 **4** 63.5 **5** 71.59, 71.62, 71.73, 71.82, 72.30, 72.36 **6** 64.44 **7** 30.5, 30.6, 40.18, 71.33, 71.34, 71.56, 71.89
Barnett, G.H. **6** 85.23
Barnett, W.E. **4** 42.5
Barnhart, D. **6** 78.7, 78.8, 78.16, 78.18, 78.19
Barnhart, D.M. **2** 69.10 **3** 83.61
Barnighausen, H. **1** 40.12 **4** 9.1 **5** 12.2 **6** 63.3
Barr, R.E. **3** 33.47
Barr, R.M. **6** 3.20, 3.21, 3.22
Barragan, I. **7** 45.42
Barrans, Y. **1** 3.78, 19.25, 52.4, 52.5 **3** 3.24, 3.33, 3.45, 3.50 **4** 3.21 **5** 3.22, 24.7 **6** 48.60, 48.67, 48.70 **7** 33.50
Barrett, A.G.M. **7** 51.56
Barrett, A.N. **1** 46.12
Barrett, E.K. **1** 35.23 **3** 28.13, 36.33 **4** 32.26, 32.29
Barrett, J.H. **1** 34.8
Barrick, J.C. **5** 4.4, 77.18 **6** 4.4, 4.5 **7** 4.6
Barrio, J.R. **6** 47.28

Barriola, A. 4 38.6
Barrow, M.J. 3 75.3, 75.28, 83.103 4 75.9, 81.41, 81.54 5 22.1 6 72.16 7 2.4, 58.26, 58.29, 73.2, 73.3, 73.18, 75.16
Bart, J.C.J. 1 19.67, 41.16, 57.1, 57.2 2 63.3, 64.69, 64.74 3 64.31, 64.32, 64.45 4 26.1, 26.2, 26.4, 71.4
Bartczak, T.J. 5 35.25 7 49.2, 64.20, 64.21
Bartindale, G.W.R. 2 71.28
Bartkowski, R.R. 5 78.3
Bartl, H. 7 14.3
Bartland, G.L. 6 35.15
Bartle, K.D. 5 24.3
Bartlett, M. 3 77.31, 86.83
Bartlett, M.W. 4 83.120
Bartolozzo, G. 3 64.42
Barton, D. 5 76.47
Barton, D.H.R. 3 58.34 7 51.56
Barton, T.J. 3 32.20 4 60.36
Barve, J.V. 4 15.4
Basak, B.S. 7 35.9
Basolo, F. 4 71.72
Bassi, I.W. 2 72.53 4 71.63 5 38.8, 72.16 6 61.8
Basson, S.S. 1 33.67 3 33.44 4 33.33 5 76.18
Bastian, J.M. 4 39.35
Batail, P. 7 16.1, 16.5
Bateman, L.R. 3 71.54
Bates, H. 6 32.38
Bates, R.B. 1 39.4 3 36.3, 53.1, 53.2, 53.19, 53.27 5 39.31, 59.25 7 58.18, 59.16
Bats, J.W. 4 60.3 5 73.15 7 16.2
Battaglia, A. 3 40.15 4 40.19
Battaglia, L.P. 5 82.1 6 76.15, 81.36, 83.66 7 66.4, 79.24, 82.6
Battelle, L. 1 42.6
Battelle, L.F. 5 73.41
Batterham, T.J. 3 35.1
Battiste, M.A. 4 38.19 7 31.7
Battisti, A. 4 39.8
Battistini, G.G. 2 79.18, 83.8, 83.9
Bau, R. 1 3.82 2 65.41 4 27.3, 72.32 5 62.6, 64.44, 71.20, 73.41 6 65.7, 71.40, 75.23 7 64.63, 64.64, 64.65, 73.41, 86.13, 86.14
Baudour, J. 1 19.30, 19.31
Baudour, J.-L. 6 32.30 7 19.4, 19.5
Baudour, J.L. 3 32.30 4 19.12 6 40.13
Bauer, H. 4 32.22
Bauer, R.A. 5 83.114
Baughman, R.H. 2 60.59
Baumann, D. 7 71.126, 71.134
Baur, W.H. 5 31.7
Bavoux, C. 5 17.1 6 17.1 7 17.2
Baxter, R.L. 6 53.17 7 59.9
Baxter, R.M. 3 39.16
Baxter, W.J. 4 59.25, 68.8
Bayer, W. 1 3.72
Beagley, B. 1 2.18, 2.33, 2.51
Beal, G. 7 84.62
Beal, J.K. 4 54.10
Beale, J.P. 3 45.41, 45.42, 86.13, 86.20 4 3.23, 3.25, 3.26, 86.5 5 17.15, 45.44, 59.9
Beall, H. 2 62.11 6 71.79
Beall, R. 3 3.49, 39.4
Bear, C.A. 3 78.2, 78.4, 78.27 5 29.4, 59.31, 86.8 6 13.18, 31.24, 60.27, 75.4, 75.5, 76.52 7 31.2, 52.3
Beard, D.W. 3 49.1, 49.7
Beard, L.N. 1 32.9
Beasley, G.H. 4 56.3
Beattie, I.R. 2 69.17
Beauchamp, A.L. 2 66.20, 66.23 5 72.7, 72.15, 83.88, 86.23 6 66.8, 86.42 7 44.10, 64.57, 64.58, 66.11, 83.7, 83.104, 85.12
Beaute, C. 3 1.7
Beaver, J.A. 5 84.16
Bechtel, F. 6 27.9, 27.10, 32.19, 32.20 7 25.3
Beck, W. 5 86.66
Becka, L.N. 1 37.4, 37.5, 37.6, 37.7, 37.8, 37.9 2 76.28 3 76.14, 79.26 5 36.3 6 16.14 7 37.5
Becker, J.W. 6 48.45
Becker, K.A. 1 20.31 2 76.7
Becker, M.R. 1 31.3
Becker, P. 4 17.3 5 17.11 6 7.3, 17.7, 17.8
Becker, T.L. 6 38.31
Beckett, R. 3 85.24 4 83.88 6 77.3, 77.6 7 85.19
Beckley, R.S. 5 31.39
Beddoes, R.L. 3 75.11, 75.13 4 78.2 5 31.9
Bednowitz, A. 1 31.4
Bednowitz, A.L. 1 1.48, 1.90, 30.27, 40.11 3 13.4, 13.5, 20.3 4 69.35 6 77.10
Beer, D.C. 5 62.9
Beer, R.J.S. 3 39.40 4 39.14 7 41.23
Beevers, C.A. 1 1.59, 21.13, 30.1, 45.30, 45.34, 45.70, 58.54 4 45.20 5 45.35, 45.36, 45.38
Begin, D. 7 3.8, 81.1
Begley, M.J. 3 59.23 4 59.18 5 51.15, 51.44, 59.29 6 38.23, 51.40, 59.16, 64.11 7 21.5, 28.9, 38.19, 59.14, 64.66, 65.7
Behforouz, M. 7 32.8
Behr, D. 7 51.48
Behrens, U. 6 72.18 7 71.72, 72.24, 72.25
Beineke, T.A. 1 27.17, 33.2 2 64.15 3 63.16, 83.107
Beintema, J. 1 24.27
Beisler, J.A. 3 51.48 4 31.25
Bejenke, V. 7 75.15
Bekiarian, P. 7 38.47
Bekoe, A. 1 45.2
Bekoe, D.A. 1 7.21, 59.66 2 67.6, 67.7
Bel'skaya, N.P. 3 68.9
Bel'skii, V.K. 7 26.3
Belford, R. 5 3.19, 71.51
Belford, R.C.E. 5 60.12 6 60.14, 77.16
Belford, R.L. 2 77.7, 77.40, 77.41, 77.43, 77.55, 81.49, 81.54 3 77.24, 81.52 5 81.29, 86.52
Belichuk, N.I. 7 76.66

Belinskij, I. **3** 76.33 **5** 83.17, 83.18
Belitskus, D. **1** 1.115
Bell, J.D. **1** 41.33 **2** 82.23, 82.36, 82.38 **5** 76.22 **7** 76.52
Bellingen, I.van **3** 60.17, 60.18
Bello, C.di **1** 34.4, 35.9
Bello, P. **7** 64.48
Bellobono, I.R. **3** 31.25
Bellon, P. **2** 60.149, 86.118 **4** 86.58 **5** 86.85 **6** 86.75
Bellon, P.L. **1** 3.41, 40.25, 40.26 **2** 71.33, 86.72, 86.74, 86.100, 86.119 **3** 3.21, 3.54, 86.22, 86.94, 86.99 **4** 86.57, 86.71 **5** 86.24, 86.48, 86.56, 86.76, 86.83 **6** 71.86, 86.69 **7** 71.118, 86.31, 86.66
Belluco, U. **3** 72.49 **4** 71.78 **7** 84.51
Bellus, D. **6** 20.6 **7** 28.3, 38.24
Belov, N.V. **2** 66.11, 66.12 **5** 2.3 **6** 76.2, 83.20, 83.21 **7** 42.3, 42.4, 67.3, 78.14, 78.15, 79.21
Belyaeva, K.F. **2** 81.43
Benci, P. **3** 37.7 **4** 37.2, 37.3 **6** 65.1, 65.2, 65.3
Bencze, W.L. **4** 24.16
Benedetti, E. **1** 1.93 **3** 1.31, 20.4, 21.5, 21.6, 21.7, 21.8, 21.10, 33.16, 33.19, 39.1, 62.16, 72.19 **4** 71.25, 71.26, 72.8 **5** 1.29, 1.32, 8.18, 13.16, 41.3, 48.42, 71.23 **6** 48.52, 72.1 **7** 1.20, 1.31, 20.4, 20.8, 38.43, 48.10
Benedicenti, C. **6** 86.69
Benes, J. **7** 3.2, 3.3
Benetollo, F. **6** 44.24
Benghiat, V. **5** 1.5, 1.6, 1.7, 1.8, 1.11, 13.1, 13.13
Bennett, A. **1** 30.26
Bennett, I. **3** 48.25
Bennett, J. **3** 86.1
Bennett, M.A. **2** 86.46 **4** 71.43, 71.46, 71.49, 71.56, 86.21 **5** 71.94, 72.40, 72.45, 72.51, 74.8, 74.9 **7** 86.35, 86.36, 86.37
Bennett, M.J. **2** 66.20, 66.23, 73.26, 73.39, 75.39, 75.66, 77.39, 77.63, 81.83, 81.88, 84.24, 85.29, 85.54, 86.28, 86.97 **3** 69.27 **4** 21.22, 28.12, 63.6, 86.25, 86.65 **5** 31.16, 71.54 **6** 62.10, 71.19, 85.31, 85.33, 85.34 **7** 69.12, 73.45
Bennett, W.E. **5** 49.11
Benno, R.H. **5** 69.7 **6** 84.24, 84.25
Beno, M.A. **7** 31.1
Benoit-Guyod, J.-L. **6** 16.18 **7** 1.21, 1.32
Benschop, H.P. **4** 52.5
Benson, J.E. **6** 35.8
Bentley, G.A. **2** 82.57 **3** 86.61 **7** 79.26
Bentollo, F. **7** 78.10
Bentrude, W.G. **3** 64.17
Benz, W. **6** 16.23
Beran, G. **3** 68.13 **5** 86.58 **7** 86.51, 86.54, 86.57
Berends, E.A. **4** 82.10
Beres, J.J. **7** 40.6
Berg, J.-E. **7** 51.48
Bergen, B. **7** 45.12
Bergendahl, T.J. **7** 84.74
Bergerhoff, G. **6** 85.4
Bergin, R. **1** 59.10, 59.11 **3** 58.2 **4** 3.18, 3.19
Bergman, J. **4** 39.16
Bergman, J.G. **2** 65.44, 81.50
Berkebile, C.A. **1** 14.14
Berking, B. **1** 21.31 **3** 51.36 **4** 36.36, 43.3, 43.5, 43.7, 45.3 **5** 45.35, 45.36, 45.38
Berkovitch-Yellin, Z. **6** 20.20 **7** 1.25
Berlin, K.D. **7** 64.49
Berlin, Yu. **4** 50.13
Berliner, L.J. **3** 12.8
Berman, H. **1** 8.8
Berman, H.M. **1** 45.44, 45.46, 45.49 **3** 32.11, 32.23, 45.5, 45.47 **4** 2.22, 36.32, 47.20 **5** 2.29, 26.5, 45.3 **6** 26.3, 45.29, 48.37, 48.79, 48.84 **7** 36.21, 64.43
Bernal, I. **1** 39.69 **2** 80.29, 85.41 **3** 11.14, 11.15, 84.20, 85.25, 85.44 **4** 33.25, 39.34, 39.41, 73.8, 73.11 **5** 3.11, 36.9, 53.14, 73.5, 73.9, 80.13, 83.113, 85.36 **6** 23.1, 23.2, 41.4, 48.33, 71.10, 73.21, 80.20, 85.19, 86.48 **7** 71.120, 73.21, 73.22, 73.58
Bernard, M.-J. **2** 83.2
Bernard, M.J. **4** 83.4
Bernard, W.H. **3** 76.77
Bernardinelli, G. **6** 28.7 **7** 38.50, 45.40
Bernauer, K. **3** 58.67
Berndt, A.F. **2** 73.8
Bernier, F. **6** 42.7
Bernotat-Wulf, H. **3** 59.9
Bernstein, J. **4** 12.9, 16.12, 16.13, 31.17, 31.18, 60.37 **5** 1.17 **6** 28.2 **7** 19.9, 20.17, 59.10
Bernstein, P.K. **5** 86.25, 86.26
Bernstein, T. **2** 73.24
Berro, C. **7** 81.4, 81.5, 81.6, 81.7
Bersted, B. **5** 73.43
Bersted, B.H. **2** 81.49
Berta, D.A. **3** 69.1, 79.9
Bertaut, E.F. **5** 81.16 **6** 81.22 **7** 81.14
Berthier, P.I.C. **7** 60.29
Berthold, H.J. **1** 11.8
Berthou, J. **1** 9.47, 31.12, 43.20, 43.21 **2** 60.28 **3** 9.18, 9.20, 11.21, 32.7 **4** 8.8 **5** 5.18, 9.16, 11.10
Berti, G. **4** 21.12
Bertin, F. **6** 67.10
Bertini, I. **4** 83.126
Bertinotti, A. **1** 16.27
Bertinotti, F. **1** 13.30, 33.1
Bertrand, G. **4** 2.9 **6** 2.5
Bertrand, J.A. **1** 46.11, 53.28, 59.5 **2** 84.69, 86.73 **3** 3.22, 77.14, 77.25, 83.65, 84.12 **4** 32.10, 77.20, 78.10, 83.21, 84.14 **5** 77.16, 78.10, 83.13, 83.42 **6** 49.14, 78.9, 78.15, 78.21, 84.10, 84.12 **7** 78.12, 84.40, 84.44
Berzjak, A. **1** 8.45
Besancon, J. **3** 74.2 **5** 74.3, 74.4, 74.5 **6** 74.4, 74.8, 74.9
Beschastnov, A.S. **3** 71.56
Bestmann, H.-J. **7** 64.59
Betkouski, M.F. **5** 38.5
Betzner, G.L. **2** 76.25

Beurskens, P. **1** 13.13
Beurskens, P.T. **1** 1.34 **2** 61.12, 80.8, 80.9, 80.34 **3** 80.11 **4** 79.10, 80.4, 80.11, 80.12, 80.16, 80.18, 85.3, 85.4 **5** 39.20, 80.24 **6** 30.8, 79.3, 80.19 **7** 30.7, 38.58, 79.9, 79.22, 80.17, 80.18, 85.8
Bevan, C.W.L. **1** 58.73 **3** 56.11
Bevan, J.A. **2** 78.6, 83.157
Beyer, H. **1** 39.1
Bezjak, A. **2** 81.60 **7** 39.11
Bezman, S.A. **5** 71.70, 71.84, 71.85, 75.36, 86.84 **6** 71.96
Bezzi, S. **1** 19.14, 39.45 **2** 83.56 **4** 19.2
Bhaduri, D. **7** 17.16, 48.30
Bhaduri, S. **6** 86.17 **7** 86.67
Bhalla, V.K. **3** 58.39
Bhandary, K.K. **5** 84.9 **7** 84.17, 84.75
Bhat, G. **7** 39.24
Bhat, H.B. **1** 59.68
Bhat, T.N. **7** 33.21, 48.31
Bhatia, S.C. **7** 78.16, 78.25
Bhattacharyya, S.C. **1** 48.19
Bhuiya, A.K. **5** 36.21
Biagini Cingi, M. **4** 84.12
Biagini, S. **1** 1.46, 40.9, 40.17, 40.18 **3** 40.6, 76.60, 83.64
Bianca, F.di **6** 66.4
Bianchi, A. **3** 86.53 **4** 76.43, 76.44 **5** 76.52, 76.53
Bianchi, R. **5** 31.12 **7** 31.20, 31.24
Bianco, V.D. **2** 86.74
Bichan, D. **4** 39.46
Bick, I.R.C. **3** 58.6 **7** 58.12
Bickelhaupt, F. **7** 67.7
Bickley, D.G. **4** 77.1
Biddlestone, M. **6** 64.45
Bideau, J.-P. **1** 43.22, 43.26, 43.30 **3** 3.33, 43.6, 43.7, 43.9 **4** 43.8 **5** 35.29 **6** 38.4, 48.59, 48.65
Bideau, J.P. **6** 43.2 **7** 43.4
Biedl, A. **3** 45.28
Biefeld, C.G. **6** 31.30, 71.84 **7** 30.5, 30.6
Bierlein, T.K. **1** 10.18
Bierrum, A.M. **3** 71.5, 71.6
Bierstedt, R. **1** 44.46
Biffar, B. **4** 33.47
Bigliardi, G. **2** 81.35, 83.10, 83.15
Bigoli, F. **2** 83.4, 83.5 **3** 83.2, 83.3 **4** 79.4, 81.2, 83.50 **5** 8.6, 84.21 **6** 83.70, 84.29 **7** 79.2, 79.3, 79.4, 79.5
Bijvoet, J.M. **1** 2.63, 2.65, 2.67, 2.70, 21.5, 48.65, 58.55
Billups, W.E. **6** 28.3
Bilofsky, H.S. **7** 39.24
Bilton, M.S. **4** 69.3
Binas, H. **2** 60.21
Bindlish, J.M. **7** 78.16, 78.25
Binnie, W.P. **1** 3.49, 3.50
Biondi, C. **1** 7.17
Birchall, T. **2** 73.122 **3** 83.137
Bird, M.J. **4** 81.44
Bird, P. **6** 86.68
Bird, P.H. **2** 65.6, 65.41, 68.5, 68.6, 75.40, 75.42, 75.43, 75.69, 75.76 **3** 58.17, 72.39, 75.32, 75.33, 75.43 **4** 63.4, 69.16, 69.17, 77.1 **5** 71.70, 77.17, 83.118, 83.119 **6** 32.18, 86.49, 86.50 **7** 38.16, 74.4, 74.5
Birker, P.J.M.W.L. **6** 79.3 **7** 79.9, 79.22
Birkett-Clews, C.J. **1** 19.59
Birknes, B. **5** 39.38 **7** 39.34
Birnbaum, G.I. **1** 54.15, 58.112, 59.32 **3** 58.49 **4** 44.23, 75.17 **5** 51.29, 51.34 **6** 51.30 **7** 50.17, 53.12, 58.10
Birnbaum, K.B. **1** 54.18 **3** 36.24 **4** 58.20, 58.27, 75.18 **5** 58.28
Birnberg, G.H. **5** 27.3
Birrell, G.B. **5** 3.5
Biryukov, B.P. **2** 69.39, 73.41, 73.49, 73.68, 73.69 **3** 69.31 **7** 45.20
Biscarini, P. **5** 84.12 **7** 84.6
Bisi, C.C. **5** 84.18 **6** 84.44
Bisi, E.C. **7** 84.19, 84.82
Bissell, E.C. **1** 3.69 **3** 49.1, 49.7 **5** 46.11, 50.15 **6** 64.22 **7** 61.2
Biswas, A.B. **1** 2.52, 2.69, 2.72, 2.85, 2.89, 48.30 **2** 82.19 **7** 81.19
Biswas, S.G. **1** 39.58 **7** 21.7, 21.8
Biushkin, V.N. **7** 76.66
Bizot, K.F. **7** 9.4, 86.34, 86.38
Bjamer, K. **1** 45.59, 54.20, 54.22
Bjornstad, H.J. **5** 7.2
Bjorvatten, T. **1** 7.8, 7.9 **2** 60.2, 60.3, 60.4, 60.53, 60.54 **3** 60.2
Bkouche-Waksman, I. **2** 84.12
Blaauw, H.J.A. **2** 80.34
Black, D.L. **7** 64.4, 64.5, 64.6
Black, K.T. **4** 63.13 **5** 9.2
Black, M. **2** 86.24
Blackmer, G.L. **5** 76.22
Blackmond, B. **7** 33.42, 33.43
Blackmore, T. **4** 71.70
Blackmore, W.R. **1** 11.65, 11.67 **2** 70.25
Blackstone, M. **2** 77.59
Bladt, S. **7** 38.17
Blain, J. **5** 2.16, 2.17
Blake, A.B. **2** 80.23 **3** 78.28, 83.84 **5** 77.13 **7** 77.10, 81.44, 83.105
Blake, C.C.F. **5** 13.8
Blakeslee, A.E. **2** 84.23
Blanchard, L. **7** 35.30
Blanck, B. **7** 66.11
Bland, W.J. **2** 86.92
Blank, G. **6** 45.16, 45.17 **7** 13.18, 44.26, 44.29, 46.2, 46.3, 46.4
Blank, G.E. **4** 46.6, 59.5 **7** 41.11
Blasio, B.di **7** 48.27
Blattmann, P. **7** 84.49
Blayden, H.E. **3** 69.6
Blaylock, B.T. **4** 16.14, 24.11
Bleidelis, Ya.Ya. **3** 69.17 **5** 27.13 **6** 69.11 **7** 16.12, 17.27, 17.28, 25.5, 25.6, 27.8, 37.8, 62.3
Blessing, R.H. **4** 32.2, 48.25 **7** 33.8, 60.3
Blickensderfer, J.R. **6** 71.70
Block, H.D. **7** 16.4, 73.9
Block, S. **1** 39.42, 39.43 **2** 62.35, 83.92 **3** 19.4

Blom, E.A. 2 69.24
Blomberg, C. 7 67.7
Bloom, A. 4 36.14, 49.22
Bloomer, A.C. 3 33.22
Bloor, J.H. 2 60.93, 60.94
Blot, J. 4 19.8, 19.9
Blount, J.F. 1 3.46, 41.33, 52.29 2 72.61, 82.24, 82.25, 83.54 3 9.19, 35.17, 35.23, 36.31, 64.67, 65.12, 82.17 4 50.15, 58.3, 58.6 5 19.20, 28.19, 32.6, 35.26, 35.27, 58.22 6 16.23, 27.21, 35.27, 38.24, 58.28, 58.34, 62.11 7 19.16, 35.31, 35.37, 39.9, 50.34, 51.44, 53.8, 64.62
Blum, J.J. 1 32.8
Blumburg, H. 3 35.24
Blundell, T.L. 2 68.22, 69.48, 86.41, 86.45 3 86.32 4 86.56 5 68.17, 86.27 7 48.46
Bo, I. 4 9.4
Boar, R.B. 6 56.5
Bobeszko, B. 6 58.54
Bobkova, R.G. 7 64.30
Bobrik, M.A. 7 85.42
Boccalon, G. 1 33.75
Bocelli, C. 6 41.16
Bocelli, G. 4 26.5, 83.125 5 9.17, 27.11, 27.12, 32.19, 32.25, 38.26, 39.2, 39.6, 39.26, 39.33, 40.11, 42.3 6 11.5, 11.6, 17.24, 31.17, 38.37, 39.1, 39.22, 39.25, 41.9, 41.11, 41.24, 72.3 7 32.21, 35.35, 35.36, 38.75, 40.9, 40.10, 41.22
Bochkov, A.F. 5 45.47
Bochkova, R.I. 7 67.3
Bock, C.M. 1 20.5 3 1.16, 3.5, 20.1, 62.11 5 16.1
Boden, N. 6 19.2
Bodner, R.L. 4 83.58
Bodor, G. 1 1.90
Bodyu, V.G. 7 79.17
Boeckman Junior, R.K. 7 50.10
Boehm, J.R. 7 9.4
Boehme, H. 5 86.66
Boelema, E. 1 39.26
Boer, B.G.de 6 62.2
Boer, F.P. 1 3.32, 11.9, 12.3, 42.1, 50.32 2 62.7, 62.22, 63.16, 63.26 3 9.1, 20.8, 20.10, 32.26, 38.18, 42.1, 63.2, 63.3, 63.10, 63.11, 69.35, 69.37, 69.38 4 27.1, 31.12, 32.12, 33.3, 46.5, 69.18, 69.19, 75.20, 75.29 5 7.5, 7.6, 33.7, 33.31 6 33.17 7 33.27, 35.8, 38.9, 38.10, 38.11, 38.20, 38.21, 38.22, 38.23, 84.23
Boer, J.de 6 10.1, 24.3
Boer, J.J.de 3 31.31, 72.22, 72.35, 72.50, 81.34 5 73.15 6 64.18 7 75.39, 75.40
Boer, J.L.de 1 16.42 2 60.122, 86.78 4 16.8, 16.9 6 73.12
Boeyens, J.C.A. 1 8.16, 30.4 2 60.9, 60.10, 60.11, 60.40, 60.41, 60.98, 60.135, 60.154, 68.7, 68.8 3 12.7, 60.6, 67.17, 77.34, 77.35 4 77.17, 77.19, 77.21, 77.22 5 77.25 7 8.24, 8.25
Boggs, R. 5 24.4 7 83.34
Boggs, R.A. 5 71.10
Bognar, R. 6 58.19
Bogucka-Ledochowska, M. 7 50.29
Bohme, R. 7 20.11
Bois d'Enghien-Peteau, M. 2 60.109 7 71.135
Bois d'Enghien, H. 1 5.27
Bois-d'Enghien-Peteau, M. 7 71.136
Bois, C. 1 17.24 3 17.3 4 17.5 5 17.7 7 5.15
Bok, L.D.C. 1 33.67 3 33.44 4 33.33 5 76.18, 77.8
Bokii, G.B. 2 72.1, 72.3, 81.1 4 72.1
Bokii, N.G. 1 9.22 2 69.4, 69.5, 69.18, 69.45 3 60.9, 64.30, 73.22, 73.38 4 69.29, 83.128, 86.36 5 41.25, 63.15, 69.3, 71.7, 71.47 6 58.8, 64.29, 69.17, 72.24 7 63.13, 63.14, 63.18, 69.10, 69.20, 69.21, 69.23, 71.18, 71.36, 71.37
Bokiy, N.K. 2 63.28
Bol'shutkin, D.N. 5 5.2
Boldrini, P. 1 8.18 2 79.2, 79.20, 83.8, 83.9 3 79.6, 79.7, 79.9 4 85.10
Boles, M.O. 7 50.24
Bolhuis, F.van 3 39.17 7 19.17
Bollinger, P. 6 53.13
Bologa, O.A. 7 83.20, 83.21, 83.32, 83.74, 83.84
Bolognesi, M. 7 27.7, 51.50
Bolton, N. 1 43.1
Bolton, W. 1 27.4, 30.30, 30.31, 43.8, 43.11, 43.13
Boman, C.-E. 6 1.5, 48.10, 81.13, 81.28, 81.29
Bombieri, G. 2 69.21, 73.19, 77.10, 78.28, 78.35 3 71.26, 72.51, 80.19, 80.20, 81.55, 83.45, 86.19, 86.25 4 71.23, 71.41, 85.26, 85.32, 86.12 5 29.3, 80.12, 81.30, 83.29, 83.30, 84.5, 84.27 6 81.9 7 33.52, 38.43, 81.22, 81.34, 82.11, 84.37, 84.43, 84.46, 84.48, 84.80, 86.93, 86.94
Bommel, A.J.van 1 2.65, 2.70
Bonamartini, A.C. 4 41.21, 41.22, 83.64 5 79.26
Bonamico, M. 1 3.70, 33.70, 33.72 2 61.13, 61.14, 68.11, 68.16, 80.13, 80.17, 80.20, 84.60, 85.72, 85.85, 85.93, 85.96 3 9.10, 68.20, 80.7, 80.8, 80.9 4 79.11, 85.17, 85.19 5 76.23, 79.23, 83.92, 84.14, 84.25, 85.6, 85.15, 85.21, 85.23, 85.30 6 85.42, 85.44 7 81.42, 84.60
Bonati, F. 2 77.2, 77.3 3 86.98
Bond, P.J. 7 47.26
Bonds Junior, W.D. 4 83.132, 86.51
Bonnemere, C. 1 59.55
Bonnet, J.-J. 3 83.46, 83.47, 83.48 5 49.8, 83.59 6 83.4
Bonnet, J.J. 4 32.8 6 32.8, 52.2 7 72.40, 86.25
Bonnett, R. 3 49.13 4 32.17 5 32.24
Bonsma, G.F. 1 58.28
Booles, S.S. 5 33.6
Boonstra, E.G. 1 15.25 3 36.32 4 32.14
Boorman, P.M. 6 85.16 7 85.36
Bootsma, G.A. 1 1.60, 1.62, 1.63
Bor, G. 6 85.38
Borah, B. 7 35.6

Bordeaux, D. **6** 33.31 **7** 33.32, 40.3
Bordner, J. **3** 9.8, 24.6, 27.11, 28.12, 30.5, 31.32, 38.17 **4** 22.10, 24.5, 27.16, 31.31, 37.15, 51.20 **5** 20.13, 26.4, 28.16, 36.19, 39.25, 51.5, 51.14, 51.36, 64.33, 68.23 **6** 30.5, 30.6, 31.9, 39.15, 51.32, 58.46, 66.6, 75.3, 75.10 **7** 3.32, 37.19, 38.66, 51.30, 51.38, 51.40, 58.9, 66.8, 83.94
Bordwell, F.G. **3** 11.26
Borel, M.M. **7** 1.4, 2.3
Borer, L.L. **6** 77.21
Borgen, B. **1** 7.10
Borgne, G.le **5** 85.16
Borisov, A.E. **6** 66.2
Borisov, S.V. **3** 76.39 **4** 76.4, 76.5 **5** 76.16, 76.19
Borkhoven, C. **1** 58.55
Born, L. **2** 64.34
Borovyak, T.E. **4** 73.10 **6** 58.8
Borowiak, T.E. **5** 41.25
Borowski, E. **7** 50.29
Bortin, O. **3** 76.9, 76.10, 76.61
Bosche, G.van den **6** 51.7
Boschi, T. **3** 71.26
Bose, N. **6** 83.72
Bosman, W.P. **6** 80.19
Bosman, W.P.J.H. **4** 80.16, 85.3 **5** 39.20
Bosnich, B. **2** 83.86
Boss, C.B. **6** 71.75
Bosshard, H.E. **6** 48.83
Boston, D.A. **7** 85.38
Bosworth, N. **7** 51.52
Botoshanskii, M.M. **7** 86.48
Bottcher, B. **3** 63.8, 63.13
Bottrill, M. **7** 71.70, 75.23
Bouas-Laurent, H. **6** 28.11
Boucher, L.J. **7** 84.34
Boucherle, A. **6** 16.18 **7** 33.38
Boulton, A.J. **4** 32.22
Bour, J.J. **4** 79.10 **7** 79.9
Boushehri, A. **7** 83.92
Boussard, G. **7** 48.50, 48.51
Boutin, H.P. **3** 34.1
Bovio, B. **1** 36.7, 36.8, 36.29 **4** 32.27, 36.16, 36.25, 40.16 **5** 40.17
Bowen, B.J. **3** 31.2, 83.23
Bowers, R.H. **5** 83.28
Bowler, S.J. **3** 83.75
Bowman, J.T. **5** 71.13
Bowman, K. **2** 80.31 **4** 83.72
Bowman, K.S. **4** 86.69
Bowman, R.M. **5** 28.15 **6** 28.22
Boyce, C.B. **1** 58.100
Boyd, D.R. **4** 40.14
Boyd, G.V. **7** 32.38, 41.31
Boyd, P.D.W. **6** 83.81
Boyko, E.R. **1** 29.2 **2** 78.29 **5** 15.2, 36.1 **7** 10.1, 78.9
Boyum, K. **4** 70.6, 70.7
Brabant, C. **6** 66.8 **7** 66.11
Bracher, B.H. **1** 1.39, 1.40 **3** 10.1
Brackett, E. **7** 83.76
Bradaczek, H. **5** 35.31
Bradford, C.W. **4** 71.71, 71.73, 71.89 **5** 71.95, 71.96, 71.99
Bradley, D.C. **2** 78.61, 83.123, 84.65, 85.4, 85.23 **3** 78.25, 83.102 **4** 80.15, 83.61, 83.85, 83.130, 83.136 **6** 84.45
Bradshaw, T.K. **6** 40.15
Bradsher, C.K. **4** 37.10
Braekman, J.C. **6** 53.12, 58.6 **7** 53.13, 54.3, 54.4
Braham, J.N. **7** 36.19
Brahde, L.B. **1** 39.38
Braibanti, A. **1** 8.28, 9.2 **2** 79.5, 81.35, 83.4, 83.5, 83.7, 83.10, 83.14, 83.15 **3** 83.2, 83.3, 83.4 **4** 8.4, 67.1, 67.2, 79.4, 81.2, 81.16, 83.50 **5** 8.5, 8.6, 41.14, 84.21 **6** 83.70, 84.29
Brain, E.G. **4** 24.13, 24.14, 24.15
Braïnina, E.M. **5** 71.47
Braithwaite, M.J. **5** 86.15
Branchi, G. **2** 69.11, 85.3
Branden, C. **1** 2.59 **2** 83.166, 85.20
Branden, C.-I. **2** 66.2, 66.6, 84.62, 84.63, 85.17, 85.21 **3** 33.43
Brandl, F. **1** 28.24, 28.25 **3** 54.8, 54.15 **4** 54.8, 54.16, 71.92 **5** 49.1
Brandon, J.K. **1** 11.5
Branfman, A.R. **6** 50.8
Brasen, S. **3** 31.16
Brassy, C. **4** 60.26 **7** 84.55, 84.56, 84.57, 84.58
Bratek-Wiewiorowska, M.D. **1** 59.32 **7** 58.13
Bratholdt, J.S. **7** 39.25
Brathovde, J.R. **1** 1.127 **2** 77.24
Bratton, W.K. **2** 81.88 **3** 81.37
Brauer, D. **5** 72.30
Brauer, D.J. **3** 68.1, 68.21 **5** 68.31, 72.19, 75.12, 75.13, 75.28, 75.32 **6** 62.9, 83.65 **7** 62.14, 69.9, 71.29, 71.44, 71.105, 72.32, 75.32
Brauer, G. **7** 83.90
Braun, A.B. **5** 5.1
Braun, P.B. **3** 21.20, 31.8, 51.8, 51.10, 51.17, 51.18, 51.21, 51.23, 51.24, 51.25 **5** 51.37, 51.38 **7** 16.14, 16.15
Braun, R.L. **2** 78.43, 78.47, 78.48
Bravic, G. **1** 43.26, 59.39 **3** 25.3 **4** 25.3, 38.22 **6** 27.9, 27.10, 27.13, 27.14, 38.34, 59.15
Bray, L.S. **3** 72.41
Breazeale, J.D. **2** 78.11
Breece, J.L. **4** 78.10 **5** 78.10 **6** 78.9, 78.15 **7** 78.12
Breen, J.J. **4** 64.14 **5** 86.42
Bregman, J. **1** 16.51, 16.52, 16.53, 23.16
Brehm, L. **1** 40.22 **3** 1.22 **4** 40.3 **5** 10.2, 40.2 **6** 40.4 **7** 40.1
Bremner, J.B. **3** 58.6 **4** 44.16 **7** 58.12
Brencic, J.V. **7** 83.64
Breneman, G.L. **3** 71.2
Brenic, J.V. **3** 3.8
Brennan, J.P. **6** 75.1
Brennan, T. **2** 80.29 **4** 26.7 **5** 47.28 **6** 47.22, 50.2
Brennan, T.F. **3** 3.39, 60.30, 60.31 **4** 60.17 **5** 11.5 **6** 80.20 **7** 51.6
Brenner, S. **1** 20.25
Brenner, S.A. **1** 1.74, 35.28

Bresciani, N. **6** 78.11 **7** 71.86, 78.7
Breton-Lacombe, M. **1** 25.1, 25.19
Breton, M. **1** 25.15 **6** 17.18 **7** 59.3
Brettle, R. **1** 33.80
Brewis, S. **3** 56.8
Brianso, J.L. **5** 9.1, 16.8 **6** 16.7, 16.10, 17.10, 17.12, 17.13, 35.12 **7** 1.8
Brice, M.D. **1** 52.17 **3** 71.35, 71.55 **4** 71.42 **5** 21.7 **6** 69.14, 73.30, 73.40
Brickenkamp, C.S. **6** 61.2, 61.3
Bridson, J.N. **4** 35.34
Briere, R. **7** 33.32
Bright, A. **4** 71.83
Bright, D. **2** 83.195, 83.200, 83.212 **3** 38.22, 67.11, 67.24, 86.47, 86.66 **4** 72.20 **5** 71.81, 73.15 **6** 10.1, 24.3, 64.18, 71.36 **7** 75.39, 75.40
Bright, W.M. **3** 11.2 **6** 37.13
Brill, R. **1** 1.9
Brill, T.B. **5** 33.9
Brimacombe, J.S. **1** 45.71 **3** 45.26 **4** 45.16
Briner, R.C. **1** 58.100
Bringeland, R. **1** 11.25
Brink-Shoemaker, C. **1** 11.14, 18.12, 49.34
Briscoe, O.V. **1** 41.33
Brisse, F. **3** 58.45, 58.60 **6** 35.25, 50.7 **7** 13.23, 13.24, 27.10, 50.12
Britnell, D. **4** 84.2 **6** 84.3, 85.5
Britton, D. **1** 7.13, 7.24, 39.49, 40.4 **2** 65.3, 69.6, 69.8, 71.18 **3** 11.4 **4** 7.1, 7.2, 7.3, 7.4, 7.9, 7.10, 40.7, 40.9, 40.10, 40.12, 63.3, 69.6, 69.7 **5** 5.9, 69.11 **6** 7.2, 7.4, 31.2, 83.3, 84.2 **7** 7.1, 7.2
Britton, R.W. **5** 59.26
Britts, K. **1** 4.18, 20.17, 20.18, 20.23, 20.25, 32.3, 35.27, 37.15, 39.57, 46.20, 52.27
Broach, R.W. **5** 65.2
Broadhurst, M.J. **4** 35.14 **6** 27.11, 27.12, 28.13, 35.17, 71.35, 75.7
Broadley, J.S. **1** 1.55
Broberg, D.E. **5** 49.11
Broch, N.C. **2** 68.14
Brock, C.P. **5** 86.36, 86.57
Brocklehurst, K. **3** 1.33
Brockway, L.O. **1** 19.40
Brodersen, K. **5** 76.2 **7** 83.1
Brodherr, N. **1** 51.49 **4** 27.19
Broggi, R. **2** 61.17, 61.19
Brondmo, N.J. **7** 11.4, 11.5
Brook, A.G. **2** 71.11, 71.12 **4** 63.12
Brook, P.R. **3** 27.8 **7** 20.23
Brookes, A. **6** 75.11
Brooks, E.H. **3** 69.27
Brooks, J.J. **5** 67.18, 67.19, 67.20
Broomhead, J.M. **1** 11.33, 11.34, 44.33, 44.35
Brossi, A. **4** 58.3, 58.6 **5** 58.22 **6** 58.34
Brot, F.E. **3** 31.10
Brotherton, P.D. **5** 79.1 **6** 79.1, 79.6 **7** 75.9, 80.8
Broussier, R. **6** 73.29
Browall, K.W. **4** 85.2 **5** 85.4, 85.5
Brown Junior, G.G. **7** 75.17
Brown, B.W. **2** 76.18, 76.32, 78.3
Brown, C.J. **1** 1.47, 2.83, 8.47, 8.48, 9.36, 9.40, 9.43, 13.25, 13.34, 13.35, 16.16, 16.35, 17.12, 17.20, 19.45, 19.57, 24.18, 26.9, 26.39, 26.40, 31.24, 32.38, 34.10, 38.1, 45.33, 45.66, 49.7, 49.10, 49.11
Brown, D. **3** 80.12, 80.13, 80.15 **4** 3.14 **7** 84.43, 84.46, 84.80
Brown, D.B. **3** 83.50
Brown, D.H. **3** 83.62 **5** 82.7
Brown, D.R. **1** 33.80
Brown, D.S. **1** 3.68 **2** 60.116, 60.133, 60.134, 76.15, 84.47, 84.48, 86.15 **7** 73.15
Brown, G.F. **6** 85.20, 85.21
Brown, G.H. **7** 17.35
Brown, G.M. **1** 3.8, 13.39, 13.40, 44.26, 45.35, 45.69 **2** 83.198, 83.199 **3** 16.14, 16.15, 21.19, 45.19, 58.21, 58.31, 73.9 **4** 73.1 **5** 5.12, 45.30, 45.35, 45.36, 45.38, 73.19 **6** 19.1, 73.38, 81.27
Brown, H.C. **6** 64.7
Brown, I. **1** 39.50
Brown, I.D. **1** 11.5 **2** 76.25, 83.6, 83.180 **6** 47.29, 77.2 **7** 77.13
Brown, J.M. **7** 72.6, 86.58, 86.62
Brown, J.N. **3** 8.8, 34.6, 64.37 **4** 37.13, 37.14, 51.1, 51.7, 81.59 **5** 10.5, 33.47, 36.11, 64.8, 64.24, 64.38, 64.40, 64.46, 84.15, 84.30 **6** 33.1, 82.21 **7** 16.7
Brown, K. **4** 59.22
Brown, K.L. **6** 3.9, 31.7, 80.12
Brown, L.D. **6** 3.24 **7** 3.27, 71.102
Brown, M.E. **5** 84.30
Brown, R. **1** 30.27 **6** 31.19
Brown, R.N. **1** 1.44
Brown, S.H. **1** 58.63
Brown, T.L. **2** 68.15
Brown, W.A.C. **1** 31.30, 50.17, 56.17 **2** 75.68
Browning, J. **5** 71.41, 72.48, 74.10
Brownlee, G.S. **4** 69.1
Broyde, S.B. **7** 47.38
Brubaker, G.R. **3** 79.16
Bruce, J.M. **6** 31.19
Bruce, M.I. **4** 71.70 **5** 71.44, 74.12, 86.79 **6** 71.74 **7** 71.131, 71.133, 71.139
Bruce, R.St.L. **6** 83.36
Bruckner, S. **2** 78.59 **3** 76.81 **5** 71.31, 76.39, 76.43, 83.41 **6** 83.10, 86.20 **7** 21.6
Brueckner, D.A. **1** 52.13
Brufani, M. **1** 35.3, 35.4, 35.7, 35.12, 35.13, 35.14, 35.15, 50.36, 50.48 **3** 50.28 **4** 1.11, 35.26, 54.19 **5** 50.13
Bruice, T.C. **7** 76.60, 85.18
Bruins, D. **3** 76.76
Brun, G. **5** 64.2
Brun, L. **2** 66.2, 84.26
Brundret, K.M. **5** 50.10
Brunel, Y. **1** 31.12
Brunelle, J.A. **7** 39.24
Brunet, J.-J. **5** 28.12, 28.14
Brunie, S. **6** 17.3, 38.40 **7** 61.5
Brunton, G.D. **3** 81.4 **4** 81.4, 81.5, 81.6, 81.7, 81.8
Brusca, R.C. **7** 17.7

Brusentsev, F.A. **2** 72.3
Brush, J.R. **5** 81.38
Brushmiller, J.G. **4** 82.10
Brusset, H. **1** 17.27 **2** 84.6, 84.12 **4** 17.3 **7** 3.15
Bruvo, M. **5** 63.1
Bryan, J.G.H. **3** 45.26
Bryan, R.F. **1** 2.91, 12.15, 12.19, 13.37, 23.2, 44.42, 44.48 **2** 69.14, 69.40, 73.31, 73.110, 73.121, 82.49, 83.62, 85.83, 86.23, 86.60, 86.91 **3** 15.3, 44.36, 44.39, 62.13, 69.22, 73.3, 73.4, 73.12, 73.13, 85.47 **4** 16.14, 24.11, 53.10, 54.4, 58.36, 62.1, 69.21, 73.33, 82.7 **5** 58.45, 58.47, 59.18, 59.19, 59.26, 59.34 **6** 36.18, 53.17, 54.2, 71.4 **7** 33.35, 38.57, 39.19, 39.20, 58.31, 59.9, 63.15, 63.16, 86.39
Bryant, M.W.R. **1** 11.58, 11.66 **3** 11.19 **4** 11.15
Bryden, J.H. **1** 8.17, 8.29, 8.33, 9.4, 9.6, 10.6, 11.32, 11.35, 32.2, 32.6, 32.16 **3** 32.21 **4** 15.5
Bryson, W.C. **4** 38.24
Bryukhova, E.V. **2** 71.1
Brzechffa, M. **3** 31.42 **5** 31.36
Bua, E. **1** 19.14
Buchanan, G.L. **1** 20.20, 38.58
Buchard, O. **5** 38.30
Buchardt, O. **1** 40.30
Buchi, G. **1** 53.40 **4** 59.27 **6** 58.52 **7** 58.28
Buchler, J.W. **6** 49.10
Buck, K.W. **1** 42.10
Buckingham, D.A. **2** 76.49, 82.54 **3** 82.12, 82.13, 82.14, 82.15 **4** 76.11, 76.17 **5** 76.31 **6** 76.16
Buckle, J. **5** 69.22
Buckles, R.E. **1** 12.20 **3** 11.13
Bucksch, R. **6** 3.5
Bugayong, R.R. **5** 48.48
Bugg, C. **1** 7.26, 33.55
Bugg, C.E. **1** 47.11 **3** 44.20, 47.15, 47.16, 47.19, 47.23 **4** 35.11, 36.31, 44.8, 44.20, 64.36, 67.6, 67.13 **5** 3.18, 44.12, 44.16, 45.24, 45.26, 45.32, 45.33, 45.39, 47.19, 48.23, 85.2 **6** 33.24, 33.25, 33.26, 35.15, 38.17, 44.13, 44.14, 45.13, 45.14, 45.36, 45.47, 47.7, 48.17, 48.58, 50.6, 58.1, 64.24 **7** 13.19, 44.1, 44.3, 44.4, 44.11, 44.12, 44.15, 44.21, 45.3, 48.19
Bukowska-Strzyzewska, M. **2** 81.7, 81.22, 81.31 **7** 64.28, 64.29
Bulgarovskaya, I.V. **5** 39.19 **7** 60.11
Bull, R.J. **6** 20.1
Bullen, G.J. **2** 62.43, 64.21, 64.46, 64.65, 77.11, 77.62 **3** 62.4, 62.17 **4** 33.23, 62.15, 64.12, 64.37 **5** 62.13, 64.4, 64.12, 64.21, 64.34, 64.45 **6** 64.5, 64.19, 64.27, 64.28, 64.45 **7** 64.55, 64.56
Bullitt, J.G. **4** 81.29 **5** 81.7
Bullock, J.I. **7** 84.61
Bullough, R.K. **1** 4.17
Bulusu, S. **6** 37.1
Bumgardner, C.L. **7** 3.32
Bunick, G. **6** 47.18
Bunnell, C.A. **7** 71.31
Buraway, A. **2** 71.16
Burbank, R.D. **1** 5.3 **3** 83.50
Burger, K. **5** 35.30, 36.25 **7** 32.15, 32.27
Burgermeister, W. **7** 48.74
Burgi, H.-B. **7** 85.2
Burgi, H.B. **1** 22.11 **3** 16.9, 16.10, 16.11 **5** 43.1 **6** 58.40
Burgmaier, G.J. **5** 31.2
Burgner II, J.W. **3** 32.11, 32.23
Burgstahler, A.W. **6** 52.1
Burke, J.M. **5** 85.34
Burke, M.E. **1** 36.5, 36.18
Burke, T.S. **6** 77.10
Burkhardt, L.A. **1** 8.17
Burkholder, P.R. **3** 58.7
Burki, H. **1** 44.5, 51.27
Burks, J.E. **7** 82.19
Burlet, P. **7** 81.14, 81.14
Burlitch, J.M. **4** 73.31 **6** 73.62
Burnell, R.H. **3** 58.17 **6** 58.14
Burnett, R.M. **4** 21.5
Burns, D.M. **1** 28.16, 29.11
Burns, J.H. **2** 65.9 **3** 77.16, 77.17 **4** 75.28 **5** 71.108, 75.39 **6** 71.52 **7** 73.30, 73.38
Burow, N.E. **1** 2.78, 2.79
Burr, A.H. **1** 16.49 **2** 64.65 **7** 64.55
Bursh, T. **5** 85.4, 85.5
Burt, J.C. **6** 75.23
Burton, C.A. **7** 83.51
Burton, D.J. **6** 64.41
Burzlaff, H. **1** 36.25, 39.63 **3** 39.15, 39.34 **6** 64.53, 64.54 **7** 20.11, 64.59
Busch, D.H. **6** 83.57
Busetta, B. **1** 36.34, 43.22 **3** 17.7 **4** 8.12, 8.13, 13.8, 17.9, 19.14, 51.8, 51.9, 51.12, 51.16, 51.18, 51.21 **5** 21.16, 30.1, 35.28, 41.40, 51.2, 51.3, 51.4, 51.11 **6** 17.20, 17.21, 17.22, 36.20, 51.3, 51.9, 59.4 **7** 17.30, 17.31, 30.8, 33.50, 36.20, 36.24, 36.28, 36.29, 50.35, 51.3, 51.9, 51.26, 59.22
Busetti, V. **1** 1.51, 38.2, 38.3, 39.8, 39.39 **3** 16.12, 32.12 **4** 39.4 **5** 42.2, 69.15, 69.25
Bush, M.A. **1** 3.57, 3.58, 3.59 **2** 66.1, 73.47, 86.17 **3** 38.15, 38.21, 73.1, 73.16, 73.17, 73.46, 73.52 **4** 10.2, 10.3, 38.26, 67.15, 67.21, 67.25, 73.2, 73.5, 76.41
Bushnell, G.W. **2** 86.15 **5** 83.22 **7** 83.26, 86.2, 86.42
Bushweller, C.H. **7** 39.24
Busing, W.R. **1** 8.21 **2** 83.199 **5** 8.2 **6** 59.18
Bussemeier, B. **4** 71.65
Butcher, R.J. **6** 81.25, 82.6
Butikova, I.K. **5** 76.6
Butler, D. **5** 32.6
Butler, I.S. **4** 86.41 **6** 86.49, 86.50
Butler, K.R. **3** 83.142 **4** 76.32, 81.24 **7** 85.4
Butler, W. **7** 63.1, 73.26, 73.47
Butler, W.M. **4** 71.1 **5** 71.8, 71.11, 71.14
Butman, L.A. **3** 77.11, 77.30 **7** 8.8, 35.10
Buttero, P.del **7** 41.22
Button, A.C. **6** 50.8

Buzell, A. **1** 7.4
Bycroft, B.W. **4** 50.8 **5** 41.24
Bye, E. **5** 48.63 **6** 1.31, 58.41 **7** 40.24, 77.12
Byram, S.K. **3** 68.3
Byrkjedal, A. **7** 48.42
Byrn, S.R. **4** 13.15, 13.16, 20.11, 20.13 **5** 37.3
Byushkin, V.N. **6** 76.2 **7** 76.32
Bywater, R.P. **3** 1.33

Cabestaing, C. **5** 27.9
Cady, H.H. **1** 15.19, 16.24, 34.1, 34.2, 40.2
Cady, R.T. **2** 83.162
Caffery, M.L. **7** 85.30, 85.31
Cagle Junior, F.W. **6** 17.14
Caglio, G. **6** 71.86, 86.69
Caglioti, L. **3** 36.21 **4** 71.41 **7** 38.43
Cahill, R. **5** 40.35
Caira, M.R. **4** 77.15 **6** 44.23, 83.74 **7** 44.9, 83.40, 83.43, 83.71, 85.13
Calabrese, J.C. **1** 13.45 **3** 1.29, 1.30, 3.38 **4** 63.1 **5** 62.10, 71.10 **6** 3.10, 31.15, 39.29, 64.47, 64.48, 65.9, 65.10, 85.11, 86.10 **7** 62.5, 71.67, 72.5, 73.42, 86.89, 86.111, 86.112
Calabresi, C. **5** 86.28, 86.32
Calas, M.-R. **1** 43.10 **4** 41.31
Calcaterra, M. **6** 46.1
Calderazzo, F. **5** 67.22
Calderon, J.L. **3** 73.20 **4** 73.6, 73.23 **6** 71.24, 73.6, 73.7
Calhoun, H.P. **6** 64.12, 64.14, 64.49, 83.67
Callahan, K.P. **5** 71.103 **6** 73.1, 73.15 **7** 71.8, 73.1, 73.16
Calleri, M. **1** 10.13, 13.53, 40.15, 40.16, 46.17 **2** 83.25 **3** 35.22, 59.4 **5** 40.24, 65.8, 83.3 **6** 40.2, 40.3, 40.6 **7** 40.2
Calleri, M.B. **3** 65.11
Calligaris, M. **1** 21.29 **2** 78.59 **3** 76.81, 78.14, 78.26, 83.150, 85.32, 85.33, 85.34, 85.40 **4** 71.40, 77.14 **5** 69.19, 71.31, 71.42, 71.57, 76.39, 76.43, 78.5, 78.6, 78.9, 83.41 **6** 71.89, 76.42, 78.11 **7** 71.32, 71.86, 78.4, 78.5, 78.7
Callow, R.K. **1** 51.38, 51.39, 51.40
Calvarin, G. **4** 73.20
Calvert, L.D. **1** 1.35
Calves, J.-Y. **7** 69.1
Calvin, M. **3** 49.6 **4** 49.8, 49.12
Calvo, C. **3** 8.2, 8.3, 81.31, 81.44 **4** 58.16, 71.51, 83.110 **5** 28.15, 40.15 **6** 4.4, 4.5, 28.22, 58.54, 64.1, 71.55 **7** 4.6, 32.26, 71.79
Calzolari, L.C. **3** 79.10
Cam, Y. **5** 41.45
Cambie, R.C. **4** 59.22
Camellini, M. **2** 83.7
Camellini, M.T. **3** 83.2 **4** 8.4, 79.4, 81.2, 81.15, 81.16 **5** 8.5, 41.14 **6** 11.12, 41.27 **7** 11.9, 11.12, 35.11, 79.4
Camerman, A. **1** 29.5, 30.6, 33.74, 45.61, 45.74, 58.79 **2** 65.19 **3** 24.4, 44.40, 52.6 **4** 13.21, 32.20, 32.25, 35.20, 48.50 **5** 33.52, 47.35 **6** 35.24, 48.73, 48.74, 64.8 **7** 17.29, 47.39, 48.44, 48.61, 48.62, 48.66, 82.14
Camerman, N. **1** 44.69, 44.73, 45.61, 47.6, 58.46, 58.79 **2** 65.1, 65.4, 65.8 **3** 44.40 **4** 32.20, 32.25, 35.20, 48.50 **5** 33.52, 47.35 **6** 48.73, 48.74, 64.8 **7** 17.29, 47.39, 48.44, 48.61, 48.62, 48.66, 82.14
Cameron, A.F. **1** 31.20 **3** 31.27, 31.28, 33.35, 51.44, 53.17, 59.16, 83.111, 86.58 **4** 3.6, 3.20, 31.24, 41.13, 83.37, 83.38, 83.41, 83.52, 83.53 **5** 52.4, 58.36, 83.52, 83.55, 83.56, 83.57 **6** 4.2, 28.15, 28.16, 31.18, 31.25, 40.1, 64.50, 83.47 **7** 36.30, 37.21, 41.33, 56.8, 58.23
Cameron, A.M. **5** 40.15
Cameron, I.R. **7** 41.33
Cameron, J.U. **4** 65.7 **7** 41.7
Cameron, T.S. **2** 60.73, 60.75 **3** 40.19, 40.20, 40.21, 64.43, 64.70 **4** 24.2, 60.15, 60.16, 64.21, 64.28, 64.29, 64.30, 71.15, 73.19 **5** 64.3, 64.21, 64.29, 64.36, 72.34, 72.53, 73.6 **6** 1.15, 16.9, 17.19, 33.44, 64.40, 73.41, 82.15, 84.1 **7** 4.8, 35.28, 35.39, 36.27, 71.24, 73.4, 73.5, 73.6, 73.7, 73.8, 73.14, 73.36
Cameroni, R. **7** 51.18
Camp, U.de la **3** 1.19, 11.11
Campbell, B.S. **5** 46.10 **7** 46.5, 64.41
Campbell, D.H. **7** 71.42, 75.21
Campbell, I.L.C. **6** 72.14 **7** 71.51, 73.11, 73.27, 73.57, 73.59
Campbell, J.A. **5** 51.6, 51.12, 77.10, 77.11
Campbell, J.C. **3** 45.31
Campbell, J.W. **5** 45.18
Campbell, M.M. **7** 41.33
Campbell, R.B. **1** 29.12, 30.14
Campbell, R.V.M. **4** 59.18
Campsteyn, H. **4** 51.28 **5** 51.17, 51.18, 51.19, 51.21 **6** 51.18, 51.19, 51.22, 51.23 **7** 51.15, 60.37
Camus, A. **6** 71.95 **7** 86.92, 86.96
Canals, E. **7** 81.4, 81.5, 81.6, 81.7
Canceill, J. **7** 28.2
Candeloro, S. **4** 83.137, 83.138
Candlin, R. **2** 82.45 **3** 82.22
Candy, H.A. **7** 59.18
Canepa, F.G. **1** 3.74 **5** 58.48
Cannas, M. **1** 1.46, 40.7, 40.8, 40.9, 40.17, 40.18 **3** 40.6, 76.48, 76.60, 83.64 **4** 76.6 **5** 76.5, 76.11, 76.38 **6** 76.7, 76.8, 83.8, 83.11 **7** 76.59, 83.63, 85.7, 85.10
Cannillo, E. **7** 84.19, 84.82
Cannilo, E. **2** 83.81
Cannon, J.F. **4** 40.14
Cannon, J.R. **6** 58.34 **7** 32.16
Cannon, R.E. **5** 68.1
Cano, F.H. **3** 13.13 **4** 71.11 **7** 40.12, 40.13, 41.1, 41.2, 41.3, 48.6, 48.73
Cantrell, C.E. **6** 38.17
Cantrell, J.S. **1** 59.22 **6** 38.30, 38.31 **7** 11.11, 17.34, 24.5
Canty, A.J. **4** 71.79, 71.88, 71.91

Capacchi, L. **1** 11.1 **2** 85.35
Capacchi, L.C. **2** 79.16, 85.14 **4** 66.8 **5** 39.14, 81.5, 81.47, 81.54
Capasso, S. **5** 36.16 **7** 36.12, 82.22, 82.23
Capiomont, A. **3** 37.3 **4** 12.13 **5** 1.37 **6** 37.6 **7** 33.33
Caramella, P. **6** 40.25
Carazzolo, G. **1** 38.2, 39.8, 39.9, 39.39
Cardin, C.J. **6** 71.53, 71.67
Cardin, D.J. **6** 71.53, 71.67
Carey, F.A. **7** 51.45, 63.15
Carfagna, P.D. **2** 86.51
Carfagno, P. **3** 79.6, 79.7
Carhart, R.E. **3** 21.17
Cariati, F. **2** 86.80 **4** 86.71
Cariello, L. **5** 36.15 **7** 36.12
Carlisle, C.H. **1** 33.10, 45.24, 46.18, 51.29, 56.13, 56.15 **3** 9.12, 9.15 **4** 39.9 **6** 41.15, 45.19, 51.42, 56.5 **7** 51.56, 64.55
Carlisle, G.O. **5** 85.3
Carlson, R.M. **6** 59.1
Carlstrom, D. **1** 8.39, 33.54, 37.26, 59.10, 59.11 **3** 35.16 **4** 3.19 **5** 17.14 **6** 63.9 **7** 20.12, 40.4
Carmichael, J.W. **2** 77.55 **3** 83.11 **4** 83.29
Carney, R.W. **4** 33.47
Caron, A. **1** 8.5, 10.19, 10.20, 48.59
Carpenter, C. **2** 75.4
Carpenter, C.L. **2** 72.47, 73.17
Carpenter, D.C. **1** 38.30
Carpenter, G.B. **1** 7.11 **2** 79.34, 79.35 **5** 15.2, 19.2
Carpenter, L.J. **7** 84.40
Carr, G. **3** 83.74
Carrabine, J.A. **3** 84.6 **4** 44.2, 44.3, 83.28, 83.40
Carrai, G. **2** 65.16
Carreck, P.W. **4** 83.25 **6** 72.8
Carrell, H.L. **2** 62.32 **3** 23.3, 31.30 **4** 2.20, 2.22, 36.26 **5** 2.27, 2.29, 26.5, 81.17, 81.18 **6** 2.21, 26.3, 36.4, 36.5, 36.15, 48.36, 48.37, 48.84 **7** 1.26, 1.27, 29.4, 29.7, 29.9, 36.21, 38.60, 38.62, 64.43, 76.60, 85.18
Carroll, P.J. **6** 64.59
Carruthers, J.R. **2** 81.27, 81.45, 81.48, 81.66, 81.81 **3** 59.33, 81.22 **4** 2.11, 81.12, 81.13, 81.14, 81.23 **6** 33.52, 33.56 **7** 76.72
Carson, J.F. **5** 48.28
Carstensen-Oeser, E. **1** 28.23 **2** 60.77 **4** 35.36 **5** 31.35, 32.16, 34.8, 34.9
Carta, G. **3** 76.48 **4** 76.6 **5** 76.5, 76.11, 76.38 **6** 76.7, 76.8, 83.8, 83.11 **7** 76.59, 83.63, 85.7, 85.10
Carter, D. **2** 83.140
Carter, D.R. **5** 7.5, 7.6, 33.7 **6** 33.17 **7** 33.27, 35.8
Carter, J.C. **2** 62.6
Carter, O.L. **2** 60.118, 73.16, 73.44, 74.6, 74.7, 74.14
Carter, V.B. **4** 7.9, 7.10
Carter, W. **2** 73.122
Carter, W.S. **7** 8.22, 33.26
Carturan, G. **7** 84.51
Carty, A.J. **2** 73.122 **3** 68.13, 78.5, 83.128, 83.137 **4** 71.61, 72.29, 83.106, 86.34 **5** 68.30, 71.90, 86.47, 86.58, 86.61 **6** 71.2, 71.5, 71.6, 71.60, 71.76, 86.43, 86.46, 86.56 **7** 71.5, 71.10, 71.98, 71.107, 71.113, 71.127, 71.138, 72.33, 72.38, 72.45, 82.2, 86.56, 86.109
Carty, P. **2** 86.70
Cartz, L. **5** 13.12 **6** 58.35, 81.40
Casagrande, G. **3** 74.3
Casagrande, G.T. **3** 74.4
Casalone, G. **1** 19.50 **3** 15.10, 19.11, 39.35 **5** 15.8 **7** 31.20, 71.60
Casalone, G.L. **1** 5.28, 5.31 **2** 60.103
Casciato, C.A. **4** 2.20, 2.21
Case, J. **3** 51.7
Case, J.A.M. **2** 72.65, 86.94
Casellato, U. **7** 78.10
Casey, A.T. **3** 83.110
Casey, C.P. **5** 71.10 **7** 71.31
Cash, G.G. **5** 75.46
Casida, J.E. **7** 52.1
Casini, G. **1** 35.3 **4** 35.26
Casinovi, C.G. **3** 59.33
Caspi, E. **4** 51.55
Cassaigne, A. **7** 64.2
Cassel, A. **7** 86.105, 86.106
Cassidy, F. **4** 24.13, 24.14, 24.15
Cassol, A. **3** 80.19, 80.20
Castain, W. **1** 9.8
Castellan, A. **5** 30.1 **7** 30.8
Castellano, E. **3** 60.29
Castellano, E.E. **3** 25.6 **4** 60.9, 60.10, 60.11 **6** 54.17
Castro, G. **6** 41.32
Castro, L. **5** 8.3
Cate, J.M.ten **7** 16.13
Catesby, C.G.C. **2** 60.6
Caticha-Ellis, S. **1** 13.54
Cattalini, L. **4** 71.41 **5** 86.21
Catterick, J. **7** 81.46
Caubere, P. **5** 28.7, 28.11, 28.12, 28.14, 31.33
Caughenour, K.N. **7** 39.30
Caughey, W.S. **1** 49.18, 49.19
Caughlan, C.N. **1** 39.93, 46.5, 46.7, 46.9, 46.10, 46.16, 46.24, 53.1, 53.3 **2** 64.25, 64.42, 69.10, 77.48, 84.22, 84.54, 84.58 **3** 46.2, 46.7, 53.4, 53.18, 64.17, 76.44, 83.61, 84.26 **4** 31.13, 53.1, 64.16, 64.19, 64.25, 64.39, 79.5, 84.15, 86.11 **5** 45.14, 46.4, 46.5, 46.8, 46.9, 59.1, 77.10, 77.11 **6** 46.5, 46.6, 46.7, 46.11, 53.14, 76.31 **7** 35.14, 71.40
Caughman, H.D. **6** 64.31, 84.19
Caulton, K.G. **2** 81.83 **4** 86.54 **5** 86.53 **6** 83.84
Cavalca, L. **1** 16.31, 39.41, 41.15 **2** 79.2, 79.5, 79.14, 79.15, 79.27, 79.33, 83.28, 84.5, 84.8, 85.3, 85.5, 85.37 **3** 16.4, 41.2, 61.8, 79.4, 79.10 **4** 9.5, 9.6, 26.5, 35.2, 39.11, 66.8, 76.12, 81.1, 83.7, 83.33 **5** 32.19, 32.25, 41.19, 42.3 **7** 32.21, 40.9, 40.10
Cavalieri, A. **6** 3.10

Cavan, M.P. **3** 30.4
Cavelier, C. **1** 43.21
Cedar, F.J. **6** 62.10
Cefalu, R. **7** 69.22
Cefola, M. **2** 77.65 **3** 33.7
Celap, M.B. **5** 82.20 **7** 81.17, 81.20, 81.21
Celotti, G. **3** 1.36
Cenini, S. **2** 86.89 **3** 86.98
Cernia, E. **7** 68.18, 68.19
Cerrini, S. **3** 32.36, 58.15, 59.33 **4** 36.34, 54.19 **5** 50.13, 56.2
Cesari, M. **3** 71.25, 78.12, 78.16 **4** 74.1 **6** 71.54, 71.78, 71.83, 72.6 **7** 32.12, 68.16, 68.17, 68.18, 68.19, 71.132, 72.22
Cesario, M. **4** 58.7 **5** 58.8, 58.9 **6** 1.17, 1.18, 1.19, 1.20, 1.21 **7** 1.28, 1.29, 1.30
Cesur, A.F. **1** 17.21, 53.18
Cetinkaya, B. **7** 71.41
Cetinkaya, E. **6** 49.8
Chacko, K.K. **4** 48.26, 48.34, 48.37 **6** 48.8, 48.61 **7** 20.1, 48.36
Chackraburtty, D.M. **5** 81.3
Chaiwasie, S. **2** 73.6
Chakraborty, D.P. **6** 58.46
Chakravorty, A. **5** 78.8
Chalyi, V.P. **7** 82.1, 82.9, 82.10
Chamberlin, J.W. **1** 50.45 **5** 50.26, 50.27, 50.28
Chan, C.-K. **7** 21.13
Chan, J. **5** 36.21
Chan, J.A. **5** 58.15
Chan, L.Y.Y. **3** 73.6 **4** 70.2, 86.3 **5** 86.14 **7** 69.12
Chan, N. **4** 83.29
Chandler, G.S. **7** 83.6
Chandler, W.D. **5** 15.10, 15.11, 15.12 **6** 13.19, 13.20, 13.21, 13.22, 13.23
Chandrasekaran, R. **1** 16.23, 20.14, 44.45, 48.81, 48.85
Chandrasekharan, G. **1** 20.14
Chandrasekharan, R. **3** 45.15, 45.16
Chandross, R.J. **6** 51.32 **7** 51.30
Chaney, M.O. **1** 48.58 **3** 29.2, 48.28, 48.45 **4** 51.10 **5** 38.23, 41.39, 50.5, 50.7, 50.26, 50.27, 50.28 **6** 47.15, 48.29, 50.10
Chang, F.C. **3** 56.2
Chang, G. **3** 86.40
Chang, K. **3** 84.1, 84.3, 84.7
Chang, M.-M.Y. **7** 11.19
Chang, S.C. **2** 83.97 **3** 81.32 **5** 81.36
Chang, S.W.-Y. **6** 71.37, 74.13 **7** 3.21, 71.97
Chang, S.W.-Y.N. **6** 71.38
Chang, T.W. **6** 67.4
Chanh, N.B. **5** 24.2 **6** 18.2, 24.1 **7** 24.2, 24.3
Chantot, J.-F. **6** 47.20
Chao-Fa, L. **1** 3.54
Chao, G.K.-J. **6** 77.14
Chao, G.Y. **2** 60.52, 60.57, 70.19
Chao, K. **4** 67.10
Chao, K.-J. **3** 67.9
Chapelle, A. **3** 58.17 **6** 58.14
Chapman, O.L. **3** 38.8 **4** 36.14, 59.16 **6** 38.22
Chapps, G.E. **7** 80.5
Chapuis, G. **6** 38.14
Charbonneau, G. **1** 19.30, 19.32
Charbonneau, G.-P. **1** 19.29, 19.33
Charlton, R.J. **2** 83.154
Charlton, T.L. **1** 15.9
Chasar, D.W. **1** 39.74
Chasseau, D. **3** 25.3, 25.5, 60.32 **4** 25.3, 35.9 **5** 7.9, 7.10, 35.10 **6** 12.1
Chastain Junior, R.V. **3** 37.6 **6** 76.19
Chastain, R.V. **2** 78.52, 83.52 **3** 8.14
Chasteen, N.D. **3** 77.24, 81.52
Chatani, Y. **1** 1.30 **3** 9.11 **6** 38.2 **7** 38.5
Chatt, J. **2** 71.38 **3** 86.72 **7** 86.100, 86.110
Chatterjee, K.K. **7** 79.8
Chatzidakis, A. **2** 64.68
Chaudhuri, B. **1** 9.38, 17.50 **4** 49.11
Chawdhury, S.A. **1** 16.46, 16.57 **3** 11.27 **7** 11.18, 32.30, 38.15
Che'ng, W. **2** 76.25
Cheeda, G.B. **3** 44.32
Cheer, C.J. **3** 21.15, 53.27 **5** 21.14, 21.15
Cheeseman, T.P. **2** 78.40, 78.50, 83.192
Chekhlov, A.N. **6** 58.20 **7** 17.13, 17.14
Chen, B.M.L. **4** 49.1
Chen, C.S. **7** 32.17
Chen, H.W. **7** 86.49
Chen, I.Y. **4** 28.10
Chen, J.-Y. **5** 86.35
Chen, K.-N. **6** 71.37
Chen, S. **5** 50.26, 50.27, 50.28
Chen, S.M. **1** 1.123
Chen, V. **5** 41.3
Chen, Y.-P. **7** 54.16
Cheney, A.J. **5** 71.86, 71.87, 71.88
Cheng-Chung Lin, Y **1** 48.48
Cheng, J.C. **1** 32.29 **3** 50.8
Cheng, P.-T. **3** 3.43, 3.44, 86.64 **4** 72.25, 86.31 **5** 72.42, 72.55 **6** 11.7, 20.18, 20.19 **7** 71.122, 71.123, 86.65
Cheng, P.T. **3** 81.54
Cherepinskii-Malov, V.D. **3** 64.30 **4** 64.17 **5** 64.32
Cherin, P. **6** 24.10
Cherkasova, T.G. **4** 83.128
Chernyshev, E.A. **4** 63.8 **7** 63.4, 63.8, 63.11, 69.10
Chesick, J.P. **4** 41.7 **5** 31.25 **6** 9.3 **7** 11.20
Chesnut, D.B. **1** 20.4 **4** 64.24, 65.9
Chetkina, L.A. **1** 5.24, 5.30, 26.1, 26.2, 26.4, 26.6, 26.7, 26.8 **4** 24.4, 24.6, 24.7 **5** 40.26, 40.27 **6** 40.16, 41.20 **7** 11.7, 26.1, 36.6, 42.10
Cheung, K.K. **1** 54.25, 59.32, 59.33, 59.34, 59.35 **2** 65.51, 79.8, 85.44 **3** 8.18, 59.16, 86.46, 86.71 **4** 75.23
Cheung, L.D. **5** 41.28, 64.43 **7** 31.30
Chevrier, B. **1** 12.5 **5** 12.3, 12.4, 66.2, 66.3, 66.4, 68.11, 68.12, 68.13 **7** 49.14
Chheda, G.B. **3** 44.29 **6** 44.16, 48.39, 48.40, 48.75 **7** 44.27, 48.45, 48.52
Chiang, C.-K. **6** 53.17
Chiang, C.C. **5** 23.1 **6** 34.2

Chiari, G. **3** 59.4 **5** 40.24 **6** 40.2, 40.3, 40.6, 40.9
Chiaroni, A. **3** 16.2, 16.7 **4** 51.23 **5** 83.96 **6** 51.11, 51.21 **7** 21.9, 51.23
Chiba, A. **1** 48.56
Chiba, T. **1** 20.1
Chidambaram, R. **1** 17.3 **3** 2.2 **5** 48.11, 48.15, 48.30, 48.48 **6** 48.32, 81.27 **7** 48.15, 48.16
Chidester, C.G. **4** 27.9, 34.3, 35.28, 35.29, 36.20, 36.33, 40.23, 45.23, 50.3, 51.46, 51.55 **5** 50.1 **6** 59.27
Chieh, P.C. **1** 37.27 **3** 1.10, 1.11, 33.27, 63.15, 69.32, 84.23 **4** 63.11, 69.32, 69.33, 78.12, 83.106 **5** 35.32, 83.86, 86.58 **6** 1.1, 71.2, 71.5, 71.6, 86.43, 86.46 **7** 71.5, 71.10, 71.98, 71.127, 71.138, 82.2, 86.109
Chiesi Villa, A. **1** 39.32
Chiesi, A. **3** 76.1, 83.10 **4** 79.12
Chih, C. **3** 67.9
Chih, H. **6** 69.5, 69.7
Chikaishi, K. **5** 60.21, 60.22
Childers, L.S. **7** 83.65
Childers, R.F. **7** 78.18
Childs, J.D. **2** 60.37
Chin-You, N. **4** 58.16
Chin, H.B. **6** 71.40 **7** 64.64, 64.65, 86.13, 86.14
Chini, P. **3** 86.72 **5** 71.35 **6** 3.11, 64.47, 64.48, 65.9, 65.10
Chioccola, G. **2** 63.29, 64.84, 86.101 **4** 71.63
Chion, B. **3** 37.3 **4** 12.13 **7** 32.9, 32.13, 32.14
Chipman, D.M. **3** 71.2 **5** 71.5
Chisholm, M. **6** 83.28
Chisholm, M.H. **7** 80.14, 80.15, 83.50
Chisholm, N.H. **2** 83.123
Chiu, C.C. **1** 1.128, 45.75
Chiu, M.F. **5** 41.37
Chiu, Y.-Y.H. **6** 38.29 **7** 50.19
Chiu, Y.H. **3** 46.1, 64.26, 64.27
Chivers, T. **6** 69.1 **7** 69.6
Chlenov, I.E. **3** 40.2
Choi, C.S. **3** 34.1 **4** 15.3 **5** 33.2, 34.3 **6** 37.1
Chomic, J. **7** 76.7
Chomyn, A. **5** 26.5
Choong, W. **7** 45.13, 81.43
Chothia, C. **1** 59.13, 59.14 **3** 20.9 **5** 58.23
Chou, B.Ch.-K. **7** 73.41
Chou, K.-D. **1** 11.19
Chow, C. **6** 65.8
Chow, W.Y. **6** 28.3
Chow, Y.L. **7** 37.9
Chow, Y.M. **3** 69.7, 69.21 **4** 63.3, 69.6, 69.7, 69.9, 69.10 **6** 83.3, 84.2
Chowdhury, B.K. **6** 58.46
Choy, C. **7** 63.19
Christensen, A. **7** 64.20
Christensen, A.T. **1** 7.20, 16.2, 51.26 **3** 27.7, 51.31, 51.40, 51.45, 51.46 **5** 51.32
Christensen, J.J. **4** 67.23 **7** 38.64, 38.65, 42.5, 42.6, 42.9
Christian, D.F. **7** 71.104
Christian, S.D. **2** 60.37 **4** 4.4
Christofferson, G.D. **2** 70.1, 70.22
Christoph, G.G. **5** 16.9, 78.20 **6** 76.30, 83.17, 83.38, 85.39 **7** 21.1, 31.1
Christophersen, C. **7** 40.7
Chu, C.T.-W. **6** 83.12
Chu, J. **5** 41.22 **6** 41.18
Chu, S. **1** 13.13
Chu, S.-C. **3** 13.1
Chu, S.C. **1** 58.101
Chu, S.S. **1** 45.42
Chu, S.S.C. **1** 18.2, 45.37, 45.41, 45.67, 45.73 **5** 39.44 **6** 39.13, 39.18, 39.20, 39.21 **7** 39.27, 41.15, 41.20
Chui, K.M. **4** 31.7, 41.27 **7** 86.73, 86.77, 86.78, 86.79, 86.80
Chung, B. **6** 39.13, 39.18, 39.20
Chupakhina, R.A. **6** 81.39
Churchill, M.R. **1** 3.82 **2** 65.41, 71.13, 71.51, 72.27, 72.41, 73.7, 73.10, 73.32, 73.67, 73.73, 73.112, 73.113, 74.19, 75.8, 75.27, 75.38, 75.40, 75.42, 75.43, 75.69, 75.76 **3** 62.14, 62.15, 64.52, 71.8, 71.16, 71.18, 71.24, 71.43, 71.60, 72.9, 72.43, 73.35, 73.40, 75.1, 75.22, 75.31, 75.32, 75.33, 75.43, 83.90, 83.93, 85.14 **4** 62.7, 62.8, 71.2, 71.35, 71.74, 71.75, 73.16, 73.18, 85.9 **5** 62.8, 71.84, 71.85, 73.39, 75.5, 75.29, 75.36, 83.68, 83.70, 86.84 **6** 62.2, 71.37, 71.38, 71.72, 71.96, 73.4, 73.18, 73.25, 74.13, 75.13, 75.25, 83.45, 83.46, 86.1, 86.70, 86.71 **7** 3.10, 3.21, 71.54, 71.97, 71.139, 73.17, 83.51, 83.89, 86.44, 86.45, 86.113
Chwang, A.K. **6** 47.14 **7** 47.25
Ciajolo, M.R. **5** 1.29, 1.32, 8.18, 13.16 **6** 48.52 **7** 1.31, 20.4
Ciampolini, M. **5** 76.42
Cian, A.de **3** 60.51, 72.21 **5** 72.14, 72.22, 72.23, 72.25, 75.6
Ciana, A. **3** 85.32, 85.33, 85.40
Ciani, G. **2** 60.149 **4** 86.57 **5** 64.20, 84.32, 86.24, 86.56, 86.79, 86.83 **6** 3.11, 86.36 **7** 86.31, 86.66, 86.107
Ciani, G.F. **1** 3.41
Cid-Dresdner, H. **1** 50.30
Ciechanowicz, M. **3** 86.74
Cihonski, J.L. **7** 71.17
Cingi, M.B. **2** 81.80 **3** 81.19 **4** 76.38, 83.55 **5** 69.14, 81.25, 81.43, 83.16, 83.36, 83.38 **7** 86.88
Ciric, J. **6** 36.16
Clapp, R.C. **1** 41.10 **3** 45.44
Clardy, J. **3** 58.24, 58.50 **4** 35.14, 36.14, 36.22, 39.17, 60.36 **5** 27.3, 27.16, 31.11, 31.39, 52.1, 56.4, 58.15, 59.21, 71.21, 86.2 **6** 27.11, 27.12, 28.13, 31.8, 35.17, 53.11, 58.33, 58.49, 58.52, 59.1, 59.11, 71.35, 71.81, 75.7, 75.16, 82.11 **7** 35.13, 50.9, 50.10, 50.13, 51.49, 54.7, 56.3, 58.3, 58.20, 58.24, 58.28, 71.137

Clardy, J.C. **3** 13.15, 31.9, 32.20, 38.8, 56.2, 58.19 **4** 13.17, 13.18, 59.16, 64.4, 64.31 **5** 31.31, 42.5, 42.8, 42.9, 76.17 **6** 38.22 **7** 64.15, 64.22, 64.23, 64.54, 86.7, 86.16, 86.19
Clark, G.L. **1** 50.11
Clark, G.R. **2** 77.17, 78.27, 83.109 **3** 39.36, 54.1, 78.8, 86.23 **4** 33.34, 58.41, 86.35 **5** 38.13, 75.2, 79.20, 79.21, 83.66, 85.33 **6** 54.4, 54.7, 83.24, 86.67 **7** 71.104, 86.68, 86.99, 86.103
Clark, H.C. **2** 69.2, 86.16
Clark, H.G. **1** 7.25
Clark, J.R. **1** 2.8, 2.34, 2.61
Clark, M.J. **6** 4.3
Clark, N.H. **3** 62.4
Clark, P.W. **5** 72.40
Clark, R. **7** 71.93
Clark, R.C. **6** 58.53
Clark, R.J.H. **3** 86.19, 86.25
Clarke, D.G. **3** 59.23
Clarke, G.R. **6** 1.8
Clarke, J.F. **7** 73.37
Clarke, P.J. **7** 32.22, 59.1, 83.28, 83.29
Clarke, P.L. **6** 69.19
Clarke, P.T. **2** 62.50
Clastre, J. **1** 17.11, 19.16, 26.34, 41.42, 48.50, 48.51 **3** 3.24
Clausen, A. **3** 76.51
Clauss, K. **4** 40.5, 44.13
Claverini, R. **5** 72.21 **6** 71.23 **7** 1.20, 20.8
Clay, R. **7** 3.6
Clay, R.M. **5** 3.8
Clayton, R. **5** 37.8
Clayton, W.R. **5** 61.4 **6** 3.32, 33.27, 86.35
Clearfield, A. **4** 35.5 **5** 83.113, 85.36 **6** 86.48 **7** 71.120, 73.21, 73.22, 83.52, 83.80
Clec'h, G. **7** 19.2
Clegg, G.A. **7** 46.1
Clegg, H.J. **3** 81.25
Clegg, W. **5** 83.106 **6** 83.58, 83.69, 84.17 **7** 45.26
Clemens, J. **4** 71.84, 71.86
Clemente, D.A. **1** 33.75 **3** 39.39, 40.15, 64.42, 69.33, 72.49 **4** 39.27, 40.19, 71.78, 78.11 **5** 69.28, 86.21 **6** 16.16, 44.24, 76.40, 84.18, 86.47 **7** 78.10, 80.2, 84.41, 84.51, 86.74, 86.88
Clerici, M.G. **6** 71.54, 71.78, 71.83
Clews, C.J.B. **1** 44.1, 44.2, 44.13, 44.14, 44.43, 44.44 **3** 19.12, 19.13, 19.14
Clippard, P.H. **4** 62.2, 62.3, 62.4
Clunie, J.S. **1** 53.17
Coates, E. **2** 77.2, 77.3
Coates, R.M. **1** 53.35
Cobbledick, R.E. **5** 13.10, 14.2 **6** 3.17, 60.12, 60.13 **7** 34.1
Coburn, R.A. **3** 45.44
Cocco, G. **1** 2.12
Cochran, T.G. **3** 52.6 **4** 13.21, 44.9, 44.18, 44.25, 50.2 **7** 58.4, 58.5
Cochran, W. **1** 2.55, 13.22, 44.1, 44.2, 44.13, 44.34, 44.44, 45.70, 48.44
Coda, A. **1** 36.13 **5** 84.18 **6** 10.2, 84.44 **7** 17.4, 17.5, 27.7, 51.50, 76.72, 84.19, 84.82
Codding, P.W. **4** 49.9 **5** 58.46, 77.23 **6** 33.41, 58.2
Cody, V. **3** 48.41, 69.34 **4** 51.2, 51.3, 51.47, 51.50, 51.51 **5** 51.35, 60.26, 60.28 **6** 38.32, 63.8 **7** 17.22, 17.25, 17.26, 24.9, 48.55, 48.60, 51.7
Coetzer, J. **3** 75.47, 79.23, 85.8 **4** 54.17, 59.25, 77.17 **5** 45.49, 58.25, 59.5 **7** 33.3, 58.14
Coffen, D.L. **7** 39.9
Coggon, P. **1** 53.7 **3** 3.48, 50.1, 53.28, 53.29, 54.7, 58.20, 59.7, 59.17, 59.34, 68.12, 78.3, 78.23, 86.1 **4** 28.10, 51.40, 51.41, 59.19, 59.26, 81.55 **5** 59.16 **6** 64.15
Coghi, L. **2** 83.28, 83.46, 84.2 **3** 79.14, 83.10 **4** 83.5, 83.12 **6** 76.45, 78.4 **7** 35.35, 35.36, 38.75, 69.13, 73.61, 73.62, 73.63, 76.22, 83.2
Cohen-Addad, C. **5** 1.25, 16.19, 16.21 **6** 1.23, 1.24, 1.25, 1.28, 16.18 **7** 1.21, 1.32
Cohen, G.H. **1** 49.24, 49.29 **2** 83.137
Cohen, M.D. **3** 1.38
Cohen, N. **6** 27.21
Coiro, V.M. **4** 8.11 **6** 51.5 **7** 48.67
Cojazzi, G. **3** 32.12 **7** 5.19
Cola, M. **5** 84.18
Colapietro, M. **1** 4.12 **2** 65.27, 80.16 **3** 32.36, 72.4 **4** 80.15 **5** 21.9
Cole-Hamilton, D.J. **7** 86.43
Cole, F.E. **3** 11.22, 48.12, 48.13, 48.39, 58.1 **4** 48.12 **5** 3.20, 33.37, 35.9 **6** 2.22, 48.21
Cole, R.J. **6** 59.1 **7** 58.20, 58.24
Cole, S.J. **2** 82.50
Colebrook, L.D. **6** 32.18
Coleman, J.M. **2** 73.117, 73.118, 85.71
Colens, A. **6** 32.36
Coles, B.F. **7** 63.9
Coletta, M. **1** 33.75
Colleter, J.-C. **1** 33.43
Colleter, J.C. **1** 33.56, 33.57 **3** 33.25 **6** 33.34, 59.4 **7** 24.8, 50.33
Collier, J.E. **4** 64.11 **7** 64.51
Collier, M.R. **4** 71.81
Collin, R.L. **1** 7.12 **5** 64.19, 84.2 **6** 46.3
Collins, A.J. **6** 64.2
Collins, D.M. **1** 58.44 **3** 49.11, 49 12, 69.29, 69.30 **4** 49.21, 49.24, 76.36 **5** 49.7 **6** 73.33, 73.34, 73.35
Collins, E. **2** 65.7
Collins, P.H. **7** 65.4, 69.8
Collins, R.K. **4** 83.76 **5** 83.54
Collman, J.P. **5** 78.13, 78.14, 86.36 **6** 69.22 **7** 49.22
Colman, P.M. **3** 60.4, 60.5 **5** 3.16 **6** 35.26
Colombo, A. **1** 29.17, 29.18, 38.27, 38.28, 38.41 **2** 61.17, 61.19 **4** 72.16 **6** 75.15
Colon, A. **6** 48.48
Colson, J. **3** 21.3, 21.4
Colson, S.D. **7** 19.1

Colton, R. **7** 86.104
Colton, R.H. **1** 2.43 **3** 60.36
Comberton, G. **4** 8.12, 51.12 **7** 51.26
Comer, F.W. **1** 53.38
Commons, C.J. **7** 86.104
Conde, A. **5** 33.32, 39.21 **6** 42.7 **7** 42.7, 45.19, 45.38
Cone, E.J. **7** 22.3
Conn, J.F. **7** 84.49
Conneely, J.A. **7** 72.6
Connelly, N.G. **3** 73.15, 73.18, 73.19 **6** 83.40
Conner, S.H. **1** 33.21, 33.24, 33.25
Connolly, J.D. **5** 56.10, 56.11 **7** 56.8
Connor, J.A. **2** 71.24
Considine, J.L. **6** 68.16
Constant, G. **4** 83.3 **5** 83.2, 83.40, 83.58, 84.1 **6** 71.14, 83.1
Conti, F. **2** 86.89 **3** 86.98
Contreras, J.G. **7** 68.15
Cook, C.D. **3** 86.64 **4** 72.25, 86.31
Cook, D.F. **7** 78.17
Cook, D.S. **1** 46.18
Cook, P.M. **5** 12.6, 12.7, 12.8, 12.9, 12.10, 77.1
Cook, R.E. **3** 39.45, 45.29, 58.29 **4** 39.47
Cook, W.J. **4** 67.6, 67.13 **5** 45.26, 45.33, 45.39 **6** 45.13, 45.14, 45.36, 65.11 **7** 44.15, 44.21
Cooke, J. **3** 64.52, 85.14
Cooke, M. **7** 75.38
Cooke, P.W. **2** 60.86
Coomes, R.M. **1** 58.52
Cooper, A. **1** 51.5, 51.7, 51.9, 51.10, 51.12, 51.13, 51.14, 51.17, 51.21, 51.22, 51.41, 51.50 **2** 60.151, 77.15, 82.53, 84.64 **3** 17.8, 51.12, 51.16 **4** 51.15, 51.17 **6** 51.33
Cooper, M.J. **3** 37.6
Cooper, M.K. **6** 72.8, 83.36, 85.23
Cooper, R.D.G. **3** 50.8
Cooper, W.F. **6** 39.6
Copeland, R.F. **1** 3.44, 33.21, 33.24, 33.25
Copeland, V.C. **4** 83.22 **5** 83.11, 83.12, 83.46 **6** 83.33
Coppens, P. **1** 1.18, 1.19, 15.20, 15.21, 17.5, 17.6, 33.3, 33.4, 33.5 **2** 64.4 **3** 33.1, 33.4, 33.7, 60.44 **4** 7.5, 22.9, 33.4, 76.15 **5** 15.1 **6** 7.3, 39.6, 74.1, 74.2, 80.7, 80.8 **7** 16.2, 33.8, 48.13, 60.3, 60.23, 60.24
Coppola, F. **1** 33.70, 33.72
Coppola, J. **1** 58.47
Coppola, J.C. **3** 46.5, 46.6, 51.42, 53.12, 53.13, 53.14, 65.5 **4** 8.9, 23.2 **5** 47.24, 47.31, 51.9, 51.22, 51.25 **6** 64.56, 65.1, 65.2, 65.3
Corao, E. **3** 79.17
Corazza, E. **7** 54.5
Corbellini, M. **7** 68.16
Corbett, M. **2** 83.213, 83.218 **3** 83.151 **5** 83.120, 83.121, 83.122
Corbridge, D.E.C. **1** 24.18 **2** 83.141
Corden, B.J. **3** 3.56 **6** 86.37, 86.54
Cordes, A.W. **1** 35.22 **3** 33.34, 83.11 **4** 64.9, 65.5, 65.8, 83.29 **5** 41.34, 65.1, 65.2, 65.5 **6** 38.20, 39.14, 41.25, 48.82, 64.9, 64.58 **7** 41.25
Corey, E.J. **3** 40.13
Corey, E.R. **2** 64.41, 64.67, 73.48, 73.96, 83.65, 84.18 **3** 69.34 **5** 69.26 **6** 63.8
Corey, J.Y. **5** 69.26
Corey, R.B. **1** 48.33, 48.60
Corfield, P.W.R. **1** 28.15, 45.18 **2** 71.23, 72.29, 72.30, 76.39, 86.16, 86.58 **3** 11.26, 21.17, 64.6, 65.7, 71.71, 77.23, 85.15, 86.16 **5** 27.6, 27.7, 45.20, 61.4, 85.1, 85.10 **6** 76.6, 76.17, 83.57, 86.60, 86.61
Corney, D.J. **4** 33.23
Corradi, A. **1** 41.21
Corradi, A.B. **4** 59.28 **5** 82.1, 82.10 **6** 40.12, 76.15, 76.41, 79.16, 81.36, 83.66, 84.34 **7** 66.4, 79.24, 82.6
Corradini, P. **1** 1.86, 1.133, 44.53 **2** 73.111, 74.24 **3** 20.4, 21.5, 21.7, 33.16, 33.19, 72.19 **5** 1.29, 1.32, 8.18, 13.16, 41.3 **7** 1.31
Corsico, A.C. **7** 27.7
Cosson, J.-P. **7** 58.32
Costamagna, J.A. **7** 84.52
Costello, W.R. **2** 85.12
Costin, C.R. **1** 1.122
Cosulich, D.B. **1** 59.50 **3** 21.13, 50.2 **4** 25.7, 41.24
Cot, L. **7** 48.5
Cotrait, M. **1** 2.93 **3** 2.23 **4** 1.10 **5** 3.22, 24.7 **6** 48.59, 48.60, 48.65, 48.66, 48.67, 48.70, 58.11 **7** 33.55
Cotton, F.A. **1** 3.63 **2** 65.44, 65.47, 66.20, 66.23, 71.44, 72.25, 73.39, 74.9, 74.17, 75.30, 75.32, 75.39, 75.57, 75.58, 75.66, 76.3, 76.44, 77.39, 77.51, 77.56, 77.57, 77.61, 77.63, 77.64, 77.67, 80.23, 81.25, 81.50, 81.83, 81.88, 81.89, 82.49, 83.1, 84.21, 84.24, 84.66, 85.29, 85.54, 85.55, 86.12, 86.59, 86.97 **3** 3.8, 11.16, 11.17, 73.20, 73.44, 75.5, 77.33, 81.37, 81.53, 83.132, 83.133, 86.57 **4** 71.10, 72.9, 73.6, 73.23, 75.6, 81.26, 81.28, 81.29, 86.25, 86.54 **5** 16.12, 71.30, 72.4, 75.9, 75.15, 75.22, 75.25, 75.31, 81.28, 81.50, 81.63, 86.45 **6** 8.4, 69.14, 71.8, 71.16, 71.31, 71.32, 71.41, 72.19, 73.24, 73.30, 73.33, 73.34, 73.35, 73.40, 73.49, 73.59, 75.6, 75.9, 75.21, 75.26, 83.9, 83.13, 83.28, 83.42, 86.15 **7** 8.18, 71.20, 71.82, 71.84, 72.15, 72.28, 73.31, 75.8, 75.13, 75.24, 81.27, 83.50, 85.3, 85.9, 85.29, 86.22, 86.61
Cotton, J.D. **7** 69.17
Cotton, S.A. **5** 71.79
Couch, D.A. **3** 83.147
Coucouvanis, D. **1** 3.69 **2** 85.89 **3** 85.31 **5** 73.10 **6** 85.1, 85.2, 85.3, 85.12, 85.13, 85.41 **7** 81.48, 85.30, 85.31, 85.40, 86.118
Couldwell, M.C. **7** 76.26

Coulter, C.L. **1** 11.21, 31.22, 47.1 **3** 47.4, 49.14 **5** 47.5 **6** 44.20, 47.4 **7** 44.22, 48.56
Countryman, R. **3** 49.12 **4** 3.17, 49.21, 71.32, 76.34 **5** 71.49, 71.92, 72.44, 86.38 **6** 67.5 **7** 19.7
Cour, T.la **6** 32.3, 32.4, 32.5, 41.1
Courseille, C. **1** 25.21 **3** 25.2 **4** 8.12, 8.13, 13.8, 17.9, 19.14, 25.2, 51.9, 51.12, 51.16, 51.18, 51.21 **5** 21.16, 30.1, 35.28, 51.2, 51.3, 51.4, 51.11 **6** 17.20, 17.21, 17.22, 18.2, 29.5, 36.20, 41.13, 51.3, 51.9 **7** 17.30, 17.31, 24.2, 24.3, 25.2, 30.8, 36.20, 36.24, 36.28, 36.29, 51.3, 51.9, 51.26, 59.3, 59.22
Courtois, A. **4** 2.9 **5** 28.7, 28.11, 28.12, 28.14, 31.33
Couvillion, J. **1** 3.55
Coville, N.J. **4** 86.41 **6** 86.49, 86.50
Cow, J.W. **2** 64.67
Cowan, D.O. **6** 60.3
Cowie, M. **6** 85.31, 85.33, 85.34
Cowley, D.J. **6** 16.9
Cowman, C.D. **6** 77.5
Cox, A. **4** 71.45
Cox, D.A. **4** 56.3
Cox, D.E. **3** 28.3, 28.4
Cox, E.G. **1** 1.8, 9.8, 39.31, 39.70, 45.11, 45.13 **2** 76.80, 83.141, 85.22 **3** 19.3
Cox, G. **1** 45.33
Cox, G.W. **1** 1.12
Cox, J.R. **1** 46.11
Cox, J.W. **2** 64.41
Cox, M.M. **2** 83.197
Cox, M.R. **1** 2.92 **4** 53.10
Cox, P.J. **4** 24.13, 24.14, 24.15 **6** 41.17 **7** 53.2, 53.3, 53.6, 53.20, 53.21, 53.22, 53.26, 53.29
Coxon, D.T. **7** 53.19
Coxon, J.M. **1** 52.17 **7** 31.7
Coyle, B.A. **3** 84.24
Cozzarelli, N.R. **6** 44.20 **7** 44.22
Crabb, T.A. **5** 40.35
Crabtree, R.H. **7** 86.110
Cradwick, E.M. **3** 83.66
Cradwick, M.E. **3** 83.12, 83.33 **4** 76.42 **5** 53.10 **6** 69.19 **7** 69.25
Cradwick, P.D. **3** 53.5, 76.50 **4** 3.13, 25.4, 25.5, 25.6, 37.12, 53.7, 76.42 **5** 53.10 **7** 58.26, 58.29, 86.110
Craig, D.C. **4** 83.113 **5** 45.16 **6** 45.15, 45.18, 45.42, 77.1, 83.61 **7** 45.13
Cramer, R. **4** 71.6
Cramer, R.E. **5** 77.22
Crane, R.I. **4** 53.3
Cras, J.A. **2** 80.8, 80.9, 80.34 **3** 80.11 **4** 80.4, 80.11, 80.18, 85.4 **5** 39.20 **6** 80.14 **7** 80.17
Cravador, A. **6** 58.6
Craven, B. **1** 1.149
Craven, B.M. **1** 1.12, 19.8, 43.5, 43.6, 43.7, 43.9, 43.15, 43.23, 43.24, 43.25, 43.27, 43.28, 43.29, 44.4, 58.56, 58.85, 59.30, 59.31 **2** 83.67, 83.68 **3** 83.115, 83.116 **4** 43.1, 43.2, 43.3, 43.6, 43.9, 60.42 **5** 3.17, 43.2, 44.1, 44.2, 46.3 **6** 8.3, 33.47, 43.1, 60.5, 60.6, 60.7, 60.8, 60.9, 60.10, 60.11, 60.16 **7** 43.1, 43.2, 43.3
Crawford, J.L. **5** 28.1
Creitz, T.C. **2** 83.32
Crescenzi, S. **5** 36.15 **7** 36.12
Cresswell, P.J. **6** 76.16
Cristini, A. **6** 83.11 **7** 76.59
Critchley, S.R. **7** 32.4, 71.24, 73.4, 73.5, 73.6, 73.7, 73.8, 73.14, 73.36, 73.49, 73.50
Croatto, U. **1** 19.14 **2** 69.21 **3** 64.42 **4** 78.11, 85.26, 85.32 **5** 84.27 **6** 76.40, 81.9
Crociani, B. **3** 71.26 **5** 69.28
Crombie, L. **3** 59.23 **4** 59.18 **5** 51.15, 51.44, 59.29 **6** 51.40
Cromer, D.T. **1** 33.58, 34.1, 34.2, 40.2 **2** 81.3, 81.4 **4** 33.5, 76.8, 83.11, 83.14 **5** 1.28 **6** 1.2 **7** 8.12, 33.1, 33.38
Cross, A.D. **1** 51.25
Cross, J.H. **3** 73.51 **4** 73.22
Cross, R.J. **4** 75.23
Crosse, B.C. **2** 85.25
Crossing, P.F. **4** 76.31
Crow, J.P. **4** 86.3
Crowder, M.M. **1** 10.26
Crowfoot, D. **1** 51.29
Crozat, M.M. **5** 63.4
Cruickshank, D.W.J. **1** 1.14, 1.55, 2.9, 19.44, 19.52, 24.14, 24.15, 37.4, 37.5, 37.6, 37.7, 37.8, 37.9, 45.13, 49.34, 52.18, 59.18 **3** 19.3, 26.4, 26.5 **4** 22.2, 46.2, 71.11, 78.2 **6** 31.11
Crumbliss, A.L. **5** 71.13 **6** 79.17
Crump, D.B. **6** 71.57
Cruse, W.B.T. **4** 54.10, 65.4
Cser, F. **1** 52.3 **7** 28.1, 38.1
Csoregh, I. **7** 83.19
Cucinella, S. **7** 68.18
Cucka, P. **1** 33.19
Cullen, D. **4** 49.19 **5** 49.10
Cullen, D.L. **3** 76.42, 83.29, 83.30 **4** 49.15 **6** 49.3, 49.4, 49.5 **7** 27.4, 53.15, 75.27
Cullen, W.R. **2** 65.10, 65.48 **4** 86.3
Cumming, H.J. **6** 3.30
Cummings, R.J. **6** 17.19
Cummins, D. **7** 77.11, 78.17
Cunningham, J.A. **2** 77.29, 77.30, 77.34 **3** 3.56, 85.51 **6** 49.21, 76.49 **7** 84.83
Cuperus, A.J. **4** 27.14
Cupper, G.L. **3** 72.14, 75.15
Curran, R. **3** 85.51
Currie Junior, J.O. **6** 27.20
Currie, M. **1** 2.32 **3** 2.12, 35.8 **4** 2.8, 2.13, 2.14, 49.3, 49.4, 49.5 **5** 2.14, 38.7, 53.19 **6** 48.44 **7** 2.4
Curry, J.D. **4** 66.7

Curry, N.A. **1** 2.32, 2.86, 2.87, 11.51, 17.15, 19.19, 19.20
Curti, R. **1** 36.7, 36.8, 36.11, 36.13
Curtin, D.Y. **1** 9.49 **4** 13.15, 13.16, 17.12, 17.13 **5** 9.18, 9.19, 9.20, 9.21 **6** 20.16 **7** 9.5, 13.1, 13.22
Curtis, N.F. **3** 81.48 **4** 83.101 **5** 77.12, 81.35 **6** 76.23, 76.24, 76.34, 76.36, 76.37, 76.38
Curtis, R.M. **1** 8.35, 39.11
Cusatis, C. **1** 43.27
Cutfield, J.F. **2** 82.55 **6** 54.4, 54.7
Cuthbertson, E. **6** 42.10 **7** 39.14
Cygler, M. **7** 8.5, 8.6
Cyrot, E. **3** 30.6
Czerinski, E.W. **4** 50.16
Czerwinski, E.W. **6** 48.83
Czugler, M. **5** 8.16 **6** 8.7 **7** 53.9

D'Addario, A. **4** 49.14
D'Aniello Junior, M.J. **6** 76.51 **7** 83.62
D'yachenko, O.A. **3** 78.22 **7** 63.4, 63.8, 63.11
Dabrowiak, J.C. **6** 76.17
Dahan, F. **6** 37.2, 71.33 **7** 37.4
Dahan, V.F. **7** 77.9
Dahl, L.F. **1** 3.46, 3.64 **2** 71.22, 72.9, 72.46, 72.48, 72.61, 73.11, 73.27, 73.58, 73.64, 73.70, 73.90, 73.91, 73.116, 73.117, 73.118, 74.4, 74.18, 75.1, 75.62, 75.63, 75.75, 75.80, 75.81, 83.203, 84.15, 84.16, 84.17, 85.25, 85.62, 85.71, 85.74, 85.82, 85.84, 86.20 **3** 3.37, 50.7, 50.9, 50.11, 64.62, 64.63, 64.64, 64.67, 65.12, 71.54, 72.12, 73.15, 73.18, 73.19, 73.27, 73.37, 75.30, 75.48, 83.31, 86.33, 86.72 **4** 3.15, 3.16, 22.7, 63.1, 64.41, 64.42, 65.11, 71.34, 73.14, 73.25, 73.27, 73.28, 85.24 **5** 12.6, 12.7, 12.8, 12.9, 12.10, 50.6, 58.10, 63.14, 73.22, 73.25, 73.28, 73.34, 73.35, 73.36, 74.6, 77.1, 81.60 **6** 3.10, 64.47, 64.48, 65.9, 65.10, 83.12, 83.40, 85.38 **7** 47.29, 73.28, 73.33
Dahl, S. **4** 23.1
Dahl, T. **1** 5.5, 21.10 **2** 60.38, 60.39, 60.58, 60.65 **3** 60.20 **4** 60.1, 60.21 **5** 60.24, 60.25 **7** 60.10
Dahlen, B. **4** 39.16, 64.28, 64.29, 64.30 **5** 1.35, 11.6, 11.7, 48.79 **6** 13.10, 42.9
Dahm, D. **3** 58.24, 58.50
Dahm, D.J. **2** 86.53 **6** 42.4
Dakin, T.W. **1** 35.26
Dale, D. **2** 83.204 **4** 49.20
Daley, L. **3** 83.60
Dalley, N.K. **3** 23.10 **4** 67.23 **5** 79.11 **7** 38.64, 38.65, 42.5, 42.6, 42.9
Daloze, D. **6** 53.12, 58.6 **7** 53.13, 54.3, 54.4
Dalton, L.R. **5** 20.8
Dalton, R.F. **3** 69.14
Daltrozzo, E. **4** 36.37
Daly, J. **4** 40.14
Daly, J.J. **1** 2.75, 4.13 **2** 63.3, 63.19, 63.29, 64.5, 64.10, 64.53, 64.57, 64.58, 64.63, 64.80, 64.82, 64.83, 64.84, 84.46, 86.101 **3** 63.7, 64.32, 64.49, 64.55, 65.1 **4** 64.6, 71.4 **5** 17.22, 64.14, 65.4, 71.77, 71.83 **6** 63.4, 64.30, 71.68, 86.21 **7** 3.29, 39.4, 63.10, 63.20, 64.38, 71.4, 71.88
Daly, J.W. **3** 58.9
Dalzell, B.C. **4** 81.20
Dalziel, W. **5** 50.10
Damiani, A. **1** 33.81 **2** 60.110, 60.111, 60.112
Damm, E. **2** 60.17, 60.18
Dammann, C.B. **5** 86.82
Dammann, L.G. **6** 47.28
Dammeier, B. **4** 13.22, 35.33
Dammeyer, R. **6** 38.12
Dance, I.G. **6** 85.11
Dancz, J.S. **6** 13.8, 13.9
Danford, M.D. **3** 77.16, 77.17
Dangoumau, J. **3** 3.45 **5** 3.22, 24.7
Danielsen, J. **1** 38.36, 59.28 **2** 84.25 **3** 39.7 **5** 1.10 **6** 29.3
Danilenko, L.P. **6** 76.5
Dann, P.E. **5** 64.4 **6** 64.5, 64.19, 64.45 **7** 64.56
Dannenberg, W. **6** 27.18, 27.19
Danno, T. **6** 3.2 **7** 18.1
Dansted, E. **1** 52.17
Dao, N.-Q. **7** 3.15
Dapporto, P. **3** 76.87, 83.88, 86.48, 86.52, 86.85 **4** 83.126 **5** 76.44, 76.52, 76.53, 86.43, 86.50 **7** 71.130, 86.75, 86.115, 86.117
Daran, J.-C. **5** 83.2, 83.40, 83.58 **6** 71.14, 83.1
Daran, J.C. **4** 83.3 **5** 84.1
Darden Junior, W.J. **4** 21.21
Dardenne, G.A. **6** 48.24
Dargay, J.M. **4** 2.22
Darlow, S.F. **1** 2.55, 2.56, 2.73, 13.47
Darnin, C. **1** 10.28
Darriet, J. **7** 64.2
Darriet, M. **7** 64.2
Dartiguenave, M. **7** 86.6
Dartiguenave, Y. **7** 86.6
Das, I.M. **4** 49.11
Das, K.C. **6** 58.46
Das, M. **6** 77.1
Dashevii, V.G. **1** 15.1
Date, T. **7** 31.15, 50.3
Datt, I.D. **1** 7.7
Datta, N. **6** 33.53
Dattagupta, J.K. **5** 16.10 **6** 33.11 **7** 17.11
Dauben, W.G. **4** 56.3 **5** 50.9, 51.20, 51.41
Dauter, Z. **7** 50.29
Davankov, V.A. **5** 82.24
Davey, G. **3** 76.20, 81.40, 81.49 **4** 81.40 **6** 1.4, 73.48, 81.7
Davidkov, K. **4** 33.21 **5** 40.9
Davidson, A.G.H. **3** 48.25
Davidson, J.L. **6** 73.61 **7** 71.22, 71.49, 71.61, 72.9
Davie, A.W. **1** 56.2

Davies, A.G. **3** 69.2
Davies, B.R. **7** 73.58
Davies, B.W. **5** 71.16 **6** 71.50 **7** 69.28, 71.35, 72.36
Davies, D.B. **7** 47.31
Davies, D.R. **1** 1.64, 32.8
Davies, G.R. **2** 72.28, 86.6, 86.52 **3** 71.57, 72.30, 73.28, 80.1, 80.2 **4** 69.36, 71.62, 71.64
Davies, J.E. **4** 78.16 **5** 78.12 **6** 76.53, 78.2, 78.5, 78.22
Davies, R.V. **6** 59.16
Davies, V.H. **4** 53.10
Davies, W.O. **1** 46.23
Davis, A.R. **7** 84.70
Davis, B.R. **2** 86.112 **3** 76.16, 86.55, 86.93 **4** 39.34, 73.11 **5** 3.11, 73.5, 73.9 **6** 41.4
Davis, J. **7** 26.4
Davis, M. **4** 41.6
Davis, P.H. **3** 79.19 **5** 86.52
Davis, P.P. **6** 19.2
Davis, R.E. **1** 45.52, 50.19, 59.38 **2** 72.68, 73.76 **3** 20.15, 27.9, 30.1, 31.29, 72.14, 72.23, 72.25, 72.26, 72.41, 72.42, 75.10, 75.12, 75.15, 75.39, 83.131 **4** 71.70, 71.84, 75.24 **5** 71.76, 71.78, 85.25 **6** 37.11, 71.25, 75.14, 75.19 **7** 28.8, 31.29, 53.24, 71.33, 71.34, 71.58, 71.131, 72.30
Davis, R.L. **6** 73.47
Davison, A. **2** 73.39 **6** 86.1
Davison, B.E. **3** 45.48
Davison, P.J. **7** 69.17
Davydova, M.A. **1** 24.2, 24.3, 24.5 **4** 24.6
Davydova, T.S. **7** 38.12
Dawans, F. **7** 72.3, 72.4
Dawson, B. **1** 48.29, 48.43
Dawson, D.J. **5** 51.39
Dawson, I.M. **1** 11.26
Dawson, J.W. **7** 86.6
Dawson, M.I. **3** 27.11 **6** 30.5, 30.6
Day, A.C. **3** 13.18 **5** 35.25
Day, F.H. **4** 37.10
Day, R.O. **5** 47.33 **6** 11.16, 27.15, 40.8, 47.30, 47.31, 49.16, 49.17, 77.8, 77.13 **7** 47.37, 71.103
Day, V.W. **2** 78.45 **3** 84.25 **4** 49.24 **6** 8.4, 11.16, 27.15, 40.8, 48.50, 49.16, 49.17, 71.11, 71.24, 72.26, 73.6, 73.7, 75.9, 75.21, 77.8, 77.13, 83.80 **7** 8.18, 71.42, 71.103, 73.20, 75.21, 77.2, 77.3, 77.6, 84.34, 86.101, 86.108
De Camp, W.H. **3** 32.28 **4** 32.3, 33.36, 33.37 **5** 33.48, 33.49 **6** 33.32 **7** 38.18, 58.21
De Perazzo, P.K. **7** 79.10
Dea, I.C.M. **6** 45.23
Deadman, W.D. **5** 31.32
Dean, C. **1** 19.8
Dean, W.K. **5** 81.60 **6** 86.10
Debaerdemaeker, T. **3** 48.44 **5** 86.54 **6** 35.21, 36.13, 41.19, 73.37 **7** 27.11, 36.8
Debaerdemaeker, T.D.J. **4** 51.6
Debbaudt, M. **4** 83.39
DeBoer, B.G. **1** 3.31 **3** 83.133 **4** 73.6, 73.23, 75.6, 81.26, 81.28 **5** 75.5 **6** 73.4, 73.25 **7** 71.54, 71.139, 73.17, 86.113
Decap, J. **1** 13.26, 16.18, 16.19, 16.22
Decazes, J.M. **5** 41.50
Deckardt, K. **7** 32.40
Declercq, J.-P. **7** 17.20, 32.24
Declercq, J.P. **6** 17.3, 36.9, 50.9 **7** 20.4, 47.33, 60.2
Declerq, J.P. **3** 67.6 **4** 32.21, 51.25, 51.30, 51.31, 51.34, 83.39 **5** 9.1, 16.8, 30.2 **6** 31.13, 31.31, 32.36, 35.32, 48.48 **7** 33.53, 59.16
Deganello, G. **5** 75.25 **7** 84.51
Degeilh, R. **1** 33.12 **2** 84.44
Degen, P.J. **6** 39.2
Degetto, S. **5** 84.5 **7** 81.22, 81.34, 84.48
Degreve, Y. **2** 72.57
Dehand, J. **2** 83.124, 83.153
Dehli, J. **3** 23.7
Dehnicke, K. **2** 73.80
DeIasi, R. **4** 78.14
Dejace, J. **1** 1.21, 19.58
DeJarnette, E. **7** 17.22, 17.26
DeJarnette, F. **4** 51.2, 51.3
DeJarnette, F.E. **4** 51.47, 51.50, 51.51
Dekaprilevich, M.O. **2** 77.33 **4** 77.12
Del Pra, A. **1** 3.85, 33.75, 34.4, 35.9, 38.3, 39.9 **4** 39.3 **5** 39.9, 69.15, 69.25 **7** 42.1
Del Re, G. **7** 48.29
DeLacy, T.P. **5** 17.21, 19.16, 19.17, 20.15, 31.6, 31.8
DeLaMatter, D. **6** 64.1
Delaplane, R.G. **1** 1.13, 1.20 **3** 33.1
Delbaere, L.T.J. **3** 73.7, 82.23 **5** 73.16, 73.17, 73.18, 82.21 **6** 47.2, 47.25 **7** 20.20, 48.38, 82.20
Delettre, J. **3** 13.17 **5** 5.18, 9.16, 41.31 **6** 51.12 **7** 51.8, 51.13, 51.25, 84.55, 84.56, 84.57, 84.58
Delise, P. **7** 71.86
Della, E.W. **6** 40.15
Dellaca, R.J. **1** 26.35 **3** 33.30, 33.31, 33.32, 71.35, 71.38, 71.59 **4** 71.39, 76.17 **5** 71.52 **6** 76.16
Deloume, J.-P. **5** 67.8 **6** 81.33
Delugeard, Y. **6** 32.30, 38.28, 40.13, 42.5 **7** 7.3
DeLuke, A.B. **4** 47.13
Demain, A.L. **6** 58.52
Demalde, A. **4** 28.5
Demarco, P.V. **3** 50.8 **5** 41.39 **6** 50.10
DeMilo, A.B. **6** 35.10
Dempsey, J.N. **2** 72.4
Denis, J.St. **7** 73.26, 73.47
Denmark, S.E. **7** 85.28, 85.46
Denne, W.A. **4** 1.2, 16.2, 16.4, 16.5, 41.6, 58.35 **5** 35.11, 58.4, 58.7 **6** 58.9, 58.44
Dennis, D. **5** 41.18, 51.39 **7** 39.30, 83.97
Denniston, M.L. **4** 62.1
Denton, B. **7** 4.8, 35.28, 35.39, 36.27, 71.24, 73.4, 73.5, 73.6, 73.7, 73.8, 73.14, 73.36
DePamphilis, B.V. **7** 85.28
Depmeier, W. **7** 45.27, 45.35

Deppisch, B. **7** 71.21
Deranleau, D.A. **6** 35.24
Derieg, M.E. **3** 9.19, 35.17
Derissen, J.L. **1** 48.25 **3** 1.18 **7** 1.2, 1.3, 13.10
Derry, J.E. **1** 36.20 **3** 34.3 **4** 22.6, 36.11
Deschamps, J. **7** 33.6, 33.12
Desiderato, R. **1** 7.16, 11.28, 35.26, 35.30 **4** 33.16, 36.1 **6** 32.7, 76.1 **7** 82.15
DeSimone, R.E. **4** 86.10 **7** 85.15
Desimoni, G. **7** 27.7
Dessy, G. **1** 3.70 **2** 68.11, 68.16, 80.13, 80.17, 84.60, 85.72, 85.85, 85.93, 85.96 **3** 68.20, 80.7, 80.8, 80.9 **4** 79.11, 85.17, 85.19 **5** 76.23, 79.23, 83.92, 84.14, 84.25, 85.6, 85.15, 85.21, 85.23, 85.30 **6** 85.42, 85.44 **7** 81.42, 84.60
DeStevens, G. **4** 33.47
Destro, R. **1** 17.25, 17.31, 17.32, 52.1 **2** 60.103 **3** 31.25 **4** 31.4 **5** 44.14 **6** 26.2, 44.15 **7** 28.5, 33.5
Desvergne, J.-P. **3** 26.9
Desvergne, M. **6** 28.11
DeTitta, G.T. **4** 60.42 **5** 46.3 **7** 59.12
Deuschl, H. **1** 32.4 **2** 74.21
Deutsch, E. **5** 39.8
Dev, S. **6** 53.9
DeVille, T.E. **1** 21.33 **4** 54.1, 54.2
Devys, M. **6** 51.41
Dewailly, J. **7** 72.3, 72.4
Dewan, J.C. **6** 80.9, 80.10 **7** 83.66, 83.101, 83.110, 86.3, 86.40, 86.41
Dewar, R. **1** 42.11, 42.12 **3** 62.18, 83.80
Dewar, R.B.K. **2** 75.26
Dewulf, B. **7** 32.28, 32.29
Dexter, D.D. **4** 3.22, 3.24 **5** 11.3
Deyrieux, R. **3** 81.2, 81.3 **7** 81.4, 81.5, 81.6, 81.7
Deyrup, J.A. **5** 38.5
Dhaneshwar, N.N. **4** 13.7 **6** 13.1, 13.14 **7** 64.47
Di Vaira, M. **7** 86.82
Diamand, R.D. **2** 77.1 **4** 59.7
Diana, G. **1** 38.59 **2** 75.37
Dias, H.W. **1** 8.42
Dichmann, K. **4** 39.46
Dichmann, K.S. **3** 36.18, 36.25, 60.34 **6** 28.27, 28.28, 73.52
Dickens, B. **2** 75.19, 75.29 **3** 12.2 **5** 34.3
Dickerhoof, D.W. **5** 12.6, 12.7, 12.8, 12.9, 12.10
Dickerson, D. **6** 3.6
Dickerson, R.E. **1** 33.51 **2** 68.15 **3** 9.8, 24.6, 27.11, 28.12, 30.5, 31.32 **4** 27.16
Dickinson, C. **1** 12.6, 15.2, 16.1, 24.6 **3** 11.18, 16.5 **4** 32.14 **5** 16.1
Dickson, R.S. **4** 71.48 **5** 71.24, 72.33
Dideberg, D. **7** 43.8
Dideberg, O. **4** 38.15, 51.28 **5** 3.15, 33.53, 41.10, 51.17, 51.18, 51.19, 51.21, 51.33, 70.7 **6** 33.15, 35.31, 51.18, 51.19, 51.22, 51.23, 70.2 **7** 32.7, 36.19, 38.53, 39.28, 43.6, 48.34, 51.15, 58.11, 58.22, 60.37
Dierks, H. **2** 75.54, 75.72
Dieterich, D.A. **7** 9.5, 13.22
Dietrich, H. **1** 10.30 **2** 67.4, 72.22, 72.77, 75.54, 75.72 **3** 20.14, 72.15, 75.19
Dijk, B.van **3** 11.23
Dijkstra, A. **2** 82.17
Dik-Edixhoven, C.J. **5** 35.4, 51.1
Dik-Edixhoven, L. **3** 33.46
Dikareva, L.M. **2** 81.56, 81.57, 81.58
Dimitrova, G.I. **7** 78.20
Dimroth, K. **2** 64.68
Dion, H.W. **7** 47.29
Dittemore, K.M. **6** 86.34
Dittmar, G. **3** 63.9
Divjakovic, V. **6** 16.11, 35.11
Dix, M.F. **5** 33.46
Dixon, F.L. **6** 49.18
Dixon, K.R. **2** 86.16 **7** 86.42
Djarmati, Z. **7** 58.21
Djerassi, C. **1** 58.63 **3** 51.41, 58.35, 58.42 **4** 58.33 **7** 51.49, 53.13, 56.3
Djuric, S. **5** 43.1 **6** 35.11 **7** 35.21
Doak, G.O. **7** 66.8
Dobinson, G.C. **2** 83.178 **3** 86.98
Dobler, M. **1** 22.12, 31.15, 37.12, 50.36, 53.40 **3** 40.7 **4** 49.6 **5** 43.1, 45.43, 50.16 **7** 38.30, 38.32, 38.33, 38.34, 38.35, 38.36, 38.37, 38.38, 38.39, 38.40, 50.22, 50.25, 50.30, 50.37, 54.11
Dobrowolska, W. **7** 64.28, 64.29
Dobrzynski, E. **6** 71.81
Dobrzynski, E.D. **7** 71.137
Dodds, T.A. **6** 71.25, 75.14, 75.19
Dodge, R. **4** 60.13
Dodge, R.P. **2** 72.34, 72.74, 72.75, 75.10, 75.79, 77.6 **3** 76.25 **4** 21.21, 38.11 **7** 35.25
Dodman, P. **6** 71.12
Dodson, G.G. **1** 58.34 **4** 76.42
Doedens, R.J. **2** 73.70, 73.116, 83.36, 83.69, 83.88, 83.203, 86.25, 86.58, 86.66, 86.79 **3** 79.11, 81.35, 83.22, 83.89 **4** 81.42, 81.60, 83.43 **5** 71.26, 71.40, 81.57, 83.33 **6** 71.74, 81.44 **7** 71.92, 81.41
Doherty, D.G. **4** 48.55, 48.56
Doherty, G.O.P. **1** 59.45
Dolcetti, G. **5** 86.36
Doleschall, G. **7** 36.13
Dolfini, J.E. **5** 59.23 **7** 59.13
Dollase, W.A. **1** 39.51 **2** 74.9
Domenicano, A. **1** 4.12 **2** 65.27, 80.16, 80.32, 80.33, 80.35, 85.9, 85.10, 85.11, 85.36, 85.69 **3** 85.1
Domiano, P. **1** 16.31, 35.8, 41.21 **2** 79.2, 79.14, 79.18, 79.27, 85.37 **3** 8.5 **4** 79.1, 83.55 **5** 8.4, 41.19, 78.2, 78.18 **6** 33.37, 41.5, 86.9 **7** 11.15, 32.10, 78.8, 78.13, 83.48
Domingos, A.J.P. **6** 71.28
Domingos, A.M. **6** 63.5, 69.3, 69.4, 69.8 **7** 69.2, 69.16
Dominguez, X.A. **4** 56.4, 59.14 **7** 53.14
Dominick, T.L. **6** 76.19
Donaldson, D.M. **1** 24.30, 30.28
Donaldson, J.D. **6** 69.2
Donaldson, P.B. **4** 86.65
Donaruma, L.G. **1** 30.27

Dondoni, A. **3** 39.39, 40.15 **4** 39.27, 40.19 **6** 16.16
Donohue, J. **1** 8.5, 8.17, 10.19, 10.20, 11.27, 15.12, 16.10, 31.11, 48.13, 48.17, 48.23, 48.33, 48.40, 48.59, 50.29 **2** 62.32, 64.11 **3** 11.25, 23.3, 31.30, 40.11, 44.6, 60.8, 65.2, 86.4 **4** 5.2, 24.1, 37.8, 41.7, 51.14, 52.4, 52.5 **5** 2.10, 24.4, 27.2, 63.5 **6** 1.14, 48.34, 59.2 **7** 11.20, 83.34, 86.12
Donovan, D.J. **7** 86.113
Donovan, W.F. **7** 58.12
Doomes, E. **3** 11.26
Doonan, D.J. **7** 71.25
Doorakian, G.A. **3** 69.35
Doorne, W.van **4** 64.9
Dori, Z. **1** 3.40 **2** 76.75, 80.31, 83.116, 85.73, 85.86 **3** 85.27, 86.96 **4** 82.13, 83.72, 86.61, 86.68, 86.69 **5** 76.36, 76.41 **6** 86.65 **7** 76.45, 82.18
Dorignac-Calas, M.-R. **4** 41.28
Dormond, A. **7** 73.60
Dornberg, M.L. **1** 2.80, 2.81
Dorosinskii, A.L. **5** 68.10 **7** 3.19
Dorset, D. **3** 1.39
Dorset, D.L. **5** 28.2
Doskotch, R.W. **4** 54.10
Dougherty, T.A. **2** 83.18
Dougill, M.W. **1** 1.8, 1.54 **2** 64.29 **6** 64.20
Douglas, B. **1** 58.100, 58.101
Douglas, P.G. **3** 86.49 **6** 86.13
Douglas, W.E. **4** 73.29
Douwes, H. **2** 64.4
Dow, J. **1** 47.15 **3** 48.29
Downie, T.C. **1** 14.5 **3** 81.8 **4** 33.15, 41.1 **5** 33.23 **6** 41.10
Downs, G.L. **2** 84.15, 84.17
Downs, H.H. **7** 84.78, 86.76
Doyle, J.R. **2** 72.47, 73.17, 75.2, 75.3, 75.4, 75.6, 75.18, 75.77 **6** 73.47, 86.34
Doyle, M.J. **7** 71.124
Doyle, P. **7** 37.21
Doyne, T.H. **1** 56.6 **2** 82.32 **3** 86.36 **7** 48.17, 51.14
Dozzi, G. **7** 68.17
Drager, M. **5** 7.4 **6** 3.19, 11.3 **7** 11.8, 41.26, 64.50, 65.2, 65.3
Drago, R.S. **6** 76.50
Dragonette, K.S. **1** 5.41, 8.41, 35.28
Drake, J.E. **2** 83.47
Drangfelt, O. **7** 16.3
Draper, M.D. **3** 33.48
Dreele, R.B.von **3** 50.17 **4** 50.9, 77.8, 77.9 **7** 48.32, 48.33
Dreiding, A.S. **5** 31.1 **7** 53.7
Dreissig, W. **2** 72.77 **4** 64.22, 64.23 **5** 29.1, 45.34, 64.25, 64.26, 64.31 **6** 45.5, 64.26 **7** 29.12
Drenth, W. **1** 19.61
Drew, J. **5** 81.48, 81.61
Drew, M.G.B. **1** 24.32, 31.43, 54.8 **2** 76.21, 76.31, 76.50, 81.46, 82.34, 83.103, 83.219, 86.21 **3** 33.9, 81.36, 82.16, 82.19, 83.5, 86.8, 86.75 **4** 3.12, 64.33, 81.36, 83.8, 83.75, 83.76, 84.2, 86.23, 86.47, 86.55 **5** 72.8, 72.12, 81.12, 83.54, 84.16, 85.22, 86.33 **6** 38.16, 71.17, 72.22, 84.3, 85.5, 85.6, 86.14 **7** 71.2, 73.37, 76.73, 81.15, 83.75, 83.108, 86.17, 86.32, 86.33, 86.53
Drew, R.E. **4** 69.5 **5** 81.24 **7** 3.8, 81.1, 81.10
Dreyfus, H. **3** 53.12, 53.13, 53.14
Driel, C.A.A.van **1** 1.70
Driver, R. **2** 82.23, 82.36, 82.38
Druding, L.F. **7** 76.5
Drummond, J. **2** 65.45 **3** 65.9 **4** 78.17
Drusiani, A. **7** 29.10
Druyan, M.E. **7** 48.56
Dryer, H.B. **2** 60.67
Dryhurst, G. **7** 43.7
Duax, W. **4** 51.2, 51.3
Duax, W.L. **3** 48.41, 51.16, 60.14, 60.15 **4** 51.26, 51.27, 51.35, 51.36, 51.47, 51.50, 51.51, 60.38, 60.39 **5** 11.11, 11.15, 50.29, 51.7, 51.23, 51.28, 51.35, 51.40, 60.28 **6** 17.23, 38.32, 51.1, 51.4, 51.8, 51.15, 51.16, 51.27, 51.31, 51.35, 51.39, 59.21, 59.22 **7** 17.17, 51.6, 51.7, 51.10, 51.12, 51.16, 51.19, 51.24, 51.28, 51.29, 51.31, 51.32, 51.33, 51.34, 51.35, 51.45, 51.51, 51.53, 59.11
Dubey, D.S. **5** 14.3, 14.4
Dubey, R.J. **3** 1.1 **6** 73.3 **7** 73.68
Dubler, E. **6** 64.46 **7** 73.24, 73.43
Duc, C.L. **7** 33.45
Duc, T.Q. **7** 81.14
Duchamp, D.J. **1** 13.41 **3** 50.6, 50.10 **4** 27.9, 34.3, 35.28, 35.29, 36.20, 36.33, 40.23, 45.23, 50.3, 50.7, 51.46, 51.55 **5** 50.1, 51.6, 51.12 **6** 50.8, 59.27
Duckworth, V.F. **2** 77.54, 83.100, 83.101 **3** 86.26, 86.49 **4** 23.5
Ducruix, A. **4** 21.16, 45.14 **6** 33.51, 45.37, 51.41 **7** 45.31, 45.33, 45.39, 45.41, 45.51, 45.53, 47.40, 54.13, 58.32
Dudakov, V.G. **7** 48.47
Dudragne, F. **7** 16.5
Dueber, J.S. **5** 69.26
Duerst, R.W. **1** 51.47
Duesler, E.M. **4** 71.50
Duesler, E.N. **4** 76.14
Duff, E. **7** 64.37
Duffield, A.M. **1** 59.41
Duffin, B. **1** 32.42 **2** 83.22, 84.1 **3** 11.28, 35.2, 35.7
Duffin, M. **6** 1.15
Duggan, D.M. **5** 76.46 **7** 76.33, 76.34, 76.63
Duisenberg, A.J.M. **7** 36.13
Duke, A.J. **3** 27.8
Duke, J.R.C. **1** 1.25 **3** 27.8, 32.3, 32.4, 32.5 **7** 33.6, 33.12
Dukefos, T. **1** 45.63
Dullforce, T.A. **1** 3.57, 3.58, 3.59 **4** 53.4, 53.9

Dulova, V.G. **2** 63.28
Dumbleton, J.H. **1** 2.97
Dumke, W.L. **1** 19.54
Dunaj-Jurco, M. **2** 83.30 **4** 76.1 **5** 61.5 **7** 76.7, 79.6
Dunand, A. **7** 76.1
Duncan, C.D. **7** 33.35
Duncan, J.A. **6** 35.26
Duncanson, F.D. **7** 56.8
Dunitz, J.D. **1** 1.11, 1.43, 1.77, 21.16, 21.18, 22.11, 22.12, 23.2, 23.4, 23.6, 23.7, 23.8, 23.10, 31.15, 34.6, 34.7, 34.11, 37.12, 38.37, 38.49, 46.21, 48.15, 49.2, 50.29, 50.44, 51.45, 53.40 **2** 73.22, 75.12, 75.53, 75.67, 83.6, 83.75, 83.77, 83.180 **3** 16.9, 16.10, 16.11, 31.6, 40.7, 83.85 **4** 23.3, 23.4, 23.6, 31.14, 31.16, 49.2, 49.3, 49.4, 49.5, 49.6, 50.11, 50.13, 83.120 **5** 5.9, 31.25, 43.1, 48.39, 83.87 **6** 23.1, 23.2, 58.40 **7** 34.2, 34.4, 34.7, 34.8, 34.9, 34.10, 34.11, 34.12, 34.13, 34.14, 38.29, 38.30, 38.31, 38.32, 38.34, 38.35, 38.36, 38.37, 38.38, 38.40, 49.3, 49.4, 49.6, 49.7, 50.25, 50.32
Dunken, H. **1** 13.9, 33.20
Dunlop, J.H. **2** 82.34, 82.37
Dunlop, R.S. **2** 73.59 **6** 2.15
Dunn, J.B.R. **5** 72.47 **6** 60.23
Dunn, R.M. **4** 2.24
Dunn, R.M.A. **4** 24.2
Dunne, T.G. **2** 71.44, 72.25
Dunphy, R.F. **4** 31.26
Duplancic, T. **7** 76.4, 76.9, 83.17, 83.31
Dupont, L. **4** 38.15, 51.28 **5** 3.15, 41.10, 51.17, 51.18, 51.19, 51.21, 51.33, 70.7 **6** 33.15, 51.18, 51.19, 51.22, 51.23, 70.2 **7** 32.7, 36.19, 38.53, 43.6, 43.8, 48.34, 51.15, 58.11, 58.22, 60.37
Dupuy Junior, A.E. **4** 32.30
Durant, F. **2** 60.25, 60.26, 60.46, 60.47, 60.64, 60.66 **3** 48.44 **4** 48.49, 59.2 **5** 39.10, 39.11, 39.12, 48.70, 48.74 **6** 33.19, 48.24, 50.9
Duranti, D. **1** 35.13
Durcanska, E. **4** 79.7
Durette, P.L. **5** 45.20 **7** 45.54
Durham, L.J. **7** 53.13
Durley, R. **6** 85.16
Durovic, S. **3** 83.37
Durr, B.G. **7** 54.11
Durr, H. **5** 31.35
Dusausoy, Y. **3** 2.5, 74.2 **5** 73.32, 73.40, 74.3, 74.4, 74.5, 75.17, 75.19, 75.20 **6** 73.29, 74.4, 74.8, 74.9 **7** 73.60, 75.7
Dustin, D.F. **6** 71.10
Dutasta, J.P. **7** 64.11
Dutkey, S.D. **4** 11.9
Dutta, B.P. **7** 8.4, 24.4
Dutta, S.N. **2** 64.26
Dutta, S.P. **6** 44.16 **7** 44.27
Duvall, L.A. **4** 67.28 **5** 16.18 **6** 16.19, 16.20, 16.21
Dvorkin, A.A. **7** 83.20, 83.21, 83.32, 83.84
Dvortsak, P. **7** 40.21
Dwek, R.A. **3** 45.31
Dwivedi, G.L. **4** 83.67 **5** 52.2, 83.45
Dwyer, M. **3** 76.38 **5** 76.31 **6** 76.16
Dwyer, P.N. **6** 49.10 **7** 49.15
Dyachenko, O.A. **7** 84.68, 85.44
Dye, J.L. **7** 40.18
Dyer, G. **2** 86.105 **4** 85.37
Dyer, J.R. **1** 59.5 **4** 32.10
Dyke, M. **1** 10.11
Dymock, K. **5** 68.30 **6** 68.11, 68.12, 68.14, 68.15 **7** 72.33, 72.38

Eaborn, C. **4** 71.81 **6** 86.32
Eachus, S.W. **6** 34.2
Ealick, S.E. **7** 31.28, 36.2, 36.3, 36.5, 36.9, 54.9, 54.10, 82.19
Earle, K.M. **3** 48.38
Earley, J.E. **4** 76.9
Eaton, G.R. **5** 49.8
Eaton, P.E. **7** 31.19
Eaton, S.S. **5** 49.8
Ebara, N. **6** 14.2
Ebner, J.R. **6** 86.15
Echols, H.M. **7** 83.97
Eck, C. **4** 53.3
Eck, J. **4** 4.1
Eckberg, R.P. **7** 83.114
Eckstein, F. **3** 47.1 **7** 47.23
Eddy, L.P. **3** 83.29
Edenharter, A. **6** 16.11
Edington, P. **6** 48.23
Edmond, E. **3** 49.13 **4** 49.23
Edmond, E.D. **7** 49.1
Edmonds, J. **3** 7.6
Edmonds, J.W. **3** 19.6, 19.7 **4** 9.3 **6** 39.6, 59.21, 59.22 **7** 59.11, 59.12
Edmunds, I.G. **1** 17.39
Edwards, A.J. **5** 83.80
Edwards, D.A. **5** 81.12
Edwards, I.A.S. **3** 3.35, 30.3
Edwards, J.M. **6** 29.1
Edwards, J.O. **3** 83.28
Edwards, S.L. **6** 32.10
Edwards, W.T. **2** 75.30, 75.32
Eeles, W.T. **1** 13.46, 58.106
Effinger, J. **1** 41.46
Efraty, A. **6** 75.27 **7** 73.48
Eger, C. **3** 60.14, 60.15 **4** 51.35, 51.36, 60.39 **6** 51.1, 51.8
Eggerding, D. **6** 85.13
Eggers, C.A. **5** 80.13
Egginton, G.M. **7** 86.33
Eglington, G. **1** 39.56
Egmond, A.B.van **5** 35.5
Egmond, J.v. **3** 22.3
Eguchi, Y. **7** 50.21
Ehrenberg, M. **1** 1.45, 28.29, 29.15, 31.47, 31.48, 31.49
Ehrenkaufer, R.L. **6** 40.14
Ehrlich, H.W.W. **1** 28.12, 30.1, 32.12 **2** 86.27
Eian, G.L. **4** 36.14
Eichelberger, H.R. **4** 81.59
Eichhorn, E.L. **1** 9.26, 33.15, 44.46
Eick, H.A. **3** 73.21 **6** 31.30, 71.84
Eickman, N. **7** 58.24

Eidawad, A.-H. **7** 84.8
Eide, G. **5** 41.2
Eijck, B.P.van **1** 1.32
Eiland, P.F. **2** 73.21
Eilbracht, P. **6** 28.8
Einhellig, K. **5** 35.30, 36.25
Einspahr, H. **3** 11.25, 86.4 **5** 2.10 **6** 1.14, 24.6, 48.17 **7** 48.19, 86.12
Einspahr, H.M. **6** 64.24 **7** 44.6
Einstein, F. **2** 70.5
Einstein, F.W. **5** 85.32
Einstein, F.W.B. **1** 3.60, 3.68 **2** 69.32, 71.10, 75.33, 83.139, 86.32 **3** 68.6, 73.6, 73.55, 83.1 **4** 68.7, 69.5, 70.2, 71.37, 73.32, 86.3, 86.9, 86.16, 86.18, 86.62 **5** 68.2, 68.19, 71.72, 71.97, 75.35, 75.48, 81.24, 83.21, 83.123, 86.14, 86.16 **6** 71.29, 73.2, 73.23 **7** 3.8, 37.9, 45.4, 68.15, 72.13, 80.1, 81.1, 81.10, 84.70, 86.23
Einstein, J.R. **1** 44.72 **3** 44.34, 85.18 **4** 30.2, 45.2, 47.4, 48.55, 48.56 **5** 83.73
Eisch, J.J. **6** 68.16
Eisenberg, R. **2** 76.75, 85.30, 85.73, 85.86, 85.97, 85.98, 86.50 **3** 3.56, 83.28, 85.27, 85.35, 85.45, 85.51, 86.31, 86.96 **4** 73.15, 85.12, 85.16, 85.39, 86.29, 86.60 **5** 85.12, 86.65, 86.75 **6** 71.82, 85.9, 85.10, 86.19, 86.37, 86.54, 86.59 **7** 71.25, 71.111, 75.10, 86.64
Eisenhut, M. **5** 64.18
Eisenhuth, W. **2** 65.5, 65.15
Eisner, U. **3** 36.29 **6** 37.13
Eiss, R. **2** 77.63, 77.64, 84.66 **3** 72.11, 75.5, 81.53, 83.148
Elcombe, M.M. **1** 8.12, 8.13, 8.14, 8.15
Elder, M. **3** 63.14, 69.20, 69.23, 69.36, 73.34, 83.146, 85.6 **4** 73.17
Elder, R.C. **2** 76.3, 77.44, 77.45, 77.51, 77.67, 83.94 **5** 38.17 **6** 38.25, 64.16, 73.27, 76.11, 76.12, 76.13, 83.73
Elding, I. **7** 81.28
Eley, D.D. **2** 60.83
Elguero, J. **1** 9.47 **3** 32.7 **4** 32.18 **6** 41.13
Eliel, E.L. **6** 39.10, 39.11
Elix, J.A. **3** 36.16
Eller-Pandraud, H.von **1** 35.41
Eller, G.von **1** 16.29
Eller, H.von **1** 39.76, 39.78 **3** 35.19
Eller, P.G. **3** 64.6, 85.15, 86.16 **5** 78.10 **6** 49.14, 78.9, 78.15, 78.21, 84.12 **7** 84.44
Ellett Junior, J.D. **5** 81.7
Elliott, I.W. **4** 37.15 **7** 37.19
Elliott, N. **4** 33.25
Ellison, R.D. **1** 2.53 **3** 1.6
Elmes, P.S. **4** 86.2 **7** 86.9, 86.18, 86.59
Elmore, N.F. **3** 33.35
Elsner, G. **7** 81.16, 81.18
Elson, C.M. **5** 83.49
Emerson, K. **2** 84.41 **3** 76.44, 83.81, 83.147
Emerson, M.T. **1** 53.3 **3** 53.18 **4** 53.1
Emery, H. **7** 51.54
Emmert, D.E. **4** 51.46
Empsall, H.D. **6** 86.25, 86.31
Ende, C.A.M.van der **7** 51.41
Ende, C.van der **7** 37.11
Endeman, H.J. **1** 48.25
Enders-Beumer, A. **5** 2.2
Endicott, J.F. **6** 83.39, 83.41 **7** 83.56, 83.61
Endres, H. **7** 71.126, 71.134
Enemark, J.H. **2** 62.36, 64.52, 85.45, 86.39, 86.85, 86.112 **3** 76.40 **4** 71.1, 71.20, 80.1, 86.30 **5** 71.8, 71.11, 71.14 **6** 76.3 **7** 9.4, 85.38, 86.34, 86.38
Engberts, J.B.F.N. **6** 1.26
Engebretson, G.R. **1** 19.55 **2** 75.24
Engel, D.W. **4** 54.12
Engel, J.F. **4** 86.17
Engel, M.R. **4** 59.16
Engel, P. **5** 38.16, 38.21 **6** 16.11 **7** 32.23
Engelen, B. **7** 4.5
Engelhardt, L.M. **5** 83.61 **6** 76.46 **7** 83.82, 83.93, 83.112
Engenieres, M.D.d' **7** 38.52
Engler, R. **6** 2.2, 3.19 **7** 39.2
Engmann, R. **1** 7.25 **2** 63.13
Ent, A.van der **6** 72.4
Entwistle, R.F. **1** 29.7, 30.11
Enwall, E. **3** 83.1 **5** 85.32
Enwall, E.L. **3** 83.81 **4** 4.4, 82.14 **7** 51.59
Epstein, E.F. **3** 75.30, 75.48, 84.20, 85.25, 85.44 **4** 73.8 **5** 85.36 **6** 73.21
Epstein, J.M. **6** 3.18 **7** 78.23, 78.24, 80.8, 83.101, 83.110
Epsztajn, J. **7** 33.6, 33.12
Eraker, J. **1** 20.13
Erasmus, C.S. **3** 77.35 **4** 77.22
Eremeyev, A.V. **7** 37.8
Erenburg, A.I. **5** 5.2
Erichsen, H.H. **1** 9.23
Erickson, J.L. **7** 33.4
Eriks, K. **1** 11.14, 12.14 **2** 83.38 **3** 46.8 **4** 12.15, 81.20 **6** 67.1 **7** 67.2
Eriksen, R. **5** 8.15
Erman, M. **7** 17.22, 17.26
Erman, M.G. **6** 51.4, 51.15, 51.35, 51.39 **7** 51.45
Ermer, O. **3** 31.6 **4** 23.4, 23.6, 31.14, 31.16 **6** 23.1, 23.2 **7** 22.4
Ern, V. **4** 36.12
Ernst, R.D. **5** 71.20
Ernst, S. **7** 17.34
Ernst, S.R. **6** 17.14
Ernster, L. **5** 50.24
Erskine, G.J. **2** 86.46
Ertl, G. **7** 63.2
Esande, A. **7** 35.5
Escande, A. **5** 32.4 **6** 35.2, 35.4
Escobar, C. **1** 41.42 **7** 43.5
Eser, H. **4** 23.3, 23.4, 23.6
Esperas, S. **5** 70.10, 70.12 **6** 64.4, 68.9 **7** 11.4, 11.5, 68.5, 70.6
Estes, E.D. **6** 3.12, 76.47 **7** 76.25, 76.51, 76.62, 84.63
Estes, W.E. **7** 76.25
Estienne, J. **4** 81.50 **6** 71.69 **7** 80.9, 80.10, 80.20

Estlin, J.A. **1** 4.18, 35.32, 39.57, 58.22 **3** 11.3, 31.34
Etienne, J. **7** 38.52
Evans, B.E. **4** 40.2, 41.3
Evans, D.A. **3** 30.5
Evans, H. **3** 49.13
Evans, J.A. **3** 71.34, 72.13, 72.16 **4** 71.83, 71.87, 85.34 **5** 71.100, 71.101
Evans, J.M. **2** 60.75
Evans, M. **4** 71.27
Evans, W.J. **6** 71.10 **7** 73.16
Everitt, G.F. **5** 71.100, 71.101
Evers, H. **1** 28.1
Evertsson, B. **2** 82.51
Evonkova, E.V. **2** 83.19
Evrard, G. **3** 20.18 **4** 20.12, 24.19, 59.2 **5** 48.70 **6** 33.19, 33.58, 35.33, 48.24, 50.9, 71.10 **7** 39.36
Ewings, P.F.R. **6** 69.18
Exner, O. **3** 10.7
Extine, M. **7** 80.14, 80.15, 83.50
Ezhkova, Z.I. **5** 26.1, 40.26 **6** 26.1, 40.16 **7** 41.17
Ezra, F.S. **6** 46.3

Faber, D.H. **6** 21.10
Fabiani, C. **4** 70.9
Fackler Junior, J.P. **3** 80.18, 85.38 **5** 77.6, 77.7, 80.20, 85.26, 85.27, 85.34 **6** 85.27, 85.36, 85.43 **7** 77.5, 84.52
Fackler, J.P. **1** 3.69 **7** 86.49
Fagherazzi, G. **2** 72.39, 72.53 **4** 46.1
Fahr, E. **5** 32.23
Fair, C.K. **4** 65.5, 83.29 **5** 65.1, 65.5 **6** 38.20
Faithful, B.D. **2** 65.37 **5** 65.7
Falgueirettes, J. **1** 9.45, 32.25 **3** 2.9, 27.5, 27.6, 32.24 **4** 9.11, 32.13 **5** 28.13, 32.4, 63.8, 63.9, 63.10 **6** 63.10
Falkenberg, G. **3** 35.16 **5** 35.16, 35.17 **6** 63.9
Falkenberg, W. **6** 41.30
Fallani, G. **4** 83.126 **5** 76.52, 76.53, 86.43, 86.50 **7** 85.37, 86.75
Faller, J.W. **2** 73.39
Fallon III, L. **6** 16.12, 27.20, 28.12, 44.1
Fallon, G.D. **7** 71.78, 76.74
Fallon, L. **7** 20.10
Falth, L. **7** 86.55, 86.105
Fanfani, L. **1** 38.43 **4** 35.10, 70.4, 83.49, 85.8 **5** 36.32, 40.25 **6** 32.13, 32.34
Fang, J.H. **7** 60.8, 60.9
Fankuchen, I. **1** 1.1, 40.11, 44.68 **2** 63.7
Farag, M.S. **1** 19.64 **4** 17.8
Fares, V. **2** 84.60, 85.93, 85.96 **3** 68.20 **4** 79.11, 85.17, 85.19 **5** 76.23, 79.23, 83.92, 84.14, 84.25, 85.6, 85.15, 85.21, 85.23, 85.30 **6** 85.42, 85.44 **7** 81.42, 84.60
Farges, C. **7** 50.35
Farhangi, Y. **6** 83.61
Farina, M. **1** 29.19 **2** 61.17, 61.19
Farkas, R. **6** 48.57
Farmer, B.L. **7** 35.20
Farney, R.F. **1** 53.35
Farnham, P.H. **5** 86.36
Farnham, W.B. **4** 52.4, 52.5
Farnsworth, N.R. **3** 59.26
Farrier, D.S. **3** 58.20
Farris, F.F. **4** 60.35
Fasiska, E.J. **3** 38.10 **5** 45.6
Faught, J.B. **3** 64.25 **5** 64.11
Faulkner, D.J. **5** 52.1 **6** 53.11
Faulstich, H. **7** 48.74
Faure, R. **4** 81.38 **6** 67.10, 81.34 **7** 81.35
Fava, G. **1** 9.3, 32.15, 39.23, 39.25 **2** 69.11, 79.9, 79.33, 84.5, 84.8, 84.9, 85.5
Fava, G.G. **5** 39.14, 40.22, 81.5, 81.47, 81.54 **6** 40.10, 40.11, 76.41, 79.16
Fawcett, J.K. **1** 28.7, 30.18, 45.55, 51.35, 51.43 **2** 65.24 **3** 20.17, 23.1, 36.18, 36.25, 46.5, 46.6, 58.47, 68.3 **7** 47.39, 48.44, 48.61, 48.66
Fawcett, K. **5** 47.35
Fay, R.C. **4** 77.8 **7** 77.3, 77.6, 84.79
Fayos, J. **3** 13.2 **4** 63.7, 83.9 **5** 35.24, 52.1, 56.4 **6** 28.13, 31.8, 45.51, 53.11, 58.33, 71.35, 71.81, 75.7, 75.16 **7** 50.10, 51.49, 54.7, 56.3, 58.20, 61.6
Fazakerley, G.V. **6** 44.23, 83.74, 83.76 **7** 83.106
Fealey, T. **4** 76.9
Featherman, S.I. **4** 64.13
Fechtig, B. **4** 33.47
Fedder, W. **4** 83.10
Fedeli, W. **1** 35.3, 35.4, 35.7, 35.12, 35.14, 35.15, 38.37, 50.48 **3** 33.45, 50.28, 58.15, 59.33 **4** 1.11, 32.22, 35.26, 36.34, 54.19 **5** 50.13, 56.2 **6** 33.52, 33.56 **7** 50.7
Fedin, E.I. **6** 66.2
Fedorov, Yu.G. **7** 60.11
Fedorov, Yu.G. **4** 24.7 **7** 36.6
Feher, F. **4** 39.7 **6** 39.12
Fehlhammer, W.P. **4** 65.11, 71.34, 73.27
Fehlmann, M. **1** 25.20, 58.122 **2** 82.54 **3** 41.5
Feil, D. **1** 8.19, 8.20, 8.22, 8.23 **2** 61.22 **4** 60.3 **6** 36.14
Felix, D. **1** 11.45
Felkin, H. **6** 73.63
Fellenberger, K. **6** 27.7
Feltham, R.D. **4** 80.1 **6** 76.3 **7** 86.34, 86.38
Felton, R.H. **6** 49.14
Fenical, W. **4** 53.2, 59.8 **5** 54.7 **6** 54.15, 59.11
Fenimore, C.P. **1** 28.27
Fenn, R.H. **1** 11.23 **2** 73.6, 81.39, 85.13 **3** 71.27, 72.27, 73.51, 83.35, 85.53, 85.55 **4** 73.22, 85.40 **5** 46.2, 72.27
Fennessey, J.P. **2** 71.51, 73.10, 73.32 **3** 58.26, 71.18, 85.14
Fenske, D. **5** 62.1, 64.15
Fenske, R.F. **5** 71.4
Fenton, D.E. **4** 67.24, 76.41 **5** 3.19, 38.25, 60.12 **6** 60.14, 67.8, 77.16

Ferguson, G. **1** 13.1, 13.2, 13.6, 19.27, 19.28, 22.6, 29.8, 31.20, 31.38, 37.28, 38.58, 52.7, 53.12, 54.3, 54.15, 54.18, 54.20, 54.22, 58.68, 59.45, 59.54, 59.58 **2** 65.30, 65.31, 65.32, 65.51, 66.16, 66.17, 73.102, 74.22, 74.23 **3** 31.27, 31.28, 39.38, 51.26, 51.30, 51.44, 53.17, 53.24, 58.54, 59.16, 59.27, 60.42, 60.55, 65.11, 86.58 **4** 35.8, 51.13, 51.22, 51.32, 58.4, 58.5, 64.44, 65.12, 65.13, 65.14, 66.4, 66.9, 85.1 **5** 20.18, 56.10, 56.11, 65.8, 71.63 **6** 64.42, 66.9, 66.10, 73.43, 74.7, 83.2 **7** 13.3, 27.9, 51.52, 54.2, 54.6, 54.14, 54.15, 59.19, 66.1, 66.2, 66.3, 71.107, 83.99, 84.1

Ferguson, G.L. **4** 3.11

Ferguson, R.B. **3** 73.7

Fergusson, J.E. **3** 83.147 **5** 83.117, 85.20

Fernando, Q. **2** 83.140, 85.47, 85.58, 85.60, 85.70, 85.81 **3** 32.32, 68.22, 77.15, 83.87, 83.94, 83.141, 85.28, 85.29 **4** 59.6, 68.10, 83.104 **6** 77.15, 81.11, 85.24

Ferrari, A. **2** 81.35, 83.10

Ferrari, M. **1** 11.1 **2** 79.16, 85.14 **3** 79.10

Ferrari, M.B. **4** 66.8 **5** 39.14, 40.19, 40.20, 40.22, 40.23, 81.5, 81.47, 81.54 **6** 40.10, 40.11, 76.41, 79.16

Ferraris, G. **1** 10.13, 13.53, 40.15, 40.16 **2** 83.25, 83.26 **3** 40.17 **4** 72.21 **5** 24.3, 83.3 **6** 29.4, 71.90, 72.12 **7** 71.108, 71.109

Ferrier, R.P. **1** 29.21, 29.22

Ferrier, W.G. **1** 29.16, 45.36, 45.39, 46.2

Ferris, J.P. **1** 58.100 **6** 58.49

Fetchin, J.A. **5** 85.26, 85.27

Field, D.S. **3** 73.13

Field, J.S. **5** 86.59 **6** 71.43, 86.17 **7** 3.8, 72.13, 80.1, 81.1

Field, R.A. **6** 65.4

Figgis, B.N. **2** 81.75 **4** 79.8 **5** 79.16, 79.17, 79.18 **7** 78.23, 78.24, 83.112

Figuiere, P. **7** 19.2

Fijuta, T. **6** 54.8

Filipenko, O.S. **7** 60.26

Filippakis, S.E. **1** 1.80, 1.100, 7.22, 38.29, 39.42, 39.43 **5** 1.23, 1.24, 1.26

Filippini, G. **4** 22.2, 31.4, 31.10 **5** 17.16, 22.1, 31.3 **6** 31.11, 38.21

Filippo, D.de **7** 85.10

Filippova, T.V. **6** 76.22 **7** 76.43

Fillion, H. **5** 11.14 **6** 11.15 **7** 11.16, 33.45

Finch, N. **3** 58.51 **4** 24.16 **7** 35.30

Finder, C.J. **5** 36.30 **6** 19.8, 36.21, 64.7

Finholt, J.E. **7** 11.10

Fink, F.H. **5** 71.108 **7** 73.30

Fink, R. **7** 31.3

Fink, W. **2** 63.19, 63.30

Finn, P.A. **7** 83.18

Finnegan, W.G. **2** 60.60, 83.27

Finney, A.J.T. **7** 58.12

Finnimore, S.R. **7** 85.17

Finocchiaro, P. **6** 62.11

Fiore, L. **5** 38.8

Fischer Junior, R.G. **3** 80.18 **5** 80.20

Fischer, E.O. **2** 71.14, 73.20

Fischer, J. **2** 84.3 **6** 58.14, 67.2, 67.6, 67.19, 84.9 **7** 80.9, 80.10, 80.20

Fischer, K. **6** 64.25 **7** 64.39, 64.40

Fischer, K.F. **7** 83.77

Fischer, M.S. **3** 48.17, 49.6 **4** 49.8, 49.12

Fischer, R.G. **6** 71.60

Fischer, W. **2** 64.68

Fischinger, A.J. **2** 81.30

Fischmann, E. **1** 18.12, 18.17, 18.18, 18.19

Fisher, C.D. **2** 60.121

Fisher, C.G. **3** 82.6

Fisher, J. **4** 51.15, 51.17

Fisher, M.S. **1** 20.11

Fitzgerald, A. **4** 64.16, 64.19 **5** 45.14, 46.4, 46.5, 46.8, 46.9, 59.1 **6** 46.6, 46.7, 46.9

Fitzwater, D.R. **1** 11.11, 11.12, 11.13, 46.8 **5** 64.5

Flaig, B.D. **4** 8.1, 44.7

Flatau, K. **5** 71.53

Fleischer, E. **3** 32.14, 32.15, 83.80

Fleischer, E.B. **1** 19.43, 31.2, 42.11, 42.12, 49.1, 49.15, 49.21, 49.22, 49.23, 49.27, 49.28, 49.30, 49.31, 49.33, 52.26 **2** 73.55, 75.26, 81.12, 83.143, 83.150, 83.184, 83.205 **3** 51.7, 62.18, 83.34, 83.134, 83.135, 83.138 **4** 49.24 **5** 16.9, 83.98, 83.99, 83.102 **6** 49.18, 69.22, 83.63 **7** 21.1

Fleming, J.E. **1** 39.7 **2** 83.161, 83.193, 84.35

Fleming, J.W. **6** 8.1

Fles, D. **7** 39.11

Fletcher, R.O.W. **1** 3.6

Fletcher, S.R. **4** 80.5, 80.14, 85.38 **5** 48.7 **6** 65.6, 71.3, 80.11, 81.26, 86.25, 86.31 **7** 71.99

Fletterick, R.J. **3** 48.26 **4** 71.9

Fliippini, G. **1** 52.1

Flippen, J. **1** 41.39

Flippen, J.L. **3** 8.17, 61.6, 61.7 **4** 37.16, 41.16, 44.26, 61.2 **5** 31.4, 34.1, 34.5, 41.16, 41.49 **6** 32.6, 41.3, 41.8, 41.22, 44.11, 45.6, 45.31, 51.28, 71.44 **7** 1.36, 31.14, 31.23, 35.19, 35.24, 35.34, 36.18, 37.16, 44.23, 44.24, 45.21

Flitcroft, N. **5** 85.32

Flood, M.T. **2** 76.81 **4** 86.39, 86.40

Flook, R.J. **5** 83.5 **6** 82.1, 82.3, 82.4, 82.12, 82.13, 82.14, 82.17, 82.18

Florencio-Sabate, F. **7** 47.34

Florencio, F. **3** 13.12 **4** 13.11 **7** 37.14, 37.15

Florian, L.R. **2** 84.18 **6** 38.25, 64.16, 76.11, 76.12, 76.13

Florian, R. **6** 49.18, 83.63

Floriani, C. **5** 67.22

Floyd, J.C. **1** 59.5

Flurry, R.L. **1** 3.44

Flynn, J.J. **1** 3.32, 11.9, 42.1, 50.32 **2** 63.26 **3** 20.10, 42.1, 63.10 **4** 21.19, 69.18, 69.19, 75.20, 75.29 **7** 21.12

Foces-Foces, C. **7** 41.2, 47.34

Foces, C. **7** 41.1, 41.3

Fodor, G. **3** 37.6 **4** 35.12
Folkers, J.B. **1** 3.67
Follner, H. **3** 2.7 **4** 2.6, 83.17, 83.18 **5** 69.17 **6** 62.5
Folting, K. **1** 19.21, 19.22, 19.26, 32.13 **2** 83.197 **6** 28.10 **7** 83.65
Fomenko, V.V. **5** 82.19 **7** 76.64, 82.5, 82.7
Font-Altaba, M. **5** 9.1, 16.8 **6** 16.7, 16.10, 17.10, 17.12, 17.13, 35.12 **7** 1.8
Fontaine, F. **7** 38.53
Fontal, B. **2** 65.41
Fontana, S. **6** 73.6, 73.7
Forder, R.A. **3** 69.8, 69.9, 69.16 **4** 10.1 **5** 68.34 **6** 10.2, 73.28 **7** 71.24, 71.74, 71.91, 73.4, 73.5, 73.6, 73.7, 73.8, 73.14, 73.23, 73.36, 73.51, 73.55, 73.65, 73.66, 73.67, 74.7
Foresti Serantoni, E. **4** 9.13, 9.14, 64.1, 64.2 **5** 38.24, 58.1 **7** 29.10, 32.6, 35.41, 35.42, 35.43, 35.44, 40.23, 51.18
Foresti, E. **3** 28.10, 32.13, 36.21 **5** 50.11
Form, G.E. **5** 81.9
Form, G.R. **5** 33.23 **6** 41.10
Forman, C. **6** 69.15, 69.16
Fornies-Marquina, J. **5** 51.13
Fornies-Marquina, J.M. **4** 13.8, 17.9, 19.14, 51.16 **6** 41.13
Forrest, J.G. **2** 81.27, 81.45, 81.48, 81.55, 81.66, 81.81
Forrest, K.P. **3** 83.111, 86.58 **4** 75.23, 83.37
Forrester, A.R. **2** 73.59
Forrester, J.D. **2** 82.59, 85.39, 85.40
Forsellini, E. **2** 78.35 **3** 71.26, 72.51, 80.19, 80.20, 83.45 **4** 71.23, 71.41, 85.26, 85.32, 86.12 **5** 80.12, 81.30, 83.29, 83.30, 84.5 **6** 81.9 **7** 33.52, 81.34, 82.11, 84.43, 84.46, 84.48
Forst, D. **2** 64.2 **7** 11.2, 11.3
Forster, D. **7** 3.29, 71.4
Forster, R. **1** 59.6
Forsythe, G.D. **3** 53.1, 53.2
Fosheim, K. **2** 70.8, 70.9
Foss, O. **1** 8.36, 8.37, 11.4, 11.25, 39.16, 39.35, 39.36, 41.6 **2** 70.2, 70.3, 70.4, 70.7, 70.8, 70.9, 70.11, 70.12, 70.13, 70.14, 70.15, 70.16, 70.24 **4** 70.6, 70.7, 70.8, 70.12 **5** 70.4, 70.5
Foster, M.S. **3** 86.33
Foster, R. **7** 40.14, 40.15
Foulon, M. **7** 52.5
Fouret, R. **7** 52.4, 52.5
Fourme, R. **1** 5.1 **4** 19.3, 19.4 **5** 5.13, 19.7, 19.8, 21.4, 21.5, 38.1, 38.2 **6** 58.10 **7** 19.2
Foust, A.S. **2** 60.155 **3** 86.33 **4** 73.13
Fouweather, C. **5** 41.24
Fowell, D.T. **3** 56.6
Fowles, G.W.A. **2** 81.46, 84.27 **3** 83.5 **4** 84.2 **6** 38.16, 84.3, 85.5
Fowlie, A.D. **3** 76.5
Fowweather, F. **1** 16.58, 16.59
Fox, M.D. **2** 78.42
Fox, M.R. **2** 78.23, 78.36
Foxman, B.M. **2** 81.89, 86.12 **3** 81.53 **4** 76.11, 83.13 **7** 72.10
Francis, J.N. **3** 62.15 **4** 72.22, 72.23, 72.24, 72.30 **5** 72.4
Franco, J.M. **5** 83.10 **6** 59.20
Frank-Kamenetskaya, O.V. **7** 12.2
Frank, G.W. **3** 47.24, 47.28, 58.1 **4** 30.5 **5** 30.3 **6** 39.2
Frank, J.A.K. **3** 71.31 **5** 31.42, 44.19
Frank, J.K. **6** 33.46, 44.26
Frank, R.W. **3** 42.2
Frankel, R.B. **7** 49.10
Frankowski, A. **5** 34.2
Franks, W.A. **2** 82.18, 82.58 **4** 82.15
Fransson, G. **5** 83.37 **7** 83.24
Franzini, M. **2** 80.3
Frasci, A. **1** 1.121
Fraser, A.J.F. **6** 86.72
Fraser, A.R. **3** 72.39 **4** 63.4, 69.16, 69.17 **5** 71.70, 77.17, 83.118, 83.119 **6** 32.18 **7** 74.4, 74.5
Fraser, G.W. **6** 64.2
Fraser, K.A. **2** 80.22, 82.25, 82.48
Fraser, L.R. **7** 77.10, 81.44, 83.105
Fraser, P.J. **5** 71.24
Frasson, E. **2** 68.9, 73.19, 77.10, 78.22, 78.31, 78.44, 83.56, 83.59, 83.60, 83.82, 83.83, 85.64, 86.38 **4** 19.2
Fraterman, H.A. **4** 21.17
Fratini, A.V. **1** 16.60, 31.45, 31.46, 52.27, 58.31, 58.33 **3** 60.52 **6** 28.24, 33.27, 63.11
Frauendorfer, E. **6** 73.6, 73.7
Fray, R.C. **4** 77.9
Fredericks, R.J. **3** 86.36
Fredette, M. **5** 77.18
Fredette, M.C. **6** 84.32
Fredin, K.S. **7** 70.3, 70.4
Freedman, H.H. **1** 22.17 **3** 69.35
Freedman, L.D. **5** 64.33
Freeman, G.F. **6** 33.24
Freeman, G.R. **3** 16.14, 16.15, 21.19 **4** 36.1 **5** 48.23 **6** 32.7, 33.25, 33.26, 35.15, 48.58, 58.1 **7** 13.19, 44.4, 44.6
Freeman, H.C. **1** 8.31, 41.33 **2** 82.5, 82.6, 82.7, 82.15, 82.23, 82.24, 82.25, 82.26, 82.27, 82.30, 82.31, 82.35, 82.36, 82.38, 82.39, 82.47, 82.54, 83.23, 83.31, 83.54, 83.173, 84.10 **3** 48.11, 76.27, 76.74, 76.75, 82.2, 82.3, 82.10, 82.11, 82.12, 82.13, 82.14, 82.15, 82.17, 83.20 **4** 82.1, 82.2, 82.8, 83.15 **5** 82.3, 82.9, 82.15, 83.5 **6** 39.3, 39.4, 69.6, 82.1, 82.3, 82.4, 82.12, 82.13, 82.14, 82.17, 82.18, 83.36
Freer, A.A. **6** 40.1 **7** 36.30, 37.21, 58.23, 73.18, 75.16
Freer, S.T. **1** 48.27, 48.28
Fregerslev, S. **1** 58.6
Frehel, D. **3** 37.6
Freiburg, C. **2** 78.19
Freimanis, Ya.F. **7** 16.12, 17.27, 25.5, 25.6, 27.8
Freiser, H. **3** 83.143 **4** 85.30

Frenz, B.A. **4** 83.133 **5** 72.4, 75.22, 75.25 **6** 71.16, 71.31, 71.32, 73.24, 73.49, 73.59, 75.9, 75.26, 83.28, 86.15 **7** 71.82, 71.84, 72.15, 72.28, 73.31, 85.3, 85.9
Fresco, J. **6** 77.5
Frew, D. **3** 39.40
Frey, A.J. **1** 58.87
Frey, M. **3** 60.7, 60.10 **4** 38.1, 84.5
Frey, M.N. **5** 48.9, 48.10, 48.59, 48.60 **6** 44.12, 48.18, 48.33, 48.57
Freyberg, D.P. **7** 84.76, 84.77
Fridrichsons, J. **1** 48.93, 50.7, 51.44, 58.23, 58.27, 58.32, 58.65, 58.77, 58.81, 58.94 **4** 83.66 **7** 58.30
Fried, J. **3** 51.7
Friedel, M.K. **3** 49.5
Friedlander, P.H. **1** 20.19, 29.9, 29.23, 30.24
Friedman, L.B. **2** 62.13, 62.36 **5** 64.16
Fries, D. **6** 46.4
Fries, D.C. **3** 45.35, 48.35, 85.38 **5** 85.26, 85.27
Frisch, P.D. **5** 73.22, 73.25, 73.28 **7** 73.13, 73.33
Frisse, M.E. **7** 49.11
Fritch, J.R. **6** 76.30
Fritchie Junior, C.J. **1** 3.28 **2** 69.35 **3** 35.24, 60.35, 60.38 **4** 71.86 **5** 36.5, 36.6, 69.7, 72.47 **6** 36.2, 60.22, 60.23, 84.24, 84.25, 84.26 **7** 36.4, 60.19, 69.18, 84.53, 84.54
Fritchie, C.J. **1** 7.30, 27.11, 32.41, 36.6, 36.23, 52.7 **2** 60.130, 73.89, 85.43
Fritz, G. **2** 63.17 **5** 63.2
Fritz, H. **3** 37.9 **6** 36.12
Fritz, H.P. **2** 85.65
Fritz, P. **6** 53.12
Frost, G.H. **2** 83.217
Frost, R. **2** 81.12
Frye, R.S. **3** 33.11
Fryer, R.I. **3** 9.19, 35.17 **6** 35.27
Fucaloro, A.F. **6** 33.1
Fuchs, B. **4** 38.5, 38.9
Fuchs, J. **6** 3.31
Fuganti, C. **3** 58.28
Fuhrmeister, C.J. **6** 66.5
Fuji, K. **1** 52.25 **6** 58.33
Fujii, K. **1** 1.108 **3** 17.5
Fujii, S. **7** 60.1
Fujiki, K. **7** 60.1
Fujimaki, H. **4** 78.1
Fujimori, K. **5** 13.6, 82.16, 83.63 **7** 83.67, 83.68
Fujimoto, H. **7** 58.19
Fujimoto, Y. **4** 48.20, 48.22
Fujita, E. **6** 54.1, 54.8 **7** 84.44
Fujita, K. **6** 59.19
Fujita, T. **4** 53.10
Fujiwara, H. **3** 27.4
Fujiwara, T. **1** 56.20, 58.93 **3** 44.12, 47.32, 56.13 **4** 56.5 **5** 4.2, 8.14, 33.34, 33.35, 33.36, 35.22, 44.5, 44.10, 47.14, 47.15 **6** 1.7, 8.5, 8.6, 35.14, 46.8, 46.10, 48.56, 58.45 **7** 1.15, 1.23, 17.10, 60.1
Fukazawa, Y. **1** 31.16, 58.111 **4** 22.3, 22.4, 22.5, 27.15 **5** 58.43 **6** 31.10, 58.21
Fukushima, R. **6** 5.4, 5.5
Fukuto, T.R. **7** 64.12
Fukuyama, K. **5** 2.28 **6** 13.7, 33.13 **7** 34.16
Fukuyo, M. **1** 13.12 **2** 65.18, 65.20 **3** 36.4 **5** 41.23, 83.107 **7** 45.1
Fukuzawa, A. **6** 59.6, 59.8
Fulke, J.W.B. **5** 56.10, 56.11 **7** 59.19
Fumagalli, M. **7** 17.4, 17.5
Funakoshi, K. **4** 48.33, 48.35
Funke, P. **5** 59.23
Furberg, S. **1** 11.56, 38.22, 44.40, 45.4, 45.5, 45.6, 45.7, 45.57, 45.58, 45.59, 45.60, 45.64, 47.9, 47.15 **3** 44.7, 44.10, 45.33 **4** 44.1 **5** 3.14, 8.13 **6** 44.17, 48.35 **7** 44.2, 44.19
Furlanetto, R. **6** 71.89
Furlani, C. **3** 85.22 **4** 85.33
Furnas Junior, T.C. **3** 49.1, 49.7
Furst, A. **4** 31.28
Furst, G.T. **6** 37.8
Furue, K. **3** 69.11
Furuhata, K. **7** 41.16
Furukawa, H. **5** 53.18 **6** 53.15 **7** 53.11
Furukawa, K. **1** 22.4, 22.5
Furukawa, M. **5** 50.25 **7** 50.36
Furusaki, A. **1** 1.108, 13.36, 22.13, 28.5, 28.14, 31.13, 50.31, 58.35, 58.120, 59.25 **3** 17.5, 33.42, 36.4, 50.5, 50.20, 53.20, 72.20, 81.24 **4** 58.1, 58.2, 81.32, 81.33, 81.35 **5** 1.20, 9.15, 50.14, 53.4, 53.5, 54.4, 54.5, 54.6 **6** 28.21, 28.26, 59.6, 59.8 **7** 28.6, 40.8, 50.11
Furuyama, S. **6** 14.2
Fushiya, S. **6** 54.14
Fusina, L. **5** 84.12 **7** 84.6
Fyfe, C.A. **6** 64.42

Gabe, E.J. **1** 2.46, 2.47, 2.76, 3.2, 4.15, 35.46, 52.22, 53.9, 58.25 **4** 26.3, 38.25 **7** 27.5, 64.13, 79.16
Gabes, W. **7** 3.18
Gabrielsen, M.V. **7** 17.18, 38.13
Gachon, P. **7** 50.35
Gadet, A. **6** 81.32, 83.49
Gadret, M. **1** 33.43, 33.56, 33.57 **3** 33.20, 33.25 **6** 33.29, 33.34, 35.1, 59.4 **7** 24.8, 50.33
Gaetani, A. **1** 35.8 **3** 41.2, 76.1
Gaetani, E. **5** 41.41
Gafner, G. **1** 19.6, 19.7, 24.1, 24.4, 58.11 **2** 60.69, 83.130 **3** 41.7, 58.5, 75.47 **4** 41.9, 41.10, 59.25, 68.8 **6** 54.3, 54.12, 54.18, 74.12 **7** 38.63, 58.17
Gagnaux, P. **2** 60.50
Gagne, R.R. **7** 49.22
Gagnon, C. **7** 44.10, 83.7, 85.12
Gaines, D.F. **5** 62.3, 62.10 **7** 38.8, 62.5
Gainsford, A.R. **5** 76.14, 76.15 **7** 76.52
Gainsford, G.J. **3** 33.30, 33.31, 33.32, 71.71 **4** 71.71, 71.73, 71.89, 73.24 **5** 71.95, 71.96, 71.99 **6** 76.16 **7** 71.115
Gal, A.W. **6** 80.14
Galantay, E. **5** 51.26, 51.27 **7** 51.17

Galbraith, A.R. **5** 83.49 **6** 86.13
Galdecki, Z. **7** 64.25, 64.26, 64.27
Gale, G.D. **7** 71.74, 73.23
Galeffi, C. **3** 59.33
Galesic, N. **4** 81.11 **7** 3.5
Galigne, J.-L. **1** 9.45, 32.25 **3** 2.9, 9.14, 32.24
Galigne, J.L. **4** 2.1, 9.11 **5** 63.9 **6** 35.2, 35.4
Gall, R.S. **6** 83.12, 83.40
Gallagher, R.T. **7** 58.12
Gallen, B. **6** 26.3, 36.15, 48.36, 48.37, 48.84 **7** 36.21
Galleri, M. **6** 40.5
Galles, C.J. **3** 71.2
Gallezot, P. **1** 44.41
Galloy, J. **7** 17.20, 32.24
Gandhi, S. **7** 67.2
Ganellin, C.R. **6** 32.17 **7** 32.4
Ganello, G.de **6** 73.49
Ganem, A. **5** 2.3
Ganguli, S.N. **6** 58.46
Ganis, P. **1** 1.78, 1.86, 1.93, 1.97, 1.110, 21.24, 21.25, 38.49, 38.59 **2** 72.38, 73.18, 75.67 **3** 1.32, 22.4, 75.36, 75.37, 83.86 **4** 1.9, 15.6, 16.6, 50.17, 72.12, 83.90 **5** 1.32, 8.18, 13.16, 72.21, 75.4 **6** 31.3, 71.23, 75.2 **7** 15.5, 15.6, 38.43, 78.1
Gantzel, P.K. **1** 11.21, 31.35 **2** 67.6, 67.7
Garaj, J. **2** 81.6, 83.30 **4** 76.1 **5** 81.4 **7** 76.7, 79.6
Garbarczyk, J. **7** 58.13
Garbassi, F. **4** 46.1 **5** 40.31 **6** 46.1
Garber, A.R. **6** 71.75
Garbuglio, C. **1** 1.51, 39.62
Garcia-Blanco, S. **1** 13.44, 59.46 **3** 13.13 **4** 13.11, 77.10 **5** 35.20, 35.24, 64.10, 83.10 **6** 13.13, 59.20 **7** 37.14, 37.15, 40.12, 40.13, 41.1, 41.2, 41.3, 47.34, 48.73
Garcia, J.F. **7** 81.13
Gardia, C. **1** 3.78
Gardner, A.P. **4** 78.8
Gardner, E.J. **5** 38.17 **6** 38.25
Gardner, J.A.F. **1** 59.61
Gardner, R.C.F. **6** 71.74
Garegg, P.J. **5** 45.19 **6** 45.25, 45.27
Garner, C.D. **7** 69.14, 83.85, 85.36
Garner, R.H. **7** 22.3
Garofalo Junior, J. **7** 49.18
Garratt, D.G. **6** 11.7
Gartland, G.L. **4** 43.9 **6** 48.58, 60.5, 60.11 **7** 43.1, 44.6, 48.19
Garton, G. **2** 86.67
Gartzke, W. **5** 71.9, 75.10, 75.11 **7** 38.2, 71.3, 71.6, 71.14, 75.12, 75.14
Garza, A. **4** 71.70 **5** 71.76, 71.78 **6** 37.11 **7** 28.8
Gasan, V.M. **5** 5.2
Gash, A.G. **4** 69.11, 69.22, 69.27, 74.11 **5** 79.24, 86.70, 86.71, 86.74 **7** 69.4, 69.5
Gasparri, G.F. **1** 41.15 **2** 79.2, 79.16, 79.18, 79.20, 79.26, 79.27, 79.31, 80.1, 83.8, 83.9, 85.2, 85.14, 85.35 **3** 8.5, 16.4, 79.10 **4** 66.8 **5** 40.19, 40.20, 40.23
Gasparrini, F. **4** 71.41
Gassmann, J. **1** 49.16, 54.24 **5** 49.1
Gastaldi, L. **4** 83.89, 85.33 **5** 79.4, 80.4 **6** 85.7, 86.23
Gateau-Olesker, A. **7** 45.52
Gatehouse, B.M. **2** 86.5 **3** 45.21, 45.22, 45.23 **4** 24.3, 43.1, 43.2, 71.79, 71.88, 71.91, 78.8, 78.16 **5** 71.24, 78.12 **6** 76.53, 78.2, 78.5, 78.22 **7** 43.1, 64.31, 71.78, 72.29, 76.74, 86.9, 86.18, 86.59
Gatilov, Yu.V. **6** 72.24
Gatteschi, D. **5** 76.44, 83.112
Gattow, G. **1** 11.3 **5** 7.4 **6** 2.2, 3.19, 11.3 **7** 39.6
Gaughan Junior, A.P. **5** 86.65 **6** 85.9, 85.10, 86.37, 86.65 **7** 71.112, 75.20, 86.83
Gaughan, A.P. **3** 86.96 **4** 82.13, 83.72, 86.61, 86.68, 86.69
Gaultier, J. **1** 11.44, 24.21, 24.24, 24.25, 25.3, 25.4, 25.5, 25.6, 25.7, 25.13, 25.16, 25.18, 25.21, 26.3, 38.32, 38.35, 59.39 **3** 25.2, 25.4, 25.5, 26.9, 60.32 **4** 21.4, 25.1, 35.9, 38.22 **5** 7.9, 7.10, 21.1, 25.1, 35.10 **6** 12.1, 21.3, 27.9, 27.10, 27.13, 27.14, 28.11, 32.19, 32.20, 38.34, 59.15 **7** 25.1, 25.3, 26.2
Gaus, P.I. **5** 71.13
Gautheron, B. **6** 73.29
Gauthier, J. **5** 41.44
Gavezzotti, A. **3** 19.11 **4** 31.20 **5** 15.8
Gavuzzo, E. **6** 38.13 **7** 48.29
Gaw, H. **4** 81.37
Gaziani, R. **7** 84.48
Gazo, J. **5** 81.4 **7** 83.23
Gearhart, R.C. **5** 33.9, 35.34 **7** 35.32
Gebala, A.E. **5** 71.56, 83.98, 83.99, 83.102
Gebreyesus, T. **4** 58.33
Geddes, A.J. **6** 17.15, 17.16, 17.17, 48.42, 48.68 **7** 76.58
Gee, R.J.D. **4** 71.68
Gehlhaus, J. **6** 38.33
Gehrer, H. **5** 5.9
Gehrz, R.C. **4** 40.10
Geiger Junior, W.E. **7** 60.8, 60.9
Geise, H.J. **1** 42.3, 46.15, 51.8, 51.30, 51.31, 51.32, 51.33, 51.34 **3** 51.37, 51.38
Geissman, T.A. **3** 53.18
Gelder, J.I. **7** 85.38
Geller, S. **1** 3.16 **3** 3.19 **5** 33.10, 33.11, 33.12, 33.13, 33.14, 33.15
Gellings, P.J. **6** 81.37 **7** 81.36
Gemenden, C.W. **3** 58.51
Gemmell, K.W. **3** 53.23, 58.67
Genet, F. **1** 3.45, 33.49, 35.19, 35.20 **2** 84.14 **4** 7.12
Gent, A.R. **4** 71.15
Gent, P.A. **1** 45.71
Gentile, P.S. **3** 79.6, 79.7 **7** 8.13, 79.7

Geoffre, S. 1 26.3 4 25.2, 51.9 6 38.34 7 17.30, 17.31, 25.2, 25.3, 35.7, 43.4, 51.9
George, J.W. 2 70.17, 70.18 7 70.6
George, T.A. 7 86.101, 86.108
Gerbeleu, N.V. 7 79.15, 79.17
Gerdil, R. 1 44.32 4 31.14, 31.16 5 34.7 6 28.7 7 38.50, 39.7, 45.40, 60.30, 60.31, 60.32, 60.33, 76.1
Gerely, N.V. 7 79.14
Gerken, V.A. 1 3.29
Gerkin, R.M. 3 40.18
Gerlach, U. 7 33.13
Gerloch, M. 2 73.29, 73.46, 78.13, 78.14, 83.89 3 78.24 6 83.81
Germain, G. 1 10.14, 10.15, 10.16, 41.26, 41.27, 41.30, 41.46 3 20.18, 48.15, 48.44, 60.17, 60.18 4 32.23, 33.43, 33.44, 40.17, 48.49, 51.25, 51.30, 51.31, 51.34 5 9.1, 16.8, 16.20, 30.2, 48.74 6 17.3, 31.13, 31.31, 32.36, 35.32, 36.9, 40.2, 40.3, 50.9 7 17.20, 20.4, 32.24, 33.53, 38.43, 47.33, 47.41, 47.42, 59.16, 60.2
Gero, S.D. 7 45.52
Gerst, K. 7 85.26
Gerstein, B.C. 7 84.3
Gerteis, R.L. 2 68.15
Gervasio, G. 4 72.21 6 71.90, 72.12 7 71.108, 71.109
Gestaut, L.J. 6 79.17
Geue, R.J. 4 83.24 5 81.38
Ghazarossian, V.E. 7 58.3
Ghelfenstein, M. 7 19.2
Ghilardi, C.A. 3 78.18, 86.44, 86.53 4 76.43, 76.44, 83.56, 83.57 5 76.50, 76.52, 76.53, 86.28, 86.32 6 86.53 7 86.81, 86.85
Ghose, S. 1 43.12
Ghotra, J.S. 6 83.54, 83.55, 84.45
Ghouse, K.M. 5 64.3, 64.36
Giacomello, G. 1 1.133, 13.30, 33.1, 33.70, 33.72, 35.3, 35.4, 35.7, 35.12, 35.13, 35.14, 35.15, 44.53, 50.48 3 50.28, 58.15
Giacomello, P. 4 8.11
Giacovazzo, C. 5 13.14 7 17.21
Giam, C.S. 7 16.7
Giannini, U. 2 72.39
Giarda, L. 4 46.1 5 40.31 6 46.1
Gibbons, C.S. 1 31.44 3 58.64 4 31.1, 31.3, 71.18 5 31.5, 39.37 6 11.8, 11.9
Gibson, C.S. 2 71.16
Gibson, D.T. 1 11.29 6 37.11
Gibson, J.G. 6 83.25
Gibson, J.W. 1 37.17 3 48.47 4 44.21
Gibson, R.E. 3 36.27, 39.32
Gidaspov, B.V. 5 1.14
Gieren, A. 1 51.49 3 49.15 4 36.37, 71.92 5 32.7, 35.30, 36.25, 36.31, 48.40, 48.69, 75.1 6 32.15, 39.12 7 4.7, 4.9, 16.9, 32.15, 32.27
Giese Junior, R.F. 3 21.3, 21.4
Giesecke, J. 6 58.16, 58.17
Giesen, M.T. 4 21.6
Giesen, W.van de 4 84.4
Gifkins, K.B. 6 59.11
Giglio, E. 1 33.81, 44.67, 59.8, 59.20 2 60.110, 60.111, 60.112 4 8.11 5 60.40, 60.41 6 38.13, 51.5 7 32.12, 48.29
Gignere, P.A. 2 60.8
Gil, E.R.de 3 72.12
Gilardi, R.D. 1 58.31, 58.33 3 51.28, 58.63 4 9.12, 37.16, 40.15, 42.6, 54.6 5 54.2, 64.42 6 35.9, 35.10, 45.6, 51.28, 58.31 7 1.36, 35.24, 35.34, 45.21
Gilbert, M.M. 5 68.2, 68.19
Giles, H.G. 6 32.18
Gilje, J.W. 5 64.7
Gill, D.F. 5 86.51
Gill, G.B. 7 38.19
Gill, J.T. 7 86.114
Gill, N.S. 3 83.119
Gillard, J. 3 58.6
Gillard, R.D. 2 77.1, 81.39, 82.28, 82.34, 82.37 3 76.63, 82.5 4 83.63
Gillean III, J.A. 6 39.14
Gilli, G. 4 78.2 6 48.86, 86.8, 86.9 7 11.15
Gillier-Pandraud, H. 1 17.7, 17.8, 17.27, 17.28 2 84.12 3 1.7 4 3.3, 17.3 5 17.9, 17.11, 17.12 6 17.8 7 45.22, 45.23
Gillier, H. 1 43.2, 43.3
Gillot, R.J.J.H. 1 39.31
Gillum, W.O. 2 83.190
Gilman, N.W. 5 35.26, 35.27
Gilmore, C.J. 3 3.52, 71.64 4 69.23, 71.77 5 58.47, 59.18, 59.19, 59.26 6 53.17, 54.2 7 53.6, 53.20, 53.21, 53.22, 58.31
Ginderow, D. 5 69.10 7 71.12, 86.5
Ginsberg, A.P. 2 73.30
Ginzburg, S.L. 3 40.2 5 40.27 6 33.30, 40.16, 41.20 7 16.8, 85.22
Giordano, F. 1 21.29 2 64.6, 84.50, 84.51 3 33.23 5 36.15, 36.16 6 51.44 7 36.12, 82.22, 82.23
Giordano, P. 7 73.48
Giraldi, G. 1 9.3, 39.23, 39.25 2 84.9
Girard, C. 4 65.6, 66.5
Girling, R.L. 2 79.19, 85.56 3 79.20, 79.24 5 45.8, 45.9, 45.10, 85.8 6 45.10, 45.11 7 79.8
Gitany, R. 3 75.34 7 53.16, 64.45
Giudice, F.L. 2 72.39
Giuliano, V. 3 75.37
Giuseppetti, G. 2 83.81
Giusta, A.D. 5 84.18 6 40.25, 84.44
Gizycki, U.v. 5 31.25
Gladkikh, E.A. 7 73.44
Gladkova, V.F. 4 9.10 5 9.11
Glanville, J.O. 2 72.83
Glaser, S.L. 6 46.5, 46.11
Glass, G.E. 2 71.18
Glass, R.S. 5 32.6
Glasser, L.S.D. 3 76.24, 85.52
Glazer, A.M. 4 36.5, 36.6, 36.7, 36.8, 36.9, 36.10
Gleason, J.G. 5 39.25 6 39.15
Gleason, W.B. 6 40.18

Glehn, M.von **1** 45.27 **3** 36.7 **4** 36.15 **5** 36.18, 50.23, 50.24 **7** 59.6
Glen, G.L. **2** 81.52
Glick, M.D. **1** 3.83, 49.24, 49.29 **2** 75.63, 76.61, 78.45, 84.15, 84.16, 84.17 **3** 30.4, 39.45, 45.29, 58.29, 64.67, 65.12, 68.18 **4** 24.12, 39.47, 81.56, 81.58, 83.77 **5** 31.20, 69.26, 77.24, 81.51, 81.52 **6** 63.8, 77.21, 83.39, 83.41, 84.20 **7** 63.1, 71.85, 73.26, 73.47, 83.56, 83.61, 84.27, 85.15
Glidewell, C. **4** 63.10 **5** 63.11
Glinskaya, E.A. **2** 85.59
Glinsukon, T. **6** 58.52
Glockling, F. **3** 73.43
Glockner, P. **3** 75.17
Gloss, R.A. **1** 4.5 **5** 64.1
Glotter, E. **4** 51.49
Glowiak, T. **5** 76.3, 76.7
Glusker, J.P. **1** 1.79, 2.76, 2.78, 2.79, 2.80, 2.81, 4.15, 44.20, 48.91 **4** 2.20, 2.21, 2.22, 26.3, 36.21, 36.32 **5** 2.26, 2.27, 2.29, 26.5, 28.5, 81.17, 81.18 **6** 26.3, 36.4, 36.5, 36.15, 48.36, 48.37, 48.84 **7** 1.26, 1.27, 29.4, 29.7, 29.9, 36.15, 36.21, 38.60, 38.62, 76.60, 79.16, 85.18
Gluskova, V.P. **1** 11.30 **2** 62.29, 62.42, 69.20
Gluvchinsky, P. **6** 78.20
Gluyas, R.E. **3** 15.1, 15.2
Gluzinski, P. **7** 83.57, 83.58
Go, K.C. **4** 32.11
Go, K.T. **1** 34.5, 41.19, 41.20 **3** 54.9 **5** 51.45, 51.46, 53.17 **6** 35.28 **7** 10.6, 10.7, 52.7
Goaman, L.C.G. **1** 21.26, 21.27 **7** 46.1
Gobillon, Y. **2** 60.23, 60.30, 60.32, 60.46
Goddard, R. **7** 71.39, 71.69, 71.70, 75.19, 75.23, 85.17
Godfrey, J.E. **6** 27.4 **7** 27.6, 84.11
Godfrey, J.M. **3** 49.13
Godwin, A.D. **6** 31.26
Goebel, C.V. **4** 81.42
Goebel, P. **1** 23.1
Goedken, V.L. **5** 83.100 **6** 71.66, 83.17, 83.22, 83.38, 83.53, 85.39 **7** 71.27, 71.28, 83.39, 83.100
Goedkoop, J.A. **1** 1.31
Goel, R.G. **4** 66.4, 66.9
Gogoi, B.N. **3** 11.7
Gokel, G.W. **5** 73.41
Gokrinsky, C. **6** 58.32
Gol'der, G.A. **1** 5.24, 5.30, 7.23, 15.3, 26.1, 26.2, 26.7, 26.8, 39.77 **3** 26.1, 26.2, 35.20 **4** 24.4, 24.6, 24.7 **5** 40.27 **7** 26.3, 36.6, 42.10
Golankiewicz, K. **3** 44.37
Gold, K. **2** 72.41 **3** 62.14, 71.60, 75.32, 75.33, 83.93 **4** 71.2 **5** 62.8
Goldberg, D.E. **7** 69.17
Goldberg, I. **4** 8.2, 38.5, 38.9, 39.18, 39.19, 69.13 **5** 60.14, 60.15, 60.16, 60.17, 60.18, 69.1, 69.2, 69.5, 69.6, 69.8 **6** 60.1 **7** 38.41, 38.55, 38.56, 60.39
Goldberg, S.Z. **4** 71.50 **6** 22.3, 84.39 **7** 71.25
Goldfarb, T.D. **2** 63.15
Goldfield, S.A. **6** 64.57
Golding, B.T. **5** 35.12 **6** 82.10 **7** 76.52
Golding, R.M. **7** 80.6, 80.7
Goldish, E. **1** 10.19, 16.10 **4** 12.9
Goldman, M. **1** 1.5
Goldschmidt, G.H. **1** 35.11
Goldschmied, E. **2** 76.34 **3** 76.30, 83.76
Goldstein, M. **4** 83.25 **6** 3.20, 3.21, 3.22
Goldstein, P. **1** 22.8, 32.5, 32.35, 32.36, 32.37, 32.39, 32.40, 32.43, 32.44 **2** 62.45, 64.66, 76.55 **6** 38.38 **7** 31.11
Golic, L. **1** 2.10, 2.11, 2.45 **4** 48.29 **5** 2.15, 2.16, 2.17, 35.19 **6** 2.4, 77.17, 81.16 **7** 2.1, 33.11, 48.22, 66.12
Golomb, M.L. **2** 82.7 **3** 82.2, 82.3, 82.10, 82.11
Golovachev, V.P. **7** 67.3
Golovastikov, N.I. **2** 66.11, 66.12
Golovchenko, L.S. **7** 71.18
Golovina, N.I. **1** 10.2, 10.3 **3** 12.1 **5** 12.1
Gomes de Mesquita, A.H. **1** 11.61, 12.14
Gonschorek, W. **7** 14.4
Good, M.L. **4** 81.59
Good, R.H. **4** 35.8
Goodacre, G.W. **2** 64.24 **3** 64.1, 64.10, 64.21 **4** 42.3
Goodall, B.L. **5** 71.44 **6** 71.74 **7** 86.90
Goodall, D.C. **2** 85.49, 86.105
Gooden, E.L. **3** 37.6
Goodgame, D.M.L. **1** 3.38, 3.39 **2** 76.21, 76.50, 83.104, 83.215 **7** 83.33
Goodman, M. **4** 15.6, 16.6 **5** 1.32, 8.18, 13.16, 41.3 **7** 15.5
Goodman, S.H. **1** 31.11
Goodwin, F.H. **1** 45.72
Goodwin, T.H. **1** 13.20, 17.29, 30.24, 45.11
Goody, R. **6** 67.16
Gopal, R. **7** 71.120, 83.52, 83.80
Gopalakrishna, E.M. **1** 15.10, 51.10, 51.12, 51.21, 51.22 **2** 60.151, 84.64 **3** 15.5, 15.6 **5** 13.12, 37.6 **6** 58.23
Gorbel, C.V. **2** 75.7
Gordon-Gray, C.G. **4** 38.21
Gordon, A.J. **6** 40.14
Gordon, J.T. **1** 56.6
Gore, E. **5** 73.16, 73.17, 73.18, 82.21
Gore, E.S. **6** 76.17
Gorio, M. **7** 84.19, 84.82
Gorres, B.T. **3** 60.33 **7** 33.20
Gorter, S. **6** 83.7
Gosling, K. **3** 68.7, 68.8
Goto, K. **1** 58.36, 58.57
Gottardi, G. **2** 65.16, 66.13
Gottlicher, S. **1** 37.19 **2** 60.77, 84.53 **3** 35.5, 37.9 **4** 37.11 **7** 10.2, 10.3, 10.4, 10.5, 76.67
Goucharov, A.N. **7** 42.4
Goucharov, A.V. **7** 42.3
Gough, T.E. **3** 77.18
Gougoutas, J.Z. **1** 17.30 **3** 13.15, 40.13, 51.39, 58.30 **4** 13.17, 13.18, 35.37, 59.27 **5** 42.5, 42.8, 42.9 **6** 35.13 **7** 13.2, 19.8, 42.11, 42.12, 59.13

Goukhberg, S.S. **5** 84.10
Gould, E.S. **1** 32.24, 39.2, 39.18
Gould, R.O. **2** 85.24 **3** 85.43 **4** 85.36 **5** 45.35, 45.36, 45.38 **6** 86.72 **7** 35.26, 45.9, 45.10
Gould, S.E.B. **7** 35.26, 45.9, 45.10
Goursolle, M. **3** 33.20, 33.25 **6** 33.29, 33.34, 35.1 **7** 24.8, 50.33
Gouterman, M. **6** 49.9
Govedic, S. **7** 76.4, 76.9, 83.17, 83.31
Gowda, D.S.S. **6** 58.35, 81.40
Graaff, R.A.G.de **4** 21.6, 27.2 **6** 51.20 **7** 51.39, 51.41
Grabowski, M. **3** 39.8
Grabowski, M.J. **7** 8.5, 8.6
Graddon, D.P. **2** 78.6, 83.101, 83.157 **6** 83.61
Grady, R.A. **5** 39.31 **7** 58.18, 59.16
Graeber, E.J. **1** 3.26 **5** 15.6 **6** 15.3, 15.4
Graf, J. **5** 62.11
Graham, A.J. **2** 81.39 **3** 71.27, 72.27 **5** 72.27
Graham, J. **4** 79.8
Graham, W.A.G. **3** 69.27 **4** 73.13 **5** 71.54 **6** 71.19 **7** 69.12
Grahn, W. **7** 36.8
Grainger, C.T. **1** 9.44 **4** 3.23, 21.14
Gram, F. **1** 9.20
Gramaccioli, C.M. **1** 17.25, 17.31, 17.32, 52.1 **2** 60.103, 82.10, 82.11 **3** 31.25 **4** 31.4, 31.6 **5** 17.16, 31.3, 31.28, 31.29
Gramlich, V. **6** 71.85 **7** 86.60
Grampel, J.C.van de **4** 27.14
Grand, A. **5** 1.25 **6** 1.23, 1.24, 1.25, 1.28 **7** 64.11
Grandjean, D. **1** 41.14 **5** 85.16 **7** 16.1, 16.5
Granoff, B. **2** 75.34, 77.21
Gransden, S.E. **7** 81.10
Grant, D.F. **1** 13.32, 21.26, 21.27, 52.22, 53.18 **2** 68.2 **4** 17.4, 53.11
Grant, I.J. **1** 56.1, 58.74
Grant, W.K. **1** 28.4
Graver, H. **7** 70.1
Gray, H.B. **1** 3.40 **2** 76.75, 76.81, 85.73, 85.86, 85.98, 86.26 **3** 85.27, 86.77 **4** 73.3, 86.39, 86.40 **5** 76.41, 86.25, 86.26 **6** 20.11, 73.8, 76.29, 77.5, 85.45 **7** 86.6
Gray, R.T. **1** 52.8
Graybeal, J.D. **3** 69.1
Graziani, M. **7** 71.117, 71.121
Graziani, R. **2** 78.28, 78.35 **3** 72.49, 72.51, 80.19, 80.20, 81.55, 83.45, 85.19, 86.19, 86.25 **4** 71.23, 71.41, 71.92, 85.26, 85.32, 86.12 **5** 80.12, 81.30, 83.29, 83.30, 84.5, 84.27 **6** 81.9 **7** 33.52, 81.34, 82.11, 84.80
Grdenic, D. **1** 8.45 **2** 71.15, 77.47, 77.49, 81.60 **3** 66.5, 83.9 **4** 83.137, 83.138 **6** 77.17, 81.5 **7** 71.9, 71.11, 76.4, 76.9, 83.17, 83.31
Greaney, T.M. **7** 77.1, 77.4
Greaves, E.O. **6** 72.7
Greaves, M.L. **3** 49.14
Green, C.D. **2** 85.58 **3** 53.19
Green, C.R. **5** 71.21
Green, D.W. **1** 44.30
Green, E.A. **7** 17.17, 47.5
Green, M. **1** 52.26 **4** 71.84, 71.86 **5** 71.3, 71.41, 74.10, 83.61 **6** 71.34, 76.46, 86.5, 86.21 **7** 71.22, 71.45, 71.49, 71.59, 71.61, 71.66, 71.70, 71.96, 72.1, 72.2, 72.9, 75.22, 75.23, 86.4, 86.10, 86.11
Green, M.J. **3** 51.7
Green, M.L.H. **3** 71.5, 71.6 **4** 73.29 **5** 68.34
Greenberg, B. **1** 9.11, 20.15
Greene, D.L. **6** 80.18
Greene, F.D. **7** 31.30
Greene, P.T. **3** 62.13, 69.22, 73.4, 73.12, 73.13, 83.18, 85.47, 86.43 **4** 62.1, 69.21, 82.7, 83.19, 83.20
Greene, R.L. **5** 28.16, 51.14 **6** 31.9
Greengrass, C.W. **6** 31.18, 31.25
Greenwood, N.N. **3** 64.46 **4** 84.9
Gregory, U.A. **3** 71.41, 71.42, 86.28, 86.29, 86.88
Greig, D.R. **7** 71.102
Grenthe, I. **3** 81.6, 81.10 **4** 81.21, 81.25 **5** 81.21, 81.22 **6** 2.16, 2.17, 2.18, 2.19, 2.20 **7** 76.17, 76.18
Gress, M.E. **6** 38.18, 73.5 **7** 45.5, 66.5
Grethe, G. **1** 41.29 **3** 35.23
Grevels, F.W. **5** 72.5
Grey, D.M. **6** 81.14
Gribenyuk, A.Yu. **7** 84.7, 85.5
Gribova, E.A. **1** 39.77 **3** 35.20
Grice, N. **4** 75.24
Grieb, R. **2** 83.188, 83.206 **5** 31.1 **7** 53.7
Griend, L.J.V. **7** 86.7
Griffe, M. **2** 60.47 **3** 48.44 **5** 39.10, 39.11, 39.12
Griffin, A.M. **7** 4.3, 41.21
Griffin, B.E. **5** 47.31
Griffin, G.W. **6** 36.16
Griffin, J.F. **5** 15.1 **7** 48.13, 51.45, 51.53
Griffin, R.G. **5** 81.7 **6** 14.1
Griffith, E.A.H. **3** 75.2, 75.8, 79.22 **4** 74.9, 75.19 **5** 15.10, 15.11, 15.12, 20.17 **6** 24.2 **7** 74.1
Griffith, E.H. **5** 79.24 **7** 44.20
Griffith, R.L. **2** 81.2
Griffith, W.P. **6** 65.6
Griffiths, A. **3** 9.7, 28.2 **4** 83.87 **6** 40.17, 41.23
Grigg, R. **3** 49.4
Grigor'eva, N.V. **1** 7.3, 10.1, 10.4, 12.1 **3** 7.1, 12.3, 12.5, 12.6 **4** 12.4, 12.7 **5** 1.14 **7** 1.10, 12.2, 12.3, 15.3
Grim, S.O. **2** 72.83 **4** 86.38 **5** 86.30, 86.31
Grimes, R.N. **4** 62.1 **5** 62.9 **6** 75.1
Grimvall, S. **1** 14.12
Grindstaff, W.K. **5** 65.5
Grob, R.L. **7** 51.14
Gronbaek, R. **1** 3.52 **2** 85.7
Groot, A.de **1** 35.42
Gross, B.von **5** 34.4, 34.6 **7** 20.19, 50.8
Grosse, G. **2** 76.7
Grosse, J. **6** 86.11

Grostic, M.F. **4** 27.9
Groth, P. **1** 1.3, 1.4, 21.8, 21.11, 21.20, 21.21, 38.46, 38.50, 38.52, 38.56, 45.32 **2** 60.12, 60.48, 60.91, 60.92 **3** 21.12, 21.14, 21.18, 23.4, 23.7, 33.17, 38.5, 38.9, 41.9, 48.46, 86.54 **4** 23.1, 33.21, 38.16, 67.19 **5** 40.7, 40.8, 40.9, 40.10, 40.28, 40.29, 41.27, 41.29 **6** 19.11, 23.3, 48.76, 48.85 **7** 23.1, 23.2, 39.29, 48.59, 48.69
Grout, R.J. **4** 35.25
Grover, S.H. **7** 53.12
Groves, D. **3** 62.3
Grubbs, R.H. **6** 71.84
Grund, N. **6** 39.5
Grundon, M.F. **6** 17.19
Grundtvig, F. **4** 39.10
Grundy, K.R. **4** 86.35
Grunwell, J.R. **7** 11.11
Grunze, H. **2** 67.2
Grutzmacher, H.-F. **3** 63.4, 63.5, 63.6
Gschwend, H.W. **5** 31.36
Gsell, R. **2** 83.32
Guastini, C. **2** 81.80 **3** 81.19 **4** 76.38, 79.2, 79.3, 83.55, 84.12 **5** 69.14, 71.27, 78.19, 81.25, 81.43, 81.44, 83.16, 83.36, 83.38, 86.37, 86.39 **6** 40.5, 40.6, 40.9, 58.27, 76.45, 78.4, 80.3 **7** 40.2, 73.61, 73.62, 73.63, 80.4, 83.2
Gubenko, N.T. **5** 72.26
Gubin, S.P. **3** 71.4, 74.9 **4** 73.10, 74.5
Gubler, B. **1** 53.40
Guder, H.-J. **7** 68.2
Guder, H.J. **6** 68.6
Guerchais, J.-E. **5** 83.80 **7** 69.1
Guerchais, J.E. **7** 77.9
Guggenberger, L.J. **1** 38.54 **2** 67.10, 69.43, 71.40 **3** 3.18, 62.9, 86.10 **4** 3.9, 36.12, 71.6, 86.39, 86.40 **5** 68.26, 71.48, 72.49, 72.50, 83.35, 86.17, 86.55, 86.62 **6** 32.12, 73.50, 73.56, 86.2, 86.64 **7** 25.4, 32.20, 71.13, 71.38, 75.29, 75.37
Guggisberg, A. **3** 58.67
Gugielmetti, L. **1** 54.7
Guha, S. **3** 48.21, 82.21 **5** 82.2 **6** 48.7, 82.8
Guilard, R. **5** 75.17, 75.19, 75.20 **7** 75.7
Guilhem, J. **1** 26.27, 26.29 **3** 58.27 **4** 51.38, 58.7 **6** 1.17, 1.18, 1.19, 1.20, 1.21, 58.55, 64.52 **7** 1.28, 1.29, 1.30, 58.32
Guillaumet, G. **5** 28.7, 28.11
Guin, H.W. **5** 27.18
Guinier-Manach, Y.le P. **6** 42.5
Gulko, A. **6** 16.6, 83.31
Gullo, V.P. **7** 56.8
Gum, P. **1** 35.27
Gundel, W.H. **7** 48.73
Gunn, P.A. **5** 56.10, 56.11
Gupta, M.P. **1** 1.50 **3** 2.15, 2.16, 2.17, 13.6 **4** 2.16, 13.10 **5** 1.13, 2.19, 14.3, 14.4, 24.5 **6** 2.14, 24.5, 28.19 **7** 2.5, 2.6, 8.4, 14.1, 19.3, 24.4, 28.4, 33.18
Gupta, N.P. **1** 1.50
Gupta, S.C. **6** 48.32
Gupta, T.N.P. **7** 2.5, 28.4
Gurr, G.E. **1** 45.22, 45.23 **2** 83.179 **3** 60.33
Gurskaya, G.V. **4** 48.41 **7** 17.8
Guseinova, M.K. **7** 33.51
Gusejnova, M.K. **2** 81.5, 81.14, 83.13
Gusev, A.I. **2** 72.72, 72.73, 72.84 **3** 40.2, 72.40, 72.48, 72.53, 75.46 **4** 63.8, 64.17, 73.10 **5** 58.6, 64.32, 72.41, 73.4, 73.7, 73.12, 73.33 **6** 64.29, 72.27, 73.55, 73.58, 81.45 **7** 42.13, 63.18, 72.44, 73.10, 73.12, 73.32, 73.46, 73.69, 81.45, 81.47, 86.72
Guss, J.M. **2** 82.6, 82.27, 82.30, 82.31, 82.39, 83.31 **4** 71.71, 71.73, 71.82, 71.89 **5** 71.80, 71.95, 71.96, 71.99, 82.15, 86.78 **6** 74.5, 74.6, 75.22
Gust, D. **6** 62.11
Gust, G.R. **7** 73.33
Gustaffson, B. **3** 3.47
Guttormson, R. **3** 54.2
Guttormson, R.J. **5** 15.3, 27.4
Guy, J.J. **5** 3.23 **6** 32.35, 33.55 **7** 3.33, 21.11, 22.1, 39.31, 41.10
Guy, M.H.P. **7** 53.6, 53.20, 53.21, 53.22
Gwinner, P.A. **6** 20.18, 20.19
Gwinup, P.D. **4** 65.8
Gyepes, E. **7** 73.29
Gyepesova, D. **5** 83.44
Gylbert, L. **6** 58.18

Haagensen, C.O. **1** 3.73, 38.36
Haas, D.J. **1** 1.74, 2.31, 8.25, 56.10 **2** 60.16, 60.31, 60.33, 60.68
Haas, R.G. **6** 66.5
Haase, J. **5** 58.13 **6** 27.1, 27.2 **7** 27.1
Haase, W. **1** 3.66, 33.20 **2** 84.28 **5** 33.22 **6** 33.10, 84.15, 84.16 **7** 84.35, 84.64, 86.69
Habenschuss, M. **7** 84.3
Habermehl, G. **1** 28.23, 37.19 **2** 60.77 **3** 37.9, 58.9 **4** 37.11
Hackbarth, J.J. **6** 86.1 **7** 71.54
Hacke, N. **7** 83.1
Hacker, M.J. **4** 85.34
Hackert, M.L. **3** 3.36, 45.4
Hackett, E.A. **6** 83.5
Hacksell, I. **7** 40.4
Haddad, S. **7** 8.13, 79.7
Hadicke, E. **3** 71.14 **4** 5.5 **5** 83.94, 83.95
Hadinec, I. **3** 53.7
Hady, S.A. **3** 2.10
Hadzi, D. **4** 2.2, 2.3
Haffner, F. **7** 3.15
Hagenbach, A. **6** 30.5, 30.6
Hager, L.P. **6** 53.16 **7** 17.7
Haget-Bouillaud, Y. **5** 24.2
Haget, Y. **6** 24.1
Hagihara, H. **2** 69.12, 69.22, 69.27, 80.4, 85.61 **4** 69.20 **5** 80.2, 80.3, 85.9
Hagihara, N. **2** 86.71, 86.81
Hahn, T. **1** 48.6
Hai-Fu Fan, Y **1** 48.48
Haigh, J.M. **4** 77.15
Haight, G.P. **7** 85.38

Haight, H.L. **2** 75.2, 75.77
Haines, A.H. **6** 45.1
Haines, R.J. **5** 73.46 **6** 73.57
Hains Junior, C.F. **5** 71.56
Hair, N.J. **3** 33.35 **4** 3.6, 3.20, 41.13 **5** 52.4 **6** 4.2, 31.18, 31.25, 64.50
Hairs, T.N.D. **6** 3.20, 3.21, 3.22
Haisa, M. **4** 33.10, 33.11 **5** 2.28, 13.2 **6** 13.7, 19.9, 33.3, 33.13, 35.5 **7** 1.11, 1.12, 17.9, 34.16, 38.14
Halasi, R. **6** 16.11 **7** 35.21
Hale, P.K. **3** 76.41
Hale, R.L. **3** 51.41
Haley, J. **3** 83.120 **4** 83.99
Hall, C.R. **7** 39.12, 39.13
Hall, D. **1** 10.7, 26.26, 50.14, 58.34 **2** 76.26, 77.16, 77.17, 78.4, 78.5, 78.7, 78.16, 78.17, 78.18, 78.21, 78.25, 78.27, 78.29, 78.40, 78.41, 78.46, 78.50, 78.58, 82.55, 83.67, 83.68, 83.109, 83.112, 83.192, 83.194 **3** 45.39, 54.4, 60.1, 60.16, 60.39, 69.20, 69.23, 76.50, 78.8, 78.29, 83.12, 83.33, 83.66, 85.6 **4** 25.4, 25.5, 25.6, 59.22, 76.42 **5** 29.4, 58.21 **6** 3.9, 3.30, 31.7, 58.22, 69.12, 76.39, 86.41 **7** 36.25
Hall, D.DeW. **3** 79.2
Hall, D.R. **5** 35.12
Hall, E.A. **2** 85.19 **3** 74.1, 74.6, 74.7, 74.12, 74.14
Hall, G.S. **2** 83.99
Hall, J.B. **5** 69.11
Hall, L.H. **2** 62.35 **3** 58.21, 73.9 **6** 73.38
Hall, M.C. **3** 86.91 **5** 86.29
Hall, R.F. **6** 60.17, 60.18
Hall, S.R. **1** 16.11, 16.12, 56.19, 58.41, 58.53 **6** 51.33 **7** 48.39, 50.17
Hall, S.S. **6** 53.11
Hallak, N. **4** 38.21
Haller, D.A. **5** 69.24
Halpern, J. **2** 83.94 **4** 81.61 **5** 86.35 **7** 86.90
Halsall, T.G. **3** 56.8, 56.11
Haltiwanger, R.C. **4** 54.4 **5** 58.47
Haluska, R.J. **1** 34.8 **4** 36.22
Haluska, V. **7** 76.7
Ham, J.T. **3** 45.38
Ham, P.J. **5** 51.15, 51.44, 56.13 **6** 51.40
Hamalainen, R. **5** 76.24 **7** 76.20, 76.29, 76.30, 76.31
Hamanaka, N. **5** 54.4, 54.5, 54.6 **6** 28.21, 28.26 **7** 28.6
Hamanaka, T. **4** 54.7
Hamano, H. **7** 17.32
Hamasaki, T. **4** 59.20 **5** 59.17
Hamburger, G. **1** 11.53, 11.54
Hamelin, M. **2** 83.53 **4** 83.27 **5** 67.4
Hamer, G. **3** 60.34
Hamer, J. **4** 40.7
Hamill, R.L. **5** 50.26, 50.27, 50.28
Hamilton, D.S. **5** 86.60
Hamilton, J. **3** 48.48
Hamilton, J.A. **1** 20.20, 48.76, 53.22, 56.1, 58.76, 59.55 **4** 58.34
Hamilton, W.C. **1** 1.28, 8.43, 8.44, 10.22, 14.14, 15.22, 28.15, 30.27 **2** 63.15, 64.22, 64.60, 64.61, 66.19, 66.21, 69.1, 77.65, 85.41 **3** 3.39, 19.6, 19.7, 72.1, 83.51 **4** 9.3, 45.10, 45.11, 48.5, 48.10, 48.29, 64.45, 76.33, 83.42, 83.132 **5** 31.2, 48.9, 48.10, 48.14, 48.21, 48.27, 48.29, 48.38, 48.47, 48.49, 48.59, 48.60, 64.47 **6** 44.12, 45.7, 45.8, 48.18, 48.79, 64.36 **7** 48.22, 64.43
Hamm, D.J. **7** 83.94
Hammer, C.F. **1** 51.42
Hammer, H. **1** 45.32
Hamodrakas, S. **6** 17.16, 48.42
Hamonniere, M. **7** 54.13
Hamor, M.J. **1** 49.19, 49.26 **2** 76.63 **7** 22.2, 27.3
Hamor, T.A. **1** 20.20, 35.6, 36.20, 42.10, 45.71, 49.18, 49.19, 49.26, 52.13, 54.13, 56.1, 58.70, 58.74, 58.76, 59.64 **2** 82.56, 84.52 **3** 34.3, 45.26 **4** 22.1, 22.6, 36.11, 42.2 **5** 3.23 **6** 31.28, 32.35, 33.55, 71.12 **7** 3.14, 3.33, 21.11, 22.1, 22.2, 27.3, 39.31, 41.10, 58.1
Hampson, G.C. **1** 9.34 **2** 71.16
Hampton, C.R.S.M. **4** 86.62
Han, K. **1** 3.51
Hanaya, K. **7** 37.9
Hanazaki, F. **1** 21.22, 21.23
Handa, B.K. **7** 48.65
Handa, R. **4** 40.11 **5** 40.13
Handlovic, M. **5** 83.85
Handy, L.B. **1** 3.64 **3** 3.37, 64.62, 64.63, 64.64 **5** 65.1, 65.2, 65.5 **7** 84.5
Haneishi, T. **7** 50.1
Haner, B.A. **1** 51.17
Hanessian, S. **6** 50.7 **7** 47.25, 50.12
Hanic, F. **1** 37.11, 53.19, 59.12 **2** 74.20, 81.78, 81.86, 83.117, 83.118 **3** 73.54 **4** 79.7 **5** 83.44, 83.67, 83.85 **7** 73.29
Hanicova, K. **2** 81.86
Hannah, R.B. **1** 7.12
Hannaway, C. **2** 73.102 **3** 51.44 **5** 58.36 **6** 28.15, 28.16
Hannoteaux, F. **6** 24.1
Hanouskova, D. **4** 31.32 **5** 59.30
Hansen, F. **3** 8.4
Hansen, F.V. **6** 62.3
Hansen, H.N. **4** 45.20
Hansen, L.K. **5** 39.5 **7** 39.10
Hansen, P.J. **2** 73.36
Hanson, A.W. **1** 17.41, 29.4, 29.25, 29.26, 29.27, 30.26, 31.21, 31.34, 36.3, 36.14, 36.15, 36.16, 36.17, 38.55, 50.6, 58.20, 58.60 **2** 60.102, 60.108, 60.115, 60.124, 60.127, 60.143, 60.152 **3** 31.20, 37.11 **4** 16.15, 39.33, 62.14 **5** 9.13, 16.22, 31.21, 31.27 **7** 13.26, 13.27, 16.10
Hanson, J.C. **3** 30.2, 83.49 **4** 62.2, 62.3, 62.4 **5** 45.28, 45.29, 45.31, 48.16, 48.24 **6** 45.35, 49.9 **7** 51.1, 51.2
Hanson, J.R. **3** 51.9 **4** 54.11, 54.14
Hanson, L.K. **6** 49.9
Hansson, E. **2** 81.41 **3** 81.14 **5** 81.19, 81.20 **6** 81.4, 81.23, 81.30, 81.31, 81.41
Hansson, I. **2** 66.7

Hanzawa, Y. **3** 36.20
Haque, M.-ul- **1** 39.93, 46.5, 46.10, 53.1 **2** 64.42, 69.10 **3** 46.2, 46.7, 53.4, 53.18, 64.16, 64.17, 64.34, 64.35, 64.39, 76.44, 83.61, 84.26 **4** 53.1, 64.25 **6** 46.11, 53.2, 53.14, 76.31 **7** 32.8, 58.7, 83.92
Harada, H. **1** 50.37, 58.86
Harada, M. **7** 1.15
Harada, S. **3** 50.22
Harada, Y. **1** 13.16 **6** 13.4, 48.26, 48.27, 48.28, 48.64
Harata, K. **5** 31.22, 60.35 **7** 24.10, 45.58
Harayama, T. **1** 58.93
Hardcastle, G.A. **1** 50.26
Hardcastle, K.I. **5** 86.45 **6** 75.9 **7** 84.8
Harder, V. **7** 73.43
Hardgrove Junior, G.L. **7** 39.25
Hardgrove, G.L. **1** 27.1, 27.2, 27.3, 51.47 **2** 73.28 **4** 40.9
Harding, A.E. **7** 56.8
Harding, M. **6** 16.15
Harding, M.M. **1** 1.147, 48.16 **2** 80.22, 82.45, 82.46, 82.48, 82.50, 85.90 **3** 48.25, 82.22, 85.43 **5** 45.18, 48.35, 48.36 **6** 48.23
Hardy, A.D. **6** 36.8, 36.17
Hardy, A.D.U. **3** 73.46, 73.52 **5** 72.9, 73.11, 73.13 **6** 42.10
Hardy, R. **1** 45.72
Hare, C.R. **2** 81.61, 81.62, 84.40, 85.31 **3** 82.29 **6** 82.19, 82.22
Harel, M. **7** 20.5
Hargis, J.H. **3** 64.17
Hargreaves, A. **1** 8.9, 11.36, 16.46, 16.57, 16.59, 17.50, 19.36, 24.19, 24.20 **3** 11.7, 11.27
Hargreaves, R.N. **3** 71.46, 71.48
Hark, T.E.M.van den **5** 80.24 **6** 22.1, 30.2, 30.8 **7** 30.7, 38.58
Harkema, S. **1** 8.19, 8.20 **4** 60.3 **5** 2.2, 35.5 **6** 81.37 **7** 81.36
Harker, D. **2** 63.8
Harle, E. **1** 54.24
Harlow, R.L. **3** 78.1 **6** 1.10, 17.5, 31.4, 33.16, 72.20, 77.7, 78.3 **7** 3.22, 3.23, 3.24, 15.4, 26.6, 26.7, 32.5, 32.37, 33.29, 38.74
Harmon, C.A. **6** 22.3
Harnik, E. **1** 28.28
Harper, F.R. **2** 65.25, 65.26
Harpp, D.N. **5** 39.25 **6** 39.15
Harreld, C.S. **4** 69.31
Harris, C.B. **2** 65.47
Harris, C.M. **2** 83.154
Harris, D.R. **1** 8.25, 47.5, 50.43, 58.52, 58.73 **2** 60.16
Harris, G.S. **2** 65.51
Harris, P.M. **3** 15.1, 15.2
Harris, R.O. **6** 86.45
Harrison, H.R. **3** 49.3, 56.8, 56.11
Harrison, I.T. **1** 55.3
Harrison, P.G. **5** 69.22 **6** 69.18 **7** 69.7, 69.19, 69.27
Harrison, P.M. **1** 48.61 **2** 83.182
Harrison, R.W. **2** 69.36
Harrison, W. **3** 81.8, 86.11, 86.45 **4** 33.15, 41.1, 48.38, 68.3, 73.9, 81.17, 83.122, 85.21 **5** 17.19, 68.6, 81.58, 83.105 **7** 73.18, 75.16
Harrod, J.F. **6** 86.68
Harrowfield, J.M.B. **6** 82.10 **7** 84.21
Harsanyi, K. **7** 40.21
Hart, F.A. **2** 83.217 **3** 84.26 **5** 71.79 **6** 76.31, 84.45
Hart, H. **2** 62.23
Hart, H.V. **6** 62.1
Hart, N.K. **5** 58.24
Hartke, K. **3** 39.15
Hartl, H. **6** 3.14, 3.31 **7** 33.10, 33.13, 33.14
Hartman, A. **1** 20.7
Hartmann, G. **2** 64.1, 64.20
Hartshorn, M.P. **1** 52.17
Hartsuck, J.A. **1** 54.26
Hartung, H. **2** 85.76 **3** 86.3
Harvan, D. **6** 59.5
Harvey, R. **7** 29.4, 29.7, 29.9, 38.62
Harvey, R.S. **7** 30.9
Haschemeyer, A.E.V. **1** 47.22, 47.23
Hasegawa, M. **4** 33.41 **5** 33.51
Hasegawa, S. **6** 9.1
Hasek, J. **5** 21.3
Hasekamp, R.C.D.E. **1** 42.2
Haser, R. **1** 18.4, 18.5
Hashimoto, M. **1** 50.2
Hass, B.S. **7** 27.4
Hass, D. **6** 85.4
Hassall, C.H. **7** 48.65
Hassanali, A. **1** 59.23
Hassel, O. **1** 1.3, 1.4, 5.5, 7.10, 21.6, 21.8, 21.9, 21.10, 21.21, 38.22 **2** 60.3, 60.12, 60.17, 60.18, 60.29, 60.35, 60.38, 60.39, 60.44, 60.45, 60.48, 60.49, 60.53, 60.58, 60.65, 60.84, 60.90, 60.91, 60.92, 60.120, 83.42 **3** 60.20 **4** 60.1
Hasselstrom, T. **1** 41.10
Hata, T. **1** 27.10, 28.20 **3** 50.20 **4** 27.5, 27.6, 64.35, 64.38 **5** 41.7, 41.9, 41.36, 44.4, 50.14, 60.4, 64.37 **6** 36.11, 44.2 **7** 47.14, 47.24, 50.1, 60.7
Hatfield, W.E. **3** 76.41 **4** 83.22 **5** 82.12, 83.11, 83.12, 83.34, 83.46, 83.75, 83.101, 85.3 **6** 76.47, 83.50, 83.51 **7** 76.25, 83.114
Hatsuda, Y. **4** 59.20 **5** 59.17
Hattori, K. **1** 50.2
Hauck, J. **6** 3.5
Hauge, S. **2** 70.13 **3** 8.12, 8.13 **4** 70.3 **5** 8.15 **7** 8.19, 8.20
Haugen, L.P. **3** 86.31
Haupt, H.-J. **5** 69.27 **7** 69.22
Haupt, H.J. **4** 69.38
Hauptman, H. **1** 51.5 **4** 51.15, 51.17, 51.27, 51.33 **5** 50.29, 51.23
Hausen, H.-D. **5** 83.27 **7** 68.2
Hausen, H.D. **6** 62.6, 68.6, 68.7 **7** 62.8, 68.6, 68.7
Hauser, D. **4** 59.10
Hauser, F.M. **3** 58.19, 58.24, 58.50
Hauser, P.J. **5** 68.23

Hauw, C. **1** 11.44, 24.21, 24.24, 24.25, 25.3, 25.4, 25.5, 25.6, 25.7, 25.13, 25.16, 25.18, 25.21, 26.3, 26.37, 38.32, 38.35, 59.39 **3** 25.2, 25.4, 25.5, 26.9, 60.32 **4** 25.1, 35.9, 38.22 **5** 7.9, 7.10, 21.6, 25.1, 35.10 **6** 12.1, 16.2, 21.3, 27.9, 27.10, 27.13, 27.14, 28.11, 32.19, 32.20, 38.34, 59.15 **7** 25.1, 25.3, 26.2
Hawkinson, S. **1** 19.43, 42.11, 42.12 **2** 83.150, 83.205 **3** 84.1, 84.7
Hawkinson, S.W. **2** 73.55, 83.143 **3** 47.4, 47.8, 49.14, 83.134, 83.135, 85.18 **7** 83.47
Hawley, D.M. **1** 53.12, 59.45 **2** 66.16, 66.17 **3** 39.38, 40.7, 53.24 **4** 50.13, 53.3 **5** 45.43 **6** 66.10
Hawthorne, M.F. **3** 62.15 **5** 71.103 **6** 71.10, 73.1, 73.14, 73.15 **7** 71.8, 73.1, 73.16
Hayafuji, Y. **6** 5.4, 5.5
Hayakawa, K. **6** 33.50
Hayashi, J. **5** 75.41, 75.42
Hayashi, M. **3** 53.20
Hayashi, S. **7** 53.27
Hayashi, T. **7** 38.48
Hayashi, Y. **5** 54.8
Hayden, T.D. **6** 67.1
Haynes, N.B. **1** 59.45
Hayter, R.G. **1** 3.64
Hazel, J.P. **5** 64.19, 84.2 **7** 51.19, 51.24, 51.28, 51.29, 51.45, 51.51
Hazell, A. **7** 29.1, 29.2
Hazell, A.C. **1** 41.8 **2** 62.21, 77.36 **3** 76.51 **4** 29.8 **5** 29.2, 70.3 **7** 4.2
Hazell, R.G. **1** 3.53 **2** 85.6, 85.8, 85.15, 85.16 **3** 8.4 **4** 19.1, 31.27, 83.60 **5** 79.2, 79.3 **6** 30.7, 62.3 **7** 84.4
Hazen Junior, E.E. **6** 8.4 **7** 8.18
Healy, J.D. **5** 64.29
Healy, M.J. **2** 82.39 **5** 82.3, 82.9
Healy, P.C. **4** 80.3, 80.9, 80.17 **5** 79.1, 80.9, 80.19 **6** 76.43, 78.1, 78.6, 78.20 **7** 80.11, 80.12, 84.76, 84.77
Hearn, R.A. **5** 3.18, 45.24, 48.23 **6** 38.17, 58.1 **7** 45.3
Heath, C. **2** 83.217 **3** 83.53, 83.129
Heath, C.E. **2** 83.123
Heath, G.A. **3** 85.24 **6** 86.63
Hecht, H.-J. **7** 69.29
Hecht, H.J. **6** 64.26
Hecht, J.K. **3** 20.10
Hecht, S. **4** 47.12, 47.15
Hecht, S.M. **4** 47.16
Hechtfischer, S. **3** 31.4, 39.24, 39.25, 39.27, 54.15 **6** 39.12
Hecker, E. **1** 54.24 **3** 54.8, 54.15 **4** 54.12
Hedberg, F.L. **4** 73.1 **5** 73.19
Hedberg, K. **2** 65.11
Hedman, B. **7** 84.12
Heeres, G.J. **1** 39.53, 39.54, 39.55
Hefelfinger, D.T. **4** 30.5 **5** 30.3
Heffer, J.P. **5** 81.39
Hegenbarth, J.J. **1** 6.2
Heger, G. **7** 3.1
Hegstrom, R. **4** 40.9
Heidner, E.G. **1** 11.7
Heijden, S.P.N.van der **6** 13.19, 13.20, 13.21, 13.22, 13.23
Heijdenrijk, D. **6** 32.2 **7** 8.2
Heijkoop, E.van **1** 51.8
Heimbach, P. **4** 71.65
Heinrich, W. **2** 85.34
Heitmann, J.A. **7** 45.47, 52.6
Heitner, H.I. **6** 85.40
Heitsch, C.W. **2** 68.13
Helgeson, R.C. **5** 38.32
Helland-Madsen, G. **3** 39.26, 64.19
Helland, S. **1** 45.4, 45.5
Heller, J. **5** 31.1
Hellerbach, J. **5** 40.33
Helling, J.F. **5** 75.46
Hellner, E. **1** 17.51, 36.25 **2** 64.68 **3** 63.9
Helm, D.van der **1** 2.78, 2.79, 2.80, 2.81, 41.7, 44.20, 53.10, 54.11 **2** 60.37, 60.95, 82.18, 82.22, 82.58, 84.44 **3** 20.5, 54.10, 82.6, 82.8, 82.12, 82.13, 82.14, 82.15 **4** 4.4, 82.14, 82.15 **5** 20.4, 82.23 **6** 44.21, 84.35 **7** 21.3, 21.4, 31.3, 31.28, 36.2, 36.3, 36.5, 36.9, 41.15, 41.20, 43.7, 51.58, 51.59, 54.9, 54.10, 64.49, 64.52, 64.53, 76.70, 82.19
Helm, F.T. **6** 84.10 **7** 84.40, 84.59
Helmholdt, R.B. **2** 72.44 **3** 75.14 **7** 5.12
Hemmerich, P. **2** 83.54
Hempel, A. **7** 50.29
Hempel, J.C. **5** 82.12
Hencken, G. **3** 67.5 **6** 73.45 **7** 68.4
Hende, J.H.van den **1** 37.23, 50.22, 50.23, 50.28, 58.110, 59.15, 59.50 **2** 64.49, 64.50, 75.50
Henderson, M.S. **5** 56.10, 56.11
Hendricks, S.B. **1** 1.6, 1.7
Hendrickson, D.N. **5** 76.46 **6** 73.8 **7** 76.33, 76.34, 76.63
Hendrickson, W.A. **3** 48.5
Hendriks, H.M. **6** 80.19 **7** 38.58
Henke, H. **4** 9.1 **6** 31.12, 31.31 **7** 38.28
Henley-Smith, P. **6** 59.16
Henn, D.E. **1** 2.43 **2** 86.7, 86.67 **3** 60.36
Hennig, H. **4** 83.127
Henning, G. **3** 3.55 **5** 83.94, 83.95
Henrick, K. **7** 86.40, 86.41
Henry, G.M. **7** 76.65
Henry, R.L. **6** 86.19
Henry, R.S. **7** 31.10
Henslee, W. **5** 85.25 **7** 28.8, 31.29, 72.30
Hepburn, S.P. **2** 73.59
Hepworth, M.A. **3** 81.8 **4** 33.15, 41.1
Herak, R. **6** 73.61, 82.7 **7** 35.21, 81.17
Herak, R.M. **7** 81.20, 81.21
Heraud, J. **1** 1.126
Herbertsson, H. **6** 1.5, 2.16, 2.18, 2.19, 48.10
Herbstein, F.H. **1** 8.16, 19.6, 19.7, 24.1, 24.4, 30.16, 30.21, 36.10 **2** 60.69, 60.135, 60.154, 68.7, 68.8, 73.24 **3** 60.45, 60.46 **4** 8.2, 29.1, 69.8, 69.13 **5** 16.15, 36.17, 69.1, 69.2, 69.5, 69.6, 69.8, 72.20 **6** 13.11, 13.12, 16.6, 83.31 **7** 38.26, 60.22, 60.25

Herceg, M. **2** 77.53 **4** 40.18, 67.12, 81.11 **5** 83.43 **6** 84.9, 84.14 **7** 3.5
Herdklotz, J. **1** 41.11 **6** 13.15
Herdklotz, J.K. **3** 3.32, 7.6, 39.4
Hergold, A. **7** 76.4, 76.9, 83.17, 83.31
Herin, M. **7** 54.3, 54.4
Herland, P. **5** 70.8, 70.9
Herlt, A.J. **7** 76.52
Herman, A. **6** 85.26
Hermann, A. **5** 80.23
Hermann, C. **1** 1.9
Hermann, H.J.A. **3** 39.32
Hermodsson, Y. **1** 3.23 **3** 60.48
Hernandez, O. **7** 63.15
Herout, V. **1** 53.14
Herpin, P. **4** 21.13 **5** 71.29
Herrin, J. **3** 50.24
Herriott, J.R. **2** 64.12 **5** 48.16, 48.24 **6** 35.24
Herskovitz, T. **5** 85.31
Herz, W. **1** 53.3 **4** 53.1 **6** 59.12 **7** 53.26
Herzog, H.L. **7** 51.4
Hesp, B. **5** 50.10 **7** 58.23
Hesper, B. **1** 10.29, 51.8 **3** 51.37, 51.38
Hess, G.G. **5** 71.2
Hess, H. **2** 62.17, 62.20, 62.46, 63.23, 64.2 **3** 68.5 **4** 62.12 **6** 62.6 **7** 9.3, 11.2, 11.3, 62.8
Hess, R.W. **7** 86.89
Hesse, M. **3** 58.67
Hesse, R. **2** 80.2, 80.37, 80.41 **3** 67.20, 68.4, 80.21, 80.22 **4** 80.20 **5** 68.18, 80.15 **6** 68.2 **7** 11.13
Hessel, L.W. **3** 81.43
Hessling, E.R. **5** 35.21
Hester Junior, J.B. **4** 36.20
Hettler, H. **3** 41.14
Hewertson, W. **3** 67.27, 72.30
Hewitt, T.G. **3** 72.22, 72.35, 72.50 **4** 77.4
Hewlins, M.J.E. **3** 64.50 **4** 44.11 **7** 85.25
Hewson, A.T. **5** 53.13
Hewson, M.J.C. **7** 64.7
Heys, P.N. **6** 86.25, 86.31
Hickel, D. **3** 50.26 **4** 50.14
Hickernell, G.L. **3** 59.32
Hidai, M. **7** 86.98
Hidaka, J. **6** 82.9
Higashi, H. **6** 1.7
Higbie, F.A. **7** 73.58
Higgs, M.A. **1** 1.29
High, D.F. **1** 51.6, 59.65
Hightower, T.C. **5** 77.16
Higson, B.M. **4** 78.13 **7** 71.83
Higuchi, T. **1** 36.26 **2** 63.9, 63.18, 73.94 **5** 1.33, 39.15 **7** 33.7, 45.1
Hijikigawa, M. **6** 5.4, 5.5
Hikino, H. **6** 54.14
Hill, N.J. **1** 3.38, 3.39
Hill, N.L. **7** 85.20
Hill, R.K. **5** 31.10 **6** 59.1
Hill, W.E. **7** 86.21
Hiller, S.A. **7** 37.8
Hillery, S.S. **3** 9.19, 35.17
Hilton, J. **2** 75.47 **5** 69.4 **6** 45.19
Hinamoto, M. **4** 83.105 **7** 20.2, 20.3
Hinazumi, H. **4** 48.14 **5** 2.21
Hinderer, A. **3** 68.5 **7** 9.3
Hine, R. **1** 48.34 **3** 9.7, 28.2
Hingerty, B. **7** 47.26, 47.38
Hipp, C.J. **6** 83.57
Hirabayashi, M. **1** 52.25
Hirai, K. **1** 50.35
Hirama, M. **4** 27.15
Hiramatsu, M. **3** 50.20
Hirao, K.-I. **7** 31.15
Hirata, Y. **1** 50.37, 54.9, 58.102, 58.114, 59.25 **3** 58.38 **4** 58.31, 58.39 **5** 36.13, 58.31, 58.35 **6** 53.8, 58.48, 58.58, 59.26 **7** 54.16
Hirayama, F. **7** 60.20
Hirayama, N. **6** 49.7
Hirobe, M. **7** 16.16
Hirokawa, S. **1** 2.84, 3.13, 3.18, 3.42, 33.63, 33.65, 33.69, 48.32, 48.41 **3** 33.36
Hirotsu, K. **2** 73.94 **3** 5.3 **4** 13.14 **5** 41.23, 54.8, 83.107 **6** 33.20, 33.21, 33.22, 36.10, 45.49, 54.6 **7** 13.13, 33.24, 45.1, 45.36, 54.12
Hirschfeld, D.R. **6** 54.15
Hirsekorn, F.J. **7** 71.103
Hirshfeld, F.L. **1** 18.24, 20.7, 23.16, 29.10, 29.24, 36.9 **3** 1.15 **7** 20.5
Hitchcock, P.B. **2** 72.16, 73.34, 86.84, 86.108 **4** 23.5, 71.43, 71.46, 71.49 **6** 46.14, 68.10, 71.39, 86.32 **7** 63.9
Hitchman, M.A. **2** 76.21, 76.50, 83.215 **3** 77.24
Hite, G. **5** 3.17 **6** 33.39, 33.40, 33.47, 33.48 **7** 17.34, 33.42, 33.43
Hjeds, H. **4** 40.3 **5** 36.26 **6** 40.4
Hjellum, O. **6** 39.30
Hjortaas, K.E. **1** 28.19 **2** 64.16
Hjortas, J. **4** 59.1 **5** 45.21, 54.1, 64.41 **6** 35.3, 45.20 **7** 45.7, 45.8, 45.45
Hlavata, D. **4** 12.12
Ho, B.Y.K. **7** 69.3
Ho, J.-Y. **1** 11.19
Hoard, J.L. **1** 3.16, 7.4, 31.1, 33.36, 49.18, 49.19, 49.24, 49.26, 49.29 **2** 76.59, 76.61, 76.63, 76.64, 76.65, 76.66, 76.69, 76.71, 76.72, 77.50, 78.45, 81.52, 81.70, 82.56, 83.137, 84.23 **3** 49.11, 49.12, 69.29, 69.30, 84.25 **4** 49.21, 49.22, 49.24, 76.34, 76.36, 81.9 **5** 49.7 **6** 48.50, 49.15, 49.21, 67.5
Hoard, L.G. **2** 64.3
Hoare, R.J. **4** 71.14 **5** 71.68, 71.69
Hobbs, J. **4** 47.1
Hobson, A.D. **1** 16.49
Hobson, R.J. **6** 84.33 **7** 84.61
Hock, A.A. **2** 72.33, 73.74
Hodder, O.J.R. **2** 61.15 **3** 13.18, 49.3, 49.13, 56.8, 56.11 **4** 2.24, 24.2, 31.11, 31.23, 40.8, 60.9, 60.10, 60.11 **5** 35.25 **6** 54.17
Hodgkin, D.C. **1** 10.28, 10.30, 48.36, 49.32, 49.34, 50.13, 50.14, 50.16, 51.45, 51.50, 59.43 **3** 49.3, 50.31, 51.43, 58.14, 82.4 **4** 49.20, 49.23, 83.137, 83.138 **7** 49.1, 49.2

Hodgson, D.J. 1 2.16 2 86.75 3 71.9, 76.41, 86.69 4 76.13, 83.22 5 33.5, 82.12, 83.11, 83.12, 83.34, 83.46, 83.75, 83.101, 86.82 6 3.12, 33.2, 38.9, 38.10, 76.47, 83.15, 83.32, 83.33, 83.50, 83.51, 83.71 7 35.1, 35.2, 35.3, 47.1, 47.9, 47.10, 47.27, 50.2, 76.25, 76.51, 76.62, 83.10, 83.69, 83.114, 84.63
Hodgson, K.O. 4 75.14, 75.15, 75.31 5 73.45, 75.26, 75.34
Hodgson, M. 3 62.11
Hodgson, P.G. 7 76.39
Hoekstra, A. 4 12.10 7 16.13, 16.14, 16.15, 16.17, 19.15
Hoel, E.L. 6 73.14
Hoffman, A.U. 4 49.24
Hoffmann, K. 3 63.4, 63.5, 63.6 5 68.3, 68.4, 68.28 7 62.12, 62.13
Hoffmann, R. 4 12.9
Hoffmann, R.W. 6 38.33
Hofman, B.M. 5 12.11
Hoge, R. 1 45.47 3 45.18, 45.20 6 56.3
Hogg, J.H.C. 6 15.5 7 39.18
Hogle, J. 6 47.9
Hoh, D.M. 5 33.25
Hohne, E. 1 38.21, 38.51, 58.105, 58.107, 58.109 3 32.17, 64.7 4 51.48, 51.56, 59.17, 85.20 5 58.39, 58.40 6 33.38, 41.30, 51.37, 51.38 7 36.14, 59.8
Holah, D.G. 3 80.12, 80.13, 80.15
Holan, G. 5 38.28
Holbrook, S.R. 7 21.3, 21.4, 64.52, 64.53, 76.70
Holden, J.R. 1 12.6, 15.2, 15.18, 16.1, 24.6 3 16.5 4 32.14 5 16.1
Holian, B.L. 3 79.13
Holland, R.V. 3 82.17
Hollander, F.J. 5 67.21, 67.25, 67.26 6 6.1, 64.33, 85.13, 85.41 7 81.48, 85.30, 85.31, 85.40, 86.118
Hollander, O. 6 3.32
Holliday, R.J. 5 65.2
Holm, R.H. 3 83.71 5 49.8, 85.31 7 49.10, 85.28, 85.42, 85.46
Holmes Junior, K.D. 6 9.2
Holmes, D.R. 1 16.41
Holmesland, O. 2 60.19, 60.20
Holst, J. 3 85.5 4 4.1, 8.7 7 4.4
Holst, J.P.C. 2 64.48
Holt Junior, S.L. 2 79.34, 79.35
Holt, E.M. 3 79.21 6 4.6, 72.13, 73.42, 82.23 7 64.46
Holt, S.L. 3 62.16, 79.21 4 78.9, 78.14 6 4.6, 72.13, 73.42, 82.23 7 64.46, 78.1
Holtedahl, B.H. 2 64.31
Holtzberg, F. 1 1.1
Holub, M. 1 53.14 3 53.7
Hon, P. 2 77.7, 77.40, 77.41, 77.43
Hon, P.-K. 3 77.24
Hon, P.K. 6 68.13, 77.11, 78.3
Honda, T. 6 58.37, 58.51
Honzl, J. 2 60.34
Hoogenboom, F.J.M. 6 51.20
Hoogschagen, G. 1 1.95
Hoogsteen, K. 1 28.26, 44.50, 44.51
Hoogzand, C. 2 72.61
Hootele, C. 6 58.6
Hooz, J. 6 62.10
Hope, H. 1 8.49, 9.27, 9.29, 11.24, 33.71, 38.25, 40.5, 40.10, 51.26 2 60.35, 60.55, 60.56, 60.120, 83.87 3 1.19, 8.7, 11.11, 39.2, 39.30, 51.31, 53.8, 70.2 4 8.5, 31.17, 31.18, 63.13 5 9.2, 37.1 6 70.1, 83.5 7 45.37, 85.20
Hopf, F.R. 7 49.12
Hopgood, D. 5 77.1
Hopkins, T.E. 2 62.33, 62.41, 65.39, 73.5 6 41.28
Hoppe, H. 1 3.66 2 84.28
Hoppe, W. 1 9.1, 28.24, 28.25, 31.6, 36.35, 49.16, 51.28, 51.49, 54.24 2 74.21, 75.61 3 31.4, 39.24, 39.25, 39.27, 49.15, 54.6, 54.8, 54.15, 59.19, 59.25, 71.14 4 13.22, 27.19, 35.33, 36.37, 51.11, 54.8, 54.12, 54.16, 71.92 5 36.31, 48.40, 48.69, 49.1, 75.1 6 39.12
Hora, J. 4 54.1
Hordvik, A. 1 11.4, 39.6, 39.16, 39.44, 39.48, 39.59, 39.60, 41.1, 41.3, 41.4, 41.5, 45.9, 45.10 3 39.16, 39.18, 39.42, 39.43 4 39.2, 39.10, 39.13, 39.30, 39.32, 39.44, 41.12, 41.26, 42.1, 45.1 5 39.5, 39.13, 39.16, 39.32, 39.38, 39.39, 39.40, 41.2, 41.30 6 39.30, 45.12 7 39.5, 39.10, 39.32, 39.34
Horhammer, L. 3 59.26
Horii, S. 5 50.2
Horii, Z. 1 58.9
Horike, M. 7 71.73, 71.90
Horn, A.S. 6 39.23 7 17.21, 29.11, 36.22, 36.23, 39.37, 39.38
Horn, P. 1 47.19
Horn, U. 5 35.12
Horne, D. 7 38.54
Horner, L. 2 64.48
Hornstra, J. 3 21.20, 31.8, 51.8, 51.10, 51.17, 51.18, 51.21, 51.23, 51.24, 51.25 5 51.37, 51.38 7 16.14, 16.15
Hornyak, G. 6 41.6
Horrocks Junior, W.DeW. 1 3.80 2 65.42, 77.46 3 76.83, 77.27, 78.6, 78.19, 79.2 4 77.24, 80.8
Horsfield, E. 2 83.131, 83.132
Horspool, W.M. 6 73.54
Horton, D. 5 45.20 7 47.40
Horvai, J.C. 5 35.15
Horvath, G. 6 39.27 7 32.2
Horvath, K. 7 40.21
Horwitz, H. 5 1.22
Hoshino, S. 1 48.9
Hoskins, B.F. 1 49.5 2 80.40, 83.90, 83.171, 83.213, 83.218 3 49.5, 80.3, 80.6, 83.78, 83.110, 83.151, 85.24 4 80.3, 83.88, 85.15 5 81.62, 83.120, 83.121, 83.122 6 77.3, 77.6 7 42.8, 70.5, 76.77, 80.21, 85.19, 85.43, 86.104
Hoskins, K. 4 71.43, 71.46, 71.49
Hosokawa, T. 4 71.51 6 71.55, 75.29 7 50.20
Hosoo, T. 3 81.17
Hosoya, S. 1 39.68, 42.13, 42.14

Hospital, M. **1** 1.75, 1.82, 1.87, 1.94, 1.98, 1.105, 1.106, 1.111, 1.112, 1.124, 1.126, 1.129, 1.134, 26.10, 36.34 **3** 1.26, 1.27, 1.28, 17.7 **4** 8.12, 8.13, 13.8, 17.9, 19.14, 51.8, 51.9, 51.12, 51.16, 51.18, 51.21 **5** 21.16, 30.1, 35.28, 51.2, 51.3, 51.4, 51.11 **6** 17.20, 17.21, 17.22, 36.20, 51.3, 51.7 **7** 17.30, 17.31, 25.2, 30.8, 33.55, 35.7, 36.20, 36.24, 36.28, 36.29, 51.3, 51.26, 59.3, 59.22
Hossain, M.B. **1** 33.10, 53.10, 54.11 **2** 82.22 **3** 54.10
Hosszu, J.L. **1** 44.72
Hosur, M.V. **7** 47.6, 47.16, 47.35
Hota, N.K. **4** 72.29
Hotsyanova, T.L. **1** 16.33, 19.38
Hotta, Y. **7** 56.7
Hou, F.L. **4** 86.3
Hough, E. **1** 59.48 **3** 58.34 **4** 38.23 **5** 45.12 **6** 35.16
House, D.A. **3** 76.5, 76.58 **4** 76.24 **5** 76.14, 76.15 **7** 76.26
Housty, J. **1** 1.75, 1.82, 1.87, 1.94, 1.98, 1.105, 1.106, 1.111, 1.112, 1.124, 1.126, 1.129, 1.134, 1.140, 19.16, 41.45, 43.22, 43.26, 43.30 **3** 43.6, 43.9 **4** 25.1 **5** 3.6, 25.1, 35.6 **7** 25.1, 35.7
Hove, M.J. **5** 12.11
Hovenkamp, S.G. **2** 64.4
Howard, G.R. **7** 38.66
Howard, J. **3** 45.45 **4** 69.26 **5** 72.17, 74.12 **6** 71.34, 71.80, 75.11 **7** 71.39, 71.93, 75.18
Howard, J.A. **6** 64.32, 64.38 **7** 66.6, 66.7, 71.96
Howard, J.A.K. **6** 75.12 **7** 71.59, 72.1, 75.1, 75.22, 75.25, 75.38
Howard, W.J. **4** 83.21
Howatson, J. **3** 76.36, 76.37 **4** 33.40, 67.16 **5** 86.81 **6** 81.14
Howe, D.B. **6** 24.7, 24.8
Howe, N.L. **3** 83.121, 83.122, 83.123, 83.124, 83.125, 83.126
Howell, P.A. **1** 39.11
Howells, M.A. **6** 64.41
Howells, R.D. **6** 64.41
Howie, R.A. **6** 69.2
Howlett, K.D. **5** 64.3, 64.21, 64.29, 64.36
Hoy, A. **5** 48.35, 48.36
Hoy, R.C. **2** 83.158
Hoy, T.G. **1** 13.49, 19.46, 19.47
Hoyano, J. **3** 69.27
Hoyano, J.K. **4** 73.13
Hrncir, D.C. **7** 36.7, 68.1
Hseu, T.-H. **6** 71.25, 75.14, 75.19
Hseu, T.H. **4** 62.11 **7** 62.4
Hsiang-Ch'i, M. **4** 66.6
Hsiung, V. **4** 45.23
Hsu, H.-Y. **7** 54.16
Hsu, I.-N. **3** 82.12, 82.13, 82.14, 82.15 **5** 20.4 **6** 60.6, 60.7, 60.8, 60.9, 60.10, 60.16 **7** 43.2, 43.3, 51.58
Hsu, I.H.-C. **3** 58.51
Htoon, S. **5** 76.10 **6** 76.9
Hu, W.-J. **6** 76.35, 85.37
Hu, W.J. **5** 76.47 **6** 28.25
Huang, C.-M. **5** 60.3, 60.9
Huang, H.-C. **5** 53.18 **7** 53.17, 53.23
Hubbard, C.R. **5** 3.12 **7** 11.1
Hubel, W. **2** 72.61, 73.116
Huber-Buser, E. **1** 23.6, 23.7
Huber, C.P. **1** 53.5 **5** 36.29
Huber, C.S. **3** 58.37 **4** 35.12, 37.5, 40.21, 58.22 **5** 37.4, 58.32 **7** 27.5, 59.20
Huber, F. **4** 69.38 **5** 69.27 **7** 69.22
Huber, M. **1** 47.18 **2** 69.29 **4** 65.6, 66.5 **5** 69.10
Huber, R. **1** 9.1, 28.24, 31.6, 51.28 **2** 75.61
Hubert, J. **5** 72.7 **6** 66.8
Huddle, B.P. **3** 10.2 **4** 33.1
Hudgens, J. **7** 11.11
Hudson, D.W. **3** 31.11 **4** 38.12 **6** 31.19
Hudson, J.M. **4** 48.53
Hudson, P. **1** 38.15, 39.22, 39.24
Huebner, C.F. **3** 31.42
Huffman, J.C. **2** 83.190 **4** 62.6, 68.2 **6** 72.10 **7** 62.1, 78.18
Hugel, R. **2** 84.66
Hughes, B. **7** 69.14
Hughes, D.L. **3** 45.45 **4** 13.9 **5** 36.7 **6** 38.26, 45.3, 67.17, 67.18 **7** 53.19
Hughes, D.O. **1** 1.24 **5** 1.1
Hughes, E.W. **1** 3.3, 8.17, 8.31, 33.8, 48.30, 48.83 **2** 60.8, 65.11 **3** 48.10
Hughes, N.A. **7** 45.26, 45.52
Hughes, R.C. **7** 81.29, 81.30
Hughes, R.E. **1** 7.2, 7.32, 37.2, 37.3 **3** 48.26, 50.17 **4** 50.5, 50.9, 71.9, 73.31 **5** 59.15 **6** 73.62 **7** 39.40, 75.6
Hughes, R.P. **7** 71.70, 72.2, 75.23
Huie, B.T. **4** 69.25 **5** 71.98 **6** 71.87, 71.94
Huisman, H.O. **3** 51.15
Huiszoon, C. **5** 35.5
Huitric, A.C. **3** 52.6 **4** 13.21
Hull, S.E. **3** 71.3, 85.7, 85.20 **4** 12.14, 28.11, 35.23 **7** 51.46, 51.47
Hullen, A. **2** 76.9 **3** 3.55
Hulme, R. **2** 60.89, 60.114, 60.136, 60.137, 60.138, 69.16, 69.17, 71.20 **3** 60.19
Hulscher, J.B. **1** 38.47 **3** 69.40 **4** 64.7 **5** 66.6
Hulshof, L.A. **5** 27.5
Huml, K. **1** 12.10, 16.42, 31.21, 36.14, 36.15, 36.16, 36.17 **3** 37.11 **5** 21.3 **6** 75.24
Hummelink-Peters, B.G.M.C. **6** 80.14 **7** 30.7
Hummelink, T.W. **4** 80.4 **5** 80.11 **6** 11.14 **7** 85.8
Humphries, A.P. **7** 71.69, 75.19
Hung, P.D. **7** 59.8
Hungerford, L. **5** 64.28
Hunt, A. **4** 65.5
Hunt, D.J. **1** 47.26, 50.12 **3** 47.13 **5** 48.54
Hunt, G.W. **4** 64.9 **5** 79.15, 79.25 **6** 64.9, 64.58, 79.5, 85.14
Hunt, M.R. **7** 42.8
Hunt, N. **6** 58.52
Hunter, D.L. **7** 71.82, 71.84, 85.3, 85.9

Hunter, F.D. **1** 45.14
Hunter, S.H. **2** 84.41
Huny, G.W. **5** 79.22
Huq, F. **3** 83.79 **4** 71.52, 71.53, 83.82 **5** 83.5
Hurley, J.C. **7** 71.40
Hursthouse, M. **4** 71.27
Hursthouse, M.B. **1** 2.92, 21.33, 50.27, 59.48 **2** 60.136, 60.137, 60.138, 78.61, 83.123, 83.217, 84.65 **3** 75.16, 78.25, 82.25, 83.53, 83.102, 83.129 **4** 32.17, 38.21, 49.16, 54.1, 54.2, 59.15, 65.1, 65.2, 80.15, 83.26, 83.61, 83.85, 83.130, 83.136 **5** 32.24, 51.43, 63.6, 71.79, 81.48, 81.61 **6** 83.54, 83.55, 84.45 **7** 81.46
Huse, G. **1** 15.4 **2** 60.86
Huse, Y. **1** 59.9, 59.21
Husebye, S. **2** 64.27, 70.6, 70.7, 70.10, 70.14, 70.15, 70.17, 70.18 **3** 39.26, 40.9, 64.18, 64.19, 70.3 **5** 40.16, 70.10, 70.11, 70.12 **6** 64.4, 68.9 **7** 11.4, 11.5, 68.5, 70.1, 70.6
Hussain, M.S. **2** 83.87 **3** 67.25, 84.13, 84.18, 84.19
Hutcheon, W.L. **4** 73.17
Hutcheon, W.L.B. **6** 47.13 **7** 47.17, 47.22
Huttner, G. **3** 71.30 **4** 71.8, 71.30, 75.1, 75.5, 75.16 **5** 71.9, 71.36, 71.39, 72.24, 74.11, 75.3, 75.10, 75.11, 75.23 **7** 38.2, 71.3, 71.6, 71.14, 73.39, 73.40, 75.12, 75.14, 75.15
Hvoslef, J. **1** 45.12, 45.17, 45.19, 45.20, 45.21 **2** 60.44, 60.45 **4** 45.17 **7** 45.2, 45.12
Hwang, K. **3** 40.11
Hwang, K.J. **3** 60.8 **4** 37.8
Hwang, Y.N. **7** 32.36
Hyatt, C. **6** 62.3
Hyatt, J.A. **7** 35.13
Hybl, A. **1** 45.76 **3** 1.39, 44.43 **4** 5.8 **5** 1.38, 28.2
Hyde, A.J. **1** 33.58
Hyde, E.M. **6** 86.25, 86.31

Iacobelli, J.A. **7** 51.44
Iball, J. **1** 26.41, 28.16, 29.7, 29.11, 29.16, 29.21, 29.22, 30.2, 30.8, 30.11, 30.17, 36.31, 36.32, 40.33, 44.36, 47.2, 47.3, 47.4 **2** 77.23, 83.181 **4** 29.5, 45.16 **6** 64.55, 73.54 **7** 26.5, 30.3, 30.4, 32.11, 40.14, 40.15, 60.27, 76.2, 76.3, 83.3, 83.4
Ibata, K. **4** 27.5, 27.6 **6** 27.8, 27.17 **7** 31.12
Ibekwe, S.D. **3** 71.41, 71.42
Ibers, J.A. **1** 1.13, 1.20, 28.15 **2** 64.52, 74.10, 75.51, 76.39, 76.40, 83.195, 83.200, 84.61, 85.30, 85.73, 85.97, 86.4, 86.16, 86.25, 86.39, 86.50, 86.58, 86.61, 86.66, 86.75, 86.76, 86.77, 86.79, 86.85, 86.87, 86.88, 86.98, 86.109, 86.112, 86.114 **3** 7.2, 7.3, 33.1, 65.7, 69.39, 71.20, 71.66, 73.2, 76.8, 76.16, 82.7, 83.22, 84.24, 85.36, 86.14, 86.21, 86.39, 86.47, 86.55, 86.56, 86.60, 86.63, 86.66, 86.68, 86.69, 86.73, 86.79, 86.89, 86.90, 86.93 **4** 71.48, 71.72, 71.93, 72.22, 72.23, 72.24, 72.30, 83.133, 86.45, 86.51, 86.59 **5** 12.11, 49.8, 68.27, 71.89, 71.102, 72.33, 85.7, 85.31, 86.36, 86.57, 86.60 **6** 32.8, 49.6, 49.19, 71.27, 71.42, 71.88, 72.30, 73.9, 86.37, 86.38, 86.39, 86.55, 86.65 **7** 3.7, 7.5, 20.15, 20.16, 49.5, 49.9, 49.10, 49.16, 71.46, 71.112, 71.117, 71.121, 72.31, 72.40, 72.43, 75.20, 83.16, 83.83, 85.28, 85.42, 85.46, 86.26, 86.27, 86.28, 86.83, 86.116
Ibrahim, E.H.M. **5** 64.21, 64.29
Ibuka, T. **7** 53.17, 53.23
Ichikawa, M. **4** 2.7 **6** 2.8
Ichikawa, T. **1** 1.66, 1.67, 48.80
Ievins, A. **2** 62.51 **7** 66.10, 85.34, 85.35
Ievins, A.F. **6** 21.1, 21.2, 27.3, 85.30
Igarashi, T. **2** 85.61
Ignat'ev, N.A. **3** 27.10
Ignatova, N.P. **7** 64.30
Iguchi, S. **7** 60.20
Ihara, M. **1** 58.45, 58.58 **3** 36.26, 58.58 **6** 58.51
Iida, K. **3** 78.10
Iijima, I. **3** 58.36
Iijima, K. **3** 78.9, 81.23
Iimura, Y. **5** 80.2 **6** 80.2
Iitaka, Y. **1** 1.66, 1.67, 21.22, 21.23, 24.33, 31.16, 41.31, 46.6, 47.13, 47.14, 48.3, 48.4, 48.42, 48.80, 50.9, 50.20, 50.21, 50.34, 50.35, 54.12, 55.1, 55.2, 55.3, 56.7, 58.90, 59.9, 59.21, 59.60 **2** 83.34 **3** 1.12, 1.17, 33.24, 36.20, 48.16, 48.30, 48.33, 53.9, 56.3, 59.29, 71.21 **4** 22.3, 22.4, 22.5, 35.38, 48.28, 50.10, 51.54, 56.6, 59.23, 71.22, 72.28 **5** 32.21, 48.17, 50.17, 50.18, 50.19, 50.22, 50.25, 67.5, 83.83 **6** 1.30, 11.10, 47.16, 48.19, 48.26, 48.27, 48.28, 48.30, 48.64, 50.3, 50.11, 53.20, 55.1, 56.4, 59.25 **7** 1.18, 16.16, 17.32, 38.48, 39.26, 40.22, 41.16, 47.30, 47.32, 50.20, 50.21, 50.36, 50.39, 53.28, 54.8, 56.7, 58.19, 58.25, 59.23, 76.37, 83.22
Iizuka, H. **6** 58.25
Ikeda, M. **6** 5.4, 5.5
Ikeda, T. **2** 80.4
Ikehara, M. **3** 47.32 **5** 47.14, 47.15
Ikekawa, T. **1** 50.9
Ikemoto, I. **2** 60.142 **3** 60.53 **4** 60.33 **5** 60.21, 60.22 **6** 60.20, 60.21, 60.26 **7** 60.16
Il'inskii, A.L. **3** 77.12 **5** 61.10, 77.21
Ilario, L.d' **7** 32.12
Iljinsky, A.L. **3** 77.11

Illingworth, S.M. **3** 69.27
Ilmaier, B. **3** 86.97
Ilyukhin, V.V. **5** 2.3 **7** 42.3, 42.4
Imado, S. **1** 58.9
Immirzi, A. **1** 21.32, 23.9, 29.20 **2** 61.17, 61.18, 61.19, 68.1, 72.20, 72.43, 72.45 **3** 21.10, 58.28, 74.3, 74.4 **4** 72.19 **5** 1.32, 8.18, 13.16, 58.30, 86.80 **6** 71.49, 71.92, 86.40 **7** 68.16, 72.11, 72.12, 81.22, 86.93, 86.94
Inaba, S. **7** 37.18
Inayama, S. **3** 53.9 **4** 53.1 **6** 53.20
Induni, G. **6** 38.21
Ingen Schenau, A.D.van **6** 83.7, 84.37, 84.38 **7** 83.44, 83.45, 83.81
Ingletto, G. **6** 3.7, 3.8, 3.29, 3.33 **7** 3.30
Inglis, F. **2** 65.51
Ingram, L. **3** 85.52
Ingri, N. **4** 83.93 **5** 83.111, 84.6
Ingrosso, G. **6** 71.49, 71.92
Innes, M. **1** 16.34, 59.62 **3** 5.4 **6** 1.22
Inouye, H. **1** 52.25
Interrante, L.V. **4** 85.2 **5** 85.4, 85.5 **7** 60.4
Intille, G.M. **5** 83.50, 83.51
Inubushi, Y. **1** 58.93
Ip, P.C. **7** 32.26
Ireland, P.R. **3** 76.58, 77.1, 83.100 **4** 71.71, 71.73, 71.89, 76.24 **5** 71.95, 71.96, 71.99
Ireland, R.E. **3** 27.11, 30.5 **4** 27.16 **5** 51.39 **6** 30.5, 30.6 **7** 75.17
Irie, H. **1** 1.108, 58.86 **3** 17.5, 58.43 **4** 58.43 **6** 58.47
Irie, T. **4** 27.8 **6** 59.6, 59.8
Irngartinger, H. **3** 29.3, 29.4, 31.37, 31.38, 31.40, 31.41 **5** 28.10, 30.4, 30.6, 31.44, 31.45, 39.36 **6** 30.1, 30.4 **7** 39.35, 71.80, 75.26, 75.35
Irreverre, F. **4** 48.20, 48.22
Irving, H.M.N.H. **1** 41.35 **5** 85.29 **6** 16.15
Irving, R.J. **5** 84.24
Isaacs, N.W. **3** 32.2, 58.6, 69.15 **4** 45.18, 67.8 **5** 47.24, 50.11, 51.9, 51.22, 51.25, 56.9 **6** 38.39, 51.6
Isakov, I.V. **1** 1.33, 16.25 **2** 83.19
Isakova, A.I. **7** 5.18
Isaksson, G. **1** 41.28
Isern, X. **6** 32.24
Ishida, M. **4** 59.20
Ishihara, N. **4** 81.33, 81.35
Ishii, H. **1** 58.93
Ishii, Y. **7** 72.40
Ishikawa, T. **5** 17.18
Ishimori, M. **4** 22.7
Ishimura, Y. **5** 82.16
Ishizuka, K. **4** 58.43 **5** 58.37
Isidor, J.L. **6** 59.1
Iske, S.D.A. **6** 73.7 **7** 86.101, 86.108
Islam, K.M.S. **1** 19.28, 29.8, 54.15, 59.54 **2** 73.102 **7** 13.3
Isler, O. **6** 54.10
Isobe, I. **4** 77.3
Isobe, T. **6** 54.6
Isono, Y. **1** 50.4
Itai, A. **1** 55.2 **4** 59.23 **5** 50.25 **6** 53.20
Ito, M. **3** 76.54 **4** 76.20, 76.21, 76.40
Ito, S. **1** 31.16, 58.111 **4** 22.3, 22.4, 22.5, 27.15 **5** 58.43 **6** 31.10
Ito, T. **2** 85.61 **3** 60.13, 82.18 **4** 69.14, 76.2, 80.19 **5** 65.3, 80.2, 83.63 **6** 32.21, 32.22, 32.23
Itoh, K. **7** 48.1, 48.2, 48.3
Ittel, S.D. **6** 72.30, 86.55 **7** 72.43, 86.26, 86.27, 86.28
Ivanov, V.I. **2** 77.33 **3** 77.12 **7** 79.18, 79.19, 79.25, 81.32
Ivanova, G.V. **7** 81.26
Ivanova, N.S. **2** 69.38, 80.21
Ivarsson, G. **5** 83.111 **6** 83.26
Ivie, R.A. **7** 53.14
Iwadare, S. **6** 58.58
Iwai, I. **5** 58.43
Iwakuma, T. **7** 31.15
Iwakura, Y. **6** 32.27
Iwamoto, R. **6** 5.9, 5.10, 5.11
Iwamoto, T. **3** 61.4 **5** 61.2 **6** 61.1, 61.5, 61.6
Iwasa, K. **6** 58.45
Iwasaki, F. **2** 60.117 **3** 21.9, 60.12 **4** 24.9 **5** 3.24 **6** 15.1, 17.4, 17.6 **7** 11.17, 48.21
Iwasaki, F.F. **1** 1.16, 1.17
Iwasaki, H. **1** 1.17 **2** 60.123, 76.35, 83.96 **4** 69.20 **5** 3.24, 80.25 **6** 71.71, 80.4, 80.16 **7** 44.13, 50.6
Iwashita, Y. **1** 7.18
Iwata, M. **2** 76.38
Iwata, Y. **7** 5.1, 45.1
Iyer, K.N. **3** 58.39 **4** 58.38
Izatt, R.M. **4** 67.23 **7** 38.64, 38.65, 42.5, 42.6, 42.9
Izawa, M. **6** 59.6
Izumi, T. **6** 13.3
Izumi, Y. **5** 33.34, 33.35, 33.36

Jack, T. **5** 72.55
Jack, T.R. **7** 71.123
Jackobs, J. **1** 39.75 **6** 45.28, 45.43
Jackson, R.B. **2** 75.9 **4** 83.109
Jacob, L.M. **4** 47.4
Jacobi, P. **1** 54.24
Jacobs, R. **5** 72.47
Jacobs, T.L. **4** 20.11, 20.13
Jacobson, R. **7** 69.11
Jacobson, R.A. **1** 33.46, 38.14, 38.54, 45.65 **2** 62.45, 64.3, 73.36, 75.34, 77.69, 86.53 **3** 3.36, 33.8, 33.10, 33.11, 37.1, 45.4, 58.24, 58.50, 71.2, 83.73, 83.107 **4** 33.20, 64.31, 83.117, 83.135, 86.8, 86.27 **5** 3.12, 33.16, 71.5, 75.24, 82.6, 82.8, 83.28, 83.84 **6** 33.33, 35.8, 38.18, 73.5, 80.13 **7** 11.1, 33.20, 66.5, 76.35
Jacobson, S. **5** 86.47 **6** 86.43, 86.46, 86.56 **7** 72.45, 86.56
Jacobus, J. **5** 19.20
Jacques, J. **7** 28.2
Jaecker, J.A. **7** 86.46, 86.97
Jaffe, A.B. **5** 9.5, 9.6, 9.7, 9.12

Jaggi, H. **1** 1.92, 13.33 **2** 83.211
Jagner, S. **7** 29.1, 29.2
Jain, P.C. **2** 76.46, 78.10, 78.38, 78.39 **3** 76.47 **6** 78.14 **7** 78.16
Jain, S.C. **1** 2.72, 2.85 **4** 39.21, 39.29, 60.43, 60.44 **7** 81.19
Jain, S.R. **6** 65.11
Jakkal, V.S. **5** 41.5
Jakle, H. **4** 71.92
Jakovenko, V.I. **1** 11.30 **2** 62.29, 69.20
James, B.R. **7** 85.6
James, J.M. **3** 83.20
James, M. **3** 45.39
James, M.N.G. **1** 44.12, 50.14, 50.40 **4** 33.38, 47.2, 54.10, 65.4 **5** 32.8, 58.2, 58.46 **6** 1.3, 2.13, 33.41, 33.50, 47.2, 47.13, 47.25, 58.2 **7** 20.20, 32.1, 32.34, 33.31, 33.46, 47.15, 47.17, 47.22
James, M.R. **5** 76.12
James, T.A. **4** 83.131
James, V. **5** 45.1, 45.2, 45.22, 45.40, 45.42
James, V.H.T. **1** 51.38, 51.39, 51.40
James, V.J. **4** 21.9, 21.10, 21.14 **5** 21.13, 21.14, 21.15, 27.15, 45.11 **6** 45.33, 45.39, 45.40, 45.41 **7** 3.12, 21.10, 45.46
Jameson, G.B. **7** 49.23
Jameson, M.B. **1** 24.26
Jamet-Delcroix, S. **2** 60.27 **4** 3.3 **5** 3.4
Jamieson, G. **4** 31.24
Jandacek, R.J. **1** 39.30 **3** 39.12, 48.31, 48.38 **4** 66.7 **5** 67.7
Janic, I. **6** 35.11
Jankowski, K. **7** 47.31
Janse-van Vuuren, P. **7** 75.6
Jansing, J.A. **3** 38.4, 42.2
Janson, V. **4** 76.17
Janusonis, G.A. **2** 83.65
Jaquier, R. **1** 9.47, 32.23 **3** 9.20
Jarchow, O.H. **1** 16.36 **7** 45.27, 45.35
Jarman Junior, C.S. **6** 66.5
Jarski, M.A. **2** 78.8
Jarvis, A.C. **7** 71.48, 71.50, 71.71
Jarvis, B.B. **4** 11.9
Jarvis, J.A.J. **1** 11.10 **2** 83.3, 83.39, 83.168, 83.207, 86.8, 86.65, 86.104 **3** 71.57, 72.3, 80.1, 80.2, 86.28, 86.29, 86.41 **4** 67.27, 69.36, 71.62, 71.64, 72.2 **5** 50.10, 71.22, 71.38, 86.29
Jarzab, T.C. **6** 82.19, 82.22
Jason, M.E. **7** 71.125, 75.41
Jaud, J. **5** 7.9, 7.10 **6** 12.1
Jaunin, A. **7** 35.23
Jaunin, R. **5** 40.33
Javora, P.H. **7** 65.1
Jayadevan, N.C. **2** 76.25 **3** 81.31, 81.44 **5** 81.3
Jayasooriya, U.A. **6** 79.2
Jayaweera, S.A.A. **3** 82.25
Jeannin, S. **5** 81.2 **7** 85.45
Jeannin, Y. **3** 83.46, 83.47, 83.48 **4** 32.8, 83.3 **5** 81.2, 83.2, 83.40, 83.58, 83.59, 84.1 **6** 52.2, 71.14, 83.1, 83.4 **7** 85.45, 86.25
Jecny, J. **2** 60.34 **6** 75.24
Jefferies, P.R. **1** 59.41
Jefferson, I.W. **7** 71.91
Jeffery, G.A. **1** 45.42
Jeffery, J.C. **7** 86.58, 86.62
Jeffery, J.W. **5** 84.28 **6** 84.40, 84.46
Jeffrey, G.A. **1** 1.8, 1.54, 1.115, 1.145, 1.149, 2.28, 5.21, 16.5, 18.2, 19.8, 19.44, 19.51, 39.12, 39.29, 39.31, 40.27, 43.9, 43.12, 43.17, 44.17, 45.8, 45.15, 45.16, 45.18, 45.25, 45.37, 45.41, 45.44, 45.45, 45.67, 45.73, 52.18, 58.47, 58.82, 58.100, 58.101 **2** 61.1, 61.5, 61.6, 61.7, 61.8, 61.11, 61.12, 61.13, 61.14, 61.22 **3** 13.1, 45.3, 45.44, 60.23, 61.2, 61.5, 81.32 **4** 21.1, 45.9, 45.10, 45.11, 45.12, 45.24, 60.22 **5** 5.14, 45.4, 45.6, 45.8, 45.9, 45.10 **6** 45.7, 45.8, 45.10, 45.11, 61.7 **7** 5.17, 19.14, 33.2, 45.5, 45.32
Jeffreys, J.A.D. **1** 58.68 **3** 58.12, 58.54 **5** 71.63, 71.64, 82.7
Jeffs, P.W. **3** 58.20
Jelen, A. **3** 69.4
Jelenic, I. **2** 81.60
Jelinski, L.W. **6** 5.8
Jellinek, F. **1** 3.9, 38.38, 40.12 **2** 72.44, 74.11, 74.12
Jeminet, G. **3** 11.21 **5** 11.10
Jenevein, R.M. **5** 33.47
Jenkins, C.M. **7** 22.1
Jenkins, P.N. **3** 58:34
Jenkins, R.A. **5** 77.1
Jenkins, R.N. **5** 64.33
Jenks, J.M. **3** 32.31 **4** 27.18
Jennings, P.W. **4** 86.11 **7** 71.40
Jennische, P. **4** 80.20 **5** 68.18, 80.15 **6** 68.2 **7** 1.7, 11.6, 11.13, 80.16
Jensen, B. **1** 38.16, 40.30 **3** 3.51, 39.11 **4** 1.13, 40.20, 40.25 **5** 59.2, 59.3 **7** 3.31
Jensen, K.G. **1** 10.24, 10.27, 40.22 **3** 10.3, 76.2, 76.3
Jensen, L.H. **1** 1.135, 1.136, 1.138, 1.155, 9.25, 12.18, 13.21, 33.74, 35.2, 35.44, 37.20, 44.9, 44.10, 44.25, 44.28, 44.40, 44.47, 46.14, 47.10, 47.15, 47.27, 48.37, 48.67, 48.68, 59.26 **2** 60.121 **3** 24.4, 33.22, 44.10, 44.31, 48.29, 52.6 **4** 13.21, 33.17, 44.16, 48.53 **5** 41.11, 45.28, 45.29, 45.31, 48.16, 48.24 **6** 45.21, 45.35, 46.9, 48.43, 48.71, 48.80, 81.3 **7** 45.48, 45.49, 48.48, 59.7
Jensen, W. **5** 46.7 **6** 3.6, 46.12, 64.13 **7** 69.11
Jensovsky, L. **2** 79.29
Jepsen, H. **4** 41.32
Jeremic, M. **5** 75.31 **7** 81.21
Jerina, D.M. **4** 40.14
Jerome-Lerutte, S. **4** 3.10
Jerslev, B. **1** 8.7, 10.27, 19.21, 19.22, 19.26, 21.1, 21.2, 38.16, 40.22, 40.23, 59.28
Jesson, J.P. **5** 83.35 **6** 86.2
Jeter, D. **6** 83.63
Jeter, D.Y. **5** 82.12, 83.34
Jezowska-Trzebiatowska, B. **5** 76.3, 76.7
Jimenez-Garay, R. **6** 38.11 **7** 45.44
Job, B.E. **2** 83.168
Job, R. **7** 76.60, 85.18

Joesten, M.D. **3** 67.25, 84.13, 84.18, 84.19 **5** 84.23, 84.29
Johans, A.W. **4** 86.55
John, K.-P. **7** 64.24, 64.42
Johns, S.R. **4** 58.35 **5** 58.24
Johnsen, J. **1** 8.36, 8.37
Johnsen, K. **1** 39.35, 39.36 **2** 70.3, 70.4
Johnson, A. **5** 71.22
Johnson, A.W. **6** 49.8
Johnson, B.F.G. **6** 71.28
Johnson, C. **5** 67.11
Johnson, C.K. **1** 2.47, 2.77, 2.78, 2.79, 2.80, 2.81, 52.21 **3** 1.6 **4** 81.4, 81.5, 81.6, 81.7, 81.8 **5** 31.14, 31.17 **6** 60.17, 60.18 **7** 39.3
Johnson, C.R. **2** 84.67 **3** 39.45
Johnson, D.A. **1** 50.26 **4** 83.29
Johnson, D.R. **3** 84.11, 84.17 **5** 82.20, 84.19
Johnson, D.W. **5** 83.21
Johnson, G. **7** 41.33
Johnson, J.E. **3** 83.107 **4** 83.117, 83.135 **5** 83.84 **6** 33.33
Johnson, J.W. **7** 73.42
Johnson, L.N. **1** 45.51
Johnson, P.L. **1** 39.83 **3** 21.15, 33.41, 39.23, 39.40, 39.41 **5** 21.14, 21.15, 27.1, 60.42 **6** 36.3, 44.25, 44.26, 48.69
Johnson, Q. **1** 3.37 **3** 3.28 **4** 21.21, 60.13 **5** 46.7 **6** 3.6, 46.12, 64.13 **7** 35.25
Johnson, R. **6** 75.27
Johnson, R.C. **5** 33.25
Johnson, R.D. **7** 17.7
Johnson, R.L. **5** 20.13
Johnson, R.N. **5** 72.51
Johnson, S.M. **1** 31.7, 34.8, 50.33, 53.35 **3** 17.11, 23.8, 23.9, 37.10, 39.21, 39.46, 44.37, 50.24, 75.7 **6** 85.1, 85.2, 85.3, 85.12
Johnson, W.S. **3** 31.10
Johnston Junior, M.D. **6** 21.9
Johnston, D.L. **3** 76.83, 77.27, 78.6 **4** 80.8
Johnston, J.P. **1** 37.28
Johnston, R.M. **6** 60.22 **7** 60.19
Johnstone, D. **5** 85.20
Jolly, P.W. **4** 71.65, 86.66
Joly, E. **6** 2.5
Jonas, K. **4** 86.66
Jones, A. **7** 36.10
Jones, C.E. **6** 86.25, 86.31
Jones, C.J. **2** 71.9 **3** 85.7, 85.20
Jones, C.R. **5** 27.8
Jones, D.D. **6** 48.33
Jones, D.S. **6** 31.1, 37.12, 37.14, 51.29
Jones, D.W. **1** 2.9 **4** 28.4 **5** 24.3 **6** 29.4 **7** 29.5, 29.6, 29.8
Jones, G. **4** 35.8, 58.4, 58.5
Jones, G.A. **7** 71.13
Jones, K. **3** 69.14
Jones, L.A. **4** 24.5 **5** 20.13, 28.16
Jones, N.D. **1** 32.29, 36.21 **2** 60.71, 73.43 **3** 29.2, 50.8 **4** 51.10, 58.42 **5** 38.23, 41.39, 50.5, 50.7, 50.26, 50.27, 50.28 **6** 47.15, 50.10
Jones, N.J. **5** 31.32
Jones, P.E. **2** 85.33
Jones, R.D.G. **3** 11.12 **4** 11.5, 68.7, 86.9, 86.16, 86.18 **5** 71.72, 75.35, 75.48, 86.16 **6** 73.23
Jones, R.E. **1** 1.26
Jones, R.O. **2** 72.35
Jong, J.G.H.de **5** 20.5, 35.21, 51.1
Jongsma, C. **3** 65.6 **4** 20.7
Jonsson, P.-G. **1** 12.4 **3** 1.4, 12.4 **4** 48.1 **5** 1.2, 1.3
Jordaan, A. **5** 45.49
Jordan, T.H. **1** 4.9 **2** 61.4, 61.5 **3** 60.21
Jose, P. **1** 48.12 **2** 82.19 **3** 77.15
Joshi, B.S. **5** 53.17 **6** 56.1
Joshi, K.K. **2** 73.33
Jost, K.-H. **2** 67.2
Jotham, R.W. **3** 71.3
Joustra, A.H. **3** 51.1
Jovanovic, B. **4** 71.19, 71.21, 71.81 **6** 71.46
Jovanovski, G. **7** 71.9, 71.11
Joys, S. **1** 41.4
Jude, R.J. **6** 13.5, 13.6, 17.2
Jukes, D.E. **1** 44.43
Julian, M.O. **6** 48.50
Julshamn, K. **4** 39.2, 39.13, 41.12, 42.1
Jumas, J.-C. **5** 64.2
Jungk, A.E. **3** 15.11, 15.12 **4** 13.23, 24.18 **5** 24.6
Jungst, R. **7** 83.111
Jurd, L. **6** 59.10 **7** 31.34, 31.35
Jurnak, F.A. **7** 59.21, 83.25

Kaas, K. **5** 83.91 **6** 32.11
Kabalkina, S.S. **1** 24.7, 26.14
Kabuto, C. **3** 58.58 **5** 12.5, 12.13, 31.19, 38.19, 75.41, 75.42 **6** 22.2, 22.4, 38.8, 53.7, 59.19 **7** 39.17
Kadonaga, M. **4** 72.27
Kadoya, S. **1** 21.22, 21.23
Kaduk, J.A. **7** 49.19
Kaesz, H.D. **2** 65.41 **5** 71.98
Kaftanat, V.N. **7** 76.66
Kaftory, M. **4** 69.8 **5** 69.1, 69.2, 69.5, 69.6 **7** 60.22, 60.25
Kagan, H.B. **5** 41.50
Kahn-Harari, A. **5** 41.38
Kahn, R. **5** 21.4, 21.5
Kahrl, J. **4** 58.33
Kahrl, J.H. **3** 48.43
Kai, G. **7** 72.35
Kai, Y. **2** 68.28 **3** 35.12, 68.19 **4** 35.39, 68.12, 71.28, 71.29, 71.58, 71.80 **5** 42.13, 66.1, 66.5, 68.22, 68.25, 68.32 **6** 66.3, 75.29 **7** 36.16, 71.62, 71.63, 71.73, 71.90
Kainosho, M. **3** 46.3
Kaiser, E.T. **1** 11.9, 42.1, 42.11, 42.12
Kaiser, J. **7** 83.90, 84.67, 85.11
Kaiser, K.L. **6** 75.28
Kajimoto, T. **4** 72.10
Kakati, K.K. **1** 9.38

Kakudo, M. **1** 2.3, 11.37, 13.3, 13.28, 13.29, 13.38, 20.1, 20.29, 20.30, 22.9, 27.13, 31.27, 33.42, 38.42, 48.18, 48.52, 48.56, 48.63, 48.84, 48.86, 48.88, 48.95, 56.12 **2** 63.2, 63.4, 63.6, 68.28, 69.31, 78.1, 78.2, 82.13, 82.29, 86.71, 86.81 **3** 13.19, 13.20, 35.12, 36.9, 41.23, 59.15, 68.19, 69.11, 69.25, 72.10, 72.18, 73.33 **4** 48.33, 48.35, 48.57, 49.13, 54.7, 60.20, 67.9, 68.12, 71.28, 71.29 **5** 17.18, 44.11, 45.15, 68.22, 68.25, 68.32, 69.12 **6** 32.9, 44.5, 48.81 **7** 47.28, 48.64, 48.68, 72.34
Kalck, P. **7** 86.25
Kalff, H.T. **1** 39.14, 39.15, 39.66
Kalinin, A.E. **6** 64.29, 73.55 **7** 63.18, 86.72
Kalinin, D.I. **7** 18.7
Kalish, R. **5** 39.43
Kalker, H.G. **7** 4.5
Kalman, A. **1** 4.6, 4.7 **3** 11.28 **4** 4.3, 11.13 **5** 4.3, 8.16, 32.12, 41.42, 42.10 **6** 8.7, 32.25, 35.11, 39.27 **7** 32.2, 32.3, 35.27, 36.13, 48.37, 53.9
Kalman, T.I. **1** 11.62 **3** 47.34
Kalra, K.L. **6** 71.72, 73.18, 73.25, 86.70, 86.71 **7** 86.44, 86.45
Kaluski, Z. **5** 58.6, 73.4 **7** 58.13, 73.46
Kaluski, Z.L. **2** 73.85, 73.87, 73.104, 73.105, 73.107, 73.115 **3** 73.45 **6** 73.55
Kalyanaraman, A.R. **4** 44.24, 48.19, 77.20, 83.21, 84.14 **5** 83.42
Kalyani, V. **1** 13.56, 13.57, 38.48
Kamat, V.N. **6** 56.1
Kamberi, B. **5** 82.20
Kamenar, B. **1** 49.5, 49.6 **2** 60.70, 60.74, 60.148, 69.15 **3** 65.10, 66.5, 83.9, 83.57 **5** 63.1, 71.6, 77.2 **6** 10.2, 32.17 **7** 48.38, 71.9, 71.11, 76.4, 76.9, 76.72, 82.20, 83.17, 83.31
Kameswaran, V. **6** 36.18
Kametani, T. **1** 58.45, 58.58 **3** 36.26, 58.58 **6** 58.37, 58.51
Kamigauchi, M. **6** 58.45
Kamijo, N. **1** 58.115 **7** 53.27
Kamijyo, N. **5** 73.23, 73.24 **7** 73.34
Kamikawa, T. **6** 54.6
Kaminskii, V.F. **5** 60.8 **7** 60.21, 60.34, 60.35, 60.36
Kaminsky, W. **7** 71.65
Kamiya, K. **1** 48.73, 50.5, 56.9, 58.35, 58.36, 58.57, 59.42, 59.52 **3** 44.42, 50.13, 50.22, 50.30 **4** 50.6, 54.3 **5** 50.2 **6** 35.29, 36.7, 40.24
Kamper, M.J. **1** 49.34
Kan-Nan, Y. **7** 83.67, 83.68
Kanamaru, F. **2** 83.156
Kanaoka, Y. **3** 31.14 **7** 37.16
Kanda, E. **1** 5.9
Kandler, H. **2** 63.25
Kane, A.R. **3** 86.10 **5** 86.17
Kane, J. **3** 3.28
Kaneda, M. **4** 51.54 **6** 55.1, 59.25
Kaneda, T. **7** 31.27
Kanehisa, N. **5** 66.1
Kaneko, H. **4** 58.43
Kaneko, M. **7** 47.24
Kaneko, T. **3** 59.1
Kannan, K.K. **1** 11.15
Kanters, J.A. **1** 1.32, 1.34, 38.47 **3** 1.8 **4** 1.1, 1.3, 1.7, 2.6, 2.15 **5** 1.19, 39.24, 63.12 **6** 43.6 **7** 1.5, 11.21
Kapadi, A.H. **4** 58.44 **7** 58.21
Kapecki, J.A. **1** 37.24, 39.61 **3** 20.12 **4** 27.7
Kapicak, L.A. **4** 38.19
Kaplan, D. **6** 42.6
Kaplan, S.F. **2** 81.8, 81.10
Kapon, M. **5** 16.15, 36.17, 69.1, 69.2, 69.5, 69.6 **6** 16.6, 83.31 **7** 38.26
Kapoor, S.K. **7** 31.9
Kapovits, I. **1** 4.7 **5** 42.10
Karagiannidis, P. **6** 85.22 **7** 80.9, 80.10, 80.20
Karanjgoakar, C.G. **6** 59.30
Karaoghlanian, B.V. **4** 39.14
Karch, N.J. **5** 13.11
Karino, S. **5** 17.18
Karipides, A. **5** 62.11, 69.24, 81.6 **6** 63.6, 69.15, 69.16 **7** 69.26
Karle, I.L. **1** 1.119, 4.18, 5.41, 8.41, 16.60, 17.44, 17.46, 20.17, 20.18, 20.23, 20.25, 31.1, 31.18, 32.3, 33.73, 35.27, 35.28, 35.32, 35.43, 37.15, 37.17, 37.25, 38.40, 39.57, 41.39, 41.49, 42.4, 44.70, 46.20, 48.69, 48.89, 51.18, 52.27, 58.22, 58.31, 58.33, 58.51, 58.66, 58.113, 58.116 **3** 8.17, 11.3, 31.14, 31.34, 37.8, 47.27, 48.14, 48.47, 48.53, 51.28, 58.9, 58.61, 59.21, 60.52, 61.6, 61.7, 67.7 **4** 9.12, 37.16, 42.6, 44.17, 44.21, 44.26, 48.20, 48.22, 48.47, 54.6, 54.9, 59.13 **5** 54.2, 54.3, 64.42 **6** 33.28, 33.49, 36.1, 37.12, 37.14, 50.13, 51.29, 58.26, 58.30, 67.22 **7** 37.16, 48.65, 48.72, 48.74, 48.75, 51.42
Karle, J. **1** 1.119, 13.51, 16.60, 20.17, 20.18, 31.1, 31.18, 35.32, 35.43, 37.17, 37.25, 38.40, 41.39, 41.49, 48.69, 48.89, 51.18, 58.22, 58.31, 58.33, 58.51, 58.66, 58.113, 58.116 **3** 37.8, 47.27, 48.5, 48.47, 59.21, 61.6, 61.7 **4** 48.20, 48.22, 54.6, 59.13, 61.2 **5** 41.16, 54.2 **7** 48.74
Karle, J.M. **4** 58.23
Karlin, K.D. **7** 76.71
Karlsson, B. **5** 53.3 **6** 31.21, 31.22, 53.1 **7** 31.13, 31.25, 51.48, 52.2, 54.1
Karlsson, R. **4** 58.8 **5** 36.14, 83.61 **6** 53.12, 58.6 **7** 53.13, 54.3, 54.4
Karolak-Wojciechowska, J. **7** 64.25, 64.26, 64.27
Karraker, D.G. **3** 77.32
Karrer, P. **3** 58.67
Kartha, G. **1** 2.66, 33.78, 48.31, 51.2, 51.3, 51.13, 56.10, 58.29, 59.68 **2** 60.151 **3** 44.14, 44.17, 44.19, 44.25, 54.9 **5** 47.6, 47.9, 47.13, 53.17, 58.29 **6** 32.14, 35.28, 38.15, 40.21, 47.12, 48.22, 48.81 **7** 10.6, 10.7, 47.2, 47.4, 47.7, 47.8, 47.18, 48.57, 48.58, 52.7

Kasai, N. **1** 31.27 **2** 63.4, 63.6, 68.28, 69.3, 69.31, 86.71, 86.81 **3** 13.19, 13.20, 35.12, 36.9, 41.23, 68.19, 69.11, 69.25, 72.10, 72.18, 73.33 **4** 35.39, 60.20, 68.12, 71.28, 71.29, 71.58, 71.80, 72.27 **5** 17.18, 42.12, 42.13, 66.1, 66.5, 68.22, 68.25, 68.32, 69.12 **6** 44.5, 66.3, 75.29 **7** 36.16, 38.67, 71.62, 71.63, 71.73, 71.90, 72.34, 72.35
Kashino, S. **4** 33.10, 33.11 **5** 2.28, 13.2 **6** 13.7, 19.9, 33.3, 33.13, 33.14, 35.5 **7** 1.11, 1.12, 17.9, 34.16, 38.14, 48.64
Kashiwagi, M. **3** 21.11 **4** 35.18, 72.10 **6** 72.11
Kashiwagi, T. **2** 86.71, 86.81 **7** 72.34
Kashman, Y. **7** 59.10
Kashyap, R.P. **7** 78.16, 78.25
Kasin, M. **7** 53.13
Kaski, B.A. **3** 58.30
Kasper, J.S. **4** 85.2 **5** 85.4, 85.5 **7** 60.4
Kastalsky, V. **2** 85.32, 85.57
Katayama, C. **1** 13.36 **3** 53.20 **5** 31.23, 31.38, 54.10 **7** 54.16
Katayama, M. **1** 1.23
Kato, K. **3** 59.6 **4** 6.2, 11.16, 11.17 **5** 11.13
Kato, N. **5** 54.10
Kato, S. **7** 43.7
Kato, T. **6** 59.19
Kato, Y. **1** 13.17, 13.18, 20.29 **5** 13.5 **6** 59.19 **7** 13.4, 13.5
Katritzky, A.R. **4** 32.22 **7** 33.6, 33.12
Katsube, Y. **1** 13.28, 48.95 **4** 59.20 **5** 13.6, 59.17, 82.16, 83.63 **7** 83.67, 83.68
Katsumoto, K. **1** 39.90 **4** 39.43
Katz, J.L. **1** 1.38
Katz, L. **1** 1.96, 38.7, 44.56, 44.58, 48.60 **2** 85.33
Katz, T.J. **2** 75.35 **3** 75.34 **4** 31.12 **5** 31.42
Kaufman, H.W. **3** 1.38 **5** 1.11 **6** 1.16
Kavcic, S. **5** 35.19
Kawada, I. **2** 64.59, 86.63 **3** 39.34 **4** 68.6
Kawaguchi, H. **3** 76.31, 76.52 **6** 77.9
Kawaguchi, S. **3** 72.10, 76.31, 76.52, 84.10 **4** 71.58, 71.80, 77.3, 77.18 **6** 76.33, 77.9 **7** 76.8
Kawaguchi, T. **4** 5.1 **6** 5.1, 5.2, 5.3, 5.4, 5.5
Kawai, K. **6** 56.4
Kawai, K.-I. **7** 59.23
Kawai, M. **3** 59.30
Kawai, S. **7** 38.48
Kawai, T. **6** 44.5
Kawamoto, I. **4** 64.35, 64.38 **5** 64.37
Kawasaki, T. **4** 54.3 **6** 59.9, 59.13
Kawase, A. **5** 41.23, 83.107
Kawase, T. **2** 79.25
Kawashima, K. **7** 83.67, 83.68
Kawashima, S. **4** 58.43
Kawazu, K. **4** 54.15
Kawazura, H. **7** 72.40
Kay, A. **3** 82.16, 82.19
Kay, H.F. **1** 23.17
Kay, J. **4** 24.12 **5** 81.51, 81.52
Kay, M.I. **1** 1.57, 1.58, 1.96, 38.7 **2** 81.3, 81.4, 81.8, 81.10 **3** 28.3, 28.4 **6** 48.2, 81.1
Kay, W.I. **1** 50.11
Kayama, Y. **5** 31.19
Kaye, P. **6** 64.55
Kayushina, R.L. **1** 48.38
Kazaryan, G.S. **7** 19.11
Kazitsyna, L.A. **1** 9.19
Kearns, D.R. **5** 27.8
Keat, R. **5** 64.3, 64.36
Keats, S. **3** 3.9
Keck, P. **5** 48.52
Keefe, W.E. **5** 4.1 **6** 24.7, 24.8
Keeler, B.T. **3** 51.39
Keeton, M. **4** 71.16, 71.17, 83.63
Keiderling, T.A. **3** 3.36
Kekka, S. **7** 1.11, 1.12
Kelecom, A. **6** 53.12
Keller-Schierlein, W. **5** 45.43
Keller, C.E. **2** 75.26
Keller, H.J. **7** 71.126, 71.134
Kellerman, M. **2** 76.20, 76.41
Kelley, B.P. **2** 80.40 **3** 85.24
Kelley, J.A. **3** 3.22, 77.14, 77.25, 83.65 **4** 78.10
Kellie, G.M. **5** 38.18
Kelly, B.P. **3** 80.3
Kelly, H.C. **4** 3.5
Kelly, J.F. **3** 42.2
Kelso, E. **4** 39.6
Kemme, A.A. **3** 27.12, 69.17 **6** 69.11 **7** 27.8, 37.8, 62.3
Kemmerich, T. **5** 86.66
Kemmitt, R.D.W. **2** 86.92 **3** 71.34 **4** 71.87, 85.34 **5** 71.100, 71.101 **7** 71.48, 71.50, 71.71
Kendall, D.S. **5** 3.10
Kende, A.S. **1** 37.23
Kendrick, N. **6** 48.83
Kennard, C.H.L. **1** 3.33, 3.35, 3.36 **2** 71.4, 71.5, 71.31, 76.66, 81.23, 81.85, 83.128, 83.171 **3** 16.6, 32.2, 60.40, 69.15, 75.9, 76.68, 83.68, 83.69, 83.136 **4** 45.18, 67.8, 68.4, 76.37, 83.73 **5** 17.21, 19.16, 19.17, 20.15, 31.6, 31.8, 32.1, 32.2, 32.10, 32.11, 35.7, 69.13, 86.49 **6** 31.5, 58.24, 81.18 **7** 33.44
Kennard, O. **1** 8.32, 51.35, 51.38, 51.39, 51.40, 56.4, 58.108 **2** 65.24, 83.208 **3** 20.17, 23.1, 46.5, 46.6, 51.42, 53.12, 53.13, 53.14, 54.13, 58.6, 58.47, 65.5 **4** 8.9, 23.2 **5** 1.18, 2.30, 20.10, 46.6, 47.24, 47.31, 50.11, 51.9, 51.20, 51.22, 51.25, 51.41, 53.13, 56.9 **6** 3.15, 30.3, 38.39, 39.23, 46.2, 51.6, 53.6, 64.56, 65.1, 65.2, 65.3 **7** 17.21, 29.11, 36.22, 36.23, 39.37, 39.38, 47.6, 47.16, 47.35, 51.36, 56.4, 64.32, 64.33
Kennedy, E.R. **6** 64.16
Kenney, M.E. **7** 63.19
Kent, P.W. **3** 45.31
Kepert, D.L. **5** 83.20, 83.32, 84.17, 84.26 **6** 65.4, 80.9, 80.10, 84.47 **7** 83.66, 83.101, 83.110, 86.3, 86.40, 86.41

Kerf, J.de **1** 41.27
Kergoat, R. **7** 77.9
Kergomard, A. **7** 50.35
Kerr, K.A. **1** 19.62, 30.7, 51.35 **2** 65.24 **3** 20.17, 23.1, 24.5, 46.5, 46.6, 58.47, 58.51 **4** 58.18 **5** 5.15, 5.16, 16.24 **6** 46.6, 46.7, 48.5, 48.6, 69.1, 85.16 **7** 69.6
Kershaw, R.A. **5** 27.6, 27.7
Kersting, G.F. **4** 12.14, 35.23
Kettle, S.F.A. **3** 71.3
Keul, H. **7** 38.28
Keulen, E. **2** 74.11 **3** 33.2, 33.3
Keve, E.T. **1** 2.44 **5** 48.7
Khadi, A. **7** 76.44
Khalifa, S. **7** 52.1
Khalil, M.F. **3** 58.17
Khalilov, A.D. **3** 77.12 **7** 33.51
Khan, A.A. **5** 31.7
Khan, M.A. **5** 83.22 **7** 83.26, 86.42
Khan, M.A.Q. **5** 31.7
Khan, M.Y. **1** 13.24
Khandkarova, V.S. **3** 74.9 **4** 74.5
Khare, G.P. **3** 85.35 **4** 85.16, 85.39 **6** 71.82 **7** 71.111, 86.90
Khattak, M.N. **5** 36.3
Khitrova, O.M. **7** 72.44
Khodashova, T.S. **7** 33.15, 35.10, 76.10
Khotsyanova, T.L. **1** 5.2, 5.8, 16.3, 19.2, 24.8 **2** 66.9 **3** 16.1, 19.2 **5** 19.1, 19.4, 19.5 **6** 73.31 **7** 39.23
Khubchandani, P.G. **1** 9.41
Khvatkina, A.N. **1** 7.6, 11.30, 41.36 **2** 62.29, 69.20, 69.38, 80.21
Kido, M. **6** 48.56
Kido, Y. **4** 71.22
Kiechel, J.R. **3** 50.19
Kiefer, E.F. **6** 5.8
Kiehl, G. **6** 36.12
Kiekbaev, I.D. **7** 81.31
Kiel, G. **6** 2.2, 11.3 **7** 11.8, 39.2, 39.6
Kiely, D.E. **6** 38.17
Kierkegaard, P. **1** 36.27 **3** 36.7, 58.4, 63.17, 80.5 **4** 45.13 **5** 50.24, 83.124 **7** 83.19
Kiers, C.T. **4** 11.7
Kietaibl, H. **7** 62.11
Kijima, K. **7** 40.22
Kikkawa, S. **2** 69.31
Kikuchi, T. **1** 58.36, 58.45, 58.57 **7** 56.6
Kikutani, N. **7** 39.26
Kilbourn, B.T. **1** 11.45, 50.44 **2** 69.48, 75.36, 83.168, 86.41, 86.62 **3** 17.10, 67.27, 71.41, 71.42, 71.57, 72.3, 73.28, 80.1, 80.2, 83.85, 86.12, 86.28, 86.29, 86.76, 86.88, 86.91 **4** 69.36, 71.62, 71.64, 72.2 **5** 71.38, 83.87, 86.29
Killean, R.C.G. **1** 45.39 **2** 63.12, 68.2 **3** 64.23 **4** 45.7, 65.7 **7** 41.7, 41.8
Killops, S.D. **7** 85.17
Kilty, P.A. **7** 83.18
Kim, H.S. **1** 45.15, 45.16, 45.45 **3** 13.1, 45.3, 45.44, 58.3, 60.23, 61.5 **4** 21.1, 60.22 **5** 16.17
Kim, J.J. **5** 47.33 **7** 84.49
Kim, J.K.S. **5** 15.2
Kim, N.E. **3** 8.7 **7** 73.56
Kim, S.-H. **3** 44.4
Kim, S.H. **1** 8.8, 44.3, 45.8, 45.18, 45.38, 45.49 **2** 60.131 **4** 47.20
Kim, Y.-H. **5** 51.42
Kim, Y.B. **6** 8.5, 8.6 **7** 1.23
Kime, N. **2** 86.87
Kimura, B.Y. **4** 71.87 **7** 71.48, 71.50, 71.71
Kimura, E. **4** 81.61 **7** 76.37, 83.22
Kimura, M. **7** 37.18
Kimura, S. **4** 59.21
Kimura, T. **3** 69.11, 69.25 **5** 69.12 **7** 36.16
Kinas, R. **7** 64.20, 64.21
Kinashi, H. **6** 50.12 **7** 50.28
Kindberg, B.L. **5** 86.69 **7** 44.7, 44.20
Kindler, H. **1** 22.7
King, A.W. **6** 60.2 **7** 38.70
King, G.S.D. **1** 1.99, 33.38, 33.41 **2** 72.32, 72.69 **3** 23.8, 23.9, 32.9, 41.10, 42.4 **4** 32.24
King, M.G. **3** 85.52
King, M.S. **7** 83.18
King, R.W. **7** 31.7
King, T.J. **1** 54.1, 59.23 **3** 49.4, 59.31, 65.8 **4** 35.25 **5** 41.24, 64.13, 69.22 **6** 64.11, 69.18 **7** 41.19, 65.6, 69.7, 69.14, 69.19, 69.27, 85.36
Kingsbury, C.A. **6** 11.16, 27.15, 40.8
Kingston, D.G.I. **7** 53.15
Kinloch, M.E. **2** 78.29
Kinoshita, T. **4** 32.5 **5** 41.7, 41.9, 41.36
Kinoshita, Y. **1** 44.17 **2** 83.114 **3** 21.11 **4** 35.18, 72.10 **6** 72.11 **7** 83.9
Kinson, P.L. **4** 29.2
Kiosse, G.A. **2** 66.11, 66.12 **5** 79.13 **7** 78.20, 79.14, 79.15, 79.17
Kirby, A.J. **1** 39.65
Kirchner, R.M. **3** 83.121, 83.122, 83.123, 83.124, 83.125, 83.126 **5** 71.102 **6** 71.27, 73.9, 78.7, 78.8, 78.16, 78.18, 78.19, 86.38, 86.39 **7** 71.46
Kirillova, N.I. **3** 72.40 **5** 73.12 **6** 72.27, 73.58 **7** 72.44, 73.10, 73.12, 73.32, 73.69
Kirin, I.S. **7** 49.20, 49.21
Kirkpatrick, J.L. **1** 58.100, 58.101
Kirksey, J.W. **6** 59.1 **7** 58.20, 58.24
Kirkwood, C.E. **3** 84.12 **5** 83.13
Kirner, J.F. **7** 49.18
Kirokawa, S. **1** 1.85
Kirschke, K. **1** 38.21
Kirson, I. **4** 51.49
Kirtley, S.W. **6** 65.7 **7** 86.13, 86.14
Kiryu, S. **2** 83.144 **4** 60.31 **7** 60.20
Kis, Z. **5** 59.27, 59.32
Kisch, H. **7** 71.100
Kishi, T. **1** 50.5 **3** 50.22, 50.30
Kishida, E. **1** 37.1
Kishida, Y. **1** 39.88 **4** 64.35, 64.38 **5** 64.37
Kispert, L.D. **1** 51.47
Kiss, B. **4** 24.16
Kiss, J. **6** 41.6 **7** 63.3, 63.6
Kissing, W. **3** 37.9

Kistenmacher, T.J. **1** 3.67 **3** 40.4 **4** 35.24, 82.12 **5** 48.54 **6** 44.4, 44.9, 44.10, 44.15, 48.25, 60.3, 76.21 **7** 44.16, 44.18, 48.8, 48.9, 76.56, 76.57, 78.11, 82.3
Kita, W.G. **6** 71.13, 72.9, 73.13
Kitagawa, I. **6** 59.29
Kitahama, K. **6** 38.2
Kitahara, Y. **1** 20.30, 38.42 **3** 58.58 **5** 12.5, 12.13, 31.19, 38.19, 75.41, 75.42 **6** 22.2, 22.4, 38.8, 53.7, 59.19 **7** 39.17
Kitaigorodskij, A.I. **1** 5.2, 5.8, 5.26, 24.7, 24.12, 26.11, 26.14, 29.28, 48.1 **2** 71.2, 71.26 **4** 33.32 **5** 19.9 **6** 24.4 **7** 17.13, 17.14
Kitano, Y. **4** 35.18, 72.10 **6** 72.11 **7** 83.9
Kitaura, Y. **6** 58.52
Kivekas, R. **7** 83.14
Kjaer, A. **3** 8.18
Kjellevold, K.E. **7** 45.2
Kjoge, H.M. **1** 39.48, 39.60 **2** 70.11, 70.12 **3** 39.18, 64.8
Klaassen, A.A.K. **6** 22.1
Klaeren, A. **4** 39.7
Klanberg, F. **2** 71.40
Klanderman, K.A. **2** 72.23, 72.24, 72.31 **3** 72.1 **7** 77.8
Klapper, H. **5** 2.20
Klar, G. **6** 10.3 **7** 66.9
Klazinga, A.H. **5** 11.9
Kleinberg, R. **6** 48.2, 81.1
Klem, E. **2** 83.184
Klement, U. **2** 84.4
Klewe, B. **1** 7.14, 7.15, 10.10, 12.13 **4** 7.6, 7.7, 7.8, 7.11, 9.4, 12.3, 12.8 **5** 7.1, 7.2, 7.3, 39.1
Kliegel, W. **7** 62.7
Klimasenko, N.L. **3** 26.1, 26.2 **6** 41.20 **7** 26.3, 42.10
Klimova, A.I. **2** 62.12
Kline, B.S. **1** 3.22
Kline, R.J. **7** 83.52, 83.80
Klinga, M. **3** 76.32 **5** 83.9 **7** 83.13
Klohs, M.W. **3** 33.48, 38.17
Kluepfel, D. **7** 50.12
Klug, A. **1** 19.12
Klug, H.P. **1** 16.28, 18.16 **2** 78.12, 80.24 **3** 16.3
Knapp Junior, F.F. **7** 51.54
Knappe, J. **1** 59.55
Knazeva, A.N. **2** 78.30
Kneale, G. **6** 17.15, 17.17
Kneale, G.G. **7** 76.58
Kneen, W.R. **4** 71.43, 71.46, 71.49
Knight, J. **4** 71.35
Knipe, A.C. **6** 17.19
Knisely, A.S. **5** 33.25
Knobler, C. **1** 38.8, 38.10, 38.11, 38.12, 42.6 **2** 60.42, 79.13 **3** 39.2, 39.30, 70.2 **5** 51.37, 51.38, 70.2 **6** 70.1
Knobler, C.B. **5** 71.98 **6** 71.70, 71.87, 71.94 **7** 71.8
Knobler, C.M. **4** 69.25
Knofel, H. **7** 28.8
Knorr, K. **7** 3.1
Know, J.R. **5** 48.52
Knowles, P.J. **6** 73.63
Knox, G.R. **3** 73.56 **6** 72.7
Knox, J.R. **2** 71.27, 73.40, 73.75, 82.14, 83.38, 85.78 **6** 45.30 **7** 45.50
Knox, K. **1** 3.69 **2** 60.99 **3** 49.1, 49.7, 85.49 **4** 49.14 **7** 63.19
Knox, S.A.R. **6** 75.11, 75.12, 75.23 **7** 71.69, 73.41, 75.1, 75.19, 75.25, 85.17
Knyazeva, A.N. **2** 77.60 **3** 77.26 **4** 77.2 **6** 77.18 **7** 78.19
Kobal, V.M. **6** 37.11
Kobalkina, S.S. **2** 83.76
Kobayashi-Tamura, H. **4** 72.17
Kobayashi, A. **5** 65.3, 76.48, 83.60 **6** 76.27, 83.35 **7** 86.71
Kobayashi, G. **6** 36.11
Kobayashi, H. **3** 3.25, 35.4 **6** 3.2, 12.3, 18.1, 19.4, 60.19 **7** 18.1, 60.14, 60.17, 60.18
Kobayashi, J. **2** 67.18, 67.19
Kobayashi, M. **4** 76.25
Kobayashi, N. **3** 59.29
Kobayashi, T. **4** 49.13, 49.18
Kobayashi, Y. **3** 36.20 **4** 71.22, 72.28 **5** 83.83 **7** 39.26
Kobelt, D. **4** 24.17, 40.5, 42.4, 44.13, 70.1, 70.5 **5** 8.19, 40.6, 69.9 **6** 41.21, 41.31 **7** 20.18, 24.11
Kobori, K. **3** 21.11 **7** 83.9
Koch, B. **1** 1.144 **4** 57.1
Koch, H.P. **1** 2.92
Koch, K.-H. **5** 32.23
Koch, M. **4** 48.49 **5** 48.74
Koch, M.H.J. **3** 48.15 **4** 33.43, 33.44 **5** 33.53, 35.33 **6** 33.58, 33.60, 35.31, 35.32, 35.33 **7** 33.53, 39.36
Koch, S. **7** 49.10
Kochi, J.K. **6** 28.10
Kocman, V. **4** 71.47
Koda, S. **2** 75.55 **3** 84.10 **4** 33.24, 77.3, 77.18 **5** 33.39, 33.40, 33.41, 36.8, 36.9 **6** 76.20 **7** 50.4
Kodama, T. **6** 32.27 **7** 60.12, 60.13, 86.98
Kodama, Y. **6** 11.10
Koenig, D.F. **1** 1.128, 28.15, 39.3, 49.17
Koenuma, M. **7** 50.28
Koer, F.J. **5** 1.12
Koerner von Gustorf, E. **5** 72.5
Koetzle, T.F. **2** 62.25 **3** 62.5 **4** 48.5, 48.10, 76.33 **5** 48.9, 48.10, 48.14, 48.21, 48.27, 48.29, 48.38, 48.47, 48.49, 48.59, 48.60 **6** 44.12, 47.17, 48.18, 48.33, 48.57 **7** 33.9, 33.25, 44.17, 44.28, 48.22, 76.49, 86.13, 86.14
Koh, L.L. **4** 12.15
Kohara, M. **6** 75.29
Kohout, J. **7** 83.23
Koizumi, H. **2** 81.77
Kojic-Prodic, B. **5** 81.10 **6** 47.11, 77.17 **7** 47.13
Kok, A.J.de **3** 1.14, 38.6 **5** 1.12, 21.8 **6** 51.36 **7** 51.55
Kokkou, S.C. **7** 32.41
Koknat, F.W. **4** 3.7 **6** 3.13
Kokotailo, G.T. **3** 85.41, 85.42 **4** 69.24, 85.28 **5** 85.18, 85.19 **6** 36.16

Kolakowski, B. 1 1.88 5 15.4 7 13.11
Kolderup, M. 4 59.3
Kolesnikov, S.P. 3 60.9
Kolobova, N.E. 2 72.72, 73.68 3 71.56 7 72.44
Kolokolov, B.N. 5 26.1 6 26.1
Kolombos, A.J. 3 67.24
Kolomnikov, I.S. 6 81.45
Kolpa, R.L. 7 64.22, 64.23, 64.54
Koltai, E. 6 39.27
Kolyadin, A.B. 7 49.20, 49.21
Komalenkova, N.G. 7 63.4, 63.8, 63.11
Komeyama, M. 6 9.1
Komiyama, Y. 2 76.4, 76.13, 76.17 3 76.31, 78.9, 78.10, 81.23 4 78.1
Komori, T. 4 54.3 6 59.9, 59.13
Komson, R.C. 4 81.48, 86.17
Kon, A.Yu. 2 83.91
Kondon, S. 7 47.30
Kondrashev, Yu.D. 1 9.30, 9.42, 44.21 7 16.6, 16.8, 17.19
Kondrashev, Yu.D. 1 9.31 4 9.10, 32.1 5 9.9, 9.10, 9.11
Konietzny, A. 7 71.101
Konig, E. 4 83.115, 83.116
Koning, R.E. 3 28.8
Koningsveld, H.van 1 5.14 3 58.10 4 21.7 5 21.11 6 21.8 7 74.3
Konishi, H. 1 48.18
Konnert, J. 1 7.13 2 71.18 3 47.27, 48.53 4 40.7, 44.17, 63.3, 69.6, 69.7
Konnert, J.H. 3 11.4
Konno, M. 5 76.29 6 12.2, 18.1, 64.34, 64.35
Kono, Y. 4 50.4
Konst, W.M.B. 3 51.15
Koo, C.H. 1 9.5, 9.18 3 3.43, 3.44, 58.3 4 72.25 5 3.3, 16.17
Koon, S.S. 5 78.13, 78.14
Kopaneva, L.I. 7 82.5
Kopecky Junior, W.J. 5 37.5
Kopecky, W.J. 4 37.9
Kopf, H. 3 73.26 4 73.8
Kopf, J. 6 10.3 7 66.9, 71.65
Kops, R.T. 5 72.11 7 20.9
Kopwillem, A. 3 80.5 5 80.10
Koran, D. 6 83.73
Korczynski, A. 2 79.11, 79.12, 79.17, 79.24 6 79.7
Koridze, A.A. 4 73.10
Koritny, E.F. 3 77.11
Kornblau, M.J. 1 7.32
Kornilov, A.N. 7 45.20
Korotkevich, A.F. 4 16.1
Korpar-Colig, B. 7 71.9, 71.11
Korvenranta, J. 3 76.57 6 76.18 7 76.21, 76.38
Korytnyi, E.F. 2 77.28 3 77.22 4 77.11
Korytnyk, W. 5 33.37 6 38.15
Koski, H.K. 7 5.5, 5.6, 5.7, 5.8, 5.9, 5.10
Kosky, C.A. 4 72.12
Kostansek, E.C. 5 8.2
Koster, J.B. 7 68.10
Koster, P.B. 3 39.17
Koster, W. 5 59.23
Kosturkiewicz, Z. 1 11.43 6 35.30 7 48.46, 50.29
Kostyanovskii, R.G. 1 32.17 3 13.7
Kosuge, T. 4 59.21
Koten, G.van 4 71.82 5 71.80 7 71.116
Kotera, K. 7 31.15, 50.3
Kothe, G. 7 45.29, 45.43
Kotick, M.P. 1 47.30
Kotsyanova, T.L. 1 19.1, 19.3
Kou, W.H. 7 44.14
Kouno, I. 6 59.9, 59.13
Kovacs, A.L. 1 37.10 2 60.97
Kovaleva, E.S. 2 79.30
Kovama, H. 2 81.74
Kowala, C. 3 58.22 5 31.37, 38.28 6 83.60 7 75.34
Koyama, G. 1 41.31, 50.20, 50.21, 50.34 5 50.3 6 50.3 7 47.30, 48.25, 51.20
Koyama, H. 1 7.28, 27.12, 33.60, 36.30, 51.20, 52.28, 58.18, 58.122 2 84.7, 85.18 3 31.12, 31.26, 38.13, 50.3, 50.4, 50.27, 51.20, 51.22, 51.33, 54.12, 56.7, 58.11, 58.16, 58.25 4 27.8, 39.37, 39.38, 48.24, 50.12 6 51.34, 53.3
Koyama, K. 7 37.16
Koyama, Y. 3 1.12
Koyano, N. 7 45.1
Koz'min, P.A. 1 33.27, 33.28, 33.30 3 81.7 4 33.2 7 81.2
Kozarek, W.J. 6 85.24
Kozerski, L. 7 31.22
Kozhin, V.M. 1 48.1
Kozlowski, D.L. 7 35.1, 35.2, 47.10, 83.69
Kozuka, M. 5 53.18 7 53.23
Krabbendam, H. 6 16.22
Krahe, E. 2 63.1, 63.17
Krahe, E.W. 5 63.2
Krajewski, A. 7 35.41, 35.42, 35.43, 35.44, 51.18
Krajewski, J. 7 50.25, 83.57, 83.58
Krakower, G.W. 3 51.39
Kramarenko, F.G. 7 82.1, 82.9, 82.10
Kramer, L. 6 64.36 7 64.43
Krannich, L.K. 6 65.11
Krasnova, T.L. 4 63.8 7 69.10
Krasochka, O.N. 3 83.17
Krass, D.K. 7 33.30
Kratky, C. 7 49.3, 49.4, 49.6, 49.7
Krause, J. 5 71.75
Krausse, J. 1 13.9 3 71.47 4 71.12 5 67.16 7 71.57
Kraut, J. 1 21.12, 31.39, 41.34, 46.1, 47.27, 48.27, 48.28, 51.6, 52.24, 52.30, 59.65
Kravchenko, V.B. 5 83.82
Kravtsov, D.N. 6 69.17 7 69.20, 69.21, 69.23, 71.18, 71.37
Krebs, B. 1 1.128, 11.3, 39.1, 39.3 6 79.9 7 12.5, 65.5, 70.2
Kreher, R. 7 32.40
Kreishman, G.P. 6 45.29
Kreisman, P. 2 86.26
Krejcaerik, G.E. 7 17.7
Kretsinger, R.H. 2 82.49
Kreutzer, P. 5 86.66

Krieg, B. **4** 71.8, 75.5 **5** 75.10, 75.11, 75.23
Krieger, M. **3** 35.24
Krieger, W. **7** 8.2
Krigbaum, W.R. **1** 1.141 **3** 9.11 **4** 13.20, 19.5 **5** 83.116 **7** 13.25
Krishna Rao, K.V. **1** 31.25, 31.26 **4** 1.4
Krishnamachari, N. **4** 83.110 **6** 58.54 **7** 32.26, 71.79
Kristensen, R. **6** 39.26
Kristoff, J.S. **6** 73.9 **7** 71.103
Kritskaya, I.I. **3** 71.49 **7** 75.33
Krivnov, V.Ya. **1** 7.6
Krogmann, K. **2** 72.15, 81.11, 81.17, 81.18, 81.19 **5** 83.27
Krogsgaard-Larsen, P. **4** 40.3 **6** 40.4 **7** 40.1
Krogsrud, S. **7** 86.26
Krom, J.C. **1** 52.19
Kromer, D.T. **1** 3.22
Kronfeld, L.R. **1** 39.67
Kroon, J. **1** 1.32, 1.34, 2.63, 2.67 **3** 1.8, 69.40 **4** 1.1, 1.3, 2.6, 2.15, 64.7 **5** 1.19, 66.6 **6** 16.22 **7** 1.5
Kroon, P.A. **3** 75.14 **6** 33.8
Kropf, R. **1** 30.27
Krstanovic, I. **7** 81.20
Krueger, C. **5** 72.30
Krueger, S.M. **3** 20.12
Kruger, C. **3** 72.17, 75.18 **4** 63.5, 71.65, 75.7, 86.46, 86.66 **5** 71.59, 71.62, 71.73, 71.82, 72.5, 72.6, 72.18, 72.19, 72.31, 72.36, 75.12, 75.13, 75.28, 75.32 **6** 29.2, 62.9, 64.44, 67.23, 83.43, 83.65, 86.44 **7** 62.14, 64.61, 68.10, 69.9, 71.29, 71.44, 71.56, 71.89, 71.100, 71.105, 72.32, 75.32
Kruger, G. **5** 51.29
Kruger, G.J. **3** 12.7, 41.7 **4** 30.1, 41.9, 41.10 **6** 54.12, 54.18, 73.57, 77.12 **7** 56.5, 58.17
Kruiisink, C.A. **1** 21.5
Krujer, G.J. **1** 30.4
Krukonis, A.P. **1** 28.21, 28.22 **3** 73.10 **4** 12.11, 73.12 **5** 19.3 **6** 7.5, 27.6, 28.17, 28.18
Kruse, F.H. **1** 11.52 **2** 70.20
Kruss, B. **6** 66.1
Kryger, L. **5** 1.10, 81.14 **6** 82.2
Kshnyakina, A.N. **1** 7.7
Ku, A.T. **4** 20.4
Kubas, G.J. **5** 73.31, 73.37
Kubina, L. **3** 51.11
Kubinyi, H. **3** 54.15
Kubota, M. **7** 86.100
Kubota, T. **1** 7.28 **5** 60.2 **6** 54.6
Kucsman, A. **1** 4.7 **3** 11.28 **4** 4.3 **5** 4.3
Kuhn, H.J. **2** 64.51
Kuhn, M. **5** 59.24
Kukina, G.A. **2** 72.1, 72.3, 80.19, 81.44, 81.67 **3** 72.32 **4** 72.1, 81.43
Kukolja, S. **5** 41.39
Kuli-Zade, T.S. **3** 72.32
Kulishov, V.I. **3** 60.9, 73.22, 73.38 **5** 71.47
Kulkarni, A.G. **7** 64.47
Kulpe, S. **1** 18.13 **2** 83.145 **4** 18.2 **5** 83.14 **7** 3.17, 3.25, 3.26, 18.4, 32.19, 33.47, 41.27
Kulshreshtha, D.K. **6** 56.4
Kulyeva, T.E. **6** 21.5, 21.6
Kumadaki, I. **7** 39.26
Kumakura, S. **2** 60.117 **5** 39.28, 39.29 **7** 60.12, 60.13
Kumar, P. **7** 33.18
Kumashiro, I. **1** 7.18
Kumra, S.K. **1** 2.73, 58.123
Kunchur, N.R. **2** 71.25, 79.10, 83.44, 83.45, 83.61, 85.4, 85.23, 85.51, 85.95 **3** 71.15, 79.12
Kung-Du, C. **1** 3.54
Kung, W.J. **7** 48.54
Kunishima, M. **7** 53.28
Kunstmann, W. **4** 24.17 **7** 24.11
Kuntsevich, T.S. **7** 73.44
Kunz, G. **5** 39.30
Kunz, V. **2** 74.13
Kuo, M.-C. **4** 71.86
Kuo, M.C. **6** 60.23
Kupchan, S.M. **3** 58.57 **4** 53.10 **5** 58.45, 59.26, 59.34 **6** 36.18, 53.17, 54.2 **7** 38.57, 59.9
Kupcik, V. **1** 53.36 **3** 83.37
Kupfer, D. **1** 33.39, 33.47, 33.48
Kupletskaya, N.B. **1** 9.19
Kupper, F.W. **2** 71.39, 71.41
Kuppers, H. **5** 2.12, 2.20 **7** 14.3, 14.4
Kupriyanova, G.N. **7** 82.1
Kurabayashi, M. **7** 59.24
Kurahashi, M. **1** 13.12 **3** 36.4 **5** 41.23, 83.107 **7** 41.14, 84.38, 84.39, 84.65, 84.66
Kurauchi, T. **6** 76.33 **7** 76.8
Kurganov, A.A. **5** 82.24 **7** 82.21
Kuribayashi, S. **1** 17.40, 48.32 **6** 62.4 **7** 5.11
Kuriyabashi, S. **1** 17.42
Kurkutova, E.N. **5** 79.5, 79.12 **7** 42.3, 79.12
Kurkutova, E.V. **7** 42.4
Kuroda, H. **2** 60.142 **3** 60.53 **4** 60.33 **5** 60.21, 60.22 **6** 60.20, 60.21, 60.26 **7** 60.15, 60.16
Kuroda, R. **5** 61.3 **6** 61.4 **7** 6.1, 76.41, 76.42, 76.76
Kuroda, T. **1** 11.37
Kuroda, Y. **7** 1.13, 1.14
Kurokawa, F. **4** 49.18
Kurosawa, E. **6** 59.6, 59.8
Kuroya, H. **2** 76.2, 76.4, 76.8, 76.10, 76.11, 76.13, 76.36, 79.25 **3** 76.23, 76.28, 76.52, 76.66, 81.17, 83.109, 84.10 **4** 33.24, 76.3, 76.10, 77.3, 77.18, 83.105 **5** 36.9, 76.4, 82.18 **6** 76.10, 76.33, 77.9 **7** 76.8
Kursanov, D.N. **2** 63.28
Kusaka, T. **1** 50.5
Kushi, Y. **2** 76.2 **3** 32.32, 68.22, 81.17 **7** 20.2, 20.3, 53.27
Kussather, E. **5** 58.13
Kuszaj, J.M. **5** 77.24 **6** 77.21, 83.39, 83.41

Kutney, J.P. **6** 58.33
Kutoglu, A. **3** 73.26 **4** 41.32 **5** 73.27, 86.54 **6** 73.11, 73.37 **7** 31.6
Kutschabsky, L. **3** 52.2 **4** 52.2, 52.3 **6** 17.9 **7** 3.17, 3.26, 33.36, 85.11, 85.33
Kutsuma, T. **3** 36.20
Kuwada, Y. **1** 59.42
Kuwano, H. **7** 59.24
Kuyper, L.F. **6** 48.82
Kuz'min, V.S. **7** 75.33
Kuz'mina, L.G. **4** 83.128, 86.36 **5** 71.7 **6** 69.20, 72.17 **7** 71.18, 71.36, 71.37, 75.5
Kuzmin, E.A. **5** 2.3
Kuznetsov, S.I. **1** 16.3 **3** 16.1 **5** 19.1, 19.4, 19.5 **6** 73.31 **7** 39.23
Kuznetsov, V.G. **1** 33.27, 33.28, 33.30 **2** 79.30 **3** 81.7
Kvick, A. **3** 12.4, 33.6 **4** 48.1, 60.5 **5** 33.6 **6** 33.6, 48.1 **7** 33.9, 33.25, 44.17, 47.13
Kvitashvili, A.I. **6** 76.5 **7** 76.11, 76.12, 76.13, 76.16, 76.19, 76.24, 76.36, 76.53, 76.54, 76.55, 84.14
Kwai, K.-I. **7** 58.25
Kwan, T. **2** 83.34
Kwiram, A.L. **5** 20.8
Kyandzhetzian, R.A. **7** 39.23
Kyogoku, Y. **1** 46.6

L'Haridon, P. **6** 84.36
L'Huillier, P.M. **5** 72.25, 75.6
La Placa, S.J. **1** 8.43, 8.44, 15.22, 28.15 **2** 64.61, 66.19, 66.21, 77.65, 86.76, 86.77, 86.109, 86.114 **3** 83.51 **4** 83.42 **5** 31.2
Laan, J. **1** 58.1
LaBar, R.A. **6** 27.20
Lacey, M.J. **6** 84.11
Lachmann, B. **5** 33.37
LaCour, T. **6** 72.19
Ladd, M.F.C. **1** 51.37, 56.13 **5** 1.27, 38.20, 38.22, 76.10 **6** 20.1, 38.23, 48.51, 76.9, 84.33
Ladell, J. **1** 1.2, 32.5, 32.35, 32.36, 32.37, 32.39, 32.40, 32.43, 32.44 **2** 64.60
Lago, J.-M. **7** 26.2
Laguna, A. **7** 71.59
Lahav, M. **6** 20.20
Lahiri, B.N. **1** 36.19
Lahuerta, P. **7** 75.13
Lai, T.F. **1** 13.27 **3** 86.40, 86.46, 86.71 **5** 47.17 **6** 59.17
Laine, R.M. **4** 72.32 **6** 75.23
Laing, K.R. **3** 86.61
Laing, M. **1** 37.16 **2** 83.131, 83.132 **3** 83.74, 85.46 **4** 31.32, 33.13, 33.14 **5** 31.41, 58.16, 59.30, 86.12, 86.20 **6** 54.13, 58.53, 86.6, 86.28 **7** 58.2, 59.18, 83.41, 83.55, 83.70, 83.86
Laing, M.B. **2** 73.72 **4** 36.2
Lajsic, S.D. **7** 58.21
Lajzerowicz-Bonneteau, J. **1** 12.8 **7** 33.32, 33.33, 40.3
Lajzerowicz, J. **3** 37.3 **4** 12.13 **6** 16.18, 33.31 **7** 32.13, 32.14
Lake, A.W. **4** 24.13, 24.14, 24.15
Lake, R.E. **6** 76.11, 76.12, 76.13
Lalancette, R. **4** 33.25
Lalancette, R.A. **2** 77.65
Lallemand, J.Y. **6** 29.5
LaLonde, R.T. **5** 58.5
Lam, S. **6** 7.4
Lam, S.Y. **3** 86.71
Lam, T.W. **1** 11.69
Lamarque, A. **1** 17.11
Lambert, J.B. **3** 21.17
Lamberton, J.A. **4** 58.35 **5** 58.24
Lami, G. **5** 38.14
Lamm, B. **1** 16.34
Lamotte-Brasseur, J. **5** 3.15
Lampin, J.P. **6** 64.37
Landa, L.I. **7** 76.6, 79.20
Landais, J. **1** 19.60
Lando, J.B. **7** 35.20
Lanfredi, A.M. **2** 81.35, 83.10
Lanfredi, A.M.M. **1** 9.2 **2** 83.4, 83.5, 83.7, 83.14, 83.15 **3** 83.3, 83.4 **4** 67.1, 67.2, 79.4 **5** 8.6 **6** 8.2 **7** 11.9, 35.11
Lang Junior, S.A. **5** 27.16, 31.11
Lange, J.J.de **1** 9.35
Lange, S. **4** 71.30 **5** 74.11
Lange, S.M. **7** 47.29
Langer, E. **4** 71.92
Langer, R. **1** 9.1
Langfelderova, H. **5** 81.4
Langford, P. **1** 42.11, 42.12
Langhoff, C.A. **3** 60.35
Langlois, N. **7** 58.32
Langridge, R. **7** 47.26, 47.38
Langs, D.A. **2** 81.61, 81.62, 84.40, 85.31 **3** 11.2, 82.29 **6** 37.13, 51.1, 51.8, 82.19, 82.22 **7** 51.19, 51.24, 51.28, 51.29, 51.51
Lapasset, J. **3** 27.5, 27.6 **4** 32.13 **5** 28.13, 32.4, 63.10 **6** 35.2
LaPlaca, S.J. **6** 45.7, 45.8 **7** 31.17
Lappaset, J. **7** 35.5
Lappert, M.F. **5** 71.38 **6** 49.8, 71.39, 71.53, 71.67 **7** 69.17, 71.41, 71.124
Lapperts, M.F. **4** 71.81
LaPrade, M.D. **2** 74.17, 75.58 **3** 64.4, 86.57 **4** 81.26, 81.28, 86.25 **5** 40.1, 75.9 **6** 14.1
Larcheres, G. **3** 82.24
Lardy, I.A. **1** 39.56
Larina, T.B. **7** 81.2
Larking, I. **3** 76.4
LaRose, R. **5** 31.31
Larrabee, R.B. **4** 69.35
Larsen, B. **6** 45.20 **7** 45.8
Larsen, E. **3** 85.4
Larsen, F.K. **3** 32.8, 60.44 **5** 29.2 **7** 60.23, 60.24
Larsen, I.K. **1** 8.7, 10.8, 10.24 **3** 10.7 **4** 13.12, 36.28, 36.29, 36.30 **5** 36.26 **7** 8.9, 8.11, 8.14, 8.15, 8.16
Larsen, K.P. **6** 16.1
Larsen, L.A. **7** 62.4
Larsen, M.L. **4** 3.1, 82.11

Larsen, S. **2** 82.2 **5** 38.30 **6** 8.4 **7** 2.8, 2.9, 8.18, 85.39
Larson, A.C. **1** 15.19, 34.1, 34.2, 40.2 **3** 51.42 **4** 23.2, 76.8, 83.11, 83.14, 84.8 **5** 1.28, 47.24, 47.31 **6** 1.2, 65.5 **7** 8.12, 33.1
Larson, S.B. **7** 38.64, 38.65, 42.5, 42.6, 42.9
Larsson, G. **1** 2.48, 5.32 **3** 60.3
Larsson, K. **1** 1.130, 1.142, 1.148, 1.154, 1.156, 1.157 **2** 63.10 **3** 1.35, 28.6 **7** 51.43
Lasheen, M.A. **1** 24.17
Latavalya, N. **7** 83.95
Lau, C.F. **5** 77.17, 83.118, 83.119
Laubereau, P.G. **4** 75.28 **5** 75.39 **6** 71.52
Laugier, R. **7** 84.69
Lauher, J.W. **6** 49.6 **7** 3.7, 20.15, 20.16, 49.16, 83.16
Laurent, A. **1** 10.9, 31.12, 33.40, 33.52 **2** 60.28, 64.38, 64.39 **3** 9.20, 11.21 **4** 8.8 **5** 5.18, 8.10, 8.11, 9.16, 11.10
Lavallee, D. **1** 49.33
Lavertue, P. **7** 44.10, 83.7, 85.12
Laves, F. **1** 36.30
Lavie, D. **4** 51.49 **5** 54.9 **6** 54.16
Lavigne, G. **5** 81.2 **7** 85.45
Lawrence, J.L. **1** 28.2, 38.20 **2** 68.2 **3** 64.23 **4** 45.7 **7** 41.7
Lawson, A.M. **1** 38.58
Lawson, J. **3** 9.2
Lawson, M.B. **4** 82.14
Lawton, D. **2** 81.59 **7** 38.76, 61.3, 61.4
Lawton, S.L. **1** 33.46 **3** 33.11, 37.1, 64.28, 85.21, 85.41, 85.42 **4** 69.24, 85.28 **5** 33.25, 85.18, 85.19 **6** 35.8, 36.16, 66.5
Layten, S.W. **5** 71.103
Layton, A.J. **6** 67.20, 67.21
Lazarev, P.I. **5** 61.10, 77.21 **6** 37.3
Lazarini, F. **6** 81.16 **7** 2.1, 33.11, 66.12
Lazarini, P. **6** 2.4
Lazzaroni, R. **4** 72.11 **6** 72.15
Le Bars, M. **5** 16.3 **6** 16.2
Le Bihan, M.-T. **2** 67.3
Le Carpentier, J.-M. **1** 12.5 **4** 12.1, 12.2, 12.5, 12.6, 33.8, 33.9, 66.2, 68.1, 83.46, 84.3 **5** 12.3, 12.4, 66.2, 66.3, 66.4, 68.11, 68.12, 68.13
Le Quesne, P.W. **4** 58.23
Leach, J.M. **5** 85.32
Leach, M. **6** 16.23, 38.24
Leander, K. **7** 51.48
Leandri, G. **3** 16.12
Leavell, K.H. **6** 28.3
Leban, I. **5** 2.15, 2.16, 2.17 **6** 2.1, 2.10, 2.11, 33.23 **7** 51.46, 51.47
Lebedda, J.D. **3** 68.12
LeBihan, M.-T. **2** 83.2 **4** 28.1, 28.2, 83.4
LeBihan, M.T. **6** 84.36
Lebioda, L. **5** 67.1 **7** 8.1, 8.3
Lechat, J. **5** 45.4
Leclaire, A. **1** 41.14 **7** 5.4
Lecomte, C. **5** 73.32, 73.40 **6** 73.29, 74.4 **7** 48.50, 48.51, 73.60
Lectard, A. **4** 21.4 **5** 21.1, 21.6 **7** 27.10
Lederer, E. **6** 51.41
Ledesert, M. **3** 60.7 **4** 38.1 **7** 1.4, 2.3
Lednicer, D. **4** 51.46
Lee, B. **2** 76.62, 76.71, 76.72 **6** 52.1 **7** 76.65
Lee, C. **4** 35.14
Lee, C.-M. **5** 31.31 **6** 35.17
Lee, C.M. **3** 31.9
Lee, E.L. **6** 11.8, 11.9
Lee, H.B. **6** 71.55
Lee, J.D. **1** 11.58, 11.66 **2** 60.88, 60.100, 64.24, 76.15, 84.47, 84.48 **3** 11.19, 11.24, 64.1, 64.10, 64.21 **4** 11.15, 42.3
Lee, K.-H. **5** 53.18 **6** 53.15 **7** 53.11, 53.17, 53.23
Lee, K.C. **7** 47.2
Lee, K.S. **1** 38.23, 41.13 **4** 41.15 **5** 48.28
Lee, L.M. **3** 86.40
Lee, M. **2** 82.8
Lee, O. **5** 3.3
Lee, S. **4** 67.10
Lee, S.O. **7** 64.9, 64.10, 64.16, 64.17
Lee, T. **2** 73.62, 73.65 **4** 67.10
Lee, T.-Y. **3** 67.9
Lee, T.C. **5** 79.14
Lee, T.J. **6** 38.35, 67.4
Lee, T.Y. **6** 67.4
Lee, Y. **2** 73.65
Leemput, P.J.H.A.M.van der **7** 85.8
Leenhouts, J.I. **3** 21.20, 31.8, 51.8, 51.10, 51.17, 51.18, 51.21, 51.23, 51.24, 51.25 **4** 38.20
Leeuwestein, C.H. **4** 27.2
Lefebvre-Soubeyran, O. **1** 32.46, 32.47, 32.48 **4** 31.29 **6** 39.7, 47.10, 58.57, 83.49
Legendre, J.-J. **4** 65.6, 66.5
Leger, J.-M. **5** 35.6, 35.29
Leger, J.M. **5** 3.6 **6** 35.1 **7** 24.8, 50.33
Legg, J.I. **4** 82.11
Leggett, J.D. **5** 76.35
LeGoff, E. **5** 31.32
Leguen, J.-L. **2** 84.14
Legzdins, P. **2** 75.66, 77.39, 77.61 **3** 73.20
Lehman, D. **7** 71.16
Lehmann, M.S. **1** 36.4 **3** 26.8, 32.8 **4** 19.1, 48.5, 48.10 **5** 29.2, 48.9, 48.10, 48.14, 48.27, 48.29, 48.38, 48.47, 48.49, 48.59, 48.60 **6** 44.12, 48.18, 48.57 **7** 48.22
Lehr, R. **4** 4.4
Lehtonen, J. **5** 76.13
Lehtonen, M. **3** 76.29 **4** 83.59
Leibnitz, P. **6** 33.38, 33.43 **7** 33.36, 33.37
Leigh, G.J. **7** 86.100
Leijonmarck, M. **3** 36.11 **4** 36.17, 36.24 **5** 83.124
Lein, M.M. **7** 39.25
Leipoldt, J.G. **1** 33.67 **3** 33.44 **4** 33.33 **5** 76.18, 77.8 **6** 80.7, 80.8

Leiserowitz, L. **1** 1.80, 1.100, 1.117, 1.118, 7.22, 16.51, 16.52, 16.53 **3** 29.3, 29.4, 31.37, 31.38, 31.40, 31.41, 38.2 **5** 1.5, 1.6, 1.7, 1.8, 1.11, 1.17, 1.23, 1.24, 1.26, 13.1, 13.13, 60.1, 60.3, 60.9 **6** 20.20 **7** 1.9, 1.16, 1.19, 1.24, 1.25, 1.34, 13.12, 24.7
Leitch, J. **6** 4.3
Leitereg, T.J. **5** 50.9
Leli, F. **7** 48.29
Leligny, H. **4** 38.1
Lemaire, H. **7** 33.32
Lemmer, F. **6** 39.12
Lempert, K. **6** 32.32, 32.33, 39.27, 41.6 **7** 36.13
Lenarda, M. **7** 71.117, 71.121
Lenhert, P.G. **1** 32.9, 49.35, 52.8 **2** 83.146 **3** 49.2, 67.25, 84.13, 84.18, 84.19 **5** 84.23, 84.29 **6** 71.59
Lentz Junior, P.J. **1** 52.31
Leonard, A. **2** 64.18, 64.19
Leonard, N.J. **3** 44.37
Leong, J. **5** 73.45
Leoni, L. **6** 72.15
Leonowicz, M.E. **7** 39.40
Lepicard, G. **1** 32.23 **2** 64.38, 64.39 **5** 5.18, 9.16, 41.31 **6** 51.12 **7** 51.8, 51.13, 51.25, 84.57
Leporati, E. **7** 79.2, 79.5
Lepore, G.C. **7** 15.5, 15.6
Lepore, U. **3** 75.36, 75.37 **4** 1.9, 83.90 **5** 75.4 **6** 75.2 **7** 15.5, 15.6
Leppard, D.G. **7** 53.7
Lerbscher, J.A. **4** 1.4, 28.7, 35.30, 69.2 **5** 39.37
Lerma, J.Lopez de **5** 35.20
Leroy, F. **1** 43.30 **3** 43.6 **5** 51.11 **6** 24.1, 29.5, 33.29, 33.34, 51.9 **7** 24.2, 24.3, 25.2, 35.7
Leroy, G. **1** 5.33
Leser, J. **4** 1.6, 13.4 **6** 28.19 **7** 13.9, 13.21, 31.4
Lesser, D.P. **7** 17.35, 43.2, 43.3
Lessinger, L. **7** 13.2, 13.17, 19.8, 42.11, 42.12
Lessor, A.E. **1** 41.7
Lester, J.E. **5** 86.36
Lester, T.E. **2** 84.27
Letcher, R. **3** 58.34
Letendre, L.J. **7** 39.24
Lethbridge, J.W. **3** 76.24
Leung, F. **1** 39.47 **4** 41.34 **5** 41.47, 41.48 **6** 45.50 **7** 45.34
Leung, Y.C. **1** 48.92
Leutzinger, E.E. **4** 44.25
Levdik, V.F. **2** 72.11
Levenson, R.A. **7** 71.17
Leverett, P. **4** 86.2 **6** 76.44
Levine, E.M. **4** 36.18
Levine, S.G. **4** 28.10, 51.20, 51.40, 51.41 **5** 51.5, 51.14, 51.36
Levisalles, J. **6** 71.33
Leviston, P.G. **7** 86.70
Levy, E.C. **6** 54.16
Levy, H.A. **1** 2.53, 2.57, 8.4, 45.35, 45.69 **3** 1.6, 16.14, 16.15, 48.9 **4** 8.1, 13.2, 36.1, 44.7, 56.3 **5** 5.12, 45.30 **7** 85.14
Lewin, A.H. **4** 83.90
Lewin, R. **2** 62.37
Lewis, A.F. **6** 50.4
Lewis, B. **7** 71.70, 75.23
Lewis, D.F. **2** 81.46, 86.21 **3** 33.9, 86.92 **4** 71.69, 80.10, 85.22 **5** 71.18 **7** 84.79
Lewis, D.L. **5** 83.75 **6** 83.15, 83.32, 83.50, 83.51, 83.71 **7** 76.71
Lewis, E.L.V. **1** 2.96
Lewis, E.S. **6** 28.3
Lewis, J. **2** 86.46 **6** 71.28
Lewis, J.C. **4** 67.7
Lewis, R.M. **4** 76.15
Lex, J. **7** 4.5
Ley, S.V. **6** 28.13, 71.35, 75.7
Li, C.-T. **1** 46.16
Li, N.C. **2** 83.97 **5** 81.36
Li, T.-I. **5** 80.5 **6** 86.73
Lianis, P.S. **7** 7.4
Libit, L. **5** 78.13, 78.14
Lichanot, A. **5** 21.2 **6** 21.3
Liebich, B.W. **7** 17.23, 17.24
Liebig, D.M. **2** 83.29, 83.43
Liebling, G. **1** 20.6
Liehr, G. **4** 85.11 **7** 83.1, 85.24
Liepa, A.J. **6** 36.18
Lightner, D.A. **4** 30.5 **5** 30.3
Liles, D.C. **7** 83.98
Liljefors, T. **7** 11.14, 32.25
Liminga, R. **1** 2.29, 2.30 **2** 60.13, 60.15, 60.24 **5** 81.10 **6** 47.11
Lin, B.T-S. **1** 31.7
Lin, C.-F. **6** 48.55, 48.63
Lin, C.-T. **2** 77.19 **6** 20.16
Lin, G.H.-Y. **3** 47.7, 47.12 **4** 8.5, 47.9, 47.10 **5** 83.15 **6** 45.28
Lin, G.H.Y. **5** 54.7, 76.35 **6** 54.15 **7** 53.18, 64.12
Lin, H.S. **7** 86.90
Lin, H.Y. **4** 50.1
Lin, K.-K.G. **6** 75.13, 75.25 **7** 83.51, 83.89
Lin, T.S. **3** 59.13
Lin, Y.C. **7** 19.13
Linck, M.H. **6** 86.3, 86.4
Lind, M.D. **2** 76.63, 76.64, 76.71, 76.72 **3** 3.19, 8.15
Lindberg, B. **4** 45.13 **5** 45.17, 45.17, 45.19 **6** 38.1
Lindberg, K.B. **6** 45.25, 45.27 **7** 45.11, 45.28
Lindblom, L.A. **4** 78.15
Linden, J.G.M.van der **3** 80.11 **5** 80.11
Linder, P.W. **6** 44.23, 83.74, 83.76 **7** 83.106
Lindgren, J. **1** 3.10, 3.17 **5** 3.2
Lindgren, O. **4** 76.1 **5** 83.85
Lindh, B. **2** 81.53
Lindley, P.F. **2** 66.1 **3** 71.32, 75.11, 75.13, 75.25, 75.27, 77.4 **6** 41.15, 56.5 **7** 32.38, 41.31, 51.56
Lindner, E. **6** 84.8 **7** 85.24

Lindner, H.J. **2** 71.39, 71.41, 84.53 **3** 28.11, 29.1, 35.9, 35.11 **4** 29.6 **5** 34.4, 34.6, 35.3 **6** 28.8, 38.33 **7** 20.19, 50.8
Lindqvist, I. **1** 11.42, 39.13 **2** 66.6, 82.3, 82.9 **3** 33.43
Lindqvist, O. **2** 70.21 **3** 69.4 **5** 41.13
Lindsay, A.R. **1** 46.2
Linek, A. **1** 53.14, 53.36 **2** 79.29 **3** 53.7
Ling, N.C. **3** 51.41, 58.35, 58.42
Lingafelter, E.C. **1** 1.127, 1.143, 3.33, 3.34, 3.35, 3.36, 8.26, 8.27, 9.25, 10.18, 29.1 **2** 76.14, 76.17, 76.18, 76.32, 76.46, 77.8, 77.12, 77.13, 78.3, 78.8, 78.9, 78.10, 78.11, 78.19, 78.23, 78.36, 78.38, 78.39, 78.43, 78.47, 78.48, 78.51 **3** 8.6, 76.42, 76.46, 76.47, 78.18, 81.25, 81.45, 81.46, 81.47, 83.29, 83.30, 83.44, 83.112, 83.118, 83.121, 83.122, 83.123, 83.124, 83.125, 83.126 **4** 8.3, 78.5, 83.56, 83.57 **5** 61.5 **6** 78.7, 78.8, 78.16, 78.18, 78.19, 83.34, 83.52
Lingier, W.R.F. **3** 41.10
Lingner, U. **6** 64.53, 64.54
Linke, K.-H. **4** 39.7 **7** 4.5
Linnros, B. **4** 36.24
Linsky, J.P. **6** 86.66
Lintvedt, R.L. **5** 77.24 **6** 77.21 **7** 84.27
Lip, H.C. **6** 35.18
Lippard, S.J. **1** 3.63 **2** 73.39, 77.39, 83.1, 83.183, 85.89, 86.69 **3** 72.34, 73.53, 85.31, 86.67, 86.92 **4** 71.69, 80.10, 85.22 **5** 71.18, 72.32, 73.10, 76.47, 80.5 **6** 71.62, 76.26, 76.35, 85.37, 85.40, 86.73 **7** 76.71, 86.114
Lippert, E.L. **2** 77.9
Lipscomb, W.N. **1** 3.1, 3.3, 4.9, 5.4, 5.11, 5.12, 7.19, 19.21, 19.22, 19.26, 22.17, 30.22, 39.11, 45.65, 52.14, 53.39, 54.26, 58.3, 58.8, 58.103, 58.125 **2** 62.1, 62.2, 62.3, 62.7, 62.9, 62.10, 62.11, 62.13, 62.15, 62.22, 62.23, 62.24, 62.25, 62.26, 62.31, 62.36, 62.37, 62.44, 64.14, 75.5, 75.19, 75.29, 85.45 **3** 3.26, 3.27, 46.1, 62.5, 64.26, 64.27 **4** 3.8 **5** 3.10 **6** 38.29, 62.1, 62.8 **7** 50.19
Liquori, A.M. **1** 13.30, 33.1, 44.67, 59.8, 59.20 **2** 60.97, 60.110, 60.111, 60.112
Lis, T. **5** 76.3, 76.7 **7** 48.7, 81.25
Lisy, J.M. **7** 35.13, 50.9, 50.13, 71.137
Little, J.L. **2** 62.39
Little, R.G. **2** 85.31 **3** 60.44 **4** 7.5, 81.60, 83.43 **5** 71.26, 71.40, 83.33 **6** 49.19, 81.44 **7** 49.5, 49.9, 60.23, 60.24, 71.92
Littleton, C.D. **1** 45.29
Liu, C.F. **3** 76.8, 82.7
Liu, C.S. **4** 63.2 **6** 67.4
Liu, S.J. **3** 50.24
Livingstone, S.E. **6** 77.1
Llabres, G. **5** 70.7 **6** 70.2
Llaguno, E.C. **4** 39.14, 39.26 **5** 39.27, 40.12, 42.4
Llewellyn, F.J. **1** 1.35, 2.26, 35.11, 45.11, 45.33 **2** 76.22, 78.26
Lloyd, D. **4** 65.13
Lloyd, D.J. **4** 24.3 **7** 86.18, 86.59
Lloyd, H.A. **7** 32.39
Lloyd, M.K. **7** 71.70, 75.23
Lo, F.Y. **7** 71.8, 73.1, 73.16
Loader, P.L. **2** 80.32, 80.33
Lobanova, G.M. **1** 19.42 **7** 17.8
Lobeeva, T.S. **6** 81.45
Lobkovskii, E.B. **5** 68.10 **7** 3.19
Lobry, M. **7** 32.18
Locchi, S. **1** 36.7, 36.8, 36.11, 36.29 **4** 32.27, 36.16, 36.25, 40.16 **5** 40.17
Lock, C.J.L. **2** 76.25 **3** 81.31, 81.44 **4** 83.110 **5** 77.18 **6** 77.2, 84.32 **7** 77.13
Locke, H.O. **1** 45.40
Lockenhoff, H.-D. **7** 16.4
Lockhart, R.W. **7** 37.9
Loehlin, J.H. **3** 83.34
Loescher, B.R. **3** 81.54
Lofving, I. **7** 62.2
Logan, N. **3** 65.8
Loghry, R.A. **7** 26.6, 26.7
Logiudice, F. **1** 9.2
Logiudice, G. **1** 29.19
Logvinenko, V.A. **5** 76.32 **7** 67.5
Lohaus, G. **4** 42.4
Lohmeyer, F.G. **6** 66.5
Lohn, J. **3** 81.5 **5** 81.15 **7** 81.8, 81.9
Lohr, L.L. **1** 4.9
Lohs, K. **3** 64.7
Loiseleur, H. **4** 81.34, 81.38 **5** 67.8, 81.42, 81.45 **6** 67.10, 81.33, 81.34 **7** 81.35
Lokensgard, D. **7** 58.20
Loliger, P. **3** 30.5
Lomborg, J.G. **4** 29.8
Lomer, T.R. **1** 1.137, 2.96, 2.97, 21.13 **4** 81.44 **7** 81.37, 81.39
Long Junior, L. **3** 45.44
Long, G.G. **6** 66.6
Long, G.J. **6** 83.78 **7** 83.79
Long, H.A. **1** 48.16 **2** 82.46
Long, J.R. **7** 85.14
Long, L. **1** 41.10
Long, R.E. **1** 7.29, 20.3
Longchambon, F. **5** 17.11 **6** 17.7 **7** 45.22, 45.23
Longoni, G. **6** 64.47, 64.48, 65.9, 65.10
Longoni, P.C.G. **6** 3.10
Longworth, J.W. **1** 44.72
Lons, J. **5** 64.15
Lonsdale, K. **1** 3.77, 3.84, 19.59, 26.19, 26.20, 26.21, 26.22, 26.23, 31.25, 31.26, 33.11
Look, W. **2** 65.35
Lopez-Castro, A. **1** 8.46, 11.41 **2** 79.23 **5** 33.32, 39.21 **6** 38.7, 38.11 **7** 45.42, 45.44
Lorenz, H. **7** 71.3, 71.6, 71.14, 73.39, 73.40
Lorenz, H.P. **6** 44.6, 44.7
Losman, D. **4** 58.8 **6** 53.12, 58.6, 76.46 **7** 53.13, 54.4
Lott, J.W. **5** 62.3, 62.10 **7** 38.8, 62.5
Lotter, H. **7** 38.17
Loub, J. **4** 81.22

Louis, R. **6** 85.18, 85.35 **7** 85.16
Lourens, G.J. **5** 45.49
Lovas, G. **7** 48.37
Love, R.A. **5** 62.6 **7** 86.13, 86.14
Love, W.E. **1** 2.80, 2.81, 44.20 **2** 64.12
Lovell, F.M. **1** 32.28 **3** 21.13, 35.14, 50.2 **4** 25.7, 35.15, 35.16, 41.24
Low, B.W. **5** 48.65, 48.66 **7** 48.70, 48.71
Low, J.N. **4** 45.16 **7** 26.5, 30.4
Low, M. **1** 21.28
Lowenstein, M.Z. **1** 39.5
Lu, C.-S. **1** 48.17 **2** 60.8
Lu, C.T. **1** 51.2, 51.3, 51.7 **6** 35.28, 48.22 **7** 10.6, 10.7
Lu, T.H. **6** 38.35
Luber, J.R. **4** 77.24
Luca, P.de **6** 51.44
Lucano, A. **6** 51.5
Luccarelli, A. **5** 86.80
Luche, J.L. **5** 41.50
Ludi, A. **3** 76.49
Ludin, A. **3** 46.8
Ludwig, M.L. **4** 31.2
Luger, P. **4** 21.2, 21.3, 71.67 **5** 29.1, 35.31, 45.34 **6** 45.5 **7** 29.12, 45.29, 45.30, 45.43, 45.54, 69.29
Lugli, G. **4** 74.1
Lugtenburg, J. **5** 21.8
Luhan, P.A. **4** 64.13 **5** 20.6, 53.18, 58.12, 58.20, 59.16 **6** 53.15 **7** 51.21, 51.22
Luk'yanova, N.A. **3** 76.67
Lukaszewicz, K. **6** 48.3, 48.4 **7** 48.7
Lumme, P. **3** 81.26
Lund, E.W. **1** 33.50
Lundberg, B.K.S. **2** 83.41 **4** 83.16, 83.93, 83.96 **5** 83.6, 83.7, 83.19, 83.37, 83.111, 84.6 **7** 83.24
Lundeland, M. **5** 70.8, 70.9
Lundell, S. **3** 76.35
Lunden, B.-M. **5** 1.36, 39.22 **6** 3.28
Lundgren, G. **1** 5.13 **3** 81.26 **5** 81.4
Lundgren, J.-O. **4** 11.1, 11.2, 15.2 **5** 11.4 **6** 11.1, 11.2 **7** 15.1
Lundin, P. **4** 11.2
Lundin, R.E. **7** 52.1
Lunsford, R.A. **7** 24.5
Lunt, E. **7** 33.6, 33.12
Luss, H.R. **4** 28.3 **5** 35.35, 35.36, 41.12 **7** 35.38
Lussier, R. **3** 83.28
Luth, H. **1** 32.21 **2** 71.10, 72.66, 79.36, 85.19 **3** 68.10, 69.12, 74.11 **4** 10.3, 69.11, 74.11 **5** 5.17
Luttke, W. **5** 39.30
Lutz, W.K. **4** 50.11
Luukkonen, E. **3** 76.29, 76.34, 76.56, 76.78 **7** 76.23, 76.27
Luxmoore, A.R. **2** 72.13
Luzzati, V. **1** 40.6, 41.12, 42.7, 47.19
Lychev, A.A. **7** 49.20, 49.21
Lydon, J.E. **2** 72.37, 83.191
Lynch, T.R. **4** 39.31
Lynton, H. **1** 39.7, 39.70 **2** 83.161, 83.193, 84.35 **3** 56.1 **4** 31.26, 83.51 **6** 31.27, 83.72 **7** 38.49
Lyssandtrae, N. **5** 70.4, 70.5
Ma, J.C.N. **7** 21.13
Ma, J.K.H. **5** 81.36
Ma, K.W. **3** 56.2
Maartmann-Moe, K. **1** 15.5, 15.6, 17.13, 17.18 **2** 70.3, 70.4 **4** 70.15 **5** 70.4, 70.5
Maas, G. **6** 64.25 **7** 64.39, 64.40
Mabbott, D.J. **7** 75.28
Mabbs, F.E. **2** 78.13, 78.14 **3** 78.3, 78.23 **4** 81.48 **6** 84.41 **7** 83.85, 85.36
Mabis, A.J. **1** 1.155, 59.22
Mabry, T.J. **7** 53.24, 59.14
Macaulay, E.W. **2** 65.30, 65.31, 65.32 **3** 51.26, 51.30, 65.11, 66.4 **4** 39.33, 62.14
Macchia, B. **4** 21.12 **5** 38.14
Macchia, F. **4** 21.12 **5** 38.14
Macdonald, A.C. **1** 31.17, 52.23, 58.69 **2** 73.86, 73.95
Macdonald, A.L. **4** 2.2, 2.3, 2.17, 2.18, 2.19 **5** 14.1, 18.1 **6** 4.1, 13.18, 48.44 **7** 59.5, 64.36
Macdonald, C.G. **6** 84.11
MacDonald, S.G.G. **1** 28.2, 30.8, 38.17, 38.20
MacDougall, M. **7** 76.2, 76.3
MacGillavry, C.H. **1** 1.31, 1.83, 1.95, 1.144, 12.14, 18.18, 18.19, 38.4, 53.27, 57.1, 57.2, 58.1 **2** 60.5, 64.4 **3** 1.37, 20.2, 37.4, 37.7 **6** 65.1, 65.2, 65.3
MacGregor, A.C. **6** 71.29, 73.2
Machajdik, D. **2** 83.118
MacInnes, D. **3** 77.27
Macintyre, W.M. **1** 6.1, 35.1, 44.19, 44.71, 47.5 **2** 84.36 **5** 31.43
Mackay, D. **5** 35.32
Mackay, I.R. **2** 60.156 **3** 51.44, 59.27, 60.42, 60.55
Mackay, K.J.H. **1** 26.41
Mackay, M. **3** 58.14
Mackay, M.F. **1** 58.16, 58.77, 59.29 **4** 16.2, 16.4, 16.5, 41.6 **5** 35.11, 58.24 **7** 45.24, 48.11, 58.30
MacKellar, F.A. **4** 34.3
MacKenzie, R.E. **5** 68.34
Mackenzie, R.K. **5** 22.3
MacLean, D.B. **4** 58.16 **6** 58.54
MacLean, H. **1** 59.61
MacNicol, D.D. **1** 22.6 **2** 61.16 **4** 61.1 **5** 22.3 **6** 42.10 **7** 39.14
Macomber, R.S. **6** 64.16
Maconochie, G.H. **1** 45.30
Macrosson, W.D.K. **1** 31.38
MacSween, D.R. **3** 83.62
Madan, O.P. **1** 46.7
Madaule-Aubry, F. **2** 83.198, 83.199, 84.6
Madden, D.P. **3** 83.128, 83.137
Madden, J.J. **3** 47.20 **5** 48.35, 48.36, 48.37
Maddox, H. **1** 20.3 **2** 60.43
Madura, P. **7** 49.15
Maeda, H. **7** 17.9
Maeda, K. **1** 41.31, 50.20, 50.21, 50.34 **5** 50.3 **7** 47.30
Maeda, S. **6** 53.7
Maeda, T. **5** 8.14

Maehr, H. **6** 16.23, 38.24
Maekawa, T. **4** 30.9
Magdesieva, N.N. **7** 39.23
Magee, R.J. **5** 81.38
Magennis, I.M. **3** 64.23
Maglio, G. **3** 72.33 **6** 72.1
Magnuson, V.R. **3** 67.12, 68.15
Magnusson, B. **5** 41.13
Magomedova, N.S. **7** 8.10
Magon, L. **6** 81.9, 84.18 **7** 82.11, 86.74, 86.88
Mague, J.T. **2** 86.59 **3** 71.68, 86.87 **4** 71.86 **6** 71.20
Maguire, J.F. **6** 73.41
Mahajan, V. **6** 3.16
Maher, R.J. **7** 39.19, 39.20, 63.16
Mahmoud, M.M. **7** 60.5
Mahon, C. **4** 7.1, 7.2, 7.4
Maier, C.A. **3** 39.21, 50.23 **4** 27.7, 29.2 **5** 31.42, 73.43 **6** 44.25
Maier, T.O. **6** 76.50
Main, P. **3** 54.2 **7** 47.33, 51.46, 51.47
Maiorana, S. **7** 41.22
Mair, G.A. **2** 86.7
Mais, R.H.B. **2** 72.28, 73.33, 75.36, 83.168, 86.6, 86.8, 86.24, 86.43, 86.52, 86.55, 86.62, 86.65, 86.104, 86.105, 86.106 **3** 67.27, 72.30, 80.1, 80.2, 86.41
Maisonnat, A. **7** 86.25
Maisseu, J. **1** 19.25
Maisto, A. **6** 48.52
Maitland, R. **1** 30.25
Maitlis, P.M. **3** 71.54 **4** 71.51 **5** 71.105 **6** 71.55, 71.61, 75.28 **7** 71.43, 71.79, 71.101, 75.28
Majeste, R. **1** 3.44, 35.10, 35.29, 35.33, 35.37 **3** 51.2, 64.37 **7** 31.30
Majeste, R.J. **3** 83.108 **4** 13.19 **5** 84.4 **6** 79.14 **7** 38.54
Majid-Hamid, A. **4** 32.22
Majumdar, S.K. **1** 1.41
Mak, T.C.W. **1** 15.11, 16.37, 28.6, 28.11, 28.13 **2** 61.3, 61.4, 61.7, 61.10, 64.78, 71.3 **3** 60.21 **7** 21.13
Makarenko, G.V. **4** 12.4, 12.7 **5** 1.14 **7** 1.10, 12.3, 15.3
Makarov, Yu.V. **6** 71.7
Makhija, R.C. **5** 86.23 **6** 86.42
Maki, A.H. **3** 60.54
Makleit, S. **6** 58.19
Malament, D.S. **3** 9.17 **5** 9.5, 9.6, 9.7, 9.12
Maldonado, J. **7** 38.52
Malik, K.M.A. **6** 84.40
Malinovskii, S.T. **7** 83.74
Malinovskii, T.I. **2** 81.32, 83.91, 83.134, 83.135, 83.136 **3** 81.30, 82.9, 83.26 **5** 79.13 **7** 76.6, 76.32, 76.66, 78.20, 79.15, 79.17, 79.18, 79.19, 79.20, 79.25, 82.16, 83.20, 83.21, 83.32, 83.74, 83.84, 86.48
Mallard, D.J.H. **7** 3.14, 58.1
Mallikarjunan, M. **1** 48.45, 48.94 **3** 45.15, 45.16 **4** 48.26 **5** 48.34 **7** 48.20, 48.24
Mallikarjunan, N. **1** 20.14
Mallinson, P.R. **2** 64.65 **3** 62.17 **4** 62.15, 64.37 **5** 38.31, 64.45 **6** 67.20, 67.21 **7** 38.71, 38.72, 38.73, 39.14
Malmborg, T. **6** 81.17
Malmros, G. **6** 28.1 **7** 38.51
Malmstrom, M.C. **4** 65.8 **5** 41.34
Malmstrom, S.M.C. **6** 41.25
Malofsky, B.M. **1** 56.16
Malone, J.F. **3** 68.16, 68.17, 75.23 **5** 68.21, 68.33 **7** 71.26
Malta, V. **3** 1.36 **7** 5.19
Mamedov, Kh.S. **6** 84.5
Mamedov, S.D. **7** 33.51
Mammi, M. **1** 1.51, 3.85, 4.16, 33.75, 34.4, 35.9, 38.2, 38.3, 39.8, 39.9, 39.39, 39.45, 39.46, 39.62 **3** 16.12, 32.12 **5** 39.9, 42.2, 69.15, 69.25 **6** 3.35
Manassero, M. **2** 60.149 **3** 86.22, 86.99 **4** 86.57, 86.71 **5** 64.20, 84.32, 86.48, 86.85 **6** 71.86, 86.36, 86.69, 86.75 **7** 71.118, 86.31, 86.66
Manchand, P.S. **5** 59.21 **7** 54.7
Mancuso, R.W. **6** 45.29
Mandava, N. **3** 37.6
Mandel, G. **4** 5.2 **5** 27.2 **6** 81.2
Mandel, G.S. **4** 24.1
Mandel, N. **3** 65.2 **4** 51.14, 52.4, 52.5 **6** 51.18
Mandyczewsky, R. **3** 86.8
Maner, R.J. **3** 11.13
Manfreddoti, A.G. **4** 79.2
Manfredotti, A. **1** 39.41 **2** 79.31
Manfredotti, A.G. **3** 85.12 **4** 8.6, 9.5, 9.6, 35.13, 76.38, 79.3, 81.1, 83.1, 83.2 **5** 69.14, 71.27, 78.19, 83.16, 83.36, 83.38, 86.37, 86.39 **6** 40.5, 40.9, 58.27, 76.45, 78.4, 80.3 **7** 73.61, 73.62, 73.63, 80.4, 83.2
Mangia, A. **1** 41.15 **2** 79.26, 83.46, 85.2 **3** 16.4, 41.2, 76.1, 79.4, 79.14, 83.10 **4** 28.5, 59.28, 76.22, 79.12, 83.5, 83.62, 83.74, 85.14 **5** 41.41, 78.22, 79.26, 79.27, 81.53, 84.12 **6** 59.3, 69.21, 83.56, 83.68 **7** 40.5, 48.63, 84.6, 84.81
Mangini, A. **3** 39.39 **4** 39.27
Mangion, M.M. **6** 70.5, 70.6, 70.7, 70.8, 70.10
Mani, A. **7** 2.2, 3.4
Mani, N.V. **1** 13.56, 53.20 **2** 64.32, 64.64 **3** 64.29
Mann, B.E. **6** 71.13, 72.9, 73.13, 75.28 **7** 71.43
Mann, F.G. **2** 65.7, 65.24, 86.99 **3** 65.5, 86.80 **6** 65.1, 65.2, 65.3
Mannan, K.A.I.F.M. **2** 86.90
Mannering, G.J. **6** 58.28
Manning, A.R. **2** 73.110, 86.23, 86.60
Mano, K. **7** 20.2, 20.3
Manohar, H. **1** 13.56, 58.75 **5** 81.37, 84.9 **7** 81.23, 83.35, 83.36, 83.37, 84.17, 84.75
Manojlovic-Muir, L. **2** 81.84 **3** 73.2, 73.46, 73.52 **4** 71.19, 71.21, 71.81, 86.19, 86.26 **5** 64.47, 86.34 **6** 71.46, 73.61, 81.38 **7** 64.18, 71.75, 71.76, 71.95, 84.26, 86.21, 86.30

Manojlovic-Muir, L.J. **6** 82.7
Manojlovic, L. **1** 2.88, 14.6, 14.13
Manojlovic, L.M. **1** 17.39
Manoli, J.-M. **7** 75.20
Manor, P.C. **3** 75.21 **5** 61.7, 61.8, 61.9 **6** 45.52 **7** 47.31, 47.41, 47.42
Manzini, G. **4** 77.14
Marangoni, G. **5** 84.5, 86.21 **7** 81.22, 81.34, 84.48
March, F.C. **3** 83.147 **4** 64.44, 65.12, 65.14, 66.4, 66.9, 85.1 **5** 83.117 **7** 66.1, 66.2, 66.3, 86.100
Marchenko, E.I. **1** 3.29
Marchessault, R.H. **3** 45.36 **6** 45.50 **7** 45.34
Marchetti, L. **7** 40.23
Marconi, W. **4** 74.1
Marcotrigiano, G. **7** 79.24
Marecek, J.F. **7** 64.43
Marek, D. **1** 43.21
Margerum, D.W. **5** 76.12, 83.114
Margolis, N.V. **1** 7.3, 10.1, 10.4, 12.1 **3** 7.1, 12.3, 12.5, 12.6 **4** 12.4, 12.7 **5** 1.14 **7** 1.10, 12.2, 12.3, 15.3
Margrave, J.L. **6** 28.3
Margulis, T.N. **1** 16.30, 20.2, 20.9, 20.11, 20.21, 20.34, 28.3, 51.42 **2** 60.119, 64.23, 75.21 **3** 20.6, 20.13 **4** 20.2, 20.6 **5** 20.8, 51.42 **6** 58.36 **7** 20.13
Marianelli, R.S. **6** 49.16, 49.17, 77.8, 77.13 **7** 77.2
Mariani, C. **1** 5.28, 5.31, 9.13, 19.50 **2** 60.103, 80.17 **3** 19.11
Mariezcurrena, R.A. **5** 45.23, 62.4 **6** 71.91
Marignan, J. **4** 9.11
Mark Junior, H.B. **6** 83.73
Mark, W. **3** 81.26, 84.16
Markila, P.L. **7** 64.14
Marko, L. **6** 85.38
Markova, T.Yu. **7** 63.11
Markova, T.Yu. **7** 63.4, 63.8
Marks, T.J. **4** 75.6 **6** 73.9, 83.80 **7** 71.94, 83.83
Markwell, A.J. **6** 3.20, 3.21, 3.22, 71.15
Marlier, M. **4** 59.2 **6** 33.19
Marly, L. **3** 43.9
Maron, L. **5** 50.24
Marongiu, G. **1** 40.7, 40.8, 40.9, 40.17, 40.18 **3** 40.6, 76.46, 76.48 **4** 76.6, 78.5 **5** 76.5, 76.11, 76.38 **6** 76.7, 76.8, 83.8, 83.11 **7** 76.59, 83.63, 85.7, 85.10
Maroy, K. **1** 4.14 **2** 70.3, 70.4, 70.7, 70.11, 70.12, 70.16 **4** 70.6, 70.7 **5** 70.8, 70.9 **6** 76.4 **7** 70.3, 70.4
Marquez, R. **5** 33.32, 39.21 **6** 38.7, 38.11, 42.7 **7** 42.7, 45.19, 45.38, 45.42, 45.44
Marr III, H.E. **5** 41.37
Marraud, M. **4** 48.23, 48.43 **5** 48.61, 48.62 **6** 37.4, 48.38, 48.47, 48.53 **7** 48.50, 48.51
Marsau, P. **1** 19.39, 41.42, 41.45 **3** 41.17 **4** 41.28, 41.31 **5** 35.6, 41.40, 41.44, 41.45 **6** 38.4, 43.2 **7** 33.50, 33.55
Marsden, C.J. **5** 85.14
Marsh, R. **1** 11.59 **2** 76.56, 86.26 **5** 86.25, 86.26
Marsh, R.A. **5** 72.17
Marsh, R.E. **1** 1.42, 11.52, 11.63, 11.64, 13.27, 13.39, 13.40, 13.41, 20.4, 20.6, 20.24, 33.12, 39.20, 39.21, 40.32, 44.15, 44.22, 44.23, 44.46, 44.62, 47.11, 48.5, 48.14, 48.23, 48.70, 48.91, 48.92 **2** 60.71, 70.20, 73.8, 73.43, 73.78, 76.81, 78.15, 82.10 **3** 44.22, 45.27, 47.19, 47.23, 76.17, 76.19, 77.9, 79.13, 86.77 **4** 29.7, 35.24, 44.8, 44.12, 76.28, 76.30, 78.15, 86.39, 86.40 **5** 2.10, 28.1, 44.14, 47.12, 47.17, 48.54, 58.10 **6** 13.8, 13.9, 13.11, 13.12, 16.6, 24.6, 28.25, 32.10, 41.12, 44.15, 47.8, 59.17, 60.22, 83.31 **7** 24.6, 33.23, 47.5, 48.8, 48.9, 48.54
Marsh, W.C. **3** 64.3, 64.9, 64.65, 83.127 **4** 35.8, 51.13, 51.22, 51.32, 58.4, 58.5, 85.21 **5** 20.18, 56.10, 56.11 **6** 64.42 **7** 50.32, 51.52, 54.2, 54.6, 54.14, 54.15
Marshak, M.L. **1** 58.73
Marshall, G.E. **5** 46.2
Marshall, G.R. **6** 48.83
Marsham, D.F. **1** 3.38, 3.39 **2** 83.215
Marsich, N. **5** 86.67 **6** 71.95
Marstokk, K.-M. **2** 60.22
Martelli, A.F. **3** 1.36
Martelli, G. **3** 32.13
Marten, D.F. **5** 71.10
Martin Junior, D.S. **5** 76.17
Martin Junior, L.L. **3** 63.16
Martin-Smith, M. **1** 39.50
Martin, D.G. **5** 50.1
Martin, D.W. **6** 78.12
Martin, G. **3** 73.53
Martin, H.A. **2** 72.44
Martin, J. **1** 31.30
Martin, J.C. **3** 17.11 **5** 11.16 **7** 42.14, 42.15
Martin, J.L. **1** 3.83 **6** 85.31, 85.33, 85.34
Martin, J.M. **5** 80.21, 80.22
Martin, J.O. **2** 83.79
Martin, L.L. **1** 38.26 **4** 50.2, 86.8 **5** 82.6, 82.8
Martin, M.A. **7** 44.10, 83.7, 85.12
Martin, R.-P. **2** 82.39
Martin, R.L. **3** 49.5, 85.24 **6** 80.6
Martin, T.I. **6** 58.54
Martinengo, S. **5** 71.35 **6** 3.10, 64.47, 64.48, 65.9, 65.10
Martinez-Carrera, S. **1** 13.44, 32.11, 43.9, 59.46 **3** 13.13 **4** 13.5, 77.10 **5** 35.20 **6** 59.20 **7** 48.6, 48.73
Martinez-Garcia, M.L. **7** 40.12, 40.13
Martinez-Ripoll, M. **6** 44.6, 44.7 **7** 48.73
Martinez, J. **1** 41.45, 43.10
Martinez, N. **4** 73.6
Martuscelli, E. **1** 1.69, 1.78, 1.93, 1.101, 1.102, 1.104, 1.110, 1.120, 1.121, 38.18 **3** 1.32, 75.36 **7** 1.17
Martynenko, L.I. **2** 81.43 **3** 68.9, 76.67, 82.9 **5** 76.34 **6** 76.22 **7** 76.43, 76.44, 76.46, 82.1
Martynov, V.P. **4** 72.5, 72.6

Marumo, F. **3** 1.23, 3.25, 35.4, 50.21, 76.54, 76.64, 76.71, 76.82, 82.18 **4** 20.1, 29.4, 50.4, 54.13, 60.32, 76.2, 76.20, 76.21, 76.25, 76.40, 81.31, 83.68 **5** 60.7, 65.3, 76.20, 76.29, 76.48, 83.60 **6** 18.1, 76.27, 82.16, 83.16, 83.35
Marusic, R. **4** 50.13
Maruyama, H. **2** 67.18, 67.19
Marx, G. **3** 71.47 **7** 71.57
Maryanoff, C.A. **7** 64.62
Marzilli, L.G. **3** 76.74 **4** 76.17 **6** 76.16 **7** 78.11
Marzilli, P.A. **2** 76.49, 82.54 **7** 78.11
Marzin, C. **3** 9.18, 9.20
Marzotto, A. **6** 44.24
Masaki, N. **1** 52.25, 58.86 **3** 58.13, 58.43, 58.52 **4** 54.15, 58.43, 59.21 **5** 17.20, 58.21, 58.37 **6** 54.8, 58.22, 58.25 **7** 56.6
Masakuni, M. **1** 3.18
Masamune, S. **4** 21.22, 28.12 **5** 31.16 **7** 20.20
Masamune, T. **4** 40.24 **7** 40.8
Mascarenhas, Y. **1** 43.6 **6** 84.42
Mascarenhas, Y.P. **5** 3.21
Masironi, L.Di M. **1** 33.76
Maskiewicz, R. **7** 85.18
Maslen, E.N. **1** 16.11, 16.12, 16.13, 16.17, 16.20, 16.21, 40.3, 44.43, 50.13, 50.16, 54.5, 54.14, 56.11, 56.19, 59.41, 59.44 **2** 72.35, 80.10, 80.14, 81.33 **3** 19.12, 19.13, 19.14, 60.40 **6** 19.7, 51.33, 80.9, 80.10 **7** 32.16, 75.9, 76.14, 77.1, 77.4, 83.82, 83.91, 83.93, 83.109, 86.3
Maslen, H.S. **1** 2.26 **6** 76.54
Mason, R. **1** 1.53, 23.11, 23.15, 26.31, 26.32, 28.18, 33.80, 36.33, 38.39 **2** 71.48, 72.16, 72.21, 72.40, 72.42, 72.59, 73.26, 73.29, 73.34, 73.35, 73.46, 73.67, 74.19, 75.8, 75.17, 75.23, 77.4, 77.62, 81.59, 82.28, 83.86, 83.178, 83.182, 86.18, 86.19, 86.28, 86.46, 86.80, 86.84, 86.89, 86.108 **3** 71.28, 71.63, 72.37, 72.38, 75.35, 82.5, 86.49, 86.70, 86.98 **4** 23.5, 71.16, 71.17, 71.43, 71.46, 71.49, 71.71, 71.73, 71.82, 71.89, 83.63 **5** 71.80, 71.95, 71.96, 71.99, 73.46, 75.30, 83.49, 86.51, 86.66, 86.78 **6** 39.3, 39.4, 39.16, 39.17, 46.14, 72.2, 74.5, 74.6, 75.22, 86.13, 86.25, 86.31, 86.63 **7** 71.60, 71.115, 73.64, 86.50, 86.100
Mason, S.A. **1** 49.12 **3** 8.9, 33.12, 33.13, 49.5
Massaux, M. **2** 83.2 **4** 83.4
Massey, A.G. **3** 75.16
Masters, C. **4** 86.14
Mastropaolo, D. **6** 69.9, 75.27, 82.24 **7** 5.17, 33.2, 41.29, 73.48
Masuda, Y. **7** 17.32
Masuko, A. **2** 83.105
Math, K.S. **3** 83.143 **4** 85.30
Math, V.B. **3** 49.13
Mather, G.G. **5** 86.1 **6** 69.10
Mathern, G. **4** 81.3, 81.19, 83.47, 83.48
Matheson, K.L. **7** 42.5, 42.6, 42.9
Mathew, M. **1** 31.33, 38.55 **2** 71.25, 83.45, 86.70 **3** 39.9, 59.3, 71.15, 78.5, 83.113, 83.137 **4** 38.19, 40.6, 60.2, 71.61, 72.29, 83.134, 84.13, 85.37 **5** 38.5, 50.8, 62.9, 71.90, 75.38, 75.46, 79.7, 79.20, 79.21, 83.109, 86.47, 86.61 **6** 41.2, 71.76, 75.31 **7** 13.16, 31.7, 71.113, 72.45, 86.51, 86.54, 86.57
Mathews, F.S. **1** 44.30, 44.39, 44.55 **2** 62.44, 75.5 **6** 48.83
Mathey, F. **6** 64.37
Mathieson, A.McL. **1** 1.151, 11.57, 45.31, 48.46, 48.47, 48.62, 48.93, 50.7, 51.44, 52.15, 58.16, 58.23, 58.27, 58.32, 58.65, 58.77, 58.81, 58.94, 59.16, 59.24, 59.29 **2** 82.42 **3** 26.3 **4** 16.2, 16.5, 58.35, 83.66 **5** 58.4, 58.7, 58.24 **6** 58.9 **7** 58.30
Mathieson, D.W. **1** 59.45
Mathieu, F. **6** 40.20
Mathisen, H. **1** 5.17, 5.18, 5.19, 5.22
Matkovic, B. **1** 3.7 **2** 77.47, 77.49, 84.49 **4** 81.11 **7** 3.5
Matkovic, P. **7** 76.4, 76.9, 83.17, 83.31
Matsuda, H. **1** 17.37, 54.9, 58.35
Matsuda, S. **2** 69.31
Matsuda, Y. **6** 36.11
Matsue, H. **7** 40.8
Matsueda, S. **3** 53.18
Matsui, K. **6** 36.10
Matsui, M. **5** 33.39, 33.40, 33.41, 36.8, 36.9, 50.14 **6** 76.33 **7** 76.8
Matsumaru, H. **1** 48.73 **3** 44.42
Matsumoto, K. **2** 76.10 **3** 76.23, 76.28, 76.52, 76.66 **4** 76.3, 76.10 **5** 76.4, 82.18 **6** 76.10, 77.9
Matsumoto, M. **5** 71.67 **6** 5.1, 5.3, 86.26 **7** 78.3
Matsumoto, T. **1** 58.120 **5** 54.4, 54.5, 54.6 **6** 28.21, 28.26 **7** 28.6, 50.11
Matsuo, A. **7** 53.27
Matsuo, K. **7** 58.28
Matsushima, M. **5** 32.8 **7** 32.1, 33.31
Matsuura, S. **7** 36.1
Matsuura, T. **3** 59.30
Matsuura, Y. **3** 73.33
Matsuzaki, T. **3** 48.33 **5** 67.5 **6** 48.87 **7** 29.3
Mattes, R. **2** 72.15, 81.11, 81.18 **5** 20.1, 62.1, 63.2, 64.15
Mattew, M. **3** 79.12
Matthews, B.W. **1** 41.24, 43.31 **3** 36.29 **5** 3.16 **6** 35.26, 37.10 **7** 32.36
Matthews, D.A. **4** 38.2, 38.3, 38.4 **6** 38.3
Matthews, J.D. **3** 73.42
Matthews, R.W. **3** 81.36 **4** 81.36
Mattia, C. **7** 82.22, 82.23
Mattson, B.M. **7** 80.19
Matzat, E. **4** 35.3
Matzuzaki, T. **6** 47.16
Mauer, F.A. **2** 83.106, 83.107
Maurin, M. **5** 64.2
Mautner, H.G. **1** 44.11, 44.12 **3** 3.29, 3.30, 3.34
Maver, F.A. **2** 62.35

Maverick, E. **4** 12.9, 20.11, 20.13 **5** 38.32 **7** 31.22
Maw Junior, C.E. **3** 83.93
Mawby, A. **4** 81.39 **5** 71.55, 73.14, 85.29, 86.40 **6** 75.17
Mawby, R.J. **2** 75.52 **6** 75.17
Maxwell, I.E. **2** 82.54 **3** 76.27, 76.38, 76.74, 76.75 **4** 76.23, 76.24, 83.83, 83.84 **6** 24.3, 76.16
May, C.J. **7** 71.123
May, F.L. **6** 42.4
Maye, C.A.de **5** 21.8
Mayer, E.F. **1** 49.2
Mayer, R.M. **1** 35.38
Mayerle, J.J. **6** 86.73 **7** 85.28, 85.46
Mayers, R.F. **1** 2.44
Mays, M.J. **4** 71.35
Mayweg, V.P. **2** 85.34
Maze-Baudet, M. **5** 17.6, 17.8, 17.17
Maze, M. **1** 17.43
Mazid, M.A. **7** 17.33, 45.6
Mazumdar, S.K. **1** 48.19, 48.71 **3** 48.21, 48.29 **6** 48.34
Mazur, Y. **5** 51.36
Mazus, M.D. **7** 76.32, 76.66
Mazwell, I.E. **4** 76.17
Mazza, F. **3** 33.45, 44.14, 44.17, 44.19, 44.25 **4** 32.22, 35.26, 36.34 **5** 50.13 **6** 33.52, 33.56, 38.13 **7** 48.29, 48.67, 50.7
Mazza, M.C. **6** 84.43 **7** 72.41, 72.42
Mazzarella, L. **1** 1.102, 37.10 **2** 60.97 **3** 33.23 **5** 36.15, 36.16, 50.4 **6** 1.13, 51.44 **7** 1.17, 36.12, 82.22, 82.23
Mazzi, F. **1** 11.18 **3** 11.5
Mazzi, U. **6** 86.47 **7** 86.74, 86.88
Mazzocchi, P.H. **5** 37.5
Mazzochi, P.H. **4** 37.9
Mazzone, G. **2** 80.20
McAdam, A. **4** 2.12, 2.13, 2.14, 72.22, 72.23, 72.24, 72.30
McAfee, E.R. **6** 35.8 **7** 33.20
McAlister, J. **6** 46.4 **7** 36.1
McArdle, J.V. **6** 86.54 **7** 71.111
McBride, D.W. **3** 73.7
McBride, H.D. **2** 68.30
McBride, J.M. **3** 9.17 **5** 9.5, 9.6, 9.7, 9.12, 13.11
McCallum, G.H. **1** 48.21
McCammon, C.S. **1** 13.52
McCandlish, L.E. **1** 3.69
McCann, S.W. **7** 80.5
McCapra, F. **1** 54.3, 54.4
McCarley, R.E. **4** 3.7 **6** 3.13 **7** 83.18
McCarty, C.G. **3** 8.14
McClelland, B.W. **4** 45.2 **5** 68.20 **6** 38.5
McCleverty, J.A. **3** 85.7, 85.20 **4** 83.131 **6** 71.13, 72.9, 73.13 **7** 73.13
McClure, M.D. **6** 75.30
McClure, R.J. **1** 53.28 **3** 53.28, 58.57 **5** 44.1, 44.2, 58.42 **6** 41.17, 43.1
McClure, W.O. **7** 17.7
McConnell, J.F. **1** 9.44, 44.16, 52.15, 52.16 **2** 78.6, 83.157, 85.32, 85.57 **3** 82.1 **4** 21.9, 21.10, 85.5 **5** 21.17 **6** 1.27, 13.17, 45.44, 45.45, 84.11
McConway, J.C. **6** 74.15
McCorkindale, N.J. **1** 59.53
McCormack, J.A. **4** 42.5
McCormick, B.J. **6** 80.18
McCormick, I.R.N. **6** 76.23, 76.24, 76.34
McCrae, L.G. **4** 65.5
McCredie, R.S. **3** 44.37
McCrindle, R. **5** 56.10, 56.11 **7** 54.2, 56.8
McCullough Junior, J.D. **4** 17.12, 17.13
McCullough, J.D. **1** 3.24, 11.20, 11.21, 11.22, 11.32, 11.35, 11.52, 11.53, 11.54, 11.63, 11.64, 39.21, 42.6 **2** 60.42, 60.43, 60.52, 60.55, 60.56, 60.57, 70.1, 70.19, 70.20, 70.22 **3** 39.2, 39.30, 70.2 **5** 70.2 **6** 70.1, 70.3 **7** 70.7
McCullough, J.J. **5** 28.15 **6** 28.22, 64.1
McDaniel, M.C. **1** 41.38
McDonald, A.N. **5** 35.25
McDonald, J. **4** 71.72
McDonald, L.J. **7** 38.70
McDonald, T.R.R. **2** 81.20 **4** 68.14
McDonald, W.S. **1** 41.35 **2** 68.32 **3** 11.10, 68.16, 68.17, 75.23 **4** 3.17, 46.2, 68.14, 71.13, 71.33, 86.28 **5** 68.21, 68.33, 71.86, 71.87, 71.88 **6** 86.7 **7** 20.23, 71.99
McDowell, J.J.H. **1** 41.40 **3** 41.20
McDowell, T.L. **6** 38.22
McEachan, C.E. **1** 53.29, 54.10
McEwen, G.K. **4** 64.4
McEwen, R.S. **2** 79.8, 84.67, 85.48, 85.75 **3** 85.39 **5** 81.55, 81.56 **6** 59.12 **7** 31.31, 53.10, 53.16, 64.45
McEwen, W.E. **2** 66.19, 66.21
McFadden, D. **6** 84.41
McFadden, D.L. **6** 71.9 **7** 83.85
McGahren, W.J. **1** 59.15
McGandy, E.L. **2** 62.5 **3** 32.11, 32.23, 32.34, 38.10, 58.23, 73.54 **4** 32.2, 48.21, 48.25 **5** 48.35, 48.36, 48.37
McGeachin, H.McD. **1** 45.34
McGeachin, S.G. **1** 50.43
McGhie, J.F. **6** 56.5
McGinnety, J. **2** 86.87
McGinnety, J.A. **2** 86.79, 86.98, 86.112 **3** 64.46, 86.89, 86.90 **4** 84.9 **6** 71.56, 72.29 **7** 71.125, 72.39, 75.41
McGrath, B.G. **6** 83.36
McGregor, A. **3** 73.43
McGregor, D.R. **1** 11.40, 14.11
McGregor, K.T. **6** 83.51
McIntosh, A.O. **1** 30.13
McIntosh, C.L. **3** 38.8
McKechnie, J. **2** 65.51
McKechnie, J.S. **1** 31.7, 32.45, 34.5, 51.24 **2** 65.46, 75.16, 75.20, 75.44, 75.78 **3** 12.11, 51.11, 59.24 **5** 73.43
McKenna, J. **1** 33.80
McKennis, J.S. **7** 71.33, 71.34
McKenzie, E.D. **2** 82.28 **3** 78.7, 78.24 **4** 78.13, 83.131 **5** 76.45, 83.64, 83.65 **6** 83.25, 83.44 **7** 71.83, 77.11, 78.17
McKenzie, T.C. **6** 28.23 **7** 39.39, 40.11, 75.17
McKillop, T.F.W. **3** 53.24 **4** 53.3
McKinley, S.V. **3** 69.35

McKinney, R.J. **5** 71.98
McKinnon, A.J. **2** 77.16, 78.4, 78.5, 78.17, 78.18, 78.46
McLachlan Junior, D. **1** 19.65
McLachlan, V.N. **3** 78.23 **6** 84.41
McLaughlin, G.M. **3** 50.19, 68.7, 68.8 **4** 54.11, 54.14 **5** 68.14, 68.15, 68.16 **6** 49.8, 49.13, 68.10, 69.10, 71.39 **7** 71.41, 71.124
McLean, S. **3** 36.18, 36.25
McLean, W.J. **2** 61.6, 61.8
McLennan, T.J. **7** 86.6
McLeod, N.J. **5** 83.120, 83.121, 83.122
McMeeking, J. **7** 71.124
McMillan, J.A. **6** 53.9, 53.16 **7** 17.7
McMillan, J.A.S. **1** 50.25, 58.2
McMillan, R.S. **7** 85.6
McMullan, R.K. **1** 13.55 **2** 61.1, 61.3, 61.5, 61.7, 61.11, 61.13, 61.14 **3** 61.2 **4** 44.22 **6** 45.51, 61.7 **7** 61.6
McNally, B.V. **7** 30.9
McOmie, J.F.W. **6** 11.16, 27.15, 40.8
McPartlin, E.M. **2** 85.50, 85.66 **4** 83.25
McPartlin, M. **2** 72.40, 86.84, 86.108 **3** 71.28 **6** 3.20, 3.21, 3.22, 71.15, 72.8, 84.21, 85.23 **7** 83.98
McPhail, A.T. **1** 13.45, 16.4, 50.39, 51.48, 53.8, 53.11, 53.22, 53.29, 53.37, 54.4, 54.10, 56.18, 58.67, 58.71, 58.87, 58.121, 59.70 **2** 60.118, 73.16, 73.44, 73.99, 74.6, 74.7, 74.14, 75.68, 84.30, 84.37, 85.12 **3** 1.29, 1.30, 3.38, 3.48, 45.48, 53.21, 53.28, 53.29, 58.20, 59.7, 59.17, 59.34, 68.12, 73.56, 78.3, 78.23, 86.1 **4** 28.10, 37.10, 51.40, 51.41, 58.32, 59.19, 59.26, 64.13, 64.14, 64.24, 65.9, 81.48, 81.55, 86.17 **5** 20.6, 27.14, 53.15, 53.18, 58.12, 58.20, 59.16, 71.13, 86.41, 86.42 **6** 39.10, 39.11, 53.15, 53.19, 59.5, 64.15, 71.9, 77.19, 79.17, 84.41 **7** 51.4, 51.21, 51.22, 53.5, 53.11, 53.17, 53.23, 53.25, 53.30, 64.9, 64.10, 64.16, 64.17, 83.85
McPherson Junior, A. **5** 76.12
McPherson, A.M. **6** 68.16 |
McPherson, W.G. **1** 3.12, 33.35, 33.44, 33.45
McQuillan, G.P. **3** 85.52
McQuillan, M.P. **6** 33.44
Meakin, P. **5** 83.35 **6** 86.2 **7** 71.38
Mealli, C. **3** 83.118, 83.121, 83.122, 83.123, 83.124, 83.125, 83.126 **4** 76.43, 76.44, 86.44 **5** 83.112 **6** 86.52 **7** 83.83, 84.43, 84.46, 86.115, 86.117
Mechlinski, W. **4** 50.17
Meck, R. **6** 53.15
Medlin, E.H. **3** 60.4, 60.5
Medrud, R.C. **1** 27.5 **2** 75.18
Meek, D.W. **3** 85.36 **4** 85.37 **5** 85.13 **6** 86.38, 86.39, 86.60, 86.61
Meer, H.van der **3** 65.6 **4** 20.7, 27.4, 36.4 **5** 16.6, 35.4, 36.28, 85.17
Meerssche, M.van **1** 5.27, 5.33, 10.14, 10.15, 10.16, 41.26, 41.27, 41.30, 41.46 **2** 60.23, 60.25, 60.26, 60.30, 60.32, 60.46, 60.64, 60.66, 60.109, 64.18, 64.19, 72.32, 72.54, 72.57, 73.38, 83.169 **3** 20.18, 60.17, 60.18, 67.3, 67.6, 71.17 **4** 20.10, 20.12, 24.19, 32.21, 32.23, 40.17, 48.4, 48.9, 48.11, 48.27, 51.25, 51.30, 51.31, 51.34, 83.39 **5** 16.20, 48.1 **6** 32.36, 36.9, 48.48 **7** 17.20, 32.24, 32.28, 32.29, 60.2, 71.135, 71.136
Meester, M.A.M. **3** 20.2 **4** 1.12 **5** 11.8
Meester, P.de **4** 83.108 **5** 81.39, 81.49, 83.24, 83.25 **6** 81.26, 83.23 **7** 83.33
Mehlsen-Sorensen, A. **6** 5.6
Mehring, M. **5** 81.7
Mehta, G. **7** 31.9
Meier, W.M. **3** 3.14, 3.15
Meij, R. **7** 71.129
Meinema, H.A. **6** 66.4
Meiners, J.H. **7** 86.16, 86.19
Meinnel, J. **1** 19.30, 19.31, 19.32 **6** 42.5
Meinwald, J. **4** 71.9 **7** 51.4, 51.21, 51.22, 75.6
Mel'nikov, V.V. **1** 7.3, 10.1, 10.4, 12.1 **3** 7.1 **4** 12.4, 12.7 **7** 1.10, 12.3, 15.3, 64.30
Mellinger, M. **6** 67.19
Mellor, I.P. **3** 3.43, 3.44 **4** 11.6, 39.24, 39.31
Melmed, K.M. **2** 86.69 **3** 86.67 **5** 73.10 **6** 86.73
Melsom, B.G. **3** 56.6
Melsom, B.G.A. **2** 76.15, 84.47, 84.48
Melson, G.A. **3** 73.3, 85.47 **5** 77.5 **6** 84.22 **7** 84.71
Melville, D. **1** 54.25
Melville, R.D. **1** 54.20
Menchetti, S. **4** 83.35 **5** 13.14
Menczel, G. **3** 9.4 **5** 9.8 **6** 41.6 **7** 63.3, 63.6
Mendel, H. **1** 48.26, 48.36
Menegus, F. **2** 68.9
Meney, J. **5** 51.42
Mercer, A. **7** 85.6, 86.15
Mercer, J.F.B. **5** 41.11
Mercer, L.F. **7** 84.62
Mercer, M. **3** 83.62 **4** 67.24, 75.23 **5** 38.25 **6** 38.36, 67.15 **7** 86.63
Merenyi, R. **3** 41.10
Meresse, A. **6** 18.2 **7** 24.2, 24.3
Mergehenn, R. **7** 84.35, 84.64
Meriani, S. **3** 85.32, 85.33, 85.40
Merk, W. **7** 71.33, 71.34
Merkel, W. **7** 41.26
Merle, A. **7** 86.6
Merlino, S. **1** 35.16, 35.18 **2** 80.7, 80.30 **4** 21.12, 72.11, 80.2 **5** 38.14, 48.75
Merrell, P.H. **7** 83.99
Merritt Junior, L.L. **1** 32.13, 33.61 **2** 83.162 **3** 35.2, 35.7 **7** 83.65
Merritt, L.L. **1** 41.7 **2** 83.197, 84.44
Mertz, K. **7** 68.6, 68.7
Merz, L. **7** 84.35
Messager, J.-C. **1** 19.30, 19.31 **4** 19.8, 19.9

Messager, J.C. **3** 32.30 **6** 38.28, 40.13
Messmer, G.G. **2** 86.4, 86.10, 86.11 **3** 40.5, 84.23, 86.62
Mester, Z.C. **7** 81.27, 85.3, 85.9
Mestvedt, B. **1** 33.62
Mesureur, C. **2** 60.14
Meszaros, Z. **7** 35.15, 35.27
Metras, F. **4** 21.4 **5** 21.1
Metras, J.C. **1** 25.2
Metz, B. **4** 67.12, 67.20 **5** 67.12, 67.13, 67.15, 69.18 **6** 67.12, 67.13, 67.14, 85.18 **7** 40.17, 40.19, 40.20, 85.21
Metzger, R.M. **6** 41.28
Meulemans, R. **3** 67.3, 67.6 **4** 48.9, 48.11, 48.27
Meunier-Piret, J. **2** 60.109, 72.32, 73.38 **3** 71.17 **4** 20.10, 32.23, 40.17 **5** 16.20, 48.1 **7** 32.28, 71.135, 71.136, 72.3, 72.4
Meunier, B. **6** 73.63
Meunier, J. **1** 41.46
Meurs, F.van **7** 74.3
Meuthen, B. **1** 17.10 **2** 78.20
Meyer Junior, E. **4** 49.19
Meyer Junior, E.F. **2** 83.189 **4** 49.2, 49.15 **5** 49.2 **6** 49.3, 49.4, 49.5
Meyer, C.D. **6** 71.82
Meyer, E. **5** 49.10
Meyer, E.F. **1** 34.11 **7** 27.4
Meyer, H.-J. **1** 3.19
Meyer, W.L. **6** 48.82
Meyerhoffer, A. **1** 37.26 **3** 37.12
Meyers, A.I. **7** 38.54
Meyers, E.A. **1** 3.12, 3.44, 33.21, 33.24, 33.25, 33.26, 33.35, 33.44, 33.45 **2** 60.150 **3** 8.8, 65.3, 83.108 **4** 13.19, 19.10, 65.3 **5** 33.21, 33.24 **6** 31.20, 60.24, 70.5, 70.6, 70.7, 70.8, 70.10 **7** 65.1
Meyers, M. **1** 11.7, 11.68
Mez, H.-C. **1** 15.26 **5** 51.24, 83.94, 83.95 **6** 20.6, 37.7, 48.34 **7** 21.2, 28.3, 28.7, 38.24, 48.23
Mez, H.C. **2** 75.53 **3** 44.6
Michalov, J. **2** 81.78
Michaud, C. **7** 16.5
Michaud, M.-C. **7** 84.58
Michejda, C.J. **6** 11.16, 27.15, 40.8 **7** 71.42, 75.21
Michel, A. **5** 48.70
Michel, P. **5** 17.2, 17.4 **7** 17.2
Michl, R.J. **4** 83.90
Micovic, I.V. **6** 33.32
Midgley, J.M. **3** 51.26, 51.30 **7** 51.52
Midollini, S. **7** 86.84, 86.87, 86.115, 86.117
Miesel, S.L. **2** 65.46
Migchelsen, T. **1** 3.61, 3.62 **4** 65.10 **5** 65.6
Migdal, S. **4** 15.6, 16.6
Mighell, A.D. **1** 38.14 **2** 83.102, 83.106, 83.107, 83.165 **3** 19.4, 83.6, 83.95 **4** 83.91, 83.92, 83.94, 83.95, 83.102, 84.11
Mignucci, G. **7** 48.67
Mikalsen, O. **7** 70.6
Mikawa, H. **3** 36.9
Mikhailov, I.E. **7** 71.36
Mikhailov, Yu. **2** 79.30
Mikhno, O.A. **5** 26.1, 40.26 **6** 26.1, 40.16 **7** 41.17
Miki, K. **7** 71.62, 71.63
Miki, M. **5** 71.67
Miklos, D. **4** 50.13 **5** 83.67
Mikolajczyk, M. **6** 64.3
Milberg, M.E. **1** 5.12
Milburn, G.H.W. **2** 60.145, 68.19 **3** 67.11, 67.24, 82.17, 82.25 **5** 82.3, 82.9 **6** 39.3, 39.4
Milburn, H. **6** 78.17 **7** 77.11
Miler-Srenger, E. **3** 41.16 **5** 41.43
Miles, M.G. **2** 71.18 **4** 83.26
Milinovic, I. **7** 39.11
Milje, L.M. **5** 41.30
Milkova, L.N. **7** 79.18, 79.19, 79.20
Millar, I.T. **2** 86.99 **3** 86.80
Millar, K. **6** 64.40
Milledge, H.J. **1** 3.77, 3.84, 19.9, 19.10, 19.11, 26.19, 26.20, 26.21, 26.22, 26.23, 31.25, 31.26 **3** 69.2 **7** 83.28, 83.29
Miller, A. **1** 21.13
Miller, C.K. **1** 49.22, 49.23, 49.28, 49.31
Miller, D.P. **4** 67.28 **5** 16.18 **6** 16.19, 16.20, 16.21
Miller, G.A. **5** 69.16
Miller, J. **4** 71.20
Miller, J.A. **6** 64.55
Miller, J.S. **5** 85.36 **6** 83.84 **7** 71.25
Miller, P.T. **5** 84.23, 84.29
Miller, R.A.L. **1** 41.10
Miller, R.C. **6** 28.22
Miller, R.S. **7** 13.1
Miller, R.W. **6** 59.5
Miller, V.R. **5** 62.9
Millington, D. **5** 64.13 **6** 64.11 **7** 86.21
Mills, H.H. **1** 2.42, 2.91, 8.25, 50.43, 58.73 **2** 60.16, 61.16, 73.59 **5** 22.3, 64.3, 64.36
Mills, J.C. **2** 71.4, 71.5, 71.31
Mills, J.F.D. **1** 38.61, 38.62, 58.84
Mills, O.S. **2** 71.14, 71.24, 71.32, 71.46, 72.14, 72.33, 72.34, 72.52, 72.64, 73.42, 73.51, 73.54, 73.74, 73.79, 73.92, 74.20, 75.28, 75.31, 75.46, 75.53, 83.170, 83.209, 83.212, 83.214, 86.47 **3** 31.11, 71.30, 71.32, 71.52, 75.3, 75.11, 75.13, 75.25, 75.27, 75.28, 83.103 **4** 22.2, 38.12, 71.14, 72.20, 75.1, 75.9, 75.16, 78.2 **5** 22.1, 31.9, 71.15, 71.68, 71.69, 71.81, 75.3, 83.69 **6** 31.11, 31.19, 71.36, 72.16
Milne, G.W.A. **7** 31.19
Milone, L. **4** 71.27 **7** 71.81
Milton, P.A. **6** 68.4
Mimum, A.S. **5** 31.29
Minacheva, L.Kh. **7** 76.15
Minacheva, L.Kh. **3** 76.15 **4** 76.7, 76.27 **5** 76.8
Minale, L. **6** 51.44
Minami, T. **3** 41.23 **4** 35.39 **5** 42.13
Minasjanc, M.X. **4** 72.14
Minasyants, M.Kh. **2** 72.7, 72.19
Minasyants, M.Kh. **7** 72.26, 72.27
Minato, H. **1** 27.12

Mingos, D.M.P. 3 71.66, 86.60, 86.63 4 86.59
Minichelli, D. 3 78.14 4 71.40
Minkin, J.A. 1 1.79, 2.76, 2.78, 2.79, 2.80, 2.81, 4.15, 44.20 4 2.21, 36.21 5 2.26 7 79.16
Minkin, V.I. 7 71.36
Minobe, M. 3 60.13
Miocquie, M. 7 38.52
Miravitlles, C. 5 9.1, 16.8 6 16.7, 16.10, 17.10, 17.12, 17.13, 35.12 7 1.8
Mironov, Yu.I. 7 3.20
Miscovic, J.D. 1 16.48, 16.50
Mishima, H. 7 59.24
Miskin, B.K. 7 64.31
Mislow, K. 1 52.26 4 52.4, 52.5 5 19.20, 31.36 6 62.11 7 19.16, 64.62
Misner, B. 6 85.16
Misra, R. 4 51.40, 51.41
Missavage, R.J. 5 37.3, 81.29
Misumi, S. 5 31.46
Mitamura, S. 7 11.17
Mitchell, D. 1 13.51
Mitchell, D.J. 1 11.39
Mitchell, J. 3 60.6
Mitchell, J.W. 7 33.6, 33.12
Mitchell, T.P. 3 76.77
Mitrofanova, N.D. 2 81.43 5 82.19 7 82.7, 82.9, 82.10
Mitscher, L.A. 1 59.15
Mitschler, A. 2 84.3 4 33.8, 33.9 6 73.63 7 80.9, 80.10, 80.20
Mitsui, T. 4 48.14, 54.7, 54.15 5 2.21, 2.22, 2.23, 47.22 7 48.1, 48.2, 48.3
Mitsui, Y. 1 48.42 7 76.37
Miura, I. 4 35.12 7 56.8
Miura, R. 6 36.10
Miwa, T. 5 1.33
Miyadera, T. 5 36.22
Miyake, Y. 7 38.5
Miyakoshi, H. 5 54.4, 54.5, 54.6
Miyashita, M. 5 58.43
Miyoshi, T. 3 61.4 5 61.2 6 61.5, 61.6 7 50.4
Mizoguchi, T. 6 75.29
Mizuno, H. 3 44.12 5 31.46, 44.5, 44.10
Mizuno, K. 1 48.73, 50.5 3 50.13
Mizushima, T. 2 83.114
Mizuta, K. 5 54.9
Mizutani, I. 2 67.18, 67.19
Mladeck, M.H. 1 23.5
Mo, F. 1 5.39 3 38.12 6 45.20, 45.21, 56.2 7 13.14, 45.48, 45.49
Moats, W.L. 3 46.2
Mocella, M.T. 6 76.51 7 83.62
Mochalin, B.V. 7 45.20
Mock, W.L. 4 31.8
Mockler, G.M. 6 78.20 7 84.76, 84.77
Modinos, A. 7 71.55
Modinos, A.G.J. 5 75.43
Moeller, T. 3 11.8, 64.25
Moers, F.G. 4 85.3
Moews, P.C. 7 45.50
Mognaschi, E.R. 6 75.15
Mohacsi, E. 6 58.28
Moise, C. 5 73.32, 73.40
Mok, K.S. 3 86.46
Mokren, J.D. 5 45.20 6 83.57
Mokuolu, J.A.A. 2 64.76, 86.56 6 86.18
Molin-Case, J. 4 81.61 5 83.100, 86.35 6 83.38
Molin-Case, J.A. 3 32.14, 32.15 4 40.22
Mollier, Y. 4 39.26 5 39.27
Moncrief, J.W. 1 45.68, 58.125, 59.56 4 80.6, 80.7 5 33.44 6 9.2, 10.4, 32.16, 32.29 7 50.5
Mongiorgi, R. 5 38.24 7 17.21, 29.10, 32.6, 35.41, 35.42, 35.43, 35.44, 40.23, 51.18
Monier, J.-C. 3 60.7 4 84.5
Moniot, J.L. 4 58.21
Montagnoli, G. 4 72.11 5 48.75
Montenero, A. 4 83.64 5 81.47, 81.54 6 3.7, 3.8, 3.29, 3.33 7 3.30
Montgomery, H. 1 39.19 2 77.8, 77.12, 77.13, 77.26, 77.27
Monti, E. 5 48.75
Monti, L. 5 38.14
Monti, S.A. 7 53.24
Mooij, J.J. 6 22.1
Mooney, J.R. 7 63.19
Mooney, R.C.L. 2 65.36
Moore, C.J. 6 82.1, 82.3, 82.4, 82.12, 82.13, 82.14, 82.17, 82.18 7 51.52
Moore, D.W. 4 38.17
Moore, E.B. 2 62.2
Moore, F.H. 2 78.25 4 2.5, 49.20, 67.8 5 21.14, 21.15 6 11.13, 19.6 7 5.16, 21.10
Moore, J.A. 5 35.34 7 35.32
Moore, J.W. 2 83.197 3 68.18 4 81.56, 81.58 5 81.51, 81.52
Moore, R.D. 4 83.131
Moore, W.J. 3 48.10
Mootz, D. 1 5.35, 20.32, 21.31, 32.18, 43.17 2 64.62, 65.35 3 11.1, 13.2, 51.36, 63.8, 63.13 4 17.2, 20.9, 36.36, 63.7, 83.9 5 8.1 6 45.51, 64.17 7 4.1, 13.6, 61.6
Mooy, J.H. 7 8.2
Moran, R.A. 6 45.48 7 52.6
Morancho, R. 6 83.1
Moras, D. 2 83.124, 83.153 4 67.12, 67.20 5 67.10, 67.13, 67.14, 67.15, 68.24 6 67.12, 67.13, 67.14 7 40.17, 40.20, 85.21
Morassi, R. 3 86.52 7 85.37
Moreau, J.M. 7 81.14
Morehouse, S.M. 2 73.39, 81.25 3 72.34 5 72.32
Moreland, J.A. 5 81.57 6 71.74, 81.44 7 81.41
Morelle, I. 3 48.25
Moreno, E. 5 8.3 7 45.19
Moreno, J. 1 16.14
Moret, C. 3 64.36
Moret, C.L. 1 32.34
Morgan, C.H. 1 47.2, 47.3, 47.4 2 77.23, 83.181
Mori, A. 5 58.43
Mori, M. 6 76.20
Moriarty, R.E. 4 72.32 5 71.20

Moriarty, R.M. **6** 71.37
Moriconi, E.J. **3** 42.2 **6** 37.8
Morikawa, K. **7** 47.32
Morimoto, C.N. **3** 8.6 **4** 8.3 **7** 27.4
Morimoto, N. **5** 31.46
Morimoto, Y. **1** 50.2 **4** 32.28 **7** 50.4
Morin, R.B. **5** 50.6
Morisaki, M. **1** 55.1
Morishima, H. **6** 1.30
Morita, K. **5** 44.5
Morita, N. **5** 38.19
Moritani, I. **6** 75.29
Moriyama, H. **7** 86.71
Morley, K.A. **1** 10.26
Morley, T. **5** 60.30, 60.31
Mornon, J.-P. **1** 24.28, 24.29, 35.36 **3** 56.4 **4** 21.13, 41.8, 60.26 **5** 5.18, 9.16, 41.15, 41.31, 41.32, 41.35 **6** 51.12 **7** 51.8, 51.13, 51.25, 84.55, 84.56, 84.57, 84.58
Mornon, J.P. **3** 13.16, 41.6
Morosi, G. **5** 31.12
Morosin, B. **1** 3.26, 3.34 **2** 67.9, 77.22, 77.24, 77.26, 77.27, 78.19 **3** 76.36, 76.37 **4** 33.40, 67.16 **5** 3.13, 15.6, 78.3, 86.81 **6** 15.3, 15.4, 33.9, 74.3, 81.14, 83.19 **7** 67.4, 81.29, 81.30, 83.27, 83.30
Morrell, D. **3** 86.98
Morris Junior, E.D. **3** 64.2
Morris, A. **3** 65.8 **6** 69.18 **7** 65.6
Morris, A.J. **6** 48.68
Morris, D.G. **1** 31.20 **4** 3.6, 3.20 **5** 52.4 **6** 4.2, 64.50
Morris, D.M. **3** 83.1
Morris, E.M. **5** 47.4
Morrison, J. **4** 12.14, 35.23
Morrison, J.D. **1** 1.55, 1.73, 1.84, 1.125
Morriss, R.H. **2** 83.158
Morrow, J.C. **2** 84.31 **3** 10.2, 77.2, 77.3 **4** 33.1 **5** 77.4 **7** 82.4
Morrow, L.R. **5** 51.36 **7** 51.38, 51.40
Mosbo, J.A. **4** 64.4 **7** 64.15
Moseley, K. **5** 71.105
Moseley, P.T. **2** 67.16 **3** 76.72, 76.73 **4** 3.14 **5** 71.1 **7** 84.80
Moss, D.S. **2** 83.174, 85.49 **6** 58.32
Moss, J.R. **5** 71.3
Moss, K.C. **4** 83.8
Moss, R.A. **7** 38.47
Mossel, A. **1** 21.7
Mostad, A. **1** 9.12, 45.6, 45.63 **3** 84.2 **4** 9.9, 48.40, 59.3 **5** 48.53, 48.57, 48.58, 48.63, 59.4 **6** 48.41 **7** 17.12, 35.17, 48.41, 48.42, 48.43
Motherwell, W.D.S. **1** 30.11, 36.31, 36.32, 40.33 **3** 51.42, 53.12, 53.13, 53.14, 58.6, 65.5 **4** 8.9, 23.2 **5** 47.24, 47.31, 50.11, 51.9, 51.25, 56.9 **6** 38.39, 64.56, 65.1, 65.2, 65.3
Motohashi, M. **2** 83.187
Mottram, M.J. **5** 13.15
Mouille, Y. **7** 33.55
Mounts, R.D. **6** 77.15, 81.11
Mourad, M.S. **5** 31.33
Moustakali, I. **6** 49.2
Mouvet, M. **3** 2.9
Mowat, W. **4** 71.52, 71.53
Muccini, U. **4** 21.12
Mudzhoyan, A.L. **7** 1.35
Mueller-Westerhoff, U. **7** 20.21
Mueller, M.H. **4** 38.2, 38.3 **5** 79.11
Muetterties, E.L. **3** 62.9, 86.10 **5** 68.26, 86.17 **7** 71.103
Mugnoli, A. **1** 5.28, 5.31, 9.13, 19.50, 22.12 **2** 60.103, 80.13 **3** 19.11, 39.35 **4** 31.20 **5** 17.16, 31.12, 31.28, 31.29 **7** 31.5, 31.16, 31.24
Muhm, H. **4** 71.92
Muir, J.A. **7** 86.91
Muir, K.M. **4** 73.30
Muir, K.W. **1** 31.40 **2** 74.22, 74.23 **3** 73.2, 86.21, 86.68, 86.79 **4** 30.8, 71.21, 71.81, 86.26 **6** 49.8, 71.46, 73.61, 82.7 **7** 2.4, 64.18, 64.19, 71.68, 71.75, 71.76, 84.26, 86.21, 86.30, 86.47, 86.52
Muir, M.M. **7** 86.91
Mukherjee, A. **7** 21.7, 21.8
Mukherjee, J. **1** 39.58
Mullen, D. **3** 60.19 **7** 3.1
Muller, B. **5** 31.35, 67.16
Muller, E. **4** 71.92 **5** 71.75
Muller, E.G. **7** 77.3, 77.6
Muller, F. **5** 36.20
Muller, F.A. **4** 47.13
Muller, H. **2** 73.80
Muller, U. **5** 12.2 **7** 27.11, 71.1, 71.30
Mullins, P. **7** 58.9
Mumme, W.G. **4** 69.15
Munakata, K. **5** 54.10
Mundy, B.P. **7** 35.14
Mundy, B.W. **2** 83.162
Munnik, J. **1** 40.12
Munnoch, P.J. **7** 60.28
Munns, A.R.I. **3** 47.17, 47.18
Muntwyler, R. **5** 50.13
Muntz, R.L. **4** 16.7
Murae, T. **7** 38.57
Murahahi, S. **1** 20.1
Murahashi, S.-I. **6** 75.29
Murakami, Y. **3** 1.17
Muralidharan, K.V. **2** 76.24 **5** 41.1, 41.4
Muraoka, Y. **1** 41.31 **7** 1.18
Murata, I. **1** 20.30, 38.42 **7** 39.17
Murata, T. **1** 56.9
Murayama, W. **1** 47.20 **3** 46.3
Murillo, C.A. **7** 72.28
Murmann, R.K. **2** 83.24 **6** 83.6, 83.29, 83.30, 84.31 **7** 76.75
Muroi, M. **1** 50.5
Murphy, D.W. **6** 69.22
Murray Junior, R.K. **4** 52.5
Murray-Rust, J. **5** 3.8 **6** 85.8 **7** 3.6, 31.10, 37.13, 40.16
Murray-Rust, P. **2** 60.146, 60.147, 81.27, 81.45, 81.48, 81.66, 81.81 **3** 33.38, 33.39, 33.40 **5** 3.8, 38.18 **6** 37.9, 45.23, 85.8 **7** 3.6, 31.10, 33.34, 37.6, 37.12, 37.13, 37.20, 40.16
Murray, A.J. **6** 71.13, 72.9
Murray, K.S. **6** 78.5 **7** 78.26
Murray, M. **7** 71.59
Murtha, D.P. **5** 77.24 **6** 77.21

Murthy, B.V.R. **7** 41.4
Murthy, N.S. **6** 45.30
Murthy, R.V.A. **7** 41.4
Murty, B.V.R. **1** 26.18
Musatti, A. **1** 16.31, 35.8, 39.41, 41.21 **2** 79.14, 79.15, 79.26, 79.31, 81.80, 85.2, 85.37 **3** 81.19 **4** 83.55, 84.12 **5** 41.19, 78.2, 78.18, 81.44, 83.36, 83.38 **6** 33.37, 41.5 **7** 32.10, 78.8, 78.13, 83.48
Muscio Junior, O.J. **4** 20.11, 20.13
Musco, A. **2** 75.37 **3** 72.33 **6** 86.40
Musher, J.I. **7** 11.19
Musso, H. **5** 31.25, 71.53 **6** 71.18
Mustafa, I.A. **7** 72.17, 72.19, 72.20, 72.21, 72.23
Mustafa, I.A.G. **5** 72.28, 72.29
Mutin, J.-C. **4** 2.9 **6** 2.5
Mutin, J.C. **3** 2.5
Muto, A. **3** 76.71
Muto, F. **3** 78.9, 81.23 **4** 78.1
Muxfeldt, H. **3** 50.17
Myasnikova, R.M. **1** 5.26 **2** 71.26 **7** 34.3, 38.12, 60.6
Myers, J.F. **4** 83.73

Nabeya, A. **6** 32.27
Nackert, M.L. **4** 33.20
Nader, F. **5** 60.1
Nader, F.W. **7** 38.27, 38.42
Nagai, S. **5** 17.18
Nagai, T. **5** 42.12
Nagai, W. **5** 1.33
Nagai, Y. **3** 36.34
Nagakura, S. **6** 19.4
Naganathan, P.S. **4** 48.13, 48.17, 48.32
Nagano, K. **3** 48.30
Nagao, R. **3** 1.23 **4** 81.31 **6** 83.16
Nagashima, N. **1** 7.18, 47.13, 47.14, 47.20 **6** 47.16, 47.27
Nagel, D.W. **7** 58.17
Nagy, J. **7** 63.5
Nahringbauer, G. **1** 11.42
Nahringbauer, I. **1** 2.2, 2.36, 2.37, 2.38, 2.39, 2.40, 2.41, 2.48 **3** 1.2, 1.3, 2.10, 2.11, 60.3
Naik, D.V. **5** 69.21 **6** 71.60, 86.56 **7** 8.21, 8.23, 79.1
Nair, P.M. **1** 59.68
Nakagawa, A. **5** 50.14
Nakagawa, M. **5** 31.23
Nakahara, A. **2** 76.8 **3** 78.9, 78.10, 81.23 **4** 78.1 **5** 82.11
Nakai, H. **3** 31.14, 38.13, 56.7, 83.109 **4** 39.37, 39.38, 83.103
Nakai, Y. **1** 2.5 **2** 81.9
Nakajima, M. **7** 86.71
Nakajima, S. **3** 60.7
Nakajima, T. **7** 37.18
Nakamua, H. **7** 1.18
Nakamura, A. **1** 7.18 **2** 73.82 **7** 78.3
Nakamura, E. **7** 54.2
Nakamura, H. **1** 41.31 **3** 36.20 **6** 1.30 **7** 47.30, 50.39, 53.28
Nakamura, K. **3** 36.9
Nakamura, M. **1** 21.14
Nakamura, N. **7** 20.20
Nakamura, T. **1** 13.38
Nakamura, Y. **3** 84.10 **4** 77.3, 77.18 **6** 76.33 **7** 76.8
Nakanishi, H. **4** 33.41 **5** 33.51
Nakanishi, K. **4** 35.12 **7** 56.8
Nakanishi, T. **1** 56.20 **4** 56.5
Nakano, E. **3** 5.3
Nakano, T. **1** 58.115 **3** 58.59
Nakao, Y. **5** 82.11
Nakata, K. **5** 13.5
Nakatsu, K. **1** 58.43 **2** 65.18, 65.20, 76.36, 76.37, 76.38, 79.25 **4** 35.31, 35.32, 38.8 **5** 1.33, 33.39, 33.40, 33.41, 36.8, 41.46, 71.67 **6** 86.26 **7** 50.18, 78.3
Nakatsu, S. **3** 36.34
Nakayama, H. **7** 6.1
Nakayama, M. **7** 53.27
Nakayama, Y. **5** 82.18 **7** 53.28
Naldini, L. **4** 86.71
Namba, Y. **1** 36.12, 36.26
Nancollas, G.H. **4** 76.15
Nappier Junior, T.E. **6** 86.38, 86.39
Naqvi, R.R. **3** 16.16, 16.17 **4** 64.1, 64.2
Narasimhamurthy, M.R. **7** 48.35
Narasimhan, K. **3** 30.4
Narayanan, P. **3** 48.1, 48.2, 48.3, 54.6, 59.25 **4** 27.19, 51.11, 54.12 **5** 48.40, 48.69 **7** 4.9, 32.15, 48.4
Nardelli, M. **1** 8.18, 9.3, 11.1, 32.15, 39.23, 39.25, 39.32 **2** 69.11, 79.5, 79.9, 79.16, 79.18, 79.20, 79.26, 79.31, 79.33, 80.1, 81.80, 83.8, 83.9, 83.28, 83.46, 84.2, 84.5, 84.8, 84.9, 85.2, 85.3, 85.5, 85.14, 85.35 **3** 8.5, 8.11, 76.1, 79.14, 81.19, 83.10, 85.12 **4** 8.6, 28.5, 35.13, 41.21, 41.22, 59.28, 76.22, 76.38, 79.12, 83.1, 83.2, 83.5, 83.12, 83.55, 83.62, 83.64, 83.74, 84.12, 85.14 **5** 40.22, 40.23, 41.41, 69.14, 78.2, 78.18, 78.22, 81.5, 81.25, 81.43, 81.44, 81.47, 81.53, 81.54, 82.1, 82.10, 83.36, 83.38 **6** 59.3, 76.41, 79.7, 79.16, 80.3, 83.56, 83.66, 83.68 **7** 48.63, 69.15, 78.13, 80.4
Nardi, N. **7** 76.61
Nardin, G. **2** 78.59 **3** 76.81, 78.14, 78.26, 83.150, 85.32, 85.33, 85.34, 85.40 **4** 71.40, 77.14 **5** 69.19, 71.31, 71.42, 71.57, 76.39, 76.43, 78.5, 78.6, 78.9, 83.41, 86.67 **6** 71.89, 71.95, 76.42, 78.11, 86.57 **7** 71.7, 71.86, 78.5, 78.7, 86.92, 86.95, 86.96
Narita, H. **4** 59.21
Naruto, S. **4** 58.43
Narwid, T.A. **7** 51.44
Nasakkala, M. **3** 76.32, 76.35 **5** 76.28
Nash, C.P. **3** 82.26
Nash, R. **7** 40.14, 40.15
Nassiff, P.J. **7** 10.1, 78.9
Nassimbeni, L.R. **4** 77.15 **5** 71.34 **6** 44.23, 71.26, 83.74, 83.76, 86.3 **7** 44.9, 51.36, 64.32, 64.33, 83.8, 83.40, 83.43, 83.71, 83.106, 83.107, 84.72, 85.13
Nasyrov, S.-M. **6** 58.20 **7** 58.16
Natarajan, S. **2** 79.6 **6** 58.35, 81.40

Natile, G. **3** 86.25
Natsuki, R. **6** 36.11
Natsume, M. **1** 54.12
Natta, G. **2** 61.17, 61.19, 72.39, 74.5, 74.25
Navaza, A. **7** 61.5
Nave, C. **6** 67.8, 72.21 **7** 34.15
Nawata, Y. **1** 50.18 **2** 83.96 **4** 50.10 **5** 50.17, 50.18, 50.19, 50.22 **6** 50.11 **7** 50.20
Naya, K. **3** 50.20, 53.20
Nayak, U.R. **6** 53.9
Naylor, A. **7** 41.23
Neel, J. **4** 48.23, 48.43 **5** 48.61, 48.62 **6** 37.4, 48.38, 48.47, 48.53
Neelakantan, L. **4** 40.22
Neely, S.C. **7** 31.3
Neeman, I. **7** 59.10
Nefedov, O.M. **3** 60.9
Negro, A.D. **5** 62.5
Neguchi, T. **2** 82.33
Neidle, S. **1** 50.27, 59.48 **3** 48.19 **4** 32.17, 38.21, 38.25, 49.16, 53.6, 59.15 **5** 32.24, 45.12, 50.10, 51.43 **7** 36.10
Neigauz, M.G. **3** 40.2 **4** 24.7 **5** 40.27 **6** 33.30, 40.16, 41.20 **7** 16.8, 18.7, 85.22
Neitzel, C.J. **7** 82.15
Nekhaev, A.I. **6** 72.17
Nelsen, S.F. **6** 31.15
Nelson, A.D. **1** 12.9 **2** 60.93, 60.94
Nelson, G.V. **4** 40.9
Nelson, N.J. **7** 73.56
Nelson, N.R. **1** 58.110, 59.50
Nelson, S.H. **7** 76.73
Nelson, S.M. **5** 72.8, 72.12 **6** 72.22
Nemiroff, M. **3** 83.86 **6** 31.3 **7** 78.1
Nepomnyashchaya, V.N. **1** 3.29
Neri, C. **3** 71.25, 78.12, 78.16
Neronova, N.N. **1** 19.34 **5** 76.32 **7** 76.50
Nesmeyanov, A.N. **2** 72.72 **3** 71.49 **6** 66.2 **7** 75.5
Nesterova, Ya.M. **1** 9.19 **4** 9.7 **5** 76.34, 76.37
Neuman, A. **5** 17.9, 17.12 **7** 45.22, 45.23
Neuman, M. **7** 1.6
Neuman, M.A. **4** 31.12, 73.28 **5** 77.5 **6** 84.22 **7** 38.9, 38.11, 38.20, 38.21, 38.22, 38.23, 84.71
Neupert-Laves, K. **7** 50.37
Neuzil, E. **7** 64.2
New, D.B. **7** 81.46
Newberry III, W.R. **7** 68.1, 68.3, 68.8, 68.9
Newing, C.W. **3** 83.102 **4** 83.61, 83.85
Newlands, M.J. **3** 73.13
Newman, B.A. **1** 23.17
Newman, P.W.G. **5** 80.1, 80.6, 80.7, 80.17, 80.21, 80.22
Newton, G. **1** 58.64 **6** 64.31
Newton, M.D. **5** 31.2
Newton, M.G. **1** 37.24, 41.38, 46.11 **2** 75.78 **3** 39.46, 54.5, 58.39, 58.40, 58.41, 58.48 **4** 42.5, 58.38, 58.44, 64.11 **5** 31.10, 36.30, 46.10 **6** 19.8, 31.16, 36.21, 64.7, 84.19 **7** 31.8, 46.5, 64.41, 64.51
Ng, T.W. **2** 73.122
Ni, S.W. **5** 73.39
Nibbering, N. **1** 38.4
Nice, J.P. **2** 73.42, 73.54
Nice, R.van **3** 84.26
Nicholas, A.F. **1** 54.11 **3** 82.6
Nicholas, H. **5** 47.33
Nicholas, K. **3** 72.41
Nichols, T.D. **7** 12.4
Nicholson, B.K. **5** 63.3
Nicholson, C. **4** 33.14
Nicholson, D.G. **6** 69.2
Nicol, A.D.I. **1** 11.33, 11.34
Nicolini, M. **5** 69.28
Niekerk, J.N.van **2** 81.21, 81.29, 81.34, 81.36, 81.37, 81.63, 81.64
Nielsen, A.T. **4** 38.17
Nielsen, J. **2** 73.116
Nielsen, K. **4** 83.60
Nielson, D.O. **3** 3.1 **4** 82.11 **6** 83.64
Nigam, G.D. **1** 26.28
Niggli, A. **1** 25.20, 58.59, 58.122 **2** 83.188, 83.206 **3** 59.9 **5** 31.1
Niimura, N. **2** 60.101
Nikolaeva, L.E. **7** 78.14, 78.15
Nilson, L. **3** 68.4, 80.21
Nilsson, A. **1** 2.30
Nilsson, B. **1** 30.36, 52.32, 53.16, 53.32, 58.21 **3** 5.4, 58.59 **6** 31.33
Nimgirawath, K. **7** 3.12
Nimrod, D.M. **1** 46.8 **5** 64.5
Nishi, Y. **1** 27.13, 56.12
Nishida, T. **3** 47.32
Nishiguchi, K. **5** 31.46
Nishihata, K. **6** 11.10
Nishikawa, M. **1** 48.73, 50.5, 56.9, 58.35, 58.36, 58.57, 59.52 **3** 44.42, 50.13, 50.22, 50.30 **4** 50.6 **5** 50.2 **6** 35.29, 36.7, 40.24
Nishikawa, Y. **4** 77.18
Nishimura, N. **6** 9.1
Nishio, K. **1** 58.93
Nishio, M. **6** 11.10
Nissen, H.M. **5** 48.58
Nitta, I. **1** 1.30, 1.108, 2.3, 2.4, 2.6, 2.84, 5.36, 5.40, 9.5, 13.36, 16.15, 17.37, 22.3, 27.8, 35.17, 37.1, 48.32, 50.37, 56.9, 58.35, 58.36, 58.57, 59.19, 59.25, 59.52 **2** 81.68, 82.4, 82.5, 83.156 **3** 17.5, 33.42, 36.4, 50.13, 50.20, 53.20 **4** 58.1, 58.2, 81.32, 81.33, 81.35 **5** 1.20
Nivellini, G.D. **5** 84.12 **7** 84.6
Niwa, M. **7** 56.6
Nixon, R.E. **1** 37.2, 37.3
Nobbs, C.L. **1** 26.26
Nockolds, C.E. **3** 83.20 **4** 49.20 **6** 39.3, 39.4
Noda, T. **1** 50.31 **3** 50.20
Noels, A.F. **7** 36.19
Nofre, C. **1** 44.41

Noguchi, M. **4** 59.21
Noguchi, T. **1** 3.18
Nojima, H. **4** 32.28
Noland, W.E. **1** 40.4 **4** 40.12
Nolte, C.R. **5** 73.46
Nolte, M.J. **5** 45.49 **6** 74.12 **7** 86.29
Noltemeyer, M. **7** 45.56, 45.57, 45.59
Noltes, J.G. **4** 71.82 **5** 71.80 **6** 66.4 **7** 71.116
Nomura, T. **2** 83.17, 83.105 **3** 76.64
Noordik, J.H. **1** 11.47 **4** 80.4, 80.11, 80.12, 85.4 **5** 80.11, 80.18 **6** 22.1, 30.2, 30.8, 80.15 **7** 30.7, 85.8
Norbury, A.H. **7** 73.15
Nord, A.G. **6** 38.1
Nordin, E. **7** 76.17, 76.18
Nordman, C.E. **1** 2.82, 30.3, 31.9, 31.10, 58.43, 58.123 **2** 62.6, 68.4, 68.13 **3** 30.2, 64.2, 64.4 **4** 38.25, 60.43 **5** 9.14 **6** 56.3 **7** 51.1, 51.2, 85.26
Norman Junior, J.G. **3** 83.132 **5** 81.28, 81.50, 81.63
Norman, N. **1** 5.17, 5.18, 5.19, 5.22, 20.16, 21.4
Norment, H.G. **1** 17.44, 17.46, 31.28 **2** 62.19
Norrestam, R. **1** 36.27, 45.27, 45.28 **3** 36.7 **4** 35.1, 36.13, 36.15, 36.19, 36.23, 60.30 **5** 36.18, 36.20, 50.23, 50.24, 82.28 **7** 59.6, 83.19
Norris, E.K. **5** 41.38 **6** 45.1 **7** 35.12, 35.16
Norris, J.M. **7** 83.100
Norris, T. **7** 32.38, 41.31
North, P.P. **3** 69.37, 69.38 **4** 27.1, 69.18 **7** 38.20, 38.21, 38.22, 38.23
Northolt, M.G. **1** 1.97, 34.9, 52.10 **2** 63.13 **3** 34.2, 34.5 **4** 31.8
Norton, D.A. **1** 51.2, 51.3, 51.5, 51.7, 51.9, 51.10, 51.12, 51.13, 51.14, 51.17, 51.21, 51.22, 51.41 **2** 60.151, 82.53, 84.64 **3** 17.8, 48.41, 51.12, 51.16, 51.19, 60.14, 60.15 **4** 51.2, 51.3, 51.15, 51.17, 51.33, 51.47, 51.50, 51.51, 60.39 **5** 50.29, 60.28 **6** 38.32
Norton, M.C. **7** 71.99
Noth, H. **5** 62.2, 75.37 **6** 86.24
Novak, C. **1** 53.36 **3** 53.7 **4** 81.22
Novakovskaya, L.A. **3** 40.2 **6** 33.30, 40.16 **7** 85.22
Novick, S. **2** 63.15
Novikov, S.S. **3** 40.2
Novotny, M. **5** 71.18
Novozhilova, N.V. **6** 76.22 **7** 76.43, 76.44, 76.46
Nowacki, M. **1** 1.92
Nowacki, W. **1** 13.33, 23.5, 44.5, 51.27, 58.28 **2** 74.13 **3** 58.26 **4** 58.42 **5** 38.15, 38.16, 38.21, 39.15 **6** 16.11 **7** 31.18, 31.32, 31.33, 32.23, 57.1, 59.2, 84.45
Nowell, I.W. **2** 86.92 **4** 86.1 **5** 71.3, 73.1, 73.2, 74.12, 75.43, 86.3, 86.4, 86.5, 86.6, 86.7, 86.9, 86.10
Nozaki, H. **7** 53.27
Nozakura, S. **1** 20.1
Nozoe, S. **1** 50.35, 55.1, 55.2 **4** 59.23
Nuber, B. **7** 71.126, 71.134, 75.2, 75.3, 75.4
Numata, S. **7** 40.8
Nunn, E.K. **5** 69.4
Nuno, M. **1** 59.42
Nunzi, A. **4** 70.4, 83.49, 85.8 **5** 36.32, 40.25 **6** 32.13, 32.34
Nuretdinov, I.A. **4** 64.17
Nuttall, R.H. **3** 83.111 **4** 83.37, 83.38, 83.41, 83.52, 83.53 **5** 83.52, 83.55, 83.56, 83.57 **6** 83.47
Nyberg, S.C. **3** 81.54
Nyburg, S.C. **1** 19.44, 38.61, 38.62, 39.47, 44.69, 44.73, 58.84 **2** 75.47 **3** 3.43, 3.44, 36.18, 36.25, 60.34, 68.3, 83.72, 86.64 **4** 11.6, 39.24, 39.31, 39.46, 41.34, 63.2, 63.12, 69.1, 72.25, 86.31 **5** 5.17, 41.47, 41.48, 72.42, 72.55 **6** 4.3, 11.7, 20.18, 20.19, 28.27, 28.28, 40.7, 86.45 **7** 71.122, 71.123, 86.65
Nygjerd, G. **6** 44.22 **7** 36.1
Nyholm, R.S. **2** 86.46 **3** 67.24, 86.97 **4** 71.43, 71.46, 71.49, 71.71, 71.73, 71.89 **5** 71.95, 71.96, 71.99, 86.15 **6** 72.8
Nyi, K. **7** 31.19
Nyitrai, J. **6** 32.32, 32.33, 39.27
Nyman, F. **2** 73.33

O'Brien, E.J. **1** 44.60, 44.61
O'Brien, J. **4** 58.3, 58.6
O'Brien, R.J. **2** 69.2 **3** 68.3 **4** 71.47
O'Brien, S. **2** 72.28
O'Brien, T.A. **2** 72.27, 73.112, 73.113 **3** 71.16, 71.43, 72.43
O'Brien, T.P. **7** 49.12
O'Bryan, N.B. **6** 76.50
O'Connell, A.M. **1** 1.139, 16.20, 16.21, 39.37, 54.14, 59.57 **6** 45.34, 59.23
O'Connor, B.H. **1** 16.17 **2** 80.14, 81.33 **4** 49.20 **6** 19.5, 19.6, 19.7
O'Connor, J.E. **2** 73.48, 73.96, 79.28, 83.65 **3** 79.5 **5** 85.8
O'Connor, M.J. **5** 81.38
O'Connor, T. **5** 71.90
O'Day, B.P. **5** 83.120, 83.121, 83.122
O'Donnell, E.A. **1** 51.37
O'Donnell, M.M. **1** 2.95
O'Flynn, K. **5** 71.86, 71.87, 71.88
Obatake, K. **6** 7.1
Obendorf, S.K. **6** 36.5 **7** 39.40
Oberhansli, W. **1** 22.6 **2** 75.62
Oberhansli, W.E. **1** 51.11 **2** 72.9, 75.80, 75.81 **3** 44.38, 58.32 **4** 31.28, 58.12, 58.17 **5** 40.33 **6** 54.10, 58.43 **7** 53.1
Obodovskaya, A.E. **3** 78.20 **6** 78.13
Oboznenko, Yu.V. **5** 80.14
Obruchnikov, I.V. **5** 45.47
Occolowitz, J.L. **6** 50.10
Ocenaskova, D. **1** 12.10, 12.11
Ochiai, A. **4** 59.21
Ochsenreiter, P. **7** 10.2, 10.3, 10.4, 10.5, 76.67

Oda, K. **3** 58.11, 58.16, 72.10, 72.18 **4** 48.24
Oda, M. **5** 12.5, 12.13, 31.19
Oda, T. **1** 36.12, 36.26
Oertel, R.P. **5** 67.3
Oeser, E. **5** 33.54, 35.38 **6** 33.45 **7** 3.28
Offereins, B. **6** 37.5 **7** 37.11
Oftedal, P. **4** 41.26
Ogawa, K. **1** 5.10 **2** 83.156 **3** 26.11 **4** 72.17, 82.6 **5** 78.1, 78.11
Ogawa, Y. **3** 50.21
Ogihara, Y. **1** 59.60 **7** 58.25
Ogilvie, R.A. **3** 24.7
Ogiso, A. **5** 59.33
Ogoshi, H. **5** 49.6 **6** 49.7 **7** 49.8
Ogura, F. **5** 31.23
Ogura, H. **7** 41.16, 53.4
Ogura, T. **3** 72.10 **4** 71.58, 71.80 **6** 81.11
Oh, P.W.T. **7** 33.44
Oh, Y.L. **1** 54.5, 56.11
Ohashi, T. **1** 2.84
Ohashi, Y. **2** 60.101, 60.123 **3** 3.25 **6** 50.1, 53.5, 53.10, 60.4, 71.58 **7** 47.28
Ohgo, Y. **6** 71.58
Ohkura, K. **5** 13.2 **6** 13.7, 19.9
Ohloff, G. **3** 53.1, 53.2
Ohno, K. **3** 58.43 **5** 58.37 **6** 58.47
Ohno, M. **5** 50.3 **7** 47.30
Ohno, T. **7** 38.48
Ohrt, J. **3** 44.29 **5** 21.12, 41.50, 58.5 **6** 44.16
Ohrt, J.M. **1** 51.13, 51.14, 51.17 **3** 52.5 **6** 48.39, 48.75 **7** 44.27, 48.45, 48.52
Ohsawa, A. **5** 83.83 **7** 39.26
Ohsawa, T. **1** 56.7
Ohshiro, Y. **3** 41.23 **4** 35.39 **5** 42.13
Okabe, K. **1** 54.17
Okada, K. **2** 81.3, 81.4 **3** 31.26
Okada, Y. **4** 38.8
Okamoto, K. **7** 72.35
Okamoto, K.-I. **6** 82.9
Okamoto, N. **6** 13.3
Okamoto, T. **3** 76.23, 76.66 **7** 16.16, 53.4
Okamoto, Y. **1** 58.36, 58.57
Okamura, W.H. **3** 85.23
Okawara, R. **2** 69.3
Okaya, Y. **1** 1.57, 1.58, 4.4, 4.8, 8.30, 9.11, 13.15, 14.9, 14.10, 21.14, 31.4, 31.14, 33.65, 33.69, 40.11, 41.17, 45.54, 48.9, 48.11, 58.42, 59.2 **2** 63.21, 63.22, 64.7, 79.13, 82.8 **3** 13.4, 13.5, 28.3, 28.4, 69.28
Okeya, S. **3** 72.10
Okuda, S. **1** 50.35, 55.2
Okuda, T. **1** 58.111 **3** 45.24
Okuno, T. **5** 54.4
Okunuki, M. **4** 60.7
Olansky, L. **5** 33.44
Olcott, R.J. **3** 83.71
Oldham, G.C. **3** 77.2, 77.3
Olejnik, S. **7** 48.7
Olekhnovich, L.P. **7** 71.36
Olie, K. **7** 3.18, 71.129
Olin, A. **5** 68.18
Oliver, A.J. **6** 71.39
Oliver, J.D. **3** 72.42 **4** 71.84 **7** 83.38
Oliver, J.E. **5** 41.16 **6** 35.10
Oliver, J.P. **3** 68.18 **7** 63.1, 73.26, 73.47
Oliveto, E.P. **5** 32.6
Ollendorf, W. **3** 81.11, 81.12, 81.13, 81.15
Ollis, C.R. **5** 85.3
Ollis, J. **7** 45.46
Olovsson, I. **1** 2.29, 2.30, 3.10, 3.17, 12.4 **3** 2.10, 12.4, 33.6 **4** 60.5
Olsen, B. **3** 85.4
Olsen, F.P. **5** 4.4 **6** 4.4, 4.5 **7** 4.6
Olson, A.J. **7** 51.1
Olson, D.H. **2** 69.23
Olthof, G.J. **7** 71.52, 73.35, 75.30, 75.36
Omae, I. **2** 69.31
Omel'chenko, Yu.A. **1** 9.31 **4** 32.1 **5** 9.9, 9.10 **7** 16.8
Omura, S. **3** 50.20 **5** 50.14
Omura, T. **7** 49.8
Onan, K.D. **6** 39.10, 39.11 **7** 53.5, 53.11, 53.17, 53.23, 53.25, 53.30, 64.9, 64.10, 64.16, 64.17
Ono, K. **3** 72.20
Onuma, K. **5** 66.5
Onuma, S. **1** 50.18 **3** 77.6
Onyszchuk, M. **4** 63.4, 69.16, 69.17
Ooi, S. **2** 76.4, 76.10, 76.11, 76.13, 79.25, 83.140, 85.70, 85.81 **3** 76.28, 77.15, 83.109, 84.10 **4** 33.24, 76.3, 76.10, 77.3, 77.18, 83.105 **5** 33.39, 33.40, 33.41, 36.8, 36.9, 82.18 **6** 76.10, 76.33 **7** 76.8
Oonishi, I. **3** 76.31, 78.9, 78.10, 81.23 **4** 78.1 **6** 82.16 **7** 81.33, 82.8
Opedal, D. **3** 8.12, 8.13 **7** 8.19, 8.20
Opferkuch, H.J. **3** 54.8
Oppolzer, W. **3** 40.12 **4** 83.112
Orama, O. **7** 83.13
Orbeladze, F.V. **7** 79.11
Orbell, J.D. **6** 83.24
Orehowsky Junior, W. **4** 2.20, 36.21
Orfanova, M.N. **2** 60.140
Organ, T.D. **3** 51.9
Orgel, L.E. **2** 73.22
Orii, S. **1** 13.38
Orio, A.A. **3** 86.77 **4** 86.39, 86.40
Orioli, P.L. **2** 76.76, 76.77, 76.78, 76.79, 78.3, 78.24, 78.32, 78.34, 78.36, 78.49, 78.53, 83.119, 83.120, 83.127, 86.49, 86.83 **3** 76.62, 78.15, 83.18, 86.44 **4** 86.44 **5** 76.42, 86.28, 86.32 **6** 35.18 **7** 76.61
Orlacchio, A. **7** 48.63
Orlandini, A. **7** 86.84, 86.86, 86.87, 86.102
Orlandini, A.B. **5** 76.50, 76.51, 86.28, 86.32, 86.64
Osaki, K. **1** 2.4, 2.5, 16.51, 17.37, 52.25, 58.36, 58.57, 58.86, 58.115 **2** 61.21, 81.9, 81.77 **3** 2.1, 45.24, 45.25, 58.43, 59.30 **4** 58.43, 62.10, 67.3 **5** 17.20, 38.12, 45.37, 62.7 **6** 38.6, 43.3, 54.1, 54.8, 58.47 **7** 1.13, 1.14
Osawa, E. **1** 31.43 **7** 38.28
Osawa, T. **6** 53.5, 53.10

Osawa, Y. **4** 51.35, 51.36, 51.47, 51.50, 51.51 **5** 51.7, 51.40 **6** 51.4, 51.15, 51.27, 51.35, 51.39 **7** 51.6, 51.12, 51.31, 51.32
Osborn, J.A. **5** 71.70, 86.84
Osborn, R.S. **4** 49.7 **6** 83.18 **7** 38.61
Oshima, E. **5** 81.32, 81.33
Oshio, H. **1** 59.52
Oskarsson, A. **2** 82.52 **4** 81.18 **6** 48.10, 48.11, 48.12, 48.13, 48.14, 48.15, 81.17, 82.20 **7** 48.12
Osterberg, R. **3** 81.51 **5** 81.40, 82.13, 82.14
Oswald, H.-R. **7** 73.24
Oswald, H.R. **4** 83.69, 83.70 **6** 64.46
Otaka, Y. **4** 5.4, 20.1, 29.4
Otake, N. **3** 50.21 **6** 50.12 **7** 50.28
Otani, S. **6** 36.10
Otsubo, T. **5** 31.46
Otsuka, S. **2** 73.82 **4** 72.17 **5** 71.67 **6** 71.42, 86.26 **7** 78.3
Ott, H. **1** 58.87
Ottenheym, H.C.J. **6** 36.1
Ottersen, T. **4** 48.40 **5** 33.3, 33.4, 33.26 **6** 3.25, 5.8, 13.16, 33.36, 48.54, 85.25 **7** 9.1, 9.2, 33.16, 33.22, 40.7, 59.15, 85.23, 85.32, 85.41
Otto, J. **5** 61.1
Otzenberger, R.D. **7** 35.14
Oughton, B.M. **1** 48.61
Oughtred, R.E. **5** 81.9
Overbeek, A.R. **5** 35.21, 39.23
Overton, K.H. **1** 54.25 **3** 54.2
Owen, J.D. **3** 64.46 **4** 84.9 **7** 20.22, 86.43
Owen, T.B. **1** 7.4, 31.1
Owens, B.B. **5** 33.10, 33.11, 33.12, 33.13, 33.14, 33.15
Owston, P.G. **1** 35.24 **2** 72.5, 72.28, 73.33, 75.36, 80.5, 83.168, 86.6, 86.8, 86.13, 86.24, 86.27, 86.43, 86.52, 86.55, 86.65, 86.86, 86.104 **3** 17.10, 72.3, 72.30, 80.1, 80.2, 86.28, 86.29, 86.41 **4** 67.27, 72.2 **5** 86.29
Oya, Kh.P. **7** 34.3
Oya, Kh.P. **7** 60.6
Oyakawa, R.T. **5** 73.41
Oyen, J.W.L.van **1** 18.21, 42.2
Oymo, I.-M.A. **7** 19.6
Oyum, P. **1** 11.56 **2** 70.24
Ozawa, T. **7** 16.16
Ozbirn, W.P. **4** 64.31
Ozeki-Minakata, K. **5** 31.38, 31.40
Ozeki, K. **1** 44.31
Ozerov, R.P. **1** 7.5, 7.7
Ozol, Ya. **2** 62.51
Ozol, Ya.K. **6** 85.30
Ozola, A. **7** 85.34, 85.35
Ozola, A.D. **6** 85.30
Ozols, J. **7** 66.10, 85.34, 85.35

Pabst, A. **2** 81.13
Pace, A.de **1** 39.62
Pachler, K. **1** 5.20
Pachler, K.G.R. **7** 58.17
Paddock, N.L. **2** 64.45 **4** 83.122 **6** 64.20, 83.67
Padilla, J.O. **1** 53.40
Padmanabhan, V.M. **1** 2.24, 2.35, 9.41 **2** 67.8, 76.24 **3** 77.8 **4** 48.3 **5** 2.24, 48.41
Padmanabhan, W.M. **7** 2.7
Padolina, W.G. **7** 53.24
Padwa, A. **1** 31.31 **4** 39.8 **7** 35.40
Page, J.E. **1** 51.38, 51.39, 51.40
Page, S.W. **3** 58.40, 58.41 **4** 58.44
Paiaro, G. **2** 75.37 **5** 71.23
Paik, H.N. **6** 71.76 **7** 71.98, 71.107, 71.113, 71.127, 72.33, 72.38
Paine, R.E. **3** 78.7
Pajunen, A. **2** 76.16, 83.50, 83.51 **3** 76.29, 76.33, 76.34, 76.43, 76.53, 76.56, 76.57, 76.78 **4** 83.59 **5** 76.13, 76.24, 76.28, 83.17, 83.18 **6** 76.18 **7** 76.20, 76.23, 76.30, 76.31, 81.38, 83.14, 84.20, 84.42
Pakhomov, V.I. **1** 3.29 **2** 71.1, 71.2, 71.8, 71.19
Pal, B.C. **3** 85.18 **4** 47.4
Palenik, G.J. **1** 10.19, 15.12, 32.1, 32.30 **2** 64.11, 65.40, 68.12, 73.122, 82.41, 83.155, 83.160, 86.68, 86.70 **3** 32.18, 32.19, 33.21, 33.27, 39.9, 39.36, 40.5, 50.14, 59.3, 68.13, 73.8, 73.14, 77.18, 77.31, 78.5, 83.113, 83.128, 83.137, 84.23, 86.23, 86.62, 86.83 **4** 15.1, 33.34, 38.19, 40.6, 58.41, 60.2, 68.11, 71.61, 72.29, 77.13, 78.12, 83.134, 84.13, 85.37, 86.34 **5** 38.5, 38.13, 50.8, 62.9, 68.30, 71.90, 75.2, 75.38, 75.46, 79.7, 79.20, 79.21, 83.109, 86.47, 86.61 **6** 41.2, 68.11, 68.12, 68.14, 68.15, 71.60, 71.76, 75.31, 79.11, 79.12, 79.13, 86.56 **7** 8.21, 8.22, 8.23, 13.16, 31.7, 33.26, 71.113, 72.33, 72.38, 72.45, 79.1, 83.87, 83.88, 83.113, 84.30, 84.31, 86.51, 86.54, 86.57
Palin, D.E. **2** 61.9
Pallister, E.T. **4** 45.22, 83.113
Palm, J.H. **1** 16.9, 21.17, 52.10 **2** 60.5
Palmer, H.T. **1** 11.70
Palmer, K.J. **1** 31.37, 38.23, 41.13 **4** 41.15, 67.7 **5** 48.28 **6** 59.10, 59.30 **7** 31.34, 31.35, 52.1
Palmer, R.A. **1** 11.70, 46.12 **3** 1.33, 68.12 **7** 17.33, 44.25, 45.6
Palmieri, C. **1** 39.32 **3** 79.4 **4** 41.21, 41.22, 83.62, 83.64
Palmieri, C.G. **5** 82.1, 82.10 **6** 40.12, 76.41, 79.16, 81.36, 83.66, 84.34
Palumbo, R. **3** 72.33 **6** 72.1
Pamphilis, B.V.de **7** 85.46
Panagiotopoulos, N.C. **6** 45.7, 45.8, 45.24
Panattoni, C. **2** 68.9, 69.21, 73.19, 77.10, 78.22, 78.28, 78.31, 78.35, 78.44, 83.59, 83.60, 83.82, 83.83, 85.64, 86.38 **3** 39.39, 40.15, 64.42, 69.33, 72.49, 72.51, 81.55, 86.19, 86.25 **4** 19.2, 39.27, 40.19, 71.78 **5** 83.29, 83.30, 84.27
Pandarese, F. **7** 17.4, 17.5
Pandraud, H. **1** 35.40, 38.53

Panetta, C.A. **1** 50.26
Pangborn, W.A. **5** 1.38
Panke, D. **2** 61.2 **3** 45.44, 61.2 **6** 61.2, 61.3
Pannan, C.D. **7** 70.5, 80.21
Pannhorst, W. **7** 81.8
Panova, G.V. **7** 78.22
Pant, A.K. **1** 13.4, 13.5
Pant, L.M. **1** 2.52, 2.69, 2.89, 3.77, 3.84, 13.8, 13.10, 13.11, 19.9, 19.10, 19.11, 48.12 **2** 82.19 **4** 13.1, 13.6, 13.7, 15.4 **6** 13.1, 13.2, 13.14 **7** 64.47
Pantaleo, D.C. **4** 80.6, 80.7
Pantini, G. **2** 72.43, 72.45 **4** 72.19
Panunzi, A. **5** 71.23, 75.4 **6** 71.23, 75.2
Paoletti, P. **3** 76.46
Paolucci, G. **7** 33.52
Paquette, L.A. **1** 34.8 **4** 35.14, 36.22, 39.17 **5** 27.3, 27.16, 31.11, 31.39 **6** 27.11, 27.12, 28.13, 35.17, 71.35, 75.7, 75.16 **7** 31.1
Pardo, J. **5** 83.10
Park, D.Y. **2** 83.97
Park, J.J. **2** 76.61 **3** 69.29, 69.30
Park, S.-W. **7** 41.26
Park, Y.-A. **6** 71.66 **7** 83.100
Park, Y.J. **3** 45.3 **4** 45.10, 45.11, 45.24
Parkanyi, L. **7** 63.5, 63.17
Parker Junior, E.B. **5** 77.4
Parker, K. **5** 38.32
Parker, L.M. **4** 68.5
Parker, R.G. **3** 33.48, 38.17 **4** 22.10
Parker, W. **3** 53.23 **4** 53.3
Parker, W.L. **1** 50.32
Parkes, A.S. **1** 7.2, 37.2, 37.3 **3** 24.7, 31.36, 75.21
Parkinson, B. **5** 27.3, 27.16
Parks, J. **5** 71.8
Parry, G.S. **1** 1.61, 2.28
Parry, R.W. **2** 62.6, 68.13
Parsons, D.G. **6** 67.20, 67.21
Parthasarathy, R. **1** 45.52, 48.35, 50.19 **3** 2.13, 2.14, 32.33, 44.29, 44.32, 52.5 **5** 3.20, 21.12, 41.50, 48.21, 58.5 **6** 2.9, 44.16, 48.16, 48.20, 48.21, 48.31, 48.39, 48.40, 48.75 **7** 32.17, 34.5, 34.6, 44.8, 44.14, 47.3, 47.36, 48.28, 48.45, 48.49, 48.52
Parthe, E. **7** 17.24
Partridge, J.M. **2** 86.13
Partridge, M.W. **4** 35.25
Pascal, Y.L. **7** 5.15
Pascard-Billy, C. **1** 25.8, 25.9, 25.10, 58.10, 58.89 **3** 43.1 **4** 21.16, 45.14, 51.23, 53.8, 60.8 **5** 36.2, 58.8, 58.9, 60.5, 83.96 **6** 19.10, 33.51, 45.37, 51.11, 51.21, 51.41 **7** 21.9, 45.52, 47.40, 51.23, 54.13, 59.4
Pascard, C. **3** 53.30 **7** 45.31, 45.33, 45.39, 45.41, 45.51, 45.53, 58.15, 58.32
Pascard, R. **4** 60.8 **5** 60.5
Paschal, J.W. **5** 41.39
Pascher, I. **1** 46.13, 59.57 **5** 48.79 **7** 51.43
Pascoe, J.D. **4** 63.12
Pasternak, R.A. **1** 1.64, 8.35, 21.3, 32.24, 48.60, 48.90
Pasynskii, A.A. **2** 72.72 **6** 72.27, 73.58 **7** 73.12, 73.32, 73.69
Patel, C.G. **3** 72.30
Patel, H.A. **3** 68.13 **4** 71.61, 72.29 **5** 86.58 **6** 71.60
Patel, M.B. **1** 58.73
Paternotte, C. **1** 41.26
Paterson, A.R.P. **7** 47.36
Patil, H. **2** 83.154
Patmore, D.J. **7** 83.72, 83.102
Paton, M.G. **1** 59.44
Paton, P.N. **1** 51.38, 51.39, 51.40
Patrick, R. **3** 1.33
Pattabhi, V. **4** 48.54 **5** 19.15 **7** 13.20, 48.18, 48.39
Pattenden, G. **7** 59.14
Patterman, S.P. **3** 67.7
Patterson, A.L. **1** 1.79, 2.76, 2.78, 2.79, 2.80, 2.81, 2.82, 4.15, 44.20
Paudler, W.W. **7** 33.30, 36.7
Paul-Roy, S. **1** 53.27
Paul, B. **7** 47.36
Paul, G.L. **3** 48.11
Paul, I.C. **1** 31.7, 32.45, 34.5, 34.8, 37.18, 37.24, 38.26, 39.61, 39.83, 41.19, 41.20, 41.29, 41.38, 50.33, 51.24, 53.35, 54.13, 59.51, 59.64 **2** 62.39, 62.48, 65.46, 73.71, 75.16, 75.20, 75.35, 75.44, 75.78, 81.49, 81.54 **3** 11.8, 12.11, 17.11, 20.12, 23.8, 23.9, 33.41, 37.10, 39.21, 39.22, 39.23, 39.40, 39.41, 39.46, 44.37, 50.23, 50.24, 51.11, 59.24, 64.25, 71.31, 75.7, 75.17, 75.29, 75.34, 77.24, 81.52 **4** 13.15, 13.16, 16.7, 17.12, 17.13, 27.7, 29.2, 32.30, 39.14, 39.26, 50.18, 71.54, 71.55 **5** 9.18, 9.19, 9.20, 9.21, 11.16, 23.1, 31.42, 37.3, 39.27, 40.12, 42.4, 44.19, 50.15, 54.9, 73.43, 81.29, 86.52 **6** 20.16, 33.46, 34.2, 44.25, 44.26, 47.28, 51.43, 53.9, 53.16, 54.16, 76.50 **7** 9.5, 11.19, 13.1, 13.22, 17.7, 41.23, 42.14, 42.15, 50.26, 50.27, 51.11, 71.19, 71.47, 83.62
Paul, S. **1** 1.56
Paulat, V. **7** 12.5, 65.5, 70.2
Pauling, L. **1** 38.30
Pauling, P. **1** 58.24, 59.6, 59.13, 59.14 **2** 65.33, 71.52, 75.12, 77.62, 83.40, 84.68, 86.120 **3** 20.9, 38.3, 83.67 **4** 16.11 **5** 58.23, 58.48 **6** 58.7
Pauling, P.J. **2** 77.4, 83.86 **3** 86.27, 86.97 **5** 28.4, 37.2, 41.21, 71.60 **6** 32.26, 33.53 **7** 32.22, 59.1
Paulsen, H. **6** 38.12 **7** 45.29, 45.30, 45.43, 45.54
Paulus, E. **2** 75.61
Paulus, E.F. **2** 73.79, 73.81, 73.92, 75.28, 85.65 **4** 24.17, 40.5, 42.4, 44.13, 70.1, 70.5 **5** 8.19, 40.6, 69.9 **6** 41.21, 41.31 **7** 20.18, 24.11, 41.12, 41.30, 42.2
Paulus, H. **3** 35.5, 83.38
Pauson, P.L. **2** 83.214 **6** 72.7
Pautler, D. **4** 7.5
Pavel, V. **5** 60.40, 60.41

Pavelcik, F. **5** 83.44
Pawley, G.S. **1** 22.17, 24.16, 27.7 **2** 62.4, 62.8, 62.27 **3** 26.8 **4** 19.1, 31.27 **6** 30.7
Pawson, B.A. **1** 52.29
Pawson, D. **6** 65.6
Payne, D.S. **2** 64.76, 86.56 **6** 86.18
Payne, N.C. **2** 82.37 **3** 76.63, 82.5, 86.73, 86.89, 86.90, 86.93 **4** 76.19 **5** 71.16, 76.25 **6** 71.45, 71.50, 71.51, 71.57 **7** 69.28, 71.35, 71.77, 72.36
Pazdernik, L.J. **7** 69.24
Peach, J.M. **3** 40.19
Pearce, H.L. **7** 75.27
Pearce, R. **5** 71.38
Pearson, R.G. **6** 71.88
Pech, L. **7** 66.10
Pechurova, N.I. **5** 76.34 **7** 76.44, 76.64
Peck, D.N. **3** 60.14, 60.15 **6** 51.1, 51.8
Pedelty, R. **6** 85.41
Pedersen, B. **1** 2.19 **2** 76.65
Pedersen, B.F. **1** 2.15, 2.17, 2.19, 2.20, 2.21, 2.27, 5.17, 5.22, 16.40 **2** 83.42 **3** 2.3 **4** 2.4
Pedersen, C.L. **5** 38.30
Pedone, C. **1** 1.93, 21.24, 21.25, 38.18, 44.53 **2** 64.13, 72.17, 72.18, 72.38 **3** 1.31, 20.4, 21.5, 21.6, 21.7, 21.8, 21.10, 33.16, 33.19, 39.1, 62.16, 72.19 **4** 71.25, 71.26, 72.8 **5** 1.32, 8.18, 13.16, 41.3, 48.42, 71.23, 72.21 **6** 1.13, 71.23, 72.1 **7** 1.20, 20.8, 48.10, 48.27
Pedretti, U. **1** 29.19 **4** 74.1
Peerdeman, A.F. **1** 2.63, 2.65, 2.67, 48.25 **2** 64.48 **3** 59.12, 69.40 **4** 64.7 **5** 66.6
Pegel, K.H. **4** 31.32 **5** 31.41 **6** 54.13
Peiren, M.A. **3** 32.9
Pelizzi, C. **3** 85.12 **4** 9.5, 9.6, 41.22, 76.22, 83.64, 83.74, 85.14 **5** 78.2, 78.18, 78.22, 81.53 **6** 33.37, 69.21, 76.41, 79.16, 83.56, 84.34 **7** 69.13, 69.15, 78.13, 83.48, 84.47, 84.81
Pelizzi, G. **1** 41.15 **2** 83.46, 85.35 **3** 16.4, 41.2, 76.1, 79.14, 83.10 **4** 28.5, 35.13, 59.28, 76.22, 79.12, 83.1, 83.2, 83.5, 83.62, 83.74, 85.14 **5** 41.41, 78.22, 79.26, 79.27, 81.53, 84.12 **6** 3.7, 3.8, 3.29, 3.33, 59.3, 69.21, 83.56, 83.68 **7** 3.30, 40.5, 48.63, 66.4, 66.12, 69.13, 69.15, 69.32, 76.22, 84.6, 84.47, 84.81
Pellerito, L. **7** 78.4
Pelletier, S.W. **1** 58.64 **3** 54.5, 58.39, 58.40, 58.41, 58.48 **4** 58.38, 58.44 **6** 33.32 **7** 38.18, 58.21
Pellinghelli, M.A. **4** 8.4, 67.1, 67.2, 81.15, 81.16, 83.7, 83.50 **5** 8.4, 40.19, 40.20, 40.22, 40.23, 41.14, 82.10, 84.21 **6** 8.2, 11.12, 40.10, 40.11, 41.27, 79.7, 83.70, 84.29 **7** 11.12, 79.2, 79.3, 79.4, 79.5
Pellisard, D. **7** 85.16
Penavic, M. **5** 71.6, 77.2
Pendergrass Junior, D.B. **5** 9.18, 9.19, 9.20, 9.21
Pendlebury, M.H. **7** 51.56
Peneloux, A. **3** 81.2, 81.3
Penfold, B.R. **1** 1.22, 7.19, 13.23, 24.26, 26.35, 33.16, 33.17, 48.44 **2** 65.43, 69.24, 69.32, 83.139 **3** 71.35, 71.38, 71.55, 71.59, 77.1, 83.146, 86.17 **4** 71.32, 71.39, 71.42 **5** 21.7, 71.41, 71.49, 71.52, 71.92, 72.44, 72.48, 74.10, 86.38 **6** 69.5, 69.7, 81.25, 82.6 **7** 76.26, 76.39
Peng, S.-M. **6** 71.66, 83.22, 83.53 **7** 71.27, 71.28, 83.39, 83.100
Penna, G. **3** 21.3, 21.4
Penrose, A.B. **4** 53.3
Penzien, K. **4** 5.5
Pepe, G. **5** 41.8, 41.17
Pepinsky, R. **1** 8.30, 14.10, 33.29, 40.1, 48.9, 58.42 **2** 73.21, 82.8, 82.32
Pera, M.-H. **7** 33.45
Perales, A. **1** 59.46 **5** 64.10 **6** 56.5
Perazzo, P.K.de **4** 38.6
Perdikatsis, B. **6** 84.8
Perdok, W.G. **1** 11.31
Perego, G. **2** 68.1, 68.20, 75.13 **3** 71.25, 78.12, 78.16 **6** 71.54, 71.78, 71.83, 72.6 **7** 32.12, 68.16, 68.17, 68.18, 68.19, 71.132, 72.22
Perera, K. **7** 81.37, 81.39
Peresie, H.J. **3** 76.45
Perez-Garrido, S. **6** 38.7 **7** 45.38
Perez-Rodriguez, M. **1** 8.46
Perez, S. **5** 3.6 **7** 13.23, 13.24
Perkins, P.G. **6** 64.2
Perl'mutter, B.M. **3** 60.9
Perloff, A. **2** 62.16, 62.35, 83.92
Pero, R.W. **6** 59.5
Perotti, A. **5** 62.5
Perozzi, E.F. **5** 11.16 **7** 42.14, 42.15
Perrin, M. **5** 17.1, 17.2, 17.4 **7** 17.6
Perrotta, A. **5** 32.6
Perrotti, E. **3** 71.25, 78.12, 78.16 **6** 71.54, 71.78, 71.83
Perry, R.W. **4** 64.9, 65.5 **5** 65.1, 65.5
Perry, S.L. **5** 64.16
Perst, H. **6** 27.18, 27.19
Pertlik, F. **6** 32.15 **7** 4.7
Perucaud-Brianso, M.-C. **6** 39.7
Perucaud, M.-C. **7** 3.9, 28.2
Perucaud, M.C. **4** 28.1, 28.2
Perutz, M.F. **2** 86.1
Petcher, T.J. **1** 58.24 **3** 38.3 **4** 16.11 **5** 36.23, 36.24, 40.32, 58.48, 59.24, 59.27, 59.32 **6** 58.3, 58.4, 58.7, 58.38, 58.42 **7** 1.37, 32.22, 35.23, 50.38
Peters Junior, J.R. **7** 66.8
Peters, C. **1** 1.9
Peters, K. **5** 20.14 **6** 63.7
Petersen, C.E.L. **4** 31.27
Petersen, C.S. **1** 45.57, 45.58, 45.59, 45.60, 45.64, 47.9 **3** 35.13 **4** 1.5, 44.1 **7** 44.19
Petersen, D.R. **1** 20.8
Petersen, J.L. **7** 73.28
Petersen, K.B. **6** 29.3
Petersen, R.B. **4** 73.31
Peterson, C.R. **2** 62.39
Peterson, E.R. **3** 3.1 **5** 3.7
Peterson, J. **1** 48.67, 48.68

Peterson, M.B. **5** 64.35
Peterson, S.W. **1** 2.57, 3.7, 3.8 **3** 3.16, 6.1, 48.9 **4** 13.2
Petraccone, V. **3** 22.4
Petracek, F.J. **3** 33.48, 38.17
Petrissans, J. **4** 21.4 **5** 21.2, 21.6 **6** 21.3
Petrova, I.M. **7** 63.18
Petrukhin, O.M. **2** 77.28 **3** 77.12, 77.22
Pettersen, R.C. **1** 31.23, 49.14, 54.15 **2** 73.25 **4** 49.10 **5** 51.20, 51.41, 53.13 **6** 51.6 **7** 53.15, 71.17, 75.27
Pettit, G.R. **4** 37.4
Pettit, L.D. **3** 11.10
Pettit, R. **2** 75.26 **3** 72.23, 72.25, 72.26, 72.41, 75.10 **4** 75.24 **7** 71.33, 71.34
Pexxanit, J.O. **7** 58.3
Peyronel, G. **2** 80.27, 80.28 **5** 29.3, 79.6, 80.16 **6** 79.4 **7** 8.17, 85.27
Pezzullo, J.C. **5** 36.1
Pfefferkorn, K. **6** 71.85
Pfister, G. **5** 51.39
Pfleiderer, W. **6** 45.38
Pflug, J.L. **3** 62.11
Pfluger, C. **7** 83.76
Pfluger, C.E. **1** 9.49, 19.23 **2** 77.7, 77.40, 77.41, 77.43 **3** 77.24, 78.1 **5** 83.50, 83.51 **6** 1.10, 17.5, 31.4, 68.13, 72.20, 77.7, 77.10, 77.11, 78.3 **7** 15.4, 71.53, 77.8
Phavanantha, P. **2** 83.215
Phelps, D.W. **6** 39.14, 48.82 **7** 41.25
Philippot, E. **5** 64.2 **7** 48.5, 83.96
Philips, J.C. **4** 39.17
Phillips II, T. **5** 47.6 **6** 40.21 **7** 47.7, 47.8
Phillips Junior, G.N. **7** 51.54
Phillips, D.C. **1** 36.22, 58.4 **3** 36.5
Phillips, F.L. **6** 65.6, 86.29
Phillips, L. **4** 31.32 **6** 54.13, 74.15 **7** 51.56
Phillips, R.K. **3** 83.12, 83.33
Phillips, R.P. **6** 75.23
Phillips, S. **7** 27.9
Phillips, S.E.V. **6** 38.26 **7** 38.44, 38.45, 38.46
Phillips, T. **2** 77.30, 77.32 **3** 44.36, 44.39 **5** 47.9
Phillips, T.E. **6** 60.3
Phipps, J.R. **4** 35.8
Phizackerley, R.P. **7** 38.30, 38.33, 38.34, 38.36, 38.38, 38.39, 38.40, 50.30
Phizackerly, R.P. **7** 38.32
Photis, J.M. **6** 75.16
Piacenza, L.P.L. **4** 31.32 **5** 31.41 **6** 54.13
Piantadosi, C. **6** 53.15
Piazzesi, A. **4** 39.3, 39.4
Piazzesi, A.M. **1** 4.16
Piciulo, P.L. **7** 49.13, 49.17
Pickard, F.H. **6** 28.27, 28.28, 40.7, 86.45
Pickworth, J. **1** 49.32
Picone, R.F. **7** 1.6
Pidcock, A. **3** 86.1, 86.40 **5** 86.1 **6** 69.10 **7** 86.2
Pierce Junior, A.G. **5** 84.15
Pierce-Butler, M. **6** 63.1 **7** 63.21, 69.30, 69.31
Pieret, A.F. **3** 48.44 **4** 48.49 **5** 39.10, 39.11, 39.12, 48.74
Piermarini, G.J. **3** 19.4 **5** 5.1
Piero, G.del **6** 71.54, 71.78, 71.83, 72.6 **7** 68.16, 68.17, 68.18, 68.19, 71.132, 72.22
Pieroni, J. **6** 37.8
Pieroni, O. **5** 48.75
Pierpont, C.G. **3** 85.45, 86.96 **4** 73.15, 85.12, 86.29, 86.60 **5** 86.65, 86.75 **6** 80.18, 84.43, 86.59, 86.66 **7** 72.41, 72.42, 76.63, 84.78, 86.76
Pierrot, M. **3** 82.24 **5** 41.8, 41.17
Pieterse, M.J. **4** 54.17
Piette, J.L. **6** 70.2
Pignataro, E. **1** 5.15 **2** 67.11
Pignedoli, A. **2** 80.27, 80.28 **5** 79.6, 80.16 **6** 79.4 **7** 8.17, 85.27
Pignolet, L.H. **3** 78.19 **7** 80.13, 80.19
Pijper, W.P. **3** 1.5
Pilati, T. **5** 17.16, 31.28 **7** 28.5, 33.5
Pilbrow, M.F. **4** 83.63
Pillai, P.M. **5** 31.20
Pilling, D. **1** 49.34
Pilot, J.F. **4** 64.25, 64.39 **5** 64.47
Pilotti, A.-M. **3** 58.44 **4** 71.37 **5** 45.48, 53.3, 59.12 **6** 31.21, 31.22, 53.1, 58.56 **7** 31.13, 31.25, 38.68, 38.69, 51.48, 52.2, 54.1
Pinel, R. **4** 39.26 **5** 39.27
Pinkerton, J.M.H. **1** 58.14
Pinkerton, M. **1** 50.45 **3** 50.25
Pinkus, J.L. **3** 40.18
Pinnavaia, T.J. **5** 77.23
Pino, P. **2** 72.39
Pioda, L.A.R. **1** 50.44
Pioli, A.J.P. **3** 71.57
Pipal, J.R. **3** 83.133 **4** 72.9, 81.26, 81.28
Pipenberg, K.J. **1** 50.11
Piquion, J. **6** 58.10
Piret, J. **2** 72.57
Piret, P. **1** 10.14, 41.26, 41.27, 41.30 **2** 60.14, 60.23, 60.25, 60.26, 60.30, 60.32, 60.46, 60.66, 72.32, 72.54, 72.57, 73.38, 83.169 **3** 20.18, 60.17, 60.18, 67.3, 67.6, 71.17 **4** 20.12, 24.19, 32.21, 40.17, 48.4, 48.9, 48.11, 48.27 **5** 16.20, 48.1 **6** 48.48 **7** 72.3, 72.4
Pirkle, W.H. **4** 16.7
Piron, J. **2** 83.169 **3** 71.17
Pitacco, G. **7** 21.6
Pitman, D.T. **1** 43.14
Pitman, I.H. **3** 44.24
Pitt, G.J. **1** 44.49
Pitts, W.D. **4** 71.90
Plaksin, P.M. **4** 83.134
Plana, F. **6** 16.7, 16.10, 17.10, 17.12, 17.13, 35.12 **7** 1.8
Planinsek, F. **3** 45.11
Plastas, H.J. **4** 86.38 **5** 86.30, 86.31
Plastas, L.A. **1** 32.32, 41.25 **4** 8.10, 33.31
Plato, K. **1** 20.31
Player, R.B. **6** 1.6, 1.9
Plessi, L. **7** 29.10
Plessius, H.A. **6** 10.1

Pletcher, J. **3** 3.47, 40.18, 41.11, 59.5 **4** 13.13, 16.17, 41.18, 46.6, 59.5 **5** 28.8, 41.22 **6** 31.23, 41.7, 41.18, 44.27, 51.2 **7** 13.18, 16.11, 41.11, 44.26, 44.29, 46.2, 46.3, 46.4, 46.6
Plettinger, H.A. **2** 81.42
Plieth, K. **1** 5.34, 16.44, 20.31, 20.32 **2** 64.51, 76.7, 76.9 **4** 21.2, 21.3, 64.22, 64.23 **5** 64.25, 64.26, 64.31 **6** 64.26
Pluth, J. **6** 83.53
Pluth, J.J. **7** 86.90
Plyasova, L.M. **7** 3.20
Plymale, D.L. **2** 86.73
Pocar, D. **7** 41.22
Podberezskaya, N.V. **3** 76.39 **4** 76.4, 76.5 **5** 76.16, 76.19
Poddar, A. **7** 17.11
Podlahova, J. **4** 81.22
Podolsky, G. **5** 77.23
Poilblanc, R. **7** 86.25
Pointer, D.J. **4** 31.7, 31.11, 31.23, 41.27 **5** 35.37
Poisson, J. **7** 54.13
Pokrywiecki, S. **4** 51.35, 51.36, 60.39 **5** 11.11, 11.15 **6** 17.23
Poland, J.S. **5** 68.34
Polder, W. **7** 64.34, 64.35
Poling, M. **3** 59.32 **6** 84.35 **7** 43.7, 64.52, 64.53
Poling, M.A. **7** 45.37
Poljak, R.J. **2** 83.62 **5** 36.3 **6** 16.14
Pollak, M. **1** 19.8
Pollard, D.R. **1** 38.58, 59.45 **3** 60.42, 60.55, 64.56, 64.57 **4** 51.45, 64.8
Pollick, P.J. **2** 85.71
Pollitz, S. **7** 11.2, 11.3
Pollock, J.J.S. **1** 40.33
Polynova, T.N. **1** 9.22 **2** 66.14, 66.18 **3** 68.9, 76.67 **5** 76.32, 76.33, 76.34, 76.37, 82.19 **6** 76.22 **7** 48.47, 67.5, 76.43, 76.44, 76.46, 76.47, 76.48, 76.50, 76.64, 81.26, 82.1, 82.5, 82.7, 82.9, 82.10
Pompa, F. **1** 40.25, 40.26 **2** 86.74
Pond, J.S. **2** 83.94
Ponomarev, V.I. **7** 60.26, 81.40
Poonia, N.S. **4** 67.24
Poore, M.C. **3** 66.2, 66.3
Popa, E.V. **6** 76.2
Popa, S.E.V. **7** 76.32
Pope, L.E. **7** 8.24, 8.25
Popov, A.I. **2** 60.93, 60.94 **4** 83.58
Popov, E.I. **5** 1.14
Popova, E.G. **7** 26.1
Popova, T.V. **5** 71.7
Popova, Z.V. **1** 33.27, 33.28 **4** 33.2
Popovich, G.A. **7** 78.20
Poppe, H. **1** 36.35
Poppleton, B.J. **1** 45.31, 59.24 **3** 45.21, 45.22, 45.23 **6** 59.24 **7** 59.17
Porai-Koshits, M.A. **1** 8.6, 9.19, 9.22 **2** 66.14, 66.18, 72.11, 76.6, 76.27, 76.29, 77.28, 77.33, 79.24, 80.19, 81.5, 81.14, 81.43, 81.44, 83.78, 83.95, 83.138, 83.175, 83.176, 83.177, 83.186, 85.59 **3** 68.9, 72.32, 76.15, 76.21, 76.67, 77.11, 77.22, 77.30, 82.9 **4** 9.7, 76.7, 76.27, 77.11, 77.12, 81.43 **5** 61.10, 76.8, 76.32, 76.33, 76.37, 77.21, 81.59, 82.19, 84.10, 84.13 **6** 37.3, 76.22, 81.39, 81.43, 84.7 **7** 8.8, 33.15, 35.10, 48.47, 61.1, 67.5, 76.10, 76.11, 76.12, 76.13, 76.15, 76.16, 76.19, 76.24, 76.36, 76.43, 76.44, 76.46, 76.47, 76.48, 76.50, 76.53, 76.54, 76.55, 76.64, 78.21, 78.22, 81.26, 81.31, 81.32, 82.1, 82.5, 82.9, 82.10, 82.16, 84.14, 84.16, 85.5
Porri, L. **2** 72.43, 72.45, 72.67 **3** 74.3, 74.4 **4** 72.19 **6** 71.49, 71.92
Porta, F. **7** 71.118
Porta, P. **2** 75.52, 85.88 **3** 72.7, 85.22, 85.50 **4** 83.89, 85.33 **5** 79.4, 80.4 **6** 85.7, 86.23
Portanova, R. **6** 84.18
Porte, A.L. **1** 1.71, 1.72
Porten, J.A. **5** 39.16
Porter, D.W. **3** 86.27
Porter, E.J. **4** 33.12
Porter, J.K. **4** 81.48
Porter, S.K. **3** 33.8, 33.10 **5** 31.11, 33.16, 71.21 **6** 82.11
Portheine, J.C. **3** 51.32 **4** 51.43, 51.44
Posey, R.G. **7** 31.7
Pospelova, L.A. **7** 8.8
Post, B. **1** 1.1, 1.2, 1.38, 1.48, 1.90, 5.15, 5.16, 8.1, 8.2, 20.15, 20.16, 39.2, 39.18, 39.71 **2** 63.7 **3** 21.7 **4** 78.9, 78.14
Post, M.L. **5** 84.24 **6** 39.23, 81.10, 81.15, 81.21 **7** 29.11, 36.22, 39.37, 39.38, 47.6, 47.16, 47.35, 81.3
Postma, H.J. **7** 19.17
Potenza, J. **6** 69.9, 75.27, 82.24 **7** 41.29, 73.48
Potenza, J.A. **2** 62.1, 62.3, 62.9, 62.10, 62.24
Potier, P. **7** 58.32
Potter, R. **1** 33.68
Potworowski, J.A. **6** 28.27, 28.28
Povet'eva, Z.P. **1** 11.30 **2** 62.29, 69.20 **4** 41.4
Povey, D.C. **5** 1.27, 38.22, 84.24 **6** 20.1, 38.23, 48.51, 84.33 **7** 84.61
Povey, D.V. **5** 38.20
Powell, D.B. **6** 79.2
Powell, H.K.J. **6** 81.25, 82.6
Powell, H.M. **1** 15.4, 16.41, 17.19, 45.2 **2** 60.86, 61.9, 61.15, 62.50, 68.22, 69.48, 71.16, 71.20, 71.21, 71.28, 73.12, 73.109, 75.52, 75.71, 86.7, 86.41, 86.44, 86.45, 86.67, 86.99, 86.105, 86.106 **3** 86.32, 86.76, 86.81, 86.82, 86.86 **4** 71.68, 86.56 **5** 68.17, 86.27 **7** 86.73, 86.77, 86.78, 86.79, 86.80
Powell, J. **5** 72.55 **7** 71.123
Powell, R.G. **7** 58.18
Power, L. **3** 41.11, 59.5 **6** 41.18

Power, L.F. **3** 11.12 **4** 2.5, 11.5, 83.66, 83.111 **6** 11.13 **7** 5.16
Powers, T.J. **7** 51.49
Powers, T.W. **4** 30.5
Poyer, N. **3** 19.16
Pozhidaev, A.I. **5** 76.32, 76.33 **7** 48.47, 67.5, 76.48, 76.50
Pra, A.del **6** 3.35
Prabhu, K.V. **3** 27.4
Pradilla-Sorzano, J. **5** 77.6, 77.7
Prakash, A. **1** 26.24, 26.25 **4** 58.16 **5** 40.15
Prasad, H.S. **4** 66.9
Prasad, N. **5** 2.19 **6** 2.14
Prasad, S.M. **3** 13.6 **4** 2.16 **6** 28.19 **7** 2.5, 19.3
Pratesi, P. **6** 3.35
Pratt, J.L. **7** 73.45
Pratt, M.R.A. **1** 35.38
Pre, S.du **1** 48.26
Precigoux, C. **4** 51.18
Precigoux, G. **5** 21.16, 51.13 **6** 51.3, 51.7, 51.9 **7** 17.30, 17.31, 33.50, 51.3, 51.9, 59.3
Predieri, G. **6** 33.37 **7** 78.8
Preer, J.R. **2** 86.26
Preez, A.L.du **6** 73.57
Preiss, H. **2** 66.8 **4** 66.1
Prelesnik, B. **3** 15.8 **6** 82.7 **7** 81.17
Prelog, V. **1** 34.6 **4** 50.13
Presley, C.T. **3** 21.2
Preston, H.S. **2** 71.4, 71.5, 71.31, 83.128 **3** 15.1, 15.2, 35.21, 83.68, 83.69, 83.136 **4** 86.38
Preston, N.W. **7** 58.12
Preston, P.N. **7** 41.19
Preuss, J. **4** 36.37 **5** 36.31 **7** 16.9
Preuss, L. **3** 39.24, 39.25, 39.27
Preut, H. **4** 69.38 **7** 69.22
Prewitt, C.T. **2** 73.89 **5** 83.35
Price, D.A. **4** 17.4
Priestley, G.M. **5** 41.11
Prigzkow, H. **7** 11.6
Prince, E. **4** 83.94, 83.95 **5** 19.12, 19.13, 33.2
Prince, R.H. **2** 83.208 **7** 71.87
Pringle, G.E. **4** 81.39 **5** 71.55, 73.14, 86.40
Prinzbach, H. **7** 28.8
Pritzkow, H. **6** 3.14 **7** 1.1, 37.1, 37.2, 37.3, 37.10
Procter, I.M. **3** 83.105
Proctor, G.R. **7** 35.31, 35.37
Prokhvatilov, A.I. **5** 5.2
Prokof'ev, A.K. **5** 69.3
Prosen, R.J. **1** 49.32 **2** 60.51
Proskina, N.N. **7** 76.6
Prota, G. **5** 36.15 **7** 36.12
Protas, J. **3** 2.5, 74.2 **4** 2.9, 48.23, 48.43 **5** 28.7, 28.11, 28.12, 28.14, 31.33, 48.61, 48.62, 73.32, 73.40, 74.3, 74.4, 74.5, 75.17, 75.19, 75.20 **6** 2.5, 37.4, 48.38, 48.47, 48.53, 73.29, 74.4, 74.8, 74.9 **7** 48.50, 48.51, 73.60
Proulx, T.W. **6** 21.9
Prout, C.K. **1** 49.5, 49.6 **2** 60.70, 60.73, 60.74, 60.75, 60.76, 60.107, 60.148, 71.27, 73.40, 73.75, 75.22, 81.27, 81.45, 81.48, 81.55, 81.66, 81.81, 82.14, 83.159, 83.167, 85.9, 85.10, 85.11, 85.36, 85.78 **3** 13.18, 25.6, 33.38, 33.39, 33.40, 35.18, 40.19, 40.20, 40.21, 45.31, 60.29, 64.43, 64.70, 65.10, 66.5, 71.5, 71.6, 81.16, 81.22, 81.42, 82.23, 83.57 **4** 2.11, 2.24, 24.2, 36.35, 40.8, 60.9, 60.10, 60.11, 60.15, 60.16, 71.15, 73.19, 73.29, 81.12, 81.13, 81.14, 81.23, 81.41, 81.49, 81.54 **5** 7.7, 60.19, 60.20, 60.27, 60.29, 60.30, 60.31, 60.32, 60.33, 60.38, 64.3, 64.21, 64.29, 64.36, 72.34, 72.53, 73.6, 73.16, 73.17, 73.18, 77.2, 82.21 **6** 10.2, 84.1
Prout, K. **5** 68.34 **6** 32.17, 60.25, 73.28, 82.15 **7** 4.8, 32.4, 35.28, 35.39, 36.27, 48.38, 71.24, 71.74, 71.91, 72.14, 72.16, 73.4, 73.5, 73.6, 73.7, 73.8, 73.14, 73.23, 73.36, 73.49, 73.50, 73.51, 73.52, 73.53, 73.54, 73.55, 73.65, 73.66, 73.67, 74.7, 76.72, 82.20
Pruet, H. **5** 69.27
Prusiner, P. **5** 35.1, 44.8 **6** 47.23, 47.24, 50.2, 50.5
Przybylska, M. **1** 17.29, 58.13, 58.15, 58.19, 58.50, 58.62, 58.80, 58.91, 58.92 **3** 54.11 **5** 31.15 **6** 58.23 **7** 31.26, 58.6
Puar, M. **6** 35.13
Puar, M.S. **5** 59.23 **7** 59.13
Puckett, R.T. **1** 9.49, 51.25 **3** 31.42, 58.55, 58.56 **4** 24.16, 33.47 **5** 31.36 **7** 35.30
Puddephatt, R.J. **5** 71.16, 71.22
Puff, H. **7** 81.16, 81.18
Puliti, R. **1** 59.8 **2** 60.110, 60.112 **5** 50.4 **6** 1.13, 51.5, 51.44
Pulsinelli, P. **6** 41.7, 44.27
Purdham, J.T. **4** 21.22, 28.12 **5** 31.16
Purdum, W.R. **7** 64.49
Purdy, R.H. **5** 51.45, 51.46
Purnell, L.G. **7** 83.10
Putkey, E. **3** 48.7
Putkey, E.F. **3** 48.6 **4** 26.7
Putten, N.van der **6** 32.2
Putzeys, J.-P. **4** 32.23, 40.17, 48.4 **7** 17.20, 32.24
Putzeys, J.P. **5** 30.2 **6** 32.36, 48.48 **7** 32.28, 32.29, 60.2
Puxley, D.C. **3** 69.2 **6** 69.2
Pyle, J.L. **7** 24.5
Pyronel, G. **2** 63.5
Pyzalska, D. **7** 43.6, 43.8

Qasba, R. **1** 28.24
Quagliano, J.V. **7** 86.21
Quagliata, C. **5** 60.40, 60.41
Quaglieri, P. **5** 81.45
Quarles, W.G. **6** 35.19, 35.20
Quastlerova-Hvastijova, M. **7** 83.23
Que Junior, L. **7** 85.42
Qui, D.T. **5** 81.16 **6** 81.22
Quick, A. **3** 82.25 **7** 35.6

Quicksall, C.O. **7** 11.1, 64.3, 64.8, 86.1
Quigley, G.J. **3** 45.36
Quin, L.D. **4** 64.13, 64.14, 86.17 **5** 86.42 **7** 64.9, 64.10, 64.16, 64.17
Quinby, M.S. **3** 76.40
Quinn, C.B. **6** 39.29
Quinn, J.D. **7** 76.68
Quinn, R.K. **6** 33.9, 83.19
Quiocho, F.A. **7** 51.54
Qureshi, I.H. **6** 29.1

Raaen, V.F. **6** 60.17, 60.18
Rabano, M. **4** 24.12
Rabaud, H. **1** 26.34
Rabczenko, A. **7** 47.31
Rabinovich, D. **1** 1.91, 18.22, 18.23, 18.24, 18.25, 18.27, 23.16 **3** 1.25, 17.6, 19.8, 19.9, 19.10, 38.2 **4** 1.6, 13.4, 51.49 **5** 1.23, 1.24, 1.26, 19.18, 36.17 **6** 1.16, 28.19 **7** 13.9, 13.21, 19.10, 19.12, 31.4, 45.22
Rabinowitz, H.N. **3** 71.31 **7** 76.71
Rabinowitz, I.N. **1** 21.12
Racanelli, P. **2** 72.43, 72.45 **4** 72.19
Racine-Weisbuch, M. **3** 19.15
Rademaker, W.J. **4** 62.1
Radlick, P. **4** 53.2, 59.8 **6** 54.15
Radonovich, L.J. **1** 3.83 **3** 30.4 **4** 49.22, 83.77 **6** 63.8, 84.20
Radoslovich, E.W. **4** 84.1 **5** 83.8
Rae, A.D. **2** 83.112, 83.194 **5** 33.46 **7** 80.6, 80.7
Rae, A.I.M. **1** 16.13, 59.41 **2** 85.42 **3** 86.70
Raeburn, U.A. **3** 82.4, 86.12
Rafferty, M. **6** 73.54
Raghavan, N.V. **5** 71.76, 71.78 **7** 71.131
Raghavan, V.K.V. **1** 59.68
Rague Schleyer, P.von **1** 31.43
Rahim, A.M.Abd.el **1** 56.15
Rahman, A. **3** 47.5 **5** 47.7
Rahman, A.-ur- **5** 58.1
Rahn, R.O. **1** 44.72
Raimondi, M. **4** 31.20 **5** 31.28
Raithby, P.R. **6** 84.45
Rajagopal, H. **5** 48.30
Rajaram, J. **5** 68.27 **6** 71.88
Ralph, B.J. **7** 80.6, 80.7
Ram, N. **5** 83.45
Ramachandra, R. **6** 31.29
Ramachandran, G.N. **1** 59.68 **3** 48.29
Ramage, R. **6** 31.18, 31.25
Ramakumar, S. **7** 37.7
Raman, S. **1** 58.8
Raman, S.V. **3** 48.1, 48.2, 48.3
Ramanadham, M. **5** 48.11, 48.15 **7** 48.15, 48.16
Ramani, R. **7** 48.54
Ramaseshan, S. **1** 58.75 **7** 2.2, 3.4
Ramaseshan, S.M. **1** 16.48, 16.50
Ramirez, F. **1** 46.7 **2** 64.61 **4** 64.25, 64.39, 64.45 **5** 64.47 **6** 46.5, 46.11, 64.36 **7** 64.43
Ramsay, W. **1** 59.6
Ramsoy, S. **4** 12.8
Rand, G.A. **7** 48.8, 48.9
Randaccio, L. **1** 21.29 **2** 78.59, 84.50, 84.51 **3** 76.81, 78.14, 78.26, 83.150, 85.32, 85.33, 85.40 **4** 71.40, 77.14 **5** 67.22, 69.19, 71.31, 71.42, 71.57, 76.39, 76.43, 78.5, 78.6, 78.9, 83.41, 86.67 **6** 69.13, 71.22, 71.89, 71.95, 76.42, 78.11, 83.10, 86.20, 86.57 **7** 71.7, 71.86, 78.2, 78.4, 78.5, 78.7, 86.92, 86.95, 86.96
Randall, E.W. **4** 71.27
Rango, C.de **5** 3.20, 28.18, 28.20, 30.2 **6** 17.3
Rannev, N.V. **1** 7.5, 7.7
Rao, A.V.R. **6** 59.30
Rao, G.S. **2** 81.47
Rao, J.K.M. **2** 82.1 **4** 82.4, 82.5 **5** 2.1
Rao, R.V.G.S. **2** 81.47
Rao, S.N. **6** 2.9, 48.16, 48.20, 48.21 **7** 34.5, 34.6
Rao, S.T. **1** 21.15, 48.94 **3** 45.35, 47.21, 47.30, 48.18, 48.34 **4** 31.19, 47.8 **5** 47.28, 48.34 **6** 48.9, 48.49, 48.62 **7** 48.20, 48.24
Rao, V.N.M. **7** 20.10
Raper, E.S. **3** 81.8 **4** 33.15, 41.1 **5** 33.23, 81.9 **6** 41.10
Raper, G. **1** 41.35 **4** 71.13, 71.33, 86.28 **6** 86.7
Raphael, R.A. **1** 22.6, 39.56 **5** 22.3
Rappard, J.C.C.W. **6** 73.60
Rasmussen, H. **6** 44.18
Rasmussen, P.G. **3** 83.49
Rasmussen, P.W. **5** 28.15
Rasmussen, S.E. **1** 3.52, 3.53, 36.4, 58.6 **2** 68.14, 76.47, 83.185, 84.25, 85.7 **3** 32.8, 85.11 **4** 83.60 **5** 1.10, 45.23, 62.4, 81.14 **6** 32.3, 32.4, 32.5, 71.91, 82.2 **7** 83.90
Rasoanaivo, P. **7** 58.32
Raspin, K.A. **2** 86.102, 86.117
Rassat, A. **1** 31.12 **7** 33.32
Rastogi, R.P. **6** 56.4
Raston, C.L. **5** 79.1, 80.6, 80.7 **6** 80.9, 80.10 **7** 41.5, 76.14, 77.1, 77.4, 78.26, 80.3, 83.6, 83.91, 83.109
Rath, J. **5** 76.2
Rau, T.F. **5** 79.5, 79.12 **7** 79.12
Raupach, M. **4** 84.1 **5** 83.8
Rausch, M.D. **7** 71.120, 73.58
Raveau, B. **3** 41.6
Ravn-Jonsen, E.J. **1** 59.28
Raymond, K.N. **2** 76.39, 76.40 **3** 76.65 **4** 71.50, 75.14, 75.15, 75.31, 76.14, 83.32 **5** 73.45, 75.26, 75.34 **6** 3.24, 22.3, 64.57, 84.39 **7** 3.27, 71.94, 71.102, 83.25
Rayner-Canham, G.W. **2** 83.104
Razakov, R. **6** 58.20
Read, L.K. **4** 36.22, 39.17 **6** 27.11, 27.12
Reay, B.R. **6** 85.15
Reck, D. **1** 1.68
Reck, G. **3** 28.1, 52.2 **4** 52.2, 52.3 **6** 33.43, 51.38 **7** 33.37, 36.14
Reddy, B.S. **4** 47.2 **7** 47.4, 47.21

Reddy, J.M. 1 58.8 2 60.99
Redhouse, A.D. 2 71.32, 71.46, 86.47 3 71.52 5 71.25, 71.44 6 73.16 7 39.16, 71.133
Redjeb, S.B. 7 5.15
Reed II, H.L. 6 60.17, 60.18
Reed, A.T. 5 81.6 6 63.6, 69.15, 69.16
Reed, C.A. 5 86.36 7 49.22
Reed, H.J. 1 41.34
Reed, J. 5 86.65 6 86.19, 86.59 7 86.64
Reed, J.W. 7 17.35
Reed, L.L. 3 76.40 4 20.8, 22.8, 41.20, 51.52 6 33.54, 48.69
Reed, T.B. 1 5.11
Reeder, S.K. 7 71.40
Reedijk, J. 4 83.81 5 83.104
Reeke Junior, G.N. 1 44.62, 58.103
Reeke, G.N. 3 69.37, 69.38
Rees, A.H. 1 58.73
Rees, B. 1 18.4, 18.5 3 18.1, 18.2, 18.3, 18.4, 33.29 6 74.1, 74.2
Rees, D.A. 7 45.9, 45.10
Rees, G.V. 4 73.29 7 71.24, 72.14, 72.16, 73.4, 73.5, 73.6, 73.7, 73.8, 73.14, 73.36, 73.49, 73.50, 73.52, 73.53, 73.54
Reese, C.B. 5 47.31
Regev, H. 4 29.1
Regitz, M. 6 64.25 7 64.39, 64.40
Regler, D. 5 71.36, 71.39, 72.24
Rehnberg, G. 4 39.1 5 42.1 7 11.14, 32.25
Reichardt, C. 7 36.8
Reichenbach, G. 6 86.8
Reichert, B.E. 7 72.29
Reichert, W.W. 6 83.28
Reichstein, T. 1 51.35
Reid, K.I.G. 3 39.22, 39.23, 39.41, 75.17, 75.29
Reid, S.T. 1 39.50
Reimann, C.W. 2 83.92, 83.102, 83.106, 83.107, 83.165 3 83.6, 83.63, 83.95 4 83.94, 83.95, 83.102, 84.11
Reimann, R. 5 86.12, 86.20 6 86.6
Reimer, K.J. 6 71.11, 72.26 7 73.20
Reimlinger, H. 3 41.10, 42.4 4 32.24
Reinheimer, H. 4 71.51
Reis Junior, A.H. 3 62.15, 83.90 4 62.7, 62.8 5 83.68, 83.70 6 83.46
Reis, A.H. 6 83.45
Reiser, B. 4 62.12
Reisner, M.G. 5 72.20
Reistad, T. 1 39.35, 39.36
Remlinger, H. 3 32.9
Remmel, R. 6 33.27
Remoortere, F.P.van 3 20.8, 32.26, 38.18, 63.11, 69.37, 69.38 4 21.19, 33.3, 69.18, 69.19 6 33.17 7 21.12, 38.9, 38.10, 38.11, 38.20, 38.21, 38.22, 38.23, 84.23
Renaud, A. 7 84.56
Renaud, M. 1 5.1 4 19.3, 19.4 5 19.7, 19.8, 21.4, 21.5
Rendall, I.F. 2 78.61 3 78.25 4 80.15
Rendle, D.F. 4 51.13, 51.22, 51.32, 64.44, 65.12, 65.13, 65.14 6 51.10, 51.13, 51.14, 68.5 7 8.22, 33.26, 51.5, 83.72, 83.73, 83.102
Reno, M.A. 6 45.43
Rentzeperis, P.J. 7 32.41
Renzi, A.de 5 71.23 6 71.23
Replogle, L.L. 1 39.90 4 39.43
Repmann, H. 1 5.34
Rerat, B. 1 16.43, 16.45, 16.54, 16.56, 40.14, 43.20, 43.21 2 60.28 3 9.18, 9.20, 11.21 4 32.18 6 24.9
Rerat, C. 1 10.9, 10.21, 17.43, 31.12, 32.23, 33.14, 33.22, 33.34, 40.14, 43.20, 43.21, 52.6, 59.1 2 60.28, 64.38, 64.39 3 9.18, 9.20, 11.21, 32.7 4 32.18
Restivo, R. 3 33.21, 50.14, 73.55, 81.44 4 68.11, 71.37, 73.32, 73.33 5 56.10, 56.11 7 71.107, 86.23
Restivo, R.J. 5 20.18, 58.45, 59.26, 59.34 6 73.43, 74.7 7 27.9, 83.99
Rettig, M.F. 3 73.24
Rettig, S. 4 48.38 5 81.58, 86.3, 86.4, 86.5, 86.6, 86.7, 86.9
Rettig, S.J. 6 62.7, 64.21, 73.53, 76.32 7 59.5, 62.6, 62.7, 68.11, 68.13, 68.14
Reuter, U. 7 11.8, 39.6
Rewicki, D. 5 29.1 7 29.12
Rey, M. 7 53.7
Reynhardt, E.C. 6 77.12
Reynolds, C.T. 4 3.14
Reynolds, W.F. 3 60.34
Rhine, W. 3 62.3 5 67.18, 67.20 7 67.6, 67.8
Rhine, W.E. 6 68.16 7 26.4, 27.2
Ribar, B. 6 16.11, 35.11 7 35.21
Ribi, M. 3 9.8
Ricard, L. 6 58.14, 71.69, 73.52, 73.63, 85.22 7 72.18, 80.9, 80.10, 80.20
Ricci Junior, J.S. 4 39.41 5 80.13, 85.36 6 64.36, 71.10, 86.48 7 64.43
Ricci, G.M.B. 3 86.94
Ricci, J. 1 39.69
Ricci, J.S. 3 11.14, 11.15 4 71.93, 86.45
Riccieri, F.M. 1 35.12, 35.14
Rice, D.A. 3 83.5 4 84.2 6 38.16, 84.3, 85.5 7 81.15
Rich, A. 1 44.3, 44.30, 44.39, 44.55, 44.56, 44.58, 44.59 2 60.131, 73.22 3 44.4, 44.9 4 60.34 5 47.33, 60.11 6 47.30, 47.31 7 3.11, 47.37, 84.49
Richard, P. 5 81.16 6 11.15, 81.22
Richards, A. 3 78.3
Richards, G.F. 2 72.47, 73.17, 75.3, 75.6 4 52.1 6 45.2, 45.48 7 45.47, 52.6
Richards, J.A. 5 69.22 7 69.7, 69.19
Richards, J.H. 2 73.43
Richards, J.P.G. 1 6.7, 13.32, 17.21 4 17.4
Richards, P.M. 6 33.9, 83.19
Richards, R. 5 81.12
Richards, R.L. 2 71.38 7 86.110
Richards, S. 2 76.65
Richardson, D.C. 2 76.44
Richardson, M.F. 2 77.29 3 77.23 4 83.45 7 76.69
Riche, C. 4 38.18 5 20.11, 38.3 6 19.10, 20.14, 38.27, 51.11, 51.21, 58.5, 58.12, 58.39 7 21.9, 45.33, 45.52, 51.23, 58.15, 58.32, 59.4

Richle, W. **5** 45.43
Richter, M.J. **5** 32.23
Richter, R. **7** 84.67, 85.11
Rickard, C.E.F. **2** 76.26 **3** 80.12, 80.13, 80.15
Rickard, R.C. **6** 79.17
Riddell, F.G. **5** 38.18
Ridley, D.R. **4** 66.4, 66.9 **6** 66.9 **7** 66.1, 66.2, 66.3
Ried, W. **7** 41.26
Riedel, E.F. **4** 86.27
Riedl, M.J. **4** 83.75 **5** 85.22
Riera, V. **6** 75.12 **7** 73.41, 75.1
Rietveld, H.M. **2** 80.10 **3** 19.12, 19.13, 19.14
Riganti, V. **1** 36.7, 36.8, 36.11, 36.13
Rigatti, G. **1** 33.75
Rigau, J.J. **3** 39.45
Rihs, G. **5** 51.24 **6** 20.6, 37.7 **7** 21.2, 28.3, 28.7, 38.24
Rij, C.van **6** 36.14
Rij, J.van **7** 7.1, 7.2
Riker-Nappier, J. **7** 86.34, 86.38
Riley, P.E. **5** 85.24 **7** 71.58
Rimala, T.S. **5** 39.13, 39.32
Rimmer, B. **1** 53.8
Rimmer, B.M. **1** 51.45
Rimsky, A. **2** 60.7 **4** 8.8
Rinehart Junior, K.L. **1** 50.33 **4** 50.18 **6** 50.8 **7** 17.7
Ringertz, H. **1** 2.71, 8.39, 8.40, 44.24 **3** 13.9, 62.1
Rinn, H.W. **7** 38.20, 38.21, 38.22, 38.23
Ripamonti, A. **1** 33.81, 59.8, 59.20 **2** 60.110, 60.111, 60.112, 64.6, 84.50, 84.51 **3** 78.26, 85.32, 85.33, 85.34, 85.40 **7** 82.22, 82.23
Ripperger, H. **1** 58.107, 58.109 **6** 51.38
Ristey, W.J. **1** 4.10
Ritter, H.L. **1** 33.58
Ritzmann, G. **6** 45.38
Riva di Sanseverino, L. **1** 5.37, 5.38, 51.38, 51.39, 51.40, 56.4, 58.108 **3** 28.10, 32.13, 36.21, 54.13, 58.47 **4** 9.13, 9.14 **5** 38.24, 50.11, 58.1 **7** 17.21, 32.6, 35.41, 35.42, 35.43, 35.44, 36.23, 40.23, 51.18
Rivarola, E. **6** 66.4
Rivera, A.J. **7** 86.91
Rivest, R. **5** 83.88, 86.23 **6** 45.4, 71.21, 86.42 **7** 83.104
Rivett, D.E.A. **6** 54.12, 54.18
Rix, C.J. **7** 86.19
Rizvi, S.H. **1** 16.46, 19.36
Rizvi, S.Q.A. **7** 33.6, 33.12
Ro, G. **4** 33.6, 33.7
Robbins, A. **7** 19.14
Robbins, A.H. **5** 43.2
Robert, F. **6** 39.7 **7** 39.15
Robert, J.-B. **6** 24.6 **7** 24.6
Robert, J.B. **7** 64.11
Roberts, D.H. **6** 68.3
Roberts, J.C. **3** 59.31
Roberts, J.D. **3** 33.48, 38.17 **6** 24.6 **7** 24.6
Roberts, J.S. **3** 53.23
Roberts, P.J. **3** 86.9, 86.17, 86.18, 86.37 **5** 2.30, 46.6, 51.22 **6** 29.2, 30.3, 38.39, 51.6 **7** 62.14, 64.61, 68.10, 71.29, 71.44, 71.105, 75.32
Robertson, B.E. **4** 83.30, 84.7 **5** 15.3, 15.10, 15.11, 15.12, 20.17, 27.4 **6** 13.19, 13.20, 13.21, 13.22, 13.23
Robertson, B.K. **1** 33.26, 33.44, 33.45
Robertson, C.G. **3** 73.56
Robertson, G.B. **1** 19.37 **2** 75.17, 77.2, 77.3, 77.4, 81.75, 82.28, 82.37, 83.40, 83.86, 84.68, 86.46, 86.80, 86.89 **3** 72.38, 75.35, 76.63, 82.5, 86.27, 86.98 **4** 71.43, 71.46, 71.49, 71.56, 86.21 **5** 71.61, 71.94, 72.40, 72.45, 72.51, 74.8, 74.9 **6** 39.3, 39.4, 39.16, 39.17, 72.23, 72.28, 74.11, 74.14, 80.6, 82.10, 85.23 **7** 50.23, 58.12, 71.110, 75.42, 75.43, 84.21, 86.35, 86.36, 86.37
Robertson, I. **2** 77.14
Robertson, I.C. **5** 71.63
Robertson, J.H. **1** 2.23, 39.22, 39.24, 45.48, 49.32, 58.54 **2** 83.29, 83.43, 85.22 **5** 72.28, 72.29 **7** 72.17, 72.19, 72.20, 72.21, 72.23
Robertson, J.M. **1** 1.11, 1.43, 1.55, 1.71, 1.72, 1.73, 1.77, 1.84, 1.125, 3.49, 3.50, 5.25, 9.34, 9.35, 11.26, 11.57, 13.20, 17.14, 17.16, 17.29, 19.40, 19.49, 19.53, 19.62, 20.19, 24.13, 24.30, 27.6, 29.12, 30.12, 30.13, 30.14, 30.19, 30.22, 30.24, 30.28, 30.29, 30.33, 30.35, 35.6, 41.10, 48.21, 49.8, 49.9, 49.13, 51.11, 52.7, 52.13, 53.8, 53.12, 53.15, 53.17, 54.13, 54.22, 54.25, 56.1, 56.2, 56.3, 58.48, 58.49, 58.70, 58.71, 58.74, 58.76, 59.45, 59.58, 59.64, 59.70 **2** 60.156, 84.36 **3** 24.5, 26.3, 51.26, 51.30, 53.17, 53.23, 53.24, 58.67, 59.16, 60.42, 60.55 **4** 30.8, 53.3 **5** 64.3, 64.36 **6** 28.15, 28.16
Robin, M.B. **2** 60.99 **3** 83.50
Robineau, M. **7** 3.9
Robins, R.K. **6** 45.29
Robinson, A.G. **4** 83.26
Robinson, B. **1** 24.20
Robinson, B.H. **3** 71.38 **4** 71.39
Robinson, B.W. **5** 80.21, 80.22 **6** 65.4
Robinson, D.J. **3** 60.40, 75.9, 76.68
Robinson, G. **2** 72.14, 72.52, 82.15 **5** 71.15
Robinson, N.J. **7** 41.19
Robinson, P.D. **7** 60.8, 60.9
Robinson, W. **7** 86.6
Robinson, W.R. **2** 71.48, 81.88 **4** 60.35, 81.37 **5** 83.114 **7** 86.46, 86.97
Robinson, W.T. **1** 26.35, 52.17 **2** 65.43, 69.24, 79.34, 79.35, 86.25 **3** 33.30, 33.31, 33.32, 71.38, 71.55, 71.66, 76.5, 76.58, 77.1, 83.100, 83.147 **4** 76.17, 76.24, 78.4, 78.6, 78.7 **5** 21.7, 63.3, 76.14, 76.15, 78.13, 78.14, 78.15, 83.93, 83.117, 85.20 **6** 76.16, 76.43, 78.1, 78.6 **7** 49.22, 49.23, 83.53, 83.54
Robson, A. **2** 75.64, 77.37
Robson, R. **4** 85.15 **5** 81.62
Roch, G. **7** 33.6, 33.12

Rochon, F. **7** 69.24
Rochon, F.D. **5** 72.15
Rodesiler, P.F. **2** 84.65 **4** 41.2, 69.11, 69.22, 69.27, 74.3, 74.8, 74.11, 75.19 **5** 63.6 **6** 81.24 **7** 69.4, 69.5
Rodewald, H. **7** 39.35
Rodgers, A. **7** 83.107, 84.72
Rodgers, A.L. **6** 83.76 **7** 44.9, 83.106
Rodgers, J. **3** 64.12, 83.73 **4** 64.33, 83.75, 83.76
Rodgers, J.R. **7** 29.11, 36.23, 39.37, 39.38
Rodig, O.R. **4** 59.6
Rodley, G.A. **2** 84.41, 84.68, 86.32 **5** 78.13, 78.14, 83.93, 86.25, 86.26 **7** 49.22, 49.23
Rodrigo, S. **1** 54.1
Rodrigues, M. **7** 13.18, 44.26
Rodrigues, M.M. **1** 43.23, 43.24, 43.25
Rodrique, L. **2** 60.23, 72.54
Roe, A.M. **3** 3.48
Roe, D.M. **5** 71.105 **7** 71.79
Roe, R.-J. **1** 1.141
Roelofsen, G. **3** 1.8 **4** 1.7 **5** 39.24 **7** 1.5, 11.21
Rogers, D. **1** 2.92, 50.12, 50.27, 53.13, 58.30, 59.48 **2** 76.21, 76.31, 76.50, 77.1, 82.34, 83.215, 86.78 **3** 10.5, 48.19, 50.15, 52.1, 53.15, 58.34, 59.8, 75.4 **4** 36.27, 38.25, 49.7, 53.6, 59.7, 59.15 **5** 45.12 **6** 35.16, 53.2, 53.14, 56.1, 83.18 **7** 38.61, 58.7
Rogers, J.B. **1** 17.30
Rogers, J.F. **5** 78.20
Rogers, M.T. **7** 1.6
Rogido, R. **3** 32.20
Rogozhin, S.V. **5** 82.24
Rohde, N.M. **6** 80.6
Rohde, W. **7** 47.19
Rohr, A. **7** 4.4
Rohrbaugh, W.J. **4** 85.28 **5** 85.18, 85.19
Rohrbaugh, W.L. **3** 76.83 **4** 80.8
Rohrer, D.C. **3** 44.11, 47.11, 47.26 **4** 45.25 **5** 45.35, 45.36, 45.38, 51.28 **6** 51.4, 51.15, 51.35, 51.39 **7** 51.10, 51.12, 51.19, 51.24, 51.28, 51.29, 51.33, 51.34, 51.35, 51.45, 51.51
Rohrl, M. **1** 54.24 **3** 54.6, 54.15, 59.19, 59.25 **4** 54.8, 54.12, 54.16 **5** 16.22, 31.21, 31.27
Rokhlina, E.M. **6** 69.17 **7** 69.20, 69.21, 69.23, 71.37
Rolfe, N. **3** 83.5 **4** 83.8
Rollett, J.S. **1** 1.107, 5.21, 46.21, 56.4, 58.108, 59.43 **7** 76.72
Romain, P. **1** 32.31
Romers, C. **1** 1.70, 10.29, 18.12, 18.17, 18.18, 18.19, 18.20, 18.21, 18.26, 21.7, 35.45, 36.36, 38.4, 38.8, 38.9, 38.10, 38.11, 38.12, 38.13, 39.14, 39.15, 39.66, 42.2, 42.3, 42.5, 51.8, 51.30, 51.31, 51.32, 51.33, 51.34 **2** 77.59 **3** 1.14, 22.3, 38.6, 51.32, 51.37, 51.38, 81.43 **4** 21.6, 21.17, 27.2, 51.43, 51.44 **5** 1.12, 21.8, 42.6, 51.37, 51.38, 67.9, 83.39 **6** 37.5, 51.20, 51.36, 84.37, 84.38 **7** 37.11, 51.39, 51.41, 51.55, 83.44, 83.45, 83.81
Romming, C. **1** 7.10, 9.12, 9.14, 9.15, 9.16, 9.20, 9.23, 20.13, 21.10, 47.9 **2** 60.17, 60.18, 60.19, 60.20, 60.49, 60.62, 60.82, 60.90, 60.139 **4** 9.2, 9.4, 9.9, 48.40, 59.3 **5** 7.3, 33.8, 48.53, 48.57, 48.58, 48.63, 59.4 **6** 16.8, 48.41 **7** 16.3, 17.12, 19.6, 35.17, 41.6, 48.41, 48.42, 48.43
Rona, S. **7** 60.2
Ronnquist, O. **1** 45.28 **3** 36.7, 36.22
Roode, H.J.G.van **6** 69.1
Roode, J.H.G.van **7** 69.6
Roos, I.A.G. **3** 85.24
Roper, W.R. **3** 86.61 **4** 86.35 **7** 71.104
Roques, R. **6** 58.10, 58.11
Ros, R. **3** 85.19
Rosa, M.de **6** 51.44
Rosalky, J.M. **5** 83.5 **7** 86.35, 86.36, 86.37
Rose-Munch, F. **6** 71.33
Rose, I.A. **1** 2.47
Rosen, L. **4** 5.8
Rosen, L.S. **3** 44.43
Rosenberg, E. **4** 71.27
Rosenberg, H. **4** 73.1 **5** 73.19
Rosenberg, J.M. **5** 47.33 **6** 47.31
Rosenblum, M. **2** 60.119, 75.21
Rosenfield Junior, R.E. **6** 48.31 **7** 48.28, 48.49
Rosenqvist, E. **3** 84.2 **4** 38.7 **5** 48.53 **6** 48.41 **7** 59.15
Rosenstein, R. **2** 82.3, 82.9
Rosenstein, R.D. **1** 32.22, 45.14, 45.15, 45.18, 45.25, 45.38, 45.44, 45.45, 45.46, 58.82 **3** 45.5, 45.6, 45.11, 45.12, 45.13, 58.23, 59.26, 60.23 **4** 37.4, 44.9, 44.18, 44.25, 48.21, 60.14 **5** 45.3, 45.25 **6** 61.7 **7** 47.5
Rosini, G. **4** 9.13, 9.14
Ross, F.K. **3** 3.39, 37.2, 83.21 **4** 33.4, 64.45 **5** 64.47 **6** 7.3 **7** 33.8
Ross, P.F. **6** 84.31
Rossell, H.J. **6** 33.4, 33.5
Rossi, G. **4** 83.35 **7** 51.50
Rossi, U. **2** 61.17, 61.19
Rossinskaya, E.R. **5** 71.104 **7** 71.128
Rossmann, M.G. **1** 11.69, 17.47, 30.22, 30.32, 52.12, 52.31, 53.39 **4** 21.5 **5** 76.12
Rossotti, F.J.C. **2** 81.27, 81.45, 81.48, 81.66, 81.81 **3** 81.16, 81.22, 81.42 **4** 2.11, 2.24, 81.12, 81.13, 81.14, 81.23, 81.41, 81.49, 81.54 **6** 82.15, 84.1
Rotaru, V.K. **7** 79.15, 79.17
Rotaru, W.K. **5** 79.13
Roth, R.J. **4** 31.12 **5** 31.42
Roth, W.L. **2** 63.8
Roth, W.R. **1** 23.1
Rothenberger, O.S. **5** 35.34
Roti, I. **4** 70.8, 70.12 **6** 70.9
Rottman, F. **7** 47.26
Rousseau, R.J. **6** 45.29, 48.79
Rousseaux, M.-P. **4** 32.23
Rousseaux, M.P. **6** 36.9
Roux, J.P. **2** 60.40, 60.41 **3** 67.17
Rovere, C. **5** 31.3

Row, T.N.G. 7 31.9
Rowe, I. 1 39.71
Rowe, J.M. 1 35.24 2 72.5, 72.10, 80.5, 86.13, 86.86
Rowley, R.J. 6 72.2
Roy, P.K. 3 2.17
Roy, P.N. 1 1.41
Roy, R.M. 5 84.28
Roychowdhury, P. 5 36.12 7 35.9
Royer, D.J. 5 83.42
Royers, E.C. 6 72.4, 73.60
Rozenberg, L.P. 2 74.16 7 60.26, 60.38
Ruban, G. 1 16.44 2 76.9 4 21.2, 21.3, 71.67
Ruben, H. 6 32.38, 33.57, 42.6 7 39.24
Rubezhov, A.Z. 3 71.4
Rubin, B. 5 83.116
Ruble, J. 6 33.48 7 33.42, 33.43, 45.32
Ruble, J.R. 4 82.3
Rubtsova, N.D. 7 84.7, 85.5
Rucklidge, J.C. 4 71.47
Rudko, A.D. 7 48.70, 48.71
Rudman, D.A. 7 78.18
Rudman, R. 1 5.16 2 63.15 3 13.3 4 64.5 5 5.10, 5.11
Rudolph, R.W. 3 62.11
Ruegg, M. 3 76.49
Ruff, J.K. 3 3.37, 64.62, 64.63, 64.64 4 3.15, 3.16, 64.41, 64.42
Rukhadze, E.G. 7 78.22
Rule, L. 2 69.17
Rumball, S.S. 3 76.5
Rundle, R.E. 1 11.11, 11.12, 11.13, 19.54, 19.55, 38.19, 45.76 2 67.1, 67.10, 67.14, 67.17, 68.3, 69.23, 71.29, 74.1, 75.24, 83.20, 83.21, 83.57, 83.58, 84.19
Runsink, J. 4 65.10 5 65.6
Rupainwar, D.C. 1 41.35
Rupprecht, G. 7 49.13, 49.17
Rush, J.J. 5 19.12, 19.13
Rush, P.E. 7 83.38
Rusholme, G.A. 4 71.10 5 71.30, 86.45 6 39.3, 39.4, 39.16, 39.17, 75.6
Russ, B.J. 2 83.49 4 83.19, 83.20
Russ, C.R. 7 75.38
Russell, D.R. 2 72.59, 83.178, 86.92 3 66.2, 66.3, 71.34, 71.41, 71.42, 72.13, 72.16 4 71.16, 71.17, 71.83, 71.87, 83.63, 85.34 5 71.100, 71.101, 73.1, 73.2 6 64.2, 64.32, 64.38 7 39.12, 39.13, 64.37, 66.6, 66.7, 71.48, 71.50, 71.71, 75.11, 75.44
Russell, G.B. 7 58.12
Russell, J.H. 1 33.66 4 33.28, 33.29, 33.30, 33.46
Russell, R.A. 7 51.56
Russell, S.E. 4 54.2
Russell, S.W. 1 21.33
Rutherford, J.S. 1 32.22 3 8.2, 8.3 5 64.21 6 64.28
Rutherford, K.W. 2 85.53
Rutten, E.W.M. 1 1.70, 38.4, 51.31, 51.34 4 21.6, 51.43, 51.44 5 51.37 6 51.20
Ruud, M. 7 83.46
Ruysink, A.F.J. 7 5.13, 5.14
Ruzic-Toros, Z. 6 47.11, 77.17 7 47.13
Ryan, R.R. 1 48.15 3 83.85 5 73.31, 73.37
Ryan, T.J. 3 9.8
Rybin, L.V. 4 86.36 5 72.26 7 75.5
Rybinskaya, M.I. 4 86.36 5 71.7, 72.26 7 75.5
Rychlewska, U. 6 35.30
Rychnovsky, V. 1 39.49
Ryderstedt-Nahringbauer, I. 1 1.146

Saakyan, G.P. 6 41.29
Saatsozov, V.V. 7 39.23
Sabacky, M.J. 3 17.11
Sabelli, C. 1 35.5, 40.19, 40.20 4 35.4, 35.10
Sabesan, M.N. 3 32.25, 48.49 4 48.52
Sabine, T.M. 1 1.12, 1.18, 1.19, 43.15 3 33.1, 48.11 5 2.4, 2.5, 2.6, 2.7, 2.8, 2.9
Sacconi, L. 2 78.22, 78.24, 78.31, 78.32, 78.34, 78.36, 78.44, 78.53, 86.49 3 76.87, 78.15, 86.43, 86.48, 86.52, 86.85 4 76.43, 76.44, 83.126, 86.44 5 76.52, 76.53, 83.112, 86.28, 86.32, 86.49, 86.50 6 86.52, 86.53 7 71.130, 86.75, 86.82, 86.84, 86.85, 86.86, 86.87, 86.102, 86.115, 86.117
Sacerdoti, M. 6 48.86, 86.8, 86.9 7 11.15
Sachdev, H.S. 3 40.13
Sackman, P. 3 3.40, 3.41, 60.30, 60.31 4 36.18, 60.17 5 11.5 6 47.22, 47.26
Sadanaga, R. 1 2.62, 19.45
Sadavoy, L.S. 6 86.45
Sadikov, G.G. 2 81.44, 81.67 4 81.43
Sadler, P.J. 4 60.9, 60.10, 60.11
Sadownick, J.A. 6 71.62
Saenger, V. 6 45.51
Saenger, W. 1 17.30 3 40.13, 41.14, 44.21, 47.1, 47.9 4 20.5, 35.37, 47.1, 47.3, 47.5, 47.7, 47.17, 47.18 5 33.42, 47.1, 47.2, 47.3, 47.21, 48.50, 61.7, 61.8, 61.9, 64.39 6 27.7, 44.8, 45.38, 45.52, 47.5, 64.3, 67.16 7 17.11, 45.56, 45.57, 45.59, 47.19, 47.23, 47.31, 47.33, 47.41, 47.42, 61.6
Saethre, L.J. 4 39.32 5 39.13, 39.32, 39.38, 39.39, 39.40, 41.2 7 39.5, 39.32, 39.34
Sager, R.S. 2 84.32, 84.34, 84.38, 84.39 3 84.8
Saha, N.N. 1 1.41, 48.19 3 48.21, 82.21 4 40.11 5 16.10, 40.13 6 33.11 7 17.11, 17.16, 48.30
Sahu, B.N. 3 2.15 4 2.16
Sahu, M. 4 13.10 5 24.5
Sahu, R.G. 3 2.16 4 2.16
Saint Clair, D. 3 62.8, 62.12 4 62.9, 71.5
Saint-Giniez-Liebig, D.de 2 64.38, 64.39
Saint-Giniez, D.de 1 32.23
Saito, A. 7 59.24
Saito, N. 3 45.43 5 59.11
Saito, T. 7 86.71

Saito, Y. **1** 1.16, 1.17, 2.3, 2.6, 50.18, 59.4 **2** 60.101, 60.117, 60.123, 76.4, 76.8, 76.35, 76.36, 76.38, 81.74, 83.17, 83.96, 83.105 **3** 1.23, 3.25, 35.4, 50.21, 60.12, 76.54, 76.64, 76.71, 76.82, 82.18 **4** 20.1, 29.4, 50.4, 54.13, 60.32, 76.2, 76.20, 76.21, 76.25, 76.40, 81.31, 83.68 **5** 60.7, 65.3, 76.20, 76.29, 76.48, 83.60 **6** 3.2, 12.2, 18.1, 19.4, 50.12, 64.34, 64.35, 76.27, 82.16, 83.16, 83.35 **7** 6.1, 50.28, 76.40, 76.41, 76.42, 76.76, 81.33, 82.8, 83.49
Sakabe, K. **5** 31.22, 31.38, 31.40, 60.35
Sakabe, N. **1** 15.23, 16.39, 44.31, 50.37, 54.17, 58.35, 58.102, 58.114, 59.25 **3** 58.36, 58.38 **5** 31.22, 31.38, 31.40, 60.35 **7** 31.27
Sakaguchi, T. **7** 40.22
Sakakibara, H. **3** 59.1
Sakakibara, J. **7** 56.7
Sakamaki, T. **5** 50.17, 50.18, 50.19, 50.22 **6** 50.11
Sakan, T. **5** 54.8
Sakanoue, S. **3** 35.12 **4** 60.20
Sakata, K. **1** 13.17, 13.18 **4** 54.15
Sakata, Y. **1** 1.30
Sakhri, A. **7** 64.57, 64.58
Sakore, T.D. **1** 13.10, 13.11 **3** 39.13, 39.14, 44.13, 44.18, 44.19, 44.25, 44.26, 44.27, 44.28 **4** 60.43 **5** 58.48 **7** 81.19
Sakurai, H. **1** 58.102 **5** 75.41, 75.42
Sakurai, K. **1** 3.15, 13.16, 13.19 **5** 13.5 **6** 13.4 **7** 13.4, 13.5
Sakurai, N. **5** 36.22, 44.4
Sakurai, T. **1** 16.5, 17.1, 17.2, 17.17, 17.35, 17.36, 18.2, 50.4 **3** 60.13 **4** 60.7, 60.12 **6** 32.21, 32.22, 32.23
Sala-Pala, J. **5** 83.80
Salah, O.M.A. **7** 71.131, 71.133, 71.139
Salamon, R. **3** 39.15
Salas-Cimingo, G. **1** 13.26, 14.7, 14.8
Saldarriaga-Molina, C.H. **7** 73.21, 73.22
Salem, G. **7** 1.24, 24.7
Sales, D.L. **2** 85.26, 85.52
Salmon, O.N. **1** 7.4
Salomon, C. **7** 86.60
Salomon, R.G. **6** 28.10
Salomone, R.A. **3** 42.2
Salvadori, P. **6** 72.15
Salzer, A. **7** 73.24
Samarskaya, V.D. **1** 5.26
Samay, G. **5** 9.8
Sammes, M.P. **7** 15.4, 33.29
Samuel, E.B. **4** 21.21
Samuel, G. **3** 40.10, 64.24, 64.38, 64.40
Samus, I.D. **6** 83.20, 83.21
Sancilio, F.D. **7** 35.31, 35.37, 76.5
Sanctis, S.C.de **5** 60.40, 60.41
Sanders, D.A. **3** 68.18
Sanders, W.W. **1** 1.70
Sandler, S. **1** 29.10, 29.24
Sandmark, C. **2** 83.166
Sandor, E. **5** 5.8 **7** 5.2, 5.3, 5.5
Sands, D.E. **1** 11.16 **2** 62.38, 77.29, 77.30, 77.32, 77.34 **3** 77.23 **5** 39.3, 77.20
Sandstrom, J. **1** 41.28 **7** 11.14, 32.25
Sankawa, U. **7** 58.19
Sanner, R.D. **7** 11.11
Sanquer, M. **6** 32.30 **7** 19.4, 19.5
Sansoni, M. **2** 71.33 **3** 3.54, 86.22, 86.99 **4** 86.58, 86.71 **5** 64.20, 71.35, 84.32, 86.85 **6** 71.86, 86.36, 86.75 **7** 71.118
Santis, P.de **1** 44.67 **2** 60.97, 60.110, 60.112
Santo, W. **4** 71.47
Santoro, A. **2** 83.102, 83.165 **3** 83.6, 83.95 **4** 83.91, 83.92, 83.94, 83.95, 83.102, 84.11
Sanz-Ruiz, F. **4** 77.10
Sanz, F. **3** 64.49, 65.1 **5** 17.22, 64.14, 65.4, 71.77, 71.83 **6** 63.4, 64.30, 71.68, 86.21 **7** 3.29, 39.4, 63.10, 63.20, 64.38, 71.88
Saperas, B. **7** 83.104
Sappa, E. **7** 71.81
Sarapu, A.C. **5** 71.4
Sarchet, C. **5** 81.42
Sargeson, A.M. **2** 76.34, 76.49, 82.2, 82.24, 82.54 **4** 76.11, 76.17 **5** 76.31 **6** 76.16, 82.10 **7** 76.52, 84.21
Sarishvili, L.P. **7** 76.16, 76.24
Sarko, A. **3** 45.36
Sarma, V.R. **1** 16.7, 16.8, 33.37, 48.94
Sartain, D. **2** 65.29, 85.91
Sartori, F. **4** 80.2 **6** 72.15
Sarynin, V.K. **7** 26.3
Sasada, Y. **1** 5.7, 11.37, 13.3, 13.28, 13.29, 13.38, 20.1, 20.29, 20.30, 22.3, 22.4, 22.5, 22.9, 27.8, 27.10, 27.13, 28.20, 31.27, 33.18, 33.42, 35.17, 38.33, 38.34, 38.42, 48.52, 48.56, 48.84, 48.86, 48.88, 48.95, 56.12, 58.7, 58.12, 58.45, 58.58 **2** 78.1, 78.2, 82.13, 82.29 **3** 31.21, 36.26, 45.43, 53.22, 58.58, 59.15 **4** 5.7, 27.5, 27.6, 33.41, 40.24 **5** 22.2, 33.51, 49.6, 58.33, 59.11 **6** 22.2, 22.4, 27.8, 27.17, 31.14, 49.7, 50.1, 53.5, 53.10, 58.37, 58.51, 71.58 **7** 20.14, 31.12, 33.7, 47.28, 48.68, 49.8
Sasaki, F. **5** 50.3
Sasaki, H. **7** 50.20
Sasaki, K. **3** 58.38 **4** 58.31 **5** 36.13, 58.31, 58.35 **6** 53.8, 58.48, 58.58, 59.26 **7** 54.16
Sasaki, M. **6** 33.3
Sasaki, N. **6** 59.19
Sasaki, Y. **3** 61.3, 61.4 **4** 58.39 **5** 61.2 **6** 61.4, 61.5, 61.6 **7** 76.76, 86.71
Sass, R.L. **1** 1.29, 1.36, 1.37, 1.76, 7.16, 7.26, 10.11, 11.28, 23.1, 33.55, 39.67, 41.11 **3** 3.32, 3.49, 7.6, 9.2, 21.2, 39.4 **6** 3.16, 13.15, 28.3 **7** 12.4
Sasse, H.-E. **6** 39.5
Sasse, H.E. **5** 72.13 **7** 75.2, 75.3, 75.4
Sasse, W.H.F. **5** 31.37
Sassmannshausen, G. **2** 64.62
Sasvari, K. **1** 21.28 **3** 45.7, 45.8 **4** 4.3, 9.8, 11.13, 21.15 **5** 4.3, 35.15, 35.18, 35.23, 42.10 **6** 45.9, 58.19 **7** 35.15, 35.18, 40.21, 63.17
Satake, S. **1** 42.14

Sato, A. **5** 59.33
Sato, M. **6** 17.4 **7** 50.18
Sato, N. **4** 40.24
Sato, S. **1** 39.88, 44.76, 48.7, 48.8 **4** 32.5 **5** 36.22, 41.7, 41.9, 41.36, 44.4, 59.33, 60.4, 64.37 **6** 36.11, 44.2, 50.12 **7** 6.1, 33.39, 33.40, 33.41, 47.14, 47.24, 50.28, 59.24, 60.7, 81.33, 82.8, 83.49
Sato, T. **1** 27.12, 40.21, 51.20, 52.28, 58.18 **2** 84.7, 85.18 **3** 31.12, 50.4, 51.20, 51.22, 51.33, 58.25 **7** 47.38
Sato, Y. **3** 51.48
Saturnino, D.J. **5** 61.4
Satyshur, K. **6** 48.49, 48.62
Satzke, L. **5** 58.24
Satzke, L.O.G. **7** 45.24
Saucy, G. **1** 52.29 **4** 27.16 **6** 27.21
Saunderson, C.P. **1** 51.46
Sauter, H. **7** 28.3, 38.24
Savage, D.S. **3** 51.44
Sawitzki, G. **6** 63.2, 63.7 **7** 84.9, 84.10, 84.28
Sawyer, J.F. **7** 62.10
Sax, M. **1** 1.145, 13.13, 13.55, 35.26, 35.30, 39.74 **3** 3.47, 40.18, 41.11, 59.5 **4** 13.13, 16.17, 41.18, 46.6, 59.5 **5** 28.8, 41.22 **6** 31.23, 41.7, 41.18, 44.27, 51.2 **7** 13.18, 16.11, 41.11, 44.26, 44.29, 46.2, 46.3, 46.4, 46.6
Sayed, K.E. **1** 26.19, 26.20, 26.21, 26.22, 26.23
Sayler, A.A. **6** 71.79
Sayre, D. **1** 29.9, 29.23 **3** 51.43
Scala, A. **3** 59.33 **7** 51.50
Scales, C.G. **3** 13.18
Scane, J.G. **2** 65.38
Scaramuzza, L. **2** 65.27, 80.16, 85.36 **3** 68.20 **4** 79.11, 85.17, 85.19 **5** 76.23, 79.23, 83.92, 84.14, 84.25, 85.6, 85.15, 85.21, 85.23, 85.30 **6** 85.42, 85.44 **7** 81.42, 84.60
Scarbrough, F.E. **2** 62.25 **4** 3.8 **7** 31.18, 31.33, 59.2
Scaringe, R.P. **7** 83.114
Scatturin, V. **1** 40.25, 40.26 **2** 86.9, 86.72, 86.74, 86.100, 86.118, 86.119
Scavnicar, S. **2** 77.53, 84.49 **5** 81.10
Schaaf, T.F. **7** 63.1
Schaap, H. **4** 85.15
Schaden, G. **5** 35.3
Schaefer, J.P. **1** 1.122, 5.23, 39.92, 50.15 **3** 21.15 **4** 20.8, 22.8, 31.9, 41.14, 41.20, 51.52 **5** 21.14, 21.15, 27.1, 60.42 **6** 33.54 **7** 36.26
Schaefer, W.P. **2** 78.15 **3** 78.30, 82.26 **4** 78.15 **5** 78.20 **6** 76.30
Schaeffer, E. **4** 81.59
Schaeffer, R. **1** 3.37 **3** 3.28 **5** 62.3 **6** 75.1
Schafer, H.L. **2** 84.31
Schaffner, C.P. **4** 50.17
Schaffrin, R. **3** 32.18, 32.19, 48.20
Schaffrin, R.M. **3** 39.29
Schairer, H.U. **1** 54.24
Schalkwyk, T.G.D.van **1** 19.18 **7** 56.5
Schantz, E.J. **7** 58.3
Schapiro, P.J. **2** 62.30
Schat, G. **7** 67.7
Schawartz, J. **7** 32.2
Scheel, H.-J. **3** 26.11
Scheidt, W.R. **3** 83.49 **4** 76.34, 76.36, 81.9 **5** 49.7, 69.21 **6** 49.10, 49.15, 49.21, 83.82, 83.83, 84.23 **7** 49.11, 49.12, 49.13, 49.15, 49.17, 49.18, 49.19
Scheie, A. **2** 70.8, 70.9
Schein, B.J.B. **1** 8.26, 8.27
Scheinbeim, J. **7** 44.5
Scheit, K.H. **3** 47.9
Schelle, S. **3** 71.30
Schempp, E. **7** 44.5
Schenk, H. **1** 53.27, 58.1 **3** 3.11, 20.2, 20.16, 37.4, 39.5, 40.14, 51.1, 51.15 **4** 1.12, 21.8, 27.17, 30.4, 37.2, 37.3, 51.19 **5** 11.8, 20.5, 35.4, 35.21, 39.23, 51.1, 72.11 **6** 32.2 **7** 20.9, 44.6
Scherer, H. **5** 69.9
Scheringer, C. **1** 17.9 **2** 78.19 **5** 25.2
Scherle, J. **7** 84.4, 84.13
Scherr, P.A. **3** 68.18
Scheuerman, R.F. **1** 1.36, 1.37, 1.76
Schevitz, R.W. **1** 52.12
Schibilla, H. **2** 67.3
Schiele, L. **5** 33.54
Schieltz, N.C. **1** 50.11
Schievelbein, V.H. **5** 83.42
Schiff, L. **2** 75.21
Schiffer, M. **1** 1.52
Schildknecht, H. **6** 39.5
Schiller, W. **6** 3.31
Schilling, J.W. **1** 30.3 **5** 9.14
Schlemper, E.O. **1** 7.24 **2** 65.3, 69.1, 69.6, 69.8, 69.37, 83.24, 83.84 **3** 83.51, 83.52 **4** 69.31, 83.42 **5** 69.16 **6** 83.6, 83.29, 83.30, 83.78, 84.31 **7** 76.75, 83.79
Schlessinger, R.H. **3** 39.13, 39.14
Schlingmann, M. **7** 83.5, 83.12
Schlodder, R. **7** 7.5, 71.117, 86.116
Schlueter, A.W. **4** 73.3 **5** 75.24 **6** 73.8
Schlupp, R. **4** 83.46, 84.3
Schlupp, R.L. **3** 22.2
Schmelter, B. **5** 35.31
Schmid, G. **5** 86.54
Schmid, G.H. **6** 11.7
Schmid, H. **3** 58.67, 59.9
Schmidt Junior, W.C. **7** 86.39
Schmidt, C. **7** 27.10
Schmidt, G.M.J. **1** 1.80, 1.91, 1.100, 1.117, 1.118, 7.22, 15.20, 16.52, 16.53, 17.5, 17.6, 18.23, 18.25, 18.27, 23.16, 28.28, 29.10, 29.24, 30.21, 36.9, 36.10, 38.29, 39.42, 39.43 **3** 1.38, 15.11, 15.12, 17.6, 19.9, 19.10, 29.3, 29.4, 31.37, 31.38, 31.40, 31.41 **4** 13.23, 16.12, 24.18, 26.1, 26.2, 26.4 **5** 1.5, 1.11, 1.23, 1.24, 1.26, 19.18, 60.3, 60.9 **6** 1.16
Schmidt, P.G. **7** 50.26, 50.27
Schmidt, W. **7** 84.67
Schmidt, W.C. **7** 33.35
Schmidt, W.H. **1** 39.85
Schmiedeknecht, K. **5** 71.75
Schmitkons, D.J. **1** 31.9, 31.10
Schmitt, H.-J. **7** 75.31

Schmitt, R.D. **3** 60.54
Schmittler-Fichtner, H. **6** 33.38, 33.43
Schmonsees, W. **5** 31.20 **7** 71.85
Schmonsees, W.G. **7** 83.56, 83.61
Schmutzler, R. **5** 64.18 **6** 86.11 **7** 64.24, 64.42
Schneemeyer, L.F. **6** 16.5
Schneider, G. **4** 59.17
Schneider, M.L. **3** 73.43 **4** 49.17, 86.41 **5** 86.13 **6** 83.2, 86.49, 86.50 **7** 84.1
Schnell, H.W. **4** 5.5
Schnering, H.G.von **2** 63.17 **4** 83.10 **5** 20.14, 33.17, 33.18, 33.19, 63.2 **6** 63.2, 63.7 **7** 84.9, 84.10, 84.28
Schnoes, H.K. **7** 58.3
Schodl, G. **3** 71.47 **4** 71.12
Schoenborn, B.P. **1** 52.15, 52.16
Schoening, F.R.L. **2** 81.21, 81.29, 81.34, 81.36, 81.37, 81.63, 81.64
Scholer, F.R. **3** 71.60, 75.1
Schollkopf, U. **6** 27.7
Scholtz, D. **3** 40.18
Schomaker, V. **1** 20.24, 48.33 **2** 72.34, 72.74, 72.75, 75.79 **3** 48.9 **4** 62.11
Schonholzer, P. **7** 53.1
Schoone, J.C. **1** 1.60, 1.62, 1.63, 58.55 **2** 82.15, 82.47
Schorno, K.S. **1** 53.28
Schousboe-Jensen, H.V.F. **5** 76.21
Schouwstra, Y. **5** 1.15, 2.25 **6** 1.11, 1.12
Schramm, V. **7** 83.77
Schrauber, H. **6** 17.9 **7** 33.47
Schrauzer, G.N. **2** 85.34 **3** 71.31, 75.17 **4** 71.54, 71.55
Schreiber, G. **4** 86.34
Schreiber, G.H. **3** 77.18 **4** 77.13
Schreiber, K. **1** 58.105, 58.107, 58.109 **4** 51.48, 51.56, 59.17 **5** 58.40 **6** 51.37, 51.38
Schreiner, A.F. **5** 68.23 **7** 83.94
Schrieke, R.R. **4** 73.24
Schrock, R.R. **7** 75.29, 75.37
Schroder, F.A. **7** 6.2, 83.90, 84.4, 84.13
Schroebler, S. **5** 20.1
Schroeder, E.D. **1** 33.61
Schroeder, L.R. **7** 45.47
Schroeder, L.W. **5** 19.12, 19.13
Schroepfer Junior, G.J. **6** 51.43 **7** 51.54
Schrumpf, G. **5** 20.3
Schubert, W.M. **2** 60.121
Schugar, H. **2** 83.183
Schugar, H.J. **6** 82.24
Schuijff, A. **1** 20.26, 29.3
Schulte-Elte, K.-H. **3** 53.1, 53.2
Schultz, A.G. **7** 39.40
Schultz, A.J. **4** 85.16 **5** 85.12, 86.65 **6** 71.82, 86.19, 86.54, 86.59 **7** 71.111
Schultz, R.J. **3** 83.8 **5** 19.14
Schulz, B. **7** 33.47
Schulz, H. **6** 81.8
Schulz, M. **1** 38.21
Schunn, R.A. **2** 69.35, 73.89
Schurr, J.M. **4** 48.53
Schuster, D.I. **3** 27.4
Schvoerer, M. **3** 25.2 **4** 25.1, 25.2 **5** 25.1, 35.10 **7** 25.1, 26.2
Schwalbe, C.H. **3** 3.26, 3.27 **4** 20.5 **5** 33.42 **6** 27.7, 67.16 **7** 35.22, 35.29, 47.41, 47.42
Schwartz, A. **4** 85.5
Schwartz, S.L. **3** 31.42
Schwarzel, W.C. **5** 50.9
Schwarzenbach, D. **1** 31.5, 33.13 **3** 81.38 **5** 81.23, 81.26, 81.37 **6** 61.7 **7** 81.23, 83.35, 83.36, 83.37
Schwarzhans, K.E. **2** 85.65
Schweikert, W.W. **2** 60.150
Schwering, H.U. **6** 68.7
Schwochau, K. **6** 3.5
Scoffer, M.D. **1** 40.11
Scordamaglia, R. **4** 71.63 **5** 38.8, 72.16 **6** 61.8
Scordari, F. **5** 13.14
Scott, A.I. **1** 50.26, 54.3, 54.4, 54.7 **6** 58.33
Scott, C.C. **1** 39.50, 58.49
Scott, H.G. **6** 33.4, 33.5
Scott, J.W. **4** 31.28
Scott, W.E. **4** 52.1 **6** 45.23 **7** 45.9, 45.10, 52.6
Scouloudi, H. **2** 76.19
Scrimgeour, S. **7** 76.2, 76.3
Scrimgeour, S.N. **6** 73.54 **7** 30.3, 32.11, 60.27
Scruton, J.C. **2** 60.89 **3** 60.19
Scudder, M.L. **5** 82.3, 82.9 **6** 82.1, 82.3, 82.4, 82.12, 82.13, 82.14, 82.17, 82.18
Scutcher, W. **7** 66.6, 66.7, 75.11
Seale, S.K. **6** 68.1, 68.3 **7** 68.12
Sears, M.C. **4** 83.51
Secaur, C.A. **7** 60.4
Seccombe, R.C. **3** 16.6 **5** 32.1, 32.2, 32.10, 32.11, 35.7, 69.13 **7** 33.44, 76.65
Seddon, D. **6** 73.13
Seebach, D. **7** 11.21
Seeden, R.P.A. **2** 84.46
Seely Junior, O. **1** 48.72, 48.79
Seely, O. **3** 48.28, 48.48
Seeman, N.C. **4** 45.3, 47.20, 48.21 **5** 47.33, 48.35, 48.36, 48.37 **6** 47.30, 47.31 **7** 47.37
Seff, K. **1** 11.7, 33.64 **2** 64.66 **4** 67.14, 82.3 **5** 64.7, 77.22, 85.24 **6** 3.25, 5.8, 13.16, 33.36, 48.54, 85.25 **7** 85.23, 85.32, 85.41
Segal, B.G. **6** 76.26
Segnitz, A. **6** 71.61, 75.28
Segrott, G.R. **3** 85.53, 85.55 **4** 85.40
Seidel, I. **4** 51.48 **5** 58.40 **6** 33.38, 51.37, 51.38 **7** 59.8
Seifert, W.J. **6** 27.18, 27.19 **7** 27.11
Seigler, D.S. **3** 20.5
Seiler, P. **7** 38.29, 38.30, 38.31, 38.32, 38.34, 38.35, 38.36, 38.37, 38.38, 38.40
Seip, H.M. **5** 39.1 **7** 19.6
Seki, M. **1** 3.18
Seki, S. **1** 37.1
Sekine, Y. **7** 39.26
Sekita, T. **3** 53.9
Sekizaki, M. **4** 83.68 **5** 83.23 **7** 82.13
Seleborg, M. **4** 78.9
Selig, W. **4** 60.13

Seligmann, O. **3** 59.26
Selinger, B.K. **1** 31.42
Selivanov, V.F. **5** 1.14
Selzer, J.O. **6** 35.26
Semeneko, K.N. **5** 68.10
Semenenko, K.N. **7** 3.19
Semeniuk, G.M. **4** 64.24, 65.9
Semin, G.K. **1** 19.1, 19.2, 19.3 **2** 71.1 **3** 16.1 **5** 19.1, 19.4, 19.5
Semion, V.A. **2** 73.53 **3** 71.4, 71.10, 71.61, 71.62 **5** 71.45
Semmingsen, D. **4** 6.1 **6** 5.7, 20.5, 21.4
Sen, D.K. **3** 36.6
Sendo, Y. **3** 58.36
Sengupta, S. **5** 41.22
Senko, M.E. **1** 8.1, 8.2, 32.7, 59.3
Senma, M. **4** 67.3 **5** 38.12, 45.37 **6** 38.6
Senoff, C.V. **4** 85.27
Senol, A. **4** 53.11
Sepulchre, A.-M. **7** 45.52
Sequeira, A. **1** 14.14 **3** 2.2 **5** 48.30, 48.48 **6** 48.32, 85.19
Serator, M. **2** 83.117
Seres, J. **5** 40.30
Serewicz, A.J. **1** 33.26
Sergeeva, G.N. **3** 71.56
Sergienko, V.S. **7** 33.15, 35.10, 76.10
Serpone, N. **4** 77.1 **7** 38.16
Sesay, L.J. **7** 38.3, 38.4, 84.22
Seshadri, T.P. **7** 47.4, 47.6, 47.16, 47.35
Sevastyanov, N.G. **1** 24.10
Sevcik, J. **3** 73.54
Seyferth, D. **3** 71.60 **5** 40.1
Seymour, J.P. **6** 52.1
Sgamellotti, A. **2** 85.88 **3** 85.22, 85.50
Sgarabotto, P. **2** 85.37 **3** 8.5, 61.8 **4** 26.5, 35.2, 39.11, 76.12, 83.7, 83.12, 83.33, 83.125, 83.129 **5** 9.17, 27.11, 27.12, 32.19, 32.25, 38.26, 39.2, 39.6, 39.26, 39.33, 40.11, 42.3 **6** 11.5, 11.6, 17.24, 31.17, 38.37, 39.1, 39.22, 39.25, 41.9, 41.11, 41.16, 41.24, 72.3 **7** 32.21, 35.35, 35.36, 38.75, 40.9, 40.10, 41.22
Shafer, J.L. **4** 83.32
Shaffner, T.J. **1** 52.8 **3** 49.2
Shahat, M. **1** 1.49
Shakked, Z. **3** 19.9, 19.10 **4** 51.49 **5** 19.18 **7** 19.10, 19.12
Shallcross, F.V. **1** 7.11
Sham, T.K. **7** 69.28
Shamala, N. **4** 48.16 **5** 33.45, 82.17
Shamma, M. **4** 58.21
Shamshin, L.N. **7** 63.4, 63.8, 63.11
Shanan-Atidi, H. **5** 1.22
Shandles, R.S. **6** 83.6
Shankar, J. **1** 9.41 **2** 67.8
Shankar, P.V. **7** 48.57, 48.58
Shannon, J.S. **6** 84.11
Shanton, K.J. **6** 38.16
Shapiro, B.L. **4** 21.11 **6** 21.9, 31.26 **7** 75.27
Shapiro, E.L. **7** 51.4
Shapiro, M.J. **7** 75.27
Shapley, J.R. **5** 71.70
Sharma, A. **7** 41.7, 41.8
Sharma, B.D. **1** 44.16, 48.30 **3** 32.1
Sharma, G.M. **3** 58.7
Sharma, V.C. **4** 45.7
Sharp, D.W.A. **3** 83.62 **6** 73.61 **7** 71.49, 72.9, 86.21
Shaver, A. **5** 75.25, 75.31 **6** 71.11, 71.16, 72.26, 75.6 **7** 73.20
Shaw, B. **4** 48.53
Shaw, B.L. **2** 71.9 **3** 86.49 **4** 71.83, 86.14 **5** 71.86, 71.87, 71.88, 83.49, 86.51 **6** 86.13, 86.25, 86.31
Shaw, B.W. **2** 72.64
Shaw, C.C. **7** 38.54
Shaw, C.M. **6** 28.24
Shaw, G. **5** 74.12, 86.79
Shaw, P.D. **7** 17.7
Shaw, R.A. **4** 64.28, 64.29, 64.30 **5** 64.21, 64.29 **6** 64.45
Shawl, E.T. **6** 64.22
Shchedrin, B.M. **3** 83.26 **6** 76.5 **7** 76.11, 76.12, 76.13, 76.16, 76.19, 76.24, 76.36, 76.53, 76.54, 76.55, 84.14
Shchipanova, E.F. **7** 69.10
Shearer, H.M.M. **1** 1.55, 5.42, 23.10, 27.6, 39.10 **2** 67.5, 67.16, 71.17, 71.23, 71.50, 71.54, 72.29, 72.30, 75.53 **3** 73.43, 76.72, 76.73 **4** 68.13 **5** 17.19, 71.1, 81.9, 86.13 **6** 73.39
Sheat-Rumball, S.V. **2** 78.58
Sheat, S.V. **2** 78.21
Sheel, H. **1** 36.30
Shefter, E. **1** 11.62, 31.31, 39.52, 42.8, 44.11, 44.12, 47.7, 47.16, 47.30, 47.31 **2** 60.104, 60.132 **3** 3.29, 3.30, 3.34, 3.39, 3.40, 3.41, 44.24, 44.33, 47.34, 60.24, 60.30, 60.31 **4** 27.13, 36.18, 40.2, 41.3, 60.17 **5** 11.5, 47.28 **6** 3.15, 47.22, 47.26, 58.40
Sheikh, Y.M. **7** 53.13
Shelby, R.M. **6** 76.29
Sheldrick, B. **1** 45.48 **6** 17.15, 17.16, 17.17, 45.26, 48.42, 48.68 **7** 2.10, 45.15, 45.16
Sheldrick, G.M. **3** 3.10, 63.1, 63.12, 68.2, 69.8, 69.9, 69.13, 69.14, 69.16 **4** 10.1, 33.12, 63.10 **5** 63.11, 85.14 **6** 33.12, 51.6, 63.5, 69.3, 69.4, 69.8, 71.28, 83.81 **7** 4.3, 35.12, 41.21, 51.36, 56.4, 63.12, 69.2, 69.16, 71.15, 71.87, 84.50, 86.67
Sheldrick, W.S. **3** 3.10, 63.1, 63.12, 68.2, 69.13, 69.14 **5** 64.18, 86.19 **6** 33.7, 64.6, 64.37, 86.11 **7** 64.1, 64.7, 64.24, 64.42, 64.44, 83.78, 86.8, 86.20
Shelestov, V.M. **3** 12.1
Shelton, G. **3** 49.4
Shelton, H.D. **6** 76.1
Shen, K. **2** 66.19, 66.21
Shen, K.-W. **5** 31.2
Shen, M.-S. **6** 33.39 **7** 38.57
Shen, M.S. **4** 45.9, 45.12 **5** 5.14 **7** 33.2, 59.9
Shenhav, H. **5** 62.3
Shepelev, Yu.F. **5** 76.6
Shepherd, J.C. **7** 83.10
Shepherd, R. **7** 64.32, 64.33
Sherfinski, J.S. **5** 47.12, 58.10 **6** 32.10, 47.8 **7** 24.6, 33.23, 36.4

Sherwood, R.C. 7 80.5
Shetty, P.S. 2 85.60 3 83.87, 83.94, 85.28
Shevyrev, A.A. 7 78.15
Shiba, T. 3 59.1
Shibaeva, R.P. 1 13.43, 32.17 2 60.140, 74.16 3 13.7, 27.12, 69.17 5 27.13, 60.8 6 33.30, 41.29, 69.11 7 27.8, 33.54, 60.21, 60.26, 60.34, 60.35, 60.36, 60.38, 71.23
Shibahara, T. 6 76.20
Shibata, M. 6 82.16
Shibata, S. 1 56.7, 59.60 3 59.29, 77.6 6 55.1, 56.4, 59.25 7 58.25, 59.23
Shibuya, I. 7 45.1
Shieh, J.J. 6 28.3
Shields, K.G. 4 67.8 5 69.13, 86.49 6 31.5, 81.18
Shigematsu, T. 6 44.9, 44.10 7 44.18
Shih, H. 5 40.1
Shikata, K. 5 47.22
Shimada, A. 1 13.12, 21.14, 22.4, 22.5, 36.26, 45.1 2 63.9, 63.18, 65.12, 65.14, 65.18, 65.20, 73.94 3 5.3, 36.4 4 13.14, 38.8 5 1.33, 41.23, 54.8, 83.107 6 33.20, 33.21, 33.22, 36.10, 44.3, 45.49, 54.6 7 13.13, 13.15, 33.7, 33.17, 33.19, 33.24, 45.1, 45.36, 54.12
Shimakawa, Y. 3 42.2
Shimanouchi, H. 1 20.30, 22.9, 27.10, 28.20, 38.42, 58.45, 58.58 3 31.21, 36.26, 45.43, 53.22, 58.58 4 5.7, 27.5, 27.6, 33.41, 40.24 5 22.2, 58.33 6 22.2, 22.4, 27.8, 27.17, 31.14, 58.37, 58.51 7 20.14, 31.12
Shimanouchi, N. 7 76.41
Shimanouchi, T. 3 1.12
Shimizu, B. 7 47.24
Shimizu, K. 2 83.34
Shimizu, S. 7 1.11, 1.12, 34.16, 38.14
Shimizu, Y. 1 47.20
Shimozawa, T. 5 39.28, 39.29
Shimura, Y. 6 82.9
Shin, H.S. 5 3.3, 16.17
Shina, M. 7 83.22
Shingu, T. 6 53.15 7 53.23
Shintani, R. 1 9.9
Shiomi, M.T. 4 72.25
Shiono, R. 1 22.2, 39.12, 45.13 3 59.26 5 17.3, 45.25 7 47.5
Shiozaki, Y. 5 2.22, 2.23
Shipley, G.G. 2 60.78, 60.79, 60.80, 60.81 6 46.14
Shirahama, H. 7 50.11
Shirell, C.D. 7 20.7
Shirley, R. 6 20.1
Shiro, M. 1 7.28, 27.12, 51.20, 52.28, 58.9, 58.18 2 76.36, 84.7, 85.18 3 31.12, 38.13, 50.4, 50.27, 51.20, 51.22, 51.33, 54.12, 83.141, 85.29 4 59.6, 68.10, 83.104 5 60.2, 82.11
Shirotani, I. 6 12.3 7 18.1
Shirrell, C,D. 6 20.8, 20.9, 39.19 7 20.6, 35.4
Shishova, T.G. 7 84.33
Shive, L. 6 83.28
Shklover, V.E. 4 63.8, 73.10 5 72.41 7 63.13, 63.14, 63.18, 69.10
Shkol'nikova, L.M. 2 77.20, 77.60, 78.30, 84.33 3 77.26, 78.20 4 77.2, 78.3 6 77.18, 78.13 7 78.19
Shmueli, U. 4 38.5, 38.9, 39.18, 39.19, 64.18 5 1.22, 60.14, 60.15, 60.16, 60.17, 60.18 6 60.1 7 59.10
Shoemaker, D.P. 1 22.7, 48.17, 48.33 3 75.21 4 67.14
Shok, L.N. 4 24.4, 24.6, 24.7
Shokhor, I.N. 1 7.3, 10.1, 10.4, 12.1 3 7.1, 12.5
Shokhor, N.I. 3 12.6
Shopron, M.V. 7 79.15
Shore, S.G. 5 61.4 6 3.32, 33.27, 86.35
Short, M.R. 4 36.22 5 27.16
Shortland, A. 4 71.53
Shrivastava, H.N. 1 14.2, 14.3, 30.34
Shriver, D.F. 2 68.4 3 7.2, 7.3 7 71.16, 73.56
Shugam, E.A. 2 77.20, 77.60, 80.11, 80.36, 84.33 3 77.5, 77.26, 78.20 4 69.28, 77.2, 78.3 5 80.14, 80.28, 80.29 6 77.18, 78.13
Shungu, T. 4 58.43
Shuto, Y. 3 71.21
Shvelashvili, A.E. 3 76.21 4 76.7 6 76.5 7 76.11, 76.12, 76.13, 76.16, 76.19, 76.24, 76.36, 76.53, 76.54, 76.55, 84.14
Shvets, A.E. 5 27.13 7 16.12, 17.27, 17.28, 25.5, 25.6, 27.8, 71.23
Shvetsov-Shilovskii, N.I. 7 64.30
Shvo, Y. 5 1.22
Sicignano, A. 7 67.2
Siddall, J.B. 1 51.25
Sieber, A.A.F. 3 9.8
Sieber, K. 7 3.17, 3.26
Sieckhaus, J.F. 6 71.79
Siegel, A. 7 73.58
Siegel, L.A. 2 64.49, 64.50
Sieker, L.C. 5 45.28, 45.29, 45.31 6 45.35
Sieler, J. 4 83.127, 85.20 6 41.30 7 84.67, 85.11
Siems, H. 5 81.15
Sievers, R. 7 81.16, 81.18
Sievers, R.E. 3 77.23 4 83.45 6 76.49 7 84.83
Siew, P.-Y. 7 38.49
Siew, P.Y. 6 31.27
Sigg, H.-P. 6 53.13
Sigg, H.P. 4 59.10
Signor, A. 5 69.25
Sigwart, C. 2 83.54
Siiman, O. 6 77.5, 85.45
Sikirica, M. 6 81.5 7 71.9, 71.11, 76.4, 76.9, 83.17, 83.31
Sikka, S.K. 1 17.3 5 48.11, 48.15 6 58.50 7 48.15, 48.16
Silver, L. 4 64.5 5 5.10, 5.11
Silverman, J. 1 26.13, 26.16, 26.38, 28.21, 28.22, 44.65 3 73.10 4 12.11, 18.3, 73.12 5 19.3 6 7.5, 27.6, 28.14, 28.17, 28.18
Silvers, S.J. 1 49.25, 58.72
Silverton, E.W. 2 81.70

Silverton, J.V. **1** 11.29, 59.70 **2** 76.65, 77.50, 81.52, 81.70 **3** 11.2, 51.48 **4** 31.25, 40.14, 76.9 **5** 59.28 **6** 29.1, 37.13, 59.29 **7** 31.19, 32.39
Sim, G.A. **1** 1.132, 3.57, 3.58, 3.59, 13.1, 13.2, 13.6, 13.20, 13.45, 16.4, 19.62, 20.20, 27.6, 31.30, 31.40, 37.21, 39.50, 39.56, 41.10, 48.21, 50.17, 50.39, 51.25, 51.36, 51.48, 52.7, 52.13, 53.6, 53.7, 53.8, 53.11, 53.22, 53.25, 53.29, 53.37, 54.3, 54.4, 54.7, 54.10, 54.13, 54.25, 56.1, 56.2, 56.3, 56.17, 56.18, 58.48, 58.49, 58.67, 58.68, 58.71, 58.74, 58.76, 58.87, 58.121, 59.33, 59.34, 59.35, 59.51, 59.58, 59.64, 59.70 **2** 60.118, 71.38, 73.16, 73.44, 73.99, 74.6, 74.7, 74.14, 74.22, 75.68, 79.8, 84.30, 84.37, 84.67, 85.12, 85.44, 85.48, 85.75 **3** 1.29, 1.30, 3.38, 8.18, 11.6, 50.19, 51.9, 53.5, 53.21, 53.23, 53.28, 53.29, 54.7, 58.51, 58.53, 58.55, 58.56, 58.57, 58.67, 59.17, 68.7, 68.8, 73.1, 73.16, 73.17, 73.46, 73.52, 73.56, 75.6, 85.39 **4** 4.2, 4.5, 24.13, 24.14, 24.15, 37.12, 53.4, 53.7, 53.9, 54.11, 54.14, 58.32, 73.2, 73.5 **5** 53.10, 53.15, 53.19, 58.42, 68.14, 68.15, 68.16, 72.9, 73.11, 73.13 **6** 41.17, 53.19, 72.7, 73.13, 75.8 **7** 53.2, 53.3, 53.6, 53.20, 53.21, 53.22, 53.26, 53.29, 58.26, 58.29, 73.2, 73.3, 73.18, 75.16
Sime, J.G. **1** 11.50, 54.15, 59.53 **2** 60.156, 82.47 **3** 2.13, 2.14 **5** 71.63 **6** 75.8
Sime, R.J. **3** 76.25 **6** 73.51, 77.14
Sime, R.L. **6** 73.51, 77.14
Simmonds, D.J. **5** 59.29
Simmons, G.L. **1** 29.1 **2** 78.19
Simms, M.L. **5** 76.9, 83.79
Simon, G.L. **5** 50.6, 73.34, 73.35, 73.36, 81.60
Simon, K. **3** 45.7, 45.8 **5** 9.8, 35.15, 35.18, 35.23 **6** 32.25, 41.6, 45.9, 58.19 **7** 32.2, 32.3, 35.15, 35.18, 35.27, 40.21, 63.5
Simon, M.S. **1** 17.30
Simon, S.L. **5** 63.14
Simon, W. **1** 50.44
Simonetta, M. **1** 5.28, 5.31, 9.13, 17.25, 17.31, 17.32, 19.50, 52.1 **2** 60.103 **3** 15.10, 19.11, 31.25 **4** 31.4, 31.6, 31.10, 31.20 **5** 15.8, 31.3, 31.12, 31.28, 31.29 **6** 26.2, 38.21 **7** 28.5, 31.5, 31.16, 31.20, 31.24, 33.5
Simonov, V.I. **7** 38.12
Simonov, Yu.A. **2** 81.32, 81.76, 81.82 **3** 81.30, 83.26 **7** 76.6, 79.18, 79.19, 79.20, 79.25, 83.20, 83.21, 83.32, 83.74, 83.84, 86.48
Simonsen, O. **4** 40.13 **6** 84.6
Simonsen, S.H. **1** 19.23, 39.30 **2** 83.189 **3** 23.10, 31.2, 31.5, 32.31, 39.12, 39.19, 39.20, 48.31, 83.23, 83.120, 83.131 **4** 27.18, 38.24, 39.12, 83.99 **5** 27.18, 76.1, 79.11 **6** 31.4, 33.16, 39.8 **7** 3.22, 3.23, 3.24, 15.4, 26.6, 26.7, 32.5, 32.37, 33.29, 38.74
Simov, D. **5** 40.9
Simpson, H.D. **3** 72.14, 75.15 **4** 75.24
Simpson, H.J. **1** 48.14
Simpson, J. **5** 63.3
Simpson, K.A. **4** 63.6
Simpson, P.G. **1** 58.63 **2** 62.15, 62.37 **3** 31.10, 48.43, 58.42 **6** 41.28, 48.45
Simpson, R. **5** 45.35, 45.36, 45.38
Simpson, R.T. **5** 86.47
Simpson, T.D. **1** 12.20 **3** 11.13
Simpson, T.H. **1** 59.45
Simpson, W.S. **1** 1.22
Sims, A.L. **6** 73.1, 73.15 **7** 73.1
Sims, J.A. **6** 54.15
Sims, J.J. **3** 20.5 **4** 53.2, 59.8 **5** 54.7 **7** 53.18
Sims, J.M. **5** 20.4
Sims, M.J. **4** 35.5 **6** 50.4
Sims, S.P. **1** 45.68
Simundza, G. **3** 44.27, 44.28
Sinclair, R.L. **2** 82.27, 82.30, 82.31, 83.31
Sinclair, V.C. **3** 26.3
Sindellari, L. **5** 83.29, 83.30 **7** 80.2
Singer, J. **1** 44.68
Singer, M.I.C. **4** 65.13
Singer, N. **3** 73.42, 83.145
Singer, P.P. **6** 58.21
Singer, S.J. **4** 71.90
Singh, C. **1** 43.16, 43.18
Singh, I. **3** 59.13 **6** 28.22
Singh, K.D. **7** 81.19
Singh, N. **3** 58.48 **4** 58.38
Singh, P. **1** 16.61, 35.1, 38.57 **2** 64.81 **4** 35.5, 83.22 **5** 33.5, 83.11, 83.34, 83.46, 83.113, 86.82 **6** 33.2, 38.9, 38.10 **7** 35.1, 35.2, 35.3, 47.1, 47.9, 47.10, 47.27, 50.2, 83.114
Singh, S. **4** 59.27 **6** 47.22, 47.26
Singh, T.P. **5** 32.17 **6** 60.15 **7** 32.33, 33.21
Singhabhandhu, A. **7** 60.8, 60.9
Singleton, E. **2** 71.9 **5** 86.12, 86.20 **6** 86.6, 86.28 **7** 86.29
Sinha, R.P. **7** 14.1
Sinha, S.P. **2** 83.174
Sinha, U.C. **3** 9.9
Sinn, E. **4** 78.4, 78.6, 78.7 **6** 76.43, 78.1, 78.6, 78.20 **7** 80.11, 80.12, 83.53, 83.54, 84.76, 84.77
Sinnreich, D. **3** 1.38
Sintes, A. **1** 1.129
Sipe III, J.P. **4** 77.24
Sipio, L.di **6** 3.7, 3.8, 3.29, 3.33 **7** 3.30
Sipski, M. **7** 83.52
Sirigu, A. **1** 1.86 **2** 64.13, 72.17, 72.18, 73.111 **3** 72.33 **4** 72.4 **5** 48.42 **7** 48.10, 48.27
Sirota, A. **7** 79.6
Sirota, M.I. **5** 19.9
Siskova, J. **2** 79.29
Sisler, H.H. **6** 65.11
Sivertsen, B.K. **3** 38.12
Sixma, F.L.J. **1** 1.95
Sjoberg, B. **3** 81.51 **5** 81.40, 82.13, 82.14
Sjolset, O. **4** 39.32 **5** 39.39
Sjostrom, L. **3** 1.35

Skakke, P.E. 3 85.11
Skapski, A.C. 1 2.44, 3.38, 3.39 2 65.17, 71.7, 71.55, 83.104, 86.78, 86.110, 86.115 3 73.36, 81.29, 83.79, 86.74, 86.95 4 71.52, 71.53, 80.5, 80.14, 83.82, 83.108, 85.38 5 16.2, 48.7, 81.39, 81.49, 83.24, 83.25 6 64.51, 65.6, 71.3, 74.15, 80.11, 81.26, 81.46, 83.23, 86.29, 86.30 7 69.1, 83.33
Skarstad, P. 7 75.6
Skarstad, P.M. 6 73.62
Skell, P.S. 7 72.7, 72.8
Skelton, B.W. 5 81.35, 83.66 7 86.99, 86.103
Skinner, J.M. 1 14.4
Sklar, N. 1 8.1, 8.2
Skolik, J. 5 58.6 7 58.13
Skolnik, S. 1 58.1
Skovlin, D.O. 7 84.8
Skoweranda, J. 7 64.28, 64.29
Skripkin, V.V. 2 73.68 5 72.41
Sky, K. 2 60.65
Slade, P.G. 4 84.1 5 83.8
Slaven, R.W. 7 72.5
Slessor, K.N. 7 45.4
Sletten, E. 1 35.2, 39.44, 39.59, 44.25 2 83.73 3 33.15, 33.18, 39.42, 39.43, 83.104 6 44.18, 44.22, 81.3 7 83.11, 83.42, 83.46
Sletten, J. 1 35.2, 39.16, 41.5, 44.25, 44.28 3 39.42, 39.43, 39.49, 44.1 4 39.23, 39.28 5 39.41 6 39.26, 39.28 7 39.33, 41.24, 41.28
Slim, D.R. 5 83.80
Slisz, E.P. 5 9.5, 9.6, 9.7, 9.12
Sljukic, M. 4 81.11 6 47.11 7 3.5
Sloan, G.J. 4 36.12
Sloan, M. 5 72.8, 72.12 6 72.22
Slogvik, S. 7 70.3, 70.4
Slomp, G. 4 34.3
Slone, C.J. 4 48.12
Slusarchyk, W.A. 5 59.23 7 59.13
Sly, W.G. 1 22.7, 27.16, 57.3 2 72.56
Smail, E.J. 6 33.12
Small, R.W.H. 1 1.24, 1.39, 1.40, 2.18, 2.33, 2.51 3 10.1 4 1.2 5 1.1, 13.8, 13.10, 14.2 6 2.3, 2.12, 3.17, 60.12, 60.13 7 8.7, 34.1
Smallwood, R.J. 4 83.130, 83.136
Smart, M.L. 1 3.38, 3.39 3 50.15, 75.4, 81.29
Smelyanskaya, E.M. 4 24.7 6 41.20 7 60.11
Smiley, I.E. 1 17.47
Smirnova, V.I. 3 19.2
Smissman, E.E. 3 3.29, 3.30, 3.34, 3.40, 3.41 4 27.13
Smit, P.H. 6 43.6 7 1.2, 1.3
Smith-Verdier, P. 6 13.13 7 37.14, 37.15, 47.34
Smith, A.E. 1 36.2 2 72.8, 85.34 3 72.36, 85.2, 85.3 4 72.13 5 39.43, 72.46, 86.46
Smith, A.J. 3 78.7 5 78.7 7 81.24
Smith, A.K. 5 74.8, 74.9
Smith, A.W. 3 77.4
Smith, B.C. 5 64.21
Smith, B.T. 7 83.33
Smith, C.H. 3 9.12, 9.15
Smith, C.P. 1 46.7 2 64.61 4 64.25
Smith, D.E. 4 67.23
Smith, D.J. 5 68.14
Smith, D.J.H. 7 39.12, 39.13
Smith, D.L. 1 35.23, 41.37 2 72.46, 73.58, 75.75 3 28.13, 36.33 4 28.3, 32.26, 32.29 5 35.35, 35.36, 41.12 7 35.38
Smith, D.W. 6 47.15
Smith, E.M. 7 51.4
Smith, G.D. 4 31.13, 64.16, 64.19, 79.5 5 77.10, 77.11 6 46.5, 46.6, 46.7, 46.11, 76.31 7 35.14, 71.40
Smith, G.M. 7 17.17
Smith, G.S. 1 3.37 2 63.11, 76.66, 76.69
Smith, G.W. 3 56.5, 56.6 5 56.5, 56.6, 56.7 7 30.1, 30.2, 56.1, 56.2
Smith, H.E. 1 52.8
Smith, H.G. 2 74.1
Smith, H.M. 2 84.22, 84.31
Smith, H.W. 1 4.9 6 3.1, 83.34, 83.52
Smith, I. 7 37.12
Smith, J.A.S. 3 19.3
Smith, J.D. 3 68.7, 68.8 4 73.24 5 68.15, 68.16 6 68.10
Smith, J.E.W.L. 2 83.23, 84.10
Smith, J.H. 6 84.12
Smith, J.S. 7 38.64, 38.65, 42.5, 42.6, 42.9
Smith, K.D. 5 71.33 6 67.9, 73.46, 84.13
Smith, M.A.R. 7 86.2
Smith, M.B. 5 64.44 7 64.64, 64.65
Smith, M.R. 3 65.3 4 39.6, 65.3 6 70.5, 70.6, 70.7, 70.10
Smith, N.E. 4 80.6, 80.7
Smith, N.O. 2 71.21
Smith, P. 3 13.12 4 13.11
Smith, P.J. 3 69.2
Smith, P.M. 4 76.9 7 39.19, 39.20, 63.16
Smith, R.A. 1 37.18 3 83.128 5 71.54 7 14.2
Smith, R.M. 4 58.36
Smith, S. 7 31.22
Smith, S.E. 5 27.17
Smits, D.W. 1 48.87
Smits, J.M.M. 6 80.15 7 79.9
Smolander, K. 5 83.17, 83.18 7 83.15, 84.20, 84.42
Smolin, Yu.I. 2 76.43 5 76.6
Smutny, E.J. 5 39.43
Snaauw, G.J. 1 5.6
Snader, K.M. 4 59.27
Snaith, R. 4 68.13
Snare, D.L. 1 16.47
Sneath, T.C. 3 36.3, 53.19, 53.27 5 39.31
Sneddon, L.G. 5 62.9 6 75.1
Sneeden, R.P.A. 5 17.22, 71.77 6 71.68 7 71.88
Snow, A.I. 2 67.1
Snow, M.R. 2 76.49, 82.5, 83.173 3 76.55, 83.142 4 71.72, 76.31, 76.32, 81.24, 83.24 5 71.89, 76.26, 76.27, 81.38, 85.7 6 79.8 7 85.4
Snyder, D.A. 3 76.6

Snyder, F.F. **5** 28.15
Snyder, R. **3** 33.7
Snyder, R.G. **2** 75.51
Snyder, R.L. **3** 32.34 **4** 60.14
Snyman, J.A. **3** 60.45, 60.46
Soares, J.R. **6** 33.40
Sobell, H.M. **1** 44.38, 44.52, 44.57, 44.59, 47.22, 47.23 **3** 39.13, 39.14, 44.13, 44.14, 44.17, 44.18, 44.19, 44.25, 44.26, 44.27, 44.28 **4** 39.21, 39.29, 60.43, 60.44 **5** 58.48
Sobolev, A.N. **7** 11.7, 36.6
Soboleva, S.V. **3** 78.22 **7** 63.4, 63.8, 63.11
Sobti, R.R. **4** 51.20 **5** 51.5, 51.14
Sodano, G. **6** 51.44
Soderberg, B. **1** 2.59
Soderberg, E. **3** 58.4 **4** 35.1, 35.6
Soderberg, R.H. **2** 60.155, 83.99, 84.21
Soderquist, R. **3** 81.51 **5** 81.40, 82.13, 82.14
Soest, T.C.van **3** 59.10, 59.11, 59.12 **6** 36.19, 72.4, 73.60
Sohn, Y.S. **6** 73.8
Sokol, V.I. **7** 8.8, 84.7, 85.5
Sokolova, Yu.A. **2** 83.78 **7** 84.32, 84.36
Sokolova, Yu.A. **4** 81.10, 84.10
Solbakk, J. **3** 44.7, 45.33 **4** 44.4 **5** 3.14, 8.13 **6** 44.17, 48.35 **7** 44.2
Soled, S. **5** 19.2
Soled, S.L. **6** 85.9, 85.10 **7** 86.64
Soleimani, E. **1** 52.2
Soleva, L.I. **5** 84.10 **6** 84.7 **7** 84.16
Solheim, J.O. **4** 70.15
Solheimsnes, S. **2** 70.8, 70.9
Soling, H. **3** 76.2, 76.3
Solodova, O.P. **2** 69.39
Solstad, P.J. **6** 85.11
Soltwisch, M. **3** 75.19
Soltzberg, L. **4** 20.2
Soltzberg, L.J. **3** 20.13 **4** 12.11
Sommer, S. **6** 32.15
Sommerville, P. **3** 85.46 **4** 31.32, 33.13 **5** 58.16, 59.30, 86.12, 86.20
Song Loong, W. **1** 8.22, 8.23
Sonin, A.S. **1** 3.29
Soos, Z.G. **7** 81.29, 81.30
Sorensen, A.M. **1** 21.1, 21.2 **2** 60.15 **3** 5.2 **5** 83.91 **7** 17.18
Soriano-Garcia, M. **6** 48.40 **7** 47.36
Soriano, M. **6** 51.4, 51.15, 51.39
Sorm, F. **3** 53.7
Sorrell, T. **6** 48.25 **7** 44.16
Sorrell, T.N. **6** 65.7
Sorum, H. **1** 5.39, 11.17 **4** 33.6, 33.7, 59.1 **7** 1.22, 1.33
Sosinsky, B.A. **7** 75.1
Sotofte, I. **2** 83.185 **3** 32.8 **4** 71.82 **5** 36.4, 40.18
Soule, F.B. **5** 2.26
Souron, C. **1** 24.25
Southwick, J. **3** 45.5
Southwick, P.L. **1** 30.27
Sowden, J.M. **7** 29.5, 29.6, 29.8
Sowerby, D.B. **5** 64.13 **6** 64.11 **7** 64.66, 65.7
Sowerby, J.D. **4** 68.13 **6** 73.39
Sowinski, F. **5** 41.33
Soylu, H. **7** 82.4
Sozzi, L. **2** 83.46
Spagna, R. **2** 85.69 **3** 32.36, 72.5, 72.6, 85.1 **4** 70.9, 72.3, 86.64 **5** 21.9, 72.1, 72.2, 72.3 **6** 72.5 **7** 4.8, 35.28, 35.39, 36.27
Sparks, R.A. **1** 7.29, 36.18, 47.31, 49.32 **2** 70.1 **3** 24.1, 26.6, 26.7 **4** 36.2
Sparrow, N. **4** 33.13
Spaulding, L.D. **6** 49.14
Speakman, J.C. **1** 2.10, 2.11, 2.32, 2.42, 2.45, 2.88, 2.91, 11.40, 14.2, 14.3, 14.4, 14.5, 14.11, 14.13, 28.4, 30.34, 46.17, 49.3 **2** 64.76, 86.56 **3** 2.12, 2.13, 2.14, 35.8, 35.22 **4** 2.2, 2.3, 2.12, 2.13, 2.14, 2.17, 2.18, 2.19 **5** 2.15, 2.16, 2.17, 14.1 **6** 2.15, 86.18 **7** 2.4
Speckamp, W.N. **3** 51.15
Spek, A.L. **5** 41.20, 84.22 **6** 66.4, 71.73, 80.5 **7** 67.7
Spencer, C. **5** 86.66
Spencer, C.B. **2** 71.17, 71.50 **3** 76.72, 76.73
Spencer, C.J. **2** 64.14
Spencer, J.B. **6** 11.1, 11.2
Spencer, J.L. **3** 71.35, 71.38 **4** 71.39 **5** 71.41, 74.10 **6** 71.34, 86.5 **7** 71.45, 71.59, 71.66, 71.96, 72.1, 75.22, 86.4, 86.10, 86.11
Spencer, R.C. **2** 83.208
Spike, T.E. **6** 51.43
Spink, J.M. **2** 81.20
Spiro, T.C. **2** 71.30
Spiro, T.G. **3** 76.65
Spittler, T.D. **7** 53.15
Spitzmiller, E.R. **5** 41.33
Spivack, B. **4** 82.13 **5** 76.36 **7** 76.45, 82.18
Spoelder, T. **4** 12.10
Spofford III, W.A. **3** 79.8, 79.24 **4** 85.10, 85.27 **5** 8.7, 79.24
Spofford, W.A. **2** 85.19, 86.51 **3** 79.6, 79.7, 79.9, 79.22
Spohn, R.J. **3** 71.60
Sprang, S. **6** 45.32
Spratley, R. **3** 72.1
Spratley, R.D. **2** 64.60
Sprecher, C.M. **4** 31.8
Spreckelmeyer, B. **5** 33.17, 33.18, 33.19
Sprenkels, A.J.J. **1** 2.64
Springer, J.P. **4** 59.16 **6** 59.1 **7** 58.3, 58.28
Sproul, G. **6** 83.48
Sproul, G.D. **5** 3.1, 36.5, 36.6
Spruijt, A.M. **4** 85.4
Squire, R.H. **5** 38.17 **6** 38.25
Srenger, E.M. **7** 41.9
Srikanta, S. **3** 2.2
Srikantha, S. **1** 2.24
Srikrishnan, T. **4** 48.42 **7** 34.8, 44.8
Srinivasan, L. **3** 83.14 **4** 83.79, 83.80
Srinivasan, R. **1** 44.45, 48.66, 48.71, 50.33 **3** 48.29 **4** 11.11, 44.24, 48.19, 48.34, 48.37, 48.42 **5** 48.26, 48.44, 48.45 **6** 16.13 **7** 31.17, 48.14, 48.40, 48.53

Srinivasan, S. **1** 9.17
Srivastava, P. **1** 13.24, 25.12
Srivastava, R.C. **1** 3.33, 3.35, 3.36, 22.7 **2** 78.10 **4** 83.36, 83.67 **5** 52.2, 78.8, 83.45 **7** 78.6
Srivastava, S.N. **1** 58.15 **3** 54.11 **5** 26.3
Srivastava, T.S. **4** 49.19, 49.24 **5** 49.10
Stackelberg, M.v. **1** 5.20, 17.10 **2** 78.20
Stadler, H.P. **1** 30.25, 30.30, 36.37 **3** 3.35, 30.3
Stahlin, W. **4** 83.69, 83.70
Stalhandske, C. **5** 16.4, 16.5 **6** 16.3
Stalhandske, C.-I. **3** 66.1
Stalick, J.K. **2** 86.88 **3** 69.39, 71.20, 86.14, 86.39, 86.56 **5** 85.13 **6** 86.60, 86.61 **7** 64.3, 64.8
Stallard, M.O. **5** 52.1
Stalley, J.C. **7** 30.1, 30.2
Stallings, W. **5** 63.5 **6** 59.2
Stalzer, R.A. **6** 38.30, 38.31
Stam-Thole, I. **4** 18.3
Stam, C.H. **1** 5.37, 5.38, 8.34, 28.1 **3** 33.46, 37.7 **4** 18.3, 84.4 **5** 19.11, 19.19, 57.1, 57.2 **6** 19.2, 40.22, 65.1, 65.2, 65.3 **7** 8.2
Stamp, L.A. **5** 2.16, 2.17
Stanfield, R.A. **1** 30.27
Stanford Junior, R.H. **1** 1.81, 41.2 **3** 24.6, 28.12, 30.5 **4** 22.10 **5** 26.4, 41.18 **6** 28.20, 28.23, 76.29 **7** 75.17
Stanford, R.H. **5** 51.39 **6** 13.8, 13.9
Stanforth, R.R. **6** 20.18, 20.19
Stanislowski, A.G. **6** 72.19 **7** 72.15
Stanko, J.A. **3** 3.53, 76.45 **4** 58.21, 64.40 **5** 3.9
Stanko, V.I. **2** 62.12, 62.14 **3** 62.2
Stankovic, S. **7** 35.21
Stanley, E. **1** 2.90, 46.23 **4** 83.30, 84.7 **5** 36.21
Starikova, Z.A. **3** 77.5 **5** 80.14 **7** 78.21, 78.22
Staron, T. **7** 50.35
Starovskii, O.V. **2** 73.14, 73.108, 74.15
Steadman, J.P. **3** 3.3
Stec, W.J. **7** 64.20, 64.21
Steele Junior, J.C.H. **5** 86.42
Steele, D. **6** 82.15, 84.1
Steele, J.C.H. **5** 86.41
Steeple, H. **1** 3.6, 41.32
Steer, I.A. **4** 65.1, 65.2
Steer, R.J. **3** 19.1
Steffen, E.D. **6** 83.27
Steffen, W.L. **7** 86.51, 86.54, 86.57
Steggarda, J.J. **4** 79.10
Steggerda, J.J. **2** 80.8, 80.9, 80.34
Steigemann, W. **3** 31.4
Stein, M.T. **4** 64.40 **6** 61.9 **7** 61.2
Steinbach, G. **6** 38.33
Steiner, E.C. **7** 38.9, 38.11, 84.23
Steinfink, H. **2** 63.7 **3** 81.4
Steinhauser, S. **3** 68.5
Steinrauf, L.K. **1** 48.58, 48.67, 48.68, 48.76, 50.45, 58.14, 59.55 **2** 77.55 **3** 48.28, 48.45, 48.48, 50.25 **4** 50.16, 58.34 **6** 48.29
Stelling, G.D. **3** 31.10
Stellman, S.D. **7** 47.38
Stelzer, O. **5** 86.19
Stempel, A. **6** 16.23, 38.24
Stempelova, D. **2** 81.86
Stemple, N.R. **1** 1.57, 1.58 **5** 59.6, 77.14 **6** 41.32 **7** 84.29, 84.62, 84.73
Sten Rouwelaar, L.N.A. **2** 81.79
Stenkamp, R.E. **5** 3.16 **6** 48.43, 48.71, 48.80 **7** 48.48
Stensland, B. **1** 36.27 **3** 63.17 **4** 35.1, 36.13 **5** 45.48, 82.28
Stenson, P. **3** 41.13, 83.130
Stepaniak, R.F. **6** 71.45, 71.51 **7** 71.77
Stephen Junior, W.F. **3** 3.40, 3.41
Stephens, D.N. **3** 36.3
Stephens, F.A. **6** 64.51, 81.46, 86.30
Stephens, F.S. **2** 64.72, 64.73, 76.51, 83.113 **3** 1.34, 60.37, 72.29, 73.39, 76.20, 76.22, 76.59, 76.69, 81.40, 81.49, 83.36, 83.105 **4** 33.23, 81.40, 83.107 **5** 75.14, 75.27, 86.18 **6** 72.14, 73.26, 73.32, 73.48, 83.62 **7** 71.51, 73.11, 73.25, 73.27, 73.57, 73.59, 82.12
Stephensen, T.A. **4** 85.36
Stephenson, N.C. **1** 3.65 **2** 76.34, 77.54, 83.33, 83.100, 83.101, 83.154, 85.50, 85.66, 86.31, 86.33, 86.34, 86.35 **3** 45.41, 45.42, 76.30, 82.1, 83.76, 86.13, 86.20, 86.26 **4** 45.22, 83.113, 86.5 **5** 45.16, 45.44 **6** 45.15, 45.18, 45.42, 77.1, 83.61 **7** 45.13, 81.43
Stepien, A. **7** 8.5, 8.6
Sterling, C. **1** 2.13, 2.22, 57.4, 57.5 **2** 69.30
Stermitz, F.R. **1** 58.52
Stern, F. **1** 1.59
Stern, P. **6** 46.5, 64.36
Sternbach, L.H. **5** 35.26, 35.27
Sternglanz, H. **5** 44.12, 44.16 **6** 44.13, 44.14, 45.47, 50.6, 64.24 **7** 44.1, 44.3, 44.4
Sterns, M. **1** 31.42 **3** 36.16, 83.119 **4** 59.9, 83.71
Stetsenko, A.N. **7** 35.10
Steuck, M.J. **3** 76.40
Stevens, C.L. **5** 31.20
Stevens, E.D. **5** 37.1 **6** 83.27
Stevens, G.N. **6** 69.6
Stevens, J.D. **3** 45.41, 45.42 **4** 45.22 **5** 21.17, 45.11, 45.16, 45.44 **6** 45.15, 45.18, 45.33, 45.39, 45.40, 45.41, 45.42, 45.44, 45.45 **7** 45.13, 45.46
Stevens, K. **7** 31.34, 31.35
Stevens, K.L. **1** 56.14
Stevenson, D.L. **2** 73.27, 86.20
Stevenson, J.L. **5** 16.2
Stevenson, R. **1** 51.42 **5** 51.42
Steward, E.G. **6** 1.6, 1.8, 1.9
Stewart Junior, R.P. **5** 71.54 **6** 71.19
Stewart, G. **1** 14.4
Stewart, J.M. **1** 8.26, 8.27, 16.1, 32.32, 41.25 **2** 72.83, 78.9, 78.11 **3** 11.18, 15.1, 15.2, 32.28, 35.21, 36.32, 39.6, 39.28 **4** 16.17, 32.3, 32.14, 39.43, 86.38 **5** 4.1, 34.3, 41.37, 86.30, 86.31 **7** 67.4
Stewart, M.V. **5** 31.10

Stewart, R.F. **1** 44.9, 44.10, 44.47 **3** 83.60 **4** 31.8
Stewart, T.G.A. **5** 76.1
Steyn, P.S. **5** 58.25 **7** 58.17
Stezowski, J.J. **3** 73.21 **4** 50.5, 73.31, 77.8 **6** 67.5 **7** 32.31, 32.32, 32.35, 33.56, 38.59, 50.14, 50.15, 50.16
Stiefel, E.I. **1** 3.40 **3** 85.27 **6** 85.20, 85.21
Stock, D. **2** 72.12
Stockel, K. **1** 51.35
Stocker, J.H. **5** 33.47
Stocklin, W. **1** 51.35
Stocks, J. **4** 85.34
Stoeckli-Evans, H. **4** 49.23 **7** 41.32, 71.106, 71.119
Stoessl, A. **7** 53.12
Stofko Junior, J.J. **7** 72.6
Stogsdill, R.M. **2** 78.51
Stokely, P.F. **3** 11.16, 11.17, 73.3 **4** 82.7
Stokes, J. **2** 85.26, 85.52
Stomberg, R. **2** 76.1, 83.35, 83.71, 83.72, 83.93 **3** 76.4, 76.11, 76.12, 81.1
Stone, A. **1** 42.11, 42.12, 49.21, 49.27
Stone, A.L. **1** 49.30 **2** 75.26
Stone, F.G.A. **4** 71.70, 71.84, 71.86 **5** 71.3, 71.41, 71.44, 74.10, 86.79 **6** 71.34, 71.74, 75.11, 75.12, 75.23, 86.5 **7** 71.22, 71.45, 71.49, 71.59, 71.61, 71.66, 71.96, 72.1, 72.9, 73.41, 75.1, 75.22, 75.25, 75.38, 86.4, 86.10, 86.11
Stone, M.E. **4** 83.30, 84.7
Stoppioni, P. **7** 86.115, 86.117
Stora, C. **3** 19.16, 45.17 **4** 17.14 **6** 24.9, 39.9
Stork-Blaisse, B.A. **4** 83.81 **5** 67.9, 83.39
Stork, G. **5** 28.19
Storr, A. **4** 68.3 **5** 68.6 **6** 68.5, 73.53 **7** 68.11, 68.13, 68.14, 83.72, 83.73, 83.102
Stosick, A.J. **1** 48.2
Stothart, P.H. **6** 47.29
Stothers, J.B. **7** 53.12
Stotter, D.A. **7** 71.87, 84.50
Stoufer, R.C. **4** 83.134
Stout, C.D. **5** 83.15
Stout, G.H. **1** 56.14, 56.16, 59.26, 59.27, 59.40, 59.65 **3** 59.13, 59.32
Stout, V.F. **1** 56.16, 59.40
Strahs, G. **1** 33.51
Strandberg, B. **2** 82.3
Strandberg, R. **4** 83.96
Strawson, C.J. **7** 58.23
Streib, W.E. **2** 62.7, 62.22, 62.28, 75.9, 83.190 **3** 62.19 **4** 62.6, 68.2, 83.109 **6** 28.10, 71.75, 72.10 **7** 78.18, 83.65
Streith, J. **5** 34.2 **6** 36.12
Strel'tsova, I.N. **1** 19.4, 19.24
Strell, I. **5** 49.1
Streurman, H.J. **3** 39.5
Strickler, P. **1** 21.16, 21.18 **2** 85.67, 85.94
Strid, K. **1** 7.1
Strieter, F.J. **1** 1.36, 1.37, 1.65 **2** 60.1
Strijk, B. **1** 21.5
Strom-Gundersen, E.K. **3** 36.25
Stromme, K.O. **1** 16.2 **2** 60.22, 60.29, 60.36, 60.84, 60.85 **6** 34.1
Strong, F.M. **7** 58.3
Strong, P. **6** 51.31
Strope, D. **7** 71.16
Stroud, R.M. **4** 39.9 **5** 35.13
Strouse, C.E. **5** 71.103 **6** 73.1, 73.14, 73.15 **7** 71.8, 73.1, 73.16
Struchkov, Yu.T. **7** 73.69
Struchkov, Yu.T. **1** 5.2, 5.8, 11.43, 15.1, 16.33, 19.4, 19.24, 19.38, 24.2, 24.3, 24.5, 24.8, 24.12, 28.8, 28.9, 28.10, 29.28 **2** 62.12, 62.14, 63.28, 65.13, 66.9, 69.4, 69.5, 69.18, 69.39, 69.45, 71.34, 72.7, 72.19, 72.72, 72.73, 72.84, 73.14, 73.41, 73.49, 73.53, 73.68, 73.69, 73.85, 73.87, 73.101, 73.105, 73.107, 73.108, 73.119, 73.120, 74.15 **3** 40.2, 60.9, 62.2, 64.30, 69.31, 71.4, 71.10, 71.49, 71.56, 71.61, 71.62, 72.40, 72.48, 72.53, 73.22, 73.29, 73.38, 73.45, 74.8, 74.9, 74.10, 74.16, 75.38, 75.46 **4** 16.1, 63.8, 64.17, 69.29, 71.38, 72.5, 72.6, 72.14, 73.10, 73.21, 74.5, 83.128, 86.36 **5** 41.25, 58.6, 63.15, 64.32, 69.3, 71.7, 71.45, 71.47, 71.104, 72.26, 72.41, 73.4, 73.7, 73.12, 73.33, 74.7, 82.24 **6** 58.8, 58.20, 64.29, 66.2, 69.17, 69.20, 71.7, 72.17, 72.24, 72.27, 73.55, 73.58, 81.45 **7** 17.13, 17.14, 18.7, 42.13, 58.16, 63.13, 63.14, 63.18, 64.30, 69.10, 69.20, 69.21, 69.23, 71.18, 71.36, 71.37, 71.128, 72.26, 72.27, 72.44, 73.10, 73.12, 73.32, 73.46, 74.6, 75.5, 75.33, 81.45, 81.47, 82.21, 86.72
Strydom, O.A.W. **6** 19.1
Stuckey, J.E. **5** 65.1
Stucky, G. **2** 67.14, 67.17 **7** 26.4, 67.1, 67.8, 83.111
Stucky, G.D. **1** 3.21, 3.27, 3.67 **2** 67.12, 67.15, 68.18 **3** 37.2, 40.4, 62.3, 67.7, 67.8, 67.12, 67.13, 67.18, 68.1, 68.11, 68.15, 68.21, 83.21 **4** 38.2, 38.3, 38.4, 67.26, 86.10 **5** 3.1, 20.7, 67.11, 67.17, 67.18, 67.19, 67.20, 68.31 **6** 20.2, 20.3, 20.4, 38.3, 62.3, 68.16, 83.48 **7** 27.2, 67.6
Studd, B.F. **2** 83.196 **3** 83.145
Stulen, G. **1** 39.91
Stults, B.R. **6** 49.16, 49.17, 71.11, 71.24, 72.26, 77.8, 77.13 **7** 77.2, 83.50
Sturdivant, J.H. **2** 71.29, 83.115
Sturm, M. **5** 48.40, 48.69, 49.1
Suares, H. **4** 58.35
Subrahmanyam, G. **7** 31.17
Subramanian, E. **1** 16.32, 31.19, 47.26, 48.64, 48.81 **3** 1.11, 44.22, 47.13, 59.2, 64.11 **4** 36.31 **5** 47.16 **7** 47.38
Subrtova, V. **1** 37.11 **4** 62.5
Suchy, M. **3** 53.7
Suck, D. **3** 44.21 **4** 47.1, 47.3, 47.7, 47.18 **5** 47.1, 47.2, 47.3, 47.21 **6** 44.8, 47.5 **7** 47.19, 47.23, 47.33, 47.41, 47.42
Sudarikov, V.S. **3** 73.22
Suddath Junior, F.L. **4** 32.10
Suddath, F.L. **5** 47.33 **7** 84.49
Suffness, M.I. **3** 58.57

Suga, H. **1** 37.1
Sugawara, I.T. **1** 5.9
Sugawara, T. **1** 2.3
Sugihara, A. **2** 82.29
Sugino, T. **3** 50.30
Suginome, H. **4** 40.24
Sugisaka, N. **3** 38.17
Sugiyama, K. **4** 5.7
Sugowdz, G. **5** 31.37
Suhara, Y. **5** 50.3
Suito, E. **4** 49.13, 49.18
Sukhanova, L.S. **3** 82.9
Sulaimankov, K. **7** 61.1
Suleyman, X. **7** 61.1
Sulligoi, L. **7** 80.6, 80.7
Sullivan, R.A.L. **1** 8.9, 16.57
Sullivan, S. **2** 60.119
Sumida, Y. **4** 33.10, 33.11
Sumner, G.G. **2** 78.12
Sumner, R.H. **2** 78.41
Sumsion, H.T. **1** 19.65
Sundaralingam, M. **1** 1.136, 12.18, 13.21, 16.5, 21.15, 25.22, 27.14, 39.75, 40.27, 46.14, 47.10, 47.28 **3** 44.11, 45.35, 47.7, 47.11, 47.12, 47.21, 47.26, 47.30, 48.6, 48.7, 48.35, 50.16, 84.6 **4** 11.3, 20.4, 26.7, 31.15, 31.19, 44.2, 44.3, 44.22, 47.8, 47.9, 47.10, 47.12, 47.15, 47.16, 50.1, 83.28, 83.40 **5** 35.1, 35.14, 44.8, 47.26, 47.28, 83.15 **6** 11.11, 13.8, 13.9, 45.28, 45.32, 45.43, 46.4, 46.9, 47.9, 47.14, 47.23, 47.24, 50.2, 50.5 **7** 36.1, 47.25
Sundarama, K. **2** 81.47
Sundararaman, P. **7** 53.10
Sundaresan, T. **4** 40.4, 60.25 **5** 33.55, 36.10, 40.5
Sundberg, M. **7** 83.13
Sundberg, R.J. **6** 71.4
Sundell, S. **7** 51.43
Sung, N. **1** 19.43
Suntzov, E.W. **5** 79.13
Surazhakaya, M.D. **3** 81.7
Surazhskaya, M.D. **7** 81.2
Surles, J.R. **7** 3.32
Susse, P. **5** 20.3, 39.30
Sussman, J.L. **4** 47.20 **6** 58.13
Susz, B.P. **2** 60.50
Sutherland, H.H. **1** 13.49, 13.50, 19.46, 19.47, 19.48, 48.10 **3** 13.14 **5** 13.15 **6** 15.5 **7** 39.18
Sutherland, S.A. **1** 50.26, 54.7, 56.3, 59.61
Sutor, D.J. **1** 1.35, 2.26, 44.54, 44.64, 44.66, 44.67, 47.21, 58.32 **2** 65.7, 65.25, 65.26
Sutton, D. **3** 83.1 **5** 71.97, 83.21, 83.123
Sutton, P.W. **2** 71.22
Suzuki, A. **5** 9.15 **6** 53.5, 53.10
Suzuki, H. **6** 59.29
Suzuki, K. **3** 36.34
Suzuki, S. **1** 50.4 **7** 50.6
Suzuki, Y. **4** 35.38
Svaeren, S.E. **5** 70.11
Svedung, D.H. **3** 83.7
Svelasvili, A.E. **2** 76.27, 76.70, 76.74
Svensson, S. **5** 45.17
Svetich, G.W. **1** 46.9, 46.24 **4** 82.10, 84.18 **6** 83.34, 83.52
Svinning, T. **7** 1.22, 1.33
Svistunov, V.S. **7** 63.14
Swahn, C.-G. **5** 45.19 **6** 45.25, 45.27
Swallow, A.G. **2** 77.58, 83.164, 83.196 **3** 73.42, 83.145
Swaminatha Reddy, B. **4** 47.9, 47.10
Swaminathan, P. **5** 48.26, 48.44, 48.45 **7** 48.14, 48.26
Swaminathan, S. **1** 9.17, 26.28 **2** 79.6 **3** 84.4 **4** 12.9 **7** 13.17
Swan-Walstra, S. **4** 65.10 **5** 65.6
Swan, E.P. **1** 59.61
Swank, D. **3** 84.1, 84.7
Swank, D.D. **1** 46.7 **2** 64.25 **4** 64.39 **6** 83.64 **7** 39.1
Swann, D.A. **4** 83.101 **5** 77.12 **6** 60.2, 76.36, 76.37, 76.38 **7** 36.25
Swanson, B. **3** 7.2, 7.3
Swanson, J. **4** 38.2, 38.3
Swanson, J.S. **3** 3.17
Swanwick, M.G. **3** 71.5, 71.6
Swardstrom, J.W. **4** 67.28 **5** 16.18
Sweeney, J. **7** 59.9
Sweeney, J.G. **7** 38.57
Sweet, R.M. **1** 44.22, 44.23 **2** 69.35 **3** 50.7, 50.9, 50.11 **5** 58.10
Swen-Walstra, S.C. **3** 31.33
Swenton, J.S. **7** 35.13
Swift, D. **6** 69.22
Swift, D.R. **3** 85.49 **5** 83.98, 83.99, 83.102
Swink, L.N. **2** 76.42
Syal, V.K. **6** 78.14
Syamal, A. **6** 76.1
Sydow, E.von **1** 1.148, 1.150, 5.32
Sygusch, J. **6** 35.25, 50.7 **7** 50.12
Szabo, W.A. **5** 38.5
Szalda, D.J. **7** 76.57, 82.3
Szary, A.C. **7** 75.25
Szmuszkovicz, J. **4** 34.3, 35.28, 35.29, 40.23
Szwarc, H. **7** 19.2
Szymanski, J.T. **2** 60.114, 82.25, 82.35 **3** 83.72 **4** 63.2, 63.12, 69.1

Tabak, J.M. **5** 28.19
Tabata, M. **5** 9.15
Tachikawa, N. **7** 60.15, 60.16
Tachikawa, R. **5** 36.22
Tachiyashiki, S. **7** 6.1
Tadini, C. **1** 11.18
Taga, T. **3** 45.24, 45.25, 58.43, 59.30 **4** 67.3 **5** 17.20, 38.12, 38.29, 45.37, 58.37 **6** 38.6, 58.47 **7** 13.25
Tagawa, H. **4** 60.12
Taggi, A.J. **7** 51.21, 51.22
Taguchi, I. **1** 16.15
Tai, C.-C. **3** 40.11
Taillandier, G. **7** 1.21, 1.32
Taira, Z. **4** 62.10 **5** 38.12, 62.7 **6** 38.6, 43.3, 54.1 **7** 1.13, 1.14
Tait, A.M. **4** 83.111
Takabatake, A. **1** 59.52

Takada, K. **7** 50.39
Takada, N. **6** 53.7
Takada, S. **1** 54.17 **4** 21.22, 28.12
Takada, Y. **5** 9.15
Takagi, S. **3** 45.13, 59.26 **5** 45.25
Takahashi, N. **7** 54.8
Takahashi, R. **4** 56.6
Takahashi, S. **2** 86.71, 86.81
Takai, M. **5** 45.15 **7** 58.25
Takaki, Y. **1** 13.16, 13.17, 13.18, 13.19, 13.38, 27.8, 33.18 **5** 13.5 **6** 13.4 **7** 13.4, 13.5
Takanashi, S. **3** 33.24
Takano, T. **1** 13.29, 20.1, 33.42 **2** 63.6
Takao, N. **6** 58.45
Takashina, K. **4** 5.1
Takats, J. **2** 75.39, 75.57 **3** 73.44 **4** 73.23 **6** 85.31, 85.33, 85.34
Takatsu, M. **6** 36.10
Takaya, H. **5** 78.13, 78.14
Takayanagi, H. **7** 41.16, 53.4
Take, T. **1** 50.31 **3** 50.20, 59.1
Takeda, T. **5** 58.33 **6** 50.1 **7** 47.28
Takei, W.J. **1** 43.7
Takemoto, K. **3** 36.34
Takemoto, T. **6** 54.14
Takenaka, A. **1** 50.31 **2** 75.55 **3** 33.42 **4** 81.32, 81.33, 81.35 **5** 1.20, 49.6, 81.32, 81.33 **6** 49.7 **7** 49.8
Takeshita, H. **4** 27.15
Takeuchi, S. **4** 50.4 **6** 71.58
Takeuchi, T. **6** 5.1, 5.2, 5.3
Takeuchi, Y. **1** 33.29, 59.4
Takigawa, T. **1** 48.84, 48.86
Takita, T. **1** 41.31 **6** 1.30 **7** 1.18, 50.39, 53.28
Takusagawa, F. **4** 13.14 **6** 33.20, 33.21, 33.22, 44.3 **7** 13.13, 13.15, 20.2, 20.3, 33.7, 33.17, 33.19, 33.24, 33.25, 45.1
Takwale, M.G. **4** 13.6
Talacki, R. **6** 36.4
Talaty, E.R. **4** 32.30
Talberg, H.J. **6** 16.8 **7** 17.3, 18.5, 41.13
Talbot, J.H. **2** 81.29
Tamamura, T. **7** 38.67
Tamburin, H.J. **4** 37.9 **5** 37.5
Tamm, C. **3** 50.19
Tamura-Kobayashi, H. **4** 82.6
Tamura, C. **1** 35.17, 37.21, 39.88, 50.1, 51.36, 58.7, 58.12 **3** 58.53 **4** 32.5, 64.35, 64.38 **5** 36.22, 41.7, 41.9, 41.36, 44.4, 59.33, 60.4, 64.37 **6** 36.11, 44.2 **7** 33.7, 33.39, 33.40, 33.41, 47.14, 47.24, 50.1, 59.24, 60.7
Tamura, G. **1** 44.76
Tamura, H. **5** 78.1, 78.11
Tamura, K. **7** 17.10
Tamura, S. **4** 54.13 **6** 53.5, 53.10
Tamura, Y. **6** 32.27
Tamura, Z. **3** 33.24 **5** 32.21
Tan, W.L. **7** 56.3
Tanaka, I. **6** 15.1
Tanaka, J. **1** 15.23, 16.39, 30.5, 44.31 **5** 31.22, 31.23, 31.38, 31.40, 60.35 **7** 24.10, 31.27
Tanaka, K. **3** 72.20 **5** 76.20 **6** 5.2
Tanaka, M. **5** 42.12
Tanaka, N. **1** 13.3, 31.27, 56.7 **3** 59.15 **4** 59.20 **5** 59.17
Tanaka, O. **1** 56.7 **3** 56.3
Tanaka, T. **4** 5.1 **5** 47.14, 47.15
Tang, C.-P. **7** 1.24
Tang, S.C. **7** 49.10
Tang, Y.-C. **1** 11.19 **2** 83.115
Tangocci, P. **1** 35.5
Tani, H. **2** 68.28 **3** 68.19
Tani, K. **7** 78.3
Tani, M.E.V. **3** 8.11 **4** 8.6, 28.5 **5** 82.1, 82.10 **6** 76.15, 83.66 **7** 66.4, 79.24, 82.6
Taniguchi, M. **7** 31.15
Taniguchi, T. **1** 13.16, 13.17, 13.18, 13.19 **6** 13.4
Tanimoto, Y. **6** 19.4
Tanimura, M. **2** 83.114 **3** 21.11 **7** 83.9
Tanisaki, S. **6** 7.1
Tapscott, R.E. **2** 81.54
Tarantelli, T. **4** 85.33
Tarkhova, T.N. **2** 81.76, 81.82, 81.87 **7** 78.14, 78.15, 79.21, 85.22
Tarnopol'skii, B.L. **3** 27.12, 33.28, 64.30 **5** 45.47
Tartakovskii, V.A. **3** 40.2
Tashiro, Y. **6** 71.58
Tashpulatov, Yu. **1** 41.18
Tasker, P.A. **3** 83.138 **5** 83.62, 83.98, 83.99, 83.102 **7** 83.98
Tasset, E.L. **6** 49.16, 49.17, 77.8, 77.13
Tatemitsu, H. **5** 31.23
Tatsch, C.E. **5** 82.23
Tatsuno, T. **7** 50.18
Tatsuno, Y. **5** 71.67 **6** 71.42
Tatsuoka, T. **7** 39.17
Taube, H. **6** 71.4
Tauer, K.J. **1** 5.4
Tauzher, G. **6** 76.42
Tavale, S.S. **1** 2.52, 2.69, 2.72, 2.85, 2.89, 13.11 **3** 44.13, 44.18 **4** 13.1 **5** 58.48 **6** 13.1, 13.2 **7** 64.47
Tavberidze, M.G. **7** 76.19, 76.36, 84.14
Taylor Junior, I.C. **7** 63.15
Taylor Junior, I.F. **3** 74.11, 74.12, 74.13, 79.24 **4** 86.24 **6** 54.5, 69.6, 71.4 **7** 39.19, 39.20, 51.57, 63.16, 74.2, 79.13, 79.23
Taylor, C.A. **1** 10.26
Taylor, D. **5** 83.20, 83.32, 84.17, 84.26 **6** 80.6, 80.9, 80.10, 84.47
Taylor, D.A.H. **3** 56.11
Taylor, D.J. **7** 50.24
Taylor, D.W. **3** 83.111 **4** 83.37, 83.38, 83.41, 83.52, 83.53 **5** 83.52, 83.55, 83.56, 83.57 **6** 83.47 **7** 73.18, 75.16
Taylor, E.C. **4** 40.2, 41.3
Taylor, H.F.W. **3** 76.24
Taylor, H.L. **4** 59.19, 59.26
Taylor, H.P. **5** 71.51
Taylor, I.F. **2** 79.3 **3** 74.7, 74.14 **7** 83.90
Taylor, J.C. **1** 8.12, 8.13, 8.14, 8.15, 59.16 **2** 83.23 **5** 2.4, 2.5, 2.6, 2.7, 2.8, 2.9
Taylor, J.H. **2** 60.83
Taylor, K.A. **2** 86.104 **3** 86.91

Taylor, M.R. **1** 2.46, 2.47, 4.15 **2** 82.26 **3** 83.14 **4** 83.79, 83.80 **5** 83.4 **6** 40.15 **7** 79.16, 83.95
Taylor, N. **4** 83.137, 83.138
Taylor, N.E. **1** 35.39, 59.43
Taylor, N.J. **6** 71.5 **7** 71.5, 71.10, 71.127, 71.138, 82.2, 86.56, 86.109
Taylor, P.J. **3** 33.35
Taylor, R. **7** 63.12, 71.87
Taylor, R.C. **4** 62.2, 62.3, 62.4 **6** 64.31, 84.19 **7** 64.4, 64.5, 64.6
Taylor, R.D. **7** 69.25
Taylor, R.M. **2** 85.24
Taylor, R.P. **2** 65.42 **3** 78.19
Taylor, S.H. **6** 86.21
Taylor, S.R. **3** 71.55
Taylor, T.G. **1** 31.39, 52.24, 52.30
Taylor, W.I. **3** 58.51
Tazzoli, V. **3** 11.5 **4** 83.35 **5** 84.18 **6** 84.44 **7** 84.19, 84.82
Teare, P.W. **1** 5.43
Tebbe, F.N. **6** 73.50, 73.56, 86.2 **7** 71.38
Tebbe, K.-F. **5** 62.1, 63.2, 64.15 **6** 63.7
Tedder, J.M. **1** 40.33
Tedenac, J.-C. **7** 83.96
Tehan, F.J. **7** 40.18
Teitel, S. **4** 58.3, 58.6 **5** 58.22
Teller, R.G. **7** 16.7
Tellgren, R. **7** 15.1
Temme III, G.H. **7** 31.19
Temple, D.M. **3** 60.40
Templeton, D.H. **1** 1.26, 1.36, 1.37, 1.65, 3.31, 3.80, 27.1, 27.2, 27.3, 31.37, 32.7, 49.36, 54.8, 59.3 **2** 60.1, 62.33, 62.41, 64.23, 65.39, 65.42, 71.30, 73.5, 73.28, 77.6, 77.46, 82.59, 83.219, 85.39, 85.40 **3** 48.17, 49.6, 62.8, 62.12, 76.25, 77.32 **4** 49.8, 49.12, 62.9, 71.5 **5** 1.4, 67.21, 67.25, 67.26 **6** 6.1, 22.3, 32.38, 33.57, 35.19, 35.20, 38.14, 42.6, 51.17, 64.33, 83.79, 84.27, 84.28 **7** 39.24, 59.21
Templeton, L.K. **1** 27.1, 27.2, 27.3 **5** 1.4
Temussi, P.A. **1** 1.86, 1.97, 21.24, 21.25
Tench, A.H. **4** 47.11 **5** 31.43
Teo, B.-K. **7** 86.111, 86.112
Teo, B.K. **4** 22.7
Teo, S.B. **5** 81.38
Terahara, A. **7** 50.1
Terao, Y. **1** 58.115
Terhaar, G. **2** 62.6
Teriz, A. **6** 45.4
Ternay Junior, A.L. **1** 39.74
Terpstra, P. **1** 11.31
Terry III, N.W. **4** 86.48 **5** 86.70, 86.71, 86.74 **6** 79.5
Terry, J.C. **4** 33.16, 36.1
Terzis, A. **3** 76.65 **5** 83.88 **6** 71.21 **7** 51.27, 51.37
Tesch, T.C. **5** 31.14, 31.17
Teufer, G. **7** 25.4
Teutsch, G. **7** 51.4
Tewari, R. **4** 83.36 **5** 78.8 **7** 78.6
Textor, M. **6** 64.46 **7** 73.24
Thackeray, M.M. **5** 71.34 **6** 54.3, 71.26 **7** 38.63, 83.8
Thakur, C.P. **5** 64.21
Than, K.A. **6** 86.68
Thanomkul, S. **6** 45.20 **7** 45.7, 45.8
Thasitis, A. **6** 84.8
Thathachari, Y.T. **6** 48.45
Thayer, N.N. **6** 33.46
Thenn, W. **7** 32.15, 32.27
Theophanides, T. **5** 72.7, 72.15 **7** 51.27, 51.37
Thewalt, U. **2** 76.56, 83.11, 83.12, 85.38 **3** 44.20, 47.16, 47.19, 47.23, 64.54, 76.17, 76.19 **4** 35.11, 44.8, 44.20, 64.36, 68.6, 76.28, 76.29, 76.30, 76.39, 83.6 **5** 44.16, 47.19, 76.30, 85.2 **6** 47.7, 65.11 **7** 44.11, 44.12, 83.5, 83.12, 85.1
Thibeault, J.C. **6** 20.11
Thich, J.A. **6** 82.24
Thiele, G. **4** 85.11 **5** 76.2 **7** 85.24
Thierry, J.-C. **3** 53.12, 53.13, 53.14
Thierry, J.C. **5** 53.7, 53.8, 53.9, 53.11, 53.16 **6** 67.3, 67.7, 85.35
Thiessen, W.E. **1** 40.5, 40.10 **3** 45.19, 51.47, 53.3, 53.8 **4** 8.1, 44.7, 56.3 **5** 51.32 **6** 41.28, 59.18
Thirase, G. **7** 71.65
Thoai, N. **3** 1.7, 30.6
Thom, E. **1** 7.15 **3** 27.7, 51.40, 51.46
Thomas II, F.D. **6** 78.7, 78.8, 78.16, 78.18, 78.19
Thomas-David, G. **6** 67.10, 81.34
Thomas, B.S. **6** 73.53
Thomas, C.K. **3** 3.53 **5** 3.9
Thomas, G. **5** 67.8, 81.45
Thomas, J. **3** 47.15
Thomas, J.O. **5** 2.11, 2.13 **6** 2.6, 2.7, 48.1
Thomas, J.T. **2** 85.22
Thomas, K.M. **5** 71.80, 75.30, 83.49, 86.51 **6** 46.14, 68.10, 72.2, 86.13, 86.25, 86.31, 86.63 **7** 69.17
Thomas, L. **5** 65.2
Thomas, M. **7** 32.9, 81.14
Thomas, P. **4** 85.20
Thomas, P.C. **1** 41.29 **2** 62.48
Thomas, R. **1** 11.14 **3** 59.8 **4** 22.9 **5** 38.20, 40.34 **7** 33.9, 33.25, 44.17
Thomas, R.H.P. **6** 63.6, 69.15, 69.16
Thompson, B. **4** 83.137, 83.138
Thompson, B.P. **1** 30.11
Thompson, C.M. **1** 19.23
Thompson, D.J. **3** 59.31
Thompson, D.T. **2** 86.6, 86.8, 86.43, 86.55 **3** 86.12, 86.41
Thompson, D.W. **4** 77.1
Thompson, G.L. **4** 39.17
Thompson, J.C. **4** 63.2
Thompson, J.E. **6** 16.9
Thompson, L.C. **1** 3.83
Thompson, L.D. **7** 10.1, 78.9
Thong, C.M. **6** 37.4
Thornley, A.S. **3** 78.3
Thornton, P. **5** 81.48, 81.61 **7** 81.46
Thorstensen, B. **7** 83.11
Thorstenson, P.C. **5** 35.34 **7** 35.32
Thorup, N. **3** 76.2, 76.3 **4** 41.11 **7** 82.17
Thozet, A. **6** 17.1 **7** 17.6
Thuijl, J.van **1** 35.45, 36.36 **2** 77.59

Thurmann-Moe, T. **2** 60.61
Thyagaraja Rao, S. **1** 48.45
Tichy, K. **1** 1.113 **3** 15.8 **7** 3.2, 3.3
Tickle, I.J. **4** 36.35 **5** 7.7, 60.19, 60.20, 60.27, 60.29, 60.30, 60.31, 60.32, 60.33, 60.38 **6** 1.26, 60.25 **7** 48.46
Tieghi, G. **4** 5.9, 35.21, 76.26 **5** 71.71 **6** 71.64, 71.65 **7** 71.64
Tiethof, J.A. **5** 85.13
Tillack, J.V. **4** 67.8 **5** 32.1 **6** 58.24 **7** 33.44
Tillberg, O. **4** 36.19, 60.30
Tillott, R.J. **7** 64.66, 65.7
Timberlake, J.W. **5** 36.11
Timewell, C.W. **7** 81.15
Timkovich, R. **3** 49.10
Timmins, P.A. **6** 51.42
Timmons, C.J. **7** 30.9
Timms, R.E. **2** 69.41, 69.47
Timofeeva, T.V. **6** 64.29
Ting, H.-Y. **4** 3.5, 56.4, 59.14
Tiripicchio Camellini, M. **1** 8.28
Tiripicchio, A. **1** 8.28, 9.2 **2** 83.4, 83.5, 83.7, 83.14, 83.15 **3** 83.2, 83.3, 83.4 **4** 8.4, 67.1, 67.2, 79.1, 79.4, 81.2, 81.15, 81.16, 83.50 **5** 8.4, 8.5, 8.6, 41.14, 84.21 **6** 8.2, 11.12, 41.27, 83.70, 84.29 **7** 11.9, 11.12, 35.11, 79.3, 79.4
Tirouflet, J. **3** 74.2 **5** 73.40, 74.3, 74.4, 74.5 **7** 73.60
Tirronen, E. **7** 83.13
Tishchenko, G.N. **2** 83.95
Tisley, D.G. **3** 86.75 **4** 80.13
Tissier, A. **7** 84.69
Titova, N.S. **3** 62.2
Titus, D.D. **3** 86.77 **4** 86.39, 86.40 **5** 76.41 **6** 64.59, 77.5
Titze, H. **3** 77.19, 77.20 **7** 77.7
Tjomsland, O. **1** 41.6
Tjornhom, T. **2** 60.82
Tkachev, V.V. **5** 73.8 **7** 84.32, 84.33, 84.36
Tkatchenko, I. **4** 71.65
Toan, T. **4** 73.25
Tobe, M.L. **2** 83.86
Tobias, R.S. **2** 71.18
Tocquer, M.C. **7** 77.9
Toda, M. **6** 58.48
Todd, G. **1** 53.15
Todd, L.J. **2** 62.39 **6** 71.75
Todd, M.J. **3** 9.8
Todd, P.F. **2** 83.168
Toeplitz, B. **6** 35.13 **7** 59.13
Toft, L. **1** 10.24
Tokuma, Y. **4** 32.28 **7** 50.4
Tokunan, H. **4** 71.58, 71.80
Tokura, N. **5** 42.12
Tokuyama, T. **3** 58.9
Toldy, L. **5** 41.42
Toledano, P. **7** 80.9, 80.10, 80.20
Tollin, P. **1** 19.48, 47.21, 47.25 **3** 47.17, 47.18 **6** 47.6 **7** 47.20
Tolpin, E.I. **6** 62.8
Toma, S. **6** 72.7
Tomalia, D.A. **1** 11.9, 42.1
Toman, K. **1** 12.10, 12.11 **2** 60.34 **5** 71.2
Tomat, G. **7** 82.11
Tomiie, Y. **1** 9.5, 37.1, 50.37, 56.9, 58.35, 58.36, 58.57, 59.19, 59.25, 59.52 **2** 67.19 **3** 50.5, 50.13, 50.20, 72.20, 81.24 **4** 58.1, 58.2 **6** 5.4, 5.5
Tomile, Y. **1** 54.9 **2** 67.18
Tominaga, Y. **6** 36.11
Tomita, K. **1** 44.42, 44.48, 44.52, 44.56, 44.58, 44.59, 56.20, 58.93 **2** 82.4, 82.5, 82.21, 83.62 **3** 44.12, 47.32, 56.13 **4** 56.5 **5** 33.34, 33.35, 33.36, 44.10, 47.14, 47.15, 48.13, 48.25 **7** 60.1
Tomita, K.-I. **1** 44.38 **5** 1.16, 4.2, 8.14, 35.22, 44.5, 48.12 **6** 1.7, 8.5, 8.6, 35.14, 48.56, 58.45 **7** 1.15, 1.23, 17.10
Tomita, M. **1** 58.36, 58.57
Tomkins, I.B. **4** 86.55 **5** 71.94, 72.45
Tomko, J. **1** 58.105 **5** 58.40
Tomlinson, A.A.G. **3** 85.22 **4** 81.1 **5** 76.23 **6** 86.23
Tomlinson, A.J. **3** 75.16
Tomlonovic, B. **5** 77.24
Tomuka, Y. **1** 48.63
Tondello, E. **3** 80.19, 80.20, 83.45 **4** 39.27 **6** 3.7, 3.8, 3.29, 3.33, 16.16, 86.47 **7** 3.30, 80.2
Toney, J. **2** 67.15 **3** 67.18 **4** 67.26 **5** 67.11
Tong, Y.C. **1** 11.9, 42.1
Toniolo, C. **1** 33.75
Toniolo, L. **7** 86.93, 86.94
Toofan, M. **7** 83.92
Toome, V. **3** 35.23 **5** 58.22
Torbjornsson, L. **1** 36.27 **5** 36.20
Torelli, L. **2** 80.35
Torii, A. **4** 72.17, 82.6
Torii, K. **3** 48.16 **4** 48.28 **6** 48.19, 48.30 **7** 47.32
Toriumi, K. **7** 76.40
Torre, L.P. **3** 83.112, 83.121, 83.122, 83.123, 83.124, 83.125, 83.126
Torti, E. **1** 21.32, 29.17, 29.18
Toth, G. **5** 41.42
Toube, T.P. **4** 38.21, 54.1
Tougard, P. **6** 47.3, 47.10, 47.19, 47.20
Toure, S. **7** 43.4
Toussaint, J. **1** 11.46, 11.48, 11.49, 13.7, 17.33, 38.24 **6** 33.15 **7** 32.7, 36.19, 39.28
Towers, C. **5** 75.43
Towl, A.D.C. **2** 86.46 **3** 71.63, 78.24 **4** 71.43, 71.46, 71.49
Town, W.G. **6** 2.12
Towns, R. **1** 35.25, 35.37
Towns, R.L. **3** 32.34, 34.6, 64.37 **4** 32.19
Towns, R.L.R. **3** 39.19, 39.20 **4** 21.11, 37.13, 37.14, 51.7 **5** 64.24 **6** 21.9, 31.26, 39.8 **7** 16.7
Townsend, L.B. **4** 44.25 **6** 50.4
Traficante, D.D. **6** 86.1
Trager, W.F. **7** 59.7
Tranqui, D. **5** 11.14 **6** 11.15 **7** 11.16, 33.38, 33.45, 84.69
Tranter, T.C. **1** 48.49
Traverso, G. **1** 39.45
Treasurywala, A.M. **6** 58.33

Treat, T.A. **5** 31.20

Trefonas, L.M. **1** 3.44, 3.55, 32.34, 35.10, 35.25, 35.29, 35.31, 35.33, 35.37, 40.21 **2** 62.44 **3** 32.34, 34.6, 36.19, 51.2, 64.36, 64.37 **4** 32.19, 35.7, 35.17, 37.13, 37.14, 40.7, 51.1, 51.7, 81.59 **5** 10.5, 33.47, 36.11, 41.28, 64.8, 64.24, 64.28, 64.38, 64.40, 64.43, 64.46, 84.4, 84.15, 84.30 **6** 33.1, 79.14, 82.21 **7** 31.30, 38.54

Treichel, P.M. **1** 3.64 **2** 73.90 **5** 81.60 **6** 86.10 **7** 71.67, 73.42, 86.89

Treppendahl, S. **7** 40.7

Tressum, L. **7** 48.41

Trigwell, K.R. **7** 86.3, 86.40, 86.41

Trinderup, P. **3** 85.4

Trinh-Toan, L **4** 73.27

Trinh-Toan, M **4** 73.28

Trippett, S. **6** 64.32, 64.38 **7** 64.37

Trofimenko, S. **5** 83.35

Trogu, E.F. **7** 85.7, 85.10

Trommel, J. **1** 48.65

Trost, B.M. **4** 29.2

Trotter, J. **1** 3.60, 13.52, 15.9, 15.11, 15.15, 15.16, 15.17, 15.24, 16.6, 16.37, 18.14, 18.15, 19.29, 19.33, 19.35, 24.9, 24.22, 24.23, 26.12, 26.15, 26.17, 26.30, 26.36, 28.6, 28.7, 28.11, 28.13, 28.17, 29.5, 29.12, 29.14, 30.6, 30.10, 30.14, 30.18, 30.23, 30.29, 30.33, 30.35, 31.17, 31.44, 32.21, 37.27, 45.3, 45.47, 45.55, 45.61, 45.74, 47.6, 51.43, 52.23, 53.38, 58.46, 58.69, 58.79 **2** 64.45, 64.78, 65.1, 65.2, 65.4, 65.8, 65.10, 65.19, 65.21, 65.22, 65.23, 65.34, 65.48, 65.49, 69.2, 69.36, 70.5, 71.3, 71.10, 73.57, 73.86, 73.95, 73.98, 75.33, 75.60 **3** 1.10, 1.11, 10.5, 27.2, 27.3, 32.18, 32.19, 35.15, 39.29, 45.18, 45.20, 45.45, 48.20, 51.49, 56.10, 58.64, 59.2, 63.15, 64.3, 64.9, 64.11, 64.20, 64.65, 69.32, 73.41, 83.32, 83.127, 86.9, 86.11, 86.17, 86.18, 86.37, 86.45 **4** 1.4, 13.9, 28.7, 31.1, 31.3, 35.30, 36.31, 48.38, 68.3, 69.2, 71.18, 73.9, 81.17, 83.122, 85.21, 86.1 **5** 17.19, 18.1, 31.5, 39.37, 68.6, 81.58, 83.105, 86.3, 86.4, 86.5, 86.6, 86.7, 86.8, 86.9, 86.10 **6** 4.1, 13.18, 31.24, 51.10, 51.13, 51.14, 53.6, 62.7, 64.12, 64.14, 64.21, 64.49, 68.5, 73.53, 75.4, 75.5, 76.32, 81.10, 81.15, 81.21, 83.67 **7** 31.2, 51.5, 52.3, 59.5, 62.6, 62.7, 64.14, 64.36, 68.11, 68.13, 68.14, 81.3, 83.72, 83.73, 83.102, 85.6, 86.15

Trotz, U.O. **3** 36.18

Troughton, P.G.H. **1** 3.38, 3.39 **2** 71.7, 71.55, 86.78, 86.110, 86.115 **3** 73.36, 86.74, 86.95 **4** 36.27 **5** 45.12

Troup, J.M. **5** 75.15, 75.22 **6** 71.8, 71.31, 71.41, 73.49, 75.9, 83.9, 83.13, 83.42 **7** 71.20, 75.8, 75.24, 83.50, 85.29, 86.61

Troxler, F. **7** 35.23, 36.11

Trueblood, K.N. **1** 7.21, 7.29, 11.7, 11.68, 15.12, 16.10, 20.3, 22.8, 31.22, 31.35, 33.64, 36.18, 47.7, 47.19, 47.31, 48.40, 49.32, 50.29, 51.45 **2** 60.51, 64.66, 67.6, 67.7, 73.72 **4** 12.9, 20.11, 20.13, 30.5, 31.17, 31.18, 36.2, 60.37 **5** 38.32 **7** 31.21, 31.22

Truelock, M.M. **4** 71.81

Truesdell, D. **5** 31.39 **6** 28.13, 71.35, 75.7

Trupcevic, D. **7** 3.5

Trus, B. **6** 30.5, 30.6

Trus, B.L. **1** 36.6 **2** 60.130 **3** 77.9 **4** 29.7, 44.12 **6** 41.12, 60.22 **7** 47.5

Truter, M.R. **1** 4.3, 8.11, 8.42, 11.2, 11.6, 11.41 **2** 60.145, 65.29, 68.19, 72.13, 72.37, 72.66, 75.64, 76.80, 77.9, 77.14, 77.36, 77.37, 77.58, 78.33, 79.10, 79.23, 79.36, 83.29, 83.164, 83.191, 85.53, 85.91 **3** 38.15, 38.21, 38.22, 67.11, 67.24, 68.10, 71.46, 71.48 **4** 10.2, 10.3, 38.26, 44.5, 44.6, 60.6, 67.15, 67.17, 67.21, 67.24, 67.25, 77.7 **5** 3.19, 36.7, 38.25, 38.31, 60.12 **6** 38.26, 38.36, 60.14, 67.8, 67.15, 67.20, 67.21, 72.21, 77.16, 78.17 **7** 34.15, 38.44, 38.45, 38.46

Tsai, C. **2** 62.28 **3** 48.26, 62.19 **4** 37.8, 50.5, 81.9

Tsai, C.-C. **5** 59.15

Tsai, M. **6** 27.21

Tsay, Y.-H. **3** 51.48 **4** 75.7, 86.46, 86.66 **5** 72.6, 72.18, 72.30, 72.31 **6** 67.23, 86.44 **7** 62.14, 68.10, 71.29, 71.44, 71.105, 75.32

Tsay, Y.H. **5** 72.5

Tschang, P.-S.W. **7** 51.4

Tscheng, P.-S.W. **6** 77.19

Tschesche, R. **1** 51.49

Tsein, L.Y. **1** 1.123

Tselinskii, I.V. **1** 7.3, 10.1, 10.4, 12.1 **3** 7.1, 12.3, 12.5, 12.6 **4** 12.4, 12.7 **7** 1.10, 12.2, 12.3, 15.3

Tsernoglou, D. **1** 44.7, 44.8

Tsintsadze, G.V. **2** 83.138 **5** 84.13 **7** 79.11

Tsivtsivadze, T.I. **7** 79.11

Tsoucaris, D. **5** 3.20

Tsoucaris, G. **1** 3.56, 33.33, 33.39, 33.47, 33.48 **3** 58.1 **5** 3.20, 28.18, 28.20, 30.2 **6** 17.3, 38.40 **7** 61.5

Tsuboi, M. **1** 48.42 **7** 47.32

Tsuchida, Y. **4** 40.24

Tsuchihashi, G. **7** 11.17

Tsuchiya, H. **4** 60.32 **5** 60.7

Tsuda, K. **1** 58.7, 58.12

Tsukada, H. **7** 20.14

Tsukihara, T. **4** 48.33, 48.35 **5** 13.6, 44.11, 59.17, 82.16, 83.63 **6** 82.9 **7** 83.67, 83.68

Tsukuda, Y. **1** 51.20 **2** 84.7 **3** 31.12, 50.3, 50.4, 51.20, 51.22, 51.33 **4** 50.12

Tsukuda, Y.M. **6** 53.3

Tsunagawa, M. **6** 59.19

Tsuneno, T. **4** 40.24

Tsutsui, M. **4** 49.19 **5** 49.10, 71.56, 72.4

Tuchscherer, C. **5** 29.1

Tuck, B. **4** 59.18

Tuck, D.G. 1 3.60, 3.68 3 68.6 5 68.2, 68.19 7 68.15
Tucker, P.A. 3 3.52 5 64.12, 64.21, 64.34 6 33.8, 64.27, 64.28, 83.62 7 16.13, 71.48, 71.50, 71.71, 75.11, 75.44
Tucker, W.F. 6 82.23
Tufte, T. 2 60.90
Tuggle, R.M. 4 75.27, 75.30 5 72.35 6 62.10, 71.19 7 73.45
Tulinsky, A. 1 39.85, 49.25, 50.28, 58.72, 59.38 2 60.93, 60.94, 67.11 3 49.10 4 49.1, 49.9 6 49.2
Tune, D.J. 6 86.32
Turano, C. 7 48.63
Turco, A. 2 86.9, 86.38
Turevskaya, E.P. 7 67.3
Turley, J.W. 1 3.32, 10.17, 40.1 2 63.16, 63.26 3 9.1, 63.2, 63.3, 76.85 4 32.12, 33.3, 40.1 5 7.6 6 33.17
Turnbull, K. 7 41.19
Turnbull, K.R. 2 82.2, 82.24 4 76.17 6 76.16
Turner, J.A. 7 31.31
Turner, J.D. 1 1.143
Turner, K. 7 71.41
Turner, K.E. 6 11.13 7 5.16
Turner, R.B. 1 23.1
Turner, R.W. 2 74.2, 74.3
Turney, T.W. 4 86.21 5 72.51
Turpeinen, E. 7 76.28
Turpeinen, U. 7 81.38
Tursch, B. 6 53.12, 58.6 7 53.13, 54.3, 54.4
Turtle, B.L. 5 71.86, 71.87, 71.88
Tustin, G.C. 3 85.23
Tuval, M. 7 1.9, 1.16, 1.19, 1.34, 13.12
Tvedten, O. 1 8.36, 8.37
Tyrrell, H.M. 7 37.17
Tyrrell, J. 1 19.27
Tysseland, M. 4 70.3 5 70.4, 70.5
Tyurk de Garcia Banos, D. 3 68.9, 76.67

Ubbelohde, A.R. 1 17.16
Ubell, E. 1 1.42, 18.27
Uchida, I. 6 54.1, 54.8
Uchida, T. 2 60.128 7 86.98
Uchida, Y. 7 86.98
Uchiyama, S. 5 32.21
Uchtman, V.A. 2 73.64 3 73.27, 73.37 5 64.1, 67.2, 67.3, 67.7, 74.6
Ucko, D.A. 4 81.26, 81.28 5 16.12
Uda, S. 6 66.3
Udel'nov, A.I. 7 69.10
Udupa, M.R. 6 79.9
Uebel, J.J. 5 77.19 6 77.20
Ueda, H. 1 15.23
Ueda, I. 1 18.1, 18.3 7 31.32, 31.33, 57.1
Uedaira, H. 7 45.58
Ueki, T. 1 48.56, 48.95 2 69.31, 78.1, 78.2, 82.13, 86.81 3 69.25, 72.18, 73.33 4 48.33, 48.35, 48.57 5 47.22 6 83.79
Uemura, D. 7 54.16
Ueno, K. 5 33.51, 59.11
Ueoka, S. 5 4.2
Uesu, Y. 2 67.18, 67.19 3 9.18, 9.20 4 32.18
Uggla, R. 3 76.32, 76.35 5 83.9 7 83.13
Ughetto, G. 5 72.3
Ugi, I. 7 64.43
Ugi, I.K. 5 73.41
Ugo, R. 2 77.2, 77.3, 86.80, 86.89 3 86.98
Ugro, J.V. 2 84.20, 86.36
Uguagliati, P. 7 84.51
Uh, H.-S. 5 27.17
Uhlemann, E. 4 85.20
Ukai, T. 7 72.40
Ukaji, T. 1 1.103
Ukei, K. 6 49.1
Ulicky, L. 6 16.4
Ulku, D. 3 10.2, 77.2, 77.3 7 82.4
Ullmann, R. 5 62.2
Ullrich, D. 7 33.10
Ulmer, S.W. 6 73.62
Ulrich, L. 3 59.9
Umans, A.J.H. 1 18.20
Umanskij, M.M. 1 7.23, 15.3, 24.10
Umezawa, H. 1 41.31, 50.9, 50.20, 50.21, 50.34 5 50.3 6 1.30, 50.3 7 1.18, 47.30, 53.28
Umino, K. 7 50.3
Umland, F. 4 83.10
Ungaretti, L. 3 11.5 5 62.5 6 40.25 7 17.4, 17.5
Unkovskii, B.V. 7 45.20
Uno, E. 7 54.16
Uno, T. 2 83.187 7 1.13, 1.14
Unsworth, W.D. 4 83.25
Urabe, K. 7 1.23
Uramoto, M. 3 50.21
Urban, L. 7 83.52
Urbanczyk-Lipkowska, Z. 7 83.57, 83.58
Urch, D.S. 3 75.16
Urdy, C.E. 2 84.57
Urry, D.W. 3 32.14, 32.15
Uskokovic, M. 1 41.29 3 35.23
Uskokovic, M.R. 7 51.44, 53.8
Ustynyuk, N.A. 6 72.24
Ustynyuk, Yu.A. 5 71.45
Utsumi-Oda, K. 6 51.34
Utsumi, H. 1 1.108 3 17.5, 33.42 4 81.32, 81.33, 81.35
Utsumi, Y. 3 50.5, 76.82
Uttech, R. 2 72.22
Uyeda, N. 4 49.13, 49.18
Uyeo, S. 1 58.86 3 58.43 4 58.43 6 58.47

Vaccaro, A.M.T. 3 59.33
Vaciago, A. 1 3.70, 4.12, 33.70, 35.3, 35.4, 35.7, 35.12, 35.14, 35.15, 50.10, 50.48 2 65.27, 80.13, 80.16, 80.17, 80.20, 80.32, 80.33, 80.35, 85.9, 85.10, 85.11, 85.36, 85.69 3 9.10, 33.45, 50.28, 58.15, 83.24, 85.1 4 32.22, 35.26, 54.19, 70.9, 80.15 5 56.2 6 33.52, 33.56
Vagg, R.S. 3 83.92 7 82.12

Vahrenkamp, H. **2** 73.91 **4** 73.14, 85.24, 86.15 **5** 62.2, 75.37, 86.11 **6** 73.10, 86.24 **7** 73.19
Vaidya, S.N. **2** 62.47
Vainshtein, B.K. **1** 48.38 **7** 17.8
Vaira, M.di **1** 3.33, 3.35, 3.36 **2** 76.76, 76.77, 76.78, 76.79, 78.24, 78.32, 78.49, 78.53, 83.119, 83.120, 83.127 **3** 76.62, 76.86, 78.15, 83.88 **5** 76.51, 86.64
Valach, F. **7** 76.7, 83.23
Valente, E.J. **7** 59.7
Vallarino, L.M. **7** 86.21
Valle, G. **1** 39.8, 39.9, 39.39 **3** 16.12, 32.12 **4** 39.3, 39.4 **5** 39.9, 42.2 **6** 3.35 **7** 42.1
Valle, M. **4** 71.27
Valle, S. **2** 71.45
Vallino, M. **3** 67.14
Valverde-Lopez, S. **3** 58.13, 58.52
Van de Velde, G.M.H. **6** 81.37
Van Soest, T.C. **4** 59.11
Vand, V. **1** 5.42, 30.13
Vandenborre, M.T. **6** 17.7, 17.8
Vanderauwera, J. **1** 41.46
VanDerveer, D.G. **6** 85.9, 85.10 **7** 75.10
Vandewalle, J.J.M. **3** 41.10
VanDoorne, W. **6** 64.58
Vane, F.M. **6** 58.28
VanEtten, R.L. **3** 32.11, 32.23, 32.34
Vannerberg, N.-G. **1** 7.1 **2** 81.71 **3** 31.16
Varenkamp, H. **2** 73.64
Varghese, J.N. **7** 41.5
Varlamova, G.L. **7** 76.64
Varshavskii, Y.S. **4** 83.128
Varughese, K.I. **5** 48.26, 48.44, 48.45 **7** 48.36
Vaska, L. **2** 86.83 **4** 86.48
Vass, G. **7** 45.52
Vassian, E.G. **3** 83.65
Vaughan, D.P. **7** 3.14, 58.1
Vaughan, P.A. **1** 29.2
Veal, J.T. **3** 71.9 **4** 76.13 **5** 82.12, 83.101 **7** 71.92
Vedavathi, B.M. **7** 2.2, 3.4
Veen, A.M.van **5** 63.12
Veen, J.M.van der **5** 48.65, 48.66
Veen, J.van der **3** 27.4
Vega, E. **7** 35.40
Veidis, M.V. **2** 86.68 **3** 32.18, 32.19, 77.18 **4** 71.74, 71.75, 73.16, 73.18 **6** 71.72
Veigel, E. **7** 68.2, 68.6
Veith, M. **6** 63.3 **7** 63.7
Vekris, J. **2** 83.47
Velde, G.M.H.van de **7** 81.36
Veldis, M.V. **2** 82.41
Velichko, F.K. **2** 71.1
Velsvik, M. **6** 39.28
Ven, C.F.W.van de **4** 27.17, 51.19
Venanzi, L.M. **2** 75.52, 75.71, 80.32, 80.33, 86.67, 86.105 **3** 72.5, 72.6 **4** 86.64
Venkataraman, K. **1** 59.68 **6** 59.30
Venkatarman, S. **7** 48.4
Venkatasubramanian, K. **5** 78.19
Venkatesan, K. **1** 23.8, 31.19, 48.94 **3** 32.25, 48.49 **4** 48.13, 48.16, 48.17, 48.32, 48.52, 48.54, 49.20 **5** 19.15, 33.45, 82.17 **6** 48.34 **7** 31.9, 37.7, 48.18, 48.35, 48.39, 48.54, 84.17, 84.75
Verbist, J. **2** 60.64 **3** 67.3 **4** 48.4 **5** 48.1
Verbist, J.F. **4** 48.10
Verbist, J.J. **5** 48.14, 48.47, 48.49 **7** 48.22
Vergamini, P.J. **4** 73.14, 85.24 **5** 73.31, 73.37
Verhoef, L.H.W. **2** 60.11, 60.98
Verhoeven, A.M. **4** 80.11
Verkade, J.G. **1** 46.8 **3** 64.12 **4** 64.4 **5** 64.5, 86.2 **7** 64.15, 64.22, 64.23, 64.54, 86.7, 86.16, 86.19
Verma, H.K.L. **3** 26.10
Veronese, A.C. **1** 33.75
Verschoor, G.C. **1** 42.2 **3** 33.2, 33.3 **4** 83.81 **5** 42.6, 67.9, 83.39, 83.104 **6** 83.7, 84.37, 84.38 **7** 83.44, 83.45, 83.81
Vesela, L. **1** 53.36
Vetter, G. **6** 10.3 **7** 66.9
Vetter, W. **4** 31.28
Vezzosi, I.M. **5** 29.3
Vialard, A.-M. **1** 38.60
Viavattene, R.L. **7** 31.30
Vicat, J. **6** 11.15
Vickery, B.L. **2** 60.145 **4** 44.5, 44.6, 60.6, 67.17, 77.7 **6** 78.17
Victor, D. **1** 9.29
Vidal, J.-P. **6** 63.10
Vidal, J.P. **5** 63.8, 63.9, 63.10
Vidali, M. **3** 85.19 **4** 78.11 **6** 76.40, 84.18 **7** 78.10
Vigato, P.A. **4** 71.41, 78.11 **6** 76.40, 84.18 **7** 78.10
Vihovde, E.H. **1** 21.9 **2** 70.2
Vijayalakshmi, B.K. **4** 11.11 **6** 16.13 **7** 48.40, 48.53
Vijayan, K. **6** 31.29 **7** 2.2, 3.4
Vijayan, M. **1** 13.57, 38.48 **2** 67.20, 67.21, 69.49 **5** 32.17 **6** 60.15 **7** 32.33, 33.21, 48.31
Villa, A. **2** 80.1
Villa, A.C. **1** 11.1 **3** 8.11, 79.4, 85.12 **4** 8.6, 35.13, 76.38, 79.2, 79.3, 81.1, 83.1, 83.2, 83.5 **5** 71.27, 78.19, 81.25, 81.43, 81.44, 83.16, 86.37, 86.39 **6** 40.6, 58.27, 76.45, 78.4, 80.3 **7** 40.2, 73.61, 73.62, 73.63, 80.4, 83.2
Villa, L. **6** 3.35
Villares, P. **7** 42.7
Villiers, J.P.R.de **3** 77.34 **4** 77.19, 77.21 **5** 77.25
Vilminot, S. **7** 48.5
Vimala, T.M. **3** 84.4 **7** 13.17
Vimba, S. **2** 62.51
Vince, D. **5** 81.62
Vincent, A.T. **4** 16.16, 64.43 **5** 15.9
Vincent, R.L. **1** 58.103
Vinciguerra, N. **2** 85.88 **3** 85.50
Vine, W.H. **6** 48.83
Viossat, C. **1** 17.27
Visscher, M.O. **5** 86.53 **6** 72.10, 83.84

Visser, G.J. **1** 3.25, 35.42, 39.26, 39.53, 39.54, 39.55, 39.91 **3** 11.23, 28.8, 31.33, 39.17 **4** 18.1, 32.16, 35.22
Visti, J. **3** 76.32, 76.35
Viswamitra, M.A. **1** 11.15 **2** 62.47, 67.20, 67.21, 69.49, 81.15 **3** 50.31 **4** 47.2, 47.9, 47.10, 82.4, 82.5, 83.137, 83.138 **5** 2.1, 47.13 **7** 47.4, 47.6, 47.16, 47.21, 47.35
Viswanathan, K.S. **2** 83.61
Viswanathan, N. **6** 56.1
Vitali, T. **5** 41.41
Viterbo, D. **1** 10.13, 13.53, 40.15, 40.16 **2** 83.25, 83.26 **3** 40.17, 59.4 **4** 24.2, 40.8 **5** 40.24, 83.3 **6** 40.2, 40.3, 40.5, 40.9 **7** 40.2
Vitulli, G. **2** 72.67 **3** 74.3, 74.4
Vizzini, E.A. **1** 43.23, 43.24, 43.25, 43.28, 43.29 **2** 79.3 **4** 43.6
Vlahov, R. **1** 53.14
Vlasse, M. **1** 45.25
Vliegenthart, J.A. **1** 1.34 **3** 1.8
Vloten, G.W.van **1** 21.5
Voblikova, V.A. **4** 78.3
Voecks, G.E. **4** 86.11
Voegel, J.C. **6** 67.3, 67.7
Voegele, J.C. **6** 67.2, 67.6
Voet, D. **2** 62.26, 62.31 **3** 44.9, 60.25 **4** 60.34 **5** 33.33, 60.10, 60.11 **6** 47.18
Voge, A.A. **4** 82.10, 84.18
Vogt Junior, L.H. **4** 33.27
Vogt, L.H. **1** 49.36
Voigt, D. **4** 51.48 **6** 51.37
Vol'pin, M.E. **2** 63.28
Vol'pin, M.R. **6** 81.45
Voll, U. **7** 64.59
Volodina, G.F. **7** 79.14
Volponi, L. **3** 86.19, 86.25
Voltattorni, C.B. **7** 48.63
Volz, K. **6** 84.27, 84.28
Volz, W.E. **7** 31.1
Vonk, C.G. **6** 73.44
Voorbergen, P. **7** 67.7
Vooren, M. **3** 37.4
Vorontsova, L.G. **1** 4.1, 11.38, 39.73 **3** 33.28 **5** 21.10, 45.47 **7** 5.18, 19.11
Vos, A. **1** 3.25, 3.61, 3.62, 11.47, 16.42, 35.42, 39.26, 39.53, 39.54, 39.55, 40.12 **2** 60.122, 64.70, 72.44 **3** 28.8, 33.1, 33.4 **4** 11.7, 12.10, 16.8, 16.9, 27.14, 32.16, 35.22 **5** 11.9, 27.5 **7** 5.13, 5.14, 16.13, 16.14, 16.15, 16.17, 19.15
Vozzennikov, V.M. **1** 11.30 **2** 62.29, 69.20
Vranka, R.G. **2** 68.10, 83.16 **3** 86.72 **4** 79.9 **6** 28.14
Vries, A.de **1** 48.31 **7** 17.35
Vukosavovich, M. **3** 3.13
Vukov, R. **5** 31.16
Vuuren, P.J.van **4** 71.9

Waal, B.W.van de **5** 35.5
Wachsman, M.A. **6** 37.8
Wachter, W.A. **6** 83.80 **7** 71.94
Wachtmeister, C.A. **7** 59.6
Wada, T. **1** 37.1
Wada, Y. **3** 50.22, 50.30 **4** 50.6, 54.3 **5** 50.2 **6** 35.29, 36.7, 40.24
Wade, K. **4** 68.13 **7** 63.21, 69.30, 69.31
Wade, T.D. **5** 36.5, 36.6 **7** 84.54
Wadhawan, V.K. **6** 58.50 **7** 2.7
Wadley, L.G.B. **4** 79.8 **5** 79.16, 79.17, 79.18
Wadsworth, W. **5** 46.7 **6** 46.12, 64.13
Waerstad, K. **1** 9.15
Wagener, R. **6** 41.30
Wagner, A. **7** 38.51, 45.28
Wagner, A.J. **2** 64.70 **3** 64.29 **4** 64.15 **5** 64.35 **7** 64.34, 64.35
Wagner, F. **6** 76.51 **7** 76.63, 83.62
Wagner, H. **3** 59.26 **7** 38.17
Wagner, H.P. **6** 54.10
Wagner, K.G. **5** 48.50
Wagner, K.P. **7** 71.67, 86.89
Wagner, R. **5** 46.7 **6** 46.12
Wagner, R.E. **6** 64.13
Wagner, W.F. **2** 77.29, 77.30, 77.32, 77.34 **5** 77.20
Wahl Junior, G.H. **4** 31.31 **5** 36.19, 39.25 **6** 31.9, 39.15
Wahlberg, A. **7** 1.7, 80.16
Waight, E.S. **4** 31.32 **6** 54.13
Wailes, P.C. **7** 75.34
Waite, M.G. **3** 11.6, 58.55, 58.56, 75.6 **4** 4.2, 4.5
Wajsman, E. **7** 8.5, 8.6
Wakabayashi, A. **6** 5.1, 5.3
Wakabayashi, K. **6** 47.16, 47.27
Wakahara, A. **5** 35.22 **6** 8.5, 8.6, 35.14, 48.56, 58.45 **7** 17.10
Wakamiya, T. **3** 59.1
Wakatsuki, Y. **6** 71.71 **7** 71.114
Wakisaka, K. **1** 58.45 **3** 36.26
Wal, H.R.van der **7** 71.52, 73.35, 75.30, 75.36
Walker, A. **2** 86.70 **4** 69.1
Walker, C. **3** 81.16, 81.42
Walker, G. **1** 2.9
Walker, G.W. **7** 83.6
Walker, J. **1** 8.32
Walker, L.A. **1** 32.13
Walker, N.W. **5** 71.12
Walker, N.W.J. **6** 71.13, 72.9 **7** 73.13
Walker, P.J.C. **6** 75.17
Walker, R. **7** 71.68, 86.30, 86.47, 86.52
Walker, W.R. **2** 82.23, 82.36, 82.38
Wall, M.E. **4** 59.19, 59.26
Wallace, J.C. **4** 45.8 **7** 45.14, 45.17, 45.18, 45.25
Wallace, J.E. **3** 78.29
Wallbridge, M.G.H. **2** 65.6, 68.5, 68.6
Walling, C. **2** 83.183
Wallwork, S.C. **1** 5.29, 17.19, 33.66, 54.1, 59.23 **2** 60.72, 60.78, 60.79, 60.80, 60.81, 60.83, 60.88, 60.100, 60.113, 60.116, 60.125, 60.126, 60.133, 60.134, 65.37, 84.1 **3** 59.34, 65.8 **4** 33.28, 33.29, 33.30, 33.46, 40.4, 60.25 **5** 33.55, 36.10, 40.5, 65.7, 69.4 **7** 30.9, 60.5, 60.29
Walser, A. **6** 35.27

Walter, R. **1** 1.128
Walter, T.J. **6** 31.16 **7** 31.8
Walter, W. **1** 4.16 **3** 85.5 **4** 4.1, 8.7 **7** 4.4
Walthers, K.K. **3** 76.40 **4** 31.9
Walton, D.R.M. **6** 86.32 **7** 63.9
Walton, R.A. **2** 83.103, 85.54, 86.21 **3** 33.9, 81.36, 86.75 **4** 80.13, 81.36, 81.37 **6** 86.15 **7** 86.46, 86.97
Wampler, D.L. **2** 83.32 **4** 8.9, 23.2 **5** 47.24, 50.11 **6** 65.1, 65.2, 65.3
Wan, C. **4** 73.31 **6** 77.2 **7** 77.13
Wan, K.Y. **4** 86.31
Wang Lund, E. **1** 21.6
Wang, A.H.-J. **4** 32.30, 50.18, 71.54, 71.55 **5** 37.3, 54.9 **6** 47.28, 51.43, 54.16, 76.51 **7** 50.26, 50.27, 51.11, 71.19, 71.47
Wang, B.-C. **3** 33.34, 78.30 **6** 81.2
Wang, B.C. **3** 83.115, 83.116 **4** 83.118, 83.119
Wang, C.-H. **2** 82.25
Wang, C.K. **5** 45.14, 46.4, 46.5, 46.8, 46.9, 59.1
Wang, F.E. **2** 62.15
Wang, J.T. **5** 81.36
Wang, M. **6** 36.2
Wang, S.Y. **5** 36.3
Wang, Y. **5** 20.7 **6** 20.2, 20.3, 20.4 **7** 33.8
Wani, M.C. **4** 59.19, 59.26
Ward, D.L. **4** 79.5, 84.15, 86.11 **6** 51.17
Wardell, J.L. **6** 69.19 **7** 69.25
Wardle, R. **4** 75.23
Warin, V. **7** 32.18
Warkentin, J. **5** 40.15 **7** 32.26
Warkusz, F. **6** 48.3, 48.4
Warmack, R.J. **7** 39.3
Warner, D. **6** 1.6, 1.8, 1.9
Warner, L.G. **6** 3.25, 13.16, 48.54, 85.25 **7** 85.23, 85.32, 85.41
Warner, P. **5** 31.31 **6** 31.8
Warr, W.A. **3** 35.18
Warren, F.L. **7** 58.2
Warren, L.F. **7** 86.35, 86.36, 86.37
Warren, R.C. **3** 82.1
Warren, S. **7** 64.32, 64.33
Warrener, R.N. **3** 36.16 **4** 44.16 **5** 41.11
Wartburg, A.von **5** 59.24
Warwicker, J.O. **1** 43.12
Waser, J. **2** 65.9, 65.11 **3** 45.27 **6** 81.2
Washecheck, D. **7** 36.2, 36.3, 36.5, 36.9
Washecheck, D.M. **7** 64.49
Wasley, R. **6** 47.12
Wasson, J.R. **3** 76.77
Wasson, S.J.S. **5** 77.20
Watanabe, E.-I. **5** 49.6 **6** 49.7
Watanabe, I. **1** 55.3
Watanabe, J.-I. **7** 76.37
Watanabe, K. **2** 76.2
Watanabe, S. **5** 45.15
Watanabe, T. **1** 2.5, 16.15, 22.4, 22.5, 33.18, 36.12, 50.31, 50.31, 58.120 **2** 61.21, 75.55, 81.9, 81.77, 82.32 **3** 50.20 **4** 5.1, 72.17, 82.6 **5** 17.20, 50.14, 53.4, 53.5, 73.23, 73.24, 81.32, 81.33 **6** 5.1, 5.2, 5.3 **7** 73.34
Watanabe, Y. **2** 69.22, 69.27 **3** 11.9, 80.4 **5** 80.3, 85.9
Watase, H. **1** 59.17, 59.19 **2** 81.68
Watase, T. **2** 63.2
Watelle-Marion, G. **4** 2.9
Watelle, G. **3** 2.5
Watenpaugh, K. **1** 47.15 **2** 77.48, 84.22, 84.54
Waterhouse, C.R. **3** 86.1
Waters, J.M. **1** 49.6 **2** 76.22, 78.46, 82.57 **3** 78.2, 78.4, 78.27, 78.29, 86.61 **4** 86.35 **5** 29.4, 59.31, 83.31, 85.33 **6** 27.4, 60.27, 76.52, 86.67 **7** 27.6, 38.70, 79.26, 84.11, 86.68
Waters, T.N. **1** 49.6 **2** 76.26, 77.16, 77.17, 78.4, 78.5, 78.7, 78.16, 78.17, 78.18, 78.21, 78.26, 78.27, 78.29, 78.40, 78.41, 78.46, 78.50, 78.58, 82.55, 82.57, 83.109, 83.112, 83.192, 83.194 **3** 54.1, 54.4, 60.1, 60.16, 60.39, 78.2, 78.4, 78.8, 78.27, 78.29 **4** 76.42, 83.101 **5** 29.4, 59.31, 77.12, 81.35, 83.66 **6** 54.4, 54.7, 60.2, 60.27, 76.23, 76.24, 76.34, 76.36, 76.37, 76.38, 76.39, 76.52, 76.54, 78.12 **7** 36.25, 38.70, 86.99, 86.103
Watkin, D. **2** 86.99
Watkin, D.J. **1** 42.10 **2** 84.52 **3** 86.81, 86.82, 86.86 **4** 22.1, 42.2 **5** 27.9 **6** 31.28
Watkins, E.D. **2** 77.30
Watkins, S.F. **3** 19.1, 69.26, 71.64, 86.2 **4** 71.90 **5** 53.14, 63.4 **6** 32.28, 32.31
Watling, R.C. **2** 78.33
Watson Junior, C.R. **7** 39.3
Watson, D.G. **1** 27.6, 44.22, 44.23, 47.21, 51.35, 56.2 **2** 65.24 **3** 20.17, 23.1, 46.5, 46.6, 51.42, 53.12, 53.13, 53.14, 54.13, 58.47, 65.5 **4** 8.9, 23.2 **5** 47.24 **6** 46.2, 65.1, 65.2, 65.3
Watson, H.C. **1** 24.19
Watson, K.J. **1** 15.7, 15.8, 50.40, 53.5, 59.44 **2** 82.2 **3** 79.21, 85.4 **4** 83.115, 83.116 **5** 10.2, 38.30, 76.31 **6** 4.6, 72.13, 82.23
Watson, R. **5** 28.16
Watson, W.H. **2** 77.19, 84.32, 84.34, 84.38, 84.39, 85.28 **3** 84.8, 84.11, 84.17 **4** 3.5, 32.11, 33.5, 56.4, 59.14, 84.8 **5** 32.9, 51.45, 51.46, 59.6, 77.14, 82.20, 84.11, 84.19 **6** 31.32, 54.5 **7** 51.57, 53.14, 53.24, 84.29, 84.62, 84.73
Watt, G.W. **7** 3.22, 3.23, 3.24
Watt, R. **4** 71.56
Watton, E.C. **2** 83.101 **3** 83.92
Watts Junior, P.H. **4** 37.9, 39.43 **5** 1.38, 37.5 **6** 64.43
Watts, P.H. **3** 39.6, 39.28
Waugh, J.S. **5** 81.7 **6** 14.1
Wawra, H. **7** 45.55
Weakley, T.J.R. **7** 83.3, 83.4
Weakliem, H.A. **2** 76.59
Weaver, D.L. **2** 84.24, 85.29, 85.55 **3** 71.44, 76.6, 76.76 **4** 75.27, 75.30 **5** 72.35 **6** 75.30
Weaver, J. **4** 71.31 **5** 73.29 **6** 73.36
Weaver, L.H. **6** 35.26, 37.10 **7** 32.36

Webb, L.E. 1 49.1, 49.22, 49.23, 49.28, 49.31 2 81.30 3 79.16 6 48.55, 48.63
Webb, M.J. 7 69.12
Webb, N.C. 1 4.5, 31.3, 59.22 2 73.78
Webb, T.R. 6 71.8 7 81.27, 86.22
Weber, A.H. 3 33.47
Weber, H.-P. 5 59.27, 59.32 6 53.13 7 50.38
Weber, H.P. 1 23.4, 34.7, 53.40 2 69.40, 73.31, 73.121, 85.83 3 40.12 4 39.35, 59.10, 83.112 5 36.23, 36.24, 40.32, 50.12, 51.26, 51.27, 58.23, 59.24 6 58.3, 58.4, 58.38, 58.42 7 6.2, 33.28, 35.23, 36.11, 37.7, 51.17
Weber, L. 5 86.54 7 51.4
Webster, M. 2 63.12 3 3.9, 69.6 4 69.3 7 65.4, 69.8
Webster, M.S. 1 10.25, 50.29
Wedegaertner, D.K. 4 21.21
Weedon, B.C.L. 1 21.33 4 54.1, 54.2
Weeks, C. 5 47.28
Weeks, C.M. 2 82.53 3 17.8, 51.12, 51.19 4 51.15, 51.17, 51.26, 51.27, 51.33 5 11.11, 11.15, 50.29 6 17.23, 51.4, 51.15, 51.16, 51.27, 51.31, 51.35, 51.39 7 51.6, 51.7, 51.10, 51.12, 51.16, 51.19, 51.24, 51.28, 51.29, 51.31, 51.32, 51.51, 59.18
Wege, D. 7 75.9
Wegner, P.A. 3 62.9
Wei, C.C. 6 58.33
Wei, C.H. 1 44.72 2 72.48, 73.90, 85.62, 85.74, 85.82, 85.84 3 44.34, 83.31 4 30.2, 30.3, 39.20, 48.55, 48.56, 83.98 5 83.71, 83.72, 83.73 6 85.38
Wei, K.-T. 7 41.23
Wei, K.-T.H. 7 11.19
Wei, L. 2 78.51
Weibel, A.T. 7 71.85
Weichert, T. 7 81.9
Weichselgartner, H. 1 49.16
Weickhardt, P.L. 3 85.24
Weidenborner, J.E. 4 69.35 5 32.23 6 41.32 7 31.17
Weidlein, J. 7 68.6, 68.7
Weigel, D. 4 73.20
Weigel, F. 3 81.11, 81.12, 81.13, 81.15
Weigelt, A. 7 29.1, 29.2
Weigold, H. 7 75.34
Weimann, G. 7 47.41, 47.42
Weimann, J. 3 1.7, 30.6
Weinblum, D. 1 44.73
Weinheimer, A.J. 7 54.9, 54.10
Weininger, M.S. 2 79.28 3 79.24 4 69.11, 69.22, 69.27, 74.11, 86.24 5 79.22, 79.25, 86.69 7 79.13, 79.23
Weinreb, S.M. 7 50.9, 50.13
Weinstein, B. 6 58.46
Weinstein, M. 7 1.24, 1.25, 24.7
Weir, C.E. 3 19.4
Weis, C. 5 86.66
Weisbach, J.A. 1 58.100, 58.101
Weiss, E. 2 73.20 3 63.4, 63.5, 63.6, 67.5 5 68.3, 68.4, 68.28 6 72.18, 73.45 7 62.12, 62.13, 68.4, 71.72, 72.24, 72.25
Weiss, G.S. 1 37.2, 37.3
Weiss, J. 2 64.1, 64.8, 64.20, 65.5, 65.15, 83.11, 83.12, 85.1, 85.38 3 64.5, 71.45 4 69.4 6 39.5 7 63.2, 64.60, 71.126, 71.134
Weiss, R. 1 12.5, 18.4, 18.5 2 83.124, 83.153, 84.3 3 33.29, 40.10, 53.12, 53.13, 53.14, 60.51, 64.24, 64.38, 64.40, 72.21 4 12.1, 12.2, 12.5, 12.6, 33.8, 33.9, 40.18, 66.2, 67.12, 67.20, 68.1, 81.3, 81.19, 81.50, 83.46, 83.47, 83.48, 84.3 5 12.3, 12.4, 53.7, 53.8, 53.9, 53.11, 53.16, 66.2, 66.3, 66.4, 67.10, 67.12, 67.13, 67.14, 67.15, 68.11, 68.12, 68.13, 68.24, 69.18, 72.14, 72.22, 72.23, 72.25, 75.6, 76.40, 83.43 6 40.20, 67.2, 67.3, 67.6, 67.7, 67.12, 67.13, 67.14, 67.19, 71.69, 73.63, 76.48, 84.14, 84.30, 85.18, 85.22, 85.35 7 31.21, 40.17, 40.19, 40.20, 49.14, 72.18, 80.9, 80.10, 80.20, 85.16, 85.21
Weiss, U. 6 29.1
Weisz, O. 2 86.1
Weitz, G. 1 58.39
Welberry, T.R. 3 31.35 6 28.4, 28.5
Welch, A.J. 4 83.61, 83.85, 83.130, 83.136 5 71.79, 81.61 6 83.55, 86.5 7 71.22, 71.45, 71.49, 71.61, 71.66, 72.2, 72.9, 86.4, 86.10, 86.11
Welch, A.L. 6 83.54
Welch, S.C. 6 30.5, 30.6
Welcker, P.S. 2 62.39 3 86.92
Weldon, A.S. 1 2.82
Wells III, W.J. 7 3.22, 3.23, 3.24
Wells, A.F. 2 66.3, 66.4, 66.5, 83.39, 86.2, 86.14
Wells, A.G. 6 45.1
Wells, J.L. 1 32.41 6 60.22
Wells, R.B. 4 38.21
Welsh, D. 6 78.7, 78.8, 78.16, 78.18, 78.19
Welsh, H.K. 1 1.151 2 82.42
Welsh, M.J. 1 59.40
Welsh, W.A. 5 33.9
Welter, A. 7 48.34
Wengelin, R.F. 1 14.12
Wentworth, R.A.D. 2 83.190 7 78.18
Wenzel, J. 7 33.36
Wenzel, J.-P. 7 33.37
Weperen, K.J.van 4 18.1
Werkema, M.S. 1 6.1, 35.1
Werner, H. 7 73.43
Werner, L.H. 7 35.30
Werner, P. 6 13.15
Werner, P.-E. 1 45.27, 45.28 3 36.7, 36.11, 36.22 4 36.17, 36.24 5 83.124
Werness, P.G. 6 28.3
Wertz, D.W. 6 75.3, 75.10
Wesselink, G.A. 6 19.2
Wessels, G.F.S. 4 33.33 5 77.8
Wessels, P.L. 7 58.17
West, B.O. 4 86.2 7 72.29, 86.9, 86.18, 86.59
West, R. 4 22.7 6 85.13 7 20.10
Wester, D. 6 79.11, 79.12, 79.13 7 83.87, 83.88, 83.113, 84.30, 84.31
Westerdahl, A. 1 1.131, 41.28
Westin, L. 5 58.19

Westley, J.W. **4** 50.15 **7** 50.34
Westover, G.F. **7** 73.58
Wet, J.F.de **2** 81.63
Wetherington, J.B. **6** 9.2, 10.4, 32.16, 32.29 **7** 50.5
Wetzel, J. **2** 66.15
Weyenberg, A.M. **4** 60.3
Whalley, W.B. **1** 2.92, 59.45 **3** 51.26, 51.30 **7** 51.52
Whang, Y.-A. **5** 83.100
Wheatley, P.J. **1** 4.2, 4.11, 4.17, 5.23, 13.42, 20.22, 22.16, 32.14, 33.6, 33.9, 41.22, 41.23, 41.47, 41.48, 44.6, 50.15 **2** 63.14, 63.30, 64.10, 64.36, 64.43, 64.57, 64.75, 64.79, 66.22, 68.21, 69.34, 85.47 **3** 16.16, 16.17 **4** 16.16, 64.1, 64.2, 64.43 **5** 15.9, 83.106, 86.59 **6** 71.43, 83.58, 83.69, 84.17, 86.17
Wheeler, A.G. **2** 60.107, 72.21, 72.42, 83.159
Wheeler, D.H. **4** 67.8
Wheeler, D.M.S. **5** 27.14
Wheeler, G.L. **6** 20.12, 33.35, 33.42, 35.6, 35.7, 35.22, 35.23, 64.43 **7** 19.1, 33.48, 33.49, 35.33
Whidden, T.K. **6** 11.8, 11.9
Whillans, F.D. **3** 83.78, 83.110 **7** 76.77, 85.43
Whimp, P.O. **3** 72.37, 75.35, 81.48 **4** 71.56, 86.21 **5** 71.61, 71.94, 72.40, 72.45, 72.51 **6** 72.23, 72.28, 74.11, 74.14, 82.10 **7** 71.110, 75.42, 75.43, 84.21
Whinnery, J.E. **4** 32.11 **5** 32.9, 84.11 **6** 31.32
White Junior, R.P. **4** 3.15, 3.16, 64.41, 64.42
White, A.H. **3** 80.6 **4** 80.3, 80.9, 80.17 **5** 79.1, 80.1, 80.6, 80.7, 80.9, 80.17, 80.19, 80.21, 80.22, 83.20, 83.32, 84.17, 84.26 **6** 3.18, 65.4, 79.1, 79.6, 80.9, 80.10, 84.47 **7** 32.16, 41.5, 75.9, 76.14, 77.1, 77.4, 78.23, 78.24, 78.26, 80.3, 80.8, 83.6, 83.66, 83.82, 83.91, 83.93, 83.101, 83.109, 83.110, 83.112, 86.3, 86.40, 86.41
White, A.J. **6** 73.24, 73.59
White, B.F. **1** 49.12
White, B.M. **2** 73.9
White, D.N.J. **3** 51.9 **4** 50.13, 53.4, 53.9, 58.32 **5** 48.39 **6** 23.4 **7** 2.4, 50.32, 53.6, 53.20, 53.21, 53.22
White, D.W. **3** 64.12
White, E. **7** 4.8, 35.28, 35.39, 36.27
White, E.H. **3** 9.8
White, E.P. **7** 58.2
White, H. **5** 71.21
White, J. **7** 79.7
White, J.C.B. **1** 13.23, 49.12 **3** 8.9, 33.12, 33.13 **4** 78.8
White, J.D. **4** 59.27 **5** 56.4, 59.21 **7** 54.7
White, J.G. **1** 3.20, 24.13, 30.9, 30.12, 30.19 **3** 38.4, 42.2 **6** 37.8
White, N.E. **1** 44.14
White, P.S. **7** 51.46, 51.47
White, R. **5** 71.70
White, R.H. **6** 53.16 **7** 17.7
Whitehead, E.V. **5** 56.5, 56.6, 56.7
Whitesides, T.H. **7** 72.5
Whitfield, H.J. **3** 85.10
Whiting, D.A. **3** 59.23 **4** 20.14, 59.18 **5** 40.35, 51.15, 51.44, 56.13, 59.29, 72.10 **6** 51.40, 59.16
Whitla, A. **2** 73.12
Whitla, W.A. **2** 75.71 **6** 11.8, 11.9
Whitlow, S.H. **1** 16.6 **2** 64.45 **3** 64.20, 83.32 **4** 2.23, 67.4 **6** 1.4, 81.7, 82.5 **7** 81.11, 81.12, 86.24
Whitt, C.D. **4** 67.18, 68.5
Whittaker, B. **7** 84.43, 84.46
Whittaker, E.J.W. **1** 17.38
Whitten, D.G. **7** 49.12
Whittle, K.R. **4** 86.35 **5** 83.31, 85.33 **6** 86.67 **7** 86.68
Whittlesey, B.R. **7** 86.13, 86.14
Wiberg, K.B. **5** 31.2 **7** 71.125, 75.41
Wickman, H.H. **7** 80.5
Wickramasinghe, J.A.F. **4** 51.55
Widdowson, D.A. **3** 58.34 **7** 51.56
Widmann, P. **6** 27.1, 27.2 **7** 27.1
Wiebenga, E.H. **1** 5.6, 19.61, 48.87, 52.19
Wiechers, A. **5** 59.5
Wiehager, A.-C. **5** 53.3 **6** 31.21, 31.22, 53.1, 58.56 **7** 31.13, 31.25, 51.48, 52.2, 54.1
Wieland, D.M. **3** 8.14
Wieland, T. **7** 48.74
Wiesner, J.R. **1** 3.33, 3.35, 3.36 **2** 76.14
Wiest, R. **5** 76.40 **6** 76.48, 84.30
Wiewiorowski, M. **1** 59.32 **5** 58.6 **7** 58.13
Wiff, D.R. **6** 63.11
Wightman, R. **7** 58.28
Wijnhoven, J.G. **4** 80.16 **5** 80.24 **6** 80.17 **7** 80.17
Wilcox, H.E. **1** 1.42
Wild, S.B. **6** 3.18 **7** 86.40, 86.41
Wildman, G.C. **4** 19.5
Wildman, W.C. **3** 58.19, 58.24, 58.50 **5** 58.15
Wildonger, K.J. **5** 36.19
Wiley, P.F. **4** 45.23
Wilford, J.B. **2** 71.21, 73.12, 73.109, 86.44 **3** 86.81, 86.82, 86.86 **4** 31.7, 31.11, 31.23, 41.27 **5** 35.37
Wilke, G. **4** 71.65
Wilkes, C.E. **2** 76.54, 77.69
Wilkes, G.R. **2** 73.90
Wilkins, C.J. **6** 81.25, 82.6
Wilkins, J.D. **5** 86.33 **6** 71.17, 85.6, 86.14 **7** 71.2, 83.75, 83.108, 86.17, 86.32, 86.33, 86.53
Wilkinson, G. **4** 71.52, 71.53 **6** 71.8, 74.15
Wilkinson, J.L. **7** 83.83
Wilkinson, J.R. **6** 71.75
Will, G. **1** 49.16 **3** 32.6
Willemse, J. **6** 80.14 **7** 80.17
Willer, R.L. **6** 39.10, 39.11
Willett, R.D. **1** 3.5, 3.11 **2** 83.20, 83.21, 84.19 **3** 3.1, 3.2, 3.3, 3.13, 84.1, 84.3, 84.7 **4** 3.1, 3.4, 82.11 **5** 3.7 **6** 65.8, 83.64 **7** 3.16, 39.1

Willey, G.R. **6** 63.1 **7** 63.21, 69.30, 69.31
Williams, B.C. **7** 32.11, 60.27
Williams, D.A. **2** 84.11 **4** 7.13 **5** 7.8
Williams, D.E. **1** 12.12, 19.54, 19.56, 38.19, 45.76 **2** 83.57, 83.58 **3** 12.10 **5** 12.12, 19.6, 19.10 **6** 20.8, 20.9, 39.19 **7** 19.13, 20.6, 20.7, 35.4
Williams, D.G. **1** 6.2, 6.3, 6.4 **3** 45.38
Williams, D.J. **1** 58.30 **2** 77.1 **3** 59.8 **5** 40.34 **6** 15.5, 56.1 **7** 35.6, 38.76, 61.3, 61.4
Williams, G. **1** 39.72
Williams, G.J.B. **4** 33.38, 47.2 **5** 58.2 **6** 1.3, 2.13, 47.17 **7** 32.34, 33.46, 44.28
Williams, H.J. **7** 26.6, 26.7
Williams, J. **7** 73.13
Williams, J.A.W. **5** 71.12
Williams, J.M. **1** 3.7, 3.8 **3** 3.16, 3.17, 6.1, 76.7 **4** 13.2 **5** 11.4 **6** 16.5, 20.3, 20.4, 47.4 **7** 11.10, 33.8, 73.28
Williams, J.R. **3** 36.29 **6** 58.34
Williams, M.D. **7** 22.3
Williams, P.A. **7** 82.12
Williams, P.P. **1** 52.11 **5** 52.3 **6** 43.4, 43.5
Williams, R.J. **2** 84.32, 85.28 **3** 84.8 **4** 33.5, 76.8, 84.8 **5** 77.14
Williams, R.M. **2** 60.72, 60.113, 60.125, 60.126
Williams, T. **1** 41.29
Williams, T.H. **6** 16.23 **7** 53.8
Williamson, D.H. **6** 71.8
Williamson, R.B. **6** 86.41
Willis, A.C. **6** 3.18 **7** 32.16, 41.5, 78.23, 78.24, 78.26, 80.8
Willis, B.T.M. **2** 73.23
Willis, C.M. **5** 71.63, 71.64
Willis, J.S. **3** 15.1, 15.2
Williston, C.S. **1** 15.15 **2** 70.5, 73.98 **3** 73.41
Willoughby, T.V. **2** 60.95 **7** 27.4
Wills, D.L. **4** 80.13
Willstadter, E. **2** 82.56
Wilson Junior, E.W. **4** 82.7
Wilson Junior, G.E. **3** 39.1
Wilson, A. **2** 60.133, 60.134
Wilson, F.B. **2** 61.16 **4** 61.1 **5** 22.3 **7** 73.18, 75.16
Wilson, H.R. **1** 44.36, 47.2, 47.3, 47.4, 47.25 **3** 47.5, 47.18 **5** 47.7 **6** 47.6 **7** 47.11, 47.20
Wilson, J.N. **1** 48.30
Wilson, R.D. **7** 64.63, 64.64, 64.65
Wilson, R.M. **5** 38.17 **6** 38.25
Wilson, S.A. **1** 19.19, 19.20
Wilson, W.S. **3** 36.16
Winarko, S. **6** 65.7
Wing, R.M. **2** 62.34, 64.56 **3** 22.2, 37.5, 60.54, 73.24, 83.148, 85.23 **4** 53.2, 59.8 **5** 27.8, 54.7, 76.35, 77.19, 80.23 **6** 54.15, 77.20, 85.26 **7** 53.18, 64.12
Wingard Junior, R.E. **4** 39.17
Wingfield, J.N. **4** 83.122
Winkler, B. **5** 78.13, 78.14
Winkler, E.K. **5** 45.43
Winkler, F. **7** 48.35
Winkler, F.K. **4** 50.11 **7** 34.2, 34.4, 34.7, 34.9, 34.10, 34.11, 34.12, 34.13, 34.14
Winkler, H. **2** 64.48
Winquist, B.H.C. **2** 86.97 **3** 77.33
Winter, G. **4** 69.15
Wirth, J.G. **4** 33.27
Wise, J.J. **2** 77.56
Wiseman, J.R. **6** 39.29
Wiseman, T.J. **2** 83.167
Wismer, R.K. **6** 80.13 **7** 76.35
Witiak, D. **5** 76.17
Witkop, B. **3** 31.14, 58.9 **4** 48.20, 48.22 **6** 36.1 **7** 37.16, 48.74
Witmore, W.G. **1** 7.20
Witt, J.R. **4** 7.1, 7.2, 7.3, 7.4
Witters, R.D. **2** 84.58
Witters, R.W. **1** 53.3
Wodak, S.J. **6** 58.13 **7** 58.8
Wogan, G.N. **6** 58.52
Wohlauer, G. **2** 83.57, 83.58
Wojcicki, A. **2** 85.71
Wojtkunski, J. **4** 33.47
Wolf, A.P. **2** 66.19, 66.21
Wolf, G.-U. **2** 67.2
Wolf, N.de **1** 42.5 **5** 42.6
Wolf, R. **7** 64.51
Wolf, W. **1** 42.8
Wolfe, G.A. **1** 39.4 **3** 53.1, 53.2
Wolfe, K. **7** 69.26
Wolfel, E. **1** 58.39 **2** 63.25
Wolff, M.E. **7** 51.7, 51.16, 51.34, 51.35
Wolfram, W. **1** 8.6
Wolodina, G.F. **2** 60.87
Wolterman, G. **7** 85.38
Wolters, A.P. **4** 3.12, 86.55 **5** 86.33 **7** 37.17, 86.32
Wolters, J. **1** 39.53, 39.54, 39.55
Wolthius-Spuy, M. **3** 1.37
Wong, C. **4** 67.10 **6** 67.4 **7** 36.7
Wong, C.-H. **1** 20.24 **2** 73.62, 73.65 **3** 67.9
Wong, C.F. **5** 58.5
Wong, C.S. **4** 81.61
Wong, F. **7** 39.9
Wong, L. **5** 35.32
Wong, P.-S. **4** 37.10 **5** 27.14
Wong, R.Y. **4** 41.15, 67.7 **5** 48.28 **6** 59.10 **7** 31.34, 31.35, 52.1
Wong, S. **6** 86.46
Wong, S.T.K. **7** 8.18
Wong, W. **4** 71.90 **6** 32.28, 32.31
Wong, Y.S. **6** 71.2, 71.5, 71.6, 86.43 **7** 71.5, 71.10, 71.98, 82.2
Woo, N. **7** 3.11
Woo, P.W.K. **7** 47.29
Wood III, J.B. **5** 59.25
Wood, A.M. **2** 73.33, 82.23, 82.36, 82.38, 86.55
Wood, C.J. **7** 45.26
Wood, J.L. **6** 28.3 **7** 35.6
Wood, J.S. **2** 65.45, 71.44, 72.25, 74.9, 77.57, 80.23, 83.47, 83.49, 83.172, 84.27 **3** 65.9, 79.19 **4** 78.17, 83.19, 83.20, 84.17 **5** 83.81 **6** 64.39 **7** 83.59, 83.60, 84.74
Wood, M. **7** 13.18, 44.29, 46.3, 46.4

Wood, M.K. **3** 31.5 **7** 16.11, 46.6, 47.12
Wood, R.G. **1** 39.72
Wood, R.H. **5** 33.9, 35.34 **7** 35.32
Woodhouse, D.I. **6** 72.7, 73.13, 75.8
Woodlock, A. **3** 33.12, 33.13
Woods, D.C. **7** 38.19
Woods, J.D. **1** 1.141
Woods, M. **4** 64.28, 64.29, 64.30 **5** 64.21
Woodward, I. **1** 5.25, 9.35, 19.49, 49.8, 49.9
Woodward, P. **2** 66.1, 73.47, 85.25, 85.26, 85.52, 86.17 **3** 3.52, 19.1, 71.64 **4** 69.23, 69.26, 71.31, 71.45, 71.77 **5** 71.51, 72.17, 73.29, 74.12 **6** 71.80, 73.36, 75.11, 75.12 **7** 71.39, 71.55, 71.69, 71.70, 71.93, 75.1, 75.18, 75.19, 75.23, 75.25, 75.38, 85.17
Woolfson, M.M. **2** 64.26 **7** 51.46, 51.47
Woolley, P.R. **7** 83.71
Worch, H. **1** 58.107
Worch, H.-H. **4** 51.56
Wormald, J. **3** 71.8, 71.24, 72.9, 73.35, 73.40, 75.22, 75.31, 85.14 **4** 71.35, 85.9 **5** 75.29, 86.84 **7** 3.10
Worsham Junior, J.E. **1** 8.4, 8.21
Worthington, C.R. **2** 67.11
Worthington, J.M. **4** 83.131 **5** 76.45 **6** 83.44
Wreford, S.S. **6** 86.1
Wright, C.E. **6** 3.30
Wright, D.A. **1** 48.70 **2** 84.11 **4** 7.13, 67.14 **5** 7.8 **7** 76.68
Wright, G.F. **1** 9.7 **2** 71.11, 71.12
Wright, H. **3** 54.5 **5** 59.21 **6** 58.49
Wright, H.E. **1** 58.64 **6** 38.22, 58.52
Wright, J.D. **2** 60.70, 60.146, 60.147, 60.148, 75.26 **5** 60.30, 60.31, 60.32, 60.33 **7** 60.28
Wright, L.H. **3** 54.5, 58.48 **4** 58.38, 58.44
Wright, N.C.A. **7** 37.21
Wright, P.E. **6** 76.39
Wright, W.B. **1** 33.38, 33.41, 48.82 **4** 19.10 **6** 60.24
Wrobel, J.T. **6** 58.54
Wu, I.B. **7** 53.24
Wu, V.Y. **5** 17.3
Wudl, F. **6** 39.6 **7** 17.17
Wulf, H. **1** 58.59
Wulff, G. **1** 51.49
Wunderlich, H. **3** 11.1 **4** 17.2, 20.9 **5** 43.3 **6** 64.17, 64.23
Wunderlich, J.A. **1** 45.65, 52.14, 58.17, 58.38 **3** 35.1, 58.22 **5** 31.37, 38.28 **6** 83.60 **7** 75.34
Wustner, D.A. **7** 64.12
Wuts, P.G.M. **5** 27.17
Wynberg, H. **1** 35.42 **5** 27.5

Yablokov, Yu.V. **7** 79.18, 79.19, 79.20, 79.25
Yacynych, A.M. **6** 76.11, 76.12, 76.13
Yadav, B. **6** 24.5 **7** 2.6
Yadav, S.R.P. **5** 1.13 **7** 19.3
Yadav, V.S. **4** 48.3
Yadava, V.S. **5** 2.24, 48.41 **7** 2.7
Yagen, B. **4** 51.55
Yagihara, M. **6** 38.8
Yagisawa, N. **7** 47.30
Yahashi, R. **6** 31.14
Yaji, H. **1** 36.26
Yakel Junior, H.L. **1** 48.83
Yakel, H.L. **1** 8.31
Yakovenko, V.I. **2** 80.25
Yakushi, K. **3** 60.53 **4** 60.33 **5** 60.21, 60.22 **6** 60.20, 60.21, 60.26 **7** 60.15, 60.16
Yale, H.L. **5** 41.33 **6** 35.13
Yamada, N. **2** 67.18, 67.19
Yamada, S. **5** 81.32, 81.33
Yamada, Y. **1** 7.18
Yamahata, K. **3** 80.4
Yamakawa, M. **1** 7.28
Yamamoto, H. **7** 37.18
Yamamoto, J. **5** 13.6
Yamamoto, S. **6** 9.1
Yamamoto, T. **4** 81.32 **5** 1.20 **6** 71.75
Yamamoto, Y. **6** 71.63
Yamamura, S. **1** 54.9 **6** 58.48
Yamane, T. **6** 32.9 **7** 38.67, 48.68
Yamasaki, K. **4** 83.68
Yamasaki, K.-I. **5** 1.20
Yamashita, S. **2** 69.12, 69.22
Yamataka, K. **4** 35.39 **5** 42.13
Yamauchi, H. **3** 56.13 **4** 56.5
Yamauchi, O. **5** 82.11
Yamauchi, T. **7** 38.5
Yamazaki, H. **4** 72.28 **6** 71.63, 71.71 **7** 71.114
Yamazaki, K. **7** 56.7
Yamazaki, M. **7** 58.19
Yamazaki, S. **4** 54.13 **7** 54.8
Yamazaki, Y. **2** 67.18, 67.19
Yanagiya, M. **1** 58.120
Yanai, H.S. **1** 58.3
Yandell, J.K. **7** 76.14
Yang, S.C. **6** 76.6
Yannoni, N.F. **1** 26.13, 26.16, 26.38, 28.21, 28.22, 44.65 **3** 73.10 **4** 12.11, 73.12 **5** 19.3 **6** 7.5, 27.6, 28.14, 28.17, 28.18
Yao, S.Y. **4** 58.21
Yarrow, D.J. **6** 71.42 **7** 71.121, 86.100
Yasaki, E. **3** 76.31
Yasin, Y.M.G. **2** 61.15
Yasuda, H. **2** 68.28 **3** 68.19
Yasuda, K. **2** 69.3
Yasuda, N. **4** 71.28, 71.29
Yasue, M. **7** 56.7
Yasuoka, N. **1** 17.40, 17.42 **2** 68.28, 69.31, 86.71, 86.81 **3** 13.19, 13.20, 35.12, 36.9, 41.23, 68.19, 69.11, 69.25, 72.10, 72.18, 73.33 **4** 35.39, 60.20, 68.12, 71.28, 71.29, 72.27 **5** 17.18, 42.12, 42.13, 68.22, 68.25, 68.32, 69.12 **6** 44.5, 66.3, 75.29 **7** 36.16, 38.67, 71.62, 71.63, 71.73, 71.90, 72.34, 72.35
Yates, P. **4** 39.46 **6** 20.18, 20.19
Yawney, D.B.W. **3** 79.11, 81.35 **4** 81.60 **5** 81.57 **6** 81.44
Yeates, D.G.R. **7** 44.25
Yeats, E.A. **1** 24.16

Yemul, S.S. **6** 59.30
Yen, T.-M. **2** 73.62
Yerkess, J. **4** 28.4 **5** 24.3 **6** 29.4
Yesinowski, J.P. **7** 71.15
Yeung, H.-N. **6** 14.1
Yevitz, M.M. **7** 72.7, 72.8
Yokota, T. **7** 54.8
Yoneda, M. **5** 47.14, 47.15
Yonehara, H. **3** 50.21 **4** 50.4 **6** 50.12
Yonemitsu, O. **3** 31.14 **7** 31.15
Yonezawa, M. **3** 76.52
Yoo, C.S. **4** 13.13, 16.17 **5** 28.8, 41.22 **6** 41.18, 51.2 **7** 16.11, 46.2
Yoshida, M. **2** 69.31
Yoshida, N. **2** 69.27
Yoshida, T. **2** 73.82 **4** 72.17 **6** 86.26
Yoshida, Z.-I. **5** 49.6 **6** 49.7 **7** 49.8
Yoshikawa, M. **7** 47.14, 60.7
Yoshimura, J. **6** 71.58
Yoshimura, M. **5** 58.43
Yoshino, A. **1** 58.90 **3** 33.24, 71.21 **5** 38.15 **7** 53.4, 84.45
Yoshioka, H. **3** 36.34 **4** 35.31, 35.32 **5** 33.39, 33.40, 33.41, 36.8, 41.46, 71.67 **6** 86.26 **7** 50.18
Yoshioka, T. **1** 44.76 **7** 33.39, 33.40, 33.41
Yoshioko, M. **5** 50.9
Yosioka, I. **6** 59.29
You-Chi, T. **1** 3.54
Young III, F.R. **7** 71.17
Young, A.E. **1** 11.9, 42.1
Young, D.W. **1** 19.48, 30.2, 30.17, 45.39, 46.2, 47.25, 48.10, 54.3, 54.4, 54.7 **3** 47.18 **5** 47.4 **6** 47.6 **7** 47.11, 47.20
Young, J.M. **3** 3.43, 3.44
Young, M. **5** 59.23
Young, R.J. **6** 74.15
Yu, A.P. **1** 23.1
Yu, M.W. **7** 69.18, 84.53
Yuan, A.S. **7** 48.17
Yuan, S.-S. **6** 58.52
Yumal', E.M. **4** 78.3
Yunusov, S.Yu. **6** 58.20
Yutal, Yu.M. **6** 77.18
Yvon, K. **6** 28.9

Zabkiewicz, J.A. **1** 22.6 **5** 22.3
Zaborsky, O.R. **1** 11.9, 42.1
Zacharias, D.E. **1** 58.82, 58.85, 58.100 **3** 58.65 **5** 28.5 **7** 1.26, 1.27, 29.4, 29.7, 29.9, 36.15, 38.60, 38.62
Zachariasen, W.H. **1** 2.1 **2** 81.42
Zacharis, H. **3** 36.19
Zacharis, H.M. **1** 35.31 **4** 35.7, 35.17
Zadock, E. **7** 59.10
Zagari, A. **7** 1.17
Zahrobsky, R.F. **1** 44.71 **2** 84.36
Zakharova, G.N. **2** 69.18 **4** 69.29 **5** 63.15
Zakrzewski, G. **3** 83.44
Zakrzewski, G.A. **3** 81.45, 81.46, 81.47 **4** 83.56, 83.57
Zalkin, A. **1** 3.31, 3.75, 3.80, 49.36, 54.8 **2** 62.33, 62.38, 62.41, 65.39, 71.30, 73.5, 77.6, 77.46, 82.59, 83.219, 85.39, 85.40 **3** 48.17, 62.8, 62.12, 76.25, 77.32 **4** 49.8, 49.12, 62.9, 71.5, 75.14, 75.15 **5** 1.4, 67.21, 67.25, 67.26 **6** 6.1, 32.38, 33.57, 35.19, 35.20, 38.14, 42.6, 51.17, 64.33, 83.79, 84.27, 84.28 **7** 39.24
Zalkow, L.H. **1** 53.28
Zalukaev, L.P. **3** 27.10
Zambonelli, L. **1** 33.70, 35.13 **2** 80.13, 80.16, 80.17, 80.20, 80.32, 80.33, 80.35 **3** 32.36, 72.4, 72.5, 72.6, 83.24 **4** 70.9, 72.3, 86.64 **5** 16.11, 21.9, 72.1, 72.2, 72.3 **6** 15.2, 72.5
Zanari, A.R. **4** 85.8
Zanazzi, P.F. **1** 3.4, 35.5, 38.43, 40.19, 40.20 **4** 35.4, 35.10, 70.4, 83.49, 85.8 **5** 36.32, 40.25 **6** 32.13, 32.34
Zand, R. **1** 20.14 **4** 48.26, 48.34, 48.37, 48.42 **6** 48.61 **7** 20.1, 48.36
Zanella, A. **4** 76.11
Zangrando, E. **6** 71.22 **7** 71.7
Zanker, V. **4** 36.37 **5** 36.31
Zannetti, R. **1** 39.46 **2** 83.59, 83.60, 83.74 **3** 1.36 **7** 5.19
Zanobini, F. **7** 85.37
Zanotti, G. **7** 42.1
Zanzari, A.R. **4** 70.4 **5** 36.32, 40.25 **6** 32.13, 32.34
Zappia, V. **7** 48.29
Zarli, B. **3** 80.19, 80.20, 81.55 **4** 85.26, 85.32 **5** 80.12, 84.27
Zaske, P. **5** 64.25
Zaslow, B. **4** 3.11 **5** 3.5
Zatko, D.A. **5** 76.9, 83.79
Zauer, K. **6** 32.32, 32.33
Zavalishin, E.I. **3** 27.10
Zavin, B.G. **7** 63.14
Zavodnik, V.E. **7** 36.6
Zazzetta, A. **3** 71.25, 78.12, 78.16 **4** 74.1 **7** 68.17, 71.132
Zbryskaya, S.G. **5** 76.37
Zechmeister, K. **3** 39.24, 39.25, 39.27, 54.6, 54.8, 54.15, 59.25 **4** 27.19, 36.37, 47.7, 51.11, 54.8, 54.12, 54.16 **5** 5.13
Zedler, A. **7** 32.19
Zee, J.J.van der **4** 67.8 **5** 86.49
Zee, S.H. **6** 38.35
Zegarski, W.J. **6** 85.27, 85.36, 85.43
Zeiger, W. **3** 30.4
Zeinstra, J.D. **6** 73.12
Zeiss, H.H. **5** 17.22, 71.77 **6** 71.68 **7** 71.88
Zelenko, B. **7** 3.5
Zelnik, R. **5** 54.9 **6** 54.16
Zelwer, C. **5** 3.20
Zenda, H. **4** 59.21
Zeppezauer, M. **3** 33.43
Zerger, R. **7** 67.1, 67.6, 67.8
Zerger, R.P. **5** 67.17
Zettler, F. **6** 62.6 **7** 11.2, 11.3, 62.8
Zhdanov, G.S. **1** 7.23, 15.3, 24.10, 39.77, 41.18 **3** 26.1, 35.20 **6** 26.1 **7** 41.17
Zieger, H.E. **5** 26.4
Ziegler, M. **2** 72.51, 85.1 **4** 69.4

Ziegler, M.F. **5** 59.26
Ziegler, M.L. **2** 64.8, 64.9 **3** 64.5, 71.45 **5** 72.13 **6** 39.5, 66.1 **7** 75.2, 75.3, 75.4, 75.31
Ziemer, B. **7** 3.25, 41.27
Ziffer, H. **3** 36.29 **6** 37.13
Zimmer, J.-C. **2** 69.29
Zimmerman, I.C. **1** 11.20
Zimmermann-Barlow, I. **1** 39.17
Zingaro, R.A. **3** 65.3 **4** 65.3 **6** 70.6 **7** 65.1
Zinnius, A. **3** 63.8, 63.13 **4** 63.7
Zins, D. **7** 3.9
Ziolkovska, B. **2** 71.26
Ziolo, R.F. **3** 86.96 **4** 86.61, 86.68 **5** 76.41 **6** 20.11, 76.29
Zirakzadeh, M. **1** 44.19
Ziser, M. **3** 44.24
Zobel, T. **1** 16.6 **2** 65.2
Zocchi, M. **2** 72.67, 83.165 **3** 83.63 **4** 5.9, 35.21, 76.26, 83.102 **5** 71.71 **6** 31.13, 40.19, 71.64, 71.65 **7** 71.64, 72.37
Zoer, H. **4** 64.15
Zol'nikova, G.P. **7** 75.33
Zolnikova, G.P. **3** 71.49
Zorky, P.M. **7** 78.21, 78.22
Zubieta, J.A. **3** 85.31 **4** 80.10, 85.22 **5** 73.46 **6** 86.13 **7** 69.3, 73.64, 86.50
Zuccaro, D.E. **1** 11.22
Zuckerman, J.J. **7** 64.54, 69.3
Zurr, Z. **4** 64.18
Zussman, J.I. **1** 47.29, 48.39
Zvonkova, Z.V. **1** 7.6, 11.30, 16.25, 41.18, 41.36 **2** 62.29, 62.42, 69.20, 69.38, 80.21, 80.25 **4** 41.4, **5** 39.19 **7** 8.10